CLIMATE CHANGE 2001: MITIGATION

Climate Change 2001: Mitigation is the most comprehensive and up-to-date scientific, technical and economic assessment of options to mitigate climate change and their costs. The report:

- Makes clear that there are strong interlinkages between climate change policy and policies towards sustainable development.
- Assesses a very large body of information on technological options to reduce greenhouse gas emissions or enhance their sinks in various economic sectors and regions.
- Analyzes which technologies and practices are available to achieve the targets of the Kyoto Protocol in the short term and to stabilize atmospheric concentrations of greenhouse gases in the longer term.
- Evaluates which technological, social, economic, cultural or institutional barriers impede the implementation of these options.
- Describes what is known about policies, measures and instruments to overcome these barriers.
- Summarizes the economic literature on the costs of the various options at the global, national and sectoral levels.
- Evaluates the other benefits that climate policy can deliver or the climate benefits of other socio-economic policies.
- Assesses how this information can be used to support climate policy decision making, acknowledging various decision-making frameworks.

This latest assessment of the IPCC will again form the standard scientific reference for all those concerned with climate change who want to be informed about ways to address this important global problem, including students and researchers in environmental economics, engineering and social sciences, and policy makers and analysts in governmental and nongovernmental organisations and the private sector worldwide.

Bert Metz is a Co-Chair of Working Group III, IPCC, and Head of the Global and European Environmental Assessment Division of the National Institute for Public Health and Environment (RIVM), The Netherlands.

Ogunlade Davidson is a Co-Chair of Working Group III, IPCC, and Director of the Energy and Development Research Centre (EDRC), University of Cape Town, South Africa.

Rob Swart is Head of the Technical Support Unit of Working Group III, IPCC.

Jiahua Pan is Economist of the Technical Support Unit of Working Group III, IPCC, and Senior Fellow, the Chinese Academy of Social Sciences, China

Climate Change 2001:
Mitigation

Edited by

Bert Metz
The Dutch National Institute for
Public Health and the Environment

Ogunlade Davidson
University of Cape Town

Rob Swart
IPCC Working Group III
Technical Support Unit

Jiahua Pan
The Chinese Academy of Social Sciences

Contribution of Working Group III to the Third Assessment Report of the Intergovernmental
Panel on Climate Change

Published for the Intergovernmental Panel on Climate Change

PUBLISHED BY THE PRESS SYNDICATE OF THE UNIVERSITY OF CAMBRIDGE
The Pitt Building, Trumpington Street, Cambridge, United Kingdom

CAMBRIDGE UNIVERSITY PRESS
The Edinburgh Building, Cambridge CB2 2RU, UK
40 West 20th Street, New York, NY 10011-4211, USA
10 Stamford Road, Oakleigh, VIC 3166, Australia
Ruiz de Alarcón 13, 28014 Madrid, Spain
Dock House, The Waterfront, Cape Town 8001, South Africa

http://www.cambridge.org

First published 2001

Printed in the United States of America

Typeface Times Roman 10/12 pt.   *System* LaTeX   [AU]

A catalog record for this book is available from the British Library.

Library of Congress Cataloging in Publication data
Climate Change 2001 : Mitigation : Contribution of Working Group III to
   the third assessment report of the Intergovernmental Panel on Climate
   Change / edited by Bert Metz ... *et al. or full list of names*

      p.   cm.
   Includes bibliographical references and index.
   ISBN 0-521-80769-7 (hb) – ISBN 0-521-01502-2 (pb)

   1. Climatic changes – Government policy.   2. Greenhouse gas
mitigation.   I. Metz, Bert.   II. Intergovernmental Panel on Climate
Change.   Working Group III.
QC981.8.C5 C511343   2001
363.738′747 – dc21
                                        2001025973

ISBN 0 521 80769 7 hardback
ISBN 0 521 01502 2 paperback

Cover Image Credit: Photo provided by Benelux Press BV.

Contents

Foreword

The Intergovernmental Panel on Climate Change (IPCC) was jointly established by the World Meteorological Organization (WMO) and the United Nations Environment Programme (UNEP) in 1988. The terms of reference include: (i) to assess available scientific and socio-economic information on climate change and its impacts and on the options for mitigating climate change and adapting to it and (ii) to provide, on request, scientific/technological/socio-economic advice to the Conference of the Parties (CoP) to the United Nations Framework Convention on Climate Change (UNFCCC). From 1990, the IPCC has produced a series of Assessment Reports, Special Reports, Technical Papers, methodologies and other products that have become standard works of reference, widely used by policymakers, scientists and other experts.

This volume, which forms part of the Third Assessment Report (TAR), has been produced by Working Group III (WGIII) of the IPCC and focuses on the mitigation of climate change. It consists of 10 chapters covering the technological and biological options to mitigate climate change, their costs and ancillary benefits, the barriers to their implementation, and policies, measures and instruments to overcome these barriers. As is usual in the IPCC, success in producing this report has depended first and foremost on the knowledge, enthusiasm and co-operation of many hundreds of experts worldwide, in many related but different disciplines. We would like to express our gratitude to all the Co-ordinating Lead Authors, Lead Authors,

Contributing Authors, Review Editors and Reviewers. These individuals have devoted enormous time and effort to produce this report and we are extremely grateful for their commitment to the IPCC process. We would like to thank the staff of the WGIII Technical Support Unit and the IPCC Secretariat for their dedication in co-ordinating the production of another successful IPCC report. We are also grateful to the governments, who have supported their scientists' participation in the IPCC process and who have contributed to the IPCC Trust Fund to provide for the essential participation of experts from developing countries and countries with economies in transition. We would like to express our appreciation to the governments of The Netherlands, Norway, Germany and South Africa, who hosted drafting sessions in their countries, to the government of Ghana, who hosted the 6th session of Working Group III for official consideration and acceptance of the report in Accra, and again to the government of The Netherlands, who funded the WGIII Technical Support Unit.

We would particularly like to thank Dr. Robert Watson, Chairman of the IPCC, for his sound direction and tireless and able guidance of the IPCC, to Dr. Sundararaman, the Secretary of the IPCC and his staff for the secretarial support, and Professor Ogunlade Davidson and Dr. Bert Metz, the Co-Chairmen of Working Group III, for their skilful leadership of Working Group III through the production of this report.

G.O.P. Obasi

Secretary-General
World Meteorological Organization

K. Töpfer

Executive Director
United Nations Environment Programme
and
Director-General
United Nations Office in Nairobi

Preface

We are pleased to present the third volume of the Third Assessment Report (TAR) of the Intergovernmental Panel on Climate Change (IPCC) prepared by Working Group III: Climate Change 2001: Mitigation. The IPCC has been established jointly by the World Meteorological Organisation (WMO) and the United Nations Environment Programme (UNEP) to assess all available factual information on the science, the impacts, and the economics of climate change and on the adaptation/mitigation options to address climate change.

At its 14th Session in 1997 in the Maldives, the IPCC agreed on the development of the TAR. Working Group III was charged to assess the scientific, technological, environmental, economic, and social aspects of the mitigation of climate change. Thus, the mandate of the Working Group was broadened in the current report from a predominantly disciplinary assessment of the Economic and Social Dimensions on Climate Change (including adaptation) in the Second Assessment Report (SAR).

This report summarizes the state of knowledge covered in previous IPCC reports, but primarily assesses information generated since the SAR, recognizing that climate change is a global issue, but emphasizing the assessment of the sectoral and regional aspects of climate change mitigation. The assessment focuses on the policy-relevant questions of today. It also draws on the IPCC Special Reports on Aviation and the Atmosphere, on Methodological and Technological Issues in Technology Transfer, on Land Use, Land-Use Change and Forestry, and Emissions Scenarios that were published in 1999 and 2000. The report, in accordance with the mandate given to Working Group III, explicitly places climate change mitigation in the broader context of development, equity, and sustainability. It draws on a broad set of literature also covering the social sciences and acknowledges different views on the linkages between climate change mitigation and sustainable development policies.

The report has been written by almost 150 lead and co-ordinating lead authors, some 80 contributing authors and 18 review editors from developed countries, developing countries, countries with economies in transition, and international organizations, who put in an enormous amount of time and effort. It has been reviewed by more than 300 experts from around the world, individually as well as through governments.

To involve experts beyond the writing teams, discuss issues which to date received insufficient attention to address current policy questions properly, and to allow for interaction with the writing teams of the other two Working Groups, Working Group III sponsored various Expert Meetings and Workshops, including:

- IPCC Regional Workshop on Integrated Assessment, Kadoma, Zimbabwe, 22-28 November 1998;
- 1st IPCC Expert Meeting on Climate Change and its Linkages with Development, Equity, and Sustainability, Colombo, Sri Lanka, 27-29 April 1999;
- Joint IPCC/TEAP Expert Meeting on Options for the Limitation of Emissions of HFCs and PFCs, Petten, Netherlands, 26-28 May 1999;
- IPCC Expert Meeting on Economic Impacts of Mitigation Measures, The Hague, Netherlands, 27-28 May 1999;
- IPCC Expert Meeting on Stabilization and Mitigation Scenarios, Copenhagen, Denmark, 2-4 June 1999;
- IPCC Expert Meeting on Costing Methodologies, Tokyo, Japan, 28 June-1 July 1999;
- IPCC Expert Meeting on Sectoral Economic Impacts, Eisenach, Germany, 14-15 February 2000;
- 2nd IPCC Expert Meeting on Development, Equity, and Sustainability, Havana, Cuba, 23-25 February 2000;
- IPCC Expert Meeting on Society, Behaviour and Climate Change Mitigation, Karlsruhe, Germany, 21-22 March 2000;
- IPCC Co-sponsored Expert Meeting on Ancillary Benefits, Washington DC, United States, 27-29 March 2000.

We are grateful to the governments of Zimbabwe, Sri Lanka, The Netherlands, Denmark, Norway, Japan, Germany and Cuba for making these meetings possible in collaboration with the local organisers. Proceedings of these meetings have been published or will be published in 2001. The writing teams of this report met four times to draft the report and discuss the results of the two consecutive formal IPCC review rounds, in Bilthoven (Netherlands, December 1998), Lillehammer (Norway, September 1999), Eisenach (Germany, February 2000) and Cape Town (South Africa, August 2000). In addition, several individual chapter team meetings, writing team teleconferences, and interactions with the UNFCCC SBSTA process contributed to the contents of the report.

According to the IPCC Procedures, the Summary for Policymakers of this report has been approved in detail by governments at the IPCC Working Group III Plenary Meeting in Accra, Ghana from 28 February to 3 March 2001. During the approval process the lead authors confirmed that the agreed text of the Summary for Policymakers is fully consistent with the underlying full report and technical summary, which has been accepted by governments, but remains the full responsibility of the authors.

We would like to acknowledge the financial support of the Netherlands' government in supporting the Technical Support Unit of Working Group III and the IPCC activities of Co-chair Dr. Bert Metz, and the IPCC Secretariat for supporting the activities of Co-chair Professor Ogunlade Davidson through the IPCC Trust Fund. This report would not have been possible without the tireless efforts of the staff of the Technical Support Unit, Dr. Rob Swart (Head of TSU), Dr. Jiahua Pan, Anita Meier, José Hesselink, Angelique Martens, Remko Ybema, Ton van Dril, Tom Kram, Jan Willem Martens, Sascha van Rooijen and Dr. Peter Kuikman. We also express our gratitude to John Ormiston, Paul Schwartzman and Ruth de Wijs for the copy-editing, reference editing and proofreading of the docu-ment and to Martin Middelburg for preparing the final lay-out and the graphics of the report.

We, as co-chairs of Working Group III, together with the other members of the Bureau of Working Group III, the Lead Authors, and the Technical Support Unit, hope that this report will assist decision-makers in governments and the private sector as well as other interested readers in the academic community and the general public to be better informed about climate change mitigation in support of appropriate response measures.

Ogunlade Davidson and Bert Metz

Co-Chairs IPCC Working Group III on Mitigation of Climate Change

SUMMARY FOR POLICYMAKERS

CLIMATE CHANGE 2001: MITIGATION

A Report of Working Group III
of the Intergovernmental Panel on Climate Change

This summary, approved in detail at the Sixth Session of IPCC Working Group III (Accra, Ghana • 28 February - 3 March 2001), represents the formally agreed statement of the IPCC concerning climate change mitigation.

Based on a draft prepared by:

Tariq Banuri, Terry Barker, Igor Bashmakov, Kornelis Blok, Daniel Bouille, Renate Christ, Ogunlade Davidson, Jae Edmonds, Ken Gregory, Michael Grubb, Kirsten Halsnaes, Tom Heller, Jean-Charles Hourcade, Catrinus Jepma, Pekka Kauppi, Anil Markandya, Bert Metz, William Moomaw, Jose Roberto Moreira, Tsuneyuki Morita, Nebojsa Nakicenovic, Lynn Price, Richard Richels, John Robinson, Hans Holger Rogner, Jayant Sathaye, Roger Sedjo, Priyaradshi Shukla, Leena Srivastava, Rob Swart, Ferenc Toth, John Weyant

CONTENTS

Introduction

1. This report assesses the scientific, technological, environmental, economic and social aspects of the mitigation of climate change. Research in climate change mitigation[1] has continued since the publication of the IPCC Second Assessment Report (SAR), taking into account political changes such as the agreement on the Kyoto Protocol to the United Nations Framework Convention on Climate Change (UNFCCC) in 1997, and is reported on here. The Report also draws on a number of IPCC Special Reports, notably the Special Report on Aviation and the Global Atmosphere, the Special Report on Methodological and Technological Issues in Technology Transfer (SRTT), the Special Report on Emissions Scenarios (SRES), and the Special Report on Land Use, Land Use Change and Forestry (SRLULUCF).

The Nature of the Mitigation Challenge

2. Climate change[2] is a problem with unique characteristics. It is global, long-term (up to several centuries), and involves complex interactions between climatic, environmental, economic, political, institutional, social and technological processes. This may have significant international and intergenerational implications in the context of broader societal goals such as equity and sustainable development. Developing a response to climate change is characterized by decision-making under uncertainty and risk, including the possibility of non-linear and/or irreversible changes (Sections 1.2.5, 1.3, 10.1.2, 10.1.4, 10.4.5).[3]

3. Alternative development paths[4] can result in very different greenhouse gas emissions. The SRES and the mitigation scenarios assessed in this report suggest that the type, magnitude,

timing and costs of mitigation depend on different national circumstances and socio-economic, and technological development paths and the desired level of greenhouse gas concentration stabilization in the atmosphere (see *Figure SPM.1* for an example for total CO_2 emissions). Development paths leading to low emissions depend on a wide range of policy choices and require major policy changes in areas other than climate change (Sections 2.2.2, 2.3.2, 2.4.4, 2.5).

4. Climate change mitigation will both be affected by, and have impacts on, broader socio-economic policies and trends, such as those relating to development, sustainability and equity. Climate mitigation policies may promote sustainable development when they are consistent with such broader societal objectives. Some mitigation actions may yield extensive benefits in areas outside of climate change: for example, they may reduce health problems; increase employment; reduce negative environmental impacts (like air pollution); protect and enhance forests, soils and watersheds; reduce those subsidies and taxes which enhance greenhouse gas emissions; and induce technological change and diffusion, contributing to wider goals of sustainable development. Similarly, development paths that meet sustainable development objectives may result in lower levels of greenhouse gas emissions (Sections 1.3, 1.4, 2.2.3, 2.4.4, 2.5, 7.2.2, 8.2.4).

5. Differences in the distribution of technological, natural and financial resources among and within nations and regions, and between generations, as well as differences in mitigation costs, are often key considerations in the analysis of climate change mitigation options. Much of the debate about the future differentiation of contributions of countries to mitigation and related equity issues also considers these circumstances[5]. The challenge of addressing climate change raises an important issue of equity, namely the extent to which the impacts of climate change or mitigation policies create or exacerbate inequities both within and across nations and regions. Greenhouse gas stabilization scenarios assessed in this report (except those where stabilization occurs without new climate policies, e.g. B1) assume that developed countries and countries with economies in transition limit and reduce their greenhouse gas emissions first.[6]

---

[1] Mitigation is defined here as an anthropogenic intervention to reduce the sources of greenhouse gases or enhance their sinks.

[2] *Climate change* in IPCC usage refers to any change in climate over time, whether due to natural variability or as a result of human activity. This usage differs from that in the UNFCCC, where *climate change* refers to a change of climate that is attributed directly or indirectly to human activity that alters the composition of the global atmosphere and that is in addition to natural climate variability observed over comparable time periods.

[3] Section numbers refer to the main body of the Report.

[4] In this report "alternative development paths" refer to a variety of possible scenarios for societal values and consumption and production patterns in all countries, including but not limited to a continuation of today's trends. These paths do not include additional climate initiatives which means that no scenarios are included that explicitly assume implementation of the UNFCCC or the emission targets of the Kyoto Protocol, but do include assumptions about other policies that influence greenhouse gas emissions indirectly.

[5] Approaches to equity have been classified into a variety of categories, including those based on allocation, outcome, process, rights, liability, poverty, and opportunity, reflecting the diverse expectations of fairness used to judge policy processes and the corresponding outcomes (Sections 1.3, 10.2).

[6] Emissions from all regions diverge from baselines at some point. Global emissions diverge earlier and to a greater extent as stabilization levels are lower or underlying scenarios are higher. Such scenarios are uncertain, do not provide information on equity implications and how such changes may be achieved or who may bear any costs incurred.

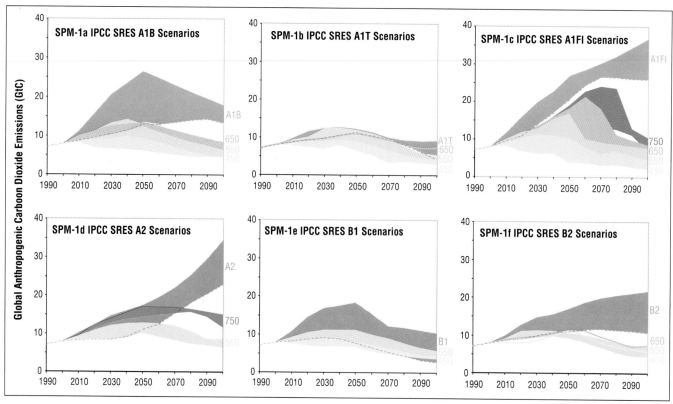

Figure SPM.1: *Comparison of reference and stabilization scenarios. The figure is divided into six parts, one for each of the reference scenario groups from the Special Report on Emissions Scenarios (SRES, see Box SPM.1). Each part of the figure shows the range of total global CO_2 emissions (gigatonnes of carbon (GtC)) from all anthropogenic sources for the SRES reference scenario group (shaded in grey) and the ranges for the various mitigation scenarios assessed in the TAR leading to stabilization of CO_2 concentrations at various levels (shaded in colour). Scenarios are presented for the A1 family subdivided into three groups (the balanced A1B group (Figure SPM.1a), non-fossil fuel A1T (Figure SPM.1b) and the fossil intensive A1FI (Figure SPM.1c)) with stabilization of CO_2 concentrations at 450, 550, 650 and 750 ppmv; for the A2 group with stabilization at 550 and 750 ppmv in Figure SPM.1d, the B1 group with stabilization at 450 and 550 ppmv in Figure SPM.1e, and the B2 group with stabilization at 450, 550 and 650 ppmv in Figure SPM.1f. The literature is not available to assess 1000 ppmv stabilization scenarios. The figure illustrates that the lower the stabilization level and the higher the baseline emissions, the wider the gap. The difference between emissions in different scenario groups can be as large as the gap between reference and stabilization scenarios within one scenario group. The dotted lines depict the boundaries of the ranges where they overlap.*

6. *Lower emissions scenarios require different patterns of energy resource development. Figure SPM.2 compares the cumulative carbon emissions between 1990 and 2100 for various SRES scenarios to carbon contained in global fossil fuel reserves and resources[7]. This figure shows that there are abundant fossil fuel resources that will not limit carbon emissions during the 21st century. However, different from the relatively large coal and unconventional oil and gas deposits, the carbon in proven conventional oil and gas reserves, or in conventional oil resources, is much less than the cumulative carbon emissions associated with stabilization of carbon dioxide at levels of 450 ppmv or higher (the reference to a particular concentration level does not imply an agreed-upon desirability of stabilization at this level). These resource data may imply a change in the energy mix and the introduction of new sources of energy during the 21st century. The choice of energy mix and associated investment will determine whether, and if so, at what level and cost, greenhouse concentrations can be stabilized. Currently most such investment is directed towards discovering and developing more conventional and unconventional fossil resources (Sections 2.5.1, 2.5.2, 3.8.3, 8.4).*

---

[7] Reserves are those occurrences that are identified and measured as economically and technically recoverable with current technologies and prices. Resources are those occurrences with less certain geological and/or economic characteristics, but which are considered potentially recoverable with foreseeable technological and economic developments. The resource base includes both categories. On top of that, there are additional quantities with unknown certainty of occurrence and/or with unknown or no economic significance in the foreseeable future, referred to as "additional occurrences" (SAR, Working Group II). Examples of unconventional fossil fuel resources include tar sands, shale oil, other heavy oil, coal bed methane, deep geopressured gas, gas in acquifers, *etc.*

Box SPM.1. The Emissions Scenarios of the IPCC Special Report on Emissions Scenarios (SRES)

A1. The A1 storyline and scenario family describes a future world of very rapid economic growth, global population that peaks in mid-century and declines thereafter, and the rapid introduction of new and more efficient technologies. Major underlying themes are convergence among regions, capacity building and increased cultural and social interactions, with a substantial reduction in regional differences in per capita income. The A1 scenario family develops into three groups that describe alternative directions of technological change in the energy system. The three A1 groups are distinguished by their technological emphasis: fossil intensive (A1FI), non-fossil energy sources (A1T), or a balance across all sources (A1B) (where balanced is defined as not relying too heavily on one particular energy source, on the assumption that similar improvement rates apply to all energy supply and end use technologies).

A2. The A2 storyline and scenario family describes a very heterogeneous world. The underlying theme is self-reliance and preservation of local identities. Fertility patterns across regions converge very slowly, which results in continuously increasing population. Economic development is primarily regionally oriented and per capita economic growth and technological change more fragmented and slower than other storylines.

B1. The B1 storyline and scenario family describes a convergent world with the same global population, that peaks in mid-century and declines thereafter, as in the A1 storyline, but with rapid change in economic structures toward a service and information economy, with reductions in material intensity and the introduction of clean and resource-efficient technologies. The emphasis is on global solutions to economic, social and environmental sustainability, including improved equity, but without additional climate initiatives.

B2. The B2 storyline and scenario family describes a world in which the emphasis is on local solutions to economic, social and environmental sustainability. It is a world with continuously increasing global population, at a rate lower than A2, intermediate levels of economic development, and less rapid and more diverse technological change than in the B1 and A1 storylines. While the scenario is also oriented towards environmental protection and social equity, it focuses on local and regional levels.

An illustrative scenario was chosen for each of the six scenario groups A1B, A1FI, A1T, A2, B1 and B2. All should be considered equally sound.

The SRES scenarios do not include additional climate initiatives, which means that no scenarios are included that explicitly assume implementation of the United Nations Framework Convention on Climate Change or the emissions targets of the Kyoto Protocol.

Options to Limit or Reduce Greenhouse Gas Emissions and Enhance Sinks

7. Significant technical progress relevant to greenhouse gas emissions reduction has been made since the SAR in 1995 and has been faster than anticipated. Advances are taking place in a wide range of technologies at different stages of development, e.g., the market introduction of wind turbines, the rapid elimination of industrial by-product gases such as N_2O from adipic acid production and perfluorocarbons from aluminium production, efficient hybrid engine cars, the advancement of fuel cell technology, and the demonstration of underground carbon dioxide storage. Technological options for emissions reduction include improved efficiency of end use devices and energy conversion technologies, shift to low-carbon and renewable biomass fuels, zero-emissions technologies, improved energy management, reduction of industrial by-product and process gas emissions, and carbon removal and storage (Section 3.1, 4.7).

Table SPM.1 summarizes the results from many sectoral studies, largely at the project, national and regional level with some at the global levels, providing estimates of potential greenhouse gas emission reductions in the 2010 to 2020 timeframe.

Some key findings are:
- Hundreds of technologies and practices for end-use energy efficiency in buildings, transport and manufacturing industries account for more than half of this potential (Sections 3.3, 3.4, 3.5).
- At least up to 2020, energy supply and conversion will remain dominated by relatively cheap and abundant fossil fuels. Natural gas, where transmission is economically feasible, will play an important role in emission reduction together with conversion efficiency improvement, and greater use of combined cycle and/or co-generation plants (Section 3.8.4).
- Low-carbon energy supply systems can make an important contribution through biomass from forestry and agricultural by-products, municipal and industrial waste to energy, dedicated biomass plantations, where suitable land and water are available, landfill methane, wind energy and hydropower, and through the use and lifetime extension of nuclear power plants. After 2010, emissions from fossil and/or biomass-fueled power plants could be reduced substantially through pre- or post-combustion carbon removal and storage. Environmental, safety, reliability and proliferation concerns may constrain the use of some of these technologies (Section 3.8.4).

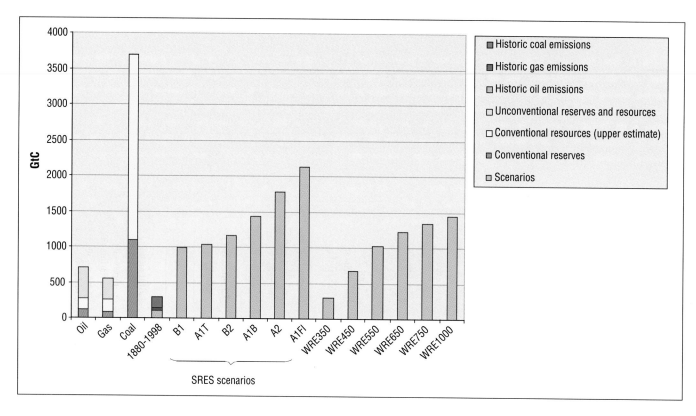

Figure SPM.2: *Carbon in oil, gas and coal reserves and resources compared with historic fossil fuel carbon emissions 1860–1998, and with cumulative carbon emissions from a range of SRES scenarios and TAR stabilization scenarios up until 2100. Data for reserves and resources are shown in the left hand columns (Section 3.8.2). Unconventional oil and gas includes tar sands, shale oil, other heavy oil, coal bed methane, deep geopressured gas, gas in acquifers, etc. Gas hydrates (clathrates) that amount to an estimated 12,000GtC are not shown. The scenario columns show both SRES reference scenarios as well as scenarios which lead to stabilization of CO_2 concentrations at a range of levels. Note that if by 2100 cumulative emissions associated with SRES scenarios are equal to or smaller than those for stabilization scenarios, this does not imply that these scenarios equally lead to stabilization.*

- In agriculture, methane and nitrous oxide emissions can be reduced, such as those from livestock enteric fermentation, rice paddies, nitrogen fertilizer use and animal wastes (Section 3.6).
- Depending on application, emissions of fluorinated gases can be minimized through process changes, improved recovery, recycling and containment, or avoided through the use of alternative compounds and technologies (Section 3.5 and Chapter 3 Appendix).

The potential emissions reductions found in *Table SPM.1* for sectors were aggregated to provide estimates of global potential emissions reductions taking account of potential overlaps between and within sectors and technologies to the extent possible given the information available in the underlying studies. Half of these potential emissions reductions may be achieved by 2020 with direct benefits (energy saved) exceeding direct costs (net capital, operating, and maintenance costs), and the other half at a net direct cost of up to US$100/t$C_{eq}$ (at 1998 prices). These cost estimates are derived using discount rates in the range of 5% to 12%, consistent with public sector discount rates. Private internal rates of return vary greatly, and are often significantly higher, affecting the rate of adoption of these technologies by private entities.

Depending on the emissions scenario this could allow global emissions to be reduced below 2000 levels in 2010–2020 at these net direct costs. Realizing these reductions involve additional implementation costs, which in some cases may be substantial, the possible need for supporting policies (such as those described in Paragraph 18), increased research and development, effective technology transfer and overcoming other barriers (Paragraph 17). These issues, together with costs and benefits not included in this evaluation are discussed in Paragraphs 11, 12 and 13.

The various global, regional, national, sector and project studies assessed in this report have different scopes and assumptions. Studies do not exist for every sector and region. The range of emissions reductions reported in *Table SPM.1* reflects the uncertainties (see *Box SPM.2*) of the underlying studies on which they are based (Sections 3.3-3.8).

Table SPM.1: *Estimates of potential global greenhouse gas emission reductions in 2010 and in 2020 (Sections 3.3-3.8 and Chapter 3 Appendix)*

Sector	Historic emissions in 1990 (MtC$_{eq}$/yr)	Historic C$_{eq}$ annual growth rate in 1990-1995 (%)	Potential emission reductions in 2010 (MtC$_{eq}$/yr)	Potential emission reductions in 2020 (MtC$_{eq}$/yr)	Net direct costs per tonne of carbon avoided
Buildings[a] CO$_2$ only	1,650	1.0	700-750	1,000-1,100	Most reductions are available at negative net direct costs.
Transport CO$_2$ only	1,080	2.4	100-300	300-700	Most studies indicate net direct costs less than US$25/tC but two suggest net direct costs will exceed US$50/tC.
Industry -energy efficiency -material efficiency CO$_2$ only	2,300	0.4	300-500 / ~200	700-900 / ~600	More than half available at net negative direct costs. Costs are uncertain.
Industry Non-CO$_2$ gases	170		~100	~100	N$_2$O emissions reduction costs are US$0-US$10/tC$_{eq}$.
Agriculture[b] CO$_2$ only Non-CO$_2$ gases	210 / 1,250-2,800	n.a	150-300	350-750	Most reductions will cost between US$0-100/tC$_{eq}$. with limited opportunities for negative net direct cost options.
Waste[b] CH$_4$ only	240	1.0	~200	~200	About 75% of the savings as methane recovery from landfills at net negative direct cost; 25% at a cost of US$20/tC$_{eq}$.
Montreal Protocol replacement applications Non-CO$_2$ gases	0	n.a.	~100	n.a.	About half of reductions due to difference in study baseline and SRES baseline values. Remaining half of the reductions available at net direct costs below US$200/tC$_{eq}$.
Energy supply and conversion[c] CO$_2$ only	(1,620)	1.5	50-150	350-700	Limited net negative direct cost options exist; many options are available for less than US$100/tC$_{eq}$.
Total	**6,900-8,400**[d]		**1,900-2,600**[e]	**3,600-5,050**[e]	

a Buildings include appliances, buildings, and the building shell.
b The range for agriculture is mainly caused by large uncertainties about CH$_4$, N$_2$O and soil related emissions of CO$_2$. Waste is dominated by landfill methane and the other sectors could be estimated with more precision as they are dominated by fossil CO$_2$.
c Included in sector values above. Reductions include electricity generation options only (fuel switching to gas/nuclear, CO$_2$ capture and storage, improved power station efficiencies, and renewables).
d Total includes all sectors reviewed in Chapter 3 for all six gases. It excludes non-energy related sources of CO$_2$ (cement production, 160MtC; gas flaring, 60MtC; and land use change, 600-1,400MtC) and energy used for conversion of fuels in the end-use sector totals (630MtC). Note that forestry emissions and their carbon sink mitigation options are not included.
e The baseline SRES scenarios (for six gases included in the Kyoto Protocol) project a range of emissions of 11,500-14,000MtC$_{eq}$ for 2010 and of 12,000-16,000MtC$_{eq}$ for 2020. The emissions reduction estimates are most compatible with baseline emissions trends in the SRES-B2 scenario. The potential reductions take into account regular turn-over of capital stock. They are not limited to cost-effective options, but exclude options with costs above US$100/tC$_{eq}$ (except for Montreal Protocol gases) or options that will not be adopted through the use of generally accepted policies.

8. *Forests, agricultural lands, and other terrestrial ecosystems offer significant carbon mitigation potential. Although not necessarily permanent, conservation and sequestration of carbon may allow time for other options to be further developed and implemented.* Biological mitigation can occur by three strategies: (a) conservation of existing carbon pools, (b) sequestration by increasing the size of carbon pools, and (c) substitution of sustainably produced biological products, e.g. wood for energy intensive construction products and biomass for fossil fuels (Sections 3.6, 4.3). Conservation of threatened carbon pools may help to avoid emissions, if leakage can be prevented, and can only become sustainable if the socio-economic drivers for deforestation and other losses of carbon pools can be addressed. Sequestration reflects the biological dynamics of growth, often starting slowly, passing through a maximum, and then declining over decades to centuries.

Conservation and sequestration result in higher carbon stocks, but can lead to higher future carbon emissions if these ecosystems are severely disturbed by either natural or direct/indirect human-induced disturbances. Even though natural disturbances are normally followed by re-sequestration, activities to manage such disturbances can play an important role in limiting carbon emissions. Substitution benefits can, in principle, continue indefinitely. Appropriate management of land for crop, timber and sustainable bio-energy production, may increase benefits for climate change mitigation. Taking into account competition for land use and the SAR and SRLU-LUCF assessments, the estimated global potential of biological mitigation options is in the order of 100GtC (cumulative), although there are substantial uncertainties associated with this estimate, by 2050, equivalent to about 10% to 20% of potential fossil fuel emissions during that period. Realization of this potential depends upon land and water availability as well as the rates of adoption of different land management practices. The largest biological potential for atmospheric carbon mitigation is in subtropical and tropical regions. Cost estimates reported to date of biological mitigation vary significantly from US$0.1/tC to about US$20/tC in several tropical countries and from US$20/tC to US$100/tC in non-tropical countries. Methods of financial analysis and carbon accounting have not been comparable. Moreover, the cost calculations do not cover, in many instances, inter alia, costs for infrastructure, appropriate discounting, monitoring, data collection and implementation costs, opportunity costs of land and maintenance, or other recurring costs, which are often excluded or overlooked. The lower end of the ranges are biased downwards, but understanding and treatment of costs is improving over time. These biological mitigation options may have social, economic and environmental benefits beyond reductions in atmospheric CO_2, if implemented appropriately (e.g., biodiversity, watershed protection, enhancement of sustainable land management and rural employment). However, if implemented inappropriately, they may pose risks of negative impacts (e.g., loss of biodiversity, community disruption and ground-water pollution). Biological mitigation options may reduce or increase non-CO_2 greenhouse gas emissions (Sections 4.3, 4.4).

9. *There is no single path to a low emission future and countries and regions will have to choose their own path. Most model results indicate that known technological options[8] could achieve a broad range of atmospheric CO_2 stabilization levels, such as 550ppmv, 450ppmv or below over the next 100 years or more, but implementation would require associated socio-economic and institutional changes.* To achieve stabilization at these levels, the scenarios suggest that a very significant reduction in world carbon emissions per unit of GDP from 1990 levels will be necessary. Technological improvement and technology transfer play a critical role in the stabilization scenarios assessed in this report. For the crucial energy sector, almost all greenhouse gas mitigation and concentration stabilization scenarios are characterized by the introduction of efficient technologies for both energy use and supply, and of low- or no-carbon energy. However, no single technology option will provide all of the emissions reductions needed. Reduction options in non-energy sources and non-CO_2 greenhouse gases will also provide significant potential for reducing emissions. Transfer of technologies between countries and regions will widen the choice of options at the regional level and economies of scale and learning will lower the costs of their adoption (Sections 2.3.2, 2.4, 2.5).

10. *Social learning and innovation, and changes in institutional structure could contribute to climate change mitigation.* Changes in collective rules and individual behaviours may have significant effects on greenhouse gas emissions, but take place within a complex institutional, regulatory and legal setting. Several studies suggest that current incentive systems can encourage resource intensive production and consumption patterns that increase greenhouse gas emissions in all sectors, e.g. transport and housing. In the shorter term, there are opportunities to influence through social innovations individual and organizational behaviours. In the longer term such innovations, in combination with technological change, may further enhance socio-economic potential, particularly if preferences and cultural norms shift towards lower emitting and sustainable behaviours. These innovations frequently meet with resistance, which may be addressed by encouraging greater public participation in the decision-making processes. This can help contribute to new approaches to sustainability and equity (Sections 1.4.3, 5.3.8, 10.3.2, 10.3.4).

---

[8] "Known technological options" refer to technologies that exist in operation or pilot plant stage today, as referenced in the mitigation scenarios discussed in this report. It does not include any new technologies that will require drastic technological breakthroughs. In this way it can be considered to be a conservative estimate, considering the length of the scenario period.

Box SPM.2.   Approaches to Estimating Costs and Benefits, and their Uncertainties

For a variety of factors, significant differences and uncertainties surround specific quantitative estimates of the costs and benefits of mitigation options. The SAR described two categories of approaches to estimating costs and benefits: bottom-up approaches, which build up from assessments of specific technologies and sectors, such as those described in Paragraph 7, and top-down modelling studies, which proceed from macroeconomic relationships, such as those discussed in Paragraph 13. These two approaches lead to differences in the estimates of costs and benefits, which have been narrowed since the SAR. Even if these differences were resolved, other uncertainties would remain. The potential impact of these uncertainties can be usefully assessed by examining the effect of a change in any given assumption on the aggregate cost results, provided any correlation between variables is adequately dealt with.

The Costs and Ancillary[9] Benefits of Mitigation Actions

11. *Estimates of cost and benefits of mitigation actions differ because of (i) how welfare is measured, (ii) the scope and methodology of the analysis, and (iii) the underlying assumptions built into the analysis. As a result, estimated costs and benefits may not reflect the actual costs and benefits of implementing mitigation actions.* With respect to (i) and (ii), costs and benefits estimates, *inter alia*, depend on revenue recycling, and whether and how the following are considered: implementation and transaction cost, distributional impacts, multiple gases, land-use change options, benefits of avoided climate change, ancillary benefits, no regrets opportunities[10] and valuation of externalities and non-market impacts. Assumptions include, *inter alia*:

- Demographic change, the rate and structure of economic growth; increases in personal mobility, technological innovation such as improvements in energy efficiency and the availability of low-cost energy sources, flexibility of capital investments and labour markets, prices, fiscal distortions in the no-policy (baseline) scenario.
- The level and timing of the mitigation target.
- Assumptions regarding implementation measures, e.g. the extent of emissions trading, the Clean Development Mechanism (CDM) and Joint Implementation (JI), regulation, and voluntary agreements[11] and the associated transaction costs.

---

[9] Ancillary benefits are the ancillary, or side effects, of policies aimed exclusively at climate change mitigation. Such policies have an impact not only on greenhouse gas emissions, but also on resource use efficiency, like reduction in emissions of local and regional air pollutants associated with fossil fuel use, and on issues such as transportation, agriculture, land-use practices, employment, and fuel security. Sometimes these benefits are referred to as "ancillary impacts" to reflect that in some cases the benefits may be negative.

[10] In this report, as in the SAR, no regrets opportunities are defined as those options whose benefits such as reduced energy costs and reduced emissions of local/regional pollutants equal or exceed their costs to society, excluding the benefits of avoided climate change.

[11] A voluntary agreement is an agreement between a government authority and one or more private parties, as well as a unilateral commitment that is recognized by the public authority, to achieve environmental objectives or to improve environmental performance beyond compliance.

- Discount rates: the long time scales make discounting assumptions critical and there is still no consensus on appropriate long-term rates, though the literature shows increasing attention to rates that decline over time and hence give more weight to benefits that occur in the long term. These discount rates should be distinguished from the higher rates that private agents generally use in market transactions.

(Sections 7.2, 7.3, 8.2.1, 8.2.2, 9.4)

12. *Some sources of greenhouse gas emissions can be limited at no or negative net social cost to the extent that policies can exploit no regrets opportunities* (Sections 7.3.4, 9.2.1):

- *Market imperfections.* Reduction of existing market or institutional failures and other barriers that impede adoption of cost-effective emission reduction measures, can lower private costs compared to current practice. This can also reduce private costs overall.
- *Ancillary benefits.* Climate change mitigation measures will have effects on other societal issues. For example, reducing carbon emissions in many cases will result in the simultaneous reduction in local and regional air pollution. It is likely that mitigation strategies will also affect transportation, agriculture, land-use practices and waste management and will have an impact on other issues of social concern, such as employment, and energy security. However, not all of the effects will be positive; careful policy selection and design can better ensure positive effects and minimize negative impacts. In some cases, the magnitude of ancillary benefits of mitigation may be comparable to the costs of the mitigating measures, adding to the no regrets potential, although estimates are difficult to make and vary widely (Sections 7.3.3, 8.2.4, 9.2.2-9.2.8, 9.2.10).
- *Double dividend.* Instruments (such as taxes or auctioned permits) provide revenues to the government. If used to finance reductions in existing distortionary taxes ("revenue recycling"), these revenues reduce the economic cost of achieving greenhouse gas reductions. The magnitude of this offset depends on the existing tax structure, type of tax cuts, labour market conditions, and method of recycling. Under some circumstances, it is possible that the economic benefits may exceed the costs of mitigation (Sections 7.3.3, 8.2.2, 9.2.1).

---

13. *The cost estimates for Annex B countries to implement the Kyoto Protocol vary between studies and regions as indicated in Paragraph 11, and depend strongly upon the assumptions regarding the use of the Kyoto mechanisms, and their interactions with domestic measures.* The great majority of global studies reporting and comparing these costs use international energy-economic models. Nine of these studies suggest the following GDP impacts[12] (Sections 7.3.5, 8.3.1, 9.2.3, 10.4.4):

Annex II countries[13]: In the absence of emissions trading between Annex B countries[14], the majority of global studies show reductions in projected GDP of about 0.2% to 2% in 2010 for different Annex II regions. With full emissions trading between Annex B countries, the estimated reductions in 2010 are between 0.1% and 1.1% of projected GDP[15]. These studies encompass a wide range of assumptions as listed in Paragraph 11. Models whose results are reported in this paragraph assume full use of emissions trading without transaction cost. Results for cases that do not allow Annex B trading assume full domestic trading within each region. Models do not include sinks or non-CO_2 greenhouse gases. They do not include the CDM, negative cost options, ancillary benefits, or targeted revenue recycling. For all regions costs are also influenced by the following factors:

- Constraints on the use of Annex B trading, high transaction costs in implementing the mechanisms, and inefficient domestic implementation could raise costs.
- Inclusion in domestic policy and measures of the no regrets possibilities[10] identified in Paragraph 12, use of the CDM, sinks, and inclusion of non-CO_2 greenhouse gases, could lower costs. Costs for individual countries can vary more widely.

The models show that the Kyoto mechanisms are important in controlling risks of high costs in given countries, and thus can complement domestic policy mechanisms. Similarly, they can minimize risks of inequitable international impacts and help to

level marginal costs. The global modelling studies reported above show national marginal costs to meet the Kyoto targets from about US$20/tC up to US$600/tC without trading, and a range from about US$15/tC up to US$150/tC with Annex B trading. The cost reductions from these mechanisms may depend on the details of implementation, including the compatibility of domestic and international mechanisms, constraints, and transaction costs.

Economies in transition: For most of these countries, GDP effects range from negligible to a several per cent increase. This reflects opportunities for energy efficiency improvements not available to Annex II countries. Under assumptions of drastic energy efficiency improvement and/or continuing economic recessions in some countries, the assigned amounts may exceed projected emissions in the first commitment period. In this case, models show increased GDP due to revenues from trading assigned amounts. However, for some economies in transition, implementing the Kyoto Protocol will have similar impact on GDP as for Annex II countries.

14. *Cost-effectiveness studies with a century timescale estimate that the costs of stabilizing CO_2 concentrations in the atmosphere increase as the concentration stabilization level declines. Different baselines can have a strong influence on absolute costs.* While there is a moderate increase in the costs when passing from a 750ppmv to a 550ppmv concentration stabilization level, there is a larger increase in costs passing from 550ppmv to 450ppmv unless the emissions in the baseline scenario are very low. These results, however, do not incorporate carbon sequestration, gases other than CO_2 and did not examine the possible effect of more ambitious targets on induced technological change[16]. Costs associated with each concentration level depend on numerous factors including the rate of discount, distribution of emission reductions over time, policies and measures employed, and particularly the choice of the baseline scenario: for scenarios characterized by a focus on local and regional sustainable development for example, total costs of stabilizing at a particular level are significantly lower than for other scenarios[17] (Sections 2.5.2, 8.4.1, 10.4.6).

---

[12] Many other studies incorporating more precisely the country specifics and diversity of targeted policies provide a wider range of net cost estimates (Section 8.2.2).

[13] Annex II countries: Group of countries included in Annex II to the UNFCCC, including all developed countries in the Organisation of Economic Co-operation and Development.

[14] Annex B countries: Group of countries included in Annex B in the Kyoto Protocol that have agreed to a target for their greenhouse gas emissions, including all the Annex I countries (as amended in 1998) but Turkey and Belarus.

[15] Many metrics can be used to present costs. For example, if the annual costs to developed countries associated with meeting Kyoto targets with full Annex B trading are in the order of 0.5% of GDP, this represents US$125 billion (1000 million) per year, or US$125 per person per year by 2010 in Annex II (SRES assumptions). This corresponds to an impact on economic growth *rates* over ten years of less than 0.1 percentage point.

[16] Induced technological change is an emerging field of inquiry. None of the literature reviewed in TAR on the relationship between the century-scale CO_2 concentrations and costs, reported results for models employing induced technological change. Models with induced technological change under some circumstances show that century-scale concentrations can differ, with similar GDP growth but under different policy regimes (Section 8.4.1.4).

[17] See *Figure SPM.1* for the influence of reference scenarios on the magnitude of the required mitigation effort to reach a given stabilization level.

15. *Under any greenhouse gas mitigation effort, the economic costs and benefits are distributed unevenly between sectors; to a varying degree, the costs of mitigation actions could be reduced by appropriate policies.* In general, it is easier to identify activities, which stand to suffer economic costs compared to those which may benefit, and the economic costs are more immediate, more concentrated and more certain. Under mitigation policies, coal, possibly oil and gas, and certain energy-intensive sectors, such as steel production, are most likely to suffer an economic disadvantage. Other industries including renewable energy industries and services can be expected to benefit in the long term from price changes and the availability of financial and other resources that would otherwise have been devoted to carbon-intensive sectors. Policies such as the removal of subsidies from fossil fuels may increase total societal benefits through gains in economic efficiency, while use of the Kyoto mechanisms could be expected to reduce the net economic cost of meeting Annex B targets. Other types of policies, for example exempting carbon-intensive industries, redistribute the costs but increase total societal costs at the same time. Most studies show that the distributional effects of a carbon tax can have negative income effects on low-income groups unless the tax revenues are used directly or indirectly to compensate such effects (Section 9.2.1).

16. *Emission constraints in Annex I countries have well established, albeit varied "spillover" effects*[18] *on non-Annex I countries* (Sections 8.3.2, 9.3).

- *Oil-exporting, non-Annex I countries: Analyses report costs differently, including, inter alia, reductions in projected GDP and reductions in projected oil revenues*[19]. The study reporting the lowest costs shows reductions of 0.2% of projected GDP with no emissions trading, and less than 0.05% of projected GDP with Annex B emissions trading in 2010[20]. The study reporting the highest costs shows reductions of 25% of projected oil revenues with no emissions trading, and 13% of projected oil revenues with Annex B emissions trading in 2010. These studies do not consider policies and measures[21] other than Annex B emissions trading, that could lessen the impact on non-Annex I, oil-exporting countries, and therefore tend to overstate both the costs to these countries and overall costs.

The effects on these countries can be further reduced by removal of subsidies for fossil fuels, energy tax restructuring according to carbon content, increased use of natural gas, and diversification of the economies of non-Annex I, oil-exporting countries.

- *Other non-Annex I countries: They may be adversely affected by reductions in demand for their exports to OECD nations and by the price increase of those carbon-intensive and other products they continue to import. These countries may benefit from the reduction in fuel prices, increased exports of carbon-intensive products and the transfer of environmentally sound technologies and know-how.* The net balance for a given country depends on which of these factors dominates. Because of these complexities, the breakdown of winners and losers remains uncertain.

- *Carbon leakage*[22]. *The possible relocation of some carbon-intensive industries to non-Annex I countries and wider impacts on trade flows in response to changing prices may lead to leakage in the order of 5%-20%* (Section 8.3.2.2). Exemptions, for example for energy-intensive industries, make the higher model estimates for carbon leakage unlikely, but would raise aggregate costs. The transfer of environmentally sound technologies and know-how, not included in models, may lead to lower leakage and especially on the longer term may more than offset the leakage.

Ways and Means for Mitigation

17. *The successful implementation of greenhouse gas mitigation options needs to overcome many technical, economic, political, cultural, social, behavioural and/or institutional barriers which prevent the full exploitation of the technological, economic and social opportunities of these mitigation options.* The potential mitigation opportunities and types of barriers vary by region and sector, and over time. This is caused by the wide variation in mitigation capacity. The poor in any country are faced with limited opportunities to adopt technologies or change their social behaviour, particularly if they are not part of a cash economy, and most countries could benefit from

---

[18] Spillover effects incorporate only economic effects, not environmental effects.

[19] Details of the six studies reviewed are found in *Table 9.4* of the underlying report.

[20] These estimated costs can be expressed as differences in GDP growth rates over the period 2000–2010. With no emissions trading, GDP growth rate is reduced by 0.02 percentage points/year; with Annex B emissions trading, growth rate is reduced by less than 0.005 percentage points/year.

[21] These policies and measures include: those for non-CO_2 gases and non-energy sources of all gases; offsets from sinks; industry restructuring (e.g., from energy producer to supplier of energy services); use of OPEC's market power; and actions (e.g. of Annex B Parties) related to funding, insurance, and the transfer of technology. In addition, the studies typically do not include the following policies and effects that can reduce the total cost of mitigation: the use of tax revenues to reduce tax burdens or finance other mitigation measures; environmental ancillary benefits of reductions in fossil fuel use; and induced technological change from mitigation policies.

[22] Carbon leakage is defined here as the increase in emissions in non-Annex B countries due to implementation of reductions in Annex B, expressed as a percentage of Annex B reductions.

innovative financing and institutional reform and removing barriers to trade. In the industrialized countries, future opportunities lie primarily in removing social and behavioural barriers; in countries with economies in transition, in price rationalization; and in developing countries, in price rationalization, increased access to data and information, availability of advanced technologies, financial resources, and training and capacity building. Opportunities for any given country, however, might be found in the removal of any combination of barriers (Sections 1.5, 5.3, 5.4).

18. *National responses to climate change can be more effective if deployed as a portfolio of policy instruments to limit or reduce greenhouse gas emissions.* The portfolio of national climate policy instruments may include - according to national circumstances - emissions/carbon/energy taxes, tradable or non-tradable permits, provision and/or removal of subsidies, deposit/refund systems, technology or performance standards, energy mix requirements, product bans, voluntary agreements, government spending and investment, and support for research and development. Each government may apply different evaluation criteria, which may lead to different portfolios of instruments. The literature in general gives no preference for any particular policy instrument. Market based instruments may be cost-effective in many cases, especially where capacity to administer them is developed. Energy efficiency standards and performance regulations are widely used, and may be effective in many countries, and sometimes precede market based instruments. Voluntary agreements have recently been used more frequently, sometimes preceding the introduction of more stringent measures. Information campaigns, environmental labelling, and green marketing, alone or in combination with incentive subsidies, are increasingly emphasized to inform and shape consumer or producer behaviour. Government and/or privately supported research and development is important in advancing the long-term application and transfer of mitigation technologies beyond the current market or economic potential (Section 6.2).

19. *The effectiveness of climate change mitigation can be enhanced when climate policies are integrated with the non-climate objectives of national and sectorial policy development and be turned into broad transition strategies to achieve the long-term social and technological changes required by both sustainable development and climate change mitigation.* Just as climate policies can yield ancillary benefits that improve wellbeing, non-climate policies may produce climate benefits. It may be possible to significantly reduce greenhouse gas emissions by pursuing climate objectives through general socio-economic policies. In many countries, the carbon intensity of energy systems may vary depending on broader programmes for energy infrastructure development, pricing, and tax policies. Adopting state-of-the-art environmentally sound technologies may offer particular opportunity for environmentally sound development while avoiding greenhouse gas intensive

activities. Specific attention can foster the transfer of those technologies to small and medium size enterprises. Moreover, taking ancillary benefits into account in comprehensive national development strategies can lower political and institutional barriers for climate-specific actions (Sections 2.2.3, 2.4.4, 2.4.5, 2.5.1, 2.5.2, 10.3.2, 10.3.4).

20. *Co-ordinated actions among countries and sectors may help to reduce mitigation cost, address competitiveness concerns, potential conflicts with international trade rules, and carbon leakage. A group of countries that wants to limit its collective greenhouse gas emissions could agree to implement well-designed international instruments.* Instruments assessed in this report and being developed in the Kyoto Protocol are emissions trading; Joint Implementation (JI); the Clean Development Mechanism (CDM); other international instruments also assessed in this report include co-ordinated or harmonized emission/carbon/energy taxes; an emission/carbon/energy tax; technology and product standards; voluntary agreements with industries; direct transfers of financial resources and technology; and co-ordinated creation of enabling environments such as reduction of fossil fuel subsidies. Some of these have been considered only in some regions to date (Sections 6.3, 6.4.2, 10.2.7, 10.2.8).

21. *Climate change decision-making is essentially a sequential process under general uncertainty.* The literature suggests that a prudent risk management strategy requires a careful consideration of the consequences (both environmental and economic), their likelihood and society's attitude toward risk. The latter is likely to vary from country to country and perhaps even from generation to generation. This report therefore confirms the SAR finding that the value of better information about climate change processes and impacts and society's responses to them is likely to be great. Decisions about near-term climate policies are in the process of being made while the concentration stabilization target is still being debated. The literature suggests a step-by-step resolution aimed at stabilizing greenhouse gas concentrations. This will also involve balancing the risks of either insufficient or excessive action. The relevant question is not "what is the best course for the next 100 years", but rather "what is the best course for the near term given the expected long-term climate change and accompanying uncertainties" (Section 10.4.3).

22. *This report confirms the finding in the SAR that earlier actions, including a portfolio of emissions mitigation, technology development and reduction of scientific uncertainty, increase flexibility in moving towards stabilization of atmospheric concentrations of greenhouse gases. The desired mix of options varies with time and place.* Economic modelling studies completed since the SAR indicate that a gradual near-term transition from the world's present energy system towards a less carbon-emitting economy minimizes costs associated with

premature retirement of existing capital stock. It also provides time for technology development, and avoids premature lock-in to early versions of rapidly developing low-emission technology. On the other hand, more rapid near-term action would decrease environmental and human risks associated with rapid climatic changes.

It would also stimulate more rapid deployment of existing low-emission technologies, provide strong near-term incentives to future technological changes that may help to avoid lock-in to carbon-intensive technologies, and allow for later tightening of targets should that be deemed desirable in light of evolving scientific understanding (Sections 2.3.2, 2.5.2, 8.4.1, 10.4.2, 10.4.3).

23. *There is an inter-relationship between the environmental effectiveness of an international regime, the cost-effectiveness of climate policies and the equity of the agreement.* Any international regime can be designed in a way that enhances both its efficiency and its equity. The literature assessed in this report on coalition formation in international regimes presents different strategies that support these objectives, including how to make it more attractive to join a regime through appropriate distribution of efforts and provision of incentives. While analysis and negotiation often focus on reducing system costs, the literature also recognizes that the development of an effective regime on climate change must give attention to sustainable development and non-economic issues (Sections 1.3, 10.2).

Gaps in Knowledge

24. *Advances have been made since previous IPCC assessments in the understanding of the scientific, technical, environmental, and economic and social aspects of mitigation of climate change. Further research is required, however, to strengthen future assessments and to reduce uncertainties as far as possible in order that sufficient information is available for policy making about responses to climate change, including research in developing countries.*

The following are high priorities for further narrowing gaps between current knowledge and policy making needs:
  - *Further exploration of the regional, country and sector specific potentials of technological and social innovation options.* This includes research on the short, medium and long-term potential and costs of both CO_2 and non-CO_2, non-energy mitigation options; understanding of technology diffusion across different regions; identifying opportunities in the area of social innovation leading to decreased greenhouse gas emissions; comprehensive analysis of the impact of mitigation measures on carbon flows in and out of the terrestrial system; and some basic inquiry in the area of geo-engineering.
  - *Economic, social and institutional issues related to climate change mitigation in all countries.* Priority areas include: analysis of regionally specific mitigation options and barriers; the implications of equity assessments; appropriate methodologies and improved data sources for climate change mitigation and capacity building in the area of integrated assessment; strengthening future research and assessments, especially in the developing countries.
  - *Methodologies for analysis of the potential of mitigation options and their cost, with special attention to comparability of results.* Examples include: characterizing and measuring barriers that inhibit greenhouse gas-reducing action; making mitigation modelling techniques more consistent, reproducible, and accessible; modelling technology learning; improving analytical tools for evaluating ancillary benefits, e.g. assigning the costs of abatement to greenhouse gases and to other pollutants; systematically analyzing the dependency of costs on baseline assumptions for various greenhouse gas stabilization scenarios; developing decision analytical frameworks for dealing with uncertainty as well as socio-economic and ecological risk in climate policy making; improving global models and studies, their assumptions and their consistency in the treatment and reporting of non-Annex I countries and regions.
  - *Evaluating climate mitigation options in the context of development, sustainability and equity.* Examples include: exploration of alternative development paths, including sustainable consumption patterns in all sectors, including the transportation sector; integrated analysis of mitigation and adaptation; identifying opportunities for synergy between explicit climate policies and general policies promoting sustainable development; integration of intra- and inter-generational equity in climate change mitigation analysis; implications of equity assessments; analysis of scientific, technical and economic implications of options under a wide variety of stabilization regimes.

TECHNICAL SUMMARY

CLIMATE CHANGE 2001: MITIGATION

A Report of Working Group III
of the Intergovernmental Panel on Climate Change

This summary was accepted but not approved in detail at the Sixth Session of IPCC Working Group III (Accra, Ghana • 28 February - 3 March 2001). "Acceptance" of IPCC reports at a session of the Working Group or Panel signifies that the material has not been subject to line-by-line discussion and agreement, but nevertheless presents a comprehensive, objective, and balanced view of the subject matter.

Lead Authors:

Tariq Banuri (Pakistan), Terry Barker (UK), Igor Bashmakov (Russian Federation), Kornelis Blok (Netherlands), John Christensen (Denmark), Ogunlade Davidson (Sierra Leone), Michael Grubb (UK), Kirsten Halsnaes (Denmark), Catrinus Jepma (Netherlands), Eberhard Jochem (Germany), Pekka Kauppi (Finland), Olga Krankina (Russian Federation), Alan Krupnick (USA), Lambert Kuijpers (Netherlands), Snorre Kverndokk (Norway), Anil Markandya (UK), Bert Metz (Netherlands), William R. Moomaw (USA), Jose Roberto Moreira (Brazil), Tsuneyuki Morita (Japan), Jiahua Pan (China), Lynn Price (USA), Richard Richels (USA), John Robinson (Canada), Jayant Sathaye (USA), Rob Swart (Netherlands), Kanako Tanaka (Japan), Tomihiro Taniguchi (Japan), Ferenc Toth (Germany), Tim Taylor (UK), John Weyant (USA)

Review Editor:

Rajendra Pachauri (India)

CONTENTS

1 Scope of the Report

1.1 Background

In 1998, Working Group (WG) III of the Intergovernmental Panel on Climate Change (IPCC) was charged by the IPCC Plenary for the Panel's Third Assessment Report (TAR) to assess the scientific, technical, environmental, economic, and social aspects of the mitigation of climate change. Thus, the mandate of the Working Group was changed from a predominantly disciplinary assessment of the economic and social dimensions on climate change (including adaptation) in the Second Assessment Report (SAR), to an interdisciplinary assessment of the options to control the emissions of greenhouse gases (GHGs) and/or enhance their sinks.

After the publication of the SAR, continued research in the area of mitigation of climate change, which was partly influenced by political changes such as the adoption of the Kyoto Protocol to the United Nations Framework Convention on Climate Change (UNFCCC) in 1997, has been undertaken and is reported on here. The report also draws on a number of IPCC Special Reports[1] and IPCC co-sponsored meetings and Expert Meetings that were held in 1999 and 2000, particularly to support the development of the IPCC TAR. This summary follows the 10 chapters of the report.

1.2 Broadening the Context of Climate Change Mitigation

This chapter places climate change mitigation, mitigation policy, and the contents of the rest of the report in the broader context of development, equity, and sustainability. This context reflects the explicit conditions and principles laid down by the UNFCCC on the pursuit of the ultimate objective of stabilizing greenhouse gas concentrations. The UNFCCC imposes three conditions on the goal of stabilization: namely that it should take place within a time-frame sufficient to "allow ecosystems to adapt naturally to climate change, to ensure that food production is not threatened and to enable economic development to proceed in a sustainable manner" (Art. 2). It also specifies several principles to guide this process: equity, common but differentiated responsibilities, precaution, cost-effective measures, right to sustainable development, and support for an open international economic system (Art. 3).

Previous IPCC assessment reports sought to facilitate this pursuit by comprehensively describing, cataloguing, and comparing technologies and policy instruments that could be used to achieve mitigation of greenhouse gas emissions in a cost-effec-

tive and efficient manner. The present assessment advances this process by including recent analyses of climate change that place policy evaluations in the context of sustainable development. This expansion of scope is consistent both with the evolution of the literature on climate change and the importance accorded by the UNFCCC to sustainable development - including the recognition that "Parties have a right to, and should promote sustainable development" (Art. 3.4). It therefore goes some way towards filling the gaps in earlier assessments.

Climate change involves complex interactions between climatic, environmental, economic, political, institutional, social, and technological processes. It cannot be addressed or comprehended in isolation of broader societal goals (such as equity or sustainable development), or other existing or probable future sources of stress. In keeping with this complexity, a multiplicity of approaches have emerged to analyze climate change and related challenges. Many of these incorporate concerns about development, equity, and sustainability (DES) (albeit partially and gradually) into their framework and recommendations. Each approach emphasizes certain elements of the problem, and focuses on certain classes of responses, including for example, optimal policy design, building capacity for designing and implementing policies, strengthening synergies between climate change mitigation and/or adaptation and other societal goals, and policies to enhance societal learning. These approaches are therefore complementary rather than mutually exclusive.

This chapter brings together three broad classes of analysis, which differ not so much in terms of their ultimate goals as of their points of departure and preferred analytical tools. The three approaches start with concerns, respectively, about efficiency and cost-effectiveness, equity and sustainable development, and global sustainability and societal learning. The difference between the three approaches selected lies in their starting point not in their ultimate goals. Regardless of the starting point of the analysis, many studies try in their own way to incorporate other concerns. For example, many analyses that approach climate change mitigation from a cost-effectiveness perspective try to bring in considerations of equity and sustainability through their treatment of costs, benefits, and welfare. Similarly, the class of studies that are motivated strongly by considerations of inter-country equity tend to argue that equity is needed to ensure that developing countries can pursue their internal goals of sustainable development–a concept that includes the implicit components of sustainability and efficiency. Likewise, analysts focused on concerns of global sustainability have been compelled by their own logic to make a case for global efficiency–often modelled as the decoupling of production from material flows–and social equity. In other words, each of the three perspectives has led writers to search for ways to incorporate concerns that lie beyond their initial starting point. All three classes of analyses look at the relationship of climate change mitigation with all three goals–development, equity, and sustainability–albeit in different and often highly complementary ways. Nevertheless, they frame the issues dif-

---

[1] Notably the Special Report on Aviation and the Global Atmosphere, the Special Report on Methodological and Technological Issues in Technology Transfer, the Special Report on Emissions Scenarios, and the Special Report on Land Use, Land-Use Change and Forestry.

ferently, focus on different sets of causal relationships, use different tools of analysis, and often come to somewhat different conclusions.

There is no presumption that any particular perspective for analysis is most appropriate at any level. Moreover, the three perspectives are viewed here as being highly synergistic. The important changes have been primarily in the types of questions being asked and the kinds of information being sought. In practice, the literature has expanded to add new issues and new tools, subsuming rather than discarding the analyses included in the other perspectives. The range and scope of climate policy analyses can be understood as a gradual broadening of the types and extent of uncertainties that analysts have been willing and able to address.

The first perspective on climate policy analysis is cost effectiveness. It represents the field of conventional climate policy analysis that is well represented in the First through Third Assessments. These analyses have generally been driven directly or indirectly by the question of what is the most cost-effective amount of mitigation for the global economy starting from a particular baseline GHG emissions projection, reflecting a specific set of socio-economic projections. Within this framework, important issues include measuring the performance of various technologies and the removal of barriers (such as existing subsidies) to the implementation of those candidate policies most likely to contribute to emissions reductions. In a sense, the focus of analysis here has been on identifying an efficient pathway through the interactions of mitigation policies and economic development, conditioned by considerations of equity and sustainability, but not primarily guided by them. At this level, policy analysis has almost always taken the existing institutions and tastes of individuals as given; assumptions that might be valid for a decade or two, but may become more questionable over many decades.

The impetus for the expansion in the scope of the climate policy analysis and discourse to include equity considerations was to address not simply the impacts of climate change and mitigation policies on global welfare as a whole, but also of the effects of climate change and mitigation policies on existing inequalities among and within nations. The literature on equity and climate change has advanced considerably over the last two decades, but there is no consensus on what constitutes fairness. Once equity issues were introduced into the assessment agenda, though, they became important components in defining the search for efficient emissions mitigation pathways. The considerable literature that indicated how environmental policies could be hampered or even blocked by those who considered them unfair became relevant. In light of these results, it became clear how and why any widespread perception that a mitigation strategy is unfair would likely engender opposition to that strategy, perhaps to the extent of rendering it non-optimal (or even infeasible, as could be the case if non-Annex I countries never participate). Some cost-effectiveness analyses had, in fact, laid the groundwork for applying this literature by

demonstrating the sensitivity of some equity measures to policy design, national perspective, and regional context. Indeed, cost-effectiveness analyses had even highlighted similar sensitivities for other measures of development and sustainability. As mentioned, the analyses that start from equity concerns have by and large focused on the needs of developing countries, and in particular on the commitment expressed in Article 3.4 of the UNFCCC to the pursuit of sustainable development. Countries differ in ways that have dramatic implications for scenario baselines and the range of mitigation options that can be considered. The climate policies that are feasible, and/or desirable, in a particular country depend significantly on its available resources and institutions, and on its overall objectives including climate change as but one component. Recognizing this heterogeneity may, thus, lead to a different range of policy options than has been considered likely thus far and may reveal differences in the capacities of different sectors that may also enhance appreciation of what can be done by non-state actors to improve their ability to mitigate.

The third perspective is global sustainability and societal learning. While sustainability has been incorporated in the analyses in a number of ways, a class of studies takes the issue of global sustainability as their point of departure. These studies focus on alternative pathways to pursue global sustainability and address issues like decoupling growth from resource flows, for example through eco-intelligent production systems, resource light infrastructure and appropriate technologies, and decoupling wellbeing from production, for example through intermediate performance levels, regionalization of production systems, and changing lifestyles. One popular method for identifying constraints and opportunities within this perspective is to identify future sustainable states and then examine possible transition paths to those states for feasibility and desirability. In the case of developing countries this leads to a number of possible strategies that can depart significantly from those which the developed countries pursued in the past.

1.3   Integrating the Various Perspectives

Extending discussions of how nations might respond to the mitigation challenge so that they include issues of cost-effectiveness and efficiency, distribution narrowly defined, equity more broadly defined, and sustainability, adds enormous complexity to the problem of uncovering how best to respond to the threat of climate change. Indeed, recognizing that these multiple domains are relevant complicates the task assigned to policy-makers and international negotiators by opening their deliberations to issues that lie beyond the boundaries of the climate change problem, *per se*. Their recognition thereby underlines the importance of integrating scientific thought across a wide range of new policy-relevant contexts, but not simply because of some abstract academic or narrow parochial interest advanced by a small set of researchers or nations. Cost-effectiveness, equity, and sustainability have all been identified as critical issues by the drafters of the UNFCCC, and they are an integral part of the charge given to the drafters of the TAR. Integration across the domains of cost-effectiveness, equity, and sustainability is therefore profoundly relevant to policy deliberations according to the letter as well as the spirit of the UNFCCC itself.

The literature being brought to bear on climate change mitigation increasingly shows that policies lying beyond simply reducing GHG emissions from a specified baseline to minimize costs can be extremely effective in abating the emission of GHGs. Therefore, a portfolio approach to policy and analysis would be more effective than exclusive reliance on a narrow set of policy instruments or analytical tools. Besides the flexibility that an expanded range of policy instruments and analytical tools can provide to policymakers for achieving climate objectives, the explicit inclusion of additional policy objectives also increases the likelihood of "buy-in" to climate policies by more participants. In particular, it will expand the range of no regrets[2] options. Finally, it could assist in tailoring policies to short-, medium-, and long-term goals.

In order to be effective, however, a portfolio approach requires weighing the costs and impacts of the broader set of policies according to a longer list of objectives. Climate deliberations need to consider the climate ramifications of policies designed primarily to address a wide range of issues including DES, as well as the likely impacts of climate policies on the achievement of these objectives. As part of this process the opportunity costs and impacts of each instrument are measured against the multiple criteria defined by these multiple objectives. Furthermore, the number of decision makers or stakeholders to be considered is increased beyond national policymakers and international negotiators to include state, local, community, and household agents, as well as non-government organizations (NGOs).

The term "ancillary benefits" is often used in the literature for the ancillary, or secondary, effects of climate change mitigation policies on problems other than GHG emissions, such as reductions in local and regional air pollution, associated with the reduction of fossil fuels, and indirect effects on issues such as transportation, agriculture, land use practices, biodiversity preservation, employment, and fuel security. Sometimes these are referred to as "ancillary impacts", to reflect the fact that in some cases the benefits may be negative[3]. The concept of "mit-

---

[2] In this report, as in the SAR, no regrets options are defined as those options whose benefits such as reduced energy costs and reduced emissions of local/regional pollutants equal or exceed their costs to society, excluding the benefits of avoided climate change. They are also known as negative cost options.

[3] In this report sometimes the term "co-benefits" is also used to indicate the additional benefits of policy options that are implemented for various reasons at the same time, acknowledging that most policies designed to address GHG mitigation also have other, often at least equally important, rationales, e.g., related to objectives of development, sustainability and equity. The benefits of avoided climate change are not covered in ancillary or co-benefits. See also Section 7.2.

igative capacity" is also introduced as a possible way to integrate results derived from the application of the three perspectives in the future. The determinants of the capacity to mitigate climate change include the availability of technological and policy options, and access to resources to underwrite undertaking those options. These determinants are the focus of much of the TAR. The list of determinants is, however, longer than this. Mitigative capacity also depends upon nation-specific characteristics that facilitate the pursuit of sustainable development – e.g., the distribution of resources, the relative empowerment of various segments of the population, the credibility of empowered decision makers, the degree to which climate objectives complement other objectives, access to credible information and analyses, the will to act on that information, the ability to spread risk intra- and inter-generationally, and so on. Given that the determinants of mitigative capacity are essentially the same as those of the analogous concept of adaptive capacity introduced in the WGII Report, this approach may provide an integrated framework for assessing both sets of options.

2 Greenhouse Gas Emissions Scenarios

2.1 Scenarios

A long-term view of a multiplicity of future possibilities is required to consider the ultimate risks of climate change, assess critical interactions with other aspects of human and environmental systems, and guide policy responses. Scenarios offer a structured means of organizing information and gleaning insight on the possibilities.

Each mitigation scenario describes a particular future world, with particular economic, social, and environmental characteristics, and they therefore implicitly or explicitly contain information about DES. Since the difference between reference case scenarios and stabilization and mitigation scenarios is simply the addition of deliberate climate policy, it can be the case that the differences in emissions among different reference case scenarios are greater than those between any one such scenario and its stabilization or mitigation version.

This section presents an overview of three scenario literatures: general mitigation scenarios produced since the SAR, narrative-based scenarios found in the general futures literature, and mitigation scenarios based on the new reference scenarios developed in the IPCC SRES.

2.2 Greenhouse Gas Emissions Mitigation Scenarios

This report considers the results of 519 quantitative emissions scenarios from 188 sources, mainly produced after 1990. The review focuses on 126 mitigation scenarios that cover global emissions and have a time horizon encompassing the coming century. Technological improvement is a critical element in all the general mitigation scenarios.

Based on the type of mitigation, the scenarios fall into four categories: concentration stabilization scenarios, emission stabilization scenarios, safe emission corridor scenarios, and other mitigation scenarios. All the reviewed scenarios include energy-related carbon dioxide (CO_2) emissions; several also include CO_2 emissions from land-use changes and industrial processes, and other important GHGs.

Policy options used in the reviewed mitigation scenarios take into account energy systems, industrial processes, and land use, and depend on the underlying model structure. Most of the scenarios introduce simple carbon taxes or constraints on emissions or concentration levels. Regional targets are introduced in the models with regional disaggregation. Emission permit trading is introduced in more recent work. Some models employ policies of supply-side technology introduction, while others emphasize efficient demand-side technology.

Allocation of emission reduction among regions is a contentious issue. Only some studies, particularly recent ones, make explicit assumptions about such allocations in their scenarios. Some studies offer global emission trading as a mechanism to reduce mitigation costs.

Technological improvement is a critical element in all the general mitigation scenarios.

Detailed analysis of the characteristics of 31 scenarios for stabilization of CO_2 concentrations at 550ppmv[4] (and their baseline scenarios) yielded several insights:

- There is a wide range in baselines, reflecting a diversity of assumptions, mainly with respect to economic growth and low-carbon energy supply. High economic growth scenarios tend to assume high levels of progress in the efficiency of end-use technologies; however, carbon intensity reductions were found to be largely independent of economic growth assumptions. The range of future trends shows greater divergence in scenarios that focus on developing countries than in scenarios that look at developed nations. There is little consensus with respect to future directions in developing regions.

- The reviewed 550ppmv stabilization scenarios vary with respect to reduction time paths and the distribution of emission reductions among regions. Some scenarios suggested that emission trading may lower the overall mitigation cost, and could lead to more mitigation in the non-OECD countries. The range of assumed mitigation policies is very wide. In general, scenarios in

which there is an assumed adoption of high-efficiency measures in the baseline show less scope for further introduction of efficiency measures in the mitigation scenarios. In part this results from model input assumptions, which do not assume major technological breakthroughs. Conversely, baseline scenarios with high carbon intensity reductions show larger carbon intensity reductions in their mitigation scenarios.

Only a small set of studies has reported on scenarios for mitigating non-CO_2 gases. This literature suggests that small reductions of GHG emissions can be accomplished at lower cost by including non-CO_2 gases; that both CO_2 and non-CO_2 emissions would have to be controlled in order to slow the increase of atmospheric temperature sufficiently to achieve climate targets assumed in the studies; and that methane (CH_4) mitigation can be carried out more rapidly, with a more immediate impact on the atmosphere, than CO_2 mitigation.

Generally, it is clear that mitigation scenarios and mitigation policies are strongly related to their baseline scenarios, but no systematic analysis has been published on the relationship between mitigation and baseline scenarios.

2.3   Global Futures Scenarios

Global futures scenarios do not specifically or uniquely consider GHG emissions. Instead, they are more general "stories" of possible future worlds. They can complement the more quantitative emissions scenario assessments, because they consider dimensions that elude quantification, such as governance and social structures and institutions, but which are nonetheless important to the success of mitigation policies. Addressing these issues reflects the different perspectives presented in Section 1: cost-effectiveness and/or efficiency, equity, and sustainability.

A survey of this literature has yielded a number of insights that are relevant to GHG emissions scenarios and sustainable development. First, a wide range of future conditions has been identified by futurists, ranging from variants of sustainable development to collapse of social, economic, and environmental systems. Since future values of the underlying socio-economic drivers of emissions may vary widely, it is important that climate policies should be designed so that they are resilient against widely different future conditions.

Second, the global futures scenarios that show falling GHG emissions tend to show improved governance, increased equity and political participation, reduced conflict, and improved environmental quality. They also tend to show increased energy efficiency, shifts to non-fossil energy sources, and/or shifts to a post-industrial (service-based) economy; population tends to stabilize at relatively low levels, in many cases thanks to increased prosperity, expanded provision of family planning, and improved rights and opportunities for women. A key impli-

---

[4] The reference to a particular concentration level does not imply an agreed-upon desirability of stabilization at this level. The selection of 550ppmv is based on the fact that the majority of studies in the literature analyze this level, and does not imply any endorsement of this level as a target for climate change mitigation policies.

cation is that sustainable development policies can make a significant contribution to emission reduction.

Third, different combinations of driving forces are consistent with low emissions scenarios, which agrees with the SRES findings. The implication of this seems to be that it is important to consider the linkage between climate policy and other policies and conditions associated with the choice of future paths in a general sense.

2.4 Special Report on Emissions Scenarios

Six new GHG emission reference scenario groups (not including specific climate policy initiatives), organized into 4 scenario "families", were developed by the IPCC and published as the Special Report on Emissions Scenarios (SRES). Scenario families A1 and A2 emphasize economic development but differ with respect to the degree of economic and social convergence; B1 and B2 emphasize sustainable development but also differ in terms of degree of convergence (see *Box TS.1*). In all, six models were used to generate the 40 scenarios that comprise the six scenario groups. Six of these scenarios, which should be considered equally sound, were chosen to illustrate the whole set of scenarios. These six scenarios include marker scenarios for each of the worlds as well as two scenarios, A1FI and A1T, which illustrate alternative energy technology developments in the A1 world (see *Figure TS.1*).

The SRES scenarios lead to the following findings:
- Alternative combinations of driving-force variables can lead to similar levels and structure of energy use, land-use patterns, and emissions.
- Important possibilities for further bifurcations in future development trends exist within each scenario family.
- Emissions profiles are dynamic across the range of SRES scenarios. They portray trend reversals and indicate possible emissions cross-over among different scenarios.
- Describing potential future developments involves inherent ambiguities and uncertainties. One and only one possible development path (as alluded to, for instance, in concepts such as "business-as-usual scenario") simply does not exist. The multi-model approach increases the value of the SRES scenario set, since uncertainties in the choice of model input assumptions can be more explicitly separated from the specific model behaviour and related modelling uncertainties.

Box TS.1. The Emissions Scenarios of the IPCC Special Report on Emissions Scenarios (SRES)

A1. The A1 storyline and scenario family describe a future world of very rapid economic growth, global population that peaks in mid-century and declines thereafter, and the rapid introduction of new and more efficient technologies. Major underlying themes are convergence among regions, capacity building, and increased cultural and social interactions, with a substantial reduction in regional differences in per capita income. The A1 scenario family develops into three groups that describe alternative directions of technological change in the energy system. The three A1 groups are distinguished by their technological emphasis: fossil intensive (A1FI), non-fossil energy sources (A1T), or a balance across all sources (A1B) (where balanced is defined as not relying too heavily on one particular energy source, on the assumption that similar improvement rates apply to all energy supply and end-use technologies).

A2. The A2 storyline and scenario family describe a very heterogeneous world. The underlying theme is self-reliance and preservation of local identities. Fertility patterns across regions converge very slowly, which results in a continuously increasing population. Economic development is primarily regionally oriented and per capita economic growth and technological change more fragmented and slower than in other storylines.

B1. The B1 storyline and scenario family describe a convergent world with the same global population, which peaks in mid-century and declines thereafter, as in the A1 storyline, but with rapid change in economic structures towards a service and information economy, with reductions in material intensity and the introduction of clean and resource-efficient technologies. The emphasis is on global solutions to economic, social, and environmental sustainability, including improved equity, but without additional climate initiatives.

B2. The B2 storyline and scenario family describe a world in which the emphasis is on local solutions to economic, social, and environmental sustainability. It is a world with continuously increasing global population, at a rate lower than in A2, intermediate levels of economic development, and less rapid and more diverse technological change than in the B1 and A1 storylines. While the scenario is also oriented towards environmental protection and social equity, it focuses on local and regional levels.

An illustrative scenario was chosen for each of the six scenario groups A1B, A1FI, A1T, A2, B1, and B2. All should be considered equally sound.

The SRES scenarios do not include additional climate initiatives, which means that no scenarios are included that explicitly assume implementation of the United Nations Framework Convention on Climate Change or the emissions targets of the Kyoto Protocol.

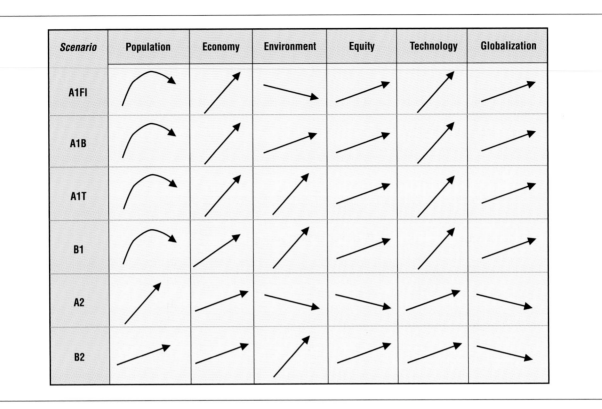

Figure TS.1: *Qualitative directions of SRES scenarios for different indicators.*

2.5   Review of Post-SRES Mitigation Scenarios

Recognizing the importance of multiple baselines in evaluating mitigation strategies, recent studies analyze and compare mitigation scenarios using as their baselines the new SRES scenarios. This allows for the assessment in this report of 76 "post-SRES mitigation scenarios" produced by nine modelling teams. These mitigation scenarios were quantified on the basis of storylines for each of the six SRES scenarios that describe the relationship between the kind of future world and the capacity for mitigation.

Quantifications differ with respect to the baseline scenario, including assumed storyline, the stabilization target, and the model that was used. The post-SRES scenarios cover a very wide range of emission trajectories, but the range is clearly below the SRES range. All scenarios show an increase in CO_2 reduction over time. Energy reduction shows a much wider range than CO_2 reduction, because in many scenarios a decoupling between energy use and carbon emissions takes place as a result of a shift in primary energy sources.

In general, the lower the stabilization target and the higher the level of baseline emissions, the larger the CO_2 divergence from the baseline that is needed, and the earlier that it must occur. The A1FI, A1B, and A2 worlds require a wider range of and more strongly implemented technology and/or policy measures than A1T, B1, and B2. The 450ppmv stabilization case requires more drastic emission reduction to occur earlier than under the

650ppmv case, with very rapid emission reduction over the next 20 to 30 years (see *Figure TS.2*).

A key policy question is what kind of emission reductions in the medium term (after the Kyoto Protocol commitment period) would be needed. Analysis of the post-SRES scenarios (most of which assume developing country emissions to be below baselines by 2020) suggests that stabilization at 450 ppmv will require emissions reductions in Annex I countries after 2012 that go significantly beyond their Kyoto Protocol commitments. It also suggests that it would not be necessary to go much beyond the Kyoto commitments for Annex I by 2020 to achieve stabilization at 550ppmv or higher. However, it should be recognized that several scenarios indicate the need for significant Annex I emission reductions by 2020 and that none of the scenarios introduces other constraints such as a limit to the rate of temperature change.

An important policy question already mentioned concerns the participation of developing countries in emission mitigation. A preliminary finding of the post-SRES scenario analysis is that, if it is assumed that the CO_2 emission reduction needed for stabilization occurs in Annex I countries only, Annex I per capita CO_2 emissions would fall below non-Annex I per capita emissions during the 21st century in nearly all of the stabilization scenarios, and before 2050 in two-thirds of the scenarios, if developing countries emissions follow the baseline scenarios. This suggests that the stabilization target and the baseline emission level are both important determinants of the

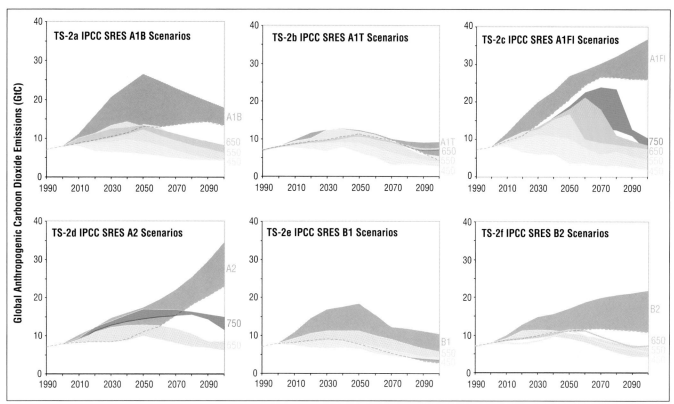

Figure TS.2: *Comparison of reference and stabilization scenarios. The figure is divided into six parts, one for each of the reference scenario groups from the Special Report on Emissions Scenarios (SRES). Each part of the figure shows the range of total global CO₂ emissions (gigatonnes of carbon (GtC)) from all anthropogenic sources for the SRES reference scenario group (shaded in grey) and the ranges for the various mitigation scenarios assessed in the TAR leading to stabilization of CO₂ concentrations at various levels (shaded in colour). Scenarios are presented for the A1 family subdivided into three groups (the balanced A1B group (Figure TS-2a), non-fossil fuel A1T (Figure TS-2b), and the fossil intensive A1FI (Figure TS-2c)) and stabilization of CO₂ concentrations at 450, 550, 650 and 750ppmv; for the A2 group with stabilization at 550 and 750ppmv in Figure TS-2d, the B1 group and stabilization at 450 and 550ppmv in Figure TS-2e, and the B2 group including stabilization at 450, 550, and 650ppmv in Figure TS-2f. The literature is not available to assess 1000ppmv stabilization scenarios. The figure illustrates that the lower the stabilization level and the higher the baseline emissions, the wider the gap. The difference between emissions in different scenario groups can be as large as the gap between reference and stabilization scenarios within one scenario group. The dotted lines depict the boundaries of the ranges where they overlap (see Box TS.1).*

timing when developing countries emissions might need to diverge from their baseline.

Climate policy would reduce per capita final energy use in the economy-emphasized worlds (A1FI, A1B, and A2), but not in the environment-emphasized worlds (B1 and B2). The reduction in energy use caused by climate policies would be larger in Annex I than in non-Annex I countries. However, the impact of climate policies on equity in per capita final energy use would be much smaller than that of the future development path.

There is no single path to a low emission future and countries and regions will have to choose their own path. Most model results indicate that known technological options[5] could achieve a broad range of atmospheric CO₂ stabilization levels, such as 550ppmv, 450ppmv or, below over the next 100 years or more, but implementation would require associated socio-economic and institutional changes..

Assumed mitigation options differ among scenarios and are strongly dependent on the model structure. However, common features of mitigation scenarios include large and continuous energy efficiency improvements and afforestation as well as low-carbon energy, especially biomass over the next 100 years and natural gas in the first half of the 21st century. Energy conservation and reforestation are reasonable first steps, but innovative supply-side technologies will eventually be required. Possible robust options include using natural gas and combined-cycle technology to bridge the transition to more

---

[5] "Known technological options" refer to technologies that exist in operation or pilot plant stage today, as referenced in the mitigation scenarios discussed in this report. It does not include any new technologies that will require drastic technological breakthroughs. In this way it can be considered to be a conservative estimate, considering the length of the scenario period.

advanced fossil fuel and zero-carbon technologies, such as hydrogen fuel cells. Solar energy as well as either nuclear energy or carbon removal and storage would become increasingly important for a higher emission world or lower stabilization target.

Integration between global climate policies and domestic air pollution abatement policies could effectively reduce GHG emissions in developing regions for the next two or three decades. However, control of sulphur emissions could amplify possible climate change, and partial trade-offs are likely to persist for environmental policies in the medium term.

Policies governing agriculture, land use and energy systems could be linked for climate change mitigation. Supply of biomass energy as well as biological CO_2 sequestration would broaden the available options for carbon emission reductions, although the post-SRES scenarios show that they cannot provide the bulk of the emission reductions required. That has to come from other options.

3   Technological and Economic Potential of Mitigation Options

3.1   Key Developments in Knowledge about Technological Options to Mitigate GHG Emissions in the Period up to 2010-2020 since the Second Assessment Report

Technologies and practices to reduce GHG emissions are continuously being developed. Many of these technologies focus on improving the efficiency of fossil fuel energy or electricity use and the development of low carbon energy sources, since the majority of GHG emissions (in terms of CO_2 equivalents) are related to the use of energy. Energy intensity (energy consumed divided by gross domestic product (GDP)) and carbon intensity (CO_2 emitted from burning fossil fuels divided by the amount of energy produced) have been declining for more than 100 years in developed countries without explicit government policies for decarbonization, and have the potential to decline further. Much of this change is the result of a shift away from high carbon fuels such as coal towards oil and natural gas, through energy conversion efficiency improvements and the introduction of hydro and nuclear power. Other non-fossil fuel energy sources are also being developed and rapidly implemented and have a significant potential for reducing GHG emissions. Biological sequestration of CO_2 and CO_2 removal and storage can also play a role in reducing GHG emissions in the future (see also Section 4 below). Other technologies and measures focus on the non-energy sectors for reducing emissions of the remaining major GHGs: CH_4, nitrous oxide (N_2O), hydrofluorocarbons (HFCs), perfluorocarbons (PFCs), and sulphur hexafluoride (SF_6).

Since the SAR several technologies have advanced more rapidly than was foreseen in the earlier analysis. Examples include

the market introduction of efficient hybrid engine cars, rapid advancement of wind turbine design, demonstration of underground carbon dioxide storage, and the near elimination of N_2O emissions from adipic acid production. Greater energy efficiency opportunities for buildings, industry, transportation, and energy supply are available, often at a lower cost than was expected. By the year 2010 most of the opportunities to reduce emissions will still come from energy efficiency gains in the end-use sectors, by switching to natural gas in the electric power sector, and by reducing the release of process GHGs from industry, e.g., N_2O, perfluoromethane (CF_4), and HFCs. By the year 2020, when a proportion of the existing power plants will have been replaced in developed countries and countries with economies in transition (EITs), and when many new plants will become operational in developing countries, the use of renewable sources of energy can begin contributing to the reduction of CO_2 emissions. In the longer term, nuclear energy technologies – with inherent passive characteristics meeting stringent safety, proliferation, and waste storage goals – along with physical carbon removal and storage from fossil fuels and biomass, followed by sequestration, could potentially become available options.

Running counter to the technological and economic potential for GHG emissions reduction are rapid economic development and accelerating change in some socio-economic and behavioural trends that are increasing total energy use, especially in developed countries and high-income groups in developing countries. Dwelling units and vehicles in many countries are growing in size, and the intensity of electrical appliance use is increasing. Use of electrical office equipment in commercial buildings is increasing. In developed countries, and especially the USA, sales of larger, heavier, and less efficient vehicles are also increasing. Continued reduction or stabilization in retail energy prices throughout large portions of the world reduces incentives for the efficient use of energy or the purchase of energy efficient technologies in all sectors. With a few important exceptions, countries have made little effort to revitalize policies or programmes to increase energy efficiency or promote renewable energy technologies. Also since the early 1990s, there has been a reduction in both public and private resources devoted to R&D (research and development) to develop and implement new technologies that will reduce GHG emissions.

In addition, and usually related to technological innovation options, there are important possibilities in the area of social innovation. In all regions, many options are available for lifestyle choices that may improve quality of life, while at the same time decreasing resource consumption and associated GHG emissions. Such choices are very much dependent on local and regional cultures and priorities. They are very closely related to technological changes, some of which can be associated with profound lifestyle changes, while others do not require such changes. While these options were hardly noted in the SAR, this report begins to address them.

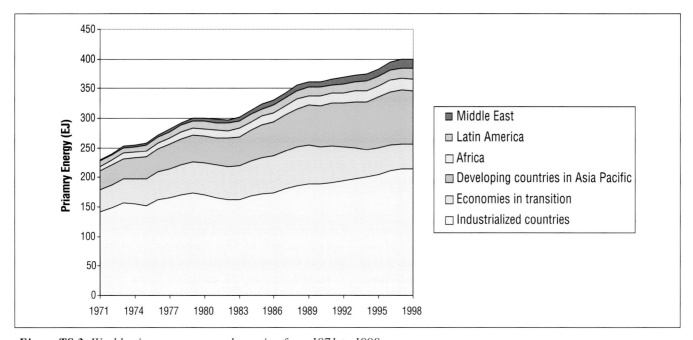

Figure TS.3: *World primary energy use by region from 1971 to 1998.*
Note: Primary energy calculated using the IEA's physical energy content method based on the primary energy sources used to produce heat and electricity.

3.2 Trends in Energy Use and Associated Greenhouse Gas Emissions

Global consumption of energy and associated emission of CO_2 continue an upward trend in the 1990s (*Figures TS.3* and *TS.4*). Fossil fuels remain the dominant form of energy utilized in the world, and energy use accounts for more than two thirds of the GHG emissions addressed by the Kyoto Protocol. In 1998, 143 exajoules (EJ) of oil, 82EJ of natural gas, and 100EJ of coal were consumed by the world's economies. Global primary energy consumption grew an average of 1.3% annually between 1990 and 1998. Average annual growth rates were 1.6% for developed countries and 2.3% to 5.5% for developing countries between 1990 and 1998. Primary energy use for the EITs declined at an annual rate of 4.7% between 1990 and 1998 owing to the loss of heavy industry, the decline in overall economic activity, and restructuring of the manufacturing sector.

Average global carbon dioxide emissions grew – approximately at the same rate as primary energy – at a rate of 1.4% per year between 1990 and 1998, which is much slower than the 2.1% per year growth seen in the 1970s and 1980s. This was in large measure because of the reductions from the EITs and structural changes in the industrial sector of the developed countries. Over the longer term, global growth in CO_2 emissions from energy use was 1.9% per year between 1971 and 1998. In 1998, developed countries were responsible for over 50% of energy-related CO_2 emissions, which grew at a rate of 1.6% annually from 1990. The EITs accounted for 13% of 1998 emissions, and their emissions have been declining at an annual rate of 4.6% per year since 1990. Developing countries in the Asia-Pacific region emitted 22% of the global total carbon dioxide, and have been the fastest growing with increases of 4.9% per year since 1990. The rest of the developing countries accounted for slightly more than 10% of total emissions, growing at an annual rate of 4.3% since 1990.

During the period of intense industrialization from 1860 to 1997, an estimated 13,000EJ of fossil fuel were burned, releasing 290GtC into the atmosphere, which along with land-use change has raised atmospheric concentrations of CO_2 by 30%. By comparison, estimated natural gas resources[6] are comparable to those for oil, being approximately 35,000EJ. The coal resource base is approximately four times as large. Methane clathrates (not counted in the resource base) are estimated to be approximately 780,000EJ. Estimated fossil fuel reserves contain 1,500GtC, being more than 5 times the carbon already released, and if estimated resources are added, there is a total of 5,000GtC remaining in the ground. The scenarios modelled

---

[6] Reserves are those occurrences that are identified and measured as economically and technically recoverable with current technologies and prices. Resources are those occurrences with less certain geological and/or economic characteristics, but which are considered potentially recoverable with foreseeable technological and economic developments. The resource base includes both categories. On top of that there are additional quantities with unknown certainty of occurrence and/or with unknown or no economic significance in the foreseeable future, referred to as "additional occurrences" (SAR). Examples of unconventional fossil fuel resources are tar sands and shale oils, geopressured gas, and gas in aquifers.

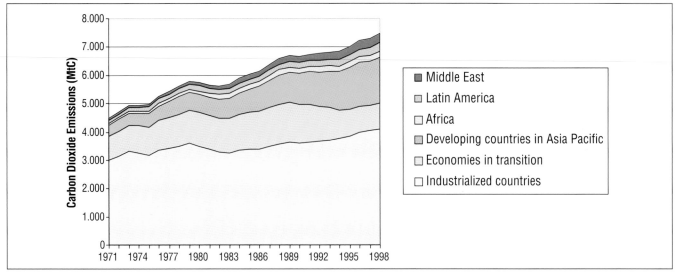

Figure TS.4: *World CO$_2$ emissions by region, 1971-1998.*

by the SRES without any specific GHG emission policies fore-see cumulative release ranging from approximately 1,000 GtC to 2,100 GtC from fossil fuel consumption between 2,000 and 2,100. Cumulative carbon emissions for stabilization profiles of 450 to 750 ppmv over that same period are between 630 and 1,300GtC (see *Figure TS.5*). Fossil-fuel scarcity, at least at the global level, is therefore not a significant factor in considering climate change mitigation. On the contrary, different from the relatively large coal and unconventional oil and gas deposits, the carbon in conventional oil and gas reserves or in conventional oil resources is much less than the cumulative carbon emissions associated with stabilisation at 450 ppmv or higher (*Figure TS.5*). In addition, there is the potential to contribute large quantities of other GHGs as well. At the same time it is clear from *Figure TS.5* that the conventional oil and gas reserves are only a small fraction of the total fossil fuel resource base. These resource data may imply a change in the energy mix and the introduction of new sources of energy during the 21st century. The choice of energy mix and associated investment will determine whether, and if so at what level and cost, greenhouse concentrations can be stabilized. Currently most such investment is directed towards discovering and developing more conventional and unconventional fossil resources.

3.3   Sectoral Mitigation Technological Options[7]

The potential[8] for major GHG emission reductions is estimated for each sector for a range of costs (*Table TS.1*). In the industrial sector, costs for carbon emission abatement are estimated to range from negative (i.e., no regrets, where reductions can be made at a profit), to around US$300/tC[9]. In the buildings sector, aggressive implementation of energy-efficient technologies and measures can lead to a reduction in CO$_2$ emissions from residential buildings in 2010 by 325MtC/yr in

developed and EIT countries at costs ranging from -US$250 to –US$150/tC and by 125MtC in developing countries at costs of –US$250 to US$50/tC. Similarly, CO$_2$ emissions from commercial buildings in 2010 can be reduced by 185MtC in developed and EIT countries at costs ranging from –US$400 to –US$250/tC avoided and by 80MtC in developing countries at costs ranging from -US$400 to US$0/tC. In the transport sector costs range from –US$200/tC to US$300/tC, and in the agricultural sector from –US$100/tC to US$300/tC. Materials management, including recycling and landfill gas recovery, can also produce savings at negative to modest costs under US$100/tC. In the energy supply sector a number of fuel switching and technological substitutions are possible at costs from –US$100 to more than US$200/tC. The realization of this potential will be determined by the market conditions as influenced by human and societal preferences and government interventions.

---

[7] International Energy Statistics (IEA) report sectoral data for the industrial and transport sectors, but not for buildings and agriculture, which are reported as "other". In this section, information on energy use and CO$_2$ emissions for these sectors has been estimated using an allocation scheme and based on a standard electricity conversion factor of 33%. In addition, values for the EIT countries are from a different source (British Petroleum statistics). Thus, the sectoral values can differ from the aggregate values presented in section 3.2, although general trends are the same. In general, there is uncertainty in the data for the EITs and for the commercial and residential sub-categories of the buildings sector in all regions.

[8] The potential differs in different studies assessed but the aggregate potential reported in Sections 3 and 4 refers to the socio-economic potential as indicated in *Figure TS.7*.

[9] All costs in US$.

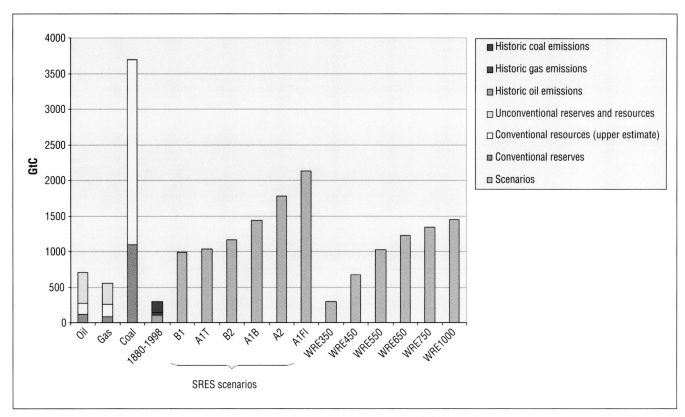

Figure TS.5: *Carbon in oil, gas and coal reserves and resources compared with historic fossil fuel carbon emissions 1860-1998, and with cumulative carbon emissions from a range of SRES scenarios and TAR stabilization scenarios up until 2100. Data for reserves and resources are shown in the left hand columns. Unconventional oil and gas includes tar sands, shale oil, other heavy oil, coal bed methane, deep geopressured gas, gas in acquifers, etc. Gas hydrates (clathrates) that amount to an estimated 12,000 GtC are not shown. The scenario columns show both SRES reference scenarios as well as scenarios which lead to stabilization of CO_2 concentrations at a range of levels. Note that if by 2100 cumulative emissions associated with SRES scenarios are equal to or smaller than those for stabilization scenarios, this does not imply that these scenarios equally lead to stabilization.*

Table TS.2 provides an overview and links with barriers and mitigation impacts. Sectoral mitigation options are discussed in more detail below.

3.3.1   *The Main Mitigation Options in the Buildings Sector*

The buildings sector contributed 31% of global energy-related CO_2 emissions in 1995, and these emissions have grown at an annual rate of 1.8% since 1971. Building technology has continued on an evolutionary trajectory with incremental gains during the past five years in the energy efficiency of windows, lighting, appliances, insulation, space heating, refrigeration, and air conditioning. There has also been continued development of building controls, passive solar design, integrated building design, and the application of photovoltaic systems in buildings. Fluorocarbon emissions from refrigeration and air conditioning applications have declined as chlorofluorocarbons (CFCs) have been phased out, primarily thanks to improved containment and recovery of the fluorocarbon refrigerant and, to a lesser extent, owing to the use of hydrocarbons and other non-fluorocarbon refrigerants. Fluorocarbon use and emission from insulating foams have declined as CFCs have

been phased out, and are projected to decline further as HCFCs are phased out. R&D effort has led to increased efficiency of refrigerators and cooling and heating systems. In spite of the continued improvement in technology and the adoption of improved technology in many countries, energy use in buildings has grown more rapidly than total energy demand from 1971 through 1995, with commercial building energy registering the greatest annual percentage growth (3.0% compared to 2.2% in residential buildings). This is largely a result of the increased amenity that consumers demand – in terms of increased use of appliances, larger dwellings, and the modernization and expansion of the commercial sector – as economies grow. There presently exist significant cost-effective technological opportunities to slow this trend. The overall technical potential for reducing energy-related CO_2 emissions in the buildings sector using existing technologies combined with future technical advances is 715MtC/yr in 2010 for a base case with carbon emissions of 2,600MtC/yr (27%), 950MtC/yr in 2020 for a base case with carbon emissions of 3,000MtC/yr (31%), and 2,025MtC/yr in 2050 for a base case with carbon emissions of 3,900MtC/yr (52%). Expanded R&D can assure continued technology improvement in this sector.

Table TS.1: *Estimations of greenhouse gas emission reductions and cost per tonne of carbon equivalent avoided following the anticipated socio-economic potential uptake by 2010 and 2020 of selected energy efficiency and supply technologies, either globally or by region and with varying degrees of uncertainty*

	Region	US$/tC avoided (−400 −200 0 +200)	2010 Potential[a]	2010 Probability[b]	2020 Potential[a]	2020 Probability[b]	References, comments, and relevant section in Chapter 3 of this report
Buildings / appliances							
Residential sector	OECD/EIT		◆◆◆◆	◇◇◇◇◇	◆◆◆◆◆	◇◇◇◇◇	Acosta Moreno et al., 1996
	Dev. cos.		◆◆◆◆	◇◇◇	◆◆◆◆	◇◇◇◇◇	Brown et al., 1998
							Wang and Smith, 1999
Commercial sector	OECD/EIT		◆◆◆◆	◇◇◇◇	◆◆◆◆	◇◇◇◇	
	Dev. cos.		◆◆◆	◇◇◇	◆◆◆	◇◇◇◇◇	
Transport							
Automobile efficiency improvements	USA		◆◆◆◆	◇◇◇◇	◆◆◆◆	◇◇◇	Interlab. Working Group, 1997
	Europe		◆◆◆◆	◇	◆◆◆◆◆	◇◇	Brown et al., 1998
							US DOE/EIA, 1998
							ECMT, 1997 (8 countries only)
	Japan		◆◆◆◆	◇◇	◆◆◆◆◆	◇◇	Kashiwagi et al., 1999
							Denis and Koopman, 1998
	Dev. cos.		◆◆◆◆	◇	◆◆◆◆◆	◇◇	Worrell et al., 1997b
Manufacturing							
CO₂ removal – fertilizer; refineries	Global		◆	◇◇◇◇	◆	◇◇◇◇	*Table 3.21*
Material efficiency improvement	Global		◆◆◆◆	◇◇◇	◆◆◆◆	◇◇◇	*Table 3.21*
Blended cements	Global		◆	◇◇◇	◆	◇◇◇	*Table 3.21*
N₂O reduction by chem. indus.	Global		◆	◇◇◇◇	◆	◇◇◇	*Table 3.21*
PFC reduction by Al industry	Global		◆	◇◇◇	◆	◇◇◇	*Table 3.21*
HFC-23 reduction by chem. industry	Global		◆◆	◇◇◇	◆◆	◇◇◇	*Table 3.21*
Energy efficient improvements	Global		◆◆◆◆	◇◇◇◇◇	◆◆◆◆	◇◇◇◇	*Table 3.19*

(continued)

Table TS.1: continued

	Region	US $/tC avoided (−400 −200 0 +200)	2010 Potential[a]	2010 Probability[b]	2020 Potential[a]	2020 Probability[b]	References, comments, and relevant section in Chapter 3 of this report
Agriculture							
Increased uptake of conservation tillage and cropland management	Dev. cos.		◆	◇◇	◆	◇◇	Zhou, 1998; *Table 3.27*; Dick *et al.*, 1998; IPCC, 2000
	Global		◆◆◆	◇◇	◆◆◆◆	◇◇◇	
Soil carbon sequestration	Global		◆◆◆	◇◇	◆◆◆◆	◇◇◇	Lal and Bruce, 1999; *Table 3.27*
Nitrogenous fertilizer management	OECD		◆	◇◇◇	◆	◇◇◇	Kroeze & Mosier, 1999; *Table 3.27*
	Global		◆	◇◇◇	◆◆◆	◇◇◇◇	OECD, 1999; IPCC, 2000
Enteric methane reduction	OECD		◆◆◆	◇◇	◆◆	◇◇◇	Kroeze & Mosier, 1999; *Table 3.27*
	USA		◆	◇◇	◆	◇◇◇	OECD, 1998
	Dev. cos.		◆	◇	◆	◇◇	Reimer & Freund, 1999; Chipato, 1999
Rice paddy irrigation and fertilizers	Global		◆◆◆	◇◇	◆◆◆	◇◇◇	Riemer & Freund, 1999; IPCC, 2000
Wastes							
Landfill methane capture	OECD		◆◆◆	◇◇◇	◆◆◆	◇◇◇◇	Landfill methane USEPA, 1999
Energy supply							
Nuclear for coal	Global		◆◆◆	◇◇	◆◆◆◆	◇◇◇◇	Totals[c] – See Section 3.8.6
	Annex I		◆◆	◇◇	◆◆◆	◇◇	*Table 3.35a*
	Non-Annex I		◆◆	◇◇◇	◆◆◆◆	◇◇◇	*Table 3.35b*
Nuclear for gas	Annex I		◆◆◆	◇	◆◆◆◆	◇	*Table 3.35c*
	Non-Annex I		◆	◇	◆◆◆	◇	*Table 3.35d*

(continued)

Table TS.1: continued

Region		US$/tC avoided (−400 to +200)	2010 Potential[a]	2010 Probability[b]	2020 Potential[a]	2020 Probability[b]	References, comments, and relevant section in Chapter 3 of this report
Gas for coal	Annex I	(cost-range bar)	◆	◇◇◇	◆◆◆◆	◇◇◇◇	Table 3.35a
	Non-Annex I	(cost-range bar)	◆	◇◇◇◇	◆◆◆	◇◇◇◇	Tables 3.35b
CO₂ capture from coal	Global	(cost-range bar)	◆	◇◇	◆◆	◇◇	Tables 3.35a + b
CO₂ capture from gas	Global	(cost-range bar)	◆	◇◇	◆◆	◇◇	Tables 3.35c + d
Biomass for coal	Global	(cost-range bar)	◆	◇◇◇◇	◆◆◆	◇◇◇◇	Tables 3.35a + b; Moore, 1998; Interlab w. gp. 1997
Biomass for gas	Global	(cost-range bar)	◆	◇	◆	◇◇◇	Tables 3.35c + d
Wind for coal or gas	Global	(cost-range bar)	◆◆	◇◇◇	◆◆◆	◇◇◇◇	Tables 3.35a - d; BTM Cons 1999; Greenpeace, 1999
Co-fire coal with 10% biomass	USA	(cost-range bar)	◆	◇◇◇	◆◆	◇◇◇	Sulilatu, 1998
Solar for coal	Annex I	(cost-range bar)	◆	◇	◆	◇	Table 3.35a
	Non-Annex I	(cost-range bar)	◆	◇	◆	◇	Table 3.35b
Hydro for coal	Global	(cost-range bar)	◆◆	◇	◆◆	◇◇	Tables 3.35a + b
Hydro for gas	Global	(cost-range bar)	◆	◇	◆◆	◇◇	Tables 3.35c + d

Notes:

[a] Potential in terms of tonnes of carbon equivalent avoided for the cost range of US$/tC given.

◆ = <20 MtC/yr   ◆◆ = 20-50 MtC/yr   ◆◆◆ = 50-100MtC/yr   ◆◆◆◆ = 100-200MtC/yr   ◆◆◆◆◆ = >200 MtC/yr

[b] Probability of realizing this level of potential based on the costs as indicated from the literature.

◇ = Very unlikely   ◇◇ = Unlikely   ◇◇◇ = Possible   ◇◇◇◇ = Probable   ◇◇◇◇◇ = Highly probable

[c] Energy supply total mitigation options assumes that not all the potential will be realized for various reasons including competition between the individual technologies as listed below the totals.

Table TS.2: Technological options, barriers, opportunities, and impacts on production in various sectors

Technological options	Barriers and opportunities	Implications of mitigation policies on sectors
Buildings, households and services: Hundreds of technologies and measures exist that can improve the energy efficiency of appliances and equipment as well as building structures in all regions of the world. It is estimated that CO_2 emissions from residential buildings in 2010 can be reduced by 325MtC in developed countries and the EIT region at costs ranging from -US$250 to -US$150/tC and by 125MtC in developing countries at costs of -US$250 to US$50/tC. Similarly, CO_2 emissions from commercial buildings in 2010 can be reduced by 185MtC in industrialized countries and the EIT region at costs ranging from -US$400 to -US$250/tC and by 80MtC in developing countries at costs ranging from -US$400 to US$0/tC. These savings represent almost 30% of buildings, CO_2 emissions in 2010 and 2020 compared to a central scenario such as the SRES B2 Marker scenario.	Barriers: In developed countries a market structure not conducive to efficiency improvements, misplaced incentives, and lack of information; and in developing countries lack of financing and skills, lack of information, traditional customs, and administered pricing. Opportunities: Developing better marketing approaches and skills, information-based marketing, voluntary programmes and standards have been shown to overcome barriers in developed countries. Affordable credit skills, capacity building, information base and consumer awareness, standards, incentives for capacity building, and deregulation of the energy industry are ways to address the aforementioned barriers in the developing world.	Service industries: Many will gain output and employment depending on how mitigation policies are implemented, however in general the increases are expected to be small and diffused. Households and the informal sector: The impact of mitigation on households comes directly through changes in the technology and price of household's use of energy and indirectly through macroeconomic effects on income and employment. An important ancillary benefit is the improvement in indoor and outdoor air quality, particularly in developing countries and cities all over the world.
Transportation: Transportation technology for light-duty vehicles has advanced more rapidly than anticipated in the SAR, as a consequence of international R&D efforts. Hybrid-electric vehicles have already appeared in the market and introduction of fuel cell vehicles by 2003 has been announced by most major manufacturers. The GHG mitigation impacts of technological efficiency improvements will be diminished to some extent by the rebound effect, unless counteracted by policies that effectively increase the price of fuel or travel. In countries with high fuel prices, such as Europe, the rebound effect may be as large as 40%; in countries with low fuel prices, such as the USA, the rebound appears to be no larger than 20%. Taking into account rebound effects, technological measures can reduce GHG emissions by 5%-15% by 2010 and 15%-35% by 2020, in comparison to a baseline of continued growth.	Barriers: Risk to manufacturers of transportation equipment is an important barrier to more rapid adoption of energy efficient technologies in transport. Achieving significant energy efficiency improvements generally requires a "clean sheet" redesign of vehicles, along with multibillion dollar investments in new production facilities. On the other hand, the value of greater efficiency to customers is the difference between the present value of fuel savings and increased purchase price, which net can often be a small quantity. Although markets for transport vehicles are dominated by a very small number of companies in the technical sense, they are nonetheless highly competitive in the sense that strategic errors can be very costly. Finally, many of the benefits of increased energy efficiency accrue in the form of social rather than private benefits. For all these reasons, the risk to manufacturers of sweeping technological change to improve energy efficiency is generally perceived to outweigh the direct market benefits. Enormous public and private investments in transportation infrastructure and a built environment adapted to motor vehicle travel pose significant barriers to changing the modal structure of transportation in many countries.	Transportation: Growth in transportation demand is projected to remain, influenced by GHG mitigation policies only in a limited way. Only limited opportunities for replacing fossil carbon - based fuels exist in the short to medium term. The main effect of mitigation policies will be to improve energy efficiency in all modes of transportation.

(continued)

Table TS.2: continued

Technological options	Barriers and opportunities	Implications of mitigation policies on sectors
	Opportunities: Information technologies are creating new opportunities for pricing some of the external costs of transportation, from congestion to environmental pollution. Implementation of more efficient pricing can provide greater incentives for energy efficiency in both equipment and modal structure. The factors that hinder the adoption of fuel-efficient technologies in transport vehicle markets create conditions under which energy efficiency regulations, voluntary or mandatory, can be effective. Well-formulated regulations eliminate much of the risk of making sweeping technological changes, because all competitors face the same regulations. Study after study has demonstrated the existence of technologies capable of reducing vehicle carbon intensities by up to 50% or in the longer run 100%, approximately cost-effectively. Finally, intensive R&D efforts for light-duty road vehicles have achieved dramatic improvements in hybrid power-train and fuel cell technologies. Similar efforts could be directed at road freight, air, rail, and marine transport technologies, with potentially dramatic pay-offs.	
Industry: Energy efficiency improvement is the main emission reduction option in industry. Especially in industrialized countries much has been done already to improve energy efficiency, but options for further reductions remain. 300 - 500MtC/yr and 700 -1,100MtC/yr can be reduced by 2010 and 2020, respectively, as compared to a scenario like SRES B2. The larger part of these options has net negative costs. Non-CO_2 emissions in industry are generally relatively small and can be reduced by over 85%, most at moderate or sometimes even negative costs.	Barriers: lack of full-cost pricing, relatively low contribution of energy to production costs, lack of information on part of the consumer and producer, limited availability of capital and skilled personnel are the key barriers to the penetration of mitigation technology in the industrial sector in all, but most importantly in developing countries.

Opportunities: legislation to address local environmental concerns; voluntary agreements, especially if complemented by government efforts; and direct subsidies and tax credits are approaches that have been successful in overcoming the above barriers. Legislation, including standards, and better marketing are particularly suitable approaches for light industries. | Industry: Mitigation is expected to lead to structural change in manufacturing in Annex I countries (partly caused by changing demands in private consumption), with those sectors supplying energy-saving equipment and low-carbon technologies benefitting and energy-intensive sectors having to switch fuels, adopt new technologies, or increase prices. However, rebound effects may lead to unexpected negative results |

(continued)

Table TS.2: continued

Technological options	Barriers and opportunities	Implications of mitigation policies on sectors
Land-use change and forestry: There are three fundamental ways in which land use or management can mitigate atmospheric CO_2 increases: protection, sequestration, and substitution[a]. These options show different temporal patterns; consequently, the choice of options and their potential effectiveness depend on the target time frame as well as on site productivity and disturbance history. The SAR estimated that globally these measures could reduce atmospheric C by about 83 to 131GtC by 2050 (60 to 87GtC in forests and 23 to 44GtC in agricultural soils). Studies published since then have not substantially revised these estimates. The costs of terrestrial management practices are quite low compared to alternatives, and range from 0 ('win-win' opportunities) to US$12/tC.	Barriers: to mitigation in land-use change and forestry include lack of funding and of human and institutional capacity to monitor and verify, social constraints such as food supply, people living off the natural forest, incentives for land clearing, population pressure, and switch to pastures because of demand for meat. In tropical countries, forestry activities are often dominated by the state forest departments with a minimal role for local communities and the private sector. In some parts of the tropical world, particularly Africa, low crop productivity and competing demands on forests for crop production and fuelwood are likely to reduce mitigation opportunities. Opportunities: in land use and forestry, incentives and policies are required to realize the technical potential. There may be in the form of government regulations, taxes, and subsidies, or through economic incentives in the form of market payments for capturing and holding carbon as suggested in the Kyoto Protocol, depending on its implementation following decisions by CoP.	GHG mitigation policies can have a large effect on land use, especially through carbon sequestration and biofuel production. In tropical countries, large-scale adoption of mitigation activities could lead to biodiversity conservation, rural employment generation and watershed protection contributing to sustainable development. To achieve this, institutional changes to involve local communities and industry and necessary thereby leading to a reduced role for governments in managing forests.
Agriculture and waste management: Energy inputs are growing by <1% per year globally with the highest increases in non-OECD countries but they have reduced in the EITs. Several options already exist to decrease GHG emissions for investments of ~US$50 to 150/tC. These include increasing carbon stock by cropland management (125MtC/yr by 2010); reducing CH_4 emissions from better livestock management (>30MtC/yr) and rice production (7MtC/yr); soil carbon sequestration (50-100MtC/yr) and reducing N_2O emissions from animal wastes and application of N measures are feasible in most regions given appropriate technology transfer and incentives for farmers to change their traditional methods. Energy cropping to displace fossil fuels has good prospects if the costs can be made more competitive and the crops are produced sustainably. Improved waste management can decrease GHG emissions by 200MtC$_{eq}$ in 2010 and 320MtC$_{eq}$ in 2020 as compared to 240MtC$_{eq}$ emissions in 1990.	Barriers: In agriculture and waste management, these include inadequate R&D funding, lack of intellectual property rights, lack of national human and institutional capacity and information in the developing countries, farm-level adoption constraints, lack of incentives and information for growers in developed countries to adopt new husbandry techniques, (need other benefits, not just greenhouse gas reduction). Opportunities: Expansion of credit schemes, shifts in research priorities, development of institutional linkages across countries, trading in soil carbon, and integration of food, fibre, and energy products are ways by which the barriers may be overcome. Measures should be linked with moves towards sustainable production methods. Energy cropping provides benefits of land use diversification where suitable land is currently under utilized for food and fibre production and water is readily available.	Energy: forest and land management can provide a variety of solid, liquid, or gaseous fuels that are renewable and that can substitute for fossil fuels. Materials: products from forest and other biological materials are used for construction, packaging, papers, and many other uses and are often less energy-intensive than are alternative materials that provide the same service. Agriculture/land use: commitment of large areas to carbon sequestration or carbon management may compliment or conflict with other demands for land, such as agriculture. GHG mitigation will have an impact on agriculture through increased demand for biofuel production in many regions. Increasing competition for arable land may increase prices of food and other agricultural products.

(continued)

Table TS.2: *continued*

Technological options	Barriers and opportunities	Implications of mitigation policies on sectors
Waste management: Utilization of methane from landfills and from coal beds. The use of landfill gas for heat and electric power is also growing. In several industrial countries and especially in Europe and Japan, waste-to-energy facilities have become more efficient with lower air pollution emissions, paper and fibre recycling, or by utilizing waste paper as a biofuel in waste to energy facilities.	Barriers: Little is being done to manage landfill gas or to reduce waste in rapidly growing markets in much of the developing world. Opportunities: countries like the US and Germany have specific policies to either reduce methane producing waste, and/or requirements to utilize methane from landfills as an energy source. Costs of recovery are negative for half of landfill methane.	Coal: Coal production, use, and employment are likely to fall as a result of greenhouse gas mitigation policies, compared with projections of energy supply without additional climate policies. However, the costs of adjustment will be much lower if policies for new coal production also encourage clean coal technology. Oil: Global mitigation policies are likely to lead to reductions in oil production and trade, with energy exporters likely to face reductions in real incomes as compared to a situation without such policies. The effect on the global oil price of achieving the Kyoto targets, however, may be less severe than many of the models predict, because of the options to include non-CO2 gases and the flexible mechanisms in achieving the target, which are often not included in the models. Gas: Over the next 20 years mitigation may influence the use of natural gas may positively or negatively, depending on regional and local conditions. In the Annex I countries any switch that takes place from coal or oil would be towards natural gas and renewable sources for power generation. In the case of the non-Annex 1 countries, the potential for switching to natural gas is much higher, however energy security and the availability of domestic resources are considerations, particularly for countries such as China and India with large coal reserves.
Energy sector: In the energy sector, options are available both to increase conversion efficiency and to increase the use of primary energy with less GHGs per unit of energy produced, by sequestering carbon, and reducing GHG leakages. Win-win options such as coal bed methane recovery and improved energy efficiency in coal and gas fired power generation as well as co-production of heat and electricity can help to reduce emissions. With economic development continuing, efficiency increases alone will be insufficient to control GHG emissions from the energy sector. Options to decrease emissions per unit energy produced include new renewable forms of energy, which are showing strong growth but still account for less than 1% of energy produced worldwide. Technologies for CO2 capture and disposal to achieve "clean fossil" energy have been proposed and could contribute significantly at costs competitive with renewable energy although considerable research is still needed on the feasibility and possible environmental impacts of such methods to determine their application and usage. Nuclear power and, in some areas, larger scale hydropower could make a substantially increased contribution but face problems of costs and acceptability. Emerging fuel cells are expected to open opportunities for increasing the average energy conversion efficiency in the decades to come.	Barriers: key barriers are human and institutional capacity, imperfect capital markets that discourage investment in small decentralized systems, more uncertain rates of return on investment, high trade tariffs, lack of information, and lack of intellectual property rights for mitigation technologies. For renewable energy, high first costs, lack of access to capital, and subsidies for fossil fuels and key barriers. Opportunities for developing countries include promotion of leapfrogs in energy supply and demand technology, facilitating technology transfer through creating an enabling environment, capacity building, and appropriate mechanisms for transfer of clean and efficient energy technologies. Full cost pricing and information systems provide opportunities in developed countries. Ancillary benefits associated with improved technology, and with reduced production and use of fossil fuels, can be substantial.	

(continued)

Table TS.2: *continued*

Technical options	Barriers and opportunities	Implications of mitigation policies on sectors
		Renewables: Renewable sources are very diverse and the mitigation impact would depend on technological development. It would vary from region to region depending on resource endowment. However, mitigation is very likely to lead to larger markets for the renewables industry. In that situation, R&D for cost reduction and enhanced performance and increased flow of funds to renewables could increase their application leading to cost reduction.

Nuclear: There is substantial technical potential for nuclear power development to reduce greenhouse gas emissions; whether this is realized will depend on relative costs, political factors, and public acceptance. |
| Halocarbons: Emissions of HFCs are growing as HFCs are being used to replace some of the ozone-depleting substances being phased out. Compared to SRES projections for HFCs in 2010, it is estimated that emissions could be lower by as much as 100MtC$_{eq}$ at costs below US\$200/tC$_{eq}$. About half of the estimated reduction is an artifact caused by the SRES baseline values being higher than the study baseline for this report. The remainder could be accomplished by reducing emissions through containment, recovering and recycling refrigerants, and through use of alternative fluids and technologies. | Barriers: uncertainty with respect to the future of HFC policy in relation to global warming and ozone depletion.

Opportunities: capturing new technological developments | |
| Geo-engineering: Regarding mitigation opportunities in marine ecosystems and geo-engineering[b], human understanding of biophysical systems, as well as many ethical, legal, and equity assessments are still rudimentary. | Barriers: In geo-engineering, the risks for unanticipated consequences are large and it may not even be possible to engineer the regional distribution of temperature and precipitation.

Opportunities: Some basic inquiry appears appropriate. | Sector not yet in existence: not applicable. |

a 'Protection' refers to active measures that maintain and preserve existing C reserves, including those in vegetation, soil organic matter, and products exported from the ecosystem (e.g., preventing the conversion of tropical forests for agricultural purposes and avoiding drainage of wetlands). 'Sequestration' refers to measures, deliberately undertaken, that increase C stocks above those already present (e.g., afforestation, revised forest management, enhanced C storage in wood products, and altered cropping systems, including more forage crops, reduced tillage). "Substitution" refers to practices that substitute renewable biological products for fossil fuels or energy-intensive products, thereby avoiding the emission of CO_2 from combustion of fossil fuels.

b Geo-engineering involves efforts to stabilize the climate system by directly managing the energy balance of the earth, thereby overcoming the enhanced greenhouse effect.

3.3.2 The Main Mitigation Options in the Transport Sector

In 1995, the transport sector contributed 22% of global energy-related carbon dioxide emissions; globally, emissions from this sector are growing at a rapid rate of approximately 2.5% annually. Since 1990, principal growth has been in the developing countries (7.3% per year in the Asia–Pacific region) and is actually declining at a rate of 5.0% per year for the EITs. Hybrid gasoline-electric vehicles have been introduced on a commercial basis with fuel economies 50%-100% better than those of comparably sized four-passenger vehicles. Biofuels produced from wood, energy crops, and waste may also play an increasingly important role in the transportation sector as enzymatic hydrolysis of cellulosic material to ethanol becomes more cost effective. Meanwhile, biodiesel, supported by tax exemptions, is gaining market share in Europe. Incremental improvements in engine design have, however, largely been used to enhance performance rather than to improve fuel economy, which has not increased since the SAR. Fuel cell powered vehicles are developing rapidly, and are scheduled to be introduced to the market in 2003. Significant improvements in the fuel economy of aircraft appear to be both technically and economically possible for the next generation fleet. Nevertheless, most evaluations of the technological efficiency improvements (*Table TS.3*) show that because of growth in demand for transportation, efficiency improvement alone is not enough to avoid GHG emission growth. Also, there is evidence that, other things being equal, efforts to improve fuel efficiency have only partial effects in emission reduction because of resulting increases in driving distances caused by lower specific operational costs.

3.3.3 The Main Mitigation Options in the Industry Sector

Industrial emissions account for 43% of carbon released in 1995. Industrial sector carbon emissions grew at a rate of 1.5% per year between 1971 and 1995, slowing to 0.4% per year since 1990. Industries continue to find more energy efficient processes and reductions of process-related GHGs. This is the only sector that has shown an annual decrease in carbon emissions in OECD economies (-0.8%/yr between 1990 and 1995). The CO_2 from EITs declined most strongly (-6.4% per year between 1990 and 1995 when total industrial production dropped).

Differences in the energy efficiency of industrial processes between different developed countries, and between developed and developing countries remain large, which means that there are substantial differences in relative emission reduction potentials between countries.

Improvement of the energy efficiency of industrial processes is the most significant option for lowering GHG emissions. This potential is made up of hundreds of sector-specific technologies. The worldwide potential for energy efficiency improvement – compared to a baseline development – for the year 2010 is estimated to be 300-500MtC and for the year 2020 700-900MtC. In the latter case continued technological development is necessary to realize the potential. The majority of energy efficiency improvement options can be realized at net negative costs.

Another important option is material efficiency improvement (including recycling, more efficient product design, and material substitution); this may represent a potential of 600MtC in the year 2020. Additional opportunities for CO_2 emissions reduction exist through fuel switching, CO_2 removal and storage, and the application of blended cements.

A number of specific processes not only emit CO_2, but also non-CO_2 GHGs. The adipic acid manufacturers have strongly reduced their N_2O emissions, and the aluminium industry has made major gains in reducing the release of PFCs (CF_4, C_2F_6). Further reduction of non-CO_2 GHGs from manufacturing industry to low levels is often possible at relatively low costs per tonne of C-equivalent (tC_{eq}) mitigated.

Sufficient technological options are known today to reduce GHG emissions from industry in absolute terms in most developed countries by 2010, and to limit growth of emissions in this sector in developing countries significantly.

Table TS.3: *Projected energy intensities for transportation from 5-Laboratory Study in the USA[a]*

Determinants	1997	2010		
		BAU	Energy efficiency	HE/LC
New passenger car l/100km	8.6	8.5	6.3	5.5
New light truck l/100km	11.5	11.4	8.7	7.6
Light-duty fleet l/100km[b]	12.0	12.1	10.9	10.1
Aircraft efficiency (seat-l/100km)	4.5	4.0	3.8	3.6
Freight truck fleet l/100km	42.0	39.2	34.6	33.6
Rail efficiency (tonne-km/MJ)	4.2	4.6	5.5	6.2

[a] BAU, Business as usual; HE/LC, high- energy/low-carbon.

[b] Includes existing passenger cars and light trucks.

3.3.4 The Main Mitigation Options in the Agricultural Sector

Agriculture contributes only about 4% of global carbon emissions from energy use, but over 20% of anthropogenic GHG emissions (in terms of MtC_{eq}/yr) mainly from CH_4 and N_2O as well as carbon from land clearing. There have been modest gains in energy efficiency for the agricultural sector since the SAR, and biotechnology developments related to plant and animal production could result in additional gains, provided concerns about adverse environmental effects can be adequately addressed. A shift from meat towards plant production for human food purposes, where feasible, could increase energy efficiency and decrease GHG emissions (especially N_2O and CH_4 from the agricultural sector). Significant abatement of GHG emissions can be achieved by 2010 through changes in agricultural practices, such as:

- soil carbon uptake enhanced by conservation tillage and reduction of land use intensity;
- CH_4 reduction by rice paddy irrigation management, improved fertilizer use, and lower enteric CH_4 emissions from ruminant animals;
- avoiding anthropogenic agricultural N_2O emissions (which for agriculture exceeds carbon emission from fossil fuel use) through the use of slow release fertilizers, organic manure, nitrification inhibitors, and potentially genetically-engineered leguminous plants. N_2O emissions are greatest in China and the USA, mainly from fertilizer use on rice paddy soils and other agricultural soils. More significant contributions can be made by 2020 when more options to control N_2O emissions from fertilized soils are expected to become available.

Uncertainties on the intensity of use of these technologies by farmers are high, since they may have additional costs involved in their uptake. Economic and other barriers may have to be removed through targetted policies.

3.3.5 The Main Mitigation Options in the Waste Management Sector

There has been increased utilization of CH_4 from landfills and from coal beds. The use of landfill gas for heat and electric power is also growing because of policy mandates in countries like Germany, Switzerland, the EU, and USA. Recovery costs are negative for half of landfill CH_4. Requiring product life management in Germany has been extended from packaging to vehicles and electronics goods. If everyone in the USA increased per capita recycling rates from the national average to the per capita recycling rate achieved in Seattle, Washington, the result would be a reduction of 4% of total US GHG emissions. Debate is taking place over whether the greater reduction in lifecycle GHG emissions occurs through paper and fibre recycling or by utilizing waste paper as a biofuel in waste-to-energy facilities. Both options are better than landfilling in

terms of GHG emissions. In several developed countries, and especially in Europe and Japan, waste-to-energy facilities have become more efficient with lower air pollution emissions.

3.3.6 The Main Mitigation Options in the Energy Supply Sector

Fossil fuels continue to dominate heat and electric power production. Electricity generation accounts for 2,100MtC/yr or 37.5% of global carbon emissions[10]. Baseline scenarios without carbon emission policies anticipate emissions of 3,500 and 4,000MtC_{eq} for 2010 and 2020, respectively. In the power sector, low-cost combined cycle gas turbines (CCGTs) with conversion efficiencies approaching 60% for the latest model have become the dominant option for new electric power plants wherever adequate natural gas supply and infrastructure are available. Advanced coal technologies based on integrated gasification combined cycle or supercritical (IGCCS) designs potentially have the capability of reducing emissions at modest cost through higher efficiencies. Deregulation of the electric power sector is currently a major driver of technological choice. Utilization of distributed industrial and commercial combined heat and power (CHP) systems to meet space heating and manufacturing needs could achieve substantial emission reductions. The further implications of the restructuring of the electric utility industry in many developed and developing countries for CO_2 emissions are uncertain at this time, although there is a growing interest in distributed power supply systems based on renewable energy sources and also using fuel cells, micro-turbines and Stirling engines.

The nuclear power industry has managed to increase significantly the capacity factor at existing facilities, which improved their economics sufficiently that extension of facility life has become cost effective. But other than in Asia, relatively few new plants are being proposed or built. Efforts to develop intrinsically safe and less expensive nuclear reactors are proceeding with the goal of lowering socio-economic barriers and reducing public concern about safety, nuclear waste storage, and proliferation. Except for a few large projects in India and China, construction of new hydropower projects has also slowed because of few available major sites, sometimes-high costs, and local environmental and social concerns. Another development is the rapid growth of wind turbines, whose annual growth rate has exceeded 25% per year, and by 2000 exceeded 13GW of installed capacity. Other renewables, including solar and biomass, continue to grow as costs decline, but total contributions from non-hydro renewable sources remain below 2% globally. Fuel cells have the potential to provide highly efficient combined sources of electricity and heat as power densities increase and costs continue to drop. By 2010, co-firing of coal with biomass, gasification of fuel wood, more effi-

[10] Note that the section percentages do not add up to 100% as these emissions have been allocated to the four sectors in the paragraphs above.

cient photovoltaics, off-shore wind farms, and ethanol-based biofuels are some of the technologies that are capable of penetrating the market. Their market share is expected to increase by 2020 as the learning curve reduces costs and capital stock of existing generation plants is replaced.

Physical removal and storage of CO_2 is potentially a more viable option than at the time of the SAR. The use of coal or biomass as a source of hydrogen with storage of the waste CO_2 represents a possible step to the hydrogen economy. CO_2 has been stored in an aquifer, and the integrity of storage is being monitored. However, long-term storage is still in the process of being demonstrated for that particular reservoir. Research is also needed to determine any adverse and/or beneficial environmental impacts and public health risks of uncontrolled release of the various storage options. Pilot CO_2 capture and storage facilities are expected to be operational by 2010, and may be capable of making major contributions to mitigation by 2020. Along with biological sequestration, physical removal and storage might complement current efforts at improving efficiency, fuel switching, and the development of renewables, but must be able to compete economically with them.

The report considers the potential for mitigation technologies in this sector to reduce CO_2 emissions to 2020 from new power plants. CCGTs are expected to be the largest provider of new capacity between now and 2020 worldwide, and will be a strong competitor to displace new coal-fired power stations where additional gas supplies can be made available. Nuclear power has the potential to reduce emissions if it becomes politically acceptable, as it can replace both coal and gas for electricity production. Biomass, based mainly on wastes and agricultural and forestry by-products, and wind power are also potentially capable of making major contributions by 2020. Hydropower is an established technology and further opportunities exist beyond those anticipated to contribute to reducing CO_2 equivalent emissions. Finally, while costs of solar power are expected to decline substantially, it is likely to remain an expensive option by 2020 for central power generation, but it is likely to make increased contributions in niche markets and off-grid generation. The best mitigation option is likely to be dependent on local circumstances, and a combination of these technologies has the potential to reduce CO_2 emissions by 350-700MtC by 2020 compared to projected emissions of around 4,00MtC from this sector.

3.3.7   *The Main Mitigation Options for Hydrofluoro-carbons and Perfluorocarbons*

HFC and, to a lesser extent, PFC use has grown as these chemicals replaced about 8% of the projected use of CFCs by weight in 1997; in the developed countries the production of CFCs and other ozone depleting substances (ODSs) was halted in 1996 to comply with the Montreal Protocol to protect the stratospheric ozone layer. HCFCs have replaced an additional 12% of CFCs. The remaining 80% have been eliminated through controlling emissions, specific use reductions, or alternative technologies and fluids including ammonia, hydrocarbons, carbon dioxide

and water, and not-in-kind technologies. The alternative chosen to replace CFCs and other ODSs varies widely among the applications, which include refrigeration, mobile and stationary air-conditioning, heat pumps, medical and other aerosol delivery systems, fire suppression, and solvents. Simultaneously considering energy efficiency with ozone layer protection is important, especially in the context of developing countries, where markets have just begun to develop and are expected to grow at a fast rate.

Based on current trends and assuming no new uses outside the ODS substitution area, HFC production is projected to be 370 kt or $170MtC_{eq}$/yr by 2010, while PFC production is expected to be less than $12MtC_{eq}$/yr. For the year 2010, annual emissions are more difficult to estimate. The largest emissions are likely to be associated with mobile air conditioning followed by commercial refrigeration and stationary air conditioning. HFC use in foam blowing is currently low, but if HFCs replaces a substantial part of the HCFCs used here, their use is projected to reach $30MtC_{eq}$/yr by 2010, with emissions in the order of 5-$10MtC_{eq}$/yr.

3.4   The Technological and Economic Potential of Greenhouse Gas Mitigation: Synthesis

Global emissions of GHGs grew on average by 1.4% per year during the period 1990 to 1998. In many areas, technical progress relevant to GHG emission reduction since the SAR has been significant and faster than anticipated. The total potential for worldwide GHG emissions reductions resulting from technological developments and their adoption amount to 1,900 to 2,600MtC/yr by 2010, and 3,600 to 5,050MtC/yr by 2020. The evidence on which this conclusion is based is extensive, but has several limitations. No comprehensive worldwide study of technological potential has yet been done, and the existing regional and national studies generally have varying scopes and make different assumptions about key parameters. Therefore, the estimates as presented in *Table TS.1* should be considered to be indicative only. Nevertheless, the main conclusion in the paragraph above can be drawn with high confidence.

Costs of options vary by technology and show regional differences. Half of the potential emissions reductions may be achieved by 2020 with direct benefits (energy saved) exceeding direct costs (net capital, operating, and maintenance costs), and the other half at a net direct cost of up to US$100/$tC_{eq}$ (at 1998 prices). These cost estimates are derived using discount rates in the range of 5% to 12%, consistent with public sector discount rates. Private internal rates of return vary greatly, and are often significantly higher, which affects the rate of adoption of these technologies by private entities. Depending on the emissions scenario this could allow global emissions to be reduced below 2000 levels in 2010-2020 at these net direct costs. Realizing these reductions will involve additional implementation costs, which in some cases may be substantial, and will possibly need supporting policies (such as those described in Section 6), increased research and development, effective technology transfer, and other barriers to be overcome (Section 5 for details).

Hundreds of technologies and practices exist to reduce GHG emissions from the buildings, transport, and industry sectors. These energy efficiency options are responsible for more than half of the total emission reduction potential of these sectors. Efficiency improvements in material use (including recycling) will also become more important in the longer term. The energy supply and conversion sector will remain dominated by cheap and abundant fossil fuels. However, there is significant emission reduction potential thanks to a shift from coal to natural gas, conversion efficiency improvement of power plants, the expansion of distributed co-generation plants in industry, commercial buildings and institutions, and CO_2 recovery and sequestration. The continued use of nuclear power plants (including their lifetime extension), and the application of renewable energy sources could avoid some additional emissions from fossil fuel use. Biomass from by-products and wastes such as landfill gas are potentially important energy sources that can be supplemented by energy crop production where suitable land and water are available. Wind energy and hydropower will also contribute, more so than solar energy because of its relatively high costs. N_2O and fluorinated GHG reductions have already been achieved through major technological advances. Process changes, improved containment and recovery, and the use of alternative compounds and technologies have been implemented. Potential for future reductions exists, including process-related emissions from insulated foam and semiconductor production and by-product emissions from aluminium and HCFC-22. The potential for energy efficiency improvements connected to the use of fluorinated gases is of a similar magnitude to reductions of direct emissions. Soil carbon sequestration, enteric CH_4 control, and conservation tillage can all contribute to mitigating GHG emissions from agriculture.

Appropriate policies are required to realize these potentials. Furthermore, on-going research and development is expected to significantly widen the portfolio of technologies that provide emission reduction options. Maintaining these R&D activities together with technology transfer actions will be necessary if the longer term potential as outlined in *Table TS.1* is to be realized. Balancing mitigation activities in the various sectors with other goals, such as those related to DES, is key to ensuring they are effective.

4 Technological and Economic Potential of Options to Enhance, Maintain and Manage Biological Carbon Reservoirs and Geo-engineering

4.1 Mitigation through Terrestrial Ecosystem and Land Management

Forests, agricultural lands, and other terrestrial ecosystems offer significant, if often temporary, mitigation potential.

Conservation and sequestration allow time for other options to be further developed and implemented. The IPCC SAR estimated that about 60 to 87GtC could be conserved or sequestered in forests by the year 2050 and another 23 to 44GtC could be sequestered in agricultural soils. The current assessment of the potential of biological mitigation options is in the order of 100GtC (cumulative) by 2050, equivalent to about 10% to 20% of projected fossil fuel emissions during that period. In this section, biological mitigation measures in terrestrial ecosystems are assessed, focusing on the mitigation potential, ecological and environmental constraints, economics, and social considerations. Also, briefly, the so-called geo-engineering options are discussed.

Increased carbon pools through the management of terrestrial ecosystems can only partially offset fossil fuel emissions. Moreover, larger C stocks may pose a risk for higher CO_2 emissions in the future, if the C-conserving practices are discontinued. For example, abandoning fire control in forests, or reverting to intensive tillage in agriculture may result in a rapid loss of at least part of the C accumulated during previous years. However, using biomass as a fuel or wood to displace more energy-intensive materials can provide permanent carbon mitigation benefits. It is useful to evaluate terrestrial sequestration opportunities alongside emission reduction strategies, as both approaches will likely be required to control atmospheric CO_2 levels.

Carbon reservoirs in most ecosystems eventually approach some maximum level. The total amount of carbon stored and/or carbon emission avoided by a forest management project at any given time is dependent on the specific management practices (see *Figure TS.6*). Thus, an ecosystem depleted of carbon by past events may have a high potential rate of carbon accumulation, while one with a large carbon pool tends to have a low rate of carbon sequestration. As ecosystems eventually approach their maximum carbon pool, the sink (i.e., the rate of change of the pool) will diminish. Although both the sequestration rate and pool of carbon may be relatively high at some stages, they cannot be maximized simultaneously. Thus, management strategies for an ecosystem may depend on whether the goal is to enhance short-term accumulation or to maintain the carbon reservoirs through time. The ecologically achievable balance between the two goals is constrained by disturbance history, site productivity, and target time frame. For example, options to maximize sequestration by 2010 may not maximize sequestration by 2020 or 2050; in some cases, maximizing sequestration by 2010 may lead to lower carbon storage over time.

The effectiveness of C mitigation strategies, and the security of expanded C pools, will be affected by future global changes, but the impacts of these changes will vary by geographical region, ecosystem type, and local abilities to adapt. For example, increases in atmospheric CO_2, changes in climate, modified nutrient cycles, and altered (either natural or human induced disturbance) regimes can each have negative or positive effects on C pools in terrestrial ecosystems.

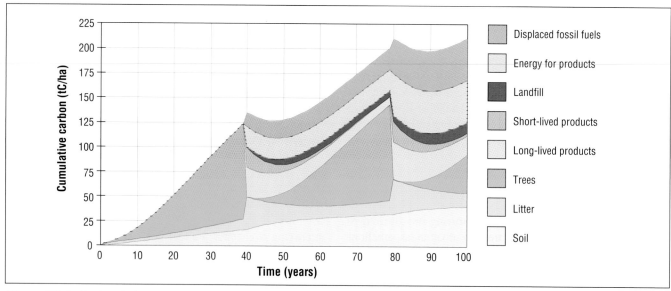

Figure TS.6: *Carbon balance from a hypothetical forest management project.*
Note: The figure shows cumulative carbon-stock changes for a scenario involving afforestation and harvest for a mix of traditional forest products with some of the harvest being used as a fuel. Values are illustrative of what might be observed in the southeastern USA or Central Europe. Regrowth restores carbon to the forest and the (hypothetical) forest stand is harvested every 40 years, with some litter left on the ground to decay, and products accumulate or are disposed of in landfills. These are net changes in that, for example, the diagram shows savings in fossil fuel emissions with respect to an alternative scenario that uses fossil fuels and alternative, more energy-intensive products to provide the same services.

In the past, land management has often resulted in reduced C pools, but in many regions like Western Europe, C pools have now stabilized and are recovering. In most countries in temperate and boreal regions forests are expanding, although current C pools are still smaller than those in pre-industrial or pre-historic times. While complete recovery of pre-historic C pools is unlikely, there is potential for substantial increases in carbon stocks. The Food and Agriculture Organization (FAO) and the UN Economic Commission for Europe (ECE)'s statistics suggest that the average net annual increment exceeded timber fellings in managed boreal and temperate forests in the early 1990s. For example, C stocks in live tree biomass have increased by 0.17GtC/yr in the USA and 0.11GtC/yr in Western Europe, absorbing about 10% of global fossil CO_2 emissions for that time period. Though these estimates do not include changes in litter and soils, they illustrate that land surfaces play a significant and changing role in the atmospheric carbon budget. Enhancing these carbon pools provides potentially powerful opportunities for climate mitigation.

In some tropical countries, however, the average net loss of forest carbon stocks continues, though rates of deforestation may have declined slightly in the past decade. In agricultural lands, options are now available to recover partially the C lost during the conversion from forest or grasslands.

4.2   Social and Economic Considerations

Land is a precious and limited resource used for many purposes in every country. The relationship of climate mitigation strategies with other land uses may be competitive, neutral, or symbiotic. An analysis of the literature suggests that C mitiga-

tion strategies can be pursued as one element of more comprehensive strategies aimed at sustainable development, where increasing C stocks is but one of many objectives. Often, measures can be adopted within forestry, agriculture, and other land uses to provide C mitigation and, at the same time, also advance other social, economic, and environmental goals. Carbon mitigation can provide additional value and income to land management and rural development. Local solutions and targets can be adapted to priorities of sustainable development at national, regional, and global levels.

A key to making C mitigation activities effective and sustainable is to balance it with other ecological and/or environmental, economic, and social goals of land use. Many biological mitigation strategies may be neutral or favourable for all three goals and become accepted as "no regrets" or "win-win" solutions. In other cases, compromises may be needed. Important potential environmental impacts include effects on biodiversity, effects on amount and quality of water resources (particularly where they are already scarce), and long-term impacts on ecosystem productivity. Cumulative environmental, economic, and social impacts could be assessed in individual projects and also from broader, national and international perspectives. An important issue is "leakage" – an expanded or conserved C pool in one area leading to increased emissions elsewhere. Social acceptance at the local, national, and global levels may also influence how effectively mitigation policies are implemented.

4.3   Mitigation Options

In tropical regions there are large opportunities for C mitigation, though they cannot be considered in isolation of broader policies in forestry, agriculture, and other sectors. Additionally, options vary by social and economic conditions: in some regions slowing or halting deforestation is the major mitigation opportunity; in other regions, where deforestation rates have declined to marginal levels, improved natural forest management practices, afforestation, and reforestation of degraded forests and wastelands are the most attractive opportunities. However, the current mitigative capacity[11] is often weak and sufficient land and water is not always available.

Non-tropical countries also have opportunities to preserve existing C pools, enhance C pools, or use biomass to offset fossil fuel use. Examples of strategies include fire or insect control, forest conservation, establishing fast-growing stands, changing silvicultural practices, planting trees in urban areas, ameliorating waste management practices, managing agricultural lands to store more C in soils, improving management of grazing lands, and re-planting grasses or trees on cultivated lands.

Wood and other biological products play several important roles in carbon mitigation: they act as a carbon reservoir; they can replace construction materials that require more fossil fuel input; and they can be burned in place of fossil fuels for renewable energy. Wood products already contribute somewhat to climate mitigation, but if infrastructures and incentives can be developed, wood and agricultural products may become a vital element of a sustainable economy: they are among the few renewable resources available on a large scale.

4.4   Criteria for Biological Carbon Mitigation Options

To develop strategies that mitigate atmospheric CO_2 and advance other, equally important objectives, the following criteria merit consideration:

- potential contributions to C pools over time;
- sustainability, security, resilience, permanence, and robustness of the C pool maintained or created;
- compatibility with other land-use objectives;
- leakage and additionality issues;
- economic costs;
- environmental impacts other than climate mitigation;
- social, cultural, and cross-cutting issues, as well as issues of equity; and
- the system-wide effects on C flows in the energy and materials sector.

Activities undertaken for other reasons may enhance mitigation. An obvious example is reduced rates of tropical deforestation. Furthermore, because wealthy countries generally have a stable forest estate, it could be argued that economic development is associated with activities that build up forest carbon reservoirs.

---

[11] Mitigative capacity: the social, political, and economic structures and conditions that are required for effective mitigation.

4.5   Economic Costs

Most studies suggest that the economic costs of some biological carbon mitigation options, particularly forestry options, are quite modest through a range. Cost estimates of biological mitigation reported to date vary significantly from US\$0.1/tC to about US\$20/tC in several tropical countries and from US\$20 to US\$100/tC in non-tropical countries. Moreover the cost calculations do not cover, in many instances, *inter alia*, costs for infrastructure, appropriate discounting, monitoring, data collection and interpretation, and opportunity costs of land and maintenance, or other recurring costs, which are often excluded or overlooked. The lower end of the ranges are biased downwards, but understanding and treatment of costs is improving over time. Furthermore, in many cases biological mitigation activities may have other positive impacts, such as protecting tropical forests or creating new forests with positive external environmental effects. However, costs rise as more biological mitigation options are exercised and as the opportunity costs of the land increases. Biological mitigation costs appear to be lowest in developing countries and higher in developed countries. If biological mitigation activities are modest, leakage is likely to be small. However, the amount of leakage could rise if biological mitigation activities became large and widespread.

4.6   Marine Ecosystem and Geo-engineering

Marine ecosystems may also offer possibilities for removing CO_2 from the atmosphere. The standing stock of C in the marine biosphere is very small, however, and efforts could focus, not on increasing biological C stocks, but on using biospheric processes to remove C from the atmosphere and transport it to the deep ocean. Some initial experiments have been performed, but fundamental questions remain about the permanence and stability of C removals, and about unintended consequences of the large-scale manipulations required to have a significant impact on the atmosphere. In addition, the economics of such approaches have not yet been determined.

Geo-engineering involves efforts to stabilize the climate system by directly managing the energy balance of the earth, thereby overcoming the enhanced greenhouse effect. Although there appear to be possibilities for engineering the terrestrial energy balance, human understanding of the system is still rudimentary. The prospects of unanticipated consequences are large, and it may not even be possible to engineer the regional distribution of temperature, precipitation, etc. Geo-engineering raises scientific and technical questions as well as many ethical, legal, and equity issues. And yet, some basic inquiry does seem appropriate.

In practice, by the year 2010 mitigation in land use, land-use change, and forestry activities can lead to significant mitigation of CO_2 emissions. Many of these activities are compatible with, or complement, other objectives in managing land. The overall effects of altering marine ecosystems to act as carbon sinks or of applying geo-engineering technology in climate change mitigation remain unresolved and are not, therefore, ready for near-term application.

5   Barriers, Opportunities, and Market Potential of Technologies and Practices

5.1   Introduction

The transfer of technologies and practices that have the potential to reduce GHG emissions is often hampered by barriers[12] that slow their penetration. The opportunity[13] to mitigate GHG concentrations by removing or modifying barriers to or otherwise accelerating the spread of technology may be viewed within a framework of different potentials for GHG mitigation (*Figure TS.7*). Starting at the bottom, one can imagine addressing barriers (often referred to as market failures) that relate to markets, public policies, and other institutions that inhibit the diffusion of technologies that are (or are projected to be) cost-effective for users without reference to any GHG benefits they may generate. Amelioration of this class of "market and institutional imperfections" would increase GHG mitigation towards the level that is labelled as the "economic potential". The economic potential represents the level of GHG mitigation that could be achieved if all technologies that are cost-effective from the consumers' point of view were implemented. Because economic potential is evaluated from the consumer's point of view, we would evaluate cost-effectiveness using market prices and the private rate of time discounting, and also take into account consumers' preferences regarding the acceptability of the technologies' performance characteristics.

Of course, elimination of all these market and institutional barriers would not produce technology diffusion at the level of the "technical potential". The remaining barriers, which define the gap between economic potential and technical potential, are usefully placed in two groups separated by a socio-economic potential. The first group consists of barriers derived from people's preferences and other social and cultural barriers to the diffusion of new technology. That is, even if market and institutional barriers are removed, some GHG-mitigating technologies may not be widely used simply because people do not like them, are too poor to afford them, or because existing social and cultural forces operate against their acceptance. If, in addition to overcoming market and institutional barriers, this second group of barriers could be overcome, what is labelled as the "socio-economic potential" would be achieved.

[12] A barrier is any obstacle to reaching a potential that can be overcome by a policy, programme, or measure.

[13] An opportunity is a situation or circumstance to decrease the gap between the market potential of a technology or practice and the economic, socio-economic, or technological potential.

Thus, the socio-economic potential represents the level of GHG mitigation that would be approached by overcoming social and cultural obstacles to the use of technologies that are cost-effective.

Finally, even if all market, institutional, social, and cultural barriers were removed, some technologies might not be widely used simply because they are too expensive. Elimination of this requirement would therefore take us up to the level of "technological potential", the maximum technologically feasible extent of GHG mitigation through technology diffusion.

An issue arises as to how to treat the relative environmental costs of different technologies within this framework. Because the purpose of the exercise is ultimately to identify opportunities for global climate change policies, the technology potentials are defined without regard to GHG impacts. Costs and benefits associated with other environmental impacts would be part of the cost-effectiveness calculation underlying economic potential only insofar as existing environmental regulations or policies internalize these effects and thereby impose them on consumers. Broader impacts might be ignored by consumers, and hence not enter into the determination of economic potential, but they would be incorporated into a social cost-effectiveness calculation. Thus, to the extent that other environmental benefits make certain technologies socially cost-effective, even if they are not cost-effective from a consumer's point of view, the GHG benefits of diffusion of such technologies would be incorporated in the socio-economic potential.

5.2   Sources of Barriers and Opportunities

Technological and social innovation is a complex process of research, experimentation, learning, and development that can contribute to GHG mitigation. Several theories and models have been developed to understand its features, drivers, and implications. New knowledge and human capital may result from R&D spending, through learning by doing, and/or in an evolutionary process. Most innovations require some social or behavioural change on the part of users. Rapidly changing economies, as well as social and institutional structures offer opportunities for locking in to GHG-mitigative technologies that may lead countries on to sustainable development pathways. The pathways will be influenced by the particular socio-economic context that reflects prices, financing, international trade, market structure, institutions, the provision of information, and social, cultural, and behavioural factors; key elements of these are described below.

Unstable macroeconomic conditions increase risk to private investment and finance. Unsound government borrowing and fiscal policy lead to chronic public deficits and low liquidity in the private sector. Governments may also create perverse microeconomic incentives that the encourage rent-seeking and corruption, rather than the efficient use of resources. Trade barriers that favour inefficient technologies, or prevent access to

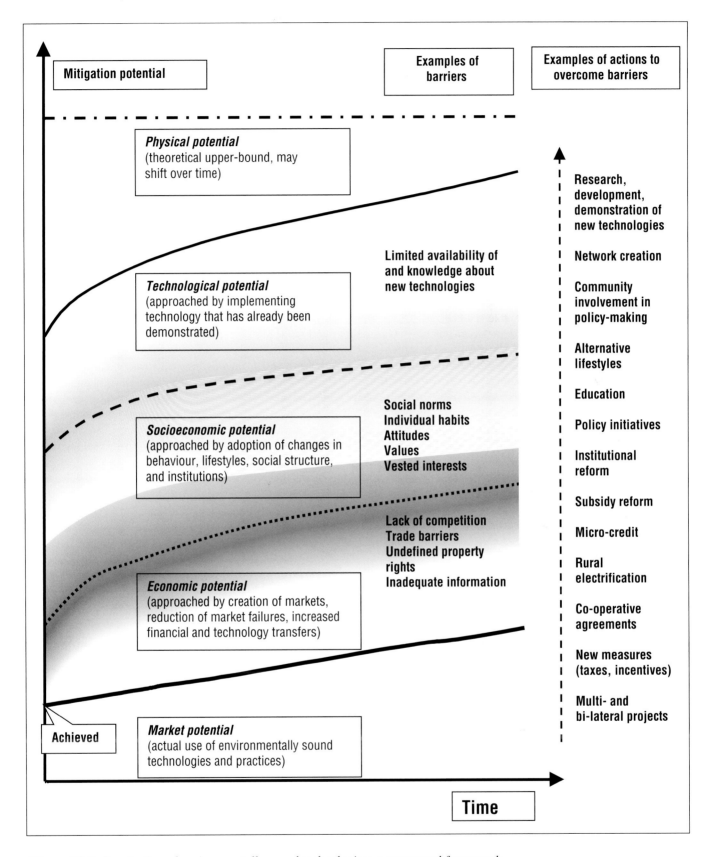

Figure TS.7: *Penetration of environmentally sound technologies: a conceptual framework.*

foreign technology, slow technology diffusion. Tied aid still dominates in official development assistance. It distorts the efficiency of technology choice, and may crowd-out viable business models.

Commercial financing institutions face high risks with developing "green" financial products. Environmentally sound technologies with relatively small project sizes and long repayment periods deter banks with their high transaction costs. Small collateral value makes it difficult to use financing instruments, such as project finance. Innovative approaches in the private sector to address these issues include leasing, environmental and ethical banks, micro-credits or small grants facilities targetted at low income households, environmental funds, energy service companies (ESCOs), and green venture capital. The insurance industry has already begun to react to risks of climate change. New green financial institutions, such as forestry investment funds, have tapped market opportunities by working towards capturing values of standing forests.

Distorted or incomplete prices are also important barriers. The absence of a market price for certain impacts(externalities), such as environmental harm, constitutes a barrier to the diffusion of environmentally beneficial technologies. Distortion of prices because of taxes, subsidies, or other policy interventions that make resource consumption more or less expensive to consumers also impedes the diffusion of resource-conserving technologies.

Network externalities can generate barriers. Some technologies operate in such a way that a given user's equipment interacts with the equipment of other users so as to create "network externalities". For example, the attractiveness of vehicles using alternative fuels depends on the availability of convenient refuelling sites. On the other hand, the development of a fuel distribution infrastructure depends on there being a demand for alternative fuel vehicles.

Misplaced incentives result between landlords and tenants when the tenant is responsible for the monthly cost of fuel and/or electricity, and the landlord is prone to provide the cheapest-first-cost equipment without regard to its monthly energy use. Similar problems are encountered when vehicles are purchased by companies for the use of their employees.

Vested interests: A major barrier to the diffusion of technical progress lies in the vested interests who specialize in conventional technologies and who may, therefore, be tempted to collude and exert political pressure on governments to impose administrative procedures, taxes, trade barriers, and regulations in order to delay or even prevent the arrival of new innovations that might destroy their rents.

Lack of effective regulatory agencies impedes the introduction of environmentally sound technologies. Many countries have excellent constitutional and legal provisions for environmental protection but the latter are not enforced. However, "informal regulation" under community pressure from, for example, non-governmental organizations (NGOs), trade unions, neighbourhood organizations, etc. may substitute for formal regulatory pressure.

Information is often considered *as a public good*. Generic information regarding the availability of different kinds of technologies and their performance characteristics may have the attributes of a "public good" and hence may be underprovided by the private market. This problem is exacerbated by the fact that even after a technology is in place and being used, it is often difficult to quantify the energy savings that resulted from its installation owing to measurement errors and the difficulty with baseline problems. Knowing that this uncertainty will prevail can itself inhibit technology diffusion.

Current lifestyles, behaviours, and consumption patterns have developed within current and historical socio-cultural contexts. Changes in behaviour and lifestyles may result from a number of intertwined processes, such as:

- scientific, technological, and economic developments;
- developments in dominant world views and public discourse;
- changes in the relationships among institutions, political alliances, or actor networks;
- changes in social structures or relationships within firms and households; and
- changes in psychological motivation (e.g., convenience, social prestige, career, *etc.*).

Barriers take various forms in association with each of the above processes.

In some situations policy development is based on a model of human psychology that has been widely criticized. People are assumed to be rational welfare-maximizers and to have a fixed set of values. Such a model does not explain processes, such as learning, habituation, value formation, or the bounded rationality, observed in human choice. Social structures can affect consumption, for example, through the association of objects with status and class. Individuals' adoption of more sustainable consumption patterns depends not only on the match between those patterns and their perceived needs, but also on the extent to which they understand their consumption options, and are able to make choices.

Uncertainty
Another important barrier is uncertainty. A consumer may be uncertain about future energy prices and, therefore, future energy savings. Also, there may be uncertainty about the next generation of equipment – will next year bring a cheaper or better model? In practical decision making, a barrier is often associated with the issue of sunk cost and long lifetimes of infrastructure, and the associated irreversibilities of investments of the non-fungible infrastructure capita.

5.3 Sector- and Technology-specific Barriers and Opportunities

The following sections describe barriers and opportunities particular to each mitigation sector (see also *Table TS.2*).

Buildings: The poor in every country are affected far more by barriers in this sector than the rich, because of inadequate access to financing, low literacy rates, adherence to traditional customs, and the need to devote a higher fraction of their income to satisfy basic needs, including fuel purchases. Other barriers in this sector are lack of skills and social barriers, misplaced incentives, market structure, slow stock turnover, administratively set prices, and imperfect information. Integrated building design for residential construction could lead to energy saving by 40%-60%, which in turn could reduce the cost of living (Section 3.3.4).

Policies, programmes, and measures to remove barriers and reduce energy costs, energy use, and carbon emissions in residential and commercial buildings fall into ten general categories: voluntary programmes, building efficiency standards, equipment efficiency standards, state market transformation programmes, financing, government procurement, tax credits, energy planning (production, distribution, and end-use), and accelerated R&D. Affordable credit financing is widely recognized in Africa as one of the critical measures to remove the high first-cost barrier. Poor macroeconomic management captured by unstable economic conditions often leads to financial repression and higher barriers. As many of several obstacles can be observed simultaneously in the innovation chain of an energy-efficient investment or organizational measure, policy measures usually have to be applied as a bundle to realize the economic potential of a particular technology.

Transport: The car has come to be widely perceived in modern societies as a means of freedom, mobility and safety, a symbol of personal status and identity, and as one of the most important products in the industrial economy. Several studies have found that people living in denser and more compact cities rely less on cars, but it is not easy, even taking congestion problems into account, to motivate the shift away from suburban sprawl to compact cities as advocated in some literature. An integrated approach to town and transport planning and the use of incentives are key to energy efficiency and saving in the transport sector. This is an area, where lock-in effects are very important: when land-use patterns have been chosen there is hardly a way back. This represents an opportunity in particular for the developing world.

Transport fuel taxes are commonly used, but have proved very unpopular in some countries, especially where they are seen as revenue-raising measures. Charges on road users have been accepted where they are earmarked to cover the costs of transport provision. Although trucks and cars may be subject to different barriers and opportunities because of differences in their purpose of use and travel distance, a tax policy that assesses the full cost of GHG emissions would result in a similar impact on CO_2 reductions in road transport. Several studies have explored the potential for adjusting the way existing road taxes, licence fees, and insurance premiums are levied and have found potential emissions reductions of around 10% in OECD countries. Inadequate development and provision of convenient and efficient mass transport systems encourage the use of more energy consuming private vehicles. It is the combination of policies protecting road transport interest, however, that poses the greatest barrier to change, rather than any single type of instrument.

New and used vehicles and/or their technologies mostly flow from the developed to developing countries. Hence, a global approach to reducing emissions that targets technology in developed countries would have a significant impact on future emissions from developing countries.

Industry: In industry, barriers may take many forms, and are determined by the characteristics of the firm (size and structure) and the business environment. Cost-effective energy efficiency measures are often not undertaken as a result of lack of information and high transaction costs for obtaining reliable information. Capital is used for competing investment priorities, and is subject to high hurdle rates for energy efficiency investments. Lack of skilled personnel, especially for small and medium-sized enterprises (SMEs), leads to difficulties installing new energy-efficient equipment compared to the simplicity of buying energy. Other barriers are the difficulty of quantifying energy savings and slow diffusion of innovative technology into markets, while at the same time firms typically underinvest in R&D, despite the high rates of return on investment.

A wide array of policies to reduce barriers, or the perception of barriers, has been used and tested in the industrial sector in developed countries, with varying success rates. Information programmes are designed to assist energy consumers in understanding and employing technologies and practices to use energy more efficiently. Forms of environmental legislation have been a driving force in the adoption of new technologies. New approaches to industrial energy efficiency improvement in developed countries include voluntary agreements (VAs).

In the energy supply sector virtually all the generic barriers cited in Section 5.2 restrict the introduction of environmentally sound technologies and practices. The increasing deregulation of energy supply, while making it more efficient, has raised particular concerns. Volatile spot and contract prices, short-term outlook of private investors, and the perceived risks of nuclear and hydropower plants have shifted fuel and technology choice towards natural gas and oil plants, and away from renewable energy, including – to a lesser extent – hydropower, in many countries.

Co-generation or combined production of power and heat (CHP) is much more efficient than the production of energy for each of these uses alone. The implementation of CHP is closely linked to the availability and density of industrial heat loads, district heating, and cooling networks. Yet, its implementation is hampered by lack of information, the decentralized character of the technology, the attitude of grid operators, the terms of grid connection, and a lack of policies that foster long-term planning. Firm public policy and regulatory authority is necessary to install and safeguard harmonized conditions, transparency, and unbundling of the main power supply functions.

Agriculture and Forestry: Lack of adequate capacity for research and provision of extension services will hamper the spread of technologies that suit local conditions, and the declining Consultative Group on International Agricultural Research (CGIAR) system has exacerbated this problem in the developing world. Adoption of new technology is also limited by small farm size, credit constraints, risk aversion, lack of access to information and human capital, inadequate rural infrastructure and tenurial arrangements, and unreliable supply of complementary inputs. Subsidies for critical inputs to agriculture, such as fertilizers, water supply, and electricity and fuels, and to outputs in order to maintain stable agricultural systems and an equitable distribution of wealth distort markets for these products.

Measures to address the above barriers include:
- The expansion of credit and savings schemes;
- Shifts in international research funding towards water-use efficiency, irrigation design, irrigation management, adaptation to salinity, and the effect of increased CO_2 levels on tropical crops;
- The improvement of food security and disaster early warning systems;
- The development of institutional linkages between countries; and
- The rationalization of input and output prices of agricultural commodities, taking DES issues into consideration.

The forestry sector faces land-use regulation and other macroeconomic policies that usually favour conversion to other land uses such as agriculture, cattle ranching, and urban industry. Insecure land tenure regimes and tenure rights and subsidies favouring agriculture or livestock are among the most important barriers for ensuring sustainable management of forests as well as sustainability of carbon abatement. In relation to climate change mitigation, other issues, such as lack of technical capability, lack of credibility about the setting of project baselines, and monitoring of carbon stocks, poses difficult challenges.

Waste Management: Solid waste and wastewater disposal and treatment represent about 20% of human-induced methane emissions. The principal barriers to technology transfer in this sector include limited financing and institutional capability, jurisdictional complexity, and the need for community involve-

ment. Climate change mitigation projects face further barriers resulting from unfamiliarity with CH_4 capture and potential electricity generation, unwillingness to commit additional human capacity for climate mitigation, and the additional institutional complexity required not only by waste treatment but also byenergy generation and supply. The lack of clear regulatory and investment frameworks can pose significant challenges for project development.

To overcome the barriers and to avail the opportunities in waste management, it is necessary to have a multi-project approach, the components of which include the following :

- Building databases on availability of wastes, their characteristics, distribution, accessibility, current practices of utilization and/or disposal technologies, and economic viability;
- Institutional mechanism for technology transfer though a co-ordinated programme involving the R&D institutions, financing agencies, and industry; and
- Defining the role of stakeholders including local authorities, individual householders, industries, R&D institutions, and the government.

Regional Considerations: Changing global patterns provide an opportunity for introducing GHG mitigation technologies and practices that are consistent with DES goals. A culture of energy subsidies, institutional inertia, fragmented capital markets, vested interests, etc., however, presents major barriers to their implementation, and may be particular issues in developing and EIT countries. Situations in these two groups of countries call for a more careful analysis of trade, institutional, financial, and income barriers and opportunities, distorted prices, and information gaps. In the developed countries, other barriers such as the current carbon-intensive lifestyle and consumption patterns, social structures, network externalities, and misplaced incentives offer opportunities for intervention to control the growth of GHG emissions. Lastly, new and used technologies mostly flow from the developed to developing and transitioning countries. A global approach to reducing emissions that targets technology that is transferred from developed to developing countries could have a significant impact on future emissions.

6   Policies, Measures, and Instruments

6.1   Policy Instruments and Possible Criteria for their Assessment

The purpose of this section is to examine the major types of policies and measures that can be used to implement options to mitigate net concentrations of GHGs in the atmosphere. In keeping within the defined scope of this Report, policies and measures that can be used to implement or reduce the costs of adaptation to climate change are not examined. Alternative policy instruments are discussed and assessed in terms of specific criteria, all on the basis of the most recent literature. There is naturally some emphasis on the instruments mentioned in the Kyoto Protocol (the Kyoto mechanisms), because they are new and focus on achieving GHG emissions limits, and the extent of their envisaged international application is unprecedented. In addition to economic dimensions, political economy, legal, and institutional elements are discussed insofar as they are relevant to these policies and measures.

Any individual country can choose from a large set of possible policies, measures, and instruments, including (in arbitrary order): emissions, carbon, or energy taxes, tradable permits, subsidies, deposit-refund systems, voluntary agreements, non-tradable permits, technology and performance standards, product bans, and direct government spending, including R&D investment. Likewise, a group of countries that wants to limit its collective GHG emissions could agree to implement one, or a mix, of the following instruments (in arbitrary order): tradable quotas, joint implementation, clean development mechanism, harmonized emissions or carbon or energy taxes, an international emissions, carbon, or energy tax, non-tradable quotas, international technology and product standards, voluntary agreements, and direct international transfers of financial resources and technology.

Possible criteria for the assessment of policy instruments include: environmental effectiveness; cost effectiveness; distributional considerations including competitiveness concerns; administrative and political feasibility; government revenues; wider economic effects including implications for international trade rules; wider environmental effects including carbon leakage; and effects on changes in attitudes, awareness, learning, innovation, technical progress, and dissemination of technology. Each government may apply different weights to various criteria when evaluating GHG mitigation policy options depending on national and sector level circumstances. Moreover, a government may apply different sets of weights to the criteria when evaluating national (domestic) versus international policy instruments. Co-ordinated actions could help address competitiveness concerns, potential conflicts with international trade rules, and carbon leakage.

The economics literature on the choice of policies adopted has emphasized the importance of interest group pressures, focusing on the demand for regulation. But it has tended to neglect the "supply side" of the political equation, emphasized in the political science literature: the legislators and government and party officials who design and implement regulatory policy, and who ultimately decide which instruments or mix of instruments will be used. However, the point of compliance of alternative policy instruments, whether they are applied to fossil fuel users or manufacturers, for example, is likely to be politically crucial to the choice of policy instrument. And a key insight is that some forms of regulation actually can benefit the regulated industry, for example, by limiting entry into the industry or imposing higher costs on new entrants. A policy that imposes costs on industry as a whole might still be supported by firms who would fare better than their competitors. Regulated firms, of course, are not the only group with a stake in regulation: opposing interest groups will fight for their own interests.

6.2   National Policies, Measures, and Instruments

In the case of countries in the process of structural reform, it is important to understand the new policy context to develop reasonable assessments of the feasibility of implementing GHG mitigation policies. Recent measures taken to liberalize energy markets have been inspired for the most part by desires to increase competition in energy and power markets, but they also can have significant emission implications, through their impact on the production and technology pattern of energy or power supply. In the long run, the consumption pattern change might be more important than the sole implementation of climate change mitigation measures.

Market-based instruments – principally domestic taxes and domestic tradable permit systems – will be attractive to governments in many cases because they are efficient. They will frequently be introduced in concert with conventional regulatory measures. When implementing a domestic emissions tax, policymakers must consider the collection point, the tax base, the variation among sectors, the association with trade, employment, revenue, and the exact form of the mechanism. Each of these can influence the appropriate design of a domestic emissions tax, and political or other concerns are likely to play a role as well. For example, a tax levied on the energy content of fuels could be much more costly than a carbon tax for equivalent emissions reduction, because an energy tax raises the price of all forms of energy, regardless of their contribution to CO_2 emissions. Yet, many nations may choose to use energy taxes for reasons other than cost effectiveness, and much of the analysis in this section applies to energy taxes, as well as carbon taxes.

A country committed to a limit on its GHG emissions also can meet this limit by implementing a tradable permit system that directly or indirectly limits emissions of domestic sources. Like a tax, a tradable permit system poses a number of design issues, including type of permit, ways to allocate permits, sources included, point of compliance, and use of banking. To be able to cover all sources with a single domestic permit regime is unlikely. The certainty provided by a tradable permit system of achieving a given emissions level for participating sources comes at the cost of the uncertainty of permit prices (and hence compliance costs). To address this concern, a hybrid policy that caps compliance costs could be adopted, but the level of emissions would no longer be guaranteed.

For a variety of reasons, in most countries the management of GHG emissions will not be addressed with a single policy instrument, but with a portfolio of instruments. In addition to one or more market-based policies, a portfolio might include standards and other regulations, voluntary agreements, and information programmes:

- Energy efficiency standards have been effective in reducing energy use in a growing number of countries. They may be especially effective in many countries where the capacity to administer market instruments is relatively limited, thereby helping to develop this administrative infrastructure. They need updating to remain effective. The main disadvantage of standards is that they can be inefficient, but efficiency can be improved if the standard focuses on the desired results and leaves as much flexibility as possible in the choice of how to achieve the results.

- Voluntary agreements (VAs) may take a variety of forms. Proponents of VAs point to low transaction costs and consensus elements, while sceptics emphasize the risk of "free riding", and the risk that the private sector will not pursue real emissions reduction in the absence of monitoring and enforcement. Voluntary agreements sometimes precede the introduction of more stringent measures.

- Imperfect information is widely recognized as a key market failure that can have significant effects on improved energy efficiency, and hence emissions. Information instruments include environmental labelling, energy audits, and industrial reporting requirements, and information campaigns are marketing elements in many energy-efficiency programmes.

A growing literature has demonstrated theoretically, and with numerical simulation models, that the economics of addressing GHG reduction targets with domestic policy instruments depend strongly on the choice of those instruments. Price-based policies tend to lead to positive marginal and positive total mitigation costs. In each case, the interaction of these abatement costs with the existing tax structure and, more generally, with existing factor prices is important. Price-based policies that generate revenues can be coupled with measures to improve market efficiency. However, the role of non-price policies, which affect the sign of the change in the unit price of energy services, often remains decisive.

6.3   International Policies and Measures

Turning to international policies and measures, the Kyoto Protocol defines three international policy instruments, the so-called Kyoto mechanisms: international emissions trading (IET), joint implementation (JI), and the Clean Development Mechanism (CDM). Each of these international policy instruments provides opportunities for Annex I Parties to fulfil their commitments cost-effectively. IET essentially would allow Annex I Parties to exchange part of their assigned national emission allowances (targets). IET implies that countries with high

marginal abatement costs (MACs) may acquire emission reductions from countries with low MACs. Similarly, JI would allow Annex I Parties to exchange emission reduction units among themselves on a project-by-project basis. Under the CDM, Annex I Parties would receive credit – on a project-by-project basis – for reductions accomplished in non-Annex I countries.

Economic analyses indicate that the Kyoto mechanisms could reduce significantly the overall cost of meeting the Kyoto emissions limitation commitments. However, achievement of the potential cost savings requires the adoption of domestic policies that allow individual entities to use the mechanisms to meet their national emissions limitation obligations. If domestic policies limit the use of the Kyoto mechanisms, or international rules governing the mechanisms limit their use, the cost savings may be reduced.

In the case of JI, host governments have incentives to ensure that emission reduction units (ERUs) are issued only for real emission reductions, assuming that they face strong penalties for non-compliance with national emissions limitation commitments. In the case of CDM, a process for independent certification of emission reductions is crucial, because host governments do not have emissions limitation commitments and hence may have less incentive to ensure that certified emission reductions (CERs) are issued only for real emission reductions. The main difficulty in implementing project-based mechanisms, both JI and CDM, is determining the net additional emission reduction (or sink enhancement) achieved; baseline definition may be extremely complex. Various other aspects of these Kyoto mechanisms are awaiting further decision making, including: monitoring and verification procedures, financial additionality (assurance that CDM projects will not displace traditional development assistance flows), and possible means of standardizing methodologies for project baselines.

The extent to which developing country (non-Annex I) Parties will effectively implement their commitments under the UNFCCC may depend, among other factors, on the transfer of environmentally sound technologies (ESTs).

6.4   Implementation of National and International Policy Instruments

Any international or domestic policy instrument can be effective only if accompanied by adequate systems of monitoring and enforcement. There is a linkage between compliance enforcement and the amount of international co-operation that will actually be sustained. Many multilateral environmental agreements address the need to co-ordinate restrictions on conduct taken in compliance with obligations they impose and the expanding legal regime under the WTO and/or GATT umbrella. Neither the UNFCCC nor the Kyoto Protocol now provides for specific trade measures in response to non-compliance. But several domestic policies and measures that might be developed and implemented in conjunction with the Kyoto Protocol could conflict with WTO provisions. International differences in environmental regulation may have trade implications.

One of the main concerns in environmental agreements (including the UNFCCC and the Kyoto Protocol) has been with reaching wider participation. The literature on international environmental agreements predicts that participation will be incomplete, and incentives may be needed to increase participation (see also Section 10).

7 Costing Methodologies

7.1 Conceptual Basis

Using resources to mitigate greenhouse gases (GHGs) generates opportunity costs that should be considered to help guide reasonable policy decisions. Actions taken to abate GHG emissions or to increase carbon sinks divert resources from other alternative uses. Assessing the costs of these actions should ideally consider the total value that society attaches to the goods and services forgone because of the diversion of resources to climate protection. In some cases, the sum of benefits and costs will be negative, meaning that society gains from undertaking the mitigation action.

This section addresses the methodological issues that arise in the estimation of the monetary costs of climate change. The focus is on the correct assessment of the costs of mitigation measures to reduce the emissions of GHGs. The assessment of costs and benefits should be based on a systematic analytical framework to ensure comparability and transparency of estimates. One well-developed framework assesses costs as changes in social welfare based on individual values. These individual values are reflected by the willingness to pay (WTP) for environmental improvements or the willingness to accept (WTA) compensation. From these value measures can be derived measures such as the social surpluses gained or lost from a policy, the total resource costs, and opportunity costs.

While the underlying measures of welfare have limits and using monetary values remains controversial, the view is taken that the methods to "convert" non-market inputs into monetary terms provide useful information for policymakers. These methods should be pursued when and where appropriate. It is also considered useful to supplement this welfare-based cost methodology with a broader assessment that includes equity and sustainability dimensions of climate change mitigation policies. In practice, the challenge is to develop a consistent and comprehensive definition of the key impacts to be measured.

A frequent criticism of this costing method is that it is inequitable, as it gives greater weight to the "well off". This is because, typically, a well-off person has a greater WTP or WTA than a less well-off person and hence the choices made reflect more the preferences of the better off. This criticism is

valid, but there is no coherent and consistent method of valuation that can replace the existing one in its entirety. Concerns about, for example, equity can be addressed along with the basic cost estimation. The estimated costs are one piece of information in the decision-making process for climate change that can be supplemented with other information on other social objectives, for example impacts on key stakeholders and the meeting of poverty objectives.

In this section the costing methodology is overviewed, and issues involved in using these methods addressed.

7.2 Analytical Approaches

Cost assessment is an input into one or more rules for decision-making, including cost-benefit analysis (CBA), cost-effectiveness analysis (CEA), and multi-attribute analysis. The analytical approaches differ primarily by how the objectives of the decision-making framework are selected, specified, and valued. Some objectives in mitigation policies can be specified in economic units (e.g., costs and benefits measured in monetary units), and some in physical units (e.g., the amount of pollutants dispersed in tonnes of CO_2). In practice, however, the challenge is in developing a consistent and comprehensive definition of every important impact to be measured.

7.2.1 Co-Benefits and Costs and Ancillary Benefits and Costs

The literature uses a number of terms to depict the associated benefits and costs that arise in conjunction with GHG mitigation policies. These include co-benefits, ancillary benefits, side benefits, secondary benefits, collateral benefits, and associated benefits. In the current discussion, the term "co-benefits" refers to the non-climate benefits of GHG mitigation policies that are explicitly incorporated into the initial creation of mitigation policies. Thus, the term co-benefits reflects that most policies designed to address GHG mitigation also have other, often at least equally important, rationales involved at the inception of these policies (e.g., related to objectives of development, sustainability, and equity). In contrast, the term ancillary benefits connotes those secondary or side effects of climate change mitigation policies on problems that arise subsequent to any proposed GHG mitigation policies.

Policies aimed at mitigating GHGs, as stated earlier, can yield other social benefits and costs (here called ancillary or co-benefits and costs), and a number of empirical studies have made a preliminary attempt to assess these impacts. It is apparent that the actual magnitude of the ancillary benefits or co-benefits assessed critically depends on the scenario structure of the analysis, in particular on the assumptions about policy management in the baseline case. This implies that whether a particular impact is included or not depends on the primary objective of the programme. Moreover, something that is seen as a GHG reduction programme from an international perspective may be seen, from a national perspective, as one in which local pollutants and GHGs are equally important.

7.2.2   *Implementation Costs*

All climate change policies necessitate some costs of implementation, that is costs of changes to existing rules and regulations, making sure that the necessary infrastructure is available, training and educating those who are to implement the policy as well those affected by the measures, etc. Unfortunately, such costs are not fully covered in conventional cost analyses. Implementation costs in this context are meant to reflect the more permanent institutional aspects of putting a programme into place and are different to those costs conventionally considered as transaction costs. The latter, by definition, are temporary costs. Considerable work needs to be done to quantify the institutional and other costs of programmes, so that the reported figures are a better representation of the true costs that will be incurred if programmes are actually implemented.

7.2.3   *Discounting*

There are broadly two approaches to discounting–an ethical or prescriptive approach based on what rates of discount should be applied, and a descriptive approach based on what rates of discount people (savers as well as investors) actually apply in their day-to-day decisions. For mitigation analysis, the country must base its decisions at least partly on discount rates that reflect the opportunity cost of capital. Rates that range from 4% to 6% would probably be justified in developed countries. The rate could be 10–12% or even higher in developing countries. It is more of a challenge to argue that climate change mitigation projects should face different rates, unless the mitigation project is of very long duration. The literature shows increasing attention to rates that decline over time and hence give more weight to benefits that occur in the long term. Note that these rates do not reflect private rates of return, which typically must be greater to justify a project, at around 10–25%.

7.2.4   *Adaptation and Mitigation Costs and the Link Between Them*

While most people appreciate that adaptation choices affect the costs of mitigation, this obvious point is often not addressed in climate policymaking. Policy is fragmented - with mitigation being seen as addressing climate change and adaptation seen as a means of reacting to natural hazards. Usually mitigation and adaptation are modelled separately as a necessary simplification to gain traction on an immense and complex issue. As a consequence, the costs of risk reduction action are frequently estimated separately, and therefore each measure is potentially biased. This realization suggests that more attention to the interaction of mitigation and adaptation, and its empirical ramification, is worthwhile, though uncertainty about the nature and timing of impacts, including surprises, will constrain the extent to which the associated costs can be fully internalized.

7.3   System Boundaries: Project, Sector, and Macro

Researchers make a distinction between project, sector, and economywide analyses. Project level analysis considers a "stand-alone" investment assumed to have insignificant secondary impacts on markets. Methods used for this level include CBA, CEA, and life-cycle analysis. Sector level analysis examines sectoral policies in a "partial-equilibrium" context in which all other variables are assumed to be exogenous. Economy-wide analysis explores how policies affect all sectors and markets, using various macroeconomic and general equilibrium models. A trade-off exists between the level of detail in the assessment and complexity of the system considered. This section presents some of the key assumptions made in cost analysis.

A combination of different modelling approaches is required for an effective assessment of climate change mitigation options. For example, detailed project assessment has been combined with a more general analysis of sectoral impacts, and macroeconomic carbon tax studies have been combined with the sectoral modelling of larger technology investment programmes.

7.3.1   *Baselines*

The baseline case, which by definition gives the emissions of GHGs in the absence of the climate change interventions being considered, is critical to the assessment of the costs of climate change mitigation. This is because the definition of the baseline scenario determines the potential for future GHG emissions reduction, as well as the costs of implementing these reduction policies. The baseline scenario also has a number of important implicit assumptions about future economic policies at the macroeconomic and sectoral levels, including sectoral structure, resource intensity, prices, and thereby technology choice.

7.3.2   *Consideration of No Regrets Options*

No regrets options are by definition actions to reduce GHG emissions that have negative net costs. Net costs are negative because these options generate direct or indirect benefits, such as those resulting from reductions in market failures, double dividends through revenue recycling and ancillary benefits, large enough to offset the costs of implementing the options. The no regrets issue reflects specific assumptions about the working and the efficiency of the economy, especially the existence and stability of a social welfare function, based on a social cost concept:

- Reduction of existing market or institutional failures and other barriers that impede adoption of cost-effective emission reduction measures can lower private costs compared to current practice. This can also reduce private costs overall.
- A double dividend related to recycling of the revenue of carbon taxes in such a way that it offsets distortionary taxes.

- Ancillary benefits and costs (or ancillary impacts), which can be synergies or trade-offs in cases in which the reduction of GHG emissions has joint impacts on other environmental policies (i.e., relating to local air pollution, urban congestion, or land and natural resource degradation).

Market Imperfections

The existence of a no regrets potential implies that market and institutions do not behave perfectly, because of market imperfections such as lack of information, distorted price signals, lack of competition, and/or institutional failures related to inadequate regulation, inadequate delineation of property rights, distortion-inducing fiscal systems, and limited financial markets. Reduction of market imperfections suggests it is possible to identify and implement policies that can correct these market and institutional failures without incurring costs larger than the benefits gained.

Double Dividend

The potential for a double dividend arising from climate mitigation policies was extensively studied during the 1990s. In addition to the primary aim of improving the environment (the first dividend), such policies, if conducted through revenue-raising instruments such as carbon taxes or auctioned emission permits, yield a second dividend, which can be set against the gross costs of these policies. All domestic GHG policies have an indirect economic cost from the interactions of the policy instruments with the fiscal system, but in the case of revenue-raising policies this cost is partly offset (or more than offset) if, for example, the revenue is used to reduce existing distortionary taxes. Whether these revenue-raising policies can reduce distortions in practice depends on whether revenues can be "recycled" to tax reduction.

Ancillary Benefits and Costs (Ancillary Impacts)

The definition of ancillary impacts is given above. As noted there, these can be positive as well as negative. It is important to recognize that gross and net mitigation costs cannot be established as a simple summation of positive and negative impacts, because the latter are interlinked in a very complex way. Climate change mitigation costs (gross and well as net costs) are only valid in relation to a comprehensive specific scenario and policy assumption structure.

The existence of no regrets potentials is a necessary, but not a sufficient, condition for the potential implementation of these options. The actual implementation also requires the development of a policy strategy that is complex as comprehensive enough to address these market and institutional failures and barriers.

7.3.3 Flexibility

For a wide variety of options, the costs of mitigation depend on what regulatory framework is adopted by national governments to reduce GHGs. In general, the more flexibility the framework allows, the lower the costs of achieving a given reduction. More flexibility and more trading partners can reduce costs. The opposite is expected with inflexible rules and few trading partners. Flexibility can be measured as the ability to reduce carbon emissions at the lowest cost, either domestically or internationally.

7.3.4 Development, Equity, and Sustainability Issues

Climate change mitigation policies implemented at a national level will, in most cases, have implications for short-term economic and social development, local environmental quality, and intra-generational equity. Mitigation cost assessments that follow this line can address these impacts on the basis of a decision-making framework that includes a number of side-impacts to the GHG emissions reduction policy objective. The goal of such an assessment is to inform decision makers about how different policy objectives can be met efficiently, given priorities of equity and other policy constraints (natural resources, environmental objectives). A number of international studies have applied such a broad decision-making framework to the assessment of development implications of CDM projects.

There are a number of key linkages between mitigation costing issues and broader development impacts of the policies, including macroeconomic impacts, employment creation, inflation, the marginal costs of public funds, capital availability, spillovers, and trade.

7.4 Special Issues Relating to Developing Countries and EITs

A number of special issues related to technology use should be considered as the critical determinants of climate change mitigation potential and related costs for developing countries. These include current technological development levels, technology transfer issues, capacity for innovation and diffusion, barriers to efficient technology use, institutional structure, human capacity aspects, and foreign exchange earnings.

Climate change studies in developing countries and EITs need to be strengthened in terms of methodology, data, and policy frameworks. Although a complete standardization of the methods is not possible, to achieve a meaningful comparison of results it is essential to use consistent methodologies, perspectives, and policy scenarios in different nations.

The following modifications to conventional approaches are suggested:

- Alternative development pathways should be analyzed with different patterns of investment in infrastructure, irrigation, fuel mix, and land-use policies.
- Macroeconomic studies should consider market transformation processes in the capital, labour, and power markets.

- Informal and traditional sector transactions should be included in national macroeconomic statistics. The value of non-commercial energy consumption and the unpaid work of household labour for non-commercial energy collection is quite significant and needs to be considered explicitly in economic analysis.
- The costs of removing market barriers should be considered explicitly.

7.5 Modelling Approaches to Cost Assessment

The modelling of climate mitigation strategies is complex and a number of modelling techniques have been applied including input-output models, macroeconomic models, computable general equilibrium (CGE) models, and energy sector based models. Hybrid models have also been developed to provide more detail on the structure of the economy and the energy sector. The appropriate use of these models depends on the subject of the evaluation and the availability of data.

As discussed in Section 6, the main categories of climate change mitigation policies include: market-oriented policies, technology-oriented policies, voluntary policies, and research and development policies. Climate change mitigation policies can include all four of the above policy elements. Most analytical approaches, however, only consider some of the four elements. Economic models, for example, mainly assess market-oriented policies and in some cases technology policies primarily those related to energy supply options, while engineering approaches mainly focus on supply and demand side technology policies. Both of these approaches are relatively weak in the representation of research and development and voluntary agreement policies.

8 Global, Regional, and National Costs and Ancillary Benefits

8.1 Introduction

The UNFCCC (Article 2) has as its ultimate goal the "stabilisation of greenhouse gas concentrations in the atmosphere at a level that will prevent dangerous anthropogenic interference with the climate system"[14]. In addition, the Convention

(Article 3.3) states that "policies and measures to deal with climate change should be cost-effective so as to ensure global benefits at the lowest possible costs"[15]. This section reports on literature on the costs of greenhouse gas mitigation policies at the national, regional, and global levels. Net welfare gains or losses are reported, including (when available) the ancillary benefits of mitigation policies. These studies employ the full range of analytical tools described in the previous chapter. These range from technologically detailed bottom-up models to more aggregate top-down models, which link the energy sector to the rest of the economy.

8.2. Gross Costs of GHG Abatement in Technology-Detailed Models

In technology-detailed "bottom-up" models and approaches, the cost of mitigation is derived from the aggregation of technological and fuel costs such as: investments, operation and maintenance costs, and fuel procurement, but also (and this is a recent trend) revenues and costs from import and exports.

Models can be ranked along two classification axes. First, they range from simple engineering-economics calculations effected technology-by-technology, to integrated partial equilibrium models of whole energy systems. Second, they range from the strict calculation of direct technical costs of reduction to the consideration of observed technology-adoption behaviour of markets, and of the welfare losses due to demand reductions and revenue gains and losses due to changes in trade.

This leads to contrasting two generic approaches, namely the engineering-economics approach and least-cost equilibrium modelling. In the first approach, each technology is assessed independently via an accounting of its costs and savings. Once these elements have been estimated, a unit cost can be calculated for each action, and each action can be ranked according to its costs. This approach is very useful to point out the potentials for negative cost abatements due to the 'efficiency gap' between the best available technologies and technologies currently in use. However, its most important limitation is that studies neglect or do not treat in a systematic way the interdependence of the various actions under examination.

[14] "The ultimate objective of this Convention and any related legal instruments that the Conference of Parties may adopt is to achieve, in accordance with the relevant provisions of the Convention, stabilization of greenhouse gas concentrations in the atmosphere at such a level that would prevent dangerous interference with the climate system. Such a level should be achieved within a timeframe sufficient to allow ecosystems to adapt naturally to climate change, to ensure that food production is not threatened, and to enable economic development to proceed in a sustainable manner."

[15] "The Parties should take precautionary measures to anticipate, prevent, or minimise the causes of climate change and mitigate its adverse effects. Where there are threats of serious irreversible damage, lack of full scientific certainty should not be used as a reason for postponing such measures, taking into account that polices and measures to deal with climate change should be cost-effective so as to ensure global benefits at the lowest possible costs. To achieve this, such policies and measures should take into account different socio-economic contexts, be comprehensive, cover all relevant sources, sinks and reservoirs of greenhouse gases and adaptation, and comprise all economic sectors. Efforts to address climate change may be carried out co-operatively by interested Parties."

Partial equilibrium least-costs models have been constructed to remedy this defect, by considering all actions simultaneously and selecting the optimal bundle of actions in all sectors and at all time periods. These more integrated studies conclude higher total costs of GHG mitigation than the strict technology by technology studies. Based on an optimization framework they give very easily interpretable results that compare an optimal response to an optimal baseline; however, their limitation is that they rarely calibrate the base year of the model to the existing non optimal situation and implicitly assume an optimal baseline. They consequently provide no information about the negative cost potentials.

Since the publication of the SAR, the bottom-up approaches have produced a wealth of new results for both Annex I and non-Annex I countries, as well as for groups of countries. Furthermore, they have extended their scope much beyond the classical computations of direct abatement costs by inclusion of demand effects and some trade effects.

However, the modelling results show considerable variations from study to study, which are explained by a number of factors, some of which reflect the widely differing conditions that prevail in the countries studied (e.g., energy endowment, economic growth, energy intensity, industrial and trade structure), and others reflect modelling assumptions and assumptions about negative cost potentials.

However, as in the SAR, there is agreement on a no regrets potential resulting from the reduction of existing market imperfections, consideration of ancillary benefits, and inclusion of double dividends. This means that some mitigation actions can be realized at negative costs. The no regrets potential results from existing market or institutional imperfections that prevent cost-effective emission reduction measures from being taken. The key question is whether such imperfections can be removed cost-effectively by policy measures.

The second important policy message is that short and medium term marginal abatement costs, which govern most of the macroeconomic impacts of climate policies, are very sensitive to uncertainty regarding baseline scenarios (rate of growth and energy intensity) and technical costs. Even with significant negative cost options, marginal costs may rise quickly beyond a certain anticipated mitigation level. This risk is far lower in models allowing for carbon trading. Over the long term this risk is reduced as technical change curbs down the slope of marginal cost curves.

8.3 Costs of Domestic Policy to Mitigate Carbon Emissions

Particularly important for determining the gross mitigation costs is the magnitude of emissions reductions required in order to meet a given target, thus the emissions baseline is a critical factor. The growth rate of CO_2 depends on the growth

rate in GDP, the rate of decline of energy use per unit of output, and the rate of decline of CO_2 emissions per unit of energy use.

In a multi-model comparison project that engaged more than a dozen modelling teams internationally, the gross costs of complying with the Kyoto Protocol were examined, using energy sector models. Carbon taxes are implemented to lower emissions and the tax revenue is recycled lump sum. The magnitude of the carbon tax provides a rough indication of the amount of market intervention that would be needed and equates the marginal abatement cost to meet a prescribed emissions target. The size of the tax required to meet a specific target will be determined by the marginal source of supply (including conservation) with and without the target. This in turn will depend on such factors as the size of the necessary emissions reductions, assumptions about the cost and availability of carbon-based and carbon-free technologies, the fossil fuel resource base, and short- and long-term price elasticities.

With no international emission trading, the carbon taxes necessary to meet the Kyoto restrictions in 2010 vary a lot among the models. Note from *Table TS.4*[16] that for the USA they are calculated to be in the range US$76 to US$322, for OECD Europe between US$20 and US$665, for Japan between US$97 and US$645, and finally for the rest of OECD (CANZ) between US$46 and US$425. All numbers are reported in 1990 dollars. Marginal abatement costs are in the range of US$20-US$135/tC if international trading is allowed. These models do not generally include no regrets measures or take account of the mitigation potential of CO_2 sinks and of greenhouse gases other than CO_2.

However, there is no strict correlation between the level of the carbon tax and GDP variation and welfare because of the influence of the country specifics (countries with a low share of fossil energy in their final consumption suffer less than others for the same level of carbon tax) and because of the content of the policies.

The above studies assume, to allow an easy comparison across countries, that the revenues from carbon taxes (or auctioned emissions permits) are recycled in a lump-sum fashion to the economy. The net social cost resulting from a given marginal cost of emissions constraint can be reduced if the revenues are targetted to finance cuts in the marginal rates of pre-existing distortionary taxes, such as income, payroll, and sales taxes. While recycling revenues in a lump-sum fashion confers no efficiency benefit, recycling through marginal rate cuts helps avoid some of the efficiency costs or dead-weight loss of existing taxes. This raises the possibility that revenue-neutral carbon taxes might offer a double dividend by (1) improving the environment and (2) reducing the costs of the tax system.

---

[16] The highest figures cited in this sentence are all results from one model: the ABARE-GTEM model.

Table TS.4: Energy Modelling Forum main results. Marginal abatement costs (in 1990 US$/tC; 2010 Kyoto target)

Model	No trading				Annex I trading	Global trading
	US	OECD-E	Japan	CANZ		
ABARE-GTEM	322	665	645	425	106	23
AIM	153	198	234	147	65	38
CETA	168				46	26
Fund					14	10
G-Cubed	76	227	97	157	53	20
GRAPE		204	304		70	44
MERGE3	264	218	500	250	135	86
MIT-EPPA	193	276	501	247	76	
MS-MRT	236	179	402	213	77	27
Oxford	410	966	1074		224	123
RICE	132	159	251	145	62	18
SGM	188	407	357	201	84	22
WorldScan	85	20	122	46	20	5
Administration	154				43	18
EIA	251				110	57
POLES	135.8	135.3	194.6	131.4	52.9	18.4

Note: The results of the Oxford model are not included in the ranges cited in the TS and SPM because this model has not been subject to substantive academic review (and hence is inappropriate for IPCC assessment), and relies on data from the early 1980s for a key parametization that determines the model results. This model is entirely unrelated to the CLIMOX model, from the Oxford Institutes of Energy Studies, referred to in *Table TS.6*.

EMF-16. GDP losses (as a percentage of total GDP) associated with complying with the prescribed targets under the Kyoto Protocol. Four regions include the USA, OECD Europe (OECD-E), Japan, and Canada, Australia and New Zealand (CANZ). Scenarios include no trading, Annex B trading only, and full global trading.

One can distinguish a weak and a strong form of the double dividend. The weak form asserts that the costs of a given revenue-neutral environmental reform, when revenues are devoted to cuts in marginal rates of prior distortionary taxes, are reduced relative to the costs when revenues are returned in lump-sum fashion to households or firms. The strong form of the double-dividend assertion is that the costs of the revenue-neutral environmental tax reform are zero or negative. While the weak form of the double-dividend claim receives virtually universal support, the strong form of the double dividend assertion is controversial.

Where to recycle revenues from carbon taxes or auctioned permits depends upon the country specifics. Simulation results show that in economies that are especially inefficient or distorted along non-environmental lines, the revenue-recycling effect can indeed be strong enough to outweigh the primary cost and tax-interaction effect so that the strong double dividend may materialize. Thus, in several studies involving European economies, where tax systems may be highly distorted in terms of the relative taxation of labour, the strong double dividend can be obtained, in any case more frequently than in other recycling options. In contrast, most studies of carbon taxes or permits policies in the USA demonstrate that recycling through lower labour taxation is less efficient than through capital taxation; but they generally do not find a strong double dividend. Another conclusion is that even in cases of no strong double-dividend effect, one fares considerably better with a revenue-recycling policy in which revenues are used to cut marginal rates of prior taxes, than with a non-revenue recycling policy, like for example grandfathered quotas.

In all countries where CO_2 taxes have been introduced, some sectors have been exempted by the tax, or the tax is differentiated across sectors. Most studies conclude that tax exemptions raise economic costs relative to a policy involving uniform taxes. However, results differ in the magnitude of the costs of exemptions.

8.4 Distributional Effects of Carbon Taxes

As well as the total costs, the distribution of the costs is important for the overall evaluation of climate policies. A policy that leads to an efficiency gain may not be welfare improving overall if some people are in a worse position than before, and vice versa. Notably, if there is a wish to reduce the income differences in the society, the effect on the income distribution should be taken into account in the assessment.

The distributional effects of a carbon tax appear to be regressive unless the tax revenues are used either directly or indi-

Table TS.5: *Energy Modeling Forum main results. GDP loss in 2010 (in % of GDP; 2010 Kyoto target)*

Model	No trading				Annex I trading				Global trading			
	US	OECD-E	Japan	CANZ	US	OECD-E	Japan	CANZ	US	OECD-E	Japan	CANZ
ABARE-GTEM	1.96	0.94	0.72	1.96	0.47	0.13	0.05	0.23	0.09	0.03	0.01	0.04
AIM	0.45	0.31	0.25	0.59	0.31	0.17	0.13	0.36	0.20	0.08	0.01	0.35
CETA	1.93				0.67				0.43			
G-CUBED	0.42	1.50	0.57	1.83	0.24	0.61	0.45	0.72	0.06	0.26	0.14	0.32
GRAPE		0.81	0.19			0.81	0.10			0.54	0.05	
MERGE3	1.06	0.99	0.80	2.02	0.51	0.47	0.19	1.14	0.20	0.20	0.01	0.67
MS-MRT	1.88	0.63	1.20	1.83	0.91	0.13	0.22	0.88	0.29	0.03	0.02	0.32
Oxford	1.78	2.08	1.88		1.03	0.73	0.52		0.66	0.47	0.33	
RICE	0.94	0.55	0.78	0.96	0.56	0.28	0.30	0.54	0.19	0.09	0.09	0.19

Note: The results of the Oxford model are not included in the ranges cited in the TS and SPM because this model has not been subject to substantive academic review (and hence is inappropriate for IPCC assessment), and relies on data from the early 1980s for a key parametization that determines the model results. This model is entirely unrelated to the CLIMOX model, from the Oxford Institutes of Energy Studies, referred to in *Table TS.6*.

rectly in favour of the low-income groups. Recycling the tax revenue by reducing the labour tax may have more attractive distributional consequences than a lump-sum recycling, in which the recycled revenue is directed to both wage earners and capital owners. Reduced taxation of labour results in increased wages and favours those who earn their income mainly from labour. However, the poorest groups in the society may not even earn any income from labour. In this regard, reducing labour taxes may not always be superior to recycling schemes that distribute to all groups of a society and might reduce the regressive character of carbon taxes.

8.5   Aspects of International Emission Trading

It has long been recognized that international trade in emission quota can reduce mitigation costs. This will occur when countries with high domestic marginal abatement costs purchase emission quota from countries with low marginal abatement costs. This is often referred to as "where flexibility". That is, allowing reductions to take place where it is cheapest to do so regardless of geographical location. It is important to note that where the reductions take place is independent of who pays for the reductions.

"Where flexibility" can occur on a number of scales. It can be global, regional or at the country level. In the theoretical case of full global trading, all countries agree to emission caps and participate in the international market as buyers or sellers of emission allowances. The CDM may allow some of these cost reductions to be captured. When the market is defined at the regional level (e.g., Annex B countries), the trading market is more limited. Finally, trade may take place domestically with all emission reductions occurring in the country of origin.

Table TS.5 shows the cost reductions from emission trading for Annex B and full global trading compared to a no-trading case.

The calculation is made by various models with both global and regional detail. In each instance, the goal is to meet the emission reduction targets contained in the Kyoto Protocol. All of the models show significant gains as the size of the trading market is expanded. The difference among models is due in part to differences in their baseline, the assumptions about the cost and availability of low-cost substitutes on both the supply and demand sides of the energy sector, and the treatment of short-term macro shocks. In general, all calculated gross costs for the non-trading case are below 2% of GDP (which is assumed to have increased significantly in the period considered) and in most cases below 1%. Annex B trading lowers the costs for the OECD region as a whole to less than 0.5% and regional impacts within this vary between 0.1% to 1.1%. Global trading in general would decrease these costs to well below 0.5% of GDP with OECD average below 0.2%.

The issue of the so-called "hot air"[17] also influences the cost of implementing the Kyoto Protocol. The recent decline in economic activity in Eastern Europe and the former Soviet Union has led to a decrease in their GHG emissions. Although this trend is eventually expected to reverse, for some countries emissions are still projected to lie below the constraint imposed by the Kyoto Protocol. If this does occur, these countries will have excess emission quota that may be sold to countries in search of low-cost options for meeting their own targets. The cost savings from trading are sensitive to the magnitude of "hot air".

Numerous assessments of reduction in projected GDP have been associated with complying with Kyoto-type limits. Most

---

[17] Hot air: a few countries, notably those with economies in transition, have assigned amount units that appear to be well in excess of their anticipated emissions (as a result of economic downturn). This excess is referred to as hot air.

economic analyses have focused on gross costs of carbon emitting activities[18], ignoring the cost-saving potential of mitigating non-CO_2 gases and using carbon sequestration and neither taking into account environmental benefits (ancillary benefits and avoided climate change), nor using revenues to remove distortions. Including such possibilities could lower costs.

A constraint would lead to a reallocation of resources away from the pattern that is preferred in the absence of a limit and into potentially costly conservation and fuel substitution. Relative prices will also change. These forced adjustments lead to reductions in economic performance, which impact GDP. Clearly, the broader the permit trading market, the greater the opportunity for reducing overall mitigation costs. Conversely, limits on the extent to which a country can satisfy its obligations through the purchase of emissions quota can increase mitigation costs. Several studies have calculated the magnitude of the increase to be substantial falling in particular on countries with the highest marginal abatement costs. But another parameter likely to limit the savings from carbon trading is the very functioning of trading systems (transaction costs, management costs, insurance against uncertainty, and strategic behaviour in the use of permits).

8.6   Ancillary Benefits of Greenhouse Gas Mitigation

Policies aimed at mitigating greenhouse gases can have positive and negative side effects on society, not taking into account benefits of avoided climate change. This section assesses in particular those studies that evaluate the side effects of climate change mitigation. Therefore the term "ancillary benefits or costs" is used. There is little agreement on the definition, reach, and size of these ancillary benefits, and on methodologies for integrating them into climate policy. Criteria are established for reviewing the growing literature linking specific carbon mitigation policies to monetized ancillary benefits. Recent studies that take an economy-wide, rather than a sectoral, approach to ancillary benefits are described in the report and their credibility is examined (Chapter 9 presents sectoral analyses). In spite of recent progress in methods development, it remains very challenging to develop quantitative estimates of the ancillary effects, benefits and costs of GHG mitigation policies. Despite these difficulties, in the short term, ancillary benefits of GHG policies under some circumstances can be a significant fraction of private (direct) mitigation costs and in some cases they can be comparable to the mitigation costs. According to the literature, ancillary benefits may be of particular importance in developing countries, but this literature is as yet limited.

The exact magnitude, scale, and scope of these ancillary benefits and costs will vary with local geographical and baseline conditions. In some circumstances, where baseline conditions involve relatively low carbon emissions and population density, benefits may be low. The models most in use for ancillary benefit estimation – the computable general equilibrium (CGE) models – have difficulty in estimating ancillary benefits because they rarely have, and may not be able to have, the necessary spatial detail.

With respect to baseline considerations most of the literature on ancillary benefits systematically treats only government policies and regulations with respect to the environment. In contrast, other regulatory policy baseline issues, such as those relating to energy, transportation, and health, have been generally ignored, as have baseline issues that are not regulatory, such as those tied with technology, demography, and the natural resource base. For the studies reviewed here, the biggest share of the ancillary benefits is related to public health. A major component of uncertainty for modelling ancillary benefits for public health is the link between emissions and atmospheric concentrations, particularly in light of the importance of secondary pollutants. However, it is recognized that there are significant ancillary benefits in addition to those for public health that have not been quantified or monetized. At the same time, it appears that there are major gaps in the methods and models for estimating ancillary costs.

8.7   "Spillover" Effects[19] from Actions Taken in Annex B on Non-Annex B Countries

In a world where economies are linked by international trade and capital flows, abatement of one economy will have welfare impacts on other abating or non-abating economies. These impacts are called spillover effects, and include effects on trade, carbon leakage, transfer and diffusion of environmentally sound technology, and other issues (*Figure TS.8*).

As to the trade effects, the dominant finding of the effects of emission constraints in Annex B countries on non-Annex B countries in simulation studies prior to the Kyoto Protocol was that Annex B abatement would have a predominantly adverse impact on non-Annex B regions. In simulations of the Kyoto Protocol, the results are more mixed with some non-Annex B regions experiencing welfare gains and other losses. This is mainly due to a milder target in the Kyoto simulations than in pre-Kyoto simulations. It was also universally found that most non-Annex B economies that suffered welfare losses under uniform independent abatement would suffer smaller welfare losses under emissions trading.

---

[18] Although some studies include multi-gas analysis, much research is needed on this potential both intertemporally and regionally.

[19] "Spillovers" from domestic mitigation strategies are the effects that these strategies have on other countries. Spillover effects can be positive or negative and include effects on trade, carbon leakage, transfer and diffusion of environmentally sound technology, and other issues.

Policies and measures \ Spillovers	Benefits from technology improvement	Impacts on energy industries activity and prices	Impacts on energy intensive industries	Resource transfers to sectors
Public R&D policies	Increase in the scientific knowledge base	↑		
"Market access" policies for new technologies	Increase in know-how through experience, learning by doing			
Standards, subsidies, Voluntary agreements	New cleaner industry/product performance standards			
Carbon taxes	Price-induced technical change and technology diffusion	Reduction of activity in fossil fuel industries Lower international prices, negative impacts for exporters, positive for importers, possibility of a "rebound effect"	Carbon leakages, positive impacts for activity, negative for envir. in receiving country	
Energy subsidy removal			Reduced distorsions in industrial competition	
Harmonized carbon taxes				
Domestic emission trading			Distorsion in competition if differentiated schemes (grandfathered vs. auctioned)	
Joint Implementation, Clean Development Mechanism				Technology transfer
International emission trading		↓		Net gain when permit price is superior (not equal) to average reduction costs

Figure TS.8: *"Spillovers" from domestic mitigation strategies are the effects that these strategies have on other countries. Spillover effects can be positive or negative and include effects on trade, carbon leakage, transfer and diffusion of environmentally sound technology, and other issues.*

A reduction in Annex B emissions will tend to result in an increase in non-Annex B emissions reducing the environmental effectiveness of Annex B abatement. This is called "carbon leakage", and can occur in the order of 5%-20% through a possible relocation of carbon-intensive industries because of reduced Annex B competitiveness in the international marketplace, lower producer prices of fossil fuels in the international market, and changes in income due to better terms of trade.

While the SAR reported that there was a high variance in estimates of carbon leakage from the available models, there has been some reduction in the variance of estimates obtained in the subsequent years. However, this may largely result from the development of new models based on reasonably similar assumptions and data sources. Such developments do not necessarily reflect more widespread agreement about appropriate behavioural assumptions. One robust result seems to be that carbon leakage is an increasing function of the stringency of the abatement strategy. This means that leakage may be a less serious problem under the Kyoto target than under the more stringent targets considered previously. Also emission leakage is lower under emissions trading than under independent abatement. Exemptions for energy-intensive industries found in practice, and other factors, make the higher model estimates for carbon leakage unlikely, but would raise aggregate costs.

Carbon leakage may also be influenced by the assumed degree of competitiveness in the world oil market. While most studies assume a competitive oil market, studies considering imperfect competition find lower leakage if OPEC is able to exercise a degree of market power over the supply of oil and therefore reduce the fall in the international oil price. Whether or not OPEC acts as a cartel can have a reasonably significant effect on the loss of wealth to OPEC and other oil producers and on the level of permit prices in Annex B regions (see also Section 9.2).

The third spillover effect mentioned above, the transfer and diffusion of environmentally sound technology, is related to induced technical change (see Section 8.10). The transfer of environmentally sound technologies and know-how, not included in models, may lead to lower leakage and especially on the longer term may more than offset the leakage.

8.8   Summary of the Main Results for Kyoto Targets

The cost estimates for Annex B countries to implement the Kyoto Protocol vary between studies and regions, and depend strongly upon the assumptions regarding the use of the Kyoto mechanisms, and their interactions with domestic measures. The great majority of global studies reporting and comparing these costs use international energy-economic models. Nine of these studies suggest the following GDP impacts[20]:

Annex II countries[21]: In the absence of emissions trading between Annex B countries[22], the majority of global studies show reductions in projected GDP of about 0.2% to 2% in 2010 for different Annex II regions. With full emissions trading between Annex B countries, the estimated reductions in 2010 are between 0.1% and 1.1% of projected GDP[23]. These studies encompass a wide range of assumptions. Models whose results are reported here assume full use of emissions trading without transaction cost. Results for cases that do not allow Annex B trading assume full domestic trading within each region. Models do not include sinks or non-CO_2 greenhouse gases. They do not include the CDM, negative cost options, ancillary benefits, or targeted revenue recycling.
For all regions costs are also influenced by the following factors:
- Constraints on the use of Annex B trading, high transaction costs in implementing the mechanisms and inefficient domestic implementation could raise costs.
- Inclusion in domestic policy and measures of the no regrets possibilities[2], use of the CDM, sinks, and inclusion of non-CO_2 greenhouse gases, could lower costs. Costs for individual countries can vary more widely.

The models show that the Kyoto mechanisms, are important in controlling risks of high costs in given countries, and thus can complement domestic policy mechanisms. Similarly, they can minimize risks of inequitable international impacts and help to level marginal costs. The global modelling studies reported

above show national marginal costs to meet the Kyoto targets from about US$20/tC up to US$600/tC without trading, and a range from about US$15/tC up to US$150/tC with Annex B trading. The cost reductions from these mechanisms may depend on the details of implementation, including the compatibility of domestic and international mechanisms, constraints, and transaction costs.

Economies in transition: For most of these countries, GDP effects range from negligible to a several percent increase. This reflects opportunities for energy efficiency improvements not available to Annex II countries. Under assumptions of drastic energy efficiency improvement and/or continuing economic recession in some countries, the assigned amounts may exceed projected emissions in the first commitment period. In this case, models show increased GDP through revenues from trading assigned amounts. However, for some economies in transition, implementing the Kyoto Protocol will have similar impacts on GDP as for Annex II countries.

Non-Annex I countries: Emission constraints in Annex I countries have well established, albeit varied "spillover" effects[24] on non-Annex I countries.
- Oil-exporting, non-Annex I countries: Analyses report costs differently, including, *inter alia*, reductions in projected GDP and reductions in projected oil revenues[25]. The study reporting the lowest costs shows reductions of 0.2% of projected GDP with no emissions trading, and less than 0.05% of projected GDP with Annex B emissions trading in 2010[26]. The study reporting the highest costs shows reductions of 25% of projected oil revenues with no emissions trading, and 13% of projected oil revenues with Annex B emissions trading in 2010. These studies do not consider policies and measures[27] other than Annex B emissions trading, that

---

[20] Many other studies incorporating more precisely the country specifics and diversity of targetted policies provide a wider range of net cost estimates.

[21] Annex II countries: Group of countries included in Annex II to the UNFCCC, including all developed countries in the Organisation of Economic Co-operation and Development.

[22] Annex B countries: Group of countries included in Annex B in the Kyoto Protocol that have agreed to a target for their greenhouse gas emissions, including all the Annex I countries (as amended in 1998) but Turkey and Belarus.

[23] Many metrics can be used to present costs. For example, if the annual costs to developed countries associated with meeting Kyoto targets with full Annex B trading are in the order of 0.5% of GDP, this represents US$125 billion (1000 million) per year, or US$125 per person per year by 2010 in Annex II (SRES assumptions). This corresponds to an impact on economic growth *rates* over ten years of less than 0.1 percentage point.

[24] Spillover effects here incorporate only economic effects, not environmental effects.

[25] Details of the six studies reviewed are found in *Table 9.4* of the underlying report.

[26] These estimated costs can be expressed as differences in GDP growth rates over the period 2000-2010. With no emissions trading, GDP growth rate is reduced by 0.02 percentage points/year; with Annex B emissions trading, growth rate is reduced by less than 0.005 percentage points/year.

[27] These policies and measures include: those for non-CO_2 gases and non-energy sources of all gases; offsets from sinks; industry restructuring (e.g., from energy producer to supplier of energy services); use of OPEC's market power; and actions (e.g. of Annex B Parties) related to funding, insurance, and the transfer of technology. In addition, the studies typically do not include the following policies and effects that can reduce the total cost of mitigation: the use of tax revenues to reduce tax burdens or finance other mitigation measures; environmental ancillary benefits of reductions in fossil fuel use; and induced technological change from mitigation policies.

could lessen the impact on non-Annex I, oil-exporting countries, and therefore tend to overstate both the costs to these countries and overall costs.

The effects on these countries can be further reduced by removal of subsidies for fossil fuels, energy tax restructuring according to carbon content, increased use of natural gas, and diversification of the economies of non-Annex I, oil-exporting countries.

- Other non-Annex I countries: They may be adversely affected by reductions in demand for their exports to OECD nations and by the price increase of those carbon-intensive and other products they continue to import. These countries may benefit from the reduction in fuel prices, increased exports of carbon-intensive products and the transfer of environmentally sound technologies and know-how. The net balance for a given country depends on which of these factors dominates. Because of these complexities, the breakdown of winners and losers remains uncertain.

- Carbon leakage:[28] The possible relocation of some carbon-intensive industries to non-Annex I countries and wider impacts on trade flows in response to changing prices may lead to leakage in the order of 5-20%. Exemptions, for example for energy-intensive industries, make the higher model estimates for carbon leakage unlikely, but would raise aggregate costs. The transfer of environmentally sound technologies and know-how, not included in models, may lead to lower leakage and especially on the longer term may more than offset the leakage.

8.9 The Costs of Meeting a Range of Stabilization Targets

Cost-effectiveness studies with a century timescale estimate that the costs of stabilizing CO_2 concentrations in the atmosphere increase as the concentration stabilization level declines. Different baselines can have a strong influence on absolute costs. While there is a moderate increase in the costs when passing from a 750ppmv to a 550ppmv concentration stabilization level, there is a larger increase in costs passing from 550ppmv to 450ppmv unless the emissions in the baseline scenario are very low. These results, however, do not incorporate carbon sequestration and gases other than CO_2, and did not examine the possible effect of more ambitious targets on induced technological change[29]. In particular, the choice of the reference scenario has a strong influence. Recent studies using the IPCC SRES reference scenarios as baselines against which to analyze stabilization clearly show that the average reduction in projected GDP in most of the stabilization scenarios

reviewed here is under 3% of the baseline value (the maximum reduction across all the stabilization scenarios reached 6.1% in a given year). At the same time, some scenarios (especially in the A1T group) showed an increase in GDP compared to the baseline because of apparent positive economic feedbacks of technology development and transfer. The GDP reduction (averaged across storylines and stabilization levels) is lowest in 2020 (1%), reaches a maximum in 2050 (1.5%), and declines by 2100 (1.3%). However, in the scenario groups with the highest baseline emissions (A2 and A1FI), the size of the GDP reduction increases throughout the modelling period. Due to their relatively small scale when compared to absolute GDP levels, GDP reductions in the post-SRES stabilization scenarios do not lead to significant declines in GDP growth rates over this century. For example, the annual 1990-2100 GDP growth rate across all the stabilization scenarios was reduced on average by only 0.003% per year, with a maximum reduction reaching 0.06% per year.

The concentration of CO_2 in the atmosphere is determined more by cumulative rather than by year-by-year emissions. That is, a particular concentration target can be reached through a variety of emissions pathways. A number of studies suggest that the choice of emissions pathway can be as important as the target itself in determining overall mitigation costs. The studies fall into two categories: those that assume that the target is known and those that characterize the issue as one of decision making under uncertainty.

For studies that assume that the target is known, the issue is one of identifying the least-cost mitigation pathway for achieving the prescribed target. Here the choice of pathway can be seen as a carbon budget problem. This problem has been so far addressed in terms of CO_2 only and very limited treatment has been given to non-CO_2 GHGs. A concentration target defines an allowable amount of carbon to be emitted into the atmosphere between now and the date at which the target is to be achieved. The issue is how best to allocate the carbon budget over time.

Most studies that have attempted to identify the least-cost pathway for meeting a particular target conclude that such as pathway tends to depart gradually from the model's baseline in the early years with more rapid reductions later on. There are several reasons why this is so. A gradual near-term transition from the world's present energy system minimizes premature retirement of existing capital stock, provides time for technology

---

[28] Carbon leakage is defined here as the increase in emissions in non-Annex B countries resulting from implementation of reductions in Annex B, expressed as a percentage of Annex B reductions.

[29] Induced technological change is an emerging field of inquiry. None of the literature reviewed in TAR on the relationship between the century-scale CO_2 concentrations and costs reported results for models employing induced technological change. Models with induced technological change under some circumstances show that century-scale concentrations can differ, with similar GDP growth but under different policy regimes (Section 8.4.1.4).

development, and avoids premature lock-in to early versions of rapidly developing low-emission technology. On the other hand, more aggressive near-term action would decrease environmental risks associated with rapid climatic changes, stimulate more rapid deployment of existing low-emission technologies (see also Section 8.10), provide strong near-term incentives to future technological changes that may help to avoid lock-in to carbon intensive technologies, and allow for later tightening of targets should that be deemed desirable in light of evolving scientific understanding.

It should also be noted that the lower the concentration target, the smaller the carbon budget, and hence the earlier the departure from the baseline. However, even with higher concentration targets, the more gradual transition from the baseline does not negate the need for early action. All stabilization targets require future capital stock to be less carbon-intensive. This has immediate implications for near-term investment decisions. New supply options typically take many years to enter into the marketplace. An immediate and sustained commitment to R&D is required if low-carbon low-cost substitutes are to be available when needed.

The above addresses the issue of mitigation costs. It is also important to examine the environmental impacts of choosing one emission pathway over another. This is because different emission pathways imply not only different emission reduction costs, but also different benefits in terms of avoided environmental impacts (see Section 10).

The assumption that the target is known with certainty is, of course, an oversimplification. Fortunately, the UNFCCC recognizes the dynamic nature of the decision problem. It calls for periodic reviews "in light of the best scientific information on climate change and its impacts." Such a sequential decision making process aims to identify short-term hedging strategies in the face of long-term uncertainties. The relevant question is not "what is the best course of action for the next hundred years" but rather "what is the best course for the near-term given the long-term uncertainties."

Several studies have attempted to identify the optimal near-term hedging strategy based on the uncertainty regarding the long-term objective. These studies find that the desirable amount of hedging depends upon one's assessment of the stakes, the odds, and the cost of mitigation. The risk premium – the amount that society is willing to pay to avoid risk – ultimately is a political decision that differs among countries.

8.10   The Issue of Induced Technological Change

Most models used to assess the costs of meeting a particular mitigation objective tend to oversimplify the process of technical change. Typically, the rate of technical change is assumed to be independent of the level of emissions control. Such change is referred to as autonomous. In recent years, the issue

of induced technical change has received increased attention. Some argue that such change might substantially lower and perhaps even eliminate the costs of CO_2 abatement policies. Others are much less sanguine about the impact of induced technical change.

Recent research suggests that the effect on timing depends on the source of technological change. When the channel for technological change is R&D, the induced technological change makes it preferable to concentrate more abatement efforts in the future. The reason is that technological change lowers the costs of future abatement relative to current abatement, making it more cost-effective to place more emphasis on future abatement. But, when the channel for technological change is learning-by-doing, the presence of induced technological change has an ambiguous impact on the optimal timing of abatement. On the one hand, induced technical change makes future abatement less costly, which suggests emphasizing future abatement efforts. On the other hand, there is an added value to current abatement because such abatement contributes to experience or learning and helps reduce the costs of future abatement. Which of these two effects dominates depends on the particular nature of the technologies and cost functions.

Certain social practices may resist or enhance technological change. Therefore, public awareness-raising and education may help encourage social change to an environment favourable for technological innovation and diffusion. This represents an area for further research.

9   Sectoral Costs and Ancillary Benefits of Mitigation

9.1   Differences between Costs of Climate Change Mitigation Evaluated Nationally and by Sector

Policies adopted to mitigate global warming will have implications for specific sectors, such as the coal industry, the oil and gas industry, electricity, manufacturing, transportation, and households. A sectoral assessment helps to put the costs in perspective, to identify the potential losers and the extent and location of the losses, and to identify the sectors that may benefit. However, it is worth noting that the available literature to make this assessment is limited: there are few comprehensive studies of the sectoral effects of mitigation, compared with those on the macro GDP effects, and they tend to be for Annex I countries and regions.

There is a fundamental problem for mitigation policies. It is well established that, compared to the situation for potential gainers, the potential sectoral losers are easier to identify, and their losses are likely to be more immediate, more concentrated, and more certain. The potential sectoral gainers (apart from the renewables sector and perhaps the natural gas sector) can only expect a small, diffused, and rather uncertain gain, spread over a long period. Indeed many of those who may gain do not exist, being future generations and industries yet to develop.

It is also well established that the overall effects on GDP of mitigation policies and measures, whether positive or negative, conceal large differences between sectors. In general, the energy intensity and the carbon intensity of the economies will decline. The coal and perhaps the oil industries are expected to lose substantial proportions of their traditional output relative to those in the reference scenarios, though the impact of this on the industries will depend on diversification, and other sectors may increase their outputs but by much smaller proportions. Reductions in fossil fuel output below the baseline will not impact all fossil fuels equally. Fuels have different costs and price sensitivities; they respond differently to mitigation policies. Energy-efficiency technology is fuel and combustion device-specific, and reductions in demand can affect imports differently from output. Energy-intensive sectors, such as heavy chemicals, iron and steel, and mineral products, will face higher costs, accelerated technical or organizational change, or loss of output (again relative to the reference scenario) depending on their energy use and the policies adopted for mitigation.

Industries concerned directly with mitigation are likely to benefit from action. These industries include renewable and nuclear electricity, producers of mitigation equipment (incorporating energy- and carbon-saving technologies), agriculture and forestry producing energy crops, and research services producing energy and carbon-saving R&D. They may benefit in the long term from the availability of financial and other resources that would otherwise have been taken up in fossil fuel production. They may also benefit from reductions in tax burdens if taxes are used for mitigation and the revenues recycled as reductions in employer, corporate, or other taxes. Those studies that report reductions in GDP do not always provide a range of recycling options, suggesting that policy packages increasing GDP have not been explored. The extent and nature of the benefits will vary with the policies followed. Some mitigation policies can lead to net overall economic benefits, implying that the gains from many sectors will outweigh the losses for coal and other fossil fuels, and energy-intensive industries. In contrast, other less-well-designed policies can lead to overall losses.

It is worth placing the task faced by mitigation policy in an historical perspective. CO_2 emissions have tended to grow more slowly than GDP in a number of countries over the past 40 years. The reasons for such trends vary but include:

- a shift away from coal and oil and towards nuclear and gas as the source of energy;
- improvements in energy efficiency by industry and households; and
- a shift from heavy manufacturing towards more service and information-based economic activity.

These trends will be encouraged and strengthened by mitigation policies.

9.2 Selected Specific Sectoral Findings on Costs of Climate Change Mitigation

9.2.1 Coal

Within this broad picture, certain sectors will be substantially affected by mitigation. Relative to the reference case, the coal industry, producing the most carbon-intensive of products, faces almost inevitable decline in the long term, relative to the baseline projection. Technologies still under development, such as CO_2 removal and storage from coal-burning plants and in-situ gasification, could play a future role in maintaining the output of coal whilst avoiding CO_2 and other emissions. Particularly large effects on the coal sector are expected from policies such as the removal of fossil fuel subsidies or the restructuring of energy taxes so as to tax the carbon content rather than the energy content of fuels. It is a well-established finding that removal of the subsidies would result in substantial reductions in GHG emissions, as well as stimulating economic growth. However, the effects in specific countries depend heavily on the type of subsidy removed and the commercial viability of alternative energy sources, including imported coal.

9.2.2 Oil

The oil industry also faces a potential relative decline, although this may be moderated by lack of substitutes for oil in transportation, substitution away from solid fuels towards liquid fuels in electricity generation, and the diversification of the industry into energy supply in general.

Table TS.6 shows a number of model results for the impacts of implementation of the Kyoto Protocol on oil exporting countries. Each model uses a different measure of impact, and many use different groups of countries in their definition of oil exporters. However, the studies all show that the use of the flexibility mechanisms will reduce the economic cost to oil producers.

Thus, studies show a wide range of estimates for the impact of GHG mitigation policies on oil production and revenue. Much of these differences are attributable to the assumptions made about: the availability of conventional oil reserves, the degree of mitigation required, the use of emission trading, control of GHGs other than CO_2, and the use of carbon sinks. However, all studies show a net growth in both oil production and revenue to at least 2020, and significantly less impact on the real price of oil than has resulted from market fluctuations over the past 30 years. *Figure TS.9* shows the projection of real oil prices to 2010 from the IEA's 1998 *World Energy Outlook*, and the effect of Kyoto implementation from the G-cubed model, the study which shows the largest fall in Organization of Oil Exporting Countries (OPEC) revenues in *Table TS.6*. The 25% loss in OPEC revenues in the non-trading scenario implies a 17% fall in oil prices shown for 2010 in the figure; this is reduced to a fall of just over 7% with Annex I trading.

Table TS.6: *Costs of Kyoto Protocol implementation for oil exporting region/countries* [a]

Model [b]	Without trading [c]	With Annex-I trading	With "global trading"
G-Cubed	-25% oil revenue	-13% oil revenue	-7% oil revenue
GREEN	-3% real income	"Substantially reduced loss"	N/a
GTEM	0.2% GDP loss	<0.05% GDP loss	N/a
MS-MRT	1.39% welfare loss	1.15% welfare loss	0.36% welfare loss
OPEC Model	-17% OPEC revenue	-10% OPEC revenue	-8% OPEC revenue
CLIMOX	N/A	-10% some oil exporters' revenues	N/A

a   The definition of oil exporting country varies: for G-Cubed and the OPEC model it is the OPEC countries, for GREEN it is a group of oil exporting countries, for GTEM it is Mexico and Indonesia, for MS-MRT it is OPEC + Mexico, and for CLIMOX it is West Asian and North African oil exporters.

b   The models all considere the global economy to 2010 with mitigation according to the Kyoto Protocol targets (usually in the models, applied to CO_2 mitigation by 2010 rather than GHG emissions for 2008 to 2012) achieved by imposing a carbon tax or auctioned emission permits with revenues recycled through lump-sum payments to consumers; no co-benefits, such as reductions in local air pollution damages, are taken into account in the results.

c   "Trading" denotes trading in emission permits between countries.

These studies typically do not consider some or all of the following policies and measures that could lessen the impact on oil exporters:

- policies and measures for non-CO_2 GHGs or non-energy sources of all GHGs;
- offsets from sinks;
- industry restructuring (e.g., from energy producer to supplier of energy services);
- the use of OPEC's market power; and
- actions (e.g., of Annex B Parties) related to funding, insurance, and the transfer of technology.

In addition, the studies typically do not include the following policies and effects that can reduce the total cost of mitigation:

- the use of tax revenues to reduce tax burdens or finance other mitigation measures;
- environmental co- or ancillary benefits of reductions in fossil fuel use; and
- induced technical change from mitigation policies.

As a result, the studies may tend to overstate both the costs to oil exporting countries and overall costs.

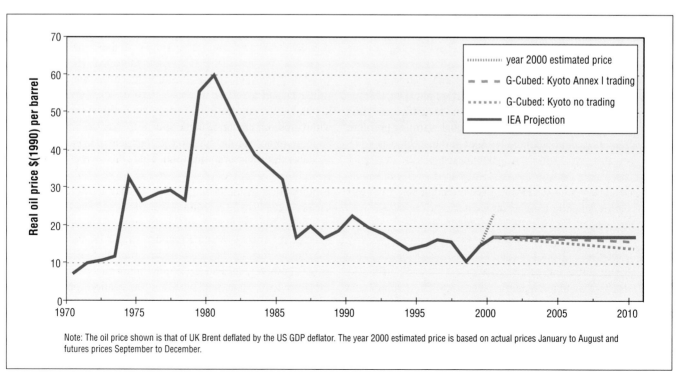

Note: The oil price shown is that of UK Brent deflated by the US GDP deflator. The year 2000 estimated price is based on actual prices January to August and futures prices September to December.

Figure TS.9: *Real oil prices and the effects of Kyoto implementation.*

9.2.3 Gas

Modelling studies suggest that mitigation policies may have the least impact on oil, the most impact on coal, with the impact on gas somewhere between; these findings are established but incomplete. The high variation across studies for the effects of mitigation on gas demand is associated with the importance of its availability in different locations, its specific demand patterns, and the potential for gas to replace coal in power generation.

These results are different from recent trends, which show natural gas usage growing faster than the use of either coal or oil. They can be explained as follows. In the transport sector, the largest user of oil, current technology and infrastructure will not allow much switching from oil to non-fossil fuel alternatives in Annex I countries before about 2020. Annex B countries can only meet their Kyoto Protocol commitments by reducing overall energy use and this will result in a reduction in natural gas demand, unless this is offset by a switch towards natural gas for power generation. The modelling of such a switch remains limited in these models.

9.2.4 Electricity

In general as regards the effects on the electricity sector, mitigation policies either mandate or directly provide incentives for increased use of zero-emitting technologies (such as nuclear, hydro, and other renewables) and lower-GHG-emitting generation technologies (such as combined cycle natural gas). Or, second, they drive their increased use indirectly by more flexible approaches that place a tax on or require a permit for emission of GHGs. Either way, the result will be a shift in the mix of fuels used to generate electricity towards increased use of the zero- and lower-emitting generation technologies, and away from the higher-emitting fossil fuels.

Nuclear power would have substantial advantages as a result of GHG mitigation policies, because power from nuclear fuel produces negligible GHGs. In spite of this advantage, nuclear power is not seen as the solution to the global warming problem in many countries. The main issues are (1) the high costs compared to alternative CCGTs, (2) public acceptance involving operating safety and waste, (3) safety of radioactive waste management and recycling of nuclear fuel, (4) the risks of nuclear fuel transportation, and (5) nuclear weapons proliferation.

9.2.5 Transport

Unless highly efficient vehicles (such as fuel cell vehicles) become rapidly available, there are few options available to reduce transport energy use in the short term, which do not involve significant economic, social, or political costs. No government has yet demonstrated policies that can reduce the overall demand for mobility, and all governments find it politically difficult to contemplate such measures. Substantial additional improvements in aircraft energy efficiency are most likely to be accomplished by policies that increase the price of, and therefore reduce the amount of, air travel. Estimated price elasticities of demand are in the range of -0.8 to -2.7. Raising the price of air travel by taxes faces a number of political hurdles. Many of the bilateral treaties that currently govern the operation of the air transport system contain provisions for exemptions of taxes and charges, other than for the cost of operating and improving the system.

9.3 Sectoral Ancillary Benefits of Greenhouse Gas Mitigation

The direct costs for fossil fuel consumption are accompanied by environmental and public health benefits associated with a reduction in the extraction and burning of the fuels. These benefits come from a reduction in the damages caused by these activities, especially a reduction in the emissions of pollutants that are associated with combustion, such as SO_2, NO_x, CO and other chemicals, and particulate matter. This will improve local and regional air and water quality, and thereby lessen damage to human, animal, and plant health, and to ecosystems. If all the pollutants associated with GHG emissions are removed by new technologies or end-of-pipe abatement (for example, flue gas desulphurization on a power station combined with removal of all other non-GHG pollutants), then this ancillary benefit will no longer exist. But such abatement is limited at present and it is expensive, especially for small-scale emissions from dwellings and cars (See also Section 8.6).

9.4 The Effects of Mitigation on Sectoral Competitiveness

Mitigation policies are less effective if they lead to loss of international competitiveness or the migration of GHG-emitting industries from the region implementing the policy (so-called carbon leakage). The estimated effects, reported in the literature, on international price competitiveness are small while those on carbon leakage appear to beat the stage of competing explanations, with large differences depending on the models and the assumptions used. There are several reasons for expecting that such effects will not be substantial. First, mitigation policies actually adopted use a range of instruments and usually include special treatment to minimize adverse industrial effects, such as exemptions for energy-intensive industries. Second, the models assume that any migrating industries will use the average technology of the area to which they will move; however, instead they may adopt newer, lower CO_2-emitting technologies. Third, the mitigation policies also encourage low-emission technologies and these also may migrate, reducing emissions in industries in other countries (see also Section 8.7).

9.5 Why the Results of Studies Differ

The results in the studies assessed come from different approaches and models. A proper interpretation of the results requires an understanding of the methods adopted and the underlying assumptions of the models and studies. Large differences in results can arise from the use of different reference scenarios or baselines. And the characteristics of the baseline can markedly affect the quantitative results of modelling mitigation policy. For example, if air quality is assumed to be satisfactory in the baseline, then the potential for air-quality ancillary benefits in any GHG mitigation scenario is ruled out by assumption. Even with similar or the same baseline assumptions, the studies yield different results.

As regards the costs of mitigation, these differences appear to be largely caused by different approaches and assumptions, with the most important being the type of model adopted. Bottom-up engineering models assuming new technological opportunities tend to show benefits from mitigation. Top-down general equilibrium models appear to show lower costs than top-down time-series econometric models. The main assumptions leading to lower costs in the models are that:

- new flexible instruments, such as emission trading and joint implementation, are adopted;
- revenues from taxes or permit sales are returned to the economy by reducing burdensome taxes; and
- ancillary benefits, especially from reduced air pollution, are included in the results.

Finally, long-term technological progress and diffusion are largely given in the top-down models; different assumptions or a more integrated, dynamic treatment could have major effects on the results.

10 Decision Analytical Frameworks

10.1 Scope for and New Developments in Analyses for Climate Change Decisions

Decision making frameworks (DMFs) related to climate change involve multiple levels ranging from global negotiations to individual choices and a diversity of actors with different resource endowments, and diverging values and aspirations. This explains why it is difficult to arrive at a management strategy that is acceptable for all. The dynamic interplay among economic sectors and related social interest groups makes it difficult to arrive at a national position to be represented at international fora in the first place. The intricacies of international climate negotiations result from the manifold often-ambiguous national positions as well as from the linkages of climate change policy with other socio-economic objectives.

No DMF can reproduce the above diversity in its full richness. Yet analysts have made significant progress in several directions since SAR. First, they integrate an increasing number of issues into a single analytical framework in order to provide an internally consistent assessment of closely related components, processes, and subsystems. The resulting integrated assessment models (IAMs) cited in Chapter 9, and indeed throughout the whole report, provide useful insights into a number of climate policy issues for policymakers. Second, scientists pay increasing attention to the broader context of climate related issues that have been ignored or paid marginal attention previously. Among other factors, this has fostered the integration of development, sustainability and equity issues into the present report.

Climate change is profoundly different from most other environmental problems with which humanity has grappled. A combination of several features lends the climate problem its uniqueness. They include public good issues raising from the concentration of GHGs in the atmosphere that requires collective global action, the multiplicity of decision makers ranging from global down to the micro level of firms and individuals, and the heterogeneity of emissions and their consequences around the world. Moreover, the long-term nature of climate change originates from the fact that it is the concentration of GHGs that matters rather than their annual emissions and this feature raises the thorny issues of intergenerational transfers of wealth and environmental goods and bads. Next, human activities associated with climate change are widespread, which makes narrowly defined technological solutions impossible, and the interactions of climate policy with other broad socioeconomic policies are strong. Finally, large uncertainties or in some areas even ignorance characterize many aspects of the problem and require a risk management approach to be adopted in all DMFs that deal with climate change.

Policymakers therefore have to grapple with great uncertainties in choosing the appropriate responses. A wide variety of tools have been applied to help them make fundamental choices. Each of those decision analysis frameworks (DAFs) has its own merits and shortcoming through its ability to address some of the above features well, but other facets less adequately. Recent analyses with well-established tools such as cost–benefit analysis as well as newly developed frameworks like the tolerable windows or safe landing approach provide fresh insights into the problem.

Figure TS.10a shows the results of a cost-effectiveness analysis exploring the optimal hedging strategy when uncertainty with respect to the long-term stabilization target is not resolved until 2020, suggesting that abatement over the next few years would be economically valuable if there is a significant probability of having to stay below ceilings that would be otherwise reached within the characteristic time scales of the systems producing greenhouse gases. The degree of near-term hedging in the above analysis is sensitive to the date of resolution of uncertainty, the inertia in the energy system, and the fact that the ultimate concentration target (once it has been revealed) must be met at all costs. Other experiments, such as those with cost-benefit models framed as a Bayesian decision analysis

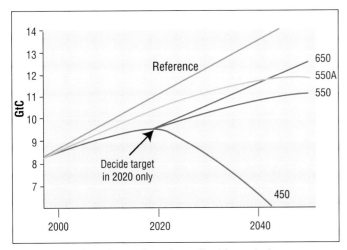

Figure TS.10a: *Optimal carbon dioxide emissions strategy, using a cost-effectiveness approach.*

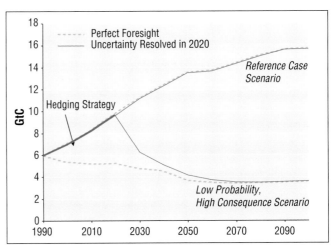

Figure TS.10b: *Optimal hedging strategy for low probability, high consequence scenario using a cost-benefits optimization approach.*

problem show that optimal near-term (next two decades) emission paths diverge only modestly under perfect foresight, and hedging even for low-probability, high-consequence scenarios (see *Figure TS.10b*). However, decisions about near-term climate policies may have to be made while the stabilization target is still being debated. Decision-making therefore should consider appropriate hedging against future resolution of that target and possible revision of the scientific insights in the risks of climate change. There are significant differences in the two approaches. With a cost-effectiveness analysis, the target must be made regardless of costs. With a cost-benefit analysis, costs and benefits are balanced at the margin. Nevertheless, the basic message is quite similar and involves the explicit incorporation of uncertainty and its sequential resolution over time. The desirable amount of hedging depends upon one's assessment of the stakes, the odds, and the costs of policy measures. The risk premium – the amount that society is willing to pay to reduce risk – ultimately is a political decision that differs among countries.

Cost-effectiveness analyses seek the lowest cost of achieving an environmental target by equalizing the marginal costs of mitigation across space and time. Long-term cost-effectiveness studies estimate the costs of stabilizing atmospheric CO_2 concentrations at different levels and find that the costs of the 450ppmv ceiling are substantially greater than those of the 750ppmv limit. Rather than seeking a single optimal path, the tolerable windows/safe landing approach seeks to delineate the complete array of possible emission paths that satisfy externally defined climate impact and emission cost constraints. Results indicate that delaying near-term effective emission reductions can drastically reduce the future range of options for relatively tight climate change targets, while less tight targets offer more near-term flexibility.

10.2 International Regimes and Policy Options

The structure and characteristics of international agreements on climate change will have a significant influence on the effectiveness and costs and benefits of mitigation. The effectiveness and the costs and benefits of an international climate change regime (such as the Kyoto Protocol or other possible future agreements) depend on the number of signatories to the agreement and their abatement targets and/or policy commitment. At the same time, the number of signatories depends on the question of how equitably the commitments of participants are shared. Economic efficiency (minimizing costs by maximizing participation) and equity (the allocation of emissions limitation commitments) are therefore strongly linked.

There is a three-way relationship between the design of the international regime, the cost-effectiveness/efficiency of climate policies, and the equity of the consequent economic outcomes. As a consequence, it is crucial to design the international regime in a way that is considered both efficient and equitable. The literature presents different theoretical strategies to optimize an international regime. For example, it can be made attractive for countries to join the group that commits to specific targets for limitation and reduction of emissions by increasing the equity of a larger agreement – and therefore its efficiency – through measures like an appropriate distribution of targets over time, the linkage of the climate debate with other issues ("issue linkage"), the use of financial transfers to affected countries ("side payments"), or technology transfer agreements.

Two other important concerns shape the design of an international regime: "implementation" and "compliance". The effectiveness of the regime, which is a function of both implementation and compliance, is related to actual changes of behaviour that promote the goals of the accord. *Implementation* refers to the translation of international accords into domestic law, policy, and regulations by national governments. *Compliance* is related to whether and to what extent countries do in fact adhere to provisions of an accord. *Monitoring, reporting,* and *verification* are essential for the effectiveness of international environmental regimes, as the systematic monitoring, assessment, and handling of implementation failures have been so far relatively rare. Nonetheless, efforts to provide "systems of implementation review" are growing, and are already incorporated into the UNFCCC structure. The challenge for the future is to make them more effective, especially by improving data on national emissions, policies, and measures.

10.3   Linkages to National and Local Sustainable Development Choices

Much of the ambiguity related to sustainable development and climate change arises from the lack of measurements that could provide policymakers with essential information on the alternative choices at stake, how those choices affect clear and recognizable social, economic, and environmental critical issues, and also provide a basis for evaluating their performance in achieving goals and targets. Therefore, indicators are indispensable to make the concept of sustainable development operational. At the national level important steps in the direction of defining and designing different sets of indicators have been undertaken; however, much work remains to be done to translate sustainability objectives into practical terms.

It is difficult to generalize about sustainable development policies and choices. Sustainability implies and requires diversity, flexibility, and innovation. Policy choices are meant to introduce changes in technological patterns of natural resource use, production and consumption, structural changes in the production systems, spatial distribution of population and economic activities, and behavioural patterns. Climate change literature has by and large addressed the first three topics, while the relevance of choices and decisions related to behavioural patterns and lifestyles has been paid scant attention. Consumption patterns in the industrialized countries are an important reason for climate change. If people changed their preferences this could alleviate climate change considerably. To change consumption patterns, however, people must not only change their behaviour but also change themselves because these patterns are an essential element of lifestyles and, therefore, of self-esteem. Yet, apart from climate change there are other reasons to do so as well as indications that this change can be fostered politically.

A critical requirement of sustainable development is a capacity to design policy measures that, without hindering development and consistent with national strategies, could exploit potential synergies between national economic growth objectives and environmentally focused policies. Climate change mitigation strategies offer a clear example of how co-ordinated and harmonized policies can take advantage of the synergies between the implementation of mitigation options and broader

objectives. Energy efficiency improvements, including energy conservation, switch to low carbon content fuels, use of renewable energy sources and the introduction of more advanced non conventional energy technologies, are expected to have significant impacts on curbing actual GHG emission tendencies. Similarly, the adoption of new technologies and practices in agriculture and forestry activities as well as the adoption of clean production processes could make substantial contributions to the GHG mitigation effort. Depending on the specific context in which they are applied, these options may entail positive side effects or double dividends, which in some cases are worth undertaking whether or not there are climate-related reasons for doing so.

Sustainable development requires radical technological and related changes in both developed and developing countries. Technological innovation and the rapid and widespread transfer and implementation of individual technological options and choices, as well as overall technological systems, constitute major elements of global strategies to achieve both climate stabilization and sustainable development. However, technology transfer requires more than technology itself. An enabling environment for the successful transfer and implementation of technology plays a crucial role, particularly in developing countries. If technology transfer is to bring about economic and social benefits it must take into account the local cultural traditions and capacities as well as the institutional and organizational circumstances required to handle, operate, replicate, and improve the technology on a continuous basis.

The process of integrating and internalizing climate change and sustainable development policies into national development agendas requires new problem solving strategies and decision-making approaches. This task implies a twofold effort. On one hand, sustainable development discourse needs greater analytical and intellectual rigor (methods, indicators, etc.) to make this concept advance from theory to practice. On the other hand, climate change discourse needs to be aware of both the restrictive set of assumptions underlying the tools and methods applied in the analysis, and the social and political implications of scientific constructions of climate change. Over recent years a good deal of analytical work has addressed the problem in both directions. Various approaches have been explored to transcend the limits of the standard views and decision frameworks in dealing with issues of uncertainty, complexity, and the contextual influences of human valuation and decision making. A common theme emerges: the emphasis on participatory decision making frameworks for articulating new institutional arrangements.

10.4   Key Policy-relevant Scientific Questions

Different levels of globally agreed limits for climate change (or for corresponding atmospheric GHG concentrations), entail different balances of mitigation costs and net damages for individual nations. Considering the uncertainties involved and future learning, climate stabilization will inevitably be an iterative process: nation states determine their own national targets based on their own exposure and their sensitivity to other countries' exposure to climate change. The global target emerges from consolidating national targets, possibly involving side payments, in global negotiations. Simultaneously, agreement on burden sharing and the agreed global target determines national costs. Compared to the expected net damages associated with the global target, nation states might reconsider their own national targets, especially as new information becomes available on global and regional patterns and impacts of climate change. This is then the starting point for the next round of negotiations. It follows from the above that establishing the "magic number" (i.e., the upper limit for global climate change or GHG concentration in the atmosphere) will be a long process and its source will primarily be the policy process, hopefully helped by improving science.

Looking at the key dilemmas in climate change decision making, the following conclusions emerge (see also *Table TS.7*):

- a carefully crafted portfolio of mitigation, adaptation, and learning activities appears to be appropriate over the next few decades to hedge against the risk of intolerable magnitudes and/or rates of climate change (impact side) and against the need to undertake painfully drastic emission reductions if the resolution of uncertainties reveals that climate change and its impacts might imply high risks;

- emission reduction is an important form of mitigation, but the mitigation portfolio includes a broad range of other activities, including investments to develop low-cost non-carbon, energy efficient and carbon management technologies that will make future CO_2 mitigation less expensive;

- timing and composition of mitigation measures (investment into technological development or immediate emission reductions) is highly controversial because of the technological features of energy systems, and the range of uncertainties involved in the impacts of different emission paths;

- international flexibility instruments help reduce the costs of emission reductions, but they raise a series of implementation and verification issues that need to be balanced against the cost savings;

- while there is a broad consensus to use the Pareto optimality[30] as the efficiency principle, there is no agreement on the best equity principle on wich to build an equitable international regime. Efficiency and equity are important concerns in negotiating emission limitation schemes, and they are not mutually exclusive. Therefore, equity will play an important role in determining the distribution of emissions allowances and/or within compensation schemes following emission trad-

ing that could lead to a disproportionately high level of burden on certain countries. Finally, it could be more important to build a regime on the combined implications of the various equity principles rather than to select any one particular equity principle. Diffusing non-carbon, energy-efficient, as well as other GHG reducing technologies worldwide could make a significant contribution to reducing emissions over the short term, but many barriers hamper technology transfer, including market imperfections, political problems, and the often-neglected transaction costs;

- some obvious linkages exist between current global and continental environmental problems and attempts of the international community to resolve them, but the potential synergies of jointly tackling several of them have not yet been thoroughly explored, let alone exploited.

Mitigation and adaptation decisions related to anthropogenically induced climate change differ. Mitigation decisions involve many countries, disperse benefits globally over decades to centuries (with some near-term ancillary benefits), are driven by public policy action, based on information available today, and the relevant regulation will require rigorous enforcement. In contrast, adaptation decisions involve a shorter time span between outlays and returns, related costs and benefits accrue locally, and their implementation involves local public policies and private adaptation of the affected social agents, both based on improving information. Local mitigation and adaptive capacities vary significantly across regions and over time. A portfolio of mitigation and adaptation policies will depend on local or national priorities and preferred approaches in combination with international responsibilities.

Given the large uncertainties characterizing each component of the climate change problem, it is difficult for decision makers to establish a globally acceptable level of stabilizing GHG concentrations today. Studies appraised in Chapter10 support the obvious expectations that lower stabilization targets involve substantially higher mitigation costs and relatively more ambitious near-term emission reductions on the one hand, but, as reported by WGII, lower targets induce significantly smaller bio/geophysical impacts and thus induce smaller damages and adaptation costs.

11 Gaps in Knowledge

Important gaps in own knowledge on which additional research could be useful to support future assessments include:

- *Further exploration of the regional, country, and sector specific potentials of technological and social innovation options, including:*
  - The short, medium, and long-term potential and costs of both CO_2 and non-CO_2, non-energy mitigation options;

---

[30] Pareto optimum is a requirement or status that an individual's welfare could not be further improved without making others in the society worse off.

Table TS.7: *Balancing the near-term mitigation portfolio*

Issue	Favouring modest early abatement	Favouring stringent early abatement
Technology development	• Energy technologies are changing and improved versions of existing technologies are becoming available, even without policy intervention. • Modest early deployment of rapidly improving technologies allows learning-curve cost reductions, without premature lock-in to existing, low-productivity technology. • The development of radically advanced technologies will require investment in basic research.	• Availability of low-cost measures may have substantial impact on emissions trajectories. • Endogenous (market-induced) change could accelerate development of low-cost solutions (learning-by-doing). • Clustering effects highlight the importance of moving to lower emission trajectories. • Induces early switch of corporate energy R&D from fossil frontier developments to low carbon technologies.
Capital stock and inertia	• Beginning with initially modest emissions limits avoids premature retirement of existing capital stocks and takes advantage of the natural rate of capital stock turnover. • It also reduces the switching cost of existing capital and prevents rising prices of investments caused by crowding out effects.	• Exploit more fully natural stock turnover by influencing new investments from the present onwards. • By limiting emissions to levels consistent with low CO_2 concentrations, preserves an option to limit CO_2 concentrations to low levels using current technology. • Reduces the risks from uncertainties in stabilization constraints and hence the risk of being forced into very rapid reductions that would require premature capital retirement later.
Social effects and inertia	• Gradual emission reduction reduces the extent of induced sectoral unemployment by giving more time to retrain the workforce and for structural shifts in the labour market and education. • Reduces welfare losses associated with the need for fast changes in people's lifestyles and living arrangements.	• Especially if lower stabilization targets would be required ultimately , stronger early action reduces the maximum rate of emissions abatement required subsequently and reduces associated transitional problems, disruption, and the welfare losses associated with the need for faster later changes in people's lifestyles and living arrangements.
Discounting and intergenerational equity	• Reduces the present value of future abatement costs (*ceteris paribus*), but possibly reduces future relative costs by furnishing cheap technologies and increasing future income levels.	• Reduces impacts and (*ceteris paribus*) reduces their present value.
Carbon cycle and radiative change	• Small increase in near-term, transient CO_2 concentration. • More early emissions absorbed, thus enabling higher total carbon emissions this century under a given stabilization constraint (to be compensated by lower emissions thereafter).	• Small decrease in near-term, transient CO_2 concentration. • Reduces peak rates in temperature change.
Climate change impacts	• Little evidence on damages from multi-decade episodes of relatively rapid change in the past.	• Avoids possibly higher damages caused by faster rates of climate change.

- Understanding of technology diffusion across different regions;
- Identifying opportunities in the area of social innovation leading to decreased greenhouse gas emissions;
- Comprehensive analysis of the impact of mitigation measures on C flows in and out of the terrestrial system; and
- Some basic inquiry in the area of geo-engineering.

- *Economic, social, and institutional issues related to climate change mitigation in all countries. Priority areas include*:
  - Much more analysis of regionally specific mitigation options, barriers, and policies is recommended as these are conditioned by the regions' mitigative capacity;
  - The implications of mitigation on equity;
  - Appropriate methodologies and improved data sources for climate change mitigation and capacity building in the area of integrated assessment;
  - Strengthening future research and assessments, especially in developing countries.

- *Methodologies for analysis of the potential of mitigation options and their cost, with special attention to comparability of results. Examples include*:
  - Characterizing and measuring barriers that inhibit greenhouse gas-reducing action;
  - Make mitigation modelling techniques more consistent, reproducible, and accessible;
  - Modelling technology learning; improving analytical tools for evaluating ancillary benefits, e.g. assigning the costs of abatement to greenhouse gases and to other pollutants;
  - Systematically analyzing the dependency of costs on baseline assumptions for various greenhouse gas stabilization scenarios;
  - Developing decision analytical frameworks for dealing with uncertainty as well as socio-economic and ecological risk in climate policymaking;
  - Improving global models and studies, their assumptions, and their consistency in the treatment and reporting of non-Annex I countries and regions.

- *Evaluating climate mitigation options in the context of development, sustainability, and equity. Examples include*:
  - More research is needed on the balance of options in the areas of mitigation and adaptation and of the mitigative and adaptive capacity in the context of DES;

- Exploration of alternative development paths including sustainable consumption patterns in all sectors, including the transportation sector, and integrated analysis of mitigation and adaptation;
- Identifying opportunities for synergy between explicit climate policies and general policies promoting sustainable development;
- Integration of inter- and intragenerational equity in climate change mitigation studies;
- Implications of equity assessments;
- Analysis of scientific, technical, and economic aspects of implications of options under a wide variety of stabilization regimes;
- Determining what kinds of policies interact with what sorts of socio-economic conditions to result in futures characterized by low CO_2 emissions;
- Investigation on how changes in societal values may be encouraged to promote sustainable development; and
- Evaluating climate mitigation options in the context of and for synergy with potential or actual adaptive measures.

- *Development of engineering-economic, end-use, and sectoral studies of GHG emissions mitigation potentials for specific regions and/or countries of the world, focusing on:*
  - Identification and assessment of mitigation technologies and measures that are required to deviate from "business-as-usual" in the short term (2010, 2020);
  - Development of standardized methodologies for quantifying emissions reductions and costs of mitigation technologies and measures;
  - Identification of barriers to the implementation of the mitigation technologies and measures;
  - Identification of opportunities to increase adoption of GHG emissions mitigation technologies and measures through connections with ancillary benefits as well as furtherance of the DES goals; and
  - Linking the results of the assessments to specific policies and programmes that can overcome the identified barriers as well as leverage the identified ancillary benefits.

1

Setting the Stage: Climate Change and Sustainable Development

Co-ordinating Lead Authors:
TARIQ BANURI (PAKISTAN), JOHN WEYANT (USA)

Lead Authors:
Grace Akumu (Kenya), Adil Najam (Pakistan), Luiz Pinguelli Rosa (Brazil), Steve Rayner (USA), Wolfgang Sachs (Germany), Ravi Sharma (India/UNEP), Gary Yohe (USA)

Contributing Authors:
Anil Agarwal (India), Steve Bernow (USA), Robert Costanza (USA), Thomas Downing (USA), Sivan Kartha (USA), Ashok Khosla (India), Ambuj Sagar (India), John Robinson (Canada), Ferenc Toth (Germany)

Review Editors:
Hans Opschoor (The Netherlands), Kirit Parikh (India)

CONTENTS

EXECUTIVE SUMMARY

This chapter places climate change mitigation, mitigation policy, and the contents of the rest of the report in the broader context of development, equity, and sustainability. This context reflects the explicit conditions and principles laid down by the UN Framework Convention on Climate Change (UNFCCC) on the pursuit of the ultimate objective of stabilizing greenhouse gas concentrations. The UNFCCC imposes three conditions on the goal of stabilization, namely, that it should take place within a time-frame sufficient to "allow ecosystems to adapt naturally to climate change, to ensure that food production is not threatened and to enable economic development to proceed in a sustainable manner" (Art. 2). It also specifies several principles to guide this process: equity, common but differentiated responsibilities, precaution, cost-effective measures, right to sustainable development, and support for an open international economic system (Art. 3).

Previous IPCC assessment reports sought to facilitate this pursuit by comprehensively describing, cataloguing and comparing technologies and policy instruments that could be used to achieve mitigation of greenhouse gas emissions in a cost-effective and efficient manner. The present assessment advances this process by including recent analyses of climate change that place policy evaluations in the context of sustainable development. This expansion of scope is consistent both with the evolution of the literature on climate change and importance accorded by the UNFCCC to sustainable development - including the recognition that "Parties have a right to, and should promote sustainable development" (Art. 3.4). It therefore goes some way towards filling the gaps in earlier assessments.

Climate Change involves complex interactions between climatic, environmental, economic, political, institutional, social, and technological processes. It cannot be addressed or comprehended in isolation from broader societal goals (such as sustainable development), or other existing or probable future sources of stress. In keeping with this complexity, a multiplicity of approaches have emerged to analyze climate change and related challenges. Many of these incorporate concerns about development, equity, and sustainability (albeit partially and gradually) into their framework and recommendations. Each approach emphasizes certain elements of the problem, and focuses on certain classes of responses, including for example, optimal policy design, building capacity for designing and implementing policies, strengthening synergies between climate change mitigation and/or adaptation and other societal goals, and policies to enhance societal learning. These approaches are therefore complementary rather than mutually exclusive.

This chapter brings together three broad classes of analysis, which differ not so much in terms of their ultimate goals as in their points of departure and preferred analytical tools. The three approaches start with concerns, respectively, about efficiency and cost-effectiveness, equity and sustainable development, and global sustainability and societal learning. The difference between the three approaches we have selected lies in their starting point, not in their ultimate goals. Regardless of the starting point of the analysis, many studies try in their own way to incorporate other concerns. For example, many analyses that approach climate change mitigation from a cost-effectiveness perspective try to bring in considerations of equity and sustainability through their treatment of costs, benefits, and welfare. Similarly, the class of studies motivated strongly by considerations of inter-country equity tend to argue that equity is needed to ensure that developing countries can pursue their internal goals of sustainable development–a concept that includes the implicit components of sustainability and efficiency. Likewise, analysts focused on concerns of global sustainability have been compelled by their own logic to make a case for global efficiency–often modelled as the decoupling of production from material flows–and social equity. In other words, each of the three perspectives has led writers to search for ways to incorporate concerns that lie beyond their initial starting point. All three classes of analyses look at the relationship of climate change mitigation with all three goals–development, equity, and sustainability–albeit in different and often highly complementary ways. Nevertheless, they frame the issues differently, focus on different sets of causal relationships, use different tools of analysis, and often come to somewhat different conclusions.

There is no presumption that any particular perspective for analysis is most appropriate at any level. Moreover, the three perspectives are viewed here as being highly synergistic. The important changes have been primarily in the types of questions being asked and the kinds of information being sought. In practice, the literature has expanded to add new issues and new tools, subsuming rather than discarding the analyses included in the other ones. The range and scope of climate policy analyses can be understood as a gradual broadening of the types and extent of uncertainties that analysts have been willing and able to address.

The first perspective on climate policy considered is Cost-effectiveness. It represents a perspective that is well represented in conventional climate policy analysis and in the First through Third Assessments. These analyses have generally been driven directly or indirectly by the question of what the most cost-effective amount of mitigation for the global economy is, starting from a particular baseline greenhouse gas (GHG) emissions scenario, reflecting a specific set of socioeconomic scenarios. Within this framework, important issues include measuring the performance of various technologies and the removal of barriers (such as existing subsidies) to the implementation of those candidate policies most likely to contribute to emissions reductions. In a sense, the focus of analysis here has been on identifying an efficient pathway through the interactions of mitigation policies and economic development, conditioned by considerations of equity and sustainability, but not primarily guided by them. At this level, policy analysis has almost always taken the existing institutions and tastes of individuals as given; assumptions that might be valid for a decade or two, but may become more questionable over many decades.

The impetus for the expansion in the scope of the climate policy analysis and discourse to include Equity considerations was to include considerations not simply of the impacts of climate change and mitigation policies on global welfare as a whole, but also of the effects of climate change and mitigation policies on existing inequalities among and within nations. The literature on equity and climate change has advanced considerably over the last two decades, but there is no consensus on what constitutes fairness. Once equity issues were introduced into the assessment agenda, though, they became important components in defining the search for efficient emissions mitigation pathways. The considerable literature that indicated how environmental policies could be hampered or even blocked by those who considered them unfair became relevant. In the light of these results, it became clear how and why any widespread perception that a mitigation strategy is unfair would likely engender opposition to that strategy, perhaps to the extent of rendering it non-optimal. Some cost-effectiveness analyses had, in fact, laid the groundwork for applying this literature by demonstrating the sensitivity of some equity measures to policy design, national perspective, and regional context. Indeed, cost-effectiveness analyses had even highlighted similar sensitivities for other measures of development and sustainability.

As mentioned, the analyses that start from equity concerns have by and large focused on the needs of developing countries, and, in particular, on the commitment expressed in Article 3.4 of the UNFCCC to the pursuit of sustainable devel-

opment. Assessing the climate challenge from a sustainable development perspective immediately reveals that countries differ in ways that have dramatic implications for scenario baselines and the range of mitigation options that can be considered. The climate policies that are feasible, and or desirable, in a particular country depend importantly on its available resources and institutions, and on its overall objectives including climate change as but one component. Moreover, although OECD centered models may give helpful first order insights into the efficacy of global scale policy interventions, their underlying assumptions may make them less useful when the heterogeneity of nations is fully incorporated. Recognizing this heterogeneity may lead to a different range of policy options than has been considered likely thus far and may ultimately feed back into policy design for Annex I. Recognizing heterogeneity among countries reveals, in short, differences in the capacities of different sectors that may also enhance appreciation of what can be done by non-state actors as well as governments to build their ability to mitigate.

While sustainability has been incorporated in the analyses in a number of ways, a class of studies takes the issue of Global Sustainability as the point of departure. One popular method for identifying constraints and opportunities within this perspective is to identify future sustainable states and then examine possible transition paths to those states for feasibility and desirability. In the case of developing countries this leads to a number of possible strategies that can depart significantly from what the developed countries pursued in the past.

The chapter closes with a discussion of preliminary attempts to integrate the information and insights that result from studies done from the three perspectives. Within this report the concept of "co-benefits" is used to capture dimensions of the response to mitigation policies from the equity and sustainability perspectives in a way that could be used to modify the cost projections produced by those working form the cost-effectiveness perspective although ancillary benefit has been more widely used in the literature. The concept of "mitigative capacity" is also introduced as a possible way to integrate results derived from the application of the three perspectives in the future.

1.1 Introduction

This chapter puts climate change mitigation and climate change mitigation policy in the broader context of development, equity, and sustainability. The ultimate objective of the United Nations Framework Convention on Climate Change (UNFCCC) "is to achieve … stabilization of greenhouse gas (GHG) concentrations in the atmosphere at a level that would prevent dangerous anthropogenic interference with the climate system. Such a level should be achieved within a timeframe sufficient to allow ecosystems to adapt naturally to climate change, to ensure that food production is not threatened and to enable economic development to proceed in a sustainable manner" (Article 2). The UNFCCC goes on to specify principles that should guide this process: equity, common but differentiated responsibilities, precaution, cost-effectiveness, the right to sustainable development, and the avoidance of arbitrary restriction on international trade (Article 3). Previous Intergovernmental Panel on Climate Change (IPCC) assessment reports sought to lay the groundwork for policymakers pursuing the UNFCCC goals by comprehensively describing, cataloguing, and comparing technologies and policy instruments that could be used to achieve the mitigation of GHG emissions.

The attention accorded in the UNFCCC to sustainable development–including the recognition that "Parties have a right to, and should promote sustainable development" (Article 3.4)–has not, however, been matched by its treatment in previous IPCC assessment reports. As a result, the present assessment seeks to address this mismatch by placing policy evaluations in the broader context of development, equity, and sustainability as outlined in the Convention. The rising stature of development, equity, and sustainability in the discussion of mitigation is, indeed, entirely consistent with the overall evolution of the scope of the literature on climate change.

In fact, the analysis of climate change policies has evolved significantly between the preparation of the First Assessment Report (FAR; IPCC, 1991), the Second Assessment Report (SAR; IPCC, 1996), and Third Assessment Report (TAR) of the IPCC. In the late 1980s, for example, the focus of policy analysis was almost exclusively on climate change mitigation through emissions reduction. GHG emissions were modelled almost exclusively in terms of carbon dioxide (CO_2) from energy use (Nordhaus and Yohe, 1983; Edmonds and Reilly, 1985); and emissions reductions were to be achieved primarily by increasing the prices of fossil fuels. Hence, it is hardly surprising that, with a few exceptions (e.g., Bradley and Williams, 1989; Parikh *et al.*, 1991), carbon taxes were overwhelmingly the most commonly analyzed policy instrument. FAR (IPCC, 1991) documents the possible ramifications of a wide range of policy instruments, but it reports that carbon taxes are again the most fully analyzed in the literature. This report, by way of contrast, demonstrates a significant enhancement in the capacity of policy analysts to consider the sources and sinks of multiple gases as well as a broader array of policy instruments to curtailing the emission of these gases into the atmosphere.

Also, little consideration was given in FAR to policies designed to enhance adaptation to climate change impacts. In TAR, though, adaptation has become a major focus of the Working Group II (WGII) report (IPCC, 2001). At the beginning of the 1990s, assessments of the capabilities of countries to achieve emissions reductions were almost exclusively based on estimates of their fossil fuel consumption. With a few exceptions (e.g., Grubb, 1991; Rayner, 1993) no explicit consideration was given to social, cultural, political, institutional, or decision-making constraints on the capacity of governments to implement climate change policies.

Consistent with the state of the policy literature on climate change, FAR (IPCC, 1991) also made no attempt to address issues of equity. Prior to the publication of *Global Warming in an Unequal World* (Agarwal and Narain, 1991a), consideration of the fairness of climate change policies (both among and within countries) received little attention from analysts and policymakers (for exceptions see Grubb, 1989; Kasperson and Dow, 1991; Parikh *et al.*, 1991). The IPCC Second Assessment Report WGIII (IPCC WG III, 1996) did, however, mention the need to extend the focus of analysis and assessment into areas that included issues not only of equity and fairness, but also of development and sustainability. Some of the studies available then did note the distributional effects of alternative policy designs and targets; and some did trace other effects into the domains of development and sustainability. The point here is not that earlier work ignored these broader issues, but that this report begins the process of making them more central in the assessment of the existing policy analyses. This report begins the task of integrating technology and policy characterizations into alternative development scenarios and policy decision-frameworks that are broadly conceived. In the same spirit, this chapter seeks to locate the work of WGIII in a broader context of development, equity, and sustainability. In the process, we draw on several themes (elaborated in subsequent chapters) to identify opportunities to enhance the capacity of regions, countries, and communities to mitigate GHG emissions while simultaneously pursuing their sustainable development goals. Neither the greenhouse gas mitigation nor the sustainable development initiative, however, eliminates the need to conduct efficiency-based assessments of the opportunity costs of mitigation and/or the enhancement of the capacity to mitigate. Instead, climate change and sustainable development both simply expand the number of objectives against which these costs need to be measured.

The expansion of IPCC's scope in this WGIII report complements that of WGII (IPCC, 2001), which addresses the impacts of continued atmospheric accumulation of GHGs and the adaptive capacity of countries to adjust to the consequences of that accumulation. The analogous concept of mitigative capacity (Yohe, in press) is offered in Section 1.5 as one tool with which policymakers and researchers alike might integrate insights drawn from the domains of cost-effectiveness, equity, and sustainability into their understanding of mitigation. Drawing attention to concepts like mitigative capacity also allows the reader to approach the complexity of mitigation within a framework that mirrors the emphasis placed on adaptive capacity by the TAR WGII Report.

The expansion of the range and scope of IPCC policy analysis, just described, can be understood as a gradual broadening of the types and extent of uncertainties that analysts have been willing and able to address. A graphic representation of this expansion of interest and capability (*Figure 1.1*) shows that the policy sciences have made significant advances since IPCC FAR. This figure simply depicts different perspectives that have been employed to examine climate policy issues and the stage at which they were incorporated into the IPCC process. Progression through the IPCC assessments displayed in *Figure 1.1* represents expansions in the scope of climate policy analyses since 1980. There is no presumption that any particular framework for analysis is most appropriate at any level. The important changes are primarily in the types of questions being asked and the kinds of information being sought. In practice, the literature has expanded to add new issues and has subsumed rather than discarded the analyses of the initial issues. With each assessment, IPCC has added to the necessary tool set without obviating the need for the tools developed in the earlier assessments.

The first concern of policy analysis to be included in IPCC assessments is labelled "Cost-effectiveness" in *Figure 1.1*. It represents the field of conventional climate policy analysis that is well represented in the First through to the Third Assessments. These analyses are generally driven directly or indirectly by the question of what is the most cost-effective amount of mitigation for the global economy starting from a particular baseline GHG emissions scenario, and reflecting a specific set of socioeconomic scenarios. Within this framework, important issues include measuring the performance of various technologies and the removal of barriers (such as existing subsidies) to the implementation of the candidate policies

most likely to contribute to emissions reductions. In a sense, the focus of such analysis is to identify an efficient pathway through the interactions of mitigation policies and economic development, in some cases conditioned by considerations of equity and sustainability, but not primarily guided by them. At this level, IPCC policy analysis has almost always taken the existing institutions and tastes of individuals as given; such assumptions might be valid for a decade or two, but may become more questionable over many decades.[1]

By introducing the issue of equity, SAR (IPCC, 1996) broadened the IPCC policy discourse; a process reflected by "Equity" in *Figure 1.1*. The impetus for this expansion in the scope of the discourse was to include considerations not simply of the impacts of climate change and mitigation policies on global welfare as a whole, but also of the effects of climate change and mitigation policies on existing inequalities among and within nations. The literature on equity and climate change has advanced considerably since SAR, but there is no consensus on what constitutes fairness. Once equity issues were introduced into the IPCC assessment agenda, though, they became important components in defining the search for efficient emissions mitigation pathways. The considerable literature that indicates how environmental policies could be hampered or even blocked by those who considered them unfair became relevant (National Academy of Engineering, 1986; Rayner and Cantor, 1987; Grubb, 1989; Weiss, 1989; Kasperson and Dow, 1991). In light of these results, it became clear how and why any widespread perception that a mitigation strategy is unfair would likely engender opposition to that strategy, perhaps to the extent of rendering it non-optimal (or even infeasible). Some cost-effectiveness analyses had, in fact, laid the groundwork for applying this literature by demonstrating the sensitivity of some equity measures to policy design, national perspective, and regional context. Indeed, cost-effectiveness analyses had even highlighted similar sensitivities for other measures of development and sustainability.

Throughout this evolution, though, the historical model of societies that industrialized in the nineteenth and twentieth centuries served as the central notion of what constitutes development in both the cost-effectiveness and equity perspectives. According to some analysts (e.g., Simon and Kahn, 1984; Beckerman, 1996) this path represents the best model for global prosperity. However, a growing parallel literature recognizes the importance of diverse development pathways in achieving an environmentally and socioeconomically sustainable world (see Section 1.4). This insight can serve as the basis of a third analytical perspective—a perspective represented in *Figure 1.1* by "Global Sustainability". As yet, however, analyses of such alternative development pathways remain largely unrealized within the framework of IPCC. Still, the first steps in this direction can be detected throughout this volume.

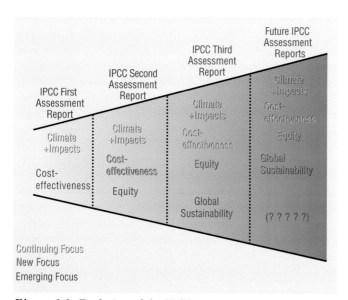

Continuing Focus
New Focus
Emerging Focus

Figure 1.1*: Evolution of the IPCC assessment process.*

---

[1] Recent work in the theory of public choice (e.g., Michaelowa and Dutschke, 1998) suggests that a more dynamic view of institutions can be incorporated into this style of analysis.

The above description of three complementary perspectives on climate change mitigation and the broad societal goals of development, equity, and sustainability bears elaboration. The rest of this chapter can be seen as a triptych, in which each section presents a particular perspective on climate change mitigation–motivated respectively by considerations of cost-effectiveness, equity, and sustainability. However, we also describe how each of the perspectives has attempted to address and incorporate concerns that lie beyond their initial starting points. For example, Section 1.2 details the Cost-effectiveness perspective; however, its two concluding sections, (1.2.5 and 1.2.6) describe how this approach has addressed concerns of equity and sustainability. Similarly, Section 1.3 is entitled "Equity and Sustainable Development" in recognition of the fact that writers examining the issue of climate change from a vantage point of global equity have generally sought to explore how developing countries could pursue their sustainable development goals. In the penultimate sub-section (1.3.4) of this section, we examine the concept of sustainable development and describe its relationship to cost-effectiveness, efficiency, and sustainability. Finally, the theme of Section 1.4 is Global sustainability; and its two main sub-sections (1.4.2 and 1.4.3) discuss issues of resource efficiency (de-coupling growth from resource flows), and values and norms that include issues of equity.

In other words, instead of forcing the literature that describes the relationship between climate change mitigation and development, equity, and sustainability into a single framework, we have tried to bring out both the commonalities and differences between alternative approaches and analytical frameworks. All three classes of analyses look at the relationship of climate change mitigation with all three goals–development, equity, and sustainability–albeit in different and often highly complementary ways. Nevertheless, they frame the issues differently, focus on different sets of causal relationships, use different tools of analysis, and often come to somewhat different conclusions. Accordingly, they are likely to be useful to decision makers in different ways.

Assessing the climate challenge with a sustainable development perspective immediately reveals that countries differ in ways that have dramatic implications for baselines and the range of mitigation options that can be considered. Moreover, although models centred on Organization of Economic Co-operation and Development (OECD) countries may give helpful first-order insights into the efficacy of global policy interventions, the underlying assumptions may make such models less useful when the heterogeneity of nations is incorporated fully. Recognition of this heterogeneity may lead to a different range of policy options than considered likely thus far, and may ultimately feed back into policy design for Annex I countries. Recognizing heterogeneity among countries reveals, in short, differences in the capacities of different sectors, which may also enhance appreciation of what can be done by non-state actors as well as governments to build their mitigative capacity.

The expansion of analytic perspectives also represents the increasing complexity of issues selected for analytic focus. On the left-hand side of *Figure 1.1*, complexity refers primarily to the analytical challenges presented by individual technologies (such as fuel cells or photovoltaics) or specific policy instruments (such as carbon taxes or tradable emissions permits). Moving from left to right across the figure, such complexities become compounded, first by interactions among technologies and policy instruments, then among mitigation and adaptation issues, and, finally among climate change issues narrowly defined and a wide range of environmental and socioeconomic issues. Finally, linkages and interactions with policy objectives for the development of the global economy come into the picture.

A major part of the complexity that must be dealt with in formulating climate policies is the uncertainties about how the world and the climate system will evolve without new policies, about what policies will be implemented now and in the future, and about the efficacy of those policies. The economist Frank Knight (1921) introduced a fundamental distinction between "risk" and "uncertainty",[2] whereby risk refers to cases for which the probable outcomes are predicted through well-established theories and methods, and with reliable data (e.g., the radiative forcing of a tonne of CO_2 or the efficiency of a gas turbine); and uncertainty to situations in which theories and methods are widely accepted, but the appropriate data are not available or are fragmentary, and probabilities and outcomes can be assessed subjectively by relevant experts. In this situation, formal decision-analytic tools can be quite useful, but only if carefully and systematically applied (Savage, 1954; Raiffa, 1968; Howard, 1980, 1988: Howard and Matheson, 1984). There is, however, a third state in the climate context, which may be called decision making under deep uncertainty (sometimes also referred to as "secondary" uncertainties; see Fischbeck, 1991). For deep uncertainty, it is not possible to specify the behaviour of major components of a system because of the absence of or contradictions in data, methods, and/or theory. Decision-analytic methods can still be applied, but the process of eliciting subjective probabilities is much more complicated. The experts must factor in assessments about the likelihood of each of the alternative theories being correct, on top of assessments of the probabilities for alternative parameter values within the methods suggested by that theory. In addition, the experts need to provide some estimate of the uncertainty in outcomes caused by factors not incorporated into any existing theory. For example, there may be discontinuities in the response of the climate or ecological systems that occur at as yet unrecognized thresholds.

---

[2] Knight defined uncertainties as either risks accessible using objective historical data or uncertainty where there is little or no data and the underlying processes are not well understood. The exposition here updates his original taxonomy to include more recent thinking on a fuller range of degrees of uncertainties.

Since they have different starting points and objectives, the three approaches to climate policy analysis have exhibited somewhat different approaches to handling uncertainty. Applications of the cost-effectiveness approach have generally ignored uncertainty completely or stayed fairly close to the traditional decision analysis approach, focusing on incorporating a limited number of subjectively accessed probabilities on key uncertainties. Applications of the equity approach have been focused on the risks climate change and climate change policies might pose to the "most vulnerable" elements of the global population and have generally employed sensitivity analyses to accomplish this objective. Studies done from the sustainability perspective have more often than not focused on the robustness of policies (and especially those designed to build climate mitigation and adaptation possibilities) across wide ranges of values for uncertain inputs and parameters.

The rest of this chapter elaborates each of the three analytic perspectives shown in *Figure 1.1*. The motivation for this elaboration is threefold. First, it is to help the reader situate each perspective in the evolution of policy science as reflected in IPCC assessments. Second, it is designed to situate the issue of GHG emissions mitigation in the context of climate policy more broadly. Third, it seeks to locate climate policy in a broader context of concerns about development, equity, and sustainability. However, it must be emphasized that *Figure 1.1* does not represent any sort of linear evolution in which one kind of analytic tool or policy focus replaces a predecessor. Rather than a hierarchy of approaches, the evolution of perspectives suggests a portfolio approach both to assessment and policy choice. Just like a personal investment portfolio, a rational global climate policy portfolio contains a flexible mix of diverse commitments consistent with different development goals, and to protect against different contingencies at various levels of uncertainty about the future.

1.2   Cost-effective Mitigation

1.2.1   Introduction

This section describes the key themes that have been pursued by the research community working from the "cost-effective mitigation" perspective (as conceptualized in *Figure 1.2*). The focus here is on the kinds of issues that the research community working from this perspective address and not on specific results.

Researchers working from a cost-effective perspective generally focus on achieving some policy objective at minimum cost. Cost minimization, in some cases, is used to compare alternative ways to meet some climate policy objective (like a specific GHG emissions or concentration target); in other cases, alternative ways to minimize the total cost of climate change and policies designed to ameliorate its impacts are considered. In the former, the policy objective is included as a constraint; but in the latter, the objective is to minimize the cost of the climate change. In either case, the policies considered are generally restricted to those that directly affect energy use or other activities with a direct impact on GHG emissions. Although equity and sustainability metrics are frequently examined in these analyses, their inclusion usually occurs after the cost-effectiveness calculations have been completed. Exceptions to this general observation include input assumptions related to discounting and utility function parameters that do represent trade-offs between the utilities of various groups and generations. Judicious use of sensitivity analysis can, however, illuminate the trade-offs implied along these dimensions, but these trade-offs are not usually the main focus of such studies. It is therefore difficult, *ex post*, to graft other policy objectives related to development or sustainability (e.g., poverty

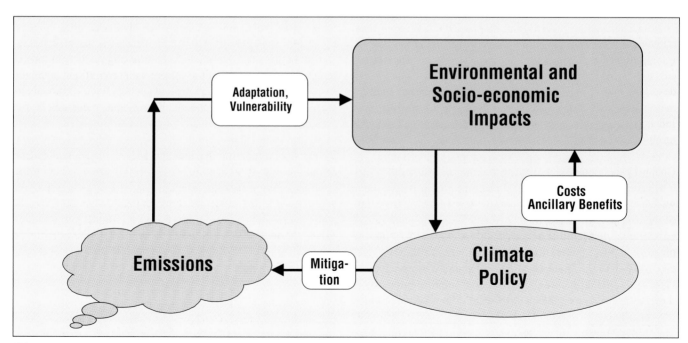

Figure 1.2: *The cost-effectiveness perspective.*

reduction, human capital development) onto a cost-effectiveness style of assessment.

1.2.2 The Costs of Climate Change Mitigation

The United Nations Framework Convention on Climate Change makes clear that cost-effectiveness is an important criterion to be used (among others) in formulating and implementing climate policies. As stated in Article 3.3 of the convention "...taking into account that policies and measures to deal with climate change should be cost-effective so as to ensure that global benefits at the lowest possible cost (UNFCC, 1992)". The impacts of climate policy can be defined as the changes that policies cause relative to some "business-as-usual" or "baseline" situation. As discussed in Chapter 2, a baseline is a scenario of how the global or regional environments, depending on the study, will evolve over time (often over 100 years or more for baselines used in climate policy studies) in the absence of climate policy intervention. Thus, a baseline is typically built upon assumptions about future population growth, economic output, and resource and technology availability, as well as upon assumptions about future non-climate environmental policies, like controls on sulphur dioxide emissions. Changes from these baselines are frequently put into categories of "benefits" and "costs". The benefits included in the calculus are estimated from avoided climate damages and other ancillary benefits that would have otherwise occurred if mitigation policies had not been introduced. The costs for mitigation and other side effects that result are estimated from economic sacrifices that might be required to mitigate climate change.

Climate change would be a relatively simple problem to overcome if it could be avoided without sacrifice and if the means to effect this avoidance were recognized widely. At present, however, there are concerns about the sacrifices that avoiding climate change might involve. A fundamental challenge in mitigation policy analysis is thus to discern how climate change can be avoided at a minimal cost or sacrifice. Chapters 3–9 describe a number of advances since WGIII SAR that identify methods to reduce the costs of climate change mitigation. Indeed, these chapters report that some degree of mitigation might be achieved at zero cost.

Chapter 7 distinguishes several cost concepts. Opportunity cost (the value of a sacrificed opportunity) constitutes a basis upon which estimates of economic cost are constructed. The extent of the costs of mitigating climate change is, from an economic perspective, measured in terms of the value of other opportunities that must be forgone (for example, the opportunity to enjoy low prices for domestic heating or other energy services). It follows that economic costs can be different when they are viewed from different perspectives. Costs of mitigation incurred by a regulated sector are, for example, generally different from economy-wide costs. Costs are sometimes measured in currency units, but they are sometimes also measured

against other metrics. In all cases, though, the underlying element of cost is the sacrifice of opportunities, goods, or services; and this element is often quite different from the overt financial outlay involved.

Chapter 7 also indicates that some notions of cost incorporate behavioural, institutional, or cultural responses that can be missed by economic analyses. In measuring opportunity costs, more specifically, economic analyses generally take personal preferences, social and legal institutions, and cultural values as given. Yet climate policies can affect (positively or negatively) the functioning of institutions. They can alter the ways in which people relate to each other; and they can influence individuals' attitudes, values, or preferences. Taking these impacts into account can alter the cost assessment. Moreover, while economic analyses (including standard benefit–cost analyses) tend to measure costs by adding up individuals' valuations of their forgone opportunities, other approaches to cost can be defined in terms that are not simple aggregations of individual measures.

As discussed below, equitable policy making brings attention to the distribution of costs as well as to their aggregate levels. There has been considerable progress since SAR in identifying ways that climate change can be avoided at lower costs. Both theoretical and modelling studies have helped to reveal the types of policies that might achieve given targets at the lowest cost. Moreover, as indicated below, models have identified certain circumstances in which at least some reductions in GHGs might be achieved at no cost.

Chapter 8 reports that the cost of mitigation can depend significantly on the selection of a designated concentration target that, typically, is assumed to be achievable within 100 or 200 years. Most model-based studies indicate that the first units of abatement are fairly inexpensive; "low-hanging fruit" is easily picked. However, most studies show that additional units of abatement require more extensive changes and involve significantly higher costs.[3] Thus, to lower the original concentration target is projected to result in a more than proportional increase in costs. Rising marginal abatement costs provide a rationale to employ broad-based, economically efficient mechanisms for GHG abatement.

The cost of mitigation depends not only upon the cumulative emissions reductions required over the next century, but on the timing of these emissions reductions as well. Chapter 8 reviews some studies that argue the most cost-effective approach to achieving a given long-term concentration target involves gradually rising abatement through time. The attraction of this approach is that it helps avoid the premature turnover of stocks of capital. In addition, deferring the bulk of abatement effort to

---

[3] It is possible for the cost curves to be very flat in certain regions, however, and technological change can shift them down significantly over time.

the future allows more discounting of abatement costs. However, other studies show potential cost advantages in concentrating more abatement towards the near term. These studies argue, in particular, that near-term abatement helps generate cost-effective "learning-by-doing", by accelerating the development of new technologies that can reduce future abatement costs. These findings are not necessarily contradictory. By introducing mitigation efforts in the near term, the process of learning-by-doing is initiated. At the same time, by increasing over time the stringency of policies (that is, the extent of abatement), nations can avoid premature capital-stock turnover and exploit the cost savings from future technological advances. Chapter 10 elaborates on these issues.

It is worth emphasizing that abatement policies (such as the introduction of national targets on carbon emissions or policies to stimulate the development of energy technologies not based on carbon, as discussed in Chapter 3) can proceed in the near term even when abatement efforts are significantly deferred to the future. The near-term introduction of policies helps to stimulate efforts to bring about new technologies, which is crucial to enable future abatement to be achieved at lower cost.

As Chapter 6 discusses, individual countries can choose from a large set of possible policy instruments to limit domestic GHG emissions. These include traditional regulatory mechanisms such as technology mandates and performance standards. They also include "market-based" instruments such as carbon taxes, energy taxes, tradable emissions permits, and subsidies to clean technologies. They also include various voluntary agreements between industries and regulators. A group of countries that wishes to limit its collective GHG emissions can agree to implement some of these policies in a co-ordinated fashion.

Chapters 6–9 reveal that the costs of achieving specified mitigation targets depend critically upon the policy instrument employed. Any given target is achieved at the lowest cost when the incremental cost of emissions reduction (abatement) is the same across all emitters. If this condition is not met, then the overall costs of emissions reduction could be reduced if firms with lower incremental costs reduced emissions a bit more, and firms with higher incremental costs pursued a bit less abatement. It follows that cost-effective emissions reductions hold the promise of allowing larger emissions reductions from any allocation of resources

While market-based instruments such as carbon taxes and tradable carbon permits have potential cost advantages, the extent to which these potential advantages are actually realized depends on whether the policy generates revenues and whether these revenues are "recycled" in the form of cuts in existing taxes. Revenue recycling is important to the costs of a carbon tax, for example. When the revenues from the carbon tax finance reductions in the rates of pre-existing taxes, some of the distortionary cost of these prior taxes can be avoided; and so the cost of mitigation is reduced. These issues are further elaborated in Chapters 6–9.

The issue of revenue recycling applies also to policies that would reduce CO_2 through carbon permits or "caps". As discussed in Chapter 6, revenues could be recycled through cuts in existing taxes if CO_2 permits are auctioned. In contrast, if the permits are distributed freely, then no revenue is collected and there is no possibility of revenue recycling. Thus, auctioning the permits has a significant potential cost advantage over free allocation.

It is also important to keep in mind that aggregate costs are not the only useful consideration in evaluating alternative policy instruments from the cost-effectiveness perspective. The distribution of these costs across businesses, regions, and individuals is important as well. Moreover, other important evaluation criteria, including administrative and political feasibility, can play a role in determining exactly how and why mitigation initiatives might emerge.

The theoretical and modelling literature also reveals that international policy co-ordination through "flexibility mechanisms" offers enormous opportunities to achieve given reductions in GHG emissions at relatively lower cost. In principle, co-ordinated policies can be designed so that cost-effectiveness is improved on a global scale. The Kyoto Protocol defines several flexibility mechanisms, including international emissions trading (IET), joint implementation (JI), and the clean development mechanism (CDM). Each of these international policy instruments provides opportunities, in theory, for Annex I Parties to fulfil their commitments cost-effectively. IET allows Annex I parties to exchange parts of their assigned amount. Similarly, JI allows Annex I parties to exchange "emission reduction units" among themselves on a project-by-project basis. Under the CDM, Annex I parties receive credit, on a project-by-project basis, for reductions accomplished in non-Annex I countries. Participation in these programmes can also increase the level of investment in clean energy technologies. International policy co-ordination in implementing climate policy also requires accounting for the "spillover" effects of mitigation in one country that can effect economic activity in other countries through international trade linkages. In general, countries that mitigate less may gain an advantage in their share of international trade over their trading partners, but can also lose market share if those trading partners control more and thus reduce their overall level of economic activity. See Chapter 8 for more on these issues.

Most studies of national or global mitigation costs focus on CO_2 from fossil energy alone (e.g., see Chapter 8), but some recent studies consider other GHGs as well. For example, Chapters 3 and 4 discuss options to reduce emissions of non-CO_2 gases and CO_2 net emissions from land-use change, respectively. Chapter 8 indicates that defining national targets in terms of a "basket" of gases (as under the Kyoto Protocol) rather than in terms of individual gases enhances flexibility and can reduce the costs of mitigating climate change. Emissions of several of the GHGs (such as methane and nitrous oxide) from some sources can, in addition, be very difficult to monitor. This practical complication raises the potential cost of mitigation over the short- to medium-term, because it highlights the need to improve the methods used to monitor these emissions.

1.2.3   The Role of Technology

The time horizon for climate change is long. The climate impacts of decisions made in the next decade or two will be felt over the next century and beyond. As a result, technology and, more specifically, improvements in the rate and direction of technological change, will play a very important role. As discussed in Chapter 2, the development and diffusion of new technologies is perhaps the most robust and effective way to reduce GHG emissions. Three aspects of technology can be distinguished: invention (the development, perhaps in a laboratory, of a new production method, product, or service), innovation (the bringing of new inventions to the market), and diffusion (the gradual adoption of new processes or products by firms and individuals). Chapter 3 indicates that hundreds of recently invented technologies can improve energy efficiency and thus reduce energy and associated GHG emissions. These technologies can yield more energy-efficient buildings and appliances and equipment used in them. There are, however, significant barriers to their innovation and diffusion. Chapter 5 (see also IPCC, 2000a) classifies these barriers and provides a framework for understanding their connections with one another. Some new low-carbon emission technologies are not adopted because their cost and performance characteristics make them unattractive relative to existing technologies. To be adopted, these technologies require tax advantages, cost subsidies, or additional cost-reducing or performance-enhancing research and development (R&D; see Chapter 6 for a discussion of the possible efficacy of such policies). Other technologies could be adopted more rapidly if market failures and other socioeconomic constraints are reduced. Market failures refers to situations in which the price system does not allocate resources efficiently (see, e.g., Opschoor, 1997). They can emerge when information is not fully disseminated or when market prices do not reflect the full social cost. So, a new technology may not be employed if potential purchasers lack information about it or if its price lies between its private value and its, potentially higher, social value.

While Chapter 3 summarizes advances in our understanding of technological options to limit or reduce GHG emissions, Chapter 4 indicates that terrestrial systems offer significant potential to capture and hold substantially increased volumes of carbon within organic material. However, the challenges associated with defining and measuring contributions to sequestration and with monitoring the performance of individual sink projects are significant. The nature of sequestration opportunities differs by region. In some regions, the least-cost method of accomplishing sequestration is to slow or halt deforestation. In others, afforestation and reforestation of abandoned agricultural lands, degraded forests, and wastelands offer the lowest-cost opportunities. The results of the IPCC (2000c) *Special Report on Land Use, Land-Use Change and Forestry*

may shed light on some of these controversies. In all cases, though, the opportunity costs associated with using terrestrial systems involve welfare implications on multiple scales.

1.2.4   The Role of Uncertainty

The uncertainties that surround climate change are vast. The connections between emissions of GHGs and climate change are not fully understood. In addition, uncertainty distorts our understanding of the impacts of climate change and the value of those impacts to humans. These uncertainties depend on scale, and become larger across the spectrum from "average" impacts across broadly defined geographical areas to specific impacts felt at a more local level.

The uncertainties that surround climate change bear on the issue of whether mitigation policies are justified. Some analysts might conclude that these uncertainties justify the postponement of significant mitigation efforts–particularly those that involve economic sacrifices–on the grounds that not enough is yet known about the problem. Proponents of this point of view argue that there is some chance that scientific inquiry will eventually reveal that the continued accumulation of GHGs will not produce significant changes in climate and/or significant associated damages. So long as the possibility exists that a "type one" error (an action that will ultimately turn out to be unnecessary) could occur, the argument goes, it is premature to undertake costly mitigation measures now.

However, uncertainty also introduces the risk that the opposite will occur. There is a significant possibility that scientific investigations will ultimately reveal that the continued accumulation of GHGs will have severe consequences for climate and substantial associated impacts. If this scenario should materialize, the cost of making this "type two" error (of taking little or no action in the near term to stem the accumulation of GHGs) could be enormous. As discussed in Chapters 8–10, it may be less costly to spread the costs of averting climate change by beginning mitigation efforts early, rather than to wait several decades and take actions after the problem has already advanced much further. Indeed, if postponing mitigation efforts allows irreversible climate impacts to occur, then no future efforts, at any cost, can undo the resultant damage.

The risks of premature (or unnecessary) action should therefore be compared with the risks of failing to take action that later proves warranted. As stated in Article 3.3 of the Framework Convention "...The parties should take precautionary measures to anticipate, prevent or minimise the causes of climate change and mitigate its adverse effects"(UNFCCC, 1992). Which risk is larger? Analyses of this issue (see Chapter 10) tend to indicate that the latter risk is sufficient to justify some mitigation efforts in the short run, despite the possibility that these efforts might ultimately prove unnecessary. These analyses depict mitigation efforts as a type of insurance against potentially serious future consequences. It is generally sensible for a person to purchase fire insurance on his or her house (despite the likelihood a fire will never occur). Likewise, it is rational for nations to insure against potentially serious damages from climate change, despite the significant chance that the most serious scenarios will not materialize.

The term precautionary principle has been employed to express the idea that it may be appropriate to take actions to prevent potentially harmful climate-change outcomes. As discussed in Chapter 10, this term has more than one meaning. A weak version of the principle is the idea that, in the presence of uncertainty, it may be prudent to engage in policies that provide insurance against some of the potential damages from climate change. Insuring against potentially serious damages can be rational simply because the costs of the insurance are less than the expected value of avoided damages. This weaker form of the precautionary principle applies even if individuals or societies are not particularly averse to risk. In its stronger form, the precautionary principle stipulates that nations should pursue whatever policies are necessary to minimize the damages under the worst possible scenario. This stronger form assumes extreme risk-aversion, since it focuses exclusively on the worst possible outcomes. It is clear, though, that there are costs associated with climate policies that could, under some circumstances, impose large costs on particular peoples and/or nations; but neither form of the precautionary principle has yet been applied to this side of the climate calculus.

Uncertainty also bears on the design of mitigation policies. As indicated in Chapters 8 and 10, the problem of climate change might be addressed most effectively through a process of sequential decision making, in which policies are adjusted over time as new scientific information becomes available and uncertainties are reduced. Moss and Schneider (2000) offer guidance on how subjective probabilities can be utilized effectively when empirical data are not available or are inconclusive. New information is valuable, and flexible policies that can make use of this information have an advantage over rigid ones that cannot. In any case, policies that help build or strengthen mitigation capacity are consistent with the insurance approach. To the extent that mitigation capacity is higher, the costs of future action can be expected to be lower.

1.2.5   *Distributional Impacts and Equity Considerations*

It is important to consider more than the aggregate (worldwide) benefits and costs of such policies in examining and evaluating mitigation options. Considerations of the national, intranational, industrial, and intergenerational distributions of the benefits and burdens of mitigation policies–as well as considerations of the historical contributions to the accumulation of GHGs–are crucial to develop equitable climate policies. The WGII report (IPCC, 2001) indicates that the impacts of climate change vary substantially across regions of the globe. Indeed, climate impacts can differ even on the scale of a few miles depending on geography, terrain, and other natural conditions. The costs

of the economic impacts of climate policies are distributed unevenly as well, although the distribution of these impacts depends on the types of mitigation policies introduced. It is important to consider the distribution of cost impacts of different potential policies across nations, socioeconomic groups, industrial sectors, and generations.

The distribution of the economic impacts of mitigation policies across economic sectors is examined in Chapter 9. Policies such as carbon taxes or carbon caps are designed to limit carbon use and are likely to cause production, output, and employment to fall in the coal and oil extraction industries. The impact on the natural gas industry is less clear. On the one hand, a carbon tax raises the cost of supplying natural gas, which tends to imply reduced demands, output, and employment in this industry. On the other hand, this tax raises the price of coal by a larger percentage, inducing shifts in demand from coal to natural gas. The impact of mitigation policies on renewable energy sources is likely to vary by resource and region but are likely to lead to larger markets for renewables. Mitigation policies are expected to lead to structural changes in manufacturing, especially in the developed countries. Sectors that supply energy-saving equipment and low-carbon technologies are likely to benefit from these policies. Sectors that rely intensively on carbon-based fuels are expected to suffer price increases and a loss of output.

Chapter 8 indicates results that concern the distribution of impacts across household income groups. According to most studies, mitigation policies that imply higher energy prices impose higher cost-burdens (relative to income) on less affluent households than on richer households. This reflects that the poor tend to spend a larger share of their income on energy. Equity considerations suggest that mitigation policies can overcome these distributional consequences by including provisions that reduce the costs they impose on the lowest-income groups.

For the most part, existing studies of the impacts across household groups (or socioeconomic groups, more broadly) apply to developed nations. There is a severe need for studies that consider the distributional impacts within developing countries. In addition, nearly all the studies lack the detail necessary to consider impacts in socioeconomic dimensions other than income. As a result, important costs to various groups within the general population may be overlooked. Important costs may also be hidden by aggregation. This is especially relevant in studies of the impacts of climate change and mitigation activity in developing countries, since existing studies may overlook major impacts to the most vulnerable individuals. Section 1.3 discusses the issue of equity in more detail and from a broader perspective.

1.2.6   *Sustainability Considerations*

Sustainability considerations are typically not the primary motivation for studies carried out from the "cost-effectiveness perspective". Besides the distributional effects of climate policies, their implications for other environmental concerns can also be calculated. For example, the implied impact of climate policies on sulphur, particulate emissions, or land uses can be calculated. Sulphur emissions in some scenarios may be so high that they have major health impacts, and the land-use requirements for a global energy industry based on a very large biomass could potentially crowd out agriculture, forestry, and the recreational use of land.

As indicated in Chapter 2, the benefits and costs from a given mitigation policy depend on the baseline circumstances to which the policy is applied. The uncertainties as to what the baseline circumstances might be are vast, in the light of which it is important to evaluate the impacts of given policies relative to a range of baseline scenarios rather than to a single baseline scenario.

Human welfare and the state of the environment (which may be a determinant of human welfare, but one that is the focus of this assessment report) depend both on the baseline path and on the policy-induced departures from the baseline. A striking conclusion from Chapter 2 is that the differences in human welfare across plausible baselines can be greater than the welfare impacts of mitigation policies. That is, the nature of the baseline–which reflects a wide range of human decisions and policies outside of the climate-policy arena–can be more important than the departures from that baseline caused by climate policy. The lower the level of baseline GHG emissions, the smaller is the effort required to achieve any specific emissions or concentration target. This does not eliminate the importance of policy actions to mitigate climate change, but it reveals the importance of developments that occur outside what is typically regarded as "climate policy".

It is not surprising that changes in the economy resulting from climate policy may be small compared to changes that may occur in response to other trends in the economy and to other policies. This is so because most the GHG emissions occur in energy production, which forms a relatively low percentage of the economy (no more than 5%–10%). In principle, rearranging energy use as one element of a mitigation strategy need not be a major shock to the economy if it is done efficiently. Important also is that the costs of mitigation are likely to vary substantially among nations because of both differences in baseline emissions trends and differences in flexibility to accomplish the emissions reductions required (see also Schneider (1998) on this subject).

Deciding what counts as "climate policy" is not always straightforward, as discussed in Chapter 2. In many policy discussions, climate-change mitigation policy is assumed to involve actions for which the primary target is a reduction in GHG concentrations. These include efforts directly aimed at reducing carbon emissions, at expanding carbon sinks, at reducing emissions of other GHGs (like methane and nitrous oxide from agriculture), and at promoting the development of

new technologies and production processes that rely less on carbon-based fuels (see Chapters 3 and 4). If this is the domain of mitigation policy, then other (anticipated) actions that do not fall in this category need to be regarded, by default, as part of the baseline. However, other activities have important consequences for climate change. For example, policies oriented towards local air pollution–such as controls on hydrocarbon emissions from automobiles–affect levels of emissions of CO_2 as well as the formation of tropospheric ozone, and thus have consequences for climate. Moreover, as discussed below, some policies, such as poverty alleviation, may ultimately have significant implications for the emissions of GHGs and are therefore extremely important to climate change.

The implications of different baseline assumptions about the future of the world reflect, in part, different assumptions about the sustainability of economic, biological, and social systems. Bringing them to bear on the analyses of mitigation opens the possibility that climate policies can be assessed within alternative worlds and that how climate policies might effect various measures of sustainability can be examined explicitly. This kind of analysis can support, though, only a limited treatment of sustainable development. A more in-depth treatment has been attempted by researchers working from the perspective of "envisioning transitions to sustainability"; their perspective is described in Section 1.4.

In addition to the direct benefits of GHG mitigation represented in terms of reductions in impacts resulting from climate change, the cost-effectiveness perspective also considers benefits from reductions in other pollutants[4] that may accompany the GHG emission reductions. Given the focus on climate change mitigation as the primary objective the term used most often is "ancillary benefits" (see also Chapter 8). The term "co-benefits" is used for situations where climate change and other environmental or socioeconomic objectives are equally important. That notion comes more naturally from the sustainability perspective and reflects that most policies designed to address GHG mitigation also have other, often at least equally important, rationales, e.g. related to development, equity and sustainability.

1.3 Equity and Sustainable Development

The above review of the literature on cost-effective GHG mitigation (including the chapters in this report) shows that elements of development, equity, and sustainability are addressed in some of the analyses. However, they generally take the form of boundary conditions, barriers, or constraints rather than the primary motivation of the analysis. There is also a large and growing volume of research that approaches mitigation directly from a concern with equity and development (*Figure 1.3*).

---

[4] In principle these ancillary benefits should be credited only to the extent of the cost of direct control of those pollutants they obviate.

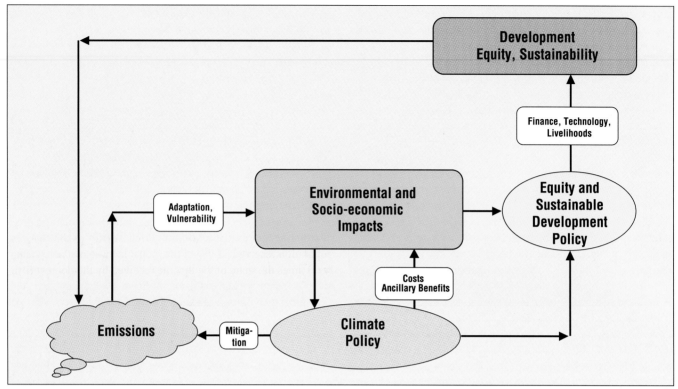

Figure 1.3: *Equity and climate change mitigation.*

While in principle, equity concerns pertain to at least three domains[5]–international, intra-country, and inter-generational–much of this literature focuses on the international dimensions of equity, and takes as its primary challenge the goal of sustainable development and poverty eradication in developing countries, (Parikh, 1992; Parikh and Parikh, 1998; Murthy, 2000).

As mentioned earlier, although this literature starts with concerns about global equity, one of its central concerns is the promotion of the prospects of sustainable development, especially in developing countries. Accordingly, we have entitled this approach, "equity and sustainable development".

An important motivation for this literature is climate change agreements in which equity–at all relevant levels (intergenerational, intragenerational, international, and intranational)–is a prominent and consistent theme. The first principle of the UNFCCC (1992, Article 3.1) states: "The Parties should protect the climate system for the benefit of present and future

generations of humankind, on the basis of equity and in accordance with their common but differentiated responsibilities and respective capabilities. Accordingly, the developed country Parties should take the lead in combating climate change and the adverse effects thereof."

The UNFCCC goes on to require developed countries to assist developing countries in coping and adapting with the impacts of climate change (Articles 4.3, 4.4, 4.5, 4.8, 4.9, 4.10), recognizes that "economic and social development and poverty eradication are the first and overriding priorities of the developing countries" (Article 4.7), and, indeed, that "Parties have a right to and should promote sustainable development" (Article 3.4). The Kyoto Protocol retained this emphasis by referring to various paragraphs of Article 4 of the UNFCCC (1992), and refrained from imposing additional commitments on developing countries (UNFCC, 1997b Article 10, preamble). It reiterated the goal of sustainable development and established the CDM to assist developing countries in achieving sustainable development while contributing to the ultimate objectives of the UNFCCC (1997b, Article 12.2; see also Jacoby *et al.*, 1998; Najam and Page, 1998; Jamieson, 2000; Agarwal *et al.*, 2000).

Finally, the issue of equity has been discussed not only with regard to the distribution of resources and burdens within and between generations, but also in terms of the role that it plays in the generation of social capital. Along with reproducible, natural, and human and intellectual capital, social capital is necessary for sustainability (Rayner *et al.*, 1999; for related

---

[5] This is an extensive and diverse literature, of which a few examples are Ramakrishna (1992), Shue (1993, 1995), Mintzer and Leonard (1994), Munasinghe (1994, 1995, 2000), Lipietz (1995), Parikh (1995), Rowlands (1995), Runnalls (1995), Jamieson (1996, 2000), Murthy *et al.* (1997), Parikh *et al.* (1997), Rajan (1997), Sagar and Kandlikar (1997), Schelling (1997), Byrne *et al.* (1998), Najam and Sagar (1998), Parikh and Parikh (1998), Tolba (1998), Agarwal *et al.* (2000).

arguments, see also Hahn and Richards, 1989; Toman and Burtraw, 1991; Rose and Stevens, 1993). Fairness is integral to the establishment and maintenance of social relations at every level, from the micro to the macro, from the local to the global.

What is fair may be the subject of disagreement, but the demand for fairness only arises because of the existence of community. It is very hard to imagine what fairness would mean if we did not live and work together in families, communities, firms, nations, and other social arrangements that persist over time (Rayner, 1995).

1.3.1 What Is the Challenge?

The challenge of climate change mitigation from an equity perspective is to ensure that neither the impact of climate change nor that of mitigation policies exacerbates existing inequities both within and across nations. The starting point for describing this challenge is the vast range of differences in incomes, opportunities, capacities, and human welfare, both between and within countries. This is combined with the fact that carbon emissions are closely correlated to income levels–both across time and across nations–which suggests that restrictions on such emissions may have strong distributional effects (Parikh *et al.*, 1991; Parikh *et al.*, 1997b; Munasinghe, 2000).

Income and consumption, as well as vulnerability to climate change, are distributed unevenly both within and between countries.[6] Concerns about the disproportionate impacts of climate change on developing countries are mirrored in similar fears with regard to poor and vulnerable communities within developing countries (Jamieson, 1992; Ribot *et al.*, 1996; Reiner and Jacoby, 1997). Similarly, issues of intergenerational equity have been raised to caution against shifting the burden of adjustment to future generations, which cannot influence political choices today (see Weiss, 1989),[7] a theme picked up in Section 1.4 below.

Academic and policy interest has focused on income distribution as well as the poverty that underlies it. Global poverty statistics are compelling. Over 1.3 billion people, or more than one-fifth of the global population, are estimated to be living at

less than US$1 per day. Other measures of poverty and vulnerability–lack of access to health, education, clean water, or sanitation–yield higher estimates of poverty. Since poverty is concentrated in non-Annex I countries–especially South Asia and Africa–whose average per capita income is less than one-quarter (in dollars of constant Purchasing Power Parity) of the average for developed countries (UNDP, 1999; World Bank, 1999), equity concerns have focused on differences between rather than within countries.

The distributional dimension of global poverty was illustrated vividly by the *Human Development Report 1989* (UNDP, 1989), in the form that has come to be known as the champagne glass (*Figure 1.4*). This representation of global income distribution shows that in 1988 the richest fifth of the world's population received 82.7% of the global income, which is nearly 60 times the share of the income received by the poorest fifth (1.4%). More recent statistics indicate that inequality has widened further since then and that in 1999 the richest quintile received 80 times the income earned by the poorest quintile (UNDP, 1999).

Besides average income levels, Annex I and non-Annex I countries differ in other ways, most importantly in terms of the capacity for collective action and access to technology and finance. Many non-Annex I countries face problems of governance because of weak administrative infrastructures, failure to invest in human and institutional capacity, lack of transparency and accountability, and a high incidence of civic, political, and regional conflicts (World Bank, 1992; UNDP, 1997; Kaufmann *et al.*, 1999; Knack, 2000; Thomas *et al.*, 2000). They also house a less than proportionate fraction of R&D infrastructure, and consequently lack access to technology and innovation. This is especially important in issues of global environmental change, which are strongly science-driven areas (Jamieson, 1992; Ramakrishna, 1992; Najam, 1995; Agarwal and Narain, 1999). Finally, many (though not all) of these countries are over-exposed to international debt–and their governments to domestic debt–and thus have less flexibility in the choice of policy options (World Bank, 1998).

Notwithstanding the diversity of initial conditions in various countries, they share a common commitment to the goal of

---

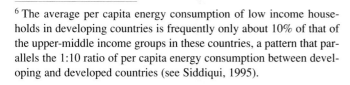

[6] The average per capita energy consumption of low income households in developing countries is frequently only about 10% of that of the upper-middle income groups in these countries, a pattern that parallels the 1:10 ratio of per capita energy consumption between developing and developed countries (see Siddiqui, 1995).

[7] Although this issue received attention in the IPCC SAR (IPCC, 1996), the discussion was framed in technical terms, namely the determination of the appropriate discount rate, which made little accommodation for philosophical, legal, and sociological perspectives on intergenerational rights and responsibilities.

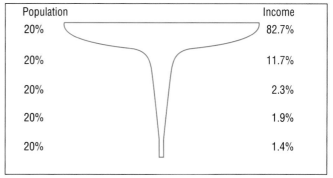

Figure 1.4: *Global distribution of income and population.*

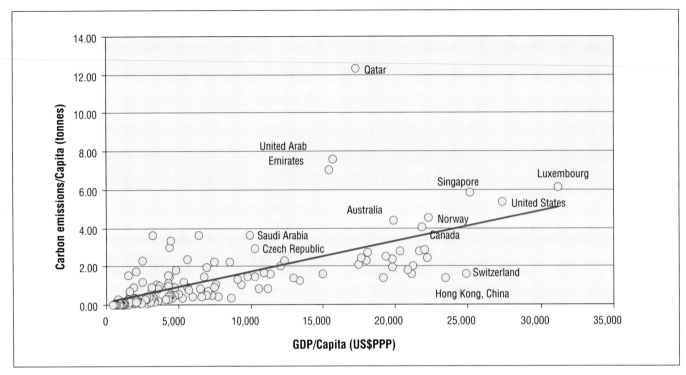

Figure 1.5: *Per capita carbon emission and income.*

economic growth, partly for its own sake and partly because it is perceived as one of the means of poverty eradication and capacity development. However, most analysts recognize that growth alone is not a solution and it needs to be combined with ancillary policies and safeguards to protect environmental and social resources. In fact, while national economic growth appears to be correlated with a reduction in poverty levels (and neutral with regard to national income distribution), over the past 50 years global income growth has been accompanied by a worsening of global income distribution (World Bank, 2000) and a persistence of poverty.[8] The concept of sustainable development has incorporated distributional aspects mainly in response to these concerns (see Lélé, 1991; Murcott, 1997). Be that as it may, economic growth continues to be the centre of government policies and plans.

This is relevant to climate change mitigation, since a fairly robust stylized fact of historical development, consistent with both cross-country and time-series data, is the close correlation between economic growth and carbon emissions. *Figure 1.5*, for example, presents cross-country data on per capita carbon emissions and income (in US$(PPP); see also *Box 1.1* on a controversy over the representation of data). The bold trend line highlights the proportionate increases (or, as in some economies in transition recently, decreases) in per capita emissions and income over time. Broadly speaking, developed

countries have per capita incomes over US$(PPP)20,000 and carbon emissions between 2 and 6 tonnes per capita. Non-Annex I countries have much lower incomes and much lower emissions, while the economies in transition fall in the middle of the range. In particular, the bulk of the world's poor live in a smaller number of non-Annex I countries, which are bunched at the bottom left corner of the graph, with incomes below US$(PPP)5,000 per capita, and emissions below 0.5tC/capita.

Useful analytical tools in this regard are various decomposition approaches[9] that represent carbon emissions as the product of three factors, carbon intensity (emissions per unit of income), affluence (income per capita), and population. The decomposition suggests that reconciliation of the goals of emissions abatement and economic growth must involve a combination of population decline and technological and managerial improvements that lead to lower carbon intensity. Some potential for improvement is evident from *Figure 1.5*, namely the large differences in per capita emissions of countries and regions at the same level of affluence (e.g., Hong Kong, Switzerland, Singapore, Japan, and the USA). This suggests the possibility of technological "leap-frogging" (see Goldemberg, 1998a, Schneider, 1998), that is the lowering of emissions by a factor of two or three without impacting income levels through investment in technological development and

---

[8] The reason for this paradox is that at the global level intercountry distributional impacts dominate over the within-country impacts (see World Bank, 2000, p. 51).

[9] See, e.g., de Bruyn *et al.* (1998) and Opschoor (1997), who develop this idea from a development perspective, and Hoffert *et al.* (1998) who uses the "Kaya Identity" to formulate decompositions from an energy economics perspective.

capacity building.[10] However, the operational and other obstacles against the realization of these possibilities have not been analyzed systematically in the literature.

In the absence of such investment, economic growth and conventional economic development are likely to remain strongly linked to the ability to emit unlimited amounts of carbon. Therefore, restrictions on emissions will continue to be viewed by many people in developing countries as yet another constraint on the development process. The mitigation challenge, therefore, is to decouple growth and economic development from emission increases.

However, mitigation policies in general, and its decoupling from economic growth in particular, have to be designed with specific contexts in mind. Policies designed for one context are generally not appropriate for another (Shue, 1993; Rahman, 1996; Jepma and Munasinghe, 1998), and identical ultimate goals–stabilization of GHG accumulation and maintenance or achievement of the quality of life–yield different priorities and strategies in Annex I and non-Annex I countries. In the former, these goals are translated as reducing emissions while improving the quality of life, and in the latter it is the other way around–improving the quality of life, *inter alia*, by maintaining the rate of economic growth, while maintaining or lowering per capita emissions.

The current global response to this situation is to exempt non-Annex I countries from climate obligations to allow them to pursue their developmental goals freely. Furthermore, UNFCCC as well as subsequent agreements stipulate the provision of financial and technological resources for voluntary mitigation actions by this group of countries. Finally, the Kyoto Protocol created the CDM to enable developing countries to contribute to emissions abatement while pursuing sustainable development.

As non-Annex I emissions continue to grow, however, this strategy may become inadequate, and more innovative mitigation efforts might be called for in non-Annex I countries. This will mean divergences of the development path of the currently developing countries from that which developed countries have displayed (Munasinghe, 1994; Jacoby *et al.*, 1998; Najam and Sagar, 1998; Barrett, 1999). As the UNDP *Human Development Report* (1998, p.7) points out, "Poor countries need to accelerate their consumption growth – but they need not follow the path taken by the rich and high-growth economies over the past half century."

Some simple calculations can help illustrate the nature of the global mitigation challenge. Current per capita carbon emissions are slightly more than 3 tonnes per year in Annex I countries and slightly less than 0.5 tonnes per year in non-Annex I countries. With about 1.3 billion people living in Annex I countries and about 4.7 billion in non-Annex I countries, total carbon emissions are in the range of $(3.1)(1.3) + (0.48)(4.7) = 6.29$ billion tonnes. Thus carbon emissions at a global scale average about 1 tonne per capita per year. The stabilization of CO_2 concentrations in the atmosphere at 450, 550, 650, and 750ppmv will require steep declines in the aggregate emissions as well emissions per capita and per dollar of gross domestic product (GDP) as illustrated in the IPCC SAR Synthesis Report (IPCC, 1996). For example, based on the SAR Synthesis Report and a recent set of calculations by Bolin and Kheshgi (2000), stabilization of CO_2 concentrations in the atmosphere at 450, 550, 650, and 750ppmv would require limiting fossil-fuel carbon emissions at about 3, 6, 9 and 12 billion tonnes, respectively, by 2100 and further reductions thereafter to less than half current global emissions. If, for example, the world population stabilized at about 10 billion people by then, an average carbon emissions per capita of 0.3, 0.6, 0.9, and 1.2 tonnes of carbon would be required to achieve the 450, 550, 650, and 750ppmv limits, respectively. We make no assumption here about how these emissions would or should be allocated globally, but simply report that the average by 2100 must work out to these levels to achieve the stabilization objectives. Thus, to achieve a 450ppmv concentration target, average carbon emissions per capita globally need to drop from about 1 tonne today to about 0.3 tons in 2100; to achieve a 650ppmv target they need to drop to 0.9 tonnes (about one-quarter of current emissions per capita in the Annex I countries) by 2100 and further thereafter. Finally, with a global economy currently producing about 25 trillion dollars of output, carbon emissions per million dollars of output are currently about 240 tonnes. If, for example, the global economy grows to 200 trillion dollars of output by 2100, the emissions per million dollars (in year 2000 dollars) would need to be limited to about 10, 25, 40, and 55 tonnes of carbon in order to achieve the 450, 550, 650, and 750ppmv CO_2 limits, respectively. If further population and economic growth continues beyond 2100 additional reductions in average emissions per capita and per unit of economic output would be required.

This framing of the mitigation challenge is central to the literature on global equity and climate change. Virtually all stabilization trajectories in the literature show an initially rising trend of aggregate global emissions, followed by a declining trend; and they also show a gradual narrowing of the gap between per capita emissions of various countries and regions. In many of these scenarios, over a finite period of time, aggregate net global emissions contract to levels consistent with the absorptive capacity of global sinks, while per capita emissions of Annex I and non-Annex I countries move towards convergence in the interest of global equity. One possible international regime to achieve stabilization would initially have only Annex I emissions decline over a period of time (to make room for the growth prospects and therefore rising emissions of non-Annex I countries). At the same time, as per capita emissions

---

[10] This possibility is also corroborated by time-series data on carbon intensity, which reveal evidence of "de-coupling" of the strong relation in some countries, including developing countries. However, the change has not been significant enough to reverse the overall trends towards increasing emissions.

Box 1.1. A Numbers Game

A persistent theme in the literature is the explicit or implicit assignment of responsibility for global warming trends. Without going into the merits of the issue, it is useful to point out that many of the arguments revolve around the appropriate way to represent the data. For example, Agarwal and Narain (1991a) criticize the uncritical use of aggregate national emissions figures, which could imply parity between developed countries and large developing countries (China, India, and Brazil) mainly because of the large populations of the latter. Instead, they recommend the use of per capita "net emissions"–that is, emissions that exceed the per capita absorptive capacity of global carbon sinks. Other analysts distinguish between "necessary" and "luxury" emissions (Agarwal *et al.*, 1999; Shue, 1993).

Another theme is the relative impact of CO_2 emissions and that of other GHGs and land-use changes, given that the latter are less strongly correlated with per capita income. Most analyses have focused on CO_2 emissions, given that it constitutes the bulk of the contribution to global warming. Others suggests that CO_2 emissions are accompanied by forced cloud changes and tropospheric aerosols, which offset their warming impact (Hansen *et al.*, 2000). There are also debates over the precision of the estimates of these associated offsets, as well as those of methane emissions in developing countries (Agarwal *et al.*, 1999). For example, Parikh *et al.* (1991) identify potentially serious problems with World Resources Institute's deforestation estimates (WRI, 1991); and Parikh (1992) shows how the IS92 IPCC scenarios may have been formulated with developed country interests hard-wired into them such that they could be very unfair to the developing countries. In response to this criticism some of the new SRES scenarios (IPCC, 2000a) explicitly explore scenarios with a narrowing income gap between the developed and developing countries.

Finally, "per capita" is not the only relevant normalization (Najam and Sagar, 1998), since emissions per unit of income can also indicate potential for efficiency improvements. Besides annual emissions, data can also be presented in terms of atmospheric concentrations, or the contribution to the global average temperature, each of which has slightly different implications for the responsibility for climate change. Given the uncertainties involved in constructing such estimates, the picture is not entirely clear. However, most estimates suggest that the developing countries may overtake Annex I countries, in terms of total annual emissions, in another 15–20 years, and in terms of the contribution to the global average temperature increase in 60–90 years (Hasselmann *et al.*, 1993; Enting, 1998; Meira, 1999; Pinguelli Rosa and Ribeiro, 2000).

of both groups decline and converge, aggregate emissions also decline–in some scenarios to close to a carbon-free situation. There are in principle many other approaches to an equitable international regime, that are discussed in Section 1.3.2.

For the purposes of this chapter, it is convenient to divide the required emissions trajectory into three segments. Phase 1, an upward sloping segment of the non-Annex I trajectory, may require only marginal deviations in baseline emissions, for which the assessment of policy options entails a central attention to the costs and benefits of mitigation. However, for options relevant for Phase 2, a downward sloping segment of non-Annex I emissions, in which deeper cuts may be called for, global equity issues will need greater attention. Finally, the policy options that can help realize Phase 3, the asymptotic segment of the trajectory, revolve to a greater extent around sustainability concerns.

1.3.2   *What Are the Options?*

These considerations have given rise to a variety of solutions, both in the evolving climate agreements and in the scholarly literature. This literature classifies options in terms of the underlying theoretical and philosophical approaches to equity. Toth (1999) constructs a useful taxonomy of perspectives on equity. We have modified this taxonomy slightly into four alternative views, based on: rights, liability, poverty, and

opportunity. A number of perspectives on equity are discussed more fully in Chapter 10.

Rights-based, that is based on equal (or otherwise defensible) rights to the global commons.[11] The earliest formulation of this approach was as a proposal for tradable permits (see, e.g., Agarwal and Narain, 1991a; Parikh *et al.*, 1991; Grubb, 1989; Ghosh, 1993). A formulation that carries this insight to its logical conclusion is that of "contraction and convergence" (Meyer, 1999), whereby net aggregate emissions decline to zero, and per capita emissions of Annex I and non-Annex I countries reach precise equality. Initial analysis assumed an equal per capita allocation of emission permits–or rights to the "atmospheric commons"–but subsequent questioning led other writers to explore equity and efficiency implications of alternative allocation formulas, including geographical area, historic use, economic activity, or some combination of these. In all this literature, the idea is that "surplus" countries or regions, namely those (mainly among non-Annex I countries) with per capita emissions below their total allocation, could sell excess

---

[11] Much of the discussion on equity invokes global commons as an organizing concept, especially with regard to the conflict between individual (or corporate) use and global community interests. This is a well-worn theme in the literature on collective action, dating back to Hardin (1968), who saw unchecked population growth as the main problem. For a recent and more nuanced view, see Ostrom (2000).

emissions rights to "deficit" countries, namely those (mainly among the Annex I countries) that exceed their quota. Besides a transfer from rich to poor countries, this scheme provided incentives to both groups to reduce their emissions–at least as long as emissions rights are a scarce commodity–to reap the financial benefits of conservation. In other words, it sought simultaneously to reward restraint, punish profligacy, provide incentives for conservation, induce a transfer from rich countries to poor ones, and thus lead to distributional equity, efficiency, and sustainability.

Liability-based, that is based on the right of people not to be harmed by others' actions without suitable compenzation (see Rayner *et al.*, 1999).[12] This literature focuses on the damage caused by overuse of the commons, and seeks to establish mechanisms through which those who cause such damage are penalized and the victims of the damage compensated. This perspective opens up possibilities of financial instruments, such as insurance, which distribute risk across society. Countries or groups that believe that the risk of harm is overstated could offer insurance to others against the liability (Sagar and Banuri, 1999). In other words, this solution is expected to lead to sustainability (incentive for restraint) and procedural (though not necessarily distributional) equity. However, broadly speaking, the climate negotiations have not taken this route in any significant manner.

Poverty-based, that is based on the need to protect the poor and vulnerable against the impact of climate change as well as climate policy. Roughly 2 billion people in the world exist at levels of consumption that, from the CO_2 emissions perspective, do not pose a threat to the climate (although their lifestyles are a threat to their own survival).[13] Unlike the high-technology sectors of the developed as well as developing countries, the poor and vulnerable communities lack the flexibility to adapt to global changes or global agreements. Options based on this approach include investment in capacity building and protection for the poor and vulnerable groups to enable them to enhance their livelihoods in an emerging climate regime, while setting aggregate emission targets for the rest of the world. This could also involve a transition to renewable energy in the developing countries, which is generally consistent with the sustainable livelihoods perspective, especially since the current menu of renewable energy technologies includes many that are small scale and appropriate for scattered and low-income populations. Elements of this solution are contained in Agenda 21, but it has not otherwise played a prominent role in discussions of global climate regimes or global governance–except for the occasional reference to intranational equity (see, e.g., Rayner and Malone, 2000).

Opportunity-based, that is based on the right of people, not to the global commons *per se*, but to the opportunity to achieve a standard of living enjoyed by those with greater access to the commons (see e.g., Najam, 2000). It has strong overlaps with the compromise solution that is emerging from the negotiations. Its exclusive focus is on the relationship between states, and it has led to agreements that place the burden of adjustment primarily on Annex I countries. It also implies a tacit consensus on such matters as:
- no large financial transfers or windfall gains;
- no sudden shocks, but a gradual approach consistent with the coping capacity of different countries;
- no financial burden on non-Annex I countries; and
- no restrictions on the space for sustainable development, particularly in the developing countries.

1.3.3   How Has Global Climate Policy Treated Equity?

Indeed, some elements of the equity agenda–primarily at the international level–have been incorporated into the emerging global climate policy regime. In particular:
- initial mitigation efforts have been concentrated in Annex I countries, resulting in a search for the most cost-effective solutions as detailed in Section 1.2;
- currently, non-Annex I countries are exempt from specific mitigation obligations;[14]
- there are agreements to provide financial resources to non-Annex I countries to cover the full cost of preliminary climate obligations (e.g., monitoring, reporting, and planning), and the incremental cost of voluntary mitigation actions;
- there are agreements and some programs to provide technical assistance and training to identify potential win–win opportunities;
- various voluntary mechanisms are being designed to induce early mitigation action in non-Annex I countries, most notably including the CDM of the Kyoto Protocol.

While the details of the CDM are still to be worked out, in broad terms it allows entities in Annex I countries to fulfil their

---

[12] In the literature cited by Rayner *et al.* (1999) see, in particular, Grubb (1995), Burtraw and Toman (1992), and Chichilnisky and Heal (1994). For a theoretical framework on accident liability, see Calabrese (1970).

[13] This group has been referred to in the literature as the "ecological refugees" (Gadgil and Guha, 1995), the "vagabonds" (Bauman, 1998), the "castaways" (Latouche, 1993), and the "excluded" (Korten, 1995). However, some writers have raised concerns that these groups impact climate change (as well as biodiversity) adversely through non-sustainable land-use practices and deforestation.

[14] However, current trends suggest that mitigation has already begun in some non-Annex I countries, even in the absence of deliberate climate policy. Reductions on fossil fuel subsidies (as a percentage of existing subsidies) have been larger in developing countries (especially China) than in OECD countries, and are leading to considerable savings in carbon emissions (International Energy Agency, 1996; Johnson *et al.*, 1996; Reid and Goldemberg, 1997).

mitigation obligations through co-operative investment in non-Annex I countries, presumably at a lower cost. It has been hailed by some analysts as an ingenious device to reconcile the goals of GHG abatement and sustainable development (see Goldemberg, 1998b; Haites and Aslam, 2000). On the other hand, it has also generated a degree of criticism. Critics fear that:

- CDM will channel investment into projects of marginal social utility (Agarwal and Narain, 1999);
- gains will not be shared fairly (Parikh *et al.*, 1991, 1997a; Parikh, 1994, 1995);
- technology transfer will not be satisfactory (Parikh, 2000);
- poorer countries (especially African countries) and vulnerable groups will be excluded (Sokona *et al.*, 1998, 1999; Goldemberg, 1998b);
- only resources for cheap mitigation options will be attracted (the so-called "low-hanging fruit"), leaving developing countries to undertake the more expensive options themselves (Agarwal *et al.*, 1999);[15]
- CDM will lead to an effective relaxation of the emission caps (Begg *et al.*, 2000; Parkinson *et al.*, 1999), and
- paradoxically, it may compromise the capacity of developing countries to pursue sustainable development (Banuri and Gupta, 2000).

Going beyond the current options, such as CDM, and to a longer time horizon raises the need to integrate mitigation goals within the broader (sustainable) development agendas of developing countries (Najam, 2000). An emerging literature has begun to explore this redefined problem (see Munasinghe, 2000). Some issues that are relevant to this discussion include:

- *Scale.* The scale of the mitigation challenge in non-Annex I countries is projected to be much broader in the long term than the short term. Instead of an exclusive reliance on financial and technological assistance, which ordinarily indicates increases in assistance levels significantly above historical trends, there is a need to invest in indigenous capacity to undertake mitigation without compromising the development agenda.
- *Timing.* To sustain the interest of both developed and developing countries in co-operative solutions, the goal must be to lower the cost of mitigation over time rather than to concentrate simply on exhausting the cheap mitigation options (the so-called "low-hanging fruit").
- *Relevance to economic growth and sustainable development.* Recent studies of the impact of foreign resource inflows demonstrate that these flows alone do

not suffice to promote economic growth or sustainable development without appropriate policy and institutional environments (World Bank, 1998). It is not clear whether financial resources alone will lead to climate mitigation and economic growth.

- *Equity and trust.* Despite consistent and repeated references to equity in climate agreements, sceptics remain wary that equity will eventually be subverted in some way and involuntary obligations imposed on non-Annex I countries (without financial compenzation) to force them to bear a disproportionate burden of mitigation (Agarwal and Narain, 1991a; Hyder, 1992; Parikh, 1992; Dasgupta, 1994; Parikh, 1995; Parikh and Parikh, 1998; Agarwal *et al.*, 1999).[16]

Some scholars propose remedies to reconcile these longer-term concerns with the more immediate goals of the existing agenda. The simplest is a proposal to restrict all co-operative measures–and thus all early and voluntary action in non-Annex I countries–to "non-carbon" projects (Agarwal and Narain, 1999). While this would exclude some legitimate mitigation options from the purview, it could channel research and entrepreneurial resources into a new market, bring down unit costs, create and strengthen technical and managerial capacities, and thus enable both developed and developing countries to engineer a transition to a carbon-free future. Renewable energy projects have been implemented at smaller scales, which make them appropriate for poor rural communities. Other proposals similarly address the potential co-benefits of the protection of primary forests (see Kremen *et al.*, 2000).

1.3.4   *Assessment of Alternatives: Sustainable Development*

While the motivating concern of the perspective described in this section is that of global equity, the literature included here has also sought to incorporate concerns of efficiency and sustainability. The main mechanism through which this has been accomplished is by using equity considerations to argue for the protection of the prospects of sustainable development in developing countries. Such an agenda is equivalent to a non-co-ordinated pursuit of sustainability in each country, as well as the formulation of policies that promote economic growth and resource efficiency.

This is analogous to the discussed in Section 1.2, in which it was shown that the cost–benefit perspective enables the assessment and comparison of alternative policy options from an efficiency standpoint. Analogously, the progression from glob-

---

[15] However, defenders of the CDM argue that the current options will disappear if not exploited immediately (for the "low-hanging fruit" will rot if not picked early), and that the early exploitation will transfer technology, capacity, and resources to developing countries and enable them to access the more expensive options later (see Haites and Aslam, 2000).

[16] For example, several authors have commented on the initiation of attempts at Kyoto to incorporate developing countries within an emissions control mandate as a retreat from the foundational principles of the UNFCCC (Cooper, 1998; Jacoby *et al.*, 1998; Schmalensee, 1998). These attempts include the call for the adoption of voluntary emissions control targets by non-Annex 1 countries (UNFCCC, 1997a).

Box 1.2. Sustainable Development

The term "sustainable development" was popularized in academic and policy circles by the Brundtland Report (WCED, 1987), although its distinctive antecedents predate the report (especially IUCN, WWF, and UNEP, 1980). The Brundtland Commission defines it as "development that meets the needs of the present without compromising the ability of future generations to meet their own needs" (WCED, 1987, p. 8). However, although the ubiquity of references to this definition suggests a degree of scholarly consensus, this is not the case. There is considerable disagreement on conceptual grounds and, perhaps most significantly, on its operationalization (see Lélé, 1991). Nevertheless, most scholars and practitioners accept a concern for economic prosperity (development), ecological integrity (sustainability), and social justice (equity) as the three pillars of sustainable development (Buitenkamp *et al.*, 1992; Opschoor, 1992; Munasinghe, 1993, 2000; Banuri *et al.*, 1994; Munasinghe and Shearer, 1995; Elkington, 1997; Carley and Spapens, 1998; Sachs *et al.*, 1998; Sachs, 1999).

Sustainable development is an integrating concept (Lélé, 1991; Perrings, 1991; Dietz *et al.*, 1992; Munasinghe, 2000) that has emerged gradually (Rayner and Malone, 1998a , 2000; Costanza, 1999; Munasinghe, 2000; Pichs-Madruga, 1999). Initially, the environmental, economic, and social domains were treated independently, and sustainability viewed as their sum or union. More recently, with the shift in emphasis towards practical and operational aspects, the literature has begun to look at synergies and trade-offs between the three goals.

al equity to sustainable development enables the comparison of policy options that emanate from concerns about global equity. This framework has evolved precisely to enable the assessment of the synergies and trade-offs involved in the pursuit of multiple goals–environmental conservation, social equity, economic growth, and poverty eradication (*Box 1.2*). These analyses touch upon many of the themes relevant to an assessment of the broad range of policy options described above–time horizon, uncertainty, and welfare.

Sustainable development is one of a series of innovative concepts–following such antecedents as human development, equitable development, or appropriate development–that seek to broaden the scope of development theory from its narrow focus on economic growth.[17] However, this evolution has not led to a radical transformation in the operational dimensions of development planning. The focus still continues to be the stock of capital–which in many ways serves as the proxy for welfare or as the index of the "real" or "permanent" income of a society (see Johnson, 1964). As such, much development policy concentrates on measures that stimulate investment and expand the stock of capital. Each innovation has served mainly to expand the definition of the capital stock.

Sustainable development, being the most recent in the series of conceptual advances, subsumes the earlier ones, and rather than meaning simply "development plus natural resource conservation", includes human development, poverty eradication, and social equity as well. Accordingly, it expands the definition of the capital stock to include human capital (skills), natural capital (natural resources and biodiversity), and, most recently, social capital.[18] In principle therefore, sustainable development

is equivalent to investment in this composite stock of capital. However, there are differences of approach rooted in the persistent controversies in development thinking. Some authors focus on investments in all relevant forms of capital, while others focus on the capacity to make such investments. Similarly, the degree of substitution that is possible between kinds of capital -- for example, between natural and human capital -- is a subject of disagreement among researchers. (see *Box 1.3*).[19]

It might appear from the above that sustainable development entails a trade-off between investment in physical capital, social capital, and natural capital, and therefore between economic growth, income distribution, and environmental conservation. However, some branches of development theory have ceased to view these as trade-offs. In particular, the goal of the research on sustainable development–especially conservation strategies and action plans–is to show that under appropriate institutional and social conditions there is a synergy rather than conflict between different goals (IUCN, WWF, and UNEP, 1980). Even earlier, development analysts had begun to question the supposed trade-off between economic growth and income distribution (World Bank, 2000; see also Kuznets, 1955; Hicks, 1979; Chenery, 1980; Fields, 1980).

These debates stem from the earliest days of development thinking, in which a distinction was made between the "bal-

---

[17] These innovations have also yielded alternative indices of welfare, including the human development index (HDI; see UNDP, 1989), basic human needs (BHN; see Streeten *et al.*, 1981), the physical quality of life index (PQLI; see Morris, 1979), and others.

[18] "Social capital" is generally taken to mean the network of social relationships, collective social capacities, and institutions (Banuri *et al.*, 1994; Clague, 1997).

[19] In the absence of detailed data that would (or, indeed, could) allow the aggregation of the different components of the capital stock into a single index, the only option is to pay attention to each component separately. The "four capitals" approach has remained largely a conceptual device rather than an operational one, even though it has often been applied at a project level to ensure that all the necessary components are accounted for.

Box 1.3. Approaches to Understanding Sustainability

Economists distinguish between four main components of the resource base: natural capital (natural resource assets), reproducible capital (durable structures or equipment produced by human beings), human capital (the productive potential of human beings), and social capital (norms and institutions that influence the interactions among humans). These are called capital because they are durable assets capable of generating flows of goods and services. In this construction, development is sustainable if some aggregate index across all forms of capital is non-decreasing.

Strong Sustainability. The strong sustainability approach of the so-called London school (Pearce, 1993) holds that different types of capital are not necessarily substitutable, so that sustainability requires the maintenance of a fixed (or minimum) stock of each component of natural capital. Under this notion, any development path that leads to an overall diminishment in the stocks of natural capital (or to a decline below the minimum) fails to be sustainable even if other forms of capital increase.

Weak Sustainability: The weak sustainability approach asserts that the different forms of capital can substitute for one another to some degree. The substitutability of different types of capital implies that the preservation of an aggregate level of capital, rather than the preservation of natural capital in particular, is crucial. The weak sustainability approach is consistent with the idea that some loss of "climate capital" could be consistent with sustainability if increases in other forms of capital could compensate for the loss.

anced growth" advocated by some writers (Rosenstein-Rodan, 1943; Nurkse, 1958), and the strategy of "unbalanced growth" advanced initially by Albert Hirschman (1958). Hirschman argued that growth is a disequilibrium process, which occurs through the efforts of economic agents to overcome bottlenecks that emerge during normal economic activity. Therefore, policy should not be restricted merely to the mobilization of financial transfers and transfer of technology, but should focus on the larger goal of creating the capacity for mobilizing and allocating such resources,[20] in effect to create conditions in which economic agents can most effectively respond to bottlenecks.

It is fair to say that the development profession has increasingly invoked themes from the latter approach. The emphasis has shifted from promoting growth towards promoting the capacity for growth. Development policy is concerned increasingly with conditions that stimulate investment–trade liberalization, structural adjustment, skill development, governance, institutional development, and market access–rather than the investment itself. This is partly because the fashion has changed from public to private investment, and partly because a large body of research shows that, while the scarcity of financial resources can inhibit the growth process, inflows do not necessarily promote it (Bauer and Yamey, 1982). For example, a recent review of cross-country experience (World Bank, 1998) discovered that the net impact of foreign resource inflows depends critically on ancillary factors–the nature of domestic policies, the fiscal stance, the institutions of governance, and the openness to international trade flows. "Successful" foreign aid led to US$2 of additional private sector investment for every dollar

of aid, while in "failed" cases foreign aid was associated with a net decline in private investment.

Similar shifts have occurred in other areas of development theory and practice. The operationalization of sustainable human development, for example, is increasingly argued to consist not of the simultaneous pursuit of several independent goals, but of investments in social capital to enable the other goals to be pursued through normal market or regulatory mechanisms (Banuri *et al.*, 1994). Poverty eradication programmes focus increasingly on institutional development rather than the creation of physical or social infrastructures. They concentrate on the fact that poor and vulnerable groups generally lack formal organizational structures and recognition as well as the capacity to respond to market opportunities.[21]

1.3.5   *Why Worry about Equity and Sustainable Development?*

While many consider equity to be a good thing in and of itself, this alone may not be reason enough to include it within the context of climate change mitigation. The literature on equity and climate change tends to argue rather that the pursuit of equity will help generate support for mitigation efforts; and that by enabling the pursuit of sustainable development within individual countries, it will lead to more effective mitigation (Lipietz, 1995; Rowlands, 1995; Runnals, 1997; Shue, 1995; Jamieson, 1996, 2000; Byrne, *et al.*, 1998; Parikh and Parikh, 1998; Tolba, 1998; Agarwal *et al.*, 1999). Given that develop-

---

[20] "Capacity" is different from "capital", although the two are related. The latter implies the availability of income-generating capacity alone, while the former suggests the freedom to make policy choices or to achieve social goals.

[21] Indeed, some analysts argue that the poor constitute a distinct "livelihood" economy, which is not well integrated into the global trading and financial system, and therefore lacks the flexibility to respond to emerging market opportunities or standard economic policies (Korten, 1990, 1995).

ing countries have a large suite of pressing social and economic concerns besides emissions control (Najam, 1995; Runnals, 1995; Tolba, 1998), they tend to be wary of mitigation policies lest they undermine other policy goals. Support for sustainable development, besides its own merits, can generate support for climate policy as well. While global climate policy seeks to push the Annex I countries towards emissions contraction, global sustainable development policy offers the opportunity to nudge the developing countries towards a potentially "convergent" trajectory.

Of course, the question is not simply of nudging and pushing countries towards an ultimately equitable path, but to arrive at a global stabilization that is both equitable and sustainable in the long run. Reaction to the Kyoto targets (Malakoff, 1997) suggests that this would require much more than just slight pushing and nudging. A growing literature suggests that this process would be helped by a the longer term focus on sustainability and the alternative development pathways that could lead to it. This is the subject of the next section.

1.4 Global Sustainability and Climate Change Mitigation

In Sections 1.2 and 1.3, we examined literature that was motivated primarily by concerns of global cost-effectiveness and global equity respectively. We now turn to a third category of literature, which is motivated largely by considerations of global sustainability. This literature views the climate problem as a component of a larger problem, namely the unsustainable lifestyles and patterns of production and consumption, and explores a broad range of options for moving the world towards a sustainable future (*Figure 1.6*).

1.4.1 Alternative Development Pathways

The modes of analysis in the studies reviewed in Sections 1.2 and 1.3 start, by and large, with existing institutions and behaviour, and examine their implications for future outcomes. The literature discussed in this section adopts a different approach. It starts with desirable outcomes and examines actions and institutions from the point of view of their compatibility with desirable outcomes. It seeks to fulfil a different objective. It aims to create shared visions of sustainable and desirable societies among the general public, and so it does not, in the first place, suggest implementation alternatives for fixed goals to decision makers (Costanza, 2000). To enlarge the range of accessible options in future decisions, authors who contribute to this line of inquiry intend to foster a process of societal learning among citizens. After all, value formation through public discussion is, as Sen (1995) suggests, the essence of democracy. In doing so, the work of these authors comple-

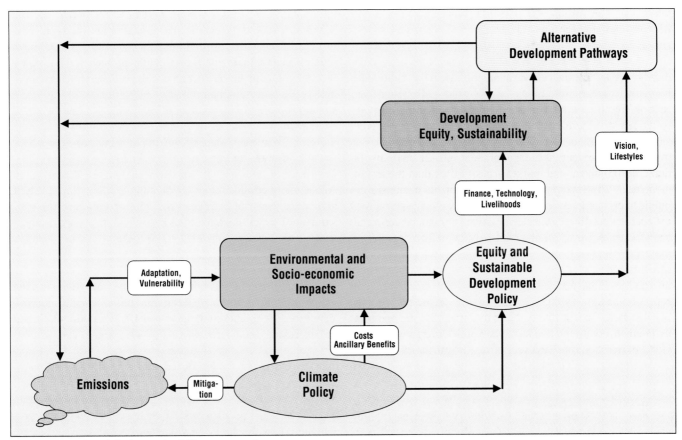

Figure 1.6: *The global-sustainability perspective.*

ments the studies discussed above by providing alternative frameworks, normative contexts, and sets of methodological tools to assess (a broader range of) policy options. Conceptually speaking, this literature takes two forms. The first offers visions of the future based on the inter-relation of various factors across a long time-scale. The second explores possible elements of future scenarios, often relying upon the extrapolation of the existing experience with sustainable practices.

The bulk of this literature starts with the recognition that long-term sustainability can imply an appropriate scale of resource flows, in society (Daly, 1997). Taking a society of appropriate physical scale as a desirable future, this literature goes on to works backwards (backcasts) through possible development paths that may lead from present-day society to a more sustainable, and in the case of concerns about climate change, low-carbon society. Authors who write from this perspective usually assume that resource availability, technology, and society move forwards in a co-evolutionary fashion (Norgaard, 1994). They work on the hypothesis that the transition to balanced and sustainable resource flows implies concomitant changes in technologies, institutions, lifestyles, and worldviews. Though this research takes a certain state of sustainability as its point of departure, it is also sensitive to the principles of equity and cost-effectiveness. It tends to view these as second-order principles that provide structure to the pursuit of sustainability, the first-order principle. In a sense, this literature can be viewed as the mirror image of the studies reviewed earlier–studies that justify the pursuit of sustainability on the grounds of efficiency and equity.

This perspective becomes relevant when it is placed in the context of concerns about unsustainability (loss of biological diversity, extinction of species, air and water pollution, deforestation, desertification, persistent poverty, and rising inequality both within and between nations, and so on). These concerns are derived from underlying pressures imposed by the growth of consumption and population and the inability of many people and communities to protect their health and livelihoods against these damages. Climate change is thus a potentially critical factor in the larger process of society's adaptive response to changing historical conditions through its choice of developmental paths (Cohen *et al.*, 1998, p. 360). Chapter 2 of this report (based on the IPCC (2000a) Special Report on Emissions Scenarios (SRES)) notes, for example, that future emissions will be determined not just by climate policy, but also and more importantly by the "world" in which we will live. Decisions about technology, investment, trade, poverty, biodiversity, community rights, social policies, or governance, which may seem unrelated to climate policy, may have profound impacts upon emissions, the extent of mitigation required, and the cost and benefits that result. Conversely, climate policies that implicitly address social, environmental, economic, and security issues may turn out to be important levers for creating a sustainable world (Reddy *et al.*, 1997, p. 6).

Backcasting from desirable future conditions can, according to Thompson *et al.* (1986), be a useful response to situations characterized by a high degree of ignorance, for which it is difficult to assess the probabilities of possible outcomes or even to know what those possible outcomes might be. Although there is a scientific consensus that anthropogenic climate change is occurring, there is considerable uncertainty about the rate of expected change and its manifestations and impacts at the regional and global levels (see IPCC, 2001, Chapter 19). Science cannot predict the climate and its impacts in Milwaukee, Mumbai, or Moscow half a century ahead very accurately, and it may never be able to do so. Moreover, these types of predictions also require scenarios of the social, economic, and technological paths that the world will follow over the same period (see Chapter 2)–knowledge that may be further beyond our reach than climate prediction. Moreover, this uncertainty increases with the time scale.

The high degree of uncertainty under which climate policy must be developed has important implications for the type of policy regimes likely to be most effective. There is a high degree of uncertainty about how ecosystems would respond to climate change in the studies reviewed here. This recognition suggests that a portfolio approach that includes a broad range of policies diversified across all the major uncertainties might be better than betting on any one particular set of outcomes. Some studies have even drawn a direct parallel between the value of biological diversity and the diversity of institutions and worldviews that contribute to the social capital necessary to maintain the sustainability of human societies (Rayner and Malone, 1998b). Stressing the relationship between risk, resilience, and governance, these authors argue that rather than seeking to anticipate and fix particular problems, the purpose of policy should be to develop coping capacity. This would both switch development and environmental management strategies more nimbly as scientific information improves and strengthen the resilience of vulnerable communities to climate impacts. Conditions of deep uncertainty make it rational for societies to focus on increasing their resilience and flexibility. Resilience in the face of unknown challenges, this research argues, may be achieved by relying on the formation of values and worldviews that embrace the goal of long-term sustainability, at least until some of the key uncertainties are resolved to the point that pursuit of a more narrowly focused policy regime can be justified.

Backcasting from a sustainable future state also supports the search for options with which certain normative goals might be achieved. For climate mitigation scenarios, such a goal might be expressed as a hypothetically acceptable stabilization threshold for GHG concentrations that may, in turn, imply certain trajectories for emission reductions. At this point, therefore, it is useful to review the historical data of global and regional carbon emissions in aggregate as well as in per capita terms (*Table 1.1*; see also *Box 1.1* on the controversy over presentation of data). In 1996, aggregate global emissions were about 6GtC, that is about 1 tonne of carbon per capita world-

Table 1.1: *Per capita income and carbon emissions in various regions*

Region/Country	History		Projections			Average annual change (%) 1996 to 2020
	1990	1996	2000	2010	2020	
North America	1550	1687	1833	2079	2314	1.3
USA	1346	1463	1585	1790	1975	1.3
Canada	126	140	151	162	182	1.1
Western Europe	936	904	947	1021	1114	0.9
Industrialized Asia	364	389	377	435	479	0.9
Japan	274	291	273	322	358	0.9
Australasia	90	99	103	113	122	0.9
Total Developed	**2850**	**2980**	**3157**	**3535**	**3907**	**1.1**
Former Soviet Union (FSU)	991	613	583	666	746	0.8
Eastern Europe (EE)	299	228	243	270	277	0.8
Total EE/FSU	**1290**	**842**	**827**	**935**	**1024**	**0.8**
Developing Asia	1065	1474	1659	2426	3377	3.5
China	620	805	930	1391	2031	3.9
India	153	230	273	386	494	3.2
Middle East	229	283	323	434	555	2.8
Africa	178	198	214	270	325	2.1
Central and South America	174	206	251	418	629	4.8
Total Developing	**1646**	**2161**	**2447**	**3547**	**4886**	**3.5**
Total World	**5786**	**5983**	**6430**	**8018**	**9817**	**2.1**

Reference case, 1990 to 2020 (MtC)

wide. Of this, the 1.2 billion people living in Annex I countries emitted roughly 64% (3.8GtC), or an average of about 3 tonnes of carbon per capita (3tC/capita). In contrast, 4.4 billion people living in non-Annex I countries were responsible for the remaining 2.1GtC, averaging only 0.5tC/capita, or about one-sixth the average for richer countries. Global emissions increased from 5.8GtC to 6GtC from 1990 to 1996, and are projected to increase to 6.4GtC in 2000 and 9.8GtC in 2020.[22] Non-Annex I emissions are growing much faster than those of Annex I countries, averaging 3.5% annual growth compared with 1% in Annex I. As a result, the Annex I share of emissions is declining–from approximately 72% in 1990 and 64% in 1995 to a projected 50% in 2020.

Table 1.2 provides long-term information by displaying aggregate emissions budgets for IPCC SRES scenarios (IPCC, 2000a) and for various stabilization goals identified in the SAR (IPCC, 1996). These goals translate into a 100-year emissions "budget" of 630GtC–13,00GtC. As discussed in section 1.3.1, the target of 450ppmv translates into a reduction (by 2100) of annual emissions to about 3GtC; that is reductions in annual emissions to half of the current level of about 6GtC. Simply stated, per capita emissions of all countries have to fall below current levels in developing countries if GHG stabilization at low levels is to be the targetted future. If these reductions were shared equally, per capita emissions of developed countries would decline by a factor of 10, while emissions from developing countries would halve[23].

These issues, as well as others with purviews beyond the confines of climate change, can provide a starting point for a variety of approaches and analyses. The studies reviewed here investigate kinds of behaviour, institutions, values, technologies, and lifestyles that would be compatible or incompatible with a "desirable" or targeted future. They argue, implicitly or explicitly, that sustainability is built on societal goals made

---

[22] EIA, *Energy Outlook*. These scenarios do not account for the impact of the recent agreements in Kyoto to curb emissions. The differences in trends in Annex I and non-Annex I are similar in other baseline scenarios. Chapter 2 discusses the change of possible scenarios and criteria for selection.

[23] While in the previous section on the equity perspective the emphasis is on an equitable distribution of greenhouse gas emissions while taking into account sustainability criteria, in this section on global sustainability the focus is on the implications of an eventual decrease of global per capita emissions taking into account equity criteria.

Table 1.2. Comparison of cumulative carbon emissions in SRES scenarios and SAR

SRES baseline scenarios	Total emissions 1990 to 2100 (GtC)
B1	989
A1T	1038
B2	1166
A1B	1437
A2	1773
A1FI	2128

IPCC SAR stabilization scenarios (Stabilization level in ppmv CO_2)	Total emissions 1990 to 2100 (GtC)
450	630–650
550	870–990
650	1030–1190
750	1200–1300

mutually supportive early in the process, when the goals and policies of society are being set, rather than downstream after the costs of unsustainable development have already been incurred (Schmidt-Bleek, 1994; Factor 10 Club, 1995). For this reason, they often adopt the industrial metabolism approach, focussing on the flow of materials and energy in modern society through the chain of extraction, production, consumption, and disposal (Ayres and Simonis, 1994; Fischer-Kowalski *et al.*, 1997; Opschoor, 1997). It is argued that the pressure the human economy exerts on the environment depends on levels and patterns of these flows between the economy and the biosphere. Within this conceptual framework, sustainability requires reductions in the overall level of resource flows, particularly the primary flow of (fossil) materials and energy at the input side. Trajectories of emissions reduction of the sort described above can, therefore, be taken as rough indicators for the order of magnitude of the changes involved in the transition to long-term sustainability. In light of this perspective, a number of studies of developed countries (Buitenkamp *et al.*, 1992; McLaren *et al.*, 1997; Carley and Spapens, 1998; Sachs *et al.*, 1998; Bologna *et al.*, 2000) have attempted to backcast a transition to a society capable of creating human welfare with a constantly diminishing amount of natural resources. Certainly, scenarios that explore such outcomes are not restricted to decarbonization or a trend toward carbon sequestration. They may, however, view policies that facilitate these trends as vehicles for nudging the world towards a sustainable future.

All of these scenarios proceed on the premise that economic growth (at least as currently measured) is not the sole goal of societies across the globe. Moreover, they assume that the relationships between economic growth and resource consumption, on the one hand, and wellbeing, on the other, are not fixed. Both should, instead, be shapable by political and social design. A given level of gross domestic product (GDP) can be

achieved with different resource flows (Adriaanse *et al.*, 1997),[24] and economic growth that takes societies beyond certain subsistence levels may not increase satisfaction, or human welfare (UNDP, 1998), or societal welfare (Cobb and Cobb, 1994; Linton, 1998). Consequently, the purpose of these visions is to explore how societies might be able to decouple economic output from resource flows (see Weizsäcker *et al.*, 1997; OECD, 1998) and wellbeing from economic output (see Robinson and Herbert, 2000). Climate change mitigation is one of the co-benefits of these decoupling processes.

1.4.2 Decoupling Growth from Resource Flows

A considerable literature has emerged recently on experiences with technologies, practices, and products that increase resource productivity and ecological efficiency, and thereby reduce the volume of resource input per unit of economic output. The ultimate hope is to shed light on ways in which economic growth and social security can be sustained while resource flows decline in developed countries and/or grow more slowly in developing countries. This literature cites macroeconomic trends with relative reductions in the intensity of resource use coupled with slight increases in absolute levels in the developed economies (Adriaanse *et al.*, 1997). It deals with issues that are central to alternative development paths that are also discussed in the SRES (IPCC, 2000a) and chapter 2. It also notes leapfrogging phases of technological development for developing economies (UNDP, 1998, p. 83). On the micro level, it identifies experiences with cleaner, more economical energy systems, and the potential for information technology to increase resource efficiency. In either case, authors

---

[24] In post-industrial economies, in particular, the resource intensity of GDP is declining.

uncover policy options that pertain mainly to support the proliferation of these trends. These options emerge from a broader conception of climate mitigation than has typically been captured in the energy supply and demand technologies represented in existing energy–economic models. Each option has the potential to reduce GHG emissions, but each needs to be carefully evaluated in terms of its impacts on economic, social, and biological systems. Moreover, each of these options needs to be evaluated alongside conventional energy supply and demand alternatives in terms of their impacts. Expanding the analysis of the set of available options in this way should make us better off, as some of the new options will be attractive upon further analysis, although others will not.

1.4.2.1   Eco-intelligent Production Systems

Many authors argue that progress in developed countries has been driven largely by the technologically based substitution of natural resources for labour. As a result, labour productivity has generally grown faster than resource productivity. Against the background of environmental scarcities, though, this pattern has and will continue to change so that innovation may increasingly be shifted away from labour-saving advances towards resource-saving technologies.

Possibilities include:
- Eco-efficient innovation, that is making products in ways that minimize resource content, utilize biodegradable materials, extend durability, and save inputs during use (Stahel, 1994; Fussler, 1996; Weaver *et al.*, 2000).
- Industrial ecology, that is moving from the nineteenth century concept of a linear throughput growth–in which materials flow through the economy as if through a straight pipe–to a closed loop economy in which industrial materials are fed back into the production cycle (Graedel *et al.*, 1995; LTI-Research Group, 1998; Pauli, 1998).
- Products to services, that is shifting the entrepreneurial focus from the sale of hardware to the direct sale of the services through leasing or renting to facilitate the full utilization of hardware, including maintenance and recycling (Deutscher Bundestag, 1995; Hennicke and Seifried, 1996; Hawken *et al.*, 1999).[25]
- Eco-efficient consumption, that is changing patterns of consumption (using new technologies) to achieve greater efficiency and to reduce waste and pollution (OECD, 1998) in sectors such as transport, food, and housing. Dematerializing consumption may go hand-in-hand with a shift from resource-intensive goods to service-intensive and knowledge-intensive goods (UNDP, 1998, p. 91).

1.4.2.2   Resource-light Infrastructures

In a complementary strand of literature attention has focused on the greater scope for a transition in developing countries by decoupling investment from resource depletion and the destruction of ecological processes. More specifically, since the physical infrastructure in developing countries is still being designed and installed, they have a better opportunity to avoid the resource-intensive trajectories of infrastructural evolution adopted by developed countries (Shukla *et al.*, 1998, p. 53; Goldemberg, 1998a). Specific examples cited in this context are efficient rail systems, decentralized energy production, public transport, grey-water sewage systems, surface irrigation systems, regionalized food systems, and dense urban settlement clusters. These can set a country on the road towards cleaner, less costly, more equitable, and less emission-intensive development patterns. The costs of such a transition are probably higher in places where considerable capital investments in infrastructures have already been made and where turnover is rather slow. For this reason, the timing of such choices is vital, as decisions about systemic technological solutions tend to lock economies onto a path with a specific resource and emission intensity.

In the context of climate policies, innovations in energy systems are of particular importance. Possible strategies advanced in the literature include a shift from expanding conventional energy supply towards emphasizing energy services through a combination of end-use efficiency, increased use of renewables, and new-generation fossil-fuel technologies (Reddy *et al.*, 1997, p. 131). Developing countries that take advantage of these sorts of innovations could follow a path that leads directly to less energy-intensive development patterns in the long run and thereby avoid large increases in energy and/or GDP intensities in the short and medium term.

In many places, renewable energy technologies seem to offer some of the best prospects for providing needed energy services while addressing the multiple challenges of sustainable development, including air pollution, mining, transport, and energy security. For instance, 76% of Africa's population relies on wood for its basic fuel needs; but research and policy design targetted to improve sustainability has been largely absent. Solar energy has a significant potential in sahelian Africa, but slow technological progress, high unit costs, and the absence of technology transfer have retarded its installation. The Brazilian ethanol programme to provide automotive fuel from renewable resources (see *Box 1.4*) is another example. Throughout the developing world the exploitation of hydro potential also remains constrained because of high capital requirements and environmental and social concerns generated by inappropriate dam building.

1.4.2.3   "Appropriate" Technologies

Development of so-called appropriate technologies could lead to environmental protection and economic security in develop-

---

[25] Most of this literature contains assessments of the economic potential of single technologies as well. For some more detail, see Chapter 3 of this report.

Box 1.4. The Brazilian Ethanol Programme

In 1974, Brazil launched a programme to shift to sugarcane alcohol (ethanol) as an automotive fuel, initially as an additive to gasoline in a proportion of about 20%. After 1979, pure alcohol-fuelled cars were produced, with the necessary technological adaptation of engines, through an agreement between the government and multinational car companies in Brazil. The conversion was driven primarily by tax policy and the regulation of fuel and vehicles. The relative prices of alcohol and gasoline were adjusted through Petrobras, the state owned oil company. In 1981 the price of alcohol was set 26% below that of gasoline, although gasoline's production cost was lower than that of alcohol (Pinguelli Rosa *et al.*, 1998).

The alcohol programme created more than 500,000 jobs in rural areas and allowed Brazil to reduce oil imports. The sales of new alcohol-powered cars grew to 30% in 1980 and to more than 90% of the total car sales after 1983 until 1987. Alcohol accounted for about 50% of car fuel consumption at that time. However, the sharp decline in world oil prices along with deregulation in the energy sector meant the abandonment of alcohol-fuelled cars. Even in 1995, though, avoided emissions through alcohol fuel use in Brazil were $24.3MtCO_2$. The cumulative avoided emissions from 1975 to 1998 can be calculated as $385MtCO_2$ (Pinguelli Rosa and Ribiero, 1998).

ing countries. The label "appropriate technologies" is used because they build upon the indigenous knowledge and capabilities of local communities; produce locally needed materials, use natural resources in a sustainable fashion, and help to regenerate the natural resource base. They may enable developing countries to keep an acceptable environmental quality within a controlled cost (Hou, 1988). Low-cost, but resource-efficient technologies are of particular importance for the rural and urban poor (see *Box 1.5*). There is a latent demand for low-cost housing, small hydropower units, low-input organic agriculture, local non-grid power stations, and biomass-based small industries. Sustainable agriculture can benefit both the environment and food production. Biomass-based energy plants could produce electricity from local waste materials in an efficient, low-cost, and carbon-free manner. Each of these options needs to be evaluated alongside conventional energy supply and demand alternatives (see Chapter 3) in terms of the impacts and contribution to sustainable development. Expanding the analysis of the set of available options in this way should make us better off, as some of the new options will be attractive upon further analysis, although others will not.

It is important, in light of these examples, to realize that the results of greater resource efficiency differ according to the performance level of the technology under consideration. Technologies devised for high eco-efficiency and intermediate performance levels consume, for example, lower absolute

amounts of resources than comparable technologies designed for high eco-efficiency and high performance levels. By design, performance levels can vary in such dimensions as level of power, speed, availability of service, yield, and labour intensity. Indeed, intermediate performance levels are often desirable because of their higher employment impact, lower investment costs, local adaptability, and potential for decentralization. For this reason, technologies that combine high eco-efficiency with appropriate performance levels hold an enormous potential for improving people's living conditions while containing the use of natural resources and GHG emissions.

1.4.2.4   Full Cost Pricing

Changing macroeconomic frameworks is often considered indispensable, in both developed and developing countries (Stavins and Whitehead, 1997), to bringing economic rationality progressively in line with ecological rationality. Economic restructuring and energy-pricing reforms both compliment and are a prerequisite for the success of many environmental policies (Bates *et al.*, 1994; TERI, 1995). As long as natural resources, including energy, are undervalued relative to labour, the tendency should be to substitute the cheaper factor for the more expensive one. Giving a boost to efficiency markets requires, first of all, the elimination of environmentally counterproductive subsidies (at least over the medium-to-long

Box 1.5. Resource-efficient Construction in India

Recent analysis shows construction-sector activities to be major drivers of Indian GHG emissions. In addition, conventional building costs place traditional construction beyond the means of an increasing fraction of rural families. A new building technology developed by an Indian non-profit organization, Development Alternatives, reverses this trend. This technology uses hand-powered rams to shape compressed earth into strong, durable, weather-resistant but unbaked bricks. The ingredients for the bricks include only locally available materials, mostly soil and water.

Building new residential and commercial structures with these rammed-earth bricks creates rural jobs and delivers structurally sound buildings with high thermal integrity and few embodied emissions of GHGs. As a result of their inherently high thermal mass, these new buildings easily incorporate passive solar design for heating and cooling. Since they use little purchased input besides human labour, their cost is well within the range of poor families.

term), as on fossil fuels, motorized transport, or pesticides, as much as concessions for logging and water extraction (Roodman, 1996; Larraín *et al.*, 1999). Reform of environmentally destructive incentives would remove a major source of price distortions. Finally, shifting the tax base gradually from labour to natural resources in a revenue-neutral manner could begin to rectify the imbalance in market prices (European Environment Agency, 1996; Hammond *et al.*, 1997). A more extensive discussion of eco-taxation, reporting a wide-ranging debate, is given in Chapter 6 of this report.

1.4.3 Decoupling Wellbeing from Production

Creating an improved, or at least a different, way of life supported by a given set of natural inputs could also enhance the overall resource productivity in society. For developed countries (and the corresponding social sections in developing countries) pursuing such an objective might start from the insight offered by some research that there is no clear link between level of GNP and quality of life (or satisfaction) beyond certain thresholds. Linton (1998) and UNDP (1998) draw this distinction clearly. Both sources argue that the quality of life is determined by subjective and non-subjective variables. On the subjective side, quality of life depends upon personal satisfaction, which in part depends on shared preferences and institutional values. On the non-subjective side, it depends upon opportunity structures, which may include access to nature, participation in community, availability of non-market goods, or public wealth, in addition to purchasing power. This literature describes situations in which GNP growth continues without a corresponding increase in human welfare as "overdevelopment" or "uneconomic growth" (Daly, 1997). For developing countries, however, the research suggests that this decoupling perspective may start from the insight that non-monetary assets (in terms of natural resources, just as in terms of community networks) need to be protected and enhanced to improve the livelihoods of the poorer and less powerful sections of society. Structures, patterns, and rates of economic growth may have to be shaped in such a way that these non-monetary assets are not diminished, but increased.

On both monetary and non-monetary accounts, a decoupling transition to sustainability implies a twin-track strategy. It may be achieved through both an intelligent reinvention of means ("efficiency") and a prudent moderation of ends ("sufficiency"; Meadows *et al.*, 1992; Sachs *et al.*, 1998) for the sake of both environmental and social sustainability. With regard to the environment, efficiency-centred strategies can have a limit; they can fail to account for the effects of continuing growth (Ayres, 1998). For instance, higher per-unit fuel efficiency of cars may not reduce total gasoline consumption in the long run if growth effects in terms of number, power, and size of cars cancel efficiency gains (see Chapter 3; Pinguelli Rosa and Tolmasquin, 1993).[26] With regard to social justice, resource consumption on the part of the rich has been shown, at times, to undermine the environmental sources of livelihood for the

poor. Frequently discussed examples are the construction of large dams for urban electricity supply, which displace large numbers of subsistence peasants, or deforestation for industrial purposes, which marginalizes indigenous people living in and from the forest. In contrast to literature that postulates a "trickling-down effect" in the long term, this school of thought is concerned about the social cost in the present. For its proponents, to secure the rights of the most vulnerable would, in many cases, imply moderation of resource extraction in terms of absolute volumes (Gadgil and Guha, 1995). In the light of these reasons, social and technological systems that combine both high eco-efficiency and intermediate performance levels may be the most likely to foster human welfare at a lower cost to the environment and to social justice.

Four dimensions–intermediate performance levels, regionalization, "appropriate" lifestyles, and community resource rights–can be distinguished in the relevant literature. Policy options identified along these four dimensions emerge from a broader concept of climate mitigation than is typically captured in the energy supply and demand technologies represented in existing energy–economic models. Each option has great potential to reduce GHG emissions, but each needs to be evaluated carefully in terms of its impacts on economic, social, and biological systems. This sort of evaluation of opportunity cost has not, however, been reported in the literature under review. Moreover, most authors are ready to admit that the conditions of public acceptance of such options are not often present at the requisite large scale; they emphasize, however, the necessity to explore these options in order to foster long-term social learning processes. Regional views on the need for or feasibility of decoupling wellbeing from production vary widely. This subsection closes with a brief review of each dimension noted here.

1.4.3.1 Intermediate Performance Levels

Most of the literature on resource-efficient technologies takes for granted that performance levels will (and should) increase. For the sake of a broader portfolio of options, however, some analysts question this assumption. It is suggested that to create resource-light economies could imply deliberately designing technologies (e.g., in construction, ventilation, refrigeration, vehicles, crop cultivation, energy delivery systems) with levels of performance that lie below the maximum feasible. These technologies are often more labour intensive. For instance, the higher speed in transportation are (efficiency gains notwithstanding) unlikely to be environmentally sustainable in the long run; moreover, it is doubtful that this trend really enhances the quality of life (Hirsch, 1976; Wachtel, 1994). Designing cars and trains with lower top speeds could give rise to a new generation of moderately motorized vehicles with much lower resource requirements. In general, renewable energy sources and locally adapted materials, it is argued, become more com-

---

[26] For a more detailed discussion of the so-called rebound effect, see the special issue of *Energy Policy*, 28 (2000), 355–495.

petitive when the performance expectations on the demand side are reduced (Meyer-Abich, 1997). Sails still drive much of ship traffic in parts of the world, as on the Niger and Nile, or the great rivers of China. And bicycles carry a substantial portion of traffic in many regions of the world. Indeed, biomass of all kinds (wood for construction and fuel, plant and animal food and fibre, medicines, dyes, etc.) has been the renewable resource base for humankind since time immemorial. However, to successfully upgrade non-carbon-based technologies, the performance level desired seems to be a critical factor for them to be technically and economically viable.

1.4.3.2 *Regionalization*

Production and lifestyles based on high volumes of long-distance transportation carry a relatively high load of energy and raw materials. Some researchers argue (Shuman, 1998; Magnaghi, 2000) that a low-input society may require that the economy evolves in a plurality of spaces, in which markets that work with "regional sourcing" and "regional marketing" can co-exist with markets that focus on "global sourcing" and "global marketing". Avoiding demand for transport rather than just optimizing the modal split between private and public means of transport is often considered the objective of sustainable policies (Whitelegg, 1993), and regionalized economies may be best suited to this objective. Moreover, solar power, which relies on the widespread but diffuse resource of sunlight, may be best developed when many operators harvest small amounts of energy, transforming and consuming the resource at close distance. A similar logic holds for biomass-centred technologies. Plant matter is widespread, available, and heavy in weight; it may be best obtained and processed in a decentralized fashion. For this reason, some analysts argue that a resource-light economy has to be, in part, a regionalized economy. On the other hand, Chapter 2 points out that regionalization may impede technology transfer, leading to higher emissions, other things being equal.

1.4.3.3 *"Appropriate" Lifestyles*

Many authors question whether the accumulation of individually owned goods beyond a certain threshold continues to increase wellbeing at the same rate. They suggest that individuals and families could be capable of enhancing their personal resource productivity–a goal which, in turn, could be defined as the ability to maintain and/or increase satisfaction with lower and/or intermediate input of resources. Some authors consider intervention in the prevailing narrative of consumption–"more (consumption) is better"–a possible strategy to interrupt the satisfaction–consumption cycle (Common, 1995; Lichtenberg, 1996; Schor, 1998). These approaches draw their motivation from the hypothesis that, ecologically, it is not only the pattern, but also the overall scale of consumption that matters. If this is correct, then social capital in its broadest sense might have to substitute for increased absolute volumes of consumption (Robinson and Herbert, 2000). Chapters 5 and 10 elaborate on the role of lifestyles as a barrier to climate change mitigation, but also as a potential opportunity.

On one level, most resource-intensive consumer goods, in effect, used for only a fraction of time because they are individually owned. Intensity of use could be increased[27] through schemes that involve co-ownership, renting, or leasing (Zukunftskommission, 1998). On another level, the marginal utility of more free time increases faster than the marginal utility of more purchasing power for the more affluent parts of society (Schor, 1998). Choosing more wealth in time rather than more wealth in goods and services can be seen as a viable option, which promises to increase freedom while containing consumption levels. Finally, under conditions of "reflexive modernization" (Beck, 1991), consumption styles might emerge that put more emphasis on quality and non-material satisfaction rather than on rising volumes of consumption (Durning, 1992). As consumption activities become reinserted into the broader contexts of human wellbeing, diverse balances may be found between satisfaction derived from the marketplace and satisfaction derived from non-monetary assets (Reisch and Scherhorn, 1999).

1.4.3.4 *Community Resource Rights*

One-third of mankind derives its sustenance directly from nature (UNDP, 1998, p. 80); and these people live, for the most part, in ecologically fragile areas. Environmental resources are valued as a source of livelihood by groups as diverse as the fisherfolk of Kerala, the forest dwellers of the Amazon, the herders of Tanzania, and the peasants of Mexico (Ghai and Vivian, 1992). In such cases, households rely on non-market goods and natural habitats for important inputs into the production system (Cavendish, 1996). Many of these communities, over the centuries, developed complex and ingenious systems of institutions and rules to regulate ownership and use of natural resources in such a way that an equilibrium between resource extraction and resource preservation could be achieved. However, particularly under the pressure of the resource needs brought forth by individuals with relatively high energy consumption, the basis of their livelihood has been undermined, degrading their dignity and sending many of them into misery (Kates, 2000). Under these circumstances, sustainable development may mean, in the first place, ensuring the rights of communities over their own resources. Properly arranged, and in concert with competitive markets and astute institutional arrangements, resource rights could make investment consistent with community values and associated positive effects on climate change mitigation. Use of ecologically sustainable resources can be made a matter of self-interest. Well-designed resource-right mechanisms permit resource users to use new information and new technology and pursue new market opportunities. Resource use by outsiders becomes a matter of negotiation or trading on more equal terms, which protects the economic security of the communities involved. Better

---

[27] This would lower the demand for capital equipment and allow larger scale more efficient equipment to be used, which in turn would lower resource use and GHG emissions.

access to resources could offer new opportunities to increase the productivity of all components of the village ecosystem–from grazing and forestlands to croplands, water systems, and animals. This may, in turn, enhance people's well-being, which in these circumstances depends on increasing and regenerating biomass in an equitable and sustainable manner. It is well known from the economics literature that the management of common property resources seems to work best when group members can draw on trust and reciprocity, have some autonomy to make their own rules, and perceive to gain benefits from their efforts (Ostrom *et al.*, 2000).

To summarize, we have examined three different perspectives that approach climate change mitigation from different vantage points–cost-effectiveness, equity, and sustainability–but converge in terms of the comprehensive set of goals to be pursued. However, the three perspectives use different analytical tools and causal relationships, and often provide different policy guidance. The main message of this chapter is that these three perspectives are complementary in nature, and can be helpful for the policymaker if used in conjunction. However, this does raise the issue of how to choose between various policy options and how to prioritize actions in the face of possibly divergent advice.

1.5   Integrating Across the Essential Domains–Cost-effectiveness, Equity, and Sustainability

To include issues of cost-effectiveness, distribution (narrowly defined), equity (more broadly defined), and sustainability adds enormous complexity to discussions on the problem of how nations can respond best to the threat of climate change. Indeed, recognition that these multiple domains are relevant complicates the task assigned to policymakers and international negotiators by opening their deliberations to issues that lie beyond the boundaries of the climate change problem, per se. Their recognition thereby underscores the need to integrate scientific thought across a wide range of new policy-relevant contexts, but not simply because of some abstract academic or narrow parochial interest advanced by a small set of researchers or nations. Cost-effectiveness, equity, and sustainability have all been identified as critical issues by the crafters of the UNFCCC, and they are an integral part of the charge given to the drafters of TAR. Integration across the domains of cost-effectiveness, equity, and sustainability is therefore profoundly relevant to policy deliberations according to the letter as well as the spirit of the Framework Convention itself.

One important preliminary step towards integration of the three perspectives that is developed in the body of this report is the use of ancillary and co-benefits, developed and assessed most fully in Chapters 7 and 8 and referred to in many of the other chapters, that could be used to augment mitigation cost estimates produced by the cost-effectiveness approach. Thus, one could add or subtract an estimate of the equivalent cost or benefits on various equity or sustainability metrics (e.g., changes

in the extent of poverty, human capital development, etc.) that would result from specific mitigation policies. Although this would be a start on a more integrated quantitative assessment of costs, it would initiate a debate on how these other metrics ought to be evaluated and aggregated. This may make it desirable to move to a broader integrating framework where multiple policies could be evaluated according to multiple metrics simultaneously. The development of the concept of "mitigative capacity" is one new, but promising, step towards the development of the systematic evaluation of mitigation options from an integrated cost-effectiveness, equity, sustainability perspective.

Yohe (2001, in press) has recently introduced mitigative capacity as an organizing tool to aid policymakers and analysts alike as they try to accomplish this integration. Briefly defined, a nation's mitigative capacity reflects its ability to diminish the intensity of the natural (and other) stresses to which it might be exposed. The list of stresses for any particular nation might include climate change and climate variability, of course. It follows that to review the diversity of the determinants of mitigative capacity from a climate perspective can help assessors who contribute to IPCC Assessments and researchers who will look to their report for guidance in setting their research agendas. These determinants can, in short, provide a framework upon which to build and through which to assess systematic and comparable representations of nations' relative capacities to cope. Mitigative capacity is therefore offered here as one means with which to integrate and to evaluate the complex issues that have emerged since the publication of SAR. There may be other means to the same end, of course, but a focus on mitigative capacity has the virtue of concentrating attention directly on the problem at hand–climate change mitigation.

1.5.1   *Mitigative Capacity–A Tool for Integration*

There are eight distinct but related determinants of mitigative capacity (Yohe, 2001, in press). Cast here in the context of a single country trying to confront its climate change mitigation challenge, they are:

- range of viable technological options for reducing emissions;
- range of viable policy instruments with which the country might effect the adoption of these options;
- structure of critical institutions and the derivative allocation of decision-making authority;
- availability and distribution of resources required to underwrite their adoption and the associated, broadly defined opportunity cost of devoting those resources to mitigation;
- stock of human capital, including education and personal security;
- stock of social capital, including the definition of property rights;
- country's access to risk-spreading processes (e.g., insurance, options and futures markets, etc.); and

- ability of decision makers to manage information, the processes by which these decision makers determine which information is credible, and the credibility of the decision makers themselves.

This section will use these determinants as organizing tools in its assessment of the degree to which current thinking, as evidenced by subsequent chapters, includes the first very preliminary steps toward a thorough integration of cost-effectiveness, equity, and sustainability on the mitigation side of the climate problem.

Mitigative capacity is the mitigation analogue of the concept of adaptive capacity introduced in Chapter 18 of the WGII report (IPCC, 2001). Indeed, adaptive capacity is offered there as a framework upon which to build systematic and comparable representations of communities' and/or countries' ability to ameliorate or exploit the impacts of the natural or social stresses that they might face. As such, adaptive capacity plays a similar organizational role for WGII in their assessment of impacts as mitigative capacity does herein. WGII authors built their assessments around the notion that a system's vulnerability to climate change is determined both by its exposure to the impacts of climate change and by its adaptive capacity. Their analysis uncovered a list of determinants for adaptive capacity that is nearly identical to the list of determinants for mitigative capacity given above. Organization of their thoughts around those determinants enabled them to integrate cost-effectiveness, equity, and sustainability into their assessments of the relative vulnerabilities of different nations, regions, and sectors.

Many of the subsequent chapters presented here offer insight into the role of the first two determinants listed above in determining the ability of various nations to mitigate climate change. Section 1.5.1.1 offers a brief introduction to these insights. An equally brief assessment of some of the related literature from which the roles of the other determinants has been gleaned is given in Section 1.5.1.2. Its coverage is more suggestive of where climate researchers and policymakers should look for aid in formulating the next round of questions; it is less indicative of where past efforts and discussion have been concentrated.

1.5.1.1   *Integrating Environmental, Social, and Economic Objectives in the Third Assessment Report*

Chapters 3 and 4 herein discuss in detail the standard technological options to mitigate climate change. Some or all of these options might be available to any country as it decides to reduce or to slow its emissions of greenhouse cases. However, each technological option must be evaluated in terms of five factors:

- its technological potential in an uncertain environment;
- its economic potential given economic uncertainty and risk;
- existence of technical and economic constraints to its adoption;

- existence of social, cultural, and political constraints to its adoption; and
- ability of key decision makers to understand and to access its potential.

Chapter 5 underlines the significance of each of these characteristics. It points out that cost and performance specifications are critical; however, a technology could be expensive in one place and relatively inexpensive in another; or it may be inexpensive when denominated in one numeraire, but expensive when denominated in another (see Schneider, 1999). Chapter 5 also highlights social and economic constraints derived from high private discount rates, market failures, closed economies, uneven allocations of resources, uneven access to decision-making processes, and other characteristics of social and cultural structures. Finally, Chapter 5 focuses considerable attention on information. Decision makers must be able to understand a technology's economic and technical potential in the context of their own countries, for which data and information may be scarce or, in cases where prices do not reflect social cost, misleading. Clearly, these observations extend the discussion beyond simply listing gadgets towards developing an understanding of how country-specific characteristics might enhance or impede decision makers' abilities to adopt mitigative technologies.

This chapter also underlines the sensitivity of acceptable policy instruments to a similar list of critical parameters that extend efficiency discussions to include equity and sustainability. These include:

- opportunity cost of their implementation, measured broadly to include their development, equity, and sustainability implications;
- sensitivity of these costs to alternative designs;
- availability of credible information and the ability to monitor critical factors in the face of uncertainty;
- definition of a wide range of policy objectives and the degree to which they complement the objective of climate mitigation;
- credibility of the policies and legitimacy of the policymakers;
- social, cultural, political, and economic constraints to their implementation, and
- the structure of the decision-making process itself.

These characteristics clearly have enormous significance when they are cast in the context of development, equity, and sustainability. Later chapters in this report show how alternative policy designs can, on average, have widely different costs and implications even if they achieve comparable results. Chapters 6 to 8 show, for example, that the cost of a policy does not depend on the specification of its targeted outcome only. It also depends on the specification of its timing, on the flexibility that it allows, and on the degree to which it is supported by the international co-ordination of similar efforts across the globe. Different policy designs for the same objective can also have different distributional impacts–different sets of winners and losers across space and time who all come to the table with different access to decision making. Moreover, the opportunity cost of any policy can be measured not only in terms of economic cost, but also in terms of non-economic metrics that measure progress or regression across a wide range of critical variables and against an equally large range of social, cultural, or political objectives (see Schneider, 1999). As a result, mitigation policies that have been successfully adopted in one country might be totally beyond the range of possibility in another.

Finally, differences in the flexibility of alternative policy designs can also mean differences in long-term sustainability from one country to another. Flexibility in response to one mitigation policy that adds efficiency and reduces costs in one place may threaten the very existence of critical systems in another. Ultimately, the goal of international agreements is to induce decision makers at various levels–national and municipal governments, corporate executives, rural communities, and individuals engaging in both production and consumption decisions–to undertake actions that lead to the mitigation of GHGs. There is, in short, a multitude of policy options and instruments available to decision makers at various levels.

Figure 1.7 illustrates this complexity in a diagrammatic form by taking the example of the Kyoto Protocol. The parties have agreed to a 5.2% reduction of Annex I emissions below 1990 levels by the first commitment period, 2008 to 2012. To realize these reductions, however, national governments in these countries have to undertake policy measures that induce corporate and other actors to modify their behaviour. As is shown present-

ly, these policy decisions cover both regulatory and market instruments. Individual economic actors will respond to these incentives through internal changes as well as domestic and international decisions. International decisions cover the innovative Kyoto mechanisms (JI, IET, and the CDM) and are relevant to non-Annex I countries, which will need to take supportive policy decisions as well. These are specifically in the area of institutional development, capacity building, project approval, project monitoring and certification, and national reporting.

Uncertainty, vulnerability to shocks, and attitudes towards risk influence the perceived legitimacy of various decision options. At a global or national level, public opinion and therefore public policy is affected by the scientific uncertainty over the range and impact of climate change. At subnational levels, such uncertainty and vulnerability lies not only in the future, but also in the present circumstances of specific groups–the poor, the communities living in fragile or threatened areas, and the ecological refugees (Gadgil and Guha, 1995). It shapes the collective experience of such groups, determines their decision objectives, and affects their choices as well as susceptibility to policy-induced changes.

Finally, the incentive situation, the nature and strength of institutions for collective action, and the quality and type of information available to decision makers affect individual decisions. All three of these factors vary from one context to another and from one level of decision making to another. Next the nature of governmental policy intervention is discussed, and then the context within which such policies are used is analyzed.

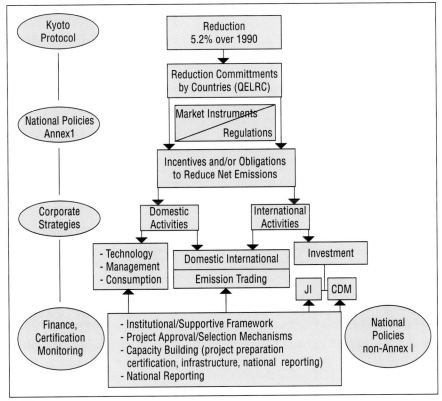

Figure 1.7: *Levels of decision making for Kyoto mechanisms.*

1.5.1.2   Expanding the Scope of Integration

Decisions that lead to the emissions of GHGs that result in global warming are made under, and generally because of, the system of incentives and institutions in place, and are based on the information available to decision makers. Influencing such decisions requires policy intervention at global, national, or local levels. Conversely, the existence of institutions and legitimacy determines the effectiveness of the menu of potential governmental policies outlined above. There is significant heterogeneity within most countries in the types of climate change impacts that might be expected and in the likely impact of GHG mitigation policies. The ability to adapt to climate change depends on the level of income and technology, as well as the capacity of the system of governance and existing institutions to cope with change. The ability to mitigate GHG emissions depends on industrial structure (the mix of industrial activities), social structure (including, e.g., the distance people must travel to work or to engage in recreational activities), the nature of governance (especially the effectiveness of government policy), and the availability and cost of alternatives. In short, what is feasible at the national level depends significantly on what can be done at the subnational, local, and various sectoral levels. However, most studies assume that the national level is the most appropriate for assessing and reacting to the externalities that result from emissions of GHGs and for negotiating international climate change agreements.

The prospect of climate change is just one of many issues of concern to governments, and in most countries climate policy is debated within a broader framework. National policymakers, therefore, have to make trade-offs in implementing climate policies within a comprehensive national and international political economy framework. Many political parties and stakeholder groups oppose climate policy because of perceived conflicts with private sector interests. They also perceive conflicts with traditional macroeconomic goals, like full employment, price stability, and international competitiveness. They also sometimes fear competition with other traditional objectives for public attention and public expenditure (e.g., health care, national security, infrastructure, and education). Likewise, some people may resist mitigation policies (regardless of who pays for them) because of the perceived adverse impacts on economic growth and poverty eradication, even though others might suggest that the implementation of such policies could provide potential opportunities for sustainable development.

Also, many countries have more than one national policy-making authority. In some cases, this diversity may reflect a separation of the executive and legislative branches of government. In others, it may simply be that separate agencies are responsible for economic policy, environment, and international affairs. These agencies will have different views regarding both the needs for climate policy and its likely impact on other goals. The decision-making process invariably reflects the relative political influence of these groups, and involves political nego-

tiations and compromises between them. As a result, O'Riordan *et al.*, 1998) argue that issues considered by governments to be on the policy periphery, like climate change, are not easily factored into consideration of issues at the policy core (such as health care, education, national economic policy, or corporate manufacturing strategy). The issue networks and policy communities around environmental ministries in most countries are weak relative to those around economic and defence ministries. Climate change is sometimes invoked to boost support for existing policy agendas, such as industrial restructuring. Nonetheless, climate change has seldom, if ever, been perceived within the powerful ministries and their policy communities as sufficiently threatening to their departmental interests to fundamentally change those agendas (O'Riordan and Jäger, 1996; Beuerman and Jäger, 1996).

There is, as well, enormous variety in the range of institutional and other conditions in various countries at the subnational level. The political decision-making process in developed countries is affected to a certain degree by powerful non-governmental institutions–including corporations and issue-based non-governmental organizations (NGOs; March and Rhodes, 1992; Sabatier and Jenkins-Smith, 1993; Smith, 1993; Michaelowa, 1998; O'Riordan *et al.*, 1998). These can be a source of resources and new ideas to address climate change as it occurs, but they can also impede the identification and response to changes because of vested interests in the current or some desired allocation of resources. In developing countries, a growing number of institutions have emerged to champion environmental agendas. These range from groups concerned with narrowly defined problems and opportunities (e.g., grassroots groups, wetlands protection groups) to broad-based rights groups (e.g. women's groups) that address a range of common problems (see, e.g., Banuri *et al.*, 1994). However, significant differences continue between the legitimacy and reach of such groups in the developed and developing regions.

The role of the informal sector can also differ between developed and developing countries (see, e.g., Cantor *et al.*, 1992). Although the term is defined somewhat loosely in the literature–often referring to urban, small-scale, non-organized economic activities, and elsewhere to activities not covered in national tax nets–estimates suggest that the informal economy may cover as much as one-third of the economic activity of some developing countries. Given its relative imperviousness to analysis as well as policy influence (indeed, its very existence is credited by some writers to its ability to escape policy influence), it is difficult to project how this sector will react to impacts from climate change or mitigation policies.

The role of information depends critically on the legitimacy of institutions that provide it. The capacity for research, analysis, and policy development is generally weak in developing countries, and especially so in terms of climate change. More importantly, this limited capacity is focused exclusively at the national level. The result is often a credibility gap between the national and local levels. In general, it is difficult to convince local actors of the significance of climate change and the need for corrective action.

More importantly, the bulk of the research and analytical capacity at the global level is concentrated in the developed countries. This is true especially of climate modelling, but also in analyses of the relationship between climate change and sustainable development. Since the late 1980s, massive investment in climate change research has taken place in the developed countries. In contrast, there is a paucity of research institutions in developing countries, with a relatively small level of research effort and investment. This is adequate neither for policy development nor for reassuring policymakers and NGOs of the developing region that the research results are unbiased (Sagar, 1999).[28]

Taken together, these insights suggest a need for investment in the research and analytical capacity of the developing countries, and for orienting the research effort in both developing and developed countries towards the local impacts of climate change and the capacity for climate change adaptation and mitigation. Section 1.5.2 indicates how approaching this complexity within the organizing concept of mitigative capacity can help to generate insights into interpreting and extending analytical exercises, integrating these exercises across multiple stresses, and using this integration to inform discussions and debates in the policy arena.

1.5.2   *Lessons from Integrated Analyses*

Integrating, organizational tools are most useful when they also provide an effective means to assess the existing literature so that new hypotheses can be articulated and new directions can be identified.

One such lesson is that to aggregate representations of mitigation across nations and/or groups may be misleading. Quite simply, the capacity to reduce emissions of GHGs can vary dramatically from nation to nation, sector to sector, region to region, group to group, and timeframe to timeframe.

Secondly, one country can easily display high adaptive capacity and low mitigative capacity simultaneously (or visa versa), even though both capacities share the same list of determinants. In a wealthy nation the damages associated with climate change may focus on a small but well-connected group of people, while the cost of a wide range of adaptation options can, through a well established tax system, be distributed across the entire population. The same country might, however, include another small group of people who would be seriously hurt by most if not all of the wide range of available mitigation options

and/or policies. The benefits of mitigation would meanwhile be marginal for most people because they would be distributed widely across the population and spread far into the future. Mitigative capacity could then be small.

Countries most vulnerable to climate change may have the smallest mitigative capacity. Vulnerability to climate change results from high exposure to climate impacts, low adaptive capacity, or both. In the high-exposure case, the opportunity cost, broadly defined, of expending resources to mitigate GHG emissions may be too high. In the case of low adaptive capacity, the factors responsible may also work to diminish mitigative capacity. And in the third case, both deleterious correlations could work to complement each other.

Enhancing any one component of mitigative capacity may (or may not) reduce the (marginal) cost of mitigation, because it either expands the set of possible mitigative options or because it reduces the constraints that stand in the way of their efficient application. Adding to the list of available technological options can, of course, lower the cost of implementing a specific policy designed to accomplish a specific objective, but the additions must be more socially acceptable than the existing alternatives, as well as structurally, socially, politically, and culturally feasible. If not, they will not be adopted. Furthermore, their informational requirements must not exceed the informational capacity of the host country.

A nation, region, or community's international position can play a significant role in determining its ability to exercise its mitigative capacity, because outside entities can influence the effectiveness of technological options and/or domestic policy alternatives. External forces can have a secondary but nonetheless significant effect on the likelihood that mitigation will occur. Section 1.2 highlights the value of international co-ordination. Trade policies, be they global or the domestic policies of significant trading partners, directly influence national incomes and their distribution. Trade also influences the degree to which a country's development plans put pressure on its stocks of social, human, and natural capital. Each of these factors subsequently affects the constraints that determine the set of feasible mitigation technologies and policies.

Developing indicators of mitigative capacity could help determine who should be expected to do what in terms of mitigation. Examining the determinants of mitigative capacity can identify weak points in the links required for countries to recognize and to act upon the need for climate mitigation This approach can organize existing information effectively as well as suggest new research directions. Specifically, attention to mitigative capacity underlines the role of instruments and targets in framing policy discussions. There are, typically, multiple targets (environmental improvement being one of them) and multiple instruments to achieve them. Contemplating the determinants of mitigative capacity suggests that there is a benefit from broadening the range of instruments used in climate policy. This may be especially so if "climate policy" is under-

---

[28] Participants in international research programme are mostly scientific experts and do not have expertise in development, equity, or sustainability issues.

stood to include mechanisms to achieve environmental goals, sustainability goals, equity goals, and development goals. In this light, mitigative capacity highlights the necessity to observe market failures, political failures, and other failures that might otherwise be overlooked. The fundamental questions are, then, ones of a broad perspective to see exactly how much public policy should be devoted to enhancing mitigative capacity in ways that can help answer questions like "Where are the payoffs clearly greater than the costs?" or "Where is the low-hanging fruit that deserves picking?"

Contemplating the complexity of mitigative capacity reveals that the sources of uncertainty in understanding mitigation extend far beyond the boundaries of the uncertainties that cloud how various technologies might be applied and how various policy designs might function. The same determinants of mitigative capacity that bring development, equity, and sustainability factors into play add to the list of these sources, just as they do on the impact side of the climate change calculus. In short, therefore, anticipating how mitigation might evolve, how much it might cost, how effective it might be, and how the costs and benefits might be distributed is just as uncertain as anticipating how systems might adapt to the impacts of climate change and climate variability.

Understanding the determinants of mitigative capacity offers a way of organizing not only the analysis of mitigation, but also the negotiations over how to meet the mitigation challenge. Indeed, enhancing mitigative capacity can be a policy objective in and of itself. The means by which this enhancement might be accomplished can be drawn directly from an understanding of how the determinants work within and across countries, how they might complement one another, and how they might conflict. Of course, the opportunity cost of enhancing mitigative capacity, measured in terms of cost of regressing against other objectives, is critical in evaluating its desirability. It is also clear, given the way in which its determinants can be expected to interact, that enhancing mitigative capacity means more than simply transferring resources from one nation to another. Weakness beyond access to adequate resources can surely impede the capacity to mitigate any stress; and so it follows that these weaknesses can undermine significantly the efficacy of offering or requesting simple financial support.

1.5.3   Mitigation Research: Current Lessons and Future Directions

Broadening the domain of analysis to include concerns of development, equity, and sustainability over multiple time scales adds enormous complexity to policy deliberations. A portfolio of strategies (not just policy instruments) that draw on efficiency and cost-effectiveness, equity, and sustainability considerations may nonetheless offer the promise of identifying new options and synergies that may make the job of implementing climate policy less disruptive to societies and

economies. In particular, it may help to broaden the range of win–win options.

Concepts like mitigative capacity can help to clarify the trade-offs within and between this expanded range of options. It can show how the assessment of climate change mitigation opportunities contained in this volume can be used and integrated to confront the problem of climate change most effectively. This is especially true when the broad lessons from WGIII herein are taken in concert with lessons drawn from the assessment provided by WGII (IPCC, 2001) on impacts and adaptation. Many of the determinants of adaptive capacity are essentially the same as those of mitigative capacity. Therefore, a portfolio of policy strategies that enhances the capacity to mitigate most effectively should also be effective in enhancing the capacity to adapt. A number of lessons and directions for future research can be enumerated.

- *Improved deliberations on appropriate climate policies in the short, medium, and long terms.*
The literature being brought to bear on the climate issue increasingly shows that policies beyond simply reducing GHG emissions from a specified baseline at minimum costs can be extremely effective in abating the emission of GHGs. Consideration of policies not directly focused on climate, such as those focused on the broader objectives of sustainable development, gives policymakers more flexibility to achieve climate policy objectives.

- *Expanded lists of tools for decision makers and analysts.*
Consideration of the objectives of development, equity, and sustainability can help buy in more participants to climate policies–beyond national and international delegations to include state, local, community, and household agents, as well as NGOs. It also expands the list of tools that can be applied to illuminate the decision-makers' deliberations, from efficiency- and/or distribution-based analytical tools to include alternative decision-analytic frameworks and the development of alternative scenarios.

- *Weighing the costs and impacts of a broader set of policies according to a longer list of objectives.*
Climate deliberations would then consider the climate ramifications of policies designed primarily to address a wide range of issues, including development, equity, sustainability, and sustainable development, as well as the likely impacts of climate policies on the achievement of these other objectives. As part of this process the opportunity costs and impacts of each instrument are measured against the multiple criteria defined by these multiple objectives.

- *A portfolio approach to policy that effectively enhances the capacity to meet the mitigation challenge as well as the capacity to adapt to climate change.*
Focusing research and policy on the determinants of mitigative and adaptive capacity simultaneously can show when, where,

and how synergies and conflicts between mitigation and adaptation might arise. Focusing research on these determinants also makes it clear that policy making in either sphere can be matched by complementary action in the other. Coping with the climate problem is not a question of mitigating and then adapting. Nor is it a question of adapting and then mitigating. It is a more holistic question of doing both at the same time; focusing attention on the common determinants of mitigative and adaptive capacities can lead productively to an understanding of exactly how to meet these coincident challenges.

- *Much additional research is needed before concepts like mitigative capacity can be used to assess the relative merits of specific options.*

Integrating concepts like mitigative capacity should prove useful as a heuristic device to integrate diverse policy instruments into a comprehensive policy portfolio, to discover the metrics with which costs and benefits should be measured, and (perhaps most immediately) to broaden the range of no regrets options.

References

Adriaanse, A. et al., 1997: *Resource Flows: The Material Basis of Industrial Economies*. World Resources Institute, Washington, DC.

Agarwal, A., and S. Narain, 1991a: *Global Warming in an Unequal World: A Case of Ecocolonialism*. Center for Science and Environment, New Delhi.

Agarwal, A., and S. Narain, 1991b: *Towards Green Villages. A Strategy for Environmentally Sound and Participatury Rural Development*. Center for Science and Environment, New Delhi.

Agarwal, A., and S. Narain , 1999: *Addressing the Challenge of Climate Change: Equity, Sustainability and Economic Effectiveness: How Poor Nations Can Help Save the World*. Center for Science and Environment. New Delhi.

Agarwal, A., N. Sunita., and A. Sharma, (eds.), 1999: *Global Environmental Negotiations I*. CSE, New Delhi.

Agarwal, A., S. Narain, and A. Sharma, 2000: *Green Politics*. Center for Science and Environment, New Delhi.

Ayres, R.U., 1998: *Turning Point: An End to the Growth Paradigm*. St Martin's, New York, NY.

Ayres, R.U., and U.E. Simonis, (eds.) , 1994: *Industrial Metabolism*. United Nations University Press, Tokyo and New York, NY.

Banuri, T., and S. Gupta, 2000: The Clean Development Mechanism and Sustainable Development: An Economic Analysis. ADB, Manila.

Banuri, T., G. Hyden, C. Juma, and M. Rivera , 1994: *Sustainable Human Development: From Concept to Operation: A Guide for the Practitioner*. UNDP, New York, NY.

Barrett, S. , 1999: Montreal versus Kyoto: International Cooperation and the Global Environment. In Inge Kaul et al. (ed.), *Global Public Goods: International Cooperation in the 21st Century*, I. Kaul et al., (eds.), OUP, New York, NY.

Bates, R. J. Cofala J., and M. Toman, 1994: Alternative Policies For The Control of Air Pollution in Poland - A World Bank Environment Paper Number 7.

Bauer, P., and B. Yamey, 1982: Foreign Aid: What is at Stake? *The Public Interest*, **Summer**, 53-69.

Bauman, Z., 1998: *Globalization: The Human Consequences*. Polity Press, Cambridge.

Beck, U., 1991: Der Konflikt der zwei Modernen. In *Politik in der Risikogesellschaft*, U. Beck,(ed.), Suhrkamp, Frankfurt, 180-195.

Beckerman, W., 1996: *Through Green Colored Glasses*. Cato Institute,Washington, DC.

Begg, K. , S. Parkinson, Y. Mulugetta, R. Wilkinson, A. Doig, and T. Anderson, 2000: *Initial Evaluation of CDM Projects in Developing Countries*. DFID, London.

Beuermann, C., and J. Jäger, 1996: Climate Change Politics in Germany: How Long Will the Double Dividend Last? In *Politics of Climate Change: A European Perspective*, T. O'Riordan, J. Jäger (eds.), Routledge, London.

Bolin, B., and H. Kheshgi, 2001: On Strategies for Reducing Greenhouse Gas Emissions. Proceeding of the National Academy of Sciences (in press).

Bologna, G. et al., 2000: *Italia capace di futuro*. EMI, Bologna.

Bradley, R.A., and E.R. Williams, (eds.), 1989: A compendium of options for government policy to encourage private sector responses to potential climate change. DOE/EH-0103 (2 vols.). Report to the Congress of the United States, US DOE.

Buitenkamp, M., H. Venner, and T. Wams, (eds.) , 1992: *Action Plan: Sustainable Netherlands*. Milieudefensie, Amsterdam.

Burtraw, D., and M.A. Toman , 1992: Equity and International Agreements for CO^2 Constraint. *Journal of Energy Engineering*, **118**(2), 122-35.

Byrne, J, W. Young-Doo, H. Lee, and J. Kim, 1998: An Equity and Sustainability-Based Policy Response to Global Climate Change. *Energy Policy*, **26** (4), 335-343.

Calabrese, G., 1970: *The Cost of Accidents*. Yale University Press, New Haven, CT.

Cantor, R., S. Henry, and S. Rayner , 1992: *Making Markets: An Interdisciplinary Perspective of Economic Exchange*. Greenwood Press, Westport, CT.

Carley, M., and P. Spapens, 1998: *Sharing the World. Sustainable Living and Global Equity in the 21st Century*. Earthscan, London.

Cavendish, W., 1996: Economics and Ecosystems: The Case of Zimbabwean Peasant Households. In *The North, The South And The Environment, Ecological Constraints and the Global Economy*, United Nations University Press, Tokyo.

Chenery, H.B., 1980: Poverty and Progress - Choices for the Developing World. *Finance and Development*, **June**, 26-30.

Chichilnisky, G., and G. Heal, 1994: Who Should Abate Carbon Emissions? An International Viewpoint. *Economic Letters*, **44**, 443-9.

Clague, C., (ed.), 1997: *Institutions and Economic Development: Growth and Governance in Less-developed Countries*. Johns Hopkins University Press, Baltimore, MD.

Cobb, C., and Cobb, J. (eds.), 1994: The Green National Product. A Proposed Index of Sustainable Economic Welfare, New York, NY.

Cohen, S., D. Demeritt, J. Robinson, and D. Rothman, 1998: Climate Change and Sustainable Development: Towards Dialogue. *Global Environment Change*, **8**(4), 341-371.

Common, M. , 1995: *Sustainability and Policy: Limits to Economics*. Cambridge University Press, Cambridge.

Cooper, R.N., 1998: Toward a Real Global Warming Treaty. *Foreign Affairs*, **77**(2), 66-79.

Costanza, R., 1999: *Ecological Sustainability, Indicators, and Climate Change*. Presented at the IPCC Expert Meeting on Development, Equity and Sustainability, Colombo, Sri Lanka, 27-29 April 1999.

Costanza, R., 2000: Visions of Alternative (Unpredictable) Futures and their Use in Policy Analysis. *Conservation Ecology*, **4** (1), 5.

Daly, H., 1997: *Beyond Growth. The Economics of Sustainable Development*. Beacon Press, Boston MA.

Dasgupta, C., 1994: The Climate Change Negotiations. In., *Negotiating Climate Change: The Inside Story of the Rio Convention*, I. Minstzer and J.A. Leonard, (eds.), Cambridge University Press, Cambridge.

De Bruyn, J.C., J.C. van den Bergh, and J.B. Opschoor, 1998: Economic Growth and Emissions: reconsidering the Empirical Basis of Environmental Kuznets Curves. *Ecological Economics*, **25** (2), 161-177.

Deutscher Bundestag, Enquete-Kommission, 1995: Protecting the Earth's Atmosphere. In *Mehr Zukunft für die Erde*, Economica, Bonn.

Dietz, F.J., U.E. Simonis, and J. van der Straaten, (eds.), 1992: *Sustainability and Environmental Policy: Restraints and Advances*. Edition Sigma, Berlin.

Durning, A., 1992: *How Much Is Enough?* Norton Publishers, Washington, DC.

Edmonds, J.A., and J. M. Reilly, 1985: *Global energy: assessing the future*. Oxford University Press, New York, NY.

Elkington, J., 1997: *Cannibals With Forks - The Triple Bottom Line of 21st Century Business*. Capstone Publications, London.

Enting, I. E., 1998: Attribution of GHG Emissions, Concentration and Radiative Forcing. Technical Paper 38, CSIRO, Australia.

European Environment Agency, 1996: *Environmental Taxes. Implementation and Environmental Effectiveness*. European Environment Agency, Copenhagen.

Factor 10 Club, 1995: *Carnoules Declaration*. Wuppertal Institute, Wuppertal, Germany.

Fields, G.S., 1980: *Poverty, Inequality, and Development*. Cambridge University Press, New York, NY.

Fischbeck, P.S., 1991: *Epistemic Uncertainty in Rational Decision Making*. PhD Thesis, Department of Industrial Engineering and Engineering Management, Stanford University, Stanford, CA.

Fischer-Kowalski, M. et al., 1997: *Gesellschaftlicher Stoffwechsel und Kolonisierung von Natur*. G+B Verlag Fakultas, Amsterdam.

Fussler, C., 1996: *Driving Eco-Innovation*. Pitman, London.

Gadgil, M., and R. Guha, 1995: *Ecology and Equity: The Use and Abuse of Nature in Contemporary India*. Penguin Books, New Delhi.

Ghai D., and J. Vivian, (eds.), 1992: *Grassroots Environmental Action - People's participation in sustainable development*. Routledge Publishers, New York, NY.

Ghosh, 1993: Structuring the Equity Issue in Climate Change. In *The Climate Change Agenda: An Indian Perspective*, A.N. Achanta, (ed.), Tata Energy Research Institute, New Delhi.

Goldemberg, J., 1998a: Leapfrog Energy Technologies. *Energy Policy*, **26**, 729-741.

Goldemberg, J., (ed.), 1998b: *Issues and Opinions: The Clean Development Mechanism.* UNDP, New York, NY.

Graedel, T., E. Graedel, and B. Allenby, 1995: *Industrial Ecology.* Englewood Cliffs.

Grubb, M., 1989: The Greenhouse Effect: Negotiating Targets. Royal Institute of International Affairs, London.

Grubb, M., 1991: Energy policies and the greenhouse effect, vol 2. Country Studies and Technical Options, RIIA/ Dartmouth, Aldershot.

Grubb, M., 1995: Seeking Fair Weather: Ethics and the International Debate on Climate Change. *International Affairs,* **71**, 463-496.

Hahn, R.W., and K.R. Richards, 1989: The Internationalization of Environmental Regulation. *Harvard International Law Journal,* **30**(2), 421-46.

Haites, E., and M. A. Aslam , 2000: *The Kyoto Mechanisms and Global Climate Change.* Pew Center on Global Climate Change, Washington, DC.

Hammond, J. et al., 1997: *Tax Waste Not Work.* Redefining Progress, San Francisco, CA.

Hansen, J., M. Sato, R.Ruedy, A. Lacis, and V. Oinas , 2000: Global Warming in the Twenty-First Century: An Alternative Scenario. *Proceedings of the National Academy of Sciences,* 10.1073, USA.

Hardin, G., 1968: The Tragedy of the Commons, *Science,* **162**, 1243-48.

Hasselmann, K., R. Sancen, E. Maier-Reimer, and R. Voss, 1993: *Climate Dynamics,* **9**(53).

Hawken, P., A. Lovins, and H.L. Lovins , 1999: *Natural Capitalism.* Little Brown, New York, NY.

Hennicke, P., and D. Seifried, 1996: *Das Einsparkraftwerk.* Birkhäuser, Basel.

Hicks, N., 1979: Growth vs. Basic Needs: Is There a Trade-off? *World Development,* **November/December.**

Hirsch, F., 1976: *Social Limits to Growth.* Harvard University Press, Cambridge, MA.

Hirschman, A.O., 1958: *The Strategy of Economic Development.* Yale University Press, New Haven, CT.

Hoffert, M.I. et al., 1998: Energy Implications of Future Stabilization of Atmospheric CO_2 Content. *Nature,* **395**, 881-84.

Hou, R., 1988: Development of appropriate technologies: A possible way forward for environmental protection of developing countries like China. *The Environmentalist,* **8** (4), 273-279.

Howard, R.A., 1980: An Assessment of Decision Analysis. *Operations Research,* **28**(1), 4-27.

Howard, R.A., 1988: Decision Analysis: Practice and Promise. *Management Science,* **34** (6), 679-695.

Howard, R.A., and J.E. Matheson, (eds.), 1984: *Readings on the Principles and Applications of Decision Analysis.* Strategic Decision Group, Menlo Park, CA.

Hyder, T.O., 1992: Climate Negotiations: The North-South Perspective. In I. Mintzer, (ed.), *Confronting Climate Change: Risks, implications and responses,* Cambridge University Press, Cambridge, UK.

IEA (International Energy Agency), 1996: *Climate Change Policy Initiatives 1995/96 Updat, Vol. II.* Organization for Economic Cooperation and Development, Paris, France.

IPCC, 1991: *The First Assessment Report of the Intergovernmental Panel on Climate Change.* Cambridge University Press, Cambridge.

IPCC, 1996: *The Second Assessment Report of the Intergovernmental Panel on Climate Change.* Cambridge University Press, Cambridge.

IPCC, 2000a: Special Report on Methodological and Technological Issues in Technology Transfer. Intergovernmental Panel on Climate Change.

IPCC, 2000b: Special Report on Emissions Scenarios. Intergovernmental Panel on Climate Change.

IPCC, 2000c: Special Report on Land Use, Land Use Change and Forestry. Intergovernmental Panel on Climate Change.

IPCC, 2001: *Climate Change 2001: Impacts and Adaptation.* O. Canziani, J. McCarthy, N. Leary, D. Dokken, K. White, (eds.), Cambridge University Press, Cambridge, MA and New York, NY.

IUCN, WWF and UNEP, 1980: *World Conservation Strategy.* International Union for the Conservation of Nature, Geneva and UNEP, Nairobi.

Jacoby, H.D., R. Prinn, and R. Schmalensee, 1998: Kyoto's Unfinished Business, *Foreign Affairs,* **77** (July/August), 54-66.

Jamieson, D., 1992: Ethics, Public Policy, and Global Warming. Science, Technology and Human Values, **17**(2), 139-153.

Jamieson, D., 1996: Ethics and International Climate Change. *Climatic Change,* **33**, 323-336.

Jamieson, D., 2000: Climate Change and Global Environmental Justice. In P. Edwards, C. Miller, (eds.), *Changing the Atmosphere: Expert Knowledge and Global Environmental Governance,* The MIT Press, Cambridge, MA.

Jepma, C., and M. Munashinghe, 1998: Climate Change Policy (68), Cambridge University Press, Cambridge, MA.

Johnson, H.G., 1964: Towards a Generalized Capital Accumulation Approach to Economic Development. In *The Residual Factor and Economic Growth,* OECD, Paris.

Johnson, T. M., J. Li, Z. Jiang, and R.P. Taylor, 1996: China: Issues and Options in Greenhouse Gas Emissions Control. World Bank Discussion Paper No. 330, World Bank, Washington, DC.

Kasperson, R., and K. Dow, 1991: *Developmental and Geographical Equity in Global Environmental Change. Evaluation Review* (Special Issue on Managing the Global Commons), **15**(1), 149-171.

Kates, R.W., 2000: Cautionary Tales: Adaptation and the Global Poor. *Climatic Change,* **45**(1), 5-18.

Kaufmann, D., A.Kraay, and P. Zoido-Lobaton, 1999: Governance Matters. World Bank Policy Research Working Paper 2196.

Knack, S. , 2000: Aid Dependance and the Quality of Governance: A Cross-Country Empirical Analysis. World Bank Policy Research Working Paper 2396.

Knight, F. , 1921: *Risk, Uncertainty and Profit.* Houghton Mifflin, Boston and New York, NY.

Korten, D., 1990: *Getting to the 21st Century: Voluntary Actions and the Global Agenda.* Kumarian Press, West Hartford, CT.

Korten, D.C., 1995: *When Corporations Rule the World.* Kumarian Press, West Hartford, CT.

Kremen, C. et al., 2000: Economic Incentives for Rainforest Conservation Across Scales, *Science,* **288**(5427).

Kuznets, S., 1955: Economic Growth and Income Inequality. *American Economic Review,* 1-28

Larraìn, S. et al., 1999: Por un Chile sustenablePropuesta, Cinda dana para el Cambio, Santiago.

Latouche, S., 1993: *In the Wake of the Affluent Society: An Exploration of Post Development.* Zed, London.

Lélé, S.M., 1991: Sustainable Development: A Critical Review. *World Development,* **19**(6), 607-21.

Lichtenberg, J., 1996: Consuming Because Others Consume. *Social Theory and Practice,* **22**, 273-297.

Linton, J., 1998: Beyond the Economics of More: the Place of Consumption in Ecological Economics. *Ecological Economics,* **25**, 239-248.

Lipietz, A., 1995: Enclosing the Global Commons: Global Environmental Negotiation in a North-South Conflictual Approach. In *The North, the South and the Environment.* Bhaskar and Glynn (eds.), Earthscan, London.

LTI-Research Group (ed.), 1998: *Long-Term Integration of Renewable Energy Sources into the European Energy System.* Physica-Verlag, Heidelberg.

Magnaghi, A., 2000: *Il progetto locale.* Bollati Boringhieri, Turin.

Malakoff, D., 1997: Thirty Kyotos Needed to Control Warming. *Science,* **278**, 2048.

March, D., and R. Rhodes, (eds.), 1992: Policy Networks in British Government.Oxford University Press, Oxford.

McLaren, D. et al., 1997: *Tomorrow's World: Britain's Share in a Sustainable Future.* Earthscan, London.

Meadows, D. and J. Randers, 1992: *Beyond the Limits.* Chelsea Green Publ., New York, NY

Meire, L. G. , 1999: Note on the Time Dependant Relationship Between Emissions of GHG and Climate Change. Presented at the IPCC Expert Meeting, Ministry of Science and Technology, Brazil.

Meyer, A., 1999: The Kyoto Protocol and the Emergence of "Contraction and Convergence" as a Framework for an International Political Solution to Greenhouse Gas Emissions Abatement. In *Man-Made Climate Change: Economic Aspects and Policy Options,* O. Hohmeyer, K. Rennings, (eds.), Physica-Verlag, Heidelberg.

Meyer-Abich, K.M. , 1997: Ist biologisches Produzieren natürlich? Leitbilder einer naturgemäßen Technik. *GALA,* **6** (4), 247-252.

Michaelowa, A., 1998: Climate Policy and Interest Groups – a Public Choice Analysis. *Intereconomics*, **33** (6), 251-259.

Michaelowa, A., and M. Dutschke, 1998: Interest Groups and Efficient Design of the Clean Development Mechanism Under the Kyoto Protocol. *International Journal for Sustainable Development*, **1** (1), 24-42.

Mintzer, I.M., and J.A. Leonard, (eds.), 1994: *Negotiating Climate Change: The Inside Story of the Rio Convention*. Cambridge University Press, New York, NY.

Morris, M.D., 1979: *Measuring the Conditions of the World's Poor: The PQLI Index*. Pergamon, Oxford.

Moss, R.H., and S.H. Schneider, 2000: Uncertainties in the IPCC TAR: Recommendation to Lead Authors For More Consistent Assessment and Reporting. In Third Assessment Report: Cross Cutting Issues Guidance Report 33-51, World Meteorological Organization, Geneva.

Munasinghe, M., 1993: *Environmental Economics and Sustainable Development*, World Bank, Washington, DC.

Munasinghe, M., 1994: *Sustainomics: A Transdisciplinary Framework for Sustainable Development*. Keynote paper, Proceedings 50th Anniversary Session of the Sri Lanka Association for the Advancement of Science (SLAAS), Colombo.

Munasinghe, M., 1995: Making Growth More Sustainable. *Ecological Economics*, **15**, 121-124.

Munasinghe, M., 2000: Development, Equity and Sustainability in the Context of Climate Change. In M. Munasingha, R. Swart (eds.), *Proceedings of the IPCC Expert Meeting on Development, Equity and Sustainability*, Colombo, 27-29 April, 1999, IPCC and World Meteorological Organization, Geneva.

Munasinghe, M. and W. Shearer (eds.), 1995: Defining and Measuring Sustainability: The Biogeophysical Foundations, UN University and World Bank, Tokyo and Washington DC, USA.

Murcott, S., 1997: Sustainable Development: A Meta review of Definitions, Principles, Criteria, Indicators, Conceptual Frameworks, Information Systems. Presented at the Annual Conference of the American Association for the Advancement of Sciences, Seattle, WA, February 13-18.

Murthy, N.S., M. Panda, and J. Parikh, 1997: Economic development, poverty reduction and carbon emissions in India. *Energy Economics*, **19**, 27-354.

Murthy, N.S., 2000: Energy, Environment and Economic Development : A Case Study of India. PhD Thesis, Indira Gandhi Institute of Development Research, Mumbai, India.

Najam, A., and A. Sagar, 1998: Avoiding a COP-out: Moving Towards Systematic Decision-Making Under the Climate Convention. *Climatic Change*, **39**(4).

Najam, A., 1995: International Environmental Negotiation: A Strategy for the South. *International Environmental Affairs*, **7**(2), 249-87.

Najam, A., 2000: The Case for a Law of the Atmosphere. *Atmospheric Environment*, **34**(23), 4047-4049.

Najam, A. and T. Page, 1998: The Climate Convention: Deciphering the Kyoto Convention, *Environmental Conservation*, **25**(3): 187-194.

NAE (National Academy of Engineering), 1986: *Hazards: Technology and Fairness*. National Academy Press, Washington, DC.

Nordhaus, W.D. and G.W. Yohe, 1983: Future carbon dioxide emissions from fossil fuels. In Changing Climate: Report of the Carbon Dioxide Assessment Committee US, NRC, National Academy Press, Washington, DC.

Norgaard, R., 1994: *Development Betrayed: The End of Progress and a Coevolutionary Revisioning of the Future*. Routledge, London.

Nurkse, R., 1958: The Conflict Between "Balanced Growth" and International Specialization. *Lectures on Economic Development*, Faculty of Economics, Istanbul University, Istanbul, 170-176.

OECD, 1998: *Sustainable Consumption and Production*. OECD, Paris.

Opschoor, J., 1992: *Environment, Economics and Sustainable Development*. Groningen, The Netherlands.

Opschoor, J.B., 1997: Industrial Metabolism, Economic Growth and Institutional Change. In *The International Handbook of Environmental Sociology*, M. Redclift, G. Woodgate, (eds.), Edgar Elgar, Cheltenham, 274-287.

O'Riordan, T., and J. Jäger, (eds.), 1996: *Politics of Climate Change: A European Perspective*. Routledge, London.

O'Riordan, T., C. Cooper, A. Jordan, S. Rayner, K. Richards, P. Runci, and S. Yoffe, 1998: Institutional Frameworks for Political Action. In *Human Choice and Climate Change*, S. Rayner, E. Malone, (eds.), Battelle Press, Columbus, OH.

Ostrom, E. et al., 2000: Revisiting the Commons: Local Lessons, Global Challenges. *Science*, **284**, 278-282.

Parikh, J., 1992: IPCC Response Strategies Unfair to the South. *Nature*, **360** (10 December), 507-508.

Parikh, J., 1994: North-South issues for climate change. *Economic and Political Weekly*, **November 5-12**, 2940-2943.

Parikh, J., 1995: North-South cooperation in climate change through joint implementation. *International Environmental Affairs*, **7**(1), 22-43.

Parikh, J., 2000: Linking Technology transfer with Clean Development Mechanism: A Developing Country Perspective. *Journal of Global Environment Engineering*, **6** (September) 1-12.

Parikh, J., and K. Parikh, 1998: Free Ride through Delay: Risk and Accountability for Climate Change. *Journal of Environment and Development Economics*, **3** (3).

Parikh, J.K., Kirit S. Parikh, Subir Gokarn, J. P. Painuly, Bibhas Saha, and Vibhooti Shukla, 1991: Consumption patterns: The driving force of environmental stress. Report prepared for United Nations Conference on Environment and Development (UNCED), IGIDR-PP-014, Mumbai.

Parikh, J., P.G. Babu, and K.S. Kavi Kumar, 1997a : Climate Change, North-South Co-operation and Collective Decision-Making Post-Rio. *Journal of International Development*, **9** (3), 403-413.

Parikh, J., M. Panda, and N.S. Murthy, 1997b: Consumption Pattern by Income Groups and Carbon dioxide Implications for India: 1990-2010. *International Journal of Global Energy Issues*, **9** (4-5).

Parkinson, S., K. Begg, P. Bailey, and T. Jackson, 1999: JI/CDM Credits under the Kyoto protocol: Does "interim period banking" help or hinder GHG emissions reduction? *Energy Policy*, **27**, 129-136.

Pauli, G., 1998: *Upsizing: The Road to Zero Emissions*. Greenleaf Publishers, Sheffield, UK.

Pearce, D.W., 1993: *Economic Values and the Natural World*. Earthscan Publications, London.

Perrings, C., 1991: *Ecological Sustainability and Environmental Control*. Centre for Resource and Environmental Studies, Australian National University.

Pichs-Madruga, R., 1999: *Climate Change and Sustainability*. Ppresented at the IPCC Expert Meeting on Development, Equity and Sustainability, Colombo, Sri Lanka, 27-29 April 1999.

Pinguelli Rosa, L., and T. Tolmasquim, 1993: An analytical model to compare energy-efficiency indices and CO_2 emissions in developed and developing countries. *Energy Policy*, 276-283.

Pinguelli Rosa, L., and S.K. Ribeiro, 1998: Avoiding Emissions of CO_2 through the Use of Fuels Derived from Sugar Cane. *Ambio*, **27**, 465-470.

Pinguelli Rosa, L., and S.K. Ribeiro, 2000: The Present, Past, and Future Contributions to Global Warming from CO_2 Emissions from Fuels. *Climate Change*, **August.**

Pinguelli Rosa, L., M.T. Tolmasquin, and M.C. Arouca, 1998: Potential for Reduction of Alcohol Production Costs in Brazil. *Energy*, **23**, 987-995.

Rahman, A., 1996: External Perspectives on Climate Change. A View from Bangla-Politics of Climate Change - An European Perspective. Routledge, London, 337-45.

Raiffa, H., 1968: *Decision Analysis*. Introductory Lectures on Choices Under Uncertainty.

Rajan, M.K., 1997: *Global Environmental Politics*. Oxford Univeristy Press, Delhi.

Ramakrishna, K. , 1992: North-South Issues, the Common Heritage of mankind and Global Environmental Change. In *Global Environmental Change and International Relations*, I.H. Rowland, M. Green, (eds.), Macmillan, London.

Rayner, S.,(ed.), 1993: National case studies of institutional capabilities to implement greenhouse gas reductions (Special issue). *Global Environmental Change*, **3** (1).

Rayner, S., 1995: A Conceptual Map of Human Values for Climate Change Decision Making. In *Equity and Social Considerations Related to Climate Change*, R. Odingo et. al., (eds.), ICIPE Science Press, Nairobi.

Rayner, S., and R. Cantor, 1987: How Fair is Safe Enough: The Cultural Approach to Technology Choice. *Risk Analysis: An International Journal,* **7**(1) 3-9.

Rayner, S., and E. Malone, 1998a: *Human Choice and Climate Change.* Batelle Press, Columbus, OH.

Rayner, S., and E. Malone, 1998b: The Challenge of Climate Change to the Social Sciences. In *Human Choice and Climate Change, Vol. 4*, S. Rayner, E. Malone, (eds.), Batelle Press, Columbus, OH.

Rayner, S., and E. Malone, 2000: Climate Change, Poverty, and Intra-Generational Equity at the National Level. In *Climate Change and its Linkages with Development, Equity, and Sustainability*, M. Munasinghe, and R. Swart, (eds.), IPCC, Geneva.

Rayner, S., E. Malone, and M. Thompson, 1999: Equity Issues in Integrated Assessment. *Fair Weather?* F. Toth, (ed.), Earthscan, London.

Reddy, A., R. Williams, and Th. Johansson, 1997: *Energy after Rio. Prospects and Challenges.* UNDP, New York, NY.

Reid, W.V., and J. Goldemberg , 1997: *Are Developing Countries Already Doing as Much as Industrialised Countries to Slow Climate Change?* World Resources Institute, **July**.

Reiner, D.M., and H.D. Jacoby, 1997: Annex I Differentiation Proposals: Implications for Welfare, Equity and Policy. MIT Joint Program on the Science and Policy of Global Change. Report No. 27, Massachusetts Institute of Technology, Cambridge, MA.

Reisch, L., and G. Scherhorn , 1999: Sustainable Consumption. In *The Current State of Economic Science, Vol. 2S,* B. Dahiya, (ed.), Spellbound Publ., Rohtak, 657-690.

Ribot, J. C., A. Najam, and G. Watson , 1996: Climate variation, vulnerability and sustainable development in the semi-arid tropics, in Ribot, J. C., A. R. Magalhaes and S. S. Pangides (eds), *Climate Variability, Climate Change and Social Vulnerability in the Semi-Arid Tropics*, Cambridge University Press, Cambridge, UK.

Robinson, J.B., and D. Herbert, 2000: Integrating Climate Change and Sustainable Development. In *Climate Change and Its Linkages with Development, Equity, and Sustainability*, M. Munasinghe, R. Swart, (eds), IPCC, Geneva, 143-161.

Roodman, D. 1996: *Paying the Piper. Subsidies, Politics, and the Environment.* World Watch Paper 133, Wordwatch Insitute, Washington, DC.

Rose, A., and B. Stevens, 1993: The Efficiency and Equity of Marketable Permits for CO_2 Emissions. *Resource and Energy Economics,* **15**, 117-46.

Rosenstein-Rodan, P. N. , 1943: Problems of Industrialization of Eastern and South-Eastern Europe. *Economic Journal,* **June-September**, 204-7.

Rowlands, I.H., 1995: *The Politics of Global Atmospheric Change.* Manchester Univeristy Press, Manchester.

Runnalls, D., 1997: The International Politics of Climate Change. *Climate Change and North-South Cooperation in Joint Implementation.* In J.K. Parikh,(ed.), Tata McGraw-Hill, New Delhi.

Sabatier, P.A., and H.C. Jenkins-Smith (eds.), 1993: Policy Change and Learning: An Advocacy Coalition Approach, Westview, Boulder, CO.

Sachs, W., 1999: *Planetary Dialectics.* Zed Books, London.

Sachs, W., 2000: Development Patterns in the North and their Implications for Climate Change. In *Climate Change and Its Linkages with Development, Equity, and Sustainability*, M. Munasinghe, R. Swart,(eds.) IPCC, Geneva, 163-176.

Sachs,W., Loske, R., and Linz, M. (eds.) , 1998: , *Greening the North. A Post-Industrial Blueprint for Ecology and Equity.* A Study of the Wuppertal Institute. London: Zed Books.

Sagar, A.D., 1999: Capacity Building and Climate Change, Policy Matters, **4**, 17-20.

Sagar, A.D., and T. Banuri, 1999: In Fairness to Current Generations: Lost Voices in the Climate Debate. *Energy Policy,* **27**(9), 509-514.

Sagar, A., and M. Kandlikar, 1997: Knowledge, Rhetoric and Power: International Politics of Climate Change. *Economic and Political Weekly,* **6 December,** 3140.

Savage, L.J., 1954: *The Foundations of Statistics.* John Wiley and Sons, New York, NY.

Schelling, T.C., 1997: The Cost of Combating Global Warming. *Foreign Affairs,* **76** (6) , 8-14.

Schmidt-Bleek, F., 1994: *Wieviel Umwelt braucht der Mensch?* Birkhaeuser, Basel.

Sen, A., 1995: Rationality and Social Choice. *American Economic Review,* **85**, 1-24.

Schmalensee, R., 1998: Greenhouse Architectures and Institutions. In *Policy Analysis for Decision-making about Climate Change.* W. Nordhaus, (ed.), Resources for the Future, Washington, DC.

Schneider, S.H., 1998: The Climate for Greenhouse Policy in the US and the Incorporation of Uncertainties into Integrated Assessments. *Energy and Environment,* **9**, 425-440.

Schneider, S., 1999: Costing of Non-Linear and/or Irreversible Events. Prepared for IPCC Expert Meeting on Costing Issues for Mitigation and Adaptation to the Climate Change, Tokyo, June 29 - July 1, 2000.

Schor, J., 1998: *The Overspent American.* Basic Books, New York, NY.

Shue, H., 1993: Subsistence Emissions and Luxury Emissions. In *LAW & POLICY,* **15**(1).

Shue, H., 1995: Equity in an International Agreement on Climate Change. In *Equity and Social Considerations Related to Climate Change*, R.Odingo et al., (eds.), ICIPE Science Press, Nairobi.

Shukla, P.R. et al., 1998: Mitigation and Adaptation Cost Assessment Concepts, Methods and Appropriate Use, Ch. 3, 58. NEP Collaborating Centre on Energy and Environment, Riso National Laboratory, Denmark.

Shuman, M., 1998: *Going Local. Creating Self-Reliant Communities in a Global Age.* Free Press, New York, NY.

Siddiqui, T.A., 1995: Energy Inequities Within Developing Countries: An Important Concern in Global Environmental Change Debate, *Global Environmental Change,* **5**: 447-54.

Simon, J., and H. Kahn, (eds.), 1984: *The Resourceful Earth: A Response to Global 2000.* Blackwell, Oxford.

Smith, M.J., 1993: *Pressure, Power, and Policy: State Autonomy and Policy Networks in Britain and the United States.* Harvester Wheatsheaf, Hemel Hempstead, UK.

Sokona, Y., S. Humphreys, and J.-P. Thomas , 1998: What Prospects for Africa? In *Issues and Options: The Clean Development Mechanism*, J. Goldemberg (ed.), UNDP, New York, NY.

Sokona, Y., S. Humphreys, and J.-P. Thomas, 1999: Sustainable Development: A Centrepiece of the Kyoto Protocol–An African Perspective. In *Towards Equity and Sustainability in the Kyoto Protocol*, SEI, Stockholm and Boston, MA.

Stahel, W., 1994: The Utilization-Focused Service Economy: Resource Efficiency and product-life Extension. In *The Greening of Industrial Ecosystems,* B. Allenby, and D. Rickards, (eds.), National Academy Press, Washington. DC.

Stavins, R.N., and B. Whitehead, 1997: Market-Based Environmental Policies. Thinking Ecologically: The Next Generation of Environmental Policy, M. Chertow, D. Esty, (eds.), Yale University Press, New Haven, CT, 105-117.

Streeten, P., S. J. Burki, M. ulHaq, N. Hicks, and F. Stewart , 1981: *First Things First: Meeting Basic Needs in Developing Countries.* Oxford University Press, New York, NY.

TERI (Tata Energy Research Institute), 1995: *Environmental Considerations and Options in Managing India's Long-Term Energy Strategy.* United Nations Environment Programme and TERI, New Delhi.

Thomas, V., M. Dailami, A. Dhareshwar, D. Kaufmann, N. Kishor, R. E. Lopez, and Y. Wang , 2000: *The Quality of Growth.* Oxford University Press, New York, NY.

Thompson, M., M. Warburton, and T. Hatley, 1986: *Uncertainty on a Himalayan Scale.* Ethnographica, London.

Tolba, M.K., 1998*: Global Environmental Diplomacy: Negotiating Environmental Agreements for the World, 1973-1992.* MIT Press, Cambridge, MA.

Toman, M., and D. Burtraw, 1991: Resolving Equity Issues: Greenhouse Gas Negotiations. *Resources,* **103**, 10-13.

Toth, F., (ed.), 1999: *Fair Weather?: Equity Concerns in Climate Change.* Earthscan, London.

UNDP (United Nations Development Programme), 1989: *Human Development Report 1989.* Oxford University Press, New York, NY.

UNDP, 1997: *Governance and Sustainable Human Development.* UNDP, New York, NY.

UNDP (United Nations Development Programme), 1998: *Human Development Report 1998*. Oxford University Press, New York, NY.

UNDP (United Nations Development Programme), 1999: *Human Development Report 1999*. Oxford University Press, New York, NY.

UNFCCC, 1992: *The United Nations Framework Convention on Climate Change*, A/AC.237/18, 9 May.

UNFCCC, 1997a: Report of the Ad-hoc Group on the Berlin Mandate on the Work of its Sixth Session, Bonn, 2-7 March 1997, Addendum. Proposals for a Protocol or Another Legal Instrument (Negotiating Text by the Chairman).

UNFCC, 1997b: Report of the Conference of Parties on its Third Session. Kyoto, 1 - 11 December, Document No. FCCC/CP/7/Add.1 18 March 1998.

Wachtel, P., 1994: *The Poverty of Affluence*. New Society Publishers, Philadelphia, PA.

WCED, 1987: Our Common Future: Report of the World Commission on Environment and Development (WCED), Oxford University Press, New York, NY.

Weaver, P., L. Jansen, G. van Grootveld, E. van Spiegel and P. Vergragt, 2000: *Sustainable Technology Development*. Greenleaf, Aizlewoods Mill, UK.

Weiss, E.B., 1989: *In Fairness to Future Generations: International Law, Common Patrimony, and Intergenerational Equity*. Transnational Publishers, Dobbs Ferry, NY.

Weizsäcker, E.U., A. Lovins, and H. Lovins, 1997: *Factor Four. Doubling Wealth - Halving Resource Use*. Earthscan, London.

Whitelegg, J., 1993: *Transport for a Sustainable Future. The Case of Europe.* London.

World Bank, 1992: *Governance and Development*. The World Bank, Washington, DC.

World Bank, 1998: . *Assessing Aid: What Works, What Doesn't, and Why.* Oxford University Press, New York, NY.

World Bank, 1999: *World Development Report 1999*. Oxford University Press and World Bank, NewYork, NY.

World Bank, 2000: *World Development Report 2000/2001*. Oxford University Press, New York, NY.

WRI (World Resources Institute), 1991: *World Resources 1990-91*. Oxford University Press, New York, NY.

Yohe, G., Mitigative Capacity: The Mirror Image of Adaptive Capacity on the Emissions Side. *Climatic Change* (in press).

Zukunftskommission der Friedrich-Ebert-Stiftung, 1998: *Wirtschaftliche Leistungsfähigkeit, sozialer Zusammenhalt, ökologische Nachhaltigkeit. Drei Ziele-ein Weg*. Dietz, Bonn.

2

Greenhouse Gas Emission Mitigation Scenarios and Implications

Co-ordinating Lead Authors:
TSUNEYUKI MORITA (JAPAN), JOHN ROBINSON (CANADA)

Lead Authors:
Anthony Adegbulugbe (Nigeria), Joseph Alcamo (Germany), Deborah Herbert (Canada), Emilio Lebre La Rovere (Brazil), Nebojša Nakicenovic (Austria), Hugh Pitcher (USA), Paul Raskin (USA), Keywan Riahi (Iran), Alexei Sankovski (USA), Vassili Sokolov (Russian Federation), Bert de Vries (Netherlands), Dadi Zhou (China)

Contributing Authors:
Kejun Jiang (China), Ton Manders (Netherlands), Yuzuru Matsuoka (Japan), Shunsuke Mori (Japan), Ashish Rana (India), R. Alexander Roehrl (Austria), Knut Einar Rosendahl (Norway), Kenji Yamaji (Japan)

Review Editors:
Michael Chadwick (UK), Jyoti Parikh (India)

CONTENTS

EXECUTIVE SUMMARY

Introduction: Summary of the Second Assessment Report and progress since this report.

This chapter reviews three scenario literatures: general mitigation scenarios produced since the Second Assessment Report (SAR), narrative-based scenarios found in the general futures literature, and mitigation scenarios based on the new reference scenarios developed in the Intergovernmental Panel on Climate Change (IPCC) Special Report on Emissions Scenarios (SRES).

Scenarios

A long-term view of a multiplicity of future possibilities is required to consider the ultimate risks of climate change, assess critical interactions with other aspects of human and environmental systems and guide policy responses. Scenarios offer a structured means of organizing information and gleaning insight into the possibilities.

Each mitigation scenario describes a future world with particular economic, social, and environmental characteristics, and therefore implicitly or explicitly contains information about development, equity, and sustainability (DES). Since the difference between reference case scenarios and their corresponding mitigation scenarios is simply the addition of deliberate climate policy, it can be the case that the differences in emissions among reference case scenarios are greater than between any one such scenario and its mitigation version.

General Greenhouse Gas Emissions Mitigation Scenarios

This chapter considers the results of 519 quantitative emission scenarios from 188 sources, mainly produced after 1990. The review focuses on 126 mitigation scenarios that cover global emissions and have a time horizon encompassing the coming century.

These mitigation scenarios include concentration stabilization scenarios, emission stabilization scenarios, tolerable windows/safe emission corridor scenarios, and other mitigation scenarios. They all include energy-related carbon dioxide (CO_2) emissions; several also include CO_2 emissions from land-use changes and industrial processes and other important greenhouse gases (GHGs).

Mitigation options used in the reviewed mitigation scenarios take into account energy systems, industrial processes, and land use, and depend on the underlying model structure. Most of the scenarios introduce simple carbon taxes or constraints on emissions or concentration levels to reflect measures that are taken to implement such options. Regional targets are introduced in the models with regional disaggregation. Emission trading is introduced in more recent work. Some models employ supply-side technology introduction, while others emphasize efficient demand-side technology options.

Allocation of emission reduction among regions is a contentious issue. Only some studies, particularly recent ones, make explicit assumptions about such allocations in their scenarios. Some studies offer global emission trading as a mechanism to reduce mitigation costs.

Technological improvement is a critical element in all the general mitigation scenarios.

Detailed analysis of the characteristics of 31 scenarios for stabilization at 550ppmv (and their respective baseline scenarios) yielded several insights[1].

There was a wide range in baselines, reflecting a diversity of assumptions, mainly with respect to economic growth and low-carbon energy supply. High economic growth scenarios tend to assume high levels of progress in the efficiency of end-use technologies; carbon intensity reductions were found to be largely independent of economic growth assumptions. The range of future trends shows greater divergence in scenarios that focus on developing countries than in scenarios that look at developed nations. There is little consensus with respect to future directions in developing regions.

The reviewed 550ppmv stabilization scenarios vary with respect to reduction time paths and the distribution of emission reductions among regions. Some scenarios show that emission trading lowers overall mitigation cost by shifting mitigation to non-OECD countries, where abatement costs are assumed to be lower. The range of assumed mitigation policies is very wide. In general, scenarios in which there is an assumed adoption of high-efficiency measures in the baseline show less scope for further introduction of efficiency measures in the mitigation scenarios. In part this is due to the structure of the models, which do not assume major technological break-

---

[1] The selection of 550ppmv scenarios is based on the relatively large number of available studies that use this level and does not imply any endorsement of this particular level of CO_2 concentration stabilization.

throughs. Conversely, baseline scenarios with high carbon intensity reductions show larger carbon intensity reductions in their corresponding mitigation scenarios. Global macroeconomic costs of mitigation in the reviewed scenarios range from 0% to 3.5% of gross domestic product (GDP), while a few simple models estimate more increase in the second half of the 21st century. No clear relationship was discovered between the GDP loss and the GDP growth assumptions in the baselines.

Only a small set of studies has reported on scenarios for mitigating non-CO_2 gases. This literature suggests that small reductions of GHG emissions can be accomplished at lower cost by including non-CO_2 gases; that both CO_2 and non-CO_2 emissions would have to be controlled in order to reduce emissions sufficiently to meet assumed mitigation targets; and that methane (CH_4) mitigation can be carried out more rapidly, with a more immediate impact on the atmosphere, than CO_2 mitigation.

In most cases it is clear that mitigation scenarios and mitigation policies are strongly related to their baseline scenarios, but no systematic analysis in this class of literature has been published on the relationship between mitigation and baseline scenarios.

Global Futures Scenarios

Global futures scenarios do not specifically or uniquely consider GHG emissions. Instead, they are more general "stories" of possible future worlds. They can complement the more quantitative emission scenario assessment because they consider dimensions that elude quantification, such as governance and social structures and institutions, but which are nonetheless important to the success of mitigation policies. Addressing these issues reflects the different perspectives presented in Chapter 1 on cost-effectiveness, equity, and sustainability.

A survey of this literature has yielded a number of insights. First, a wide range of future conditions has been identified by futurists, ranging from variants of sustainable development to collapse of social, economic, and environmental systems. Since the underlying socio-economic drivers of emissions may vary widely in the future, it is important that climate policies should be designed so that they are resilient against widely different future conditions.

Second, the global futures scenarios that show falling GHG emissions tend to show improved governance, increased equity and political participation, reduced conflict, and improved environmental quality. They also tend to show increased energy efficiency, shifts to non-fossil energy sources, and/or shifts to a post-industrial economy. Furthermore, population tends to stabilize at relatively low levels, in many cases as a result of increased prosperity, expanded provision of family planning, and improved rights and opportunities for women. A key implication is that sustainable development policies can make a significant contribution to emission reduction.

Third, different combinations of driving forces are consistent with low emission scenarios. The implication of this would seem to be that it is important to consider the linkage between climate policy and other policies and conditions associated with the choice of future paths in a general sense.

Special Report on Emission Scenarios

Six new GHG emission reference scenario groups (not including specific climate policy initiatives), organised into 4 scenario "families", were developed by the IPCC and published as the Special Report on Emission Scenarios (SRES). Scenario families A1 and A2 emphasize economic development but differ with respect to the degree of economic and social convergence; B1 and B2 emphasize sustainable development but also differ in terms of degree of convergence. In all, six models were used to generate 40 scenarios that comprise the six scenario groups. In each group of scenarios, which should be considered equally sound, one illustrative case was chosen to illustrate the whole set of scenarios. These six scenarios include marker scenarios for each of the scenario families as well as two scenarios, A1FI and A1T, which illustrate alternative energy technology developments in the A1 world.

The SRES scenarios lead to the following findings:
- Alternative combinations of driving-force variables can lead to similar levels and structure of energy use, land-use patterns and emissions.
- Important possibilities for further bifurcations in future development trends exist within each scenario family.
- Emissions profiles are dynamic across the range of SRES scenarios. They portray trend reversals and indicate possible emissions cross-over among different scenarios.
- Describing potential future developments involves inherent ambiguities and uncertainties. One and only one possible development path (as alluded to for instance in concepts such as "business-as-usual scenario") simply does not exist. The multi-model approach increases the value of any scenario set, since uncertainties in the choice of model input assumptions can be more explicitly separated from the specific model behaviour and related modelling uncertainties.

Review of Post-SRES Mitigation Scenarios

Recognizing the importance of multiple baselines in evaluating mitigation strategies, recent studies analyze and compare mitigation scenarios using as their baselines the new SRES scenarios. This allows for the assessment in this report of 76 "Post-SRES Mitigation Scenarios" produced by nine modelling teams.

These mitigation scenarios were quantified on the basis of storylines for each of the six SRES scenarios which describe the relationship between the kind of future world and its capacity for mitigation.

Quantifications differ with respect to the baseline scenario including assumed storyline, the stabilization target, and the model that was used. The post-SRES scenarios cover a very wide range of emission trajectories but the range is clearly below the SRES range. All scenarios show an increase in CO_2 reduction over time. Energy reduction shows a much wider range than CO_2 reduction, because in many scenarios a decoupling between energy use and carbon emissions takes place as a result of a shift in primary energy sources.

In general, the lower the stabilization target and the higher the level of baseline emissions, the larger the CO_2 divergence from the baseline that is needed, and the earlier that it must occur. The A1FI, A1B, and A2 worlds require a wider range and more strongly implemented technology and/or policy measures than A1T, B1, and B2. The 450 ppmv stabilization case requires very rapid emission reduction over the next 20 to 30 years.

A key policy question is what kind of emission reductions in the medium term (after the Kyoto protocol commitment period) would be needed. Analysis of the post-SRES scenarios (most of which assume developing country emissions to be below baselines by 2020) suggests that stabilization at 450ppmv will require emissions reductions in Annex I countries after 2012 that go significantly beyond their Kyoto Protocol commitments. It also suggests that it would not be necessary to go much beyond the Kyoto commitments for Annex I countries by 2020 to achieve stabilization at 550ppmv or higher. However, it should be recognized that several scenarios indicate the need for significant Annex I emission reductions by 2020 and that none of the scenarios introduces other constraints such as a limit to the rate of temperature change.

An important policy question already mentioned concerns the participation of developing countries in emission mitigation. A preliminary finding of the post-SRES scenario analysis is that, if it is assumed that the CO_2 emission reduction needed for stabilization would occur in Annex I countries only, Annex I per capita CO_2 emissions would fall below non-Annex I per capita emissions during the 21st century in nearly all of the stabilization scenarios, and before 2050 in two-thirds of the scenarios. This suggests that the stabilization target and the baseline emission level are both important determinants of the timing when developing countries' emissions might need to diverge from their baseline.

Climate policy would reduce per capita final energy consumption in the economy-emphasized worlds (A1FI, A1B, and A2), but not in the environment-emphasized worlds (B1 and B2).

The reduction in energy use caused by climate policies would be larger in Annex I than in non-Annex I. However, the impact of climate policies on equity in per capita final energy use would be much smaller than that of the future development path.

No single measure will be sufficient for the timely development, adoption, and diffusion of mitigation options to stabilize atmospheric GHGs. Instead, a portfolio based on technological change, economic incentives, and institutional frameworks could be adopted. Combined use of a broad array of known technological options has a long-term potential which, in combination with associated socio-economic and institutional changes, is sufficient to achieve stabilization of atmospheric CO_2 concentrations in the range of 450–550ppmv or below.

Assumed mitigation options differ among scenarios and are strongly dependent on the model structure. However, common features of mitigation scenarios include large and continuous energy efficiency improvements and afforestation as well as low-carbon energy, especially biomass, over the next one hundred years and natural gas in the first half of the 21st century. Energy conservation and reforestation are reasonable first steps, but innovative supply-side technologies will eventually be required. Possible robust options include using natural gas and combined-cycle technology to bridge the transition to more advanced fossil fuel and zero-carbon technologies, such as hydrogen fuel cells. Solar energy along with either nuclear energy or carbon removal and storage would become increasingly important for a higher emission world or lower stabilization target.

Integration between global climate policies and domestic air pollution abatement policies could effectively reduce GHG emissions in developing regions for the next two or three decades; however, control of sulphur emissions could amplify possible climate change, and partial trade-offs are likely to persist for environmental policies in the medium term.

Policies governing agriculture and land use and energy systems need to be linked for climate change mitigation. Supply of biomass energy as well as biological CO_2 sequestration would broaden the available options for carbon emission reductions, although the post-SRES scenarios show that they cannot provide the bulk of the emission reductions required. That has to come from other options.

2.1   Introduction: Summary of the Second Assessment Report and Progress since this Report

Various options for mitigating climate change, which constitute the basis of this Working Group III report, depend on societal visions of the future. These visions largely define the decision analytical frameworks used (see Chapter 10) and form the basis for evaluating options. As this chapter will make clear, existing visions of the future are very different in scope and scale, in time horizons, in constituents and uncertainties, and cover different areas of human activities, natural conditions, etc. Whereas some authors explore the future by extrapolating trends, others aim at a more desirable future state.

Many visions of the future can be modified into scenarios through the systematization of data and other available information, using various modelling techniques, and thereby leading to quantitative interpretations of the future. The spectrum of scenarios can be as broad as that of visions, however, articulating a scenario can provide a more detailed picture of the framework for decisions and the associated limitations for decision-making processes and policy interventions in any particular area.

Climate change and its impacts have a long history in the existing scenario literature, while mitigation scenarios that explore policy options to be implemented are of more recent origin. In the Second Assessment Report (SAR) of the Intergovernmental Panel on Climate Change (IPCC), greenhouse gas (GHG) mitigation scenarios were reviewed. Since that time, there has been considerable development of such scenarios, focussing on issues of the timing, location, and extent of responses required to stabilize atmospheric concentrations at various levels. These new mitigation scenarios are reviewed in this chapter.

Another literature, consisting of more narrative-based scenarios of alternative global futures, is also reviewed in this chapter. These more general scenarios provide a basis for contextualizing the more traditional emissions scenarios, and providing a link to development, equity, and sustainability (DES).

In addition, in 1996, the IPCC commissioned a new report on emissions scenarios (the Special Report on Emissions Scenarios, or SRES), in which new scenarios were developed (Nakicenovic *et al.*, 2000). During 1999 and 2000 various modellers used these new reference scenarios as the basis of new mitigation and stabilization analyses. This post-SRES work is also reviewed in this chapter.

Section 2.2 provides a background of scenarios in general, and emission and mitigation scenarios in particular, and discusses the link between scenarios and DES. Section 2.3 reviews general mitigation scenarios produced since the SAR. Section 2.4 discusses global futures scenarios, which are narrative-based scenarios found in the general futures literature. Section 2.5 provides a review of the SRES and discusses post-SRES mitigation scenarios. Finally, Section 2.6 provides recommendations for future research.

2.2   Scenarios

2.2.1   Introduction to Scenarios

Climate change assessment addresses a highly complex set of interactions between human and natural systems, a scientific challenge that is compounded by the cumulative and long-term character of the phenomenon. While the world of many decades from now is indeterminate, scenarios offer a structured means of organizing information and gleaning insight into the possibilities. Scenarios can draw on both science and imagination to articulate a spectrum of plausible visions of the future and pathways of development. Some scenarios are assumed to evolve gradually and continuously from current social, economic, and environmental patterns and trends; others deviate in fundamental ways. A long view of a multiplicity of future possibilities is required to consider the ultimate risks of climate change, assess critical interactions with other aspects of human and environmental systems, and guide policy responses.

The term "scenarios" appears in two distinct streams of inquiry, one based on qualitative narrative and the other on mathematical models. Qualitative scenarios are primarily literary exercises, aimed at holistic and integrated sketches of future visions and compelling accounts of a progression of events that might lead to those futures. Quantitative, formal models seek mathematical representation of key features of human and/or environmental systems in order to represent the evolution of the system under alternative assumptions, such as population, economic growth, technological change, and environmental sensitivity. Qualitative scenarios have a greater power to posit system shifts, to explore the implications of surprise, and to include critical factors that defy quantification, such as values, cultural shifts, and institutional features. On the other hand, qualitative scenarios may appear arbitrary, idiosyncratic, and weakly supported. Model-based scenarios are useful for examining futures that result from variations of quantitative-driving variables, and they offer a systematic and replicable basis for analysis.

A first wave of global assessments began in the 1970s. Ambitious global modelling exercises aimed to forecast the behaviour over many decades of development, resource, and environmental systems, and to assess resource constraints (Meadows *et al.*, 1972; Mesarovic and Pestel, 1974). The Latin American world model stressed social and political concerns, rather than physical limits, by positing a normative egalitarian future to examine the actions required to achieve it (Herrera *et al.*, 1976). A second wave of integrated global scenario analyses responded to new concerns about sustainable development and the future (WCED, 1987). Many of these were in the qualitative tradition (Svedin and Aniansson, 1987; Toth *et al.*, 1989; Milbrath, 1989; Burrows *et al.*, 1991; Kaplan, 1994; Gallopin *et al.*, 1997; WBCSD, 1997; Bossel, 1998). In addition, stimulated largely by the climate issue, there have been a number of new models that quantitatively link energy and other human activities to atmospheric, oceanic, and terrestrial systems (e.g., Rotmans and de Vries, 1997). Finally, scenario studies have begun recently to synthesize the modelling and qualitative approaches, in order to blend structured quantitative analysis with textured and pluralistic scenario narratives (Raskin *et al.*, 1998; Nakicenovic *et al.*, 2000).

IPCC GHG emission scenarios were prepared for the first assessment report of 1990. These initial scenarios were updated and extended, and led to the publication in 1992 of alternative emissions scenarios for the period 1990 through 2100 (Leggett *et al.*, 1992; Pepper *et al.*, 1992). These so-called IS92 emission scenarios were used by the IPCC to assess changes in atmospheric composition and climate over this time horizon. Analysts have used the IS92 scenarios, and particularly IS92a, as the preferred reference scenarios for mitigation and stabilization studies. A subsequent IPCC evaluation of the IS92 scenarios (Alcamo *et al.*, 1995) found that for the purposes of driving atmospheric and climate models, the carbon dioxide (CO_2) emissions trajectories of the IS92 scenarios provided a reasonable reflection of variations found in the open literature. However, the review found that these scenarios should not be used for evaluating the consequences of interventions to reduce GHG emissions since the scenarios have insufficient sectoral and regional detail for careful analyses. This review also took into account criticism by Parikh (1992) who suggested the need for a more coherent approach and scenarios that show improved equity between the developed and the developing countries.

The 1995 review also emphasized the need for analysts to consider the full range of IS92 emissions scenarios, rather than a single "business-as-usual" reference scenario. The uncertain-

ties in long-range future assumptions make the assignation of a most-probable trajectory problematic.

In 1996, the IPCC initiated a process for establishing a new set of reference emissions scenarios. The new scenarios are described in the IPCC Special Report on Emissions Scenarios (Nakicenovic *et al.*, 2000). These are designed to be non-mitigation or reference scenarios, that is, scenarios in which additional policy initiatives aimed specifically at lowering GHG emissions are assumed to be absent.

Owing to fundamental uncertainties, it is impossible to predict or forecast the long-range global future, even with the most sophisticated methods. Long-range indeterminism implies that probabilities cannot be rigorously assigned for either a given set of driving assumptions or the likelihood of structural shifts in societies and natural systems. Consequently, instead of a single "business as usual" scenario, multiple baseline scenarios are needed to scan a spectrum of plausible possibilities in order to guide the formulation of robust policies that are not geared to an overly rigid sense of where the world is heading.

To account for the wide variety of possible futures and the large uncertainties involved in such forward projections, the SRES team opted for a multiple baseline approach.[2] It also decided to fuse a qualitative, narrative approach with a more formal approach with different models, to guarantee structural variance and methodological diversity in the scenarios. As such, the SRES-scenarios combine elements from both the more story-like scenarios discussed in Section 2.4 below, and the more model-based scenarios discussed in Section 2.3. The relationship between these three kinds of scenarios is shown in *Figure 2.1*.

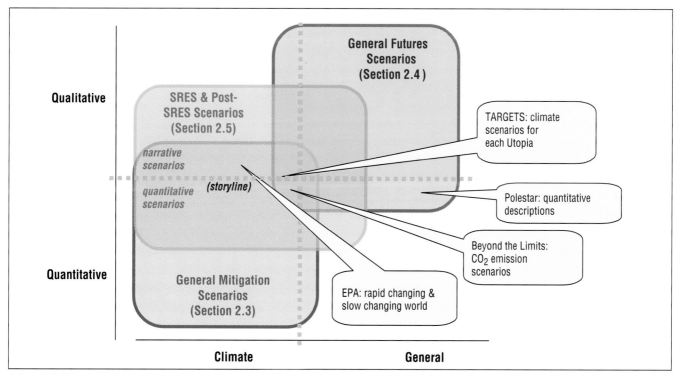

Figure 2.1: *Relationship among the three groups of literature reviewed in Chapter 2.*

2.2.2   *Mitigation and Stabilization Scenarios*

Mitigation scenarios are usually defined as a description and a quantified projection of how GHG emissions can be reduced with respect to some baseline scenario. They contain new emission profiles as well as costs associated with the emission reduction. Stabilization scenarios are mitigation scenarios that aim at a pre-specified GHG reduction target. Usually the target is the concentration of CO_2 or the CO_2-equivalent concentration of a "basket" of gases by 2100 or at some later date when atmospheric stabilization is actually reached.

There are two common difficulties associated with the formulation and quantification of mitigation scenarios. First, in certain cases there is not a clear-cut distinction between intervention and non-intervention scenarios, that is, scenarios with or without explicit climate policy. This is discussed in detail in *Box 2.1*. The second important problem regarding mitigation scenarios has to do with the difference between top-down and bottom-up models. Whereas the latter focus on engineering trends and technology costs, the former view resource development from a macroeconomic price-mediated perspective. Although, as discussed in the SAR (IPCC, 1995), the differ-

Box 2.1. Differentiating Between Climate Policy and No-climate-policy Scenarios

Recent discussions among IPCC experts and reactions from reviewers of this report and the SRES report revealed the need to clarify differences between various types of GHG emission scenarios, in particular, between climate policy scenarios (CP scenarios) and scenarios without climate policies (NCP scenarios) but with low emissions.

CP scenarios (also known as climate intervention or climate mitigation scenarios) are defined in this report as those that: (1) include explicit policies and/or measures, the primary[3] goal of which is to reduce GHG emissions (e.g., carbon tax) and/or (2) mention no climate policies and/or measures, but assume temporal changes in GHG emission sources or drivers required to achieve particular climate targets (e.g., GHG emission levels, GHG concentration levels, temperature increase or sea level rise limits).[4]

CP scenarios are often, but not always, constructed with reference to a corresponding reference or baseline scenario that is similar to the CP scenario in every respect except the inclusion of climate mitigation measures and/or policies. In fact, climate policy analysis often starts with the construction of such a reference scenario, to which is added climate policy to create the CP scenario.

Another type of CP scenario is not originally built around such "no-policy" baselines. Developers of such scenarios envision future "worlds" that are internally consistent with desirable climate targets (e.g., a global temperature increase of no more than 1°C by 2100), and then work "backwards" to develop feasible emission trajectories and emission driver combinations leading to these targets. Such scenarios, also referred to as "safe landing" or "tolerable windows" scenarios, imply the necessary development and implementation of climate policies, intended to achieve these targets in the most efficient way.

The general definition of CP scenarios provided here enables one to effectively discriminate between CP scenarios and other scenarios with low emissions (e.g., IS92c, SRES-B1). Unlike the former, NCP scenarios have low emissions but do not assume any explicit emission abatement measures or policies, nor are they designed specifically to achieve certain climate targets. NCP scenarios by themselves may explore a wide variety of alternative development paths, including "green" or "dematerialization" futures.

Confusion can arise when the inclusion of "non-climate-related" policies in a NCP scenario has the effect of significantly reducing GHG emissions. For example, energy efficiency or land use policies that reduce GHG emissions may be adopted for reasons that are not related to climate policies and may therefore be included in a NCP scenario. Such a NCP scenario may have GHG emissions that are lower than some CP scenarios.

The root cause of this potential confusion is that, in practice, many policies can both reduce GHG emissions and achieve other goals. Whether such policies are assumed to be adopted for climate or non-climate policy related reasons in any given scenario is determined by the scenario developer based on the underlying scenario narrative. While this is a problem in terms of making a clear distinction between CP and NCP scenarios, it is at the same time an opportunity. Because many decisions are not made for reasons of climate change alone, measures implemented for reasons other than climate change can have a large impact on GHG emissions, opening up many new possibilities for mitigation. Chapters 7, 8, and 9 discuss ancillary benefits of climate mitigation and the co-benefits of policies integrating climate mitigation objectives with other goals.

---

[2] It is perhaps worth noting in this connection that, in a similar way, the IPCC had originally recommended that climate and other modellers use the full set of IS92 scenarios but, in practice, this advice has not been followed by most researchers who have focussed primarily on the "central" IS92 case, thereby potentially contributing to an unjustified sense of probability or accuracy.

[3] Some climate polices have multiple benefits. For example, a particular policy designed to reduce methane leaks from natural gas systems may also increase the operating company's profitability and improve safety. However, if this policy was originally developed to reduce emissions it should be classified as a climate policy, not as a policy to increase profitability or improve safety.

[4] Such targets may be reached without specific additional climate policies, e.g., by pursuing particular development pathways.

ences between these approaches are continiously narrowing as each incorporates elements of the other, there is still quite a difference in their formulation of emission reduction strategies. This suggests the importance of including multiple methodological approaches in scenario analysis.

2.2.3 Scenarios and "Development, Equity, and Sustainability (DES)"

The climate issue is embedded in the larger question of how combined social, economic, and environmental subsystems interact and shape one another over many decades. There are multiple links. Economic development depends on maintenance of ecosystem resilience; poverty can be both a result and a cause of environmental degradation; material-intensive lifestyles conflict with environmental and equity values; and extreme socio-economic inequality within societies and between nations undermines the social cohesion required for effective policy responses.

It is clear that climate policy, and the impacts of climate change, will have significant implications for sustainable development at both the global and sub-global levels. In addition, policy and behavioural responses to sustainable development issues may affect both our ability to develop and successfully implement climate policies, and our ability to respond effectively to climate change. In this way, climate policy response will affect the ability of countries to achieve sustainable development goals, while the pursuit of those goals will in turn affect the opportunities for, and success of, climate policy responses.

In this report and its Working Group II companion report, climate change impacts, mitigation, and adaptation strategies are discussed in the broader context of DES (see Munasinghe, 1999).

The issues raised by a consideration of DES are of particular relevance to the scenarios discussed in this chapter. Because they are necessarily based upon assumptions about the socio-economic conditions that give rise to emissions profiles, mitigation and stabilization scenarios implicitly or explicitly contain information about DES. In principle, each stabilization or mitigation scenario describes a particular future world, with particular economic, social, and environmental characteristics. Given the strong interactions between development, environment, and equity as aspects of a unified socio-ecological system and the interplay between climate policies and DES policies, emissions scenarios are viewed in this report as an aspect of broad sustainable development scenarios.

The allocation of emissions in a scenario is coupled closely to an important policy question in climate negotiations: the fair distribution of future emission rights among nations, or "burden sharing". For example, an egalitarian formulation of the rights of developing countries to future "climate space" is often

expressed in terms of equal per capita emissions allocations. Alternative assumptions on burden sharing have important implications for equity, sustainable development, and the economics of emissions abatement. However, it is noteworthy that this critical conditioning variable is usually not explicitly treated in mitigation scenarios in the literature (see section 2.3 below). Indeed, documentation of scenarios generally does not address the implications of the scenarios for equity and burden sharing. In rare cases, mitigation scenarios have been developed which explicitly impose the simultaneous co-constraints of climate and equity goals (e.g., Raskin *et al.*, 1998).

In this and other ways scenario analysis could become an important way of linking DES issues to climate policy considerations. However, as discussed in more detail in section 2.4, many quantitative mitigation and stabilization scenarios have not been designed with this purpose in mind. As a result, it is not always easy to draw out the DES implications of particular stabilization and mitigation scenarios.

Although this chapter focuses on mitigation and stabilization scenarios, it is important to note that DES issues are also implicit in the base case or reference scenarios that underlie mitigation and stabilization scenarios. Since the difference between reference case scenarios and stabilization and mitigation scenarios is simply the addition of deliberate climate policy, it can be the case that the DES differences among different reference case scenarios are greater than between any one such scenario and its stabilization or mitigation version. This is of particular relevance in the discussion below in section 2.5.2 of scenarios based on the baselines produced in the IPCC's SRES (Nakicenovic *et al.*, 2000).

2.3 Greenhouse Gas Emissions: General Mitigation Scenarios

This chapter reviews three scenario literatures, which span a range from more quantitative scenario analysis to analysis that is based more on narrative descriptions (see *Figure 2.1*). At the quantitative end of the spectrum are the "general mitigation scenarios" reviewed in this section, which consist mainly of quantitative descriptions of driving forces and emission profiles.

2.3.1 Overview of General Mitigation Scenarios

More than 500 emission scenarios have already been quantified, including non-mitigation (non-intervention) scenarios and mitigation (intervention) scenarios that assume policies to mitigate climate change. These scenarios have been published in the literature or reported in conference proceedings, and many of them were collected in the IPCC SRES database (Morita & Lee, 1998a) and made available through the Internet (Morita & Lee, 1998b). Using this database, a systematic review of non-mitigation scenarios has already been reported in the SRES

(Nakicenovic *et al.*, 2000). However, several mitigation and other scenarios were missing from this database and new emission scenarios have been quantified since the SRES review. Accordingly, the missing scenarios and new scenarios were collected and the database revised for this new review of mitigation scenarios (Rana and Morita, 2000).

The current database collection, covered in this report, contains the results of a total of 519 scenarios from 188 sources. These scenarios were mainly produced after 1990. Two questionnaires were sent to representative modellers in the world, and sets of scenarios from the International Energy Workshop (IEW) and Energy Modelling Forum (EMF) comparison programmes were collected. The database is intended to include only scenarios that are based on quantitative models. Therefore, it does not include scenarios produced using other methods; for example, heuristic estimations such as Delphi.

Of the 519 scenarios, a total of 380 were global GHG emission scenarios, most of which were disaggregated into several regional emission profiles. Of these 380 global emission scenarios, a total of 150 were mitigation (climate policy) scenarios. This review focuses on mitigation scenarios that cover global emissions and also have a time horizon encompassing the coming century. Of the 150 mitigation scenarios, a total of 126 long-term scenarios that cover the next 50 to 100 years were selected for this review. 24 scenarios were excluded on the basis of their short time coverage.

Table 2.1 presents an outline of several representative scenarios in this review; these scenarios exemplify the modelling literature. Columns 1 and 2 of the table show the main identifiers of the scenarios, namely, the model name and source and the policy scenario name, as given by the modellers. The third and fourth columns show the policy scenario type and specific scenario assumptions. The remaining columns contain additional important features of the policy scenarios, including reduction time-paths and burden sharing, GHGs analyzed, policy options and approaches, and feedback. Only five studies among the selected sources of *Table 2.1* have detailed policies. Most of the other scenarios assume very simple policy options such as carbon taxes and simple constraints.

Based on the type of mitigation, the scenarios can be classified into four categories: concentration stabilization scenarios, emission stabilization scenarios, safe emission corridor (tolerable windows/safe landing) scenarios, and other mitigation scenarios.

Scenarios for concentration stabilization account for a large proportion of the mitigation scenarios, with 47 of the 126 mitigation scenarios being classified into this type. Many scenarios of this type were quantified in the process of the EMF comparison (Weyant and Hill, 1999) where a systematic guideline was prepared for stabilization quantification. Of the 47 scenarios, two-thirds are intended to stabilize atmospheric concentrations of CO_2 at 550ppmv. The concentration of 550ppmv was

used as a benchmark for stabilization in the previous studies on mitigation scenarios. This number may be related to the frequent references made to it in political discussions. The adoption by the European Union of a maximum increase in global average temperature of 2°C above pre-industrial levels is roughly equivalent to a stabilization level of 550ppmv CO_2 equivalent or 450ppmv CO_2. It does not imply an agreed-upon desirability of stabilization at this level. In fact, environmental groups have argued for desirable levels well below 550ppmv, while other interest groups and some countries have questioned the necessity and/or feasibility of achieving 550ppmv. Scenarios with levels of concentration stabilization other than 550ppmv are contained in IPCC (1990), Manne *et al.* (1995), Alcamo and Kreileman (1996), Ha-Duong *et al.* (1997), Manne and Richels (1997), and Fujii and Yamaji (1998).

The emission stabilization scenarios account for 20 of the 126 mitigation scenarios. Most scenarios of this type are intended to stabilize at 1990 emission levels in Annex I or the Organization for Economic Co-operation and Development (OECD) countries. Some scenarios have emissions stabilizing at other levels, for example, the emissions stabilization scenario of DICE (Nordhaus, 1994) aims at a level of 8GtC/yr of CO_2 and chlorinated fluorocarbons (CFCs) by 2100. Other stabilization scenarios, namely the "Safe Emissions Corridor" or "Tolerable Windows" (WBGU, 1995; Alcamo and Kreileman, 1996; Matsuoka *et al.*, 1996) and "Climate Stabilization" (Nordhaus, 1994) scenarios, determine the upper limit of emissions based on a constraint of some natural threshold, such as global mean temperature increase rate. Only a few studies are based on such scenarios.

Other scenarios based on DICE (Nordhaus, 1994), MERGE (Manne and Richels, 1997) and MARIA (Mori and Takahashi, 1998) determine the level of emission reduction based on net benefit maximization, which is estimated as the benefit produced by climatic policy minus the policy implementation cost. In addition to the above, the low CO_2-emitting energy supply system (LESS) constructions should be noted. These scenarios were developed on the basis of detailed assessments of technological potentials, and can therefore be distinguished from many other mitigation scenarios (see Box 2.2).

Of the remaining mitigation scenarios, a total of 50 adopt other criteria to reduce GHGs. Some of these scenarios assume the introduction of specific policies such as a constant carbon tax, while others assume the Kyoto Protocol targets for Annex I countries up to 2010 and a stabilization of their emissions thereafter at 2010 levels.

While all the scenarios deal necessarily with energy-related CO_2 emissions that have the most significant influence on climate change, several models include CO_2 emissions from land use changes and industrial processes (e.g., IPCC, 1992; Nakicenovic *et al.*, 1993; Matsuoka *et al.*, 1995; Alcamo and Kreileman, 1996). Some of them include other important GHGs in their calculations, such as methane (CH_4) and nitrous

Table 2.1: Overview of mitigation scenarios: the main futures of representive scenarios from 26 sources

#	Model name and source	Policy scenario name	Policy scenario type	Specific scenario assumptions[2]	Reduction time paths and burden sharing	GHGs analyzed	Sectors in which policies are introduced	Feedbacks[3]
1	ASF	RCWP	Emission stab.	475ppm	Based on policy scenario	CO_2, CO, CH_4, N_2O	Detailed policy scenario	EP to M
	EPA (1990)	RCWR	Emission stab.	350ppm		NO_x, CFCs	Energy supply; End use	EP to M
2	ASF/ IMAGE	Control policy (2x CO_2 by 2090)	Conc. stab.	540ppm	Based on policy scenario	CO_2, CO, CH_4, N_2O	Detailed policy scenario	EP to M
	IPCC (1990)	Accelerated control (< 2x CO_2)	Conc. stab.	465ppm		NO_x, CFCs	Energy supply; End use	
3	ASF	IS92b	Other mitigation[1]	18.6BtC (CO_2 emissions)		CO_2, CO, CH_4, N_2O, VOC. SO_x, CFCs, NO_x	Energy supply; Industrial processes	EP to M
	IPCC (1992)							
4	MESSAGE	ECS'92 +	Other mitigation[1]			CO_2	Energy supply; Industrial processes; End use	
	Nakicenovic et al. (1993)							
5	DICE Nordhaus (1994)	Optimal policy;	Other mitigation[1]		Utility maximization	CO_2, CFCs	Energy	C to M
		10-yr delay of optimal policy	Other mitigation[1]		Utility maximization	Other GHGs are		
		emission stabilization	Emission stab.	8BtC/yr (CO_2+CFCs)	Based on policy scenario	Exogenous		I to M
		20% emission cut	Other mitigation[1]	6BtC/yr (CO_2+CFCs)	Based on policy scenario			C to M
		Geoengineering	Other mitigation		Based on policy scenario			I to M
		Climate stabilization	Slow global temp. increase	0.2°C/decade	Based on policy scenario			
6	CETA	"Selfish" case	Emissions cont. by OECD		Cost minimization (regional)	CO_2, CO, CH_4, N_2O,	Energy	C to M
	Peck and Tiesberg (1995)	"Altruistic" case	Emissions cont. by OECD		Cost minimization (global)	CFCs		
		"Optimal" case	Emissions cont. by both		Cost minimization (global)			

(continued)

Table 2.1: *continued*

	Model name and source	Policy scenario name	Policy scenario type	Specific scenario assumptions[2]	Reduction time paths and burden sharing	GHGs analyzed	Sectors in which policies are introduced	Feedbacks[3]
7	LESS (IPCC, 1996)	LESS Constructions	Other mitigation[1]			CO_2	Energy supply	
8	Manne et al. (1995) MERGE	Delayed tax; Early tax / Emission stab. / Conc. stab.	Other mitigation[1] / Emission stab. / Conc. stab.	750 ppm; 540 ppm / 540 ppm / 415 ppm	Utility maximization / Utility maximization / Utility maximization	CO_2, CH_4, N_2O	Energy	C to M
9	MESSAGE	Case C	Other mitigation[1]	430 ppm	Based on policy scenario	CO_2, CO, CH_4, N_2O, SO_x, CFCs NO_x, VOC.	Energy supply	
	WEC (1995)	Ecologically driven					End use	
10	WBGU (1995) (German Adv. Council)	Tolerable temp. window	Safe corridor temp. rise constraint / Temperature rise const.	deltaT = 1°C (upper limit)	Temp. rise constraint	CO_2		
11	AIM/Top-down Matsuoka et al. (1996)	Negotiable safe emiss. corridor	Safe corridor	deltaT = 1-2°C	Temp. rise constraint	CO_2	Energy	EP to M
12	DICE/RICE	Cooperative RICE	Other mitigation[1]		Global welfare optimization	CO_2	Energy; Land use	C to M
	Nordhaus and Yang (1996)	Non-cooperative RICE	Other mitigation[1]	Regional welf. optimization				I to M
13	IMAGE 2 Alcamo and Kreileman (1996)	Stab 350–650 ppm / Stab yr 1990 / St2000-a - St2000-e Safe emissions corridor	Conc. stab. / Conc. stab. / Other mitigation[1] Safe corridor	367–564 ppm / 354 ppm / 633–433 ppm deltaT = 1-2°C deg	Temp. rise constraint	CO_2, CH_4, N_2O	Energy supply; Industrial processes; Land use	
14	MiniCAM Edmonds et al.	Adv. tech (5 Cases using	Other mitigation[1]		Based on policy scenario	CO_2, CH_4, N_2O, SO_x, aerosols, Halocarbons	Energy supply	EP to M

(coninud)

Table 2.1: continued

Model name and source	Policy scenario name	Policy scenario type	Specific scenario assumptions[2]	Reduction time paths and burden sharing	GHGs analyzed	Sectors in which policies are introduced	Feedbacks[3]
15 YOHE Yohe and Wallace (1996)	Stabilization	Conc. stab.		Based on policy scenarios	CO_2	Energy	C to M
16 DIAM Ha-Duong, *et al.* (1997)	450A-D; 550A-D; 650A	Conc. stab.		Cost minimization	CO_2	Energy	
17 FUND 1.6 Tol (1997)	Non-cooperative optimum Cooperative optimum	Other mitigation[1] Other mitigation[1]		Regional welf. optimization Generational welf. optim.	CO_2, CH_4, N_2O		
18 MERGE 3.0 Manne and Richels (1997)	Range of scenarios 350 to 750 ppm	Conc. stab.	350 to 750ppm depending on scenario	Utility maximization (non-Annex I begin limit in 2030)	CO_2	Energy	C to M
19 SGM Edmonds *et al.* (1997)	M1990 ; M1990+10%; M1990−10%; M1995	Other mitigation[1]			CO_2, CO, CH_4, N_2O, NO_x, VOC, SO_x.	Energy	EP to M
20 ABARE/GTEM Tulpule *et al.* (1998)	Independent abatement; Annex B trading; Double bubble	Other mitigation[1]	Kyoto targets		CO_2	Energy	C to M
21 AIM/Top-down Kainuma *et al.* (1998)	No trading; Annex I Trading; Global trading; Double bubble; Annex I + Chn&Ind; No trading 5% offset	Other mitigation[1]	Kyoto targets	Based on policy scenario			C to M

(continued)

Table 2.1: continued

Model name and source	Policy scenario name	Policy scenario type	Specific scenario assumptions[2]	Reduction time paths and burden sharing	GHGs analyzed	Sectors in which policies are introduced	Feedbacks[3]
22 G-CUBED McKibbin (1998)	Annex I trading; Double bubble Global permit trade	Other mitigation[1]			CO_2	Energy	C to M
23 MARIA Mori and Takahashi (1998)	Case B	Emission stab.	1990 level			Energy supply; End use; Land use	C to M
24 NE21 Fujii and Yamaji (1998)	Conc. regulation	Conc. stab.	Below 550ppm	Cost minimization	CO_2	Energy	C to M
25 WorldScan Bollen *et al.* (1996)	No Trade; Full trade; Clubs Restricted trade; CDM	Other mitigation[1]	Kyoto targets		CO_2		EP to M
26 FUND 1.6 Tol (1999)	EMF-14 scenarios (WRE/WGI+ 450/550/650+ NC/C)	Conc. stab.	Various	Various			

Notes:

1 Other mitigation means emission reduction not necessarily leading to stabilization.

2 Specific scenario assumption indicates year 2100 level of emissions/conc. unless specified otherwise.

3 EP: Energy price; M: Macro economy; C: Cost; I: Impact.

Box 2.2. Review of Low Carbon Dioxide Emitting Energy Supply System (LESS) Constructions from the Second Assessment Report

The LESS constructions described in the IPCC's SAR, Working Group II (IPCC, 1996, Ch19), were probably the only constructions akin to mitigation "scenarios" taken up in SAR. They are similar to the mitigation scenarios reviewed in this chapter in that they also explore alternative paths to energy futures in order to achieve mitigation of carbon dioxide.

A number of technologies with potential for reducing CO_2 emissions exist or are in a state of possible commercialization. The LESS constructions illustrate the potential for reducing emissions by using energy more efficiently and by using various combinations of low CO_2-emitting energy supply technologies, including shifts to low-carbon fossil fuels, shifts to renewable and nuclear energy sources, and decarbonization of fuels. The assumed technological feasibility and costs of each of the technologies included in these variants is based on an extensive literature review.

Both bottom-up and top-down approaches were used in the LESS constructions. For the reference cases in the bottom-up analyses, the energy demand projections for the high economic growth variant of the "Accelerated Policies" scenarios developed by the Response Strategies Working Group (RSWG, 1990) were adopted.

The five variants constructed in the bottom-up analyses were (1) BI: biomass intensive, (2) NI: nuclear intensive, (3) NGI: natural gas intensive, (4) CI: coal intensive, and (5) HD: high demand. The BI variant explores the potential for using renewable electricity sources in power generation. Both intermittent renewables (wind, photovoltaics, and solar thermal-electricity technologies) and advanced biomass electricity-generating technologies (biomass-integrated gasifier and/or gas turbine technologies through 2025 and biomass-integrated gasifier and/or fuel-cell technologies through 2050 and beyond) were applied. The NI variant involves a revitalization of the nuclear energy option and deployment of nuclear electric power technology worldwide. In the NGI variant, the emphasis is on natural gas. Any natural gas in excess of that for the reference cases is used to make methanol (CH_4O) and hydrogen (H_2). These displace CH_4O and H_2 produced from plantation biomass. In the CI variant, the strategy for achieving deep reductions involves using coal and biomass for CH_4O and H_2 production, along with sequestration of the CO_2 separated out at synthetic fuel production facilities. Finally, in the HD variant the excess demand is met by providing an extra supply of fuels with low emissions. To illustrate the possibilities, the HD variant is constructed with all of the incremental electricity provided by intermittent renewables.

A top-down exercise was carried out to test the robustness of the bottom-up energy supply analyses by incorporating performance and cost parameters for some of the key technologies in the BI variant. Six technology cases were modelled using the Edmonds–Reilly–Barns (ERB) model. The results for CO_2 emissions in two cases (cases 5 and 6) were comparable to the bottom-up LESS variants, but the energy end-uses were different owing to different assumptions.

The central finding of the LESS construction exercise is that deep reductions of CO_2 emissions from the energy sector are technically possible within 50 to 100 years, using alternative strategies. Global CO_2 emissions could be reduced from about 6GtC in 1990 to about 2GtC in 2100, in many combinations of the options analyzed. Cumulative CO_2 emissions, from 1990 to 2100, would range from about 450 to about 470GtC in the alternative LESS constructions. Higher energy efficiency is underscored in order to achieve deep reductions in CO_2 emissions, increase the flexibility of supply-side combinations, and reduce overall energy system costs.

oxide (N_2O) (e.g., EPA, 1990; IPCC, 1990; Manne *et al.*, 1995; Tol, 1997), and a few go even further to include sulphates, volatile organic compounds (VOCs), and halocarbons (e.g. IPCC, 1992; WEC, 1995; Edmonds *et al.*, 1996, 1997). With respect to the policy options used in the scenario quantifications, three fields are taken into account in the reviewed studies: energy systems (including both supply and demand), industrial processes (including cement and metal production), and land use (including agriculture and forest management).

Since most of the modelling exercises have been carried out to study the CO_2 emissions from human activities linked to the use of energy, energy supply and end-use are naturally the areas where policy is applied. Energy supply options include natural gas, renewable energy, and commercial biomass; introduction of new technologies; and so on. End-use options

chiefly pertain to increased energy efficiency in industry, transport, and residential and/or commercial applications.

The policy instruments analyzed depend on the underlying model structure. Most of the scenarios introduce policies such as simple carbon taxes or a constraint on emissions or concentration levels for achieving the desired reduction or stabilization. How the constraint is imposed varies from scenario to scenario. Among the models with regional disaggregation, a few regional targets have been introduced (e.g., Nordhaus, 1994; Tol, 1999). Regional disaggregation also allows modellers to let the regions trade in emission permits. Permit trading is introduced in more recent work, especially just before and after the Third Conference of the Parties to the United Nations Framework Convention on Climate Change in Kyoto (December 1997). Some studies offer permit trading as a

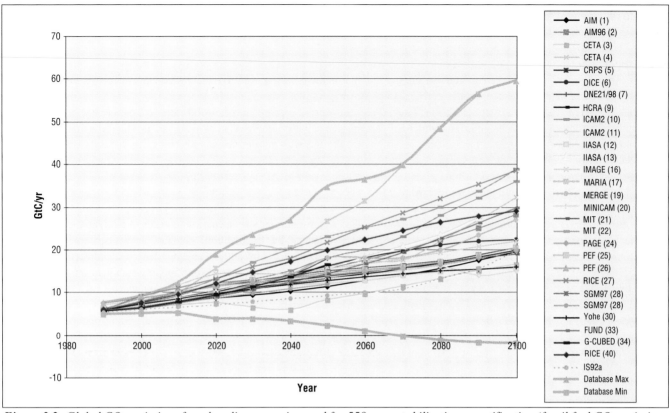

Figure 2.2: *Global CO₂ emissions from baseline scenarios used for 550ppmv stabilization quantification (fossil fuel CO₂ emissions over the period 1990 to 2100 with the maximum and minimum numbers of the database of scenarios). This figure excludes the SRES scenarios (for legend details see Appendix 2.1).*

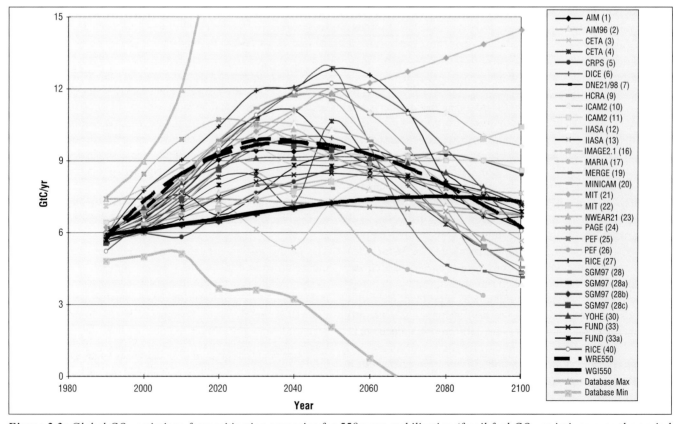

Figure 2.3: *Global CO₂ emissions from mitigation scenarios for 550ppmv stabilization (fossil fuel CO₂ emissions over the period 1990 to 2100 with the maximum and minimum numbers of the database of scenarios). This figure excludes the post-SRES scenarios (for legend details see Appendix 2.1).*

mechanism to reduce the overall costs of abatement. Much of the work done in the early 1990s led to the development of detailed scenarios for introducing such policies (EPA, 1990; IPCC, 1990, 1992). Some models employ policies of supply-side technology introduction (Nakicenovic *et al.*, 1993; Edmonds *et al.*, 1996; Fujii and Yamaji, 1998), while other models emphasize the introduction of efficient demand-side technology (EPA, 1990; Kainuma *et al.*, 1999a).

The issue of burden sharing among regions is a contentious one and it was sparsely treated in the first half of the 1990s. Most discussions about burden sharing are of a qualitative and partial nature and are not related to model-based mitigation scenarios. A few studies (most notably Rose and Stevens, 1993; Enquete Commission, 1995; and Manne and Richels, 1997) present a set of burden-sharing rules in their scenarios. Of late, the EMF exercises looking at the Kyoto scenarios have treated this issue better than in the past (Weyant, 1999).

The time-paths of emission reduction are determined in three ways in the reviewed studies. First, the emission trajectories are determined by policy scenarios that have been designed in detail for regions over the time frame (EPA, 1990; IPCC, 1990; WEC, 1995; Edmonds *et al.*, 1996; Yohe and Wallace, 1996; Kainuma *et al.*, 1998). Second, dynamic optimization models automatically determine these reduction time-paths by global cost minimization over time (e.g., Peck and Tiesberg, 1995; Fujii and Yamaji, 1998) or economic welfare maximization (Nordhaus, 1994; Manne *et al.*, 1995). Third, mitigation scenarios of tolerable windows/safe landing, or safe emission corridors, can fix the time series of emission reduction by introducing a specific constraint of the rate of change in natural systems including the global temperature change rate (e.g., Alcamo and Kreileman, 1996).

Finally, there are differences in the treatment of feedback to the macro-economy in the models. While most bottom-up models have no feedback from cost to the macro-economy, top-down models allow for the feedback of energy prices to the macro-economy. The MERGE (Manne *et al.*, 1995) and CETA (Peck and Tiesberg, 1995) models also have feedback from impacts to the macro-economy.

Technological improvement is a critical element in all the general mitigation scenarios. This is apparent when the detailed policy options are studied, where such literature is available. For instance, Nakicenovic *et al.* (1993) (using MESSAGE) incorporated policies of dematerialization and recycling, efficiency improvements and industrial process changes, and fuel-mix changes in the industrial sector; fuel efficiency improvements, modal split changes, behavioural change, and technological change in the transport sector; and efficiency improvements of end-use conversion technologies, fuel-mix changes, and demand-side measures in the household and services sector. It should be noted that efficiency improvement through technological advancement is emphasized in all sectors. Similar policies leading to efficiency improvement were also underlined in

earlier modelling studies such as EPA (1990), IPCC (1990), and IPCC (1992).

2.3.2 *Quantitative Characteristics of Mitigation Scenarios*

From the large number of mitigation scenarios, a selection must be made in order to clarify in a manageable way the quantitative characteristics of mitigation scenarios. One of the efficient ways to analyze them is to focus on a typical mitigation target. As the most frequently studied mitigation target is the 550ppmv stabilization scenario, a total of 31 stabilization scenarios adopting that target were selected along with their baseline (reference or non-intervention) scenarios in order to analyze the characteristics of the stabilization scenarios as well as their baselines[5]. *Figure 2.2* shows these baseline scenarios, and *Figure 2.3* shows the mitigation scenarios for 550ppmv stabilization. (The sources and scenario names are noted in *Appendix 2.1*).

2.3.2.1 *Characteristics of Baseline Scenarios*

In order to analyze the characteristics of stabilization scenarios, it is very important to identify the features of the baseline scenarios that have been used for mitigation quantification. Although the general characteristics of non-intervention scenarios have already been analyzed in the SRES (Nakicenovic *et al.*, 2000), more specific analyses are conducted here, focusing on the baseline scenarios that have been used for 550ppmv stabilization quantification.

First, it is clear that the range of CO_2 emissions in baseline scenarios used for 550ppmv stabilization quantification is very wide at the global level, as shown in *Figure 2.2*. The maximum levels of CO_2 emissions represent more than ten times the current levels, while the minimum level represents four times current levels. The range of baseline scenarios covers the upper half of the total range of the database, and most of them were estimated to be larger than IS92a (IPCC 1992 scenario "a"). This means that the baseline scenarios used for the 550ppmv stabilization analyses have a very wide range and are high relative to other studies.

This divergence can be explained by the Kaya identity (Kaya, 1990), which separates CO_2 emissions into three factors: gross domestic product (GDP), energy intensity, and carbon intensity[6]:

---

[5] This closer look at 550ppmv CO_2 stabilization scenarios is solely based on the frequency of their occurrence in the literature, which in turn has been influenced by frequent reference to this level in the policy area (e.g., it has been selected as a long-term target by the European Union). The discussion in this chapter does not imply any endorsement of this particular level as a policy target. There is a need for analysis of the feasibility and implications of stabilization levels other than 550 ppmv.

[6] The usual form of the Kaya identity separates the GDP term into population × GDP/capita. However, population assumptions were not provided for most scenarios and thus the GDP term was not disaggregated.

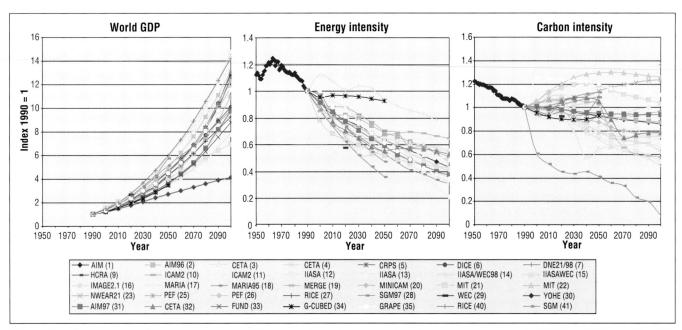

Figure 2.4: *Range of baseline assumptions in GDP, energy intensity, and carbon intensity over the period 1990 to 2100 used for 550 ppmv stabilization analyses (indexed to 1990 levels), with historical trend data for comparison (for legend details see Appendix 2.1).*

CO_2 emissions = GDP * Energy intensity * Carbon intensity
= GDP * (energy/GDP) * (emissions/energy)

Figure 2.4 shows these factors. For comparability of the factors, which were not harmonized to be the same number among models in the base year of 1990, all the values are indexed to 1990 levels. CO_2 emissions are mostly determined by energy consumption. This, in turn, is determined by the levels of GDP, energy intensity, and carbon intensity. However, the ranges of GDP and of carbon intensities in the scenarios are larger than the range of energy intensities. This suggests that the large range of CO_2 emissions in the scenarios is primarily a reflection of the large ranges of GDP and carbon intensity in the scenarios. Thus, the assumptions made about economic growth and energy supply result in huge variations in CO_2 emission projections.

These characteristics are also observed in regional scenarios. For example, in both the OECD and non-OECD scenarios, CO_2, GDP, energy intensity, and carbon intensity have wide ranges, and in particular, the range among scenarios for the non-OECD nations is wider than the range among scenarios for OECD nations. In addition, the growth of CO_2 emissions in non-OECD nations is generally larger than the growth of emissions in OECD nations. This is mainly caused by higher GDP growth in the non-OECD countries.

With regard to regional comparisons, it is very difficult to come to any general conclusions, as the ranges involved in the regional scenarios are extraordinarily large. Moreover, with the exception of the USA, Europe, the Former Soviet Union (FSU) and China, the number of available scenarios is limited. However, some general trends can be identified that are associated with the medium ranges of the scenarios: for Asian

countries, GDP growth is the most significant factor, resulting in high levels of energy use and CO_2 emissions; energy efficiency improvements are the most significant factor in the scenarios for China; and carbon intensity reductions are very high in Africa, Latin America, and Southeast Asia, because of drastic energy mix changes.

Other interesting characteristics at the global level can be identified in the relationships among GDP, energy intensity, and carbon intensity. *Figure 2.5* shows a scatter plot of GDP growth rate versus energy intensity reduction from the baseline scenarios. As might be expected, the energy intensity reduction is higher with a higher GDP growth rate, while a lower energy intensity reduction is associated with a lower GDP growth rate. This relationship suggests that high economic growth scenarios assume high levels of progress in end-use technologies.

Unlike energy intensity reductions, carbon intensity reductions in the models are apparently seen as largely independent of economic growth and consequently are a function of societal choices, including energy and environmental policies. The scenarios do not show any clear relationship between energy intensity reduction and carbon intensity reduction. The values depend on regional characteristics in energy systems and technology combinations. Energy intensity reduction can include many measures other than fuel shifting. Most of the efficiency measures will result in lower carbon emissions, and fuel shifts from high-carbon to low- or non-carbon fuels can increase the efficiency of energy systems in many cases. However, carbon intensity reductions can also lead to reduced efficiency in energy systems, as in the case of shifts to biomass gasification or liquefaction, or result in increased energy consumption, as in the case of industrial carbon sequestration.

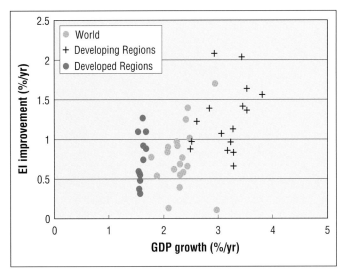

Figure 2.5: *Scatter plot of GDP growth versus energy intensity reduction in baseline scenarios (including world and regional data).*

2.3.2.2 Characteristics of Stabilization Scenarios

The stabilization scenarios that were estimated based on the above baselines also have a very wide range, as shown in *Figure 2.3*. This wide range is caused by several factors, including differences in emission time-paths for the stabilization, differences in timing of the stabilization at 550 ppmv, and different carbon cycle models used to assess the stabilization.

The divergence in reduction time-path has been discussed based on two sets of popular scenarios. One is a set of IPCC Working Group (WG) I scenarios (Houghton *et al.*, 1996) which is sometimes referred to as "early action scenarios" and denoted as "WGI"; the other is a set of scenarios published by Wigley *et al.* (1996), sometimes referred to as "delayed action scenarios" and denoted "WRE". Chapter 8 explains that these terms are misleading, since WRE scenarios may not assume early emissions reductions, but do assume early actions to facilitate such reductions later. *Figure 2.3* compares the 550 ppmv stabilization scenarios of these two scenario sets with the reviewed scenarios, and it shows that scenarios reviewed here cover a wider range than that of the WGI and WRE scenarios. While the RICE and MERGE scenarios show late reduction (WRE type) trajectories, the CETA, MARIA and MIT scenarios show more severe reduction (WGI type) trajectories.[7] A few scenarios, for example ICAM2, show no drastic reduction even in the latter half of the 21st century. Most of the scenarios have emissions trajectories that lie in between.

The reduction time-path of emissions is a controversial point, which is closely related to the intergenerational equity issue. However, no conclusion can be drawn from such global trajec-

tories, since behind them lies a distribution between countries and the political, technical, economic, and social acceptability of this distribution would depend on how the equity concerns are sorted out.

Figures 2.6 and *2.7* show energy-related CO_2 reduction at the global and the non-OECD levels, respectively, which were estimated for each scenario source by subtracting stabilization scenario emissions (*Figure 2.3*) from baseline scenario emissions (*Figure 2.2*). These figures show that the range of reduced CO_2 emissions for 550ppmv stabilization is also very wide both at the global and the non-OECD levels. This wide range is apparently caused by the divergent baseline scenarios shown in *Figure 2.2*, while other factors such as differences in emission time-path, in timing of stabilization and in the carbon cycle model used also tend to increase the range.

Figures 2.6 and *2.7* show the simulation results of models, assuming that non-OECD countries would participate in mitigation. The distribution of mitigation among the countries is based on different approaches, such as the introduction of emission caps, or the assumption of the same rate of emission reduction for all countries, or global emission trading. The results show that emission trading may lower the mitigation cost, and could lead to more mitigation in the non-OECD countries.

The regional allocation of reductions is a controversial and highly political issue from the equity viewpoint. Mostly, modellers do not explicitly state the burden-sharing rule. Nevertheless, the emission reduction from baseline by the non-Annex I countries is a good indicator of when it is assumed that these countries start sharing the reductions. The data set used in this analysis is limited in the sense that models have different regional specifications; it was therefore difficult to obtain a large number of data points to analyze non-Annex I emissions. As a proxy, emission reduction from the baseline by the non-OECD region is used, which includes Russia and Eastern Europe. This is shown in *Figure 2.7*. In part of the AIM, MiniCAM, FUND, and PEF scenarios, introduction of climate policy in the non-OECD region is assumed not to begin by 2010. Although Russia and Eastern European countries are included in the Kyoto Protocol, the models do assume that because of the decreased emissions in these countries since 1990, actual climate policies would not be needed until 2010. Some scenarios show that non-OECD regions may not have to significantly reduce emissions before 2030. However, there are still other scenarios that show an opposite picture. The RICE, MERGE, MIT, and MARIA scenarios show a very steep increase in emission reduction from baseline levels in the non-OECD region starting very early in the 21st century.

One of the ways to explain this divergence in reduction time series is to differentiate the assumptions about trade in these scenarios. Some scenarios assume trade in emission credits, which are allotted initially to each country or region. This allows some countries to purchase emission rights from other countries to minimize the cost of meeting their emission tar-

---

[7] For a more detailed discussion of the WRE and WGI trajectories, see Chapters 8 and 10 of this report.

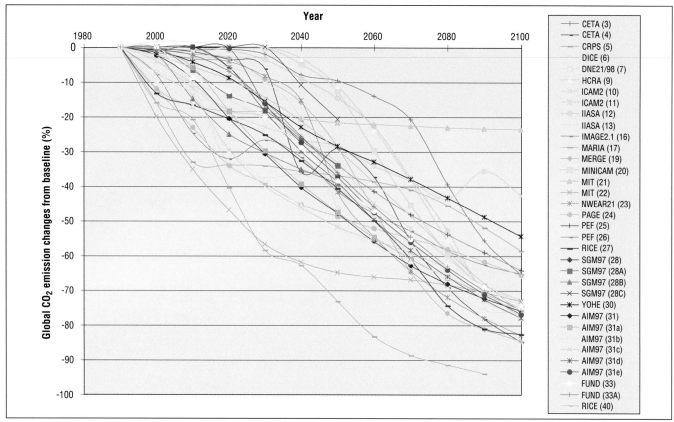

Figure 2.6: *Global CO₂ emission reduction from baseline for 550ppmv stabilization scenarios, estimated for each scenario source as baseline emissions minus emissions in the 550ppmv stabilization scenario (for legend details see Appendix 2.1).*

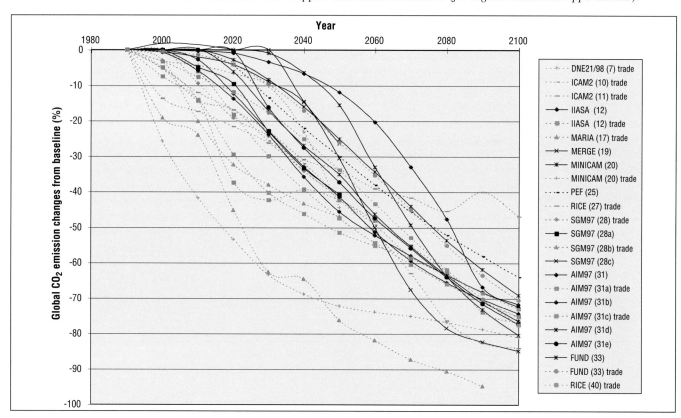

Figure 2.7: *Non-OECD CO₂ emission reduction for 550ppmv stabilization, estimated for each scenario source as baseline emissions minus emissions in the 550ppmv stabilization scenario divided by baseline emissions. Dotted lines show the scenarios which assume carbon credit trading between the OECD and developing regions (for legend details see Appendix 2.1).*

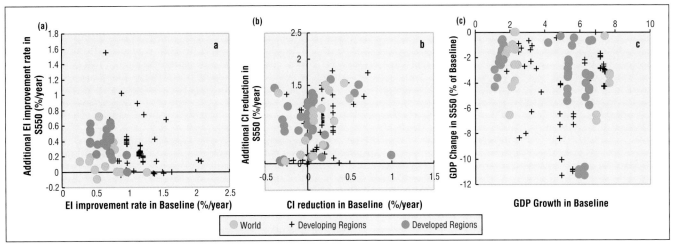

Figure 2.8: *Scatter plots to analyze the relationships between baseline scenario assumptions and mitigation scenario outputs in Energy Intensity (a), Carbon Intensity (b), and GDP growth (c).*

gets. The dotted lines in *Figure 2.7* show the scenarios that assume trade in emission credits between the Annex I and non-Annex I countries. The scenarios that show an early reduction of emissions in the non-OECD region are included in the trade scenarios, and they assume the OECD region would transfer funds to the non-OECD region via emission credit trading. Most of the other scenarios assume that the non-OECD region would start to introduce reduction policies after 2010.

With regard to overall mitigation, the range of assumed policies is very wide, resulting in a wide range of emission reductions. The additional increase in energy efficiency improvement from the baseline ranges between minus 0.04 and 1.56% per year within the sampled data, while the additional reduction in carbon intensity from the baseline is between zero and 3.76% per year. Although it is difficult to identify detailed policy assumptions from the database, the range of these factors suggests divergent policy options among scenarios. These policy options are dependent not only on the level of CO_2 reduction, but also on the baseline scenarios that have been used for 550 ppmv stabilization quantification.

Figure 2.8 (a) shows the relationship between the effects of efficiency improvement policy in mitigation scenarios and the energy intensity reduction assumption in baseline scenarios. This figure suggests an inverse relationship between them. The implication of this is that scenarios in which there is an assumed adoption of high-efficiency measures in the baseline usually would have less scope for further introduction of efficiency measures in the mitigation scenarios, as compared to scenarios that have a lower level of efficiency improvement in their baseline.[8] As a result, the additional reduction of energy

intensity in mitigation scenarios over the base cases would be lower when the assumed energy intensity reduction is high in the base case, and vice versa. In the case of unanticipated technological breakthroughs, of course, this relationship may not hold and one could expect further energy efficiency improvements, even when the baseline has a fair amount of energy efficiency built into it.

Figure 2.8 (b) shows the relationship between the effects of decarbonization policies and the carbon intensity reductions assumed in the baseline scenarios. This figure suggests that baseline scenarios with high carbon intensity reductions show larger carbon intensity reductions in their mitigation scenarios, while those with low carbon intensity reductions in the base case show smaller reductions in carbon intensity in their corresponding stabilization cases. This is somewhat counterintuitive and difficult to explain simply on the basis of the results available. One might expect that high carbon intensity reductions in the base case might "use up" decarbonization potential, giving rise to lower additional reduction of carbon intensity in mitigation scenarios. On the other hand, increased investment in low-carbon energy technology in the base case could increase the resource base of low-carbon energy, thereby providing more opportunity to reduce CO_2 emissions in the stabilization case. The mitigation potential in this direction depends not only on the technology but also, and perhaps more, on the economics and social acceptance of the technology. A closer and more careful analysis of which particular mitigation policies were assumed in constructing the scenario than was possible on the basis of the available information, would reveal the underlying reasons for such a pattern.

Finally, *Figure 2.8 (c)* shows the relationship between macroeconomic costs[9] in the mitigation scenarios and GDP growth assumptions in the baseline scenarios. No clear relationship is visible, but it can be observed that macroeconomic costs for the world as a whole are estimated to range between 0% and 3.5% of GDP in 2100, while a few simple models estimate more

---

[8] In part this is an artefact of the structure of the models, which cannot easily account for changes in social and technological structure such as significant changes in consumption patterns, land use, or urban form.

Box 2.3. Non-CO$_2$ Mitigation Scenarios

Since the publication of IPCC's SAR, the literature on mitigation scenarios has continued to focus on the reduction of CO$_2$ emissions rather than on other GHGs. This is unfortunate because non-CO$_2$ emissions make up a significant fraction of the total "basket of gases" that must be reduced under the Kyoto Protocol. However, a small set of papers has reported on scenarios for mitigating non-CO$_2$ gases, especially CH$_4$ and N$_2$O. In one such paper, Reilly *et al.* (1999) compared scenarios for achieving emission reductions with and without non-CO$_2$ emissions in Annex B countries (those countries that are included in emission controls under the Kyoto Protocol). Scenarios that omitted measures for reducing non-CO$_2$ gases had 21% higher annual costs in 2010 than those that included them. Tuhkanen *et al.* (1999) and Lehtilä *et al.* (1999) came to similar conclusions — in a scenario analysis for 2010, they found that including CH$_4$ and N$_2$O in mitigation strategies for Finland reduced annual costs by 20% in the year 2010 relative to a baseline scenario. The general conclusion of these papers is that small reductions of GHG emissions, for example of the magnitude required by the Kyoto Protocol, can be accomplished at a lower cost by taking into account measures to reduce non-CO$_2$ gases, and that a small reduction of non-CO$_2$ gases can produce large impacts at low cost because of the high global warming potential (GWP) of these gases.

In another type of scenario analysis, Alcamo and Kreileman (1996) used the IMAGE 2 model to evaluate the environmental consequences of a large set of non-CO$_2$ and CO$_2$ mitigation scenarios. They concluded that non-CO$_2$ emissions would have to be controlled along with CO$_2$ emissions in order to slow the increase of atmospheric temperature to below prescribed levels. Hayhoe *et al.* (1999) pointed out two additional benefits of mitigating CH$_4$, an important non-CO$_2$ gas. First, most CH$_4$ reduction measures do not require the turnover of capital stock (as do CO$_2$ measures), and can therefore be carried out more rapidly than CO$_2$ reduction measures. Second, CH$_4$ reductions will have a more immediate impact on mitigating climate change than CO$_2$ reductions because the atmosphere responds more rapidly to changes in CH$_4$ than to CO$_2$ concentrations.

increase in the second half of the 21st century. The GDP loss may or may not be related to the GDP growth assumptions in baselines. For instance, high baseline economic growth would lead to higher emissions of GHGs, which would lead to increased GHG reduction costs compared to the corresponding mitigation scenario for a low-growth baseline. On the other hand, high economic growth could provide increased funds for research and development (R&D) of advanced technologies, which would decrease the cost of GHG reduction. The net cost would depend on the relative strengths of these effects. Another aspect is that the costs are also dependent upon the structure of economies, i.e., economies with high fossil fuel dependence, via either exports or domestic consumption, are likely to experience higher costs compared with economies with relatively lower fossil fuel dependence.

2.3.3   Summary of General Mitigation Scenario Review

Many mitigation as well as stabilization scenarios have already been quantified and published. Most assume very simple policy options for their mitigation scenarios, and only some of them have detailed policy packages. These policy options have a very wide range in their level, which is apparently caused by the divergent baseline scenarios and GHG reduction targets,

with other factors such as differences in models and reduction time-paths also acting to increase the range. Allocations of emission reductions between OECD and non-OECD countries also vary widely, and are affected by policy assumptions and model structures.

The mitigation scenarios under review were quantified based on a wide range of baselines that reflect a diversity of assumptions, mainly with respect to economic growth and low-carbon energy supply. The range of future trends shows greater divergence in scenarios that focus on developing countries than in scenarios that consider developed nations. There is little consensus with respect to future directions among the existing disaggregated scenarios in developing regions.

Some general conclusions about the relationships between baseline scenarios and mitigation policies are suggested by this review: an assumption of high economic growth in the baseline tends to be associated with more technological progress; the additional improvement of energy efficiency in mitigation scenarios tends to be lower when the energy efficiency improvement is high in the base case; and baseline scenarios with high carbon intensity reductions lead to mitigation scenarios with relatively more carbon intensity reduction. The counterintuitive nature of some of these conclusions suggests that the relationship between economic growth and the macroeconomic cost of emission reduction is very complicated.

Most generally, it is clear that mitigation scenarios and mitigation policies are strongly related to their baseline scenarios, but no systematic analysis has been published on the relationship between mitigation and baseline scenarios.

---

[9] The macroeconomic cost is defined here as the reduction of GDP caused by GHG emission reduction in comparison to baseline GDP. It should be noted that these costs do not take into account the benefits that would occur from avoiding climate change-related damages or any co-benefits. See also Chapters 7 and 8 for a discussion of these issues.

2.4   Global Futures Scenarios

2.4.1   The Role of Global Futures Scenarios

In contrast to the GHG emission scenarios discussed in sections 2.3 and 2.5 of this chapter, "global futures" scenarios do not specifically or uniquely consider GHG emissions. Instead, they are more general "stories" of possible future worlds. Global futures scenarios can complement the more quantitative emission scenario assessments, because they consider several dimensions that elude quantification, such as governance, social structures, and institutions, but which are nonetheless important to the success of mitigation (and adaptation) policies and, more generally, describe the nature of the future world.

In this assessment, the global futures scenario literature was reviewed to achieve three objectives. First, it was consulted in order to determine the range of possible future worlds that have been identified by futurists. This aids climate change policy analysis by providing a range of potential futures against which the robustness of policy instruments may be assessed.

Second, global futures scenarios were analyzed to determine whether they displayed any relationships between the various scenario dimensions and GHG emissions. Although these relationships are often based entirely on qualitative analysis, they might nonetheless yield insights about the relationships between some dimensions, especially those that are difficult to quantify, and emissions.

Third, global futures scenarios may provide a link between the more quantitative emission scenarios and sustainable development issues. Global futures scenarios generally provide good coverage of sustainable development issues, while the quantitative emission scenarios generally provide only limited coverage of these issues. Linking the global futures scenarios with the quantitative emission scenarios therefore might also provide a link between the latter and sustainable development issues.

2.4.2 Global Futures Scenario Database

An extensive review of the futures literature was conducted and, from this review, a database of scenarios was constructed. This database contains 124 scenarios from 48 sources.[10]

Scenarios were selected which were global[11], long-term, and multidimensional in scope. The scenarios consider timelines that run from the base year to anywhere between 2010 and 2100. Most scenarios are detailed and comprehensive depictions of possible future worlds, with descriptions of the social, economic, and environmental characteristics of these worlds. Others are less detailed but still describe more than one characteristic of the future world. Some scenarios are derived from the authors' judgement about most likely future conditions. Others are part of sets of possible futures, usually posited as alternatives to a reference case. Still others are normative scenarios, in that they describe the authors' visions of desirable future worlds.

In general, the global futures scenarios provide few quantified projections, although there are some notable exceptions such as CPB (1992), Meadows *et al.* (1992), Duchin *et al.* (1996), Gallopin *et al.* (1997), OECD (1997), Rotmans and de Vries (1997), Glenn and Gordon (1998), Nakicenovic *et al.* (1998), and Raskin *et al.* (1998). Several scenarios explicitly consider energy use, GHG emissions, and/or future climate change, but not all of these provide numerical estimates of the relevant variables. These quantified scenarios are different from the scenarios in the previous section since they present quantifications of primarily narrative scenarios. The basis of the scenarios in the previous section is a purely quantitative analysis of emissions profiles without narrative description.

2.4.3   Global Futures Scenarios: Range of Possible Futures

The global futures scenarios vary widely along different demographic, socio-economic, and technological dimensions, as shown in *Table 2.2*. Scenarios range from economic collapse to virtually unlimited economic prosperity; from population collapse (caused by famine, disease, and/or war), to stabilization near current levels, to explosive population growth. Governance systems range from decentralized, semi-autonomous communities with a form of direct democracy to global oligarchies. Some scenarios posit large improvements in income and social equality, within and among nations, while others foresee a widening of the income gap. Many scenarios envisage a future world that is high-tech, with varying rates of diffusion, but some envisage a world in which a crisis of some kind leads to a decline in technological development and even a loss of technological capability. Most scenarios are pessimistic with respect to resource availability; some are more

---

[10] See Barney, 1993; Bossel, 1998; Coates and Jarratt, 1990; Coates, 1991, 1997; Cornish, 1996; Costanza, 1999; CPB, 1992; Duchin *et al.,* 1994; Gallopin *et al.,* 1997; GBN, 1996; Glenn and Gordon, 1997, 1998; Henderson, 1997; Hughes, 1997; IDEA Team, 1996; Kahane, 1992; Kinsman, 1990; Linden, 1998; Makridakis, 1995; McRae, 1994; Meadows *et al.,* 1992; Mercer, 1998; Millennium Project, 1998; Nakicenovic *et al.,* 1998; OECD, 1997; Olson, 1994; Price, 1995; Ramphal, 1992; Repetto, 1985; Rotmans and de Vries, 1997; Schindler and Lapid, 1989; Schwartz, 1991, 1995; Schwartz and

Leyden, 1997; Science Advisory Board, 1995; Shinn, 1982; Stokke *et al.,* 1991; Sunter, 1992; Svedin and Aniansson, 1987; Toffler, 1980; van den Bergh, 1996; Wallerstein, 1989; WBCSD, 1997; 1998; Wilkinson, 1995; World Bank, 1995; WRI, 1991.

[11] The literature contains a great many scenarios that focus on specific countries or regions. However, time and space limitations precluded including these scenarios in this review.

Table 2.2: Descriptive statistics for global futures scenario dimensions

	Number of scenarios	Range	Most common (mode)	Number of scenarios showing changes (compared to current situation)		
				Declining	Same	Rising
Total Scenarios	**124**					
Size of Economy	102	collapse to high growth	Rising	24	13	65
Population Size	84	collapse to high growth	Rising	10	5	69
Level of Technology	98	stagnation & decline to very high	Rising	4	9	85
Degree of Globalization	84	isolated communities to global civilization	More global	22	1	61
Government Intervention in Economy	76	laissez-faire to strong regulation	Declining	36	9	31
Pollution	85	very low to very high	Rising	34	3	48
International Income Equality	99	very low to very high	Rising	32	16	50
Intranational Income Equality	53	very low to very high	Rising	24	0	29
Degree of Conflict	76	peace to many wars/world war	Rising	26	14	36
Fossil Fuel Use	49	virtually zero to high		24	1	24
Energy Use	51	low to high	Rising	14	0	37
GHG Emissions	45	low to high	Rising	11	1	33
Climate Change (yes/no)	0	no climate change to severe climate change				
Structure of Economy	50	agrarian/subsistence to "quaternary" (leisure)	Increasingly post-industrial	4	6	40
Percentage of Older Persons in Population	11	primarily young population to ageing population	Rising	2	0	9
Migration	30	low to high	Rising	10	0	20
Human Health	38	worsening to improving	Improving	13	3	22
Degree of Competition	41	low to high	Rising	14	0	27
Citizen Participation in Governance	56	autocracy to meaningful participation	Rising	14	14	28
Community Vitality	42	breakdown to very strong	Rising	12	0	30
Responsiveness of Institutions	75	irrelevant to very responsive/citizen-driven	Improving	21	16	38
Social Equity	38	low to high		19	1	18
Security Activity	30	low to high	Rising	13	0	17
Conflict Resolution	30	inadequate to successful	Improving	10	1	19
Technological Diffusion	58	low to high	Improving	9	13	36
Rate of Innovation	45	low to high	Rising	3	14	28
Renewable Resource Availability	28	low to high	Declining	19	1	8
Non-renewable Resource Availability	35	low to high	Rising	15	4	16
Food Availability	45	low to high	Rising	16	4	25
Water Availability	18	low to high	Declining	12	0	6
Biodiversity	33	low to high	Declining	21	2	10
Threat of Collapse	26	unlikely to likely	Rising	9	1	16

optimistic, pointing to the ability of technology and demand changes to alleviate scarcity. Most scenarios also project increasing environmental degradation; more positively, many of these scenarios portray this trend reversing in the long-term, leading to an eventual improvement in environmental quality. The sustainable development scenarios, on the other hand, describe a future in which environmental quality improves throughout the scenario.

The scenarios were grouped together according to their main distinguishing features and were combined into four groups, according to whether they described futures in which, accord-

Table 2.3: *Global futures scenario groups*

Scenario group	Scenario subgroups	Number of scenarios
1. Pessimistic Scenarios	Breakdown: collapse of human society	5
	Fractured World: deterioration into antagonistic regional blocs	9
	Chaos: instability and disorder	4
	Conservative: world economic crash is succeeded by conservative and risk-averse regime	2
2. Current Trends Scenarios	Conventional: no significant change from current and/or continuation of present-day trends	12
	High Growth: government facilitates business, leading to prosperity	14
	Asia Shift: economic power shifts from the West to Asia	5
	Economy Paramount: emphasis on economic values leads to deterioration in social and environmental conditions	9
3. High-Tech Optimist Scenarios	Cybertopia: information & communication technologies facilitate individualistic, diverse and innovative world	16
	Technotopia: technology solves all or most of humanity's problems	5
4. Sustainable Development Scenarios	Our Common Future: increased economic activity is made to be consistent with improved equity and environmental quality	21
	Low Consumption: conscious shift from consumerism	16

ing to the scenario authors, conditions deteriorate (group 1), stay the same (group 2), or improve (groups 3 and 4). These groups are summarized in Table 2.3.

The scenarios in group 1 describe futures in which conditions deteriorate from present. Some of these scenarios describe a complete breakdown of human society, because of war, resource exhaustion, or economic collapse. Other scenarios describe a future in which the world is fractured into antagonistic blocs or in which society deteriorates into chaos. Still others describe futures in which the global economic system crashes and is succeeded by a conservative, risk-averse regime.

The scenarios in group 2 describe futures in which conditions do not change significantly from the present, or in which current trends continue. Many of these scenarios are "reference" scenarios, which are used by their authors to contrast other alternative future scenarios. In general, these scenarios are pessimistic; they describe futures in which many current problems get worse, although there may be improvement in some areas. This is particularly true of the "Economy Paramount" scenarios, which describe futures in which an emphasis on economic over other values leads to deteriorating environmental and social conditions. Other scenarios in group 2 describe a more optimistic future in which government and business co-operate to improve market conditions (generally through market liberalization and free trade), leading to an increase in prosperity. Several of the group 2 scenarios foresee a shift in economic power from the West to Asia.

The group 3 scenarios could be characterized as "High-Tech Optimist" scenarios. They describe futures in which technology and markets combine to produce increased prosperity and opportunity. Many of these scenarios describe "Cybertopias"

in which information and communication technologies enable a highly individualistic, diverse, and innovative global community. Other group 3 scenarios describe worlds in which technological advances solve all or most of the problems facing humanity, including environmental problems.

The scenarios in group 4 are "Sustainable Development" scenarios. In general these scenarios envisage a change in society towards improved co-operation and democratic participation, with a shift in values favouring environment and equity. These scenarios can be subdivided into two subgroups. The first subgroup might be described as "Our Common Future" scenarios in which economic growth occurs, but is managed so that social and environmental objectives may also be achieved. The second subgroup could be characterized as "Low Consumption" sustainable development scenarios. They describe worlds in which economic activity and consumerism considerably decline in importance and, usually, population is stabilized at relatively low levels. Many of these scenarios also envisage increasing regional autonomy and self-reliance.

These groups correspond quite closely with the scenario archetypes that have been developed by the Global Scenarios Group (see Box 2.4). They also roughly correspond with the 4 new emission scenario "families" that were developed in the IPCC SRES (see Section 2.5.1 below) and the scenarios developed by the World Business Council for Sustainable Development (WBCSD, 1997).

2.4.4 Global Futures Scenarios, Greenhouse Gas Emissions, and Sustainable Development

Of the 124 global futures scenarios in the database, 35 provide some kind of projection of future GHG (usually CO_2) emis-

Box 2.4. The Global Scenarios Group: Scenarios and Process

A few organizations have been developing futures scenarios that incorporate both narrative and quantitative elements, including, for example, the Dutch Central Planning Bureau (CPB, 1992), the Millennium Project (Glenn and Gordon, 1998), and the Global Scenario Group (Gallopin *et al.*, 1997). The latter is discussed here as an illustration of this kind of approach to scenario development.

The Global Scenario Group (GSG) was convened by the Stockholm Environment Institute in 1995 as an international process to illuminate the requirements for a transition to global sustainability. It is a continuing and interdisciplinary process involving participants from diverse regional perspectives, rather than a single study. The GSG scenarios are holistic, developed both as narratives — accounts of how human values, cultural choices, and institutional arrangements might unfold — and detailed quantitative representations of social conditions such as level of poverty, economic patterns, and a wide range of environmental issues.

The GSG framework includes three broad classes of scenarios for scanning the future — "Conventional Worlds", "Barbarization", and "Great Transitions" — with variants within each class. All are compatible with current patterns and trends, but have very different implications for society and the environment in the 21st century (Gallopin *et al.*, 1997). In "Conventional Worlds" scenarios, global society develops gradually from current patterns and dominant tendencies, with development driven primarily by rapidly growing markets as developing countries converge towards the development model of advanced industrial ("developed") countries. In "Barbarization" scenarios, environmental and social tensions spawned by conventional development are not resolved, humanitarian norms weaken, and the world becomes more authoritarian or more anarchic. "Great Transitions" explore visionary solutions to the sustainability challenge, which portray the ascendancy of new values, lifestyles, and institutions.

"Conventional Worlds" is where much of the policy discussion occurs, including most of the analysis of climate mitigation. The integrated GSG approach situates the discussion of alternative emission scenarios in the context of sustainable development, by making poverty reduction an explicit scenario driver, and highlighting the links between climate and other environment and resource issues (Raskin *et al.*, 1998). The regional distribution of emissions becomes an explicit consideration in scenario design that is linked to poverty reduction, equity, and burden sharing in environmentally-sound global development. By underscoring the interactions between environmental and social goals, the policy strategies for addressing climate are assessed for compatibility and synergy with a wider family of actions for fostering sustainable development.

sions. These projections range from narrative descriptions (e.g., "emissions continue to rise") to numerical estimates. *Figure 2.9* shows global carbon dioxide emissions projections from the scenarios that provide numerical estimates.

Most (22) of these scenarios project increased emissions, but several (13) foresee declining emissions. All but one of the latter scenarios are Sustainable Development scenarios in which there is a concerted policy effort towards emission reduction, innovation in energy development towards improved efficiency and conservation, and/or alternatives to fossil fuels. The exception is a High-Tech Optimist scenario in which energy efficiency technologies and a shift to low- and non-fossil fuels bring about declining emissions.

The Sustainable Development scenarios that project declining emissions are in general characterized by increased co-operation and political participation; many assume that there is strong international agreement on the environment and development in general and climate change in particular. There is improved environmental quality and equity and, in several scenarios, increased material affluence globally (although some scenarios indicate a decline in consumerism). Population continues to grow but at slower rates and stabilizes at relatively low levels. In most scenarios significant developments of energy efficiency, energy conservation, and alternative energy tech-

nologies are key to emission reduction; a number of scenarios assume a tax on fossil fuels.

Table 2.4 summarizes the apparent relationships between emissions and scenario dimensions. It is important to note that there is considerable variety among the scenarios; *Table 2.4* therefore shows relationships that were in the majority, but not necessarily all, of the scenarios. It should also be noted that the relationships shown in *Table 2.4* do not by themselves prove causation; they simply reflect what the majority of scenarios with rising and falling GHG emissions, respectively, indicate for each scenario dimension.

What is clear from *Table 2.4* is that there are no strong patterns in the relationship between economic activity and GHG emissions. Growth in economic activity is compatible, across this set of scenarios, with both increasing and decreasing GHG emissions. In the latter case, mediating factors include increased energy efficiency, shifts to non-fossil energy sources, and/or shifts to a post-industrial (service-based) economy. Similarly, population growth is present in scenarios with rising emissions as well as scenarios with falling emissions, although in the latter group of scenarios, population tends to stabilize at relatively low levels, in many cases owing to increased prosperity, expanded provision of family planning, and improved rights and opportunities for women.

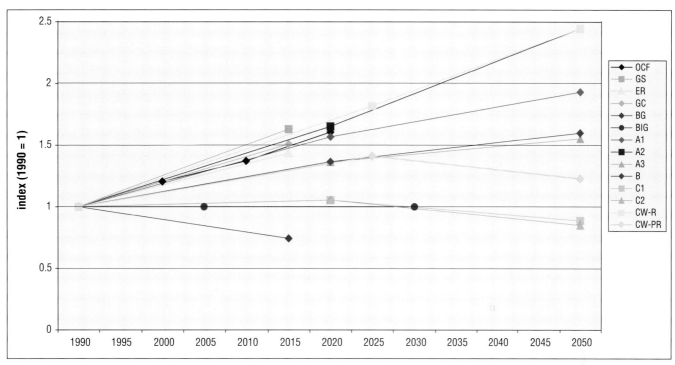

Figure 2.9: *CO_2 Emissions in Global Futures Scenarios (narrative scenarios). Acronyms: OCF, the "Our Common Future" scenario from Duchin et al., 1994; GS, the "Global Shift"; ER, the "European Renaissance"; GC, the "Global Crisis"; and BG the "Balanced Growth" scenarios from the Central Planning Bureau of the Netherlands (CPB, 1992); A1, A2, A3, B, C1 and C2, scenarios from Nakicenovic et al., 1998; CW-R, "Conventional Worlds – Reference"; and CW-PR, "Conventional Worlds – Policy Reform" from Gallopin et al., 1997 and Raskin et al., 1998. Note that this figure shows emission projections from a subset of the Global Futures Scenarios which discuss emissions, and a slightly higher proportion of scenarios in this larger group foresee declining emissions (13 of 35 scenarios, compared to 4 of 14 scenarios shown in the figure).*

Table 2.4: *Factors associated with changing GHG emissions in global futures scenarios*

Factor	Rising GHGs	Falling GHGs
Economy	Growing, post-industrial economy with globalization, (mostly) low government intervention, and generally high level of competition	Some scenarios show rising GDP, others show economic activity limited to ecologically sustainable levels; generally high level of government intervention
Population	Growing population with high level of migration	Growing population that stabilizes at relatively low level; low level of migration
Governance	No clear pattern in governance	Improvements in citizen participation in governance, community vitality, and responsiveness of institutions
Equity	Generally declining income equality within nations and no clear pattern in social equity or international income equality	Increasing social equity and income equality within and among nations
Conflict/ Security	High level of conflict and security activity (mostly), deteriorating conflict resolution capability	Low level of conflict and security activity, improved conflict resolution capability
Technology	High level of technology, innovation, and technological diffusion	High level of technology, innovation, and technological diffusion
Resource Availability	Declining renewable resource and water availability; no clear pattern for non-renewable resource and food availability	Increasing availability of renewable resources, food and water; no clear pattern for non-renewable resources
Environment	Declining environmental quality	Improving environmental quality

The major visible difference has to do with environmental impacts. As might be expected, pollution and the risk of ecological collapse are generally high in scenarios which show rising GHG emissions, and low in scenarios which show falling GHG emissions. Water availability and biodiversity decline in the scenarios with rising GHG emissions, and rise or stay the same in the scenarios with falling GHG emissions.

On a different front, in the scenarios with rising GHG emissions, conflict and security activity are generally high, while government intervention in the economy and income equality (within nations) are generally low. The reverse is true in the scenarios with falling GHG emissions, which also show improving equity between North and South. This would be expected from the fact that all but one of these scenarios are Sustainable Development scenarios.

Chapter 3 of the SRES discusses the relationships between GHG emissions and a number of driving forces, including population, economic and social development (including equity), and technology. What is clear from that discussion, which is consistent with the evidence summarized in *Table 2.4*, is that the impacts on GHG emissions of changes in these underlying driving forces are complex.

These complex relationships suggest that the choice of future "world" is more fundamental than the choice of a few driving forces in determining GHG emissions. The wide range of emissions in the various SRES baseline scenarios also demonstrates this point. Choices about DES are crucial, not just for the underlying conditions which give rise to emissions, but also for the nature and severity of climate change impacts, and the success of particular mitigation and adaptation policies. This finding is consistent with the discussion in Chapter 1, which suggests the central importance of DES issues in any consideration of climate change.

It is important therefore that emission scenarios consider qualitative aspects that are potentially important for future GHG emissions and mitigation policies. One way to do this is to link these scenarios with the broader global futures scenarios. However, this will be difficult because there are few areas of overlap, as a result of the very different natures of the two kinds of scenarios. Perhaps a more fruitful way of incorporating qualitative dimensions into quantitative scenarios, already pursued by the Global Scenarios Group and others, as well as in the SRES, is to develop quantitative estimates of key variables based on qualitative descriptions of future worlds.

2.4.5   Conclusions

A survey of the global futures literature has yielded a number of insights that are relevant to GHG emission scenarios and sustainable development. First, a wide range of future conditions has been identified by futurists, ranging from variants of sustainable development to collapse of social, economic, and

environmental systems. Since future values of the underlying socio-economic drivers of emissions may vary widely, it is important that GHG emission scenarios in particular, and climate change analysis in general, not limit themselves to a narrow range of possible futures, but consider the implications for mitigation of quite different sets of future conditions. In turn, climate policies should be designed so that they are resilient against widely different future conditions.

Second, the global futures scenarios describe a wide range of worlds, from pessimistic to optimistic, that are consistent with rising GHG emissions and a smaller range of (generally optimistic) worlds that are consistent with falling emissions. Scenarios that show falling emissions tend to show improved governance, increased equity and political participation, reduced conflict, conditions supportive of lower birth rates, and improved environmental quality. Scenarios with rising emissions generally show reduced environmental quality and equity within nations and increased conflict, and are more mixed with respect to governance and international equity. Both types of scenarios generally indicate continued technological development. The Sustainable Development scenarios suggest that sustainable development approaches are feasible, and can lead to futures characterized by relatively low emissions. A key implication is that sustainable development policies, taken generally, can make a significant contribution to emission reduction.

Third, scenarios do not all show a positive relationship between emissions and economic and population growth, as is commonly assumed (see also the discussion of the Kaya identity in Section 2.3.2.1 of this chapter). This is largely because, in the scenarios with declining emissions and rising population and economic activity, policy, lifestyle choices, and technological development act to reduce emissions through efficiency improvements, energy conservation, shifts to alternative fuels, and shifts to post-industrial economic structures. This suggests that different combinations of driving forces are consistent with low emission scenarios, which agrees with the SRES findings. The implication of this would seem to be that it is important to consider the linkage between climate policy and other policies and conditions associated with the choice of future paths in a general sense. In other words, low emission futures are associated with a whole set of policies and actions that go beyond the development of climate policy itself.

In general, the global futures scenarios provide more comprehensive coverage of the issues relevant to sustainable development than the general mitigation scenarios described in section 2.3. They therefore represent an important complement to the quantitative emission scenarios. However, there are significant difficulties involved in trying to connect the mainly narrative-based scenarios discussed in this section with the more quantitatively oriented scenarios discussed earlier. In this connection, the work of the Global Scenarios Group, the SRES, and others in linking narrative scenarios addressing social, environmental, and economic elements of sustainable development with model "quantifications" appears to point the way to the type of work needed to better assess the implications of GHG mitigation for sustainable development and vice versa. Section 2.5 below discusses the SRES scenarios and process, as well as mitigation scenarios that were developed on the basis of the SRES baseline scenarios.

2.5 Special Report on Emissions Scenarios (SRES) and Post-SRES Mitigation Scenarios

This section reviews two scenario literatures. One is the SRES, which reports on the development of multiple GHG emissions baselines based on different future world views, and the other is the post-SRES literature, which involves the quantification of mitigation scenarios based on the new SRES baseline scenarios.

2.5.1 Special Report on Emissions Scenarios: Summary and Differences from TAR

2.5.1.1 IPCC Emissions Scenarios and the SRES Process

First, the reference scenarios are reviewed, namely the SRES GHG emissions scenarios. These are "reference" scenarios in the sense that they describe future emissions in the absence of specific new policies to mitigate climate change. The new scenarios are published as the Special Report on Emissions Scenarios (SRES) by the IPCC (Nakicenovic *et al.*, 2000).

A key feature of the SRES process was that different methodological approaches and models were used to develop the scenarios. Another was that an "open process" was used to develop the scenarios through which researchers and other interest groups throughout the world could review and comment on the SRES scenarios as they were being developed. The SRES also aimed at improving the process of scenario development by extensively documenting the inputs and assumptions of the SRES scenarios; by formulating narrative scenario storylines; by encouraging a diversity of approaches and methods for deriving scenarios; by making the scenarios from different groups more comparable, and by assessing their differences and similarities; by expanding the range of economic-development pathways, including a narrowing of the income gap between developing and industrially developed countries; by incorporating the latest information on economic restructuring throughout the world; and by examining different trends in and rates of technological change.

2.5.1.2 SRES Approach to Scenario Development

The basic approach of the SRES writing team was to construct scenarios that were both qualitative and quantitative. The process involved first the formulation of the qualitative scenario characteristics in the form of narrative storylines and then their quantification by six different modelling approaches. The

qualitative description gives background information about the global setting of the scenarios, which can be used to assess the capability of society to adapt to and mitigate climate change, and for linking the emission scenarios with DES issues. The quantitative description of emission scenarios can be used as input to models for computing the future extent of climate change, and for assessing strategies to reduce emissions.

The relation between qualitative and quantitative scenarios can be characterized in terms of *Figure 2.10*.

The SRES writing team developed four scenario "families" (see *Box 2.5* for an explanation of terminology used in the SRES), because an even number helps to avoid the impression that there is a "central" or "most likely" case. The scenarios cover a wide range – but not all possible futures. In particular, there are no "global disaster" scenarios. None of the scenarios include new explicit climate policies.

Each family has a unifying theme in the form of a "storyline" or narrative that describes future demographic, social, economic, technological, and policy trends. Four storylines were developed by the whole writing team that identified driving forces, key uncertainties, possible scenario families, and their logic. Six global modelling teams then quantified the storylines. The quantification consisted of first translating the storylines into a set of quantitative assumptions about the driving forces of emissions (for example, rates of change of population and size of the economy and rates of technological change). Next, these assumptions were input to six integrated, global models that computed the emissions of GHGs and sulphur dioxide (SO_2). As a result, a total of 40 scenarios were produced for the four storylines. The large number of alternative scenarios showed that a single storyline could lead to a large number of feasible emission pathways.

In all, six models were used to generate the 40 scenarios that comprise the four scenario families. Six of these scenarios,

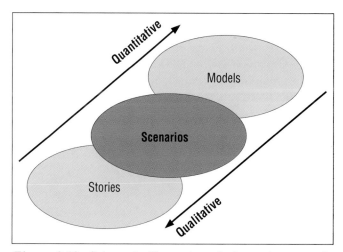

Figure 2.10: *Schematic illustration of alternative scenario formulations ranging from narrative storylines to quantitative formal models (source: Nakicenovic et al., 2000).*

Box 2.5. IPCC SRES Scenario Terminology (Source: Nakicenovic *et al.*, 2000)

Model: a formal representation of a system that allows quantification of relevant system variables.

Storyline: a narrative description of a scenario (or a family of scenarios) highlighting the main scenario characteristics, relationships between key driving forces, and the dynamics of the scenarios.

Scenario: a description of a potential future, based on a clear logic and a quantified storyline.

Family: scenarios that have a similar demographic, societal, economic, and technical-change storyline. Four scenario families comprise the SRES: A1, A2, B1, and B2.

Group: scenarios within a family that reflect a variation of the storyline. The A1 scenario family includes three groups designated by A1T, A1FI, and A1B that explore alternative structures of future energy systems. The other three scenario families consist of one group each.

Category: scenarios are grouped into four categories of cumulative CO_2 emissions between 1990 and 2100: low, medium–low, medium–high, and high emissions. Each category contains scenarios with a range of different driving forces yet similar cumulative emissions.

Marker: a scenario that was originally posted on the SRES website to represent a given scenario family. A marker is not necessarily the median or mean scenario.

Illustrative: a scenario that is illustrative for each of the six scenario groups reflected in the Summary for Policymakers of this report. They include four revised "scenario markers" for the scenario groups A1B, A2, B1, and B2, and two additional illustrative scenarios for the A1FI and AIT groups. See also "(Scenario) Groups" and "(Scenario) Markers".

Harmonized: harmonized scenarios within a family share common assumptions for global population and GDP while fully harmonized scenarios are within 5% of the population projections specified for the respective marker scenario, within 10% of the GDP and within 10% of the marker scenario's final energy consumption.

Standardized: emissions for 1990 and 2000 are indexed to have the same values.

Other scenarios: scenarios that are not harmonized.

which should be considered equally sound, were chosen to illustrate the whole set of scenarios. They span a wide range of uncertainty, as required by the SRES Terms of Reference. These encompass four combinations of demographic change, social and economic development, and broad technological developments, corresponding to the four families (A1, A2, B1, B2), each with an illustrative "marker" scenario. Two of the scenario groups of the A1 family (A1FI, A1T) explicitly explore energy technology developments, alternative to the "balanced" A1B group, holding the other driving forces constant, each with an illustrative scenario. Rapid growth leads to high capital turnover rates, which means that early small differences among scenarios can lead to a large divergence by 2100. Therefore, the A1 family, which has the highest rates of technological change and economic development, was selected to show this effect.

To provide a scientific foundation for the scenarios, the writing team extensively reviewed and evaluated over 400 published scenarios. Results of the review were published in the scientific literature (Alcamo and Nakicenovic, 1998), and made available to the scientific community in the form of an Internet scenario database. The background research by the six modelling teams for developing the 40 scenarios was also published in the scientific literature (Nakicenovic, 2000).

2.5.1.3   A Short Description of the SRES Scenarios

Since there is no agreement on how the future will unfold, the SRES tried to sharpen the view of alternatives by assuming that individual scenarios have diverging tendencies — one emphasizes stronger economic values, the other stronger environmental values; one assumes increasing globalization, the

other increasing regionalization. Combining these choices yielded four different scenario families (*Figure 2.11*). This two-dimensional representation of the main SRES scenario characteristics is an oversimplification. It is shown just as an illustration. In fact, to be accurate, the space would need to be multi-dimensional, listing other scenario developments in many different social, economic, technological, environmental, and policy dimensions.

The titles of the four scenario storylines and families have been kept simple: A1, A2, B1, and B2. There is no particular order among the storylines; they are listed in alphabetical and numerical order:

- The A1 storyline and scenario family describes a future world of very rapid economic growth, global population that peaks in mid-century and declines thereafter, and the rapid introduction of new and more efficient technologies. Major underlying themes are convergence among regions, capacity building, and increased cultural and social interactions, with a substantial reduction in regional differences in per capita income. The A1 scenario family develops into three groups that describe alternative directions of technological change in the energy system. The three A1 groups are distinguished by their technological emphasis: fossil intensive (A1FI), non-fossil energy sources (A1T), or a balance across all sources (A1B).[12]

- The A2 storyline and scenario family describes a very heterogeneous world. The underlying theme is self-

---

[12] Balanced is defined as not relying too heavily on one particular energy source, on the assumption that similar improvement rates apply to all energy supply and end-use technologies.

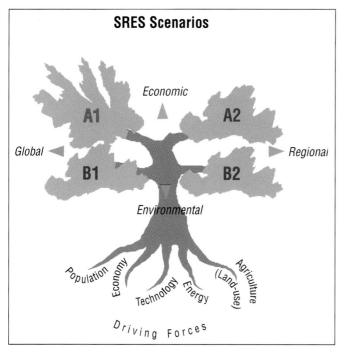

SRES Scenarios

Figure 2.11. *Schematic illustration of SRES scenarios. The four scenario "families" are shown, very simplistically, for illustrative purposes, as branches of a two-dimensional tree. The two dimensions shown indicate global and regional scenario orientation, and development and environmental orientation, respectively. In reality, the four scenarios share a space of a much higher dimensionality given the numerous driving forces and other assumptions needed to define any given scenario in a particular modelling approach. The schematic diagram illustrates that the scenarios build on the main driving forces of GHG emissions. Each scenario family is based on a common specification of some of the main driving forces.*

reliance and preservation of local identities. Fertility patterns across regions converge very slowly, which results in continuously increasing global population. Economic development is primarily regionally oriented and per capita economic growth and technological change are more fragmented and slower than in other storylines.

• The B1 storyline and scenario family describes a convergent world with the same global population that peaks in mid-century and declines thereafter, as in the A1 storyline, but with rapid changes in economic structures towards a service and information economy, with reductions in material intensity, and the introduction of clean and resource-efficient technologies. The emphasis is on global solutions to economic, social, and environmental sustainability, including improved equity, but without additional climate initiatives.

• The B2 storyline and scenario family describes a world in which the emphasis is on local solutions to economic, social, and environmental sustainability. It is a world with a continuously increasing global population at a rate lower than in A2, intermediate levels of economic development, and less rapid and more diverse technological change than in the B1 and A1 storylines. While the scenario is also oriented towards environmental protection and social equity, it focuses on local and regional levels.

In all, six models were used to generate the 40 scenarios that comprise the four scenario families. They are listed in *Table 2.5*. These six models are representative of emissions scenario modelling approaches and different integrated assessment frameworks in the literature, and include so-called top-down and bottom-up models.

Table 2.5: *Models used to generate the SRES scenarios*

Model	Source	Reference
Asian Pacific Integrated Model (AIM)	National Institute of Environmental Studies in Japan	Morita *et al.*, 1994 Kainuma *et al.*, 1998, 1999a, 1999b
Atmospheric Stabilization Framework Model (ASF)	ICF Consulting in the USA	EPA 1990; Pepper *et al.*, 1992
Integrated Model to Assess the Greenhouse Effect (IMAGE), used in connection with the WorldScan model	IMAGE: RIVM and WorldScan: CPB (Central Planning Bureau), The Netherlands	IMAGE: Alcamo 1994; Alcamo *et al.*,1998; de Vries *et al.*, 1999 WorldScan: CPB Netherlands, 1999
Multiregional Approach for Resource and Industry Allocation (MARIA)	Science University of Tokyo in Japan	Mori and Takahashi, 1998
Model for Energy Supply Strategy Alternatives and their General Environmental Impact (MESSAGE)	IIASA in Austria	Messner *et al.*, 1996; Riahi and Roehrl, 2000
The Mini Climate Assessment Model (MiniCAM)	PNNL in the USA	Edmonds *et al.*, 1996

2.5.1.4   Emissions and Other Results of the SRES Scenarios

Figure 2.12 illustrates the range of global energy-related and industrial CO_2 emissions for the 40 SRES scenarios against the background of all the 400 emissions scenarios from the literature documented in the SRES scenario database. The six scenario groups are represented by the six illustrative scenarios. *Figure 2.12* also shows a range of emissions of the six scenario groups next to each of the six illustrative scenarios.

Figure 2.12 shows that the four marker and two illustrative scenarios by themselves cover a large portion of the overall scenario distribution. This is one of the reasons that the SRES Writing Team recommended the use of all four marker and two illustrative scenarios in future assessments. Together, they cover most of the uncertainty of future emissions, both with respect to the scenarios in the literature and the full SRES scenario set. *Figure 2.12* also shows that they are not necessarily close to the median of the scenario family because of the nature of the selection process. For example, A2 and B1 are at the upper and lower bounds of their scenario families, respective-

ly. The range of global energy-related and industrial CO_2 emissions for the six illustrative SRES scenarios is generally somewhat lower than the range of the IPCC IS92 scenarios (Leggett *et al.*, 1992; Pepper *et al.*, 1992). Adding the other 36 SRES scenarios increases the covered emissions range. Jointly, the SRES scenarios cover the relevant range of global emissions, from the 95th percentile at the high end of the distribution all the way down to very low emissions just above the 5th percentile of the distribution. Thus, they only exclude the most extreme emissions scenarios found in the literature – those situated out in the tails of the distribution. What is perhaps more important is that each of the four scenario families covers a sizable part of this distribution, implying that a similar quantification of driving forces can lead to a wide range of future emissions. More specifically, a given combination of the main driving forces is not sufficient to uniquely determine a future emission path. There are too many uncertainties. The fact that each of the scenario families covers a substantial part of the literature range also leads to an overlap in the emissions ranges of the four families. This implies that a given level of future emissions can arise from very different combinations of dri-

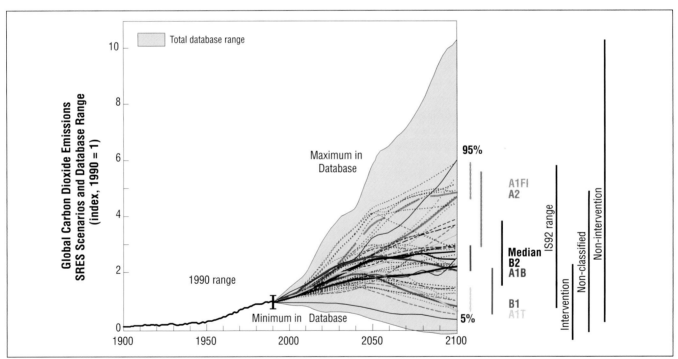

Figure 2.12: *Global CO_2 emissions from energy and industry, historical development from 1900 to 1990 and in 40 SRES scenarios from 1990 to 2100, shown as an index (1990 = 1). The range is large in the base year 1990, as indicated by an "error" bar, but is excluded from the indexed future emissions paths. The dashed time-paths depict individual SRES scenarios and the blue shaded area the range of scenarios from the literature (as documented in the SRES database). The median (50th), 5th, and 95th percentiles of the frequency distribution are shown. The statistics associated with the distribution of scenarios do not imply probability of occurrence (e.g., the frequency distribution of the scenarios in the literature may be influenced by the use of IS92a as a reference for many subsequent studies). The 40 SRES scenarios are classified into six groups. Jointly the scenarios span most of the range of the scenarios in the literature. The emissions profiles are dynamic, ranging from continuous increases to those that curve through a maximum and then decline. The coloured vertical bars indicate the range of the four SRES scenario families in 2100. Also shown as vertical bars on the right are the ranges of emissions in 2100 of IS92 scenarios, and of scenarios from the literature that apparently include additional climate initiatives (designated as "intervention" scenarios emissions range), those that do not ("non-intervention"), and those that cannot be assigned to either of these two categories ("non-classified").*

ving forces. This result is of fundamental importance for assessments of climate change impacts and possible mitigation and adaptation strategies.

An important feature of the SRES scenarios obtained using the SAR methodology is that their overall radiative forcing is higher than the IS92 range despite comparatively lower GHG emissions (Wigley and Raper, 1992; Wigley *et al.*, 1994; Houghton *et al.*, 1996; Wigley, 1999; Smith *et al.*, 2000; IPCC, 2001). This results from the loss of sulphur-induced cooling during the second half of the 21st century. On one hand, the reduction in global sulphur emissions reduces the role of sulphate aerosols in determining future climate, and therefore reduces one aspect of uncertainty about future climate change (because the precise forcing effect of sulphate aerosols is highly uncertain). On the other hand, uncertainty increases because of the diversity in spatial patterns of SO_2 emissions in the scenarios. Future assessments of possible climate change need to account for these different spatial and temporal dynamics of GHG and sulphur emissions, and they need to cover the whole range of radiative forcing associated with the scenarios.

In summary, the SRES scenarios lead to the following findings:
- Alternative combinations of driving-force variables can lead to similar levels and structure of energy use and land-use patterns, as illustrated by the various scenario groups and scenarios. Hence, even for a given scenario outcome, for example, in terms of GHG emissions, there are alternative combinations and alternative pathways that could lead to that outcome. For instance, significant global changes could result from a scenario of high population growth, even if per capita incomes would rise only modestly, as well as from a scenario in which a rapid demographic transition (low population levels) coincides with high rates of income growth and affluence.
- Important possibilities for further bifurcations in future development trends exist within one scenario family, even when adopting certain values for important scenario driving force variables to illustrate a particular possible development path.
- Emissions profiles are dynamic across the range of SRES scenarios. They portray trend reversals and indicate possible emissions crossover among different scenarios. They do not represent mere extensions of a continuous increase of GHGs and sulphur emissions into the future. This more complex pattern of future emissions across the range of SRES scenarios reflects the recent scenario literature.
- Describing potential future developments involves inherent ambiguities and uncertainties. One and only one possible development path (as alluded to for instance in concepts such as "business-as-usual scenario") simply does not exist. And even for each alternative development path described by any given scenario, there are numerous combinations of driving forces and numerical values that can be consistent with

a particular scenario description. This particularly applies to the A2 and B2 scenarios that imply a variety of regional development patterns that are wider than in the A1 and B1 scenarios. The numerical precision of any model result should not distract from the basic fact that uncertainty abounds. However, in the opinion of the SRES writing team, the multi-model approach increases the value of the SRES scenario set, since uncertainties in the choice of model input assumptions can be more explicitly separated from the specific model behaviour and related modelling uncertainties.
- Any scenario has subjective elements and is open to various interpretations. While the SRES writing team as a whole has no preference for any of the scenarios, and has no judgement about the probability or desirability of the scenarios, the open process and reactions to SRES scenarios have shown that individuals and interest groups do have such judgements. This will stimulate an open discussion in the political arena about potential futures and choices that can be made in the context of climate change response. For the scientific community, the SRES scenario exercise has led to the identification of a number of recommendations for future research that can further increase understanding about potential development of socio-economic driving forces and their interactions, and associated GHG emissions.

2.5.2 Review of Post-SRES Mitigation Scenarios

2.5.2.1 Background and Outline of Post-SRES Analysis

The review of general mitigation scenarios shows that mitigation scenarios and policies are strongly related to their baselines, and that there has been no systematic comparison of the relationship between baseline and mitigation scenarios. Modellers participating in the SRES process recognized the need to analyze and compare mitigation scenarios using as their baselines the new IPCC scenarios, which quantify a wide range of future worlds. Consequently, they participated (on a voluntary basis) in a special comparison programme to quantify SRES-based mitigation scenarios (Morita *et al.*, 2000a; 2000b). These SRES-based scenarios are called "Post-SRES Mitigation Scenarios".

The process of the post-SRES analysis was started by a public invitation to modellers. A "Call for Scenarios" was sent to more than one hundred researchers in March 1999 by the Co-ordinating Lead Authors of this chapter and the SRES to facilitate an assessment of the potential implications of mitigation scenarios based on the SRES cases, which report was developed in support of the Third Assessment Report. Modellers from around the world were invited to prepare quantified stabilization scenarios for two or more concentrations of atmospheric CO_2, based on one or more of the six SRES scenarios. Concentration ceilings include 450, 550 (minimum require-

Table 2.6: *Post-SRES participants and quantified scenarios (indicated by CO_2 stabilization target in ppmv)*

Baseline scenarios	A1B	A1FI	A1T	A2	B1	B2
AIM (NIES and Kyoto University, Japan)	450, 550, 650	550		550	550	550
ASF (ICF Corporation, USA)				550, 750		
IMAGE (RIVM, Netherlands)	550				450	
LDNE (Tokyo University, Japan)	550	550	550	550	550	550
MARIA (Science University of Tokyo, Japan)	450, 550, 650		450, 550, 650		450, 550	450, 550, 650
MESSAGE-MACRO (IIASA, Austria)	450, 550, 650	450[*], 550[*] 650[*], 750[*]	450, 550	550, 750		550
MiniCAM (PNNL, USA)	550[*]	450, 550, 650, 750		550	450, 550	550 [*]
PETRO (Statistics Norway, Norway)	450, 550, 650, 750		450, 550 650, 750			
WorldScan (CPB, Netherlands)	450 [**], 550[**]			450, 550[**]	450[**], 550	450[**], 550

Notes: [*] High and low baselines were used; [**] An early action and a delayed response were quantified.

ment), 650, and 750ppmv, and harmonization with the SRES scenarios was required by tuning reference cases to SRES values for GDP, population, and final energy demand.

Nine modelling teams participated in the comparison programme, including six SRES modelling teams and three other teams: AIM team (Jiang *et al.*, 2000), ASF team (Sankovski *et al.*, 2000), IMAGE team, LDNE team (Yamaji *et al.*, 2000), MESSAGE-MACRO team (Riahi & Roehrl, 2000), MARIA team (Mori, 2000), MiniCAM team (Pitcher, 2000), PETRO team (Kverndokk *et al.*, 2000) and WorldScan team (Bollen *et al.*, 2000). *Table 2.6* shows all the modelling teams and the stabilized concentration levels which were adopted as stabilization targets by each one. Most of the modelling teams covered more than two SRES baseline scenarios, and half of them developed multiple stabilization cases for at least one baseline, so that a systematic review can be conducted to clarify the relationship between baseline scenarios and mitigation policies and/or technologies.

While all baselines were analyzed, the A1B baseline was most frequently used. Across baselines, the stabilization target of 550ppmv seemed to be the most popular. Because of time constraints involved in quantifying the stabilization scenarios, the modelling teams mostly focused their analyses on energy-related CO_2 emissions. However, about half of the modelling teams, notably the AIM, IMAGE, MARIA, and MiniCAM teams, have quantified mitigation scenarios in non-energy CO_2 emissions as well as in non-CO_2 emissions. The modelling teams that did not estimate non-energy CO_2 emissions intro-

duced scenarios of them from outside of their models for estimating atmospheric concentrations of CO_2.

In order to check the performance of CO_2 concentration stabilization for each post-SRES mitigation scenario, a special "generator" (Matsuoka, 2000) was used by the modelling teams to convert the CO_2 emissions into CO_2 concentration trajectories. In addition, the generator was used by them to estimate the eventual level of atmospheric CO_2 concentration by 2300, based on the 1990 to 2100 CO_2 emissions trajectories from the scenarios. This generator is based on the Bern Carbon Cycle Model (Joos *et al.*, 1996), which was used in the IPCC SAR (IPCC, 1996) and TAR (IPCC, 2001). Using this generator, each modelling team adjusted their mitigation scenarios so that the interpolated CO_2 concentration reached one of the alternative fixed target levels at the year 2150 within a 5% error. The year 2150 was selected based on Enting *et al.* (1994) who gave a basis for stabilization scenarios of the IPCC SAR (IPCC, 1996).[13] A further constraint imposed was that the

---

[13] Enting *et al.* (1994) selected the timings to reach alternative target levels in 2100 year for 450ppmv, 2150 year for 550 ppmv, 2200 year for 650ppmv, and 2250 year for 750ppmv. Post-SRES modellers selected only the year 2150 for all the stabilization targets; this decision was a consequence of the tight time constraints the modelling teams faced for preparation of the scenarios. As a result, 450ppmv stabilization scenarios of post-SRES require slightly more reductions of CO_2 than those of IPCC (1995), while 650 and 750ppmv stabilization scenarios of post-SRES require slightly less reductions than those of IPCC (1995), both during the period from now to 2150.

interpolated emission curve should be smooth after 2100, the end of the time-horizon of the scenarios. This adjustment played an important role in the post-SRES analyses for harmonizing emissions concentrations levels across the stabilization scenarios. The key driving forces of emissions such as population, GDP, and final energy consumption were harmonized in baseline assumptions specified by the six SRES scenarios.

2.5.2.2 Storylines of Post-SRES Mitigation Scenarios

The procedure for creating post-SRES mitigation scenarios was similar to the SRES process, even though the period for the post-SRES work was much shorter than that for the SRES and, in contrast to the SRES process, the exercise was voluntary and not mandated by the IPCC. The storyline approach of SRES indicates that different future worlds will have different mitigative capacities (cf. Chapter 1). Hence, the first step of the post-SRES scenario work was to create storylines for the mitigation scenarios.

In general, mitigation scenarios are defined relative to a baseline scenario. If mitigation strategies are formulated and implemented in any of the future worlds as described within SRES, a variety of aspects of that world will determine the capacity to formulate and implement carbon reduction policies, for instance:

- The availability and dissemination of relevant knowledge on emissions and climate change;
- The institutional, legal, and financial infrastructure to implement mitigation policies and measures;
- Entrepreneurial and/or governmental policies for generating innovation and encouraging the penetration of new technologies; and
- The mechanisms by which consumers and entrepreneurs respond to changing prices and new products and processes.

In the post-SRES process, it was difficult for the modelling teams to consider all of these aspects with relation to the SRES future worlds, because of their inherent complexity and the amount of time available for the work. However, some aspects were considered by some modelling teams and these were reflected in the quantification assumptions. The rest of this section illustrates these major points in the form of storylines for each of the six SRES scenarios, which describe the relationship between the kind of future world on the one hand and the capacity for mitigation on the other.

The A1 world is well equipped to formulate and implement mitigation strategies in view of its high-tech, high-growth orientation and its willingness to co-operate at a global scale, provided the major actors acknowledge the need for mitigation. There will be good monitoring and reporting on emissions and climate change, and possible signs of climate change will be detected early and become part of the international agenda. Market-oriented policies and measures will be the preferred response. Least-cost options will be searched for and implemented through international negotiation and mechanisms with the support of governments and multinational companies. New emission reduction technologies from developed countries will enable developing countries to respond more rapidly and effectively if barriers to technology transfer can be overcome. In this high-growth world, the economic costs associated with the response to climate change are likely to be bearable. In the A1B scenario, where mitigation strategies may hit the limits of renewable energy supply, and in the A1FI scenario, carbon removal and storage as well as higher end-use energy efficiency will become major emission reduction options. In the A1T scenario, technology developments are such that mitigation policies and measures only require limited additional efforts.

Developing and implementing climate change mitigation measures and policies in the A2 world can be quite complicated. This is a result of several features embedded in the scenario storyline: rapid population growth, relatively slow GDP per capita growth, slow technological progress, and a regional and partially "isolationist" approach in national and international politics. Because of all these serious challenges, the abatement of GHG emissions in the A2 world becomes plausible only in the situation when the negative effects of climate change become imminent and the associated losses "outweigh" the costs of mitigation. The same features that make the A2 world "non-receptive" to worldwide mitigation policies may exacerbate the climate change effects and prompt nations to act. Measures such as a rapid shift towards high-tech renewable energy or deep-sea carbon storage will be highly improbable in the A2 world as a consequence of technology limitations. Instead, such relatively low-tech measures as limiting energy consumption, and capturing and using methane from natural gas systems, coal mining, and landfills better fit the A2 world's economic and technological profile. The lack of global co-operation may cause rather large regional variations in the feasibility and cost of mitigation policies and measures.

The B1 world is also well equipped to formulate and implement mitigation strategies, in view of its high economic growth and willingness to co-operate at a global scale. In comparison with the A1 world, however, it will be confronted with higher marginal abatement costs, although total costs are much lower than in A1B or A1FI. This is because baseline carbon emissions are lower in the B1 world compared to the A1 world, a consequence of the emphasis on sustainable development in B1. There will be intense monitoring and reporting of emissions and climate change. The precautionary principle informs international agenda setting and policy formulation, with governments taking responsibility for climate change-related preventive and adaptive action. Tightening international standards generates incentives for further innovation towards energy-efficiency and low- and zero-carbon options. Educational campaigns are another important instrument. Developed regions support the less developed regions in a variety of ways, including transfer of energy-efficiency and renewable-energy related technologies. Carbon taxes are introduced; an elaborate phase-in mechanism for less developed regions is negotiated and implemented. A part of the carbon tax revenue is used to compensate some fossil-fuel exporters and for a fund to compensate those affected by climate change.

In the B2 scenario actions to reduce GHG emissions are taken mainly at a local or regional scale in response to climate change impacts. Environmentally aware citizens of the B2 world will increasingly attribute damages to human-induced climate change. High-income countries, which are generally less vulnerable to climate change impacts, will increasingly see the need for climate policy action as a consequence of cost-benefit analyses. With increasing costs of damage, counter-measures challenge existing energy sector policies and institutional frameworks. Generally high educational levels promote both development and environmental protection. Resource availability, economic development, and technical change are uneven over regions. In relative terms, R&D expenditures are expected to stay constant, but they will be more targeted towards cleaner and less carbon-intensive energy technologies. Existing bilateral trade links will foster bilateral technology transfer from OECD countries to some developing countries. This is because rapidly increasing energy and, in particular,

electricity demand in developing countries present business opportunities no longer available in OECD countries. Therefore, there exist a number of incentives for bilateral environmental policy co-operation between R&D intensive countries in the North and developing countries of the South. Energy trade links, first for oil and later for natural gas and methanol, will play an important seed role for new environmental bilateral co-operation, leading to a regionally heterogeneous approach to GHG reduction.

2.5.2.3 Comparison of Quantified Stabilization Scenarios

Based on the storylines, 76 stabilization scenarios were quantified as shown in *Table 2.6*. The assessment of the post-SRES work in this section is restricted to the analysis of CO_2 emissions and energy use in the different model runs. The detailed comparison of macroeconomic costs of reducing CO_2 emissions costs is not dealt with here: Chapter 8 addresses this aspect of stabilization.

Figure 2.13 shows the CO_2 emission trajectories of the 76 post-SRES mitigation scenarios along with the ranges of SRES and

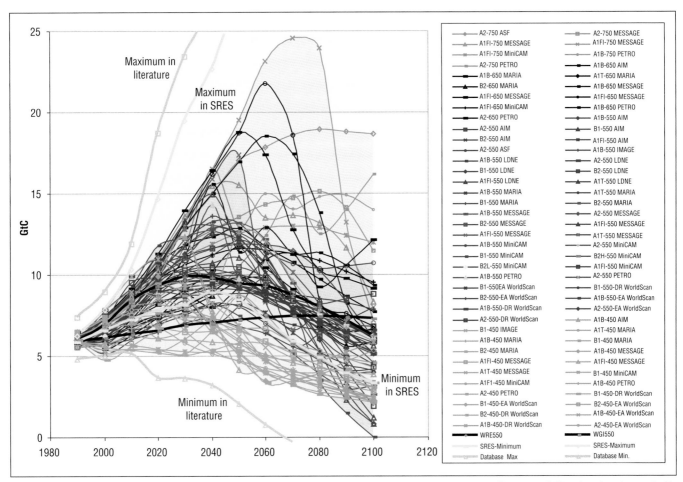

Figure 2.13: *The 76 post-SRES stabilization scenarios of world fossil fuel CO_2 emissions. Different stabilization levels are indicated by colour, with 750ppmv in red, 650ppmv in black, 550ppmv in blue, and 450ppmv in green. For comparison, the minimum and maximum of the ranges of scenarios from the literature (grey) and the SRES (yellow) as well as the WRI and WRE 550ppmv stabilization scenarios (bold black) are also shown.*

other published scenarios. Quantifications differ with respect to the baseline scenario including assumed storyline, the stabilization target, and the model that was used. As shown in *Figure 2.13*, the post-SRES scenarios cover a very wide range of emission trajectories, but the range is relatively below the SRES range, and they are apparently classified into groups according to the different stabilization targets. The figure shows the WRE late-response scenario and WGI early-action scenario for 550ppmv stabilization to compare with post-SRES scenarios, and it shows that the post-SRES range covers a much wider range than that between WRE and WGI.

Figure 2.14 shows the comparison of SRES and post-SRES scenario ranges in total global CO_2 emissions. The post-SRES ranges are estimated based on the selected scenarios quantified by SRES participants in order to compare the formal SRES ranges in Nakicenovic *et al.* (2000). It is shown clearly in the figure that concentration stabilization requires much more reduction of CO_2 emissions under development paths with high emissions such as A1FI and A2 than under development paths such as B1 and B2. These differences in reduction requirements result in selection of different technology and/or policy measures and, as a consequence, different costs to stabilize concentrations even at the same level. In the A1 scenario family, with its different scenarios in technological development (A1B, A1FI, and A1T), technological change is also a key component in bringing down the costs of mitigation options and their contribution to the emissions reduction. The A1FI stabilization scenarios, which are based on the highest baseline emissions, require much larger emission reductions than the A1T stabilization scenarios. The role of technology has been found to be crucial in the A1 scenario variants.

Morita *et al.* (2000a) compared all the stabilization variants in detail and found several common characteristics among these scenarios. These findings are as follows:

- Comparing the CO_2 emissions reductions from SRES baselines, the models have many points in common, but there are also some clear differences. All models show an increase in CO_2 reduction over time. This reflects the strong constraint of atmospheric CO_2 concentration. There is a considerable range in reductions among models in early years. However, most models achieve a similar proportional reduction from the baseline over the observation period.

- For achieving stabilization at 550ppmv, the highest reductions in CO_2 emissions compared to the baseline are observed in the A2 family. B1 shows the lowest reductions. CO_2 reduction at the end of the 21st century ranges in A2 between 75% and 80%, A1B between 50% and 75%, B2 between 40% and 70%, and B1 between 5% and 40%.

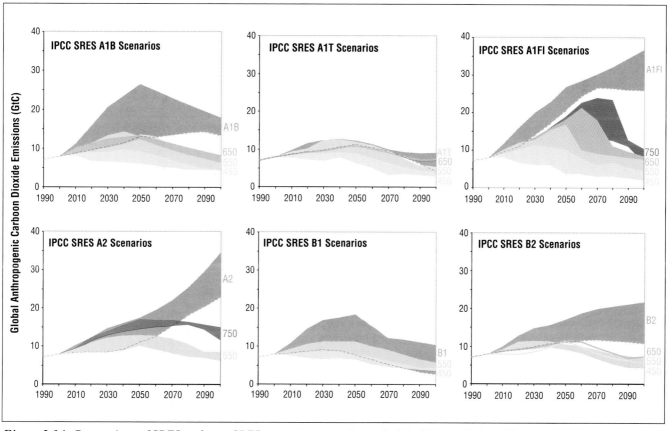

Figure 2.14: *Comparison of SRES and post-SRES scenario ranges in total global CO_2 emissions. The post-SRES ranges are estimated based on the selected scenarios quantified by SRES participants in order to compare SRES ranges.*

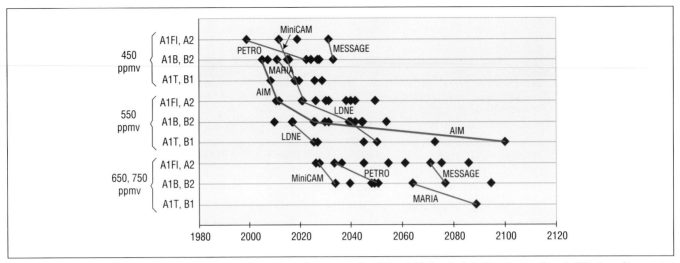

Figure 2.15: *Timing when the stabilization scenarios achieve a reduction of 20% of global energy-related CO$_2$ baseline emissions, compared across stabilization targets as well as baselines. Slanted lines join scenarios quantified by the same model.*

- The target stabilization level also significantly affects the CO$_2$ reduction, even when based on the same baseline scenario. In the 450ppmv stabilization case, the reduction reaches 70% to 100%[14] of A1B baseline emissions at the end of the 21st century.
- Energy consumption reductions are more complicated among models. There is no strong relationship between the level of energy consumption and the stabilization level.
- Different baselines lead to different macroeconomic costs in order to reach a stabilization target. In spite of the wide range among models, A2 would be the most expensive case while B1 would require the lowest cost for stabilization at 550ppmv. The GDP loss in B1 would be less than one-tenth that in the A1B case, and less than one-twentieth that in the A2 case.
- The CO$_2$ reduction and macroeconomic costs are also significantly affected by the target stabilization level, even when based on the same baseline scenario. The economic cost for 450ppmv stabilization would be around three times that for 550ppmv, and six to eight times that for 650ppmv. These relationships can be observed at both the global and regional levels.
- Different stabilization targets also require different timing for the introduction of reduction policies. The 450ppmv stabilization case requires drastic emission reductions that occur earlier than under the 650ppmv case. Very rapid increases in emission reduction over 20 to 30 years are also observed in the 450ppmv stabilization case.

In order to compare the scenarios in further detail, several indices were calculated for this review.

First, a CO$_2$ reduction index was compared among stabilization levels as well as among SRES worlds. This index is calculated by subtracting baseline emissions from mitigation scenario emissions. In general, the lower the stabilization level that is required, as well as the higher the level of baseline emissions caused by the selected development path, the larger the CO$_2$ divergence from the baseline that is needed in all the regions. However, it does not simply follow from the larger divergence in emissions that there is an earlier divergence from the baseline.

The impact on the timing of emission reduction of both the stabilization level and the baseline level of emissions is further elaborated in *Figure 2.15*. This figure shows when the reduction in energy-related CO$_2$ emissions in each stabilization scenario would reach 20% of baseline emissions. This figure indicates that more stringent stabilization targets require earlier emission reductions from baseline levels. Higher emission worlds such as A1F1 and A2 also require earlier reduction than lower emission worlds such as A1T and B1.

A key policy question is what kind of emission reductions would be needed in the medium term, after the commitment period of the Kyoto Protocol (assuming that it will be implemented). *Figure 2.16* shows the percent reduction in energy-related CO$_2$ emissions in Annex I countries from 1990 for the various stabilization cases. Since the first commitment period of the Kyoto Protocol ends in 2012, this can give some indication of the extent to which emission reduction commitments after 2012 would be needed to achieve the various stabilization levels. It should be noted that about two thirds of the scenarios assume that developing countries have already diverged from their baseline emission trajectories in 2020. Another point is that the post-SRES scenarios were not developed specifically to include the Kyoto targets, so there is a range of Annex I emission reductions (from 1990 levels) in 2010, 2020 and 2030. The mid-course scenarios are indicated in *Figure 2.16* as

---

[14] The 100% reduction scenario based on LDNE assumes the large scale introduction of carbon sequestration technologies.

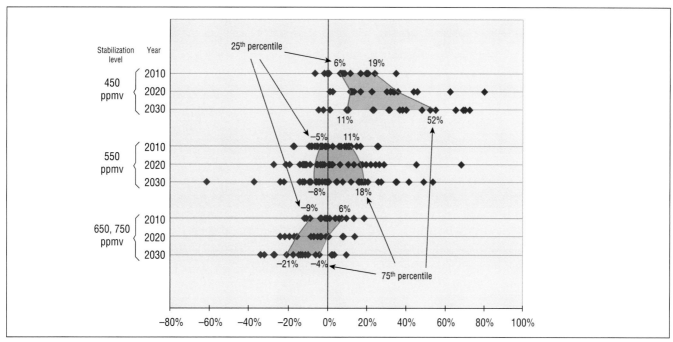

Figure 2.16: *The reduction of energy-related CO_2 emissions from 1990 levels in Annex I countries for stabilization at 450ppmv, 550ppmv, and 650–750ppmv. For each stabilization level, emission reductions are shown for the years 2010 (upper lines), 2020 (middle lines), and 2030 (lower lines). Shaded areas show the range between the 25th and 75th percentiles of the frequency distribution of the scenarios.*

the range between the 25th and 75th percentiles of the frequency distribution of the scenarios.

Figure 2.16 shows that:

• In the 450ppmv stabilization scenarios, the middle range (between the 25th and 75th percentiles) of Annex I emissions in 2010 lies between the Kyoto target and a 19% reduction from 1990 levels. This range increases after 2010, as does the decrease in Annex I emissions that would be needed to achieve stabilization at 450 ppmv. The percent reduction from 1990 levels in the middle range of scenarios is 13%–34% in 2020 and 11%–52% in 2030;

• In the 550ppmv stabilization scenarios, the middle range of Annex I emissions in 2010 is around the Kyoto target (from an 11% decrease to a 5% increase from 1990 levels); in 2020, the middle range of emissions lies between a 17% decrease and an 8% increase from 1990 levels; and in 2030, it lies between an 18% decrease and an 8% increase from 1990 levels. The average level of emissions slightly decreases after 2010; and

• The 650 or 750ppmv stabilization scenarios show similar changes in emission levels in 2010 compared to 1990, and few of them show any additional reduction in Annex I emissions after 2010. The middle range of emissions lies between an increase of 1%–17% from 1990 levels in 2020, and an increase of 4%–21% from 1990 levels in 2030.

This suggests that achievement of stabilization at 450ppmv will require emissions reductions in Annex I countries by 2020 that go significantly beyond their Kyoto Protocol commitments for 2008 to 2012.[15] It also suggests that it would not be necessary to go much beyond the Kyoto commitments for Annex I countries (assuming as indicated that developing countries diverge from their baselines by 2020) to achieve stabilization at 550ppmv or higher. However, it should be recognized that several scenarios do indicate the need for significant emission reductions by 2020 in order to achieve these stabilization levels. These findings should be interpreted in light of the facts that CO_2 concentrations are assumed to reach one of the alternative fixed target levels in the year 2150, and unlike "emission corridor" analyses, these scenarios do not introduce other conditions such as a constraint on the rate of temperature increase.

Another important policy question concerns the participation of developing countries in emission mitigation. As a first step in addressing this question, the post-SRES scenarios were evaluated according to when per capita CO_2 emissions in Annex I countries would fall below per capita emissions in non-Annex I countries, assuming that all CO_2 emission reduction necessary for stabilization would occur in Annex I countries and that non-Annex I countries would emit CO_2 without

---

[15] It should be noted, however, that a few scenarios show the possibility of achieving 450ppmv stabilization even if the initial Kyoto commitments are not met, provided that emissions decline sufficiently by 2020.

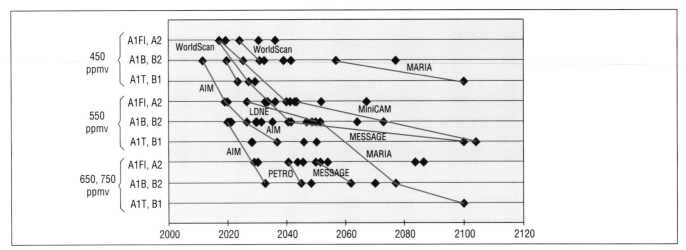

Figure 2.17: *Timing of when per capita CO_2 emissions in Annex I countries would fall below per capita CO_2 emissions in non-Annex I countries, assuming that all CO_2 emission reduction necessary for stabilization would occur in Annex I countries and that non-Annex I countries would emit CO_2 without any controls.*

any controls. This hypothetical assumption permits the analysis of one of the determinants of when non-Annex I emissions might begin to diverge from baseline levels. The results are shown in *Figure 2.17* for each stabilization level and for three groups of SRES baselines.

Figure 2.17 shows that:

*   Assuming that all the CO_2 reductions for concentration stabilization are undertaken in Annex I countries, most of the post-SRES scenarios indicate that per capita Annex I emissions would fall below per capita non-Annex I emissions in the 21st century. This situation occurs before 2050 in two-thirds of the scenarios. Only in the A1T or B1 worlds would per capita CO_2 emissions in developing countries remain below those of developed countries in the 21st century.
*   These timings are significantly affected by the time series of emission reductions in the scenarios, and consequently they diverge in the scenarios. However, comparison within individual models suggests that the lower the stabilization level, the earlier that Annex I per capita emissions fall below non-Annex I per capita emissions. Stabilization scenarios based on higher emission worlds such as A1FI and A2 also tend to show earlier timing for Annex I to fall below non-Annex I per capita emissions compared to scenarios based on the lower emission worlds of B1 or A1T. This suggests that the stabilization target and the baseline emission level are both important determinants of the timing when developing countries' emissions might need to diverge from their baseline emissions.

In order to assess priority setting in energy intensity reduction or in carbon intensity reduction, a "response index" was calculated for all stabilization variants of post-SRES scenarios for the years 2020, 2050, and 2100, as shown in *Figure 2.18*. This index relates the impact on CO_2 emission reduction of switch-

ing towards low-carbon or carbon free energy to the impact of energy intensity reduction. The response index is the ratio of the change in carbon intensity to the change in primary energy intensity[16].

When energy intensity reduction is relatively larger than carbon intensity reduction, the index shows more than 1.0, and less than 1.0 in the opposite case.

It is clear from *Figure 2.18* that the priority of response to reduce CO_2 emissions would change over time. Energy intensity reduction would be relatively larger than carbon intensity reduction in the beginning of 21st century, but these would be of equal weight by the middle of the century. The impact of energy intensity reduction would be saturated towards the end of the 21st century, and the use of low-carbon or carbon-free energy would become relatively much larger. This pattern is generally consistent across the stabilization levels. The lower the stabilization target, the higher the relative importance of energy intensity reduction in the beginning of the 21st century, and the higher the relative importance of low-carbon or carbon free energy towards the end of the 21st century.

These trends are important, but it is necessary at the same time to understand the model assumptions behind them. Most of the models do not accommodate very well structural and consumption-pattern-related efficiency measures (e.g., advanced dematerialization, major structural change, and changes in consumption patterns and lifestyles). A few cases which incorpo-

---

16

$$R = \frac{(Ci_{t,MS}/CI_{t,BS})}{(EI_{t,Ms}/EI_{t,BS})}$$

In this expression, CI denotes carbon intensity and EI energy intensity. The indices BS and MS refer to the baseline and mitigation scenario, respectively; the time is given by t.

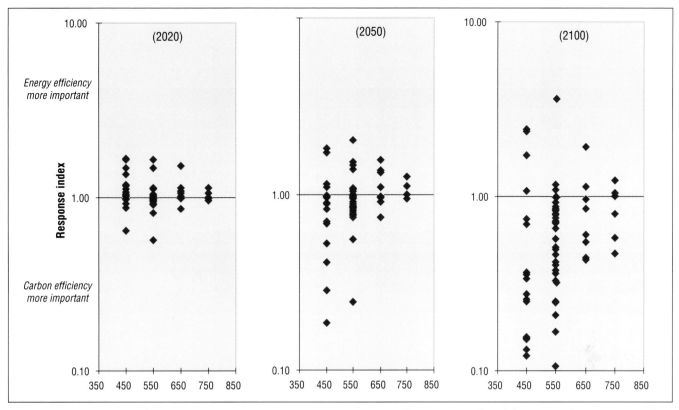

Figure 2.18: *Response index to assess priority setting in energy intensity reduction (more than 1.0) or in carbon intensity reduction (less than 1.0) for all stabilization variants of post-SRES scenarios in 2020, 2050, and 2100.*

rate drastic changes in social structure (e.g., some of the scenarios based on AIM and WorldScan) give relatively high priority to energy efficiency improvement even in the latter half of 21st century.

A per capita final energy index was calculated in order to analyze equity between North and South. Since one of the weak points of quantified scenario analysis concerns equity or "burden sharing", the comparison of this kind of index is very important. Even though the per capita income is the most popular index to analyze equity, this index was not estimated by all the modelling teams. Therefore, final energy consumption per person in each region was adopted as an appropriate index for the equity analysis, because this index is closely related to per capita economic welfare. *Figure 2.19* shows this index (in GJ/capita) among the OECD, EFSU, ASIA and ALM regions[17] for all post-SRES and SRES variants over the period 1990 to 2100.

As shown in this figure, some interesting trends can be observed:

- In the development-emphasized worlds (A1B and A2) climate policy would reduce per capita final energy in both the Annex I and non-Annex I countries, while in the environment-emphasized worlds (B1 and B2) climate policy would have little effect on energy use. These impacts would slightly improve equity in per capita final energy use between the Annex I and non-Annex I countries, because the reduction in energy use caused by climate policies would be larger in Annex I than in non-Annex I.

- However, the impact of climate policies on equity in per capita final energy use would be much smaller than that of the future development path. The differences among the various SRES baseline conditions have the largest impact upon whether per capita energy use values converge between Annex I and non-Annex I countries, with the highest degree of convergence occurring in the A1B and B1 worlds. This can be seen in *Figure 2.19* by comparing the (smaller) change in energy use between regions within each of the four columns (i.e., between the baseline and the 55ppmv stabilization scenario for each world) with the (much larger) change between regions across each of the two rows (i.e., across the baseline or across the 550ppmv stabilization scenario).

---

[17] These regional aggregations were defined by Nakicenovic *et al.* (2000). OECD: OECD member countries as of 1990; EFSU: the East and Central European countries and Former Soviet Union; ASIA: all non-Annex I countries in Asia (excluding the Middle East); ALM: Africa, Latin America, and the Middle East, and the rest of the world.

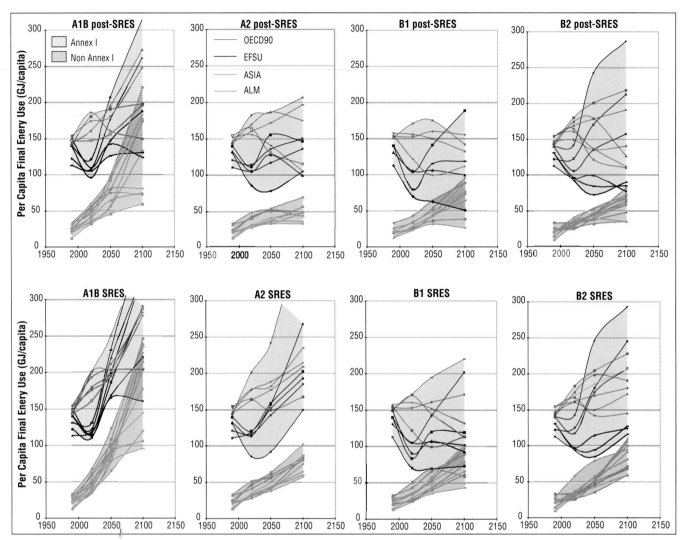

Figure 2.19: *An equity index to compare per capita final energy use (GJ/capita) between the Annex I (pink) and non-Annex I (blue) countries for all post-SRES (550ppmv stabilization) and SRES (baseline) variants from 1990 to 2100. Climate policy has a much smaller impact on equity in energy use than does choice of development path. This can be seen by comparing the change in energy use within each of the four worlds (i.e., between the baseline and the 550ppmv stabilization for each world) with the change among the worlds (i.e., across the baselines or across the 550ppmv stabilization).*

Though the analyses described above mainly focus on CO_2 emissions from energy consumption, it is also important to consider non-CO_2 emissions as well as non-energy-related CO_2 emissions. However, very few scenarios that include these emissions have been quantified and therefore it was not possible to include this additional review in this report. Some of the nine modelling approaches used here do include other radiatively active gases. However, the mitigation and/or stabilization scenarios include explicit limitations only on CO_2 emissions, and hence the reductions in other gases are indirect results (or ancillary benefits) of the CO_2 reduction measures.

2.5.2.4   Comparison of Technology and/or Policy Measures and Assessment of Robustness

Assumed technology and/or policy options differ among models (Morita *et al.*, 2000a). These differences are strongly

dependent on the model structure. MESSAGE-MACRO, LDNE, and MARIA are dynamic optimization-type models that incorporate detailed supply-side technologies; once a constraint on CO_2 emission or concentration is imposed, the optimal set of technology and/or policy measures (focusing on energy supply) is automatically selected in the model. AIM and IMAGE are recursive simulation-type models which integrate physical and land-use modules rather than focus on energy demand, so that highly detailed technology and/or policy measures are assumed for each region and time as exogenous scenarios. ASF, MiniCAM, PETRO, and WorldScan are other types of integrated models focusing on the economics of energy systems. In these models, only a carbon tax is used for the post-SRES analyses.

In order to reduce CO_2 and other GHG emissions, each modelling team assumed specific technology and/or policy measures

Table 2.7: *Sources of emissions reduction for 550ppmv stabilization across the nine post-SRES models. Minimum-maximum and (median) at 2100 (GtC)*

	A1B	**A1FI**	A1T	**A2**	**B1**	B2
Substitution among fossil fuels	-0.1 – 2.2 (0.97)	0.2 – 11.8 (1.82)	0.1 – 0.1 (0.09)	2.4 – 5.4 (2.95)	0.0 – 0.2 (0.09)	0.6 – 2.7 (1.35)
Switch to nuclear	0.3 – 6.4 (0.55)	-2.4 – 1.9 (1.20)	0.0 – 2.0 (1.03)	0.3 – 1.7 (1.18)	0.0 – 3.1 (0.02)	-0.2 – 5.1 (2.28)
Switch to biomass	-0.8 – 1.5 (1.03)	-0.2 – 5.5 (2.50)	-0.2 – 0.3 (0.07)	1.1 – 3.8 (1.84)	0.0 – 4.3 (0.04)	-1.9 – 1.5 (0.63)
Switch to other renewables	0.1 – 2.5 (1.51)	0.6 – 15.1 (2.70)	-0.1 – 0.0 (-0.05)	2.2 – 6.7 (3.33)	0.1 – 0.3 (0.28)	0.1 – 3.2 (2.07)
CO_2 scrubbing and removal	0.0 – 4.7 (0.00)	0.0 – 23.8 (0.39)	0.5 – 1.6 (1.06)	0.0 – 5.8 (0.00)	0.0 – 1.1 (0.00)	0.0 – 3.0 (0.63)
Demand reduction	0.5 – 6.6 (0.94)	1.9 – 17.7 (10.4)	0.0 – 0.2 (0.11)	5.2 – 15.6 (10.21)	0.1 – 0.3 (0.08)	0.7 – 3.5 (1.64)
TOTAL reduction	*7.1 – 11.9 (9.16)*	*21.7 – 30.5 (21.1)*	*0.3 – 4.4 (2.31)*	*21.7 – 26.9 (22.81)*	*0.2 – 9.6 (0.39)*	*6.0 – 10.6 (8.14)*

Note: Emission reductions are estimated by subtracting the mitigation value (in GtC) from the baseline value (in GtC) of each scenario.

for its scenario quantification. The main reduction measures are:

- demand reductions and/or efficiency improvements;
- substitution among fossil fuels;
- switch to nuclear energy;
- switch to biomass;
- switch to other renewables;
- CO_2 scrubbing and removal; and
- afforestation.

Table 2.7 summarizes the contribution of these emission mitigation options and/or measures for the post-SRES scenarios. The table shows the emission reduction (in GtC) between the baseline and the mitigation and/or stabilization cases, corresponding to the first six points of the list above. For simplicity, the total ranges as well as the median value in 2100 are shown only for the 550ppmv stabilization case. As shown in *Table 2.7*, no single source will be sufficient to stabilize atmospheric CO_2

concentrations. Across the scenarios, the contributions of demand reduction, substitution among fossil fuels, and switching to renewable energy are all relatively large. The contributions of nuclear energy, of CO_2 scrubbing and removal differ significantly among the models and also across the post-SRES scenarios.

With respect to the role of biofuels, it should be noted that the models assume trade in biofuels across regions; hence, biomass produced in Africa and/or South America can satisfy the fuel needs of Asia. In all mitigation scenarios, the additional role of biomass, as a mitigation option, is limited and the world supply never exceeds 400EJ/yr; this is possible because the other options (solar and/or wind, nuclear, and CO_2 removal and storage) also play a key role in mitigation strategies. *Table 2.8* shows the ranges in primary biomass use in 2050 in the post-SRES scenarios.

Table 2.8: *Ranges of primary use of biomass in 2050 in the post-SRES scenarios (EJ)*

Stabilization target	A1B	A1FI	A1T	A2	B1	B2
450ppmv	246 - 328	226 - 246	137 - 246	128	96 - 186	127 - 189
550ppmv	76 - 228	78 - 217	74 - 217	22 - 232	36 - 176	27 - 157
650ppmv	0 -180	143 -184	133	(*)		121
750ppmv	(*)	131		25 - 63		

Note: As the PETRO model does not separate biomass energy from primary energy, no number is filled in (*).

To contribute to a synthesis of findings, each modelling team was asked to respond to the following questions about the policy implications of the scenarios:

- How do technology and/or policy measures vary among different baselines for a given stabilization level?
- How does the stabilization level affect the technology and/or policy measures used in the scenarios?
- What packages of technology and/or policy measures are robust enough to be effected in the different baseline worlds?

As shown in *Table 2.7*, high emission worlds such as A1FI, A2, and A1B require a larger introduction of energy demand reduction, switching to renewable energy, and substitution among fossil fuels, in comparison to other SRES worlds. The contribution of CO_2 scrubbing and removal is largest in the A1FI stabilization scenarios, while mitigation measures in the A1T world depend mainly on a switch to nuclear power as well as carbon scrubbing and removal. Biomass energy steadily contributes across the SRES worlds and also across stabilization targets.

The following summarizes more detailed differences in technology and/or policy measures across the regions as well as the different SRES worlds:

- The timing and the pace of the emissions reduction are particularly influenced by the region's resource availability. Regions with large amounts of cheap fossil fuel reserves and resources (ASIA: coal; EFSU: natural gas) rely comparatively longer on fossil fuel-based power generation. In the long run the emissions mitigation measures are predominantly the result of the technology assumptions consistent with the scenario storylines. In the fossil-intensive A2 scenario, emissions reduction for 2100 in ASIA and EFSU are mainly a result of shifts to advanced fossil technologies in combination with carbon scrubbing and/or removal and increasing shares of solar-photovoltaic, and advanced nuclear technologies. For the B2 scenario, the shift towards non-fossil fuels in ASIA and EFSU is more complete, and hence, scrubbing plays a less important role. In A2 and B2, synthetic fuel production from biomass plays a key role in the ALM region. In both scenarios the emissions mitigation in the OECD region is because of shifts to wind, solar-photovoltaic, biomass, and nuclear technologies. In the OECD, fossil fuels contribute roughly 30% to the power generation, which comes predominantly from fuel cells (MESSAGE-MACRO team: Riahi and Roehrl, 2000);
- In the 550ppmv cases, the composition of primary energy is diversified, with increased shares of various renewable energy sources, nuclear power, and natural gas. Among the renewable energy sources, photovoltaics (PV) seem to be the most promising abatement measure in the A1 and A2 scenarios, where the final energy demands grow quite substantially, while CO_2 recovery and disposal measures play a very important

role in the B1 and B2 scenarios. In the case of A1B and A2, PV would increase rapidly especially in the Middle East and North Africa (ALM) where PV panels could be set in wide desert areas. For the entire SRES world, methanol would be made from hydrogen (H_2) and carbon monoxide (CO) through gas splitting mainly in the Former Soviet Union and Eastern Europe (EFSU) where there are plenty of natural gas resources. Wind energy production would play an important role in North America (LDNE team: Yamaji *et al.*, 2000);

- In the A1B and A1T worlds, expansion of biomass utilization is the major strategy, rather than nuclear power, for carbon emission control in OECD and EFSU. In the latter, biomass mainly substitutes for natural gas in public and other sectors, and a shift from coal to natural gas in the industry sector is also observed. Nuclear power is mainly used in the Asia-Pacific and ALM regions. In contrast, the B1 scenarios give very similar figures among regions, except for a small increase of biomass in the OECD region. Carbon sequestration is implemented in all regions for the purpose of carbon emission control. B2 scenarios are basically similar to those of the A1 family, except that nuclear energy and biomass are introduced in the OECD region (MARIA team: Mori, 2000);
- In the A1 and B1 families, technology transfer to developing countries would occur with respect to renewable energy production, unconventional oil and gas exploitation, and nuclear power generation. In these worlds, there would be a large increase in biomass use in the Asia and ALM regions. Coal is mainly produced in the Asia-Pacific region. Nuclear technology is widely used in developing regions. In the A2 and B2 worlds, energy supply and use heavily depend on local energy resources because of international trade barriers. The Asia-Pacific region will rely on nuclear energy and coal, while ALM may use much renewable energy. The OECD region makes much use of advanced end-use technology and modern renewable energy technologies. Large gas resources in the EFSU region can satisfy much of the energy demand in that region (AIM team: Jiang *et al.*, 2000);
- The allocation of carbon "taxes" across regions based on their per capita GDP levels leads to substantial differences in levels of CO2 reductions relative to the baseline. The largest relative reductions are implemented in regions with relatively high per capita GDP growth (e.g., OECD) and regions with a relatively low cost of renewable energy (Latin America). The lowest relative reductions are achieved in regions with low per capita GDP and a relatively high cost of renewable energy (e.g., Africa) (ASF team: Sankovski *et al.*, 2000); and
- Assuming that there are no constraints on fuel trade, the Middle East and later the Commonwealth of Independent States (CIS) will still be major fossil fuel exporters; their revenues may decline significantly by

the middle of the 21st century as a consequence of carbon mitigation measures. Parts of Africa and South America may develop into important biofuel exporters. High-income regions with limited fossil fuel resources, such as Europe and the USA, will probably be among the first to introduce high-efficiency and non-carbon technologies. This results over time in sizeable cost reductions, enabling less industrialized regions to replace their indigenous coal use by these relatively capital-intensive supply side options.

One of the major results of the post-SRES analyses is the identification of "robust climate policy options" across the different SRES worlds as well as across different stabilization targets. Most of the modelling teams have identified several such options based on their simulations. The following list summarizes the major findings:

- Robust policies include technological efficiency improvements for both energy use technology and energy supply technology, social efficiency improvements such as public transport introduction, dematerialization promoted by lifestyle changes and the introduction of recycling systems, and renewable energy incentives through the introduction of energy price incentives such as a carbon tax (AIM, IMAGE, MARIA, MiniCAM (Pitcher, 2000), PETRO (Kverndokk *et al.*, 2000), and WorldScan teams);
- It would be reasonable to start with energy conservation and reforestation to cope with global warming. However, innovative supply-side technologies will eventually be required to achieve stabilization of atmospheric CO_2 concentration (AIM, ASF, IMAGE, and LDNE teams);
- Robust options across the SRES worlds are natural gas and the use of biomass resources. Innovative transitional strategies of using natural gas as a "bridge" towards a carbon-free hydrogen economy (including CO_2 sequestration) are at a premium in a possible future world with low emissions (MESSAGE-MACRO, AIM, MARIA, and MiniCAM teams);
- In all mitigation scenarios, gas combined-cycle technology bridges the transition to more advanced fossil (fuel) and zero-carbon technologies. The future electricity sector is not dominated by any single dominant technology, however, hydrogen fuel cells are assumed to be the most promising technology among all stabilization cases (MESSAGE-MACRO, IMAGE, and MiniCAM teams);
- Climate stabilization requires the introduction of natural gas and biomass energy in the first half of the 21$_{ST}$ century, and either nuclear energy or carbon removal and storage in the latter half of the century as the cost effective pathways. Carbon removal and storage has a role to play in high emission worlds such as A1FI and A1B for the serious or moderate targets (LDNE, MiniCAM, and MARIA teams);
- Even in the B1 world there are very difficult decisions

to be made and these may well imply the need to significantly further redirect the energy system (MiniCAM and WorldScan teams); and
- Energy systems would still be dependent on fossil fuels at more than 20% of total primary energy over the next century, even with the stabilization of CO_2 concentration (LDNE and WorldScan teams).

The post-SRES analyses supplied several other findings from individual model simulations. The AIM and the MESSAGE-MACRO teams as well as other teams found that technological progress plays a very important role in stabilization, and that knowledge transfer to developing countries is a key issue in facilitating their participation in early CO_2 emission reduction. With respect to policy integration, the AIM team found that integration between climate policies and domestic policies could effectively reduce GHGs in developing regions from their baselines, especially for the next two or three decades. On the other hand, the MESSAGE-MACRO team estimated that regional air pollution control with respect to sulphur emissions tends to: (1) amplify global climate change in the medium-term perspective, and (2) accelerate the shift towards less carbon (and sulphur) intensive fuels such as renewables. The MiniCAM team concluded that agriculture and land use and energy system controls need to be linked, and that failure to do this can lead to much larger than necessary costs.

The above results are found with robust technology and/or policy measures across the SRES worlds and across different stabilization targets, and many of them are common among different modelling teams. A part of these common results can be tested by more detailed analyses of emission reduction sources, shown in *Table 2.7*. This table as well as time series analyses of the contribution of sources clearly show that:

- Large and continuous energy efficiency improvements are common features of mitigation scenarios in all the different SRES worlds;
- Introduction of low-carbon energy is also a common feature of all scenarios, especially biomass energy introduction over the next one hundred years and natural gas introduction in the first half of the 21st century;
- Solar energy and other renewable energy sources could play an important role in climate stabilization in the latter half of the 21st century, especially for higher emission baselines or lower stabilization levels; and
- Mitigation scenarios with reduced fossil fuel use will further decrease regional sulphur emissions and hence open up the possibility of earlier and larger climate change effects.

2.5.2.5 *Summary of Post-SRES Scenario Review*

A new type of policy assessment has been conducted by the post-SRES activities, with nine modelling teams quantifying various simulation cases. Even though stabilization scenarios show a range among the models, several common trends and characteristics can be observed.

The different SRES baseline worlds require different technology and/or policy measures to stabilize at the same level. The A1F1, A1B, and A2 worlds require a wider range of stronger technology and/or policy measures than A1T, B1, and B2. For example, energy efficiency improvements in all sectors, the introduction of low-carbon energy, and afforestation would all be required in the A1F1, A1B, and A2 worlds in the first half of the 21st century, with the additional introduction of advanced technologies in renewable energy and other energy sources in the second half of the 21st century. The level of technology and/or policy measures in the beginning of this century would be significantly affected by the choice of development path over the next one hundred years. Higher emission worlds such as A1F1 and A2 require earlier reduction than low emission worlds such as A1T and B1.

The stabilization level chosen also significantly affects technology and/or policy measures and the timing of their introduction. More stringent stabilization targets require earlier emission reductions from baseline levels. The post-SRES scenario analysis suggests that stabilization at 450ppmv will require emissions reductions in Annex I countries that go significantly beyond the Kyoto Protocol commitments. It also suggests that maintaining emissions at the level of the Kyoto commitments may be adequate for achieving stabilization at 550ppmv or higher, although it should be recognized that several scenarios do indicate the need for significant emission reductions by 2020 in order to achieve these stabilization levels.

With respect to the important policy question of the role of developing countries in GHG emission mitigation, a preliminary finding of the post-SRES scenario analysis is that, assuming that the CO_2 emission reduction needed for stabilization occurs in Annex I countries only, per capita CO_2 emissions in Annex I countries would fall below per capita emissions in non-Annex I countries during the 21st century except in some of A1T and B1 stabilization scenarios, and this occurs before 2050 in two-thirds of the scenarios. This suggests that, especially for more stringent stabilization targets and/or worlds with relatively high baseline emissions, there is a need for emissions to diverge from baseline levels in developing countries. The stabilization target and the baseline emission level were both important determinants of the timing when developing countries emissions might need to diverge from their baseline emissions.

No single measure will be sufficient for the timely development, adoption, and diffusion of mitigation options to stabilize atmospheric GHGs. Rather, a portfolio based on technological change, economic incentives, and institutional frameworks might be adopted. Large and continuous energy efficiency improvements and afforestation are common features of mitigation scenarios in all the different SRES worlds. Introduction of low-carbon energy is also a common feature of all scenarios, especially biomass energy introduction over the next one hundred years, as well as natural gas introduction in the first half of the 21st century. Reductions in the carbon intensity of ener-

gy have a greater mitigation potential than reductions in the energy intensity of GDP in the latter half of the 21st century, while energy intensity reduction is greater than carbon intensity reduction in the beginning of the century. This result appears to be robust across the storylines and stabilization levels, if drastic social changes are not assumed for energy efficiency improvement. In an A1B or A2 world, either nuclear power or carbon sequestration would become increasingly important for GHG concentration stabilization, the more so if stabilization targets are lower. Solar energy could play an important role in climate stabilization in the latter half of the 21st century, especially for a higher emission baseline or lower stabilization levels.

Robust policy and/or technological options include technological efficiency improvements for energy supply and use, social efficiency improvements, renewable energy incentives, and the introduction of energy price incentives such as a carbon tax. Energy conservation and reforestation are reasonable first steps, but innovative supply-side technologies will eventually be required to achieve stabilization of atmospheric CO_2 concentration. Possibilities include using natural gas and combined-cycle technology to bridge the transition to more advanced fossil (fuel) and zero-carbon technologies such as hydrogen fuel cells. However, even with emissions control, some modellers found that energy systems would still be dependent on fossil fuels over the next century.

Integration between global climate policies and domestic air pollution abatement policies could effectively reduce GHG emissions in developing regions for the next two or three decades. However, control of sulphur emissions could amplify possible climate change, and partial trade-offs are likely to persist for environmental policies in the medium term.

Policies governing agriculture and land use and energy systems need to be linked for climate change mitigation. Failure to do this can lead to much larger than necessary costs. At tight levels of control, even some ability to acquire additional emissions capacity from land sequestration can have major cost-reducing impacts. Moreover, a high potential supply of biomass energy would ameliorate the burden of carbon emission reductions.

2.6   Recommendations for Future Research

- Rigorous techno-economic analysis of multiple mitigation measures for each baseline and mitigation target;
- More explicit analysis of policy instruments leading to mitigation;
- Inclusion of other GHGs in addition to CO_2;
- Analysis of the feasibility and costs of stabilizing atmospheric concentrations at levels other than 550ppmv CO_2;
- Explicit cost-benefit analysis of the impacts of timing and burden sharing on mitigation costs and targets;

- Quantitative analysis of linkages between DES targets (e.g., international equity) and climate change mitigation costs and benefits;
- More extensive attempts to link qualitative narrative-based scenarios analysis with quantitative modelling work; and
- Capacity building for scenario analyses in developing countries.

References

Alcamo, J. (ed., co-author), 1994: *IMAGE 2.0: Integrated Modeling of Global Climate Change.* Kluwer Academic Press, Dordrecht/Boston, 314 pp.

Alcamo, J., and G.J.J. Kreileman, 1996: Emission Scenarios and Global Climate Protection. *Global Environmental Change*, **6**(4), 305-334.

Alcamo, J., and N. Nakicenovic (eds.), 1998: Long-Term Greenhouse Gas Emissions Scenarios and Their Driving Froces. *Mitigation and Adaptation Strategies for Clobal Change*, **3**(2-3), 95-466.

Alcamo, J., and N. Nakicenovic, 2000: Mitigation and Adaptation Strategies for Global Change. *Technological Forecasting and Social Change*, **63**(2&3).

Alcamo, J., J. A. Bouwman, J. Edmonds, A. Grubler, T. Morita, and A. Sugandhy, 1995: An Evaluation of the IPCC IS92 Emission Scenarios. In *Climate Change 1994, Radiative Forcing of Climate Change and An Evaluation of the IPCC IS92 Emission Scenarios*. Cambridge University Press, Cambridge.

Alcamo, J., G.J.J. Kreileman, and R. Leemans (eds., co-authors), 1998: *Global Change Scenarios of the 21st Century.* Pergamon/Elsevier Science, Oxford, 296 pp.

Barney, G.O., 1993: Global 2000 Revisited: *What Shall We Do? The Critical Issues of the 21st Century.* Millennium Institute, Arlington, VA.

Bollen, J.C., A.M.C. Toet, and H.J.M. de Vries 1996: Evaluating Cost-Effective Strategies for Meeting Regional CO_2 Targets. *Global Environmental Change — Human and Policy Dimensions*, **6**(4), 359-373.

Bollen, J., T. Manders, and H. Timmer, 2000: The Benefits and Costs of Waiting - Early action versus delayed response in the post-SRES stabilization scenarios. *Environmental Economics and Policy Studies*, **3**(2).

Bossel, H. 1998: *Earth at a Crossroads: Paths to a Sustainable Future.* Cambridge University Press, Cambridge.

Burrows, B., A. Mayne, and P. Newbury, 1991: *Into the 21st Century: A Handbook for a Sustainable Future.* Adamantine, Twickenham.

CPB, 1992: *Scanning the Future: A Long-Term Scenario Study of the World Economy 1990-2015.* SDU Publishers, The Hague.

CPB, 1999: *WorldScan — the Core version.* Bureau for Economic Policy Analysis (CPB), The Hague, **December**.

Coates, J.F., 1991: Factors Shaping and Shaped by the Environment: 1990-2010. *Futures Research Quarterly*, **7**(3), 5-55.

Coates, J.F., 1997: Long-term technological trends and their implications for management. *International Journal of Technology Management*, **14**(6-8), 579-595.

Coates, J.F. and J. Jarratt, 1990: What Futurists Believe: Agreements and Disagreements. *The Futurist*, **XXIV**(6).

Coates, J., J. Mahaffie, and A. Hines, 1997: *2025: Scenarios of US and Global Society Reshaped by Science and Technology.* Oakhill Press, Greensboro, NC.

Cornish, E., 1996: 92 Ways Our Lives Will Change by the Year 2025. *The Futurist*, **30**(1).

Costanza, R., 1999: Four Visions of the Century Ahead: Will It Be Star Trek, Ecotopia, Big Government or Mad Max? *The Futurist*, **33**(2), 23-29.

De Vries, H.J.M., M. Janssen, and A. Beusen, 1999: Perspectives on Global Energy Futures - Simulations with the TIME model. *Energy Policy*, **27**, 477-494.

Duchin, F., G.-M. Lange, K. Thonstad, and A. Idenburg, 1994: *The Future of the Environment: Ecological Economics and Technological Change.* Oxford University Press, New York, NY.

Edmonds, J., M. Wise, H. Pitcher, T. Wigley, and C.N. MacCracken, 1996: *An Integrated Assessment of Climate Change And The Accelerated Introduction of Advanced Energy Technologies.* Pacific Northwest National Laboratory, Washington, DC.

Edmonds, J., S.H. Kim, C.N. MacCracken, R.D. Sands, and M. Wise, 1997: *Return to 1990: The Cost of Mitigating United States Carbon Emissions in the Post-2000 Period.* Report No. PNNL-11819, Pacific Northwest National Laboratory, Washington, DC.

Enquete Commission, 1995: *Mehr Zukunft für die Erde.* Final Report of the Enquete Commission on "Protecting the Earth's Atmosphere" of the 12th Session of the German Bundestag. Economica Verlag, Bonn.

Enting, I.G., T.M.L. Wigley, and M. Heimann, 1994: *Future emissions and concentrations of carbon dioxide: Key ocean/atmosphere/land analyses.* Technical Paper No.31, CSIRO Division of Atmospheric Research, 120pp.

EPA (Environmental Protection Agency), 1990: *Policy Options for Stabilizing Global Climate.* EPA, Washington, DC.

Fujii, Y., and K. Yamaji, 1998: Assessment of Technological Options in the Global Energy System for Limiting the Atmospheric CO_2 Concentration. *Environmental Economics and Policy Studies*, **1**(2), 113-139.

Gallopin, G., A. Hammond, P. Raskin, and R. Swart, 1997: *Branch Points: Global Scenarios and Human Choice.* Polestar Series, Report no. 7, Stockholm Environment Institute, Boston, MA.

GBN, 1996: *Twenty-First Century Organizations: Four Plausible Prospects.* GBN, Emeryville, CA (as quoted in Ringland, 1998).

Glenn, J.C., and T.J. Gordon, 1997: *1997 State of the Future: Implications for Action Today.* American Council for the United Nations University, Washington, DC.

Glenn, J.C., and T. J. Gordon, 1998: *1998 State of the Future: Issues and Opportunities.* American Council for the United Nations University, Washington, DC.

Ha-Duong, M., M.J. Grubb, and J.-C. Hourcade, 1997: The Influence of Socioeconomic Inertia on Optimal CO_2 Abatement. *Nature*, **390**(20 November).

Hayhoe, K., A. Jain, H. Pitcher, C. MacCracken, M. Gibbs, D. Wuebbles, R. Harvey, and D. Kruger, 1999: *Science*, **286**, 905-906.

Henderson, H., 1997: Looking Back from the 21st Century. *Futures Research Quarterly*, **13**(3), 83-98.

Herrera, A., H. Scolnik, G. Chichilnisky, G. Gallopin, J. Hardoy, D. Mosovich, E. Oteiza, G. de Romero Brest, C. Suarez and L. Talavera, 1976: *Catastrophe or New Society? A Latin American World Model.* International Development Research Centre, Ottawa, Canada.

Houghton, J.T., L.G. Meira Filho, B.A. Callander, N. Harris, A. Kattenberg and K. Maskell, (eds.), 1996: *Climate Change, The Science of Climate Change.* Contribution of Working Group 1 to the Second Assessment Report of the Intergovernmental Panel on Climate Change, Cambridge University Press, Cambridge.

Hughes, B.B., 1997: Rough Road Ahead: Global Transformations in the 21st Century. *Futures Research Quarterly*, **13**(2), 83-107.

IDEA (Innovators of Digital Economy Alternatives) Team, 1996: *Creating the Future: Scenarios for the Digital Economy.* Simon Fraser University, Vancouver.

IPCC (Intergovernmental Panel on Climate Change), 1990: *Report of the Expert Group on Emission Scenarios.* World Meteorological Organization and United Nations Environment Programme, New York

IPCC, 1992: *Climate Change 1992: The Supplementary Report to the IPCC Scientific Assessment.* J.T. Houghton, B.A. Callander, S.K. Varney (eds.), WMO and IPCC, Cambridge University Press, Cambridge.

IPCC, 1995: *Climate Change 1995: Impacts, Adaptations and Mitigation of Climate Change: Scientific-Technical Analyses.* Contribution of Working Group II to the Second Assessment Report of the Intergovernmental Panel on Climate Change, R.T. Watson, M.C. Zinyowera, R.H. Moss (eds.), Cambridge University Press, Cambridge.

IPCC, 2001: *Climate Change 2001: Impacts and Adaptation.* O. Canziani, J. McCarthy, N. Leary, D. Dokken, K. White (eds.), Cambridge University Press, Cambridge, MA and New York, NY.

Jiang, K., T. Morita, T. Masui, and Y. Matsuoka, 2000: Global Long-term GHG Mitigation Emission Scenarios based on AIM. *Environmental Economics and Policy Studies*, **3**(2).

Joos, F., M. Bruno, R. Fink, T.F. Stocker, U. Siegenthaler, C. Le Quéré, and J. L. Sarmiento., 1996: An efficient and accurate representation of complex oceanic and biospheric models of anthropogenic carbon uptake. *Tellus*, **48B**, 389-417.

Kahane, A. 1992: Scenarios for Energy: Sustainable World vs. Global Mercantilism. *Long Range Planning*, **25**(4), 38-46.

Kainuma, M., Y. Matsuoka, and T. Morita, 1998: Analysis of Post-Kyoto Scenarios: The AIM Model. In *Economic Modeling of Climate Change: OECD Workshop Report*, Organisation for Economic Co-operation and Development, Paris.

Kainuma, M., Y. Matsuoka, T. Morita, and G. Hibino, 1999a: Development of an End-Use Model for Analyzing Policy Options to Reduce Greenhouse Gas Emissions. *IEEE Transactions on Systems, Man and Cybernetics, Part C: Applications and Reviews*, 29(3), 317-324.

Kainuma, M., Y. Matsuoka, and T. Morita, 1999b: Analysis of Post-Kyoto Scenarios: the Asian-Pacific Integrated Model. In *The Costs of the Kyoto Protocol: A Multi-Model Evaluation, Special Issue of The Energy Journal*. J.P. Weyant (ed.), pp 207-220.

Kaplan, R.D, 1994: The coming anarchy. *The Atlantic Monthly*, 273(2), 44-76.

Kaya, Y., 1990: Impact of carbon dioxide emission control on GNP growth: Interpretation of proposed scenarios. Paper presented to the IPCC Energy and Industry Subgroup, Response Strategies Working Group, Paris (photocopy).

Kinsman, F., 1990: *Millennium: Towards Tomorrow's Society*. W H Allen, London.

Kverndokk, S., L. Lindholt, and K.E. Rosendahl, 2000: Stabilisation of CO_2 concentrations: Mitigation scenarios using the Petro model. *Environmental Economics and Policy Studies*, 3(2), 195-224.

Leggett, J., W. Pepper, and R. Swart, 1992: Emissions Scenarios for IPCC: An Update. In *Climate Change 1992. The Supplementary Report to the IPCC Scientific Assessment*, J. Houghton. B. Callander, and S. Varney (eds.), Cambridge University Press, Cambridge.

Lehtilä, A., S. Tuhkanen, and I. Savolainen, 1999. Cost-effective Reduction of CO_2, CH_4 and N_2O. Emissions in Finland. In Riemer, P., B. Eliasson, A. Wokaun (eds.), *Greenhouse Gas Control Technologies*, Elsevier Science, pp. 1129-1131.

Linden, E. 1998: *The Future in Plain Sight: Nine Clues to the Coming Instability*. Simon & Schuster, New York, NY.

Makridakis, S., 1995: The Forthcoming Information Revolution: Its Impact on Society and Firms. *Technological Forecasting and Social Change*, 27(8), 799-821.

Manne, A., and R. Richels, 1997: On Stabilizing CO_2 Concentrations — Cost-Effective Emission Reduction Strategies. *Environmental Modeling and Assessment*, 2, 251-265 .

Manne, A., R. Mendelsohn, and R. Richels, 1995: MERGE: A Model For Evaluating Regional and Global Effects of GHG Reduction Policies. *Energy Policy*, 23(1), 17-34.

Matsuoka, Y., 2000: Development of a Stabilization Scenario Generator for Long-term Climatic Assessment. *Environmental Economics and Policy Studies*, 3 (2).

Matsuoka, Y., M. Kainuma, and T. Morita, 1995: Scenario Analysis of Global Warming Using the Asian Pacific Integrated Model (AIM). *Energy Policy*, 23(4-5), 357-371.

Matsuoka, Y., T. Morita, Y. Kawashima, K. Takahashi, and K. Shimada, 1996: An Estimation of a Negotiable Safe Emissions Corridor Based on AIM Model (unpublished).

McKibbin, W.J., 1998: Greenhouse Abatement Policy: Insights From the G-Cubed Multi-Country Model. *Australian Journal of Agricultural and Resource Economics*, 42(1), 99-113.

McRae, H., 1994: *The World in 2020: Power, Culture and Prosperity*. HarperCollins Publishers, London.

Meadows, D.H., D.L. Meadows, J. Randers, and W. W. Behrens III, 1972: *The Limits to Growth*. Earth Island Press, London.

Meadows, D.H., D.L. Meadows, and J. Randers, 1992: *Beyond the Limits*. Chelsea Green Publishing Company, Post Mills, VT.

Mercer, D., 1998: *Future Revolutions: A Comprehensive Guide to the Third Millennium*. Orion Business Books, London.

Mesarovic, M., and E. Pestel, 1974: *Mankind at a Turning Point*. Dutton, New York, NY.

Messner, S., A. Golodnikov, and A. Gritsevskii, 1996: *A Stochastic Version of the Dynamic Linear Programming Model MESSAGE III*. RR-97-002, International Institute for Applied Systems Analysis, Laxenburg, Austria.

Milbrath, L.W. 1989: *Envisioning a Sustainable Society: Learning Our Way Out*. SUNY Press, Albany, NY.

Millennium Project, American Council for the United Nations University (online: www.geocities.com/CapitolHill/Senate/4787/millennium/scenarios/explor-s.html)

Mori, S. 2000: Effects of Carbon Emission Mitigation Options under Carbon Concentration Stabilization Scenarios. *Environmental Economics and Policy Studies*, 3(2).

Mori, S., and M. Takahashi, 1998: An Integrated Assessment Model for New Energy Technologies and Food Production — An Extension of the MARIA Model. *International Journal of Global Energy Issues*, 11(1-4), 1-17.

Morita, T., and H.-C. Lee, 1998a: *IPCC SRES database, Version 1.0, Emissions Scenario Database prepared for IPCC Special Report on Emission Scenarios* (online: http:/ www-cger.nies.go.jp/cger-e/db/ipcc.html).

Morita, T., and H. Lee, 1998b: IPCC Emission Scenarios Database. *Mitigation and Adaptation Strategies for Global Change*, 3(2-4), 121-131.

Morita, T., Y. Matsuoka, M. Kainuma, and H. Harasawa, 1994: AIM - Asian Pacific integrated model for evaluating policy options to reduce GHG emissions and global warming impacts. In *Global Warming Issues in Asia*, S. Bhattacharya *et al.* (eds.), AIT, Bangkok, pp. 254-273

Morita, T., N. Nakicenovic and J. Robinson, 2000a: Overview of Mitigation Scenarios for Global Climate Stabilization based on New IPCC Emission Scenarios (SRES). *Environmental Economics and Policy Studies*, 3(2).

Morita, T., N. Nakicenovic, and J. Robinson, 2000b: *The Relationship between Technological Development Paths and the Stabilization of Atmospheric Greenhouse Gas Concentrations in Global Emissions Scenarios*. CGER Research Report (CGER-I044-2000), Center of Global Environmental Research, National Institute for Environmental Studies.

Munasinghe, M., 1999: *Development, Equity and Sustainability (DES) in the Context of Climate Change*. In Proceedings of the IPCC Expert Meeting in Colombo, Sri Lanka, M. Munasinghe, R. Swart, (eds.), 27-29 April 1999, World Bank, Washington, DC, pp. 13-66.

Nakicenovic, N. (ed.), 2000: Global Greenhouse Gas Emissions Scenarios: Five Modeling Approaches. *Technological Forecasting and Social Change*, 63(1-2), 105-371.

Nakicenovic, N., A. Grubler, A. Inaba, S. Messner, S. Nilsson, Y. Nishimura, H-H. Rogner, A. Schafer, L. Schrattenholzer, M. Stubegger, J. Swisher, D. Victor , and D. Wilson, 1993: Long-Term Strategies For Mitigating Global Warming. *Energy*, 18(5), 401-609.

Nakicenovic, N., A. Grubler, and A. McDonald, 1998: *Global Energy Perspectives*. Cambridge University Press, Cambridge.

Nakicenovic, N., J. Alcamo, G. Davis, H.J.M. de Vries, J. Fenhann, S. Gaffin, K. Gregory, A. Grubler, T.Y. Jung, T. Kram, E.L. La Rovere, L. Michaelis, S. Mori, T. Morita, W. Papper, H. Pitcher, L. Price, K. Riahi, A. Roehrl, H-H. Rogner, A. Sankovski, M. Schlesinger, P. Shukla, S. Smith, R. Swart, S. van Rooijen, N. Victor, and Z. Dadi, 2000: *Special Report on Emissions Scenarios*. Intergovernmental Panel on Climate Change, Cambridge University Press, Cambridge.

Nordhaus, W.D., 1994: *Managing the Global Commons: The Economics of the Greenhouse Effect*. MIT Press, Cambridge, MA.

Nordhaus, W.D., and Z.L. Yang, 1996: A Regional Dynamic General-Equilibrium Model of Alternative Climate-Change Strategies. *American Economic Review*, 86(4), 741-765.

Olson, R.L., 1994: Alternative Images of a Sustainable Future. *Futures*, 26(2), 156-169.

OECD (Organisation for Economic Development and Cooperation), 1997: *The World in 2020: Towards a New Global Age*. OECD, Paris.

Parikh, J., 1992: IPCC Response Strategies Unfair to the South. *Nature*, **360** (10 December), 507-508.

Peck, S.C., and T. J. Tiesberg, 1995: International CO_2 Emissions Control: An Analysis Using CETA. *Energy Policy*, 23(4-5), 297-208.

Pepper, W., J. Legett, R. Swart, J. Wasson, J. Edmonds, and I. Mintzer, 1992: Emissions Scenarios for the IPCC. An Update: Asssumptions, Methodology, and Results. In *Climate Change 1992. The Supplementary Report to the IPCC Scientific Assessment*, J. Houghton. B. Callander, S. Varney (eds.), Cambridge University Press, Cambridge.

Pitcher, H.M. 2000: An Assessment of Mitigation Options in a Sustainable Development World. *Environmental Economics and Policy Studies*, 3(2).

Price, D., 1995: Energy and Human Evolution. *Population and Environment*, **16**(4), 301-319.

Ramphal, S., 1992: *Our Country, The Planet: Forging a Partnership for Survival*. Island Press, Washington, DC.

Rana, A., and T. Morita, 2000: Scenarios for Greenhouse Gas Emissions Mitigation: A Review of Modeling of Strategies and Policies in Integrated Assessment Models. *Environmental Economics and Policy Studies*, 3(2).

Raskin, P., G. Gallopin, P. Gutman, A. Hammond, and R. Swart, 1998: *Bending the Curve: Toward Global Sustainability*. Stockholm Environment Institute, Stockholm.

Reilly, J., R. Prinn, J. Harmisch, J. Fitzmaurice, H. Jacoby, D. Kicklighter, J. Melillo, P. Stone, A. Sokolov, and C. Wang, Multiple Gas Assessment of the Kyoto Protocol, 1999: *Nature*, **401**, 549-555.

Repetto, R., 1985: *The Global Possible*. Yale University Press, New Haven, CT.

Riahi, K., and R.A. Roehrl, 2000: Robust Energy Technology Strategies for the 21st Century - Carbon dioxide mitigation and sustainable development. *Environmental Economics and Policy Studies*, 3(2).

Ringland, G., 1998: *Scenario Planning: Managing for the Future*. Wiley, New York.

Rose, A., and B. Stevens, 1993: The Efficiency and Equity of Marketable Permits for CO_2 Emissions. *Resource and Energy Economics*, **15**(1), 117-146.

Rotmans, J., and H.J.M. de Vries (eds.), 1997: *Perspectives on Global Change: The TARGETS Approach*. Cambridge University Press, Cambridge.

RSWG, 1990: *Emissions Scenarios*. Appendix of the Expert Group on Emissions Scenarios (Task A: under the RSWG Steering Committee), US Environmental Protection Agency, Washington, DC.

Sankovski, A., W. Barbour, and W. Pepper, 2000: Climate Change Mitigation in Regionalized World. *Environmental Economics and Policy Studies*, 3(2).

Schindler, C., and G. Lapid, 1989: *The Great Turning: Personal Peace, Global Victory*. Bear & Company Publishing, Santa Fe, NM.

Schwartz, P., 1991: *The Art of the Long View*. Doubleday, New York.

Schwartz, P., 1995: The New World Disorder. *Wired*, **3**(11), 104-107.

Schwartz, P., and P. Leyden, 1997: The long boom: A history of the future 1980-2020. *Wired*, **5**(7), 115.

Science Advisory Board (SAB), 1995: *Beyond the Horizon: Using Foresight to Protect the Environmental Future*. Report No. EPA-SAB-EC-95-007/007A, Science Advisory Board, US EPA, Washington, DC.

Shinn, R.L., 1982: *Forced Options: Social Decisions for the 21st Century*. Harper & Row Publishers, San Francisco, CA.

Smith, S.J., T.M.L. Wigley, N. Nakicenovic, and S.C.B. Raper, 2000: Climate implications of greenhouse gas emissions scenarios. *Technological Forecasting & Social Change*, 65(3).

Stokke, P.R., T.A. Boyce, W.K. Ralston, and I.H. Wilson, 1991: Visioning (and Preparing for) the Future: The Introduction of Scenarios-Based Planning into Statoil. *Technological Forecasting and Social Change*, **40**(2), 131-150.

Sunter, C., 1992: *The New Century: Quest for the High Road*. Human and Rousseau (Pty) Ltd./Tafelberg Publishers Ltd, Cape Town.

Svedin, U., and B. Aniansson, 1987: *Surprising Futures: Notes from an International Workshop on Long-Term World Development*. Swedish Council for Planning and Coordination of Research, Friibergh Manor, Sweden.

Toffler, A., 1980: *The Third Wave*. Morrow, New York, NY.

Tol, R.S.J., 1997: On the Optimal Control of Carbon Dioxide Emissions: an Application of FUND. *Environmental Modelling and Assessment*, **2**, 151-163.

Tol, R.S.J., 1999: Spatial and Temporal Efficiency in Climate Policy: Application of FUND. *Environmental and Resource Economics*, **14** (1), 33-49.

Toth, F.L., E. Hizsnyik, and W. Clark, (eds.) 1989: *Scenarios of Socioeconomic Development for Studies of Global Environmental Change: A Critical Review*. International Institute for Applied Systems Analysis, Laxenburg, Austria.

Tuhkanen, S., A. Lehtilä, and I. Savolainen, 1999: The Role of CH_4 and N_2O Emission Reductions in the Cost-Effective Control of the Greenhouse Gas Emissions from Finland. *Mitigation and Adaptation Strategies for Global Change*, **4**, 91-111.

Tulpule, V., S. Brown, J. Lim, C. Polidane, H. Pant, and B.S. Fisher, 1998: An Economic Assessment of the Kyoto Protocol using the Global Trade and Environment Model. In *Economic Modeling of Climate Change*. OECD Workshop Report, Organisation for Economic Co-operation and Development, Paris.

Van den Bergh, M., 1996: Charting a course — preparing for the oil and gas business of the 21st century. Speech presented at the State University of Groningen (as quoted in Ringland, 1998).

Wallerstein, I., 1989: The Capitalist World Economy — Middle-Run Prospects. *Alternatives: Social Transformation and Humane Governance*, **14**(3), 279-288.

WBGU (German Advisory Council on Global Change), 1995: *World in Transitions: Way Towards Global Environmental Solutions*. Annual Report, Springer.

Weyant, J. (ed.), 1999: The Costs of the Kyoto Protocol: a Multi-Model Evaluation, *The Energy Journal*, (Special Issue).

Weyant, J.P., and H. Hill, 1999: Introduction and Overview. *The Energy Journal* (Special issue on the Costs of the Kyoto Protocol: a Multi-Model Evaluation) **vii-xliv**.

Wigley, T.M.L., 1999: The Science of Climate Change. *Global and U.S. Perspectives*, PEW Center on Global Climate Change, Arlington, VA, 50 pp.

Wigley, T.M.L., and S.C.B. Raper, 1992: Implications for Climate and Sea Level of revised IPCC Emission Scenarios. *Nature*, **357**, 293-300.

Wigley, T.M.L., Solomon, M. and Raper, S.C.B., 1994: Model for the Assessment of Greenhouse-gas Induced Climate Change Version 1, 2. Climate Research Unit, University of East Anglia, UK

Wigley, T.M.L, R. Richels, and J.A. Edmonds, 1996: Economic and Environmental Choices in the stabilization of Atmospheric CO_2 Concentrations. *Nature*, **379**, 240-243.

Wilkinson, L., 1995: How To Build Scenarios. *Wired* (Special Edition - Scenarios: 1.01), **September**, 74-81.

World Bank, 1995: *World Development Report 1995 — Workers in an Integrating World*. World Bank, Washington, DC.

WBCSD (World Business Council for Sustainable Development), 1997: *Exploring Sustainable Development: WBSCD Global Scenarios 2000-2050 Summary Brochure*. World Business Council for Sustainable Development, London.

WBCSD, 1998: *Exploring Sustainability 2000-2050*. WBCSD, Geneva.

WCED (World Commission on Environment and Sustainable Development), 1987: *Our Common Future*. Oxford University Press, Oxford.

WEC (World Energy Council), 1995: *Global Energy Perspectives to 2050 and Beyond*. WEC and IIASA, Austria.

WRI (World Resources Institute), 1991: *The Transition to a Sustainable Society*. World Resources Institute, Washington, DC.

Yamaji, K., J. Fujino, and K. Osada, 2000: Global Energy System to Keep the Atmospheric CO_2 Concentration at 550 ppm. *Environmental Economics and Policy Studies*, 3(2).

Yohe, G., and R. Wallace, 1996: Near Term Mitigation Policy for Global Change Under Uncertainty: Minimizing the Expected Cost of Meeting Unknown Concentration Thresholds. *Environmental Modeling and Assessment*, **1**, 47-57.

Chapter 2 Appendix

Appendix 2.1: Details of scenarios from IPCC-SRES database in legends of *Figures 2.2, 2.3, 2.4, 2.6,* and *2.7*

Legend Key	Baseline scenario name	Stabilzation scenario name	Legend key name	Baseline scenario name	Stabilzation scenario name
AIM (1)	Standard Ref	Stblz ppm/STD	PEF (25)	Modeler's Ref	Stblz ppm/MOD
AIM96 (2)	Standard Scenario	Scenario_3	PEF (26)	Standard Ref	Stblz ppm/STD
CETA (3)	Modeler's Ref	Stblz ppm/MOD	RICE (27)	Modeler's Ref	Stblz ppm/MOD
CETA (4)	Standard Ref	Stblz ppm/STD	SGM97 (28)	Reference	MID550 (full trade)
CRPS (5)	Standard Ref	Stblz ppm/STD	SGM97 (28a)	--	MID550 (partial trading)
DICE (6)	Modeler's Ref	Stblz ppm/MOD	SGM97 (28b)	--	WGI550 (trade)
DNE21/98 (7)	Ref	550ppmv	SGM97 (28c)	--	WRE550 (trade)
HCRA (9)	Standard Ref	Stblz ppm/STD	WEC (29)	--	C
ICAM2 (10)	Modeler's Ref	Stblz ppm/MOD	YOHE (30)	Modeler's Ref	Stblz ppm/MOD
ICAM2 (11)	Standard Ref	Stblz ppm/STD	AIM97 (31)	--	Stblz ppm/MOD
IIASA (12)	Modeler's Ref	Stblz ppm/MOD	AIM97 (31a)	--	MID550 (full trade)
IIASA (13)	Standard Ref	Stblz ppm/STD	AIM97 (31b)	--	MID550 (no trade)
IIASA/WEC98 (14)	--	C1	AIM97 (31c)	--	WRE550 (full trade)
IIASAWEC (15)	--	C1	AIM97 (31d)	--	WRE550 (no trade)
IMAGE2.1 (16)	Baseline-A	Stab 550 All	AIM97 (31e)	--	WGI550 (no trade)
MARIA (17)	Standard Ref	Stblz ppm/STD	CETA (32)	--	550_stab
MARIA95 (18)	--	A	FUND (33)	Modeler's Reference	Kyoto+Min.Cost 550ppm
MERGE (19)	Standard Ref	Stblz ppm/STD	FUND (33a)	--	Min. Cost 550ppm
MINICAM (20)	Standard Ref	Stblz ppm/STD	G-CUBED (34)	Modeler's Reference	Stblz ppm
MIT (21)	Modeler's Ref	Stblz ppm/MOD	GRAPE (35)	--	Stblz ppm
MIT (22)	Standard Ref	Stblz ppm/STD	RICE (40)	Modeler's Reference	Min. Cost 550ppm
NWEAR21 (23)	--	Stblz ppm/MOD	SGM (41)	--	WRE550 (trade)
PAGE (24)	Standard Ref	Stblz ppm/STD			

Note: The scenario names are taken from the IPCC scenario database (Morita & Lee, 1998a)

3

Technological and Economic Potential of Greenhouse Gas Emissions Reduction

Co-ordinating Lead Authors:
WILLIAM R. MOOMAW (USA), JOSE ROBERTO MOREIRA (BRAZIL)

Lead Authors:
Kornelis Blok (Netherlands), David L. Greene (USA), Ken Gregory (UK), Thomas Jaszay (Hungary), Takao Kashiwagi (Japan), Mark Levine (USA), Mack McFarland (USA), N. Siva Prasad (India), Lynn Price (USA), Hans-Holger Rogner (Germany), Ralph Sims (New Zealand), Fengqi Zhou (China), Peter Zhou (Botswana)

Contributing Authors:
Frank Ackerman (USA), Erik Alsema (Netherlands), Harry Audus (IEA GHG), Jeroen de Beer (Netherlands), Ranjan K. Bose (India), John Davison (IEA GHG), Paul Freund (IEA GHG), Jochen Harnisch (Germany), Gilberto de M. Jannuzzi (Brazil), Anja Kollmuss (Switzerland), Changsheng Li (USA), Evan Mills (USA), Kiyoyuki Minato (Japan), Steve Plotkin (USA), Sally Rand (USA), A. Schafer (USA), David Victor (USA), Arnaldo C. Walter (Brazil), John J. Wise (USA), Remko Ybema (Netherlands)

Review Editors:
Ramon Pichs-Madruga (Cuba), Hisashi Ishitani (Japan)

CONTENTS

EXECUTIVE SUMMARY

The technological and economic potential to reduce greenhouse gas (GHG) emissions is large enough to hold annual global greenhouse gas emissions to levels close to or even below those of 2000 by 2010 and even lower by 2020. Realization of these reductions requires combined actions in all sectors of the economy including adoption of energy-efficient technologies and practices, increased fuel switching toward lower carbon fuels, continued growth in the use of efficient gas turbines and combined heat and power systems, greater reliance on renewable energy sources, reduced methane emissions through improved farm management practices and ruminant methane reduction strategies, diversification of land use to provide sinks and offsets, increased recovery of landfill methane for electricity production and increased recycling, reduction in the release of industrial gases, more efficient vehicles, physical sequestration of CO_2, and improving end-use efficiency while protecting the ozone layer. Countervailing socioeconomic and behavioural trends that cause greenhouse gas emissions to increase also exist, including increased size of dwelling units, increased sales of heavier and more powerful vehicles, growing vehicle kilometers travelled, reduced incentives for efficient use of energy or the purchase of energy efficiency technologies as a result of low real retail energy prices, increased consumption of consumer goods, and stimulated demand for energy-consuming products as a result of increased electrification.

A number of new technologies and practices have gained importance since the Second Assessment Report (SAR). As a result, greater opportunities for energy efficiency are available, often at lower cost than was expected. Annual growth in global consumption of primary energy and related carbon dioxide emissions dropped to 1.3% and 1.4%, respectively, between 1990 and 1998 after experiencing much higher growth rates of 2.4% and 2.1% between 1971 and 1990. This decrease in growth rate is because of the combined effects of improved energy efficiency technologies, increased fuel switching and adoption of renewable energy sources, and the dramatic decrease in emissions of countries with economies in transition (EITs) as a result of economic changes (*Figures 3.1* and *3.2*).

Sustained progress in the development and adoption of technologies and practices to reduce greenhouse gas emissions requires continued efforts in the areas of research and development, demonstration, dissemination, policies, and programmes. There has been a reduction in both public and private resources devoted to research and development to develop and implement new technologies that will reduce greenhouse gas

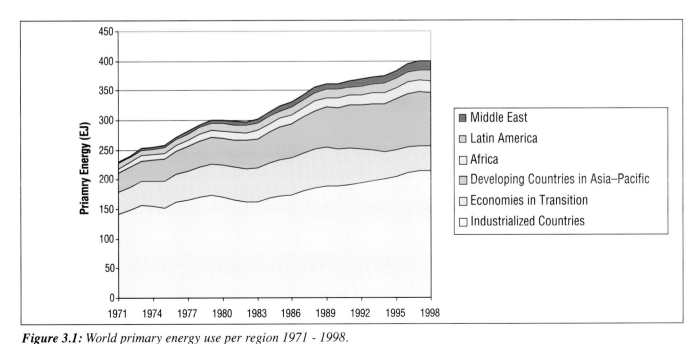

Figure 3.1: *World primary energy use per region 1971 - 1998.*
Note: Primary energy calculated using the IEA's physical energy content method based on the primary energy sources used to produce heat and electricity (IEA, 2000).

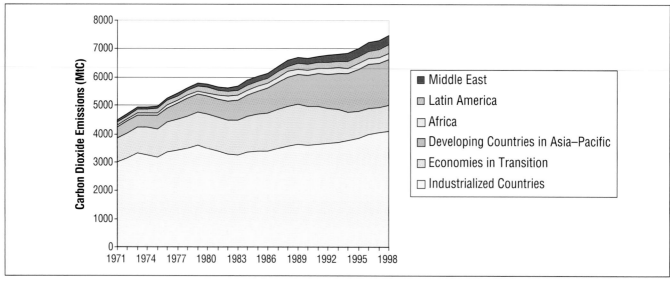

Figure 3.2: *World CO$_2$ emissions by region, 1971 - 1998.*

emissions. Despite the development of new, efficient technologies, current rates of energy efficiency improvements alone will not be sufficient to significantly reduce greenhouse gas emissions in the near term. In addition, policies or programmes to increase energy efficiency and promote renewable energy technology are lacking in many countries.

Technological innovation and change are influenced by the differing needs of different economies and sectors. A large percentage of capital is invested in a relatively small number of technologies that are responsible for a significant share of the energy supply and consumption market (automobiles, electric power generators, industrial processes, and building heating and cooling systems). There is a tendency to optimize these few technologies and their related infrastructure development, gaining them advantages and locking them into the economy. That makes it more difficult for alternative, low-carbon technologies to compete. For example, a particular technological configuration such as road-based automobiles has become "locked-in" as the dominant transportation mode. In industrial countries, technologies are developed as a result of corporate innovation or government-supported R&D, and in response to environmental regulations, utility deregulation, energy tax policies, or other incentives. In many developing countries, where electric power capacity and much end-use demand is growing most rapidly, there is often greater emphasis on getting technology such as electric power generation established in order to enhance economic development, with less concern for environmental and other issues. Capital flows and differing types of technology transfer may also determine technology choices. It is important to recognize that often values other than energy efficiency or greenhouse gas emissions are the dominant shapers of technological choice and innovation.

This chapter describes technologies and practices to reduce greenhouse gas emissions in the end-use sectors of the economy as well as through changes in energy supply. The end-use

sectors addressed are buildings, transport, industry, agriculture, and waste. Energy supply includes non-renewable resources, renewable resources, and physical carbon dioxide sequestration. In addition, options for reducing global warming contributions from substitutes for ozone depleting substances are discussed in the Appendix to this chapter.

The buildings sector contributes about 31% of global energy-related carbon dioxide emissions and these emissions grew at an average annual rate of almost 2.0% between 1971 and 1995. Growth in emissions varied significantly by region; between 1990 and 1995 the largest annual increases were experienced in developing countries (around 5.0% per year), moderate growth was seen in developed countries (around 1.0% per year), and emissions declined in the EITs (–3.0% per year). The growth in emissions in the developing and developed countries is largely caused by the increased amenity that consumers demand – in terms of increased purchase and use of appliances, larger dwellings, and the modernization and expansion of the commercial sector – as economies grow. Technology has continued on an evolutionary trajectory with incremental gains during the past decade in windows, lighting, insulation, space heating, refrigeration, air conditioning, building controls, passive solar design, and infiltration reduction. Although CFCs have been eliminated in developed countries as working fluids in heat pumps, air conditioners, and refrigerators, and as foam blowing agents for insulation, research and development (R&D) has been able to continue to improve energy efficiency of refrigerators and cooling and heating systems. Integrated building design has demonstrated very large reductions in energy use and greenhouse gas emissions. Expanded R&D is needed to assure continued technology improvement, but implementation policies remain the major hurdle to their rapid introduction.

The transport sector contributes 22% of carbon dioxide emissions; globally, emissions from this sector are growing at a rate of approximately 2.5% annually. Between 1990 and 1995, growth was highest in the developing countries (7.3% per year in the developing countries of Asia–Pacific and 4.6% per year in the remaining developing countries), moderate in the developed countries (1.9% per year) and is actually declining at a rate of –5.0% per year for the EITs. Technology improvements may generate operational cost reductions that have a rebound effect that stimulates further personal transportation use. These issues show the necessity of both policies and behavioural changes to lower emissions from the transport sector. Hybrid gasoline-electric vehicles have been introduced on a commercial basis with fuel economies 50% to 100% better than that of comparably sized four-passenger vehicles. The development of extremely low-polluting engines may reduce the incentive for hybrid and battery electric vehicles that were previously thought to encourage the adoption of vehicles that would also reduce greenhouse gases. Lightweight materials have the potential to improve fuel economy for all land transport. Fuel cell powered vehicles are developing rapidly, and could be introduced to the market sometime during the coming decade. Substantial potential for improving the fuel economy of heavy-duty trucks seems feasible. Only incremental improvements of the order of 1%/yr are expected for aircraft over the next several decades. There appears to be little attention being given to rail or public transportation systems, but waterborne transport of freight is already highly efficient, and has potential for additional gains.

Industrial emissions account for over 40% of carbon dioxide emissions. Global industrial sector carbon dioxide emissions grew at a rate of 1.5% per year between 1971 and 1995, slowing to 0.4% per year between 1990 and 1995. This is the only sector that has shown an annual decrease in carbon emissions in industrial economies (–0.8% per year between 1990 and 1995) as well as in the EITs (–6.4% per year between 1990 and 1995). Emissions from this sector in developing countries, however, continue to grow (6.3% per year in developing countries of Asia–Pacific and 3.4% per year in the remaining developing countries). Substantial differences in the energy efficiency of industrial processes between countries exist. Improvement of energy efficiency is the most important emission reduction option in the short term. However, industries continue to find new, more energy efficient processes which makes this option also important for the longer term. The larger part of the energy can be saved at net negative costs. In addition, material efficiency improvement (including more efficient product design, recycling, and material substitution) can greatly contribute to reducing emissions. For many sources of non-CO_2 emissions, like those from the aluminium industry, and adipic acid and HCFC-22 production, substantial emission reductions are possible or are already being implemented.

The agricultural sector has the smallest direct CO_2 emissions, contributing 4.0% of total global emissions. Growth in these emissions between 1990 and 1995 was greatest in the developing countries (6.0% per year in the developing countries of Asia–Pacific and 9.3% per year in the remaining developing countries), modest in the developed countries (1.3% per year),

and declined at a rate of –5.4% per year in the EIT. However, methane and nitrous oxide emissions dominate the agricultural sector, which contributes over 20% of global anthropogenic greenhouse gas emissions in terms of CO_2 equivalents. Reductions can be made by improved farm management practices such as more efficient fertilizer use, better waste treatment, use of minimum tillage techniques, and ruminant methane reduction strategies. Biotechnology and genetic modification developments could provide additional future gains and also lead to reduced energy demand, but the conflict between food security and environmental risk is yet to be resolved. Mitigation solutions exist overall for 100–200MtC$_{eq}$/yr but farmers are unlikely to change their traditional farming methods without additional incentives. Diversification of land use to energy cropping has the technical potential to provide both carbon sinks and offsets in regions where suitable land and water are available. Transport biofuel production costs remain high compared with oil products, but do provide additional value in the form of oxygenates and increased octane (ethanol). Because of market liberalization policies, the potential for biofuels has declined, though there is a growing demand for biodiesel in Germany. Improvements in biofuel conversion routes, such as the enzymatic hydrolysis of lignocellulosic material to ethanol, may help narrow the cost disadvantage versus fossil fuels.

Greenhouse gas emissions are being lowered substantially by increased utilization of methane from landfills and from coal beds for electric power generation. Significant energy-related greenhouse gas reductions are identified for improved waste recycling in the plastics and carpet industries, and through product remanufacturing. A major discussion is taking place over whether the greater reduction in lifecycle CO_2 emissions occurs through paper recycling or by utilizing waste paper as a biofuel in waste to energy facilities. In several developed countries, and especially in Europe and Japan, waste-to-energy facilities have become more efficient with lower air pollution emissions.

Abundant fossil fuel reserves that are roughly five times the total carbon already burned are available. The electric power sector accounts for 38% of total CO_2 emissions. Low cost, aero-derivative, combined cycle gas turbines with conversion efficiencies approaching 60% have become the dominant option for new electric power generation plants, wherever adequate natural gas supply and infrastructure are available. With deregulation of the electric power sector, additional emission reductions have occurred in most countries through the utilization of waste heat in combined heat and power systems that are capable of utilizing 90% of the fossil fuel energy. Low carbon-emitting technologies such as nuclear power have managed to significantly increase their capacity factor at existing facilities, but relatively few new plants are being proposed or built because of public concern about safety, waste storage, and proliferation. There has also been rapid deployment of wind turbines and smaller, but expanding markets for photovoltaic solar power systems. The annual growth rate from a small base for both wind and solar currently exceeds 25% per year, and together with an increasing number of bioenergy plants, accounts for around 2% of global electricity generation. Modern biomass gasification is increasing the opportunities for this renewable resource. There remains additional hydropower potential in some locations, but most large sites have already been developed in many regions of the world. Fuel cells appear to be a promising combined heat and electric power source as part of evolving distributed generation systems.

Further analysis since the SAR suggests that physical sequestration of CO_2 underground in aquifers, in depleted gas and oil fields, or in the deep ocean is potentially a viable option. Technical feasibility has been demonstrated for CO_2 removal and storage from a natural gas field, but long-term storage and economic viability remain to be demonstrated. Environmental implications of ocean sequestration are still being evaluated. The utilization of hydrogen from fossil fuels, biomass, or solid waste followed by sequestration appears particularly attractive. Along with biological sequestration, physical sequestration might complement current efforts at improving energy efficiency, fuel switching, and the further development and implementation of renewables, but it must compete economically with them.

Hydrofluorocarbon (HFCs) and perfluorocarbon (PFCs) use is growing as CFCs and, to a much lesser extent HCFCs, are eliminated. There is a variety of uses for these substances as alternatives in refrigeration, mobile and stationary air-conditioning, heat pumps, in medical and other aerosol delivery systems, insulating plastic foams, and for fire suppression and solvents. The replacement of ozone-depleting substances with HFCs and PFCs has been about one-tenth on a mass basis, with the difference being attributed to improved containment, recovery of fluids, and the use of alternative substances. The importance of considering energy efficiency simultaneously with ozone layer protection is discussed in the Appendix, especially in the context of developing countries.

This chapter concludes with a quantification of the potential for reducing greenhouse gas (GHG) emissions in the various end-use sectors of the economy and through changes in energy supply. It is found that sufficient technological potential exists to stabilize or lower global greenhouse gas emissions by 2010, and to provide for further reductions by 2020. The quantification is based on sector-specific analyses and, thus, caution should be taken when adding up the various estimates resulting from interactions between different types of technologies. These sector-based analyses can be used to provide further understanding of the results of global mitigation scenarios, such as those presented in Chapter 2, which account for inter-sectoral interactions, but typically do not provide estimates of sectoral level GHG emissions reduction potential or costs.

Some of the costs associated with sector specific options for reducing GHG emissions may appear high (for example US\$300/tC_{eq}). However, we estimate that there is technological potential for reductions of between 1,900 and 2,600MtC_{eq}/yr in 2010 and 3,600 to 5,050MtC_{eq}/yr in 2020. Half of these reductions are achievable at net negative costs (value of energy saved is greater than capital, operating and maintenance costs), and most of the remainder is available at a cost of less than US\$100tC_{eq}/yr. The continued development and adoption of a wide range of greenhouse gas mitigation technologies and practices will result not only in a large technical and economic potential for reducing greenhouse gas emissions but will also provide continued means for pursuing sustainable development goals.

3.1   Introduction

Technologies and measures to reduce greenhouse gas emissions are continuously being developed (Nadel *et al.*, 1998; National Laboratory Directors, 1997; PCAST, 1997; Martin *et al.*, 2000). Many of these technologies focus on improving the efficiency of fossil fuel use since more than two-thirds of the greenhouse gas emissions addressed in the Kyoto Protocol (in carbon dioxide equivalents) are related to the use of energy. Energy intensity (energy consumed divided by gross domestic product (GDP)) and carbon dioxide intensity (CO_2 emitted from burning fossil fuels divided by the amount of energy produced) have been declining for more than 100 years in developed countries without explicit government policies for decarbonization and both have the potential to decline further. Non-fossil fuel energy sources are also being developed and implemented as a means of reducing greenhouse gas emissions. Physical and biological sequestration of CO_2 can potentially play a role in reducing greenhouse gas emissions in the future. Other technologies and measures focus on reducing emissions of the remaining major greenhouse gases - methane, nitrous oxide, hydrofluorocarbons (HFCs), perfluorocarbons (PFCs), and sulphur hexafluoride (SF_6)(see Section 3.5 and Appendix to this Chapter).

Table 3.1 shows energy consumption in the four end-use sectors of the global economy – industry, buildings, transport, and agriculture–over time[1]. Data are displayed for six world regions – developed countries, countries with economies in transition (EITs), developing Asia-Pacific countries, Africa, Latin America and the Middle East. Comparing global annual average growth rates (AAGRs) for primary energy use in the period 1971 to 1990 and 1990 to 1995 a significant decrease is noticed– from 2.5% in the first period to about 1.0% in the latter, due almost entirely to the economic crisis in the EITs. Overall, growth averaged about 2.0% per year from 1971 to 1995. *Table 3.1* also shows carbon dioxide emissions from energy consumption for four world regions. The AAGR of global carbon dioxide emissions from the use of energy also declined (from 2% to 1%) in the same periods. A different picture emerges if the countries with economies in transition are excluded. In this case, growth in world energy use averaged about 2.5% per year in both the 1971 to 1990 and 1990 to 1995 periods, while average annual growth in carbon dioxide emissions was 2.0% and 2.6% during the same time periods, respectively.

Uncertainty in *Table 3.1* arises in a number of areas. First, the quality of energy data from the International Energy Agency

(IEA) is not homogeneous because of the use of various reporting mechanisms and "official" sources of national data (IEA, 1997a; IEA, 1997b; IEA, 1997c)[2]. Second, for the economies in transition, primary energy use data and carbon dioxide data are from two different sources (BP, 1997; IEA, 1997a; IEA, 1997b; IEA, 1997c). There are inconsistencies between the two sources, and no analysis has yet been done to resolve them. Third, IEA statistics report sectoral data for the industrial and transport sectors, but not for buildings and agriculture, which are reported as "other". These sectors have been estimated using an allocation scheme described in Price *et al.* (1998)[3]. In general, the most uncertainty is associated with data for the economies in transition region, and for the commercial and residential sub-categories of the buildings sector in all regions.

It is likely that total commercial energy production and demand estimates will be known accurately for most developed countries (within one or a few per cent), relatively accurately for some developing countries (with an uncertainty of 1% to 5%), and less accurately for developing countries with poorly functioning data gathering and statistical systems. Converting the energy data into carbon emissions introduces some increased uncertainty – primarily as a consequence of the fraction of natural gas that leaks to the atmosphere and the fraction of all fossil fuels that are left uncombusted – the uncertainty in carbon emissions is greater than that of energy use. Uncertainties in non-CO_2 greenhouse gas emissions are greater than those for carbon emissions.

In general, energy supply statistics, and their disaggregation into fuel types, are more reliable than statistics for energy demand. In particular, the estimates of sectoral energy demand (buildings, industry, transportation, agriculture) and the further disaggregation into subsectors (e.g., residential and commercial buildings; auto transportation; specific industries), and then into end uses has relatively high levels of uncertainty for at least two reasons. First, the full data to perform these disaggregations are rarely gathered at the national level, so that assumptions and approximations need to be made. Second, the conventions vary among different countries as to what energy use belongs to which sector or subsector (e.g., the distinction

---

[1] The data in this table differ slightly from the data presented in *Figures 3.1* and *3.2* because those figures are based on IEA data alone while the data in the table represents a combination of IEA and British Petroleum data (further described in the next paragraph of the text). Also, in *Figure 3.1*, primary energy was calculated using IEA's physical energy content method which is based on the primary energy sources used to produce heat and electricity (IEA, 2000) while in *Table 3.1*, primary energy was calculated using a standard electricity conversion efficiency of 33% (Price *et al.*, 1998).

[2] The IEA explains: "Countries often have several 'official' sources of data such as a Ministry, a Central Bureau of Statistics, a nationalized electricity company, etc. Data can also be collected from the energy suppliers, the energy consumers or the customs statistics. The IEA tries to collect the most accurate data, but does not necessarily have access to the complete data set that may be available to national experts calculating emissions inventories for the UNFCCC" (IEA, 1997c).

[3] The results of this allocation scheme were compared to the Lawrence Berkeley National Laboratory (LBNL) sectoral energy data for a number of developed countries. In general, the sectoral energy consumption values based on allocated IEA data compare favourably to LBNL data for total buildings and agriculture for most countries. Larger discrepancies were seen between the LBNL data and the allocated IEA data at the level of commercial and residential buildings.

Table 3.1: World carbon dioxide emissions and primary energy use by sector and region – 1971 to 1995 (Price et al., 1998, 1999)

Buildings sector	Carbon dioxide emissions (MtC)						Average annual growth rate			Primary energy use (EJ)						Average annual growth rate		
	1971	1975	1980	1985	1990	1995	1971-1990	1990-1995	1971-1995	1971	1975	1980	1985	1990	1995	1971-1990	1990-1995	1971-1995
Developed Countries	790	836	886	887	915	958	0.8%	0.9%	0.8%	44.4	48.9	52.3	56.8	62.3	68.5	1.8%	1.9%	1.8%
Residential	522	543	549	537	539	560	0.2%	0.8%	0.3%	28.3	30.5	33.0	34.6	36.7	40.6	1.4%	2.0%	1.5%
Commercial	268	293	336	350	377	398	1.8%	1.1%	1.7%	16.1	18.4	19.3	22.2	25.6	27.9	2.5%	1.7%	2.3%
Economies in Transition	240	296	362	381	373	320	2.3%	-3.0%	1.2%	10.7	13.0	18.2	21.0	23.0	16.2	4.1%	-6.8%	1.7%
Residential	164	213	266	290	279	256	2.9%	-1.8%	1.9%	8.1	9.8	12.9	14.3	15.1	10.4	3.3%	-7.2%	1.1%
Commercial	76	83	97	92	94	64	1.1%	-7.3%	-0.7%	2.6	3.2	5.3	6.7	7.9	5.8	6.0%	-6.1%	3.4%
Dev. countries Asia-Pacific	67	88	131	179	232	292	6.7%	4.7%	6.3%	3.6	4.6	5.6	7.9	10.2	12.9	5.7%	4.8%	5.5%
Residential	57	75	110	145	180	210	6.2%	3.1%	5.6%	3.0	3.9	4.6	6.3	7.9	9.3	5.2%	3.4%	4.8%
Commercial	10	14	21	33	51	81	9.0%	9.7%	9.1%	0.6	0.8	1.0	1.6	2.3	3.6	7.8%	9.0%	8.1%
Africa	15	18	23	30	38	48	5.0%	5.1%	5.0%	0.6	0.8	1.1	1.4	1.9	2.5	6.0%	5.4%	5.9%
Residential	11	12	16	22	29	39	5.1%	6.0%	5.2%	0.5	0.6	0.8	1.1	1.5	2.0	6.1%	6.0%	6.0%
Commercial	3	6	7	8	8	9	4.8%	1.7%	4.2%	0.1	0.2	0.3	0.3	0.4	0.5	5.8%	3.1%	5.2%
Latin America	18	21	24	24	30	34	2.6%	2.8%	2.7%	1.7	2.1	2.8	3.3	4.1	5.0	4.9%	4.1%	4.7%
Residential	14	16	19	19	22	24	2.5%	1.6%	2.3%	1.3	1.6	2.0	2.3	2.9	3.4	4.2%	3.3%	4.0%
Commercial	4	5	6	5	8	10	3.2%	5.9%	3.7%	0.3	0.5	0.8	0.9	1.2	1.6	7.0%	5.8%	6.8%
Middle East	9	13	23	41	58	80	10.5%	6.7%	9.7%	0.4	0.7	1.2	2.2	4.1	4.6	12.3%	2.8%	10.3%
Residential	7	10	17	30	41	59	10.0%	7.6%	9.5%	0.4	0.7	1.1	2.0	3.4	3.7	11.5%	1.8%	9.4%
Commercial	2	4	6	12	17	21	12.0%	4.3%	10.3%	0.0	0.0	0.1	0.2	0.7	1.0	20.2%	7.0%	17.3%
Rest of World	42	52	71	95	125	162	6.0%	5.3%	5.8%	2.7	3.7	5.1	6.9	10.1	12.1	7.1%	3.8%	6.4%
Residential	32	38	52	70	92	122	5.7%	5.8%	5.7%	2.2	2.8	3.9	5.4	7.8	9.1	6.8%	3.2%	6.0%
Commercial	10	14	19	25	33	40	6.7%	4.0%	6.2%	0.5	0.8	1.2	1.5	2.3	3.0	8.5%	5.7%	7.9%
World	1140	1273	1450	1542	1646	1732	2.0%	1.0%	1.8%	61.5	70.3	81.3	92.6	105.6	109.8	2.9%	0.8%	2.4%
Residential	775	869	977	1042	1091	1148	1.8%	1.0%	1.6%	41.7	47.1	54.5	60.6	67.4	69.4	2.6%	0.6%	2.2%
Commercial	364	404	473	500	555	584	2.2%	1.0%	2.0%	19.8	23.2	26.8	31.9	38.2	40.3	3.5%	1.1%	3.0%

(continued)

Table 3.1: continued

	Carbon dioxide emissions (MtC)						Average annual growth rate			Primary energy use (EJ)						Average annual growth rate		
	1971	1975	1980	1985	1990	1995	1971-1990	1990-1995	1971-1995	1971	1975	1980	1985	1990	1995	1971-1990	1990-1995	1971-1995
Transport sector																		
Developed Countries	494	554	612	636	743	816	2.2%	1.9%	2.1%	26.2	29.4	32.5	33.8	39.4	43.3	2.2%	1.9%	2.1%
Economies in Transition	69	75	77	83	87	67	1.2%	-5.0%	-0.1%	6.0	7.3	8.0	9.2	10.0	7.3	2.7%	-6.0%	0.8%
Dev. Countries Asia-Pacific	51	54	69	87	122	173	4.7%	7.3%	5.2%	2.0	2.4	3.3	4.3	6.0	8.7	5.9%	7.6%	6.2%
Africa	17	21	26	30	31	33	3.2%	1.5%	2.9%	0.8	1.0	1.3	1.5	1.6	1.7	3.6%	1.6%	3.2%
Latin America	33	44	53	55	62	79	3.4%	5.0%	3.7%	2.2	2.9	3.8	3.9	4.5	5.5	3.9%	4.2%	4.0%
Middle East	7	13	24	34	33	46	8.6%	6.6%	8.2%	0.4	0.7	1.3	1.8	1.7	2.4	8.6%	6.5%	8.1%
Rest of World	57	77	104	119	126	159	4.3%	4.6%	4.4%	3.3	4.6	6.3	7.2	7.8	9.6	4.6%	4.2%	4.5%
World	672	760	862	925	1078	1215	2.5%	2.4%	2.5%	37.5	43.6	50.1	54.4	63.3	69.0	2.8%	1.7%	2.6%
Industrial sector																		
Developed Countries	932	911	970	859	887	852	-0.3%	-0.8%	-0.4%	48.6	49.3	55.0	52.3	54.3	56.8	0.6%	0.9%	0.7%
Economies in Transition	416	494	597	615	621	447	2.1%	-6.4%	0.3%	26.0	31.6	34.0	36.9	38.0	26.0	2.0%	-7.3%	0.0%
Dev. Countries Asia-Pacific	223	287	384	483	632	859	5.6%	6.3%	5.8%	8.8	11.5	15.5	20.0	26.1	34.8	5.9%	5.9%	5.9%
Africa	35	43	57	63	68	66	3.5%	-0.4%	2.6%	1.4	1.8	2.5	2.8	3.1	3.1	4.5%	0.0%	3.5%
Latin America	33	41	55	53	58	70	3.1%	3.8%	3.2%	2.5	3.4	4.8	5.7	6.3	7.4	5.1%	3.4%	4.7%
Middle East	13	19	31	38	28	45	4.0%	10.2%	5.3%	0.7	1.1	1.6	2.1	1.5	2.4	3.9%	9.6%	5.1%
Rest of World	81	104	143	154	154	182	3.4%	3.4%	3.4%	4.6	6.2	8.9	10.5	11.0	13.0	4.7%	3.5%	4.5%
World	1653	1796	2094	2110	2293	2340	1.7%	0.4%	1.5%	88.0	98.5	113.5	119.8	129.4	130.8	2.1%	0.2%	1.7%
Agricultural sector																		
Developed Countries	35	33	38	45	48	51	1.7%	1.3%	1.6%	1.8	1.8	2.1	2.6	2.7	3.0	2.2%	1.6%	2.0%
Economies in Transition	44	53	72	88	96	72	4.2%	-5.4%	2.1%	1.3	1.6	1.8	2.4	3.0	1.7	4.5%	-10.6%	1.1%
Dev. Countries Asia-Pacific	17	23	36	38	51	68	5.9%	6.0%	5.9%	0.9	1.2	1.6	1.7	2.3	3.0	4.8%	5.6%	5.0%
Africa	2	3	4	4	4	7	2.9%	11.3%	4.6%	0.1	0.1	0.2	0.2	0.2	0.3	3.1%	9.8%	4.5%
Latin America	3	4	6	6	7	10	4.1%	7.2%	4.7%	0.2	0.3	0.5	0.6	0.6	0.8	4.6%	6.2%	5.0%
Middle East	1	2	4	5	5	8	7.6%	10.2%	8.1%	0.0	0.0	0.1	0.0	0.1	0.5	11.3%	35.7%	16.0%
Rest of World	7	10	13	15	16	25	4.6%	9.3%	5.5%	0.4	0.5	0.7	0.8	0.9	1.6	4.7%	12.6%	6.3%
World	103	120	159	186	210	217	3.8%	0.6%	3.1%	4.4	5.1	6.1	7.5	8.9	9.3	3.8%	0.8%	3.1%

(continued)

Table 3.1: continued

World Total	Carbon dioxide emissions (MtC)						Average annual growth rate			Primary energy use (EJ)						Average annual growth rate		
	1971	1975	1980	1985	1990	1995	1971-1990	1990-1995	1971-1995	1971	1975	1980	1985	1990	1995	1971-1990	1990-1995	1971-1995
Developed Countries	2252	2334	2506	2426	2593	2678	0.7%	0.6%	0.7%	121.0	129.3	141.8	145.5	158.8	171.7	1.4%	1.6%	1.5%
Economies in Transition	770	918	1108	1167	1177	907	2.3%	-5.1%	0.7%	44.0	53.5	62.0	69.5	74.0	51.3	2.8%	-7.1%	0.6%
Dev. Countries Asia-Pacific	358	453	620	787	1036	1392	5.8%	6.1%	5.8%	15.4	19.7	26.0	33.9	44.7	59.5	5.8%	5.9%	5.8%
Africa	70	85	110	127	141	155	3.8%	2.0%	3.4%	2.9	3.8	5.1	5.9	6.9	7.7	4.6%	2.3%	4.1%
Latin America	87	110	138	139	157	194	3.1%	4.3%	3.4%	6.5	8.7	11.8	13.4	15.5	18.8	4.7%	3.9%	4.5%
Middle East	30	48	83	118	124	179	7.7%	7.7%	7.7%	1.6	2.5	4.2	6.1	7.4	10.0	8.5%	6.0%	8.0%
Rest of World	187	243	330	383	422	528	4.4%	4.6%	4.4%	11.0	14.9	21.1	25.4	29.8	36.4	5.4%	4.1%	5.1%
World	3567	3948	4565	4763	5227	5504	2.0%	1.0%	1.8%	191.4	217.5	251.0	274.2	307.2	318.8	2.5%	0.7%	2.1%

Notes: Emissions from energy use only; does not include feedstock or CO_2 from calcination in cement production. Biomass = no emissions. Rest of World = Africa, Latin America, Middle East. Primary energy use and CO_2 emissions for Economies in Transition are from different sources and thus cannot be compared to each other.
Primary energy calculated using a standard 33% electricity conversion rate.

between residential and commercial buildings; the issue of whether energy use in industrial buildings counts as industrial or building energy use).

The least accurate data are for non-commercial energy use, especially in developing countries – dung, plant or forest waste, logs, and crops used for energy. Energy use from these sources is generally estimated from surveys, and is known very poorly. Because of uncertainty about whether these sources are used in sustainable ways and, even more importantly, because the release of products of incomplete combustion – which are potent greenhouse gases – are poorly characterized, the overall contribution of non-commercial energy sources to greenhouse gas emissions is only somewhat better than an educated guess at this time.

An important observation from *Table 3.1* is the high AAGR in the transport sector for energy and carbon emission. AAGR is not only the greatest for the transport sector, but it has slowed only slightly since 1960 despite significant improvements in technology. Because of the increase in the number of vehicles, and the recent decline in energy efficiency gains as vehicles have become larger and more powerful, transportation now is responsible for 22% of CO_2 emission from fuel use (1995). Unlike electricity, which can be produced from a variety of fuels, air and road transport is almost entirely fuelled with petroleum, except for ethanol and biodiesel used in a few countries. Biomass-derived fuels and hydrogen production from fossil fuels with carbon sequestration technology, in parallel with improved fuel efficiency conversion, are some of the few more promising alternatives for reducing significantly carbon emissions in the transport sector for the next two decades. The accelerated introduction of hybrid and fuel cell vehicles is also promising, but these gains are already being offset by increased driving, and the rapid growth of the personal vehicle market worldwide.

Oil, gas, and coal availability is still recognized to be very extensive. Fossil fuel reserves are estimated to be approximately five times the carbon content of all that have been used since the beginning of the industrial revolution. The possibility of using gas hydrates and coal bed methane as a source of natural gas has increased since the SAR.

Greenhouse gas (GHG)-reducing technologies for energy systems for all sectors of the economy can be divided into three categories – energy efficiency, low or no carbon energy production, and carbon sequestration (Acosta Moreno *et al.*, 1996; National Laboratory Directors, 1997). Even though progress will continue to be made in all categories, it is expected that energy efficiency will make a major contribution in the first decade of the 21st century. Renewable technologies are expected to begin to be significant around 2010, and pilot plants for the physical carbon sequestration from fossil fuels[4]

will be the last mitigation option to be adopted because of cost (National Laboratory Directors, 1997). Nevertheless, with appropriate policies, economic barriers can be minimized, opening possibilities for all the three categories of mitigation options. Considering the large number of available technologies in all categories and the still modest results obtained to date (see *Table 3.1*), it is possible to infer that their commercial uses are being constrained by market barriers and failures as well as a lack of adequate policies to induce the use of more costly mitigation options (see Chapters 5 and 6). This should not be interpreted as a reason to reduce R&D efforts and funding, since technological advances always help to cut costs and consequently reduce the amount and intensity of policies needed to overcome the existing economic barriers. Implementing new technological solutions could start soon by establishing policies that will encourage demand for these devices and practices. Complex technological innovations advance through a non-linear, interactive innovation process in which there is synergy between scientific research, technology development, and deployment activities (OTA, 1995a; Branscomb *et al.*, 1997; R&D Magazine, 1997). Early technology demand can be stimulated through well-placed policy mechanisms.

In this chapter numerous technologies are discussed that are either already commercialized or that show a probable likelihood to be in the commercial market by the year 2020, along with technologies that might possibly contribute to GHG abatement by 2010. For the quantification of the abatement capacity of some of the technologies a horizon as far as 2050 must be considered since the capital stock turnover rate, especially in the energy supply sector, is very low.

A number of new technologies and practices have gained importance since the preparation of SAR, including:

Buildings
- Off-grid building photovoltaic energy supply systems;
- Integrated building design for greater efficiency.

Transportation
- Hybrid electric vehicles;
- Fuel cell vehicles.

Industry
- Advanced sensors and controls for optimizing industrial processes;
- Large reductions in process gases such as CF_4, N_2O and HFCs through improved industrial processes;
- Reduced energy use and CO_2 emissions through improvements in industrial processing, remanufacturing, and use of recycled materials;
- Improved containment and recovery of CFC substitutes, the use of low Global Warming Potential (GWP) alternatives, and the use of alternative technologies.

Agriculture
- Biotechnology development for crop improvements (including energy crops), alternative fuels other than biomass, carbon cycle manipulation/sequestration, bio-

---

[4] Biological carbon sequestration is discussed in Chapter 4.

processing for fuels and chemicals and biological/bio-chemical hydrogen production;
- Minimum tillage practices in agriculture to reduce energy requirements and soil erosion, and improved management systems that lower N_2O emissions.

Energy
- Grid-connected Alternating Current (AC) solar panels;
- Combined cycle gas turbines for standard electric power production;
- Distributed combined heat and power systems;
- Fuel cells for distributed power and low temperature heat applications;
- Conversion of cellulosic materials for production of ethanol;
- Wind-based electricity generation;
- Carbon sequestration in aquifers and depleted oil and gas wells;
- Increased coal bed methane and landfill gas use;
- Replacement of grid connected electricity by PV;
- Nuclear plants life extension.

Cost data are presented in this chapter for many mitigation options. They are derived from a large number of studies and are not fully comparable. However, in general, the following holds for the studies quoted in this Chapter. The specific mitigation costs related to the implementation of an option are calculated as the difference of levelized costs[5] over the difference in greenhouse gas emissions (both in comparison to the situation without implementation of the option). Costs are generally calculated on a project basis (for a definition see Chapter 7, Section 7.3.1). The discount rates used in the cost calculation reflect real public sector discount rates (for a discussion of discount rates, see Chapter 7, Section 7.2.4). Generally, the discount rates in the quoted studies are in the range of 5%–12% per year. It should be noted that the discount rates used here are lower than those typically used in private sector decision making. This means that options reported in this chapter to have negative net costs will not necessarily be taken up by the market. Furthermore, it should be noted that in some cases even small specific costs may form a substantial burden for companies.

3.2   Drivers of Technological Change and Innovation

Reduction of greenhouse gas emissions is highly dependent upon both technological innovation and practices. The rate of introduction of new technologies, and the drivers for adoption are, however, different in industrial market economies, economies in transition and developing countries.

In industrial countries, technologies are developed as a result of corporate innovation or government-supported R&D, and in

response to environmental regulations, energy tax policies, or other incentives. The shift of electric and gas utilities from regulated monopolies to competing enterprises has also played a major role in the strong shift to combined cycle gas turbines, often with utilization of the waste heat in the electric power sector.

The most rapid growth in the electric power sector and many energy intensive industries is now occurring in developing countries, which have come to rely heavily upon technology transfer for investments in energy infrastructure. Capital for investment flows from industrial countries to developing countries through several pathways such as multilateral and bilateral official development assistance (ODA), foreign direct investments (FDI), commercial sales, and commercial and development bank lending. During the period 1993 to 1997, ODA experienced a downward trend with an increase in 1998, while FDI has increased substantially by a factor of five (see *Figure 3.3*) (OECD, 1999; Metz *et al.*, 2000). This shift is a consequence of the many opportunities that have opened for private capital in developing countries, and a reluctance by some industrial countries to increase ODA. The energy supply sector of developing countries is also undergoing deregulation from state to private ownership, increasing the role of the private sector in technology innovation.

A large percentage of capital is invested in a relatively small number of technologies that are responsible for a significant share of the energy supply and consumption market (automobiles, electric power generators, and building heating and cooling systems). There is a tendency to optimize these few technologies and their related infrastructure development, gaining them advantages that will make it more difficult for subsequent competing technologies to catch up. For example, a particular technological configuration such as road-based automobiles can become "locked-in" as the dominant transportation mode. This occurs because evolution of technological systems is as important as the evolution of individual new technologies. As their use expands their development becomes intertwined with the evolution of many other technologies and institutional and social developments. The evolution of technologies for oil exploration and extraction and for automobile production both affect and are affected by the expansion of infrastructures such as efficient refineries and road networks. They also affect and are affected by social and institutional developments, such as political and military power and settlement patterns, and business adaptation to changed transportation options, respectively.

Lock-in effects have two implications. First, early investments and early applications are extremely important in determining which technologies will be most important in the future. Second, learning and lock-in make technology transfer more difficult. Learning is much more dependent on successful building and using technology than on instruction manuals. Furthermore, technological productivity is strongly dependent upon complementary networks of suppliers, repair persons and

---

[5] Levelized costs include capital costs, operation and maintenance costs, fuel costs, *etc.*

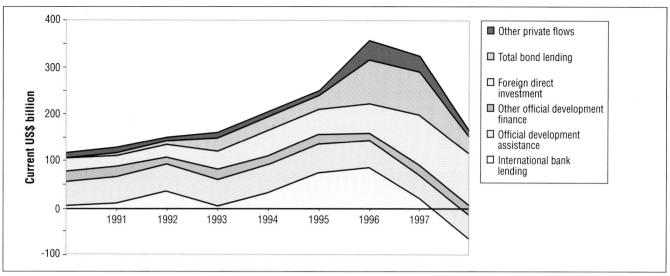

Figure 3.3: *Total net resource flows to aid recipient countries.*

training which is difficult to replicate in another country or region (IIASA/WEC, 1998; Unruh, 1999, 2000).

There are multiple government-driven pathways for technological innovation and change. Through regulation of energy markets, environmental regulations, energy efficiency standards, and market-based initiatives such as energy and emission taxes, governments can induce technology changes and influence the level of innovations. Important examples of government policies on energy supply include the Clean Air Act in the USA, the Non Fossil Fuel Obligation in the UK, the Feed-in-Law in Germany, the Alcohol Transport Fuel Program in Brazil, and utility deregulation that began in the UK and has now moved to the USA, Norway, Argentina, and many other countries. Voluntary agreements or initiatives implemented by the manufacturing industry, including energy supply sections, can also be drivers of technological change and innovation.

In the energy-consuming sector, major government actions can promote energy efficient use and the replacement of high (like coal) to lower carbon fuels (like natural gas and renewables). Energy efficiency standards for vehicles, appliances, heating and cooling systems, and buildings can also substantially encourage the adoption of new technologies. On the other hand, continued subsidies for coal and electricity, and a failure to properly meter electricity and gas are substantial disincentives to energy efficiency gains and the uptake of renewable and low carbon technologies. Government-supported R&D has also played a significant role in developing nuclear power, low carbon technologies such as gas turbines, and carbon-free energy sources including wind, solar, and other renewables. Such government actions in the energy-consuming sector can ensure increasing access to energy required for sustainable development.

While regulation in national energy markets is well established, it is unclear how international efforts at GHG emission

regulation may be applied at the global level. The Kyoto Protocol and its mechanisms represent opportunities to bring much needed energy-efficient practices and alternative energy to the continuously growing market of developing countries and in reshaping the energy markets of the economies in transition.

Important dimensions and drivers for the successful transfer of lower GHG technologies to developing countries and economies in transition are capacity building, an enabling environment, and adequate mechanisms for technology transfer (Metz *et al.*, 2000). Markets for the use of new forms of energy are often non-existent or very small, and require collaboration among the local government and commercial or multilateral lending banks to promote procurement. It may also be necessary to utilize temporary subsidies and market-based incentives as well. Because energy is such a critical driver of development, it is essential that strategies to reduce GHG emissions be consistent with development goals. This is true for all economies, but is especially true for developing countries and economies in transition where leap-frogging to modern, low emitting, highly efficient technologies is critical (Moomaw *et al.*, 1999a; Goldemberg, 1998).

Non-energy benefits are an important driver of technological change and innovation (Mills and Rosenfeld, 1996; Pye and McKane, 2000). Certain energy-efficient, renewable, and distributed energy options offer non-energy benefits. One class of such benefits accrues at the national level, e.g. via improved competitiveness, energy security, job creation, environmental protection, while another relates to consumers and their decision-making processes. From a consumer perspective, it is often the non-energy benefits that motivate decisions to adopt such technologies. Consumer benefits from energy-efficient technologies can be grouped into the following categories: (1) improved indoor environment, comfort, health, safety, and productivity; (2) reduced noise; (3) labour and time savings; (4)

improved process control; (5) increased reliability, amenity or convenience; (6) water savings and waste minimization; and (7) direct and indirect economic benefits from downsizing or elimination of equipment. Such benefits have been observed in all end-use sectors. For renewable and distributed energy technologies, the non-energy benefits stem primarily from reduced risk of business interruption during and after natural disasters, grid system failures or other adverse events in the electric power grid (Deering and Thornton, 1998).

Product manufacturers often emphasize non-energy benefits as a driver in their markets, e.g. the noise- and UV-reduction benefits of multi-glazed window systems or the disaster-recovery benefits of stand-alone photovoltaic technologies. Of particular interest are attributes of energy-efficient and renewable energy technologies and practices that reduce insurance risks (Mills and Rosenfeld, 1996). Approximately 80 specific examples have been identified with applications in the buildings and industrial sectors (Vine *et al.*, 1998), and insurers have begun to promote these in the buildings sector (Mills, 1999). The insurance sector has also supported transportation energy efficiency improvements that increase highway safety (reduced speed limits) and urban air quality (mass transportation) (American Insurance Association, 1999). Insurance industry concern about increased natural disasters caused by global climate change also serves as a motivation for innovative market transformation initiatives on behalf of the industry to support climate change adaptation and mitigation (Mills 1998, 1999; Vellinga *et al.*, 2000; Nutter, 1996). Market benefits for industries that adopt low carbon- emitting processes and products have also been increasingly recognized and documented (Hawken *et al.*, 1999; Romm, 1999).

3.3   Buildings

3.3.1   Introduction

This section addresses greenhouse gas emissions and emissions reduction opportunities for residential and commercial (including institutional) buildings, often called the residential and service sectors. Carbon dioxide emissions from fossil fuel energy used directly or as electricity to power equipment and condition the air (including both heating and cooling) within these buildings is by far the largest source of greenhouse gas emissions in this sector. Other sources include HFCs from the production of foam insulation and for use in residential and commercial refrigeration and air conditioning, and a variety of greenhouse gases produced through combustion of biomass in cookstoves.

3.3.2   Summary of the Second Assessment Report

The Second Assessment Report (SAR) reviewed historical energy use and greenhouse gas emissions trends as well as mitigation options in the buildings sector in Chapter 22,

Mitigation Options for Human Settlements (Levine *et al.*, 1996a). This chapter showed that residential and commercial buildings accounted for 19% and 10%, respectively, of global carbon dioxide (CO_2) emissions from the use of fossil fuels in 1990. More recent estimates increase this percentage to 21% for residential buildings and 10.5% for commercial buildings, both for 1990 and 1995, as shown in *Table 3.1*. Globally, space heating is the dominant energy end-use in both residential and commercial buildings. Developed countries account for the vast majority of buildings-related CO_2 emissions, but the bulk of growth in these emissions over the past two decades was seen in developing countries. The SAR found that many cost-effective technologies are available to reduce energy-related CO_2 emissions, but that consumers and decision-makers often do not invest in energy efficiency for a variety of reasons, including existing economic incentives, levels of information, and conditions in the market. The SAR concluded that under a scenario with aggressive adoption of energy-efficiency measures, cost-effective energy efficiency could likely cut projected baseline growth in carbon emissions from energy use in buildings by half over the next two decades.

3.3.3   Historic and Future Trends

CO_2 from energy use is the dominant greenhouse gas emitted in the buildings sector, followed by HFCs used in refrigeration, air conditioning, and foam insulation, and cookstove emissions of methane and nitrous oxide (see *Table 3.2*). Developed countries have the largest emissions of CO_2 and HFCs, while developing countries have the largest emissions of greenhouse gases from non-renewable biomass combustion in cookstoves (Smith *et al.*, 2000). It is noted, however, that the biomass energy source is being replaced with non-renewable carbon-based fuels (Price *et al.*, 1998). This trend is expected to continue.

Energy use in buildings exhibited a steady growth from 1971 through 1990 in all regions of the world, averaging almost 3% per year. Because of the decline in energy use in buildings in the former Soviet Union after 1989, global energy use in buildings has grown slower than for other sectors in recent years. Growth in commercial buildings was higher than growth in residential buildings in all regions of the world, averaging 3.5% per year globally between 1971 and 1990. Energy-related CO_2 emissions also grew during this period. By 1995, CO_2 emissions from fuels and electricity used in buildings reached 874MtC and 858MtC, respectively, for a total of 1732MtC, or 98% of all buildings-related GHG emissions. Growth in these CO_2 emissions was slower than the growth in primary energy in both the developed countries and the rest-of-world region, most likely the result of fuel switching to lower carbon fuels in these regions. In contrast, growth in energy-related CO_2 emissions in the developing countries — Asia Pacific region — was 6.3% per year between 1971 and 1995, greater than the 5.5% per year growth in primary energy use, reflecting a growing reliance on more carbon-intensive fuels in this region.[6]

Table 3.2: *Overview of 1995 greenhouse gas emissions in the buildings sector (in MtC) by region* (Price *et al.*, 1998, 1999; Smith *et al.*, 2000).

Greenhouse gas source	Developed Countries	Countries with Economies in Transition	Developing Countries in Asia-Pacific	Rest of World	Total
Fuel CO_2	397	235	167	75	874
Electricity CO_2[a]	561	85	125	87	858
Refrigeration, A/C, foam insulation HFCs					45[b]
Biomass cookstove CH_4					40[c]
Total					1817

a CO_2 emissions from production of electricity.

b Based on an estimated range of 47 to 50MtC in the year 2000 (see Appendix to this Chapter).

c Based on an estimate of global annual emissions of 7 Tg of CH_4. Estimates for N_2O emissions from biomass cookstoves are not available (Smith *et al.*, 2000).

Non-CO_2 greenhouse gas emissions from the buildings sector are hydrofluorocarbons (HFCs)[7] used or projected to be used in residential and commercial refrigerators, air conditioning systems, and in open and closed cell foam for insulation. HFCs in the building sector were essentially zero in 1995, but are projected to grow as they replace ozone-depleting substances (see Appendix to this chapter). In addition, methane (CH_4), nitrous oxide (N_2O), carbon monoxide (CO), and nitrogen oxides (NO_x) (along with CO_2) are produced through combustion of biomass in cookstoves (Levine *et al.*, 1996b; Smith *et al.*, 2000). It is estimated the biomass cookstoves emit about 40MtC_{eq}, 2% of total buildings-related GHG emissions (Smith *et al.*, 2000). These emissions are concentrated in developing countries, where biomass fuels can account for more than 40% of the total energy used in residences (UNDP, 1999).[8]

Key drivers of energy use and related GHG emissions in buildings include activity (population growth, size of labour force, urbanization, number of households, per capita living area, and persons per residence), economic variables (change in GDP and personal income), energy efficiency trends, and carbon intensity trends. These factors are in turn driven by changes in consumer preferences, energy and technology costs, settlement patterns, technical change, and overall economic conditions.

Urbanization, especially in developing countries, is clearly associated with increased energy use. As populations become more urbanized and commercial fuels, especially electricity, become easier to obtain, the demand for energy services such as refrigeration, lighting, heating, and cooling increases. The number of people living in urban areas almost doubled between 1970 and 1995, growing from 1.36 billion, or 37% of the total, in 1970 to 2.57 billion, or 45% of the total, in 1995 (UN, 1996).

Driving forces influencing the use of HFCs include both its suitability as a replacement for CFCs and HCFCs, as well as an awareness of the contribution of HFCs to global climate change. It is expected that this awareness will continue to drive decisions to use HFCs only in highest value applications. Some countries have enacted regulations limiting emissions of HFCs while others have established voluntary agreements with industry to reduce HFC use (see Appendix to this chapter).

Global projections of primary energy use for the buildings sector show a doubling, from 103EJ to 208EJ, between 1990 and 2020 in a baseline scenario (WEC, 1995a). The most rapid growth is seen in the commercial buildings sector, which is projected to grow at an average rate of 2.6% per year. Increases in energy use in the EITs are projected to be as great as those in the developing countries, as these countries recover from the economic crises and as the growth in developing countries begins to slow. Under a scenario where state-of-the-art technology is adopted, global primary energy consumption in the buildings sector will only grow to about 170EJ in 2020. A more aggressive "ecologically driven/advanced technology" scenario, which assumes an international commitment to energy efficiency as well as rapid technological progress and widespread application of policies and programmes to speed the adoption of energy-efficient technologies in all major regions of the world, results in primary energy use of 140EJ in 2020 (WEC, 1995a).

The IPCC's IS92a scenario projected baseline global carbon dioxide emissions from the buildings sector to grow from 1900 MtC to 2700MtC between 1990 and 2020. An analysis of the

---

[6] Trends in primary energy use and CO_2 emissions in the EIT region cannot be compared because these values are from two different data sources (see Price *et al.*, 1998).

[7] HFCs are used as a replacement gas for chlorofluorocarbons (CFCs) which are being phased out globally under the Montreal Protocol on Substances that Deplete the Ozone Layer.

[8] For example, traditional fuels based on biomass account for a large share of residential energy consumption in Nicaragua (43%), El Salvador (44%), Honduras (50%), Paraguay (51%), Guatemala (61%) and Haiti (87%) (UNDP, 1999).

potential reductions from implementation of energy-efficient technologies found that annual global carbon dioxide emissions from the buildings sector could be reduced by an estimated 950MtC in 2020 compared to the IS92a baseline scenario (Acosta Moreno *et al.*, 1996). Over 60% of these projected savings are realized through improvements in residential equipment and the thermal integrity of buildings globally. Carbon dioxide emissions from commercial buildings grow from 37% to 41% of total buildings emissions between 1990 and 2020 as a result of expected increases in commercial floor space (which implies increases in heating, ventilation, and air conditioning systems (HVAC)) as well as increased use of office and other commercial sector equipment (Acosta Moreno *et al.*, 1996; WEC, 1995a).

The B2 scenario from the IPCC's Special Report on Emissions Scenario projects buildings sector carbon dioxide emissions to grow from 1,790MtC in 1990 to 3,090MtC in 2020. The most rapid growth is seen in the developing countries, which show an average growth in buildings-related carbon dioxide emissions of over 3% per year. In contrast, this scenario envisions that the emissions from buildings in the EIT region continue to decline, at an average annual rate of –1.3% (Nakicenovic *et al.*, 2000).

3.3.4   New Technological and Other Options

There are myriad opportunities for energy efficiency improvement in buildings (Acosta Moreno *et al.*, 1996; Interlaboratory Working Group, 1997; Nadel *et al.*, 1998) (see *Table 3.3*). Most of these technologies and measures are commercialized but are not fully implemented in residential and commercial buildings, while some have only recently been developed and will begin to penetrate the market as existing buildings are retrofitted and new buildings are designed and constructed.

A recent study identified over 200 emerging technologies and measures to improve energy efficiency and reduce energy use in the residential and commercial sectors (Nadel *et al.*, 1998). Individual country studies also identify many technologies and measures to improve the energy efficiency and reduce greenhouse gas emissions from the buildings sector in particular climates and regions.[9] For example, a study for South Africa discusses 15 options for the residential sector and 11 options for the commercial sector (Roos, 2000). Examples of other studies that identify energy efficiency or greenhouse gas mitigation options for the buildings sector include those for Brazil (Schaeffer and Almeida, 1999), Bulgaria (Tzvetanov *et al*, 1997), Canada (Bailie *et al.*, 1998); China (Research Team of China Climate Change Country Study, 1999); Czech Republic (Tichy, 1997), the European Union (Blok *et al.*, 1996; van

Velsen *et al.*, 1998), India (Asian Development Bank, 1998), Indonesia (Cahyono Adi *et al.*, 1997), Mexico (Mendoza *et al.*, 1991), Poland (Gaj and Sadowski, 1997); Ukraine (Raptsoun and Parasyuk, 1997), and the US (Interlaboratory Working Group, 1997; National Laboratory Directors, 1997; STAPPA/ALAPCO, 1999). Below examples are given of three new developments out of many that could be cited: integrated building design, reducing standby power losses in appliances and equipment, and photovoltaic systems for residential and commercial buildings. These examples focus on options for reducing greenhouse gas emissions from the buildings sector in which there has been significant recent research: improving the building shell, improving building equipment and appliances, and switching to lower carbon fuels to condition the air and power the equipment and appliances in buildings. In addition, recent developments in distributed power generation for buildings are briefly described (see also Section 3.8.5.3).

3.3.4.1   Integrated Building Design

Integrated building design focuses on exploiting energy-saving opportunities associated with building siting as well as synergies between building components such as windows, insulation, equipment, and heating, air conditioning, and ventilation systems. Installing increased insulation and energy-efficient windows, for example, allows for installation of smaller heating and cooling equipment and reduced or eliminated ductwork.[10] Most importantly, it will become possible in the future to design a building where operation can be monitored, controlled, and faults detected and analyzed automatically. For large commercial buildings, such systems (which are currently under development) have the potential to create significant energy savings as well as other operational benefits. Two recent projects that used integrated building design for residential construction found average energy savings between 30% and 60% per cent (Elberling and Bourne, 1996; Hoeschele *et al.*, 1996; Parker *et al.*, 1996), while for commercial buildings energy savings have varied between 13% and 71% (Piette *et al.*, 1996; Hernandez *et al.*, 1998; Parker *et al.*, 1997; Thayer, 1995; Suozzo and Nadel, 1998). Assuming an average savings of 40% for integrated building design, the cost of saved energy for residential and commercial buildings has been calculated to be around US$3/GJ (the average cost of energy in the US buildings sector is about US$14/GJ) (Nadel *et al.*, 1998; US DOE/EIA, 1998).

3.3.4.2   Reducing Standby Power Losses in Appliances and Equipment

Improving the energy efficiency of appliances and equipment can result in reduced energy consumption in the range of 10 to 70%, with the most typical savings in the 30% to 40% range (Acosta Moreno *et al.*, 1996; Turiel *et al.*, 1997).

---

[9] Many countries provide a discussion of these technologies and measures in their National Communications to the UNFCCC (http://www.unfccc.int/text/resource/natcom/).

[10] It is noted that production of these efficient products increases energy use for materials production in the industrial sector (Gielen, 1997).

Table 3.3: *Overview of opportunities for energy efficiency improvement in buildings*
(Acosta Moreno *et al.*, 1996; Interlaboratory Working Group, 1997; Nadel *et al.*, 1998; Suozzo and Nadel, 1998).

End use	Energy efficiency improvement opportunities
Insulation	Materials for buildings envelopes (e.g., walls, roofs, floors, window frames); materials for refrigerated spaces/cavities; materials for highly heated cavities (e.g., ovens); solar reflecting materials; solar and wind shades (e.g., vegetation, physical devices); controls; improved duct sealing
Heating, ventilation, and air conditioning (HVAC) systems	Condensing furnaces; electric air-source heat pumps; ground-source heat pumps; dual source heat pumps; Energy Star residential furnaces and boilers; high efficiency commercial gas furnaces and boilers; efficient commercial and residential air conditioners; efficient room air conditioners; optimization of chiller and tower systems; desiccant coolers for supermarkets; optimization of semiconductor industry cleanroom HVAC systems; controls (e.g., economizers, operable windows, energy management control systems); motors; pumps; chillers; refrigerants; combustion systems; thermal distribution systems; duct sealing; radiant systems; solar thermal systems; heat recovery; efficient wood stoves
Ventilation systems	Pumps; motors; air registers; thermal distribution systems; air filters; natural and hybrid systems
Water heating systems	High efficiency electric resistance water heaters; water heaters; air-source heat pump water heaters; exhaust air heat pump water heaters; integrated space/water heating systems; integrated gas-fired space/water heating systems, high efficiency gas water heaters; instantaneous gas water heaters; solar water heaters; low-flow showerheads
Refrigeration	Efficient refrigerators; high efficiency freezers; commercial refrigeration technologies
Cooking	Improved biomass stoves; efficient wood stoves; Turbochef combination microwave/convection oven; high efficiency gas cooking equipment
Other appliances	Horizontal axis washing machine; increase washing machine spin speed; heat pump clothes dryer; efficient dishwashers; consumer electronics with standby losses less than 1 watt; consumer electronics with efficient switch-mode power supplies
Windows	Double and triple-glazed windows; low-emittance windows; spectrally selective windows; electrochromic windows
Lighting systems	Compact fluorescents (including torchères); halogen IR lamps; electronic ballasts; efficient fluorescents and fixtures; HIDs, LED exit signs; LED traffic lights; solid state general purpose lighting (LEDs and OLEDs); lighting controls (including dimmers); occupancy controls; lighting design (including task lighting, reducing lighting levels);daylighting controls; replacement of kerosene lamps
Office equipment	Efficient computers; low-power mode for equipment; LCD screens
Motors	Variable speed drives; high efficiency motors; integrated microprocessor controls in motors; high quality motor repair practices
Energy management	Buildings energy management systems; advanced energy management systems; commercial building retro-commissioning
Design	Integrated building design; prefabricated buildings; solar design (including heat or cold storage); orientation; aspect ratio; window shading; design for monitoring; urban design to mitigate heat islands; high reflectance roof surfaces
Energy sources	Off-grid photovoltaic systems; cogeneration systems

Implementation of advanced technologies in refrigerator/freezers, clothes washers, clothes dryers, electric water heaters, and residential lighting in the US is estimated to save 3.35EJ/yr by 2010, reducing energy use of these appliances by nearly 50% from the base case (Turiel *et al.*, 1997).

A number of residential appliances and electronic devices, such as televisions, audio equipment, telephone answering machines, refrigerators, dishwashers, and ranges consume electricity while in a standby or off mode (Meier *et al.*, 1992; Herring, 1996; Meier and Huber, 1997; Molinder, 1997; Sanchez, 1997). These standby power losses are estimated to consume 12% of Japanese residential electricity, 5% of US residential electricity, and slightly less in European countries (Nakagami *et al.*, 1997; Meier *et al.*, 1998). Metering studies have shown that such standby losses can be reduced to one watt in most of these mass-produced goods (Meier *et al.*, 1998). The costs of key low-loss technologies, such as more efficient switch-mode power supplies and smarter batteries, are low (Nadel *et al.*, 1998) and a recent study found that if all US appliances were replaced by units meeting the 1-watt target, aggregate standby losses would fall at least 70%, saving the USA over US$2 billion annually (Meier *et al.*, 1998).

Table 3.4: *Buildings sector 1995 fuel, electricity, primary energy, CO₂ emissions, and average annual growth rates (AAGRs) for 1971 to 1990 and 1990 to 1995 by region* (Price *et al.*, 1998, 1999).

	Fuels			**Electricity**			**Primary energy**			**CO₂ emissions**		
	1995 Energy (EJ)	AAGR 1971-1990	AAGR 1990-1995	1995 Energy (EJ)	AAGR 1971-1990	AAGR 1990-1995	1995 Energy (EJ)	AAGR 1971-1990	AAGR 1990-1995	1995 CO₂ (MtC)	AAGR 1971-1990	AAGR 1990-1995
Developed Countries	25.45	-0.7%	0.7%	14.21	4.5%	2.7%	68.51	1.8%	1.9%	958.46	0.8%	0.9%
Countries with Economies in Transition	11.98	3.4%	-6.4%	1.39	6.6%	-7.9%	16.19	4.1%	-6.8%	319.83	2.3%	-3.0%
Developing Cos. In Asia-Pacific	7.34	4.3%	1.5%	1.85	10.4%	10.5%	12.93	5.7%	4.8%	291.62	6.7%	4.7%
Rest of World	5.14	6.2%	0.5%	2.31	8.2%	6.7%	12.15	7.1%	3.8%	162.32	6.0%	5.3%
World	49.91	1.3%	-1.2%	19.76	5.3%	2.6%	109.78	2.9%	0.8%	1732.23	2.0%	1.0%

Note: Data sources are IEA, 1997a; IEA, 1997b, IEA, 1997c and BP, 1997. For the EIT region only, energy data from British Petroleum were used instead of IEA data. Thus, primary energy and CO₂ emissions for the EIT region cannot be compared. For a more detailed description of the data, see Price *et al.*, 1998, 1999.

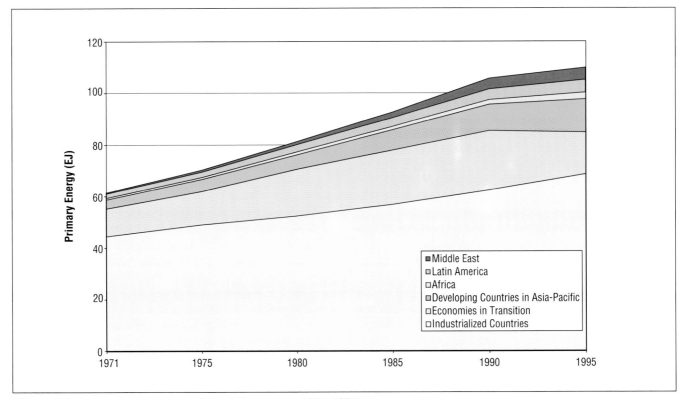

Figure 3.4: *Primary energy use in the buildings sector, 1971-1995.*

3.3.4.3 *Photovoltaic Systems for Buildings*

Photovoltaic systems are being increasingly used in rural off-grid locations, especially in developing countries, to provide electricity to areas not yet connected to the power infrastructure or to offset fossil fuel generated electricity. These systems are most commonly used to provide electricity for lighting, but are also used for water pumping, refrigeration, evaporative cooling, ventilation fans, air conditioning, and powering various electronic devices. In 1995, more than 200,000 homes worldwide depended on photovoltaic systems for all of their electricity needs (US DOE, 1999a). Between 1986 and 1998, global PV sales grew from 37MW to 150MW (US DOE, 1999b). Rural electrification programmes have been established in many developing countries. In Brazil, more than 1000 small stand-alone systems that provide power for lighting, TVs, and radios were recently installed in homes and schools, while two hybrid (PV-wind-battery) power systems were installed in the Amazon Basin to reduce the use of diesel generators that supply power to more than 300 villages in that area (Taylor, 1997). Similar projects have been initiated in South Africa (Arent, 1998), Egypt (Taylor and Abulfotuh, 1997), India (Stone and Ullal, 1997; US DOE, 1999b), Mexico (Secretaria de Energia, 1997), China, Indonesia, Nepal, Sri Lanka, Vietnam, Uganda, Solomon Islands, and Tanzania (Williams, 1996). Recent developments promoting increased adoption of photovoltaic systems include the South African Solar Rural Electrification Project (Shell International, 1999), the US Million Solar Roofs Initiative (US DOE, 1999a), the effort to install 5000MW on residences in Japan by 2010 (Advisory Committee for Energy, 1998), and net metering, which allows the electric meters of customers with renewable energy generating facilities to be reversed when the generators are producing energy in excess of residential requirements (US DOE, 1999b).

3.3.4.4 *Distributed Power Generation for Buildings*

Distributed power generation relies on small power generation or storage systems located near or at the building site. Several small scale (below 500kW), dispersed power-generating technologies are advancing quite rapidly. These technologies include both renewable and fossil fuel powered alternatives, such as photovoltaics and microturbines. Moving power generation closer to electrical end-uses results in reduced system electrical losses, the potential for combined heat and power applications (especially for building cooling), and opportunities to better co-ordinate generation and end-use, which can together more than compensate for the lower conversion efficiency and result in overall energy systems that are both less expensive and emit less carbon dioxide than the familiar central power generating station. The likelihood of customer sites becoming net generators will be determined by the configuration of the building and/or site, the opportunities for on-site use of cogenerated heat, the availability and relative cost of fuels, and utility interconnection, environmental, building code, and other regulatory restrictions (NRECA, 2000).

3.3.5 *Reginal Differences*

There are significant regional differences in levels of energy use and related GHG emissions from the buildings sector. *Table 3.4* presents 1995 buildings sector's fuels, electricity, primary energy, and CO_2 emissions and historical growth rates for the 1971 to 1990 and 1990 to 1995 periods for four regions (Price *et al.*, 1998, 1999). *Figure 3.4* provides a graphical presentation of the data on primary energy use in buildings, with the fourth region (Rest of World) desegregated into Middle East, Latin America, and Africa. Three very important trends are apparent:

- Developed countries have by far the largest CO_2 emissions from the buildings sector and have exhibited a relatively steady long-term trend of annual primary energy growth in the 1.8% to 1.9% range (with lower growth through 1985 and higher growth thereafter).
- Since the late 1980s, energy use and related CO_2 emissions from buildings in the developing countries, particularly in the Asia–Pacific region, have grown about five times as fast as the global average (and more than twice as fast as in developed countries).
- The growth rate of buildings' energy use globally has declined since 1990 because of the economic crisis in the EITs. The world other than the EITs continued its long-term trend (1971-1995) of annual energy growth in the 2.8% to 2.9% range.

The average annual increase in urban population was nearly 4.0% per year in Asia and Rest of World regions. This increased urbanization led to increased use of commercial fuels, such as kerosene and liquefied petroleum gas (LPG), for cooking instead of traditional biomass fuels. In general, higher levels of urbanization are associated with higher incomes and increased household energy use, including significantly increased purchase and use of a variety of household appliances (Sathaye *et al.*, 1989; Nadel *et al.*, 1997, Sathaye and Ketoff, 1991). Wealthier populaces in developing countries exhibit consumption patterns similar to those in developed countries, where purchases of appliances and other energy-using equipment increase with gains in disposable income (WEC, 1995a).

Between 1971 and 1990, global primary energy use per capita in the buildings sector grew from 16.5GJ/capita to 20GJ/capita. Per capita energy use in buildings varied widely by region, with the developed and EIT regions dominating globally. Energy use per capita is higher in the residential sector than in the commercial sector in all regions, although average annual growth in commercial energy use per capita was higher during the period, averaging 1.7% per year globally compared to 0.6% per year for the residential sector.

Energy consumption in residential buildings is strongly correlated with household income levels. Between 1973 and 1993, increases in total private consumption translated into larger homes, more appliances, and an increased use of energy services (water heating, space heating) in most developed countries (IEA, 1997d). In developed countries, household floor area increased but household size dropped from an average of 3.5 persons per household in 1970 to 2.8 persons per household in 1990. These trends led to a decline in energy use per household but increased residential energy use per capita (IEA, 1997d).

In the commercial sector, the ratio of primary energy use to total GDP as well as commercial sector GDP fell in a number of developed countries between 1970 and the early 1990s. This decrease, primarily a result of increases in energy efficiency, occurred despite large growth in energy-using equipment in commercial buildings, almost certainly the result of improved equipment efficiencies. Growth in electricity use in the commercial sector shows a relatively strong correlation with the commercial sector GDP (IEA, 1997d).

Space heating is the largest end-use in the developed countries as a whole and in the EIT region (Nadel *et al.*, 1997), although not as important in some developed countries with a warm climate. The penetration of central heating doubled from about 40% of dwellings to almost 80% of dwellings in many developed countries between 1970 and 1992 (IEA, 1997d). District heating systems are common in some areas of Europe and in the EIT region. Space heating is not common in most developing countries, with the exception of the northern half of China, Korea, Argentina, and a few other South American countries (Sathaye *et al.*, 1989). Residential space heating energy intensities declined in most developed countries (except Japan) between 1970 and 1992 because of reduced heat losses in buildings, lowered indoor temperatures, more careful heating practices, and improvements in energy efficiency of heating equipment (IEA, 1997d; Schipper *et al.*, 1996).

Water heating, refrigeration, space cooling, and lighting are the next largest residential energy uses, respectively, in most developed countries (IEA, 1997d). In developing countries, cooking and water heating dominate, followed by lighting, small appliances, and refrigerators (Sathaye and Ketoff, 1991). Appliance penetration rates increased in all regions between 1970 and 1990. The energy intensity of new appliances declined over the past two decades; for example, new refrigerators in the US were 65% less energy-intensive in 1993 than in 1972, accounting for differences in size or performance (IEA, 1997d; Schipper *et al.*, 1996). Electricity use and intensity (MJ/m^2) increased rapidly in the commercial buildings sector as the use of lighting, air conditioning, computers, and other office equipment has grown. Fuel intensity (PJ/m^2) declined rapidly in developed countries as the share of energy used for space heating in commercial buildings dropped as a result of thermal improvements in buildings (Krackeler *et al.*, 1998). Fuel use declined faster than electricity consumption increased, with the result that primary energy use per square meter of commercial sector floor area gradually declined in most developed countries.

The carbon intensity of the residential sector declined in most developed countries between 1970 and the early 1990s (IEA, 1997d). In the service sector, carbon dioxide emissions per square meter of commercial floor area also dropped in most developed countries during this period in spite of increasing carbon intensity of electricity production in many countries (Krackeler *et al.*, 1998). In developing countries, carbon intensity of both the residential and commercial sector is expected to continue to increase, both as a result of increased demand for energy services and the continuing replacement of biomass fuels with commercial fuels (IEA, 1995).

3.3.6   *Technological and Economic Potential*

An estimate of the technological and economic potential of energy efficiency measures was recently prepared for the IPCC (Acosta Moreno *et al.*, 1996).[11] This analysis provides an estimate of energy efficiency potential for buildings on a global basis. Using the B2 Message marker scenario (Nakicenovic *et al.*, 2000) as the base case,[12] the analysis indicates an overall technical and economic potential for reducing energy-related CO_2 emissions in the buildings sector of 715MtC/yr in 2010 for a base case with carbon emissions of 2,600MtC/yr (27%), of 950MtC/yr in 2020 for a base case with carbon emissions of 3,000MtC/yr (31%), and of 2,025MtC/yr in 2050 for a base case with carbon emissions of 3,900MtC/yr (52%) (see Table 3.5).[13] It is important to note that the availability of technologies to achieve such savings cost-effectively depends critically on significant R&D efforts.

Estimates of the ranges of costs of carbon reductions are based on a synthesis of recent studies of costs (Brown *et al.*, 1998); these estimates are similar to those provided in an International Energy Agency Workshop on Technologies to Reduce Greenhouse Gas Emissions (IEA, 1999a). The qualitative rankings for the reductions in carbon emissions follow the results of the IPCC Technical Paper (Acosta Moreno *et al.*, 1996). In general, it is assumed that costs are initially somewhat higher in developing countries because of the reduced availability of advanced technology and the lack of a sufficient delivery infra-

---

[11] The review of more recent information bearing on the technical and economic potential of energy efficiency measures gives no reason to change the earlier estimate in the IPCC technical paper (see, for example, Brown *et al.*, 1998; de Almeida and Fonseca, 1999; Jochem, 1999; Kainuma *et al.*, 1999a; Lenstra, 1999; Levine *et al*, 1996; Schaeffer and Almeida, 1999; Sheinbaum *et al.*, 1998; Urge-Vorsatz and Szesler, 1999; Zhou, 1999).

[12] The original analysis was based on the IS92a scenario. From the set of IPCC SRES scenarios, for the period covered in this chapter (up to 2020), scenario B2 most resembles baseline scenarios with the low levels of technology introduction used in the literature assessed here.

[13] Of the efficiency measures that are technically and economically feasible, the IPCC report estimated that between 35% and 60% could be adopted in the market through known and established policy approaches.

Table 3.5: *Technical and economic potential for reducing energy-related carbon dioxide emissions from the buildings sector* (Acosta Moreno *et al.*, 1996).

	Projected emissions reductions (MtC)			Share of projected total emissions		
	2010	**2020**	**2050**	**2010**	**2020**	**2050**
Developed Countries + EIT Region						
Residential	325	420	660	30%	35%	54%
Commercial	185	245	450	32%	38%	68%
Total	510	665	1110	31%	36%	59%
Developing Countries						
Residential	125	170	515	20%	21%	39%
Commercial	80	115	400	24%	26%	57%
Total	205	285	915	21%	23%	45%
World	715	950	2025	27%	31%	52%

Note: Projected total emissions based on B2 Message marker scenario (standardized) (Nakicenovic *et al.*, 2000).

structure. However, depending upon conditions in the country or region, these high costs could be offset by the fact that there are many more low-cost opportunities to improve energy efficiency in most developing countries.

These studies show that with aggressive implementation of energy-efficient technologies and measures, CO_2 emissions from residential buildings in 2010 can be reduced by 325MtC in developed countries and the EIT region at costs ranging from −US$250 to −US$150/tC saved and by 125MtC in developing countries at costs of −US$200 to US$50/tC saved. Similarly, CO_2 emissions from commercial buildings in 2010 can be reduced by 185MtC in developed countries and the EIT region at costs ranging from −US$400 to −US$250/tC saved and by 80MtC in developing countries at costs ranging from -US$400 to US$0/tC saved.

3.3.7 Conclusions

Energy demand in buildings worldwide grew almost 3% per year from 1971 to 1990, dropping slightly after that as a consequence of the significant decrease in energy use in the EIT region. Growth in buildings energy use in all other regions of the world continued at an average rate of 2.5% per year since 1990. This growth has been driven by a wide variety of social, economic, and demographic factors. Although there is no assurance that these factors will continue as they have in the past, there is also no apparent means to modify most of the fundamental drivers of energy demand in residential and commercial buildings. However, there is considerable promise for improving the energy efficiency of appliances and equipment used in buildings, improving building thermal integrity, reducing the carbon intensity of fuels used in buildings, reducing the emissions of HFCs, and limiting the use of HFCs to those areas

where appropriate. There are many cost-effective technologies and measures that have the potential to significantly reduce the growth in GHG emissions from buildings in both developing and developed countries by improving the energy performance of whole buildings, as well as reducing GHG emissions from appliances and equipment within the buildings.

3.4 Transport and Mobility

3.4.1 Introduction

This section addresses recent patterns and trends in greenhouse gas (GHG) emissions by the transport sector, and the technological and economic potential to reduce GHG emissions. The chapter focuses on areas where important developments have occurred since the SAR. It does not attempt to comprehensively present mitigation options for transport, as was done there (Michaelis *et al.*, 1996). For a discussion of barriers and market potential with respect to advanced transportation technologies, the reader is referred to Chapter 5, especially Section 5.4.2. For a discussion of policies, measures and options, including behavioural strategies, the reader is referred to Chapter 6.

Recent successes with key future technologies for motor vehicles such as fuel cell power trains and advanced controls for air pollutants (carbon monoxide, hydrocarbons, oxides of nitrogen, and particulate matter) seem to promise dramatic changes in the way the transport sector uses energy and in its impacts on the environment. At the same time, the rapid motorization of transport around the world, the continued availability of low-cost liquid fossil fuels, and the recent trend of essentially constant fuel economy levels caused by demand for larger, more powerful vehicles, all point towards steadily increasing GHG emissions from transport in the near future (e.g., WEC,

1998a; Ogawa *et al.*, 1998). These are challenges that must be met by the evolution of policies and institutions capable of managing environmentally beneficial change in an increasingly global economy.

3.4.2   Summary of the Second Assessment Report

The SAR's chapter 21, Mitigation Options in the Transportation Sector (Michaelis *et al.*, 1996), provides an overview of global trends in transportation activity, energy intensities, and GHG emissions, along with a comprehensive review of economic, behavioural, and technological options for curtailing GHG emissions from the global transport sector. It concludes with an assessment of transport policies and their effects on GHG emissions. Its review of mitigation options for transportation demand management, modal structure, and alternative fuels, and its analysis of transport policies are still essentially up to date and are not repeated in this section.

Historically, transportation energy use and GHG emissions have increased because reductions in energy intensities have not kept pace with increasing transport activity. The world's motor vehicle fleet grew at an average annual rate of 4.5% from 1970 to 1990. Over the same period, light-duty vehicle fuel economy improved by 2% per year or less. Increases in vehicular fuel economy have also been accompanied by declining vehicle occupancy rates. It is noted below that the fuel economy of road passenger transport vehicles has levelled off since the publication of the SAR, and no longer appears to be improving. Air travel and truck freight activity have also grown more rapidly than energy intensities (energy use per passenger km) have declined. Since 1970, transport energy use and GHG emissions have grown at an average annual rate of 2.4%.

The SAR concluded that by 2010 it might be technically feasible to reduce energy intensities for new transport vehicles by 25% to 50% without reduction of performance or quality, by adopting a variety of fuel economy technologies. It noted that the economic potential would likely be smaller. The adoption of energy efficiency improvements throughout the sector was estimated to be able to reduce transportation energy use in 2025 by one-third versus projected levels.

The SAR also extensively reviewed the life cycle GHG emissions from alternative fuels and concluded that only fuels derived from biomass or electricity generated from substantially non-fossil sources could reduce life cycle GHG emissions by more than 20% versus conventional gasoline internal combustion engine vehicles. Compressed or liquefied natural gas and liquefied petroleum gases are capable of reducing full fuel cycle GHG emissions by 10% to 20% over gasoline-powered light-duty vehicles, but emissions would actually increase if these fuels were used to replace diesel engines in heavy-duty vehicles.

3.4.3   Historic and Future Trends

Since the publication of the SAR, important advances have been achieved in several areas of automotive technology. Among the most significant are: (1) two global automotive manufacturers are now selling hybrid automobiles 5-10 years ahead of what was anticipated just 5 years ago; (2) dramatic reductions have been made in fuel cell cost and size, such that several manufacturers have announced that they will introduce fuel cell vehicles by 2005, 10-20 years ahead of what was previously anticipated; and (3) improvements in fuels, engine controls, and emissions after-treatment led to the production of a gasoline internal combustion engine vehicle with virtually zero emissions of urban air pollutants. This achievement, combined with regulations requiring low-sulphur fuels, may foreshadow the development of acceptable emissions control systems for more energy efficient direct injection engines, although significant hurdles remain. It may also reduce the incentive for adopting alternative fuel vehicles, such as battery electric and natural gas vehicles, which can also have lower greenhouse gas emissions. These developments could have profound effects on future GHG emissions from road, rail, marine, and pipeline transport. Also, since the publication of the SAR, the IPCC has released a comprehensive report on the impacts of aviation on the global atmosphere (Penner *et al.*, 1999) that includes a projection of expected progress in reducing energy intensity and GHG emissions from commercial air transport, and adds greatly to the information about aviation's effects on climate.

Worldwide, transport produces roughly 20% of carbon emissions and smaller shares of the other five greenhouse gasses covered under the Kyoto Protocol. According to IEA statistics, the transport sector's share of world GHG emissions increased from about 19% in 1971 to 22% in 1995 (Price *et al.*, 1998) and 23% in 1997 (IEA, 1999c, p. II.67). Excluding emissions from vehicle air conditioners (described in the Appendix), CO_2 from combustion of fossil fuels is the predominant GHG

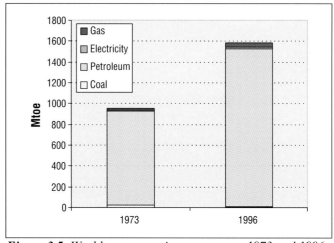

Figure 3.5: *World transportation energy use, 1973 and 1996.*

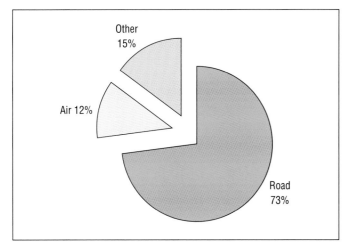

Figure 3.6: *World transport energy by mode, 1996.*

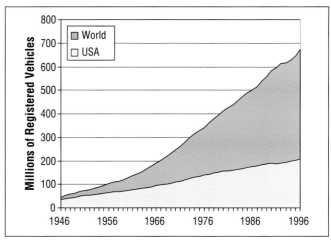

Figure 3.7: *Growth of world motor vehicle population, 1946-1996.*

produced by transport, accounting for over 95% of the annual global warming potential produced by the sector. Nitrous oxide produced by vehicles equipped with catalytic converters, and methane emitted by internal combustion engines account for nearly all the remainder. Almost all of the carbon comes from petroleum fuels. Between 1973 and 1996, world transportation energy use, of which petroleum-derived fuels comprise over 95%, increased by 66% (*Figure 3.5*). Alternative energy sources have not played a significant role in the world's transport systems. Despite two decades of price upheavals in world oil markets, considerable research and development of alternative fuel technologies, and notable attempts to promote alternative fuels through tax subsidies and other policies, petroleum's share of transport energy use has not decreased (94.7% in 1973 and 96.0% in 1996) according to IEA statistics (IEA, 1999c).

On a modal basis, road transport accounts for almost 80% of transport energy use (*Figure 3.6*). Light-duty vehicles alone comprise about 50%. Air transport is the second largest, and most rapidly growing mode, with about 12% of current transport energy use according to International Energy Agency estimates (IEA, 1999c).

The growth of transport energy use, its continued reliance on petroleum and the consequent increases in carbon emissions are driven by the long-term trends of increasing motorization of world transport systems and ever-growing demand for mobility. Immediately after World War II, the world's motor vehicle fleet numbered 46 million vehicles, and 75% of the world's cars and trucks were in the USA. In 1996, there were 671 million highway vehicles worldwide, and the US share stood at just over 30% (*Figure 3.7*). Since 1970, the US motor vehicle population has been growing at an average rate of 2.5% per year, but the population of vehicles in the rest of the world has been increasing almost twice as rapidly at 4.8% per year (AAMA, 1998, p. 8). The same patterns of growth are discernible in statistics on vehicle stocks (ECMT, 1998).

Transport achieved major energy efficiency gains in the 1970s and 1980s, partly because of an economic response to the oil price increases of 1973 to 1974 and 1979 to 1980, and partly as a result of government policies inspired by the oil price shocks. Driven principally by mandatory standards, the average fuel economy of new passenger cars doubled in the USA between 1974 and 1984 (e.g., Greene, 1998). In Europe, similar improvements were achieved by a combination of voluntary efficiency agreements and higher taxes on motor fuels. From 1980 to 1995 the average sales-weighted fuel consumption rates of passenger cars sold in Europe and Japan fell by 12%, from 8.3 l/100km to 7.3 l/100km (Perkins, 1998). All of the decrease, however, occurred between 1980 and 1985 (*Figure 3.8*). Since 1985, the fuel economies of light-duty vehicles sold in the USA and Europe have remained essentially constant.

Energy efficiency improvements in other modes have also slowed or stagnated over the past 10-15 years. Average energy

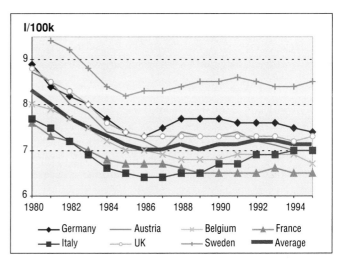

Figure 3.8: *Weighted average fuel consumption of new passenger cars.*

Table 3.6: *Modal energy intensities, 1973 to 1994*
(IEA, 1997d, Table 3.1).

Energy intensity	Europe-8	Europe-8	USA	USA	Japan	Japan
MJ per pass-km	1973	1993	1973	1994	1973	1994
Average, all modes	1.46	1.56	3.10	2.52	1.22	1.67
Car	1.65	1.73	3.10	2.59	2.20	2.46
Bus	0.58	0.71	0.79	1.03	0.54	0.73
Rail	0.58	0.48	1.81	2.15	0.17	0.19
Air	4.55	2.78	4.92	2.46	3.49	2.13

use per passenger-kilometre in Europe and Japan actually increased between 1973 and 1993/4, but declined by almost 20% in the USA (*Table 3.6*). Bus and rail modal energy intensities generally increased, with the exception of rail travel in Europe. The energy intensity of commercial air travel, however, has declined consistently, achieving a 40%-50% reduction over the last 25 years.

On the freight side, trucking's share of tonne km increased in every OECD country, included in a recent analysis of energy trends by the IEA (1997d, *Figure 4.6*), leading to an overall increase in the energy intensity (MJ/t-km) of freight movements. Unlike passenger modes, for freight, changes in modal structure tend to dominate changes in modal energy intensities in determining overall energy intensity (IEA, 1997d, p. 127).

The slowing of energy efficiency improvements in recent years has occurred despite the fact that new technologies with the potential to increase energy efficiency continue to be adopted. In Europe, the market share of diesel cars increased from 7% in 1980 to 17% in 1985 and 23% in 1995, due in part to lower diesel fuel taxes (Perkins, 1998). In the USA, emissions and fuel economy standards increased the use of multipoint fuel injection from 16% of new light-duty vehicles in 1985 to 100% in 1999, and installation of 4- and 5-valve engines increased from zero to 40% over the same period (Heavenrich and Hellman, 1999, *Table 4*). Manufacturers also continued to substitute lighter weight materials such as high-strength steel and aluminium, and to reduce aerodynamic drag and tyre rolling resistance. Yet fuel economy stagnated because vehicles were made larger and much more powerful. Between 1988 and 1999, the average mass of a new US light-duty vehicle

increased from 1381 kg to 1534 kg. At the same time, power per kg increased 29% (Heavenrich and Hellman, 1999). In Europe, the average power per car increased by 27% between 1980 and 1995, from 51 to 65 kW (Perkins, 1998).

Because of the slowing down of energy efficiency gains, world transportation energy use is now increasing at just slightly less than the rate of growth in transportation activity. Given the relatively close correlation between economic growth and the demand for transport (*Table 3.7*; see WEC, 1995b, Ch. 3.2 for further details), it is reasonable to expect continued strong growth of transport energy use and carbon emissions, unless significant, new policy initiatives are undertaken. The following paragraphs review several studies of future transportation demand and energy use. A common theme of these and many others is strong growth in transport energy use and the challenges it poses to reducing greenhouse gas emissions from the sector.

Projections of future transport energy use under baseline assumptions reflect an expectation of robust growth in transport activity, energy demand, and carbon emissions through 2020. The World Energy Council (WEC, 1995b) considered three alternative scenarios for transport energy demand through 2020: (1) "markets rule", (2) "muddling through", and (3) "green drivers". Of these, markets rule reflects a high-growth baseline future (2.8%/yr in the OECD, 5.2% in the rest of the world), muddling through a lower growth one (2.2%/year in OECD, 4.2% elsewhere). In the markets rule scenario, world transport energy consumption grows 200% in the quarter century from 1995 to 2020. In the muddling through scenario, transport energy use grows by 100% by

Table 3.7: *Annual growth in GDP and transport in OECD countries, 1975-1990*
(WEC, 1995b, Table 3.2.1).

	GDP	Freight traffic	Passenger traffic
OECD Europe	2.6%	2.8%	2.8%
USA	2.8%	2.6%	2.3%
Japan	4.2%	3.6%	2.6%

Table 3.8: *Energy information administration projections of global transport energy use to 2020* (US DOE/EIA, 1999b, Tables E1, E7, E8, E9).

	1996	2010	2020	Average annual
		Millions of barrels per day		percent change
Road	25.5	37.4	45.2	2.4
Air	4.0	6.6	9.6	3.7
Other	5.1	5.8	6.5	1.0
TOTAL	34.6	49.9	61.3	2.4

2020, with most of the shortfall from the markets rule scenario occurring after 2010. In the green drivers scenario, transport energy use is nearly constant as a result of much higher energy taxes and comprehensive environmental regulation. In all three scenarios, growth in freight transport and air travel far outpace the growth of passenger vehicle travel, so that the passenger car's share of total transport energy use falls from about 50% in 1995 to 30% by 2020.

A more recent WEC (1998a) report foresaw considerably slower growth in transport energy use through 2020: 55% in a base case with an 85% increase in a higher economic growth case. In both cases, light-duty vehicles continued to dominate through 2020, accounting for 44% of global transport energy demand in the base case. Still road freight and air travel gained on highway passenger vehicles. Road freight increased from 30% of transport energy demand in 1995 to 33% in 2020. Air transport's share grew from 8% to almost 13%. Global carbon emissions from transport were expected to grow by 56% in the base case, from 1.6GtC in 1995 to 2.5GtC in 2020.

The US DOE and US Energy Information Administration's (EIA's) International Energy Outlook (1999b, p.115) foresees transportation's share of world oil consumption climbing from 48% in 1996 to 53% by 2010 and 56% by 2020. The EIA expects a 77% increase in total world transport energy use by 2020, an average annual global growth rate of 2.4% (*Table 3.8*). Road dominance of energy use is maintained by the rapid increase in vehicle stocks outside of the OECD. The world motor vehicle population is projected to surpass 1.1 billion vehicles in 2020. The SAR (Michaelis *et al.*, 1996, *Table 21-3*) presented projections of future global vehicle stocks ranging from 1.2 to 1.6 billion by 2030, rising to 1.6 to 5.0 billion by 2100.

Projections of passenger travel, energy use, and CO_2 emissions to 2050 by Schafer and Victor (1999) show carbon emissions rising from 0.8GtC in 1990 to 2.7GtC in 2050, driven by an increase in travel demand from 23 trillion passenger-kilometres in 1990 to 105 trillion p-km in 2050. The model used is based on constant travel budgets for time and money, so that as incomes and travel demand grow, passenger travel must shift to faster modes in order to stay within time budget limits. As a result, automobile travel first increases, and then eventually

declines as travel shifts to high-speed rail and air. The projections assume that car, bus and conventional rail systems maintain their energy intensities at approximately 1990 levels through 2050. Energy intensity of the air mode (which by the authors' definition includes high-speed rail) is assumed to decrease by 70% by 2050, substantially more than the Penner *et al.* (1999)-report estimates. No change in the average carbon content of transportation fuels is assumed.

Projections such as these suggest that it will be very difficult to attain a goal such as holding transport's carbon emissions below 1990 levels by 2010. Lead times for introducing significant new technologies, combined with the normal lifetimes for transportation equipment on the order of 15 years, imply that sudden, massive changes in the trends and outlooks described above can be achieved only with determined effort. At the same time, dramatic advances in transport energy technology have been achieved over just the past 5 years, and the potential for further advances is very promising. By 2020 and beyond the world may see revolutionary changes in energy sources and power plants for new transport equipment, provided that appropriate policies are implemented to accelerate and direct technological changes towards global environmental goals.

3.4.4   *New Technology and Other Options*

Significant energy efficiency technologies that less than ten years ago were thought too "long-term" to be considered in an assessment of fuel economy potential through 2005 (NRC, 1992), are now available for purchase in at least some OECD countries. The US Partnership for a New Generation of Vehicles (PNGV), the European "Car of Tomorrow" and Japanese Advanced Clean Energy Vehicle programmes have helped achieve these striking successes. In December 1997, a commercial hybrid electric vehicle was introduced in Japan, demonstrating a near doubling of fuel economy over the Japanese driving cycle for measuring fuel economy and emissions. In 1998, a practical, near zero-emission (considering urban air pollutants) gasoline-powered passenger car was developed, and demonstrated. This achievement established the possibility that modern emissions control technology, combined with scientific fuel reformulation, might be able to achieve virtually any desired level of tailpipe emissions at reasonable cost using conventional fossil fuel resources. Emissions problems now limit the application of lean-burn fuel economy technologies such as the automotive diesel engine. Advanced technologies and cleaner fuels may achieve similar results for lean-burn gasoline and diesel engines in the near future. Such advances in urban air pollutant emissions controls for fossil fuel burning engines reduce the environmental incentives for curbing fossil fuel use by road vehicles. Automotive fuel cells also realized order of magnitude reductions in size and cost, and dramatic improvements in power density. The status of these key technologies is reviewed below.

3.4.4.1   Hybrid Electric Vehicles

A hybrid electric vehicle combines an internal combustion engine or other fuelled power source with an electric drivetrain and battery (or other electrical storage device, e.g., an ultracapacitor). Potential efficiency gains involve: (1) recapture of braking energy (with the motor used as generator and captured electricity stored in the battery); (2) potential to downsize the engine, using the motor/battery as power booster; (3) potential to avoid idling losses by turning off the engine or storing unused power in the battery; and (4) increasing average engine efficiency by using the storage and power capacity of the electric drivetrain to keep engine operation away from low efficiency modes. Toyota recently introduced a sophisticated hybrid subcompact auto, the Prius, in Japan and has since introduced a version into the US market. Honda also began selling in model year 2000 its Insight hybrid, a two seater. Ford, GM, Daimler/Chrysler and several others have hybrids in advanced development. The most fuel-efficient hybrid designs can boost fuel economy by as much as 50% at near-constant performance under average driving conditions. The added complexity of the dual powertrain adds significantly to the cost of hybrids, and this could hinder their initial market penetration in countries with low fuel prices, unless policies are adopted to promote them.

Hybrids attain their greatest efficiency advantage—potentially greater than 100%—over conventional vehicles in slow stop-and-go traffic, so that their first applications might be urban taxicabs, transit buses, and service vehicles such as garbage trucks. An assessment of the potential for hybridization to reduce energy consumption by medium-sized trucks in urban operations concluded that reductions in l/100km of 23% to 63% could be attained, depending on truck configuration and duty cycle (An *et al.*, 2000).

Testing the Toyota Prius under a variety of driving conditions in Japan, Ishitani *et al.*,(2000) found that the hybrid electric design gave 40%–50% better fuel economy at average speeds above 40 km/h, 70%–90% better in city driving at average speeds between 15 and 30 km/h and 100%–140% better fuel economy under highly congested conditions with average speeds below 10 km/h. Actual efficiency improvements achieved by hybrids will depend on both design of the vehicle and driving conditions. Much of the efficiency ben-

efit of hybrids is lost in long-distance, constant high-speed driving.

3.4.4.2   Lower Weight Structural Materials

Mass reduction via materials substitution is a potentially important strategy for improving light-duty vehicle fuel economy, because it permits synergistic reductions in engine size without loss of performance. The use of alternative materials to reduce weight has been historically restrained by cost considerations, manufacturing process technology barriers, and difficulty in meeting automotive requirements for surface finish quality, predictable behaviour during crash tests, or repairability. The past few years have seen significant developments in space frame structures, advanced new manufacturing technology for plastics and aluminium, and improved modelling techniques for evaluating deformability and crash properties. Ford has displayed an advanced lightweight prototype that is a mid-size car with a weight of only 900 kg, as compared to vehicles weighing 1450 kg today. Even if some of the more exotic weight-saving materials from Ford's prototype were discarded, a weight reduction of 30% or more appears possible. With engine downsizing to maintain a constant ratio of kW/kg, this should produce a 20% fuel economy improvement. Some aluminium-intensive luxury cars have already been introduced (for example, the Audi A8 and the new Volkswagen Lupo with 3l/100km consumption), and Ford is known to be considering the introduction of such a vehicle in the mass market.

According to Bouwman and Moll (1999), 85% of life cycle vehicle energy use occurs in the vehicle use phase, with about 15% accounted for in vehicle production and about 3% recovered in recycling. Mass reductions of 30% to 40% via extensive substitution of aluminium for steel have been incorporated in the designs of advanced, high fuel economy prototypes, improving fuel economy by 20% to 25%. Because the production of aluminium requires more energy than production of steel, and the recycling of aluminium auto bodies is more difficult given current recycling technology, the benefits of substituting aluminium for steel must be assessed by a life cycle analysis of greenhouse gas emissions (efforts are being made to improve aluminium recycling technology, however). Analyses have shown that accounting for life cycle impacts diminishes, but does not eliminate GHG emission reductions caused by the use of aluminium for mass reduction in motor vehicles (*Figure 3.9*). The amount of reduction, however, is sensitive to several key assumptions. Considering the total life cycle emissions for a typical passenger car in the USA, Das (2000) concluded that higher net emissions in the production plus recycling stages would reduce the potential GHG benefits of aluminium in the vehicle use stage by 6.5% versus conventional steel auto bodies, but by 15.8% versus advanced, ultralight steel body (ULSAB) designs.

Because the increased emissions come first in the production stage, there is a "recovery" period before net emissions reductions are realized. Das (2000) found a recovery period of four

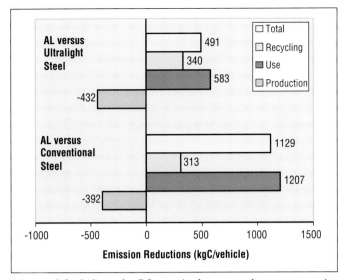

Figure 3.9: *Life cycle CO$_2$-equivalent greenhouse gas emission estimates for automobile body materials.*

years versus steel but 10 years versus ultra-light steel auto-bodies (ULSAB) for an aluminium-intensive vehicle. An analysis by Clark (1999) of aluminium versus conventional steel, assuming fewer lifetime kilometres, found a cross-over point at approximately eight years for a single vehicle, but at 15 years for an expanding fleet of aluminium-intensive vehicles. In comparison to ULSAB, the car fleet crossover point was found to be at 33 years. In other OECD countries where lifetime vehicle kilometres may be one-half, or less, the levels of the USA, the cross-over points would be even farther in the future. Sensitivity analyses have shown that the results depend strongly on key assumptions, especially the sources of energy for aluminium production and lifetime vehicle miles.

Bouwman and Moll (1999) obtained similar results in scenarios based on the growing Dutch passenger car fleet. A scenario in which aluminium vehicles were introduced in 2000achieved lower energy use than a steel scenario after 2010. By 2050, the aluminium scenario energy use was 17% below that of the all steel scenario.

3.4.4.3   Direct Injection Gasoline and Diesel Engines

Direct injection lean-burn gasoline engines have already been introduced in Japan and Europe, but have been restricted in North America by a combination of tight emission standards and high sulphur content in gasoline. Fuel sulphur levels will be drastically reduced in Europe and North America over the next 10 years. The US EPA, for example, has proposed regulations that would set caps on sulphur content of 30 ppm for gasoline and 15 ppm for diesel fuel (Walsh, 2000). While planned reductions in the sulphur content of fuels to the range of 10 to 30 ppm will allow direct injection gasoline engines to be introduced, it is not yet clear that the full fuel efficiency benefits can be retained at lower NO$_x$ levels. Preliminary evaluations suggest that benefits may be in the 12% to 15% range

rather than the 16% to 20% range available in Japan and Europe, but even this assumes some advances in after treatment technology. Engine costs, however, seem quite moderate, in the range of US$200 to US$300 more than a conventional engine.

Direct injection (DI) diesel engines have long been available for heavy trucks, but recently have become more competitive for automobiles and light trucks as noise and emission problems have been resolved. These new engines attain about 35% greater fuel economy than conventional gasoline engines and produce about 25% less carbon emissions over the fuel cycle. In light-duty applications, DI diesels may cost US$500 to US$1000 more than a comparable gasoline engine. Tightening of NO$_x$ and particulate emissions standards presents a challenge to the viability of both diesel and gasoline lean-burn engines, but one that it may be possible to overcome with advanced emissions controls and cleaner fuels (e.g., Martin *et al.*, 1997; Gerini and Montagne, 1997; Mark and Morey, 1999; Greene, 1999). Further improvements in diesel technology also offer substantial promise in heavy-duty applications, especially heavy trucks but also including marine and rail applications. Current research programmes are aiming to achieve maximum thermal efficiencies of 55% in heavy-duty diesels (compared to current peak efficiencies of about 40%-45%), with low emissions.

3.4.4.4   Automotive Fuel Cells

Fuel cells, which have the potential to achieve twice the energy conversion efficiency of conventional internal combustion engines with essentially zero pollutant emissions, have received considerable attention recently, with most major manufacturers announcing their intentions to introduce such vehicles by the 2005 model year. The recent optimism about the fuel cell has been driven by strong advances in technology performance, including rapid increases in specific power that now allow a fuel cell powertrain to fit into a conventional vehicle without sacrificing its passenger or cargo capacities. While fuel cell costs have been reduced by approximately an order of magnitude, they are still nearly 10 times as expensive per kW as spark ignition engines. Recent analyses project that costs below US$40/kW for complete fuel cell drivetrains powered by hydrogen can be achieved over the next ten years (Thomas *et al.*, 1998). Hydrogen is clearly the cleanest and most efficient fuel choice for fuel cells, but there is no hydrogen infrastructure and on-board storage still presents technical and economic challenges. Gasoline, methanol or ethanol are possible alternatives, but require on-board reforming with consequent cost and efficiency penalties. Mid-size fuel cell passenger cars using hydrogen could achieve fuel consumption rates of 2.5 gasoline equivalent l/100 km in vehicles with lightweight, low drag bodies; comparable estimates for methanol or gasoline-powered fuel cell vehicles would be 3.2 and 4.0 l/100 km (gasoline equivalent), respectively. While gasoline is relatively more difficult to reform, it has the benefit of an in-place refuelling infrastructure, and progress has been made in reformer technology (NRC, 1999a).

The fuel economy of hydrogen fuel cell vehicles is projected to be 75% to 250% greater than that of conventional gasoline internal combusiton engine (ICE) vehicles, depending on the drive cycle (Thomas *et al.*, 1998). Primarily as a result of energy losses in reforming, comparable estimates of the fuel economy benefit of methanol-powered fuel cells range from 25% to 125%. The GHG reduction potential of hydrogen or methanol fuel cells, however, requires a "well-to-wheels" analysis to measure the full fuel cycle impacts. Both sources cited here include emissions of all significant greenhouse gases produced in the respective processes. Assuming hydrogen produced by local reforming of natural gas, Thomas *et al.* (1998, *Figure 8*) estimated roughly a 40% reduction in well-to-wheels GHG emissions for a direct hydrogen fuel cell vehicle versus a conventional gasoline ICE vehicle getting 7.8 l/100km (about 150 g CO_2 equivalent per km, versus 250). Wang (1999a, p. 4) concluded that direct hydrogen fuel cell vehicles, with hydrogen produced at the refuelling station by reforming natural gas, would reduce full fuel cycle GHG emissions by 55% to 60% versus a comparably sized 9.8 l/100km gasoline vehicle. Hydrogen could also be produced from methane in large-scale centralized facilities. This could create opportunities for sequestering carbon but would also require an infrastructure for hydrogen transport. Hydrogen produced via electrolysis was estimated to produce 50% to 100% *more* full fuel cycle

GHG emissions, depending on the energy sources used to generate electricity. Methanol produced from natural gas was estimated to give a 50% reduction in full fuel cycle GHG emissions. Wang (1999b, *Table 4.4*) projected direct hydrogen fuel cell vehicles to be 180% to 215% more energy efficient, and methanol fuel cell vehicles to be 110% to 150% more efficient. These analyses attempt to hold other vehicle characteristics constant but, of course, that is never entirely possible.

3.4.4.5   Fuel Cycle Emissions

In considering the impacts of advanced technologies and alternative fuels on emissions of greenhouse gases, it is important to include the full fuel cycle, since emissions in feedstock and fuel production can vary substantially. The same fuel can be produced from several feedstocks, and this too has important implications for greenhouse gas emissions. Finally, as Ishitani *et al.* (2000) have demonstrated, the use of different drive cycles as a basis for comparison can also change the ranking of various advanced technologies. Hybrid vehicles, for example, will perform relatively better under congested, low-speed driving conditions. *Table 3.9* shows a sample of results obtained by Wang (1999a) based on US assumptions for passenger car technologies expected to be available in the year 2010. In all cases, carbon dioxide is the predominant GWP-weighted

Table 3.9: *GHG emissions from advanced automotive technologies and alternative fuels* (Wang, 1999, App. B-II).

	CO₂-equivalent grams per km						
	Fuel cycle stage			**Greenhouse gas**			
	Feedstock	**Fuel**	**Operation**	**CO_2**	**N_2O**	**CH_4**	**Total**
Gasoline (reformulated)	15.6	52.7	228.9	282.2	5.7	9.4	295.6
Gasoline direct injection (DI)	12.6	42.1	184.3	225.6	5.7	7.7	237.6
Propane (from natural gas)	19.0	13.6	197.6	217.5	5.5	7.3	228.9
Compressed natural gas (CNG)	30.7	21.3	174.6	206.2	3.1	17.3	225.3
Diesel DI	10.6	27.2	161.6	191.7	3.3	4.5	198.4
20% biodiesel DI	11.7	32.7	132.7	169.1	3.7	4.3	176.1
Grid-Hybrid (RFG)	9.8	63.5	88.8	152.7	4.1	5.3	161.2
Hybrid (RFG)	8.6	27.5	123.3	148.3	5.7	5.4	158.5
Electric vehicle (EV, US mix)	12.3	145.2	0.0	152.1	0.6	4.8	156.6
Fuel Cell (Gasoline)	7.8	26.1	112.6	140.8	1.4	4.4	145.7
Hybrid. CNG	19.1	13.2	110.5	127.6	2.7	12.5	142.0
Fuel cell (methanol. NG)	8.1	17.9	83.1	105.0	1.2	3.0	108.5
Fuel cell (H_2 from CH_4))	11.0	97.3	0.0	103.1	0.2	5.0	107.62
EV (CA mix)	10.4	51.1	0.0	58.5	0.2	2.8	61.1
Fuel cell (solar)	0.0	20.3	0.0	18.9	0.2	1.2	20,2

100-year global warming potentials		
CO_2	**N_2O**	**CH_4**
1	310	21

greenhouse gas. Advanced direct injection gasoline engines appear to achieve nearly the same greenhouse gas emissions reductions as spark-ignition engine vehicles fuelled by propane or compressed natural gas. Direct-injection diesel vehicles show a reduction of one-third over advanced gasoline vehicles. The gasoline hybrid achieves almost a 50% reduction, while the grid-connected hybrid does no better because of the large share of coal in the US electricity generation mix. The dependence of electric vehicle (EV) emissions on the power generation sector is illustrated by the very large difference between EVs using California versus US average electricity. Fuel cell vehicles using gasoline are estimated by Wang (1999a) to achieve a 50% reduction in emissions, but hybrid vehicles fuelled by compressed natural gas (CNG) do slightly better. Fuel cells powered by hydrogen produced by reforming natural gas locally at refuelling outlets are estimated to reduce fuel cycle greenhouse gas emissions by almost two thirds, while those using hydrogen produced from solar energy achieve more than a 90% reduction. Clearly, Wang's (1999b) estimates differ substantially from those of Thomas *et al.* (1998) as noted above. Such differences are common, as a result of differences in the many assumptions that must be made in fuel cycle analyses.

3.4.4.6   Use of Biofuels

Liquid and gaseous transport fuels derived from a range of biomass sources are technically feasible (see Section 3.8.4.3.2). They include methanol, ethanol, di-methyl esters, pyrolytic oil, Fischer-Tropsch gasoline and distillate, and biodiesel from vegetable oil crops (Section 3.6.4.3). Ethanol is commercially produced from sugar cane in Brazil and from maize in the USA where it has been sold neat or blended for more than a decade. Ethanol is blended with gasoline at concentrations of 5-15%, thereby replacing oxygenates more typically used in North America such as methyl-t-butylether (MTBE) and ethyl-t-butylether (ETBE) additives.   ETBE production from bio-ethanol is also a promising market in Europe but the production costs by hydrolysis and fermentation from cereals or sweet sorghum crops remain high (Grassi, 1998).

In Brazil the production of ethanol-fuelled cars achieved 96% market share in 1985 but declined to 3.1% in 1995 and 0.1% in 1998. Since the government approved a higher blend level (26%) of ethanol in gasoline the production of ethanol has continued to increase achieving a peak of 15,307m^3 in the 1997/98 harvesting season. This represented 42.73% of the total fuel consumption in all Otto cycle engines giving an annual net carbon emission abatement of 11% of the national total from the use of fossil fuels (IPCC, 2000).

National fuel standards are in place in Germany for biodiesel and many engine manufacturers such as Volkswagen now maintain warranties (Schindlbauer, 1995). However, energy yields (litres oil per hectare) are low and full fuel cycle emissions and production costs are high (see Section 3.8.4.3.2).

3.4.4.7   Aircraft Technology

Several major technologies offer the opportunity to improve the energy efficiency of commercial aircraft by 40% or more (*Table 3.10*). The Aeronautics and Space Engineering Board of the National Research Council (NRC, 1992, p. 49) concluded that it was feasible to reduce fuel consumption per seat mile for new commercial aircraft by 40% by about 2020. Of the 40%, 25% was expected to come from improved engine performance, and 15% from improved aerodynamics and weight. A reasonable preliminary goal for reductions in NO$_x$ emissions was estimated to be 20%–30%.

An assessment of breakthrough technologies by the US National Research Council (1998) estimated that the blended wing body concept alone could reduce fuel consumption by 27% compared to conventional aircraft, assuming equal engine efficiency. The NRC report also identified a number of breakthrough technologies in the areas of advanced propulsion systems, structures and materials, sensors and controls, and alternative fuels that could have major impacts on aircraft energy use and GHG emissions over the next 50 years.

Noting that the energy efficiency of new production aircraft has improved at an average rate of 1-2% per year since the dawn of the jet era, the IPCC Special Report on *Aviation and the Global Atmosphere* concluded that the fuel efficiency of new production aircraft could improve by 20% from 1997 to 2015 (*Table 3.11*), as a result of a combination of reductions in aerodynamic drag and airframe weight, greater use of high-bypass engines with improved nacelle designs, and advanced, "fly-by-light" fibre optic control systems (Penner *et al.*, 1999, Ch. 7). Advanced future aircraft technologies including laminar flow concepts, lightweight materials, blended wing body designs, and subsystems improvements were judged to offer 30%-40% to 40%-50% efficiency improvements by 2050, with the lower range more likely if reducing NO$_X$ emissions is a high priority. The purpose of these scenarios was not to describe the technological or economic potential for efficiency improvement and emissions reductions, but rather to provide a "best judgement" scenario for use in assessing the impacts of aviation on the global atmosphere through 2050. A

Table 3.10: Energy information administration aircraft technology estimates

Technology	Year of introduction	% gain in seat-km per kg
Ultra-high bypass engine	1995	10
Propfan engine	2000	23
Hybrid laminar flow	2020	15
Advanced aerodynamics	2000	18
Material substitution	2000	15
Engine thermodynamics	2010	20

Table 3.11: *Historical and future improvements in new production aircraft energy efficiency (%)* (Lewis and Niedzwiecki, 1999, Table 7.1).

Time period	Airframe	Propulsion	Total	percent per year
1950 to 1997	30	40	70	1.13
1997 to 2015	10	10	20	1.02
1997 to 2050	25	20	45	0.70

number of alternatives to kerosene jet fuel were considered. None were considered likely to be competitive with jet fuel without significant technological breakthroughs. On a fuel cycle basis, only liquid methane and hydrogen produced from nuclear or renewable energy sources were estimated to reduce greenhouse gas emissions relative to jet fuel derived from crude oil.

In operation, aircraft seat-km per kg is also influenced by aircraft size, and overall passenger-km per kg efficiencies depend on load factors as well. Industry analysts (Henderson, 1999) have forecasted an increase in global load factors to 73% by 2018, but foresee only a small potential for increasing aircraft size, however, since most additional capacity is expected to be supplied by increased flight frequencies. If average aircraft size could be increased, perhaps as a strategy for reducing airport congestion, further reductions in energy intensity could be achieved.

3.4.4.8   Waterborne Transport

Opportunities for reducing energy use and GHG emissions from waterborne transport were not covered in the SAR. The predominant propulsion system for waterborne transport is the diesel engine. Worldwide, 98% of freighters are powered by diesels. Although the 2% powered by steam electric drive tend to be the largest ships and account for 17% of gross tonnage, most are likely to be replaced by diesels within the next 10 years (Michaelis, 1997). Still, diesel fuel accounted for only 21% of international marine bunker fuel consumed in 1995 (Olivier and Peters, 1999). Modern marine diesel engines are capable of average operating efficiencies of 42% from fuel to propeller, making them already one of the most efficient propulsion systems. The best modern low-speed diesels can realize efficiencies exceeding 50% (Farrell *et al.*, 2000).

Fuel cells might be even more efficient, however, and might possibly be operated on fuels containing less carbon (Interlaboratory Working Group, Appendix C, 1999). Design studies suggest that molten carbonate fuel cell systems might achieve energy conversion efficiencies of 54%, and possibly 64% by adding a steam turbine bottoming cycle. These studies do not consider full fuel cycle emissions, however. Farrell *et al.* (2000) estimated the cost of eliminating carbon emissions from marine freight by producing hydrogen from fossil fuel, sequestering the carbon, and powering ships by solid oxide or molten

carbonate fuel cells at US$218/tC, though there is much uncertainty about costs at this time.

A number of improvements can be made to conventional diesel vessels in, (1) the thermal efficiency of marine propulsion (5%–10%); (2) propeller design and maintenance (2%–8%); (3) hydraulic drag reduction (10%); (4) ship size; (5) speed (energy use increases to the third power of speed); (6) increased load factors; and (7) new propulsion systems, such as underwater foils or wings to harness wave energy (12%–64%) (CAE, 1996). More intelligent weather routing and adaptive autopilot control systems might save another 4%–7% (Interlaboratory Working Group, Appendix C, 1999).

3.4.4.9   Truck Freight

Modern heavy trucks are equipped with turbo-charged direct-injection diesel engines. The best of these engines achieve 45% thermal efficiency, versus 24% for spark-ignited gasoline engines (Interlaboratory Working Group, 1997). Still, there are opportunities for energy efficiency improvements and also for lower carbon alternative fuels, such as compressed or liquified natural gas in certain applications. By a combination of strategies, increased peak pressure, insulation of combustion chambers, recovery of waste heat, and friction reduction, thermal efficiencies of 55% might be achievable, though there are unresolved questions about nitrogen oxide emissions (US DOE/OHT, 1996). For medium-heavy trucks used in short distance operations, hybridization may be an attractive option. Fuel economy improvements of 60%-75% have been estimated for smaller trucks with 5-7 litre engines (An *et al.*, 1999). With drag coefficients of 0.6 to 0.9, heavy trucks are much less aerodynamic than light-duty vehicles with typical drag coefficients of 0.2 to 0.4. Other potential sources of fuel economy improvement include lower rolling resistance tyres and reduced tare weight. The sum total of all such improvements has been estimated to have the potential to improve heavy truck fuel economy by 60% over current levels (Interlaboratory Working Group, 2000).

3.4.4.10   Systems Approaches to Sustainability

Recognizing the growing levels of external costs produced by the continuing growth of motorized transport, cities and nations around the world have begun to develop plans for achieving sustainable transport. A recent report by the ECMT

(1995) presents three policy "strands", describing a progression of scenarios intended to lead from the status quo to sustainability. The first strand represents "best practice" in urban transport policy, combining land-use management strategies (such as zoning restrictions on low-density development and parking area controls) with advanced road traffic management strategies, environmental protection strategies (such as tighter pollutant emissions regulations and fuel economy standards), and pricing mechanisms (such as motor fuel taxes, parking charges, and road tolls). Even with these practices, transport-related CO_2 emissions were projected to increase by about one-third in OECD countries over the next 20 years and by twice that amount over the next 30 to 40 years. A second strand added significant investment in transit, pedestrian, and bicycle infrastructure to shape land use along with stricter controls on development, limits on road construction plus city-wide traffic calming, promotion of clean fuels and the setting of air quality goals for cities, as well as congestion pricing for roads and user subsidies for transit. The addition of this strand was projected to reduce the growth in CO_2 emissions from transport to a 20% increase over the next 20 years. The third strand added steep year-by-year increases in the price of fuel, full-cost externality pricing for motor vehicles (estimated at 5% of GDP in OECD countries), and ensuring the use of high-efficiency, low-weight, low-polluting cars, vans, lorries, and buses in cities. Addition of the third strand was projected to reduce fuel use by 40% from 1995 to 2015.

3.4.5 Regional Differences

Technical and economic potentials for reducing greenhouse gas emissions will vary by region according to differences in geography, existing transportation infrastructure, technological status of existing transport equipment, the intensity of vehicle use, prevailing fuel and vehicle fiscal policies, the availability of capital, and other factors. Differences in spatial structure, existing infrastructure, and cultural preferences also influence the modal structure and level of transport demand.

Many developing countries and countries with economies in transition are experiencing rapid motorization of their transport systems but are not yet locked into a road-dominated spatial structure. In addressing the transport problems of these economies, the World Bank (1996) has emphasized the importance of combining efficient pricing of road use (including external costs) with co-ordinated land use and infrastructure investment policies to promote efficient levels of transport demand and modal choice. Without providing specific GHG emission reduction estimates, the World Bank study notes that non-highway modes such as rail can reduce energy requirements by two-thirds versus automobiles and 90% versus aircraft, in situations where the modes provide competitive services.

Studies of transport mitigation options in Africa and Asia have emphasized behavioural, operational, and infrastructure mea-

Table 3.12: Estimated costs of greenhouse gas mitigation options in Southern Africa (Zhou, 1999; UNEP/Southern Centre, 1993).

Measure	Cost (US$/tC)
Paved roads	−41.42
Road freight to rail	−31.47
Petroleum and product pipelines	−18.91
Fuel pricing policies	0.00
Vehicle inspections	0.20
Rail electrification	111.94
Compressed natural gas	1.37
Ethanol	−186.5

sures in addition to technology. In Africa, in particular, options that have been examined include: the reduction of energy intensity through expanding mass transit systems (e.g., modal shifts from road to rail), vehicle efficiency improvement through maintenance and inspection programmes, improved traffic management, paving roads, and the installation of fuel pipelines (e.g., modal shift from road or rail to pipeline), provision of infrastructure for non-motorized transport, and decarbonization of fuels through increased use of compressed natural gas or biomass ethanol (Baguant and Teferra, 1996; Zhou, 1999). Mass movements of goods, passengers, and fuel become more cost-effective as the volumes and load factors increase, and for most African countries this is likely to be achievable only after 2010 (Zhou, 1999). In studies conducted for East and Southern Africa, these options were found to be implementable at little or no cost per tC (*Table 3.12*). Zhou (1999) has estimated that investments in paving roads, rail freight systems and pipelines could reduce greenhouse gas emissions in Botswana at negative cost (*Table 3.12*). Vehicle inspection programmes, as well as fuel decarbonization by use of compressed natural gas and biomass ethanol were all estimated to be no cost to low-cost options. Bose (1999a) notes that in developing countries mass transport modes and demand management strategies are an essential complement to technological solutions because of three factors: (1) lack of leverage in global vehicle markets to influence the development of appropriate transport technologies; (2) the relatively greater importance of older, more polluting vehicles combined with slower stock turnover; and (3) the inability to keep pace with rapid motorization in the provision of infrastructure.

3.4.6 Technological and Economic Potential

This section addresses the technological potential to cost-effectively increase energy efficiency in transport and thereby reduce GHG emissions. Most studies concentrate on light-duty vehicles because of their 50% share of energy use and GHG emissions, and on technology or fuel pricing policies. Technical efficiency improvements, in the absence of complementary fiscal policies, are subject to a "rebound effect" in that they reduce the fuel cost of travel. Rebound effects in the USA amount to about 20% of the potential GHG reductions (Greene, 1999). In Europe, where fuel prices are higher, rebound effects may be as large as 40% (Michaelis, 1997). Most assessments take the rebound effect into account when estimating technical efficiency impacts. Fewer studies address policies such as land use planning, investment in or subsidy of particular transport modes, or information.

An Asian four-country study of the technological and economic potential to reduce GHG emissions considered five types of options for GHG mitigation in transport: (1) improving fuel efficiency, (2) improving transportation system efficiency, (3) behavioural change, (4) modal split changes, and (5) technological change (Bose, 1999b). The Indian study concluded that abatement costs for transport were high relative to options available in other sectors, and projected little change in transport for emissions constraints less than a 20% reduction from the baseline. The Bangladesh study, using a different methodology, concluded that a wide array of near-term technology options had no net cost, but that the cost of 4-stroke engines for 3-wheeled vehicles fell between US$48 and US$334/tC reduced, depending on the application. The Thailand study found that lean-burn engines would improve efficiency by 20% at a negative net cost of US$509/tC. The Korean study also concluded that several "no regrets" options were available, including use of continuously variable transmissions, lean-burn engines, and exclusive bus lanes.

Recognizing that transportation energy consumption and CO_2 emissions increased by 16% from 1990 to 1995, and that carbon emissions may be 40% higher in 2010 than in 1990 if measures are not taken, the government of Japan has strengthened energy efficiency standards based on a "Front Runners" approach,

which sets standards to meet or exceed the highest energy efficiency achieved among products currently commercialized (MITI/ANRE, 1999). These require a 22.8% improvement over 1995 new gasoline car fuel economy in 1/km by 2010, and a 13.2% improvement for gasoline light-duty freight vehicles (Minato, 1998). For diesel-fuelled vehicles the corresponding requirements are 14.9% and 6.5% by 2005. Technological improvements in other modes are expected to produce efficiency improvements of 7% for railways, 3% for ships, and 7% for airlines over the same period (Minato, 1998). Cost-effective technical potentials have also been reported by Kashiwagi *et al.* (1999), who cite 27.7 PJ of energy savings in Japan's transport sector achievable at US$0.044/kWh, or less.

There are significant barriers to the kinds of fuel economy improvements described above, and substantial policy initiatives will be needed to overcome them. In Europe, for example, the European automobile manufacturers' association, ACEA, and the European Union have agreed to voluntary standards to reduce carbon emissions from new passenger cars by 25% over the next 10 years. The European standards will require reducing average fuel consumption of new cars from 7.7 to 5.8 1/100 km, creating a strong incentive to adopt advanced fuel economy technologies. A survey of 28 European countries identified 334 separate measures countries were taking to reduce CO_2 emissions from transport (Perkins, 1998).

At least nine recent studies have assessed the economic potential for technology to improve light-duty vehicle fuel economy (Weiss *et al.*, 2000; Greene and DeCicco, 1999; Michaelis, 1997). The conclusions of eight of the studies are summarized in the form of quadratic fuel economy cost curves describing incremental purchase cost versus the improvement in fuel economy over a typical 8.4 1/100 km passenger car (*Figure 3.10*). Most of the technology potential curves reflect a short-

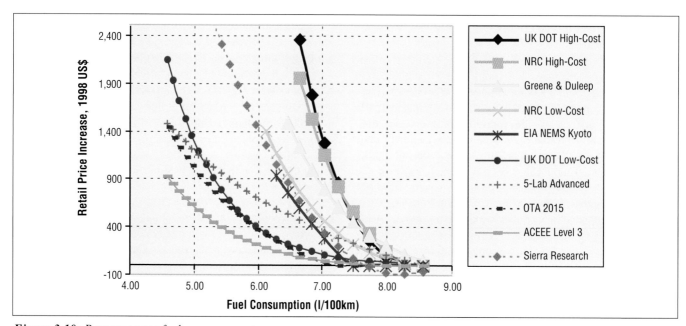

Figure 3.10: *Passenger car fuel economy cost curves.*

Table 3.13: *Estimated technological potential for carbon emissions reductions in the US transportation sector*
(Brown *et al.*, 1998).

	1990	2010	2020	2030
Business as usual (MtC)	432	598	665	741
Technology potential (%)		7–12	15–17	27–40

	1990	2010		
		Baseline	Efficiency	High efficiency[a]
Transport emissions (MtC)	432	616	543	513
Reduction (%)			12	17

[a] Includes US$50/tC permit cost.

run perspective, considering what can be achieved using only proven technologies over a 10-year period. The two most pessimistic (which reflect a 1990 industry view of short-term technology potential) indicate that even a reduction from 8.4 to 6.5 l/100 km would cost nearly US$2000. The curves labelled "ACEEE Level 3" and "UK DOT Low-Cost" are limited to proven technologies, but allow substantial trade-offs in performance, transmission-management and other features that may affect customer satisfaction. The curves labelled "5-lab" and "OTA 2015" include the benefits of technologies in development, but not yet commercialized (NRC, 1992; DeCicco and Ross, 1993; US DOE/EIA, 1998). The most optimistic of these suggest that an improvement to less than 5.9 l/100 km is possible at an incremental cost of less than US$1000 per vehicle (1998 US$). The Sierra Research (Austin *et al.*, 1999) curve is intended to pertain to the year 2020, but reflects industry views about technology performance, and excludes certain key technologies such as hybrids and fuel cell vehicles that could have dramatic impacts over the next 20 years.

Three of the studies (OTA, 1995b; DeCicco and Ross, 1993; National Laboratory Directors, 1997) considered more advanced technologies such as those described above (e.g., direct-injection engines, aluminium-intensive designs, hybrid vehicles, fuel cells). These concluded that by 2015, consumption rates below 4.7 l/100 km could be attained at costs ranging from under US$1000 to US$1500 per vehicle. These long-run curves span a range similar to fuel consumption/cost curves for European passenger cars reported by Denis and Koopman (1998, *Figure 3*), except that the base fuel consumption rate is 7 l/100 km as opposed to 8.5 in the USA, and improvements to the range of 4 to 5 l/100 km were judged achievable at incremental costs of 2000 to 700 ECU, respectively (1990 ECU).

A lifecycle analysis of the greenhouse gas impacts of nine hybrid electric and fuel cell vehicles was compared to a 1996 vehicle and an "evolved 2020" baseline vehicle for the year 2020 by Weiss *et al.* (2000). The study concluded that a hybrid

vehicle fuelled by compressed natural gas could reduce GHG emissions by almost two-thirds relative to the 1996 reference vehicle, and by 50% compared with an advanced 2020 internal combustion engine vehicle. Other technologies capable of 50%, or greater lifecycle GHG reductions versus the 1996 reference vehicle included: gasoline and diesel hybrids, battery-electric, and hydrogen fuel cell vehicles.

A recent study by five of the US Department of Energy's (DOE's) National Laboratories (Interlaboratory Working Group, 1997) assessed the economic market potential for carbon reductions, using the EIA's National Energy Modelling System. Transport carbon emissions were projected to rise from 487 MtC in 1997 to 616 MtC by 2010 in the baseline case. In comparison to the baseline case, use of cost-effective technologies reduced carbon emissions by 12% in 2010 in an "Efficiency" case (*Table 3.13*). More optimistic assumptions about the success of R&D produced a reduction of 17% by 2010. The authors noted that lead times for cost-effectively expanding manufacturing capacity for new technologies and the normal turnover of the stock of transport equipment significantly limited what could be achieved by 2010. Efficiency improvements in 2010 for new transportation equipment were substantially greater (*Table 3.14*). New passenger car efficiency increased by 36% in the "Efficiency" case and by 57% in the more optimistic case (Brown *et al.*, 1998).

Eleven of the US DOE's National Laboratories completed a comprehensive assessment of the technological potential to reduce GHG emissions from all sectors of the US economy (National Laboratory Directors, 1997). This study intentionally made optimistic assumptions about R&D success, and did not explicitly consider costs or other market factors. The study concluded that the technological potential for carbon emissions reductions from the US transport sector was 40–70 million metric tons of carbon (MtC) by 2010, 100–180MtC by 2020 and 200–300MtC by 2030. These compare to total US transportation carbon emissions of 473MtC in 1997 (note that this

Table 3.14: *Projected transportation efficiencies of 5-Laboratory Study* (Interlaboratory Working Group, 1997).

Determinants	1997	2010		
		Baseline	Efficiency	HE/LC[a]
New passenger car l/100 km	8.6	8.5	6.3	5.5
New light truck l/100 km	11.5	11.4	8.7	7.6
Light-duty fleet l/100 km[b]	12.0	12.1	10.9	10.1
Aircraft efficiency (seat-l/100 km)	4.5	4.0	3.8	3.6
Freight truck fleet l/100 km	42.0	39.2	34.6	33.6
Rail efficiency (tonne-km/MJ)	4.2	4.6	5.5	6.2

[a]   HE/LC, high-energy/low-carbon.
[b]   Includes existing passenger cars and light trucks

Table 3.15: *Assumptions and results of three European studies*

	Dutch	Hanover	EU
Base and target years (length of scenario in years)	1995, 2020 (25 years)	1990, 2010 (20 years)	1990, 2000 (10 years)
CO_2 emissions in target year: baseline (Mt)	36.6–43.3	1.9	649.8
Annual percentage growth in baseline emissions (Mt)	0.4% to 1.4% per year	0.6% per year	1.7% per year
Solution scenario	(I) Best technical means, (II) Intensifying current policy, (III) Non-conventional local transport technologies	(A) Local/regional, (B) National	(R) Reasonable restrictive, (T) Target orientation
Base and target years (length of scenario, in years)	1995, 2020 (25 years)	1990, 2010 (20 years)	1990, 2000 (10 years)
CO_2 emission reduction (transport sector - Mt)	(I) 11–13, (II) 3–11, (III) 18	(A) 0.16 and (B) 0.34	(R) 84 and (T) 177
Reduction of total transport emissions (including non-road transport) relative to baseline in target year	(I) 30%, (II) 8%–25%, (III) 42%	(A) 8% and (B) 18%	(R) 13% and (T) 25%
Economic evaluation			
Net annual costs	Not quantified, though asserted to be <€0 /tC	Not quantified	Not quantified

base year estimate differs from that for the Interlaboratory Working Group). The report suggested the following technological potentials for carbon emissions reductions by mode of transport over the next 25 years: (1) light-duty vehicles with fuel cells, 50%–100%; (2) heavy trucks via fuel economy improvements, 20%–33%; and (3) air transport, 50%. It is difficult to interpret the practical implications of these conclusions, however, since no attempt was made by this study to estimate achievable market potentials.

Three European studies of the technical-economic potential for energy savings and CO_2 reduction were reviewed by van Wee and Annema (1999). Generally, the studies focused on technological options, such as improving the fuel efficiencies of conventional cars and trucks, promotion of hybrid vehicles, switching trucks and buses to natural gas, and electrifying buses, delivery trucks, and mopeds. Only the study for Hanover included investment in improved public transport as a major policy option. The results, summarized in *Table 3.15*, suggest that emissions reductions of 8% to as much as 42% over business-as-usual projections may be possible.

The effects of a variety of fiscal and regulatory policies on CO_2 emissions from road passenger vehicles have been estimated for Europe over a 15-year forecast horizon (Jansen and Denis, 1999; Denis and Koopman, 1998). These studies, both using the EUCARS model developed for the European Commission, concluded that CO_2 reductions on the order of 15% over a baseline case could be achieved in the 2011 to 2015 time period at essentially zero welfare loss. Among the more effective policies were fuel taxes based on carbon content, fuel consumption standards requiring proportional increases for all cars, and the combination of fuel-consumption based vehicle sales taxes with a fuel tax. When reductions in external costs and the benefit of raising public revenues are included in the calculation of social welfare impacts, the feebate (a policy combining subsidies for fuel efficient vehicles and taxes on inefficient ones) and fuel tax policy combination was able to achieve CO_2 reductions of 20% to 25% in the 2011 to 2015 time period at zero social cost (Jansen and Denis, 1999).

3.4.7   Conclusions

Over the past 25 years, transport activity has grown at approximately twice the rate of energy efficiency improvements. Because the world's transportation system continued to rely overwhelmingly on petroleum as an energy source, transport energy use and GHG emissions grew in excess of 2% per year. Projections to 2010 and beyond reviewed above reflect the belief that transport growth will continue to outpace efficiency improvements and that without significant policy interventions, global transport GHG emissions will be 50%–100% greater in 2020 than in 1995. Largely as a result of this anticipated growth, studies of the technical and economic potential for reducing GHG emissions from transport generally conclude that while significant reductions from business-as-usual pro-

jections are attainable, it is probably not practical to reduce transport emissions below 1990 levels by the 2010–2015 time period. On the other hand, the studies reviewed generally indicate that cost-effective reductions on the order of 10%–20% versus baseline appear to be achievable. In addition, more rapid than expected advances in key technologies such as hybrid and fuel cell vehicles, should they continue, hold out the prospect of dramatic reductions in GHG emission from road passenger vehicles beyond 2020. Most analyses project slower rates of GHG reductions for freight and air passenger modes, to a large extent reflecting expectations of faster rates of growth in activity.

Assessing the total global potential for reducing GHG emissions from transportation is hindered by the relatively small number of studies (especially for non-OECD countries) and by the lack of consistency in methods and conventions across studies. Not all studies shown in Table 3.16 cover the entire transportation sector, even of the countries included in the study. Most consider a limited set of policy options, (e.g., only motor vehicle fuel economy improvement). In general, the studies do not report marginal costs of GHG mitigation, but rather average costs versus a base case. Keeping all of these limitations in mind, *Table 3.16* summarizes the findings of several major studies. For 2010, the average low GHG reduction estimate is just under 7% of baseline total transport sector emissions in 2010, with the higher estimates averaging a 17% reduction. There is, however, considerable dispersion around both numbers, indicative both of uncertainty and differences in methodology and assumptions. For studies looking ahead to 2020, the average low estimate is 15% and the average high estimate is 34% of baseline 2020 transport sector emissions. Estimated (average rather than marginal) costs are generally negative (as much as -US$200/tC), indicating that fuel savings are expected to outweigh incremental costs. There are some positive cost estimates as high as US$200/tC, however. The majority of the studies cited in *Table 3.16* are based on engineering-economic analyses. Some argue that this method tends to underestimate welfare costs because trade-offs between CO_2 mitigation and non-price attributes (e.g., performance, comfort, reliability) are rarely explicitly considered (Sierra Research, Inc., 1999).

3.5   Manufacturing Industry

3.5.1   Introduction

This section deals with greenhouse gas emissions and greenhouse gas emission reduction options from the sector *manufacturing industry*[14]. Important are the energy intensive (or heavy) industries, including the production of metals (especially iron and steel, and aluminium), refineries, pulp and

---

[14] NACE codes 15 to 37.

Table 3.16: Estimates of the costs of reducing carbon emissions from transport based on various studies, 2010-2030 (Brown *et al.*, 1998; ECMT, 1997; US DOE/EIA, 1998; DeCicco and Mark, 1998; Worrell *et al.*, 1997b; Michaelis, 1997; Denis and Koopman, 1998)

Study	Year of publication	Application	Year of scenario	Years in future	Country	Quantity		Reduction		Cost in US$/MtC	
						Low (MtC)	High (MtC)	Low (%)	High (%)	Low	High
OECD Working Paper 1	1997	Light-duty road vehicle efficiency	2010	13	OECD	50	150	2.5	7.5	US$0	US$0
US National Academy of Sciences	1992	Vehicle efficiency	2010	18	USA	20	79	3.2	12.7	-US$275	-US$77
	1992	System efficiency	2010	18	USA	3	13	0.5	2.1	-US$183	US$18
US DOE 5-Lab Study	1997	Transport sector	2010	13	USA	82	103	13.2	16.6	-US$157	US$6
US Energy Information Administration	1998	Transportation sector	2010	12	USA	41	55	6.6	8.9	-US$121	US$163
Tellus Institute	1997	Transportation efficiency	2010	13	USA	90	90	14.5	14.5	-US$465	-US$465
	1997	Transportation demand reduction	2010	13	USA	61	61	9.8	9.8	US$0	US$0
ACEEE	1998	Transport sector	2010	12	USA		125		22.6	-US$139	
US DOE, Clean Energy Futures	2000	Transport sector	2010	10	USA	20	66	3.2	10.5	-US$280	-US$144
European Council of Ministers of Transport	1997	Transport sector	2010	13	Austria	2			8.3		
	1997	Transport sector	2010	13	Belgium	4			13.3		
	1997	Transport sector	2010	13	Czech R.	6			57.1		
	1997	Transport sector	2010	13	Netherlands	11			37.2		
	1997	Transport sector	2010	13	Poland	5			12.8		
	1997	Transport sector	2010	13	Slovak R.	1			16.3		
	1997	Transport sector	2010	13	Sweden	4			23.2		
	1997	Transport sector	2010	13	UK	22			14.3		
Summary for 2010					**Minimum/maximum**			**0.5**	**57.1**	**-US$465**	**US$163**
					average			**6.7**	**16.9**	**-US$153**	**-US$62**
Denis and Koopman	1998	Road pricing	2015	17	EU				25.0		
	1998	CO$_2$ tax	2015	17	EU				13.0		
	1998	Purchase subsidy + CO$_2$ tax	2015	17	EU				14.0	US$0	US$0
US Congress OTA	1991	Transportation efficiency	2015	24	USA		195		29.2	-US$180	US$195
Summary for 2015					**Minimum/maximum**			**13.0**	**29.2**	**-US$180**	**US$195**
					average				**20.3**		
US DOE, Clean Energy Futures	2000	Transport sector	2020	20	USA	58	163	8.3	23.4	-US$234	-US$153
ACEEE	1998	Transport sector	2020	22	USA		260		42.4	-US$164	
United Nations	1997	Transport sector	2020	23	Industrialized	153	423	14.9	41.2		
	1997	Transport sector	2020	23	Transitional	72	126	18.2	31.8		
	1997	Transport sector	2020	23	Developing	297	450	28.4	43.1		
OECD Working Paper 1	1997	Light-duty road vehicle efficiency	2020	23	OECD	100	500	4.3	21.7	US$0	US$0
Summary for 2020					**Minimum/maximum**			**4.3**	**43.1**	**-US$234**	
					average			**14.8**	**34.0**		
ACEEE	1998		2030	32	USA		401		58.8	-US$192	

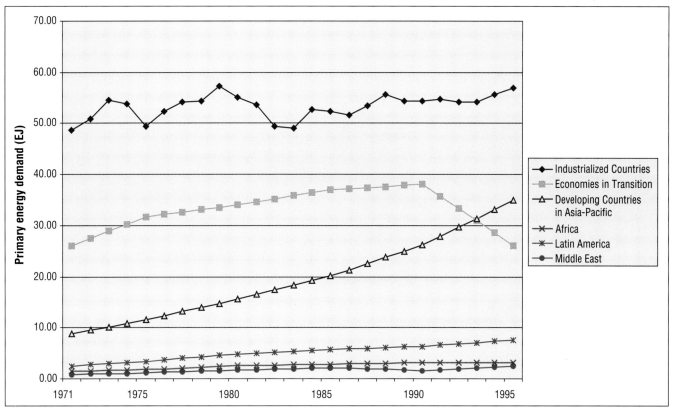

Figure 3.11: *Development of industrial energy use in terms of primary energy (direct fuel use and indirect fuel use in power plants) in the different world regions. Data from Price et al. (1998, 1999).*

paper, basic chemicals (important ones are nitrogen fertilizers, petrochemicals, and chlorine), and non-metallic minerals (especially cement). The less energy intensive sectors, also called *light industry*, are among others, the manufacture of food, beverages, and tobacco; manufacturing of textiles; wood and wood products; printing and publishing; production of fine chemicals; and the metal processing industry (including automobiles, appliances, and electronics). In many cases these industries each produce a wide variety of final products. Non-CO_2 gases emitted from the manufacturing sector include nitrous oxide (N_2O), hydrofluorocarbons (HFCs), perfluorocarbons (PFCs) and sulphur hexafluoride (SF_6). Adipic acid, nitric acid, HCFC-22 and aluminium production processes emit these gases as unintended by-products. A number of other highly diverse industries, including a few sectors replacing ozone-depleting substances, use these chemicals in manufacturing processes[15].

All direct emissions from manufacturing are taken into account, plus emissions in the electricity production sector, as far as they are caused by electricity consumption by manufacturing industry firms.

Kashiwagi *et al.* (1996) dealt with industry emission reduction options in IPCC (1996). In that chapter, processes, energy consumption, and a range of emission reduction options (mainly for CO_2) have been described on a sector-by-sector basis. For the TAR, these options are summarized (see Section 3.5.3) and estimates of potentials and costs for emission reduction are quantified. The scope of TAR has been expanded to also include greater detail on non-CO_2 greenhouse gases and the differences in regional emission profiles and emission reduction opportunities.

3.5.2 Energy and GHG Emissions

Emissions of carbon dioxide are still the most dominant contribution of manufacturing industry to total greenhouse gas emission. These emissions are mainly connected to the use of energy. In *Figure 3.11* an overview is given of the energy consumption of the manufacturing industry (see also *Table 3.1*). Energy use is growing in all regions except in the economies in transition, where energy consumption declined by 30% in the period 1990 to 1995. This effect is so strong that it nearly offsets growth in all other regions. In industrialized countries energy use is still growing at a moderate rate; electricity consumption grows faster than fuel consumption. The strongest growth rates occur in the developing countries in the Asia-Pacific region. All developing countries together account for 36% of industrial energy use. However, industry in industrial-

---

[15] Chapter 3 Appendix *Options to Reduce Global Warming Contributions from Substitutes for Ozone Depleting Substances* elaborates on the sectors that would be affected by both the Montreal Protocol and the Kyoto Protocol.

Table 3.17: Overview of greenhouse gas emissions by manufacturing industry (in MtC_{eq}) in 1990 (1995 for the fluorinated gases). Note that the accuracy is much less than 1 MtC_{eq}
Sources: see notes.

Source	OECD	EIT	Asia-Pacific DCs	Other DCs	Total	Trends after 1990 (% per year)
Fuel CO_2[f]	546	454	461	105	1567	Stable (90-95)
Electricity CO_2[f]	341	167	170	66	726	+1.2% (90-95)
CO_2 from cement[a]	51	25	60	19	155	
CH_4[a]					8	
N_2O[b]	34	13	13	4	65	
HFC-23[c]	19	~1	~2	~1	22	+2% (90-97)
PFCs[d]	>11	>4	>4		31	Decreasing
SF_6[e]	26	6	7		40	+4% (90-96)
Total					2614	

[a] Olivier *et al.*, 1996.

[b] Total N_2O emissions are estimated to be 489 ktonnes ($65MtC_{eq}$) (Olivier *et al.*, 1999). Main industrial process that lead to emissions of N_2O are the production of adipic acid ($38MtC_{eq}$) and nitric acid ($23MtC_{eq}$).

[c] At present, the main HFC source from industrial processes is the emission of HFC-23 (trifluoromethane, with an estimated GWP of 11,700) as an unintended by-product of HCFC-22 (chlorodifluoromethane) production. The weight percentage by-product is estimated to be 4%, 3%-5% (March Consulting, 1998) or 1.5%–3% (Branscome and Irving, 1999) of the HCFC-22 production. Some abatement takes place, but the fraction for 1995 is not known. Atmospheric measurements of HFC-23 suggest an *emitted* by-product fraction of 2.1% (Oram *et al.*, 1998). This leads to the reported $22MtC_{eq}$ These are not inconsistent with reported US -23 emissions of $9.5MtC_{eq}$ in 1990 and $7.4 MtC_{eq}$ in 1995 (US EPA, 1998) and for Europe of $9.5MtC_{eq}$. Regional breakdown and trend from Olivier (2000). For other HFC emissions see the Appendix to this Chapter.

[d] Perfluorocarbons (PFCs) have the general chemical formula C_xF_{2x+2}. The manufacturing industry is thought to be responsible for all PFC emissions, mainly CF_4 and C_2F_6. On the basis of recent atmospheric concentration data, Harnisch (1998) estimates emissions of 10,500 tonnes and 2000 per year respectively ($20 MtC_{eq}$). Most of these emissions are the by-product of aluminium smelting; a smaller but growing contribution is from plasma etching in semi-conductor manufacturing and use as solvent 1.4 - 4 MtC_{eq} (Victor and McDonald, 1999; Harnisch *et al.*, 1998). Some applications for higher carbon PFCs have also been identified and may become significant. C_3F_8 ($1.4MtC_{eq}$) is emitted as a result of various activities, like plasma etching, fire extinguishers and as an additive to the refrigerant R-413a. Emissions of c-C_4F_8 ($4MtC_{eq}$) may result from the pyrolysis of fluoropolymers, whereas C_6F_{14} originates from use of this substance as a solvent (5 MtC_{eq}) (Harnisch *et al.*, 1998; Harnisch, 2000). Regional breakdown is based on Victor and McDonald (1999) and is only for CF_4 and C_2F_6.

[e] Maiss and Brenninkmeijer (1998) estimate the following breakdown of 1995 emissions (in tonnes SF_6): switchgear manufacturers: 902; utilities and accelerators: 3476; magnesium industry: 437; electronics industry: 327; "using adiabatic properties": 390; other uses: 498; total 6076. The regional breakdown is extrapolated from Victor and MacDonald (1999).

[f] Price *et al.*, 1999.

ized countries on a per capita basis uses about 10 times as much energy as in developing countries.

The CO_2 emissions by the industrial sector worldwide in 1990 amounted to 1,250MtC. A breakdown of 1990/1995 emissions is given in *Table 3.17*. However, these emissions are only the direct emissions, related to industrial fuel consumption. The indirect emissions in 1990, caused by industrial electricity consumption, are estimated to be approximately 720MtC (Price *et al.*, 1998 and Price *et al.*, 1999). In the period 1990 to 1995 carbon emissions related to energy consumption have grown by 0.4% per year.

Note that the energy-related CO_2 in a number of sectors are partly process emissions, e.g., in the refineries and in the production of ammonia, steel, and aluminium (Kashiwagi *et al.*,

1996). However, the statistics often do not allow us to make a proper separation of these emissions.

Olivier *et al.*, (1996) also report 91MtC of non-energy use (lubricants, waxes, etc.) and 167MtC for feedstock use (naphtha, *etc.*). Further work on investigating the fact of these carbon streams is necessary; knowledge about emission reduction options is still in an early stage (Patel and Gielen, 1999; Patel, 1999).

An overview of industrial greenhouse gas emissions is given in *Table 3.17*. The manufacturing industry turns out to be responsible for about one-third of emissions of greenhouse gases that are subject to the Kyoto Protocol. Non-CO_2 greenhouse gases make up only about 6% of the industrial emissions.

Underlying Causes for Emission Trends

Unander *et al.* (1999) have analysed the underlying factors for the development of energy consumption in OECD countries in the period 1990 to 1994. Generally, the development of energy use can be broken down into three factors: volume, structure and energy efficiency. In the period examined, development of production volume differed from country to country, ranging from a 2.0% growth per annum in Norway to a 1.4% per annum decline in Germany. The second factor is structure: this is determined by the shares that the various sectors have in the total industrial production volume. A quite remarkable result is that in nearly all countries, structural change within the manufacturing industry has an increasing effect on energy use, i.e. there is a shift towards more energy-intensive industrial sectors. This is a contrast with earlier periods. Finally, Unander *et al.* (1999) found – with some exceptions – a continuing decline in energy intensity within sectors, be it at a lower pace than in the period 1973 to 1986. For more results see *Table 3.18*.

In the paper by Unander *et al.* (1999), energy intensity is measured in terms of energy use per unit of value added. An indicator more relevant to the status of energy efficiency in a country is the specific energy consumption, corrected for structural differences. Also, such an indicator shows a continuous downward trend, as can be seen in *Figure 3.12*. Similar results were obtained for the iron and steel industry (Worrell *et al.*, 1997a).

A substantial part of industrial greenhouse gas emissions is related to the production of a number of primary materials. Relevant to this is the concept of dematerialization (the reduction of society's material use per unit of GDP). For most individual materials and many countries dematerialization can be observed. Cleveland and Ruth (1999) reviewed a range of studies that show this. They suggest that it cannot be concluded to be due to an overall decoupling of economy and material inputs, among other reasons because of the inability to measure aggregate material use. Furthermore, they note that some analysts observe relinking of economic growth and material use in more recent years. They warn against "gut" feeling that technical change, substitution, and a shift to the "information age" inexorably lead to decreased materials intensity and reduced environmental impact.

3.5.3 New Technological and Other Options for CO_2 and Energy

3.5.3.1 Energy Efficiency Improvement

Energy efficiency improvement can be considered as the major option for emission reduction by the manufacturing industry. A wide range of technologies is available to improve energy efficiency in this industry. An overview is given in *Table 3.19*. Note that the total technical potential consists of a larger set of

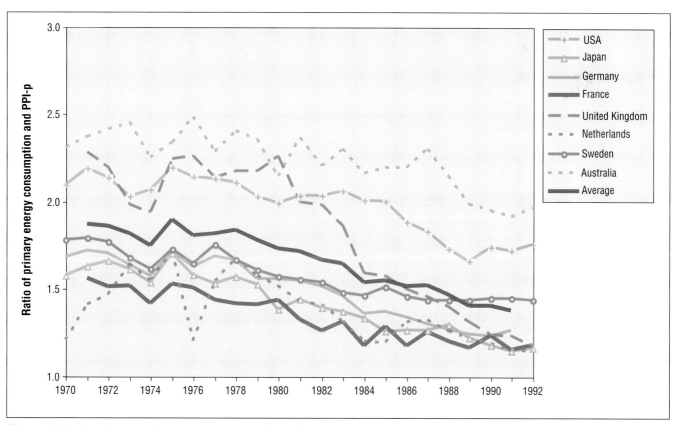

Figure 3.12: *Development of the primary energy demand per unit of production in the pulp and paper industry (PPI-p) in OECD countries.*

Table 3.18: *Average annual rates of change in manufacturing energy use, and the degree to which changes in volume, structure and energy intensity contribute to such change.* Source: Unander et al. (1999).

Country	Development in energy use			Effect of volume development on energy use			Effect of structural change in industry on energy use			Effect of energy intensity changes within sectors on energy use		
	1973-1986	1986-1990	1990-1994	1973-1986	1986-1990	1990-1994	1973-1986	1986-1990	1990-1994	1973-1986	1986-1990	1990-1994
Australia	0.3%	3.3%	0.8%	1.1%	3.2%	1.9%	0.0%	0.6%	-0.4%	-1.2%	-2.1%	0.1%
Canada	N/A	0.7%	0.8%	2.0%	1.7%	1.4%	N/A	-0.1%	0.4%	N/A	-0.8%	-1.0%
Denmark	-1.1%	-3.3%	1.5%	2.1%	-0.6%	0.9%	-0.3%	-0.1%	0.0%	-2.9%	-2.6%	0.7%
Finland	1.7%	3.3%	1.8%	2.9%	3.2%	1.6%	-0.1%	0.3%	1.6%	-2.0%	-0.2%	-1.5%
France	-2.3%	1.3%	0.7%	1.2%	3.2%	-0.5%	-0.2%	0.1%	0.0%	-3.3%	-2.0%	1.2%
Germany	-1.8%	0.6%	-0.5%	1.1%	2.7%	-1.4%	-0.4%	-0.5%	1.0%	-2.6%	-1.6%	-0.1%
Italy	-1.8%	3.8%	-0.7%	3.4%	4.0%	0.2%	0.0%	0.2%	0.4%	-5.2%	-0.4%	-1.4%
Japan	-1.8%	3.5%	-0.1%	3.2%	6.3%	-0.4%	-2.0%	-0.2%	0.1%	-3.0%	-2.6%	0.2%
Netherlands	-4.0%	4.4%	0.0%	1.8%	2.8%	0.6%	1.1%	-0.4%	0.8%	-6.9%	2.0%	-1.5%
Norway	0.1%	-0.9%	1.5%	0.5%	-1.3%	2.0%	0.6%	2.2%	0.8%	-1.1%	-1.8%	-1.3%
Sweden	-1.4%	0.0%	0.0%	1.3%	1.5%	1.3%	-0.4%	0.3%	2.8%	-2.2%	-1.9%	-4.1%
UK	-3.6%	0.0%	-2.4%	-0.7%	3.9%	-0.2%	-0.4%	-0.3%	-0.5%	-2.6%	-3.6%	-1.6%
USA	-1.9%	2.9%	1.9%	2.0%	3.0%	1.8%	-1.1%	-0.5%	0.1%	-2.8%	0.5%	1.6%

Technological and Economic Potential of Greenhouse Gas Emissions Reduction

Table 3.19: *Overview of important examples of industrial energy efficiency improvement technologies and indications of associated emission reduction potentials and costs. For an explanation see the legend below. Note that the scale is not linear. Cost may differ from region to region. This overview is not meant to be comprehensive, but a representation of the most important options.*
Sources: Kashiwagi *et al.* (1996), De Beer *et al.* (1994), ETSU (1994), WEC (1995a), IEA Greenhouse Gas R&D Programme (2000a), Martin *et al.* (2000).

Sector	Technology	Potential in 2010	Emission reduction costs	Remarks
All industry	Implementation of process control and energy management systems	■■■■	-	Estimate: 5% saving on primary energy demand worldwide
	Electronic adjustable speed drives	■■	++	In industrial countries ~30% of industrial
	High-efficiency electric motors	■■	+ *	electricity demand is for electric drive systems
	Optimized design of electric drive systems, including low-resistance piping and ducting	■■	+++	Not known for developing countries.
	Process integration, e.g., by applying pinch technology	■■■■	+	Savings vary per plant from 0%-40% of fuel demand; costs depend on required retrofit activity.
	Cogeneration of heat and power	■■■■	-	
Food, beverages and tobacco	Application of efficient evaporation processes (dairy, sugar)	■	+	
	Membrane separation	■	++	
Textiles	Improved drying systems (e.g., heat recovery)	■	++	
Pulp and paper	Application of continuous digesters (pulping)	■	+	Applicable to chemical pulping only; energy generally supplied as biofuels
	Heat recovery in thermal mechanical pulping	■	+++	Energy generally supplied as biofuels
	Incineration of residues (bark, black liquor) for power generation	■	+	
	Pressing to higher consistency, e.g., by extended nip press (paper making)	■	-	Not applicable to all paper grades
	Improved drying, e.g., impulse drying or condensing belt drying	■■	-	Pre-industrial stage; results in a smaller paper machine (all paper grades)
	Reduced air requirements, e.g., by humidity control in paper machine drying hoods	■	+	
	Gas turbine cogeneration (paper making)	■■	-	
Refineries	Reflux overhead vapour recompression (distillation)	■	+	
	Staged crude preheat (distillation)	■	+	
	Application of mechanical vacuum pumps (distillation and cracking)	■■	+	
	Gas turbine crude preheating (distillation)	■■	-	Applicable to 30% of the heat demand of refineries
	Replacement of fluid coking by gasification (cracking)	■	+	
	Power recovery (e.g., at hydrocracker)	■	-	
	Improved catalysts (catalytic reforming)	■■	+	

(continued)

Table 3.19: *continued*

Sector	Technology	Potential in 2010	Emission reduction costs		Remarks
Fertilizers	Autothermal reforming	■	-	*	
	Efficient CO_2 separation (e.g., by using membranes)	■	+	*	Saving depends strongly on opportunities for process integration of old and new techniques.
	Low pressure ammonia synthesis	■	+	*	Site-specific: an optimum has to be found between synthesis pressure, gas volumes to be handled, and reaction speed
Petrochemicals	Mechanical vapour recompression (e.g., for propane/propene splitting)	■	+		
	Gas turbine cogeneration	■	-		Not yet demonstrated for furnace heating
	De-bottlenecking	■■	-		Estimate: 5% saving on fuel demand
	Improved reactors design, e.g., by applying ceramics or membranes	■	+		Not yet commercial
	Low pressure synthesis for methanol	■	+	*	Site-specific: an optimum has to be found between synthesis pressure, gas volumes to be handled, and reaction speed
Other chemicals	Replacement of mercury and diaphragm processes by membrane electrolysis (chlorine)	■	+	*	In some countries, e.g., Japan, membrane electrolysis is already the prevailing technology
	Gas turbine cogeneration	■■	-		
Iron and steel	Pulverized coal injection up to 40% in the blast furnace (primary steel)	■■	-		Maximum injection rate is still topic of research
	Heat recovery from sinter plants and coke ovens (primary steel)	■■	+		
	Recovery of process gas from coke ovens, blast furnaces and basic oxygen furnaces (primary steel)	■■■	-		
	Power recovery from blast furnace off-gases (primary steel)	■	+		
	Replacement of open-hearth furnaces by basic oxygen furnaces (primary steel)	■■■	-	*	Mainly former Soviet Union and China
	Application of continuous casting and thin slab casting	■■■	-	*	Replacement of ingot casting
	Efficient production of low-temperature heat (heat recovery from high-temperature processes and cogeneration)	■	++		Heat recovery from high temperature processes is technically difficult
	Scrap preheating in electric arc furnaces (secondary steel)	■	+		
	Oxygen and fuel injection in electric arc furnaces (secondary steel)	■	-		
	Efficient ladle preheating	■			
	Second-generation smelt reduction processes (primary steel)	■■■	-		First commercial units expected after 2005
	Near-net-shape casting techniques	■■	-		Not yet commercial
Aluminium	Retrofit existing Hall-Héroult process (e.g., alumina point-feeding, computer control)	■	-/+		
	Conversion to state-of-the-art PFBF technology	■	+		
	Wettable cathode	■	+++		Not yet commercial
	Fluidized bed kilns in Bayer process	■	++		
	Cogeneration integrated in Bayer process				

(continued)

Table 3.19: *continued*

Sector	Technology	Potential in 2010	Emission reduction costs		Remarks
Cement and other non-metallic minerals	Replacement of wet process kilns	■■	-/+	*	
	Application of multi-stage preheaters and pre-calciners	■	+		No savings expected in retrofit situations
	Utilization of clinker production waste heat or cogeneration for drying raw materials	■■	-		
	Application of high-efficiency classifiers and grinding techniques	■	+		
	Application of regenerative furnaces and improving efficiency of existing furnaces (glass)	■	+		Costs of replacing recuperative furnaces by regenerative furnaces are high (++)
	Tunnel and roller kilns for bricks and ceramic products	■	-	*	
Metal processing and other light industry	Efficient design of buildings, air conditioning and air treatment systems, and heat supply systems	■	-	*	
	Replacement of electric melters by gas-fired melters (foundries)	■	-	*	
	Recuperative burners (foundries)	■	-	*	
Cross-sectoral	Heat cascading with other industrial sectors	■■	+		
	Waste heat utilization for non-industrial sectors	■■	+		

Legend

Potential: ■ = 0-10MtC; ■■ = 10-30MtC; ■■■ = 30-100MtC; ■■■■ > 100MtC.

Annualized costs at discount rate of 10%: - = benefits are larger than the costs; + = US$0-US$100/tC ; ++ = US$100-US$300/tC; +++ > US$300/tC

An asterisk (*) indicates that cost data are only valid in case of regular replacement or expansion.

options and differs from country to country (see Section 3.5.5). Especially options for light industry are not worked out in detail. An important reason is that these sectors are very diverse, and so are the emission reduction options. Nevertheless, there are in relative terms probably more substantial savings possible than in heavy industry (see, e.g., De Beer *et al.*, 1996). Examples of technologies for the light industries are efficient lighting, more efficient motors and drive systems, process controls, and energy saving in space heating.

An extended study towards the potential of energy efficiency improvement was undertaken by the World Energy Council (WEC, 1995a). Based on a sector-by-sector analysis (supported by a number of country case studies) a set of scenarios is developed. In a baseline scenario industrial energy consumption grows from 136EJ in 1990 to 205EJ in 2020. In a state-of-the-art scenario the assumption is that replacement of equipment takes place with the current (1995 in this case) most efficient technologies available; in that case industrial primary energy requirement is limited to 173EJ in 2020. Finally, the *ecologically driven/advanced technology* scenario assumes an international commitment to energy efficiency, as well as rapid technological progress and widespread application of policies

and programmes to speed up the adoption of energy efficient technologies in all major regions of the world. In that case energy consumption may stabilize at 1990 levels. The difference between baseline and ecologically driven/advanced technology is approx. 70EJ, which is roughly equivalent to 1100 MtC. Of this reduction approx. 30% could be realized in OECD countries; approx. 20% in economies-in-transition, and approximately 50% in developing countries. The high share for developing countries can be explained by the high production growth assumed for these countries and the currently somewhat higher specific energy use in these countries.

Apart from these existing technologies, a range of new technologies is under development. Important examples are found in the iron and steel industry. Smelt reduction processes can replace pelletizing and sinter plants, coke ovens, and blast furnaces, and lead to substantial savings. Near net shape casting techniques for steel avoids much of the energy required for rolling (De Beer *et al.*, 1998). Other examples are black liquor gasification in the pulp industry, improved water removal processes for paper making, e.g., impulse drying and air impingement drying, and the use of membrane reactors in the chemical industry. A further overview is given in Blok *et al.*

(1995). Although some of these options already can play a role in the year 2010 (see *Table 3.19*), their full implementation may take some decades. De Beer (1998) carried out an in-depth analysis for three sectors (paper, steel and ammonia). He concludes that new industrial processes hold the promise to reduce the current gap between industrial best practice and theoretical minimum required energy use by 50%.

3.5.3.2   *Fuel Switching*

In general not much attention is paid to fuel switching in the manufacturing industry. Fuel choice to a large extent is sector dependent (coal for dominant processes in the iron and steel industry, oil products in large sectors in the chemical industry). Nevertheless, there seems to be some potential. This may be illustrated by the figures presented in *Table 3.20* where – per sector – the average carbon intensity of fuels used in industry is compared to the country with the lowest carbon intensity. This indicates that fuel switching within fossil fuels can reduce CO_2 emissions by 10%–20%. However, it is not clear whether the switch is feasible in practical situations, or what the costs are. However, there are specific options that combine fuel switching with energy efficiency improvement. Examples are: the replacement of oil- and coal-fired boilers by natural-gas fired combined heat and power (CHP) plant; the replacement of oil-based partial oxidation processes for ammonia production by natural-gas based steam reforming; and the replacement of coal-based blast furnaces for iron production by natural-gas based direct reduction. Daniëls and Moll (1998) calculate that costs of this option are high under European energy price conditions. In the case of lower natural gas prices this option may be more attractive.

3.5.3.3   *Renewable Energy*

See Section 3.8.4.3 for an extensive assessment of renewable energy technology.

3.5.3.4   *Carbon Dioxide Removal*

Carbon dioxide recovery from flue gases is feasible from industrial processes that are operated on a sufficiently large scale. Costs are comparable with the costs of recovering CO_2 from power plant flue gases. See the discussion of these options in Section 3.8.4.4.

However, there are a number of sectors where cheaper recovery is possible. These typically are processes where hydrogen is produced from fossil fuels, leaving CO_2 as a by-product. This is the case in ammonia production (note that some of the CO_2 is already utilized), and increasingly in refineries. Costs can be limited to those of purification, drying and compression. They can be on the order of about US$30/tC avoided (Farla *et al.*, 1995). Another example of carbon dioxide recovery connected to a specific process is the recovery of CO_2 from the calcination of sodium bicarbonate in soda ash production. The company Botash in Botswana recovers and reuses 70% of the CO_2 generated this way (Zhou and Landner, 1999). There are several industrial gas streams with a high CO_2 content from which carbon dioxide recovery theoretically is more efficient than from flue gas (Radgen, 1999). However, there are no technical solutions yet to realize this (Farla *et al.*, 1995).

3.5.3.5   *Material Efficiency Improvement*

In heavy industry most of the energy is used to produce a limited number of primary materials, like steel, cement, plastic, paper, etc. Apart from process changes that directly reduce the CO_2 emissions of the processes, also the limitation of the use of these primary materials can help in reducing CO_2 emissions of these processes. A range of options is available: material efficient product design (Brezet and van Hemel, 1997); material substitution; product recycling; material recycling; quality cascading; and good housekeeping (Worrell *et al.*, 1995b). A review of such options is given in a report for the UN (1997).

Table 3.20: Specific carbon-emission factors for fossil fuel use in manufacturing industry
The figures are calculated on the basis of the IEA Energy Balances

Sector	Specific carbon emission (kg/GJ)	Lowest specific carbon emission found (kg/GJ)
Iron and steel industry	23.6	19.8[a]
Chemical industry	19.1	15.3
Non-ferrous metals industry	19.2	15.3
Non-metallic minerals industry	20.4	16.7
Transportation equipment industry	17.3	15.3
Machine industry	17.7	15.5
Food products industry	18.4	15.6
Pulp and paper industry	18.5	15.3
Total industry	20.1	18.1

[a] Excludes Denmark (no primary steel production)

An interesting integral approach to material efficiency improvement is the suggestion of the "inverse factory" that does not transfer the ownership of goods to the consumers, but just gives the right of use, taking back the product after use for the purpose of reuse or recycling (Kashiwagi *et al.*, 1999).

Some quantitative studies are available on the possible effects of material efficiency improvement. For the USA, Ruth and Dell' Anno (1997) calculate that the effect of increased glass recycling on CO_2 emissions is limited. According to these authors, light-weighting of container glass products may be more promising. In addition, Hekkert *et al.* (2000) show that product recycling of glass bottles (instead of recycling the material to make new products) is also a promising way to reduce CO_2 emissions.

For packaging plastics it is estimated that more efficient design (e.g., use of thinner sheets) and waste plastic recycling could lead to savings of about 30% on the related CO_2 emissions. Hekkert *et al.* (2000) found a technical potential for CO_2 emission reduction for the *total* packaging sector (including paper, wood, and metals) of about 50%.

Worrell *et al.* (1995c) estimate that more efficient use of fertilizer by, e.g., improved agricultural practices and slow release fertilizer, in the Netherlands may lead to a reduction of fertilizer use by 40%.

Closed-loop cement recycling is not yet technically possible (UN, 1997). A more important option for reducing both energy-related and process emissions in the cement industry is the use of blended cements, where clinker as input is replaced by, e.g., blast furnace slag or fly ash from coal combustion. Taking into account the regional availability of such inputs and maximum replacement, it is estimated that about 5%–20% of total CO_2 emissions of the cement industry can be avoided. Costs of these alternative materials are generally lower than those of clinker (IEA Greenhouse Gas R&D Programme, 1999). Note that these figures are based on a static analysis for the year 1990 (Worrell *et al.*, 1995a).

Some integral approaches give an overview of the total possible impact of changes in the material system. Gielen (1999) has modelled the total Western European materials and energy system, using a linear optimization model (Markal). In a baseline scenario emissions of greenhouse gases in the year 2030 are projected to be 5000 MtC_{eq}. At a cost of US$200/tC 10% of these emissions can be avoided through "material options"; at a cost of US$800/tC this increases to 20%. Apart from "end-of-pipe" options, especially material substitution is important, e.g., replacement of petrochemical feedstocks by biomass feedstocks (see also Chapter 4); steel by aluminium in the transport sector; and concrete by wood in the buildings sector. At higher costs, waste management options (energy recovery, plastics recycling) are also selected by the model. Gielen (1999) notes that in his analysis the effect of material efficiency of product design is underestimated.

A study for the UN (1997) estimates that the effect of material efficiency improvement in an "ecologically-driven/advanced technology" scenario in the year 2020 could make up a difference of 40 EJ in world primary energy demand (approximately 7% of the baseline energy use), which is equivalent to over 600 Mt of carbon emissions.

3.5.4 Emission Reduction Options for Non-CO_2 Greenhouse Gases

Non-CO_2 gases from manufacturing (HFCs, PFCs, SF_6, and N_2O) are increasingand. Furthermore, PFCs and SF_6 have extremely long atmospheric lifetimes (thousands of years) and GWP values (thousands of times those of CO_2) resulting in virtually irreversible atmospheric impacts. Fortunately, there are technically-feasible, low cost emission reduction options available for a number of applications. Since the SAR, implementation of major technological advances have led to significant emission reductions of N_2O and the fluorinated greenhouse gases produced as unintended by-products. For the case of fluorinated gases being used as working fluids or process gases, process changes, improved containment and recovery, and use of alternative compounds and technologies have been adopted. On-going research and development efforts are expected to further expand emission reduction options. Energy efficiency improvements are also being achieved in some refrigeration and foam insulation applications, which use fluorinated gases. Emission reduction options by sector are highlighted below. The Chapter 3 Appendix reviews use and emissions of HFCs and PFCs being used as substitutes for ozone-depleting substances.

3.5.4.1 *Nitrous Oxide Emissions from Industrial Processes*

Adipic acid production. Various techniques, like thermal and catalytic destruction, are available to reduce emissions of N_2O by 90% – 98% (Reimer *et al.*, 2000). Reimer *et al.* (2000) report costs of catalytic destruction to be between US$20 and US$60/tN_2O, which is less than US$1/$tC_{eq}$. Costs of thermal destruction in boilers are even lower. The inter-industry group of five major adipic acid manufacturers worldwide in 1991 to 1993 have agreed on information exchange and on a substantial emission cut before the year 2000. These major producers probably will have reduced their joint emissions by 91%. It is estimated that emissions from the 24 plants producing adipic acid worldwide will be reduced by 62% in the year 2000 compared to 1990 (Reimer *et al.*, 2000).

Nitric acid production. Concentrations of N_2O in nitric acid production off-gases are lower than in the case of adipic acid production. Catalytic destruction seems to be the most promising option for emission reduction. Catalysts for this purpose are under development in a few places in the world. Oonk and Schöffel (1999) estimate that emissions can be reduced to a large extent at costs between US$2 and US$10/tC_{eq}.

3.5.4.2   PFC Emissions from Aluminium Production

The smelting process entails electrolytic reduction of alumina (Al_2O_3) to produce aluminium (Al). The smelter pot contains alumina dissolved in an electrolyte, which mainly consists of molten cryolite (Na_3AlF_6). Normal smelting is interrupted by an "anode effect" that is triggered when alumina concentrations drop; excess voltages between the anode and alumina bath result in the formation of PFCs (CF_4 and C_2F_6) from carbon in the anode and fluorine in the cryolite (Huglen and Kvande, 1994; Cook, 1995; Kimmerle and Potvin, 1997). Several processes for primary aluminium production are in use, with specific emissions ranging from typically 0.15 to 1.34 kg CF_4 per tonne Al[16] depending on type of technology (determined by anode type and alumina feeding technology) (IAI, 2000). Measurements made at smelters with the best available technology (point feed prebake) indicate an emissions rate as low as 0.006 kg CF_4 per tonne Al (Marks *et al.*, 2000). Worldwide average emissions for 1995 are estimated to range from 0.26 to 0.77 kg CF_4 per tonne Al (Harnisch *et al.*, 1998; IEA, 2000). Manufacturers have carried out two surveys on the occurrence of anode effects and associated PFC-emissions (IPAI, 1996; IAI, 2000). Based on 60% coverage of world production (no data on Russia and China) they estimated a mean emission value of 0.3 kg CF_4 per tonne Al in 1997. Emission reductions were achieved from 1990 to 1995 by conversion to newer technologies, retrofitting existing plants, and improved plant operation. Industry-government partnerships also played a significant role in reducing PFC emissions. As of November 1998, 10 countries (which accounted for 50% of global aluminium production in 1998) have undertaken industry-government initiatives to reduce PFC emissions from primary aluminium production (US EPA, 1999d). It has been estimated that emissions could be further reduced via equipment retrofits, such as the addition or improvement of computer control systems (a minor retrofit) and the conversion to point-feed systems (a major retrofit). One study estimated 1995 emissions could be reduced an additional 10%–50% (depending on technology type and region) with maximum costs ranging from US$110/t$CO_{2eq}$ for a minor retrofit to nearly US$1100/t$CO_{2eq}$ for a major retrofit (IEA, 2000). A second study estimates that 1995 emissions could be reduced by 40% at costs lower than US$30/t$C_{eq}$, by 65% at costs lower than US$100/t$C_{eq}$ and by 85% at costs lower than US$300/t$C_{eq}$ (Harnisch *et al.*, 1998; 15% discount rate, 10 year amortization).

The development of an inert, non-carbon anode is being pursued through governmental and industrial research and development efforts. A non-carbon anode would remove the source of carbon for PFC generation, thereby eliminating PFC emissions (AA, 1998). A commercially viable design is expected by 2020.

3.5.4.3   PFCs and Other Substances used in Semiconductor Production

The semiconductor industry uses HFC-23, CF_4, C_2F_6, C_3F_8, c-C_4F_8, SF_6 and NF_3 in two production processes: plasma etching thin films (etch) and plasma cleaning chemical vapour deposition (CVD) tool chambers. These chemicals are critical to current manufacturing methods because they possess unique characteristics when used in a plasma that currently cannot be duplicated by alternatives. The industry's technical reliance on high GWP chemicals is increasing as a consequence of growing demand for semiconductor devices (15% average annual growth), and ever-increasing complexity of semiconductor devices.

Baseline processes consume from 15%-60% of influent PFCs depending on the chemical used and the process application (etch or CVD). PFC emissions, however, vary depending on a number of factors: gas used, type/brand of equipment used, company-specific process parameters, number of PFC-using steps in a production process, generation of PFC by-product chemicals, and whether abatement equipment has been implemented. Semiconductor product types, manufacturing processes, and, consequently, emissions vary significantly across worldwide semiconductor fabrication facilities.

PFC use by the semiconductor industry began in the early 1990s. Global emissions from semiconductor manufacturing have been estimated at 4 MtC_{eq} in 1995 (Harnisch *et al.*, 1998). Options for reducing PFC emissions from semiconductor manufacture include process optimization, alternative chemicals, recovery and/or recycling, and effluent abatement. A number of emission reduction options are now commercially available. For plasma-enhanced CVD chamber cleans, switching to PFCs that are more fully dissociated in the plasma or installing reactive fluorine generators upstream of the chamber is favoured. For etch tools, PFC abatement is currently available (Worth, 2000). However, the size of wafers being processed and the design and age of the fabrication facility have a major impact on the applicability of PFC emission reduction technology. A recent study for the EU (Harnisch and Hendriks, 2000) estimated that 60% of projected emissions from this sector could be abated through the use of NF_3 in chamber cleaning at US$110/t$C_{eq}$. According to the same study another 10% are available through alternative etch chemistry at no costs and about 20% through oxidation of exhausts from etch chambers at US$330/t$C_{eq}$. The remaining emissions from existing systems are assumed to be currently virtually unabatable.

Through the World Semiconductor Council, semiconductor manufacturers in the EU, Japan, Korea, Taiwan (China), and the USA have set a voluntary emission reduction target to lower PFC emissions by at least 10% by 2010 from 1995 (1997 for Korea and 1997/1999 average for Taiwan (China) baselines (World Semiconductor Council, 1999). Members of the World Semiconductor Council represent over 90% of global semiconductor manufacture.

---

[16] C_2F_6 emissions typically are 10% of CF_4 emissions.

3.5.4.4   HFC-23 Emissions from HCFC-22 Production

HFC-23 is generated as a by-product during the manufacture of HCFC-22 and emitted through the plant condenser vent. There are about 20 HCFC-22 plants globally. Additional new plants are expected in developing countries as CFC production plants are converted to comply with the Montreal Protocol and demand for refrigeration grows. Although HCFC-22 is an ozone-depleting chemical and production for commercial use will be phased out between 2005 and 2040, production as a feedstock chemical for synthetic polymers will continue.

Technologies available to reduce emissions of HFC-23 have been reviewed by the Research Triangle Institute (RTI, 1996; Rand *et al.*, 1999) and March Consulting Group (March Consulting, 1998). Two emission reduction options were identified.

- Optimization of the HCFC-22 production process to minimize HFC-23 emissions. This technology is readily transferable to developing countries. Process optimization is relatively inexpensive and is demonstrated to reduce emissions of fully optimized plants to below 2% of HCFC-22 production. Nearly all plants in developed countries have optimized systems.

- Thermal destruction technologies are available today and can achieve emissions reductions of as high as 99%, although actual reductions will be determined by the fraction of production time that the destruction device is actually operating. Cost estimates are 7 ECU/tC for the EU (March Consulting, 1998, 8% discount rate).

3.5.4.5   Emissions of SF_6 from the Production, Use and Decommissioning of Gas Insulated Switchgear

SF_6 is used for electrical insulation, arc quenching, and current interruption in electrical equipment used in the transmission and distribution of high-voltage electricity. SF_6 has physical properties that make it ideal for use in high-voltage electric power equipment, including high dielectric strength, excellent arc quenching properties, low chemical reactivity, and good heat transfer characteristics. The high dielectric strength of SF_6 allows SF_6-insulated equipment to be more compact than equivalent air-insulated equipment. An SF_6-insulated substation can require as little as 10% of the volume of an air-insulated substation. Most of the SF_6 used in electrical equipment is used in gas-insulated switch gear and circuit breakers. SF_6 in electric equipment is the largest use category of SF_6 with global estimates of over 75% of SF_6 sales going to electric power applications (SPS, 1997). Options to reduce emissions include upgrading equipment with low emission technology, and improved handling during installation maintenance and/decommissioning (end-of-life) of SF_6-insulated equipment, which includes the avoidance of deliberate release and systematic recycling. Guidelines on equipment design to allow ease of gas recycling, appropriate gas handling and recycling procedures, features of gas handling and recycling equipment, and the impact of voluntary emission reduction programmes

are contributing to the reduction of emissions from this sector (Mauthe *et al*, 1997; Causey, 2000).

Significant emissions may also occur during the manufacturing and testing of gas-insulated switch gear when the systems are repeatedly filled with SF_6 and re-evacuated (Harnisch and Hendriks, 2000). Historically these emissions have been in the range of 30%-50% of the total charge of SF_6. The existence and appropriate use of state-of-the art recovery equipment can help to reduce these emissions down to at least 10% of the total charge of SF_6.

3.5.4.6   Emissions of SF_6 from Magnesium Production and Casting

In the magnesium industry, a dilute mixture of SF_6 with dry air and/or CO_2 is used as a protective cover gas to prevent violent oxidation of the molten metal. It is assumed that all SF_6 used is emitted to the atmosphere. 7% of global SF_6 sales is estimated to be for magnesium applications (SPS, 1997). Manufacturing segments include primary magnesium production, die casting, gravity casting and secondary production (i.e., scrap metal recycling). Because of differing production processes and plant scale, emission reduction potential varies across manufacturing segments. Emissions of SF_6 in magnesium casting can potentially be reduced to zero by switching to SO, a highly toxic and corrosive chemical used over20 years ago as a protective cover gas. Harnisch and Hendriks (2000) estimate that net costs of switching from SF_6 to SO_2-based cover gas systems are about US\$1/tC_{eq}, but as a result of the high toxicity and corrosivity of SO_2 much more careful handling and gas management is required. In many cases the specific usage of SF_6 can be reduced by operational changes, including moderate technical modifications (Maiss and Brenninkmeijer, 1998). Companies may also reduce SF_6 emissions and save money by carefully managing the concentration and application of the cover gas (IMA, 1998). A study is currently beingundertaken to identify and evaluate chemical alternatives to SF_6 and SO_2 for magnesium melt protection (Clow and Hillls, 2000).

3.5.4.7   Some Smaller Non-CO_2 Emission Reduction Options

There are a number of small emission sources of SF_6, some of which are considered technically unnecessary. For example, SF_6 has been used as a substitute for air, hydrogen or nitrogen in sport shoes and luxury car tyres to extend the lifetime of the pressurized system. SF_6 in sport shoes has been used by a large global manufacturer for over a decade under a patented process. Soundproof windows have been manufactured with SF_6 in several countries in Europe.

Small quantities of SF_6 are used as a dielectric in the guidance system of radar systems like the airborne warning and control system (AWACS) aircraft and as a tracer gas for pollutant dispersion studies. Small quantities of PFCs and SF_6 are used in medical applications such as retina repair, collapsed lung expansion, and blood substitution (UNEP, 1999).

Table 3.21: *Overview of greenhouse gas emission reduction options in industry (excludes energy efficiency improvement, see Table 3.19). Note that the scales are not linear.*

Sector	Technology	Potential in 2010	Emission reduction costs	Remarks
All industry	Fuel switching	■■■	?	Rough estimate
Fertilizer, refineries	Carbon dioxide removal	■■	+	Excludes carbon dioxide removal from flue gases
Basic materials industries	Material efficiency improvement	■■■■	–/+/++	First estimate of potentials; option is not yet worked out in detail
Cement industry	Application of blended cements	■■■	–	
Chemical industry	Nitrous oxide emission reduction	■■	+	Excludes emission reduction measures taken before the year 2000
Aluminium industry	PFC emission reduction	■	+/–	
Chemical industry	HFC-23 emission reduction	■■	+	

Legend

Potential: ■ = 0-10MtC; ■■ = 10-30MtC; ■■■ = 30-100MtC; ■■■■ > 100MtC

Annualized costs at discount rate of 10%:

– = benefits are larger than the costs; + = US$0-100/tC; ++ = US$100-300/tC; +++ > US$300/tC

3.5.4.8   Summary of Manufacturing Industry GHG Emission Reduction Options

An overview of greenhouse gas emission reduction options in manufacturing industry due to fuel switching, carbon dioxide removal, material efficiency improvements, and reduction of non-CO_2 greenhouse gases emissions practices is presented in *Table 3.21*, which complements information from *Table 3.19*.

3.5.5   Regional Differences

Differences in emission reduction potential are mainly caused by differences in specific energy consumption of industrial processes. In recent years attention was paid to developing methods to compare energy efficiency levels on a physical basis (Phylipsen *et al.*, 1998a), in addition to methods that compare energy efficiency levels on a monetary basis (see, e.g., Schipper and Meyers, 1992). Energy efficiency indicators on a physical basis start from the level of energy consumption per unit of product (e.g., expressed in GJ/t). However, countries may differ in the production structure per sector (i.e. differences in product mix and associated differences in feedstocks used). Correction for such differences can take place by relating the specific energy consumption level for each product to a best practice level, resulting in a so-called energy efficiency index. The more efficient the aggregate of processes in a sector in a country, the lower the energy efficiency index. The energy efficiency indices are scaled in such a way that if all processes were operated at the best-practice level, the index would be 100.

Results up to now are presented in *Figure 3.13*. Apart from correction for structure, international comparison of energy efficiency requires correction for statistical errors. A common source of error in the process industries is the double counting of fuels (e.g., in the iron and steel industry double counting of coke input to the blast furnace and blast furnace gas). After correction for such errors the energy efficiency indicators – like those presented in *Table 3.19* – show a typical uncertainty of 5% (Farla, 2000).

Despite the remaining uncertainties, some conclusions can be drawn from these data. In general Japan and South Korea and countries in Western Europe show the lowest energy efficiency index (i.e., they are most efficient). Developing countries, economies in transition and some OECD countries (like the USA and Australia) show higher levels of this index. However, there are certainly exceptions; for instance, some developing countries show fairly low levels of the energy efficiency index for some sectors. This may be explained by the fact that these countries are developing at a high rate, and hence apply relatively young and modern technology. In general the countries with the highest energy efficiency index will have the highest technical potential for energy efficiency improvement. The dif-

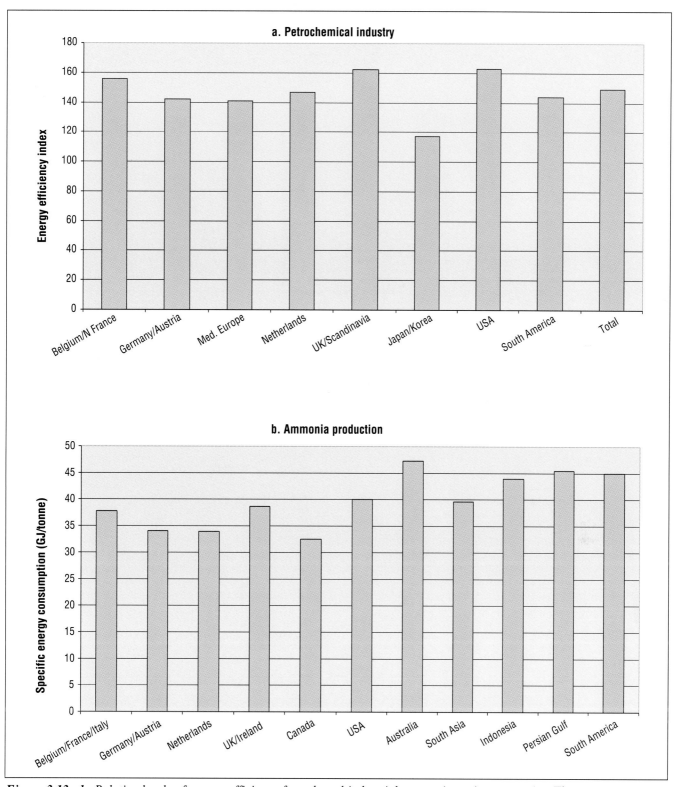

Figure 3.13a-b: *Relative levels of energy efficiency for selected industrial sectors in various countries. The aggregate energy efficiency index (EEI) is calculated as:*

$$EEI = (\Sigma_i P_i \cdot SEC_i)/(\Sigma_i P_i \cdot SEC_{i,BP}),$$

where P_i is the production volume of product i; SEC_i is the specific energy consumption for product i, and $SEC_{i,BP}$ is a best-practice reference level for the specific energy consumption for product i. By applying this approach a correction is made in order to account for structural differences between countries in each of the tracked industrial sectors. A typical statistical uncertainty for these figures is 5%. Because of statistical errors higher uncertainties may occur in individual cases.

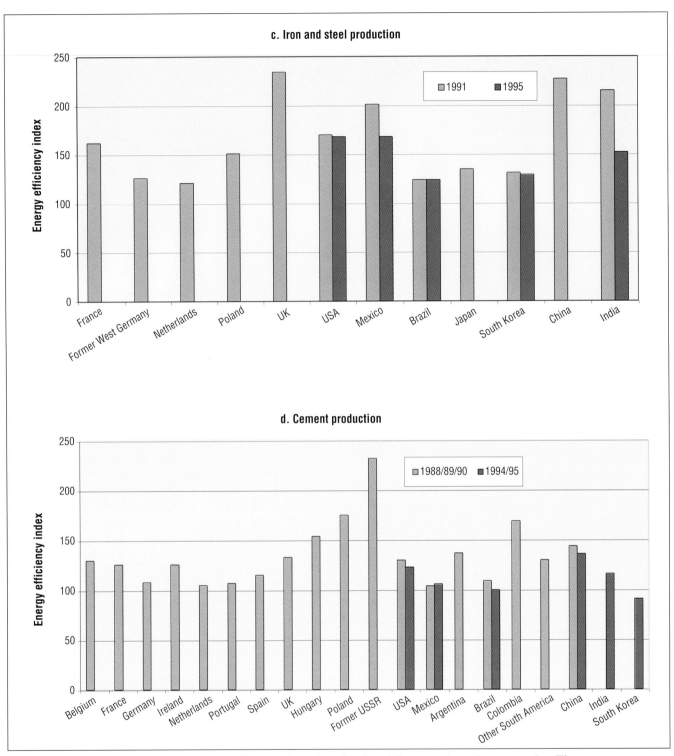

Figure 3.13c - d: *Relative levels of energy efficiency for selected industrial sectors in various countries. The aggregate energy efficiency index (EEI) is calculated as:*

$$EEI = (\Sigma_i \, P_i \cdot SEC_i)/(\, \Sigma_i \, P_i \cdot SEC_{i,BP}),$$

where P_i is the production volume of product i; SEC_i is the specific energy consumption for product i, and $SEC_{i,BP}$ is a best-practice reference level for the specific energy consumption for product i. By applying this approach a correction is made in order to account for structural differences between countries in each of the tracked industrial sectors. A typical statistical uncertainty for these figures is 5%. Because of statistical errors higher uncertainties may occur in individual cases.

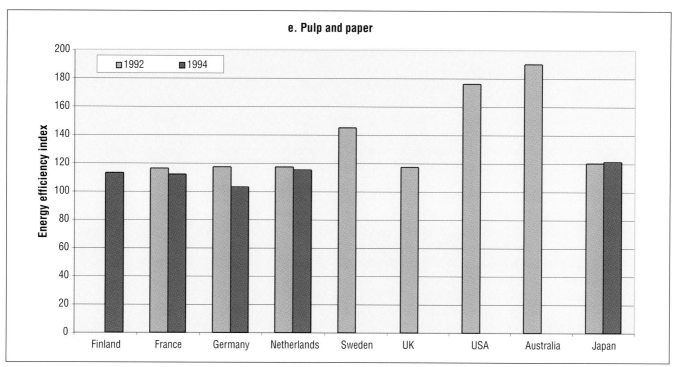

Figure 3.13e: *Relative levels of energy efficiency for selected industrial sectors in various countries. The aggregate energy efficiency index (EEI) is calculated as:*

$$EEI = (\Sigma_i\, P_i \cdot SEC_i)/(\,\Sigma_i\, P_i \cdot SEC_{i,BP}),$$

where P_i is the production volume of product i; SEC_i is the specific energy consumption for product i, and $SEC_{i,BP}$ is a best-practice reference level for the specific energy consumption for product i. By applying this approach a correction is made in order to account for structural differences between countries in each of the tracked industrial sectors. A typical statistical uncertainty for these figures is 5%. Because of statistical errors higher uncertainties may occur in individual cases.

ferences in economic potential may be smaller, as a consequence of the lower energy prices that often occur in the less efficient countries. In this section a number of regional studies – mainly into energy efficiency in industry – are reviewed.

3.5.5.1 China

Industry is responsible for 75% of commercial energy end-use in China (IEA, 1997d). The period from 1980 to 1996 has seen a strong economic growth and growth of industrial production, but also a substantial decline of the energy/GDP ratio of about 4% per year (China Statistical Yearbook). The share of energy efficiency and structural change in this decline is uncertain, but it is clear that substantial energy efficiency improvement was obtained (Zhou and Hu, 1999; Sinton, 1996). Nevertheless, Chinese industry is still substantially less energy efficient than most OECD countries (Wu and Wei, 1997), see also *Figure 3.13*. Within industry, the steel industry is most important, consuming 23% of industrial energy use in 1995 (IEA, 1997d). Zhou and Hu (1999) analysed the differences between the Chinese and the efficient Japanese iron and steel industry and identified a range of measures to improve the specific energy consumption of the Chinese steel industry. Important measures are the recovery of residual gases (2.7GJ/t steel); boiler modification and CHP (2.1 GJ/t); improved feedstock quality (2.1GJ/t); wider application of continuous casting (1.0GJ/t); and others (2.0GJ/t). The total

leads to a reduction of 25% compared to the present average of 35.6GJ/t (Zhou and Hu, 1999). An analysis of future prospects by Worrell (1995) shows that, in the case steel production grows from 93Mt in 1995 to 140Mt in 2020, energy consumption in the Chinese steel industry is likely to grow. But the growth can be very moderate if modern technologies, like smelt reduction and near-net-shape casting, are adopted. Also for two other important sectors, the building materials industry and the chemical industries, substantial technical saving potentials are reported (Zhou and Hu, 1999). Liu *et al.* (1995) report for the cement industry – consuming 10% of industrial energy use in 1995 – a potential for reduction of the specific energy consumption of 32% in the period 1990 to 2000; associated investments are estimated at 105 billion yuan (~US$13 billion). Important economically viable options are comprehensive retrofit of vertical kilns (e.g., improving refractory lining) and wet kilns, and kiln diameter enlargement and retrofit. Similar savings can be reached when adding a pre-calciner to the kilns, which is, however, the most expensive option. All cost-effective measures add up to a 20% reduction of primary energy consumption compared to the base line energy use in 2010 (Sinton and Yang, 1998).

3.5.5.2 Japan

In Japan, industry accounts for nearly half of the final energy demand. Industrial energy demand is stabilizing, mainly because of the shift from heavy industry to sectors like electrical machinery, precision instruments, and motor vehicles.

Substantial energy efficiency improvements have been obtained, and Japan is now one of the most efficient countries in the world (see also *Figure 3.13*). Nevertheless, there are still energy efficiency improvement potentials. Current technical potential is 10%–12% in the iron and steel industry. Under the influence of a carbon tax, the potential is 8% in the cement industry and 10% in the chemical industry. Costs of saving energy are in the majority of the cases lower than energy purchase costs at a 5% discount rate (Kashiwagi *et al.*, 1999). Kainuma *et al.* (1999) have carried out an analysis of various policies using the AIM model and find maximum absolute reductions of industrial CO_2 emissions of 15% (in the base case the absolute emission reduction is 3%). The increasing concern about the climate change issue has required setting a new higher target to curb energy use to the FY 1996 level in FY 2010, which requires an energy savings of approximately 10% of final demand in the industrial sector by the revision of the Energy Conservation Law put into force in April 1999 (MITI, 1999).

3.5.5.3   Latin America

In Latin American countries, industry consumes about 30% of final energy use. Energy intensity has increased, partly because of a deterioration of the energy efficiency in the heavy industries. Substantial energy efficiency improvement potentials are reported, see *Table 3.22*.

As an example it is useful to give some information on industrial electricity use in Brazil. Industry accounts for 48% of electricity consumption in Brazil, about half of this is for electric motors. Geller *et al.* (1998) report low-cost saving possibilities of 8%–15%. The use of energy-efficient motors is more costly (typically 40% more investment than conventional), but still simple payback times range from 1 to 7 years. Such motors could save about 3% of industrial electricity use. In addition, variable speed controls may save 4% of industrial electricity use (Moreira and Moreira, 1998).

3.5.5.4   USA and Canada

The manufacturing industry is responsible for one-third of total USA energy use and for nearly half of total Canadian energy use. A set of studies is available regarding possible developments of carbon dioxide emissions in this sector. A comparison of three of these studies was presented by Ruth *et al.* (1999); see *Table 3.23*. All three studies do not present a technical or economic potential, but take into account incomplete penetration of available technologies. The outcomes in the *policy case* for the USA range from a 2% carbon dioxide emission growth to a strong decline. The two studies for the USA rely on the same model, but differ in the extent to which technologies are implemented. Furthermore, there are differences in assumed structural development and the treatment of combined generation of heat and power.

Table 3.22: *Potential energy savings in energy intensive industries in Latin America. The table shows the percentage reduction of average specific energy consumption that can be achieved with additional investments (Pichs, 1998).*

	Short term/ small investments	Long term/ medium size investments
Steel	5 - 7	5 – 13
Aluminium	2 - 4	10 – 15
Oil	7 - 12	15 – 25
Fertilizer	2 - 5	20 – 25
Glass	10 - 12	15 – 20
Construction	10 - 15	15 – 20
Cement	10 - 20	10 – 30
Pulp and paper	10 - 15	10 – 16
Food	8 - 18	12 – 85
Textile	12 - 15	15 – 17

For the USA a series of studies have determined the static potentials for three energy-intensive sectors. A study of the iron and steel industry concludes that steel plants are relatively old. A total of 48 cost-effective measures were identified that can reduce carbon dioxide emissions per tonne of steel from this sector by 19% (Worrell *et al.*, 1999). For the cement industry a cost-effective potential of 5% excluding blending (30 technologies) and 11% including blending was calculated (Martin *et al.*, 1999). For the pulp and paper industry the cost-effective potential is 14% (16% including paper recycling) and the technical potential 25% (37% including recycling) (Martin *et al.*, 2000).

For the important Canadian pulp and paper industry for 2010 (compared to 1990) a technical potential for reduction of specific energy consumption of 38% was found; the cost-effective potential is 9% (Jaccard, 1996). All these cost-effective potentials are calculated from the business perspective (e.g., for the USA a pay-back criterion of 3 years is used).

3.5.5.5   Africa

Typically the industry in Africa is characterized by slow replacement of equipment like motors, boilers, and industrial furnaces. Small and medium enterprises are the most affected as a result of limited financial resources and skills. Greenhouse gas emission mitigation opportunities identified in past national studies in Southern Africa (UNEP/Southern Centre, 1993; CEEZ, 1999; Zhou, 1999) are centred on retrofitting boilers and motors, cogeneration using waste process heat, and introduction of high efficiency motors on replacement. The costs for implementing these measures are in the range of negative to low per tonne of carbon.

Table 3.23: *Change in carbon emissions from the industrial sector, 1990 to 2010, base and policy cases.*
Source: Ruth *et al.* (1999)

		USA – I (Interlaboratory Working Group, 1997)	USA – II (Bernow *et al.*, 1997)	ERG (Bailie *et al.*, 1998)
Base case, 2010 emissions relative to 1990	*fuel*	+20%	+20%	+25%
	Electricity	+28%	+24%	+50%
	Total	+22%	+23%	+29%
Policy case, 2010 emissions relative to 1990	Fuel	+7%	-13%	+7%
	Electricity	-6%	-54%	+28%
	Total	+2%	-28%	+11%

Table 3.24: *Energy efficiency improvement potential in terms of reduction of aggregate specific energy consumption compared to frozen efficiency. In the figures for Germany combined generation of heat and power is not included, in the Netherlands it is included. (*see Blok *et al.*, 1995).

	Germany (BMBF, 1995; Jochem and Bradke, 1996) (discount rate 4%)	The Netherlands (De Beer *et al.*, 1996) (discount rate 10%)	United Kingdom (discount rates vary by sector)
Technical potential	1995/2005: 20% 1995/2020: 25%	1990/2000: heavy industry: 25% light industry: 40%	1990/2010 high-temperature industries: 45% low-temperature industries: 32% horizontal technologies (excluding CHP): 15% (ETSU, 1994)
Economic potential	1995/2005: 7%.to 13% 1995/2020: 16% to 20%	1990/2000: heavy industry: 20% light industry: 30%	1990/2000 all industry 24% of CO_2 (ETSU, 1996)

3.5.5.6   Western Europe

Industry in Western Europe is relatively efficient, as was shown in *Figure 3.13*. For some countries results of detailed studies into the technical and economic potential for energy efficiency are shown in *Table 3.24*. These studies show that the economic potential for energy efficiency improvement typically ranges from 1.4%–2.7% per year, whereas the technical potential may be up to 2.2%–3.5% per year[17].

Assessment of total potential for energy efficiency improvement

The previous overview gives results for a range of studies carried out for a variety of countries. It should be noted that the studies differ in starting points, methods of analysis, and completeness of the analysis. Some studies give technical or economic potentials, others take into account implementation rates in an accelerated policy context.

Nevertheless, it may be concluded that in all world regions substantial potentials for energy efficiency improvement exist. This is also the case for regions like Western Europe and Japan that – according to *Figure 3.13* – were already fairly efficient. For the other regions energy efficiency improvement potentials generally are higher, although both detailed sector studies and comprehensive overviews are lacking for most countries.

In order to make an estimate of the worldwide potential of enhanced energy efficiency improvement a number of assumptions are made. It is assumed growth of industrial production in physical terms to be 0.9% per annum in the OECD region; 1.0% per annum in economies in transition; 3.6% per annum in the Asian developing countries; 3.9% per annum in the rest of

---

[17] The 1995 to 2020 potentials for Germany are lower on an annual basis, but this may be due to the long time-frame underestimating the potential.

the world. Autonomous energy efficiency improvement is assumed to lead to a reduction of specific energy use by 0.5%–1.0% per year (assumption for the average: 0.75%). The total is equivalent to the outcomes in terms of CO_2 emissions in the SRES-B2 scenario. For calculating the potential of industrial energy efficiency improvement, it is assumed that from the year 2000 the enhanced energy efficiency improvement is 1.5%–2.0% per year in the OECD countries (average); and 2%–2.5% per year in the other world regions. Starting from the energy use and emission figures quoted in section 3.5.2, a potential of 300–500MtC is calculated for the year 2010 and 700–900MtC for the year 2020. These figures are consistent with earlier estimates, e.g. WEC, 1995a).

3.5.6   Conclusions

It once again becomes clear that enhanced energy efficiency improvement remains the main option for emission reduction in the manufacturing industry. There are substantial differences in the level of energy efficiency between countries and also potentials differ. For most OECD countries and for a number of developing countries extended inventories of emission reduction options in industry exist. However, the focus is still very much on the heavy industrial sector. The total potential of energy efficiency improvement for the year 2010 can be estimated to be 300–500MtC for the year 2010. It seems possible to develop new technologies to sustain energy efficiency improvement in the longer term; if such innovations materialize the potential can be 700 - 900MtC for the year 2020. The larger part of these emission reductions can be attained at net negative costs.

A category of options to which only limited attention was paid in relation to greenhouse gas emission reduction is material efficiency improvement. It is clear that substantial technical potentials exist. These may be sufficient to attain emission reductions on the order of 600MtC in the year 2020 (UN, 1997). However, a significant effort is needed in selection, development, and implementation of such options. For the shorter term the potential will be substantially smaller (e.g., 200MtC), because of the complexity of introducing these options.

For virtually all sources of non-CO_2 greenhouse gases in the manufacturing industry, options are available that can reduce emissions substantially, in some sectors to near zero. However, the total contribution to the emission reduction is limited: approximately 100MtC$_{eq}$ emission reduction is possible at a cost less than US$30/tC$_{eq}$.

3.6   Agriculture and Energy Cropping

3.6.1   Introduction

Agriculture contributes to over 20% of global anthropogenic greenhouse gas emissions as a result of:

- CO_2 (21%–25% of total CO_2 emissions) from fossil fuels used on farms, but mainly from deforestation and shifting patterns of cultivation;
- CH_4 (55%–60% of total CH_4 emissions) from rice paddies, land use change, biomass burning, enteric fermentation, animal wastes;
- N_2O (65%–80% of total N_2O emissions) mainly from nitrogenous fertilizers on cultivated soils and animal wastes (OECD, 1998).

Direct emissions of greenhouse gases occur during agricultural production processes from soils and animals and as a result of meeting demands for heat, electricity, and tractor and transport fuels. In addition, indirect N_2O emissions are induced by agricultural activities (Mosier *et al*, 1998b) and CO_2 also results from the manufacturing of other essential inputs such as machinery, inorganic fertilizers, and agro-chemicals. Emissions occur at various stages of the production chain and full life cycle analyses are necessary to identify their extent.

In developing countries such as India, emissions mainly arise from ruminant methane, field burning of agricultural residues, and paddy cultivation. Mitigation is difficult to achieve but research into more frequent draining of paddy fields, reduction in the use of nitrogenous fertilizers, and improved diets of cattle is ongoing. Cattle numbers are expected to increase 50% by 2020, which would largely offset any methane avoidance.

As for energy inputs, in many developing countries traditional agriculture still depends on human labour and animal power together with firewood for cooking. Modern agriculture in industrialized countries relies on direct fossil fuel inputs together with embedded energy in fertilizers, and for transport to markets. In the USA each food item purchased has been transported an average of over 2500km (Resources for the Future, 1998) and even further in Europe and Australasia. Recent data for OECD countries suggest the embodied energy in food and drink is 42 GJ per person per year, being 10 times the energy content of the food (Treloar and Fay, 1998).

Increasing energy inputs to meet the growing needs for food and fibre are shown in *Figure 3.14*. Demand has declined in EITs, increased only slightly in Latin America and Africa in spite of population increases, and increased significantly elsewhere. In developing countries the provision and uptake of "leapfrog" technologies to enable human energy to be replaced by non-fossil fuel energy could be stimulated (Best, 1998).

Primary production methods used by farmers, foresters, and fisheries are not energy intensive compared with the industrial and transport sectors, so carbon dioxide emissions are comparatively small, being 217MtC in 1995 (*Table 3.1*) from an annual energy demand of around 3% of total consumer energy (*Table 3.25*).

The worldwide trend towards energy intensification (GJ/ha) of food and fibre production grown on arable land continues.

Table 3.25: *Energy use in the agricultural sector in 1995 and annual growth rates in the preceding periods* (Price *et al.*, 1998).

	Fuel			**Electricity**			**Primary energy use**		
	Annual growth rate '71-'90	Annual growth rate '90-'95	Cons. 1995 (EJ)	Annual growth rate '71-'90	Annual growth rate '90-'95	Cons. 1995 (EJ)	Annual growth rate '71-'90	Annual growth rate '90-'95	Cons. 1995 (EJ)
OECD Countries	2.3%	1.4%	2.36	1.8%	2.3%	0.20	2.2%	1.6%	2.97
EIT	4.4%	-14.1%	1.04	4.7%	-2.7%	0.22	4.5%	-10.6%	1.71
DCs Asia-Pacific	2.8%	2.4%	1.25	7.9%	8.2%	0.59	4.8%	5.6%	3.03
Rest of the World	3.5%	13.1%	1.05	8.2%	11.6%	0.17	4.7%	12.6%	1.56
World	3.2%	-1.4%	5.70	5.3%	4.6%	1.18	3.8%	0.8%	9.28

China, for example, began its "socialism marketing system" recently with the aim of changing agriculture from traditional to more modern production methods (Zhamou and Yanfei, 1998). As a result, total food production on the same land area is projected to rise by around 15% and the standard of living for farmers will be higher but also associated with higher risk. Without greater access to modern energy sources, food and fibre production is unlikely to increase (FAO, 1995). The energizing of the food production chain in terms of quantity and quality is necessary for the attainment of global food security to meet demand for more than one year. To meet the targets of the World Food Summit to reduce the undernourished population to half the current level by 2015, a 4 to 7 fold increase in current commercial energy inputs into agriculture, particularly in developing countries, is anticipated (Best, 1998). In order for agricultural production to be undertaken in a more sustainable manner, one can use husbandry methods and management techniques to minimize the inputs of energy, synthetic fertilizers, and agro-chemicals on which present industrialized farming methods depend. Any method of reducing these inputs in both developed and developing countries using new technologies must be considered.

Integrated assessment methodologies, which include both direct and indirect energy inputs, have been developed for

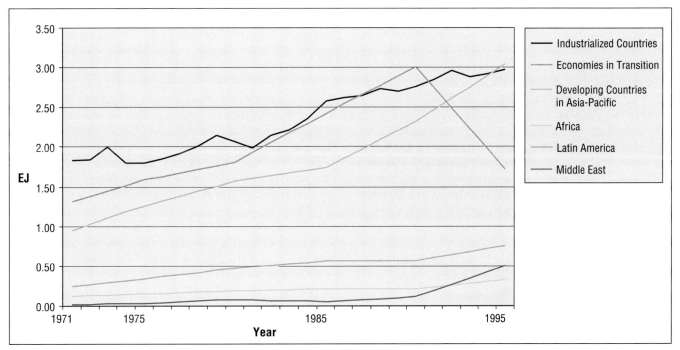

Figure 3.14: *Energy use in the agricultural sector from 1971 to 1995.*

Table 3.26: *Major sources of methane and nitrous oxide by region in 1995 (MtC$_{eq}$/yr).*
(Adapted from OECD, 1998)

Source	Canada	USA	Europe	Japan	EIT	Oceania	World
CH$_4$ animals	2.9	23.6	39.5	1.2	34.6	21.0	438.1
CH$_4$ animal wastes	0.8	11.9	18.7	2.5	12.3	1.6	84.0
N$_2$O fertilizer	2.4	21.0	15.0	0.7	18.9	4.1	112.4
N$_2$O animal wastes	0.8	4.8	8.2	0.6	7.2	1.6	85.6

crops by Kramer *et al.* (1999) and for milk production by Wells (1999). Both studies analysed the complete production chain up to the "farm gate" and both identified fertilizer inputs as being the major contributor of carbon emissions from the system. For example, manufacturing nitrogenous fertilizers in Germany has specific cumulative energy inputs of around 59MJ/kg of fertilizer (having been reduced by energy efficiency methods from 78MJ/kg in 1970), whereas in the USA they remain at a higher level and have slightly increased (Scholz, 1998).

Spedding (1992) expounded the view that if abundant renewable energy supplies were to become available, then energy would be the only important natural resource since all other natural resources could be generated and all waste streams neutralized. In theory even soil could be considered to be a dispensable resource since crops could be grown hydroponically in nutrient solutions, though in practice this would not be feasible. Agricultural industries can contribute to a more sustainable energy future by providing biomass products. Surplus crop and animal waste products (where not used for soil amendments or fertilizers) can be used as bioenergy sources. Growing crops for energy is well understood, though usually only economically viable where some form of government incentive exists or the environmental benefits are fully recognized (Sims, 1997). Possible conflicts of land use for sustainable food production, soil nutrient depletion, water availability, and biodiversity need to be addressed.

Farming, fishing, and forestry continue to grow in energy intensity to meet the ever-increasing global demands for food and fibre. The present challenge is to offset this trend by introducing more efficient production methods and greater adoption of new technologies and practices. Whilst reducing energy intensity, agriculture must also become more sustainable in terms of reduced nutrient inputs, lower environmental impacts, and with zero depletion of the world's natural resources such as fish and topsoil. This can only be successfully achieved if practical support is received from primary producers, and this will only occur if other benefits are perceived (Section 3.6.4.5).

As to methane and nitrous oxide, accurate measurement of these anthropogenic greenhouses gas emissions poses challenges as they arise from diffuse sources and wide ranges are quoted. Methane arises from conversion of tropical rainforests to pasture (120-480MtC/yr); rice paddies (120-600MtC/yr); ruminants (390-600MtC/yr); and animal wastes (60-160MtC/yr). Nitrous oxide mainly arises from use of mineral nitrogenous fertilizers (140-200MtC/yr); use of organic fertilizers (140-200 MtC/yr); and deforestation by burning and subsequent cultivation (200-260MtC/yr) (Ahlgrimm, 1998). Another estimate for total agricultural N$_2$O emissions exceeded 840MtC$_{eq}$/yr (6.3 TgN/yr) but also included manure storage, animal droppings on pasture (Oenema *et al.,* 1997), and cultivation of organic soils (Kroeze and Mosier, 1999). OECD regional sources give lower estimates (*Table 3.26*). Thus there remains a high degree of uncertainty concerning the actual levels of N$_2$O emissions, since many countries have not yet adopted the IPCC revised 1996 guidelines for national greenhouse gas inventories for emissions from agricultural systems.

Strategies for reducing methane emissions from paddy rice and ruminant animals are being evaluated (Yagi *et al.*, 1997), as are techniques to reduce N$_2$O emissions by better treatment of wastes, improved pasture and animal management, and improved use of nitrogenous fertilizers. Reducing N$_2$O emissions has to be achieved in areas of intensive agriculture by reducing the N surplus of the system. Improvements in modelling nitrogen and carbon fluxes for agricultural ecosystems have recently been developed (see, for example, Li, 1998, 1999) and applied on the county level for the USA (Li 1995, Li *et al.* 1996) and for China (Li *et al.*, 1999). By considering the specific interaction between agricultural management with climate and soil conditions, the model simulations have demonstrated large potentials for mitigating N$_2$O and other greenhouse gas emissions by changing management practices. These include adjusting fertilizer use in poor or rich soils; altering timing of fertilizer or manure applications based on rainfalls; and altering timing and depth of tillage. Based on the modelled results for the USA and China, the most effective way to reduce agricultural N$_2$O emissions and to ensure adequate crop yields is to optimize fertilizer use in arable soils, particularly those which contain soil organic carbon greater than 3%. However, farmers will need to first accept this management practice if it is to be implemented. A full discussion on the complexities of soil carbon is provided in the Special Report on *Land Use, Land Use Change, and Forestry* (IPCC, 2000).

There is debate whether such process-based, spatially and temporally integrated models are ready to be used for country inventories and would be better than the IPCC methodology (Frolking, 1998). Considerable uncertainty remains as a result of the sensitivity of underlying assumptions (as discussed at the European Federation of Clean Air and Environmental Protection Associations' Second International Symposium on Non-CO_2 Greenhouse Gases, Noordwijkerhout, Netherlands, 8-10 September, 1999). Validation by independent atmospheric budget measurements is needed.

Land clearance activities are covered in Chapter 4; the transportation of products from the "farm gate" to the market or processing plant in Section 3.4; and processing of the agricultural, horticultural, forest, or fish products in Section 3.5.

3.6.2   Summary of the Second Assessment Report

Little has changed in the industrialized agricultural sector during the past few years apart from the continuing trends towards genetically modified crops and animals, and reduced chemical input production methods (Section 3.6.4.2). Farming systems remain a major contributor of anthropogenic emissions, not just from energy inputs but mainly from methane from ruminants (cattle, goats and sheep), livestock wastes, rice paddy fields, and nitrous oxide from the application of nitrogenous fertilizers, and circulation of N within crop and livestock production. Land use change activities, such as the clearance of forests by burning and cultivation to provide more land for agricultural production with subsequent soil degradation, are also major contributors (Chapter 4). Carbon sequestration by soils continues to be quantified and estimates refined further, including the effects from reducing organic matter losses by changing to minimum tillage techniques. Improved farm management can result in lower emissions of CH_4 and N_2O and increased soil carbon uptake.

New energy saving technologies, such as ice bank refrigeration for milk cooling (CAE, 1996), continue to be developed, but need to be widely implemented if they are going to have any significant effects on global greenhouse gas emissions. Methods to reduce emissions by the agricultural sector are outlined in Section 3.6.4. Energy crop production continues to provide a possible alternative land use where suitable land is available and markets exist for the products.

3.6.3   Historic and Future Trends

Although on-farm energy intensity (GJ/ha) continues to increase, energy inputs per unit of production (GJ/t) have tended to decline in modern intensive industrialized agricultural systems, mainly caused by increasing crop yields (IPCC, 1996). The current trend in OECD countries is towards less intensive farming systems because of public concerns for animal welfare, reduced chemical inputs, and increasing demand for organically grown food. If this demand continues it could lead to reduced GJ/ha and also to lower GJ/t if yields can be maintained, (though this is generally not the case for low input farming). Conversely, in developing countries energy use (both GJ/ha and GJ/t) is increasing in attempts to increase yields (t/ha) by substituting machinery for manual labour, developing irrigation schemes, and improving crop storage systems to reduce losses. For example, Indian agricultural production has increased threefold since 1970 to 200Mt of food in 1998, whilst during this period animal energy/ha declined 35%, and diesel and electricity inputs increased by over 15 times (Prasad, 1999).

The development and introduction of biotechnology and gene technology could offer new chances to accelerate and support the traditional plant and animal breeding procedures. However, the conflict between food security and environmental risks is yet to be resolved. In developing countries uptake of transgenic technologies would require support of the farmers in adopting non-traditional techniques. If, following a comprehensive risk assessment, public confidence in producing genetically modified organisms is obtained, it could ultimately result in:

- yield increases per hectare of food and fibre crops;
- improved performance efficiency of livestock animals;
- reduced inputs of agrio-chemicals and fertilizers because of new resistant cultivars; and
- development of low input cultivars with improved nutrient and water use efficiency.

Energy inputs per unit of product could be reduced by around 10%-20% as a result, and methane and nitrous oxide emissions also lowered. For energy crops, improvements through traditional plant breeding techniques have barely begun. Opportunities for developing high yielding crops more suitable for energy purposes (such as high erucic acid oilseed rape) using genetic engineering techniques by transferring genes through recombinant DNA technology also have potential (Luhs and Friedt, 1998).

More difficult to predict are changes in diet and food consumption based on availability, quality, health, and environmental decisions. New protein sources will also impact on land use as will the growing of new crops to provide biomaterials specifically for manufactured products.

Social problems in rural areas continue to occur as does urban drift, particularly in developing countries, resulting from unemployment caused by substitution of manual labour by fossil fuel powered tractors and machinery. Rural economies are also struggling financially (even in developed countries where farm subsidies are available) because of current surpluses of many food and fibre commodities leading to low prices. Therefore, limited funds are available for investment in more modern, less energy intensive equipment. Further evaluation of combining traditional manual techniques with modern crops and scientific knowledge to improve sustainability is required.

Possible removal of agricultural subsidies in Europe and the USA is a further threat to the future profitability of these rural regions (which may then need to be stimulated through a government development plan), but could have beneficial effects in terms of reducing greenhouse gas emissions (Storey, 1997). However, it will also be an opportunity for unsubsidized energy crops to compete more successfully with the more traditional uses of the land.

3.6.4 New Technological Options and Social and Behavioural Issues

3.6.4.1 Uptake of Management Techniques

These include use of conservation tillage techniques, improved soil, pasture, and livestock management, paddy field management, careful use of nitrogenous fertilizers, better tractor operation, and irrigation scheduling as outlined in *Table 3.27*.

Crop production in heated greenhouses is particularly energy intensive, and in many cases intended to satisfy luxury demands for vegetables grown out of season or cut flowers (Japan Resources Association, 1994). A range of options exist to reduce the energy inputs (CAE, 1996).

3.6.4.2 Uptake of New Technologies

There is potential for improving yields of food, fibre, and energy crops yet reducing inputs by using genetic selection or modification. Animals can also be bred to convert feed more efficiently. Transgenetic technologies will be difficult to implement unless publicly supported. Following careful scientific research, including life cycle assessment analyses, and stringent government controls over the release of genetically modified organisms into the environment, then it may be possible that future agricultural production systems will involve lower inputs of nutrients and energy. The extent of the uptake of such developments will be largely based on assessments of risks, benefits, and public perceptions and is hard to predict.

Options to increase soil carbon levels are given in *Table 3.27*. Emissions of soil carbon of around 0.2–3tC/ha resulting from cultivation can be reduced by using zero or minimum tillage techniques. However, a reverse of land use activities would soon lose any accumulated soil carbon. In Canada a group of 7 energy companies are paying farmers (through an insurance company acting as an aggregator of credits), CAN$1.50–13/ha/yr to change to zero tillage so they can claim the resulting carbon credits for the effective accumulation period (Ag Climate, 1999). The return to farmers depends on the recruiting and support programme costs, scientific proof of higher carbon gains, and the extent to which other on-farm carbon emission reduction activities are implemented.

3.6.4.3 Energy Cropping

Other than traditional forest crops (see Chapter 4), a number of annual and perennial species have been identified as having high efficiency properties when converting solar energy into stored biomass which can then be converted into heat, electricity or transport fuels with zero or very low carbon emissions (Veenendal *et al.*, 1997). (Conversion of biomass is described in Section 3.8.4.3.2 and biofuels for transport in Section 3.4.4).

High yielding short rotation forest crops or C4 plants (e.g., sugar cane and sorghum) can give stored energy equivalents of over 400 GJ/ha/yr at the commercial scale, leading to very positive input/output energy balances of the overall system (El Bassam, 1996). Ethanol production from maize and other cereals in the USA, from sugar cane in Brazil, and biodiesel from oilseed rape in Europe, are being commercially produced but are subject to commodity price fluctuations and government support. The relatively low energy yields per hectare for many oil crops (around 60 to 80GJ/ha/yr for oil) compared with crops grown for cellulose or starch/sugar (200–300GJ/ha/yr), has led to the US National Research Council advising against any further research investment (NRC, 1999b).

Liquid biofuels (see Section 3.4.4.6) when substituted for fossil fuels will directly reduce CO_2 emissions. Therefore, a combination of bioenergy production with carbon sink options can result in maximum benefit from mitigation strategies. This can be achieved by planting energy crops such as short rotation coppice into arable or pasture land, which increases the carbon density of that land, while also yielding a source of biomass. Converting the accumulated carbon in the biofuels for energy purposes, and hence recycling it, alleviates the critical issue of maintaining the biotic carbon stocks over time as for a forest sink. Increased levels of soil carbon may also result from growing perennial energy crops (IEA Bioenergy, 1999), but a detailed life cycle assessment is warranted for specific crops and regions.

Land needed to grow energy crops competes directly with food and fibre production unless grown on marginal or degraded land, or unless surplus land is available. For the USA and Europe. Hall and Scrase (1998) calculated there to be sufficient land physically available to grow crops to supply all human needs of food, fibre, and energy at current population levels, though the study did not include social, economic, and logistical constraints. Sufficient labour, water, and nutrients must also be available if a sustainable and economic bioenergy industry is to be developed. On marginal lands, such as the increasingly saline soils of Australia, growing short rotation eucalyptus in strips between blocks of cereal crops can help lower the water table and hence, under certain circumstances, reduce the soil saline levels to bring back the natural fertility. However, water demands can be high for short rotation forest crops so the resulting overall effects are yet to be determined. An additional benefit is that the decentralization of energy production using energy crops to supply local conversion plants creates

Table 3.27: Uptake of management techniques and new technologies to reduce greenhouse gas emissions in the agricultural sector

Management techniques	Techniques and technologies to be considered	References
Conservation tillage	Conventional tillage consumes 60% of the tractor fuel used in industrialized crop production and decreases soil carbon. Minimum and zero cultivation techniques save tractor fuel, conserve soil moisture, and reduce soil erosion. Uptake is continuing worldwide. Greater chemical weed control may be required. Benefits need to be achieved without reducing crop yields which is more likely under dry conditions as a result of moisture conservation. Animal powered versions of conservation tillage used in developing countries can also reduce the manual drudgery. Cost of uptake in Botswana is around US$31 – 38/tC saved. Globally 150-175MtC/yr sequestration is possible.	Allmaras and Dowdy (1995) Derpsch, 1998 UNEP/Southern Centre (1993) Zhou (1999)
Soil carbon uptake	Typical agricultural soils contain 100-200tC/ha to 1m depth. Overuse of soils leads to degradation, salinization, erosion, and desertification, and will lead to lower organic matter contents with consequent carbon emissions. A change of land use of intensively cultivated soils could result in increased organic matter and carbon sequestration till the soil finds a new balance. Total sequestration potential of world cropland is around 750- 1000MtC/yr for 20-50 years from: erosion control (80-120MtC/yr), restoration (20-30MtC/yr), conservation tillage and crop residue management (150-170MtC/yr), reclamation of saline soils (20-40MtC/yr), improved cropping (180-240MtC/yr) and C offsets through energy crop production (300-400MtC/yr).	Lal and Bruce (1999) Takahashi and Sanada (1998) Batjes (1998) IPCC (2000)
Paddy rice	Estimates have been corrected downwards to around 360MtC/yr. Emissions can be reduced by intermittent flooding and greater use of inorganic fertilizers, but these benefits will be offset by increasing areas grown to meet increasing food demand.	Ahlgrimm (1998) Neue (1997) Mosier *et al.* (1998a)
Nitrogenous fertilizers	Anthropogenic agricultural nitrous oxide emissions (over 800MtC/yr) released after application of N fertilizers as a result of nitrification and denitrification and from animal wastes, exceed carbon emissions from fossil fuels used in agriculture. Measuring emissions is difficult (±85%) because of soil variability. Reductions resulting from use of N fertilizer strategies, slow release fertilizers, organic manures and nitrification inhibitors, could tentatively cut emissions by 30% on a global scale. Costs would be between US$0 – 14/tC in Europe for 3-4MtC/yr. Genetically engineered leguminous plants may have further potential.	Augustin *et al.* (1998) Hendriks *et al.* (1998) Kramer *et al.* (1999) Kroeze and Mosier (1999)
Tractor operation and selection	Correct operation of tractors and size matching to machinery can save fuel, improve tyre life, reduce soil compaction, and save time. Behavioural change by driver education is required but with cheap diesel fuel there is little incentive.	Sims *et al.* (1998)
Irrigation scheduling	Applying water only as needed saves both water and energy for pumping. Cheap and accurate field soil moisture sensors are necessary but not yet available.	Schmitz and Sourell (1998)

(continued)

Table 3.27: continued

New technologies	Techniques and technologies to be considered	References
Ruminant enteric methane	Average methane emissions of grazing animals in temperate regions are 76.8 kg/head/yr for dairy cattle; beef cattle, 67.5kg; deer, 30.6kg; goats, 16.5kg; and sheep, 15.1kg. Reduction is by either improving the productivity of the animal or reducing emissions by chemical, antibiotic control (vaccines) or biological methods (bacteriocins) without affecting animal performance. Poor animal diet in developing countries produces higher methane per unit of production. A range of options are being researched, but limited economic analysis of mitigation opportunities has been conducted other than in Europe (15MtC/yr at US$0-14/tC). Selective breeding and magnesium licks may be cheap options. The reduction in ruminant livestock numbers caused by reduced demand for meat, milk (for health reasons) and wool products may continue. Since the sources of emissions are dispersed, they will be difficult to measure, and therefore challenging to include within an enforceable trading regime.	Storey (1999) Ullyatt *et al.* (1999)
Postharvest crop losses	A reduction in postharvest crop losses could make a significant impact on energy use, particularly in developing countries such as India, where average losses for cereals average 10% up to 25% loss of the harvested perishables including fruit, meat, milk, and fish. Solar drying on the ground leads to vermin and pest losses. Storage in sealed buildings with natural ventilation and solar heated air will reduce losses for minimal energy inputs. For fresh crops, refrigeration and heat pumps are used to maintain the cool chain but energy inputs can be significant. Solar panels on refrigerated truck roofs are technically feasible but not economic.	Prasad (1999)
Global positioning systems	Commercially available GPS and GIS systems are available to map then monitor the position of working tractors to enable strategic applications of fertilizers and chemicals to be applied depending on crop yields and soil types. Plantation forest mapping is also used to plan roads and harvests. Energy inputs can be saved as a result.	Oliver (1999)
Controlled environment	Crops grown in greenhouses can use less energy per production unit if the available growing area is increased and better control of heating and ventilation occurs. The effects on energy inputs of producing fish by aquacultural methods rather than sea trawling needs investigation.	CAE (1996)

employment in rural areas (Grassi, 1998; El Bassam *et al.*, 1998; Moreira and Goldemberg, 1999).

Certain woody crops and also perennial grasses grown to produce biomass have theoretically high dry matter yields, but commercial yields are often lower than expected from those produced in small plot research trials. In Sweden, for example, where 16,000 ha of coppice *Salix* species have been planted, around 2000ha were harvested for the first time during the winters of 1996 to 1998 to yield only 4.2 oven dry t/ha/yr on average (Larsson *et al.*, 1998). With better management, genetic selection, and grower experience once viable markets for the product are established, it had been anticipated that commercial yields closer to 10 oven dry t/ha/yr would result.

Correct species selection to meet specific soil and climatic site conditions is necessary in order to maximize yields in terms of MJ/ha/yr (Sims *et al.*, 1999). For example, the saccharose yield of Brazilian sugar cane has increased 10% to 143kg/t of fresh cane (70% moisture content wet basis) since 1990. Methods of identifying appropriate species based on non-destructive yield measurements and species fuelwood characteristics have been developed (Senelwa and Sims, 1998). Energy balance ratios for each unit of energy input required to produce solid fuels from short rotation forest crops are up to 1: 30, and can be even higher when crop residues are also utilized (Scholz, 1998). Woody crops normally require less energy inputs per hectare than food crops.

Forest sinks are covered in Chapter 4 and also in the Special Report on *Land Use, Land-Use Change and Forestry* (IPCC, 2000), but there is a link between these low cost sinks and eventually using some of the biomass grown for energy purposes. Once the limited area of available land is covered in forest sinks, no more planting will be possible and recycling of the carbon to displace fossil fuels may then become feasible. Economic mechanisms to link a forest sink project with a biofuel project have been suggested (Read, 1999).

3.6.4.4 Crop and Animal Wastes

Crop residues such as straw, bagasse, and rice husks, if not returned to the land for nutrient replenishment and soil conditioning, could be used more in the future for heat and power generation, at times in co-combustion with coal, and in appropriate conversion equipment now that the technology is well proven. Wood residues used in small-scale biomass gasifiers will become reliable and more cost effective in time, but at present have some operational risk attached, particularly under developing country conditions (Senelwa and Sims, 1999).

Animal manures and industrial organic wastes are currently used to generate biogas. For example, in Denmark there are 19 decentralized community scale biogas plants for electricity generation (Nielsen *et al.*, 1998). Biogas can also be used for cogeneration, direct heating or as a transport fuel.

3.6.4.5 Behavioural Changes

Many farmers in both developed and developing countries will remain unlikely to change their traditional production methods in the short term unless there are clear financial incentives to do so. Behavioural changes as a result of advisors educating members of farming communities to adopt new measures have rarely succeeded to date. Cultural factors have a strong influence on the general unwillingness to accept inappropriate development and hence new ideas. Changing attitudes are unlikely to occur unless farmers can also perceive personal co-benefits such as increased profitability, time saving, cost reductions, improved animal health, increased soil fertility, and less arduous tasks. Regulations in some form are the alternative (OECD, 1998a) but would probably be difficult to monitor, particularly in developing countries. Education of local extension officers is needed to encourage the uptake of new methods and more rapid implementation into the field. These barriers are discussed in section 5.4.5.

Dietary changes from meat to fish or vegetables could help reduce emissions by $55MtC_{eq}$ in Europe alone (Gielen *et al*, 1999) and possibly release land for energy cropping.

3.6.5 Regional Differences

Comparative regional studies of agricultural emissions per unit of GDP or per hectare or per capita would show significant differences but as a result of local farming systems, climate, and management techniques employed, a useful comparison between regions is not possible. Standard methods for measuring and reporting of agricultural emissions are being developed and will enable more accurate and useful comparisons to be made between alternative production systems in the future (Kroeze and Mosier, 1999).

Developing countries are slowly moving towards using modern food and fibre production techniques. Economies in transition are also implementing modern production methods encouraged by foreign investors but many challenges remain. From a sustainability point of view, traditional methods may well be preferable.

3.6.6 Technological and Economic Potential

A summary of the technical and market potential for reducing greenhouse gas emissions from the agricultural industry is given in *Table 3.36*. If agricultural production per hectare in developing countries could be increased to meet the growing food and fibre demand as a result of a greater uptake of new farming techniques, modern technologies and improved management systems, then there would be less incentive for deforestation to provide more agricultural land.

3.7   Waste

3.7.1   Summary of the Second Assessment Report

The major emphasis in the SAR was on the reduction of greenhouse gases associated with industrial recycling in the metals, glass and paper industries. These topics are addressed in the industrial Section 3.5 of this chapter. There was a less systematic account of the methane emissions from landfills, or of the consumer dimension of recycling, both of which will be emphasized here.

3.7.2   Historic and Future Trends

Waste and waste management affect the release of greenhouse gases in five major ways: (1) landfill emissions of methane; (2) reductions in fossil fuel use by substituting energy recovery from waste combustion; (3) reduction in energy consumption and process gas releases in extractive and manufacturing industries, as a result of recycling; (4) carbon sequestration in forests, caused by decreased demand for virgin paper; and (5) energy used in the transport of waste for disposal or recycling. Except for the long-range transport of glass for reuse or recycling, transport emissions of secondary materials are often one or two orders of magnitude smaller than the other four factors (Ackerman, 2000).

3.7.2.1   Landfills

Worldwide, the dominant methods of waste disposal are landfills and open dumps. Although these disposal methods often have lower first costs, they may contribute to serious local air and water pollution, and release high GWP landfill gas (LFG). LFG is generated when organic material decomposes anaerobically. It comprises approximately 50%-60% methane, 40%-45% CO_2 and the traces of non-methane volatile organics and halogenated organics. In 1995, US, landfill methane emissions of 64 MtC_{eq} slightly exceed its agricultural sector methane from livestock and manure.

Methane emission from landfills varies considerably depending on the waste characteristics (composition, density, particle size), moisture content, nutrients, microbes, temperature, and pH (El-Fadel, 1998). Data from field studies conducted worldwide indicate that landfill methane production may range over six orders of magnitude (between 0.003-3000g/m²/day) (Bogner *et al.*, 1995). Not all landfill methane is emitted into the air; some is stored in the landfill and part is oxidized to CO_2. The IPCC theoretical approach for methane estimation has been complemented with more recent, site-specific models that take into account local conditions such as soil type, climate, and methane oxidation rates to calculate overall methane emissions (Bogner *et al.*, 1998).

Laboratory experiments suggest that a fraction of the carbon in landfilled organic waste may be sequestered indefinitely in

landfills depending upon local conditions. However, there are no plausible scenarios in which landfilling minimizes GHG emissions from waste management. For yard waste, GHG emissions are roughly comparable from landfilling and composting; for food waste, composting yields significantly lower emissions than landfilling. For paper waste, landfilling causes higher GHG emissions than either recycling or incineration with energy recovery (US EPA, 2000).

3.7.2.2   Recycling and Reuse

Recycling involves the collection of materials during production or at the end of a product's useful lifetime for reuse in the manufacturing process. The degree of treatment varies from simple remelting of glass, aluminium, or steel, to the breaking apart and reconstitution of paper or other fibres (e.g., textiles or carpets), to depolymerization of plastics and synthetic fibres to monomers, which are then used instead of petrochemicals to synthesize new polymers.

In many cases, manufacturing products from recycled materials is less energy intensive and associated with fewer GHG emissions than making products from virgin materials. This is especially true for aluminium and steel, which are energy intensive and release significant process GHGs during production (CO_2 and PFCs). A US EPA analysis finds lower GHG emissions over the product life cycle from recycling than from virgin production and disposal of paper, metals, glass, and plastics under typical American conditions (US EPA, 2000).

Overall energy consumption is lower for recycled paper than for virgin paper, yet there is some debate over life cycle GHG emissions between paper recycling (Blum *et al.*, 1997; Finnveden and Thomas, 1998; US EPA, 1998) and paper consumption with energy recovery (Bystroem and Loennstedt, 1997; Ruth and Harrington, 1998; IIED, 1996). These conflicting analyses make different underlying assumptions concerning the fuel displaced by energy released from paper incineration, the energy source for the electricity used in paper production, how the recycled paper is utilized, and how much carbon sequestration can be credited to uncut forests because of recycling. In all studies, landfilling of paper clearly releases more GHGs than either recycling or incineration.

The life cycle environmental impact and GHG emissions from recycling are usually higher than reusing products. This may not hold true if the used materials have to be transported over long distances. To address this issue, some countries such as Germany, Norway, Denmark, and other European countries have standardized bottles for local reuse.

3.7.2.3   Composting and Digestion

Composting refers to the aerobic digestion of organic waste. The decomposed residue, if free from contaminants, can be used as a soil conditioner. As noted above under landfilling, GHG emissions from composting are comparable to landfilling for yard waste, and lower than landfilling for food waste. These estimates do not include the benefits of the reduced need for synthetic fertilizer, which is associated with large CO_2 emissions during manufacture and transport, and N_2O releases during use. USDA research indicates that compost usage can reduce fertilizer requirements by at least 20% (Ligon, 1999), thereby significantly reducing net GHG emissions (see Section 3.6).

Composting of yard waste has become widespread in many developed countries, and some communities compost food waste as well. Small, low-technology facilities handling only yard waste are inexpensive and generally problem-free. Some European and North American cities have encountered difficulties implementing large-scale, mixed domestic, commercial and industrial bio-waste collection and composting schemes. The problems range from odour complaints to heavy metal contamination of the decomposed residue. Also, large-scale composting requires mechanical aeration which can be energy intensive (40-70 kW/t of waste) (Faaij *et al.*, 1998). However, facilities that combine anaerobic and aerobic digestion are able to provide this energy from self-supplied methane. If 25% or more of the waste is digested anaerobically the system can be self-sufficient (Edelmann and Schleiss, 1999).

For developing countries, the low cost and simplicity of composting, and the high organic content of the waste stream make small-scale composting a promising solution. Increased composting of municipal waste can reduce waste management costs and emissions, while creating employment and other public health benefits.

Anaerobic digestion to produce methane for fuel has been successful on a variety of scales in developed and developing countries. The rural biogas programmes based upon manure and agricultural waste in India and China are very extensive. In industrial countries, digestion at large facilities utilizes raw materials including organic waste from agriculture, sewage sludge, kitchens, slaughterhouses, and food processing industries.

3.7.2.4 Incineration

Incineration is common in the industrialized regions of Europe, Japan and the northeastern USA where space limitations, high land costs, and political opposition to locating landfills in communities limit land disposal. In developing countries, low land and labour costs, the lack of high heat value materials such as paper and plastic in the waste stream, and the high capital cost of incinerators have discouraged waste combustion as an option.

Waste-to-energy (WTE) plants create heat and electricity from burning mixed solid waste. Because of high corrosion in the boilers, the steam temperature in WTE plants is less than 400 degrees Celsius. As a result, total system efficiency of WTE plants is only between 12%–24% (Faaij *et al.*, 1998; US EPA, 1998; Swithenbank and Nasserzadeh, 1997).

Net GHG emissions from WTE facilities are usually low and comparable to those from biomass energy systems, because electricity and heat are generated largely from photosynthetically produced paper, yard waste, and organic garbage rather than from fossil fuels. Only the combustion of fossil fuel based waste such as plastics and synthetic fabrics contribute to net GHG releases, but recycling of these materials generally produces even lower emissions.

3.7.2.5 Waste Water

Methane emissions from domestic and industrial wastewater disposal contribute about 10% of global anthropogenic methane sources (30-40Mt annually). Industrial wastewater, mainly from pulp and paper and food processing industries, contributes more than 90% of these emissions, whereas domestic and commercial wastewater disposal contributes about 2 Mt annually. Unlike methane emissions from solid waste, most of the methane from wastewater is believed to be generated in non-Annex I countries, where wastewater is often untreated and stored under anaerobic conditions (SAR).

3.7.3 New Technological and Other Options

3.7.3.1 Landfill Management

LFG capture and energy recovery is a frequently applied landfill management practice. There have been many initiatives during the past few years to capture and utilize LFG in gas turbines; a number of such facilities are currently generating electricity. US regulations now require capture of an average of 40% of all landfill methane nationwide. Yet even after compliance with those regulations, it remains profitable (at a carbon price of zero or negative cost) to capture 52% of the landfill methane. At a price of US$20/tC$_{eq}$ (in 1996 dollars), an additional 19% of the methane could be captured, an amount that approaches the estimated maximum practical attainable level (US EPA, 1999a). Official estimates suggest that approximately half, or 35MtC$_{eq}$, of landfill methane could be recovered by 2000.

Other studies have found that the methane yield from landfills is about 60-170 l/kg of dry refuse (El-Fadel *et al.*, 1998). Some landfills produce electricity from LFG by installing cost effective gas turbines or technologically promising, but still expensive fuel cells (Siuru, 1997). Later reports dispute this claim (US EPA, 2000).

One study suggests that landfilling of branches, leaves and newspaper sequesters carbon even without LFG recovery, whereas food scraps and office paper produce a net increase in GHGs, even from landfills with methane recovery (US EPA, 1998).

3.7.3.2 Recycling

Many programmatic initiatives and incentives can boost the rate of recycling. The potential gains are quite large: if everyone in the USA increased from the national average recycling rate to the per capita recycling rate achieved in Seattle, Washington, the result would be a reduction of 4% of total US GHG emissions (Ackerman, 2000). While often associated with affluent countries, recycling is also an integral part of the informal economy of developing countries; innovative approaches to recycling have been adopted in poor neighbourhoods of Curitiba, Brazil, and in other cities.

The literature on techniques for increasing the rate of recycling is too extensive for adequate citation here (see, for example, Ackerman (1997) and numerous sources cited there). One much-discussed initiative is the use of variable rates, or pay-per-bag/per-can charges for household solid waste collection. This provides a clear financial incentive to the householder to produce less waste, particularly when accompanied by free curbside recycling (Franke *et al.*, 1999). Strict packaging and lifetime product responsibility laws for manufacturers in Germany have brought about innovations in the manufacture and marketing of a wide range of products. Other market incentives such as repayable deposits on glass containers, lead acid batteries, and other consumer products have led to major gains in recycled materials in many countries. Voluntary recycling programmes have met with a mixed range of success, with commercial and institutional recycling of office paper and cardboard, and curbside recovery of mixed household materials generally having higher recycling rates. Countries such as Austria and Switzerland successfully require separation of household waste into many disaggregated categories for high value recovery.

3.7.3.3   Composting

Increased composting of household food waste would reduce GHG emissions, but may be difficult to achieve in developed countries, where an additional separation of household waste would be required. In low-income developing countries, the high proportion of food waste in household and municipal waste makes composting attractive as a primary waste treatment technology.

Other new opportunities involve composting or anaerobic digestion of agricultural and food industry wastes. Livestock manure management accounts for 10% of US methane emissions; capture of about 70% of the methane from livestock manure appears technologically feasible. Some 20% of the feasible methane capture is profitable under existing conditions, with a carbon price of zero; 28% can be recovered at US$20/tC$_{eq}$ and 61% at US$50/tC$_{eq}$(US EPA, 1999a).

Biogas facilities intentionally convert organic waste to methane; use of the resulting methane can substitute for fossil fuels, reducing GHG emissions. High ammonia content (e.g., in swine manure) can inhibit conversion of organic waste to methane. This problem can be avoided by mixing agricultural waste with other, less nitrogenous wastes (Hansen *et al.*, 1998). Wastes with high fat content can, on the other hand, enhance

and increase methane output. In Denmark, a number of biogas facilities have been running successfully, accepting livestock manure as well as wastes from food processing industries (Schnell, 1999). In Germany and Switzerland, pilot projects compress the methane from biogas plants and supply it to natural gas vehicles. Canadian engineers have completed a pilot project using a mixture of waste-activated sludge, food waste, industrial sludge from potato processing, and municipal waste paper. Methane production reached 50 l/kg of total solids, and heavy metal contamination was found to be far below regulatory levels (Oleszkiewicz and Poggi-Varaldo, 1998). Woody waste with high lignin content cannot be converted to methane, and yard waste is better handled by composting.

3.7.3.4   Incineration

New combustion technologies with higher efficiencies of energy production and lower emissions are currently being developed:

- Fluidized bed combustion (FBC) is a very efficient and flexible system that can be used for intermittent operation, and can run with solid, liquid, or gaseous fuels. Despite high operating costs, this low pollution combustion technology is increasingly used in Japan, and has also been used in Scandinavia and the USA (NEDO, 1999; http://www.residua.com/wrftbfbc.html).

- Gasification (partial incineration with restricted air supply) and pyrolysis (incineration under anaerobic conditions) are two technologies that can convert biomass and plastic wastes into gas, oil, and combustible solids. Gasification of biomass produces a gas with a heating value of 10%-15% that of natural gas. When integrated with electricity production, it can prove economically and environmentally attractive; it appears best suited for clean biomass, such as wood wastes. Pilot projects are now using pyrolysis for plastic wastes, and for mixed municipal solid waste (MSW); they potentially have very high energy efficiency (Faaij *et al.*, 1998). Combined pyrolysis and gasification (Thermoselect) and combined pyrolysis and combustion (Schwelbrenn-Verfahren) have also been developed and implemented.

- Co-incineration of fossil fuel jointly with waste leads to improved energy efficiency. Stringent emission standards in some countries may limit the extent to which co-incineration is possible (Faaij *et al.*, 1998). In other countries, emission standards for industrial combustion processes are less tight than those for incinerators, leading some to fear that co-incineration might produce higher emissions of air pollutants (Kossina and Zehetner, 1998).

3.7.3.5   Wastewater Treatment

Conventional sewage collection is very water intensive. Vacuum toilets, using less than 1 litre per flush, have long been used on ships and have now been installed in the new ICE trains in Germany. Human waste collected in this way can then be anaerobically digested. This process reduces GHG emissions and water usage is minimal. Acceptance of this technology has been slow because of cost (Schnell, 1998).

Modular anaerobic or aerobic systems are available (Hairston *et al.*, 1997). Anaerobic digestion has the advantage of generating methane that can be used as a fuel, yet many sewage treatment plants simply flare it. The potential for energy generation is clearly very large. New York City's 14 sewage plants, for example, generate 0.045 billion cubic metres of methane every year, most of which is flared. Cities such as Los Angeles sell methane to the local gas utility, and one New York plant and the Boston Harbor facility were equipped with fuel cells in 1997. This new technology successfully provides needed electricity and heat, but is still expensive.

Because of concerns about contamination of sewage sludge by heavy metals, policies in many countries now encourage incineration rather than soil application. However, the energy needed to dry the sludge for incineration leads to a net increase in GHGs. Alternatives to sludge incineration are anaerobic digestion, gasification, wet oxidation, and co-incineration with coal. These technologies are under development and yield improved energy efficiencies and low GHG emissions (Faaij *et al.*, 1998).

3.7.4   Regional Differences

Individual countries have adopted different strategies and innovations in waste management that reduce GHG emissions. It is not possible to provide a comprehensive description in this chapter, but a sampling of different national and regional strategies is summarized below.

3.7.4.1   Germany

Germany promotes recycling through the world's most stringent return requirements for packaging and many other goods, including automobiles; materials management is the responsibility of the manufacturer through the end of product life, including ultimate disposal or reuse of the materials from which it is made. This has led to high recycling rates, but also to high monetary costs, prompting ongoing controversy in Germany and elsewhere.

Every year Germany generates about 30 million tonnes of solid municipal waste. German landfills emit yearly 1.2-1.9Mt of methane, accounting for 25%-35% of Germany's methane emissions and about 3%-7% of national GWP. To meet the provisions of a 1993 law requiring that by 2005 all wastes disposed in landfills have to have a total organic carbon content of

less than 5% will require incineration. Under this law methane emissions are projected to drop by two-thirds by 2005, and by 80% by 2015 (Angerer and Kalb, 1996).

3.7.4.2   USA

The USA produces about 200 million tonnes of municipal waste each year. In 1997, 55% was landfilled, 28% was recycled or composted, and 17% was incinerated (US EPA, 1999b). The 11.6Mt of methane emitted by landfills accounts for 37% of anthropogenic methane emissions, or about 4% of national GHG emissions. US regulations now require the largest landfills to collect and combust LFG, which is projected to reduce emissions to 9.1Mt in 2010 (US EPA, 1999a). There are more than 150 LFG-to-energy projects in operation, and 200 more in development, promoted by government technical support and tax incentives (Kerr, 1998; Landfill, 1998).

If all the material currently recycled in the USA were instead landfilled, national GHG emissions would increase by 2%, even with the new LFG regulations (Ackerman, 2000). More than 9,000 municipal recycling programmes collect household materials, and numerous commercial enterprises also recycle material. Many innovative uses of recycled materials are reducing emissions in manufacturing; for example, remanufacture of commercial carpet from recovered fibres lowers energy inputs by more than 90%, and some products are now said to have zero net GHG impacts (Hawken *et al.*, 1999).

3.7.4.3   Japan

With a large waste stream and very limited land area, Japan relies heavily on both recycling and incineration as alternatives to landfilling. Widespread participation in recycling recovers not only easily recycled materials such as metals and glass, but also large quantities of unconventional recycled materials, such as aseptic packaging (juice boxes).

Japan has approximately 1,900 waste incineration facilities of which 171 produce electric power with a capacity of 710MW. A major new commitment to create high efficiency waste to energy facilities has been announced by the Japanese government. In 1998 a corrosion resistant, high temperature, fluidized bed WTE facility achieved 30% conversion efficiency to electricity with low dioxin and stack gas emissions. The facility can accept mixed municipal and industrial waste including plastics and recovers ash for road foundations and recyclable metals (NEDO, 1999).

3.7.4.4   India

Recycling is a very prevalent part of Indian society. Unskilled labourers, working in the informal economy, collect newspapers, books, plastic, bottles, and cans and sell them to commercial recyclers. In recent years a shift from collecting for reuse to collecting for recycling has taken place. Because of changing lifestyles and increased consumption of goods, the use of recyclables has increased dramatically over the past few years (from 9.6% in 1971 to 17.2% in 1995). Paper accounts for 6% and ash and fine earth for 40%. Total compostable matter is over 42% of the waste stream.

Plastic in the waste stream increased from 0.7% in 1971 to 4%-9% in 1996, and is expected to grow rapidly. Though current consumption is 1.8 kg/capita/yr compared to a world average of 18 kg and a US average of 80 kg, India recycled between 40-80% of its plastics, compared to 10%-15% in developed nations. There are about 2000 plastic recycling facilities in India, which often cause serious environmental harm as a result of outdated technology. Current per capita paper consumption is 3.6 kg, compared to a world average of 45.6 kg. Paper consumption is projected to increase to 8 kg by 2021. India imports approximately 25% of its paper fibre as waste paper from the US and Europe.

Almost 90% of solid waste is deposited in low-lying dumps and is neither compacted nor covered; 9% is composted. In 1997, landfill emissions were India's third largest GHG contributors, equivalent to burning 11.6Mt of coal (Gupta *et al.*, 1998).

3.7.4.5   China

China generated 108 million tonnes of municipal waste in 1996, an amount that is increasing every year by 8%-10%. In 1995, the GEF approved an action plan and specific projects for methane recovery from municipal waste (Li, 1999).

According to a survey of ten cities, the per capita waste generation averages 1.6kg/day, but in some rapidly developing cities in southern China, per capita waste production is almost as high as in developed countries (e.g., Shenzhen, 2.62 kg/day). Between 60%-90% of Chinese municipal solid waste is high moisture organic material with a low heat value. The composition of waste is changing, with cinder and soil content decreasing while plastic, metal, glass and organic waste are increasing. Kitchen waste has replaced coal cinder as the largest component, raising the water content. By the end of 1995, incineration treatment capacity was 0.9% of total MSW.

Estimates are that in 2010 China will produce 290 million tonnes of MSW. If 70% is disposed of in landfills with methane collection, the landfill gas recovered could be equivalent to 40 to 280 billion m^3 of natural gas (Li, 1999).

3.7.4.6   Africa

The average annual solid waste generation in Africa is estimated to be about 0.3 to 0.5t/ capita and for a population for Africa of about 740 million in 1997, the total continent's annual generated waste could be as much as 200 million tonnes. It is estimated that anything from 30%–50% if the waste is not subjected to proper disposal, presenting severe health and environmental hazards (INFORSE, 1997). With few financial

resources, and population increasing at 3% per annum, with the most rapid growth in urban regions from migration, this poses a serious challenge for waste management in the future.

An analysis of energy content of MSW generated in South Africa alone indicates that if one-third were utilized for combustion energy it would be equivalent to 2.6% of the total electricity distributed in 1990 (529Million GJ) by the country's largest utility, ESKOM. Technologies are not yet available on the continent to make this a reality.

Mitigating CH_4 through extraction of landfill gas for energy use has been estimated to cost below US$10/t$C_{eq}$ in Africa (Zhou, 1999). Both incineration of MSW and extraction of landfill gas have significant potential to reduce emissions of methane in Africa, and will provide the co-benefit of addressing the severe waste management problem on the continent.

3.7.5   *Technological and Economic Potential*

Economic analysis of waste management strategies yields widely varying results, with far less reliable standard cost estimates than in fields such as energy production. In the USA, the most successful communities report that ambitious waste reduction, recycling, and composting programmes cost no more than waste disposal, and often cost significantly less (US EPA, 1999c). Overall, average recycling costs appear to be slightly above landfill disposal costs (Ackerman, 1997). Not all waste management strategies have been fully analyzed for their economic potential or distributional cost and benefit implications. The waste hierarchy (reduce, reuse, recycle, incinerate, landfill) on which many countries' waste policies are based has not been comprehensively evaluated on a country and materials specific basis (Bystroem and Loennstedt, 1997).

Integrated waste management that considers environmental protection, economic efficiency, social acceptability, flexibility, transparency, market-oriented recovery and recycling, appropriate economies of scale, and continuous improvement is being developed throughout Europe (Franke *et al.*, 1999).

Considering only GHG emissions, the most favourable management options are those that reduce fossil fuels use in manufacturing as does recycling, or replace them as does incineration with energy recovery. There is, however, disagreement over the most ecological waste disposal method. Some argue for incineration of all solid waste in modern, energy recovering incinerators (Pipatti and Savolainen, 1996; Aumonier, 1996); others advocate increased composting and anaerobic digestion of organic wastes (Ackerman, 1997; Dehoust *et al.*, 1998; Finnveden and Thomas, 1998; Ligon, 1999). The estimated GHG emissions for different scenarios depend heavily on the parameter assumptions made in each model. If GHGs from waste disposal are the only concern, incineration with energy recovery is the most favourable solution. If economic and other environmental factors (e.g., emissions of heavy metals) are taken into account the answer is less clear. Also, if the whole life cycle and not just the disposal of the material is considered, recycled materials usually are associated with lower GHG emissions than virgin materials. Numerous technologies appropriate to differing national needs are available at a range of technological complexities for reducing GHGs from waste. Many options are highly cost effective, and can lead to significant reductions on the order of several per cent of national greenhouse gas emissions. Source reduction is indisputably the most environmentally sound and cost effective tool to reduce GHG emissions from solid waste.

3.8   Energy Supply, Including Non-Renewable and Renewable Resources and Physical CO_2 Removal

3.8.1   *Introduction*

This section reviews the major advances in the area of GHG mitigation options for the electricity and primary energy supply industries that have emerged since IPCC (1996). The global electricity supply sector accounted for almost 2,100MtC/yr or 37.5% of total carbon emissions. Under business-as-usual conditions, annual carbon emissions associated with electricity generation, including combined heat and power production, is projected to surpass the 4,000MtC mark by 2020 (IEA, 1998b). Because a limited number of centralized and large emitters are easier to control than millions of vehicle emitters or small boilers, the electricity sector is likely to become a prime target under any future involving GHG emission controls and mitigation.

3.8.2   *Summary of the Second Assessment Report*

Chapter 19 of the IPCC Second Assessment Report (1996) gave a comprehensive guide to mitigation options in energy supply (Ishitani and Johansson, 1996). The chapter described technological options for reducing greenhouse gas emissions in five broad areas:

- *More efficient conversion of fossil fuels.* Technological development has the potential to increase the present world average power station efficiency from 30% to more than 60% in the longer term. Also, the use of combined heat and power production replacing separate production of power and heat, whether for process heat or space heating, offers a significant rise in fuel conversion efficiency.
- *Switching to low-carbon fossil fuels and suppressing emissions.* A switch to gas from coal allows the use of high efficiency, low capital cost combined cycle gas turbine (CCGT) technology to be used. Opportunities are also available to reduce emissions of methane from the fossil fuel sector.
- *Decarbonization of flue gases and fuels, and CO_2 storage.* Decarbonization of fossil fuel feedstocks can be used to make hydrogen-rich secondary fuel for use in

fuel cells in the longer term. CO_2 can be stored, for example, in depleted gas fields.
- *Increasing the use of nuclear power.* Nuclear energy could replace baseload fossil fuel electricity generation in many parts of the world if acceptable responses can be found to concerns over reactor safety, radioactive waste transport, waste disposal, and proliferation.
- *Increasing the use of renewable sources of energy.* Technological advances offer new opportunities and declining costs for energy from renewable sources which, in the longer term, could meet a major part of the world's demand for energy.

The chapter also noted that some technological options, such as CCGTs, can penetrate the current market place, whereas others need government support by improving market efficiency, by finding new ways to internalize external costs, by accelerating R&D, and by providing temporary incentives for early market development of new technologies as they approach commercial readiness. The importance of transferring efficient technologies to developing countries, including technologies in the residential and industrial sectors and not just in power generation, was noted.

The Energy Primer of the IPCC Second Assessment Report (Nakicenovic *et al.*, 1996) gave estimates of energy reserves and resources, including the potential for various nuclear and renewable technologies which have since been updated (WEC, 1998b; Goldemberg, 2000; BGR, 1998). A current version of the estimates for fossil fuels and uranium is given in *Table 3.28a*. The potential for renewable forms of energy is discussed later.

A variety of terms are used in the literature to describe fossil fuel deposits, and different authors and institutions have various meanings for the same terms which also vary for different fossil fuel sources. The World Energy Council defines resources as "the occurrences of material in recognisable form" (WEC, 1998b). For oil and gas, this is essentially the amount of oil and gas in the ground. Reserves represent a portion of these resources and is the term used by the extraction industry. British Petroleum notes that proven reserves of oil are "generally taken to be those quantities that geological and engineering information indicates with reasonable certainty can be recovered in the future from known reservoirs under existing economic and operating conditions" (BP, 1999). Resources, therefore, are hydrocarbon deposits that do not meet the criteria of proven reserves, at least not yet. Future advances in the geosciences and upstream technologies – as in the past – will improve knowledge of and access to resources and, if demand exists, convert these into reserves. Market conditions can either accelerate or even reverse this process.

The difference between conventional and unconventional occurrences (oil shale, tar sands, coalbed methane, clathrates, uranium in black shale or dissolved in sea water) is either the nature of existence (being solid rather than liquid for oil) or the geological location (coal bed methane or clathrates, i.e., frozen

Table 3.28a: *Aggregation of fossil energy occurrences and uranium, in EJ*

	Consumption 1860-1998	1998	Reserves	Resources[a]	Resources base[b]	Additional occurrences
Oil						
Conventional	4,854	132.7	5,899	7,663	13,562	
Unconventional	285	9.2	6,604	15,410	22,014	61,000
Natural gas[c]						
Conventional	2,346	80.2	5,358	11,681	17,179	
Unconventional	33	4.2	8,039	10,802	18,841	16,000
Clathrates						780,000
Coal	5,990	92.2	41,994	100,358	142,351	121,000
Total fossil occurrences	13,508	319.3	69,214	142,980	212,193	992,000
Uranium – once through fuel cycle[d]	1,100	17.5	1,977	5,723	7,700	2,000,000[e]
Uranium – reprocessing & breeding[f]			120,000	342,000	462,000	>120,000,000

[a.] Reserves to be discovered or resources to be developed as reserves
[b.] Resources base is the sum of reserves and resources
[c.] Includes natural gas liquids
[d.] Adapted from OECD/NEA and IAEA, 2000. Thermal energy values are reactor technology dependent and based on an average thermal energy equivalent of 500 TJ per t U. In addition, there are secondary uranium sources such as fissile material from national or utility stockpiles, reprocessing former military materials, and from re-enriched depleted uranium
[e.] Includes uranium from sea water
[f.] Natural uranium reserves and resources are about 60 times larger if fast breeder reactors are used (Nakicenovic *et al.*, 1996)

Table 3.28b: *Aggregation of fossil energy occurrences, in GtC*

	Consumption 1860-1998	1998	Reserves	Resources[a]	Resources base[b]	Additional occurrences
Oil						
Conventional	97.1	2.7	118	153	271	
Unconventional	5.7	0.2	132	308	440	1,220
Natural gas[c]						
Conventional	35.9	1.2	82	179	261	
Unconventional	0.5	0.1	123	165	288	245
Clathrates	-	-	-	-	-	11,934
Coal	156.4	2.4	1,094	2,605	3,699	3,122
Total fossil occurrences	295.6	6.5	1,549	3,410	4,959	16,521

- Negligible volumes
[a],[b] and [c] see *Table 3.28a*

ice-like deposits that probably cover a significant portion of the ocean floor). Unconventional deposits require different and more complex production methods and, in the case of oil, need additional upgrading to usable fuels. In essence, unconventional resources are more capital intensive (for development, production, and upgrading) than conventional ones. The prospects for unconventional resources depend on the rate and costs at which these can be converted into quasi-conventional reserves.

3.8.3   *Historic Trends and Driving Forces*

Table 3.28a categorizes fossil deposits into reserves, resources and additional occurrences for both conventional and uncon-

ventional oil and gas deposits. The categories reflect the definitions of reserves and resources given above, with the exception that resources are further disaggregated into resources and occurrences so as to better reflect the speculative nature associated with their technical and economic feasibility (Rogner, 1997, 2000a).

Table 3.28b presents the global fossil resource data of *Table 3.28a* in terms of their respective carbon content. Since the onset of the industrial revolution, almost 300GtC stored in fossil fuels have been oxidized and released to the atmosphere. The utilization of all proven conventional oil and gas reserves would add another 200GtC, and those of coal more than 1,000

GtC. The fossil fuel resource base represents a carbon volume of some 5,000GtC indicating the potential to add several times the amount already oxidized and released to the atmosphere during the 21st century. To put these carbon volumes into perspective, cumulative carbon emissions associated with the stabilization of carbon dioxide at 450ppm are estimated to be at 670GtC. *Figure SPM.2* combines the reserve and resource estimates with cummulative emissions for various reference and stabilization scenarios, taken from other chapters and the IPCC WGI report.

Potential coal reserves are large – of that there is little doubt. However, there is an active debate on the ultimate size of recoverable oil reserves. The pessimists see potential reserves as limited, pointing to the lack of major new discoveries for 25 years or so (Laherrere, 1994; Hatfield, 1997; Campbell, 1997; Ivanhoe and Leckie, 1993). They see oil production peaking around 2010. The optimists point to previous pessimistic estimates being wrong. They argue that "there are huge amounts of hydrocarbons in the Earth's crust" and that "estimates of declining reserves and production are incurably wrong because they treat as a quantity what is really a dynamic process driven by growing knowledge" (Adelman and Lynch, 1997; Rogner, 1998a). They further point to technological developments such as directional drilling and 3D seismic surveys which are allowing more reserves to be discovered and more difficult reserves to be developed (Smith and Robinson, 1997). The optimists see no major supply problem for several more decades beyond 2010.

Estimates of gas reserves have increased in recent years (IGU, 2000; Rogner, 2000a; Gregory and Rogner, 1998) as there is much still to be discovered, often in developing countries that have seen little exploration to date. The problem in the past has been that there needed to be an infrastructure to utilize gas before it could have a market, and without an infrastructure, exploration appeared unattractive. The development of CCGT power stations (discussed below) means that a local market for gas can more readily be found which could encourage wider exploration. In the longer term, it is estimated that very substantial reserves of gas can be extracted from the bottom of deep oceans in the form of methane clathrates, if technology can be developed to extract them economically

With uranium, there has only been very limited exploration in the world to date but once more is required, new exploration is likely to yield substantial additional reserves (Gregory and Rogner, 1998; OECD-NEA and IAEA, 2000) (see *Table 3.28a)*.

The other major supply of energy comes from renewable sources, which meet around 20% of the global energy demand, mainly as traditional biomass and hydropower. Modern systems have the potential to provide energy services in sustainable ways with almost zero GHG emissions (Goldemberg, 2000).

The following sections focus on energy supply and conversion technologies in which there have been developments since the

Second Assessment Report and which may be key to achieving substantial reductions in greenhouse gas emissions in the coming decades.

On a global basis, in 1995 coal had the largest share of world electricity production at 38% followed by renewables (principally hydropower) at 20%, nuclear at 17%, gas at 15%, and oil at 10%. On current projections, electricity production is expected to double by 2020 compared to 1995 and energy used for generation to increase by about 80% as shown in *Table 3.29* (IEA, 1998b).

- Coal is projected to retain the largest share with a 90% increase in use from strong growth in countries such as India and China reflecting its importance there, steady growth in the USA but a decline in Western Europe.
- Gas is projected to grow strongly in many world regions reflecting the increasing availability of the fuel, with an overall increase of 160%.
- Nuclear power is projected to decline slightly on a global basis after 2010. Capacity additions in developing countries and in economies in transition roughly balance the capacity being withdrawn in OECD countries. Few new power stations will be built in many countries without a change in government policies. IAEA projections for 2020 cover a range from a 10% decline to an optimistic 50% increase in nuclear generating capacity (IAEA, 2000a).
- Hydropower is projected to grow by 60%, mainly in China and other Asian countries.
- New renewables have expanded substantially, in absolute terms, throughout the 1990s (wind 21% per year, solar PV more than 30% per year); these are projected to grow by over tenfold by 2020, but they would still supply less than 2% of the market.

3.8.4 *New Technological Options*

3.8.4.1 *Fossil Fuelled Electricity Generation*

3.8.4.1.1 Pulverized Coal

In a traditional thermal power station, pulverized coal (or fuel oil or gas) is burned in a boiler to generate steam at high temperature and pressure, which is then expanded through a steam turbine to generate electricity. The efficiencies of modern power stations can exceed 40% (lower heating value (LHV)), although the average efficiency, worldwide, of the installed stock is about 30% (Ishitani and Johansson, 1996). The typical cost of a modern coal- fired power station, with SO_2 and NO_x controls, is US$1,300/kW (Ishitani and Johansson, 1996). These costs vary considerably and can be more than 50% higher depending on location. Less efficient designs with fewer environmental controls are cheaper.

Table 3.29: *Past and projected global electricity production, fuel input to electricity production and carbon emissions from the electricity generating sector*
(Source: IEA, 1998b)

Global electricity generation (TWh)

	1971	1995	2000	2010	2020
Oil	1,100	1,315	1,422	1,663	1,941
Natural gas	691	1,932	2,664	5,063	8,243
Coal	2,100	4,949	5,758	7,795	10,296
Nuclear	111	2,332	2,408	2,568	2,317
Hydro	1,209	2,498	2,781	3,445	4,096
Renewables	36	177	215	319	433
Total	5,247	13,203	15,248	20,853	27,326

Fuel input (EJ)

	1971	1995	2000	2010	2020
Oil	11	13	14	15	18
Natural Gas	10	24	29	43	62
Coal	26	57	65	85	106
Nuclear	1	25	26	28	25
Hydro	4	9	10	12	15
Renewables	0	1	2	3	5
Total	53	129	146	187	230

CO_2 emissions (MtC)

	1971	1995	2000	2010	2020
Oil	224	258	273	307	350
Natural gas	158	362	443	662	946
Coal	668	1,471	1,679	2,185	2,723
Nuclear	0	0	0	0	0
Hydro	0	0	0	0	0
Renewables	0	0	0	0	0
Total	1,050	2,091	2,395	3,155	4,019

Average emissions per kWh

	1971	1995	2000	2010	2020
gC/kWh	200	158	157	151	147

The development of new materials allows higher steam temperatures and pressures to be used in "supercritical" designs. Efficiencies of 45% are quoted in the Second Assessment Report, although capital costs are significantly higher at around US$1,740/kW (Ishitani and Johansson, 1996). More recently, efficiencies of 48.5% have been reported (OECD, 1998b) and with further development, efficiencies could reach 55% by 2020 (UK DTI, 1999) at costs only slightly higher than current technology (Smith, 2000).

3.8.4.1.2 Combined Cycle Gas Turbine (CCGT)

Developments in gas turbine technology allow for higher temperatures which lead to higher thermodynamic efficiencies. The overall fuel effectiveness can be improved by capturing the waste heat from the turbine exhaust in a boiler to raise steam to generate electricity through a steam turbine. Thus in such a CCGT plant, electricity is generated by both the gas and steam turbines driving generators. The efficiency of the best available natural gas fired CCGTs currently being installed is now around 60% (LHV) (Goldemberg, 2000) and has been improving at 1% per year in the past decade. Typical capital costs for a power station of 60% efficiency are around US$450-500/kW, including selective catalytic reduction (for NO_x), dry cooling, switchyard, and a set of spares. Costs can be higher in some regions, especially if new infrastructure is required. These costs have been falling as efficiencies improve (IIASA-WEC, 1998). Together with high availability and short construction times, this makes CCGTs highly favoured by power station developers where gas is available at reasonable prices. Developments in the liquefied natural gas markets could further expand the use of CCGTs. Further improvements might allow electricity generating efficiencies of over 70% to be achievable for CCGTs within a reasonable period (Gregory and Rogner, 1998).

3.8.4.1.3   Integrated Gasification Combined Cycle (IGCC)

IGCC systems utilize the efficiency and low capital cost advantages of a CCGT by first gasifying coal or other fuel. Gasifiers are usually oxygen blown and are at the early commercial stage (Goldemberg, 2000). Coal and difficult liquid fuels such as bitumens and tar can be used as feedstocks. Biomass fuels are easier to gasify (Section 3.8.4.3.2), which may reduce the cost and possibly the efficiency penalty as an oxygen plant is not required (Lurgi GmbH, 1989). Gas clean-up prior to combustion in the gas turbine, which is sensitive to contaminants, is one of the current areas of development. The potential efficiency of IGCCs is around 51%, based on the latest CCGTs of 60% efficiency (Willerboer, 1997). Vattenfall, using a GE Frame 6 gas turbine, indicated a net efficiency of 48% in trials (Karlsson *et al.*, 1998), and an efficiency of 50%-55% was claimed to be achievable by using the latest gas turbine design. With continuing development in hot gas cleaning and better heat recovery as well as the continuing development of CCGTs, commercially available coal- or wood-fired IGCC power stations with efficiencies over 60% may be feasible by 2020.

In addition to the potential high efficiencies, IGCC offers one of the more promising routes to CO_2 capture and disposal by converting the gas from the gasifier into a stream of H_2 and CO_2 via a shift reaction. The CO_2 can then be removed for disposal before entering the gas turbine (see Section 3.8.4.4). The resultant stream of H_2 could be used in fuel cells and not just in a gas turbine.

3.8.4.1.4   Cogeneration

Combined heat and power (CHP) generation can yield fuel energy utilization rates of up to 90% and can therefore be an effective GHG mitigation option. CHP is possible with all heat machines and fuels (including nuclear, biomass and solar thermal) from a few kW-rated to 1000MW steam-condensing power plants. At the utility level the employment of CHP is closely linked with industrial heat loads as well as the availability or development of district heating and/or cooling net-

works. These are energy transmission systems suited for the distribution of heat and/or cooling within areas with sufficiently high heat/cooling load densities (Kalkum *et al.*, 1993; Rogner, 1993). The expanded use of natural gas may provide a basis for increased dispersed cogeneration. Industrial CHP utilizes temperature differentials between the heat source and the process temperature requirements for electricity generation. More recently, in some countries electricity market deregulation has made it easier for large industrial users to generate their own electricity as well as heat by being more easily able to sell any surplus electricity (see Section 3.8.5.1). Conversely, following deregulation in Germany and elsewhere, large grid CHP has suffered market loss as a consequence of existing surplus generating capacity and independent generation. There is good potential for cogeneration of biomass including bagasse in developing countries such as India, where the market potential is 3500MW. However, a heat demand is necessary for CHP plants to be implemented successfully.

3.8.4.1.5   Fuel Cells

Several types of fuel cell compete for early entry into a variety of prospective markets (Gregory and Rogner, 1998). Proton exchange membrane, phosphoric acid, fuel cells (PAFCs) and solid oxide fuel cells are the current technology options. Each type has its own distinctive characteristics such as operating temperatures, efficiency ranges, fuel use, markets and costs[18]. The potential advantages of fuel cells over gas turbines include smaller unit sizes at similar efficiencies, the potential of a low or quasi zero GHG emission technology at the point of use, lower maintenance costs, less noise and, eventually, better economic performance.

The internal fuel is hydrogen, but some fuel cell types can use fuels such as CO, methanol, natural gas or even coal if externally converted to hydrogen at the plant via gasification and steam reforming or partial oxidation. Alternatively, some fuel cell designs perform the hydrogen conversion step internally as an integral part of the technology.

Hydrogen production from hydrocarbon fuels generates some airborne emissions (NO_x, CO, CO_2 and NMVOCs) but these are – with the exception of CO_2 - orders of magnitude lower than those associated with combustion cycles. The electrochemistry of most fuel cells demands the use of sulphur-free natural gas, hence no SO_2 emissions occur. CO_2 emissions are a function of the electrical efficiency and as such are comparable with the efficient CCGT. Non fossil-derived hydrogen, e.g., by way of solar powered electrolysis or from methanol derived from biomass, can be used with virtually zero GHG emissions.

---

[18] Although primarily targeted for electricity generation in less densely populated rural areas (distributed generation), fuel cell manufacturers have always had high aspirations for the transport market.

Proton exchange membrane (PEM) conversion efficiencies are currently at 45%-48%[19] using hydrogen as the onboard fuel, and are expected to approach 55%-60% in the near term future. The joint venture between the leading PEM fuel cell producer and major automobile manufacturers aimed at mass producing PEMs for vehicle propulsion targets to bring down costs to less than US\$500/kW$_e$ by 2000 and to less than US\$250/kW$_e$ by 2010 (Rogner, 1998b). In the very long term, costs comparable to current internal combustion engines of approximately US\$50/kW$_e$ have been suggested (Lovins, 1996).

Phosphoric acid fuel cells (PAFCs) operate at around 200°C and pressures up to 8 bar. At present, PAFCs, in particular the 200 kW ONSI PC25, are commercially the most advanced fuel cells in the market place and have accumulated more than one million hours of operating experience world-wide. Stationary PAFC applications include the world's largest fuel cell power plant of 11 MW in Goi, Japan, which was in test operation from 1991 to 1997, commissioned by Tokyo Electric Power Company (TEPCO). Natural gas to electricity conversion was 36 to 38% efficient. Overall fuel effectiveness with waste heat utilization can be as high as 80%. Capital costs for an integrated system, including a fuel processor based on natural gas, are expected to decline to about US\$1,500/kW$_e$ by 2000. Long-term costs will depend on manufacturing volumes, but industry experts project costs below US\$700/kW$_e$ by 2010 (Tauber and Jablonski, 1998) at electrical efficiencies of 50% and overall fuel effectiveness of 90% for cogeneration.

Molten carbonate fuel cells (MCFC) operating at 650°C open the possibility of using carbonaceous fuels and internal reforming. With steam turbines in a bottoming cycle, the overall electrical efficiency could be as high as 65%. There are still major technical problems associated with MCFCs such as electrode corrosion, the sintering of the structural fuel cell material, and sensitivity to fuel impurities.

The operating temperature of 1000°C of solid oxide fuel cells (SOFCs) allows internal reforming and produces high quality by-product heat for cogeneration or for use in a bottoming cycle. The development of suitable materials and the fabrication of ceramic structures are presently the key technical challenges facing SOFCs. They are currently being demonstrated in a 100-kilowatt plant and are expected to be competitive with traditional fossil-fired generation early in the 21st century. Installation costs will eventually reach about US\$700/kW$_e$ (EPRI, 1997) though other sources report US\$1,620/kW$_e$ for the period 2005-2010 (OECD, 1998b).

Hybrid SOFC/CCGT systems have projected efficiencies of 72 to 74%, and, depending on R&D progress, would represent the ultimate fossil fuel based electricity generation (Federal Energy Technology Center, 1997). Typical plant sizes would be 1 to 100 MW and, fuelled with natural gas, would produce the lowest emissions of all fossil fuel electricity generating options of about 75–80gC/kWh.

3.8.4.2   Nuclear Power

3.8.4.2.1 Present Situation

Nuclear power is a mature technology with 434 nuclear reactors operating in 32 countries in 1999, with a total capacity of around 349GW$_e$ generating 2,398 TWh or some 16% of global electricity generation in 1999 (IAEA, 2000b). In general, the majority of current nuclear power plants worldwide are competitive on a marginal cost basis in a deregulated market environment[20].

The life cycle GHG emissions per kWh from nuclear power plants are two orders of magnitude lower than those of fossil-fuelled electricity generation and comparable to most renewables (EC, 1995; Krewitt *et al.*, 1999; Brännström-Norberg *et al.*, 1996; Spadaro *et al.*, 2000). Hence it is an effective GHG mitigation option, especially by way of investments in the lifetime extension of existing plants.

Whether or not nuclear power would be accepted in the market place depends on new capacities becoming economically competitive and on its ability to restore public confidence in its safe use.

3.8.4.2.2 Nuclear Economics

Where gas supply infrastructures are already in place, new nuclear power plants at US\$1700–US\$3100/kW$_e$ (Paffenberger and Bertel, 1998) cannot compete against natural gas-fuelled CCGT technology at current and expected gas prices (OECD, 1998b). Nuclear power can be competitive versus coal and natural gas, especially if coal has to be transported over long distances or natural gas infrastructures are not in place. Discount rates are often critical in tilting the competitive balance between nuclear power and coal. A study (OECD, 1998b) surveyed the costs of nuclear, coal, and natural gas-fuelled electricity generation in 18 countries for plants that would go into operation in 2005. The results, estimated for both 5% and 10% discount rates, showed that nuclear power is the least cost option in seven countries at a 5% discount rate (generating cost range US\$0.025–0.057/kWh), but only in two countries at a 10% discount rate (generating cost range US\$ 0.039–0.080/kWh). In fully deregulated markets such as the UK's, rates of return in excess of 14% have been required at which level new nuclear plant construction would not be competitive at current fossil fuel market prices[21].

---

[19] For fuel cells all efficiencies cited are based on higher heating values (HHV).

[20] Because of low operating costs and the fact that many nuclear power plants are already fully depreciated.

[21] Rising coal and gas market prices such as observed for oil in 2000 would change the competitive position of nuclear substantially.

3.8.4.2.3 Waste Disposal

Technological approaches for safe and long-term disposal of high-level radioactive waste have been extensively studied (Posiva Oy, 1999; EC, 1999). One possible solution involves deep geological repositories, however, no country has yet disposed of any spent fuel or high-level waste in such a repository because of public and political opposition (NEA, 1999). Several countries are actively researching this issue. Long-term disposal of radioactive wastes should not be an intractable problem from a technical perspective, because of the small quantities of storage space required (Goldemberg, 2000; Rhodes and Beller, 2000). Radioactive waste storage density limits defined for storing light water reactor (LWR) fuel at Yucca Mountain are about 41 m^2/MW_e of nuclear generating capacity for a power plant over its expected 30 years of operating life[22] (Kadak, 1999). High level waste volumes can be further reduced if spent fuel is reprocessed so that most of the plutonium and unused uranium is extracted for reuse. The remaining high-level waste is compacted and "vitrified" (melted with other ingredients to make a glassy matrix), and placed into canisters that are appropriate for long-term disposal. However, reprocessing of spent fuel and the separation of plutonium are often viewed as potentially opening the door for nuclear weapons proliferation. For this and economic reasons, several countries therefore prefer once-through fuel cycles and direct disposal of spent reactor fuel.

Because of the low waste volumes, it may be plausible to accumulate high level radioactive wastes in a few sites globally rather than every country seeking national solutions (Goldemberg, 2000) These international repositories would be operated and controlled by an international organization which would also assume the responsibility of safeguarding these sites (McCombie, 1999a; 1999b; McCombie *et al.*, 1999; Miller *et al.*, 1999). For the time being, most governments remain committed to identifying suitable high-level waste disposal or interim storage solutions within their own national territories.

In the longer run, fundamentally new reactor configurations may need to be developed that are based on innovative designs that integrate inherent operating safety features and waste disposal using previously generated radioactive waste as fuel and, by way of transmutation, convert nuclear waste or plutonium to less hazardous and short-lived isotopic substances (Rubbia, 1998).

Present technology can be used to reduce the growth of the plutonium stocks by use of mixed plutonium/uranium oxide fuels (MOX) in thermal reactors. Belgium, France, Germany, and Switzerland use MOX fuels in existing reactors. Japan also has been progressing its MOX utilizing programme.

3.8.4.2.4 New Reactor Technologies

The future of nuclear power will depend on whether it can meet several objectives simultaneously – economics, operating safety, proliferation safeguards, and effective solutions to waste disposal. While present new nuclear power plants already incorporate unprecedented levels of safety based on in-depth designs, their economics need further improvement to be competitive in most markets. Safe waste disposal for approximately 1 million years is technically feasible (Whipple, 1996) and would add US$0.0002/kWh to generating costs (Kadak, 1999; Goldemberg, 2000). Disposal cost estimates for Sweden are higher, i.e., US$0.0013. Proliferation is a political issue primarily, but can also be addressed by technology. Evolutionary technology improvements of existing designs are important elements for the near-term viability of nuclear power but may not be sufficient to meet all the objectives optimally. For example, smaller grid sizes in developing countries demand smaller unit size reactors. Therefore, new technology that addresses these objectives by integral design holds the key to the future of nuclear power.

Building on more than 40 years of experience with LWR technology, major nuclear reactor vendors have now developed modified LWRs that offer both improved safety and lower cost (CISAC, 1995; Kupitz and Cleveland, 1999). These evolutionary development efforts resulted in standardized designs for which there can be a high degree of confidence that performance and cost targets will be met. All employ active but simplified safety systems, and some have some passive safety features.

One reactor in this category is the Westinghouse AP600, a 600 MW_e pressurized water reactor (PWR). The design is simpler than existing PWRs and modular, with about half the capacity of most existing PWRs—which allows some components to be factory built and assembled faster onsite at lower cost than for plants that are entirely field constructed. The AP600 is expected to be safer than existing PWRs, constructed in three years, and costs about 15% less than existing PWRs of the same capacity (NPDP, 1998)[23].

Other examples include the ABB/Combustion Engineering System 80+ and the GE Advanced Boiling Water Reactor (ABWR)[24]. The System 80+ is a large (1,350 MW_e) unit for which the estimated core damage frequency is two orders of magnitude lower than for its predecessor. The ABWR has as a design objective stepped-up operating safety and a target capital cost that is 20% less than for BWRs previously built in Japan (NPDP, 1998). Two ABWRs are now operating in Japan. Two more are under construction in Japan and also in Taiwan/China.

[23] In late 1999 the AP600 received Design Certification from the US Nuclear Regulatory Commission.

[24] Both designs received Design Certification from the US Nuclear Regulatory Commission in 1997.

[22] The actual high-level waste accumulated over 30 years amounts to approximately 1 m^3/MWe.

In Europe, a Framatome/Siemens joint venture has developed the European pressurized water reactor (EPR), a 1,450 to 1,750 MW_e system designed to specifications endorsed by utilities in Europe—with hoped-for economies of scale at this large unit size. The EPR is being offered on the international market.

One of the innovative designs is the pebble bed modular reactor (PBMR) developed by the South African utility ESKOM. The fundamental concept of the design is to achieve a plant that has no physical process, however unlikely, that could cause a radiation-induced hazard outside the site boundary. This is principally achieved in the PBMR[25].

The current ESKOM assessment is that the capital cost of a production of 1000MW_e block of 10 modules will be US$1,000/$kW_e$ (US$1,200/$kW_e$ for the prototype). The low capital costs are the result of a much lower energy density of the reactor core than present reactor technologies; the elimination of heat exchangers; use of a direct helium turbine; the shift of the containment from the plant periphery to the pebble fuel; a high in-shop manufacturing component of the plant, and a short construction period of 2 years. This would produce attractive generating costs with unprecedented safety aspects. As a base load station with a depreciation period of 20 years and at a 10% discount rate, the expected cost of power would be approximately US$0.018/kWh including the full fuel cycle and decommissioning. These costs are low indeed and in part the result of engineering optimism. Other studies based on the less advanced modular high temperature reactor (HTR) designs conclude that generating costs may range from US$0.020 to US$0.034/kWh (Lako, 1996). Kadak (1999) estimates the unit capital cost for a PBMR plant with some different design characteristics at twice the ESKOM value, i.e., US$2,090/$kW_e$ which results in generating costs of US$0.033/kWh but still less than average present technology.

3.8.4.2.5 GHG Mitigation Potential

Increased performance and lifetime extension of the currently existing nuclear reactors often present a zero-costs greenhouse gas mitigation option. However, given current market conditions, new nuclear power capacity is a least-cost alternative in only some countries usually characterized by limited indigenous fossil resources or by large distances between resource location and consumption centres. Under such circumstances nuclear power is a zero cost mitigation option. If the optimistic PBMR generating costs can be accomplished, this would certainly imply negative mitigation costs. For new nuclear plants of state-of-the-art designs, a value of US$0 to 40/tC avoided would put nuclear at par with coal-fired electricity, while it

would take a value of a US$100 to 250/tC for nuclear power to break even with natural gas combined cycle electricity (IEA, 1998a; Rogner, 2000b). Based on the current global electricity mix, nuclear power avoids some 600MtC of carbon per year (Rogner, 1999).

The most recent projection of the International Atomic Energy Agency (IAEA) is that in the absence of any policies with regard to climate change, capacity in 2020 could be in the range 300GW_e to 520GW_e (from 349GW_e in 1999). This means that the share of nuclear power in total power generation could decline to 6% - 8% by 2020 (IAEA, 2000a). Compared to the reference case of the IEA World Energy Outlook (IEA, 1998b), the higher projection of IAEA would avoid the emission of 87MtC in 2010 and 281MtC in 2020 over and above the reference case assuming that nuclear power displaces coal-fired electricity[26]. Maintaining the past nuclear share in global electricity generation would avoid annually 280MtC in 2010 and 550MtC in 2020 (again above the IEA reference case of approximately 600GtC).

3.8.4.3 Renewable Energy Conversion Technologies

Natural energy flows vary from location to location, and make the techno-economic performance of renewable energy conversion highly site-specific. Intermittent sources such as wind, solar, tidal, and wave energy require back-up if not grid connected, while large penetration into grids may eventually require storage and/or back-up to guarantee reliable supply. Therefore, it is difficult to generalize costs and potentials.

3.8.4.3.1 Hydropower

Hydroelectricity remains the most developed renewable resource worldwide with global theoretical potential ranges from 36,000 to 44,000TWh/yr (World Atlas, 1998). Approximately 65% of the technical hydro potential has been developed in Western Europe, and 76% in the USA. This indicates a limit caused by societal and environmental barriers. For many developing countries the total technical potential, based on simplified engineering and economic criteria with few environmental considerations, has not been fully measured. The economic potential resulting from detailed geological and technical evaluations, but including social and environmental issues, is difficult to establish because these parameters are strongly driven by societal preferences inherently uncertain and difficult to predict. A rate of utilization between 40% and 60% of a region's technical potential is therefore a reasonable assumption and leads to a global economic hydro-electricity potential of 7,000 to 9,000TWh/yr (see *Table 3.30*).

---

[25] The integrated heat loss from the reactor vessel exceeds the decay heat production in the post accident condition, and the peak temperature reached in the core during the transient is below the demonstrated fuel degradation point and far below the temperature at which the physical structure is affected. The prospect of a "core melt" scenario is therefore zero.

[26] The nuclear mitigation potential is based on the following assumption: between 2000 and 2020 nuclear capacities increase to the 520GW_e of the high IAEA projection and substitute incremental coal capacities with global average efficiencies of 39.4% and 44.1% in 2010 and 2020, respectively (compared to 33.1% and 35.1% in the IEA World Energy Outlook (IEA, 1986)). The nuclear share in global electricity generation would then amount to 13%.

Table 3.30: *Annual large hydroelectric development potential (TWh/yr)*

	Theoretical potential		Technological potential		Economic potential	
	TWh[a]	TWh[b]	TWh[a]	TWh[b]	TWh[a]	TWh[b]
Africa	3,307	3,633	1,896	1,589	815	866
North America	5,817	5,752	1,509	1,007	912	957
Latin America	7,533	8,800	2,868	3,891	1,198	2,475
Asia (excluding former USSR)	15,823	14,138	4,287	4,096	1,868	2,444
Australasia	591	592	201	206	106	168
Europe	3,128	3,042	1,190	942	774	702
Former USSR	3,583	3,940	1,992	2,105	1,288	1,093
World	**39,784**	**39,899**	**13,945**	**13,839**	**6,964**	**8,708**

[a] World Atlas, 1999

[b] International Water Power & Dam Construction, 1997

Numerous small (<10MW), mini (<1MW) and micro (<100kW) scale hydro schemes with low environmental impacts continue to be developed globally. The extent of this resource, particularly in developing countries such as Nepal, Oceania, and China, is unknown but likely to be of significance to rural communities currently without electricity.

Large-scale hydropower plant developments can have high environmental and social costs such as loss of fertile land, methane generation from flooded vegetation, and displacement of local communities (Moomaw *et al*., 1999b). At the 18,200 MW Three Gorges dam under construction in China, 1.2 million people have been moved to other locations. Another limitation to further development is the high up-front capital investment which the recently privatized power industries are unlikely to accept because of the low rates of return.

The remote locations of many potential hydro sites result in high transmission costs. Development of medium (<50MW) to small (<10MW) scale projects closer to demand centres will continue. In countries where government or aid assistance is provided to overcome the higher investment costs/MW at this scale, power generation costs around US$0.065/kWh will result (UK DTI, 1999). Mini- and micro-hydro low head turbines are under development but generating costs at this scale are likely to remain high, partly as a result of the cost of the intake structure needed to withstand river flood conditions. Even at this small scale, environmental and ecological effects often result from taking water from a stream or small river and discharging it back again, even after only a short distance.

3.8.4.3.2 Biomass Conversion

Globally, biomass has an annual primary production of 220 billion oven-dry tonnes (odt) or 4,500 EJ (Hall and Rosillo-Calle, 1998a). Of this, 270 EJ/yr might become available for bioener-

gy on a sustainable basis (Hall and Rosillo-Calle, 1998a) depending on the economics of production and use as well as the availability of suitable land. In addition to energy crops (Section 3.6.4.3), biomass resources include agricultural and forestry residues, landfill gas and municipal solid wastes. Since biomass is widely distributed it has good potential to provide rural areas with a renewable source of energy (Goldemberg, 2000). The challenge is to provide the sustainable management, conversion and delivery of bioenergy to the market place in the form of modern and competitive energy services (Hall and Rao, 1994).

At the domestic scale in developing countries, the use of firewood in cooking stoves is often inefficient and can lead to health problems. Use of appropriate technology to reduce firewood demand, avoid emissions, and improve health is a no-regrets reduction opportunity (see Section 4.3.2.1).

Agricultural and forest residues such as bagasse, rice husks, and sawdust often have a disposal cost. Therefore, waste-to-energy conversion for heat and power generation and transport fuel production often has good economic and market potential, particularly in rural community applications, and is used widely in countries such as Sweden, the USA, Canada, Austria, and Finland (Hall and Rosillo-Calle, 1998b; Moomaw *et al*., 1999b; Svebio, 1998). Energy crops have less potential because of higher delivered costs in terms of US$/GJ of available energy.

Harvesting operations, transport methods, and distances to the conversion plant significantly impact on the energy balance of the overall biomass system (CEC, 1999; Moreira and Goldemberg, 1999). The generating plant or biorefinery must be located to minimize transport costs of the low energy density biomass as well as to minimize impacts on air and water use. However, economies of scale of the plant are often more sig-

Table 3.31: *Projection of technical energy potential from biomass by 2050*
(Derived from Fischer and Heilig, 1998; D'Apote, 1998; IIASA/WEC, 1998)

Region	Population in 2050	Total land with crop production potential	Cultivated Land in 1990	Additional cultivated land required in 2050	Available area for biomass production in 2050	Max. Additional amount of energy from biomass[a]
	Billion	Gha	Gha	Gha	Gha	EJ/yr
Developed[b]	-	0.820	0.670	0.050	0.100	30
Latin America						
Central & Caribbean	0.286	0.087	0.037	0.015	0.035	11
South America	0.524	0.865	0.153	0.082	0.630	189
Africa						
Eastern	0.698	0.251	0.063	0.068	0.120	36
Middle	0.284	0.383	0.043	0.052	0.288	86
Northern	0.317	0.104	0.04	0.014	0.050	15
Southern	0.106	0.044	0.016	0.012	0.016	5
Western	0.639	0.196	0.090	0.096	0.010	3
China[c]	-	-	-	-	-	2
Rest of Asia						
Western	0.387	0.042	0.037	0.010	-0.005	0
South –Central	2.521	0.200	0.205	0.021	-0.026	0
Eastern	1.722	0.175	0.131	0.008	0.036	11
South –East	0.812	0.148	0.082	0.038	0.028	8
Total for regions above	8.296	2.495	0.897	0.416	1.28	396
Total biomass energy potential, EJ/yr						**441**[d]

[a] Assumed 15 odt/ha/yr and 20GJ/odt

[b] Here, OECD and Economies in Transition

[c] For China, the numbers are projected values from D'Apote (1998) and **not** maximum estimates.

[d] Includes 45 EJ/yr of current traditional biomass.

nificant than the additional transport costs involved (Dornburg and Faaij, 2000). The sugar cane industry has experience of harvesting and handling large volumes of biomass (up to 3Mt/yr at any one plant) with the bagasse residues often used for cogeneration on site to improve the efficiency of fuel utilization (Cogen, 1997; Korhonen *et al.*, 1999). Excess power is exported. In Denmark about 40% of electricity generated is from biomass cogeneration plants using wood waste and straw. In Finland, about 10% of electricity generated is from biomass cogeneration plants using sawdust, forest residues, and pulp liquors (Pingoud *et al.*, 1999; Savolainen, 2000). In other countries biomass cogeneration is utilized to a lesser degree as a result of unfavourable regulatory practices and structures within the electricity industry (Grohnheit, 1999; Lehtilä *et al.*, 1997).

Land used for biomass production will have an opportunity cost attributed to it for the production of food or fibre, the value being a valid cost which can then be used in economic analyses. *Table 3.31* shows the technical potential for energy crop production in 2050 to be 396EJ/yr from 1.28Gha of available land[27]. By 2100 the global land requirement for agriculture is estimated to reach about 1.7Gha, whereas 0.69-1.35Gha would then be needed to support future biomass energy requirements in order to meet a high-growth energy scenario (Goldemberg, 2000). Hence, land-use conflicts could then arise.

Several developing countries in Africa (e.g., Kenya) and Asia (e.g., Nepal) derive over 90% of their primary energy supply

---

[27] Practical/technical constraints on the use of land for bioenergy such as the distance of a proposed biomass production site from energy demand centres, the power distribution grid, or sources of labour are not considered here. Hence the estimate exceeds the 270EJ of Hall and Rosillo-Calle (1998a).

from traditional biomass. In India it currently provides 45% and in China 30%. Modern bioenergy applications at the village scale are gradually being implemented, leading to better and more efficient utilization which, in many instances, complement the use of the traditional fuels (FAO, 1997) and provide rural development (Hall and Rosillo-Cale, 1998b). For example, production of liquids for cooking, from biomass grown in small-scale plantations, using the Fischer-Tropsch process (modified to co-produce electricity by passing unconverted syngas through a small CCGT), is being evaluated for China using corn husks (Larson and Jin, 1999). Biomass and biofuel were identified by a US Department of Energy study (Interlaboratory Working Group, 1997) as critical technologies for minimizing the costs of reducing carbon emissions. Co-firing in coal-fired boilers, biomass-fuelled integrated gasification combined-cycle units (BIGCC) for the forest industry, and ethanol from the hydrolysis of lignocellulosics were the three areas specifically recognized as having most potential. Estimates of annual carbon offsets from the uptake of these technologies in the USA alone ranged from 16-24Mt, 4.8Mt, and 12.6-16.8Mt, respectively, by 2010. The near term energy savings from use of each of these technologies should cover the associated costs (Moore, 1998), with co-firing giving the lowest cost and technical risk.

Woody biomass blended with pulverized coal at up to 10%–15% of the fuel mix is being implemented, for example, in Denmark and the USA, but may be uneconomic as a consequence of coal being cheaper than biomass together with the costs of combustion plant conversion (Sulilatu, 1998). However, major environmental benefits can result including the reduction of SO_2 and NO_x emissions (van Doorn *et al.*, 1996).

Gasification of biomass

Biofuels are generally easier to gasify than coal (see Section 3.8.4.1.3), and development of efficient BIGCC systems is nearing commercial realization. Several pilot and demonstration projects have been evaluated with varying degrees of success (Stahl and Neergaard, 1998; Irving, 1999; Pitcher and Lundberg, 1998). Capital investment for a high pressure, direct gasification combined-cycle plant of this scale is estimated to fall from over US$2,000/kW at present to around US$1,100/kW by 2030, with operating costs, including fuel supply, declining from 3.98c/kWh to 3.12c/kWh (EPRI/DOE, 1997). By way of comparison, capital costs for traditional combustion boiler/steam turbine technology were predicted to fall from the present US$1,965/kW to US$1,100/kW in the same period with current operating costs of 5.50c/kWh (reflecting the poor fuel efficiency compared with gasification) lowering to 3.87c/kWh.

A life cycle assessment of the production of electricity in a BIGCC plant showed 95% of carbon delivered was recycled (Mann and Spath, 1997). From the energy ratio analysis, one unit of fossil fuel input produced approximately 16 units of carbon neutral electricity exported to the grid.

Liquid biofuels

Ethanol production using fermentation techniques is commercially undertaken in Brazil from sugar cane (Moreira and Goldemberg, 1999), and in the USA from maize and other cereals. It is used as a straight fuel and/or as an oxygenate with gasoline at 5%-22% blends. Enzymatic hydrolysis of lignocellulosic feedstocks such as bagasse, rice husks, municipal green waste, wood and straw (EPRI/DOE, 1997) is being evaluated in a 1t/day pilot plant at the National Renewable Energy Laboratory and is nearing the commercial scale-up phase (Overend and Costello, 1998). Research into methanol from woody biomass continues with successful conversion of around 50% of the energy content of the biomass at a cost estimate of around US$0.90/litre (US$34/GJ) (Saller *et al.*, 1998). In Sweden production of biofuels from woody biomass (short rotation forests or forest residues) was estimated to cost US$0.22/litre for methanol and US$0.54/litre for ethanol (Elam *et al.*, 1994). However, the energy density (MJ/l) of methanol is around only 50% that of petrol and 65% for ethanol. Using the available feedstock for heat and power generation might be a preferable alternative (Rosa and Ribeiro, 1998).

Commercial processing plants for the medium scale production of biodiesel from the inter-esterification of triglycerides have been developed in France, Germany, Italy, Austria, Slovakia, and the USA (Austrian Biofuels Institute, 1997). Around 1.5 million tonnes is produced each year, with the largest plant having a capacity of 120,000 tonnes. Environmental benefits include low sulphur and particulate emissions. A positive energy ratio is claimed with 1 energy unit from fossil fuel inputs giving 3.2 energy units in the biodiesel (Korbitz, 1998). Conversely, other older studies suggest more energy is consumed than produced (Ulgiati *et al.*, 1994).

Biodiesel production costs exceed fossil diesel refinery costs by a factor of three to four because of high feedstock costs even when grown on set-aside land (Veenendal *et al.*, 1997), and they are unlikely to become more cost effective before 2010 (Scharmer, 1998). Commercial biodiesel has therefore only been implemented in countries where government incentives exist. Biofuels can only become competitive with cheap oil if significant government support is provided by way of fuel tax exemptions, subsidies (such as for use of set-aside surplus land), or if a value is placed on the environmental benefits resulting.

3.8.4.3.3 Wind Power

Wind power supplies around 0.1% of total global electricity but, because of its intermittent nature and relatively recent emergence, accounts for around 0.3% of the global installed generation capacity. This has increased by an average of 25% annually over the past decade reaching 13,000MW by 2000, with estimates of this increasing to over 30,000MW capacity operating by 2005 (EWEA, 1999). The cost of wind turbines

Table 3.32: *Assessment of world wind energy potential on land sites with mean annual wind speeds greater than 5.1m/s* (Grubb and Meyer, 1993)

Region	Percent of land area	Population density	Gross electric potential	Wind energy potential	Estimated second order potential	Assessed wind energy potential
	%	capita/km^2	TWh ×10^3/yr	EJ/yr[a)	TWh ×10^3/yr	EJ/yr[a]
Africa	24	20	106	1,272	10.6	127
Australia	17	2	30	360	3.0	36
North America	35	15	139	1,670	14.0	168
Latin America	18	15	54	648	5.4	65
Western Europe	42	102	31	377	4.8	58
EITs	29	13	106	1,272	10.6	127
Asia	9	100	32	384	4.9	59
World	23	-	498	5,976	53.0	636

[a]   The energy equivalent in TWh is calculated on the basis of the electricity generation potential of the referenced sources by dividing the electricity generation potential by a factor of 0.3 (a representative value for the efficiency of wind turbines including transmission losses) resulting in a primary energy estimate.

continues to fall as more new capacity is installed. The trend follows the classic learning curve and further reductions are projected (Goldemberg, 2000). In high wind areas, wind power is competitive with other forms of electricity generation.

The global theoretical wind potential is on the order of 480,000TWh/yr, assuming that about $3×10^7$ km^2 (27%) of the earth's land surface is exposed to a mean annual wind speed higher than 5.1 m/s at 10 metres above ground (WEC, 1994). Assuming that for practical reasons just 4% of that land area could be used (derived from detailed studies of the potential of wind power in the Netherlands and the USA), wind power production is estimated at some 20,000 TWh/yr, which is 2.5 times lower than the assessment of Grubb and Meyer (1993) (see *Table 3.32*). The *Global Wind Energy Initiative*, presented by the wind energy industry at the 4th Conference of Parties meeting in Buenos Aires (BTM Consult, 1998), demonstrated that a total installed capacity of 844GW by 2010, including offshore installations, would be feasible. A report by Greenpeace and the European Wind Energy Association estimated 1,200GW could be installed by 2020 providing almost 3,000TWh/yr or 10% of the global power demand assumed at that time (Greenpeace, 1999).

Many of the turbines needed to meet future demand will be sited offshore, exceed 2MW maximum output, and have lower operating and maintenance costs, increased reliability, and a greater content of local manufacture. Shallow seas and planning consents may be a constraint.

Various government-enabling initiatives have resulted in the main uptake of wind power to date occurring in Germany, Denmark, the USA, Spain, India, the UK and the Netherlands. Typically turbines in the 250 – 750kW range are being installed (Gipe, 1998). Significant markets are now emerging in China, Canada, South America, and Australia.

Denmark aims to provide 40%-50% of its national electricity generation from wind power by 2030 and remains the main exporter of turbine technology (Krohn, 1997; Flavin and Dunn, 1997). China and India, based on recent wind survey programmes, have a high technical wind potential of 250–260GW and 20–35GW respectively, and are major turbine importers (Wang, 1998; MNES, 1998). However, following various government incentives, both China and India now manufacture their own turbines with export orders in place (Wang, 1998; AWEA, 1998).

Wind power continues to become more competitive, and commercial development is feasible without subsidies or any form of government incentives at good sites. In 1999, for example, a privately owned 32MW wind farm constructed in New Zealand on a site with mean annual wind speed of greater than 10m/s was competing at below US$0.03/kWh in the wholesale electricity market (Walker *et al.*, 1998). The rapidly falling price of wind power is evidenced by the drop in average prices (adjusted for short contract lengths). Over successive rounds of the British NFFO (non-fossil-fuel obligation), average tendered kWh prices declined from 7.95p in 1990 to 2.85p (US$0.043/kWh) in 1999 (Mitchell, 1998; UK DTI, 1999). These confirm the estimate of Krohn (1997) that wind generated electricity costs from projects >10MW would decline to US$0.04/kWh on good sites. The global average price is expected to drop further to US$0.027–0.031/kWh by around 2020 as a result of economies of scale from mass production and improved turbine designs (BTM consult, 1999). EPRI/DOE (1997) predicted the installed costs will fall from US$1,000 to US$635/kW (with uncertainty of +10% -20%), and operating costs will fall from 0.01c/kWh to 0.005c/kWh. However, on poorer sites of around 5m/s mean annual wind speed, the generating costs would remain high at around US$0.10-0.12/kWh (8% discount rate).

Table 3.33a: Key assumptions for the assessment of the solar energy potential

Region	Assumed annual clear sky irradiance[a] kW/m^2		Assumed annual average sky clearance[b], %	
	Min	Max	Min	Max
NAM (North America)	0.22	0.45	0.44	0.88
LAM (Latin America and the Caribbean)	0.29	0.46	0.48	0.91
AFR (Sub-Saharan Africa)	0.31	0.48	0.55	0.91
MEA (Middle East and North Africa)	0.29	0.47	0.55	0.91
WEU (Western Europe)	0.21	0.42	0.44	0.80
EEU (Central and Eastern Europe)	0.23	0.43	0.44	0.80
FSU (Newly independent states of the former Soviet Union)	0.18	0.43	0.44	0.80
PAO (Pacific OECD)	0.28	0.46	0.48	0.91
PAS (Other Pacific Asia)	0.32	0.48	0.55	0.89
CPA (Centrally planned Asia and China)	0.26	0.45	0.44	0.91
SAS (South Asia)	0.27	0.45	0.44	0.91

[a] The minimum assumes horizontal collector plane; the maximum assumes two-axis tracking collector plane

[b] The maxima and minima are as found for the relevant latitudes in Table 2.2 of WEC (1994).

Since wind power is intermittent the total costs will be higher if back-up capacity has to be provided. In large integrated systems it has been estimated that wind could provide up to 20% of generating capacity without incurring significant penalty. In systems that have large amounts of stored hydropower available, such as in Scandinavia, the contribution could be higher. The Denham wind (690kW)/diesel(1.7MW) system in Western Australia uses a flywheel storage system and new power station controller software to displace around 70% of the diesel used in the mini-grid by wind (Eiszele, 2000).

3.8.4.3.4 Solar Energy

An estimation of solar energy potential based on available land in various regions (*Tables 3.33a* and *3.33b*) gives 1,575 to 49,837 EJ/yr. Even the lowest estimate exceeds current global energy use by a factor of four. The amount of solar radiation intercepted by the earth may be high but the market potential for capture is low because of:

(1) the current relative high costs;

(2) time variation from daily and seasonal fluctuations, and hence the need for energy storage, the maximum solar flux at the surface is about 1 kW/m^2 whereas the annual average for a given point is only 0.2 kW/m^2;

(3) geographical variation, i.e. areas near the equator receive approximately twice the annual solar radiation than at 60° latitudes; and

(4) diffuse character with low power such that large-scale generation from direct solar energy can require significant amounts of equipment and land even with solar concentrating techniques.

Photovoltaics

The costs of photovoltaics are slowly falling from around US$5,000/kW installed as more capacity is installed in line with the classical learning curve (Goldemberg, 2000). Present generating costs are relatively high (20 – 40c/kWh), but solar power is proving competitive in niche markets, and has the potential to make substantially higher contributions in the future as costs fall. Photovoltaics can often be deployed at the point of electricity use, such as buildings, and this can give a competitive advantage over power from central power stations to offset higher costs.

Conversion technology continues to improve but efficiencies are still low. Growing markets for PV power generation systems include grid connected urban building integrated systems; off-grid applications for rural locations and developing countries where 2 billion people still have no electricity; and for independent and utility-owned grid-connected power stations. The size of the annual world market has risen from 60MW in 1994 to 130MW in 1997 with anticipated growth to over 1000MW by 2005 (Varadi, 1998). This remains small compared with hydro, wind, and biomass markets. Industrial investment in PV has increased with Shell and BP-Solarex establishing new PV manufacturing facilities with reductions in the manufacturing costs anticipated (AGO, 1999).

Conversion efficiencies of silicon cells continue to improve with 24.4% efficiency obtained in the laboratory for monocrystalline cells and 19.8% for multicrystalline (Green, 1998; Zhao *et al.*, 1998), though commercial monocrystalline-based mod-

Table 3.33b: *Assessment of the annual solar energy potential*

Region	Unused land (Gha)	Assumed for solar energy[d] (Mha)		Solar energy potential[e] (EJ/yr)	
	Available[c]	Min	Max	Min	Max
NAM (North America)	0.5940	5.94	59.4	181.1	7,410
LAM (Latin America and the Caribbean)	0.2567	2.57	25.7	112.6	3,385
AFR (Sub-Saharan Africa)	0.6925	6.93	69.3	371.9	9,528
MEA (Middle East and North Africa)	0.8209	8.21	82.1	412.4	11,060
WEU (Western Europe)	0.0864	0.86	8.6	25.1	914
EEU (Central and Eastern Europe)	0.0142	0.14	1.4	4.5	154
FSU (Newly independent states of the former Soviet Union)	0.7987	7.99	79.9	199.3	8,655
PAO (Pacific OECD)	0.1716	1.72	17.2	72.6	2,263
PAS (Other Pacific Asia)	0.0739	0.74	7.4	41.0	994
CPA (Centrally planned Asia and China)	0.3206	3.21	32.1	115.5	4,135
SAS (South Asia)	0.1038	1.04	10.4	38.8	1,339
World total	3.9331	39.33	39.33	1575.0	49,837
Ratio to the current primary energy consumption (425 EJ/yr[f])	-	-	-	3.7	117
Ratio to the primary energy consumption projected[g] for 2050 (590-1,050 EJ/yr)	-	-	-	2.7 - 1.5	84 - 47
Ratio to the primary energy consumption projected[g] for 2100 (880-1,900 EJ/yr)	-	-	-	1.8 - 0.8	57 - 26

[c] The "other land" category from FAO (1999)

[d] The maximum corresponds to 10% of the unused land; the minimum corresponds to 1% of the unused land

[e] The minimum is calculated as (9) = (2)×(4)×(7)× 315 EJ/a, where numbers in parentheses are column numbers in *Tables 3.33a* and *3.33b*, 315 is a coefficient of unit conversion; the maximum (10) is (3)×(5)×(8)× 315 EJ/yr

[f] Source: IEA (1998b)

[g] Source: IIASA/WEC (1998)

ules are obtaining only 13%-17% efficiency and multicrystalline 12%-14%. Modules currently retail for around US$4,000 – 5,000/kW peak with costs reducing as predicted by the Worldwatch Institute (1998) as a result of manufacturing scale-up and mass production techniques. Recent studies showed a US$660M investment in a single factory producing 400MW (5 million panels) a year would reduce manufacturing costs by 75%. KPMG (1999) and Neij (1997) calculated a US$100 billion investment would be needed to reach an acceptable generating level of US$0.05/kWh.

Thin film technologies are less efficient (6%-8%) but cheaper to produce, and can be incorporated into a range of applications including roof tile structures. Further efficiency improvements are proving difficult, whereas both cadmium telluride and copper indium gallium selenide cells have given 16%-18% efficiencies in the laboratory (Green *et al.*, 1999) and are close to commercial production. New silicon thin film technology using multilayer cells, which combine buried contact technology with new silicon deposition and recrystallization techniques, enables manufacture to be automated. A commercially

viable product now appears to be feasible with an efficiency of around 15% and cost of around US$1500/kW (Green, 1998). Recycling of PV modules is being developed at the pilot scale for both thin film and crystalline silicon modules (Fthenakis *et al.*, 1999).

Advances in inverters (including incorporation into the modules to give AC output) and net metering systems have encouraged marketing of PV panels for grid-connected building integration projects either in government sponsored large scale installations (up to 1MW) or on residential buildings (up to 5kW) (IEA, 1998c; Moomaw *et al.*, 1999b; IEA, 1999b; Schoen *et al.*, 1997). Japan aims to install 400 MW on 70,000 houses by 2000 (Flavin and Dunn, 1997) and 5000MW by 2010. Simple solar home systems with battery storage and designed for use in developing countries are being installed and evaluated in South Africa and elsewhere by Shell International Renewables with funding from the World Bank. Integrated building systems and passive solar design is covered in Section 3.3.4.

A promising low-cost photovoltaic technology is the photo-sensitization of wide-band-gap semiconductors (Burnside *et al.*, 1998). New photosensitizing molecules have been developed in the laboratory, which exhibit an increased spectral response, though at low efficiencies of <1%. Arrays of large synthetic porphyrin molecules, with similar properties to chlorophyll, are being developed for this application (Burrell *et al.*, 1999).

Solar Thermal

In Europe 1 million m² of flat plate solar collectors were installed in 1997, anticipated to rise to 5 million m² by 2005 (ESD, 1996). Combined PV/solar thermal collectors are under development with an anticipated saving in system costs, though these remain high at US$0.18-0.20/kWh at 8% discount rate and 10 year life (Elazari, 1998). High temperature solar thermal power generation systems are being developed to further evaluate technological improvements (Jesch, 1998). The Californian "power tower" pilot project has been successful at the 10 MW scale and is now due to be tested at 30MW with 100MW the ultimate goal (EPRI/DOE, 1997). Dish systems giving concentration ratios up to 2000 and therefore performing at temperatures up to 1,500°C can supply steam directly to a standard turbo-generator (AGO, 1998). Capital costs are projected to fall from US$4,000/kW to US$2,500 by 2030 (Moomaw *et al.*, 1999b) with other estimates much lower (AGO, 1998).

3.8.4.3.5 Geothermal

Geothermal energy is a heat resource used for electricity generation, district heating schemes, processing plants, domestic heat pumps, and greenhouse space heating, but is only "renewable" where the rate of depletion does not exceed the heat replenishment.

The geothermal capacity installed in 20 countries was 7,873 MW$_e$ in 1998: this provided 0.3% (40TWh/yr) of the total world power generation (Barbier, 1999). Geothermal direct heat use was an additional 8,700 MW$_{th}$. This energy resource could be increased by a factor of 10 in the near term with much of the resource being in developing countries such as Indonesia (Nakicenovic *et al.*, 1998).

3.8.4.3.6 Marine Energy

The potential for wave, ocean currents, ocean thermal conversion, and tidal is difficult to quantify but a significant resource exists. For example, resources of ocean currents greater than 2 m/s have been identified, and in Europe alone the best sites could supply 48TWh/yr (JOULE, 1993). Technical developments continue but several proposed schemes have met with economic and environmental barriers. Many prototype systems have been evaluated (Duckers, 1998) but none have yet proved to be commercially viable (Thorpe, 1998).

Several ocean current prototypes of 5 to 50kW capacity have been evaluated with estimated generating costs of around US$0.06-0.11/kWh (5% discount rate) depending on current speed, though these costs are difficult to predict accurately (EECA, 1996). The economics of tidal power schemes remain non-viable, and there have been environmental concerns raised over protecting wetlands and wading birds on tidal mudflats.

3.8.4.4   Technical CO$_2$ Removal and Sequestration

Substantial reductions in emissions of CO$_2$ from fossil fuel combustion for power generation could be achieved by use of technologies for capture and storage of CO$_2$. These technologies have become much better understood during the past few years, so they can now be seriously considered as mitigation options alongside the more well established options, such as the improvements in fossil fuel systems described in Section 3.8.4.1, and the substitutes for fossil fuels discussed in Sections 3.8.4.2 and 3.8.4.3. Strategies for achieving deep reductions in CO$_2$ emissions will be most robust if they involve all three types of mitigation option.

The potential for generation of electricity with capture and storage of CO$_2$ is determined by the availability of resources of fossil fuels plus the capacity for storage of CO$_2$. Fossil fuel resources are described in Section 3.8.2 and published estimates of CO$_2$ storage capacity are discussed below. These show that capacity is not likely to be a major constraint on the application of this technology for reducing CO$_2$ emissions from fossil fuel combustion.

The technology is available now for CO$_2$ separation, for piping CO$_2$ over large distances, and for underground storage. This technology is best suited to dealing with the emissions of large point sources of CO$_2$, such as power plant and energy-intensive industry, rather than small, dispersed sources such as transport and heating. Nevertheless, as is shown below, it could have an important role to play in reducing emissions from all of these sources.

3.8.4.4.1 Technologies for Capture of CO$_2$

CO$_2$ can be captured in power stations, either from the flue gas stream (post-combustion capture) or from the fuel gas in, for example, an integrated gasification combined cycle process (pre-combustion capture). At present, the capture of CO$_2$ from flue gases is done using regenerable amine solvents (Audus, 2000; Williams, 2000). In such processes, the flue gas is scrubbed with the solvent to collect CO$_2$. The solvent is then regenerated by heating it, driving off the CO$_2$, which is then compressed and sent to storage. This technology is already in use for removing CO$_2$ from natural gas, and for separating CO$_2$ from flue gases for use in the food industry.

The concentration of CO$_2$ in power station flue gas is between about 4% (for gas turbines) and 14% (for pulverized-coal-fired plant). These low concentrations mean that large volumes of gas have to be handled and powerful solvents have to be used, resulting in high energy consumption for solvent regeneration. Research and development is needed to reduce the energy consumption for solvent regeneration, solvent degradation rates, and costs. Nevertheless, 80%-90% of the CO_2 in a flue gas stream could be captured by use of such techniques.

In pre-combustion capture processes, coal or oil is reacted with oxygen, and in some cases steam, to give a fuel gas consisting mainly of carbon monoxide and hydrogen. The carbon monoxide is reacted with steam in a catalytic shift converter to give CO_2 and more hydrogen. Similar processes can be used with natural gas but then air may be preferred as the oxidant (Audus *et al.*, 1999). The fuel gas produced contains a high concentration of CO_2, making separation easier, so a physical solvent may be better suited for this separation; the hydrogen can be used in a gas turbine or a fuel cell. Similar technology is already in use industrially for producing hydrogen from natural gas (e.g., for ammonia production). The integrated operation of these technologies for generating electricity whilst capturing CO_2 has no major technical barriers but does need to be demonstrated (Audus *et al.*, 1999).

The concentration of CO_2 in a power station flue gas stream can be increased substantially (to more than 90%) by using oxygen for combustion instead of air (Croiset and Thambimuthu, 1999). Then post-combustion capture of CO_2 is a very easy step, but the temperature of combustion must be moderated by recycling CO_2 from the exhaust, something which has been demonstrated for use with boilers but would require major development for use with gas turbines. Currently, the normal method of oxygen production is by cryogenic air separation, which is an energy intensive process. Development of low-energy oxygen separation processes, using membranes, would be very beneficial.

Other CO_2 capture techniques available or under development include cryogenics, membranes, and adsorption (IEA Greenhouse Gas R&D Programme, 1993).

After the CO_2 is captured, it would be pressurized for transportation to storage, typically to a pressure of 100 bar. CO_2 capture and compression imposes a penalty on thermal efficiency of power generation, which is estimated to be between 8 and 13 percentage points (Audus, 2000). Because of the energy required to capture and compress CO_2, the amount of emissions avoided is less than the amount captured. The cost of CO_2 capture in power stations is estimated to be approximately US$30-50/t CO_2 emissions avoided (US$110-180/tC), equivalent to an increase of about 50% in the cost of electricity generation.

3.8.4.4.2 Transmission and Storage of CO_2

CO_2 can be transported to storage sites using high-pressure pipelines or by ship. Pipelines are used routinely today to transport CO_2 long distances for use in enhanced oil recovery

(Stevens *et al.*, 2000). Although CO_2 is not transported by ship at present, tankers similar to those currently used for liquefied petroleum gas (LPG) could be used for this purpose (Ozaki, 1997).

CO_2 exists in natural underground reservoirs in various parts of the world. Potential sites for storage of captured CO_2 are underground reservoirs, such as depleted oil and gas fields or deep saline reservoirs. CO_2 injected into coal beds may be preferentially absorbed, displacing methane from the coal; sequestration would be achieved providing the coal is never mined. Another possible storage location for captured CO_2 is the deep ocean, but this option is at an earlier stage of development than underground reservoirs; so far only small-scale experiments for preliminary investigation have been carried out (Herzog *et al.*, 2000); the deep ocean is chemically able to dissolve up to 1800GtC (Sato, 1999). An indication of the global capacities of the major storage options is given in *Table 3.34*. The capacities of these reservoirs are subject to substantial uncertainty, as purposeful exploration has only been conducted in some parts of the world so far. Other published estimates of the global capacity for storage in underground aquifers range up to 14,000GtC (Hendriks, 1994). Other methods of CO_2 storage have been suggested but none are competitive with underground storage (Freund, 2000).

Substantial amounts of CO_2 are already being stored in underground reservoirs:

- Nearly 1Mt/yr of CO_2 is being stored in a deep saline reservoir about 800m beneath the bed of the North Sea as part of the Sleipner Vest gas production project (Baklid and Korbol, 1996). This is the first time CO_2 has been stored purely for reasons of climate protection.
- About 33Mt/yr of CO_2 is used at more than 74 enhanced oil recovery (EOR) projects in the USA. Most of this CO_2 is extracted from natural CO_2 reservoirs but some is captured from gas processing plants. Much of this CO_2 remains in the reservoir at the completion of oil production; any CO_2 produced with the oil is separated and reinjected. An example of an EOR scheme which will use anthropogenic CO_2 is the Weyburn project in Canada (Wilson *et al.*, 2000). In this project, 5,000 t/d of CO_2 captured in a coal gasification plant in North Dakota, USA will be piped to the Weyburn field in Saskatchewan, Canada.
- At the Allison unit in New Mexico, USA, (Stevens *et al.*, 1999), over 100,000 tonnes of CO_2 has been injected over a three-year period to enhance production of coal bed methane. The injected CO_2 is sequestered in the coal (providing it is never mined). This is the first example of CO_2-enhanced coal bed methane production.

The cost of CO_2 transport and storage depends greatly on the transport distance and the capacity of the pipeline. The cost of transporting large quantities of CO_2 is approximately US$1-3/t

Table 3.34: *Some natural reservoirs which may be suitable for storage of carbon dioxide*
(Freund, 1998; Turkenburg, 1997)

Reservoir type	Storage option	Global capacity (GtC)
Below ground		
	Disused oil fields	100
	Disused gas fields	400
	Deep saline reservoirs	>1000
	Unminable coal measures	40
Above ground		
	Forestry	1.2 GtC/yr
Ocean		
	Deep ocean	>1000

per 100km (Ormerod, 1994; Doctor *et al.*, 2000). The cost of underground storage, excluding compression and transport, would be approximately US$1-2/t$CO_2$ stored (Ormerod, 1994). The overall cost of transport and storage for a transport distance of 300 km would therefore be about US$8/t$CO_2$ stored, equivalent to about US$10/t of emissions avoided (US$37/tC).

If the CO_2 is used for enhanced oil recovery (EOR) or enhanced coal bed methane production (ECBM), there is a valuable product (oil or methane, respectively) which would help to offset the cost of CO_2 capture and transport. In some EOR or ECBM projects, the net cost of CO_2 capture and storage might be negative. Other ideas for utilizing CO_2 to make valuable products have not proved to be as useful as sequestration measures, because of the amount of energy consumed in the process and the relatively insignificant quantities of CO_2 which would be used.

If no valuable products were produced, the overall cost of CO_2 capture and storage would be about US$40-60/t CO_2 emissions avoided (150-220/tC). As with most new technologies, there is scope to reduce these costs in the future through technical developments and wider application.

3.8.4.4.3 Other Aspects

If CO_2 is to be stored for mitigation purposes, it is important that the retention time is sufficient to avoid any adverse effect on the climate. It is also important to avoid large-scale accidental releases of CO_2. It is expected that these goals will be achievable with underground storage of CO_2 and may be achievable with ocean storage. Oil and gas fields have remained secure for millions of years, so they should be able to retain CO_2 for similar timescales, providing extraction of oil or gas or injection of CO_2 does not disrupt the seal. Deep saline reservoirs are generally less well characterized than oil and gas reservoirs because of their lack of commercial importance to

date. Their ability to contain CO_2 for the necessary timescales is less certain, but research is underway to improve understanding of this aspect (Williams, 2000).

If CO_2 storage were to be used as a basis for emissions trading or to meet national commitments under the UNFCCC, it would be necessary to establish the quantities of CO_2 stored in a verifiable manner. Most verification requirements for geologically stored CO_2 can be achieved with technology available today. Validation of CO_2 storage in the ocean would be more difficult, but it should be possible to verify quantities of CO_2 stored in concentrated deposits on the seabed.

3.8.4.4.4 Applicability of Capture and Storage in Other Industries

The possibility of using CO_2 capture and storage in the manufacturing industry is described in Section 3.5.3.4, including capture of CO_2 during production of hydrogen from fossil fuels. Hydrogen is widely used for ammonia production and in oil refineries, but it can also be used as an energy carrier. Applications would be for small, dispersed, and/or mobile energy users where capture of CO_2 after combustion would not be feasible. Particular examples are in transport e.g., cars and aircraft, and small-scale heat and power production. Production of hydrogen from fossil fuels with CO_2 storage could be an attractive transition strategy to enable the wide-scale introduction of hydrogen as an energy carrier (Turkenburg, 1997; Williams, 1999).

3.8.4.4.5 The Role of CO_2 Capture and Storage in Mitigation of Climate Change

Some CO_2 could be captured from anthropogenic sources and stored underground at little or no overall cost, for example where the CO_2 is already available in concentrated form, such as in natural gas treatment plants or in hydrogen or ammonia manufacture. If the captured CO_2 were to be stored in depleted hydrocarbon reservoirs, such as in enhanced oil recovery schemes, or through enhanced production of coal bed methane, the income produced would also help to offset the costs. Such opportunities are available for early action to combat climate change using technology which is available today.

Substantial quantities of CO_2 from fossil fuel combustion could be captured in the future and sequestered in natural reservoirs (Williams, 2000). Potentially, this approach could achieve deep reductions in emissions of CO_2. Edmonds *et al.* (2000) have considered various possible strategies to achieve stabilization of CO_2 concentrations around 550-750 ppmv. It has been shown that inclusion of the option of capture and storage of CO_2 offers significant reduction in overall cost compared with strategies which do not include this option.

3.8.4.5 *Emissions from Production, Transport, Conversion, and Distribution*

Methane can be released during the production, transport and use of coal, oil, and gas. Various techniques can be used to reduce these emissions, some of which can be captured for use as an energy resource (Williams, 1993; IEA Greenhouse Gas R&D Programme, 1996, 1997; IGU, 1997, 2000; US EPA, 1993, 1999a).

With coal, methane is trapped in coal seams and surrounding strata in varying amounts and is released as a result of mining. Coal mines are ventilated to dilute the methane as it is released to prevent an explosive build up of the gas. The diluted methane is then normally released to the atmosphere. The emissions can be reduced substantially by capturing some of the methane in a more concentrated form in areas of old workings in a mine or by drilling into the coal seams to release the methane prior to mining, then using it as an energy source. Around 50% of emissions from coal mining could be prevented at costs in the range US$1-4/tC$_{eq}$ (IEA Greenhouse Gas R&D Programme, 1996).

With natural gas, leakage of methane occurs at exploration, through transportation to final use. In North America and Europe, the major source is from fugitive emissions, often leaked from above ground installations or old cast iron or steel pipelines that were originally installed for coal- and oil-derived town gas. Vulnerable networks can be replaced with polyethylene pipes. In Russia, the main sources of leakage are from exploration, compressors, pneumatic devices, and fugitive emissions. Techniques are being applied to reduce emissions including replacing seals, increasing compressor efficiencies, and replacing gas-operated pneumatic devices. Around 45% of global emissions from gas could be eliminated and produce a saving of US$5 billion (IEA Greenhouse Gas R&D Programme, 1997). A further 10% of gas emissions could be reduced at costs up to US$108/tC$_{eq}$, and a further 15% at costs up to US$135/tC$_{eq}$. However, the cost of emission reduction for old distribution systems remains very high, and

the reduction potential will be reached mainly as networks are replaced.

Methane and other gases often occur with oil in the ground and are brought to the surface during extraction. If there was no market for the gas, it was normally vented to the air, but since the1960s methane has increasingly been utilized, compressed and reinjected into the oil field to aid oil production or flared rather than vented. Emissions can be reduced typically by 98% by such methods and these are now common. In Nigeria, where venting is still practised, Shell has made a commitment to end continuous venting by 2003 and continuous flaring by 2008, and has started to liquefy the gas for export.

Sulphur hexafluoride, SF_6, used as an insulator in electrical transmission equipment, is covered in Section 3.5.

3.8.5 *Regional Differences*

3.8.5.1 *Privatization and Deregulation of the Electricity Sector*

In many countries, state owned or state regulated electricity supply monopolies have been privatized and broken up to deregulate markets such that companies compete to generate electricity and to supply customers. These moves affect the types of power station favoured. Traditional, large power stations (> 600MW) have had high capital costs and construction periods of 4-7 years, which have led to high interest payments during construction and the need for higher planning margins. Under the new circumstances, the new power generators use higher discount rates, seek lower overall costs, and try to minimize project risks by preferring plants of smaller unit size. They thus favour projects with low capital costs, rapid construction times, use of proven technology, high plant reliability/availability, and low operating costs. CCGTs meet all of these new criteria, and are favoured by generators where gas is available at acceptable costs. This could point the way for the development of new designs for other types of power station, which need to be smaller with modular designs that are largely factory built rather than site built. Economies of scale then come from replication on an assembly line rather than through size (see also Section 6.2.1.3).

Community ownership of distributed renewable energy projects, particularly wind turbines and biogas plants, is becoming common in Denmark (Tranaes, 1997) and more recently in the UK (UK DTI, 1999). The trend towards privately owned distributed power supply systems, either independent or grid connected, is likely to continue as a result of growing public interest in sustainability and technical improvements in controls and asynchronous grid connections.

In countries where privatization of transmission line companies is occurring, there is no longer any commercial rationale to construct and maintain lines only to service a small demand. This has historically often been a social investment by governments and aid agencies. Where grid connections are already in place, it is possible that disconnections may occur in the future where the lines are uneconomic. Then existing residents will have to choose between installing independent domestic-scale systems or establishing community-owned co-operative schemes.

State owned utilities have been able to cross-subsidize otherwise non-competitive projects including nuclear and renewable technologies. Privatization of these utilities requires new methods of supporting technology implementation objectives.

In some cases, electricity tariffs and regulatory systems may need to be amended to include the benefits and costs of embedded generation. This would enable renewable energy projects to be sited on the distribution network at nodes where they would bring most benefit to quality of supply (see Mitchell, 1998, 1999; and Chapter 6). One detrimental impact could be an increase of fossil fuel electricity generation caused by the increased need to operate in load-following mode.

3.8.5.2 Developing Country Issues

In the past there has been little incentive to explore for gas in developing countries unless there was an existing infrastructure to utilize it. The development of CCGT technology now means that, if electricity generation is required, an initial market for the gas can be developed quite rapidly and this market extended to other sectors as the infrastructure is built.

Developing countries have a large need for capital to meet the development of hospitals, schools, and transport and not just for energy in general or electricity in particular. In such circumstances, cheaper power stations are often built at lower efficiency than might otherwise be the case, for example 30%-35% efficiency for an old coal-fired design rather than 40%+ for a modern design. The low price of fuel in some of these countries can also make a cheaper, less efficient design economically more attractive. In India, coal-fired power station design has been standardized at 37.5% efficiency and capacity of 250 and 500MW. Capital costs are US$884/kW whereas a 40% efficiency station would cost around US$977/kW. The coal price is US$25 - US$37/t, depending on location. Even at the higher price, the increased capital costs for the higher efficiency power station outweigh the economic benefits from its lower fuel demand and hence lower emissions.

Technology transfer of advanced power generation technologies including CCGT, nuclear, clean coal, and renewable energy would lead to emission reduction and could be encouraged through the Kyoto mechanisms (see Chapter 6). In addition to limited capital resources that can make advanced technologies unaffordable, many developing countries face skill shortages that can impede the construction and operation of such technology. This is discussed more fully in Chapter 5.

Electricity plants and boilers are sometimes not operated as efficiently as possible in developing countries. In some cases,

incremental investment in such a plant will yield benefits but, more often, it is investment in training the operators that is lacking and that will yield substantial gains. The extension of grids in regions such as India and Africa could allow better use to be made of efficient power stations in order to displace less efficient local units. In India, one trading scheme by three electricity companies resulted in an emissions reduction of 2MtC (Zhou, 1998), and there are similar possibilities in southern and east Africa (Batidzirai and Zhou, 1998). The same study shows that there is a large scope in the subregion for exploiting hydropower, sharing of natural gas resources for power generation, and utilization of wind power along the coastal areas. These measures can displace coal-based generation which currently emits 30–40MtC in southern Africa alone.

An alternative to the extension of grids in developing countries is to increase development of efficiently distributed power generation. This is discussed further in the section below.

3.8.5.3 Distributed Systems

Distributed power comprises small power generation or storage systems located close to the point of use and/or controllable load. Worldwide, these include more than 100MW of existing compressed ignition and natural gas-fired spark ignition engines, small combustion turbines, smaller steam turbines, and renewables. Emerging distributed power technologies include cleaner natural gas or biodiesel engines, microturbines, Stirling engines and fuel cells, small modular biopower and geopower packaged as cogeneration units, and wind, photovoltaics and solar dish engine renewable generation. Increased integration of distributed power with other distributed energy resources could further enhance technology improvement in this sector.

Interest is growing in generating power at point of use using independent or grid-connected systems, often based on renewable energy. These could be developed, owned, and operated by small communities. The European "Campaign for Take-off" target for 100 communities to be supplied by 100% renewable energy and become independent of the grid by 2010 will require a hybrid mix of technologies to be used depending on local resources (Egger, 1999). Local employment opportunities should result and the experience should aid uptake in developing countries.

For small grid-connected embedded generation systems, power supply companies could benefit from improved power quality where the distributed sites are located towards the end of long and inefficient transmission lines (Ackermann *et al.*, 1999). Expensive storage would be avoided where a grid system can provide back-up generation.

3.8.6 *Technological and Economic Potential*

Several studies have attempted to express the costs of power generation technologies on a comparable basis (US DOE/EIA,

2000; Audus, 2000; Freund, 2000; Davison, 2000; Goldemberg, 2000; OECD, 1998b). The OECD data are for power stations that are mainly due for completion in 2000 to 2005 in a wide cross section of countries, and these show that costs can vary considerably between projects, because of national and regional differences and other circumstances. These include the need for additional infrastructure, the trade-off between capital costs and efficiency, the ability to run on baseload, and the cost and availability of fuels. The costs of reducing greenhouse gas emissions will similarly vary both because of variability in the costs of the alternative technology and because of the variability in the costs of the baseline technology. Because of this large variation in local circumstances, the generating costs of studies can rarely be generalized even within the boundaries of one country. Consequently, costs (and mitigation potentials) are highly location dependent. The analysis in this section uses two principle sources of data, the OECD (1998b) data and the US DOE/EIA (2000) data. The latter data are for a single country and may reduce some of the variability in costs seen in multi-country studies.

Tables 3.35a-d are derived from the OECD (1998b) survey which gives data on actual power station projects due to come on stream in 2000 to 2005 from 19 countries including Brazil, China, India, and Russia, together with a few projects for 2006 to 2010 based on more advanced technologies. Data from other sources have been added where necessary and these are identified in the footnote to the tables. The tables present typical costs per kWh and CO_2 emissions of alternative types of generation expected for 2010. *Tables 3.35a* and *3.35b* use a baseline pulverized coal technology for comparative purposes. *Table 3.35a* contains data for Annex I countries (as defined in the UN Framework Convention on Climate Change) in the OECD dataset, and *Table 3.35b* contains data for non-Annex I countries. In addition to coal, the table gives projected costs for gas, nuclear, CO_2 capture and storage, PV and solar thermal, hydro, wind, and biomass. In the baseline, costs and carbon emissions are an average of the coal-fired projects in the OECD database for Annex I/non-Annex I countries respectively, with flue gas desulphurization (FGD) included in all Annex I cases and in around 20% of the non-Annex I cases. Other technologies are then compared to the coal baseline using cost data from the OECD database and other sources. In *Tables 3.35c* and *3.35d*, the baseline technology is assumed to be CCGT burning natural gas, and costs and emissions are similarly calculated for Annex I and non-Annex I countries.

In the tables, the first column of data gives the generation costs in USc/kWh and the emissions of CO_2 in grams of carbon per kWh (gC/kWh) for the baseline technology and fuel, coal, and gas, respectively. The subsequent columns give a range of possible generation options, and the costs and emissions for alternative technologies that could be used to reduce C emissions over the next 20 years and beyond. Additionally, it might be noted that the non-Annex I baseline coal technology is cheaper than that for Annex I countries (both based on the costs of power stations under construction) and that CO_2 emissions

(expressed as gC/kWh) are higher. This reflects the lower efficiencies of power stations currently being built in non-Annex I countries. The costs of reducing greenhouse gas emissions in the mitigation options varies both because of variability in the costs of the alternative technology and because of the variability in the costs of the baseline technology.

Tables 3.35a-d also present estimates of the CO_2 reduction potential in 2010 and 2020 for the alternative mitigation options. Baseline emissions of CO_2 are used, derived from projections of world electricity generation from different energy sources (IEA, 1998b). The IEA projections essentially are enveloped by the range of SRES marker scenarios for the period up to 2020. The IEA projections were used as the baseline because of their shorter time horizon and higher technology resolution. In the tables, it is assumed that a maximum of 20% of new coal baseline capacity could be replaced by either gas or nuclear technologies during 2006 to 2010 and 50% during 2011 to 2020. Similarly, it is assumed that a maximum of 20% of new gas capacity in 2006 to 2010 and 50% in 2011 to 2020 could be displaced by mitigation options. These assumptions would allow a five-year lead-time (from the publication of this report) for decisions on the alternatives to be made and construction to be undertaken. It is assumed that the programme would build up over several years and hence the maximum capacity that could be replaced to 2010 is limited. After 2010 it is assumed that there will be practical reasons why half the new coal capacity could not be displaced. The rate of building gas or nuclear power stations that would be required using these assumptions should not present problems. For nuclear power, the rate of building between 2011 and 2020 would be less than that seen at the peak for constructing new nuclear plants. For gas, the gas turbines are factory made, so no problems should arise from increasing capacity, and less would be required in terms of boilers, steam turbines, and cooling towers than the coal capacity being replaced. For renewables such as wind, photovoltaics (PV) and biomass, maximum penetration rates were derived from the Shell sustainable growth scenario (Shell, 1996) and applied to replace new coal or gas capacities. For wind and PV, these penetration rates imply substantial growth, but are less than what could be achieved if the industries continued to expand at the current rate of 25% per year until 2020. For biomass, most of the fuel would be wood process or forest waste. Some non-food crops would also be used. The introduction of CO_2 capture and storage technology would require similar construction processes as for a conventional power plant. The CO_2 separation facilities would need additional equipment but, in terms of physical construction, involve no more effort than, say, the establishment of a similar scale of biomass gasification plant. CO_2 storage facilities would be constructed using available oil/gas industry technology and this is not seen to be a limiting factor. Storage would be in saline aquifers of depleted oil and gas fields. For CO_2 capture and storage, it is assumed that pilot plants could be operational before 2010, and the mitigation potential is put at 2–10MtC each for coal and gas technologies. It is assumed, arbitrarily, that these would be in Annex I countries. For 2020, the total mitigation potential is put at 40–200MtC, split equally between coal and gas, and between Annex I and non-Annex I countries. Again, this is somewhat arbitrary, but reflects, on the one hand, the potential to move forward with the technology if no major problems are encountered, and, on the other, the potential for more extended pilot schemes. It is assumed, for simplicity, that fuel switching, from coal to gas or vice versa, would not occur in addition to CO_2 capture and storage, although this would be an extra option.

The tables show that the reduction potential in 2020 is substantially higher than in 2010, which follows from the assumptions used and reflects the time taken to take decisions and, especially in the case of renewables and CO_2 capture and storage, to build up manufacturing capacity, to learn from experience, and to reduce costs. The tables show that each of the mitigation technologies can contribute to reducing emissions, with nuclear, if socio-politically desirable, having the greatest potential. Replacement of coal by gas can make a substantial contribution as can CO_2 capture and storage. Each of the renewables can contribute significantly, although the potential contribution of solar power is more limited. The potential reductions within each table are not addable. The alternative mitigation technologies will be competing with each other to displace new coal and gas power stations. On the assumption about the maximum displacement of new coal and gas power stations (20% for 2006 to 2010, 50% for 2011 to 2020), the maximum mitigation that could be achieved would be around 140MtC in 2010 and 660MtC in 2020. These can be compared with estimated and projected global CO_2 emissions from power stations of around 2400MtC in 2000, 3150MtC in 2010 and 4000MtC in 2020 (IEA, 1998b).

In practice, a combination of technologies could be used to displace coal and natural gas fired generation and the choice will often depend on local circumstances. In addition to the description in the tables, oil-fired generation could also be displaced and, on similar assumptions, there is a further mitigation potential of 10MtC by 2010 and 40MtC by 2020. Furthermore, in practice not all of the mitigation options are likely to achieve their potential for a variety of reasons – unforeseen technical difficulties, cost limitations, and socio-political barriers in some countries. The total mitigation potential for all three fossil fuels from power generation, allowing for potential problems, is therefore estimated at about 50-150MtC by 2010 and 350-700MtC by 2020.

In contrast to the OECD data which span a wide range reflecting local circumstances, *Table 3.35e* presents costs for the USA, mainly based on data used in the *Annual Energy Outlook* of the US Energy Information Agency (US DOE/EIA, 2000). By and large, the mitigation costs fall in the range of costs given in *Tables 3.35a-d*. The electricity generating costs are based on national projections of utility prices for coal and natural gas, while capital costs and generating efficiencies are dynamically improving depending on their respective rates of market penetration. The table indicates that once sufficient

capacities have been adopted in the market place, coal-fired integrated gasification combined cycle power stations would have similar costs but lower emissions than the pulverized fuel (pf) power station (because of its higher efficiency). In many places, gas-fired CCGT power stations offer lower cost generation than coal at current gas prices and produce around only half the emissions of CO_2. Data on CO_2 capture and storage have been taken from IEA Greenhouse Gas R & D Programme studies (Audus, 2000; Freund, 2000; Davison, 2000). This could reduce emissions by about 80% with additional costs of around 1.5c/kWh for gas and 3c/kWh for coal pf and 2.5c/kWh for coal IGCC. In the EIA study, nuclear power is more expensive than coal-fired generation, but generally less than coal with carbon capture and storage. Wind turbines can be competitive with conventional coal and gas power generation at wind farm sites with high mean annual speeds. Biomass can also contribute to GHG mitigation, especially where forestry residues are available at very low costs (municipal solid waste even at negative costs). Where biofuel is more costly, either because the in-forest residue material used requires collection and is more expensive or because purpose grown crops are used, or where wind conditions are poorer, the technologies may still be competitive for reducing emissions. Photovoltaics and solar thermal technologies appear expensive against large-scale power generation, but will be increasingly attractive in niche markets or for off-grid generation as costs fall.

Table 3.35e also gives estimated CO_2 emissions and mitigation costs compared to either a coal-fired pf power station or a gas-fired CCGT. For the coal base-case, it is projected that in 2010 under assumptions of improved fossil fuel technologies, an IGCC would offer a small reduction in emissions at positive or negative cost. A gas-fired CCGT has generally negative mitigation costs against a coal-fired pf baseline, reflecting the lower costs of CCGT in the example used. CO_2 capture and storage would enable deep reductions in emissions from coal-fired generation but the cost would be about US$100-150/tC depending on the technology used. Gas-fired CCGT with CO_2 capture and storage appears attractive, but this is principally because switching to CCGT is attractive in itself. Nuclear power mitigation costs are in the range US$50-100/tC when coal is used as the base for comparison. It is uncertain whether there would be sufficient capacity available for wind or biomass to deliver as much electricity as could be produced by fossil fuel-fired plants, but certainly not at the low costs shown in *Table 3.35e*.

If a gas-fired baseline is assumed, most of the mitigation options are found to be more expensive. CO_2 capture and storage appears relatively attractive, achieving deep reductions in emissions at around US$150/tC avoided. Wind, biomass, and nuclear could be attractive options in some circumstances. Other options show higher costs. PV and solar thermal are again expensive mitigation options, and, as noted above, are more suited to niche markets and off-grid generation.

Table 3.35a: Estimated costs of alternative mitigation technologies in the power generation sector compared to baseline coal-fired power stations and potential reductions in carbon emissions to 2010 and 2020 for Annex I countries

Technology	pf+FGD, NO_x, etc	IGCC and Super-critical	CCGT	pf+FGD+ CO_2 capture	CCGT+ CO_2 capture	Nuclear	PV and thermal solar	Hydro	Wind turbines	BIGCC
Energy source	Coal	Coal	Gas	Coal	Gas	Uranium	Solar radiation	Water	Wind	Biofuel
Generating costs (c/kWh)	4.90	3.6-6.0	4.9-6.9	7.9	6.4-8.4	3.9-8.0	8.7-40.0	4.2-7.8	3.0-8	2.8-7.6
Emissions (gC/kWh)	229	190-198	103-122	40	17	0	0	0	0	0
Cost of carbon reduction (US$/tC)	Baseline	-10 to 40	0 to 156	159	71 to 165	-38 to 135	175 to 1400	-31 to 127	-82 to 135	-92 to 117
Reduction potential to 2010 (MtC/yr)	Baseline	13	18	2-10	-	30	2	6	51	9
Reduction potential to 2020 (MtC/yr)	Baseline	55	103	5-50	-	191	20	37	128	77

see notes on page 258

Table 3.35b: Estimated costs of alternative mitigation technologies in the power generation sector compared to baseline coal-fired power stations and potential reductions in carbon emissions to 2010 and 2020 for non-Annex I countries

Technology	pf+FGD, NO_x, etc	IGCC and Super-critical	CCGT	pf+FGD+ CO_2 capture	CCGT+ CO_2 capture	Nuclear	PV and thermal solar	Hydro	Wind turbines	BIGCC
Energy source	Coal	Coal	Gas	Coal	Gas	Uranium	Solar radiation	Water	Wind	Biofuel
Generating costs (c/kWh)	4.45	3.6-6.0	4.45-6.9	7.45	5.95-8.4	3.9-8.0	8.7-40.0	4.2-7.8	3.0-8	2.8-7.6
Emissions (gC/kWh)	260	190-198	103-122	40	17	0	0	0	0	0
Cost of carbon reduction (US$/tC)	Baseline	-10 to 200	0 to 17	136	62 to 163	-20 to 77	164 to 1370	-10 to 129	-56 to 137	-63 to 121
Reduction potential to 2010 (MtC/yr)	Baseline	36	20	0	-	36	0.5	20	12	5
Reduction potential to 2020 (MtC/yr)	Baseline	85	137	5-50	-	220	8	55	45	13

see notes on page 258

Table 3.35c: *Estimated costs of alternative mitigation technologies in the power generation sector compared to gas-fired CCGT power stations and the potential reductions in carbon emissions to 2010 and 2020 for Annex I countries*

Technology	CCGT	pf+FGD+ CO$_2$ capture	CCGT+ CO$_2$ capture	Nuclear	PV and thermal solar	Hydro	Wind turbines	BIGCC
Energy source	Gas	Coal	Gas	Uranium	Solar	Water	Wind radiation	Biofuel
Generation costs (c/kWh)	3.45	7.6-10.6	4.95	3.9-8.0	8.7-40.0	4.2-7.8	3.0-8	2.8-7.6
Emissions (gC/kWh)	108	40	17	0	0	0	0	0
Cost of carbon reduction (US$/tC)	Baseline	610 to 1050	165	46 to 421	500 to 3800	66 to 400	-43 to 92	-60 to 224
Reduction potential to 2010 (MtC/yr)	Baseline	-	2-10	62	0.8	3	23	4
Reduction potential to 2020 (MtC/yr)	Baseline	-	5-50	181	9	18	61	36

see notes on page 258

Table 3.35d: *Estimated costs of alternative mitigation technologies in the power generation sector compared to gas-fired CCGT power stations and the potential reductions in carbon emissions to 2010 and 2020 for non-Annex I countries*

Technology	CCGT	pf+FGD+ CO_2 capture	CCGT+ CO_2 capture	Nuclear	PV and thermal solar	Hydro	Wind turbines	BIGCC
Energy source	Gas	Coal	Gas	Uranium	Solar	Water	Wind radiation	Biofuel
Generation costs (c/kWh)	3.45	6.9-8.7	4.95	3.9-8.0	8.7-40.0	4.2-7.8	3.0-8	2.8-7.6
Emissions (gC/kWh)	108	40	17	0	0	0	0	0
Cost of C reduction (US$/t)	Baseline	507-772	165	46 to 421	500 to 3800	66 to 400	-43 to 92	-60 to 224
Reduction potential to 2010 (MtC/yr)	Baseline	-	0	10	0.2	9	5	1
Reduction potential to 2020 (MtC/yr)	Baseline	-	5-50	70	4	26	21	6

Notes to Tables 3.35a-d

Costs are derived from OECD (1998b) for coal, gas and nuclear. The additional costs of CO_2 capture and storage are derived from IEA Greenhouse Gas R & D Programme (Audus, 2000; Freund, 2000; Davison, 2000). For coal, the costs for CO_2 capture and storage are given for pf power stations. Photovoltaic costs are taken from the World Energy Assessment (Goldemberg, 2000) and are based on today's costs of US$5,000 per peak kilowatt capacity at the high end, and US$1,000 per peak kilowatt capacity projected after 2015 at the low end. Data for hydropower are taken from Goldemberg (2000) and Ishitani and Johansson (1996). Data on wind power are based on today's costs (Walker *et al.*, 1998) and projected future costs are also given in Ishitani and Johansson (1996). Finally, for biofuels, the technology used is BIGCC with capital and operating costs assumed to be similar to those for coal IGCC. Fuel costs are set at US$0-2.8/GJ, based on either wood process and forest residues or specially grown crops (from *Table 3.35e*). A 10% discount rate was used throughout. IEA (1998b) projections of electricity generation were used to estimate the potential CO_2 reductions for each mitigation technology.

For *Tables 3.35a* and *3.35b*, if gas infrastructures are in place and CCGT is a lower cost option than coal, it is assumed that natural gas is the fuel of choice. Hence no negative cost reduction options exist for natural gas to displace coal.

In estimating the maximum CO_2 mitigation each technology could achieve, it was assumed that only newly planned or replacement of end-of-life coal or gas power stations could be displaced – no early retirements of existing plant and equipment were assumed.

As a general rule, increasing the amount of abatement to be achieved will require moving to higher cost options. This will apply particularly to renewables as additional capacity might require moving to sites with less favourable conditions.

pf+FGD: pulverized fuel + flue gas desulphurization.

Table 3.35e: *Estimated costs of coal and gas baselines and alternative mitigation technologies in the US power generation sector*

Technology	pf+FGD, NOx, etc.	IGCC	CCGT	pf+FGD+ CO$_2$ capture	IGCC + CO$_2$ capture	CCGT+ CO$_2$ capture	Nuclear	PV and thermal solar	Wind turbines	Biomass	Biomass
Energy source	Coal	Coal	Gas	Coal	Coal	Gas	Uranium	Solar radiation	Wind	Forestry residues	Energy crops
Generating costs (c/kWh)	3.3-3.7	3.2 – 3.9	2.9-3.4	6.3-6.7	5.7-6.4	4.4-4.9	5.0-6.0	9.0 – 25.0	3.3-5.4	4.0-6.7	6.4-7.5
Emissions (gC/kWh)	247 - 252	190-210	102-129	40	37	17	0	0	0	0	0
Cost of C reduction compared to coal pf (US$/tC)	Baseline	-80 to 168	-53 to 8	141-145	93-148	30-70	52-102	210 – 880	-16 to 85	12-138	107-170
Cost of C reduction compared to gas CCGT (US$/tC)			Baseline	326-613	250-538	134-176	124-304	434 –2167	-8 to 245	47-373	233-450

Notes to *Table 3.35e:* Data derived from US DOE/EIA (2000) except for additional costs of CO$_2$ capture and disposal which are from IEA Greenhouse Gas R & D Programme (Audus, 2000; Freund, 2000; Davison, 2000). Additionally, coal costs of US$1.07/GJ were used for coal-fired power stations; and gas costs of US$3.07/GJ were used for gas-fired power stations (from US DOE/EIA, 1999a). The alternative wind costs come from high and low cases in the US DOE/EIA (2000) study. The costs of biomass-derived electricity used a fuel cost of US$2.8/GJ for purpose grown crops, whereas the fuel costs for biofuel wastes were US$0-2/GJ. A 10% discount rate was used. It should be noted that the costs of renewables given in the table apply to specific conditions of availability of sunlight, wind, or crops. As production from renewables increases, it may be necessary to move to sites with less favourable conditions and higher costs. The large-scale reliance on intermittent renewables may require the construction of back-up capacity or energy storage at additional costs.

3.8.7   *Conclusions*

The section on energy sources indicates that there are many alternative technological ways to reduce GHG emissions, including more efficient power generation from fossil fuels, greater use of renewables or nuclear power, and the capture and disposal of CO_2. There are also opportunities to reduce emissions of methane and other non-CO_2 gases associated with energy supply. In general, this new review reinforces the conclusions reached in the SAR, as discussed in Section 3.8.2.

3.9   Summary and Conclusions

The analysis in this chapter is based upon a review of existing and emerging technologies, and the technological and economic potential that they have for reducing GHG emissions. In many areas, technical progress relevant to GHG emission reduction since the SAR has been significant and faster than anticipated. A broad array of technological options have the combined potential to reduce annual global greenhouse gas emission levels close to or below those of 2000 by 2010 and even lower by 2020.

Estimates of the technical potential, an assessment of the range of potential costs per metric tonne of carbon equivalent (tC_{eq}), and the probability that a technology will be adopted are presented in *Table 3.36* by sector. Specific examples and the estimation methodologies are discussed more fully in the chapter for each sector.

Available estimates of the technological potential to reduce greenhouse gas emissions and its costs suffer from several important limitations:

- There are no consistent estimates of technological and economic potential covering all the major regions of the world;
- Country- and region-specific studies employ different assumptions about the future progress of technologies and other key factors;
- Studies make different assumptions about the difficulty of overcoming barriers to the market penetration of advanced technologies and the willingness of consumers to accept low-carbon technologies;
- Most studies do not describe a range of costs over a domain of carbon reduction levels, and many report average rather than marginal mitigation costs; and
- Social discount rates of 5%-12% are commonly used in studies of the economic potential for specific technologies which are lower than those typically used by individuals and in industry.

A summary of the estimates of the potential for worldwide emission reductions is given in *Table 3.37*. Overall, the total potential for worldwide greenhouse gas emissions reductions resulting from technological developments and their adoption are estimated to amount to 1,900-2,600MtC/yr by 2010[28] and 3,600–5,050MtC/yr by 2020.

In the scenarios that were constructed within the SRES emissions of the six Kyoto Protocol greenhouse gases develop as follows (in MtC_{eq}, rounded numbers):

1990:	9,500
2000:	10,500
2010:	11,500 – 13,800
2020:	12,000 – 15,900

It was not possible to calculate the emission reduction potential of the short-term mitigation options presented in this Chapter on the basis of the SRES scenarios, mainly because of lack of technological detail in the SRES. In order to come to a comprehensive emission reduction estimate, it has been ensured that for all the sectors the estimates are compatible with one of the scenarios, i.e. the B2-Message (standardized) scenario. The emission reductions presented in *Table 3.37* total 14% - 23% of baseline emissions in the year 2010 and to 23% - 42% of baseline emissions in the year 2020.[29] If these percentages also apply to the other scenarios - there is no obvious reason why this would not be the case – it is concluded that *in most situations* the annual global greenhouse gas emission levels can be reduced to a level close to or below those of 2000 by 2010 and even lower by 2020.

The evidence on which this conclusion is based is extensive, but is subject to the limitations outlined above. Therefore, the estimates as presented in the table should be considered to be indicative only. Nevertheless, the main conclusion presented above can be drawn with a high degree of confidence.

Costs of options vary by technology, sector and region (see cost discussion in *Table 3.37*). Based upon the costs in a majority of the studies, approximately half of the potential for emissions reductions cited above for 2010 and 2020 can be achieved at net negative costs (value of energy saved exceeds capital, operating and maintenance costs) using the social discount rates cited. Most of the remainder can be achieved at a cost of less than US$100/$tC_{eq}$.

The overall rate of diffusion of low emission technologies is insufficient to offset the societal trend of increasing consumption of energy-intensive goods and services, which results in increased emissions. Nevertheless, substantial technical progress has been made in many areas, including the market introduction of efficient hybrid engine cars, the demonstration of underground carbon dioxide storage, the rapid advancement of wind turbine design, and the near elimination of N_2O emissions from adipic acid production.

---

[28] For comparison: the total commitment of Annex I countries according to the Kyoto Protocol is estimated to be 500MtC (SRES-scenario B2 as reference).

[29] Some double-counting in the emission reduction estimates occur, especially between electricity saving options and options in the electricity production sector. However, further analysis shows that the effect of double-counting is just noise within the uncertainty range.

Table 3.36: *Estimations of greenhouse gas emission reductions and cost per tonne of carbon equivalent avoided following the anticipated socio-economic potential uptake by 2010 and 2020 of selected energy efficiency and supply technologies, either globally or by region and with varying degrees of uncertainty*

	Region	US$/tC avoided (−400 −200 0 +200)	2010 Potential[a]	2010 Probability[b]	2020 Potential[a]	2020 Probability[b]	References; comments; relevant section of this chapter
Buildings / appliances							
Residential sector	OECD/EIT	*(bar)*	◆◆◆◆	◇◇◇◇◇	◆◆◆◆	◇◇◇◇	Acosta Moreno *et al.*, 1996; Brown *et al.*, 1998; Wang and Smith, 1999
	Dev. cos.	*(bar)*	◆◆◆	◇◇◇	◆◆◆◆	◇◇◇◇◇	
Commercial sector	OECD/EIT	*(bar)*	◆◆◆◆	◇◇◇◇	◆◆◆◆◆	◇◇◇◇◇	
	Dev. cos.	*(bar)*	◆◆◆	◇◇◇	◆◆◆◆	◇◇◇◇	
Transport							
Automobile efficiency improvements	USA	*(bar)*	◆◆◆◆	◇◇◇◇	◆◆◆◆◆	◇◇◇	Interlab. Working Group, 1997; Brown *et al.*, 1998; US DOE/EIA, 1998; ECMT, 1997 (8 countries only); Kashiwagi *et al.*, 1999; Denis and Koopman, 1998; Worrell *et al.*, 1997b
	Europe	*(bar)*	◆◆◆◆	◇◇	◆◆◆◆	◇◇	
	Japan	*(bar)*	◆◆◆◆	◇◇	◆◆◆◆	◇◇	
	Dev. cos.	*(bar)*	◆◆◆◆	◇◇	◆◆◆◆◆	◇◇	
Manufacturing							
CO$_2$ removal – fertilizer; refineries	Global	*(bar)*	◆	◇◇◇◇	◆	◇◇◇◇	*Table 3.21*
Material efficiency improvement	Global	*(bar)*	◆◆◆◆	◇◇◇	◆◆◆◆	◇◇◇	*Table 3.21*
Blended cements	Global	*(bar)*	◆	◇◇◇	◆	◇◇◇	*Table 3.21*
N$_2$O reduction by chem. indus.	Global	*(bar)*	◆	◇◇◇◇	◆	◇◇◇	*Table 3.21*
PFC reduction by Al industry	Global	*(bar)*	◆	◇◇◇	◆	◇◇◇	*Table 3.21*
HFC-23 reduction by chem. industry	Global	*(bar)*	◆◆	◇◇◇	◆◆	◇◇◇	*Table 3.21*
Energy efficient improvements	Global	*(bar)*	◆◆◆◆◆	◇◇◇◇◇	◆◆◆◆◆	◇◇◇◇	*Table 3.19*

(continued)

Table 3.36: continued

	Region	US $/tC avoided (−400 −200 0 +200)	2010 Potential[a]	2010 Probability[b]	2020 Potential[a]	2020 Probability[b]	References; comments; relevant section of this chapter
Agriculture							
Increased uptake of conservation tillage and cropland management	Dev. cos.		◆	◇◇	◆	◇◇	Zhou, 1998; *Table 3.27* / Dick *et al.*, 1998
	Global		◆◆◆	◇◇	◆◆◆◆	◇◇◇	IPCC, 2000
Soil carbon sequestration	Global		◆◆◆	◇◇	◆◆◆◆	◇◇◇	Lal and Bruce, 1999 / *Table 3.27*
Nitrogenous fertilizer management	OECD		◆	◇◇◇	◆	◇◇◇	Kroeze & Mosier, 1999 / *Table 3.27*
	Global		◆	◇◇◇	◆◆◆	◇◇◇◇	OECD, 1999; IPCC, 2000
Enteric methane reduction	OECD		◆◆	◇◇	◆◆	◇◇◇	Kroeze & Mosier, 1999 / *Table 3.27*
	USA		◆	◇◇	◆	◇◇	OECD, 1998 / Reimer & Freund, 1999
	Dev. cos.		◆	◇	◆	◇	Chipato, 1999
Rice paddy irrigation and fertilizers	Global		◆◆◆	◇◇	◆◆◆◆	◇◇◇	Riemer &Freund, 1999 / IPCC, 2000
Wastes							
Landfill methane capture	OECD		◆◆◆◆	◇◇◇	◆◆◆◆	◇◇◇◇	Landfill methane USEPA, 1999
Energy supply							
Nuclear for coal	Global		◆◆◆	◇◇	◆◆◆◆◆	◇◇◇◇	[c]**Totals** – See Section 3.8.6
	Annex I		◆◆	◇◇	◆◆◆	◇◇	*Table 3.35a*
	Non-Annex I		◆◆	◇◇◇	◆◆◆◆◆	◇◇◇	*Table 3.35b*
Nuclear for gas	Annex I		◆◆◆	◇	◆◆◆	◇	*Table 3.35c*
	Non-Annex I		◆	◇	◆◆◆	◇	*Table 3.35d*

(continued)

Table 3.36: continued

Region		US$/tC avoided (−400 −200 0 +200)	2010 Potential[a]	2010 Probability[b]	2020 Potential[a]	2020 Probability[b]	References; comments; relevant section of this chapter
Gas for coal	Annex I		◆	◇◇◇	◆◆◆◆	◇◇◇◇	*Table 3.35a*
	Non-Annex I		◆	◇◇◇◇	◆◆◆◆	◇◇◇◇	*Tables 3.35b*
CO₂ capture from coal	Global		◆	◇◇◇	◆◆	◇◇	*Tables 3.35a + b*
CO₂ capture from gas	Global		◆	◇◇	◆◆	◇◇	*Tables 3.35c + d*
Biomass for coal	Global		◆	◇◇◇◇	◆◆◆	◇◇◇◇	*Tables 3.35a + b* Moore, 1998; Interlab w. gp. 1997
Biomass for gas	Global		◆	◇◇	◆	◇◇◇	*Tables 3.35c + d*
Wind for coal or gas	Global		◆◆	◇◇◇	◆◆◆◆◆	◇◇◇◇	*Tables 3.35a - d* BTM Cons 1999;Greenpeace, 1999
Co-fire coal with 10% biomass	USA		◆	◇◇◇	◆◆	◇◇◇	Sulilatu, 1998
Solar for coal	Annex I		◆	◇◇	◆	◇	*Table 3.35a*
	Non-Annex I		◆	◇◇	◆	◇	*Table 3.35b*
Hydro for coal	Global		◆◆	◇◇	◆◆◆	◇◇	*Tables 3.35a + b*
Hydro for gas	Global		◆	◇◇	◆◆	◇◇	*Tables 3.35c + d*

Notes:

[a] Potential in terms of tonnes of carbon equivalent avoided for the cost range of US$/tC given.

◆ = <20 MtC/yr	◆◆ = 20-50 MtC/yr	◆◆◆ = 50-100MtC/yr	◆◆◆◆ = 100-200MtC/yr	◆◆◆◆◆ = >200 MtC/yr

[b] Probability of realizing this level of potential based on the costs as indicated from the literature.

◇ = Very unlikely   ◇◇ = Unlikely   ◇◇◇ = Possible   ◇◇◇ = Probable   ◇◇◇◇ = Highly probable

[c] Energy supply total mitigation options assumes that not all the potential will be realized for various reasons including competition between the individual technologies as listed below the totals.

Table 3.37: Estimates of potential global greenhouse gas emission reductions in 2010 and in 2020.

Sector		Historic emissions in 1990 (MtC_{eq}/yr)	Historic C_{eq} annual growth rate in 1990-1995 (%)	Potential emission reductions in 2010 (MtC_{eq}/yr)	Potential emission reductions in 2020 (MtC_{eq}/yr)	Net direct costs per tonne of carbon avoided
Buildings[a]	CO_2 only	1650	1.0	700-750	1000-1100	Most reductions are available at negative net direct costs.
Transport	CO_2 only	1080	2.4	100-300	300-700	Most studies indicate net direct costs less than US$25/tC but two suggest net direct costs will exceed US$50/tC.
Industry -energy efficiency -material efficiency	CO_2 only	2300	0.4	300-500 ~200	700-900 ~600	More than half available at net negative direct costs. Costs are uncertain.
Industry	Non-CO_2 gases	170		~100	~100	N_2O emissions reduction costs are US$0-$10/tC_{eq}.
Agriculture[b]	CO_2 only Non-CO_2 gases	210 1250-2800		150-300	350-750	Most reductions will cost between US$0-100/$tC_{eq}$ with limited opportunities for negative net direct cost options.
Waste[b]	CH_4 only	240	1.0	~200	~200	About 75% of the savings as methane recovery from landfills at net negative direct cost; 25% at a cost of US$20/$tC_{eq}$.
Montreal Protocol replacement applications	Non-CO_2 gases	0		~100		About half of reductions due to difference in study baseline and SRES baseline values. Remaining half of the reductions available at net direct costs below US$200/$tC_{eq}$.
Energy supply and conversion[c]	CO_2 only	(1620)	1.5	50-150	350-700	Limited net negative direct cost options exist; many options are available for less than US$100/$tC_{eq}$.
Total		**6,900-8,400[d]**		**1,900-2,600[e]**	**3,600-5,050[e]**	

[a] Buildings include appliances, buildings, and the building shell.

[b] The range for agriculture is mainly caused by large uncertainties about CH_4, N_2O, and soil-related emissions of CO_2. Waste is dominated by landfill methane and the other sectors could be estimated with more precision as they are dominated by fossil CO_2.

[c] Included in sector values above. Reductions include electricity generation options only (fuel switching to gas/nuclear, CO_2 capture and storage, improved power station efficiencies, and renewables).

[d] Total includes all sectors reviewed in Chapter 3 for all six gases. It excludes non-energy related sources of CO_2 (cement production, 160MtC; gas flaring, 60MtC; and land use change, 600-1,400MtC) and energy used for conversion of fuels in the end-use sector totals (630MtC). Note that forestry emissions and their carbon sink mitigation options are not included.

[e] The baseline SRES scenarios (for six gases included in the Kyoto Protocol) project a range of emissions of 11,500-14,000MtC_{eq} for 2010 and of 12,000-16,000MtC_{eq} for 2020. The emissions reduction estimates are most compatible with baseline emissions trends in the SRES-B2 scenario. The potential reductions take into account regular turnover of capital stock. They are not limited to cost-effective options, but exclude options with costs above US$100/$tC_{eq}$ (except for Montreal Protocol gases) or options that will not be adopted through the use of generally accepted policies.

Hundreds of technologies and practices exist to reduce greenhouse gas emissions from the buildings, transport, and industrial sectors. These energy efficiency options are responsible for more than half of the total emission reduction potential of these sectors. Efficiency improvements in material use (including recycling) will also become more important in the longer term.

The energy supply and conversion sector will remain dominated by cheap and abundant fossil fuels but with potential for reduction in emission caused by the shift from coal to natural gas, conversion efficiency improvement of power plants, the adoption of distributedcogeneration plants, and carbon dioxide recovery and sequestration. The continued use of nuclear power plants (including their lifetime extension) and the application of renewable energy sources will avoid emissions from fossil fuel use. Biomass from by-products, wastes, and methane from landfills is a potentially important energy source which can be supplemented by energy crop production where suitable land and water are available. Wind energy and hydropower will also contribute, more so than solar energy because of the latter's relatively high costs.

N_2O and some fluorinated greenhouse gas reductions have already been achieved through major technological advances.

Process changes, improved containment, recovery and recycling, and the use of alternative compounds and technologies have been implemented. Potential for future reductions exists, including process-related emissions from insulated foam and semiconductor production, and by-product emissions from aluminium and HCFC-22. The potential for energy efficiency improvements connected to the use of fluorinated gases is of a similar magnitude to reductions of direct emissions.

Agriculture contributes 20% of total global anthropogenic emissions, but although there are a number of technology mitigation options available, such as soil carbon sequestration, enteric methane control, and conservation tillage, the widely diverse nature of the sector makes capture of emission reductions difficult.

Appropriate policies are required to realize these potentials. Furthermore, on-going research and development is expected to significantly widen the portfolio of technologies to provide emission reduction options. Maintaining these R&D activities together with technology transfer actions will be necessary if the longer term potential as outlined in *Table 3.37* is to be realized. Balancing mitigation activities in the various sectors with other goals such as those related to development, equity, and sustainability is the key to ensuring they are effective.

References

AA (US Aluminum Association), 1998: *The Inert Anode Roadmap.* Washington DC.

AAMA (American Automobile Manufacturers Association), 1998: *World Motor Vehicle Data.* Washington DC.

Ackerman, F., 1997: *Why Do We Recycle? Markets, Values, and Public Policy.* Island Press, Washington DC.

Ackerman, F., 2000: Waste Management and Climate Change. *Local Environment,* 5(2), 223-229.

Ackermann, T., K. Garner, and A. Gardiner, 1999: *Wind Power Generation in Weak Grids – economic optimisation and power quality simulation.* Proceedings, World Renewable Energy Congress, Perth, Australia, February.

Acosta Moreno, R., R. Baron, P. Bohm, W. Chandler, V. Cole, O. Davidson, G. Dutt, E. Haites, H. Ishitani, D. Kruger, M.D. Levine, L. Zhong, L. Michaelis, W. Moomaw, J.R. Moreira, A. Mosier, R. Moss, N. Nakicenovic, L. Price, N.H. Ravindranath, H-H. Rogner, J. Sathaye, P. Shukla, L. Van Wie McGrory, and T. Williams, 1996: Technologies, Policies and Measures for Mitigating Climate Change. In *IPCC Technical Paper 1.* R.T. Watson, M.C. Zinyowera, and R.H. Moss (eds.), Intergovernmental Panel on Climate Change, Geneva.

Adelman, M.A., and M.C. Lynch, 1997: Fixed View of Resources Limits Creates Undue Pessimism. *Oil and Gas Journal,* **April**, 56-60.

Advisory Committee for Energy, 1998: Long Term Energy Supply/Demand Outlook. The Energy Supply and Demand Subcommittee of the Advisory Committee for Energy (an advisory body to the Minister of International Trade and Industry). Advisory Committee for Energy, Japan.

Ag Climate, 1999: Canadian energy companies to pay US farmers for carbon credits. *Wall Street Journal,* 21 October.

AGO (Australian Greenhouse Office), 1998: *Renewable energy showcase projects.* Australian Greenhouse Office, Canberra, Australia, December [http://www.greenhouse.gov.au/renewable/renew3.html].

AGO, 1999: *Renewable energy commercialization program, RECP round 1 grants.* Australian Greenhouse Office, Canberra, Australia. April [http://www.greenhouse.gov.au/renewable/renew3.html].

Ahlgrimm, H-J., 1998: Trace gas emissions from agriculture. In *Sustainable Agriculture for Food, Energy and Industry.* James and James, Ltd., London, pp. 74-78.

Allmaras, R.R., and R.H. Dowdy, 1995: Conservation tillage systems and their adoption in the United States. *Soil and Tillage Research,* 5, 197-222.

American Insurance Association, 1999: *Property-Casualty Insurance and the Climate Change Debate: A Risk Assessment.* American Insurance Association, Washington DC [http://www.aiadc.org/media/press/april/pr41999cas.htm]

An, F., F. Stodolsky, and J.J. Eberhardt, 1999: *Fuel and Emission Impacts of Heavy Hybrid Vehicles.* SAE Technical Paper 99 CPE015, Society of Automotive Engineers, Warrandale, PA.

An, F., F. Stodolsky, A. Vyas, R. Cuenca, and J.J. Eberhardt, 2000: *Scenario Analysis of Hybrid Class 3-7 Heavy Vehicles.* SAE Technical Paper, Society of Automotive Engineers, Warrendale, PA.

Angerer, G., and H. Kalb, 1996: Abatement of Methane Emissions from Landfills - the German Way. *World Resource Review,* **8**(3), 311.

Arent, D., 1998: *Rural Electrification in South Africa - Renewables for Sustainable Village Power: Project Brief.* National Renewable Energy Laboratory, Boulder, CO [http://www.rsvp.nrel.gov].

Asian Development Bank, 1998: *Asia Least-cost Greenhouse Gas Abatement Strategy: India.* Asian Development Bank, Global Environmental Facility, United Nations Development, Manila.

Audus, H., 2000: *Leading Options for the Capture of CO_2 at Power Stations.* Proceedings of the Fifth International Conference on Greenhouse Gas Control Technologies, Cairns, Australia, 13-16 August 2000.

Audus, H., O. Kaastad, and G. Skinner, 1999: *CO_2 capture by precombustion decarbonisation of natural gas.* Proceedings of the 4th International Conference on Greenhouse Gas Control Technologies, 30 August-2 September, Interlaken, Switzerland.

Augustin, J., W. Merbach, J. Kading, W. Schmidt, and G. Schalitz, 1998: Nitrous oxide and methane fluxes of heavily drained and reflooded fen sites in northern Germany. In *Sustainable Agriculture for Food, Energy and Industry.* James and James Ltd., London, pp. 69-73.

Aumonier, S., 1996: The Greenhouse Gas Consequences of Waste Management - Identifying Preferred Options. *Energy Conservation Management,* 37 (6-8), 1117-1122.

Austin, T.C., R.G. Dulla, and T.R. Carlson, 1999: *Alternative and Future Technologies for Reducing Greenhouse Gas Emissions from Road Vehicles.* Sierra Research, Inc., Sacramento, CA.

Austrian Biofuels Institute, 1997: *Biodiesel development status worldwide.* Report to the International Energy Agency, Vienna.

AWEA (American Wind Energy Association), 1998: *World wind industry grew by record amount in 1997.* American Wind Energy Association [http://www.igc.apc.org/awea/news/news9801intl.html].

Baguant, J., and M. Teferra, 1996: *Transport Energy in Africa.* M.R. Bhagavan (ed.), African Energy Policy Network/Zed Books, London.

Bailie, A., B. Sadownik, A. Taylor, M. Nanduri, R. Murphy, J. Nyboer, M. Jaccard, and A. Pape, 1998: *Cost Curve Estimation for Reducing CO_2 Emissions in Canada: An Analysis by Province and Sector.* Prepared for Natural Resources Canada (Energy Research Group, School of Resource and Environmental Management Programme, Simon Fraser University), Vancouver BC.

Baklid A., and R. Korbøl, 1996: *Sleipner Vest CO_2 disposal, CO_2 injection into a shallow underground aquifer.* SPE 36600, Society of Petroleum Engineers, Richardson, TX.

Barbier E., 1999: Geothermal energy - a world overview. *Renewable Energy World,* 2(4), 148-155.

Batidzirai, and P.P. Zhou, 1998: Inventory of GHG Emissions from the Power Sector. In *Options for Greenhouse Gas Mitigation in SADC's Power Sector.* R.S. Maya (ed.), Southern Centre Publication, Harare, Zimbabwe.

Batjes, N.H., 1998: Mitigation of atmospheric CO_2 concentrations by increased carbon sequestration in the soil. *Biological Fertility Soils,* 27, 230-235.

Best, G., 1998: Energizing the food production chain for the attainment of food security. In *Sustainable Agriculture for Food, Energy and Industry.* James and James Ltd., London, pp.1123-1129.

BGR (Bundesanstalt für Geowissenschaften und Rohstoffe), 1998: *Reserven, Ressourcen und Verfügbarkeit von Energierohstoffen.* BGR, Hannover, Germany.

Blok, K., H. Bradke, A. Haworth, and J. Vis, 1999: *Economic-engineering studies for Western-Europe - A Review.* Proceedings of the International Workshop on Technologies to Reduce Greenhouse Gas Emissions, International Energy Agency/US Department of Energy [http://www.iea.org/acti.htm].

Blok, K., W.C. Turkenburg, W. Eichhammer, U. Farinelli, and T.B. Johansson (eds.), 1995: *Overview of Energy RD&D Options for a Sustainable Future.* European Commission, Directorate-General XII, Brussels, Belgium.

Blok, K., D. van Vuuren, A. van Wijk, and L. Hein, 1996: *Policies and Measures to Reduce CO_2 Emissions by Efficiency and Renewables: A Preliminary Study for the Period to 2005.* Department of Science, Technology and Society, Utrecht University, Utrecht, The Netherlands.

Blum, L., R. Denison, and J. Ruston, 1997: A Life-Cycle Approach to Purchasing and Using Environmentally Preferable Paper: A Summary of the Paper Task Force Report. *Journal of Industrial Ecology,* 1(3), 15-46.

BMBF (Bundesministerium für Bildung, Wissenschaft, Forschung und Technologie), 1995: *IKARUS – Instrumente für Klimagas-Reduktionsstrategien.* BMBF, Bonn, Germany.

Bogner, J., K. Spokas, E. Burton, R. Sweeney, and V. Corona, 1995: Landfills as Atmospheric Methane Sources and Sinks. *Chemosphere,* 31(9), 4119-4130.

Bogner, J., M. Meadows, and E. Repa, 1998: New Perspective: Measuring and Modeling of Landfill Methane Emissions. *Waste Age,* 29 (6), 8.

Bose, R.K., 1999a: *Growing Automobiles and Emissions: Mitigation Strategies for Developing Countries.* Tata Energy Research Institute, New Delhi.

Bose, R.K., 1999b: *Engineering-Economic Studies of Energy Technologies to Reduce Carbon Emissions in the Transport Sector.* Presented at the IEA international workshop on Technologies to Reduce Greenhouse Gas Emissions: Engineering- Economic Analyses of Conserved Energy and Carbon, Tata Energy Research Institute, New Delhi.

Bouwman, M., and H. Moll, 1999: *Status quo and Expectations Concerning the Material Composition of Road Vehicles and Consequences for Energy Use.* University of Groningen, Groningen, The Netherlands.

BP (British Petroleum), 1997: *BP Statistical Review of World Energy 1997.* BP, London [http://www.bp.com].

BP, 1999: BP *Statistical Review of World Energy 1999.* British Petroleum, London [http://www.bp.com].

Branscomb, L., M. Lewis, and J.H. Keller (eds.), 1997: *Investing in Innovation: Creating a Research and Innovation Policy That Works.* MIT Press, Cambridge, MA.

Branscome, M., and W.N. Irving, 1999: *HFC-23 emissions from HCFC-22 production.* Background paper prepared for the IPCC/OECD/IEA Programme on National Greenhouse Gas Inventories, Washington, DC, January 1999.

Brännström-Norberg, B.M., U. Dethlefsen, R. Johansson, C. Setterwall, and S. Tunbrant, 1996: *Life-Cycle Assessment for Vattenfall's Electricity Generation.* Summary Report, 1996, pp. 12-20, Stockholm.

Brezet, J.C., and C. van Hemel, 1997: *Ecodesign; a Promising Approach to Sustainable Production and Consumption.* UNEP, Paris.

Brown, M.A., M.D. Levine, J.P. Romm, A.H. Rosenfeld, and J.G. Koomey, 1998: Engineering-Economic Studies of Energy Technologies to Reduce Greenhouse Gas Emissions: Opportunities and Challenges. *Annual Review of Energy and Environment*, **23**, 287-385.

BTM Consult, 1998: *Global wind energy initiative.* Report presented at COP4, Buenos Aires, November, 1998, http://home4.inet.tele.dk/btm-cwind.

BTM Consult, 1999: *International wind energy development – world market update 1998 + forecast 1999-2003.* BTM Consult ApS, Denmark. April.

Burnside, S.D., K. Brooks, A.J. Mcevoy, and M. Gratzel, 1998: Molecular photovoltaics and nanocrystalline junctions. *Chimia, * **52** (10), 557-560.

Burrell, A.K., W.M. Campbell, D.K. Officer, D.C.W. Reid, S.M. Scott, and K.Y. Wild, 1999: *Porphyrin dyes for new photovoltaic technology.* In Proceedings World Renewable Energy Congress, Murdoch University, Perth, Australia, pp. 403-406.

Bystroem, S., and L. Loennstedt, 1997: Paper recycling: environmental and economic impact. *Resources, Conservation and Recycling, * **21**, 109-127.

CAE (Centre for Advanced Engineering), 1996: *Energy efficiency – a guide to current and emerging technologies, Vol. 2. Industry and Primary production: Transport.* Centre for Advanced Engineering, University of Canterbury, Christchurch, New Zealand, 488 pp.

Cahyono A., C.L. Malik, A. Nurrohim, R.T.M. Sutamihardja, M. Nur Hidajat, I.B. Santoso, Amirrusidi, and A. Suwarto, 1997: Mitigation of Carbon Dioxide from Indonesia's Energy System. *Applied Energy*, **56** (3/4), 253-263.

Campbell, C.J., 1997: Better Understanding Urged for Rapidly Depleting Reserves. *Oil and Gas Journal*, **April**, 51-54.

Causey, W., 2000: Reduce SF_6 - and save money. *Electrical World*, **September/October**, 50-52.

CEC, 1999: *Evaluation of biomass ' to' ethanol.* Californian Energy Commission Report P500-99-011, August.

CEEZ (Centre for Energy, Environment and Engineering), 1999: *Climate Change Mitigation in Southern Africa.* Zambia Country Study, UNEP Centre Riso National Laboratory, Riso, Denmark.

China Statistical Yearbook, various years. State Statistical Bureau, Beijing.

Chipato, C., 1999: *Ruminant methane in Zimbabwe.* Ruminant Workshop, Global Livestock Group, Washington DC, http://www.theglg.com/newsletter1662.

CISAC (Committee on International Security and Arms Control of the National Academy of Sciences), 1995: *Management and Disposition of Excess Weapons Plutonium: Reactor-Related Options.* National Academy of Sciences Press, Washington DC.

Clark, J., 1999: *Fleet-Based LCA: Comparative CO_2 Emission Burden of Aluminium and Steel Fleets.* Materials Systems Laboratory, Massachusetts Institute of Technology, Cambridge, MA.

Cleveland, C.J., and M. Ruth, 1999: Indicators of Dematerialization and the Materials Intensity of Use. *Journal of Industrial Ecology*, **2**, 15-50.

Cogen, 1997: *European Cogeneration Review.* A Study co-financed by the SAVE Programme of the European Commission, Brussels, Cogen Europe, 182 pp.

Cook, E., 1995: *Lifetime Commitments: Why Climate Policy-Makers Can't Afford to Overlook Fully Fluorinated Compounds.* World Resources Institute, Washington DC.

Croiset, E., and K.V. Thambimuthu, 1999: *Coal combustion with flue gas recirculation for CO_2 recovery.* Proceedings of the 4th International Conference on Greenhouse Gas Control Technologies, 30 August - 2 September, Interlaken, Switzerland.

D'Apote, S.L., 1998: IEA Biomass Energy Analysis and Projections. In *Biomass Energy: Data, Analysis and Trends.* Proceedings of OECD/IEA Conference in Paris, March 23-24, OECD/IEA, Paris.

Daniëls, B.W., and H.C. Moll, 1998: *The Base Metal Industry: Technological Descriptions of Processes and Production Routes - Status Quo and Prospects.* Interfacultaire Vakgroep Energie en Milieukunde, Groningen University, Groningen, The Netherlands.

Das, S., 2000: *Life Cycle Impacts of Aluminum Body-in-White Automotive Material.* Report to the Office of Transportation Technologies, U.S. Department of Energy, Oak Ridge National Laboratory, Oak Ridge, TE.

Davison, J., 2000: *Comparison of CO_2 abatement by use of renewable energy and CO_2 capture and storage.* Proceedings of the Fifth International Conference on Greenhouse Gas Control Technologies, Cairns, Australia, 13-16 August 2000.

De Almeida, A., and P. Fonseca, 1999: Carbon Savings Potential of Energy-Efficient Motor Technologies in Central and Eastern Europe. Paper presented at the International Energy Agency 'International Workshop on Technologies to Reduce Greenhouse Gas Emissions: Energy-Engineering Analysis of Conserved Energy and Carbon', 5-7 May, Washington DC.

De Beer, J.G., 1998: *Potential for Industrial Energy-efficiency Improvement in the Long Term.* Ph.D. Thesis, Utrecht University, Utrecht, The Netherlands.

De Beer, J.G., M.T. van Wees, E. Worrell, and K. Blok, 1994: ICARUS-3 - *The Potential of Energy Efficiency Improvement in the Netherlands up to 2000 and 2015.* Department. of Science, Technology and Society, Report no. 94013, Utrecht University, Utrecht, The Netherlands.

De Beer, J.G., E. Worrell, and K. Blok, 1996: Sectoral Potentials for Energy Efficiency Improvement in the Netherlands. *International Journal of Global Energy Issues,* **8**, 476-491.

De Beer, J.G., E. Worrell, and K. Blok, 1998: Future technologies for energy efficient iron and steel making. *Annual Review of Energy and Environment,* **23**, 123-205.

DeCicco, J.M., and M. Ross, 1993: *An Updated Assessment of the Near-Term Potential for Improving Automotive Fuel Economy.* American Council for an Energy-Efficient Economy, Washington DC.

DeCicco, J., and J. Mark, 1998: Meeting the Energy and Climate Challenge for Transportation in the United States. *Energy Policy,* **26**(5), 395-412.

Deering, A., and J.P. Thornton, 1998: *Solar Technology and the Insurance Industry: Issues and Applications.* NREL/MP 520-25866, National Renewable Energy Laboratory, Golden, CO, 80401-3393, 14 pp.

Dehoust, G., H. Stahl, P. Gebhard, S. Gaertner, D. Bunke, W. Jenseit, and R. Esposito, 1998: *Systemvergleich unterschiedlicher Verfahren der Restabfallbehandlung für die Stadt Münster.*Oeko-Institut e.V., Darmstadt, Germany.

Denis, C., and G.J. Koopman, 1998: *EUCARS: A Partial Equilibrium Model of EUropean CAR Emissions (Version 3.0).* II/341/98-EN, Directorate General II, European Commission, August.

Derpsch, R., 1998: Historical review of no-tillage cultivation of crops. *FAO International Workshop, Conservation Tillage for Sustainable Agriculture.* J. Benites *et al.* (eds.), pp. 205-218 [http://www.fao.org/waicent/faoinfo/agricult/ags/AGSE/MENU.htm]

Dick, W.A., R.L. Blevins, W.W. Frye, S.E. Peters, D.R. Christenson, F.J. Pierce, and M.L. Vitosh, 1998: Impacts of agricultural management practices on C sequestration in forest-derived soils of the eastern Corn Belt. *Soil & Tillage Research*, **47**, 235-244.

Doctor, R.D., J.C. Molburg, and N.F. Brockmeier, 2000: *Transporting Carbon Dioxide Recovered from Fossil-Energy Cycles.* Proceedings of the Fifth International Conference on Greenhouse Gas Control Technologies, Cairns, Australia, 13-16 August 2000.

Dornburg, V., and A. Faaij, 2000: *System analysis of biomass energy system efficiencies and economics in relation to scale.* Proceedings of the 1st World Conference and Exhibition on Biomass for Energy and Industry, CARMEN, Sevilla.

Duckers, L., 1998: Power from the waves. *Renewable Energy World*, **1**(3), 52-57.

EC (European Commission), 1995: *Externalities of Energy.* ExternE Report, Rep. EUR-16522, EC, Brussels.

EC, 1999: *Communication and Fourth Report from the Commission on Present Situation and Prospects for Radioactive Waste Management in the European Union.* COM(98)799, Brussels.

ECMT (European Conference of Ministers of Transport), 1995: *Urban Travel and Sustainable Development.* OECD, Paris.

ECMT, 1997: *CO$_2$ Emissions from Transport.* OECD, Paris.

ECMT, 1998: *Statistical Trends in Transport 1965-1994.* OECD Publications Service, Paris.

Edelmann, W., and K. Schleiss, 1999: *Energetic and Economic Comparison of Treating Biogenic Wastes by Digesting, Composting, or Incinerating* (http://www.biogas.ch/arbi/ecobalan.htm)

Edmonds, J.A., P. Freund, and J.J. Dooley, 2000: *The Role of Carbon Management Technologies in Addressing Atmospheric Stabilization of Greenhouse Gases.* Proceedings Fifth International Conference on Greenhouse Gas Control Technologies, Cairns, Australia, 13-16 August 2000.

EECA (Energy Efficiency and Conservation Authority), 1996: *New and emerging renewable energy opportunities in New Zealand.* EECA, Wellington, New Zealand, 266 pp.

Egger, C. 1999: *100 Communities, 100% Renewable Energy.* European Federation of Regional Energy and Environment Agencies [http://www.fedarene.org]

Eiszele, D.R., 2000: Western Power Corporation - a quarter of a century of renewable technology innovation. In *Proceedings World Renewable Energy Congress VI Perth, Australia*, A.A.M. Sayigh (ed.), Pergamon, London, pp. 25-32.

El Bassam, N., 1996: *Renewable energy – potential energy crops for Europe and the Mediterranean region.* FAO Regional Office for Europe, REU Technical series 46, Rome.

El Bassam, N., M. Graef, and K. Jakob, 1998: Sustainable energy supply for communities from biomass. *Sustainable Agriculture for Food, Energy and Industry*, James and James, Ltd., London, pp. 837-843.

Elam N., C. Ekstrom, and A. Ostman, 1994: *Methanol och etanol ur trädrå-vara – Huvudrapport.* ISSN 1100-5130.

Elazari, A., 1998: A multi-solar system combining PV and solar thermal. *Renewable Energy World*, **1**(3), 71-72.

Elberling, L., and R. Bourne, 1996: ACT2 Project Results: Maximizing Residential Energy Efficiency. In *Proceedings of the 1996 ACEEE Summer Study on Energy Efficiency in Buildings*, American Council for an Energy-Efficient Economy, Washington DC.

El-Fadel, M., N. Angelos, J. Findikakis, and O. Leckie, 1998: Estimating and Enhancing Methane Yield from Municipal Solid Waste. *Hazardous Waste & Hazardous Materials, *13(3), 309.

EPRI (Electric Power Research Institute), 1997: *Solid Oxide Fuel Cells.* EPRI, Palo Alto, CA [http://www.epri.com/srd/sofc]

EPRI/DOE, 1997: *Renewable energy technology characterizations.* Electric Power Research Institute and US Department of Energy report EPRI TR-109496, December.

ESD, 1996: *Energy for the future -meeting the challenge: TERES II (The European Renewable Energy Study).* Report and CD ROM prepared for Directorate General XVII – Energy, European Commission by ESD Ltd, Corsham, Wiltshire, UK.

ETSU (Energy Technology Support Unit), 1994: *An Appraisal of UK Energy Research, Development, Demonstration and Dissemination.* HMSO, London.

ETSU, 1996: *Energy and Carbon Dioxide Supply Curves for UK Manufacturing Industry.* ETSU, Oxfordshire, UK.

EWEA (European Wind Energy Association), 1999: *Wind Energy – the Facts.* Study prepared by the European Wind Energy Association for the Directorate-General XVII - Energy, European Commission.

Faaij, A., M. Hekkert, E. Worrell, and A. van Wijk, 1998: Optimization of the final waste treatment system in the Netherlands. *Resources Conservation and Recycling*, **22,** 47-82.

FAO (Food and Agriculture Organization)**,** 1995: *Future energy requirements for Africa's agriculture.* Food and Agriculture Organization Report, Rome, 96 pp.

FAO, 1997: *Regional study on wood energy today and tomorrow in Asia.* FAO Field Document no. 50, Bangkok.

FAO, 1999: *Statistical Databases on the Internet.* Food and Agriculture Organization of the United Nations, Rome.

Farla, J.C.M., C.A. Hendriks, and K. Blok, 1995: Carbon Dioxide Recovery from Industrial Processes. *Climatic Change*, **29,** 439-461.

Farla, J.C.M., 2000: *Physical Indicators of Energy Efficiency.* PhD Thesis (Chapter 6), Utrecht University, Utrecht, The Netherlands.

Farrell, A., D.W. Keith, and J.J. Corbett, 2000: *True Zero-Emission Vehicles: Fuel Cells and Carbon Management in the Transportation Sector.* Conference Proceedings of the 21st Annual North American Conference, International Association for Energy Economics, Cleveland, OH [http://www.usaee.org]

Federal Energy Technology Center: 1997: *Fuel Cell Overview.* U.S. Department of Energy/EIA.

Finnveden, G., and E. Thomas, 1998: Life-cyle assessment as a decision-support tool – the case of recycling versus incineration of paper. *Resources, Conservation and Recycling*, **24,** 235-256.

Fischer G., and G.K. Heilig, 1998: *Population Momentum and the Demand on Land and Water Resources.* Report IIASA-RR-98-1, International Institute for Applied Systems Analysis (IIASA), Laxenburg, Austria.

Flavin C., and S. Dunn, 1997: *Status report on renewable energy technologies and policies in the United States and Europe.* Report for NEDO (New Energy and Industrial Technology Development Organisation, Japan), Worldwatch Institute, Washington DC.

Franke, M., F. McDougall, and F. Sher, 1999: *Integrated Waste Management in Europe – an Analysis of 11 Case Studies.* Presented at the UN-China: Mayors Seminar on Municipal Solid Waste Management and Landfill Gas Utilization, Nanjung, China, March 1999.

Freund, P., 1999: The IEA Greenhouse Gas R&D Programme: International Initiative to Combat Climate Change. In *Proceedings of the 4th International Conference on Greenhouse Gas Control Technologies*, B. Eliasson, P. Riemer, A. Wokaun (eds.), 30 August-2 September 1998, Elsevier Science, Amsterdam (http://www.ieagreen.org.uk/pfghgt4b.htm).

Freund, P., 2000: *Progress in understanding the potential role of CO$_2$ storage.* Proceedings Fifth International Conference on Greenhouse Gas Control Technologies, Cairns, Australia, 13-16 August 2000.

Frolking K., 1998: Comparisons of N$_2$O emissions from soils at temperate agricultural sites; simulations of year round measurement by four models. *Nutrient Cycling in Agroecosystems*, **52,** 77-105.

Fthenakis, V.M., K. Zweibel, and P.D. Moskowitz, 1999: *Report on the Workshop Photovoltaics and the Environment 1998.* Brookhaven National Laboratories, Keystone, CO, 23-24 July 1998.

Gaj, H., and M. Sadowski, 1997: Climate Change Mitigation: Case Studies from Poland. Prepared for the US Environmental Protection Agency, Pacific Northwest National Laboratory, Washington, DC.

Geller, H., G. de Martino Jannuzzi, R. Schaeffer, and M. Tiomno Tolmasquin, 1998: The efficient use of electricity in Brazil: progress and opportunities. *Energy Policy*, **26,** 859-872.

Gerini, A., and X. Montagne, 1997: Automotive Direct Injection Diesel Sensitivity to Diesel Fuel Characteristics. SAE Technical Paper 972963. In *Combustion & Emissions in Diesel Engines.* SP-1299, Society of Automotive Engineers, Warrendale, PA, pp. 45-56.

Gielen, D.J., 1997: *Building materials and CO$_2$ – Western European emission reduction strategies.* Report ECN-C—97-065, Petten, The Netherlands.

Gielen, D.J., 1999: *Materialising Dematerialisation.* PhD Thesis, Delft University of Technology, Delft, The Netherlands.

Gielen, D.J., A.J.M. Bos, M.A.P.C. de Feber, and T. Gerlagh, 1999: Reduction de l'emission de gaz a effet de serre en agriculture et forestiere (greenhouse gas emission reduction in agriculture and forestry: a western European systems engineering perspective). *Comptes rendus de lácademie d'agriculture de France,* **85**(6), 344-359.

Gipe P., 1998: *Overview of worldwide wind generation.* Proceedings 36th Annual conference of the Australian and New Zealand Solar Energy Society, Christchurch, New Zealand, November, pp. 631 – 636.

Goldemberg, J., 1998: Leapfrog Energy Technologies. *Energy Policy,* **26**(10), 729-741.

Goldemberg, J. (ed.), 2000: *World Energy Assessment: Energy and the Challenge of Sustainability.* United Nations Development Programme (UNDP), United Nations Department of Economic and Social Affairs (UN-DESA), and World Energy Council (WEC), New York, NY.

Grassi, G., 1998: Modern bioenergy in the European Union. *Sustainable Agriculture for Food, Energy and Industry,* James and James Ltd., London, pp. 829–836.

Green, M.A., 1998: *Solar electricity: from the laboratory to the rooftops of the world.* Photovoltaics Special Research Centre, University of New South Wales, Sydney.

Green, M.A., K. Emery, K. Buecher, D.L. King and S. Igari, 1999: Solar Cell Efficiency Tables (Version 14). *Progress In Photovoltaics: Research and Applications,* **7**(4), 321-326.

Greene, D.L., 1998: Why CAFE Worked. *Energy Policy,* **26**(8), 595–614.

Greene, D.L., 1999: *An Assessment of Energy and Environmental Issues Related to the Use of Gas-to-Liquid Fuels in Transportation,* ORNL/TM-1999/258. Oak Ridge National Laboratory, Oak Ridge, TN, November.

Greene, D.L., and J.M DeCicco, 1999: *Engineering-Economic Analyses of Automotive Fuel Economy Potential in the United States.* International Energy Agency, Paris.

Greenpeace, 1999: *Wind Force 10: a blueprint to achieve 10% of the world's electricity from wind power by 2020.* Greenpeace and European Wind Energy Association Report, Greenpeace, Amsterdam, The Netherlands.

Gregory K., and H.-H. Rogner, 1998: Energy Resources and Conversion Technologies for the 21st Century. *Mitigation and Adaptation Strategies for Global Change,* **3**, 171-229.

Grohnheit, P.E., 1999: *Energy Policy Responces to the Climate Change Challenge: The Consistency of European CHP, Renewables and Energy Efficiency Policies.* Risoe-R-1147 (EN), Risoe National Laboratory, Denmark, 148pp.

Grubb, M.J., and N.I. Meyer, 1993: Wind Energy: Resources, Systems and Regional Strategies. In *Renewable Energy: Sources for Fuels and Electricity.* T.B. Johansson, H. Kelly, A.K.N. Reddy, and R.H. Williams (eds.), Island Press, Washington DC.

Gruppelaar H., J.L. Kloosterman, and R.J.M. Konings, 1998: *Advanced technologies for the reduction of nuclear waste.* ECN-R—008, Netherlands Energy Research Foundation ECN, Petten, The Netherlands.

Gupta, S., K. Mohan, P. Rajkumar, and A. Kansal, 1998: Solid Waste management in India: Options and opportunities. *Resources, Conservation and Recycling,* **24**, 137-154.

Hairston, D., D.G. Robinson, P.E. James, White, and A.J. Callier, 1997: Aerobic versus Anaerobic Wastewater Treatment. *Chemical Engineering,* **104**(4), 102.

Hall, D.O., and K.K. Rao, 1994: *Photosynthesis, 5th Edition.* Cambridge University Press, Cambridge.

Hall, D.O., and F. Rosillo-Calle, 1998a: *Biomass Resources Other than Wood.* World Energy Council, London.

Hall, D.O., and F. Rosillo-Calle, 1998b: *The role of bioenergy in developing countries.* In Proceedings of the 10th European conference on Biomass for Energy and Industry, C.A.R.M.E.N., Würzburg, pp. 52-55.

Hall, D.O., and Scrase J.I., 1998: Will biomass be the environmentally friendly fuel of the future? *Biomass and Bioenergy,* **15**, 357-367.

Hansen, K.H., I. Angelidaki, and B.K. Ahring, 1998: Ammonia Control Aids Anaerobic Digestion. *Waste Treatment Technology News,* **13**, 5.

Harnisch, J., 2000: Atmospheric perfluorocarbons: sources and concentrations. In *Non-CO_2 Greenhouse Gases - Scientific Understanding, Control and Implementation.* J. van Ham, A.P.M. Baede, L.A. Meyer, and R. Ybema (eds.), Kluwer Academic Publishers, Dordrecht.

Harnisch, J., and C.A. Hendriks, 2000. *Economic Evaluation of Emission Reductions of HFCs, PFCs and SF$_6$ in Europe.* Ecofys Energy and Environment, Cologne, Germany.

Harnisch, J., I.S. Wing, H.D. Jacoby, and R.G. Prinn, 1998: *Primary Aluminum Production: Climate Policy, Emissions and Costs.* Report no. 44, MIT Joint Program on the Science and Policy of Global Change, Cambridge, MA.

Hatfield, C.B., 1997: Oil Back on the Global Agenda. *Nature,* **387**, 121.

Hawken, P., A. Lovins, and H. Lovins, 1999: *Natural Capitalism.* Little Brown and Company, Boston, MA.

Heavenrich, R.M., and K.H. Hellman, 1999: *Light-Duty Automotive Technology and Fuel- Economy Trends Through 1999.* EPA420-R-99-018, Air and Radiation, US Environmental Protection Agency, Ann Arbor, MI.

Hekkert, M.P., L.A.J. Joosten, E. Worrell, and W. Turkenburg, 2000: Reduction of CO_2 emissions by management of material and product use, the case of primary packaging. *Resources, Conservation and Recycling,* **29**, 33-64.

Henderson, S., 1999: *Load Factors.* Project Director, Airline Industry Analysis, Boeing Commercial Airplanes, Seattle, WA, 19 April 1999, (personal communciation).

Hendriks, C.A., 1994: *Carbon Dioxide Removal from Coal-Fired Plants.* Ph.D thesis, Department of Science, Technology, and Society, Utrecht University, The Netherlands.

Hendriks, C.A., D. de Jager, and K. Blok, 1998: *Emission reduction potential and costs for methane and nitrous oxide in the EU-15.* Interim report M714 for the DGXI, European Commission, Ecofys, Utrecht, The Netherlands.

Hernandez, G., G. Brohard, and E. Kolderup, 1998: Evaluation Results from an Energy-Efficient New Office Construction Project. In *Proceedings of the 1998 ACEEE Summer Study on Energy Efficiency in Buildings.* American Council for an Energy-Efficient Economy, Washington DC.

Herring, H. 1996: *Standby Power consumption in brown goods: A case study of hi-fi equipment.* EERU Report No. 73, The Open University, Milton Keynes, UK.

Herzog, H., E. Adams, M. Akai, G. Alendal, L. Golmen, P. Haugan, S. Masuda, R. Matear, S. Masutani, T. Ohsumi, and C.S. Wong, 2000: Update on the international experiment on CO_2 ocean sequestration. In *Proceedings of the Fifth International Conference on Greenhouse Gas Control Technologies.* Cairns, Australia, 13-16 August 2000.

Hoeschele, M., D. Springer, and J. Kelly, 1996: Implementation and Operation of an 'Integrated Design' Desert House. In *Proceedings of the 1996 ACEEE Summer Study on Energy Efficiency in Buildings.* American Council for an Energy-Efficient Economy, Washington DC.

Huglen, R., and H. K van de, 1994: Global Considerations of Aluminium Electrolysis on Energy and the Environment. In *Light Metals.* U. Mannweiler (ed.), The Minerals, Metals and Materials Society, Warrendale, PA, pp. 373-380.

IAEA (International Atomic Energy Agency), 2000a: *Energy, Electricity, and Nuclear Power Estimates for the Period Up to 2020.* IAEA-Reference Data Series No.1, July 2000 Edition, Vienna.

IAEA 2000b: *Nuclear Power Reactors in the World.* AEA-Reference Data Series No.2, April 2000 Edition, Vienna.

IAI (International Aluminium Institute [formerly the International Primary Aluminium Institute]), 2000: *Anode Effect Survey 1994-1997 and Perfluorocarbon Compounds Emissions Survey 1990-1997.* IAI, London.

IEA (International Energy Agency), 1995: *World Energy Outlook,* 1995 Edition. IEA/OECD, Paris.

IEA, 1997: *Energy Statistics and Balances of Non-OECD countries 1994-1995.* IEA, Paris.

IEA, 1997a: *Energy Balances of OECD Countries, 1960-1995.* IEA/OECD, Paris.

IEA, 1997b: *Energy Balances of Non-OECD Countries, 1960-1995.* IEA/OECD, Paris.

IEA, 1997c: *CO_2 Emissions from Fuel Combustion: Annex II Countries, 1960-1995; Non-Annex II Countries, 1971-1997.* IEA/OECD, Paris.

IEA, 1997d: *Indicators of Energy Use and Efficiency: Understanding the Link Between Energy and Human Activity.* IEA/OECD. Paris.

IEA, 1998a: *Nuclear Power: Sustainability, Climate Change and Competition.* IEA/OECD, Paris.

IEA, 1998b: *World Energy Outlook - 1998 Update.* IEA/OECD, Paris.

IEA, 1998c: 1 million solar roofs for the USA. *PV Power: Newsletter of the IEA PVPS programme*, **9**, 8.

IEA, 1999a: *Engineering-Economic Analyses of Conserved Energy and Carbon.* International Workshop on Technologies to Reduce Greenhouse Gas Emissions, Washington DC, 5-7 May 1999, IEA, Paris.

IEA, 1999b: Building integrated PV plant. *PV Power: Newsletter of the IEA PVPS Programme*, **11**, 4.

IEA, 1999c: *Total Final Consumption by Sector.* http://www.iea.org/stats/files/keystats/p_0303.html, June 3, 1999.

IEA, 1999d: *Energy Balances of Non-OECD Countries: 1996-1997.* IEA/OECD, Paris.

IEA, 1999e: CO_2 *Emissions from Fuel Combustion: 1971-1997*, 1999 Edition. IEA/OECD, Paris.

IEA, 2000: *Greenhouse Gases from Major Industrial Sources – IV The Aluminium Industry.* Report Number PH3/23, Prepared by ICF Consulting, April.

IEA Bioenergy, 1998a: The role of bioenergy in greenhouse gas mitigation, Position Paper Task 25. *IEA Bioenergy*, **T25**(2) [http://www.forestresearch.cri.nz/ieabioenergy/home.htm]

IEA Bioenergy, 1998b: Towards a standard methodology for greenhouse gas balances of bioenergy systems in comparison with fossil fuel energy systems. *Biomass and Bioenergy*, **13**(6), 359-375.

IEA Bioenergy, 1999: *Greenhouse Gas Balances of Bioenergy Systems, Task 25.* Proceedings of the Workshop: Bioenergy for mitigation of CO_2 emissions: the power, transportation, and industrial sectors, 27-30 September, Gatlinburg, TN [http://www.joanneum.ac.at/iea-bioenergy-task25]

IEA Greenhouse Gas R&D Programme, 1993: *Capture of carbon dioxide from power stations.* IEA Greenhouse Gas R&D Programme, Cheltenham, UK.

IEA Greenhouse Gas R&D Programme, 1996: *Methane Emission from Coal Mining.* IEA Greenhouse Gas R&D Programme, Stoke Orchard, Cheltenham, UK.

IEA Greenhouse Gas R&D Programme, 1997: *Methane Emission from the Oil and Gas Industry.* IEA Greenhouse Gas R&D Programme, Stoke Orchard, Cheltenham, UK.

IEA Greenhouse Gas R&D Programme, 1999: *The Reduction of Greenhouse Gas Emissions from the Cement Industry.* IEA Greenhouse Gas R&D Programme, CRE Group, Cheltenham, UK.

IEA Greenhouse Gas R&D Programme, 2000a: *Greenhouse Gases from Major Industrial Sources – III, Iron and Steel Production.* Report Number PH3/30, Prepared by Ecofys, The Netherlands, IEA Greenhouse Gas R&D Programme, CRE Group, Cheltenham, UK.

IEA Greenhouse Gas R&D Programme, 2000b: *Greenhouse Gases from Major Industrial Sources – IV The Aluminium Industry.* Report Number PH3/23. Prepared by ICF Consulting (USA), IEA Greenhouse Gas R&D Programme, CRE Group, Cheltenham, UK.

IGU (International Gas Union), 1997: *Gas and the Environment – Methane Emissions.* Report of IGU Task Force I, 20th World Gas Conference Proceedings, Copenhagen.

IGU, 2000: *Emisson of Methane from the Gas Industry.* 21st World Gas Conference Proceedings, Nice.

IIASA/WEC, 1998: *Global Energy Perspectives.* N. Nakicenovic, A. Grübler, and A. McDonald (eds.), International Institute for Applied Systems Analysis (IIASA) and World Energy Council (WEC), Cambridge University Press, Cambridge, UK.

IIED (International Institute for Environment and Development), 1996: Towards a Sustainable Paper Cycle. IIED, York, UK [http://www.iied.org/bookshop/pubs/x136.htm]

IMA (International Magnesium Association), 1998: *Recommended Practices for the Conservation of Sulfur Hexafluoride in Magnesium Melting Operations.* Technical Committee Report, McLean, VA.

INFORSE, 1999: *International Network for Sustainable Energy Sustainable Energy News,* No. 17 (May), Copenhagen.

Interlaboratory Working Group, 1997: *Scenarios of U.S. Carbon Reductions: Potential Impacts of Energy Technologies by 2010 and Beyond.* Lawrence Berkeley National Laboratory, Berkeley, CA, and Oak Ridge National Laboratory, Oak Ridge,TN (LBNL-40533 and ORNL-444, resp.).

International Water Power & Dam Construction, 1997: *Water Power & Dam Construction Yearbook* [http://www.uvigo.es/webs/servicios/biblioteca]

IPAI (International Primary Aluminium Institute), 1996: *Anode effect and PFC emission survey 1990-1993.* IPAI, London.

IPAI, 2000: *Perfluorocarbon compounds emissions survey 1990-1997.* IPAI, London.

IPCC (Intergovernmental Panel on Climate Change), 1996: Climate Change 1995: Impacts, Adaptations and Mitigation of Climate Change: Scientific-Technical Analyes, Cambridge University Press, Cambridge, UK.

IPCC, 2000: *Land Use, Land Use Change and Forestry.* R.T. Watson, I.R. Noble, B. Bolin, N.H. Ravindranath, D.J. Verardo, and D.J. Dokken (eds.), A special report of the IPCC, Cambridge University Press, Cambridge, UK, 377 pp.

Irving, J., 1999: McNeil Plant Manager, Burlington, Vermont (personal communication).

Ishitani, H., 2000: *Drive Cycle Dependence of Hybrid Vehicle Greenhouse Gas Emissions.*

Ishitani, H., and T.B. Johansson, 1996: Energy Supply Mitigation Options. In *Climate Change 1995: Impacts, Adaptations and Mitigation of Climate Change: Scientific-Technical Analysis.* R.T. Watson, M.C. Zinyowera, and R.J. Moss (eds.), Contribution of Working Group II to the Second Assessment Report of the Intergovernmental Panel on Climate Change, Cambridge University Press, Cambridge, pp. 585-647.

Ishitani, H., Y. Baba and O. Kobayashi, 2000: Well to wheel energy efficiency of fuel cell electric vehicles by various fuels. Japan Society of Energy and Resources, *Energy and Resources*, **21**(5), 417-425.

Ishitani, H., Y. Baba, and R. Matsuhashi, 2000: *Evaluation of Energy Efficiency of a Commercial HEV, Prius at City Driving.* Department of Geosystems Engineering, University of Tokyo, Japan.

Ivanhoe, L.F., and G.G. Leckie, 1993: Global Oil, Gas Fields, Sizes Tallied, Analyzed. *Oil and Gas Journal*, **15 February**, 87-91.

Jaccard, M., 1996: *Industrial Energy End-Use Analysis & Conservation Potential in Six Major Industries in Canada.* MK Jaccard and Associates and Willis Energy Services Ltd., Vancouver.

Jansen, H., and C. Denis, 1999: A Welfare Cost Assessment of Various Policy Measures to Reduce Pollutant Emissions from Passenger Road Vehicles. *Transportation Research-D*, **4D**(6), 379–396.

Japan Resources Association, 1994: *Life cycle energy in domestic life.* Anhorume Co Ltd., Japan.

Jesch, L., 1998: Solar thermal power. *Renewable Energy World*, **1**(2), 52-53.

Jochem, E.K., 1999: *Policy Scenarios 2005 and 2020 – Using Results of Bottom-Up Models for Policy Design in Germany.* Presented at the International Energy Agency International Workshop on Technologies to Reduce Greenhouse Gas Emissions: Energy-Engineering Analysis of Conserved Energy and Carbon, 5-7 May 1999, Washington DC.

Jochem, E., and H. Bradke, 1996: *Energie-effizienz, Strukturwandel und Produktionsentwicklung der deutschen Industrie.* Fraunhofer Institut Systemtechnik und Innovationsforschung, Karlsruhe, Germany.

Joule, 1993: *Marine currents energy extraction: resource assessment.* Final report, European Union JOULE contract JOU2-CT93-0355.

Kadak, A.C., 1999: *The Politically Correct Reactor.* Nuclear Engineering Department, Massachusetts Institute of Technology (MIT), Cambridge, MA.

Kainuma, M., Y. Matsuoka, T. Morita, and G. Hibino, 1999: Development of an End-Use Model for Analyzing Policy Options to reduce Greenhouse Gas Emissions. *IEEE-SMCC* Part C, **29**.

Kalkum B., Z. Korenyl, and J. Caspar, 1993: *Utilization of District Heating for Cooling Purposes. An Absorption Chilling Project in Mannheim, Germany.* Mannheimer Versorgungs- und Verkehrsgesellschaft mbH, Germany, 21 pp.

Karlsson, G., C. Ekstrom, and L. Liinanki, 1998: *The Development of a Biomass IGCC for Power and Heat Production.* Vattenfall Utveckling AB, Stockholm.

Kashiwagi, T., J. Bruggink, P.-N. Giraud, P. Khanna, and W.R. Moomaw, 1996: Industry. In *Climate Change 1995 – Impacts, Adaptations and Mitigation of Climate Change: Scientific-Technical Analysis.* R.T. Watson, M.C. Zinyowera, and R.H. Moss (eds.), Cambridge University Press.

Kashiwagi, T., B.B. Saha, D. Bonilla, and A. Akisawa, 1999: Energy Efficiency and Structural Change for Sustainable Development and CO_2 Mitigation. Contribution to the IPCC Expert Meeting "Costing Methodologies", Dept. of Mechanical Systems Engineering, Tokyo University of Agriculture and Technology, Tokyo, June.

Kerr, T., 1998: Developing Landfill Gas Projects Without Getting Credit. *World Wastes,* **5,** 41-50.

Kimmerle, F.M., and G. Potvin, 1997: Measured versus Calculated Reduction of the PFC Emissions from Prebaked Hall Héroult Cells. In *Light Metals 1997.* R. Huglen (ed.), The Minerals, Metals and Materials Society, Warrendale, PA, pp. 165-171.

Korbitz, W., 1998: Biodiesel – from the field to the fast lane. *Renewable Energy World,* **1**(3), 32-37.

Korhonen, J., M. Wihersaari, and I. Savolainen, 1999. Industrial ecology of a regional energy supply system - the case of Jyväskylä. *Greener Management International,* **26**, 57-67.

Kossina, I., and G. Zehetner, 1998: *Energetische Verwendung von Abfällen in Industrieanlagen. Rechtliche und konzeptionelle Bedingungen für die Republik Österreich.* Umweltbundesamt Oesterreich, Vienna (http://www.ubavie.gv.at).

KPMG, 1999: *Solar energy – from perennial promise to competitive alternative.* Project number 562, KPMG Bureau voor Economische Argumentatie, Hoofddorp, Netherlands, 61 pp.

Krackeler, T., L. Schipper, and O. Sezgen, 1998: Carbon dioxide emissions in OECD service sectors: the critical role of electricity. *Energy Policy* **26**(15), 1137-1152. See also Krackeler, T., L. Schipper, and O. Sezgen, 1998: *The Dynamics of Service Sector Carbon Dioxide Emissions and the Critical Role of Electricity Use: A Comparative Analysis of 13 OECD Countries from 1973-1995.* Berkeley, CA, Lawrence Berkeley National Laboratory, LBNL-41882.

Kramer, K.J., H.C. Moll, and S. Nonhebel, 1999: Total greenhouse gas emissions related to the Dutch crop production system. *Agriculture, Ecosystems and Environment,* **72**(1), 9-16.

Krewitt, W., T. Heck, A. Truckenmueller, and R. Friedrich, 1999: Environmental damage costs from fossil electricity generation in Germany and Europe. *Energy Policy,* **27,** 173-183.

Kroeze C., and A. Mosier, 1999: New estimates for emissions of nitrous oxides. In *Second International Symposium on Non-CO_2 Greenhouse Gases (8-10 September 1999).* J. van Ham, A.P.M. Baede, L.A. Meyer and R. Ybema (eds.), Kluwer Academic Publishers, Dordrecht, The Netherlands.

Krohn S., 1997: *Danish wind turbines, an industrial success story.* Danish Wind Turbine Manufacturers Association [http://www.windpower.dk/articles/success.htm]

Kupitz, J., and J. Cleveland, 1999: *Overview of global development of advanced nuclear power plants, and the role of the IAEA.* International Atomic Energy Agency, Vienna.

Laherrere, J., 1994: Published Figures and Political Reserves, *World Oil,* January 1994, p. 33.

Lako, P., 1996: *Economie van nieuwe concepten van de modulaire Hoge Temperatuur Reactor (Economics of New High Temperature Reactor Concepts).* Netherlands Energy Research Foundation ECN, ECN-C—96-043, Petten, The Netherlands.

Lal, R., and J.P. Bruce, 1999: The potential of world cropland soils to sequester C and mitigate the greenhouse effect. *Environmental Science and Policy,* **2**(2), 177- 186.

Landfill, 1998: Landfill Management Foster Use of Landfill Gas Throughout Renewables Program, SWANA Says. *Solid Waste Report,* **16,** April 29.

Larson, E.D., and H. Jin, 1999*: A preliminary assessment of biomass conversion to Fischer-Tropsch cooking fuels for rural China.* Proceedings of the 4th Biomass Conference of the Americas, Elsevier Science Ltd., Oakland, UK.

Larsson S., G. Melin, and H. Rosenqvist, 1998: *Commercial harvest of willow wood chips in Sweden.* Proceedings of the 10th European Conference on Biomass for Energy and Industry, C.A.R.M.E.N., Würzburg, pp. 200-203.

Lehtilä, A., I. Savolainen, and S. Tuhkanen, 1997 *: Indicators of CO_2 emissions and energy efficiency. -comparison of Finland with other countries.* Technical Research Centre of Finland, Espoo, Publication 328, 80 pp.

Lenstra, J., 1999: The ICARUS Database and Dutch Climate Policy. Presented at the International Energy Agency International Workshop on Technologies to Reduce Greenhouse Gas Emissions: Energy-Engineering Analysis of Conserved Energy and Carbon, 5-7 May, 1999, Washington DC.

Levine, M.D., H. Akbari, J. Busch, G. Dutt, K. Hogan, P. Komor, S. Meyers, H. Tsuchiya, G. Henderson, L. Price, K.R. Smith, and L. Siwei, 1996a: Mitigation Options for Human Settlements. In *Climate Change 1995: Impacts, Adaptations and Mitigation of Cimate Change: Scientific-Technical Analyses.* R.T. Watson, M.C. Zinyowera, and R.H. Ross (eds.), Contribution of Working Group II to the Second Assessment Report of the Intergovernmental Panel on Climate Change, Cambridge University Press, London.

Levine, M.D., L. Price, and N. Martin, 1996b. Mitigation Options for Carbon Dioxide Emissions from Buildings. *Energy Policy,* 24(10/11), 937-949.

Lewis, J.S., and R.W. Niedzwiecki, 1999: Aircraft Technology and Its Relation to Emissions. In *Aviation and the Global Atmosphere.* Intergovernmental Panel on Climate Change, Oxford University Press, Oxford.

Li, C.S., 1995: Impact of agricultural practices on soil C storage and N2O emissions in 6 states in the US. In *Advances in Soil Science - Soil Management and Greenhouse Effect.* R. Lal, J. Kimble, E. Levine, and B.A. Steward (eds.), Lewis Publishers, Boca Raton, FL, pp. 101-112.

Li, C.S., 1998: *The DNDC Model.* Available as a CD ROM from the Institute for the Study of Earth, Oceans and Space, University of New Hampshire, Durham, NH 03824, USA.

Li, C.S., 1999: The challenges of modeling nitrous oxide emissions, In *Reducing nitrous oxide emissions from agroecosystems.* R. Desjardins, J. Keng, and K. Haugen-Kozyra (eds.), International N_2O Workshop, Banff, Alberta, Canada, March 3-5.

Li, C.S., V. Narayanan, and R. Harriss, 1996: Model estimates of nitrous oxide emissions from agricultural lands in the United States. *Global Biogeochemical Cycles,* **10**, 297-306.

Li, C.S., Y.H. Zhuang, M.Q. Cao, P.M. Crill, Z.H. Dai, S. Frolking, B. Moore III, W. Salas, W.Z. Song, and X.K. Wang, 1999: Developing a national inventory of N_2O emissions from arable lands in China using a process-based agro-ecosystem model. *Nutrient Cycles in Agro-ecosystems.*

Ligon, P., 1999: Sustainable Solid Waste Management in Developing Countries. Tellus Institute, Boston, MA (unpublished).

Liu, F., M. Ross, and S. Wang, 1995: Energy efficiency of China's cement industry. *Energy - The International Journal,* **20**, 669-681.

Lovins, A.B., 1996: Negawatts: Twelve transitions, eight improvements and one distraction. *Energy Policy,* 24(4), 331-343.

Luhs, W., and W. Friedt, 1998: Recent developments in industrial rapeseed breeding. In *Sustainable Agriculture for Food, Energy and Industry.* James and James, Ltd., London, pp. 156-164.

Lundberg, H., M. Morris, and E. Rensfelt, 1998: Biomass gasification for energy production. In *The World Directory of Renewable Energy,* James and James, Ltd., London, pp. 75-82.

Lurgi GmbH, 1989: *Gasification of Lignite and Wood in the Lurgi Circulating Fluidized-Bed Gasifier.* Report prepared for the Electric Power Research Institute, EPRI GS-6436.

Maiss, M., and C.A.M. Brenninkmeijer, 1998: Atmospheric SF_6: Trends, Sources and Prospects. *Environmental Science and Technology,* **32**, pp. 3077-3086

Mann, M.K., and P.L. Spath, 1997: *Life cycle assessment of a biomass gasification combined-cycle system.* National Renewable Energy Laboratory report for the US Department of Energy, NREL/TP-430-23076, 94 pp. [http://www.eren.doe.gov/biopower/life_cycle.html]

March Consulting Group, 1998*: Opportunities to Minimize Emissions of Hydrofluorocarbons (HFCs) from the European Union.* March Consulting Group, Manchester, UK, 1998.

Mark, J., and C. Morey, 1999: *Diesel Passenger Vehicles and the Environment.* Union of Concerned Scientists, Berkeley, CA.

Marks, J., R. Roberts, V. Bakshi, and E.Dolin, 2000: *Perfluorocarbon Generation during Primary Aluminum Production, Light Metals 2000.* Proceedings of the 2000 TMS Annual Meeting and Exhibition, The Minerals, Metals and Materials Society, Warrendale, PA, pp. 365-372.

Martin, B., P. Aakko, D. Beckman, N. Del Giacomo, and F. Giavazzi, 1997: Influence of Future Fuel Formulations on Diesel Engine Emissions – A Joint European Study, SAE Technical Paper 972966. In *Combusion & Emissions in Diesel Engines*. SP-1299, Society of Automotive Engineers, Warrendale, PA, pp. 97—109

Martin, N., N. Anglani, D. Eistein, M. Khrushch, E. Worrell, and L.K. Price, 1999: *Opportunities to Improve Energy Efficiency and Reduce Greenhouse Gas Emissions Reduction in the U.S. Pulp and Paper Industry.* Lawrence Berkeley National Laboratory, LBNL-46141, Berkeley, CA.

Martin, N., E. Worrell, and L.K. Price, 1999: *Energy Efficiency and Carbon Dioxide Emissions Reduction Opportunities in the U.S. Cement Industry.* Lawrence Berkeley National Laboratory, LBNL-44812, Berkeley, CA.

Martin, N., E. Worrell, L. Price, M. Ruth, N. Elliott, N., A. Shipley, and J. Thorne, 2000: *Emerging Energy-Efficient Industrial Technologies.* American Council for an Energy-Efficient Technology, Washington DC, and Lawrence Berkeley National Laboratory, Berkeley, CA.

Mauthe, G., B.M. Ptyor, L. Niemeyer, R. Probst, J. Poblotzki, H.D. Morrison, P. Bolin, P. O'Connell, and J. Henriot, 1997: *SF_6 recycling guide: Reuse of SF_6 gas in electrical power equipment and final disposal.* CIGRE WG 23, 10 Task Force 01, CIGRE-ELECTRA 173, Paris.

McCombie, C., 1999a: *Multinational repositories: a win-win disposal strategy.* Proceedings of the ENS Topseal '99 Conference, Antwerp, Belgium, October.

McCombie, C., 1999b: *A prospective global solution for the disposal of unwanted nuclear materials.* ICEM Conference, Nagoya, Japan.

McCombie, C., G. Butler, M. Kurzeme, D. Pentz, J. Voss, and P. Winter, 1999: *The Pangea International Repository: A technical overview.* Proceedings of the WM99 Conference, Tucson, March 1999, Tucson, AZ.

Meier, A., and W. Huber. 1997. *Results from the investigation on leaking electricity in the USA.* First International Conference on Energy Efficiency in Household Appliances, November 10-12, Florence, Italy.

Meier, A., W. Huber, and K. Rosen, 1998: *Reducing Leaking Electricity to 1 Watt.* Proceedings of the 1998 ACEEE Summer Study on Energy Efficiency in Buildings, Washington DC, American Council for an Energy-Efficient Economy and Berkeley, CA.

Meier, A., L. Rainer, and S. Greenberg, 1992: Miscellaneous electrical energy use in homes. *Energy,* **17,** 509.

Mendoza, Y., O. Masera, and P. Macias, 1991: Long-term energy scenarios for Mexico. *Energy Policy,* December, pp. 962-969.

Metz, B., O. Davidson, J-W. Martens, S.N.M. van Rooijen, and L. Van Wie McGrory (eds.), 2000: *Methodological and Technological Issues in Technology Transfer.* A special report of IPCC Working Group III, Cambridge University Press, Cambridge, UK.

Michaelis, L., 1997: CO_2 Emissions from Road Vehicles. Working Paper 1, Annex I, Expert Group on the United Nations Framework Convention on Climate Change, Organisation for Economic Co-operation and Development, Paris.

Michaelis, L., D. Bleviss, J-P. Orfeuil, R. Pischinger, J. Crayston, O. Davidson, T. Kram, N. Nakicenovic, and L. Schipper, 1996: Mitigation Options in the Transportation Sector. In *Climate Change 1995: Impacts, Adaptation and Mitigation of Climate Change: Scientific-Technical Analyses.* Intergovernmental Panel on Climate Change, Cambridge University Press, Cambridge, UK.

Miller, I., J. Black, C. McCombie, D. Pentz, and P. Zuidema, 1999: *High-isolation sites for radioactive waste disposal: a fresh look at the challenge of locating safe sites for radioactive repositories.* Proceedings of the WM99 Conference, March 1999, Tucson, AZ.

Mills, E., 1998: The Coming Storm - Global Warming and Risk Management. *Risk Management,* May, pp. 20-27.

Mills, E., 1999:*The Insurance and Risk Management Industries: New Players in the Delivery of Energy-Efficient Products and Services.* Proceedings of the ECEEE 1999 Summer Study, European Council for an Energy-Efficient Economy, May 31-June 4, 1999, Mandelieu, France, and for the United Nations Environment Programme's 4th International Conference of the Insurance Industry Initiative, Natural Capital at Risk: Sharing Practical Experiences from the Insurance and Investment Industries, 10-11 July 1999, Oslo, Norway.

Mills, E., and A. Rosenfeld, 1996: Consumer Non-Energy Benefits as a Motivation for Making Energy-Efficiency Improvements. *Energy -The International Journal,* **21**(7/8), 707-720. (Also in Proceedings of the 1994 ACEEE Summer Study of Energy Efficiency in Buildings, pp. 4.201-4.213.)

Minato, K., 1998: *Road Transportation and Global Environmental Problems.* Japan Automobile Research Institute, Ibaraki, Japan.

Mitchell, C., 1998: *Renewable energy in the UK.* Report for the Council for the Protection of Rural England, SP Research Unit, Sussex University, UK.

Mitchell, C., 1999: *The value of renewable electricity.* EC Report.

MITI, Energy and Resource Agency, Energy Conservation Division, 1999: *Pledge and Review: Japan's Target to Save Energy, 56 Million kl by FY 2010.* Proceedings of APEC/EWG Meetings, New Zealand, November 18-19.

MITI/ANRE, Japanese Government, 1999: *Energy in Japan—Facts and Figures.* February.

MNES (Ministry of Non-conventional Energy Sources), 1998: *Annual Report.* Ministry of Non-conventional Energy Sources, India.

Molinder, O., 1997: Study on miscellaneous standby power consumption of household equipment. Prepared for the European Union under contract 4.1031/E96-008 (EU-DG XVII, June 25), Brussels, Belgium.

Moomaw, W., K. Ramakrishna, K. Gallagher, and T. Freid, 1999a: The Kyoto Protocol: A Blueprint for Sustainability. *Journal of Environment and Development,* **8,** 82-90.

Moomaw, W., A. Serchuk, G. Unruh, J. Sawin, and F. Sverrison, 1999b: Renewable energy in a carbon limited world. *Advances in Solar Energy 3.* D.Y. Goswami (ed.), American Solar Energy Society, Boulder. CO, pp. 68-137.

Moore, T., 1998: Electrification and global sustainability. *EPRI Journal,* **January/February,** pp. 43-52.

Moreira, J.R., and J. Goldemberg, 1999: The Alcohol Program. *Energy Policy,* **27,** 229-245.

Moreira, J.R., and G. Moreira, 1998: Energy Conservation in Brazil. Prepared for the Minister of Science and Technology, Brasilia, Brazil.

Mosier, A.R., J.M. Duxbury, J.R. Freney, O. Heinemeyer, K. Minami, and D.E. Johnson, 1998a: Mitigating agricultural emissions of methane. *Climate Change,* **40,** 39-84.

Mosier, A.R., C. Kroeze, C. Nevison, O. Oenema, S. Seitzinger and O. van Cleemput, 1998b: Closing the global N2O budget: nitrous oxide emissions through the agricultural nitrogen cycle. *Nutrient Cycling in Agroecosystems,* **52,** 225–248.

Nadel, S.M., D. Fridley, J. Sinton, Y. Zhirong, and L. Hong, 1997: *Energy Efficiency Opportunities in the Chinese Building Sector.* American Council for an Energy-Efficient Economy,Washington DC.

Nadel, S., L. Rainer, M. Shepard, M. Suozzo, and J. Thorne, 1998: *Emerging Energy-Saving Technologies and Practices for the Buildings Sector.* American Council for an Energy-Efficient Economy, Washington DC.

Nakagami, H., A. Tanaka, and C. Murakoshi, 1997: *Standby Electricity Consumption in Japanese Houses.* Jyukanko Research Institute, Japan.

Nakicenovic, N., M. Aman, and G. Fischer, 1998: *Global energy supply and demand and their environmental effects.* Mimeo, International Institute for Applied Systems Analysis, Laxenburg, Austria.

Nakicenovic, N., A. Gruebler, H. Ishitani, T. Johansson, G. Marland, J.R. Moreira, and H-H. Rogner, 1996: Energy Primer. In *Climate Change 1995, Impacts, Adaptation and Mitigation of Climate Change: Scientific-Technical Analyses.* R.T. Watson, M.C. Zinyowera, and R.H. Moss (eds.), Intergovernmental Panel on Climate Change, Cambridge University Press, Cambridge, UK.

Nakicenovic, N., A. Grubler, and A. McDonald, 1998: *Global energy perspectives.* Cambridge University Press, Cambridge.

Nakicenovic, N., J. Alcamo, G. David, B. de Vries, J. Fenhann, S. Gaffin, K. Gregory, A. Grubler, T.Y. Jung, T. Kram, E.L. La Rovere, L. Michaelis, S. Mori, T. Morita, W. Pepper, H. Pitcher, L. Price, K. Riahi, A. Roehrl, H-H. Rogner, A. Sandkovski, M. Schlesinger, P. Shukla, S. Smith, R. Swart, S. van Rooijen, N. Victor, and D. Zhou, 2000: *Special Report on Emission Scenarios.* A Special Report of Working Group II of the Intergovernmental Panel on Climate Change, Cambridge University Press, Cambridge, UK.

National Laboratory Directors, 1997: *Technology Opportunities to Reduce U.S. Greenhouse Gas Emissions*. Prepared for the U.S. Department of Energy. Oak Ridge National Laboratory, Oak Ridge, TN, http://www.ornl.gov/climate_change

NEA (Nuclear Energy Agency), 1999: *Progress towards geological disposal of radwaste. Where do we stand? An international assessment*. Nuclear Energy Agency of the Organsation for Economic Co-Operation and Development (OECD-NEA), Paris, France, p. 25.

NEDO, 1999: Fluidised bed combustion. Technical Brief from *Residua & Warmer Bulletin* [http://www.residua.com/wrftbfbc.html]

Neij, L., 1997: Use of experience curves to analyze the prospects for diffusion and adoption of renewable energy technology. *Energy Policy, 23*, 1099-1107.

Neue, H.U., 1997: Fluxes of methane from rice fields and potential for mitigation. *Soil Use and Management,* **13**, 258-267.

Nielsen, P.S., K. Karlsson, and J.B. Holm-Nielsen, 1998: The role of transportation and co-fermentation in the CO_2 balance for utilisation of biogas for energy. *Sustainable Agriculture for Food, Energy and Industry*, James and James Ltd., London, pp. 959-963.

NPDP (Nuclear Power Plant Design Project), 1998: *A Response to the Environmental and Economic Challenge of Global Warming, Phase I: Review of Options and Selection of Technology of Choice*, MIT, January.

NRC (National Research Council), 1992: *Automotive Fuel Economy*. Committee on Fuel Economy of Automobiles and Light Trucks, Energy Engineering Board, National Academy Press, Washington, DC.

NRC, 1999a: *Research Program of the Partnership for a New Generation of Vehicles, Fifth report*. National Academy Press, Washington DC.

NRC, 1999b: *Review of the Research Strategy for biomass-derived transportation fuels*. National Academy Press, Washington DC, 60 pp.

NRECA (National Rural Electric Cooperative Association), 2000: *White Paper on Distributed Generation* [http://www.nreca.org/leg_reg/DGWhitepaper.pdf]

Nutter, F.W., 1996: Insurance and the Natural Sciences: Partners in the Public Interest. *Journal of Society of Insurance Research*, Autumn.

OECD, 1998a: *Agriculture and Forestry: identification of options for net greenhouse gas reduction*. Working Paper, Vol. VI, No. 67, OECD/GD(97)74.

OECD, 1998b: *Projected Costs of Generating Electricity: Update 1998*. Organization for Economic Cooperation and Development (OECD), International Energy Agency (IEA) and Nuclear Energy Agency (NEA), Paris, France.

OECD, 1999: *New Release: Financial Flows to Developing Countries in 1998. Rise in Aid; Sharp Fall in Private Flows*. OECD, Paris, France, 10 June.

OECD-NEA and IAEA, 2000: *Uranium 1999: Resources, Production and Demand*. A joint report of the OECD Nuclear Energy Agency (NEA) and the International Atomic Energy Agency (IAEA), Paris, France.

Oenema O., G.L. Velthof, S. Yamulki, and S.C. Jarvis, 1997. Nitrous oxide emissions from grazed grasslands. *Soil Use and Management,* **13**, 288-295.

Ogawa, J., B. Bawks, N. Matsuo, and K. Ito, 1998: *The Characteristics of Transport Energy Demand in the APEC Region*. Asia Pacific Energy Research Center, Tokyo.

Oleszkiewicz, J.A., and H.M. Poggi-Varaldo, 1998: Anaerobic Digestion of Mixed Solid Waste. *Waste Treatment Technology News,* **13**(4).

Oliver, M.A., 1999: Exploring soil spatial variation geostatistically. In *Proceedings of the 2nd European Conference on Precision Agriculture*, J.V. Stafford (ed.), Odense, Denmark, Sheffield Academic Press, ISBN 1-84172-042-3, pp. 3-18.

Olivier, J., 2000: Personal Communication, October 2000. National Institute for Public Health and Environment (RIVM), Bilthoven, The Netherlands..

Olivier, J.G.J., and J A.H.W. Peters, 1999: *International marine and aviation bunker fuel: trends, ranking of countries and comparison with national CO_2 emissions*. RIVM report 773301 002, National Institute of Public Health and the Environment, Bilthoven, The Netherlands.

Olivier, J.G.J., A.F. Bouwman, C.W.M. van der Maas, J.J.M. Berdwoski, C. Veldt, J.P.J. Bloos, A.J.H. Visschedijk, P.Y.J. Zandveld, and J.L. Haverlag, 1996: *Description of EDGAR Version 2.0*. National Institute of Public Health and Environment (RIVM), Bilthoven, The Netherlands.

Oonk, H., and K. Schöffel, 1999: Catalytic Abatement of Nitrous Oxide from Nitric Acid Production. *In Greenhouse Gas Control Technologies*. B. Eliasson, P.W.F. Riemer, and A. Wokaun (eds.), Elsevier Science, Oxford.

Oram, D.E., W.T. Sturges, S.A. Penkett, A. Mc Cullough, and P.J. Fraser, 1998: Growth of Fluoroform (CHF_3, HFC-23) in the background atmosphere. *Geophysical Research Letters,* **25**, 35-38.

Ormerod, B., 1994: *The Disposal of Carbon Dioxide from Fossil Fuel Fired Power Stations*. IEA Greenhouse Gas R&D Programme report IEAGHG/SR3, Cheltenham, UK, ISBN 1 898373 02 7.

OTA (Office of Technology Assessment), 1995a: *Innovation and Commercialization of Emerging Technologies*. U.S. Government Printing Office, OTA-BP-ITC-165, Washington DC, September.

OTA, 1995b. *Advanced Automotive Technology: Visions of a Super-Efficiency Family Car*. OTA-ETI-638, U.S. Congress, Washington DC.

Overend, R., and R. Costello, 1998: Bioenergy in North America: an overview of liquid biofuels, electricity and heat. In *Proceedings of the 10th European Conference on Biomass for Energy and Industry*. Würzburg, C.A.R.M.E.N., pp. 59 – 61.

Ozaki, M., 1997: CO_2 Injection and Dispersion in Mid-Ocean Depth by Moving Ship. *Waste Management,* **17**(5/6), 369-373.

Paffenbarger, J.A., and E. Bertel, 1998: Results from the OECD report on international projects of electricity generating costs. In *Proceedings of IJPGC 98: International Joint Power Generation Conference and Exhibition,* 24-26 August.

Parker, D., S. Barkaszi, J. Sherwin, and C. Richardson, 1996: Central Air Conditioner Usage Patterns in Low-Income Housing in a Hot and Humid Climate: Influences on Energy Use and Peak Demand. In *Proceedings of the 1996 ACEEE Summer Study on Energy Efficiency in Buildings,* American Council for an Energy-Efficient Economy, Washington DC.

Parker, D., P. Fairey, and J. McIlvaine, 1997: Energy-Efficient Office Building Design for Florida's Hot and Humid Climate. *ASHRAE Journal,* **April**.

Patel, M., 1999: *Closing Carbon Cycles*. Ph.D. Thesis, Utrecht University, Utrecht, The Netherlands.

Patel, M., and D.J. Gielen (eds.), 1999: *Non-energy Use CO_2 Emissions*. Proceedings of the first NEU-CO_2 Workshop, Netherlands Energy Research Foundation (ECN), Petten, The Netherlands.

PCAST (President's Committee of Advisors on Science and Technology – Panel on Energy Research and Development), 1997: Report to the President on *Federal Energy Research and Development for the Challenges of the Twenty First Century*. Executive Office of the President of the United States, Washington, DC.

Penner, J.E., D.H. Lister, D.J. Griggs, D.J. Dokken, and M. McFarland, 1999: *Aviation and the Global Atmosphere*. Intergovernmental Panel on Climate Change, Cambridge University Press, Cambridge, UK.

Perkins, S., 1998: Focus on Transport Emissions Needed If Kyoto's CO_2 Target Are To Be Met. *Oil and Gas Journal,* **96**(3), 36–39.

Phylipsen, G.J.M., K. Blok, and E. Worrell, 1998a: *Handbook on International Comparisons of Energy Efficiency in the Manufacturing Industry*. Department of Science, Technology and Society, Utrecht University, The Netherlands.

Phylipsen, G.J.M., K. Blok, and E. Worrell, 1998b: *Benchmarking the Energy Efficiency of the Dutch Energy-intensive Industry*. Department of Science, Technology and Society, Utrecht University, The Netherlands.

Phylipsen, G.J.M., L.K. Price, E. Worrell, K. Blok, 1999. Industrial Energy Efficiency in Light of Climate Change Negotiations: Comparing Major Developing Countries and the U.S. In *Proceedings of the 1999 American Council on Energy Efficiency Summer Study on Energy Efficiency in Industry*. Washington DC: ACEEE.

Pichs, R., 1998: *Tecnología Energía y Medio Ambiente. Potencialidades y limitaciones internacionales para una restructuración energética sostenible y retos para México*. Universidad National Autónoma de México (UNAM), México D.F. The Table is cited by Pichs from E. Videla, 1994: Panorama del potencial de ahorro en el sector industrial de America Latina. In SEMIP-IEA-CONAE, Eficiencia Energética en América Latina.

Piette, M., B. Nordman, O. deBuen, and R. Diamond, 1996: Over the Energy Edge: Results from a Seven Year New Commercial Buildings Research and Demonstration Project. In *Proceedings of the 1996 ACEEE Summer Study on Energy Efficiency in Buildings*. American Council for an Energy-Efficient Economy, Washington DC.

Pingoud, K., A. Lehtilä, and I. Savolainen, 1999: Bioenergy and forest industry in Finland after the adoption of the Kyoto Protocol. *Environmental Science and Policy, 2*, 153-163.

Pipatti, R., and I. Savolainen, 1996: Role of Energy Production in the Control of Greenhouse Gas Emissions from Waste Management. *Energy Conservation Management, 37*(6-8), 1105-1110.

Pitcher, K., and H. Lundberg, 1998: The development of a wood fuel gasification plant utilizing short rotation coppice abd forestry residues - Project Arbre. In *Making a Business from Biomass*. Pergamon Press, Oxford, pp. 1367–1378.

Posiva Oy, 1999: An Overall Description of the Facility for Final Disposal of Spent Nuclear Fuel. A document related to the application for a decision in principle filed to the council of State, Posiva Oy, Helsinki, Finland, June 1999.

Prasad, S.S., 1999: Trends in Indian Agriculture from 1970 till 1998. In *Survey of Indian Agriculture 1999*. A compilation of research papers, The Hindu.

Price, L., L. Michaelis, E. Worrell, and M. Khrushch, 1998: Sectoral Trends and Driving Forces of Global Energy Use and Greenhouse Gas Emissions. *Mitigation and Adaptation Strategies for Global Change, 3*, 263–319.

Price, L., E. Worrell, and M. Khrushch, 1999: *Sector Trends and Driving Forces of Global Energy Use and Greenhouse Gas Emissions: Focus on Buildings and Industry*. Lawrence Berkeley National Laboratory, LBNL-43746, Pergamon Press, Berkeley, CA.

Pye, M., and A. McKane, 2000: Making a Stronger Case for Industrial Energy Efficiency by Quantifying Non-Energy Benefits. *Resources, Conservation and Recycling, 28*(3-4), 171-183.

R&D Magazine, 1997: Basic Research White Paper: Defining Our Path to the Future. *R&D Magazine*. A. Cahners Publication, Des Plainis, Ill.

Radgen, P., 1999: Abscheidung, Nutzung und Entsorgung von CO_2 aus Energie- und Stoffumwandelnden Prozessen. VDI Bericht Nr. 1457, pp. 423-436, VDI Verlag, Düsseldorf, Germany.

Rand, S., D. Ottinger, and M. Branscome, 1999: Opportunities for the Reduction of HFC-23 Emissions from the Production of HCFC-22. In *Proceedings of the Joint IPCC/TEAP Expert Meeting on Options for the Limitation of Emission of HFCs and PFCs*, Netherlands Energy Research Foundation ECN, Petten, The Netherlands.

Raptsoun, N.V., and N.V. Parasyuk, 1997: Assessment of GHG Mitigation Measures in Ukraine. *Applied Energy, 56*(3/4), 367-380.

Read, P., 1999. Comparative static analysis of proportionate abatement obligations – a market based instrument for responding to global warming. *New Zealand Economic Papers, 33*(1), 137-147.

Reimer, P., and P. Freund, 1999: *Technologies for reducing methane emissions*. IEA Greenhouse Gas Programme [http://www.ieagreen.org.uk/prghgt43.htm]

Reimer, R.A., C.S. Slaten, M. Sepan, T.A. Koch, and V.G. Triner, 2000: Adipic Acid Industry – N_2O Abatement: Implementation of Technologies for Abatement of N_2O Emissions Associated with Adipic Acid Manufacture. In *Non-CO_2 Greenhouse Gases - Scientific Understanding, Control and Implementation*. J.van Ham, A.P.M. Baede, L.A. Meyer, and R. Ybema (eds.), Kluwer Academic Publishers, Dordrecht, The Netherlands.

Research Team of China Climate Change Country Study, 1999. *China Climate Change Country Study*. Tsinghua University Press, Beijing.

Resources for the Future, 1998: *Transport of foodstuffs*. Resources for the Future, Washington, USA.

Rhodes, R., and D. Beller, 2000: The Need for Nuclear Power. *Foreign Affairs*, **70**(1), January-February.

Rogner, H-H., 1993: Clean energy services without pain: district energy systems. *Energy Studies Review*, **5**(2), 114-120.

Rogner, H-H., 1997: An Assessment of World Hydrocarbon Resources. *Annual Review Energy Environment, 22*, 217–62.

Rogner, H-H., 1998a: Climate Change Assessments: Technology Learning and Fossil Sources – How Much Carbon Can Be Mobilized? In: *The assessment of climate change damages*. IEA Greenhouse Gas R&D Programme (SR6), Cheltenham, UK, pp. 49-65.

Rogner, H-H., 1998b: Hydrogen Technologies and the Technology Learning Curve. *International Journal Hydrogen Energy, 23*(9), 833-840.

Rogner, H-H., 1999: *Sustainable Energy Development – Economics and Externalities*. Proceedings of the Scientific Forum of the 43rd General Conference, International Atomic Energy Agency (IAEA), Vienna, Austria.

Rogner, H-H., 2000a: Energy Resources. In *World Energy Assessment: Energy and the Challenge of Sustainability*. J. Goldenberg (ed.), United Nations Development Programme (UNDP), United Nations Department of Economic and Social Affairs (UN-DESA) and World Energy Council (WEC), New York.

Rogner, H-H., 2000b: Rethinking the Options: Kyoto's Flexible Mechanisms & Nuclear Power. *IAEA Bulletin, 42*(2).

Romm, J., 1999: *Cool Companies: How the Best Businesses Boost Profits and Productivity by Cutting Greenhouse Gas Emissions*. Island Press, Washington DC.

Roos, G., 2000: Ancillary Costs and Benefits of Mitigation Options in the Households and Tertiary Sectors. In *Sectoral Economic Costs and Benefits of GHG Mitigation*. Proceedings of the IPCC Expert Meeting held in Eisenach, Germany, 14-15 February 2000, RIVM, Bilthoven, The Netherlands.

Rosa L.P., and S.K. Ribeiro, 1998: Avoiding emissions of carbon dioxide through the use of fuels derived from sugar cane. *Ambio, 27*, 465-470.

RTI (Research Triangle Institute), 1996: *The Reduction of HFC-23 Emissions from the Production of HCFC-22*. RTI, Research Triangle Park, USA.

Rubbia, C., 1998: *Accelerator Driven Energy Systems*. Presentation at the Technical University of Vienna, Austria.

Ruth, M., and P. Dell'Anno, 1997: An Industrial Ecology of the U.S. Glass Industry. *Resources Policy, 23*(3), 109-124.

Ruth, M., and T. Harrington Jr., 1998: Dynamics of Material and Energy Use in U.S. Pulp and Paper Manufacturing. *Journal of Industrial Ecology,* **1**, 147-168.

Ruth, M.B., S. Bernow, G. Boyd, R.N. Elliot, and J. Roop, 1999: Analytical Approaches to Measuring the Potential for Carbon Dioxide Emission Reductions in the Industrial Sectors of the United States and Canada. In *Proceedings of the International Workshop on Technologies to Reduce Greenhouse Gas Emissions*. International Energy Agency / US Department of Energy.

Saller, G., G. Funk, and W. Krumm, 1998: Process chain analysis for production of methanol from wood. In *Proceedings of the 10th European Conference on Biomass for Energy and Industry*. C.A.R.M.E.N., Würzburg, pp. 131 –133.

Sanchez, M., 1997: *Miscellaneous Electricity Use in U.S. Residences*. Lawrence Berkeley National Laboratory, LBNL Report 40295, UC-1600, Berkeley, CA.

Sathaye, J., and A. Ketoff, 1991: CO_2 Emissions from Major Developing Countries: Better Understanding the Role of Energy in the Long Term. *The Energy Journal, 12*(1), 161-196.

Sathaye, J., A. Ketoff, L. Schipper, and S. Lele, 1989: *An End-Use Approach to Development of Long-Term Energy Demand Scenarios for Developing Countries*. Lawrence Berkeley National Laboratory, LBL-25611, Berkeley, CA.

Sato, T., 1999: Can we make the Biological Impacts of CO_2 Negligible by Dilution? In *Proceedings of the 2nd International Symposium on Ocean Sequestration of Carbon Dioxide*. NEDO, 21-22 June 1999, Tokyo, Japan.

Savolainen, I. 2000: Corporate Greenhouse Gas Management in Paper Industry. *World Resource Review, 12*(1), 111-125

Schafer, A., and D. Victor, 1999: Global Passenger Travel: Implications for Carbon Dioxide Emissions. *Energy, 24*(8), 657–679.

Schaeffer, R., and M.A. Almeida, 1999: *Potential for Electricity Conservation in the Residential Sector in Brazil*. Presented at the International Energy Agency International Workshop on Technologies to Reduce Greenhouse Gas Emissions: Energy-Engineering Analysis of Conserved Energy and Carbon, 5-7 May 1999, Washington DC.

Scharmer, K., 1998: Biodiesel from set-aside land. In *Sustainable Agriculture for Food, Energy and Industry*. James and James Ltd., London, pp. 844-848.

Schindlbauer, H., 1995: *Standardization and analysis of biodiesel*. Proceedings of the first international conference, FICHTE/Technical University, Vienna, Austria.

Schipper, L., R. Hass, and C. Sheinbaum, 1996: Recent Trends in Residential Energy Use in OECD Countries and Their Impact on Carbon Dioxide Emissions: A Comparative Analysis of the Period 1973-1992. *Journal of Mitigation and Adaptation Strategies for Global Change*, **1**, 167-196.

Schipper, L.J., and S. Meyers, 1992: *Energy Efficiency and Human Activity*. Cambridge University Press, Cambridge.

Schmitz, M., and H. Sourell, 1998: Efficient use of water for irrigation. *Sustainable Agriculture for Food, Energy and Industry*, James and James Ltd., London, pp. 311-314.

Schnell, R., 1999: *Mit der Bio-gas Technik zu ökologisch verträglichen und ökonomisch tragfähigen Lösungen für den ländlichen Raum - Regenerative Energie aus Gülle, Abfall und Abwasser* ;http://www.members.aol.com/biogasde/texte/anwend.htm]

Schoen, T., E. ter Horst, J. Cace, and F. Vlek, 1997: Large-scale Distributed PV Projects in The Netherlands. *Progress in Photovoltaics: Research and Applications*, **5**(3), 187-194.

Scholz, V., 1998: Energy balance of solid biofuels. *Sustainable Agriculture for Food, Energy and Industry*, James and James Ltd., London, pp. 861-866.

Secretaria de Energia, 1997: *Balance Nacional de Energia*. Secretaria de Energia, Mexico DF.

Senelwa, K., and R.E.H. Sims, 1999: Opportunities for small-scale biomass electricity systems in Kenya. *Biomass and Bioenergy*, **17**(3), 239 – 255.

Senelwa, K., and R.E.H. Sims, 1998: Determination of yield indices for short rotation forestry species grown for fuelwood in New Zealand. *Sustainable Agriculture for Food, Energy and Industry*, James and James Ltd., London, pp. 773 –779.

Sheinbaum, C., I. Jauregui, and V. Rodriguez, 1998: Carbon Dioxide Emission Reduction Scenarios in Mexico for Year 2005: Industrial Cogeneration and Efficient Lighting. *Mitigation and Adaptation Strategies for Global Change*, **2**, 359-372.

Shell, 1996: *The evolution of the world's energy systems*. Shell International Limited, Shell Centre, London, 7 pp.

Shell International, 1999: Lighting Up Rural South Africa. *The Shell Report 1999* ;http://www.shell.com/shellreport]

Sims, R.E.H., 1997: Energy sources from agriculture. *Sustainable Agriculture for Food, Energy and Industry*, James and James Ltd., London, pp. 748–752.

Sims, R.E.H., G.A. Martin, and R.W. Young, 1988: Tractor efficiency and fuel conservation. *Agricultural Engineering*, **43**(1), 12.

Sims, R.E.H., K. Senelwa, T. Maiava, and B. Bullock, 1999: Eucalyptus species for biomass energy in New Zealand –Part I growth screening trials at first harvest. *Biomass and Bioenergy*, **16**, 199-205.

Sinton, J.E., 1996: *Energy Efficiency in Chinese Industry: Positive and Negative Influences of Economic System Reform*. Ph.D. Thesis, University of California, Berkeley, CA.

Sinton, J.E., and F. Yang, 1998: *Sectoral Energy Efficiency Opportunities in China*. Lawrence Berkeley National Laboratory, Energy Analysis Program.

Siuru, B., 1997: Fuel Cells: Garbage in, Electricity out. *World Waste*, **5**(40), 7.

Smith, D., 2000: Horizontal boiler make 700ºC steam economic. *Modern Power Systems*, **20**(5), 37-41.

Smith, N.J., and G.H. Robinson, 1997: Technology Pushes Reserves 'Crunch' Date Back in Time. *Oil and Gas Journal*, **7 April**, 43-50.

Smith, K.R, R. Uma, V.V.N. Kishore, K. Lata, V. Joshi, J. Zhang, R.A. Rasmussen, and M.A.K. Khalil, 2000: *Greenhouse Gases from Small-Scale Combustion Devices in Developing Countries. Phase IIA: Household Stoves in India.Phase 2A.*. U.S. EPA, Washington, DC.

Spadaro, J., L.M. Langlois, and B. Hamilton, 2000: Assessing the Difference: Greenhouse Gas Emissions of Electricity Generating Chains. *IAEA Bulletin*, **42**(2), 19-24.

Spedding, C.R.W., 1992: Constraints and pressures on livestock farming systems. In *Sustainable Livestock Farming into the 21st Century*. B.J. Marshall (ed.), CAS paper 25, Reading University, UK, pp. 51-60.

SPS (Science and Policy Services), 1997: *Sales of Sulfur Hexafluoride by End-Use Applications, Annual Sales for 1961-1996, Sales Projection for 1997 through 2000*. Washington DC.

Stahl, K., and M. Neergaard, 1998: Experiences from the biomass fuelled IGCC plant at Varnamo. *Proceedings of the 10th European Conference on Biomass for Energy and Industry*, C.A.R.M.E.N., Würzburg, pp. 291-294.

STAPPA/ALAPCO, 1999: *Reducing Greenhouse Gases and Air Pollution: A Menu of Harmonized Options. Final Report*. State and Territorial Air Pollution Program Administrators (STAPPA) and Association of Local Air Pollution Control Officials (ALAPCO), Washington, DC.

Stevens, S.H., C.E. Fox, and S.L. Melzer, 2000: McElmo Dome and St. Johns Natural CO_2 Deposits: Analogs for Geologic Sequestration. In *Proceedings of the Fifth International Conference on Greenhouse Gas Control Technologies*, Cairns, Australia, 13-16 Aug. 2000.

Stevens, S.H., D. Spector, and P. Riemer, 1999: Enhanced coal bed methane recovery by use of CO_2. *Journal of Petroleum Technology*, **October 1999**, 62.

Stone, J.L., and H.S. Ullal, 1997: *The Ramakrishna Mission PV Project – A Cooperation Between India and the United States. Renewables for Sustainable Village Power: Project Brief*. National Renewable Energy Laboratory, Boulder, CO, http://www.rsvp.nrel.gov

Storey, M., 1997: The climate implications of agricultural policy reform; policies and measures for common action. Working Paper 16, Annex I, Expert Group of the UNFCCC.

Storey, M., 1999: *Ruminant methane: strategic issues and opportunities*. Ministry of Agriculture and Forestry, Wellington, New Zealand.

Sulilatu, W.F., 1998: *Co-combustion of biofuels*. Bioenergy Agreement report T13:combustion:1998:06, International Energy Agency, 67 pp.

Suozzo, M., and S. Nadel, 1998: *Selecting Targets for Market Transformation Programs: A National Analysis*. American Council for an Energy-Efficient Economy, Washington DC.

Svebio, 1998: *Environmental and energy policies in Sweden and the effects on bioenergy development*. Swedish Bioenergy Association [http://www.sve-bio.se/environment/env_contents.html]

Swithenbank, J., and V. Nasserzadeh, 1997: Co-Incineration, New Developments and Trends. Presented at the Workshop on the Co-incineration of Waste, October, ISPRA, Italy.

Takahashi, M., and E. Sanada, 1998: *Carbon distribution of the earth and carbon in forest soils*. Report, Forest and Forest Products Research Institute, Hokkaido, Japan.

Tauber, C., and R. Jablonski, 1998: Brennstoffzellen aus Sicht eines überregionalen EVU. *Energiewirtschaftliche Tagesfragen*, **48**(1/2), 68-74.

Taylor, R., 1997: *Joint U.S./Brazilian Renewable Energy Rural Electrification Project. Renewables for Sustainable Village Power: Project Brief*. National Renewable Energy Laboratory, Boulder, CO [http://www.rsvp.nrel.gov]

Taylor, R., and F. Abulfotuh, 1997: *Photovoltaic Electricity in Egypt. Renewables for Sustainable Village Power: Project Brief*. National Renewable Energy Laboratory, Boulder, CO [http://www.rsvp.nrel.gov]

Thayer, B., 1995. Daylighting and Productivity at Lockheed. *Solar Today*, **May/June**.

Thomas, C.E., B.D. James, F.D. Lomax Jr., and I.G. Kuhn Jr., 1998: *Societal Impacts of Fuel Options for Fuel Cell Vehicles*. SAE Technical Paper 982496, Society of Automotive Engineers, Warrendale, PA.

Thorpe, T.W., 1998: *Overview of wave energy technologies*. ETSU report for the Department of Trade and Industry, London, UK.

Tichy, M., 1997: GHG Emission Mitigation Measures and Technologies in the Czech Republic. *Applied Energy*, **56**(3/4), 309-324.

Tranaes, F., 1997: Danish wind co-operatives. Danish Wind Turbine Owners Association [http://www.windpower.dk.articles/coop.htm]

Treloar, G., and R. Fay, 1998: *The embodied energy of living*. Deakin University, Geelong, Australia.

Turiel, I., B. Atkinson, S. Boghosian, P. Chan, J. Jennings, J. Lutz, J. McMahon, S. Pickle, and G. Rosenquist, 1997: Advanced Technologies for Residential Appliance and Lighting Market Transformation. *Energy and Buildings*, **26**, 241-252.

Turkenburg, W.C., 1997: Sustainable development, climate change and carbon dioxide removal. *Energy Conversion and Management*, **38**, S3-S12.

Tzvetanov, P., M. Ruicheva, and M. Denisiev, 1997: Scenarios of Energy Demand and Efficiency Potential for Bulgaria. *Applied Energy*, **56**(3/4), 287-297.

UK DTI, 1999: *New and emerging renewable energy-prospects for the 21ˢᵗ century*. Report DTI/Pub 4024/3K/99/NP URN 99/744, Department of Trade and Industry, London, UK.

Ulgiati, S., S. Bastianoni, L. Nobili, and E. Tiezzi, 1994: A thermodynamic assessment of biodiesel production from oil seed crops - Energy analysis and environmental loading. In *27ᵗʰ ISATA proceedings for the Dedicated Conferences on Electric hybrid and alternative fuel vehicles and supercars*, ENEA, Italy, pp. 477-489.

Ullyatt, M., K. Betteridge, J. Knapp, and R.L. Baldwin, 1999: Methane production by New Zealand ruminants. In: *Global Change: Impacts on Agriculture and Forestry*. W.M. Williams (ed.), Bulletin 30, Royal Society of New Zealand, Wellington. pp. 89-93.

Unander, F., S. Karbuz, L.J. Schipper, M. Khrushch, and M. Ting, 1999: Manufacturing Energy Use in OECD Countries: Decomposition of Long Term Trends. In: *New Equilibria in the Energy Markets, Energy Policy*, **27**(13), 769-778.

UN (United Nations), 1996: *World Population Prospects:1996 Revision*. United Nations, New York.

UN, 1997: *Potentials and Policy Implications of Energy and Material Efficiency Improvement*. United Nations, Department for Policy Coordination and Sustainable Development, New York.

UNDP (United Nations Development Programme), 1999: *Human Development Report 1999*. Oxford University Press, Oxford, UK.

UNEP (United Nations Environment Programme), 1999: *The Implications to the Montreal Protocol of the Inclusion of HFCs and PCFs in the Kyoto Protocol*. Technology and Economic Assessment Panel, Nairobi, Kenya.

UNEP/Southern Centre, 1993: *UNEP Greenhouse gas Abatement Costing Studies. Zimbabwe Country Study Phase II*. Risoe National Laboratory, Denmark.

Unruh, G., 1999: *Avoiding Carbon Lock-in*. PhD. Dissertation, The Fletcher School of Law and Diplomacy, Tufts University, Medford, MA, USA.

Unruh, G., 2000: Understanding carbon lock-in. *Energy Policy, 28*(12), 817-830.

Urge-Vorsatz, D., and A. Szesler, 1999: Assessment of CO_2 Emission Mitigation by Technology Improvement in Central and Eastern Europe: Case Studies from Hungary, Poland, and Estonia. Presented at the International Energy Agency International Workshop on Technologies to Reduce Greenhouse Gas Emissions: Energy-Engineering Analysis of Conserved Energy and Carbon, 5-7 May 1999, Washington DC.

US DOE (U.S. Department of Energy), 1999a: *About Photovoltaics*. U.S. DOE Photovoltaics Program, http://www.eren.doe.gov/pv/

US DOE, 1999b: *Photovoltaic Energy Program Overview: Fiscal Year 1998*. Washington. DC.

US DOE/EIA (U.S. Department of Energy, Energy Information Administration), 1998: *Impacts of the Kyoto Protocol on U.S. Energy Markets and Economic Activity*. SR/OIAF/98-03, Washington, DC.

US DOE/EIA, 1999a: *International Energy Annual 1997*. DOE/EIA-0219(97), Washington DC.

US DOE/EIA, 1999b: *International Energy Outlook*, DOE/EIA-0484(99), Washington DC.

US DOE/EIA, 2000: *Assumptions to the Annual Energy Outlook 2000*. Report no. DOE/EIA-0554(2000), Washington DC.

US DOE/OHT (U.S. Department of Energy, Office of Vehicle Technologies), 1996: *Multiyear Program Plan for 1996-2000*. Office of Transportation Technologies, Washington DC.

US EPA (U.S. Environmental Protection Agency), 1993: *Options for Reducing Methane Emissions Internationally, Volume II: International Opportunities for Reducing Methane Emissions*. Report to Congress, EPA 430-R-93-006 B, Washington DC.

US EPA, 1998: *Greenhouse Gas Emissions From Management of Selected Materials in Municipal Solid Waste*. Washington DC.

US EPA, 1999a: *U.S. Methane Emissions 1990-2020: Inventories, Projections, and Opportunities for Reduction*. EPA 430-R-99-013, US EPA Office of Air and Radiation, Washington DC.

US EPA, 1999b. *Characterization of Municipal Solid Waste in the United States: 1998 Update*. Washington DC.

US EPA, 1999c: *Cutting the Waste Stream in Half: Community Record-Setters Show How*. EPA 530-R-99-013, Washington DC.

US EPA, 1999d: *International Efforts to Reduce Perfluorocarbon (PCF) Emissions from Primary Aluminum Production*. EPA 430-R-001, Washington DC.

US EPA, 2000: *Greenhouse Gas Emissions From Management of Selected Materials in Municipal Solid Waste; a revision of US EPA, 1998*. Washington, DC.

Van Doorn, J., P. Bruyn, and P. Vermeij, 1996: Combined combustion of biomass, municipal sewage sludge and coal in an atmospheric fluidized bed installation. In *Proceedings of the 9th European Biomass Conference for Energy and the Environment*, Copenhagen, pp. 1007-1012.

Van Velsen, A.F.M., O. Stobbe, K. Blok, A.H.M. Struker, 1998: *Building Regulations as a Means of Requiring Energy Saving and Use of Renewable Energies*. Project No. EP/IV/STOA/98/0503/01, Ecofys, Utrecht, The Netherlands.

Van Wee, B., and J.A. Annema, 1999: *Transport, Energy Savings and CO_2 Emission Reductions: Technical-economic Potential in European Studies Compared*. IEA, Paris.

Varadi, P., 1998: PV, why are we waiting? *Renewable Energy World*, **1**(3), 12–19.

Vellinga, P., G. Berz, S. Huq, L. Kozak, E. Mills, J. Paulikof, B. Schanzenbacher, and G. Soler, 2000: Financial Services. In *Climate Change: Impacts, Adaptation and Vulnerability. Contribution of Working Group II to the Third Assessment Report of the Intergovernmental Panel on Climate Change*. Cambridge University Press, Cambridge.

Veenendal, R., U. Jorgensen, and C. Foster, 1997: European energy crops overview project – synthesis report. *Biomass and Bioenergy*, Summer /autumn special issue.

Victor, D.G., and G.J. MacDonald, 1999: A model for estimating future emissions of sulfur hexafluoride and perfluorocarbons. *Climatic Change, 42*, 633-662.

Vine, E., E. Mills, and A. Chen, 1998: *Energy-Efficiency and Renewable Energy Options For Risk Management and Insurance Loss Reduction: An Inventory of Technologies, Research Capabilities, and Research Facilities at the U.S. Department of Energy's National Laboratories*. Lawrence Berkeley National Laboratory Report No. 41432, Berkeley, CA. A briefer version is published in *Energy, 25*(2000), 131-147.

Walker, D., P. Botha, and G. White, 1998: Tararua wind farm. In *Proceedings of the 36th Annual Conference of the Australian and New Zealand Solar Energy Society*, Christchurch, November, pp. 637 – 641.

Walsh, M., 2000: *EPA Proposal Regarding Heavy Duty Engines & Diesel Fuel* and *Highlights of the Tier 2/Gasoline Sulfur Progra*. [http://walshcarlines.com]

Wang, M., 1999a: *GREET 1.5 – Transportation Fuel-Cycle Model, Volume 2: Appendices of Data and Results*. ANL/ESD-39, vol. 2, Center for Transportation Research, Argonne National Laboratory, Argonne, Illinois.

Wang, S., 1998: *Wind power making a promising contribution to our world*. China Renewable Energy Information Network [http://www.crein.org.ch/newpage21.htm]

Wang, M.Q.,1999b: *A Full Fuel-Cycle Analysis of Energy and Emissions Impacts of Transportation Fuels Produced from Natural Gas*. ANL/ESD-40, Center for Transportation Research, Argonne National Laboratory, Argonne, IL.

Wang, X., and K.R. Smith, 1999: *Near-Term Health Benefits of Greenhouse Gas Reductions: A Proposed Assessment Method and Application in Two Energy Sectors of China*. World Health Organization, Geneva.

Watson, R.T., I.R. Noble, B. Bolin, N.H. Ravindranath, D.J. Verardo, and D.J. Dokken (eds.), 2000: *Land Use, Land-Use Change, and Forestry*. A Special Report of the Intergovernmental Panel on Climate Change, Cambridge University Press, Cambridge, UK.

WEC (World Energy Council), 1994: *New Renewable Energy Resources*. World Energy Council, London, UK.

WEC, 1995a: *Efficient Use of Energy Using High Technology – An Assessment of Energy Use in Industry and Buildings*. M.D. Levine, N. Martin, L. Price, and E. Worrell (eds.), World Energy Council, London, UK.

WEC, 1995b: *Global Transport Sector Energy Demand Towards 2020*. World Energy Council, London.

WEC, 1998a: *Global Transport and Energy Development: The Scope for Change*. World Energy Council, London.

WEC, 1998b: *Survey of Energy Resources, 18th Edition*. World Energy Council, London.

Weiss, M.A., J.T. Heywood, E.M. Drake, A. Schafer, and F.F. AuYeung, 2000: *On the Road in 2020*. MIT EL 00-003, Massachusetts Institute of Technology, Energy Laboratory, Cambridge, MA.

Wells, C., 1999: *Total energy indicators of agricultural sustainability: dairy farming case study*. Ministry of Agriculture report, Wellington, New Zealand.

Whipple, C.G., 1996: Can nuclear waste be stored safely at Yucca Mountain? *Scientific American*, **274**(6).

Willerboer, W., 1997: *Future IGCC concepts*. Demkolec BV, Moerdijk, The Netherlands

Williams, A. (ed.), 1993: *Methane Emissions*. The Watt Committee on Energy, Elsevier, London.

Williams, N., 1996: Photovoltaic Rural Electrification. In *Proceedings of the International Workshop on Greenhouse Gas Mitigation Technologies and Measures*, Beijing, 12-15 November 1996, Lawrence Berkeley National Laboratory, Berkeley, CA, LBNL-39686.

Williams, R.H., 1999: Hydrogen production from coal and coal bed methane, using byproduct CO_2 for enhanced methane recovery and sequestering the CO_2 in the coal bed. In *Greenhouse Gas Control Technologies: Proceedings of the 4th International Conference on GHG Control Technologies, Interlaken, Switzerland, 30 August – 2 September*. B. Eliasson, P. Riemer, and A. Wokaun (eds.), Pergamon, Amsterdam, 1999, pp. 799-804.

Williams, R.H., 2000: Advanced Fossil Technologies. In *World Energy Assessment: Energy and the Challenge of Sustainability*. J. Goldemberg (ed.), United Nations Development Programme (UNDP), United Nations Department of Economic and Social Affairs (UN-DESA) and World Energy Council (WEC), New York, USA.

Wilson, M., R. Moburg, B. Stewart, and K. Thambimuthu, 2000: CO_2 Sequestration on Oil Reservoirs – A Monitoring and Research Opportunity. In *Proceedings of the Fifth International Conference on Greenhouse Gas Control Technologies*. Cairns, Australia, 13-16 August 2000.

World Atlas, 1998: 1998 World Atlas & Industry Guide. *The International Journal on Hydropower & Dams*. Aqua-Media International, Sutton, UK.

World Atlas, 1999: World Atlas & Industry Guide 1999-2000. *The International Journal on Hydropower & Dams*, Aqua- Media International, Sutton, UK.

World Bank, 1996: *Rural energy and development: improving energy supplies for two billion people*. World Bank Industry and Energy Department, Report no. 1512 GLB, Washington DC.

World Semiconductor Council, 1999: *Position Paper Regarding PFC Emissions Reduction Target*. World Semiconductor Council, Fuiggi, Italy.

Worldwatch Institute, 1998: *Vital signs 1998*. Worldwatch Institute, W.W. Norton & Co., New York, pp. 58-61.

Worrell, E., R. Smit, G.J.M. Phylipsen, K. Blok, F. van der Vleuten, and J. Jansen, 1995a: International Comparison of Energy Efficiency Improvement in the Cement Industry. In *Proceedings of the ACEEE 1995 Summer Study on Energy Efficiency in Industry*, ACEEE, Washington DC.

Worrell, E., A.P.C. Faaij, G.J.M. Phylipsen, and K. Blok, 1995b: An Approach for Analysing the Potential For Material Efficiency Improvement. *Resources, Conservation and Recycling*, **13**, 215-232.

Worrell, E., B. Meuleman, and K. Blok, 1995c: Energy Savings by Efficient Application of Fertilizer. *Resources, Conservation and Recycling*, **13**, 233-250.

Worrell, E., 1995. Advanced Technologies and Energy Efficiency in the Iron and Steel Industry in China. *Energy for Sustainable Development*, **2**, 27-40.

Worrell, E., L.K. Price, N. Martin, J.C.M. Farla, and R. Schaeffer, 1997a: Energy Intensity in the Iron and Steel Industry: a Comparison of Physical and Economic Indicators. *Energy Policy*, **25**, 727-744.

Worrell, E., M. Levine, L. Price, N. Martin, R. van den Broek, and K. Blok, 1997b: *Potentials and Policy Implications of Energy and Material Efficiency Improvement*. Department for Policy Coordination and Sustainable Development, United Nations, New York, table 2.15.

Worth, W.F., 2000: International SEMATECH, Reducing PFC Emissions: A Technology Update. *Future Fab International*, **9**, 57-62.

Wu, Z., and Z. Wei 1997: Mitigation Assessment Results and Priorities for China's Energy Sector. *Applied Energy*, **56**, 237-251.

Yagi, K.H., H. Tsuruta, and K. Minami, 1997: Possible options for mitigating methane emission from rice cultivation. *Nutrient cycling in Agroeco*, **49**, 213-220.

Zhamou, W., and W. Yanfei, 1998: Sustainable agriculture development in China. *Sustainable Agriculture for Food, Energy and Industry*, James and James Ltd., London, pp. 1275 - 1276

Zhao, J., A. Wang, and F. Ferrazza, 1998: Novel 19.8% efficient "honeycomb" textured multicrystalline and 24.4% monocrystalline solar cell. *Applied Physics Letters*, **73**, 1991-1993.

Zhou, F., and X. Hu, 1999: *Energy Efficiency and Structural Change in End-Use Energy in China*. Energy Research Institute of SDPC, Beijing. This publication relies to a large part on: Hu, X., 1997: Energy System Network Chart and Energy Efficiency Study in China, Energy Research Institute of SDPC, Beijing.

Zhou, P.P., 1998: *Energy Efficiency for Climate Change – Opportunities and Prerequisites for Southern Africa*. International Network for Sustainable Energy (Inforse), No. 20, Copenhagen, Denmark

Zhou, P.P., 1999: *Climate change mitigation in Southern Africa. Botswana Case Study*. UNEP Collaborating Centre for Energy and Environment, RISOE National Laboratory, Denmark, ISBN 87-550-2429-9.

Zhou, P.P., and L. Landner, 1999: *Business and Environment in Botswana - An Assessment of Business Sector Interests in Environmental Resource Management*. International Union for Conservation of Nature (IUCN), Botswana.

Chapter 3 Appendix

Options to Reduce Global Warming Contributions from Substitutes for Ozone-Depleting Substances

Lead Authors:

Stephen O. Andersen (USA), Suely Maria Machado Carvalho (Brazil), Sukumar Devotta (India), Yuichi Fujimoto (Japan), Jochen Harnisch (Germany), Barbara Kucnerowicz-Polak (Poland), Lambert Kuijpers (Netherlands), Mack McFarland (USA), William Moomaw (USA), Jose Roberto Moreira (Brazil)

Contributing Authors:

Paul Ashford (UK), Paul J. Atkins (UK), James A. Baker (USA), Kornelis Blok (Netherlands), Denis Clodic (France), Abid Merchant (USA), E. Thomas Morehouse Jr. (USA), Sally Rand (USA), Robert J. Russell (USA), Remko Ybema (Netherlands)

Review Editors:

Ramon Pichs-Madruga (Cuba), Hisashi Ishitani (Japan)

CONTENTS

EXECUTIVE SUMMARY

Hydrofluorocarbons (HFCs) and to a lesser extent perfluorocarbons (PFCs) have been introduced to replace ozone-depleting substances (ODSs) that are being phased out under the Montreal Protocol on Substances that Deplete the Ozone Layer. HFCs and PFCs have a significant global warming potential (GWP) and are listed in the Kyoto Protocol. This Appendix estimates consumption and emissions and assesses alternative practices and technologies to reduce emissions. Emissions as by-products of manufacturing are treated in the main part of Chapter 3.

In the absence of the Montreal Protocol the use of chlorine-containing compounds and especially CFCs would have expanded significantly. However, because of this treaty, developed countries replaced about 8% of projected chlorofluorocarbon use with HFCs, 12% with HCFCs, and eliminated the remaining 80% by controlling emissions, specific use reductions, or by using alternative technologies and fluids including ammonia, hydrocarbons, carbon dioxide, water, and not-in-kind options.

In 1997, the production of HFCs was about 125 kilotons (50MtC$_{eq}$), and the production of PFCs amounted to 5 kilotons (12MtC$_{eq}$). The production of HFCs in 2010 is projected to be about 370 kilotons or 170MtC$_{eq}$ and less than 12MtC$_{eq}$ for PFCs, assuming current trends in use and regulations, substantial investment in new HFC production capacity, and success of voluntary agreements. Since most of the HFCs and some of the PFCs are contained in equipment or products, annual emissions lag production when use is growing.

Refrigeration, air conditioning, and heat pumps are the largest source of emissions of HFCs. Improved design, tighter components, and recovery and recycling during servicing and disposal can reduce lifetime HFC emissions at moderate to low costs. Non-HFC alternatives include hydrocarbons, ammonia, and carbon dioxide, or alternative technologies. Lifecycle climate performance (LCCP) analysis of the entire system, including direct fluid emissions and indirect emissions from carbon dioxide resulting from energy use by the device, provides a means of assessing the net contribution of a system to global climate change. The LCCP calculations are very system specific and can be used to make relative rankings. However, since the LCCP approach involves regional climate conditions and local energy sources, the results cannot be generalized in order to make globally valid comparisons.

Insulating foams are anticipated to become the second largest source of HFC emissions and HFC use is expected to grow rapidly as CFCs and HCFCs are replaced with HFC-134a, HFC-245fa, and HFC-365mfc. Alternative blowing agents including the different pentanes and carbon dioxide have lower direct climate impact from direct emissions. However, they also have lower insulating values than CFCs and HCFCs, and hence may have higher indirect emissions from energy use if the foam thickness is not increased to offset the higher conductivity. Non-foam insulation alternatives such as mineral fibres are also used, and vacuum panels may play a role in the future.

Other sources of HFC and PFC emissions are industrial solvent applications, medical aerosol products, other aerosol products, fire protection, and non-insulating foams. A variety of options are available to reduce emissions including increased containment, recovery, destruction, and substitution by non-fluorocarbon fluids and not-in-kind technologies. There are no zero- or low-GWP alternatives for some medical and fire protection applications.

A3.1   Introduction

Alternatives and substitutes for HFCs, perfluorocarbons (PFCs), and ozone depleting substances (ODSs) have recently been extensively evaluated. The Montreal Protocol Technology and Economic Assessment Panel (TEAP) and its technical committees published a comprehensive assessment (UNEP, 1999b). Furthermore, reports were published within the framework of the joint IPCC/TEAP workshop (IPCC/TEAP, 1999) and the second non-CO_2 greenhouse gases conference (van Ham *et al.*, 2000).

The HFCs that are projected for large volume use have global warming potentials (GWPs) which are generally lower than those of the ODSs they replace. The GWP of HFCs replacing ODSs range from 140 to 11,700. HFC-23 with a GWP of 11,700 is used as a replacement for ODSs to only a very minor extent. However, there are relatively large emissions of HFC-23 from the HCFC-22 manufacturing process. The majority of HFCs have GWPs much lower than that of HFC-23. PFCs

have GWPs that are generally higher than those of the ODSs they replace, ranging from 7,000 to 9,200 (IPCC, 1996). *Table A3.1* lists the atmospheric properties of the HFCs and HFC blends considered in this Appendix.

Most HFCs are used for energy-consuming applications such as refrigeration, air conditioning and heat pumps, and building and appliance insulation. Life cycle climate performance (LCCP) analysis is being used to estimate the net contribution to climate change. It includes all direct greenhouse gas emissions and indirect emissions related to energy consumption associated with the design and the operational modes of systems (UNEP, 1999b; Papasavva and Moomaw, 1998). The LCCP is a very system specific parameter that can be used to make relative rankings. However, LCCP analysis involves regional differences – including different fuel sources – and the related equipment operating conditions; the results can therefore not be generalized in order to make globally valid comparisons.

Table A3.1: Atmospheric properties (lifetime, global warming potential (GWP)) for the HFC chemicals described in the Appendix (IPCC, 1996; WMO, 1999)

Compound	Chemical formula	Lifetime (yr) (IPCC, 1996)	GWP (100 yr) (IPCC, 1996)	Lifetime (yr) (IPCC, 2000)	GWP (100 yr) (IPCC, 2000)
HFC-23	CHF_3	264	11,700	260	12,000
HFC-32	CH_2F_2	5.6	650	5.0	550
HFC-125	CHF_2CF_3	32.6	2,800	29	3,400
HFC-134a	CH_2FCF_3	14.6	1,300	13.8	1,300
HFC-143a	CH_3CF_3	48.3	3,800	52	4,300
HFC-152a		1.5	140	1.4	120
HFC-227ea	CF_3CHFCF_3	36.5	2,900	33	3,500
HFC-245fa[a]	$CF_3CH_2CHF_2$	-	-	7.2	950
HFC-365mfc[a]	$CF_3CH_2CF_2CH_3$	-	-	9.9	890
HFC-43-10mee	$CF_3CHFCHFCF_2CF_3$	17.1	1,300	15	1,500
R-404A (44% HFC-125, 4% HFC-134a, 52% HFC-143a)			3,260		
R-407C (23% HFC-32, 25% HFC-125, 52% HFC-134a)			1,525		
R-410A (50% HFC-32, 50% HFC-125)			1,725		
R-507 (50% HFC-125, 50% HFC-143a)			3,300		

[a] No lifetime or GWP listed in IPCC (1996)

Note: GWP values to be used by Parties for reporting any emissions and for any other commitments under the Kyoto Protocol are the 100 year GWP values from IPCC (1996) (decision taken at CoP 3, 1997)

The energy efficiency of equipment and products can be expressed in at least three ways: theoretical maximum efficiency, maximum efficiency achievable with current technology, and actual efficiency for commercial scale production (often expressed as a range of values). Systems optimized for a new refrigerant have been compared to sub-optimum systems with other refrigerants. Furthermore, appliance sizes and features that influence energy performance vary between studies and test conditions, and methodologies are often significantly different. These factors have led to a wide range of energy efficiency claims in technical reports and commercial publications. Ultimately, the performance and cost effectiveness of specific products from commercial scale production must be directly compared. Furthermore, costs reported in this appendix might not always be comparable because of differing estimation methods, including estimates based on both consumer and producer costs.

A3.1.1 Past Trends

Unlike anthropogenic greenhouse gases emitted as an immediate consequence of the burning of fossil fuels to generate energy, most HFCs and PFCs are contained within equipment or products for periods ranging from a few months (e.g., in aerosol propellants) to years (e.g., in refrigeration equipment) to decades (e.g., in insulating foams). Thus, emissions significantly lag consumption and, because HFC systems are relatively new, emissions will continue to grow after 2010.

Both the quantities used and patterns of use of ODSs, HFCs, and PFCs are changing (see *Figure A3.1*) as ODSs are phased out under the Montreal Protocol (IPCC/TEAP, 1999; McFarland, 1999). In 1986, less than half of total ODS use

was in insulating foams, fire protection, refrigeration, air conditioning, and heat pumps, with more than half as aerosol product propellants, non-insulating foam, solvent, and specialized applications. However, by 1997, the global consumption of fluorocarbons (CFCs, HCFCs and HFCs) had decreased by about 50% as solvent, aerosol product, and non-insulating foam applications switched to alternatives other than fluorocarbons. Refrigeration, air conditioning, and insulating foam accounted for about 85% of the remaining total fluorocarbon use. 80% of projected chlorofluorocarbon demand was avoided by reducing emissions, redesign, and use of non-fluorocarbon technologies. As CFCs, halons, and HCFCs are phased out globally, the quantities of fluorocarbons are expected to continue to decline in the short term, but are expected to grow in the longer term.

A3.1.2 Projections

Future global HFC and PFC consumption and/or emissions as substitutes for ODSs have been separately estimated by IPCC (1995), Midgley and McCulloch (1999), and UNEP (1998a). Midgley and McCulloch (1999) projected carbon-equivalent emissions of HFCs and PFCs (excluding unintended chemical by-product emissions) at $60MtC_{eq}$ in 2000, $150MtC_{eq}$ in 2010 and $280MtC_{eq}$ in 2020. Projected consumption data for 2000 and 2010 are primarily based on UNEP reports (UNEP, 1998f, 1999b) and are shown in *Table A3.2*. Considering that emissions lag consumption by many years, the Midgley and McCulloch figures are much larger than the UNEP figures. This discrepancy is consistent with the Midgley and McCulloch scenario which was constructed to represent plausible upper limits to future emissions (McFarland, 1999). HFC emissions in the SRES scenarios (IPCC, 2000) are $54MtC_{eq}$ in

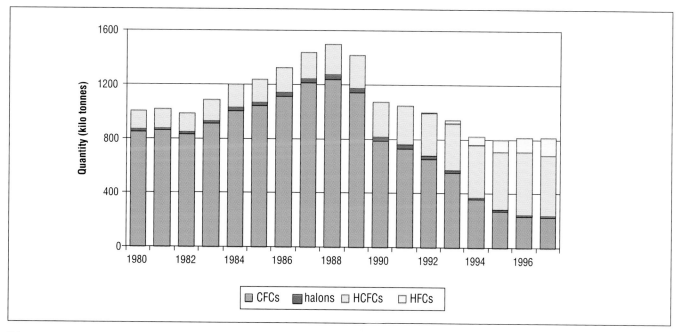

Figure A3.1: *Estimated global consumption of CFCs, halons, HCFCs, and HFCs (McFarland, 1999).*

Table A3.2: Estimated and projected global HFC consumption and emission for different sub-sectors for 2000 and 2010

Sub-sector	2000				2010			
	HFC consumption kt/yr	HFC consumption MtC$_{eq}$/yr	HFC emission kt/yr	HFC emission MtC$_{eq}$/yr	HFC consumption kt/yr	HFC consumption MtC$_{eq}$/yr	HFC emission kt/yr	HFC emission MtC$_{eq}$/yr
Refrigeration & A/C [a,b]	102-112	47-50	40-44	18-19	195-255	106-139	82-124	42-64
Mobile A/C [c]	64-74	23-26	31-35	11-12	58-79	21-28	37-54	13-19
Domestic refrigeration [f]	7	2.5	0.9	0.3	15-17	5.5-6.4	3.5-4.5	1.3-1.7
Comm. refrigeration [d,e,f]	19	15	5	4.5	46.5-64	39-54	19.5-31	16-26
Cold storage [d]	4.5	3	1.2	0.8	9-12	6-8	3-4	2-2.5
Industrial refrigeration [d]	1.5	1	0.3	0.2	3-4	2-2.7	0.6-0.8	0.4-0.5
Chiller A/C	2.5	1	0.2	0.1	3.5-4.5	2.3-3	0.5-0.7	0.3-0.5
Transport refrigeration [d,e,f]	3.3	1	1.3	0.7	17-23.5	8.5-12	10-14.5	5-7
Unitary air conditioning	-	-	-	-	43-51	22-25	8-14	4-7
Insulating foams [g]	4+	1.5+	<1	<0.5	115	29.5	20-40	5-10
Solvents/ cleaning [h]	<2	<9	<2	<9	>2	<9	>2	<9
Med. aerosol [h]	1	<1	1	<1	<9	<4	<9	<4
Other aerosol [h]	<15	<4	<15	<4	<20	<5	<20	<5
Fire protection [a,b,i]	1.0 - 1.6 0.6 - 0.9	0.8 - 1.3 0.5 - 0.8	0.2 - 0.4	0.2 - 0.3	1.6 - 2.0	1.3 - 1.7		
TOTAL	125-136	63-66	59-62	32-33	343-403	155-189	134-196	66-93

a Consumption and emission estimates are based on information contained in (UNEP, 1998a, 1999b); A/C = Air Conditioning

b The average growth has been estimated as 2.5% annually over the period 2000-2010

c See text and *Table A3.3* for explanations

d The mix of refrigerants (both pure HFCs and HFC blends) is estimated based on information in (UNEP, 1998a, 1999b) for commercial, cold storage, industrial and transport applications

e 2010 emission factors have been defined as 7%-10% for commercial refrigeration, 12%-18% for transport refrigeration;

f Equipment life for domestic refrigeration is assumed to be 15 years or longer implying that no emissions at disposal will occur by the year 2010. Equipment life for commercial and transport refrigeration is assumed to be 10 years with between 60% and 80% emitted (20%-40% recovery) at disposal

g Emissions for insulating foam were based on a methodology described in Gamlen et al., 1986

h Emissions of HFCs used as solvents and medical and other aerosol propellants occurs within one year of consumption (Gamlen et al., 1986)

i Emissions for the fire protection sector for 2000 are estimated to be 5% of the installed base (the same level as the average halon emission in the recent decade); for the year 2010 they are assumed to be 2.5% of the installed base, as a result of improved design and service practices (IPCC/TEAP, 1999; UNEP, 1999b).

2000 and 130-136MtC$_{eq}$ in 2010. These values are higher than those presented in *Table A3.2* because of the top-down approach used in SRES that does not adequately account for delay between use and emissions in the 2000 to 2010 timeframe. Considering this fact and given the options for substitution, containment, etc., it is estimated that emissions in 2010 could well be about 100MtC$_{eq}$ below the SRES forecast at a marginal cost lower than US$200/tC$_{eq}$. None of the scenarios have considered the implications of new uses of HFCs or PFCs other than as substitutes for ODSs.

Projected consumption and emission estimates for HFCs by sub-sector for 2000 and 2010 are summarized in *Table A3.2*.

A3.2   Refrigeration, Air Conditioning, and Heat Pumps

Most current and projected HFC consumption and emissions is in this sector. HFC consumption in refrigeration and mobile and stationary air conditioning in 1997 was, on a mass basis, about 30% of projected developed country CFC consumption in the absence of the Montreal Protocol (McFarland, 1999). Most of the remaining 70% of projected consumption has been eliminated by reducing leaks, reduced charge per application, and improved service practices; the substitution by other fluids and new technologies played a lesser role (some substitution – a few per cent – by HCFCs also took place). Globally, there is still a huge potential to further reduce HFC emissions. Estimated consumption and emissions of HFCs for this sector for 2000 and 2010 are shown in *Table A3.2*. Emissions significantly lag consumption because HFC systems are relatively new so emissions will occur well after 2010.

The primary options for limiting HFC emissions are the use of alternative refrigerants and technologies, reduced refrigerant charge, improved containment, recovery with recycling, and/or destruction. There are no globally representative estimates of the cost effectiveness of improved containment and recovery. In developed countries, recovery during servicing of small domestic refrigerators captures a relatively insignificant proportion of HFCs, while end-of-life recovery is significant. For medium-sized devices such as commercial units with substantial leakage rates, recovery during both multiple servicing and at the end of useful life is both significant. For very large units recovery both during servicing and at end of life is frequently done already because of the high economic value associated with the large quantities of recovered fluids.

In developing countries, where low cost is important, the quality of equipment is often poor, resulting in high failure rates. Since the service sector in developing countries is normally not equipped with the tools for recycling, the emissions of refrigerants during servicing and product disposal form a significant portion of the overall emissions.

Recovery at the end of equipment life is likely to exhibit a poor cost-effectiveness for smaller units. For these units, the intro-

duction of economic incentives will be necessary, probably together with voluntary agreements and/or government regulations (as already exist in some countries) to achieve significant reductions in this sector.

Carbon dioxide emissions associated with energy consumption by refrigeration, air conditioning, and heat pump equipment are usually the largest contributions to global warming associated with cooling equipment (AFEAS, 1991; Papasavva and Moomaw, 1998). Japanese manufacturers estimate that energy-related CO_2 emissions represent an even larger fraction of lifetime emissions for their low leakage rate, small charge appliances. Thus, improvements in equipment energy efficiency are often a cost-effective way to reduce greenhouse gas emissions and to lower costs to consumers (March, 1998). Proper equipment design, component performance, and the selection of the most appropriate refrigerant fluid are the most important factors contributing to energy efficiency. Examination of the LCCP of the system will determine which combination of operating efficiency and fluid choice yields the lowest overall contribution to global warming.

Hydrocarbons, carbon dioxide, and to a lesser extent, ammonia are the most likely alternatives to HFC refrigerants. No ammonia vapour compression units have capacity less than 50kW. Since both hydrocarbons and ammonia are flammable and ammonia is toxic, their acceptance will depend on cultural norms and specific regulations in each country. Hydrocarbons are currently being used in about 50% of the refrigerators manufactured in Europe and in some manufactured in Asia and Latin America; their use in these products as well as in other refrigeration and air conditioning systems could increase. Large charges can present a safety concern, and globally standardized mechanical and electrical safety standards are being established.

If safety is a concern, secondary loops containing a heat transfer fluid can be used. For modest cooling, such as water chilling for residential air conditioning or industrial process chilling, there is no energy penalty from using a secondary loop. For medium temperature applications in food processing and commercial refrigeration, secondary loops permit the safe use of ammonia and hydrocarbons, or enable minimization of an HFC refrigerant charge, generally with a modest energy penalty. If safety concerns require a secondary loop for low temperature applications in food processing and cold storage, in which normally the refrigerant is used as the direct heat transfer fluid, a substantial energy penalty may ensue.

Where they are required, the estimated cost of utilizing secondary loops with ammonia and hydrocarbons to replace HFCs is estimated to exceed US$100/tC$_{eq}$ (Harnisch and Hendriks, 2000). Secondary loop systems designed to achieve comparable efficiency and demonstrated in Europe have up to a 15% higher cost.

An optimal transition strategy from ODSs to alternatives can substantially lower costs and better meet development goals

Table A3.3: Estimated global HFC-134a emissions for vehicle air conditioning
Source: Baker (1999).

Year	Vehicles w/134a A/C (million)	Recycle at service/disposal (%)	HFC-134a use (kt)	HFC-134a emissions (kt)
2000	214-247	60	64-74	31-35
2010	464-530	60	58-79	37-54

for developing countries, especially in the refrigeration and air conditioning sectors (Papasavva and Moomaw, 1997). The Montreal Protocol Multilateral Fund (MLF) and the Global Environment Facility (GEF) have just begun to coordinate financing of ozone and climate protection (IPCC/TEAP, 1999). To date, one project has been jointly funded by the MLF and the GEF, which addresses energy efficiency in the replacement of CFCs. Energy use forms a major problem for the stressed energy supply system of capital-strapped developing countries. Since the greatest growth in refrigeration and air conditioning is projected to occur in developing countries, it is important that they select the most effective (in terms of costs and energy efficiency) non-ODS technology. Currently, customers in developing countries make purchase decisions based on initial cost with little consideration of energy consumption.

A3.2.1 Mobile Air Conditioning

HFC-134a replaced CFC-12 in virtually all vehicle air conditioners produced after 1993/94. Motor vehicle air conditioning uses HFC-134a refrigerant in an integrated system of components that provide cooling, heating, defrosting, demisting, air filtering, and humidity control. It is technically and economically feasible to significantly reduce emissions of HFC-134a refrigerants: by recovery and recycling of refrigerant during servicing and vehicle disposal; by using high quality components with low leakage rates, hoses with lower permeation rates, and improved connections; and by minimizing refrigerant charge. Efficiency improvements and smaller, lighter units can further reduce energy-related CO_2 emissions. New systems using alternative refrigerants –carbon dioxide or hydrocarbons– are being developed as described below (see Section A3.2.1.3).

A3.2.1.1 Estimates of Global HFC-134a Emissions

Globally, 65%–75% of air-conditioned vehicles in service in 2000 have HFC-134a air conditioners and it is predicted that between 2000 and 2010, 70%–80% of all new vehicles will have HFC-134a air conditioners. This projection assumes a continuation of current trends of mobile air conditioning installed in vehicles in normally cool climates where air conditioning may not be necessary. When air conditioning systems were redesigned to use HFC-134a, vehicle manufacturers used

a smaller refrigerant charge and reduced leakage rates. Typical direct HFC emissions over a 10-year period in the USA are 1.4 kg if recycling is undertaken during service and disposal and 3.2 kg without recycling (Baker, 2000). Estimates for HFC-134a air conditioner emissions are included in *Table A3.3*.

A3.2.1.2 Strategies for Reducing Emissions and Improving Energy Efficiency of HFC-134a Systems

Vehicle manufacturers and their suppliers are working to increase the energy efficiency and reduce the emissions of HFC-134a systems. Typical CO_2 exhaust emissions resulting from air-conditioner operation are in the range of 2% to 10% of total vehicle CO_2 emissions (SAE, 2000). Comparison of reduced emissions HFC-134 systems and CO_2 systems have been published (Petitjean *et al.*, 2000; March, 1998). HFC-134a systems can be redesigned for higher energy efficiency and smaller refrigerant charge within 2–4 years, and manufacturers of replacement parts could supply high-quality components within 2–4 years (SAE, 2000). SAE (2000) estimates that improved HFC-134a systems can be introduced faster and at lower incremental cost than carbon dioxide, hydrocarbon, and HFC-152a systems.

Lowering the demand for cooling and humidity control can reduce indirect emissions from fuel consumption and could allow smaller air conditioning systems having reduced refrigerant charges. This is accomplished by increasing thermal insulation and decreasing thermal mass in the passenger compartment, by sealing the vehicle body against unwanted air infiltration, by minimizing heat transfer through window glass, and by controlling the compressor to minimize over-cooling and subsequent re-heating of air.

A3.2.1.3 Strategies for Developing Efficient Alternative Air Conditioning Systems

Considerable activity is underway to develop alternatives to HFC-134a air conditioning systems for vehicles. Prominent efforts are the European "Refrigeration and Automotive Climate Systems under Environmental Aspects (RACE) Project" (Gentner, 1998), and the Society of Automotive Engineers/US EPA/Mobile Air Conditioning Society Worldwide "Mobile Air Conditioning Climate Protection Partnership" (SAE, 1999, 2000).

Two categories of alternative refrigerant candidates have emerged for new systems: 1) transcritical carbon dioxide systems and 2) hydrocarbon or HFC-152a systems.

1. Transcritical CO_2 systems require substantial new engineering, reliability, and testing efforts. These carbon dioxide systems have potential energy efficiency that is comparable or better than HFC-134a systems and the lowest direct global warming emissions of any candidate refrigerant. Prototype systems from several European vehicle manufacturers provided comparable passenger cooling comfort in medium-sized vehicles, and one reported improved efficiency over HFC systems at the Scottsdale Symposium (SAE, 1999). A CO_2 system in a small vehicle was less efficient, especially during idling (Kobayashi *et al.*, 1998). With a higher heat rejection temperature compared to HFC-134a cycles, carbon dioxide systems can also efficiently operate in reverse mode to heat vehicle interiors. New equipment and technician training will be required to safely repair systems with operating pressures up to 6 times higher than systems with HFC-134a. The first CO_2 systems could be commercially available within 4-7 years (SAE, 2000).

2. Hydrocarbon and HFC-152a systems, with secondary cooling loops to mitigate flammability risk, are under study and development by several manufacturers in co-operation with suppliers (Baker, 2000; Ghodbane 1999; Dentis *et al.*, 1999; SAE, 1999a). One prototype achieved a cooling performance at the 1999 Phoenix Forum comparable to HFC-134a systems (Baker, 2000; Gentner, 1998; Ghodbane, 1999; SAE, 2000). Systems using flammable refrigerants will require additional engineering and testing, development of safety standards and service procedures, and training of manufacturing and service technicians before commercialization, but would require fewer technical breakthroughs than carbon dioxide systems. If proven safe to Original Equipment Manufactures (OEMs), it is estimated that systems with flammable refrigerants could be commercially implemented in the first vehicles in as little as 4-5 years (SAE, 2000).

Highly efficient air conditioning and heating systems are particularly important to the commercial success of electric, hybrid, fuel cell, and other low-emission vehicles to help overcome the limited power of such vehicles.

A3.2.1.4 Cost-Effectiveness of Reducing Emissions from Vehicle Air Conditioning

Recovery and Recycle
It is estimated that recycling rates can be increased from 60% to 90% within one to two years in developed countries (SAE, 2000). Recovery and recycling of HFCs can reduce emissions by more than 10 kt annually (Baker, 1999). About 50% of the global fleet of HFC-134a air conditioned vehicles are in the USA where recycling is mandatory, and 25% are in Japan, where voluntary programme achieve a substantial recycling rate. The remainder are in Europe where recycling ranges from zero in some countries to near 100% participation in others, and

in developing countries, where a wide range of recovery practices is found. A UNDP survey of 1300 Brazilian garages found one-quarter of garages recycling HFC-134a (UNDP, 1999).

The current market value of HFC-134a recovered during service or disposal in the USA more than pays for the cost of labour, equipment, and maintenance for shops servicing more than 6 vehicle air-conditioning systems per week. By 2002 to 2003 it is technically feasible to reduce system charge and leakage rates significantly. Recovery of 0.33 kg of HFC-134a will cost US$0.70 in large shops and US$1.50 in small shops. For large shops, recovery costs for improved, low-charge vehicles are estimated at less than US$3.50/t$C_{eq}$. Even for small shops, the cost-effectiveness per tonne of carbon equivalent can then be calculated in the range of US$1.18-12.81/t$C_{eq}$ depending upon the size of the charge (EPA, 1998).

Reduced Charge and Improved Containment
By 2002 to 2003, it is technically feasible to reduce system charge and leakage rates worldwide. It is estimated that the vehicle charge in the US can be reduced from 0.9kg to 0.8kg and that annual vehicle leakage could be reduced from 0.07 kg/yr to 0.04kg/yr (UNEP, 1998a; Baker, 1998; Sand *et al.*, 1997; Wertenbach and Caesar, 1998). In Europe, refrigerant charges average about 0.7 kg per vehicle (Clodic, 1999). For the USA it is estimated (Baker, 2000) that emissions can be reduced from 8% to 5% per year for a 10-year reduction of 1.2 kg/vehicle without recovery and recycling or 1.0kg with recovery and recycling. Two studies (Harnisch and Hendriks, 2000; March, 1998) estimate that, in Europe, the cost per vehicle to reduce leakage rates from 10% to 4%-5%/yr is only US$11-US$13.

Alternative Systems
Three authors have published estimates of the cost of emission reductions achieved through alternative vehicle air conditioning using carbon dioxide as the refrigerant (March, 1998; Baker, 1999, 2000; Harnisch and Hendriks, 2000). These studies reported widely diverging results on the specific abatement costs of HFC emissions for the use of transcritical CO_2 systems (from US$90 to >US$1000/tC_{eq}). Differences in cost estimates can be traced back to a number of factors among which two are most important: (1) the use of producer-costs versus consumer costs and (2) differing assumptions about the existing degree of recovery of HFC-134a during servicing and at the end of life. Of lesser importance were differing assumptions on the average fluid charge of an HFC air conditioning system, annual leakage rates, relative differential costs, and applied discount rates. Once the results are normalized to common assumptions on the major factors, the abatement costs differ by only a factor of two or less (see *Table A3.4*).

As reported in *Table A3.4*, costs of avoiding HFC emissions through alternative air conditioning systems vary between US$20 and US$2100/tC_{eq} depending on the emission characteristics of the reference HFC system (and on whether consumer or producer prices are used). Consequently in countries

Table A3.4: *Abatement costs [a] of avoiding HFC emissions by using alternative systems (based on CO_2 or secondary hydrocarbons) relative to different baseline HFC emission scenarios*

Alternative system compared against current HFC-134a systems				Alternative system compared against improved HFC-134a systems with reduced leakage rate			
With recycling, 1.4 kg 10-year emission baseline[b] (US$/tC$_{eq}$)		without recycling, 3.2 kg 10-year emission baseline (US$/tC$_{eq}$)		With recycling, 0.4 kg 10-year emission baseline[b] (US$/tC$_{eq}$)		without recycling 2.0 kg 10-year emission baseline (US$/tC$_{eq}$)	
Producer cost	Consumer cost	Producer cost	Consumer cost	Producer cost	Consumer cost	Producer cost	Consumer cost
102-173	306-519	21-53	63-159	460-711	1380-2133	59-109	177-327

a   Assuming:(i) equivalent energy efficiency for conventional and alternative systems, (ii) an increase of producer cost by US$60-90 per vehicle relative to current HFC-systems, (iii) a discount rate of 4% per year, and (iv) a factor of 3 between consumer cost and producer cost (Crain, 1999).

b   Incremental costs for the improved HFC system and for establishing and enforcing a recovery system are not included but assumed to be small compared to additional costs of alternative systems.

where systems already exist to ensure HFC recycling during servicing and at the end of life, alternative air conditioning systems will need to exhibit significantly reduced indirect emissions in order to be cost-effective in abating greenhouse gas emissions.

A3.2.2   Domestic Appliances

In developed countries, the only replacements for the fluid CFC-12 in refrigerators and freezers have been HFC-134a and isobutane (R-600a). Developing countries have chosen the same replacements, but some still utilize CFC-12; here, the complete conversion of new equipment from CFC-12 is not expected until 2001-2002. Globally, in 1996 to isobutane was used in about 8% of new appliances (UNEP, 1998a). Isobutane accounts for a much higher and growing percentage in Northern European countries such as Germany, where it is used in virtually all new domestic appliances. It is estimated that isobutane currently is the coolant used in 45%-50% of domestic refrigerator and freezer sold in Western Europe. Projected use and emissions of HFC-134a are shown in *Table A3.2*.

HFC emission reductions achieved during servicing and through recovery of the refrigerant upon disposal of appliances are costly. Next to economic incentives, regulations (as already exist for CFC-containing appliances in several countries) would probably be required to obtain significant emissions reductions through HFC recovery (March, 1998). One study (Harnisch and Hendriks, 2000) reports a value of US$334/tC$_{eq}$ for the recovery of HFCs from refrigerators; the larger part of this is the cost for the transport and collection scheme.

Product liability, export market opportunities, and regulatory differences among regions are likely to be significant factors in determining the choice between isobutane and HFC-134a systems. Isobutane may well account for over 20% of domestic appliances globally by the year 2010. Published estimates suggest that isobutane systems are US$15 to US$35 more expensive than HFC-134a systems (Juergensen, 1995; Dieckmann *et al.*, 1999). These costs would translate into a cost effectiveness of US$600/tC$_{eq}$ due to the relatively small refrigerant charge (about 120 g of HFC-134a).

A3.2.3   Commercial Refrigeration

The primary refrigerants used in this sector are R-404A and HFC-134a; usage of R-407C and R-507 is relatively small. Hydrocarbons are being applied in smaller direct expansion systems and in both small and large systems with a secondary loop, whereas ammonia is mainly applied in larger systems with secondary loops (UNEP, 1998a). Projected consumption and emissions of HFCs are shown in *Table A3.2*.

Historical emission rates of CFC refrigerants from the commercial refrigeration sector were 30% or more of the system charge per year. Regulations have resulted in improved system designs and service practices with significantly lower emissions in many countries (UNEP, 1998a; IEA, 1998). These practices are being carried over to HFC systems and the emissions savings are reflected in the projections shown in *Table A3.2* (UNEP, 1999b). March (1998) estimated that refrigerant emissions could be further reduced through better containment and recovery by an additional 30% to 50% in 2010 for Europe. In many developing countries, the supermarket refrigeration units are often produced by small and medium enterprises to lower quality standards, leading to considerable emissions of HFCs. The existing stock of supermarket refrigerators continues to operate with CFC-12 and HCFC-22.

The use of hydrocarbons and ammonia as refrigerants in this sector is growing from a small base. Several large commercial refrigeration manufacturers are developing systems using carbon dioxide which are expected to enter the market shortly. The HFC projections shown in *Table A3.2* are based upon the assumption that less than 10% of the systems will use ammonia, hydrocarbons, and carbon dioxide in 2010.

A3.2.4 *Residential and Commercial Air Conditioning and Heating*

Most existing residential air conditioning and heating systems (unitary systems) currently use HCFC-22 as the refrigerant; in the manufacturing of new systems HCFC-22 is being displaced by HFC blends, and to a lesser extent, by propane in some systems. In developed countries, the Montreal Protocol and more stringent national regulations are leading to a replacement of HCFC-22 in virtually all new equipment, ultimately by 2010. The leading HFC alternatives are R-407C and R-410A (UNEP, 1998a), the latter particularly for smaller units in the developed countries at present. In developing countries, HCFC-22 will be available for many more years and the use of HFC blends may remain small. Split HC based air conditioning equipment is produced by some smaller European manufacturers; production of these units is being announced by others. Estimated consumption and emission amounts for 2010 are shown in *Table A3.2*.

In small water chillers, applying a variety of compressor types, there is emphasis on the use of R-407C. For large water chillers that apply centrifugal compressors, the primary alternatives to CFCs are HFC-134a and HCFC-123. HCFC-123 is used in virtually all low-pressure chillers since it has a very high energy-efficiency and so far no highly efficient, low-pressure non-ODS alternative has become available (Wuebbles and Calm, 1997). Certain existing high-pressure HFC equipment or new low pressure HFCs may take over the low-pressure market gradually in the near future (IEA, 1998). Ammonia chillers form an important replacement and they are already in use in some regions. In large chillers, there is some use of water as a refrigerant, particularly in Northern Europe, where the water can also be used – in ice slurry form – as the cooling agent in the secondary loop. Use of hydrocarbon refrigerants for chillers is growing from a small base. Estimated consumption and emissions of HFCs are shown in *Table A3.2*.

Continued improvement in emissions reductions is anticipated. In 1994, the annual emission rates from low-pressure CFC chillers were estimated at 7% and for high pressure CFC-12 chillers at 17% (UNEP 1998a); for current new low (HCFC-123) and high pressure chillers the emissions are estimated at less than 2% and 8%, respectively.

A3.2.5 *Food Processing, Cold Storage, and Other Industrial Refrigeration Equipment*

Owing to their long lifetime, three out of four CFC systems are still in use in cold storage and food processing. The main non-CFC refrigerants used are ammonia, HCFC-22, HFC-134a, HFC blends, and hydrocarbons, with significant regional differences (see Section A3.4).

In the industrial sub-sector all types of refrigerants are used, with HCFCs and ammonia currently representing the majority of the refrigerant volume. Hydrocarbons hold a significant market share in industrial sub-sectors that handle flammable fluids. Since industrial refrigeration does not pose risks to the public, efficient ammonia and hydrocarbon systems are often used. The majority of the larger CFC systems used for cold storage and food processing are still in operation and may keep operating until 2010 to 2015.

Ammonia has traditionally been used in the cold storage sector because of its low cost and high efficiency. It has increased its importance in Europe and Australia. In the USA it is estimated to have a 90% market share in systems of 100kW cooling capacity and above; however, the market share of ammonia in industrial systems is much lower. In the developing countries, ammonia and HCFC-22 are expected to remain the most important alternatives. Unfortunately, many of these systems exhibit low efficiency due to poor system design.

HFC-134a has not been used much since the use of CFC-12 was traditionally small relative to HCFC-22. R-404A and R-507 are currently the most commonly used HFCs in these sub-sectors. However, their efficiencies are low compared to ammonia and HCFC-22 if the equipment is not very well designed. R-410A is well suited for industrial applications, with an insignificant market share at present, but it is estimated to grow significantly during the next decade (UNEP, 1998a). The HFCs are currently used in about 10% of new systems in Europe and in 20% in other developed countries. The demand for HFCs is expected to grow by about 40% between 2000 and 2010. It is expected that recovery and re-use will be cost effective in this sector. Rough estimates are that the emission rates are currently 6% per annum for new HFC systems, and are expected to decrease further over the next decade.

A3.2.6 *Transport Refrigeration*

Transport refrigeration relates to reefer ships, containers, railcars, and road transport. The majority of reefer ships currently use HCFC-22; the vast majority of new containers are equipped with HFC-134a or R-404A, and also R-410A.

For road transport, new equipment uses HFC-134a, R-404A, and still a considerable amount (estimated 25%) of HCFC-22. Owing to the mechanically and thermally harsh operating envi-

Table A3.5: *Alternative refrigerant options for the specific refrigeration and air conditioning sub-sectors: buildings (domestic and commercial refrigeration, residential and commercial air conditioners, chillers), industry (food processing and cold storage, other industrial processes), transport (transport refrigeration and mobile air conditioning)*

Refrigerant options	HFC-134a	R-404A	HFC blends	HCs	NH₃	Absorp-tion	HCFC	CO₂	H₂O
Domestic refrigeration	*			*					
Comm. refrigeration									
Small (< 5 kW)	*	*	*	*			*		
Other (> 5 kW)		*	*	* S	* S		*	P	
Residential A/C									
Unitary A/C (<20 kW)	*		*	*			*	P	
Commercial A/C									
Unitary A/C (>20 kW)			*	O/S					O
Chillers									
Centrifugal	*			O	*	*	*		O
Industrial									
Food processing	*	*	*	*	*		*		
Cold storage		*	*		*		*	P	
Other industrial	*	*		*	*	*	*		P
Transport refrigeration	*	*	*	*			*	O/P	
Mobile A/C	*			P				P	

Note: (*) indicates current practice, (O) small number installed, (P) prototype installed, (S) includes secondary loop.

ronment, the emissions estimated from transport refrigeration are significant and exceed 25% of the charge annually in many applications. One German manufacturer produces trucks equipped with refrigeration systems using propane.

Although the use of ammonia on ships has a reasonable potential, its proliferation has not been significant. Carbon dioxide has been tested for cooling containers; however, its future market share is difficult to predict and cost indications are lacking.

HFC-134a and, in the near future, R-410A are forecast to be the most important refrigerants for transport refrigeration. It is almost certain that all reefer ships will utilize R-410A. The fraction of equipment using HFCs in the mid-1990s was about 15% (UNEP, 1998a) and that fraction is expected to grow, e.g. one study (Harnisch and Hendriks, 2000) estimates the fraction in Europe to be 70% by 2010 and 100% by 2030.

In the developing countries, the use of HCFC-22 could continue until phase-out is required in 2040, after which HFC-134a or other options developed by then may take over the market.

A3.2.7 *Summary of Alternative Refrigerant Use*

An overview of the current pattern of refrigerant fluids by sub-sector and technology is provided in *Table A3.5*.

A3.3 Foams

A3.3.1 *Insulating Foams*

The global market for thermal insulation materials is large, complex, and has substantial regional variation. Since the prime purpose of insulation in addition to energy conservation is to maintain appropriate ambient conditions within a defined space, insulation use is affected most by external climatic conditions. However, in developing countries per-capita use is often lower than local climatic conditions would predict. Increasing insulation use can therefore often go towards improving comfort levels as well as saving energy use and resultant carbon dioxide emissions.

Climatic conditions, space constraints, local building code requirements, and construction costs can all influence the choice of insulation material. In mass markets polymeric foams offer the best insulation performance at higher unit cost. The thermal efficiency of foam is influenced by the choice of blowing agent, and HFCs promise to yield a performance similar to that of previously used HCFCs (UNEP, 1998d).

In Europe, where construction applications dominate, the market for insulation foams, polyurethane, extruded polystyrene, and phenolic resins accounts for about 13% of the total insulation market. Mineral fibres and expanded polystyrene have historically been the dominant materials in terms of volume and mass (roughly 80%), primarily on grounds of lower unit cost. However, performance characteristics are becoming an increasingly important factor in material selection to meet the demands of greater prefabrication.

In North America, the timber frame method of construction has contributed to a more widespread use of polyurethane (PU), polyisocyanurate (PIR), and extruded polystyrene foams. The PU and PIR systems also have better production economics than in Europe because of higher line speeds and less stringent thickness tolerance criteria. In Japan, the market is shaped by strict fire codes and much of the construction is based around concrete. PU spray foams have done particularly well in enclosed spaces and use of phenolic foam is preferred in some exposed applications because of its lower flammability compared to other alternatives. In developing countries, the use of foam for cold storage applications predominates.

Where HFCs are used, they will be emitted during the manufacturing and over the life of the foam (25–50 years). Retention in the foam at end-of-life will generally depend on the thickness of the foam and the facings used.

A3.3.2 *Insulating Foams in Appliances*

Appliance foams are currently produced with either hydrocarbons or HCFC-141b. Foams produced with HCFC-141b generally provide 5%-15% more insulation per unit of thickness than those produced with other blowing agents. HFCs (primarily "liquid" HFCs such as HFC-245fa and HFC-365mfc) are anticipated to partly replace HCFCs because they produce foams with similar insulating properties. The contribution of the foam to the overall energy efficiency of the appliance is important since the energy used to operate the appliance accounts for the majority of the global warming impacts in most cases.

Where HFCs are selected, options to reduce emissions include the use of formulations that minimize the amount and GWP of the blowing agent used, and the end-life destruction of the HFC. The latter is particularly important, since it is technically possible to recover and destroy over 90% of the HFC blowing agent at an estimated cost-effectiveness of between US$30 and US$100/tC_{eq} (AFEAS, 2000).

A recent study for the European Commission (Harnisch and Hendriks, 2000) estimates that in Europe about 70% of all

polyurethane foams for appliances will be blown with hydrocarbons and about 30% with HFCs by 2010. HFCs are more likely to be selected where more stringent energy standards exist. In contrast, the investment related to the introduction of a new blowing agent might play a determining role in developing countries. However, in practice, this effect has been broadly offset by the supporting activities of the multilateral fund. Significant concern about hydrocarbon use exists in North America and Japan, related to product liability and process safety costs.

Vacuum panels may partly replace insulation foam in the future but the cost-effectiveness of this option is uncertain. A few domestic and commercial applications already use vacuum insulation panels in combination with polyurethane foams. These systems have up to 20% lower energy consumption than those using CFC or HCFC blown foam insulation systems (UNEP, 1998d).

A3.3.3 *Insulating Foams in Residential Buildings*

The use of HFC blown foams in the residential sector is expected to be relatively limited because of the high cost-sensitivity of this market. However, it is preferable to base the choice of insulation for all buildings on a proper consideration of the LCCP, including the comparative energy saving impacts of alternative insulation materials, the potential emissions of blowing agents, and the embodied energies of the insulating materials themselves. Where HFCs are used, the cost to destroy the HFC will be determined primarily by the cost of separating the construction materials. There are trends towards prefabricated construction and requirements for recycling of building materials in some regions that could lower these costs in time. Emissions of HFCs partially offset the benefits of low energy consumption arising from their use.

The use of hydrocarbon and carbon dioxide as blowing agents for polyurethane and extruded polystyrene insulation foams is expanding. A recent European study (Harnisch and Hendriks, 2000) estimated that by 2010 about 50% of all polyurethane and extruded polystyrene foams in this sector will be blown by hydrocarbons and carbon dioxide, respectively. It is estimated that substituting the remaining HFCs by hydrocarbon use in the mass markets of polyurethane foam production would cost between US$90 and US$125/tC_{eq}. There is some concern about the use of flammable hydrocarbons in the residential environment and indoor air quality could also be affected. The replacement of HFCs by CO_2/water blown polyurethane spray systems is estimated to be available at a cost-effectiveness of about US$80/t$C_{eq}$ (Harnisch and Hendriks, 2000). In Europe, one major producer is converting its extruded polystyrene production lines to use CO_2 as the blowing agent. The cost-effectiveness of the use of CO_2 as the blowing agent for extruded polystyrene is estimated at US$40/t$C_{eq}$ (March, 1998) and at US$25/t$C_{eq}$ (Harnisch and Hendriks, 2000) for the remaining manufacturers in Europe.

Fibrous insulation materials and expanded polystyrene are used extensively for residential construction in most parts of the world. The increased thickness required to achieve a desired energy efficiency can cost more; however, builders have been willing to increase the cavity wall size substantially since the 1970s to comply with increasing insulation standards in some regions.

A3.3.4   Insulating Foams in Commercial Buildings

For commercial buildings, the choice of foam type and facings is more likely to be based on lifetime costs (performance related) than on initial cost. An additional factor is the increased use of prefabricated building techniques, particularly in Europe. Both aspects suggest that HFC blown foams could penetrate the commercial and industrial sectors to a greater extent than the residential sector previously discussed. Harnisch and Hendriks (2000) estimate that avoiding HFCs in most mass applications by switching to hydrocarbon systems would cost in the region of US$90 to US$125/tC$_{eq}$. For the switch from HFC to CO$_2$ use in extruded polystyrene, one study estimates US$40/tC$_{eq}$ (March, 1998), whilst another (Harnisch and Hendriks, 2000) estimates US$25/tC$_{eq}$ for the remaining European manufacturers.

Fibrous insulation materials and expanded polystyrene are used extensively for commercial construction and are expected to play a significant role in the future. However, whether this role will expand technically seems in doubt.

A3.3.5   Insulating Foams in Transportation

Hydrocarbon blown foams and vacuum insulation panels are alternative options. Hydrocarbon blown foams have a somewhat lower insulating value per unit of thickness than HFC blown foams, and the vacuum insulating panels currently cost substantially more. Insulating performance is crucial in this sub-sector and serious thickness constraints exist, limiting the available options.

A3.3.6   Other Insulating Foams

Another application of HFCs for insulating foams will be in industrial process applications, where an estimated 2500 tonnes will be used – primarily in process pipework. Owing to high foam densities in this sector the differences in insulation performance between different blowing agents are small. For Europe it is estimated (Harnisch and Hendriks, 2000) that in 2010 hydrocarbons will have a market share of 50% of the pipe insulation production.

Both the building industry and the do-it-yourself market use one-component foams in a variety of applications, including sound and thermal insulation applications. The thermal conductivity of the foam, however, is not a critical requirement.

HFC-134a and HFC-152a, hydrocarbons, propane, butane, and dimethyl ether (DME) are all technically suitable and in use. These are frequently used in blends; for example, a blend of HFC-134a/DME/propane/butane is widely used in Europe (UNEP, 1998d). Some replacement of HFC use in this sector is likely although concerns over the flammability of mixtures may delay this process in some regions.

A3.3.7   Non-Insulating Foams

Non-insulation HFC blown foams are expected to be used only in those applications where product or process safety are paramount, for example, integral skin foams for safety applications. Harnisch and Hendriks (2000) project that HFCs will not be required for the production of non-insulation foams in Europe. However, in view of different product specifications elsewhere in the world, liquid HFCs could replace a significant part of the current small use of HCFCs.

A3.4   Solvents and Cleaning Agents

Less than 3% of projected demand for CFCs solvents has been replaced by HFCs and PFCs (McFarland, 1999).The high cost of fluorocarbons, regulatory prohibitions on HFC and hydrofluoroethers and hydrofluoroesters (HFE) solvents, and investment in emission reduction measures are expected to maintain carbon equivalent use and emissions in 2010 to current baseline levels. Annual PFC solvent emissions are estimated at 3,000–4,000 tonnes (UNEP, 1999b; Harnisch *et al.*, 1999) and HFC emissions are estimated to be 1,000–2,000 tonnes (UNEP, 1999b). These values convert to less than 7.5MtC$_{eq}$ for PFCs and less than 1 MtC$_{eq}$ for HFCs.

Perfluorocarbons (PFCs such as C$_5$F$_{12}$, C$_6$F$_{14}$, C$_7$F$_{16}$, and C$_8$F$_{18}$) were introduced in the early 1990s as substitutes for ozone-depleting CFC-113 solvents and are also used in some applications where ODS solvents were never used. HFC-43-10mee and its azeotropic blends with alcohol, hydrochlorocarbons, and hydrocarbons were introduced in the mid-1990s to replace CFC-113 and PFCs. HFE solvents became commercially available in the late 1990s to replace PFCs, CFCs, HCFCs, and HFCs.

HFCs and HFEs are used in specialized cleaning of delicate materials, oxygen systems, and precision parts; as a flush fluid for particulate removal in precision cleaning; as a rinsing agent in a co-solvent process for cleaning printed circuit boards and mechanical components; and to dry electronics and precision parts after aqueous or semi-aqueous processing. In some circumstances, HFC drying may have a lower LCCP than thermal drying. HFCs and HFEs are also replacing PFCs and CFC-113 as carrier fluids for specialized fluorocarbon lubricants, as dielectric and heat transfer fluids, in developing latent fingerprints off porous surfaces, in rain repellent sprays for aircraft windshields, and in other applications demanding unique solvency properties (UNEP, 1998e, 1999b).

The four emission reduction options are: (1) changing production processes and product designs to avoid the need for fluorocarbon solvents (e.g., "no-clean" soldering and aqueous cleaning); (2) switching to lower GWP fluorocarbon or non-fluorocarbon solvents; (3) reducing emissions through process improvements (UNEP, 1999b); and (4) utilizing solvent recovery and recycling where possible. Progress is being made in each of these options.

One source estimates that process improvements could reduce fluorocarbon solvent emissions in the European Union by 20% by 2010 at a cost effectiveness of about US$160/t$C_{eq}$ and that an 80% reduction could be achieved at about US$330/t$C_{eq}$ (March, 1998).

A3.5   Aerosol Products

A3.5.1   Medical Applications

Metered dose inhalers (MDIs) form a reliable and effective therapy for asthma and chronic obstructive pulmonary disease (COPD).

There are estimated to be 300 million patients with asthma and COPD worldwide. Approximately 450-500 million MDIs are used annually worldwide with asthma prevalence increasing as urbanization of developing countries continues. It is estimated that 10,000 metric tons (tonnes) of CFC and 1000 tonnes of HFCs were used in MDIs worldwide in 1998 (UNEP, 1998b, 1999b). HFC-based MDIs are essential for the near-term CFC phaseout, because other available options, including dry powder inhalers (DPIs) (single or multi-dose), nebulizers (hand held or stationary), orally administered drugs (tablets, capsules, or oral liquids), and injectable drugs, which are alternatives for not using CFCs or HFCs, cannot currently replace CFC products for all patients (UNEP, 1999b). The transition to HFC MDIs began in 1995, and approximately 5% in 1998 and 10% in 1999 contain HFC (UNEP, 1999b). HFC-based MDIs and DPIs are expected to help minimize the use of CFCs by 2005 in developed countries, while providing essential medication for patients. Important factors in the conversion to DPIs will include their acceptance by doctors, patients, insurance companies, and medical authorities.

Assuming the complete phase-out of CFC MDIs and a continued growth rate in demand for asthma and COPD treatment of 1.5%–3.0%/yr, it is estimated that HFC consumption and emissions will be 7,500 to 9,000t/yr – about 3–3.6MtC_{eq} in 2010 (UNEP, 1999b).

DPIs have been formulated successfully for many anti-asthma drugs. Dry powder inhalers are an immediately available alternative free of CFCs and HFCs; however, they are not a satisfactory alternative to the pressurized MDIs for some patients with very low inspiratory flow (e.g., some small children and elderly people, patients) with acute asthma attacks or with severe respiratory diseases, and emergency-room patients. Use is likely to accelerate, particularly as they may be more suitable for young children than the older DPIs (UNEP, 1999b). In Scandinavian countries, government policies have led to greater use of DPIs than of MDIs (IPCC/TEAP, 1999; UNEP, 1999b; March, 1998).

The abatement cost estimates to reduce future HFC emissions by replacing MDIs with DPIs depend on the price of DPIs. The cost per equivalent dose varies between products and countries, with some CFC-free MDIs being more expensive than CFC-based MDIs and some DPIs more expensive than both CFC- and HFC-based MDIs (ARCF, 2000). In Europe, prices are less as much as US$4 higher for a DPI than for a comparable MDI (Harnisch and Hendriks, 2000). It is estimated that, by 2010, the EU can reduce HFC emissions by 30% at a cost of about US$460/t$C_{eq}$ and 50% at about US$490/t$C_{eq}$ (March, 1998), which translates to a differential cost of US$4 over MDIs; for one country in Europe there is no differential cost (Harnisch and Hendriks, 2000). It is not currently medically feasible to replace MDIs by DPIs completely because approximately 25% of MDI use is for patients who require medication be forced into their respiratory system (Öko-Recherche, 1999).

A3.5.2   Cosmetic, Convenience, and Technical Aerosol Propellants

Global 1998 consumption and emissions of HFCs in non-medical aerosol products was less than 15,000 tonnes (UNEP, 1998d) with two-thirds HFC-134a and one-third HFC-152a – less than 4MtC_{eq}. Emissions of HFCs are projected to not exceed 20,000 tonnes in 2010 (IPCC/TEAP, 1999) or about 5MtC_{eq} (calculated assuming equal emissions of HFC-134a and HFC-152a). HFCs have replaced only about 2% of the aerosol product market that would have used CFCs had there not been the Montreal Protocol (McFarland, 1999). Hydrocarbon, dimethyl ether (DME), carbon dioxide, nitrogen propellants, and not-in-kind alternative products have replaced the remaining 98% of projected demand.

HFCs are used in aerosol products primarily to comply with technical requirements or environmental regulations. HFC-134a is the propellant of choice for products that must be completely non-flammable. An example of HFC use based on a technical requirement is non-flammable, far-reaching insecticide products used on high-voltage power lines and transformers where workers cannot escape from wasps and hornets. HFC-152a is the propellant of choice to replace hydrocarbon aerosol propellants restricted in Southern California and in some applications where hydrocarbons and dimethyl ether are too flammable but the flammability of HFC-152a is acceptable. HFC-134a and HFC-152a are the propellants of choice for laboratory, analytical, and experimental uses where chemical properties are important and flammability may be a concern.

One source estimates that about 45% of HFC emissions from cosmetic and convenience applications where flammability is an issue could be eliminated at a cost of US$70/tC$_{eq}$ and about 70% could be eliminated at a cost of about US$130/tC$_{eq}$ (March, 1998).

The aerosol product industry has every incentive to minimize HFC use. HFCs cost more than other propellants and unnecessary HFC use has the potential to re-ignite consumer boycotts like the CFC boycotts in the early 1970s that led to national bans on certain cosmetic products. Boycotts could threaten sales of all aerosol products because consumers may not be able to distinguish targeted HFC products from acceptable hydrocarbon products (UNEP, 1999b).

A "self-chilling beverage can" was designed to achieve refrigeration through the physics of expanding and emitting approximately 35–75g of HFC-134a directly to the atmosphere for every beverage can chilled. The inventing company pledged not to manufacture or license the technology and to discourage its use, the US government banned the use of HFCs in self-chilling beverage cans (US Federal Register, 1999), and a number of HFC producers have stated publicly that they will not supply such an application. However, self-chilling cans using HFC-134a are marketed in at least one country and it is estimated that even a small market penetration could substantially increase emissions of greenhouse gases (US Federal Register, 1999).

The UNEP/TEAP HFC and PFC Task Force (UNEP, 1999b) developed principles to guide the use of HFCs for aerosol products:

- recommend HFCs be used only in applications where they provide technical, safety, energy, or environmental advantage that are not achieved by not-in-kind alternatives; and
- select the HFC compound with the smallest GWP that still meets the application requirements.

Application of these principles justifies the use of HFCs for some products in some circumstances but these "responsible use" criteria are not satisfied when not-in-kind alternatives are technically and economically suitable. The above-mentioned study (UNEP, 1999b) includes detailed evaluation of alternatives and substitutes for aerosol safety products (insecticides, boat horns, noise-makers), cosmetic products (deodorants, hair sprays, shaving creams), convenience products (room fresheners, dust blowers, tyre inflators, foam caulk, and insulation), and novelty products (foam party streamers, pneumatic pellet and bait guns).

A3.6 Fire Protection

A range of alternatives to halon with no or low GWP, such as water-based technologies, dry powders, inert gases, and carbon dioxide, have displaced about 75% of previous halon use in countries classified as developed under the Montreal Protocol.

About 5% of the existing and new halon applications are considered critical, with no technically or economically feasible alternatives. These critical uses include military vehicles, civil and military aircraft, and other high-risk explosion scenarios involving unacceptable threat to humans, the environment, or national security. Recovered and recycled halon is being used to meet these needs (IPCC/TEAP, 1999).

Relatively small, but important, quantities of HFCs and PFCs are being used as substitutes for halon in fire protection. About 20% of the systems that would have used halons in the absence of the Montreal Protocol currently use HFCs and only about 1% use PFCs (UNEP, 1998c, 1999b).

Growth in HFC use is limited by high cost compared to other choices. PFCs are not technically necessary as halon replacements except in rare and special circumstances (UNEP, 1999b). However, relatively strong growth of HFC/PFC use in developing countries and countries with economies in transition is being driven by aggressive marketing, and is producing a new dependency that could lead to a rapidly growing market in applications where other alternatives are available. Awareness campaigns involving fire protection experts and their customers could help limit uses that are not technically justified (UNEP, 1999b).

The Montreal Protocol prompted various improvements in the management of halons and their replacements, resulting in a fourfold decrease in annual emissions. Testing and training with halon and HFC was eliminated and the unintended discharges of systems were greatly reduced through intensified maintenance and operational improvements. With only 20% of new fire protection systems using HFCs and with the fourfold decrease in emissions, HFC emissions are 5% compared to those from halon systems before the Montreal Protocol (UNEP, 1999b; McFarland, 1999).

Emissions of HFC from the installed bank of fire protection equipment, including necessary emissions to suppress fires, are estimated to be about 4%–6% per year (UNEP, 1999b). These emissions could be reduced by up to 50% through continued improvements to eliminate unnecessary discharges and by increased recycling of the HFCs (IPCC/TEAP, 1999). There are no estimates of the cost-effectiveness of such measures.

A3.7 Developing Countries and Countries with Economies in Transition

Developing countries have until 2010 to phase out CFCs, whereas some countries with economies in transition (EIT) have largely met the more stringent schedules of the developed countries. However, both country groups are concerned that any potential future restrictions on the use of HFCs in the developed countries might reduce the availability of these substances to developing countries and EITs. This could limit the possibilities for them to comply with their Montreal

Protocol obligations. Possible impacts are anticipated in the refrigeration, air conditioning, and foam sectors. It will be clear that the availability of HFC supplies to those developing countries and EITs that have selected HFC technologies is essential for manufacturing if supplies and service to customers are to be maintained.

It will be advantageous for both developing countries and countries with economies in transition if they develop and prioritize consistent strategies that simultaneously address the protection of the ozone layer and the mitigation of climate change. Such strategies utilized to date include emission reductions, the selection of zero ODP and low GWP solutions wherever possible, as well as the optimization of the energy efficiency of products in conversion projects by the Multilateral Fund (MLF) and the Global Environment Facility (GEF). The mechanisms that have guided developing countries towards a successful implementation of the Montreal Protocol should be studied within the framework of mechanisms that are being negotiated for the Kyoto Protocol. It can be emphasized here that capacity building is seen as at least as important for the implementation of the Kyoto Protocol as it is for the Montreal Protocol.

A3.7.1   Technology Selection

Certain non-ODP substitutes and alternative technologies to CFCs and HCFCs have become available in recent years for many applications. The selection of the substitute or alternative technology is based on a balance of maturity, availability, cost-effectiveness, energy-efficiency, safety, and safety costs. The selection is also influenced by local circumstances, preferences of enterprises, accessibility and cost-effectiveness of certain technologies, joint venture partners and customers, availability of training, and regulatory compliance. This implies that developing countries need access to the newest information and need to be part of an adequate technical review process so that they can assess the choice of the most appropriate and integrated environmental solutions. In addition, those developing countries that receive financial assistance from the MLF for the conversion process and select HCFC-based technologies must submit a thorough justification as to why these are preferred. This is because the countries have to take into account the decisions by the Montreal Protocol Parties that state that certain fluorocarbon-based technologies should be avoided if more environmentally friendly and acceptable technologies are available, as well as the guidelines developed by the MLF Executive Committee for the implementation of these technologies.

A3.7.2   Impact of Replacement Technology Options in Montreal Protocol MLF Projects

HCFC- and HFC-based technologies have not been significant alternative choices in the phase-out of ODSs in aerosols and in solvent applications, or in the fire extinguishing sector. However, HCFCs and HFCs have been selected as significant alternatives to ODSs in the foam, refrigeration, and air conditioning sectors. *Table A3.6* shows the quantities of controlled substances (CFCs) that have been (or are in the process of being) phased out in developing countries through projects approved under the Montreal Protocol's multilateral fund (to date, over US$1 billion has been used to support these phase-out activities). *Table A3.6* also shows the replacement technology selected in the different refrigeration and foam sectors and sub-sectors. *Table A3.6* presents data for projects, approved by the Executive Committee of the MLF and listed under the Inventory of Approved Projects of the MLF Secretariat as of March 1999, see UNEP (1999b).

A3.7.2.1   Foams Sector

The present contribution of HFCs as a direct replacement technology for ODSs in projects approved under the MLF in the foam sector is much less than 1% of the total tonnage of ODS replaced in this sector. *Table A3.6* presents the breakdown of ODS replaced in each of the foam sub-sectors. The contribution of hydrocarbons is significant. The overall contribution of zero-ODP and low-GWP technologies selected to replace ODSs is close to 27,000 ODP tonnes or about 75% (see *Table A3.6*).

Wherever application of zero-ODP technologies was not feasible because of availability, safety, and safety-related costs, or for energy-efficiency reasons, HCFCs (HCFC-22, -141b, and -142b) have been selected as a transitional replacement in all foam sub-sectors. In the medium term HCFCs are expected to be replaced by zero-ODP and low-GWP substitutes, such as water, carbon dioxide or hydrocarbons, except in certain parts of the rigid polyurethane foam sub-sector where HFC alternatives are expected to play an important role in the medium to long term. *Table A3.6* also presents the HCFC contribution; it amounts to about 25% of the total tonnage of ODSs replaced.

While several mid-size and large domestic and commercial refrigeration companies have switched to hydrocarbons in the rigid foam sector, most small and medium-sized enterprises (SMEs) in the developing countries have had more difficulties in this selection of hydrocarbons because of safety concerns and related higher manufacturing costs. Next to large companies, many of these SMEs have selected HCFC-141b as a transitional substance. All these companies will have to switch to the use of other, non-ODP substances when HCFC availability cannot be guaranteed or HCFCs will be phased out according to Montreal Protocol schedules. It is expected that a large part of this SME sector will convert to HFC alternatives in the medium to long term. With regard to HCFCs, questions on HCFC availability after 2003 are of serious concern to developing countries; these will be evaluated by the Technology and Economic Assessment Panel, at the request of the Parties to the Montreal Protocol. With regard to HFCs, it should be mentioned that enterprises are uncertain whether their businesses

Table A3.6: Replacement technology options in multilateral Fund-approved projects in developing countries (UNEP, 1999b)

Use sector (# projects)	ODS	Impact (ODP t)	ODP tonnes to be eliminated according to technology selected							
			HCFC Type	(t)	HFC Type	(t)	Hydrocarbons Type	(t)	Other Type	(t)
1-Refrigeration										
a-Domestic (168)										
Foam	CFC-11	16,589	HCFC-141b	4,379		0	Cyclopentane	12,188		0
Refrigerant	CFC-12	5,241		0	HFC-134a/152a/blends	4,553	Isobutane	688		0
b-Commercial (161)										
Foam	CFC-11	2,432	HCFC-141b	1,648		0	Cyclopentane	784		0
Refrig (plus chillers)	CFC-12	1,136	HCFC-22	4	HFC-134a	1,132		0		0
	R-502	1	HCFC-22	1						
c-Insulation foam (34)	CFC-11	1,998	HCFC-141b/blends	636		0	Cyclopentane	849	H_2O/CO_2	513
	CFC-12	8	HCFC-141b	8		0		0		0
2-Foam										
a-Flexible molded (12)	CFC-11	450	HCFC-141b	66		0		0	H_2O/CO_2/Me-Cl	384
b-Flexible slabstock (159)	CFC-11	11,934	HCFC-141b	35		0		0	H_2O/CO_2/Me-Cl	11,899
c-Integral skin (84)	CFC-11	2,573	HCFC-141b	597		0	Hexane/pentane	345	H_2O/CO_2	1,631
d-Polystyr./polyethyl. (63)	CFC-11	1,204	HCFC-141b	0		0	Butane/ isobutane/ LPG/ Pentane / isopentane	980	CO_2/CO_2-butane blend	224
e-Rigid foam (238)	CFC-12	6,280	HCFC-22/-142b	196		0	Butane/LPG/pentane	6,084		0
	CFC-114	40		0		0	LPG	40		0
	CFC-11	10,938	HCFC-141b/-22/-142b	7,144	HFC-134a	58	Cyclopentane	3,003	H_2O/CO_2	733
f-Multiple sub-sector (30)	CFC-11	1,829	HCFC-141b	556		0	Butane	200	H_2O/ CO_2/ Me-Cl/ LCD	1,073

Note: Data have been reproduced from the data presented in UNEP (1999b) which were directly taken from the internal report "Inventory of Approved Projects", as published by the Multilateral Fund Secretariat, Montreal, March 1999.

will be impacted if, in the near future, certain developed countries decide to put certain (national) restrictions on the use of HFCs, influencing their availability for the developing countries.

A3.7.2.2   Refrigeration Sector

There are only a limited number of options to replace ODSs in this sector. HCFCs have been selected as an interim replacement technology for ODSs, where non-ODP alternatives could not be applied, and their share represents about 24% of the total tonnage of ODSs replaced in the sector as a whole (see *Table A3.6*).

In refrigeration products the foam considered is exclusively rigid polyurethane foam. As a direct replacement for ODS blowing agents in the foam, the contribution of HFCs is negligible in the projects approved by the Multilateral Fund. In contrast to this very small contribution, hydrocarbons have accounted for 53% of the total ODS replacement in the sector, which includes both the refrigeration and the foam part; their share is about 66% in the replacement of ODS foam blowing agents. In these projects, zero-ODP and low-GWP alternatives could meet the requirements on availability, safety and safety related costs, and the stringent energy efficiency.

As shown in *Table A3.6*, the contribution of HFC-based technology as a direct refrigerant replacement technology is close to 21% of the total ODS replacement in the sector, if both the refrigeration and the insulating foam part are included. Where it concerns the refrigeration part, for both domestic and commercial refrigeration, HFCs constitute about 89% of the refrigerant replacement. The conversion of refrigeration components and the refrigeration manufacturing plants is to a large extent determined by market availability and by market forces (compressor suppliers); of course, there is also a direct relation to manufacturing and safety costs.

A3.7.3   General Concerns and Opportunities

For the developing countries, financial assistance is available for agreed incremental costs associated with the ODS phase-out through the multilateral fund under the Montreal Protocol. Likewise, financial assistance from the GEF is available for countries with economies in transition. GEF financing is currently available to improve energy efficiency and other reductions of greenhouse gas emissions. The Clean Development Mechanism (CDM), guidelines of which are still being negotiated within the Kyoto Protocol framework, might also provide opportunities to reduce HFC emissions.

Further opportunities exist in those parts of the refrigeration and air conditioning sector in which large emissions of HFCs occur, and for equipment that will need thorough maintenance; this particularly applies to mobile air conditioning, commercial, and transport refrigeration. Where emission reductions are possible, best-practices training is needed (UNDP *et al.*, 1999).

Under the Multilateral Fund, enterprises are eligible for financial assistance for only one conversion. This makes it crucial for an enterprise to choose a technology that is cost effective, environmentally acceptable, and globally sustainable. It is very important that developing countries and countries with economies in transition examine opportunities for consistent strategies to simultaneously protect the ozone layer and to mitigate climate change. Such opportunities, *inter alia*, may be in the field of emission reductions, the direct transition to non-fluorocarbon low GWP alternatives where possible, as well as in the field of enhancing energy efficiencies. It would be advantageous if assistance given by the multilateral fund could be expanded to extra assistance from the GEF in terms of addressing the energy efficiency optimization aspect.

To date, when funds were available, manufacturers in the developing countries have responded rapidly to the goals of the Montreal Protocol, and to regulations in the developed countries that prohibited import of products made with or containing ODSs. Uncertainties regarding the availability of HCFCs and regarding the impact of possible restrictions on the use of HFCs in certain developed countries may delay the implementation of the Montreal Protocol in EITs and developing countries; this aspect can be considered as an interlinkage between the Montreal and Kyoto Protocols and it is the subject of further study.

References

AFEAS/DOE (Alternative Fluorocarbon Environmental Acceptability Study and U.S. Department of Energy), 1991: *Energy and Global Warming Impacts of CFC Alternative Technologies.* S.K. Fischer, P.J. Hughes, and P.D. Fairchild (eds.), Washington.

AFEAS (Alternative Fluorocarbon Environmental Acceptability Study), 2000: *Development of a Global Emission Function for Blowing Agents Used in Closed Cell Foam.* Submitted to AFEAS by Caleb Management Services, September.

ARCF (American Respiratory Care Foundation), 2000: Consensus Statement: Aerosol and Delivery Devices. *Respiratory Care Journal,* **45**(6), June 2000, from: Conference Proceedings of the American Respiratory Care Foundation, 24-26 September 1999).

Baker, J.A., 1998: *Mobile Air Conditioning and Global Warming.* Proceedings of the Phoenix Alternate Refrigerant Forum 1998, Suntest Engineering, Scottsdale, AZ, USA.

Baker, J.A., 1999: Mobile Air Conditioning: HFC-134a Emissions and Emission Reduction Strategies. In *Proceedings of the Joint IPCC/TEAP Expert Meeting on Options for the Limitation of Emissions of HFCs and PFCs.* L. Kuijpers, and R. Ybema (eds.), Energieonderzoek Centrum Nederland (ECN), Petten, Netherlands, 15 July 1999, ECN-RX-99-029.

Baker, J.A., 2000: *Vehicle Performance of Secondary Loop Air Conditioning Systems (using flammable refrigerants).* Proceedings of the MACS 2000 Technical Conference and Trade Show, Las Vegas, NV, USA, January 2000.

Clodic, D., L. Palandre, and A.M. Pougin (eds.), 1999: *Inventory and Provision of Emissions of HFCs Used as Refrigerants in France According to Different Scenarios.* Final Report ADEME/ARMINES 98.74061, m016.07775.

Clodic, D., Y.S. Chang, and A.M. Pougin, 1999: *Evaluation of Low-GWP refrigerants for Domestic Refrigeration, Commercial Refrigeration, Refrigerated Transport and Mobile Air Conditioning.* Final Report for the French Ministry of Environment, Centre d'Energétique, Ecole des Mines, Paris.

Crain, K.E., 1999: *Automotive News '99 Market Data Book.* Crain Communications Inc., Detroit, MI, pp. 75-88.

Dentis, L., A. Mannoni, and M. Parrino, 1999: *HC Refrigerants: An Ecological Solution of Automotive A/C Systems.* Vehicle Thermal Management Systems (VTMS4) Conference Proceedings, Professional Engineering Publishing Ltd, London, UK.

Dieckmann, J., A.D. Little and H. Magid (eds.), 1999: *Global Comparative Analysis of HFC and Alternative Technologies for Refrigeration, Air Conditioning etc.* A.D. Little, Inc.k, Cambridge, ref. nr. 49468.

EPA (Environmental Protection Agency), 1998: *Regulatory Impact Analysis, the Substitutes Recycling Rule.* Section 608 and 609 of the Clean Air Act Amendments of 1990, 18 May 1998.

Gamlen, P.H., B.C. Lane, P.M. Midgley, and J.M. Steed, 1986: The Production and Release to the Atmosphere of CCl_3F and CCl_2F_2 (Chlorofluorocarbons CFC-11 and CFC-12). *Atmospheric Environment,* 20, 1077-1085.

Gentner, H., 1998: *RACE (Refrigeration and Automotive Climate Systems under Environmental Aspects) Synthesis Report for Publication. Contract No: BRE2-CT94-0555, Project No. 7282.* Proceedings of the Phoenix Alternate Refrigerant Forum 1998. Suntest Engineering, Scottsdale, AZ, USA, Report funded by the European Community (Brite Euram III), BMW AG, München, Germany.

Ghodbane, M., 1999: *An Investigation of R-152a and Hydrocarbon Refrigerants in Mobile Air Conditioning.* Society of Automotive Engineers (ed.), Warrendale, PA. Paper No. 1999-01-0874 presented at The International Congress and Exposition, Detroit, MI, 1-4 March 1999.

Harnisch, J., I.S. Wang, H.D. Jacoby, and R.G. Prinn, 1999: Primary Aluminum Production: Climate Policy, Emissions and Costs. In *Proceedings of the Joint IPCC/TEAP Expert Meeting on Options for the Limitation of Emissions of HFCs and PFCs.* L. Kuijpers and R. Ybema (eds.), Energieonderzoek Centrum Nederland (ECN), Petten, Netherlands, 15 July 1999, ECN-RX-99-029.

Harnisch, J., and C. Hendriks, 2000: Economic Evaluation of emission reduction of HFCs, PFCs and SF6 in Europe. Contribution to the Report *Economic Evaluation of Sector Targets for Climate Change.* Ecofys Energy & Environment, Utrecht, The Netherlands, for the DG Environment of the European Commission, also available at a website: http://europa.eu.int/comm/environment/enveco/climate_change/sectoral_objectives.htm)

IEA, 1998: *Guidelines for Design and Operation of Compression Heat Pump, Air Conditioning and Refrigeration Systems with Natural Working Fluids.* J. Stene (ed.), IEA Heat Pump Programme, Annex 22, Report HPP-AN22-4, SINTEF Energy Research, Trondheim.

IPCC (Intergovernmental Panel on Climate Change), 1995: *Climate Change 1994: Radiative Forcing of Climate Change and an Evaluation of the IPCC IS92 Emission Scenarios.* J.T. Houghton, L.G. Meira Filho, J. Bruce, H. Lee, B.A. Callander, E. Haites, N. Harris, and K. Maskell (eds.), Cambridge University Press, Cambridge, UK.

IPCC (Intergovernmental Panel on Climate Change), 1996: *Climate Change 1995: The Science of Climate Change.* J.T Houghton, L.G. Meira Filho, B.A. Callander, N. Harris, A. Kattenberg, and K. Maskell (eds.), Cambridge University Press, Cambridge, UK.

IPCC (Intergovernmental Panel on Climate Change), 2000: *Emissions Scenarios.* N. Nakicenovic, J. Alcamo, G. Davis, B. de Vries, J. Fenhann, S. Gaffin, K. Gregory, A. Grübler, T.Y. Jong, T. Kram, E. Lebre La Rovere, L. Michaelis, S. Mori, T. Morita, W. Pepper, H. Pitcher, L. Price, K. Riahi, A. Roehrl, H-H. Rogner, A. Sankovski, M. Schlesinger, P. Shukla, S. Smith, R. Swart, S. van Rooijen, N. Victor, and D. Zhou (eds.), Special Report of WGIII of the IPCC, Cambridge University Press, Cambridge, UK, ISBN 0-521-80493-0.

IPCC/TEAP (Intergovernmental Panel on Climate Change and Technology and Economic Assessment Panel), 1999: Proceedings of the *Joint IPCC/TEAP Expert Meeting on Options for the Limitation of Emissions of HFCs and PFCs.* L. Kuijpers, and R. Ybema (eds.), Energieonderzoek Centrum Nederland (ECN), Petten, Netherlands, 15 July 1999, ECN-RX-99-029.

Juergensen, H., 1995: Experiences with Hydrocarbon Use in Household Refrigerators and Freezers. In *Proceedings of the Conference on Hydrocarbons and other progressive answers to refrigeration, Washington DC, 22 October 1995.* Greenpeace International (ed.).

Kobayashi, N., T. Matsuno, and T. Hirata, 1998: *Concerns of CO_2 A/C System for Compact Vehicles.* Proceedings of the 1998 Earth Technologies Forum. Washington D.C., October 1998.

March, 1998: *Opportunities to Minimise Emissions of Hydrofluorocarbons (HFCs) from the European Union.* March Consulting Group, UK.

McFarland, M., 1999: *Applications and Emissions of Fluorocarbon Gases: Past, Present and Prospects for the Future.* In *Proceedings of the NCGG-2 Symposium 'Non-CO_2 Greenhouse Gases - Scientific Understanding, Control and Implementation'.* J. van Ham, A.P.M. Baede, L.A. Meyer, and R. Ybema (eds), Kluwer Academic Publishing, Dordrecht, The Netherlands, 2000.

Midgley, P.M., and A. McCulloch, 1999: Properties and Applications of Industrial Halocarbons. *The Handbook of Environmental Chemistry,* **4**, Part E, Springer-Verlag, Berlin, Heidelberg.

Öko-Recherche, 1999: *Emission and reduction potentials for HFCs, PFCs and SF_6 in Germany.* A study commissioned by the German Environmental Protection Agency, October 1999, Germany.

Papasavva, S., and W.R. Moomaw, 1997: Adverse Implications of the Montreal Protocol Grace Period for Developing Countries. *Journal of International Environmental Affairs,* **9**(3), 219 -231.

Papasavva, S., and W.R. Moomaw, 1998: Life Cycle Global Warming Impact of CFCs and CFC Substitutes for Refrigeration. *Journal of Industrial Ecology,* 1, 71-91.

Petitjean, C., G. Guyonvarch, M. BenYahia, and R. Beauvis, 2000: TEWI Analysis for Different Automotive Air Conditioning Systems. In *Proceedings of The Future Car Congress, April 2000, Crystal City, VA, USA.* Society of Automotive Engineers.

SAE (Society of Automotive Engineers), 1999: *SAE Alternate Refrigerants Symposium,* Scottsdale, AZ. Sponsored by the Society of Automotive Engineers, Warrendale, PA, USA.

SAE, 2000: *Technical Options for Motor Vehicle Air Conditioning Systems.* S.O. Andersen, W. Atkinson, J.A. Baker, S. Oulouhojan, and J.E. Phillips (eds.), Prepared for the Society of Automotive Engineers Mobile Air Conditioning Climate Protection Partnership, March 2000, http: www.sae.org/misc/aaf99/index.html.

Sand, J.R., S.K. Fischer, and V.D. Baxter, 1997**:** *Energy and Global Warming Impacts of HFC Refrigerants and Emerging Technologies.* Prepared by Oak Ridge National Laboratory, sponsored by AFEAS and the U.S. Department of Energy, Oak Ridge, TN, USA, 214 pp.

UNDP, UNIDO and World Bank, 1999: Options for Reduction of Emissions of HFCs in Developing Countries. F. Pinto, S.M. Si-Ahmed, and S. Gorman. In *Proceedings of the Joint IPCC/TEAP Expert Meeting on Options for the Limitation of Emissions of HFCs and PFC.* L. Kuipers and R. Ybema (eds.), Energieonderzoek Centrum Nederland (ECN), Petten, Netherlands, 15 July 1999, ECN-RX-99-029.

UNDP, 1999: *Brazil MAC Survey, First Phase Report – Automobiles.* Submitted by the United Nations Development Programme, Montreal Protocol Unit, SEED/BDP, on behalf of the Brazilian Government to the Multilateral Fund, October 1999.

UNEP (United Nations Environment Programme), 1998a: UNEP Refrigeration, Air Conditioning and Heat Pumps Technical Options Committee (RTOC). *1998 Report of the Refrigeration, Air Conditioning and Heat Pumps Technical Options Committee.* 1998 RTOC Assessment Report, United Nations Environment Programme, Nairobi, December 1998, 285 pp., ISBN 92-807-1731-6.

UNEP, 1998b: UNEP Aerosols, Sterilants, Miscellaneous Uses and Carbon Tetrachloride Technical Options Committee (ATOC). *1998 Report of the Aerosols, Sterilants, Miscellaneous Uses and Carbon Tetrachloride Technical Options Committee.* 1998 ATOC Assessment Report, United Nations Environment Programme, Nairobi, March 1999, 98 pp., ISBN 92-807-1726-X.

UNEP, 1998c: UNEP Halons Technical Options Committee (HTOC). *1998 Report of the Halons Technical Options Committee.* 1998 HTOC Assessment Report, United Nations Environment Programme, Nairobi, April 1999, 210 pp., ISBN 92-807-1729-4.

UNEP, 1998d: UNEP Flexible and Rigid Foams Technical Options Committee (FTOC). *1998 Report of the Flexible and Rigid Foams Technical Options Committee.* 1998 FTOC Assessment Report, United Nations Environment Programme, Nairobi, March 1999, 80 pp., ISBN 92-807-1728-6.

UNEP, 1998e: UNEP Solvents, Coatings and Adhesives Technical Options Committee (STOC). *1998 Report of the Solvents, Coatings and Adhesives Technical Options Committee.* 1998 STOC Assessment Report, United Nations Environment Programme, Nairobi, April 1999, ISBN 92-807-1732-4.

UNEP, 1998f: UNEP Technology and Economic Assessment Panel (TEAP). *1998 Report of the Technology and Economic Assessment Panel.* 1998 TEAP Assessment Report, United Nations Environment Programme, Nairobi, December 1998, 286 pp., ISBN 92-807-1725-1.

UNEP, 1999: UNEP HFC and PFC Task Force of the Technology and Economic Assessment Panel (TEAP). *The Implications to the Montreal Protocol of the Inclusion of HFCs and PFCs in the Kyoto Protocol.* 1999 HFC/PFC Task Force Report, United Nations Environment Programme, Nairobi, October 1999, 202 pp.

US Federal Register, 1999: 40 CFR Part 82: *Protection of Stratospheric Ozone - Listing of Substitutes for Ozone-Depleting Substances – Final Rule,* **64**(41), 3 March 1999.

Van Ham, J., A.P.M. Baede, L.A. Meyer, and R. Ybema, 2000: *Non-CO$_2$ Greenhouse Gases: Scientific Understanding, Control and Implementation.* Kluwer Academic Publishing, Dordrecht, The Netherlands.

Wertenbach, J., and R. Caesar, 1998**:** An Environmental Evaluation of an Automobile Air-Conditioning System with CO$_2$ versus HFC-134a as Refrigerant. In *Proceedings of the Phoenix Alternate Refrigerant Forum 1998.* Suntest Engineering, Scottsdale, AZ, USA.

WMO, 1999: *Scientific Assessment of Ozone Depletion: 1998.* World Meteorological Organisation Global Ozone Research and Monitoring Project, Report No. 44, Geneva, pp. 10.18-10.32.

Wuebbles, D.J., and J.M. Calm, 1997: An environmental rationale for retention of endangered chemicals. *Science,* **278**, 1090-1091.

4

Technological and Economic Potential of Options to Enhance, Maintain, and Manage Biological Carbon Reservoirs and Geo-engineering

Co-ordinating Lead Authors:
PEKKA KAUPPI (FINLAND), ROGER SEDJO (USA)

Lead Authors:
Michael Apps (Canada), Carlos Cerri (Brazil), Takao Fujimori (Japan), Henry Janzen (Canada), Olga Krankina (Russian Federation/USA), Willy Makundi (Tanzania/USA), Gregg Marland (USA), Omar Masera (Mexico), Gert-Jan Nabuurs (Netherlands), Wan Razali (Malaysia), N.H. Ravindranath (India)

Contributing Authors:
David Keith (USA), Haroon Kheshgi (USA), Jari Liski (Finland)

Review Editors:
Eduardo Calvo (Peru), Birger Solberg (Norway)

CONTENTS

EXECUTIVE SUMMARY

Terrestrial ecosystems offer significant potential to capture and hold carbon at modest social costs. The Intergovernmental Panel on Climate Change (IPCC) Second Assessment Report estimated that about 60 to 87GtC could be conserved or sequestered in forests by the year 2050 and another 23 to 44 GtC could be sequestered in agricultural soils. In this chapter, we describe and assess biological mitigation measures in terrestrial ecosystems, focusing on the physical mitigation potential, ecological and environmental constraints, economics, and social considerations. Also the so-called geo-engineering options are discussed.

The mitigation costs through forestry can be quite modest, US$0.1–US$20/tC in some tropical developing countries, and somewhat higher (US$20–US$100/tC) in developed countries. The costs of biological mitigation, therefore, are low compared to those of many other alternative measures. The costs would be expected to rise, however, if large areas of land were taken from alternative uses. The technologies for preserving existing terrestrial C and enhancing C pools, while using biomass in a sustainable way, already exist and can be further improved.

Increased carbon pools from management of terrestrial ecosystems can only partially offset fossil fuel emissions. Moreover, larger C stocks may pose a risk for higher carbon dioxide (CO_2) emissions in the future, if the C-conserving practices are discontinued. For example, abandoning fire control in forests or reverting to intensive tillage in agriculture may result in rapid loss of at least part of the C accumulated during previous years. However, using biomass as a fuel or wood to displace more energy-intensive materials in products can provide permanent carbon mitigation benefits. It is useful to evaluate terrestrial sequestration opportunities alongside emission reduction strategies as both approaches will likely be required to control atmospheric CO_2 levels.

Carbon reservoirs in most ecosystems eventually approach some maximum level. Thus, an ecosystem depleted of carbon by past events may have a high potential rate of carbon accumulation, while one with a large carbon pool tends to have a low rate of carbon sequestration. As ecosystems eventually approach their maximum carbon pool, the sink (i.e., the rate of change of the pool) will diminish. Although both the sequestration rate and pool of carbon may be relatively high at some stages, they cannot be maximized simultaneously. Thus, management strategies for an ecosystem may depend on whether the goal is to enhance short-term accumulation or to maintain the carbon reservoirs through time. The ecologically achievable balance between the two goals is constrained by disturbance history, site productivity, and target time frame. For example, options to maximize sequestration by 2010 may not maximize sequestration by 2020 or 2050; in some cases, maximizing sequestration by 2010 may lead to higher emissions in later years.

The effectiveness of C mitigation strategies, and the security of expanded C pools, will be affected by future global changes, but the impacts of these changes will vary by geographic region, ecosystem type, and local abilities to adapt. For example, increases in atmospheric CO_2, changes in climate, modified nutrient cycles, and altered disturbance regimes can each have negative or positive effects on C pools in terrestrial ecosystems.

In the past, land management has often resulted in reduced C pools, but in many regions like Western Europe, C pools have now stabilized and are recovering. In most countries in temperate and boreal regions forests are expanding, although current C pools are still smaller than those in pre-industrial or pre-historic times. While complete recovery of pre-historic C pools is unlikely, there is potential for substantial increases in carbon stocks. The Food and Agriculture Organization (FAO) and the UN Economic Commission for Europe (ECE)'s statistics suggest that the average net annual increment has exceeded timber fellings in managed boreal and temperate forests in the early 1990s. For example, C stocks in the live tree biomass has increased by 0.17billion tonnes (gigatonnes = Gt) C/yr in the USA and 0.11GtC/yr in Western Europe, absorbing about 10% of global fossil CO_2 emissions for that time period. Though these estimates do not include changes in litter and soils, they illustrate that land surfaces play a significant and changing role in the atmospheric carbon budget and, hence, provide potentially powerful opportunities for climate mitigation.

In some tropical countries, however, the average net loss of forest carbon stocks continues, though rates of deforestation may have declined slightly in the last decade. In agricultural lands, options are now available to recover partially the C lost during the conversion from forest or grasslands.

Land is a precious and limited resource used for many purposes in every country. The relationship of climate mitigation strategies with other land uses may be competitive, neutral, or symbiotic. An analysis of the literature suggests that C mitigation strategies can be pursued as one element of more comprehensive strategies aimed at sustainable development, where increasing C stocks is but one of many objectives. Often, measures can be adopted within forestry, agriculture, and other land uses to provide C mitigation and, at the same time, also advance other social, economic, and environmental goals. Carbon mitigation can provide additional value and income to land management and rural development. Local solutions and targets can be adapted to priorities of sustainable development at national, regional, and global levels.

A key to making C mitigation activities effective and sustainable is to balance C mitigation with other ecological and/or environmental, economic, and social goals of land use. Many biological mitigation strategies may be neutral or favourable for all three goals and become accepted as "no regrets" or "win–win" solutions. In other cases, compromises may be needed. Important potential environmental impacts include effects on biodiversity, effects on amount and quality of water resources (particularly where they are already scarce), and long-term impacts on ecosystem productivity. Cumulative environmental, economic, and social impacts could be assessed within individual projects and also from broader, national and international perspectives. An important issue is "leakage" – an expanded or conserved C pool in one area leading to increased emissions elsewhere. Social acceptance at the local, national, and global scale may also influence how effectively mitigation policies are implemented.

In tropical regions, there are large opportunities for C mitigation, though they cannot be considered in isolation from broader policies in forestry, agriculture, and other sectors. Additionally, options vary by social and economic conditions: in some regions, slowing or halting deforestation is the major mitigation opportunity; in others, where deforestation rates have declined to marginal levels, improved natural forest management practices and, afforestation and reforestation of degraded forests and wastelands are the most attractive opportunities.

Non-tropical countries also have opportunities to preserve existing C pools, enhance C pools, or use biomass to offset fossil fuel use. Examples of strategies include fire or insect control, forest conservation, establishing fast-growing stands, changing silvicultural practices, planting trees in urban areas, ameliorating waste management practices, managing agricultural lands to store more C in soils, improving management of grazing lands, and re-planting grasses or trees on cultivated lands.

Wood and other biological products play several important roles in carbon mitigation: they act as a carbon reservoir; they can replace construction materials that require more fossil fuel input; and they can be burned in place of fossil fuels for renewable energy. Wood products already contribute somewhat to climate mitigation, but if infrastructures and incentives can be developed, wood and agricultural products may become vital elements of a sustainable economy: they are among the few renewable resources available on a large scale.

A comprehensive analysis of carbon mitigation measures would consider:

- potential contributions to C pools over time;
- sustainability, security, resilience, permanence, and robustness of the C pool maintained or created;
- compatibility with other land-use objectives;
- leakage and additionality issues;
- economic costs;
- environmental impacts other than climate mitigation;
- social, cultural, and cross-cutting issues as well as issues of equity; and
- the system-wide effects on C flows in the energy and materials sector.

Activities undertaken for other reasons may enhance mitigation. An obvious example is reduced rates of tropical deforestation. Furthermore, because wealthy countries generally have a stable forest estate, it could be argued that economic development is associated with activities that build up forest carbon reservoirs in the long run.

Marine ecosystems may also offer possibilities for removing CO_2 from the atmosphere. The standing stock of C in the marine biosphere is very small, however, and efforts could focus not only on increasing biological C stocks, but also on using biospheric processes to remove C from the atmosphere and transport it to the deep ocean. Some initial experiments have been performed, but fundamental questions remain about the permanence and stability of C removals, and about possible unintended consequences of the large-scale manipulations required to have significant impact on the atmosphere. In addition, the economics of such approaches have not yet been determined.

Geo-engineering involves efforts to stabilize the climate system by directly managing the energy balance of the earth, thereby overcoming the enhanced greenhouse effect. Although there appear to be possibilities for engineering the terrestrial energy balance, human understanding of the system is still rudimentary. The likelihood of unanticipated consequences is large, and it may not even be possible to engineer the regional distribution of temperature, precipitation, etc. Geo-engineering raises scientific and technical questions as well as many ethical, legal, and equity issues. And yet, some basic inquiry does seem appropriate.

In practice, by the year 2010 mitigation in land use, land-use change, and forestry activities can lead to significant mitigation of CO_2 emissions. Many of these activities are compatible with, or complement, other objectives in managing land. The overall effects of altering marine ecosystems to act as carbon sinks or of applying geo-engineering technology in climate change mitigation remain unresolved and are not, therefore, ready for near-term application.

4.1 Introduction

Land is used to raise crops, graze animals, harvest timber and fuel, collect and store water, create the by-ways of travel and the foundations of commerce, mine minerals and materials, dispose of our wastes, recreate people's bodies and souls, house the monuments of history and culture, and provide habitat for humans and the other occupants of the earth. Can land, and water, also be managed to retain more carbon, and thereby mitigate the increasing concentration of atmospheric carbon dioxide (CO_2)? This chapter examines the present scientific thinking on this question.

The atmosphere now contains about 760 billion tonnes (gigatonnes = Gt) of carbon as CO_2, an amount that has increased by an average of 3.3 ± 0.2GtC each year throughout the 1990s, mostly from combustion of fossil fuels (IPCC, 2000a). Atmospheric C represents only a fraction (~ 30%) of the C in terrestrial ecosystems; vegetation contains nearly 500GtC, while soils contain another 2000 GtC in organic matter and detritus (Schimel, 1995; WBGU, 1998) as cited in Intergovernmental Panel on Climate Change (IPCC) Special Report on Land Use, Land-Use Change and Forestry (LULUCF) (IPCC, 2000a). *Table 4.1* provides estimates of the carbon stocks in terrestrial ecosystems now.

The Second Assessment Report (SAR) of the IPCC (1996) suggested that 700Mha of forestland might be available for carbon conservation globally – 138Mha for slowed tropical deforestation, 217Mha for regeneration of tropical forests, and 345Mha for plantations and agroforestry. The IPCC suggested that by 2050 this area could provide a cumulative mitigation impact of 60 to 87GtC, of which 45 to 72GtC in the tropics. Towards the end of this time interval, the mitigation impact could approach a maximum rate of 2.2GtC/yr. The cost of mitigation (excluding land and other transaction costs) was envisioned to be about 2 to 8US\$/tC. The SAR (IPCC, 1996) further suggested

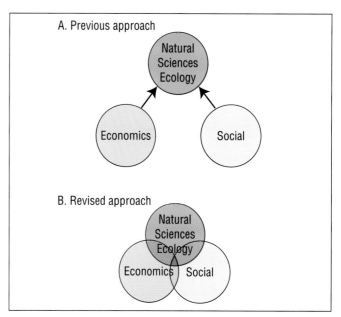

Figure 4.1: *Evolution of approaches to carbon sequestration in terrestrial ecosystems. Previous assessments (e.g., IPCC, 1996) tended to focus on ecological processes and potentials, and treated economic and social factors as constraints (A). A slightly different viewpoint considers the three dimensions as mutually reinforcing and seeks to maximize the overlaps (B).*

that, over the next 50 years, an additional 0.4 and 0.8GtC could be sequestered per year in agricultural soils, with the adoption of appropriate management practices.

The current report, while supporting many of these earlier findings, provides a broader evaluation of the potential for management of C stocks in the biosphere (*Figure 4.1*). Recent studies, for example, suggest that costs may often be higher than estimated earlier, particularly when opportunity costs of the

Table 4.1: *Estimates of global carbon stocks in vegetation and soils to 1 m depth (from Bolin et al., 2000; based on WGBU, 1998).*

Biome	Area (million km²)	Carbon stocks (GtC)		
		Vegetation	**Soils**	**Total**
Tropical forests	17.6	212	216	428
Temperate forests	10.4	59	100	159
Boreal forests	13.7	88	471	559
Tropical savannas	22.5	66	264	330
Temperate grasslands	12.5	9	295	304
Deserts and semideserts	45.5	8	191	199
Tundra	9.5	6	121	127
Wetlands	3.5	15	225	240
Croplands	16.0	3	128	131
Total	**151.2**	**466**	**2,011**	**2,477**

land are included. In addition, the issue of "leakage" (where actions at one site influence actions elsewhere, a problem not considered by the SAR) is examined. This report considers forests, grasslands, croplands, and wetlands, and, where possible, examines all C pools within them. Carbon mitigation is evaluated as one of many services provided by ecosystems. The objectives of this chapter are to review progress made since the IPCC-SAR, and to evaluate prospects for storing more carbon in ways that ensure the continued provision of other goods and services from the varied and finite land resources.

The aim of this chapter is not to assess specifically the implications of the Kyoto Protocol (UNFCCC, 1997), a mandate assigned to the IPCC Special Report on LULUCF (IPCC, 2000a). Rather, it seeks to provide a broader scientific view of the prospects and problems of land management for carbon sequestration, unconstrained by the limited scope of the Kyoto Protocol.

This chapter begins by describing the current state of land use, the history of land use, ongoing changes in land use, pressures driving these changes, and potential competition among demands for land (Section 4.3). It then considers opportunities for enhanced C stocks, especially in forestry and agriculture (Section 4.4). Having identified possible C conservation measures, the physical, environmental, social, and economic impacts of these measures are examined; and assessment is made of how they augment or compete with other services provided by land (Sections 4.5 - 4.7). How these options might be evaluated and, where appropriate, encouraged (Sections 4.8 and 4.9) is also considered. Finally, the prospects for managing marine ecosystems to increase carbon sequestration, and the possibility of managing the global ecosystem by 'geo-engineering' of the earth's energy balance (Section 4.10) are considered.

Land-use changes and the pressures that influence them vary widely, especially between tropical and non-tropical regions. Both of these regions are addressed.

4.2 Land Use, Land-Use Change, and Carbon Cycling in Terrestrial Ecosystems

Terrestrial ecosystems provide an active mechanism (photosynthesis) for biological removal of CO_2 from the atmosphere. They act as reservoirs of photosynthetically-fixed C by storing it in various forms in plant tissues, in dead organic material, and in soils. Terrestrial ecosystems also provide a flow of harvestable products that not only contain carbon but also compete in the market place with fossil fuels, and with other materials for construction (such as cement), and for other purposes (such as plastics) that also have implications for the global carbon cycle.

Human activities have changed terrestrial carbon pools. The largest changes occurred with the conversion of natural ecosys-

tems to arable lands. Such disruptions typically result in a large reduction of vegetation biomass and a loss of about 30% of the C in the surface 1 metre of soil (Davidson and Ackermann, 1993; Anderson, 1995; Houghton, 1995a; Kolchugina *et al.*, 1995). Globally, conversion to arable agriculture has resulted in soil C losses of about 50GtC (Harrison *et al.*, 1993; Scharpenseel and Becker-Heidmann, 1994; Houghton, 1995a; Cole *et al.*, 1996; Paustian *et al.*, 2000), and total emissions of C from land use change, including that from biomass loss, have amounted to about 122 ± 40GtC (Houghton, 1995b; Schimel, 1995). Most of the soil C losses occur within a few years or decades of conversion, so that in temperate zones, where there is little expansion of agricultural lands now, losses of C have largely abated (Cole *et al.*, 1993; Anderson, 1995; Janzen *et al.*, 1998; Larionova *et al.*, 1998). Tropical areas, however, remain an important source of CO_2 because of widespread clearing of new lands and reduced duration of "fallow" periods in shifting agriculture systems (Paustian *et al.*, 1997b; Scholes and van Breemen, 1997; Woomer *et al.*, 1997; Mosier, 1998).

The competition for land varies among countries and within a country. Land-use and forestry policies for C management may be most successful when climate mitigation is considered alongside other needs for land, including agriculture, forestry, agroforestry, biodiversity, soil and water conservation, and recreation. Forest fires, for example, are controlled, in many parts of the world, not as a measure for carbon mitigation, but simply because fire threatens areas of human settlement and the habitats of living organisms.

Similarly, biodiversity and landscape considerations have motivated protection of old-growth stands in temperate, boreal, and tropical rain forests from commercial logging. In many cases such decisions have prevented C release into the atmosphere, even though C mitigation was not the initial intent (Harmon *et al.*, 1990). The impact of harvest restrictions on C pool in old-growth forests may be affected by "leakage". If one ecosystem is protected but timber demand remains constant, logging may simply be shifted to another, similar ecosystem elsewhere, perhaps to a country where conservation priorities are lower.

4.2.1 Historical Land-Use Change in the Tropics

4.2.1.1 Trends in Land Use and Changes in Carbon Stocks

Tropical forests were largely intact until colonial times, when large tracts were removed to provide raw materials for railroads, ships, etc., in the period following the industrial revolution. The loss of tropical forests escalated in the second half of the 20th century. According to the UN Food and Agriculture Organization (FAO, 1996), about 15.4 million ha of natural tropical forests are lost each year. Of this, 42% occurs in Latin America, 31% in Africa, and 27% in Asia. Brunner *et al.* (1998) estimated tropical deforestation at 19.1 million ha/yr during the period 1990 to 1995. There has, how-

Box 4.1. Stocks and Flows

The global carbon cycle consists of the various stocks of carbon in the earth system and the flows of carbon between these stocks. It is discussed at length in IPCC WG I (Prentice *et al.*, 2001) and IPCC Special Report on LULUCF (IPCC, 2000a) and is illustrated in *Figure 4.2*.

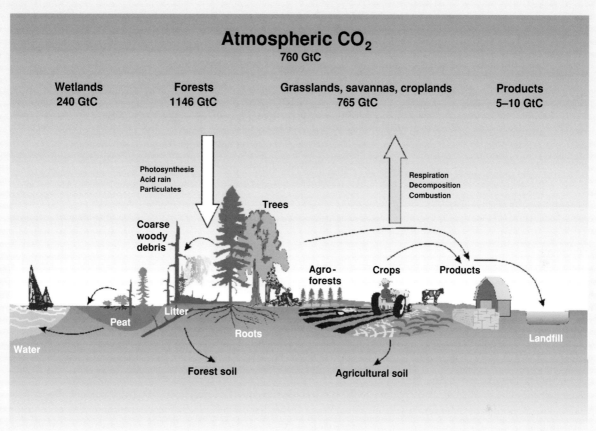

Figure 4.2: *Different ecosystems, their components, and human activities. The carbon stocks associated with the different ecosystems are stored in aboveground and belowground biomass, detrital material (dead organic matter), and soils. Carbon is withdrawn from the atmosphere through photosynthesis (vertical down arrow), and returned by oxidation processes that include plant respiration, decomposition, and combustion (vertical up arrow). Carbon is also transferred within ecosystems and to other locations (horizontal arrows). Both natural processes and human activities affect carbon flows. Mitigation activities directed at one ecosystem component generally have additional effects influencing carbon accumulation in, or loss from, other components. Estimates of ecosystem and atmospheric C stocks are adapted from Bolin et al. (2000). Values for C stocks in some ecosystems are still very uncertain. Not shown are estimates of C stocks in tundra (127GtC), deserts and semi-deserts (199GtC), and oceans (approx. 39,000GtC) (numbers are taken from Special Report on LULUCF, Fig 1-1, page 30; IPCC, 2000a).*

A consequence of the conservation of mass is that the net of all of the flows (measured as a rate variable in units such as tC/yr) into and out of a given reservoir or stock (measured in units such as tC) during a period of time must equal the change in the stock (tC) in that period. Conversely, a change in stock of a reservoir during a given period must exactly equal the integrated net difference in C flows into and out of that reservoir during that period. Elsewhere in this text the word "pool" is sometimes used to represent the various reservoirs of carbon in the global carbon cycle. The word "sink" is used to indicate the net positive flow of carbon into a terrestrial carbon pool.

The maximum rate of net ecosystem carbon uptake cannot occur at the same time as the maximum ecosystem carbon stock (see *Figure 4.3*). An ecosystem depleted of carbon by past events may have much higher rates of carbon accumulation than a comparable one in which carbon stocks have been maintained. Ecosystems eventually approach some maximum carbon stock – a carrying capacity – at which time the flows into the carbon pool are balanced by flows out of the carbon pool. Because C sink and C stock in ecosystems cannot be maximized simultaneously, mitigation activities aimed at enhancing the sink and maintaining the biological carbon stock coincide only partially (IGBP, 1998).

(continued)

Box 4.1. continued

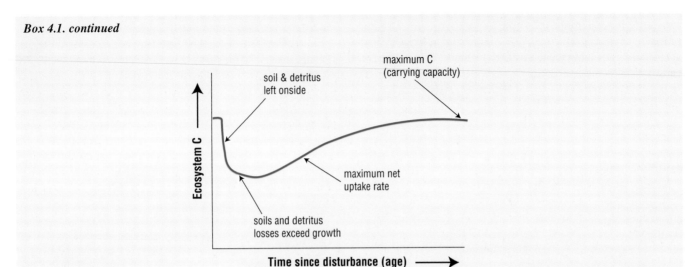

Figure 4.3: *An example of net changes in ecosystem carbon stocks over time. Changes in individual ecosystem components take place at different rates, but it is the net of the changes in all interconnected pools that determines the net flow to or from the atmosphere. In the example, the accumulation of biomass initially is at a lower rate than the decomposition of the dead organic matter stock so the stock of ecosystem C declines. Later in the cycle, dead organic matter stocks may increase, although other components have reached a steady state. Maximum ecosystem stocks (highest value of ecosystem C) occur at a later time than the maximum rate of net carbon uptake (steepest slope of the ecosystem C line).*

Similarly, the maximum rate of C substitution cannot occur at the same time as maximum C conservation. High rates of carbon substitution, through use of forest products or biofuels, generally require high productivity and efficient manufacture and use of derived products.

Carbon taken up by the biosphere may also accumulate in offsite pools – as products or in landfills – but it continues to oxidize at rates that depend on the conditions of those pools. It is the net of many flows that defines the changes in carbon stocks of off-site pools as well as of on-site pools. Carbon accumulation in off-site pools is an often overlooked, but a potentially important, form of sequestration.

ever, been a large increase in area devoted to forest plantations. By 1990, there were 61.3 million ha under plantations and the rate of establishment is now about 3.2 million ha/yr (FAO, 1996).

As pointed out by the IPCC (IPCC, 1996) global estimates of C emissions from deforestation have remained highly uncertain and show high geographical variability. The magnitude of forest regeneration (particularly secondary forest regrowth and regrowth of abandoned lands) and forest degradation processes is not well documented. Improving the accuracy of these estimates remains an urgent and challenging task (Houghton *et al.*, 2000).

Estimates of C emissions from land-use change and forestry activities in the tropics during the1990s range from 1.1 to 1.7GtC/yr, with a best estimate of 1.6GtC/yr (Brown *et al.*, 1996b; Melillo *et al.*, 1993; Bolin *et al.*, 2000). These estimates may change with improved information on biomass densities and land-use conversion. Detailed studies for major tropical countries in the early 1990s, studies that include forest regeneration and afforestation, show lower net emissions

for most countries than those from aggregate estimates (Makundi *et al.*, 1998).

A review of scenarios of future land-use changes in the tropics, and their implications for greenhouse gas (GHG) emissions, shows a wide range of estimates, particularly for the first part of the 21st century, where estimates differ by a factor of 14 (Alcamo and Swart, 1998). These disparities reflect a lack of agreement on the definition of deforestation, and a lack of knowledge and agreement on the estimation of C emissions (Alcamo and Swart, 1998). These scenarios can be divided into two groups: in one group emissions decline smoothly after 1990; in the other group emissions increase for a few decades after 1990.

4.2.1.2   Driving Forces for Land-use Change

The rates and causes of land-use change vary by region and scale (Kaimowitz and Angelsen, 1998). Deforestation is often considered a one way process, but the landscape is a dynamic mosaic of land uses and vegetation types, with transitions both to and away from forest (Houghton *et al.*, 2000). Natural factors, such as forest fires and pests, as well as socio-economic processes, many of which are not seen at the local level, interact in complex ways, complicating analysis. Understanding the causes of this mosaic of land-use and/or land-cover transitions in order to understand and predict the net effect on deforestation rates and C emissions remains a key research challenge.

Conversion of forests to pasture and cropland has been the most important proximal cause of tropical deforestation. Non-sustainable logging has been the leading factor in parts of Southeast Asia, whereas excessive harvest of wood fuel has been important only in specific sub-country regions and in some African countries (Kaimowitz and Angelsen, 1998). According to Bawa and Dayanandan (1997), the causes (correlates) of deforestation are many and varied, with complex interactions. Overall, Bawa and Dayanandan found that population density, cattle density, and external debt were the key factors. In Africa, the most important factors were extraction of fuelwood and charcoal and demand for cropland; in Asia, it was cropland; and in Latin America, it was cattle density.

Most analyses of land-use change and forestry have concentrated on proximal reasons for land-use and/or land-cover change; that is, on land uses such as agriculture, pasture, and timber extraction that replace forests. But Meyer and Turner (1992) have identified six "underlying" forces: (1) population, (2) level of affluence, (3) technology, (4) political economy, (5) political structure, and 6) attitudes and values. The influence of each varies by region and country.

The rate of population growth is now apparently declining, but the population, and hence the demand for food and other land services, is still growing (Roberts, 1999). Population growth has been widely cited as a major cause of deforestation (Myers, 1989), but the relationship between population and deforestation is not simple. Population growth exerts increasing pressure on resources, but whether these pressures lead to forest degradation or to positive changes (e.g., afforestation, improved forest management, and better technology) depends largely on social structure. Extensive migration may also lead to deforestation and soil erosion. Simplistic assumptions about population and deforestation also do not apply where high population densities and/or growth rates are accompanied by forest conservation and reforestation programmes. In India, for example, deforestation rates have declined since 1980, despite population growth, owing to effective forest conservation legislation (Ravindranath and Hall, 1994).

Patterns that affect land-use are changed by economic development. Affluence usually increases consumption, but it does not necessarily decrease terrestrial C stocks. The maintenance of ecosystems tends to improve with increasing and better distribution of wealth, as well as with proper institutional structures and sound development strategies. The demand for and interest in forests and their services is the driving force for the technological and economic capacity to maintain forests. Also, wealthy societies tend to be urbanized and this may reduce

destructive pressures on forests. Technological development provides efficient tools for land-use change and for high-value, alternative uses. Technology can also limit encroachment. As seen by the "green revolution" in agriculture, technological development can increase productivity on intensively managed land, thereby releasing other land areas from agriculture (Waggoner, 1994). Nevertheless, there is always the risk of leakage (*i.e.*, tendencies to transfer destructive operations from the developed to less developed areas and countries), or the possibility that technology development and transfer will have positive spillover effects (Brown *et al.*, 2000; Noble *et al.*, 2000)

In many countries, especially those seeking development of frontier areas, subsidies are provided for activities promoting economic development. Land clearing may be subsidized directly or by providing property rights to cleared land. Frontier development is often considered desirable for security or where there is a disputed area.

Land-use change is driven largely by efforts perceived as "best and highest" use of the land. But benefits of the land that are non-market and/or external to the direct user (e.g., watershed protection, biodiversity, and carbon mitigation) may be ignored by land managers. For example, the decision to convert forest-land to agriculture may ignore the many external and non-market benefits lost. Moreover, where long-term land rights are insecure, lands may be used to generate short-term benefits, with disregard for long-term benefits.

Factors related to social structure and political economy have not been studied widely, but studies at the country and regional levels suggest that deforestation is favoured by the following factors: growing landlessness and persistent inequalities in access to land, insecure land tenure, land speculation, rising external debt, large-scale expansion in commercial agriculture, erosion of traditional systems of resource management and community control, and widespread migration of impoverished people to ecologically fragile areas (Hecht, 1985; Palo and Uusivuori, 1999; Tole, 1998).

4.2.2   Land Use in the Temperate and Boreal Zones

4.2.2.1   Historical and Present Land Use in the Temperate and Boreal Zones

The temperate zone is the most populated zone of the world, while the boreal zone is quite sparsely populated. For thousands of years forest area has diminished, particularly in the temperate zone, as forests were cleared for agriculture and pasture. Clearing of the European Mediterranean region began *ca* 5000 years ago; in Central Europe and in China deforestation occurred in early Medieval times; in parts of Russia and Mongolia forest clearing occurred in late Medieval times; and in North America clearing occurred mainly in the 19th century (Mather, 1990, see *Figure 4.4*). Since the mid 20th century the net forest area of the temperate zone has no longer decreased but has instead increased (Kauppi *et al.*, 1992). The inner parts of the boreal zone in Siberia, Alaska, and Canada have not been subject to significant land-use management. The opportunities present to store carbon in terrestrial ecosystems in the boreal and temperate zones are thus very much determined by historical land-use change and the associated losses of carbon (Kurz and Apps, 1999).

Understanding the historic and current net sink of C in the temperate and boreal zones is important to assessing the potential of present and future management options. In general, estimates of C flows have been based on a variety of methods and data, resulting in a wide range of reported values for C flows per region. The confidence level in each separate value is therefore low. For example, for European forests the estimates of the present C sink vary from almost 0 to 0.5 GtC/yr

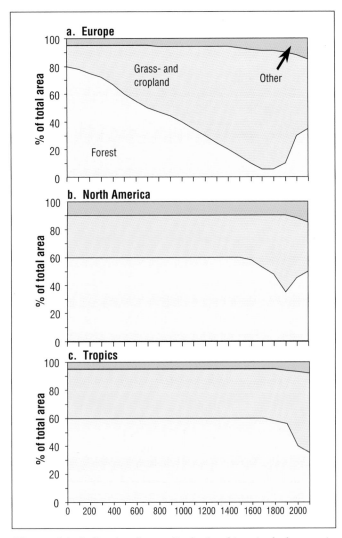

Figure 4.4: *Indicative figure displaying historical changes in land use in three world regions. The presented values should not be taken as absolute, because the historical evidence is often only anecdotal (Mather, 1990; Kauppi et al., 1992; Palo and Uusivuori, 1999; Farrell et al., 2000).*

(Nabuurs *et al.*, 1997; Martin *et al.*, 1998; Valentini *et al.*, 2000; Schulze, 2000). For Canada, early estimates, based on a static assessment, indicated a net sink of 0.08GtC/yr for the mid-1970s (Kurz *et al.*, 1992); whereas subsequent analyses, accounting for changes in forest disturbances over time (see section 4.2.3), indicated that Canadian forests became a small net source of C (–0.068GtC/yr) by the early 1990s (Kurz and Apps, 1999). Estimates of carbon accumulation in woody biomass for the USA also show a large uncertainty. While the average rate for the USA C sink ranges from 0.020 to 0.098GtC/yr for the 1980s and 1990s (Birdsey and Heath, 1995; Turner *et al.*, 1995; Houghton *et al.*, 1999), atmospheric inversion models applied to the North American continent suggest a sink of 1.7 ± 0.5GtC/yr, largely south of 51°N (Fan *et al.*, 1998), but with very low levels of confidence (Bolin *et al.*, 2000).

In the less intensively managed forests of Russia and Canada, changes in mortality associated with natural disturbances appear to dominate over management influences (see Section 4.2.4). In European Russia, managed forest ecosystems were estimated to be a sink of 0.051GtC/yr between 1983 and 1992, but the less actively managed Siberian forest was a net source of 0.081–0.12GtC/yr (Shepashenko *et al.*, 1998). The available estimates for Siberia differ even more than for the other regions mentioned above, and their confidence level may be "low" (Schulze *et al.*, 1999).

Recent FAO statistics on 55 countries in the temperate and boreal zones indicate a general increase in the forest carbon stock (trees only) of 0.88GtC/yr (UN-ECE/FAO, 2000). Changes in forest management and changes in the environment have contributed to this trend. In Europe, the trend is consistent with the observation of increased growth in individual stands noted by Spiecker *et al.* (1996). The FAO statistics indicate that between the 1980s and 1990s both net annual increment and timber fellings increased, but that the rate of change was lower for fellings than for growth, resulting in a substantial increase in the carbon sink from the 1980s to the 1990s (Kuusela, 1994; Kauppi *et al.*, 1992; Sedjo, 1992; Dixon *et al.*, 1994; UN-ECE/FAO, 2000). The carbon sink in live woody vegetation was on the order of 10% of the fossil fuel CO_2 emissions in the USA and in western Europe, and higher in the 1990s than in the 1980s (c.f. Kauppi *et al.*, 1992).

These relatively high sequestration rates are not a result of active policies aimed at climate mitigation, but less rather appear to be related to general trends in land use and land-use change. In the USA, Schimel *et al.* (2000) and Houghton *et al.* (1999) estimate that the observed sink is a result mainly of changes in land use and land management, rather than a response to changes in the environment. The latest observations, based on forest inventory data (UN-ECE/FAO, 2000), are reflected in the Special Report on LULUCF (IPCC, 2000a). The IPCC (2000a) estimates that the total global terrestrial biospheric sink in the 1990s amounted to 0.7GtC/yr, despite a source from land-use change in the tropics of 0.9GtC/yr.

4.2.2.2 Driving Forces for Land-Use Change

Land management decisions are influenced by many factors. In the temperate zone, and in the European parts of the boreal zone, these are mainly technological and economic. Agricultural production is, for example, heavily influenced by evolving technologies, economic opportunities, subsidies, and restrictions on international trade. Forestry practices are similarly influenced by economic returns, trade, and pressures from society (Clawson, 1979; Waggoner, 1994; Wernick *et al.*, 1998). It is within these pressures and opportunities that carbon mitigation possibilities may be found, and preferably they would be region specific. *Table 4.2* gives an overview of some of the specific issues of importance in the temperate and boreal zone of the world.

Competition for land between forestry and agriculture has become less severe. Forest area is increasing in many regions of the boreal and temperate zone, partly because agricultural yields have improved or because the profitability of marginal agriculture has declined. The ability to produce agricultural goods has grown faster than demand, resulting in a downwards trend in prices (Alig *et al.*, 1990; Waggoner, 1994). Much abandoned agricultural land has reverted to forest, either naturally or through deliberate planting. Superimposed over these land conversions is a transition in forestry from a foraging and gathering operation, dependent upon primary forest, through a stage of more intensively managed forest, to total forest ecosystem management. The latter occurs when urbanized societies press for nature-oriented forest management. Continuously improving technologies allow low-cost establishment and higher productivity from planted and plantation forests (Sedjo, 1983; 1999a). In agriculture, also, practices are changing towards maintaining site fertility or decreasing the risk of erosion.

Silvicultural practices have increased forest growth in many boreal and temperate regions. The increasing concentration of atmospheric CO_2 may also have contributed to the enhanced growth of forests.

Incentives for planting forests are provided by a combination of market factors and public policy. Remaining wild forests, such as the public forests in the US National Forest System and in British Columbia, are becoming less accessible and have increased harvesting restrictions. Subsidies to harvesting of natural forests are also being withdrawn elsewhere. For example, large subsidies for harvesting Russian forests were prevalent during the Soviet era, largely through subsidized transportation, but have now disappeared. The economic structures are in transition and industrial production has declined. As a result, harvests have fallen dramatically in Russia since the 1990s (Nilsson and Shvidenko 1998).

Market forces, reflecting industrial needs for wood, have provided financial incentives for expansion of commercial forests (Sedjo and Lyon, 1990). This is a trend expected to continue, because of growing demand for industrial wood and low profitability in agriculture (Sohngen *et al.*, 1999). Early analyses suggested that economic returns from plantations (in the tropics as well as in the temperate and boreal zone) justify investment in a number of regions (Sedjo, 1983). Recent studies confirm that forest plantations are being established at a rate of 600,000ha/yr (Pandey, 1992; Postel and Heise, 1988; UN-ECE/FAO, 2000). However, industrial plantation forestry is new in many tropical areas and yields vary considerably across ecosystems. In many locations where plantations have only recently been established, little is known about the potential capabilities for increasing productivity as well as the potential problems that may limit yields.

4.2.3 Forest Disturbance Regimes

The concept of "forest disturbance" refers to events such as forest fire, harvesting, wind-throw, insect and disease outbreak (epidemics), and forest flooding that cause large pulses of CO_2 to be released into the atmosphere through combustion or decomposition of resulting dead organic matter. Stand-replacing disturbances, such as crown fires and wind-throw, are associated with the sudden death of large cohorts of trees near one another (Pickett and White, 1985; Kurz *et al.*, 1995a, 1995b; Kurz and Apps, 1999, see *Box 4.2*). Some disturbance agents, such as pollution and some insects and disease outbreaks, may result in large areas with productivity decline but only local mortality (Hall and Moody, 1994). Disturbances play an important natural part in the lifecycle and succession dynamics of many forest systems. In boreal systems large-scale, natural, stochastic forces tend to dominate the ecosystem dynamics, even when direct human influences are considered (Kurz *et al.*, 1995b). The return interval of these disturbances, their intensity, and their specific impacts are referred to as the disturbance regime (Weber and Flanigan, 1997). Kurz *et al.* (1995b) and Price *et al.* (1998) (having compiled insect, fire, and harvest data) showed that the disturbance regime of Canadian forests changed over the last quarter of the 20th century from about 2.5Mha/yr prior to 1970 to 4Mha/yr between 1970 and 1990. Using these data, Kurz and Apps (1999) showed that these changes in the disturbance regime resulted in a switch of Canadian forests from being a net sink of C to a small net source of C to the atmosphere.

Disturbances, both human-induced and natural, are major driving forces that determine the transition of forest stands, landscapes, and regions from carbon sink to source and back. The current pattern of forest vegetation and its role in carbon cycling reflects the combined effects of anthropogenic and natural disturbances over a range of time scales. For C stocks with very slow turnover rates (such as soils and peat) the effects of past disturbances on carbon cycling may reverberate for centuries and millennia (*Figure 4.5*). For example, carbon continues to accumulate in young soils (such as those associated with the isostatic uplift following deglaciation in Canada and Finland), which appear to be actively accruing carbon (Harden

Table 4.2: *Overview of biological carbon mitigation issues and opportunities in selected countries/regions*
(Based, in part, on Sedjo and Lyon, 1990; Fujimori, 1997; Nilsson and Shvidenko, 1998; De Camino *et al.*, 1999; Sohngen *et al.*, 1999; Zhang, 1996)

Region	Issues	Options to store carbon arising from the issues
USA/Canada	• Primary forest based forestry and second rotation forestry • High tech forest industry • Fierce environmental debates • Large impacts of natural disturbances • Agriculture under pressure (excess agricultural land)	• Fire management • Afforestation • Efficient use of wood products • Bioenergy • Farming practices (e.g., reduced tillage) that restore soil C
Europe	• Agriculture under pressure, afforestation of agricultural lands • Changing ownership • Forest health problems • Move towards nature-oriented forest management • High tech forest industry • In eastern Europe, privatization of forest ownership	• Nature-oriented forest management • Nature reserves • Afforestation • Efficient use of wood products • Bioenergy • Farming practices (e.g., reduced tillage) that restore soil C
Russia	• Transition to market economy • Bad financial situation of forest service • Large impacts of natural disturbances • Low levels of fellings	• Natural regeneration on abandoned agricultural land • Fire management • Capacity building • Farming practices that restore soil C
Japan	• Plantation-based forestry and managed secondary forestry • High tech forest industry • Forest health problems • Move towards nature-oriented management	• Efficient use of wood • Nature-oriented forests • Reserves • Bioenergy
China	• Transition to market economy • Transition from non-wood fibre sources to using wood fibre • Floods resulting from loss of forest	• Afforestation with plantations • Protecting primary forests • Flood protection • Farming practices (e.g., reduced tillage) that restore soil C
Australia/ New Zealand	• Plantation-based forestry and some primary forest based forestry • High tech forest industry • Afforestation of agricultural lands	• Fire management • Afforestation with plantations • Efficient use of wood products • Bioenergy • Halting deforestation • Farming practices (e.g., more forages) that enhance soil C
Argentina, Chile, Brazil	• Plantation-based forestry and some primary forest based forestry • High tech forest industry developing • Plantations are not able to reduce deforestation because they provide different set of products and services.	• Afforestation with plantations • Efficient use of wood products • Bioenergy • Halting deforestation • Farming practices (e.g., reduced tillage) that enhance soil C
Mexico	• Forestry largely based on native forests • Large deforestation rates • Economic incentives favour agriculture/cattle over forestry • Afforestation of degraded lands mostly for restoration	• Halting deforestation • Sustainable forest management of native forests • Social forestry • Afforestation with local species • Bioenergy

---

Box 4.2. Disturbance, Age-class Distribution, and their Implications for Forest Carbon Dynamics

At the stand scale, disturbance events (both natural and anthropogenic) have three main impacts on the carbon budget (Apps and Kurz, 1993). First, they redistribute the existing carbon by transferring carbon from living material, above and below ground, to the dead organic matter pools. Second, they transfer some of the carbon out of the ecosystem (e.g., into the atmosphere as combustion products, in the case of fire, and/or into the forest product sector as raw feedstock, in the case of harvest). Third, by opening the forest canopy, the disturbance changes the site micro-environment and restarts the successional cycle for new stand development.

At the scale of forests (typically comprising many stands), the disturbance regime determines the age-class structure (e.g., the even-age structure associated with stand-replacing disturbance regimes or the uneven-age structures associated with individual tree mortality and gap-phase replacement), and age-class structure of stands and trees making up the forest. The C stocks in a forest landscape, and the changes in these stocks over time, are strongly influenced by the age-class distribution (Kurz *et al.*, 1995b; Turner *et al.*, 1995; MacLaren, 1996; Apps *et al.*, 2000; Bhatti *et al.*, 2001). In managed plantation forests, the age-class distribution is controlled by the management regime and harvest cycle (Heath and Birdsey, 1993; MacLaren, 1996), while in natural forests other mortality agents play a major role. See Heath *et al.* (1996) and Kurz *et al.* (1995a) for examples.

Box 4.3. The Reduced Impact Logging Project, Carbon Sequestration Through Reduced Impact Logging

The RIL (reduced impact logging) project developed by Innoprise, a Malaysian company with forestry activities, and the New England Power Company, USA, aims to save CO_2 already stored in forest biomass by reducing damage to vegetation and soils during harvesting. The hope is to reduce damage by 50% compared to that of conventional harvesting. The techniques employed are modifications of conventional bulldozer harvesting techniques; including pre-felling climber cutting, directional felling, skid trail design, and post harvest operations such as rehabilitation of log landings. Today the total project area amounts to 2,400 ha. The pilot research project has quantified the carbon implications and costs on 1,415ha. They found that avoided emissions amounted to 65–90MgC/ha and that the associated costs were US$3.55/MgC (Wan Razali and Tay, 2000).

---

et al., 1992). In these soils, losses from decomposition of accumulated organic matter are exceeded by the inputs of fresh organic debris (Liski *et al.*, 1999). Human influences on the disturbance regime include both direct effects, such as harvesting or inducing and/or suppressing natural disturbances (fires, insects, flooding, etc.), and indirect influences from altering the forest environment. Indirect influences include both climate change and atmospheric pollution, and their effects on tree health and survival.

The different types of disturbances are often linked. For example, in some forest types the probability of fire may increase following insect outbreaks because of increases in available fuel (litter). In some cases salvage logging (recovering the usable timber following a disturbance) can reduce the total area of living forest that is disturbed in a given year by all agents combined. It is common to try to replace natural disturbances (such as wildfires) with commercial harvesting, using a combination of protection and scheduled logging. In Sweden and Finland, for example, logging has become the main disturbance type; and large-scale natural disturbances resulting from wildfire, insect outbreaks, or storms have been almost non-existent for half a century (Lähde *et al.*, 1999).

Disturbances affect the carbon stocks of all components of forested ecosystems. During and following a disturbance, carbon is transferred from living material, above and below ground, to the dead organic matter pools (*Figure 4.2*). In the

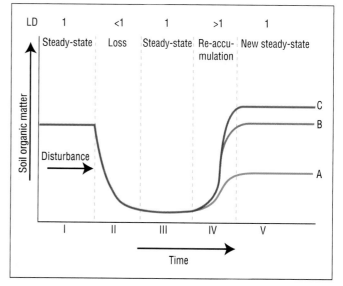

Figure 4.5: *Conceptual model of soil organic matter decomposition and accumulation following disturbance (after Johnson, 1995; IPCC, 2000a). At steady state (I), carbon (C) inputs from litter (L) equal C losses via decomposition (D) (i.e., L/D = 1). After a disturbance, D often exceeds L resulting in loss of C (II), until a new, lower steady state is reached (III). Adoption of new management, where L exceeds D results in a re-accumulation of C (IV) until a new, higher steady state is reached (V). The eventual steady state (A, B, or C) depends on the new management adopted.*

case of a forest fire, part of the ecosystem carbon is released immediately into the atmosphere as combustion products. Disturbed forest stands continue to release carbon into the atmosphere as the enlarged pools of dead organic matter tend towards a new steady-state condition (Bhatti *et al.,* 2001). Regrowth follows, but maximum uptake may not be achieved for some time (decades or more), and during much of this period decomposition of dead organic matter may exceed vegetative uptake. The corresponding re-sequestration of carbon through regrowth can last 50 to 200 years or more.

Management of natural disturbance regimes can provide significant C mitigation opportunities, e.g., through activities to prevent or suppress disturbances. Such measures can significantly enhance the strength of C sinks (Kurz *et al.,* 1995a; Apps *et al.,* 2000; Bhatti *et al.,* 2001) and maintain existing C stocks, but only as long as the programmes are maintained. Other factors being equal, during periods of reduced disturbance (e.g., with increasing suppression effort), C stocks tend to increase as biomass accumulates and litter production (in all forms) increases: forests act as a sink for atmospheric C (Bhatti *et al.,* 2001). In contrast, with increasing disturbance (e.g., with reduction in suppression effort), the net losses of C from forest ecosystems can exceed inputs from photosynthesis (*Figures 4.2 and 4.3*) and the forests could become a net source of C. We note that all forms of disturbance, not just highly visible fires, play a role in these dynamics. In a changing climate, the control of new pathogens and immigrant herbivores (especially insects and disease), to which local forest ecosystems may be maladapted, may be critical to avoid emissions and maintain existing forest C stocks.

Disturbances affect the carbon stocks in vegetation, in soil, and in dead organic matter. All these stocks vary over time as a function of the history of disturbances (MacLaren, 1996; Bhatti *et al.,* 2001; Kurz and Apps, 1999). With an increase of widespread disturbance events the carbon stocks of living vegetation decrease and the age-class distribution of the forest shifts to younger stands containing less carbon. If forests are disturbed at regular intervals (i.e., an unchanging, disturbance regime), the carbon stock of large tracts of forest can be relatively stable.

4.2.4   Changes in Global Climate and Other Indirect Human Effects

Evaluating the long-term outcome of carbon mitigation activities will require estimating how carbon reservoirs will change in the future. Carbon stocks sequestered through mitigation activities today may be more or less secure, depending on how the environment changes and how society adapts to those changes. Estimating future C stocks in ecosystems is complicated by our inability to predict the magnitude and impact of impending changes in the environment. Some of the possible changes favour larger C stocks; others would lead to smaller stocks. The impact of global climate change on future C stocks

is particularly complex. These changes may result in both positive and negative feedbacks on C stocks (Houghton *et al.,* 1998). For example, increases in atmospheric CO_2 are known to stimulate plant yields, either directly or via enhanced water-use efficiency, and thereby to enhance the amount of C added to soils (Schimel, 1995; Woodwell *et al.,* 1998). Higher CO_2 concentrations may also suppress decomposition of stored C, because C/N ratios in residues may increase and because more C may be allocated below ground (Owensby, 1993; Morgan *et al.,* 1994; Van Ginkel *et al.,* 1996; Torbert *et al.,* 1997). Predicting the long-term influence of elevated CO_2 concentrations on the C stocks of forest ecosystems remains a research challenge (Bolin *et al.,* 2000; Prentice *et al.,* 2001).

Where plant growth is now limited by nitrogen (N) deficiencies, increased deposition of N associated with intensified production of bio-available N (Schindler and Bayley, 1993; Vitousek *et al.,* 1997) may accelerate plant growth. This may, eventually, enhance the carbon stock of the soil (Wedin and Tilman 1996). Nadelhoffer *et al.* (1999) caution, however, that the global impact of N deposition may be comparatively small. Moreover, where the N fertilization effect increases growth, especially in the N-deficient northern forests, it also delays the hardening-off process, resulting in increased winter damage, and thus negating some of the growth enhancement (Makipaa *et al.,* 1999).

Increased soil temperatures associated with increased atmospheric CO_2 have long been expected to result in increased soil respiration (Schimel, 1995; Townsend and Rastetter, 1996; Woodwell *et al.,* 1998). Data recently reported by Giardina and Ryan (2000), however, suggest that decomposition of organic carbon in mineral soil layers is relatively insensitive to changes in air temperature. Modelling studies by Liski *et al.* (1998) suggest similar results. Nevertheless, IPCC reviews (Bolin *et al.,* 2000; Prentice *et al.,* 2001) conclude that existing terrestrial C sinks may gradually diminish over time, in part because of increasing losses via respiration.

Over the long term, as climate gradually changes, the time scales for adaptation of ecosystems to climatic conditions will become important. Vegetation types (and other organisms) have adapted to the combination of site conditions, including climate, where they now occur. It cannot be assumed that tree growth will increase with climate change, or that the plant populations will remain optimally adapted to their current sites. Analysis of provenance (seed source) data, in the light of global change, indicates either no net increase in growth rate as a result of warming or small decreases in growth rate. Trees may be under more stress in a changed climate, leaving them more susceptible to insects and diseases.

The various processes of environmental change may occur over different time periods and with varying intensity at different locations. Ecosystems that initially absorb C in response to higher atmospheric CO_2 will become "saturated" or even later release CO_2 if increasing temperatures lead to enhanced decomposition and respiration (Cao and Woodward, 1998; Scholes *et al.*, 1999). Fires and other disturbances could increase in frequency and intensity if temperatures increase and precipitation patterns change. The net impact of these, and other global changes, is an area of active research (e.g., Hungate *et al.*, 1997; Kauppi *et al.*, 1997; Norby and Cotrufo, 1998; Woodwell *et al.*, 1998).

The effects of climate change on mitigation activities in the terrestrial biosphere are difficult to anticipate, as they are dependent on the timing and the specific spatial character and distribution of changes. Present climate scenarios are neither spatially nor temporally very precise, and averages over the scale of typical global circulation climate models are inadequate for estimating impacts on very specific, localized mitigation activities. Moreover, the responses of ecosystems are dependent on the ecological mechanisms, the climate change imposed, and the management responses to these factors. For example, planting of species adapted to present conditions may be inappropriate for future conditions and the species might grow more slowly under chronic climate change. Conversely, species planted for an anticipated future climate may not be able to survive current variations.

Climate change can also affect the economic and social dimensions of land use and forestry. Currently, productive lands may become less productive and less attractive for food and fibre production. The current patterns of land use and disturbance could change. Model results reported by Darwin *et al.* (1995, 1996) and others suggest, for example, that conversion from forestland to cropland is a significant adaptive response to climate change in some regions. Protection from fire or insect and/or disease predation, in boreal regions especially, may become increasingly hard to maintain. Reliable estimates of risks to, or enhancements of, mitigation activities carried out today will require increased understanding of the interactions between the important ecological, economic, and social impacts of climate change. As described in this chapter, the carbon stocks in terrestrial ecosystems respond to a combination of ecological, economic, and social drivers. That will not change even if the global environment changes.

4.3 Processes and Practices that Can Contribute to Climate Mitigation

4.3.1 System Constraints and Considerations

In terrestrial ecosystems the carbon cycle exhibits natural cyclic behaviour on a range of time scales. Most ecosystems, for example, have a diurnal and seasonal cycle. Often this means that the ecosystem functions as a source of C in the winter and a sink for C in the summer, and this shows up in fluctuations at the global scale, as shown by the annual oscillations in the global atmospheric CO_2 concentration. Large-scale fluctuations occur at other temporal scales as well, ranging from decades (Braswell *et al.*, 1997; Turner *et al.*, 1997; Karjalainen

et al., 1998; Kurz and Apps, 1999; Bhatti *et al.*, 2001) to several centuries (Campbell *et al.*, 2000) and longer (Harden *et al.*, 1992).

The net balance of C flows between the atmosphere and the terrestrial biosphere also undergoes management-induced cycles that occur over long time scales (decades to millennia), and that can cause the transition of terrestrial systems from sink to source and back (Harden *et al.*, 1992). Of relevance for C mitigation are the human-induced changes that occur on an annual to centennial time scale. This would include the harvest cycle of managed, production forests.

The intent of any mitigation option is to reduce atmospheric CO_2 relative to that which would occur without implementation of that option. Biological approaches to curb the increase of atmospheric CO_2 can occur by one of three strategies (IPCC, 1996):

- conservation: conserving an existing C pool, thereby preventing emissions to the atmosphere;
- sequestration: increasing the size of existing carbon pools, thereby extracting CO_2 from the atmosphere; and
- substitution: substituting biological products for fossil fuels or energy-intensive products, thereby reducing CO_2 emissions.

The benefits of these strategies show contrasting temporal patterns. Conservation offers immediate benefits via prevented emissions. Sequestration impacts often follow an S-curve: accrual rates are often highest after an initial lag phase and then decline towards zero as C stocks approach a maximum (e.g., *Figure 4.3*). Substitution benefits often occur after an initial period of net emission, but these benefits can continue almost indefinitely into the future (*Figure 4.6*).

This section deals primarily with carbon conservation and sequestration in the terrestrial biosphere, but acknowledges the complementarity and trade-offs among the three strategies. Carbon sequestration in forest products is included here and the substitution benefits of forest products are treated briefly. The role of energy cropping is treated in greater depth in Chapter 3 (Section 3.6.4.3) and in the IPCC Special Report on LULUCF (IPCC, 2000a). Here the discussion is restricted to the secondary use of biomass products for energy (e.g., waste products) and non-commercial uses (e.g., domestic heating, cooking, *etc.*).

The general goal of sequestration activities is to maintain ecosystems in the sink phase. However, if the system is disturbed (a forest burns or is harvested, or land is cultivated), a large fraction of previously accumulated C may be released into the atmosphere through combustion or decomposition (*Figure 4.2*). When the system recovers from the disturbance, it re-enters a phase of active carbon accumulation. Thus, the disturbance history of terrestrial ecosystems involves in large C loss-

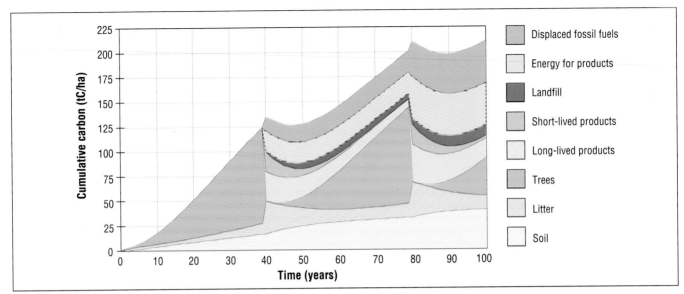

Figure 4.6: *Cumulative carbon changes for a scenario involving afforestation and harvest. These are net changes in that, for example, the diagram shows savings in fossil fuel emissions with respect to an alternative scenario that uses fossil fuels and alternative, more energy-intensive products to provide the same services (adapted from Marland and Schlamadinger, 1999).*

es in the past (Houghton *et al.*, 1999; Kurz and Apps, 1999), but opportunities for C sequestration in the present.

A comprehensive systems analysis is useful to fully evaluate mitigation options. Factors to be considered may include: ecosystem C stocks and sinks; sustainability, security, resilience, and robustness of the C stock maintained or created; temporal patterns of C accumulation; other land-use goals and related C flows in the energy and materials sector; and effects on other non-CO_2 GHGs. For example, one option might have both a high maximum C stock and a high or more sustained rate of sequestration, yet be incompatible with other demands placed on the land. A second option may have a high maximum C stock, but reach that level only very slowly. Still another option may offer high short-term sequestration, but reach maximum C stocks very quickly. Yet another option might manage production systems to maximize the flow of harvested carbon into products, thus maximizing the displacement of alternate, energy-intensive products. Thus, while a wide array of practices may be technically possible, options that meet all criteria may be much fewer, and a combination of complementary options may best accomplish C mitigation goals. Although scientists now recognize the value of system-wide analyses (Cohen *et al.*, 1996; Alig *et al.*, 1997), rarely have mitigation options been subjected to such comprehensive evaluations.

An upper bound for the technical potential for global C mitigation in the terrestrial biosphere, a physical upper limit, can be estimated for conservation, sequestration, and substitution measures. The technical potential for conservation measures would equal the current existing C stock of the world's ecosystems. This assumes that all ecosystems are threatened, but all could be conserved by implementing protection measures. The technical potential for sequestration would roughly equal the

carbon stocks lost in deforestation, desertification, and other human-induced changes in land cover and land use over centuries and millennia. The theoretical upper limit would thus correspond to the full recovery of lost biomass in ecosystems, and to a steady state at the natural carrying capacity for biomass on earth. The technical potential for substitution is related to the sustainable production of harvestable biomass and its substitution for fossil fuels and energy-intensive products. Clearly, each of these upper limits violates in practice the ideals of development, equity, and sustainability. And yet, they help to appreciate that there are bounds on the role that managing the biosphere might play in carbon mitigation.

4.3.2   Opportunities in Forests

Many silvicultural and forest management practices have been reported to enhance carbon mitigation (Lunnan *et al.*, 1991; Hoen and Solberg, 1994; Karjalainen, 1996; Row, 1996; Binkley *et al.*, 1997; Price *et al.* 1998; Birdsey *et al.*, 2000; Fearnside, 1999; Anonymous, 1999; Nabuurs *et al.*, 2000). Measures suggested for forests include: protecting against fires; protecting from disease, pests, insects, and other herbivores; changing rotations; controlling stand density; enhancing available nutrients; controlling the water table; selecting useful species and genotypes; using biotechnology; reducing regeneration delays; selecting appropriate harvest methods such as reduced-impact logging; managing logging residues; recycling wood products; increasing the efficiency with which forest products are manufactured and used; and establishing, maintaining, and managing reserves.

Sampson *et al.* (2000) provide an overview of the potential impacts of some different management alternatives on carbon mitigation, and examine both additional benefits and some possible unintended, negative effects of these practices. They estimate that 10% of the global forest area could be technically available by the year 2010, and that the global potential of forest management practices could be 0.17GtC/yr. These opportunities rise to 50% of the global forest area and 0.7GtC/yr by the year 2040. Sampson *et al.* (2000) emphasize win-win situations, but also indicate the low level of certainty associated with their estimates and the possibility for certain negative impacts.

Nabuurs *et al.* (2000) also estimate the potential of a broad range of forest-related activities (including protection from natural disturbance, improved silviculture, savannah thickening, restoration of degraded lands, and management of forest products) at 0.6GtC/yr over six regions in the temperate and boreal zone (Canada, USA, Australia, Iceland, Japan, and EU, *Figure 4.7*). According to their estimates, alternative forest management for C sequestration is technically feasible on 10% (on average) of the forest area in each region examined. *Figure 4.8* shows that the relative importance of the different practices for the various regions depends on the current situation in the respective regions.

The analyses of Sampson *et al.* (2000) and Nabuurs *et al.* (2000) estimate that the hectare-scale effectiveness of these activities ranges from 0.02tC/ha/yr for forest fertilization to 1.2tC/ha/yr for several practices combined in Loblolly pine stands. However, they show that the impact of most practices is in the range of 0.3–0.7tC/ha/yr.

Forest management and protection offer high mitigation potential in some countries. For example, additional pools of 40-160tC/ha and 215tC/ha may be possible in Cameroon and the Philippines, respectively (Sathaye and Ravindranath, 1998). Afforestation or plantation forest options have the potential to increase carbon stocks by 70–100tC/ha in many places, and the potentials for some commercial plantations may be even higher: 165tC/ha for timber estates in Indonesia, 120tC/ha for timber forestry in India, and 236tC/ha for long rotation forestry in the Philippines (Sathaye and Ravindranath, 1998). The suggested opportunities for mitigation potential in 12 developing countries are summarized in *Table 4.3*.

The study of Sathaye and Ravindranath (1998) suggests that, in 10 tropical and temperate countries in Asia, about 300Mha may be available for mitigation options: 40Mha for conservation, protection, and management; 79Mha of degraded forest land for regeneration; and 181Mha of degraded land for plantation forestry and, hence, for C sequestration (Sathaye and Ravindranath, 1998). A further 172Mha was estimated to be available in these countries for agroforestry. These esti-

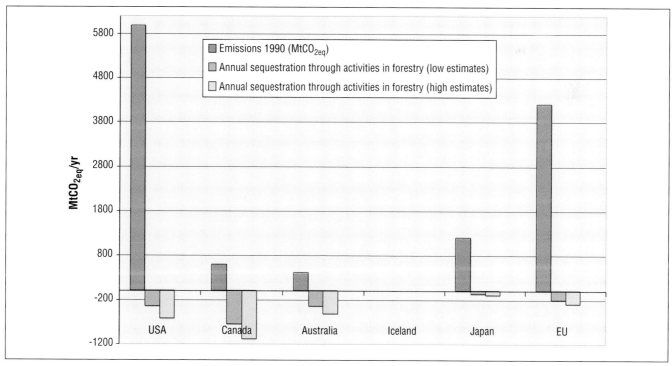

Figure 4.7: *Indications of the magnitude of the carbon sink in case study countries for a set of forest management measures (MtCO$_{2eq}$, adapted after Nabuurs et al. 2000). The values for the three bars for Iceland are 2.6, 2.8, and 2.9, respectively. The figure is based on the forest part of the model "Access to Country Specific Data" (ACSD). It was designed to provide insight into the potential magnitude of carbon sequestration that may be achieved when alternative sets of management measures are adopted. Therefore, the exact numbers provided in this figure result from the assumptions chosen for a certain set of measures. The estimates in this figure are tentative and only illustrative. In these studies all forestry activities under discussion were included, but applied on average on some 10% of mostly the exploitable forest area.*

Table 4.3: *Mitigation options, mitigation potential, and investment cost per tonne of carbon (US$/tC) abated in selected countries* (Sathaye and Ravindranath, 1998)

Mitigation option	Mitigation potential (tC/ha)	Investment cost[1] (US$/tC)	Mitigation option	Mitigation potential (tC/ha)	Investment cost[1] (US$/tC)
ASIA					
China			**India**		
North & North West			Natural regeneration[2]	62.0	1.5
Assisted natural regeneration[2]	13.0	1.3	Enhanced natural regeneration[2]	87.5	2.5
Plantation	55.0	1.3	Agroforestry	25.4	1.6
Agroforestry	15.0	16.3	Community woodlot	75.8	5.6
South, South West & North East			Softwood forestry	80.1	7.3
Assisted natural regeneration	13.9	3.5	Timber forestry	120.6	3.3
Plantation[3]	71.0	5.0			
Agroforestry	6.0	9.8			
Indonesia			**South Korea**		
Timber estate	165.0	1.9	Improved management of natural forest	99.4	6.0
Social forestry	94.0	1.1	Urban forestry	299.0	9.2
Reforestation[4]	214.0	0.9	Enhanced regeneration of		
Private forests	99.0	2.1	*L. leptolepsis*	123.0	13.8
Afforestation	106.0	0.6	*P. koraiensis*	85.0	21.0
Mongolia			**Pakistan**		
Private forests	99.2	0.8	Intensified forest management		
Natural regeneration	67.5	0.6	- *Conifer forest – protection*	41.6	0.1
Agroforestry	9.8	0.8	- *Conifer forest – natural regeneration (enhanced)*	33.8	8.8
Bioenergy	80	-	Reforestation[4]	39.1	19.3
Shelter belt	101.7	0.9	Riverain forest plantation	32.9	40.6
			Commercial forest plantation	54.6	40.6
			Watershed management	26.7	34.8
			Agroforestry	29.7	1.6
			Plantation on agricultural land[2]	7.5	0.7
			Rangeland management.	20.0	17.4
Philippines			**Thailand**		
Forest protection plus sustainable management	215.0	1.3	Short rotation in:		
Forest protection – total log ban	215.0	0.5	- *Managed forests*	185.5	2.5
Long rotation forestry	236.0	2.1	- *Non protected areas*	158.9	2.9
Urban forestry	90.0	5.3	Long rotation in community managed forests	169.0	3.2
			Medium rotation in non protected areas	112.5	4.3
			Forest protection and rotation forestry for conservation in		
			- *Protected area*	38.6	7.5
			- *Community managed forests*	38.1	10.7
Vietnam			**Myanmar**		
Forest protection	106.9	0.1	Natural regeneration	33.0	0.1
Degraded forest protection	64.3	0.2	Reforestation long[4]	155.0	0.8
Natural regeneration (enhanced)	57.1	0.8	Forest protection	47.0	1.6
Scattered trees	64.0	0.9	Reforestation short[4]	55.0	3.8
Reforestation short[4]	43.0	2.2	Bio electricity	78.0	21.4
Reforestation long	68.2	1.7			

(continued)

Tabel 4.3: *continued*

Mitigation option	Mitigation potential (tC/ha)	Investment cost[1] (US$/tC)	Mitigation option	Mitigation potential (tC/ha)	Investment cost[1] (US$/tC)
AFRICA					
Ghana			**Cameroon**		
Evergreen forest			Evergreen forest		
- agroforestry	13-88	1-6	*- agroforestry*	16-58	1-5
- slowing deforestation	35-140	1-2	*- slowing deforestation.*	40-160	1-2
Deciduous forest			*- forestation[5]*	73-195	1-19
-slowing deforestation	35-140	1-2	Deciduous forest		
- forestation[5]	31-154	1-27	*- forestation[5]*	27-169	21-19
Savannah			Savannah		
- agroforestry	29-61	4-12	*- forestation[5]*	36-170	1-31

[1] Investment costs (US$/tC): This largely includes forest or plantation establishment costs incurred during the initial 2-3 years; discounted for only the initial 2-3 year period. For forest protection, the costs include expenditure on erecting barriers for protection, training, and other organizational costs incurred during the initial 2-3 year period. Mitigation potential is in pertuity, assuming one full cycle; rotation length for mitigation option subject to harvesting (such as short and long rotation) and for others 40 years.

[2] Natural regeneration of forest is increasing the biomass density to that of closed forests on partially degraded open forest areas; assisted or enhanced natural regeneration would involve planting a (few) trees and/soil and water conservation activity to assist or enhance natural regeneration.

[3] Plantations involve planting of one or more species at high densities.

[4] Reforestation in a short rotation has a 5 to 15 year harvest cycle, reforestation in a long rotation has a 30-100 year harvest cycle.

[5] Forestation includes both afforestation and reforestation.

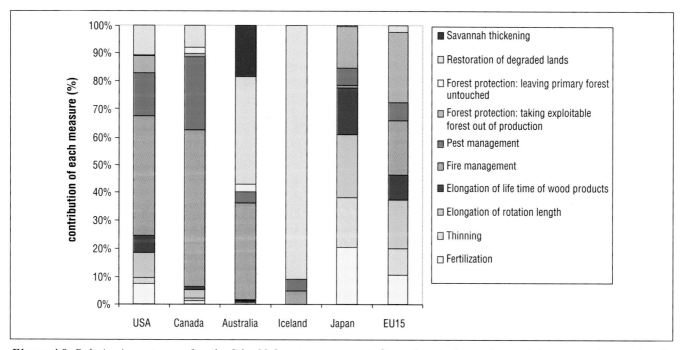

Figure 4.8: *Relative importance of each of the 10 forest management alternatives in the total potential sequestration as given in Figure 4.7. These data give an indication of opportunities and do not necessarily represent national plans. For example, silvicultural practices in Japan generally do not accompany fertilization and the figure for Japan is probably an overestimate. Nevertheless it shows that opportunities vary among countries because of both the national situation, the mix of current forestry practices, and/or the historic management. One common recommendation of which measures would yield the largest carbon sequestration can therefore not be given (adapted from Nabuurs et al., 2000).*

Table 4.4: *Land categories and extent of availability for mitigation in selected developing countries* (Sathaye and Ravindranath, 1998)

Country	Forest land for conservation, protection, and management (Mha)	Degraded forest land for regeneration (Mha)	Degraded land for plantation forestry (Mha)	Agroforestry (Mha)	Others (Mha)	Total geographic area (Mha)	Area under forests (Mha)
Asia							
China		19.2	105.2	75.9		932.6	134.0
India		36.9	41.3	96.0		329.0	63.3
Indonesia			30.5			193.0	144.7
Mongolia		2.4	1.6			156.6	17.5
Myanmar	3.3	6.9				65.8	49.3
Pakistan	0.5	0.3	2.6	1.2		77.1	3.7
Philippines	6.6	2.5			0.60	29.8	6.5
South Korea	0.7	0.3			0.05	9.9	6.5
Thailand	17.8	4.4				51.1	14.0
Vietnam	10.5	6.0			2.50	32.5	19.0
Total	39.4	78.9	181.2	171.9	3.15	1877.4	458.5
Africa							
Cameroon	1.6		7.3	1.6		46.0	36.0
Ghana	0.9		0.3	2.5		23.0	18.0
Total	2.5	0	7.6	4.1	0	69.0	54.0
Total (12 countries)	41.9	78.9	188.8	176	3.15	1946.4	512.5

mates are much larger than those in IPCC (1996) (*Table 4.4*).

Current estimates suggest that the cumulative C mitigation potential of forests in 10 Asian countries is about 26.5GtC, suggesting that the SAR estimates for the tropical region were conservative. China (9.7GtC) and India (8.7GtC) have particularly large mitigation potentials in the forestry sector (Sathaye and Ravindranath, 1998).

Latin America, which accounts for 51% of the global area of tropical forests (FAO, 1997), has an estimated mitigation potential of at least 9.7GtC, an estimate based on analyses of Mexico, Venezuela, and partly Brazil (*Table 4.5*). This total includes native forest management, protected areas, commercial plantations, agroforestry, and restoration plantations. The technical potential C mitigation in forestry is estimated at about 4.8GtC for Mexico, 1.4GtC for Venezuela, and 3.5GtC for Brazil (Da Motta *et al.*, 1999, *Table 4.5*). The feasible mitigation potential, which is largely constrained by land tenure policies and socio-economic pressures (land availability), is, however, often much lower than this technical potential. The feasible socio-economic mitigation potential is about 50% less than the technical potential in Mexico and about 44% lower than the technical potential in India (Ravindranath and Somashekar, 1995).

Deforestation in the Brazilian Amazon is a significant source of CO_2 and, with 90% of the originally forested area still uncleared, Brazil remains a large potential source of future emissions. The deforestation rate in Amazonia was estimated to be 1.38 million ha/yr in 1990, corresponding to an emission of 251MtC/yr (Fearnside, 1997). The rate of deforestation has increased in recent years, to 2.91Mha/yr in 1995 and 1.82Mha/yr in 1996 (Fearnside, 1998). Reducing the deforestation rate by 50% would conserve 125MtC/yr. Thus, Brazil alone offers a large potential for mitigation through slowing of deforestation.

What is the permanence of C sequestered by forest management activities? Clearly, tree plantations that are harvested and not re-established do not contribute to long-term carbon sequestration, though they may reduce atmospheric C in the short term. But, if a new forest is maintained so that harvest equals net growth, the forest can both be a source of wood products and still retain the captured C. In other words, the sequestration phase may be finite, lasting only a few decades, but the conservation phase need not be finite. Although there is an exchange of carbon between the atmosphere and the biomass, a considerable pool of carbon can be permanently stored in the steady-state biomass while wood products continue to be produced. This C pool remains withdrawn from the atmosphere as long as the forest exists. The substitution phase, which

Table 4.5: *Biological GHG mitigation potential in Latin America*

Option	Land available in 2030 (Mha)		Unit C sequestration (tC/ha)	Total C sequestration (MtC)		Unit cost[b] (US$/tC)	Total cost	Reference
	Technical potential	Economic potential		Technical potential	Economic potential			
Native forest management								
Mexico	18.7	13.2	132	2465	1550	0.1-4	4930	Masera, 1995; Masera et al., 1997a; Masera and Ordóñez, 1997
Venezuela	9.8		75	735		9	6615	Bonduki and Swisher, 1995
Brazil	60		18	735		1.8	1323	Da Motta et al., 1999
Protected areas								
Mexico	6	4.9	89	535	470	1-6	1872.5	Masera, 1995; Masera et al., 1997a, Masera et al. 1995 and 1997b
Venezuela	4		94	376		4	1504	Bonduki and Swisher, 1995
Brazil	151		18	2718		3	7650	Da Motta et al., 1999
Restoration plantations								
Mexico	4.2	2.5	76	320	200	7	2240	Masera, 1995; Masera et al., 1997a, Masera and Ordóñez, 1997
Commercial plantations (includes energy plantations)								
Mexico	6.6	2.4	208	1375	1075	5-7	8250	Masera, 1995; Masera et al., 1997a, Masera et al. 1995 and 1997b
Venezuela	4.9		52-62	295		17	5015	Bonduki and Swisher, 1995
Brazil degraded land								
- Pulp			24			1.4		Da Motta et al., 1999
- Charcoal			180			0.7		Da Motta et al., 1999
- Timber			43			–9.5		Da Motta et al., 1999 '–' means profitable
Agroforestry								
Mexico	1.9	1.5	53	100	80	2-11	650	Masera, 1995; Masera et al., 1997a; Masera and Ordóñez, 1997; De Jong et al., 1995
Venezuela	1		27	27		20	540	Bonduki and Swisher, 1995
Total								
Mexico	37.4	24.5		4795	3375		17943	
Venezuela	19.7			1433			13674	
Brazil[a]	211			3453			8973	

a   Unit carbon sequestration considers the difference between sustainable and unsustainable logging. Unit price is NPV(net of present value of benefits minus present value of costs)

b   Unit cost US$/tC is NPV.

begins at the onset of the first harvest, can be sustained. Each timber crop, in a cumulative manner, can substitute for fossil-fuel resources. The forest thus offers a sustainable alternative to the unsustainable use of fossil-fuel resources (Schlamadinger and Marland, 1996).

Land owners are unlikely to manage their forest resources for C sequestration alone. In the absence of financial incentives, any C sequestration will likely be incidental, or have the role of a by-product in the management of forests to produce valued goods and services (ITTA, 1983, 1994). In the tropical biome, the optimal mix of management strategies will likely reflect a balance between various forest management systems and agricultural production. Existing policies for forest and agricultural land management, however, do not yet reflect economic incentives for C management and probably are not optimal (see for example Poore *et al.*, 1989).

The effectiveness of various strategies for C sequestration will depend on the initial status of the forest ecosystems. For lands without tree cover, afforestation permits large C gains per hectare (Dyson, 1977; Sedjo and Solomon, 1989). Industrial plantation forests are already being created on a large scale and expansion of this area for C sequestration is possible (Sedjo and Sohngen, 2000). The establishment of forest plantations is generally the most reliable silvicultural method for afforestation, reforestation, and sustainable regeneration (regeneration soon after cutting). Plantation establishment can enhance productivity if desired species are planted on suitable sites. Plantations can reduce the pressures to degrade natural forests (Sedjo and Botkin, 1997). However, following the harvest of a mature or old-growth forest, the land can remain a source of carbon for many decades, even when it is regenerated (Hoen and Solberg, 1994; Cohen *et al.*, 1996; Schlamadinger and Marland, 1996; Bhatti *et al.*, 2001). Therefore, for primary and mature forests, conserving and protecting the existing C pools is often the only mitigation option that yields near-term benefits.

Because of the diversity in the current global forest status and socio-economic situation, the optimal mix of mitigation strategies will vary with country and region, in both the tropics and the non-tropics. For many countries, slowing or halting deforestation is a major opportunity for mitigation (e.g., Brazil: Fearnside, 1998, and Mexico: Masera, 1995). In countries such as India, where deforestation rates have declined to marginal levels, afforestation and reforestation in the degraded forest and non-forest lands offer large mitigation opportunities (Ravindranath and Hall, 1995). Ravindranath and Hall (1995) have shown the potential of using this degraded land and small biomass gasifiers to sustainably produce electricity from woody biomass and displace 40 million tonnes of C annually. In Africa an important opportunity for mitigation is in conserving wood fuel and charcoal through improved efficiencies of stoves and charcoal kilns (Makundi, 1998). The selection of mitigation strategies or projects in tropical countries, particularly, will be determined by economic development priorities,

changing pressures on land use, and resource constraints. In many industrialized countries, adjusting forest management regimes and material flows in the forest products sector (including substitution) appears most promising (Hoen and Solberg, 1994; Binkley *et al.*, 1997).

To quantify accurately the effects of changes in forest management on the net transfer of C to the atmosphere, the whole system could be considered (see *Box. 4.1*). Many earlier studies focused on the immediate results of forest management measures, e.g. the higher biomass growth rate following a silvicultural treatment or the protected stock of C if wildfire or logging is prevented. Global assessments based on these studies (e.g., Dixon *et al.*, 1994; Brown *et al.*, 1996b) have limitations. Estimates, in terms of tC/ha or tC/ha/yr, leave unanswered the critical questions of the timing, security, and sustainability of these effects. Also, recent, more comprehensive studies indicate the importance of complete accounting for all the C flows in and out of the system and the analysis of long-term patterns. For example, Schlamadinger and Marland (1996) showed that the positive effect of short-rotation plantations for fossil fuel substitution is less than implied by the simple substitution of fossil fuels, because of the continued input of fossil fuels needed to operate the system. While the limitations of earlier studies are now evident, data for comprehensive analysis at the global scale are not yet available. This, in part, explains why global-level estimates of the potential for C mitigation in forestry remain unchanged from those in SAR.

4.3.2.1   Wood Products

Wood products are an integral part of the managed forest ecosystem and the forest sector C cycle. They play three roles in the forest sector carbon cycle: (1) a physical pool of carbon, (2) a substitute for more energy-intensive materials and, (3) a raw material to generate energy (Burschel *et al.*, 1993; Nabuurs and Sikkema, 1998; Harmon *et al.*, 1996; Karjalainen, 1996; Matthews *et al.*, 1996; Marland and Schlamadinger, 1997; Apps *et al.*, 1999).

Wood removed from a forest by harvest, whether by thinning or clear-cut, can be viewed as a replacement for the natural mortality that would otherwise occur eventually (albeit at a faster rate). Harvested wood provides renewable raw material for use as fuel, fibre, and building materials; as well as income and employment for rural populations (Glück and Weiss, 1996). Globally, about 3.4 billion m^3 of wood are harvested per year, excluding wood that is burned on site (FAO, 1997). Harvest rates are expected to increase at 0.5% per year (Solberg *et al.*, 1996). Of the total harvest, about 1.8 billion m^3 is for fuelwood, used mainly in the tropics. The total fuelwood consumption in tropical countries increased from 1.3 to 1.7 billion m^3 during the period 1990 to 1995 (FAO, 1997; Nogueira *et al.*, 1998).

If the fossil fuel based energy required to produce and transport forest products is less than that needed for alternative products, then CO_2 emissions will be avoided by the use of forest products. Buchanan and Levine (1999) show, for example, that when wood is used for building construction in place of brick, aluminium, steel, and concrete, there can be net savings in CO_2 emissions. For construction of small buildings in New Zealand, the carbon substitution effect was larger than the direct carbon storage in wood building products (Buchanan and Levine, 1999). Forest products can also substitute in the marketplace for alternative materials, such as cement, that involve carbon emissions in their manufacture.

A systems approach has been used recently to recognize interdependencies among products and sectors. For example, Adams (1992) and Alig *et al.* (1997) examined the effects of sequestering C in forests in the USA on the availability of agricultural land, and Sedjo and Sohngen (2000) used a sectoral approach that explicitly recognized interrelations among various wood investment decisions, and between wood investment and C sequestration activities. The systems approach also recognizes the joint product nature of industrial wood and carbon sequestration. In a study in Argentina, for example, Sedjo (1999b) found that timber alone does not generate sufficient returns to justify plantation investment, but the simultaneous sequestration of C can justify investment above some threshold C price. The models do not yet incorporate a potential increase in demand for wood as a fuel to displace fossil fuels.

In the developing world most fuelwood and charcoal use is devoted to satisfying energy needs for cooking (Makundi, 1998). The potential for conservation of fuelwood is significant, both through improved cooking stoves and by substitution with liquefied or gasified biofuels. India, China, and some African countries have large programmes for the distribution of more efficient wood stoves. In India alone 28 million improved stoves have been disseminated (Ravindranath and Hall, 1995). The carbon mitigation costs of improved wood stoves in India range from US\$0.10/tC abated (Luo and Hulscher, 1999) to US\$12/tC abated (Ravindranath and Somashekar, 1995). A review of case studies in Asia showed an average mitigation cost of US\$0.8/tC abated in Thailand to US\$1.7/tC in India, through programmes to encourage use of improved wood stoves (Hulscher *et al.*, 1999). The experience with wood stoves shows that – when appropriately designed, implemented, and monitored – efficient stove programmes can provide substantial benefits to local residents. There are no estimates of the global potential for carbon conservation via this option, however, in India alone it is estimated that 20MtC could be saved annually (Ravindranath and Hall, 1995).

There is also a significant potential for saving fuelwood and charcoal in a large number of small industries. Charcoal making, brick making, pottery making, bakeries, etc. use fuelwood as their primary energy source in many areas. Fuelwood and charcoal consumption in tropical countries is projected to increase from 1.34 billion m^3 in 1991 to 1.81 billion m^3 in 2010 (FAO, 1993).

Most of the forest harvest in the boreal and temperate zone is for industrial roundwood (*i.e.*, cut logs). About one-half to two thirds of the roundwood finds its way into final products, and the rest is used for energy or ends up as decomposing residues (e.g., Apps *et al.*, 1999). The annual production of roundwood, according to FAO (1997) statistics, corresponds to a harvest flux of about 1.6 billion m^3, resulting in about 0.9 billion m^3 in final products. This represents a C flux of about 0.3GtC/yr into the product pool.

According to the SAR (IPCC, 1996), the current global stock of C in forest products is about 4.2GtC and the net sink is 0.026GtC/yr. Other sources suggest a stock of 10-20GtC (Sampson *et al.*, 1993; Brown *et al.*, 1996b) and a global sink of 0.139GtC/yr (Winjum *et al.*, 1998). There is a large uncertainty in the estimates. Even if the high end of the range is correct, the C sink in wood products appears small compared to the current rate of C sequestration in boreal and temperate forest ecosystems. Whether the physical pool of carbon in wood products in use acts as a sink depends on the relative rates of input and output from the product pool, *i.e.*, the difference between the production of new products and the decay of the C stock in existing products (Apps *et al.*, 1999).

Options to increase physical sequestration of carbon in wood products include:
- Increasing consumption and production of wood products;
- Improving the quality of wood products;
- Improving processing efficiency; and
- Enhancing recycling and re-use of wood and wood products.

Several studies have been carried out on the impacts of these measures on the amount of carbon sequestered in wood products. These studies generally conclude that the sink potential is quite small at the national or global level (Karjalainen, 1996; Nabuurs, 1996; Marland and Schlamadinger, 1997).

Use of wood as a fuel reduces CO_2 emissions from fossil fuels (Hall *et al.*, 1991; Brown *et al.*, 1996a; Nabuurs, 1996; Marland and Schlamadinger, 1997). Where the costs of growing biofuels on agricultural lands are higher than the costs of using fossil fuel, some form of incentive may be required to generate significant shifts to biofuels (Sedjo, 1997). The use of abandoned forest products for energy rather than disposal as waste can provide additional opportunities for displacing use of fossil fuels (Apps *et al*, 1999). Chapters 3 and 6 provide further discussion of the use of bioenergy within the energy sector.

Micales and Skog (1997) estimate that of the total amount of carbon-based products disposed of in the USA in 1993, as either paper or wood products, 28TgC (out of a total domestic harvest of approximately 123TgC/yr) will remain stored in landfills. Heath *et al.* (1996) and Karjalainen *et al.* (1994) emphasize the increasing role of landfills as a store of C.

Production of methane through anaerobic decomposition deserves to be considered when evaluating the mitigation potential.

While C sequestration in wood products can reach saturation, the C benefits of materials substitution can be sustained. Assuming a material substitution effect of 0.28tC/m³ of final wood product (Burschel *et al.*, 1993), and a flux corresponding to a roundwood volume of 0.9 billion m³ annually, the substitution impact of industrial wood products may be as large as 0.25GtC/yr. Although this estimate is highly uncertain, it is possible that for wood products the substitution impact is larger than the sequestration impact. This substitution is additional to the sinks in wood products mentioned before.

4.3.2.2 Managing Wetlands

Globally, wetlands contain large reserves of organic carbon - about 300 to 600GtC (Gorham, 1991; Eswaren *et al.*, 1993; Scharpenseel, 1993; Kauppi *et al.*, 1997). A major portion of this carbon is found in peat-forming wetlands (peatlands), often associated with forests, in both northern (302Mha, 397GtC) and tropical (50Mha, 144GtC) biomes (Zoltai and Martikainen, 1996). Over the long term, peatlands gradually accumulate additional carbon, because decomposition is suppressed under flooded conditions (Harden *et al.*, 1992; Mitsch and Wu, 1995; Rabenhorst, 1995; Zoltai and Martikainen, 1996; Kasimir-Klemedtsson *et al.*, 1997). The beneficial effect of this carbon accumulation, however, is at least partially offset by release of methane, which is also a GHG (Gorham, 1995).

There are few opportunities to augment the accumulation of carbon in wetlands by improved management. Drainage of forested peatlands, largely concentrated in boreal regions, can enhance tree growth significantly, but the net ecosystem carbon changes are less clear – some studies report large net gains while others indicate large net losses of carbon to the atmosphere (see review by Zoltai and Martikainen, 1996). A more important mitigation measure, from the perspective of atmospheric CO_2, is the preservation of the vast carbon reserves already present (van Noordwijk *et al.*, 1997) in peatlands. Drainage of wetlands for agricultural or other uses results in rapid depletion of stored C (Kasimir-Klemedtsson *et al.*, 1997).

4.3.3 Opportunities in Agricultural Lands

Most ecosystems, under constant conditions, eventually approach a steady-state C stock that is dictated by management, climate, and soil properties. But changes imposed on the ecosystem can alter the balance of C inputs and losses, shifting the ecosystem, eventually, to a new steady state (Paustian *et al.*, 1997c). For example, after conversion of forests or grasslands to arable agriculture, losses of C often exceed inputs temporarily, resulting in a net loss of C to the atmosphere until a new, lower equilibrium level is reached (Balesdent *et al.*, 1998;

Huggins *et al.*, 1998; Solomon *et al.*, 2000). At least a portion of C lost, however, can often be recovered by adopting management practices that again favour higher C stocks (Cole *et al.*, 1997). The accumulation of C in soil can continue until a new steady state is reached, often after several or more decades. Most of the additional C is stored in the soil as organic matter. Apart from agroforests, agricultural lands store very little carbon in plant biomass (*Table 4.1*).

There are two general ways of increasing C stocks in agricultural lands: by changing management within a given land use (e.g., cropland, rice land, grazing land, or agroforests) or by changing from one land use to another (e.g., cropland to grassland or cropland to forest) (Sampson *et al.*, 2000). In this section, we review briefly the possible ways of increasing C stocks in agricultural lands, first within a land use and then by a change in land use. We then review recent estimates of the potential for increasing C stocks in agricultural lands globally. A more detailed assessment of management practices and corresponding rates of C accrual is reported in the IPCC Special Report on LULUCF (IPCC, 2000a).

Croplands, as referred to here, are lands devoted, at least periodically, to the production of arable crops (wetland rice, because of its unique features, is discussed separately). Soil C in these lands can often be preserved or enhanced by using farming systems with reduced tillage intensity, thus slowing the rate at which soil organic matter decomposes (Bajracharya *et al.*, 1997; Feller and Beare, 1997; Rasmussen and Albrecht, 1997; Dick *et al.*, 1998). Another way to promote higher soil C is to increase crop yields. This can be done by applying organic amendments, by effective use of fertilizers, by using improved crop varieties, or by irrigating. These practices help replenish soil organic matter by increasing the amount of crop residues returned to the soil (Raun *et al.*, 1998; Huggins *et al.*, 1998; Paustian *et al.*, 1997b; Lal *et al.*, 1998; Smith *et al.*, 1997; Fernandes *et al.* 1997; Izac 1997). Further, soil C can often be increased by using practices that extend the duration of C fixation by photosynthesis; for example, cover crops, perennial forages in rotation, and avoiding bare fallow tend to increase organic C returns to soil (Lal *et al.*, 1997; Singh *et al.*, 1997a; Smith *et al.*, 1997; Carter *et al.* 1998; Tiessen *et al.*, 1998; Tian *et al.*, 1999; Paustian *et al.*, 1997a, 2000). Farming techniques that reduce erosion (e.g., terracing, windbreaks, and residue management) maintain productivity and also prevent loss of C from agricultural soils. The net effect of soil erosion on atmospheric CO_2 is still uncertain, however, because the C removed may be deposited elsewhere and at least partially stabilized (van Noordwijk *et al.*, 1997; Lal *et al.*, 1998; Stallard, 1998).

Rice land, as the term is used here, refers to areas that are at least periodically flooded for wetland rice production. Carbon stocks in these systems can be preserved or enhanced by the addition of organic amendments (Singh *et al.*, 1997b; Kumar *et al.*, 1999) and nutrient management (Yadav *et al.*, 1998). Rice lands, however, are an important source of methane and, from the standpoint of overall radiative forcing, management effects on methane emissions may be more important than effects on C storage (Greenland, 1995; Sampson *et al.*, 2000). Methane emissions can be suppressed to some extent by soil amendments, altered tillage practices, water management, crop rotation, and cultivar selection (Minami, 1995; Kern *et al.*, 1997; Neue, 1997; Yagi *et al.*, 1997; Van der Gon, 2000). For more information on CH_4 and N_2O emissions from land use, see Section 3.6.

Grazing lands refer to natural grasslands, intensively managed pastures, savannas, and shrublands used, at least periodically, to graze livestock. One way to increase C stocks in these lands is to introduce new plant species. For example, the introduction of N-fixing legumes increases productivity, thereby favouring C storage (Fisher *et al.*, 1997; Conant *et al.*, 2001). Large increases in soil C have been also reported from the introduction of deep-rooted grasses in South American savannas (e.g., Fisher *et al.*, 1994), though the area over which these findings apply is still uncertain (Davidson *et al.*, 1995). Other management practices that can affect C storage include: changing grazing intensity and frequency (Manley *et al.*, 1995; Ash *et al.*, 1996; Burke *et al.*, 1997, 1998); adding nutrients, especially phosphorus (Barrett and Gifford, 1999); controlling fire (Burke *et al.*, 1997; Kauffman *et al.*, 1998); and irrigation (Conant *et al.*, 2001).

Agroforests include trees on farms as part of the agricultural landscape (Sampson *et al.*, 2000). Unlike most other agricultural systems, agroforests store C in the above and below ground vegetation as well as in soil organic matter (Fernandes *et al.*, 1997; Woomer *et al.*, 1997). Examples of practices that can enhance C stocks include: integrated pest management, optimum tree densities, superior tree or crop cultivars, and better nutrient management (Sampson *et al.*, 2000).

Land-use conversion involves transferring a given land area from one use to another. Where the shift is to a land use with higher potential C storage, the conversion can result in increased C stocks. For example, conversion of cropland to grassland often increases soil C (e.g., Paustian *et al.*, 1997b; Reeder *et al.*, 1998; Potter *et al.*, 1999; Post and Kwon, 2000). Carbon stocks may also be enhanced by conversion of cropland to forests (reforestation, afforestation) or to agroforests (e.g., Fernandes *et al.*, 1997; Woomer *et al.*, 1997; Falloon *et al.*, 1998; Post and Kwon, 2000). In some cases, cultivated lands can be restored as wetlands (Paustian *et al.*, 1998; Lal *et al.*, 1999), resulting in carbon gains, though this practice may also result in higher net CH_4 emissions (Willison *et al.*, 1998; Batjes, 1999; Sampson *et al.*, 2000).

Another form of land-use conversion is the rehabilitation of severely degraded lands. Severely degraded lands are those where previous management has caused a drastic decline or disruption of productivity. Large areas of degraded lands occur on lands previously used for agriculture; lands abandoned after excessive erosion, over-grazing, desertification, or salinization (Oldeman, 1994; Lal and Bruce, 1999). Often the degradation was caused by social and economic pressures, and land reha-

bilitation may depend on the amelioration of the underlying causes of degradation. Specific rehabilitation practices include: introduction of new species (e.g., reforestation), addition of nutrients, and organic amendments (e.g., Lal and Bruce, 1999; Lal *et al.*, 1998; Izaurralde *et al.*, 1997).

Various attempts have been made to estimate potential C storage by improved management of agricultural lands. In the IPCC Second Assessment Report, Cole *et al.* (1996) estimated the potential for C storage in agricultural soils from improved management of existing croplands, restoration of degraded lands, and conversion to grass or forestlands. By assuming that one-half to two-thirds of the estimated historic C loss from cultivated soils could be recovered in 50 years, they proposed potential soil C increases of about 0.4 to 0.6GtC/yr from better management of existing agricultural soils. According to their estimates, additional C could be stored by set-aside of surplus upland soils (0.015 to 0.03GtC/yr), restoration of wetlands (0.006 to 0.012GtC/yr), and restoration of degraded lands (0.024 to 0.24GtC/yr), yielding a combined potential of about 0.44 to 0.88GtC/yr over a 50-year period. Later studies have provided similar estimates. Lal and Bruce (1999), using rates of soil C gain from the literature, estimated global C storage potentials of 0.43 to 0.57GtC/yr in the next 20-50 years, from erosion control, soil restoration, conservation tillage and residue management, and improved cropping practices. Batjes (1999), based partly on C gains estimated by Bruce *et al.* (1999), proposed that an additional 14GtC (±7) could be stored in agricultural soils over the next 25 years by improved management of "degraded" and "stable" agricultural lands. Including "extensive grasslands" and "regrowth forests" increased the estimate to 20GtC (±10), corresponding to an average rate of 0.58 to 0.80GtC/yr.

Sampson *et al.* (2000) recently completed a comprehensive assessment of potential net C storage from land management as part of the IPCC Special Report on LULUCF (IPCC, 2000a). According to their estimate, improved management within a land use could result in global rates of C gain, in 2010, of 0.125GtC/yr for cropland, <0.008GtC/yr for rice paddies, 0.026GtC/yr for agroforestry, and 0.237GtC/yr for grazing land. Potential rates of C gain in 2010 for land use conversion were 0.391 GtC/yr for conversion of unproductive cropland and grasslands to agroforests, <0.004GtC/yr for restoring severely degraded land, 0.038GtC/yr for conversion of cropland to grassland, and 0.004GtC/yr for conversion of drained land back to wetland. Corresponding rates of potential C gains for 2040 were consistently higher than those for 2010, often by a factor of about 2, though confidence in these values was lower. Sampson *et al.* (2000) cautioned that their estimates "are approximations, based on interpretation of available data" and that, "for some estimates of potential carbon storage, the uncertainty may be as high as ±50%".

Most of these estimates assume widespread, concerted adoption of C-conserving practices, and all have high uncertainty, stemming in part from the difficulty of predicting adoption of

C-conserving practices. The various estimates, furthermore, cannot always be compared directly because of differences in practices, scope, time-frame, and underlying assumptions. Most of the more recent estimates, however, are within the same order of magnitude as those presented in the SAR (Cole *et al.*, 1996).

Increases in soil carbon content in response to improved practices cannot continue indefinitely. Eventually, soil C storage will approach a new equilibrium where C gains equal C losses (Paustian *et al.*, 2000). This new equilibrium will depend on the management practices adopted, as well as on soil type and climatic conditions. Consequently, rates of C gain will diminish with time, and estimates for a given year cannot be extrapolated far into the future.

Once soils reach a new equilibrium, there is little further accumulation of C. And if the C-conserving practice is discontinued (e.g., reversion from no-tillage to intensive tillage), much of the previously gained C may be lost back to the atmosphere as CO_2 (Dick *et al.*, 1998; Stockfisch *et al.*, 1999). Consequently, the C stocks stored in soils are not necessarily permanent and irreversible.

4.4   Environmental Costs and Ancillary Benefits

4.4.1   *Environmental Costs and Ancillary Benefits in Forests*

Forests serve many environmental functions aside from carbon mitigation. Natural forests with various stages of stand development, including old-growth forests with snags and fallen logs, provide diverse habitats necessary for biodiversity (Harris, 1984; Franklin and Spies, 1991). Stopping or slowing deforestation and forest degradation, therefore, not only maintains carbon stocks but also preserves biodiversity, as shown by studies in Belize (EPA/USIJI, 1998) and Paraguay (Dixon *et al.*, 1993).

Although plantations usually have lower biodiversity than natural forests (Yoshida, 1983: Kurz *et al.*, 1997; Frumhoff and Losos, 1998), they can reduce pressure on natural forests, leaving greater areas to provide for biodiversity and other environmental services (Sedjo and Botkin, 1997). Plantations can negatively affect biodiversity if they replace biologically rich native grassland or wetland habitat, but non-permanent plantations of exotic or native species can be designed to enhance biodiversity by stimulating restoration of natural forests (Keenan *et al.*, 1997; Lugo, 1997; Parrotta *et al.*, 1997a, 1997b). Measures to promote biodiversity of intensively managed plantations include the adoption of longer rotation times, reduced or eliminated clearing of understory vegetation, use of native tree species, and reduced chemical inputs (Allen *et al.*, 1995; Da Silva Jr *et al.*, 1995; Fujimori, 1997).

Preserving forests conserves water resources and prevents flooding. For example, the flood damage in Central America following hurricane Mitch was apparently enhanced by loss of forest cover. By reducing runoff, forests control erosion and salinity. Consequently, maintaining forest cover can reduce siltation of rivers, protecting fisheries and investment in hydro-electric power facilities (Chomitz and Kumari, 1996).

Afforestation and reforestation, like forest protection, may also have beneficial hydrological effects. After afforestation in wet areas, the amount of direct runoff initially decreases rapidly, then gradually becomes constant, and baseflow increases slowly as stand age increases towards a mature stage (Kobayashi, 1987; Fukushima, 1987), suggesting that reforestation and afforestation help reduce flooding and enhance water conservation. In water-limited areas, afforestation, especially plantations of species with high water demand, can cause significant reduction of streamflow, affecting inhabitants in the basin (Le Maitre and Versfeld, 1997). The hydrological benefits of afforestation may need to be evaluated site by site.

Forest protection may, however, have negative social effects, such as displacement of local populations, reduced income, and reduced flow of subsistence products from forests. Conflicts between protection of natural ecosystems and their other functions, such as production of food, fuelwood, and roundwood, can be minimized by appropriate land use on the landscape (Boyce, 1995; Forman, 1995) and appropriate stand management.

In arid and semi-arid regions, where deforestation is advancing (Kharin, 1996) and leading to carbon loss (Duan *et al.*, 1995), restoring forests by afforestation and proper management of existing secondary forests can help combat desertification (Cony, 1995; Kuliev, 1996). Afforestation of desertified lands may be limited, however, by costs and insufficient knowledge of ecology, genetics, and physiology (Cony, 1995). In relatively arid regions, fuelwood plantations may reduce pressure on natural woodlands, thereby retarding deforestation (Kanowski *et al.*, 1992).

Agroforestry can both sequester carbon and produce a range of economic, environmental, and socioeconomic benefits. For example, trees in agroforestry farms improve soil fertility through control of erosion, maintenance of soil organic matter and physical properties, increased N, extraction of nutrients from deep soil horizons, and promotion of more closed nutrient cycling (Young, 1997). Thus, agroforestry systems improve and conserve soil properties (Nair, 1989; MacDicken and Vergara, 1990; Wang and Feng, 1995). Examples of mitigation projects that promote soil conservation through agroforestry include the AES Thames Guatemala project, and the Profator project in Ecuador (Dixon *et al.*, 1993; FACE Foundation, 1997).

We note that decisions to protect or enlarge forest cover on a large scale could also have secondary climate consequences through their feedbacks on the earth's albedo, the hydrological cycle, cloud cover, and the effect of surface roughness on air movements (see, for example, Pielke and Avissar, 1990; Nobre *et al.*, 1991; Garratt, 1993). Analyses by Bonan and Shugart (1992) suggest that large-scale changes in vegetative cover in the boreal zone may be especially important, with potentially global-scale impacts. In the boreal zone the albedo contrast between forested and unforested land during the winter is particularly large (differences as large as 40%). Indications are that the nature, magnitude, and even direction of climate changes driven by changes in surface vegetative cover will depend on the nature, location, hydrological setting, etc. of the vegetative change.

4.4.2   *Environmental Costs and Ancillary Benefits in Agricultural Lands*

Management strategies that conserve C in agricultural soils may have ancillary benefits quite apart from atmospheric CO_2 removal. Foremost among these is a favourable effect on soil productivity. Numerous studies have shown a strong link between the organic C content of a soil and its quality for crop production (e.g., Carter *et al.*, 1997; Christensen and Johnston, 1997; Herrick and Wander, 1997). Consequently, a gain in soil C may promote crop yields, and preserve or enhance future soil productivity (Cole *et al.*, 1997; Rosenzweig and Hillel, 2000). For example, application of fertilizers to agro-pastoral systems in parts of South America may not only induce soil C accumulation, but also enhance agricultural productivity (Fisher *et al.*, 1997). Many of the practices advocated for soil C conservation – reduced tillage, more vegetative cover, greater use of perennial crops – also prevent erosion, yielding possible benefits for improved water and air quality (Cole *et al.*, 1993). As a result of these benefits, adoption of practices that promote C conservation in agricultural lands is often justified even without the additional benefits arising from CO_2 mitigation.

Soil carbon sequestration, however, may sometimes have some potential adverse effects on the emission of other GHGs, notably nitrous oxide (N_2O). Where the C accumulation requires addition of higher amounts of N as fertilizer or manure, it carries the risk of increased N_2O emissions (Cole *et al.*, 1993; Batjes, 1998). Furthermore, some C-conserving practices like reduced tillage may increase N_2O emissions by favouring higher soil moisture content (Cole *et al.*, 1993; MacKenzie *et al.*, 1997; Ball *et al.*, 1999), though this effect is not always observed (e.g., Jacinthe and Dick, 1997; Lemke *et al.*, 1999). Because the radiative forcing of N_2O is about 310 times that of CO_2 (kg per kg), when calculated over a 100-year time frame (IPCC, 1996), even a small increase in N_2O emissions, if confirmed, can significantly offset gains from C sequestration.

Carbon sequestration strategies may also have an effect on energy use and, hence, CO_2 emission from fossil fuel use. Changes in fertilizer use, pesticides, and agricultural machinery may enhance or offset any gains in soil C because of CO_2 released from fossil fuel. For example, roughly 1 kgC (or

more) is released into the atmosphere as CO_2 per kgN used (Flach *et al.*, 1997; Janzen *el al.*, 1998; Schlesinger, 1999). In tropical areas where shifting cultivation is now practiced, intensification of crop production may maintain higher C stocks, by leaving more land under natural forest, but additional fossil fuel may have to be used to compensate for the fuelwood previously collected from the fallow period (van Noordwijk *et al.*, 1997). In some cases, the adoption of C-conserving practices may reduce energy use. For example, using less intensive tillage may not only favour soil C gains, but also permits savings in CO_2 emission from fossil fuel combustion (Kern and Johnson, 1993). An evaluation of the net benefit of a C-sequestering practice, therefore, must consider energy use in addition to changes in C stocks. Whereas the duration of soil C gain in response to improved management may be finite, savings in CO_2 emissions from energy use continue indefinitely (Cole *et al.*, 1997).

Aside from their secondary effects on GHG emissions, practices that sequester soil C may also have other potential adverse effects, at least in some regions or conditions. Possible effects include enhanced contamination of groundwater with nutrients or pesticides via leaching under reduced tillage (Cole *et al.*, 1993; Isensee and Sadeghi, 1996), and possible environmental effects from widespread application of manures or sludges (Batjes, 1998). These possible negative effects, however, have not been widely confirmed nor quantified, and the extent to which they may offset the environmental benefits of C sequestration is uncertain.

4.5   Social and Economic Considerations

4.5.1   *Economics*

The method of calculating costs for forestry and agricultural projects differs. Forestry almost always looks at private market costs. However, many, if not most, forestry projects have positive externalities (or ancillary benefits) in the form of erosion control, water protection, flora and fauna habitat, non-timber forest products, water protection, and so forth (Makundi, 1997; Frumhoff *et al.*, 1998; Trexler and Associates, 1998). For agricultural projects the approach is typically tied to the idea that the carbon-sequestering projects are essentially productivity enhancing and therefore can be viewed as "no regrets" activities; these are actions that have benefits in themselves aside from climate mitigation, which make the project socially desirable even without its carbon benefits. Such "no regrets" activities generally take the form of soil management activities, which both generate increased sequestered carbon and improve agricultural productivity.

There are basically three different ways of estimating the costs of sequestration of forestry projects – point estimates, i.e., cost for a particular level of output; partial equilibrium estimates, e.g., a cost function construction with the prices of inputs being held constant; and more general equilibrium types of approaches, e.g., a market equilibrium model in which some other prices, such as the prices of land inputs and the relative price of all other goods, are allowed to change owing to market forces. Additionally, economic models can incorporate changing climate conditions to estimate changes in economic variables as the climate and ecosystem change. Early studies tended to look at individual projects, relating the private costs of establishing a project to the cumulative carbon sequestered over the life of the project (see Sedjo *et al.*, 1995). Many of the point estimate type studies provide undiscounted private market cost point estimates of the carbon sequestration in afforestation projects. However, this approach usually reveals little about how costs might change if the project were expanded to involve truly large land areas, as they do not recognize rising costs required to increasingly bid land away from alternative uses. These types of estimates tended to be biased downwards, partly because the opportunity costs of the land (land rents) were often ignored.

Point Estimates: The cost estimates of actually sequestering carbon obtained in point estimate type of studies tend to be quite low; in the SAR (IPCC, 1996) a range was given of US$3-US$7 per tonne of carbon. Additionally, a large number of more recent point estimate country studies reported most unit abatement costs in this low range, or lower. The earlier IPCC estimates for SAR were that of an investment of US$ 168-220 billion required to mitigate 45-72GtC in the tropical regions. More recent work provides estimates that the cumulative investment required for mitigating 26.53GtC to be US$63.6 billion at an overall cost of US$2.4/tC (Sathaye and Ravindranath, 1998). The unit cost given in *Table 4.3* shows that the investment cost of mitigation is generally quite low for carbon conservation options in selected developing countries and South Korea (e.g., US$0.10/tC in Vietnam, and US$1- 2/tC in Cameroon and Ghana). The mitigation cost is lower than US$2/tC for the majority of the options in Indonesia, the Philippines, Vietnam, and Mongolia.

Partial Equilibrium: Partial equilibrium involves a more complete estimation of a static cost function that estimates rising costs (e.g., as a result of land price increases as one moves to lands with higher opportunity costs) associated with increased sequestration activities. These studies generate marginal cost functions that tend to suggest most costs are higher than those of the simple point estimates. This is because, for example, they include in the cost estimates the opportunity costs of the land, and they recognize rising costs associated with additional planning activity and, for some, because they apply a discount rate to future physical carbon sequestered. The costs for modest amounts of carbon sequestered in specific areas are generally in the US$20-US$100/tC range (Moulton and Richards, 1990; Adams *et al.*, 1995; Parks and Hardie, 1997;Stavins, 1999; Plantinga *et al.,* 1999). Costs tend to depend on the forest growth rates anticipated and the opportunity costs of the land. Where projects are small, land prices would be expected to be stable. However, in regions where projects are large, land prices, and hence sequestration costs, will tend to rise.

Market Equilibrium Models: This approach incorporates sectoral and general equilibrium interrelationships. It recognizes that expanding the forest for carbon sequestration purposes has implications for current and future industrial forest production and prices, and for agricultural production and prices. These price and production changes then generate feedbacks through the market to the forest and agricultural sector behaviour. Alig *et al.* (1997), for example, examine the effects on welfare costs of meeting alternative carbon sequestration targets by land re-allocations between agriculture and forestry in the USA. This model explicitly treats agriculture and wood production as interrelated. Allocating more land to trees to capture carbon has implications on the price and quantity of agricultural products, as well as on timber. Thus, the costs of carbon plantations are found both in the price of establishing the plantations and in the higher agricultural prices, and thus involve welfare shifts across sectors. A different approach, also recognizing sectoral interrelationships, is that of Sedjo and Sohngen (2000). This approach expands on earlier global timber supply models by explicitly incorporating the interrelations between the industrial wood sector and carbon plantations by recognizing the joint product nature of industrial wood and carbon. This approach finds that tree planting carbon sequestration activities tend to have a somewhat more modest effect than anticipated, since the tree planting for carbon purposes leads to an expected increase in future timber supplies and a corresponding decrease in expected future prices. Through the effects of price expectations on the timber market, carbon activities may discourage industrial timber investments and thereby lose some of the carbon gains made from the initial project. This is a form of leakage not often recognized.

Climate Feedback Models: These market equilibrium models incorporate the impact of the climate-driven changing ecology into their assessment of the potential and costs. Perez-Garcia *et al.* (1997) examine the effects of climate change, using a global trade model (CGTM). This approach imposes a global circulation model (GCM) and a terrestrial ecosystem model on the world's industrial wood economy, and estimates the welfare effects on forest owners and forest consumers of such changes. Sohngen and Mendelsohn (1998) use a timber model of the USA to estimate the changes in the forest market sector that would be expected to occur with a climate warming using GCM and terrestrial ecosystem models. However, neither study considers the impacts of increased fuelwood demand to replace fossil fuels.

In summary, most studies, of all methodologies, suggest that there are many opportunities for relatively low-cost carbon sequestration through forestry. Estimates of the private costs of sequestration range from about US$0.10-US$100/tC, which are modest compared with many of the energy alternatives (see *Table 3.9* and *Figure 4.9*). Additionally, it should be noted that most forest projects have positive non-market benefits, thus increasing their social worth. However, as the studies have become more sophisticated, incorporating both the full private

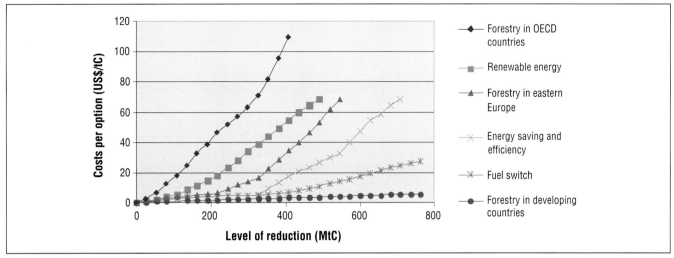

Figure 4.9: *Indicative curves of costs (US\$/tC, cost of US\$28/tC is equivalent to US\$100 per tonne of CO_2) of emission reduction or carbon sequestration by level of total reduction. The curves display how comparable options vary in costs between world regions. However, costs per option are also reported to vary widely at comparable total levels of reduction. This is mainly because cost studies have not been carried out in the same way. In some options net monetary profit may occur as well (i.e., costs may be negative as well) (Brown et al., 1996a, Hol et al., 1999; Jepma et al., 1997; Sedjo et al., 1995).*

opportunity costs of the land and market effects on land and resource prices, estimates of carbon sequestration costs have tended to rise. The cost estimates tend to vary for regions, with high costs generally associated with high opportunity costs for land. In the many regions that have low opportunity costs for land, including many subtropical regions, the costs tend to remain low.

4.5.2 Institutional Structures and Equity Issues

In order to realize the mitigation potential in part or in full, it would be helpful to have a set of institutions to translate the policies and measures into avoided emissions or carbon sequestration. In the United Nations Conference on Environment and Development (UNCED) held in 1992, the importance of sustainable forest management was emphasized under the "Forest Principles." Subsequently, the formulation of criteria and indicators was worked out under the Helsinki and Montreal Processes, in which the maintenance and enhancement of forest resources to contribute to the global carbon cycle is described. The same is a criterion under the United Nations (UN-ECE/FAO, 2000). The three main types of necessary institutions are global and/or regional, national and local, and/or community based (IPCC, 2000b). At the global level, there exist government-based multilateral institutions such as the United Nations Food and Agriculture Organization (FAO), United Nations Environment Programme (UNEP), the World Bank, and the International Tropical Timber Organization (ITTO). All of these institutions are involved in natural resource management, and can play a significant role in integrating mitigation objectives in tropical forest management. Also, a wide array of non-governmental organizations (NGOs) active in resource conservation and sustainable utilization, as

well as bilateral aid organizations, could play a more important role in incorporating mitigation in their policy objectives. For example, investment promotion agencies could be created to assist in the co-ordination of investment into carbon projects (e.g., see Moura-Costa *et al.*, 1999). Additionally, global private enterprises could be encouraged to include climate mitigation measures in their plans. Financial incentives may be required to achieve broad participation.

In tropical countries, forestry is dominated by government-based institutions, mostly the departments of forestry and agriculture and/or those involved in environmental management (WRI, 1987). These departments may need support and new insight in order to effectively incorporate mitigation policies and measures in their resource management activities. At a national level, there also exist some institutions involving NGOs that focus on conservation and forest expansion, as well as those dedicated to encouraging sustainable agriculture. Such institutions may also include umbrella organizations involved in developmental activities such as gender, poverty alleviation, etc. A few institutions, including non-governmental and especially those involved in nature conservation and environmental services, e.g., game reserves, tourism companies, and large-scale agricultural production, could also incorporate mitigation considerations in their efforts.

At the local level, effective institutions include community leaderships, religious institutions, schools, traditional organizations, and indeed the family. These institutions are essential with regard to natural resource management and agricultural practices, as well as for introducing mitigation-type activities that do not contravene their basic needs to use their land and natural resources for sustenance.

Public, NGO, and private institutions, at each spatial level where they exist, can focus on including GHG mitigation as one of their considerations, while they oversee the use of forest and land resources to meet the developmental aspirations of those in tropical countries and elsewhere. For example, a recent study on sustainable livelihoods and carbon management discussed arrangements to facilitate the involvement of small-scale farmers and rural communities in carbon trading (Bass *et al.*, 2000). An optimal mix of conservation, sequestration, and substitution will be incidental or arise from the optimal management of forest resources for producing desired goods and services as shown under various tropical forest management stipulations (ITTA, 1983, 1994). In the tropical biome, the most likely use of the optimal mix of management strategies will be based on optimal management of forestry and agricultural resources in each country. For example, balance between forest conservation, afforestation, reforestation, and multiple land use of the forest and agricultural areas will predetermine the extent of utilization of the land-use sectors for mitigation activities. However, the existing policies in managing forest and agricultural resources have been criticized as non-optimal (see, for example, Poore *et al.*, 1989). Optimal levels of substitution will be determined by the energy and industrial policies of these countries, rather than by carbon sequestration criteria.

The so-called "no regrets options" can be identified and pursued (see Chapters 7 and 8 for a discussion of no regrets options). Analysis has suggested that adequately designed and implemented GHG mitigation options in forestry and agriculture could help advance the countries' own development priorities, at the same time providing significant carbon sequestration (see Sheinbaum and Masera (2000) for analysis at the country level).

4.6   Market and Non-market Options to Enhance, Maintain, and Manage Carbon Pools

4.6.1.   Introduction, Taxes, and Quotas

There are a host of market and non-market options to manage carbon pools in the terrestrial biosphere. Some of the most relevant questions related to carbon sequestration deal with the types of instruments, policies, and mechanisms that could play a role in promoting increased sequestration and how the various arrangements would actually affect outcomes. Market mechanisms could be important in promoting or discouraging carbon sequestration. Potential mechanisms might include taxes or subsidies for activities that affect carbon directly or that affect activities with large carbon implications. The UK, for example, has proposed a "Climate Change Levy" to be adopted by the UK's 2001 Budget. The Kyoto Protocol introduces flexible mechanisms allowing joint implementation, emissions trading, and the clean development mechanism. When dealing with terrestrial systems any policies that influence land use can affect carbon sequestration. Finally, there is

the question of how the various instruments and policies are likely to influence leakage of carbon flows outside the targetted system.

Agricultural subsidies are common in many, if not most, countries. Agricultural subsidies and absent forestry subsidy policies can often be viewed as discouraging forest production and thus, inadvertently, discouraging some possibilities for carbon sequestration. Similarly, tax policies can promote or discourage certain types of land use. In some countries, however, subsidies do promote afforestation and reforestation. The movement of land from agriculture to forests generally leads to gains in the forest sector and losses in the agricultural sector. The cost of any additional carbon storage can involve a change in welfare across two sectors. Lower taxes for agricultural lands and subsidies for forest clearing may be part of the package of instruments to promote development.

To reach objectives for carbon sequestration, market mechanisms are important, but an appropriate institutional setting is also useful. In some tropical countries the profitability of maintaining forests could be improved in order to prevent conversion to alternative uses of the land. Success could entail revising policies that directly or indirectly subsidize cattle ranching (as has been historically the case in Latin America) or agriculture. Success in C sequestration could also entail technical and financial training and capacity building at the local level. It should be recognized, however, that in many tropical countries, particularly within Asia and Africa, forests are harvested and used according to the subsistence needs of local communities. In these cases, some have argued that approaches based on market mechanisms will not be effective. Also, non-timber forest products are an important component of the total demand for forest products and could be considered.

It is clear that some measures aimed at sequestering C in the biosphere have relatively low cost compared to other approaches for mitigating the atmospheric increase of CO_2 (Section 4.5). However, to date only a small number of projects involving a small and varied group of stakeholders has been initiated in the terrestrial biosphere. These projects (forest expansion, forest management, soil carbon management, community forestry, and agroforestry) have covered, worldwide, an area of 3.6 to 6.4 million hectares in 1999 (for an overview see Brown *et al.*, 2000). Incentives that would create projects aimed at carbon sequestration in the biosphere on a large scale are not yet in place.

An important change in motivating carbon sequestration has been the creation of the Kyoto Protocol, in December 1997. Although few countries have yet to ratify the Kyoto Protocol, it introduces ceilings and/or quotas for CO_2 emissions for Annex B countries. In addition, the Kyoto Protocol explicitly recognizes afforestation, reforestation, and deforestation (ARD) as having carbon implications, and it provides credits (and debits) for these activities in meeting carbon-emissions targets. This arrangement has contributed to pressures to find ways to give sequestered carbon value in the market place. A detailed explanation of how the Kyoto Protocol might influence management of C stocks is given in the IPCC Special Report on LULUCF (IPCC, 2000a).

Through setting emissions targets and introducing taxes on CO_2 emission in some countries, carbon gains monetary value and could become a new product for the forestry sector. From existing emissions taxes this value is estimated at US\$200/tCO_2 in Norway and US\$100/tCO_2 in the Netherlands (Solberg, 1997; Nabuurs, 1998). In the case of the Netherlands, this carbon value is equivalent to US\$17.5/m^3 of roundwood, more than the stumpage value of wood as a raw material. However, in the first trades of certified carbon credits, Moura-Costa and Stuart (1998) found that prices ranged between US\$5–US\$10/tC. More generally, Moura-Costa and Stuart (1998) found that the average price for carbon credits for carbon sequestered in developing countries ranged from US\$0.19 to US\$12/tC, and that these differences are very much linked to uncertainty about long term policy.

The Dutch Government is considering the introduction of CO_2 certificates as part of a test of CO_2 emissions trading. In this system, each economic sector and each firm could achieve its targets partly through certificates. Funds generated from these certificates would be used to establish forests.

4.6.2   *Carbon Offsets, Tradable Permits, and Leakage*

Markets created for carbon credits from management of the biosphere, of course, will be heavily influenced by the many other commodities produced by the biosphere (food, wood, etc.). Food security may, for example, be a reason for a government to continue its policy of agricultural subsidies in the absence of forestry (carbon) subsidies. On the other hand, some studies (e.g., Callaway and McCarl, 1996) have shown that when diverting agricultural subsidies to carbon payments, the net impact on the national budget could be zero. In tropical countries, the institutions and subsidies for forest clearing may remain as part of the package to promote economic development. Only if the monetary value of carbon stocks and sinks is recognized and paid for will markets be efficient in encouraging C sequestration. Some developing countries see markets for C offsets as providing resources to facilitate capital inflows to finance conservation and other activities.

An emerging instrument that is likely to have a large effect on carbon sequestration is the tradable emissions permit. Tradable permits to deal with environmental pollutants have precedents in other areas. In the USA, for example, there is an active market for sulphur emissions permits (Burtraw, 2000). Firms with excess emissions permits can trade these to firms in need of additional permits. Thus, incremental emissions are no longer free, but incur additional costs to the firm. Firms that have excess permits can either sell those permits or forego the opportunity of receiving a payment – an opportunity cost. Such

an approach allows the market to reallocate emission permits, and thus emissions, to the users that receive the highest return from the permits, thereby distributing carbon emissions permits to the most efficient users. This approach is beginning to be contemplated in addressing the problem of increasing atmospheric carbon and is endorsed in the Kyoto Protocol.

Currently, there are a series of brokers prepared to trade carbon credits in the USA and Europe, e.g., Natsource and Canto Fitzgerald (Stuart and Moura-Costa, 1998), and the Sydney Futures Exchange in Australia is planning to begin trading in the latter part of 2000[1]. In addition to tradable carbon emission permits, the door is open for consideration of an analogous instrument, tradable "carbon offsets". Activities, such as planting and protecting forests, could provide carbon sequestration services that could be sold or traded.

To date there is only limited experience with certified carbon offset instruments. In the USA, the electrical power industry, through the Edison Electric Institute (EEI - an association of private electrical power companies), has formed the Utility Carbon Management Tree Program whereby the various member companies invest money into a project fund to develop or purchase carbon offset credits (Sedjo, 1999a). Another market approach has been created, the Certified Tradable Offsets, issued by the Costa Rican government, and the first carbon-backed securities worldwide (Stuart and Moura-Costa, 1998). These offsets are like JI or CDM as defined in the Kyoto Protocol, but would be tradable.

A potentially serious problem with carbon offsets is that there may be carbon leakage. Leakage refers to the situation in which a carbon sequestration activity (e.g., tree planting) on one piece of land inadvertently, directly or indirectly, triggers an activity which, in whole or part, counteracts the carbon effects of the initial activity. It can be shown that most of these types of problems arise from differential treatment of carbon in different regions and circumstances, and the problem is not unique to carbon sequestration activities but pervades carbon mitigation activities in the energy sector as well.

In land use, leakage can occur from either protection or planting activities. Suppose, for example, that a forest or wetland that was to be cleared is instead protected. Protection of one such forest or wetland may simply deflect the pressure to another piece of land that is not protected and will be cleared instead. Leakage can occur across both spatial and temporal boundaries. Additionally, a forest protected in one year is subject to the possibility of clearing in subsequent years.

A similar situation may also exist with activities such as tree planting. Trees provide at least two services: producing industrial wood and sequestering carbon. Trees planted for carbon sequestration, because they may eventually be used for wood,

---

[1] See *International Herald Tribune*, 31/08/1999.

can affect expectations about future industrial wood prices, thereby influencing the planting decisions of forest products companies. If carbon credits are provided to carbon forests but not to industrial forests, and if some carbon forests are anticipated to enter future timber markets, then forest industrial firms may reduce investments in new forests. Such a reduction would partly offset carbon sequestered in the newly planted carbon forest, thereby reducing the net total carbon that would have accumulated by both industrial and carbon forests (Sedjo and Sohngen, 2000). This leakage effect would not occur if both industrial and carbon forests could expect to receive payment for both their carbon and their wood.

Leakage from industrial forests, resulting from forests established for carbon purposes, has been estimated by Sohngen and Sedjo (1999) to be about 40%, globally, assuming that all carbon forests are made available to the timber market. This compares with estimated leakages in the energy sector of about 5%–20%. No estimates of leakage generated from protection activities are available, but it is suggested that it may vary by country and site, unlike planted forests that are linked through the global timber market.

The leakage problem may be addressed reasonably well within nations by caps imposed on total emissions, but leakage of emissions across national boundaries may still occur in the absence of global coverage.

Conceptually, a permanent net carbon offset should be equivalent to a tradable emissions permit. If a new activity permanently reduces net atmospheric carbon by one tonne, the climatic implications are the same as if the tonne of carbon was never released. Thus, a carbon-offset credit would be equivalent to a tradable emission credit. However, since carbon offset can quickly be liquidated, offset credits have greater liability problems. One approach might be treated on an annual (or decadal) basis as the rental of (perhaps temporary) carbon sequestering services. Although different from carbon emissions permits, they nevertheless would expand the number of "credits" available, and thus have a mitigating effect on the market price of the credits. A discussion of some of the options is presented in IPCC (2000a).

4.6.3   *Risks, Rights, and Practical Economics*

Protecting forestlands, grasslands, and other natural ecosystems is often proposed as the best way to maintain large carbon reservoirs at lowest cost. The cost of such an approach, however, may in fact be significant, although low in comparison with many of the options in the energy sector and attempts at forest protection have failed in many parts of the world. The incentives to convert often far outweigh the incentives to protect. This problem is often exacerbated by the absence of well-defined, enforceable property rights, either private or public, and the absence of other necessary institutions. In an open access situation the incentives are to "use it or lose it", since

there are no certain claims on the future use of the resource. Because there is no long-term claim on the resource in the future, the result is that resources may be used wastefully in excess of their economic optimum. Thus, deforestation and land clearing are a form of the open access problem (Hardin, 1968).

The costs of carbon management may not be distributed in the same way as the benefits. Carbon management options in developing countries may have low market costs but high local social costs in land commitments, and the benefits that arise may not be shared with local peoples. Analysis of a forest protection project in Madagascar suggests that there are financial benefits for local inhabitants and social benefits for the global community, but short-term debits at the national level (Kremen *et al.*, 2000). Formal adoption of markets for forest carbon could increase incentives for forest protection, especially if mechanisms assure that local peoples share in the benefits. Similarly, costs and benefits may be realized at different times; future benefits are often weighted against current costs. How communities value present and future costs depends on wealth, culture, and economic and environmental priorities.

International consensus on carbon management begins to have important implications for national sovereignty and personal property rights, an issue brought to prominence by recent turmoil regarding international trade agreements (see Chapter 6 for a detailed discussion on policies, measures and instruments).

4.7   Biological Uptake in Oceans and Freshwater Reservoirs, and Geo-engineering

The net primary production of marine ecosystems is roughly the same as for terrestrial ecosystems (50GtC/yr for marine ecosystems and 60GtC/yr for terrestrial ecosystems), and there are opportunities to increase the net carbon flow into the marine biosphere. There are fundamental differences between the two systems, however, as the marine biosphere does not include large stores of carbon in the living and dead biomass. There are some 3 GtC in marine biota versus nearly 2500GtC in terrestrial vegetation and soils (*Table 4.1*). The key to increasing the carbon stocks in ocean ecosystems is thus to move carbon through the small reservoir of the marine biota to the larger reservoirs of dissolved inorganic carbon (the "biological pump") in ways that will isolate the carbon and prevent its prompt return to the atmosphere. The biological pump serves to move carbon from the atmosphere to the deep oceans, as organisms take up CO_2 by photosynthesis in the surface ocean, and release the carbon when the organic material sinks and is oxidized at depth.

Several researchers have suggested that ocean productivity in major geographical regions is limited by the availability of primary or micronutrients, and that productivity could be increased substantially by artificially providing the limiting nutrients. This might involve providing nitrogen or phosphorus in large quantities, but the quantities to be supplied would be much smaller if growth were limited by a micronutrient. In particular, there is evidence that in large areas of the Southern Ocean productivity is limited by availability of the micronutrient iron. Martin (1990, 1991) suggested that the ocean could be stimulated to take up additional CO_2 from the atmosphere by providing additional iron, and that 300,000 tonnes of iron could result in the removal of 0.8GtC from the atmosphere. Other analyses have suggested that the effect may be more limited. Peng and Broecker (1991) examined the dynamic aspects of this proposal and concluded that, even if the iron hypothesis was completely correct, the dynamic issues of mixing the excess carbon into the deep ocean would limit the magnitude of the impact on the atmosphere. Joos *et al.* (1991) reported on a similar model experiment and found the ocean dynamics to be less important, the time path of anthropogenic CO_2 emissions to be very important, and the maximum potential effect of iron fertilization to be somewhat greater than reported by Peng and Broecker (1991).

Some of the concepts of iron fertilization have now been tested with 2 small-scale experiments in the equatorial Pacific Ocean. In experiment IronEX 1 (November, 1993) 480 kg of iron were added over 24 hours to a 64 km^2 area of the equatorial Pacific. In IronEX 2 (May/June, 1995) a similar 450 kg of iron (as acidic iron sulphate) were added over a 72 km^2 area, but the addition occurred in 3 doses over a period of one week.

The IronEx 1 experiment showed unequivocally that there was a biological response to the addition of iron. However, although plant biomass doubled and phytoplankton production increased fourfold, the decrease in CO_2 fugacity (in effect the partial pressure of CO_2 decreased by 10 micro atm) was only about a tenth of that expected (Martin *et al.*, 1994; Watson *et al.*, 1994; Wells, 1994). In the IronEX 2 experiment the abundance and growth rate of phytoplankton increased dramatically (by greater than 20 and twice, respectively), nitrate decreased by half, and CO_2 concentrations were significantly reduced (the fugacity of CO_2 was down 90μatm on day 9). Within a week of the last fertilization, however, the phytoplankton bloom had waned, the iron concentration had decreased below ambient, and there was no sign that the iron was retained and recycled in the surface waters (Monastersky, 1995; Coale *et al.*, 1996; Cooper *et al.*, 1996; Frost, 1996).

These two experiments have demonstrated that week-long, sustained additions of iron to nutrient-rich, but iron-poor, regions of the ocean can produce massive phytoplankton blooms and large drawdowns of CO_2 and nutrients. While the results of these two experiments cannot be uncritically extrapolated, they suggest a very important role for iron in the cycling of carbon (Cooper *et al.*, 1996). The consequences of larger, longer-term introductions of iron remain uncertain. Concerns that have been expressed relate to the differential impact on different algal species, the impact on concentrations of dimethyl sulphide in surface waters, and the potential for creating anoxic regions at depth (Coale *et al.*, 1996; Frost, 1996; Turner *et al.*, 1996). There is much to be learned of the ecological consequences of large-scale fertilization of the ocean.

Jones and Young (1998) suggest that the addition of reactive nitrogen in appropriate areas, perhaps in conjunction with trace nutrients, would increase production of phytoplankton and could both increase CO_2 uptake and provide a sustainable fishery with greater yield than at present.

Chemical buffering of the oceans to decreases in pH associated with uptake of CO_2 leads to an increase in dissolved inorganic carbon that does not rely on alteration of the biological pump. Buffering of the oceans is enhanced by dissolution of alkaline minerals. Dissolution of alkaline materials in ocean sediments with rising pH occurs in nature, but does so on a time-scale of thousands of years or more (Archer *et al.*, 1997). Intentional dissolution of mined minerals has been considered, but the quantity (in moles) of dissolved minerals would be comparable to the quantity of additional carbon taken up by the oceans (Kheshgi, 1995).

Stallard (1998) has shown that human modifications of the earth's surface may be leading to increased carbon stocks in lakes, water reservoirs, paddy fields, and flood plains as deposited sediments. Burial of 0.6 to 1.5GtC/yr may be possible theoretically. Although Stallard (1998) does not suggest intentional manipulation for the purpose of increasing carbon stocks, it is clear that human activities are likely leading to carbon sequestration in these environments already, that there are opportunities to manage carbon via these processes, and that the rate of carbon sequestration could be either increased or decreased as a consequence of human decisions on how to manage the hydrological cycle and sedimentation processes.

The term "geo-engineering" has been used to characterize large-scale, deliberate manipulations of earth environments (NAS, 1992; Marland, 1996; Flannery *et al.*, 1997). Keith (2001) emphasizes that it is the deliberateness that distinguishes geo-engineering from other large-scale, human impacts on the global environment; impacts such as those that result from large-scale agriculture, global forestry activities, or fossil fuel combustion. Management of the biosphere, as discussed in this chapter, has sometimes been included under the heading of geo-engineering (e.g., NAS, 1992) although the original usage of the term geo-engineering was in reference to a proposal to collect CO_2 at power plants and inject it into deep ocean waters (Marchetti, 1976). The concept of geo-engineering also includes the possibility of engineering the earth's climate system by large-scale manipulation of the global energy balance. It has been estimated, for example, that the mean effect on the earth surface energy balance from a doubling of CO_2 could be offset by an increase of 1.5% to 2% in the earth's albedo, i.e. by reflecting additional incoming solar radiation back into space. Because these later concepts offer a potential approach for mitigating changes in the global climate, and because they are treated nowhere else in this volume, these additional geo-engineering concepts are introduced briefly here.

Summaries by Early (1989), NAS (1992), and Flannery *et al.* (1997) consider a variety of ways by which the albedo of the earth might be increased to try to compensate for an increase in the concentration of infrared absorbing gases in the atmosphere (see also Dickinson, 1996). The possibilities include atmospheric aerosols, reflective balloons, and space mirrors. Most recently, work by Teller *et al.* (1997) has re-examined the possibility of optical scattering, either in space or in the stratosphere, to alter the earth's albedo and thus to modulate climate. The latter work captures the essence of the concept and is summarized briefly here to provide an example of what is envisioned. In agreement with the 1992 NAS study, Teller *et al.* (1997) found that $\sim 10^7$ t of dielectric aerosols of ~ 100 nm diameter would be sufficient to increase the albedo of the earth by $\sim 1\%$. They showed that the required mass of a system based on alumina particles would be similar to that of a system based on sulphuric acid aerosol, but the alumina particles offer different environmental impact. In addition, Teller *et al.* (1997) demonstrate that use of metallic or optically resonant scatterers can, in principle, greatly reduce the required total mass of scattering particles required. Two configurations of metal scatterers that were analyzed in detail are mesh microstructures and micro-balloons. Conductive metal mesh is the most mass-efficient configuration. The thickness of the mesh wires is determined by the skin-depth of optical radiation in the metal, about 20 nm, and the spacing of wires is determined by the wavelength of scattered light, about 300nm. In principle, only $\sim 10^5$t of such mesh structures are required to achieve the benchmark 1% increase in albedo. The proposed metal balloons have diameters of ~ 4 mm and a skin thickness of ~ 20nm. They are hydrogen filled and are designed to float at altitudes of ~ 25km. The total mass of the balloon system would be $\sim 10^6$t. Because of the much longer stratospheric residence time of the balloon system, the required mass flux (e.g., tonnes replaced per year) to sustain the two systems would be comparable. Finally, Teller *et al.* (1997) show that either system, if fabricated in aluminium, can be designed to have long stratospheric lifetimes yet oxidize rapidly in the troposphere, ensuring that few particles are deposited on the surface.

One of the perennial concerns about possibilities for modifying the earth's radiation balance has been that even if these methods could compensate for increased GHGs in the global and annual mean, they might have very different spatial and temporal effects and impact the regional and seasonal climates in a very different way than GHGs. Recent analyses using the CCM3 climate model (Govindasamy and Caldeira, 2000) suggest, however, that a 1.7% decrease in solar luminosity would closely counterbalance a doubling of CO_2 at the regional and seasonal scale (in addition to that at the global and annual scale) despite differences in radiative forcing patterns.

It is unclear whether the cost of these novel scattering systems would be less than that of the older proposals, as is claimed by Teller *et al.* (1997), because although the system mass would be less, the scatterers may be much more costly to fabricate. However, it is unlikely that cost would play an important role in the decision to deploy such a system. Even if we accept the higher cost estimates of the NAS (1992) study, the cost may be very small compared to the cost of other mitigation options (Schelling, 1996). It is likely that issues of risk, politics (Bodansky, 1996), and environmental ethics (Jamieson, 1996) will prove to be the decisive factors in real choices about implementation. The importance of the novel scattering systems is not in minimizing cost, but in their potential to minimize risk. Two of the key problems with earlier proposals were the potential impact on atmospheric chemistry, and the change in the ratio of direct to diffuse solar radiation, and the associated whitening of the visual appearance of the sky. The proposals of Teller *el al.* (1997) suggest that the location, scattering properties, and chemical reactivity of the scatterers could, in principle, be tuned to minimize both of these impacts. Nonetheless, most papers on geo-engineering contain expressions of concern about unexpected environmental impacts, our lack of complete understanding of the systems involved, and concerns with the legal and ethical implications (NAS, 1992; Flannery *et al.*, 1997; Keith, 2000). Unlike other strategies, geo-engineering addresses the symptoms rather than the causes of climate change.

4.8   Future Research Needs

This chapter suggests a host of future research needs. A combination of statistical, ecological, and socio-economic research would be helpful to better understand the situation of the land, the forces of land-use change and the dynamic of forest carbon pools in relation to human activities and natural disturbance. More precise information is needed about degradation or improvement of secondary and natural forests throughout the world, but particularly in developing countries.

Some specific examples are:
- assessment of land available for mitigation options based on socio-economic pressures and land tenure policies. Furthermore, it would be beneficial if the impact of market price of carbon mitigated on land available for mitigation opportunities in different countries was understood;
- implications of financial incentives and mechanisms on LULUCF sector mitigation potential in different countries;
- comparative advantage (mitigation cost, ancillary benefits, etc.) of LULUCF sector mitigation options over energy sector opportunities;
- development and assessment of different approaches to developing baselines for LULUCF activities and comparison with other sectors; and
- socio-economic and environmental costs and benefits of implementing LULUCF sector mitigation options in developing countries, including issues such as property rights and land tenure.

Finally, an important consideration is the problem of leakages. Research would help to determine the conditions under which leakage is likely to be a serious problem and when it may be less so. Estimates of the degree of leakage under varying circumstances could be made so that appropriate adjustments in carbon credits can be made.

References

Adams, D.M., 1992: Long-term timber supply in the United States: an analysis of resources and policies in transition. *Journal of Business Administration*, **20**, 131-156.

Adams, D.M., R.J. Alig, J.M. Callaway, B.A. McCarl, and S.M. Winnett, 1995: *The forest and agricultural sector optmization model (FASOM): Model structure and policy implications*. Research paper PNW-RP USDA-Forest Service, Pacific Northwest Research Station, USA, 166 pp.

Alcamo, J., and R. Swart, 1998: Future trends of land use emissions of major greenhouse gases. *Mitigation and Adaptation Strategies to Global Change*, **3**, 343-381.

Alig R., D. Adams, B. McCarl, J.M. Callaway, and S. Winnett, 1997: Assessing Effects of Mitigation Strategies for Global Climate Change with an Intertemporal Model of the U.S. Forest and Agricultural Sectors. In *Economics of Carbon Sequestration in Forestry*. R.A. Sedjo, R.N. Sampson and J. Wisniewski (eds.), Lewis Publishers, Boca Raton, pp. 185-193.

Alig, R.J., W.G. Hohenstein, B.C. Murray, and R.G. Haight, 1990: *Changes in area of timberland in the United States, 1952-2040, by ownership, forest type, region and state*. General Technical Report SE 64, USDA-Forest Service, Southeastern Forest Experiment Station, 34 pp.

Allen, R.K., K. Platt, and S. Wiser, 1995: *Biodiversity in New Zealand plantation*. New Zealand Forestry, 26-29 February.

Anderson, D.W., 1995: Decomposition of organic matter and carbon emissions from soils. In: *Soils and Global Change*. R. Lal, J. Kimble, E. Levine, and B. A. Stewart (eds.), Boca Raton, CRC Lewis Publishers, pp. 165-175.

Anonymous, 1999: *Land-use, land-use change and forestry in Canada and the Kyoto Protocol. Canada's National Climate Change Process*. Sinks Table Options Paper, 169 pp., http://www.nccp.ca/html/index.htm

Apps, M.J., and W.A. Kurz, 1993: The Role of Canadian Forests in the Global Carbon Balance. In *Carbon Balance on World's Forested Ecosystems: Towards a Global Assessment*. M. Kannien (ed.), Proceedings Intergovernmental Panel on Climate Change Workshop, Joensuu, Finland, 11-15 May 1992, Publications of the Academy of Finland, Helsinki, pp. 14-28.

Apps, M.J., J.S. Bhatti, D. Halliwell, H. Jiang, and C. Peng, 2000: Ch 7: Simulated carbon dynamics in the boreal forest of central Canada under uniform and random disturbance regimes. In *Global Climate Change and Cold Regions Ecosystems*. R. Lal, J.M. Kimble, and B.A. Stewart (eds.), Advances in Soil Science, Lewis Publishers, Boca Raton, pp. 107—121.

Apps, M.J., W.A. Kurz, S.J. Beukema, and J.S. Bhatti, 1999: Carbon budget of the Canadian forest product sector. *Environmental Science and Policy*, **2**, 25-41.

Ash, A.J., S.M. Howden, and J.G. McIvor, 1996: Improved rangeland management and its implications for carbon sequestration. In *Proceedings of the Fifth International Rangeland Congress*, Salt Lake City, Utah, USA, 23-28 July 1995, **1**, 19-20.

Bajracharya, R.M., R. Lal, and J.M. Kimble. 1997: Long-term tillage effects on soil organic carbon distribution in aggregates and primary particle fractions of two Ohio soils. In *Management of Carbon Sequestration in Soil*. R. Lal, *et al.* (eds.), CRC Press, Boca Raton, pp. 113-123.

Balesdent, J., E. Besnard, D. Arrouays, and C. Chenu, 1998: The dynamics of carbon in particle-size fractions of soil in a forest-cultivation sequence. *Plant and Soil*, **201** 49-57.

Ball, B.C., A. Scott, and J.P. Parker, 1999: Field N_2O, CO_2 and CH_4 fluxes in relation to tillage, compaction and soil quality in Scotland. *Soil and Tillage Research*, **53**, 29-39.

Barrett, D.J., and R.M. Gifford, 1999: Increased C-gain by an endemic Australian pasture grass at elevated atmospheric CO_2 concentration when supplied with non-labile inorganic phosphorus. *Australian Journal of Plant Physiology*, **26**, 443-451.

Bass, S., J. Ford, O. Dubois, P. Moura-Costa, C. Wilson, M. Pinard, and R. Tipper, 2000: *Rural Livelihoods and Carbon Management. An Issues Paper*. International Institute for Environment and Development, Forestry and Land Use, London, UK.

Batjes, N.H., 1998. Mitigation of atmospheric CO_2 concentrations by increased carbon sequestration in the soil. *Biology and Fertility of Soils*, **27**, 230-235.

Batjes, N.H., 1999: *Management options for reducing CO_2-concentrations in the atmosphere by increasing carbon sequestration in the soil*. Report 410 200 031, Dutch National Research Programme on Global Air Pollution and Climate Change, Project executed by the International Soil Reference and Information Centre, Wageningen, The Netherlands, 114 pp.

Bawa, K.S., and S. Dayanandan, 1997: Socioeconomic factors and tropical deforestation. *Nature*, **386**, 562-563.

Bhatti, J. S., M.J. Apps and H. Jiang, 2001: Examining the Carbon Stocks of Boreal Forest Ecosystems at Stand and Regional Scales. In *Assessment of Methods for Soil C Pools*. R. Lal, M. Kimble, R.F. Follett, and B.A. Stewart (eds.), *Advances in Soil Science*. Lewis Publishers, Boca Raton, pp. 513-532.

Binkley, C.S., M.J. Apps, R.K. Dixon, P.E. Kauppi, and L.O. Nilsson, 1997: Sequestering carbon in natural forests In *Economics of carbon sequestration in Forestry*. R.A. Sedjo, R.N. Sampson, and J. Wisniewski (eds.). Critical Reviews in Environmental Science and Technology, **27**, 23-45.

Birdsey, R.A., and L.S. Heath, 1995: Carbon changes in US forests. *Productivity of America's Forests and Climate Change*, **271**, 56-70.

Birdsey, R.A., R. Alig, and D. Adams, 2000: Mitigation activities in the forest sector to reduce emissions and enhance sinks of greenhouse gases. In *The Impact of Climate Change on America's Forests*. L. Joyce and R. Birdsey (eds.), U.S. Department of Agriculture, Rocky Mountain Research Station, Fort Collins, CO, USA, 133 pp.

Bodansky, D., 1996: May We Engineer the Climate? *Climatic Change*, **33**, 309-321.

Bolin, B., R. Sukumar, P. Ciais, W. Cramer, P. Jarvis, H. Kheshgi, C. Nobre, S. Semenov, and W. Steffen, 2000: Global perspective. In *Land Use, Land Use Change and Forestry*. R.T. Watson, I.R. Noble, B. Bolin, N.H. Ravindranath, D.J. Verardo, and D.J. Dokken (eds.), A Special report of the IPCC, Cambridge University Press, pp. 23-51.

Bonan, G.B., and H.H. Shugart, 1992: Soil temperature, nitrogen mineralization and carbon source sink relationships in boreal forests. *Canadian Journal of Forest Research*, **22**, 629-639.

Bonduki, Y., and J.N. Swisher, 1995: Options for mitigating greenhouse gas emissions in Venezuela's forest sector: A general overview. *Interciencia*, **20**(6), 380-387.

Boyce, S.G., 1995: *Landscape forestry*. John Wiley & Sons, Inc., New York, pp. 67-166.

Braswell, B.H., D.S. Schimel, E. Linder, and B. Moore III, 1997: The response of global terrestrial ecosystems to interannual temperature variability. *Science*, **278**, 870-872.

Brown, S., J. Sathaye, M. Cannell, and P.E. Kauppi, 1996a: Mitigation of carbon emissions to the atmosphere by forest management. *Commonwealth Forestry Review*, **75**, 80-91.

Brown, S., J. Sathaye, M. Cannell, and P.E. Kauppi, 1996b: Management of forests for mitigation of greenhouse gas emissions. In *Climate Change 1995 - Impacts, Adaptations and Mitigation of Climate Change: Scientific-Technical Analyses*. R.T. Watson, M.C. Zinyowera, R.H. Moss, and D.J. Dokken (eds.), Contribution of Working Group II to the Second Assessment Report of the Intergovernmental Panel on Climate Change, Cambridge University Press, Cambridge, UK, pp. 773-797.

Brown, S., O. Masera, and J. Sathaye, 2000: Project based activities. In *Land Use, Land Use Change and Forestry*. R.T. Watson, I.R. Noble, B. Bolin, N.H. Ravindranath, D.J. Verardo, and D.J. Dokken (eds.), A Special report of the IPCC, Cambridge University Press, pp.283-338.

Bruce, J.P., M. Frome, E. Haites, H. Janzen, R. Lal, and K. Paustian, 1999: Carbon sequestration in soils. *Journal of Soil and Water Conservation*, **54**, 381-389.

Brunner, J., K. Talbot, and C. Elkin, 1998: *Logging Burma's Frontier Forests: Resources and the Regime*. The World Resources Institute, Washington DC, USA.

Buchanan, A.H., and S.B. Levine, 1999: Wood based building materials and atmospheric carbon emissions. *Environmental Science and Policy*, **2**, 427-437.

Burke, I.C., W.K. Lauenroth, and D.G. Milchunas. 1997: Biogeochemistry of managed grasslands in central North America. In *Soil Organic Matter in Temperate Agroecosystems. Long-Term Experiments in North America.* E.A. Paul, E.T. Elliott, K. Paustian, and C.V. Cole (eds.), CRC Press, Boca Raton, pp. 85-102.

Burke, I.C., W.K. Lauenroth, M.A. Vinton, P.B. Hook, R.H. Kelly, H.E. Epstein, M.R. Aguiar, M.D. Robles, M.O. Aguilera, K.L. Murphy, and R.A. Gill, 1998: Plant-soil interactions in temperate grasslands. *Biogeochemistry*, **42**, 121-143.

Burschel, P., E. Kuersten, and B.C. Larson, 1993: *Die Rolle von Wald und Forstwirtschaft im Kohlenstoffhaushalt. Eine Betrachtung für die Bundesrepublik Deutschland.* (Role of Forests and forestry in the carbon cycle; a try-out for Germany). Forstliche Forschungsberichte München, Schriftenreihe der Forstwissenschaftlichen Fakultät der Universität München und der Bayerischen Forstlichen Versuchs- und Forschungsanstalt, **126**, 135 pp.

Burtraw, D., 2000: *Innovation Under the Tradable Sulfur Dioxide Emission Permits Program in the U.S. Electricity Sector.* RFF Discussion Paper 00-38, Resources for the Future, Washington DC, USA, 28 pp.

Callaway, J.M., and B.A. McCarl, 1996: The economic consequences of substituting carbon payments for crop subsidies in U.S. agriculture. *Environmental and Resource Economics*, **7**, 15-43.

Campbell, I., C. Campbell, Z. Yu, D. Vitt, and M.J. Apps, 2000:. Millennial-scale rhythms in peatlands in the western interior of Canada and in the global carbon cycle. *Quaternary Research*, **54**, 155-158.

Cao, M., and F.I. Woodward, 1998:. Dynamic reponses of terrestrial ecosystem carbon cyling to global climate change. *Nature*, **393**, 249-252.

Carter, M.R., E.G. Gregorich, D.A. Angers, R.G. Donald, and M.A. Bolinder, 1998: Organic C and N storage, and organic C fractions, in adjacent cultivated and forested soils of eastern Canada. *Soil & Tillage Research*, 47,253-261.

Carter, M.R., E.G. Gregorich, D.W. Anderson, J.W. Doran, H.H. Janzen, and F.J. Pierce, 1997: Concepts of soil quality and their signific ance. In *Soil Quality for Crop Production and Ecosystem Health.* E.G. Gregorich, and M.R. Carter (eds.), Elsevier, Amsterdam, The Netherlands, pp. 1-19.

Chomitz, K.M., and K. Kumari, 1996: *The domestic benefits of tropical forests: a critical review emphasizing hydrological functions.* Policy Research Working Paper, World-Bank, No. WPS1601, 41 pp.

Christensen, B.T., and A.E. Johnston, 1997: Soil organic matter and soil quality – Lessons learned from long-term experiments at Askov and Rothamsted. In *Soil Quality for Crop Production and Ecosystem Health.* E.G. Gregorich, and M.R. Carter (eds.), Elsevier, Amsterdam, The Netherlands, pp. 399 –430.

Clawson, M., 1979: Forests in the Long Sweep of American History. *Science*, **204**, 1168-1174.

Coale, K.H., K.S. Johnson, S.E. Fitzwater, R.M. Gordon, S. Tanner, F.P. Chavez, L. Ferioli, C. Sakamoto, P. Rogers, F. Millero, P. Steinberg, P. Nightingale, D. Cooper, W.P. Cochran, M.R. Landry, J. Constantinou, R. Rollwagen, A. Trasvina, and R. Kudela, 1996: A massive Phytoplankton Bloom Induced by an Ecosystem-Scale Iron Fertilization Experiment in the Equatorial Pacific Ocean. *Nature*, **383**, 495-501.

Cohen, W.B., M.E. Harmon, D.O. Wallin, and M. Fiorella, 1996: Two decades of carbon flux from forests of the Pacific Northwest. *BioScience,* **46**(11), 836-844, 20-33.

Cole, C.V., J. Duxbury, J. Freney, O. Heinemeyer, K. Minami, A. Mosier, K. Paustian, N. Rosenberg, N. Sampson, D. Sauerbeck, and Q. Zhao, 1996: Agricultural options for mitigation of greenhouse gas emissions. In *Climate Change 1995 - Impacts, Adaptations, and Mitigation of Climate Change: Scientific-Technical Analyses.* R.T. Watson, M.C. Zinyowera, R.H. Moss, and D.J. Dokken (eds.), Cambridge University Press, pp. 745-771.

Cole, C.V., J. Duxbury, J. Freney, O. Heinemeyer, K. Minami, A. Mosier, K. Paustian, N. Rosenberg, N. Sampson, D. Sauerbeck, and Q. Zhao, 1997: Global estimates of potential mitigation of greenhouse gas emissions by agriculture. *Nutrient Cycling in Agroecosystems*, **49**, 221-228.

Cole, C.V., K. Flach, J. Lee, D. Sauerbeck, and B. Stewart, 1993: Agricultural sources and sinks of carbon. *Water Air and Soil Pollution,* **70**, 111-122.

Conant, R.T., K. Paustian, and E.T. Elliot, 2001: Grassland management and conversion into grassland: Effects on soil carbon. *Ecological Applications* (in press).

Cony, M.A., 1995: Rational afforestation of arid and semiarid lands with multipurpose trees. *Interciencia*, **2**(5), 249-253.

Cooper, D.J., A.J. Watson, and P.D. Nightingale, 1996: Large Decrease in Ocean Surface CO2 Fugacity in Response to in situ Iron Fertilization. *Nature*, **382**, 511-513.

Da Motta, S., C. Young, and C. Ferraz, 1999: *Clean Development Mechanism and Climate Change: Cost Effectiveness and Welfare Maximization in Brazil.* Report to the World Resources Institute, Institute of Economics, Federal University of Rio de Janeiro, Brazil, 42 pp.

Da Silva, F. Jr., S. Rubio, and F. de Souza, 1995: Regeneration of an Atlantic forest formation in the understory of a *Eucalyptus grandis* plantation in south-eastern Brazil. *Journal of Tropical Ecology*, **11**, 147-152.

Darwin, R., M. Tsigas, J. Lewandrowski, and A. Raneses, 1995: *World Agriculture and climate change: economic adaptations.* Agricultrural Economic Report No. 703, US Department of Agriculture, Economic Research Service, Washington DC, USA, 86 pp.

Darwin, R., M. Tsigas, J. Lewandrowski, and A. Raneses, 1996. Land Use and Cover in Ecological Economics. *Ecological Economics*, **17**(3), 157-181.

Davidson, E.A., D.C. Nepstad, C. Klink, and S.E. Trumbore, 1995: Pasture soils as carbon sink. *Nature*, **376**, 472-473.

Davidson, E.A., and I.L. Ackerman, 1993: Changes in soil carbon inventories following cultivation of previously untilled soils. *Biogeochemistry*, **20**, 161-164.

Dick, W.A., R.L. Blevins, W.W. Frye, S.E. Peters, D.R. Christenson, F.J. Pierce, and M.L. Vitosh, 1998: Impacts of agricultural management practices on C sequestration in forest-derived soils of the eastern Corn Belt. *Soil & Tillage Research*, **47**, 235-244.

Dickinson, R.E., 1996: Climate Engineering - A Review of Aerosol Approaches to Changing the Global Energy Balance. *Climatic Change*, **33**, 279-290.

Dixon, R.K., S. Brown, R.A. Houghton, A.M. Solomon, M.C. Trexler, and J. Wisniewski, 1994: Carbon pools and flux of global forest ecosystems. *Science,* **263**, 185-190.

Dixon, R.K., J.K. Winjum, and P.E. Schroeder, 1993: Conservation and sequestration of carbon: the potential of forest and agroforest management practices. *Global Environmental Change*, **3**(2), 159-173.

Duan, Z., X. Liu., and J. Qu, 1995: Effect of land desertification on CO_2 content of atmosphere in China. *Agricultural Input and Environment 1995,* 279-302.

Dyson, F.J., 1977: Can we control the carbon dioxide in the atmosphere ?. *Energy*, **2**, 287-291.

Early, J.T., 1989: Space-based Solar Screen to Offset the Greenhouse Effect. *Journal of the British Interplanetary Society,* **42**, 567-569.

EPA/USIJI (US Environmental Protection Agency / US Initiative on Joint Implementation), 1998: *Activities Implemented Jointly: Third Report to the Secretariat of the UN Framework Convention on Climate Change.* 2 volumes, EPA-report 236-R-98-004. US Environmental Protection Agency, Washington DC, USA, p.19 (Vol. I) and p. 607 (Vol. II).

FACE Foundation, 1997: *Annual report 1996.* Foundation FACE, Arnhem, The Netherlands.

Falloon, P.D., P. Smith, J.U. Smith, J. Szabo, K. Coleman, and S. Marshall, 1998: Regional estimates of carbon sequestration potential: linking the Rothamsted Carbon Model to GIS databases. *Biology and Fertility of Soils*, **27**, 236-241.

Fan, S., M. Gloor, J. Mahlman, S. Pacala, J. Sarmiento, T. Takahashi, and P. Tans, 1998: A large terrestrial carbon sink in North America implied by atmospheric and oceanic carbon dioxide data and models. *Science*, **282**, 442-446.

FAO (Food and Agriculture Organization), 1993: *Forestry Statistics Today and tomorrow.* FAO, Rome, Italy.

FAO, 1996: *Forest resources assessment 1990: Survey of tropical forest cover and study of change processes.* FAO Forestry paper 130. Food and Agriculture Organization of the United Nations, Rome, Italy.

FAO, 1997. *State of the World's Forests.* Food and Agriculture Organization of the United Nations, Rome, Italy, 200 pp.

Farrell, E.P., E. Führer, D. Ryan, F. Andersson, R. Hüttl, and P. Piussi, 2000: European forest ecosystems: building the future on the legacy of the past. *Forest Ecology and Management*, **132**, 5-20.

Fearnside, P.M., 1997: Greenhouse gases from deforestation in Brazilian Amazonia: net committed emissions. *Climatic Change*, **35**(3), 321-360.

Fearnside, P.M., 1998: Plantation forestry in Brazil: projections to 2050. *Biomass and Bioenergy*, **15**, 437-450.

Fearnside, P.M., 1999: Forests and global warming mitigation in Brazil: opportunities in the Brazilian forest sector for responses to global warming under the "clean development mechanism". *Biomass-and-Bioenergy*, **16**, 171-189.

Feller, C., and M.H. Beare, 1997. Physical control of soil organic matter dynamics in the tropics. *Geoderma*, **79**, 69-116.

Fernandes, E.C.M., P.P. Motavalli, C. Castilla, and L. Mukurumbira, 1997: Management control of soil organic matter dynamics in tropical land-use systems. *Geoderma*, **79**, 49-67.

Fisher, M.J., I.M. Rao, M.A. Ayarza, C.E. Lascano, J.I. Sanz, R.J. Thomas, and R.R. Vera, 1994: Carbon storage by introduced deep-rooted grasses in the South American savannas. *Nature*, **371**, 236-238.

Fisher, M.J., R.J. Thomas, and I.M. Rao, 1997: Management of tropical pastures in acid-soil savannas of South America for carbon sequestration in the soil. In *Management of Carbon Sequestration in Soil*. R. Lal *et al.* (eds.), CRC Press, Boca Raton, pp. 405-420.

Flach, K.W., T.O. Barnwell, Jr., and P. Crosson, 1997: Impacts of agriculture on atmospheric carbon dioxide. In *Soil Organic Matter in Temperate Agroecosystems. Long-Term Experiments in North America*. E.A. Paul, E.T. Elliott, K. Paustian, and C.V. Cole (eds.), CRC Press, Boca Raton, pp. 3-13.

Flannery, B.P., H. Kheshgi, G. Marland, M.C. MacCracken, H. Komiyama, W. Broecker, H. Ishitani, N. Rosenberg, M. Steinberg, T. Wigley, and M. Morantine, 1997: Geoengineering Climate. In *Engineering Response to Global Climate Change*. R.G. Watts (ed.), CRC/Lewis Publishers, Boca Raton, Louisiana, USA, pp. 379-427.

Forman, R.T.T., 1995: *Land Mosaics - The ecology of landscapes and regions*. Cambridge University Press, New York, 632 pp.

Franklin, J.F., and T.A. Spies, 1991: *Composition, function, and structure of old-growth Douglas-fir forests. Wildlife and vegetation of unmanaged Douglas-fir forests*. USDA Forest Service General Technical Report, PNW-GTR-285, pp. 71-80.

Frost, B.W., 1996: Phytoplankton Bloom on Iron Rations. *Nature*, **383**, 475-476.

Frumhoff, P.C., and E.C. Losos, 1998: *Setting Priorities for Conserving Biological Diversity in Tropical Timber Production Forests*. Union of Concerned Scientists, Cambridge, MA, USA, 14 pp.

Frumhoff, P.C., D.C. Goetze, and J.J. Hardner, 1998: *Linking Solutions to Climate Change and Biodiversity Loss Through the Kyoto Protocol's Clean Development Mechanism*. Union of Concerned Scientists, Cambridge, MA, USA, 14 pp.

Fujimori, T., 1997: Overview of forest resources and forestation activities in Japan. Environmentally friendly tree products and their processing technology. In *Proceedings of the Japanese/Australian workshop on environmental management*. Japan Science and Technology Agency and Forestry and Forest Products Research Institute, pp. 21-27.

Fukushima, Y., 1987: Influence of forestation on mountainside at granite highlands. *Water Science*, **177**, 17-34.

Garratt, J.R., 1993: Sensitivity of Climate Simulations to Land-Surface and Atmospheric Boundary-Layer Treatments – A Review. *Journal of Climate*, **6**, 419-449.

Giardina, C.P., and M.G. Ryan, 2000: Evidence that decomposition rates of organic carbon in mineral soils do not vary with temperature. *Nature*, **404**, 858-861.

Glück, P., and G. Weiss, 1996: *Forestry in the Context of Rural Development: Future Research Needs*. Proceedings of the COST seminar 'Forestry in the Context of Rural Development', Vienna, Austria, 15-17 April 1996, EFI Proceedings No. 15, European Forest Institute Joensuu, Finland, 173 pp.

Gorham, E., 1991: Northern peatlands: role in the carbon cycle and probable response to climate warming. *Ecological Applications*, **1**(2), 182-195.

Gorham, E., 1995: The biogeochemistry of northern peatlands and its possible responses to global warming. In *Biotic Feedback in the Global Climatic System. Will the Warming Feed the Warming?* G.M. Woodwell and F.T. McKenzie (eds.), Oxford University Press, New York, pp. 169-187.

Govindasamy, B., and K. Caldeira, 2000: Geoengineering Earth's Radiation Balance to Mitigate CO_2-Induced Climate Change. *Geophysical Research Letters*, **27**, 2141-2144.

Greenland, D.J, 1995: Land use and soil carbon in different agroecological zones. Soils and Global Change. In *Soil Management and the Greenhouse Effect*. R. Lal, J. Kimble, E. Levine, and B.A. Stewart (eds.), Lewis Publishers, pp. 9-24.

Hall, P.J., and B. Moody, 1994: *Forest depletions caused by insects and diseases in Canada 1982-1987*. Information Report ST-X-8, Natural Resources Canada, Canadian Forest Service, Ottawa, 14 pp.

Hall, D.O., H.E. Mynick, and R.H. Williams, 1991: Cooling the Greenhouse with bioenergy. *Nature*, **353**, 11-12.

Harden, J.W., E.T. Sundquist, R.F. Stallard, and R.K. Mark, 1992: Dynamics of soil carbon during deglaciation of the Laurentide ice sheet. *Science*, **258**, 1921-1924.

Hardin, G., 1968: The tragedy of the commons. *Science*, **162**, 1243-1248.

Harmon, M.E., J.M. Harmon, W.K. Ferell, and D. Brooks, 1996: Modelling carbon stores in Oregon and Washington forest products: 1900 - 1992. *Climatic Change*, **33**, 521-550.

Harmon, M.E., W.K. Ferel, and J.F. Franklin, 1990: Effects on carbon storage of conversion of old-growth forests to young forests. *Science*, **247**(4943), 699-703.

Harris, L.D., 1984: *The fragmented forest: Island biogeography theory and the preservation of biotic diversity*. The University of Chicago, 211pp.

Harrison, K.G., W.S. Broecker, and G. Bonani, 1993: The effect of changing land use on soil radiocarbon. *Science*, **262**, 725-726.

Heath, L.S., and R.A. Birdsey, 1993: Carbon trends of productive temperate forests of the conterminous United States. *Water, Air, and Soil Pollution*, **70**, 279-293.

Heath, L.S., R.A. Birdsey, C. Row, and A.J. Plantinga, 1996: Carbon pools and fluxes in U.S. forest products. In *Forest Ecosystems, Forest Management, and the Carbon Cycle*. M.J. Apps, and D.T. Price (eds.), NATO ASi Series, **140**, 271-278.

Herrick, J.E., and M.M. Wander, 1997: Relationships between soil organic carbon and soil quality in cropped and rangeland soils: the importance of distribution, composition, and soil biological activity. In: *Soil Processes and the Carbon Cycle*. R. Lal, J.M. Kimble, R.F. Follett, and B.A. Stewart (eds.), Boca Raton, CRC Press, pp. 405-425.

Hoen, H.F., and Solberg, B. 1994: Potential and economic efficiency of carbon sequestration in forest biomass through silvicultural management. *Forest Science*, **40**, 429-451.

Hol, P., R. Sikkema, E. Blom, P. Barendsen, and W. Veening, 1999: *Private investments in sustainable forest management. I. Final Report*. Form Ecology Consultants and Netherlands Committee for IUCN, The Netherlands.

Houghton, R.A., 1995a: Changes in the storage of terrestrial carbon since 1850. In *Soils and Global Change*. R. Lal, J. Kimble, E. Levine, and B.A. Stewart (eds.), Lewis Publishers, Boca Raton, pp. 45-65.

Houghton, R.A., 1995b: Effects of land-use change, surface temperature, and CO_2 concentration on terrestrial stores of carbon. In *Biotic Feedbacks in the Global Climate Systems*. G.M. Woodwell, and E.T. MacKenzie (eds.), Oxford University Press, New York, pp. 333-350.

Houghton, R.A., E.A. Davidson, and G.M. Woodwell, 1998: Missing sinks, feedbacks, and understanding the role of terrestrial ecosystems in the global carbon balance. *Global Biogeochemical Cycles*, **12**, 25-34.

Houghton, R.A., J.L. Hackler, and K.T. Lawrence, 1999: The U.S. carbon budget: Contributions from land-use change. *Science*, **285**, 574-578.

Houghton, R.A., D.L. Skole, C.A. Nobre, J.L. Hackler, K.T. Lawrence, and W.H. Chomentowski, 2000: Annual fluxes or carbon from deforestation and regrowth in the Brazilian Amazon. *Nature*, **403**(6767), 301-304.

Huggins, D.R., G.A. Buyanovsky, G.H. Wagner, J.R. Brown, R.G. Darmody, T.R. Peck, G.W. Lesoing, M.B. Vanotti, and L.G. Bundy, 1998: Soil organic C in the tallgrass prairie-derived region of the corn belt: effects of long-term crop management. *Soil & Tillage Research*, **47**, 219-234.

Hulscher, W.S., Z. Luo, and A. Koopmans, 1999: Stoves on the carbon market. *Wood Energy News*, **14**(3), 20-21.

Hungate, B.A., E.A. Holland, R.B. Jackson, F.S. Chapin, H.A. Mooney, and C.B. Field, 1997: The fate of carbon in grasslands under carbon dioxide enrichment. *Nature*, **388**, 576-579.

IGBP, 1998. The terrestrial carbon cycle: Implications for the Kyoto Protocol. *Science*, **280**, 1393-1394.

IPCC (Intergovernmental Panel on Climate Change), 1996: *Climate Change 1995: Impacts, Adaptations and Mitigation of Climate Change: Scientific-Technical Analyses.* Contribution of working group II to the Second Assessment Report of the Intergovernmental Panel on Climate Change, R. Watson, M.C. Zinyowera, and R. Moss (eds.), Cambridge University Press, Cambridge, UK, 880 pp.

IPCC, 2000a: *Land Use, Land Use Change and Forestry.* R.T. Watson, I.R. Noble, B. Bolin, N.H. Ravindranath, D.J. Verardo, and D.J. Dokken (eds.), A Special report of the IPCC, Cambridge University Press, Cambridge, UK, 377 pp.

IPCC, 2000b: *Methodological and Technological Issues in Technology Transfer.* B. Metz, O.R. Davidson, J.-W. Martens, S.N.M. van Rooijen, and L. van Wie-McGrory (eds.), A Special report of the IPCC, Cambridge University Press, Cambridge, UK, 466 pp.

Isensee, A.R., and A.M. Sadeghi, 1996: Effect of tillage reversal on herbicide leaching to groundwater. *Soil Science*, **161**, 382-389.

ITTA (International Tropical Timber Agreement), 1983: *1983 United Nation Conference on Trade and Development.* Geneva.

ITTA (International Tropical Timber Agreement), 1994: *1994 United Nations Conference on Trade and Development.* Geneva.

Izac, A-M.N., 1997: Developing policies for soil carbon management in tropical regions. *Geoderma*, **79**, 261-276.

Izaurralde, R.C., M. Nyborg, E.D. Solberg, H.H. Janzen, M.A. Arshad, S.S. Malhi, and M. Molina-Ayala, 1997: Carbon storage in eroded soils after five years of reclamation techniques, In *Soil Processes and the Carbon Cycle.* R. Lal, J.M. Kimble, R.F. Follett, and B.A. Stewart (eds.), CRC Press, Boca Raton, pp. 369-385.

Jacinthe, P.A,. and W.A. Dick, 1997: Soil management and nitrous oxide emissions from cultivated fields in southern Ohio. *Soil and Tillage Research,* **41**, 221-235.

Jamieson, D., 1996: Ethics and Intentional Climate Change. *Climatic Change,* **33**, 323-336.

Janzen, H.H., C.A. Campbell, R.C. Izaurralde, B.H. Ellert, N. Juma, W.B. McGill, and R.P. Zentner, 1998: Management effects on soil C storage on the Canadian prairies. *Soil and Tillage Research,* **47**, 181-195.

Jepma, C.J., S. Nilsson, M. Amano, Y. Bonduki, L. Lonnstedt, J. Sathaye, and T. Wilson, 1997: Carbon sequestration and sustainable forest management: common aspects and assessment goals. In *Economics of cabon sequestration in Forestry.* R.A. Sedjo, R.N.Sampson, and J. Wisniewski (eds.), Critical Reviews in Environmental Science and Technology, **27**, 83-96.

Johnson, M.G., 1995: The role of soil management in sequestering soil carbon. In *Soil Management and Greenhouse Effect.* R. Lal, J. Kimble, E. Levine and B.A. Stewart, Boca Raton, CRC Lewis Publishers, pp. 351-363.

Jones, I.S.F., and H.E. Young, 1998: Enhanced Oceanic Uptake of Carbon Dioxide – an AIJ Candidate. In: *Greenhouse Gas Mitigation: Technologies for Activities Implemented Jointly.* P.W.F. Riemer, A.Y. Smith, and K.V. Thambimuthu (eds.), Elsevier Science Ltd., Oxford, UK, pp. 267-272.

Jong, B.H.J. de, G. Montoya-Gomez, K. Nelson, and L. Soto-Pinto, 1995: Community forest management and carbon sequestration: A feasibility study from Chiapas, Mexico. *Interciencia*, **20**, 409-416.

Joos, F., J.L. Sarmiento, and U. Siegenthaler, 1991: Estimates of the Effect of Southern Ocean Iron Fertilization on Atmospheric CO2 Concentrations. *Nature*, **349**, 772-774.

Kaimowitz, D., and A. Anglesen, 1998: *Economic Models of Tropical Deforestation: a Review.* Center for International Forestry Research (CIFOR), Bogor, Indonesia, 139 pp.

Kanowski, P.J., P.S. Savill, P.G. Adlard, J. Burley, J. Evans, J.R. Palmer, and P.J. Wood, 1992: Plantation forestry. In *Managing the World's Forests.* N.P. Sharma (ed.), Kendall-Hunt, Dubuque, Iowa, USA, pp. 375-401.

Karjalainen, T., A. Pussinen, S. Kellomäki, and R. Mäkipää, 1998: The History and Future Carbon Dynamics of Carbon Sequestration in Finland's Forest Sector. In *Carbon Mitigation in Forestry and Wood Industry.* G.H. Kohlmaier, M. Weber, and R.A. Houghton (eds.), Springer-Verlag, Berlin, pp. 25-42.

Karjalainen, T., 1996: Dynamics and potentials of carbon sequestration in managed stands and wood products in Finland under changing climatic conditions. *Forest Ecology and Management*, **80**, 113-132.

Karjalainen, T., S. Kellomaeki, and A. Pussinen, 1994: Role of wood based products in absorbing atmospheric carbon. *Silva Fennica*, **28**(2), 67-80.

Kasimir-Klemedtsson, A., L. Klemedtsson, K. Berglund, P. Martikainen, J. Silvola, and O. Oenema. 1997: Greenhouse gas emissions from farmed organic soils: a review. *Soil Use and Management*, **13**, 245-250.

Kauffman, J.B., D.L. Cummings, and D.E. Ward, 1998: Fire in the Brazilian Amazon. 2. Biomass, nutrient pools and losses in cattle pastures. *Oecologia*, **113**, 415-427.

Kauppi, P.E., K. Mielkainen, and K. Kuusela, 1992: Biomass and carbon budget of European forests, 1971-1990. *Science*, **256**, 70-74.

Kauppi, P.E., M. Posch, P. Hänninen, H.M. Henttonen, A. Ihalainen, E. Lappalainen, M. Starr, and P. Tamminen, 1997: Carbon reservoirs in peatlands and forets in the boreal regions of Finland. *Silva Fennica,* **31**(1), 13-25.

Keenan, R., D. Lamb, O. Woldring, T. Irvine, and R. Jensen, 1997: Restoration of plant biodiversity beneath tropical tree plantations in Northern Australia. *Forest Ecology and Management*, **99**, 117-131.

Keith, D.W., 2000: Geoengineering the climate: History and prospect. *Annual Reviews Energy and the Environment,* **25**, 245-284.

Keith, D.W., 2001: Geoengineering. In *Oxford Encyclopedia of Global Change: Environmental Change and Human Society.* A.S. Goudie (ed.), Oxford University Press, New York, 1440 pp.

Kern, J.S., G. Zitong, Z. Ganlin, Z. Huizhen, and L. Guobao, 1997: Spatial analysis of methane emissions from paddy soils in China and the potential for emissions reduction. *Nutrient Cycling in Agroecosystems*, **49**, 181-195.

Kern, J.S., and M.G. Johnson, 1993: Conservation tillage impacts on national soil and atmospheric carbon levels. *Soil Science Society America Journal,* **57**, 200-210.

Kharin, N., 1996: Strategy to combat desertification in Central Asia. *Desertification Control Bulletin,* **29**, 29-34.

Kheshgi, H., 1995: Sequestering Carbon Dioxide by Increasing Ocean Alkalinity. *Energy*, **20**, 915-922.

Kobayashi, K., 1987: Hydrologic effects of rehabilitation treatment for bare mountain slops. *Bulletin of the Forestry and Forest Products Research Institute*, **300**, 151-185.

Kolchugina, T.P., T.S. Vinson, G.G. Gaston, V.A. Rozhkov, and A.Z. Shwidenko, 1995: Carbon pools, fluxes, and sequestration potential in soils of the former Soviet Union. In *Soil Management and Greenhouse Effect.* R. Lal, J. Kimble, E. Levine, and B.A. Stewart (eds.), CRC Lewis Publishers, Boca Raton, pp. 25-40.

Kremen, C., J.O. Niles, M.G. Dalton, G.C. Daily, P.R. Ehrlich, J.P. Fray, D. Gerwal, and R.P. Guillery, 2000: Economic incentives for rain forest conservation across scales. *Science*, **288**, 1828-1832.

Kuliev, A., 1996: Forests - an important factor in combatting desertification. *Problems of desert development*, **4**, 29-31.

Kumar, V., B.C. Ghosh, and R. Bhat, 1999: Recycling of crop wastes and green manure and their impact on yield and nutrient uptake of wetland rice. *Journal of Agricultural Science*, **132,** 149-154.

Kurz, W.A., M.J. Apps, B.J. Stocks, and W.J.A. Volney, 1995a: Global climatic change: disturbance regimes and biospheric feedbacks of temperate and boreal forests. In *Biotic feedbacks in the global climatic system: Will the warming speed the warming?* G.M. Woodwell, and F.T. Mackenzie (eds.), Oxford University Press, New York, pp. 119-133.

Kurz, W.A., and M.J. Apps, 1999: A 70-year retrospective analysis of carbon fluxes in the Canadian forest sector. *Ecological Applications*, **9**, 526-547.

Kurz, W.A., M.J. Apps, S.J. Beukema, and T. Lekstrum, 1995b: Twentieth century carbon budget of Canadian forests. *Tellus,* **47B**, 170-177.

Kurz, W.A., M.J. Apps, T. Webb, and P. MacNamee, 1992: *The Carbon Budget of the Canadian Forest Sector: Phase 1.* ENFOR Information Report NOR-X-326, Forestry Canada Northwest Region, Edmonton, Alberta, Canada, 93 pp.

Kurz, W.A., S.J. Beukema, and M.J. Apps, 1997: Carbon budget implications of the transition from natural to managed disturbance regimes in forest landscapes. *Mitigation and Adaptation Strategies for Global Change,* **2**, 405-421.

Kuusela, K., 1994: *Forest resources in Europe 1950-1990*. Research Report 1, European Forest Institute, Joensuu, Finland, 154 pp.

Lähde, E., O. Laiho, and Y. Norokorpi, 1999: Diversity-oriented silviculture in the boreal zone of Europe. *Forest Ecology and Management*, **118**, 223-243.

Lal, R., and J.P. Bruce, 1999: The potential of world cropland soils to sequester C and mitigate the greenhouse effect. *Environmental Science and Policy*, **2**, 177-185.

Lal, R., J. Kimble, and R. Follett, 1997: Land use and soil C pools in terrestrial ecosystems. In: *Management of Carbon Sequestration in Soil*. R. Lal, J.M. Kimble, R.F. Follett, and B.A. Stewart. (eds.), CRC Press, Boca Raton, pp. 1-10.

Lal, R., J.M. Kimble, R.F. Follett, and C.V. Cole, 1998: *The Potential of U.S. Cropland to Sequester Carbon and Mitigate the Greenhouse Effect*. Sleeping Bear Press, Inc., Ann Arbor Press, Chelsea, MI.

Lal, R., R.F. Follett, J. Kimble, and C.V. Cole, 1999: Managing U.S. cropland to sequester carbon in soil. *Journal of Soil and Water Conservation*, First Quarter 1999: 374-381.

Larionova, A.A., A.M. Yermolayev, S.A. Blagodatsky, L.N. Rozanova, I.V. Yevdokimov, and D.B. Orlinsky, 1998: Soil respiration and carbon balance of gray forest soils as affected by land use. *Biology and Fertility of Soils*, **27**, 251-257.

Le Maitre, D.C., and D.B. Versfeld, 1997: Forest evaporation models: relationships between stand growth and evaporation. *Journal of Hydrology*, **193**, 240-257.

Lemke, R.L., R.C. Izaurralde, M. Nyborg, and E.D. Solberg, 1999: Tillage and N source influence soil-emitted nitrous oxide in the Alberta Parkland region. *Canadian Journal of Soil Science*, **79**, 15-24.

Liski, J., H. Ilvesniemi, A. Makela, and M. Starr, 1998: Model analysis of the effects of soil age, fires and harvesting on the carbon storage of boreal forest soils. *European Journal of Soil Science*, **49**, 397-406.

Liski, J.; H. Ilvesniemi, A. Makela, and C.J. Westman, 1999: CO2 emissions from soil in response to climatic warming are overestimated - the decomposition of old soil organic matter is tolerant of temperature. *Ambio*, **28**(2), 171-174.

Lugo, A., 1997: The apparent paradox of reestablishing species richness on degraded lands with tree monocultures. *Forest Ecology and Management*, **99**, 9-19.

Lunnan, A., S. Navrud, P.K. Rorstad, K. Simensen, and B. Solberg, 1991: *Skog og skogproduksjon i norge som virkemiddel mot CO₂ -opphopning i atmosfaeren*. Forest and wood products in Norway as a mean to reduce CO2-accumulation in the atmosphere. Skogforsk 6-1991, Norsk Institut for skogforskning og institut fro skogfag, Norges Landbrukshogskole, As Norway, 86 pp.

Luo, Z., and W. Hulscher, 1999: Woodfuel emissions and costs. *Wood Energy News*, **14**(3), 13-15.

MacDicken, K.G., and N.T. Vergara, 1990: Introduction to agroforestry. In *Agroforestry: Classification and Management*. K.G. Macdicken, and N.T. Vergara (eds.), John Wiley and Sons, New York, USA, pp. 1-30.

MacKenzie, A.F., M.X. Fan, and F. Cadrin, 1997: Nitrous oxide emission as affected by tillage, corn-soybean-alfalfa rotations and nitrogen fertilization. *Canadian Journal of Soil Science*, **77**, 145-152.

MacLaren, J.P., 1996: Plantation forestry - its role as a carbon sink: conclusions from calculations based on New Zealand's planted forest estate. In *Forest Ecosystems, Forest Management and the Global Carbon Cycle*. M.J. Apps, and D.T. Price (eds.), NATO ASI Series I (Global Environmental Change), Vol. I 40, Springer-Verlag Academic Publishers, Heidelberg, Germany, pp. 257-270.

Makipaa, R., T. Karjalainen, A. Pussinen, and S. Kellomaki, 1999: Effects of climate change and nitrogen deposition on the carbon sequestration of a forest ecosystem in the boreal zone. *Canadian Journal of Forest Research*, **29**, 1490-1501.

Makundi, W.R., 1997: Global Climate Change Mitigation and Sustainable Forest Management – The Challenge of Monitoring and Verification. *Journal of Mitigation and Adaptation Strategies for Global Change*, **2**, 133-155.

Makundi, W.R., 1998: Mitigation Options in Forestry, Land-use Change and Biomass Burning in Africa. *In Climate Change Mitigation in Africa*. G. Mackenzie, J. Turkson, and O. Davidson (eds), Proceedings of UNEP/SCEE Workshop, Harare, May 1998.

Makundi, W.R., W. Razali, D. Jones, and C. Pinso, 1998: Tropical Forests in the Kyoto Protocol: Prospects for Carbon Offset Projects after Buenos Aires. *Tropical Forestry Update*, **8**(4), 5-8, International Tropical Timber Organization (ITTO), Yokohama, Japan.

Manley, J.T., G.E. Schuman, J.D. Reeder, and R.H. Hart, 1995: Rangeland soil carbon and nitrogen responses to grazing. *Journal of Soil and Water Conservation*, **50**, 294-298.

Marchetti, C., 1976: *On Geoengineering and the CO₂ Problem*. Research Memorandum RM-76-17, International Institute for Applied Systems Analysis, Vienna, Austria

Marland, G., 1996: Geoengineering. In *Encyclopedia of climate and weather, volume 1*. S.H. Schneider (ed.), Oxford University Press, New York, 338-339.

Marland, G., and B. Schlamadinger, 1997: Forests for carbon sequestration or fossil fuel substitution? A sensitivity analysis. *Biomass and Bioenergy*, **13**(6), 389-397.

Marland, G., and B. Schlamadinger, 1999: The Kyoto Protocol could make a difference for the optimal forest-based CO₂ mitigation strategy: some results from GORCAM. *Environmental Science and Policy*, **2**, 111-124.

Martin, J.H., 1990: A New Iron Age, or a Ferric Fantasy. *US JGOFS Newsletter*, **1**(4), 5-6.

Martin, J.H., 1991: Iron, Liebig's Law, and the Greenhouse. *Oceanography*, **4**, 52-55.

Martin, J.H., K.H. Coale, K.S. Johnson, S.E. Fitzwater, 1994: Testing the Iron Hypothesis in Ecosystems of the Equatorial Pacific Ocean. *Nature*, **371**, 123-130.

Martin, P.H., R. Valentini, P. Kennedy, and S. Folving, 1998: New estimate of the carbon sink strength of EU forests integrating flux measurements, field surveys, and space observations: 0.17-0.35 Gt(C). *Ambio*, **27**(7), 582-584.

Masera, O.R., 1995: Carbon Mitigation Scenarios for Mexican Forests: Methodological Considerations and Results. *Interciencia*, **20**, 388-395.

Masera, O.R., and A. Ordóñez, 1997: Forest Management Mitigation Options. In *Final report to the USAID-Support to the National Climate Change Plan for Mexico*. C. Sheinbaum (Coord.), Instituto de ingeniería, National University of Mexico (UNAM), Report 6133, UNAM, Mexico City, pp. 77-93.

Masera, O.R., M. Bellon, and G. Segura, 1995: Forest management options for sequestering carbon in Mexico. *Biomass and Bioenergy*, **8**(5), 357-368.

Masera, O.R., M.R. Bellon, and G. Segura, 1997b: Forestry Options for Sequestering Carbon in Mexico: Comparative Economic Analysis of Three Case Studies. *Critical Reviews in Environmental Science and Technology*, **27**, 227-244.

Masera, O.R., M.J. Ordoñez, and R. Dirzo, 1997a: Carbon emissions from Mexican Forests: Current Situation and Long-term Scenario. *Climatic Change*, **35**, 265-295.

Mather, A.S., 1990: *Global forest resources. Chapter 3. Historical perspectives on forest resource use*. Timber Press, Portland, OR, USA, pp. 30-57.

Matthews, R., G.J. Nabuurs, V. Alexeyev, R.A. Birdsey, A. Fischlin, J.P. MacLaren, G. Marland, and D. Price, 1996: WG3 Summary: Evaluating the role of forest management and forest products in the carbon cycle. In *Forest ecosystems, forest management and the global carbon cycle*. M.J. Apps, and D.T. Price (eds.), NATO Advanced Science Institute Series, NATO-ASI Vol I 40, Berlin, Heidelberg, Proceedings of a workshop held in September 1994 in Banff, Canada, pp. 293-301.

Melillo, J.R., A.D. McGuire, D.W. Kicklighter, B. Moore, C. J. Vorosmarty, and A.L. Schloss, 1993: Global climate change and terrestrial net primary production. *Nature*, **363**, 234-240.

Meyer, W.B., and B.L. Turner II, 1992: Human population growth and global land-use / cover change. *Annual Review Ecology Systematics*, **23**, 39-61.

Micales, J.A., and K.E. Skog, 1997: The decomposition of forest products in landfills. *International Biodeterioration and Biodegradation*, **39**, 145-158.

Minami, K., 1995: The effect of nitrogen fertilizer use and other practices on methane emission from flooded rice. *Fertilizer Research*, **40**, 71-84.

Mitsch, W.J., and X. Wu, 1995: Wetlands and global change. In *Soil Management and Greenhouse Effect*. R. Lal, J. Kimble, E. Levine, and B.A. Stewart (eds.), Boca Raton, CRC Lewis Publishers, pp. 205-230.

Monastersky, R., 1995: Iron versus the Greenhouse. *Science News*, **148**, 220-222.

Morgan, J.A., W.G. Knight, L.M. Dudley, and H.W. Hunt, 1994: Enhanced root system C-sink activity, water relations and aspects of nutrient acquisition in mycotrophic *Bouteloua gracilis* subjected to CO_2 enrichment. *Plant and Soil*, **165**, 139-146.

Mosier, A.R., 1998: Soil processes and global change. *Biology and Fertility of Soils*, **27**, 221-229.

Moulton, R., and K. Richards, 1990: *Costs of Sequestrating Carbon Through Tree Planting and Forest Management in the United States.* GTR WO-58, USDA Forest Service, Washington DC, USA.

Moura-Costa, P., and M. Stuart, 1998: Forestry based greenhouse gas mitigation: a story of market evolution. *Commonwealth Forestry Review*, **77**, 191-202.

Nabuurs, G.J., 1996: Significance of wood products in forest sector carbon balances. In *Forest ecosystems, forest management and the global carbon cycle.* M.J. Apps, and D.T. Price (eds.), NATO Advanced Science Institute Series, Vol I 40, Berlin, Heidelberg, Proceedings of the workshop held in September 1994 in Banff, Canada, 245-256.

Nabuurs, G.J., 1998: Bos wordt meer geld waard – Dutch forests become more valuable. *Nederlands Bosbouwtijdschrift*, **70**(2), 69.

Nabuurs, G.J., and R. Sikkema, 1998: *The role of harvested wood products in national carbon balances - an evaluation of alternatives for IPCC guidelines.* IBN Research report 98/3, Institute for Forestry and Nature Research, Institute for Forest and Forest Products, 53 pp.

Nabuurs, G.J., A.V. Dolman, E. Verkaik, P.J. Kuikman, C.A. van Diepen, A. Whitmore, W. Daamen, O. Oenema, P. Kabat, and G.M.J. Mohren, 2000: Article 3.3 and 3.4. of the Kyoto Protocol - consequences for industrialised countries' commitment, the monitoring needs and possible side effect. *Environmental Science and Policy*, **3**(2/3) 123-134.

Nabuurs, G.J., R. Paeivinen, R. Sikkema, and G.M.J. Mohren, 1997: The role of European forests in the global carbon cycle - a review. *Biomass and Bioenergy*, **13**(6), 345-358.

Nadelhoffer, K.J., B.A. Emmett, P. Gundersen, O.J. Kjønaas, C.J. Koopmans, P. Schleppi, A. Tietema, and R.F. Wright, 1999: Nitrogen deposition makes a minor contribution to carbon sequestration in temperate forests. *Nature*, **398**, 145-148.

Nair, P.K.R., 1989: The role of trees in soil productivity and protection. In *Agroforestry systems in the tropics.* P.K.R. Nair (ed.), Kluwer Academic Publishers, Dordrecht, The Netherlands, pp. 567-589.

NAS (National Academy of Sciences), 1992: *Policy Implications of Greenhouse Warming: Mitigation, Adaptation, and the Science Base.* Panel on Policy Implications of Greenhouse Warming, U.S. National Academy of Sciences, National Academy Press, Washington DC, USA.

Neue, H.U., 1997: Fluxes of methane from rice fields and potential for mitigation. *Soil Use and Management*, **13**, 258-267.

Nilsson, S., and A. Shvidenko, 1998: *Is sustainable development of the Russian forest sector possible?* IUFRO-Occasional-Paper, 1998, No. 11, 76 pp.

Noble, I., M. Apps, R. Houghton, D. Lashof, W. Makundi, D. Murdiyarso, B. Murray, W. Sombroek, and R. Valentini, 2000: Implications of different definitions and generic issues. In: *Land Use, Land Use Change and Forestry.* R.T. Watson, I.R. Noble, B. Bolin, N.H. Ravindranath, D.J. Verardo, and D.J. Dokken (eds.), A Special report of the IPCC, Cambridge University Press, UK, pp. 55-126.

Nobre, C.A., P.J. Sellers, and J. Shukla, 1991: Amazonian Deforestation and Regional Climate Change. *Journal of Climate*, **4**, 957-988.

Nogueira, L.A.H., M.A. Trossero, L. Couto, 1998: A discussion of the relationship between wood fibre for energy supply and overall supply of wood fibre for industry. *Unasylva*, **49**, 51-56.

Norby, R.J., and M.F. Cotrufo, 1998: A question of litter quality. *Nature*, **396**, 17-18.

Oldeman, L.R., 1994: The global extent of soil degradation. In *Soil Resilience and Sustainable Land Use.* D.J. Greenland, and I. Szaboles (eds.), CAB International, Wallingford, UK, pp. 99-118.

Owensby, C.E., 1993: Potential impacts of elevated CO_2 and above- and belowground litter quality of a tallgrass prairie. *Water Air and Soil Pollution*, **70**, 413-424.

Palo, M., and J. Uusivuori, 1999: Globalization of Forests, Societies and Environments. In *World Forests, Society and Environment.* M. Palo, and J. Uusivuori (eds.), Kluwer, pp. 3- 14.

Parks, P.J., and I.W. Hardie, 1995: Least-cost forest carbon reserves: cost effective subsidies to convert marginal agricultural land to forests. *Land Economics*, **71**, 122-136.

Paustian, K., E.T. Elliot, and K. Killian, 1997a: Modeling soil carbon in relation to management and climate change in some agroecosystems in central North America. In *Soil Processes and the Carbon Cycle.* R. Lal, J.M. Kimble, R.F. Follett, and B.A. Stewart (eds.), CRC Press, Boca Raton, pp. 459-471.

Paustian, K., O. Andren, H.H. Janzen, R. Lal, P. Smith, G. Tian, H. Tiessen, M. van Noordwijk, and P.L. Woomer, 1997b: Agricultural soils as a sink to mitigate CO_2 emissions. *Soil Use and Management,* **13**, 230-244.

Paustian, K., H.P. Collins, and E.A. Paul, 1997c: Management controls on soil carbon. Soil Organic Matter in Temperate Agroecosystems. In *Long-Term Experiments in North America.* E.A. Paul, E.T. Elliott, K. Paustian, and C.V. Cole (eds.), CRC Press, Boca Raton, pp. 15-49.

Paustian, K., C.V. Cole, D. Sauerbeck, and N. Sampson, 1998: CO_2 mitigation by agriculture: an overview. *Climatic Change*, **40**, 135-62.

Paustian, K., J. Six, E.T. Elliott, and H.W. Hunt, 2000: Management options for reducing CO_2 emissions from agricultural soils. *Biogeochemistry*, **48**, 147-163.

Peng, T.H., and W.S. Broecker, 1991: Dynamical Limitations on the Antarctic Iron Fertilization Strategy. *Nature*, **349**, 227-229.

Perez-Garcia, J., L.A. Joyce, C.S. Binkley, and A.D. McGuire, 1997: Economic Impacts of Climate Change on the Global Forest Sector. In *Economics of Carbon Sequestration in Forestry.* R.S. Sedjo (ed.), Lewis Publishers, Boca Raton.

Pickett, S.T.A., and P.S. White, 1985: *The Ecology of Natural Disturbance and Patch Dynamics.* Academic Press Inc., San Diego, USA, 472 pp.

Pielke, R.A., and R. Avissar, 1990: Influence of Landscape Structure on Local and Regional Climate. *Landscape Ecology*, **4**, 133-155.

Plantinga, A.J., T. Mauldin, and D.J. Miller, 1999: An econometric analysis of the costs of sequestering carbon in forests. *American Journal of Agricultural Economics*, **81**(4), 812-824.

Poore, D., P. Burges, J. Palmer, S. Rietbergen, and T. Synnott, 1989: *No Timber Without Trees: Sustainability in the Tropical Forest.* Earthscan Publications, London.

Post, W.M., and K.C. Kwon, 2000: Soil carbon sequestration and land-use change: processes and potential. *Global Change Biology*, **6**, 317-327.

Postel, S., and L. Heise, 1988: *Reforesting the Earth.* Worldwatch Institute, Washington DC, USA, 55 pp.

Potter, K.N., H.A. Torbert, H.B. Johnson, and C.R. Tischler, 1999: Carbon storage after long-term grass establishment on degraded soils. *Soil Science*, **164**, 718-725.

Prentice, I.C., G.D. Farquhar, M.J.R. Fasham, M.L. Goulden, M. Heimann, V.J. Jaramillo, H.S. Kheshgi, C. Le Quéré, R.J. Scholes, D.W.R. Wallace, 2001: The Carbon Cycle and Atmospheric CO_2. In *Climate Change 2001: The Scientific Basis.* Contribution of Working Group I to the IPCC Third Assessment Report, Cambridge University Press.

Price, D.T., M.J. Apps, and W.A. Kurz, 1998: Past and possible future dynamics of Canada's Boreal Forest Ecosystems. In *Carbon Dioxide Mitigation in Forestry and Wood Industry.* G.H. Kohlmaier, M. Weber, and R.A. Houghton (eds.), Springer-Verlag, Berlin, Germany, pp. 63-88.

Rabenhorst, M.C. 1995: Carbon storage in tidal marsh soils. In *Soils and Global Change.* R. Lal, J. Kimble, E. Levine, and B.A. Stewart, CRC Lewis Publishers, Boca Raton, pp. 93-103.

Rasmussen, P.E., and S.L. Albrecht, 1997: Crop management effects on organic carbon in semi-arid Pacific Northwest soils. In *Management of Carbon Sequestration in Soil.* R. Lal, J.M. Kimble, R.F. Follett, and B.A. Stewart, CRC Press, Boca Raton, pp. 209-219.

Raun, W.R., G.V. Johnson, S.B. Phillips, and R.L. Westerman, 1998: Effect of long-term N fertilization on soil organic C and total N in continuous wheat under conventional tillage in Oklahoma. *Soil and Tillage Research*, **47**, 323-330.

Ravindranath, N.H., and B.S. Somashekhar, 1995: Potential and Economics of Forestry Options for Carbon Sequestration in India. *Biomass and Bioenergy*, **8**, 323-336.

Ravindranath, N.H., and D.O. Hall, 1994: Indian forest conservation and tropical deforestation. *Ambio*, **23**(8), 521-523.

Ravindranath, N.H., and D.O. Hall, 1995: *Biomass, Energy and Environment. A Developing Country Perspective from India.* Oxford University Press, Oxford, UK, 376 pp.

Reeder, J.D., G.E. Schuman, and R.A. Bowman, 1998: Soil C and N changes on conservation reserve program lands in the Central Great Plains. *Soil and Tillage Research*, **47**, 339-349.

Roberts, L., 1999: *World Resources: 1998-99.* A joint publication by The World Resources Institute, The United Nations Environment Programme, The United Nations Development Programme and The World Bank, Oxford University Press, New York.

Rosenzweig, C., and D. Hillel, 2000: Soils and global climate change: Challenges and opportunities. *Soil Science*, **165**, 47-56.

Row, C., 1996: Effects of selected forest management options on carbon storage. In *Forests and Global Change. Vol. 2. Forest Management Opportunities for Mitigating Carbon Emissions.* N. Sampson, and D. Hair (eds.), American Forests, Washington DC, USA, pp. 27-58.

Sampson, R.N., M. Apps, S. Brown, C.V. Cole, J. Downing, L.S. Heath, D.S. Ojima, T.M. Smith, A.M. Solomon, and J. Wisniewski, 1993: Workshop Summary Statement - Terrestrial Biospheric Carbon Fluxes - Quantification of Sinks and Sources of CO_2. *Water Air and Soil Pollution*, **70**(1-4), 3-15.

Sampson, R.N., R.J. Scholes, C. Cerri, L. Erda, D.O. Hall, M. Handa, P. Hill, M. Howden, H. Janzen, J. Kimble, R. Lal, G. Marland, K. Minami, K. Paustian, P. Read, P.A. Sanchez, C. Scoppa, B. Solberg, M.A. Trossero, S. Trumbore, O. Van Cleemput, A. Whitmore, and D. Xu, 2000: Additional Human-induced Activities – Article 3.4. In *Land Use, Land-use Change, and Forestry.* In R.T. Watson, I.R. Noble, B. Bolin, N.H. Ravindranath, D.J. Verardo, and D.J. Dokken (eds.), A Special Report of the Intergovernmental Panel on Climate Change, Cambridge University Press, Cambridge, UK, 377 pp.

Sathaye, J., and N.H. Ravindranath, 1998: Climate change mitigation in the energy and forestry sectors of developing countries. *Annual Review of Energy and Environment*, **23**, 387-437.

Scharpenseel, H.W., 1993: Major carbon reservoirs of the pedosphere; source-sink relations; potential of d14C and d13C as supporting methodologies. *Water Air and Soil Pollution*, **70**, 431-442.

Scharpenseel, H.W., and P. Becker-Heidmann, 1994: Sustainable land use in the light of resilience/elasticity to soil organic matter fluctuations. In *Soil Resilience and Sustainable Land Use.* D.J. Greenland, and I. Szabolcs (eds.), CAB International, Wallingford, pp. 249-264.

Schelling, T.C., 1996: The Economic Diplomacy of Geoengineering. *Climatic Change*, **33**, 303-307.

Schimel, D., J. Melillo, H. Tian, A.D. McGuire, D. Kicklighter, T. Kittel, N. Rosenbloom, S. Running, P. Thornton, D. Ojima, W. Parton, R. Kelly, M. Sykes, R. Neilson, and B. Rizzo, 2000: Contribution of increasing CO_2 and climate to carbon storage by ecosystems in the United States. *Science*, **287**, 2004-2006.

Schimel, D.S., 1995: Terrestrial ecosystems and the carbon cycle. *Global Change Biology*, **1**, 77-91.

Schindler, D.W., and S.E. Bayley, 1993: The biosphere as an increasing sink for atmospheric carbon: Estimates from increased nitrogen deposition. *Global Biogeochemical Cycles*, **7**, 717-733.

Schlamadinger, B., and G. Marland, 1996: The role of forest and bioenergy strategies in the global carbon cycle. *Biomasss and Bioenergy*, **10**, 275-300.

Schlesinger, W.H., 1999: Carbon sequestration in soils. *Science*, **284**, 2095.

Scholes, R.J., and N. van Breemen, 1997: The effects of global change on tropical ecosystems. *Geoderma*, **79**, 9-24.

Scholes, R.J., E.D. Schulze, L.F. Pitelka, and D.O. Hall, 1999: Biogeochemistry of terrestrial ecosystems. In *The Terrestrial Biosphere and Global Change: Implications for Natural and Managed Ecosystems.* B. Walker, W. Steffen, J. Canadell, and J. Ingram (eds.), Cambridge University Press, Cambridge, pp. 271-303.

Schulze, E.D., 2000: Carbon and nitrogen cycling in European forest ecosystems. *Ecological Studies*, Vol. 142, Springer, Heidelberg, Germany, 500 pp.

Schulze, E.D., J. Lloyd, F.M. Kelliher, C. Wirth, C. Rebmann, B. Luhker, M. Mund, A. Knohl, I.M. Milyukova, W. Schulze, W. Ziegler, A. Varlargin, A.F. Sogachev, R. Valentini, S. Dore, S. Grigoriev, O. Kolle, M.I. Panfyorov, N. Tchebokova, and N.N. Vygodskaya, 1999: Productivity of forests in the Eurosiberian boreal region and their potential to act as a carbon sink – a synthesis. *Global Change Biology*, **5**, 703-722.

Sedjo, R.A., 1983: *The Comparative Economics of Plantation Forestry.* Johns Hopkins Press, Resources for the Future, Baltimore, MD, USA, 161 pp.

Sedjo, R.A., 1992: Forest Ecosystem in the Global Carbon Cycle. *Ambio*, **21**(4), 274-277.

Sedjo, R.A., 1997: The economics of forest-based biomass supply. *Energy Policy*, **25**(6), 559-566.

Sedjo, R.A., 1999a: Land Use Change and Innovation in U.S. Forestry. In *Productivity in Natural Resources Industries: Improvement Through Innovation.* David Simpson (ed.), Resources for the Future, Washington DC, USA, pp. 141-174.

Sedjo, R.A., 1999b: *Potential for Carbon Forest Plantations in Marginal Timber Forests: The Case of Patagonia, Argentina.* RFF Discussion Paper 99-27, Resources for the Future, Washington DC, 20 pp.

Sedjo, R.A., and D. Botkin, 1997: Using Forest Plantations to Spare natural Forests. *Environment*, **30**(10), 14-20 and 30.

Sedjo, R.A., and K.S. Lyon: 1990: *The Long-Term Adequacy of World Timber Supply.* Resources for the Future, Washington DC, USA, 230 pp.

Sedjo, R.A., and B. Sohngen, 2000: *Forestry Sequestration of CO2 and Markets for Timber.* RFF Discussion Paper 00-35, Resources for the Future, Washington DC, USA, 83 pp.

Sedjo, R.A., and A.M. Solomon, 1989: Climate and Forests. In *Greenhouse Warming : Abatement and Adaptation.* N.J. Rosenberg, W.E. Easterling, P.R. Crosson, and J. Darmstadter (eds.), Resources for the Future, Washington DC, pp. 105-120.

Sedjo, R.A., J. Wisniewski, A.V. Sample, and J.D. Kinsman, 1995: The economics of managing carbon via forestry: assessment of existing studies. *Environmental and Resource Economics*, **6**(2), 139-165.

Sheinbaum, C., and O.R. Masera, 2000: Mitigating Carbon Emissions while Advancing National Development Priorities - The Case of Mexico. *Climatic Change*, **47**(3), 259-282.

Shepashenko, D., A. Shvidenko, and S. Nilsson, 1998: Phytomass (live biomass) and carbon of Siberian forests. *Biomass and Bioenergy*, **14**(1), 21-31.

Singh, B.R., T. Borresen, G. Uhlen, and E. Ekeberg, 1997a: Long-term effects of crop rotation, cultivation practices, and fertilizers on carbon sequestration in soils in Norway. *In: Management of Carbon Sequestration in Soil.* R. Lal, J.M. Kimble, R.F. Follett, and B.A. Stewart (eds.), CRC Press, Boca Raton, pp.195-208.

Singh, B., Y. Singh, M.S. Maskina, and O.P. Meelu, 1997b: The value of poultry manure for wetland rice grown in rotation with wheat. *Nutrient Cycling in Agroecosystems*, **47**, 243-250.

Smith, P., D.S. Powlson, M.J. Glendining, and J.U. Smith, 1997: Opportunities and limitations for C sequestration in European agricultural soils through change in management. In *Management of Carbon Sequestration in Soil.* R. Lal, J.M. Kimble, R.F. Follett, and B.A. Stewart (eds.), CRC Press, Boca Raton, pp. 143-152.

Sohngen, B., and R. Mendelsohn, 1998: Valuing the impact of large-scale ecological change in a market: The effect of climate change on US timber. *American Economic Review*, **88**(4), 686-710.

Sohngen, B., and R.A. Sedjo, 1999: *Carbon Sequestration by Forestry – Effects of Timber Markets.* Report number PH3/10, IEA Greenhouse Gas R&D Programme, UK, 74 pp.

Sohngen, B., R. Mendelsohn, and R.A. Sedjo, 1999: Forest management, Conservation, and Global Timber Markets. *American Journal of Agricultural Economics*, **81**, 1-13.

Solberg, B., 1997: Forest biomass as carbon sink – economic value and forest management/policy implications. *Critical reviews in Environmental Science and Technology*, **27**, S323-333.

Solberg, B., D. Brooks, H. Pajuoja, T.J. Peck, and P.A. Wardle, 1996: *Long-term trends and prospects in world supply and demand for wood and implications for sustainable forest management.* EFI research report 6, European Forest Institute/Norwegian Forest Research Institute, A contribution to the CSD Ad Hoc Intergovernmental Panel on Forests, 31 pp.

Solomon, D., J. Lehmann, and W. Zech, 2000: Land use effects on soil organic matter properties of chromic luvisols in semi-arid northern Tanzania: carbon, nitrogen, lignin and carbohydrates. *Agriculture Ecosystems and Environment*, **78**, 203-213.

Spiecker, H., K. Mielikainen, M. Kohl, and J.P. Skovsgaard, 1996: *Growth Trends in European Forests*. Springer-Verlag, Heidelberg, Germany, 372 pp.

Stallard, R.F., 1998: Terrestrial sedimentation and the carbon cycle: Coupling weather and erosion to carbon burial. *Global Biogeochemical Cycles*, **12**, 231-257.

Stavins, R., 1999: The Costs of Carbon Sequestration: A Revealed Preference Approach. *The American Economic Review*, **89**, 994-1009.

Stockfisch, N., T. Forstreuter, and W. Ehlers, 1999: Ploughing effects on soil organic matter after twenty years of conservation tillage in Lower Saxony, Germany. *Soil and Tillage Research*, **52**, 91-101.

Stuart, M.D., and P.H. Moura-Costa, 1998: Greenhouse gas mitigation: a review of international policies and initiatives. In *Policies that Work for People*, Series no. 8, International Institute of Environment and Development, London, UK, pp. 27-32.

Teller, E., L. Wood, and R. Hyde, 1997: *Global Warming and Ice Ages: I. Prospects for Physics-Based Modulation of Global Change*. UCRL-JC-128715, Lawrence Livermore National Laboratory, Livermore, California, USA, 20 pp.

Tian, G., G.O. Kolawole, F.K. Salako, and B.T. Kang, 1999: An improved cover crop-fallow system for sustainable management of low activity clay soils of the tropics. *Soil Science*, **164**, 671-682.

Tiessen, H., C. Feller, E.V.S.B. Sampaio, and P. Garin, 1998: Carbon sequestration and turnover in semiarid savannas and dry forest. *Climatic Change*, **40**, 105-117.

Tole, L., 1998: Source of deforestation in tropical developing countries. *Environmental Management*, **22**, 19-33.

Torbert, H.A., H.H. Rogers, S.A. Prior, W.H. Schlesinger, and G.B. Runions, 1997: Effects of elevated atmospheric CO_2 in agro-ecosystems on soil carbon storage. *Global Change Biology*, **3**, 513-521.

Townsend, A.R., and E.B. Rastetter, 1996: Nutrient constraints on carbon storage in forested ecosystems. In *Forest Ecosystems, Forest Management and the Global Carbon Cycle*. M.J. Apps, and D.T. Price (eds.), NATO ASI Series I (Global Environmental Change), Vol. I 40, Springer-Verlag Academic publishers, Heidelberg, Germany, pp. 35-46.

Trexler and Associates, Inc., 1998: *Final Report of the Biotic Offsets Assessment Workshop, Baltimore, USA*. Prepared for the US Environmental Protection Agency, Washington DC, USA, 107 pp.

Turner, D.P., G.J. Koerper, M.E. Harmon, and J.L. Lee, 1995: A carbon budget for forests of the conterminous United States. *Ecological Applications*, **5**, 421-436.

Turner, S.M., P.D. Nightingale, L.J. Spokes, M.I. Liddicoat, and P.S. Liss, 1996: Increased Dimethyl Sulphide Concentrations in Sea Water From in situ Iron Enrichment. *Nature*, **383**, 513-517.

Turner, D.P., J.K. Winjum, T.P. Kolchugina, and M.A. Cairns, 1997: Accounting for biological and anthropogenic factors in national land-based carbon budgets. *Ambio*, **26**, 220-226.

UN-ECE/FAO (UN Economic Committee for Europe / Food and Agricultural Organisation), 2000: *Forest resources of Europe, CIS, North America, Australia, Japan and New Zealand*. Geneva Timber and Forest Study papers No 17, United Nations Economic Committee for Europe, Food and Agricultural Organisation, Geneva, Switzerland, 445 pp.

UNFCCC (UN Framework Convention on Climate Change), 1997: The Kyoto Protocol to the United Nations Framework Convention on Climate Change. Document FCCC/CP/1997/7/Add.1. http://www.unfccc.de/

Valentini, R., A. Matteucci, A.J. Dolman, and E.D. Schulze, 2000: Respiration as the main determinant of carbon balance of European forests. *Nature*, **404**, 861-865.

Van der Gon, H.D., 2000: Changes in CH_4 emission from rice fields from 1960 to 1990s. 1- Impacts of modern rice technology. *Global Biogeochemical Cycles*, **14**, 61-72.

Van Ginkel, J.H., A. Gorissen, and J.A. van Veen, 1996: Long-term decomposition of grass roots as affected by elevated atmospheric carbon dioxide. *Journal of Environmental Quality*, **25**, 1122-1128.

Van Noordwijk, M., C. Cerri, P.L. Woomer, K. Nugroho, and M. Bernoux, 1997: Soil carbon dynamics in the humid tropical forest zone. *Geoderma*, **79**, 187-225.

Vitousek, P.M., H.A. Mooney, J. Lubchenco, and J.M. Melillo, 1997: Human domination of earth's ecosystems. *Science*, **277**, 494-499.

Waggoner, P., 1994: *How Much Land Can Ten Billion People Spare for Nature?* Task Force Report No. 121, Council for Agricultural Science and Technology, Ames, IA, 64 pp.

Wan Razali, W.M., and J. Tay, 2000: Forestry Carbon Emission Offset and Carbon Sinks Project: Examples of Opportunity in Carbon Sequestration and its Implications from Kyoto Protocol. International Workshop on *The response of tropical forest ecosystems to long term cyclic climate change*, Science and Technology Agency, Japan, National Research Council of Thailand, 24-27 January, 2000, Kanchanaburi, Thailand, 10 pp.

Wang, X., and Z. Feng, 1995: Atmospheric carbon sequestration through agroforestry in China. *Energy*, **20**(2), 117-121.

Watson, A.J., C.S. Law, K.A. Van Scoy, F.L. Millero, W. Yao, G.E. Friedrich, M.I. Liddicoat, R.H. Wanninkhof, R.T. Barber, and K.H. Coale, 1994: Minimal Effect of Iron Fertilization on Sea-Surface Carbon Dioxide Concentrations. *Nature*, **371**, 143-145.

WBGU (Wissenschaftlicher Beirat der Bundesregierung Globale Umweltveränderungen), 1998: *Die Anrechnung biologischer Quellen und Senken im Kyoto-Protokoll: Fortschritt oder Rückschlag für den globalen Umweltschutz?* Sondergutachten 1998, WBGU, Bremerhaven, Germany, 76 pp., (available in English).

Weber, M.G., and M.D. Flanningan, 1997: Canadian boreal forest ecosystem structure and function in a changing climate: impact on fire regimes. *Environmental Reviews*, **5**, 145-166.

Wedin, D.A., and D. Tilman, 1996: Influence of Nitrogen Loading and Species Composition on the Carbon Balance of Grasslands. *Science*, **274**, 1720-1723.

Wells, M.L., 1994: Pumping Iron in the Pacific. *Nature*, **368**, 295-296.

Wernick, I.K., P.E. Waggoner, and J.H. Ausubel, 1998: Searching for Leverage to Conserve Forests. The Industrial Ecology of Wood Products in the United States. *Journal of Industrial Ecology*, **1**, 125-145.

Willison, T.W., J.C. Baker, and D.V. Murphy, 1998: Methane fluxes and nitrogen dynamics from a drained fenland peat. *Biology and Fertility of Soils*, **27**, 279-283.

Winjum, J.K., S. Brown, and B. Schlamadinger, 1998: Forest harvests and wood products: sources and sinks of atmospheric carbon dioxide. *Forest Science*, **44**(2), 272-284.

Woodwell, G.M., F.T. Mackenzie, R.A. Houghton, M. Apps, E. Gorham, and E. Davidson, 1998: Biotic Feedbacks in the Warming of the Earth. *Climatic Change*, **40**, 495-518.

Woomer, P.L., C.A. Palm, J.N. Qureshi, and J. Kotto-Same, 1997: Carbon sequestration organic and resource management in African smallholder agriculture. In *Management of Carbon Sequestration in Soil*. R. Lal, J.M. Kimble, R.F. Follett, and B.A. Stewart (eds.), CRC Press, Boca Raton, pp. 153-173.

WRI (World Resources Institute), 1987: *Tropical Forestry Action Plan: Recent Developments*. The World Resources Institute, Washington DC, USA.

Yadav, R.L., D.S. Yadav, R.M. Singh, and A. Kumar, 1998: Long term effects of inorganic fertilizer inputs on crop productivity in a rice-wheat cropping system. *Nutrient Cycling in Agroecosystems*, **51**, 193-200.

Yagi, K., H. Tsuruta, and K. Minami, 1997: Possible options for mitigating methane emission from rice cultivation. *Nutrient Cycling in Agroecosystems*, **49**, 213-220.

Yoshida, K., 1983: Heterogeneous environmental structure in a moth community of Tomakomai Experiment Forest, Hokkaido University. *Japan Journal of Ecology*, **33**, 445-451.

Young, A., 1997: *Agroforestry for Soil Management* (2nd edition). CAB International, Oxford, UK, 320 pp.

Zhang, Z.X., 1996: Some economic aspects of climate change. *International Journal Environment and Pollution*, **6**, 185-195.

Zoltai, S.T., and P.J. Martikainen, 1996: Estimated extent of forested peatlands and their role in the global carbon cycle. In *Forest Ecosystems, Forest Management and the Global Carbon Cycle*. M.J. Apps, and D.T. Price (eds.), NATO ASI Series, Series I: Global Environmental Change, Vol. 40, pp. 47-58.

5

Barriers, Opportunities, and Market Potential of Technologies and Practices

Co-ordinating Lead Authors:
JAYANT SATHAYE (USA), DANIEL BOUILLE (ARGENTINA)

Lead Authors:
Dilip Biswas (India), Philippe Crabbe (Canada), Luis Geng (Peru), David Hall[†](UK), Hidefumi Imura (Japan), Adam Jaffe (USA), Laurie Michaelis (UK), Grzegorz Peszko (Poland), Aviel Verbruggen (Belgium), Ernst Worrell (Netherlands/USA), F. Yamba (Zambia)

Contributing Authors:
Mauricio Tolmasquim (Brazil), Henry Janzen (Canada)

Review Editors:
Michael Jefferson (UK), R. T. M. Sutamihardja (Indonesia)

[†] Professor David Hall, a close colleague, passed away in August 1999. He inspired us all through his vigorous support for bioenergy, and its just uses in the developing world.

CONTENTS

EXECUTIVE SUMMARY

The transfer of technologies and practices that have the potential to reduce greenhouse gas (GHG) emissions is often hampered by barriers[1] that slow their penetration. The opportunity[2] to mitigate GHG concentrations by removing or modifying barriers to the spread of technology may be viewed within a framework of different potentials for GHG mitigation (*Figure 5.1*). The "market potential" indicates the amount of GHG mitigation that might be achieved under forecast market conditions, with no changes in policy or implementation of measures whose primary purpose is the mitigation of GHGs. The market potential can be close to zero as a result of extreme poverty, absence of markets, and remoteness of communities. The inability of the poor or isolated communities to access modern energy services reflects this situation. Because interventions to address poverty fall outside the immediate scope of this chapter, they receive only limited treatment here despite the intrinsic general importance of the subject.

In addition to the market potential, there is also the economic potential and the socioeconomic potential to be considered. Eliminating imperfections of markets, public policies, and other institutions that inhibit the diffusion of technologies that are (or are projected to be) cost-effective for consumers (evaluated using consumers' private rate of time discounting and prices) without reference to any GHG benefits they may generate would increase GHG mitigation to the level defined as the "economic potential". The "socioeconomic" potential consists of barriers derived from people's individual habits, attitudes and social norms, and vested interests in the diffusion of new technology. This potential represents the level of GHG mitigation that would be achieved if technologies that are cost effective from a societal perspective are implemented.

Finally, some technologies might not be widely used simply because they are too expensive from a societal perspective. This leads to the level of the "technical potential", which can be improved upon by solving scientific and technological problems. Policies to overcome this category of barriers must be aimed at fostering research and development (R&D).

Technological and social innovation is a complex process of research, experimentation, learning, and development that can contribute to GHG mitigation. Several theories and models have been developed to understand its features, drivers, and implications. New knowledge and human capital may result from R&D spending, through learning by doing, and/or in an evolutionary process. Most innovations require some social or behavioural change on the part of users. Rapidly changing economies, as well as social and institutional structures offer opportunities for locking-in to GHG-mitigative technologies that may lead countries on to sustainable development pathways. The pathways will be influenced by the particular socioeconomic context that reflects prices, financing, international trade, market structure, institutions, the provision of information, and social, cultural and behavioural factors; key elements of which are described below.

Unstable Macroeconomic Conditions
Such conditions increase risk to private investment and finance. Unsound government borrowing and fiscal policy lead to chronic public deficits, reducing the availability of credit to the private sector. Trade barriers that favour inefficient technologies, or prevent access to advanced knowledge and hardware, can slow the diffusion of mitigation options.

Commercial Financing Institutions
These institutions face high risks when developing "green" financial products. Innovative approaches in the private sector to address this and other issues include leasing, environmental and ethical banks, micro-credits or small grants facilities targeted at low income households, environmental funds, energy service companies (ESCOs), and green venture capital.

Distorted or Incomplete Prices
The absence of a market price for certain impacts, such as environmental harm, can constitute a barrier to the diffusion of environmentally beneficial technologies. Distortion of prices arising from taxes, subsidies, or other policy interventions that make resource consumption more or less expensive to consumers can also impede the diffusion of resource-conserving technologies.

Information as a Public Good
Generic information regarding the availability of different kinds of technologies and their performance characteristics has the attributes of a "public good" and hence may be underprovided by the private market.

---

[1] A barrier is any obstacle to reaching a potential that can be overcome by a policy, programme, or measure.

[2] An opportunity is a situation or circumstance to decrease the gap between the market potential of a technology or practice and the economic, socioeconomic, or technological potential.

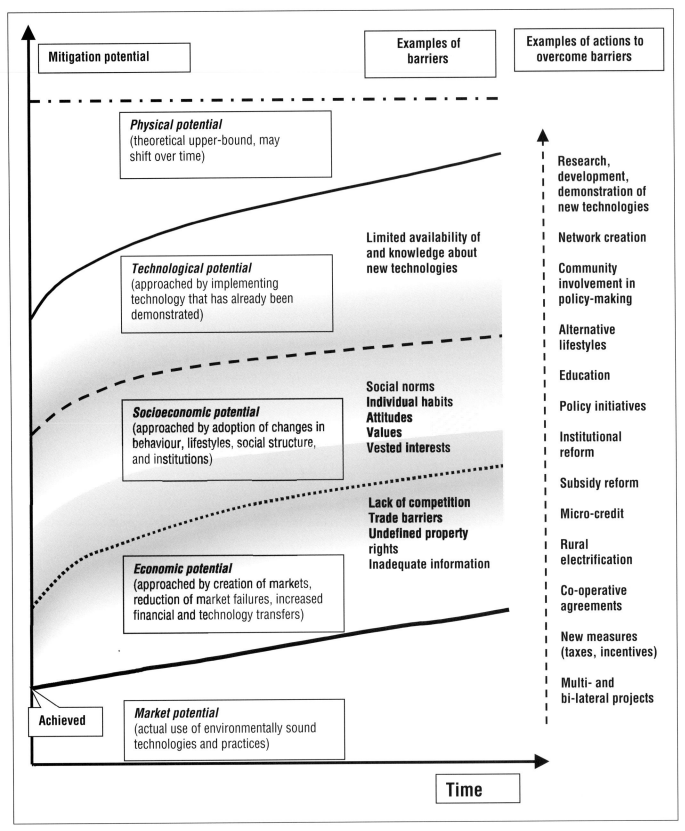

Figure 5.1: *Penetration of environmentally sound technologies: a conceptual framework. Various barriers prevent the different potentials from being realized. Opportunities exist to overcome barriers through innovative projects, programmes and financing arrangements. An action can address more than one barrier. Actions may be pursued to address barriers at all levels simultaneously. Their implementation may require public policies, measures and instruments. The socioeconomic potential may lie anywhere in the space between the economic and technological potential.*

Lack of Effective Regulatory Agencies

Many countries have on their books excellent constitutional and legal provisions for environmental protection but the latter are not enforced. However, "informal regulation" under community pressure may substitute for formal regulatory pressure.

Lifestyles, Behaviours, and Consumption Patterns

These have developed within current and historical socio-cultural contexts. Changes in behaviour and lifestyles may result from a number of intertwined processes. Barriers take various forms in association with each of the above processes.

Conventional Policy Development

This type of development is based on a model of human psychology, where people are assumed to be rational welfare-maximizers, that has been widely criticized. Such a model does not explain processes, such as learning, habituation, value formation, or the bounded rationality observed in human choice.

Buildings

The poor in every country are affected far more by barriers in this sector than the rich, because of inadequate access to financing, low literacy rates, adherence to traditional customs, and the need to devote a higher fraction of income to satisfy basic needs, including fuel purchases.

Measures to overcome these barriers that have been implemented include voluntary programmes, building efficiency standards, equipment efficiency standards, state market transformation programmes, financing, government procurement, tax credits, accelerated R&D, and a carbon cap and trade system.

Transport

The low relative cost of fuel, split incentives, a perception that the car is more convenient or economical than alternatives, are some of the barriers that slow the use of mitigation technologies in this sector. The car has also become charged with significance as a means of freedom, mobility and safety, a symbol of personal status and identity, and as one of the most important products in the industrial economy. A combination of policies protecting road transport interests, rather than any single policy, poses the greatest barrier to change.

Industry

Barriers include the high transaction costs for obtaining reliable information, the use of capital for competing investment priorities, high-hurdle rates for energy efficiency investments, lack of skilled personnel for small and medium-sized enterprises (SMEs), and the low relative cost of energy. Information programmes, environmental legislation, and voluntary agreements have been used and tested in developed countries with varying rates of success in reducing barriers.

Energy Supply

The increasing deregulation of energy supply has raised particular concerns. Volatile spot and contract prices, short-term out-look of private investors, and the perceived risks of nuclear and hydropower plants have shifted fuel and technology choice towards natural gas and oil plants, and away from hydro in many countries. Co-generation is hampered by lack of information, the decentralized character of the technology, the hostile attitude of grid operators, the terms of grid connection, and lack of policies that foster long-term planning. Firm public policy and regulatory authority are necessary to install and safeguard harmonized conditions, transparency, and unbundling of the main power supply functions.

Agriculture and Forestry

Adoption of new technology is limited by small farm size, credit constraints, risk aversion, lack of access to information and human capital, inadequate rural infrastructure and tenurial arrangements, and unreliable supply of complementary inputs. Subsidies for critical inputs to agriculture, such as fertilizers, water supply, and electricity and fuels, and to outputs in order to maintain stable agricultural systems and an equitable distribution of wealth distort markets for these products. In relation to climate change mitigation, other issues such as lack of technical capability, lack of credibility about setting project baselines, and monitoring of carbon stocks pose difficult challenges.

Waste Management

The principal barriers to technology transfer include limited financing and institutional capability, jurisdictional complexity, and the need for community involvement. Climate change mitigation projects face further barriers owing to the unfamiliarity with methane (CH_4) capture and potential electricity generation, unwillingness to commit additional human capacity for climate mitigation, and the involvement of diverse institutions at all levels.

Regional Considerations

Changing global patterns provide an opportunity for introducing GHG mitigation technologies and practices that are consistent with development, equity, and sustainability (DES) goals. A culture of energy subsidies, institutional inertia, fragmented capital markets, vested interests, etc., however, presents major barriers to their implementation in the developing countries and those with economies in transition (EIT). Situations in these two groups of countries call for a more careful analysis of trade, institutional, financial, and income barriers and opportunities; distorted prices and information gaps. In the developed countries, other barriers such as the current carbon-intensive lifestyle and consumption patterns, social structures, network externalities, and misplaced incentives offer opportunities for intervention to control the growth of GHG emissions. Lastly, new and used technologies mostly flow from the developed to developing and transitioning countries. A global approach to reducing emissions that targets technology being transferred from developed to developing countries could have a significant impact on future emissions.

5.1   Introduction

Technology transfer comprises a broad set of processes covering the flows of know-how, experience, and equipment for mitigating and adapting to climate change among different stakeholders such as governments, private sector entities, financial institutions, non-governmental organizations (NGOs), and research and/or education institutions (IPCC, 1996; IPCC, 2000b). The term transfer encompasses diffusion of technologies and technology co-operation across and within countries. It comprises the process of learning to understand, utilize, and replicate the technology, including the capacity to choose and adapt it to local conditions, and integrate it with indigenous technologies.

The previous chapters (Chapters 3 and 4) have discussed the characteristics of different technologies and practices, and their potential and costs for the mitigation of climate change. Chapter 3 has identified numerous negative cost or "no regrets" options whose full implementation is prevented by various types of barriers. The focus of this chapter, thus, is on the various barriers that inhibit the process of technology transfer, but not on technology programmes, which are covered in Chapter 3. A "barrier" is any obstacle to reaching a potential that can be overcome by a policy, programme, or measure (*Figure 5.1*). This chapter describes the barriers that lie below the "socioeconomic potential" line in *Figure 5.1*. Barriers to technology transfer may also be viewed as opportunities for intervention by the aforementioned stakeholders so that technologies can reach their full potential. An "opportunity" is thus any situation or circumstance to decrease the gap between the "market potential" of a technology and the economic, socioeconomic, or technical potential. Barriers and opportunities tend to be context-specific, and can change over time and vary across countries. Policies, programmes, and measures may be used to take advantage of the opportunities to help overcome the barriers. The interventions are largely described and assessed in Chapter 6, although some types of interventions at the sectoral level are illustrated in Section 5.4 of this chapter.

Opportunities for climate change mitigation exist both in reducing the intensity of greenhouse gas (GHG) emissions and the level of activities that cause these emissions. Reducing the level of an activity, for instance vehicle travel, need not reduce the services associated with it if a substitute like telecommuting can satisfy the same need. GHG mitigation can thus be achieved without sacrificing consumer welfare. Opportunities for such changes are equally important and need to be actively sought out. The interventions needed for achieving changes in the level of activity, however, can encompass the broad array of macro and micro policies that affect consumers and producers alike. In this chapter, the barriers, opportunities, and sectoral interventions for both the GHG intensity and "activity" changes are discussed. The broader macro-interventions are discussed in Chapter 6.

An element that lies largely unexplored is the connection between poverty and climate change mitigation. A large proportion of the world's population lives in poverty, often outside a cash economy, and does not have access to modern fuels. Even when the poor are part of a cash economy, they are often deprived of access to financial instruments that require collateral. The literature on barriers and opportunities to address their need for fuels, and the consequent GHG emissions, is relatively sparse. In this chapter, the limited material on barriers, opportunities, and interventions associated with the provision of energy services to the poor is reviewed primarily in the sections on finance (Section 5.3.3), energy use in buildings (Section 5.4.1) and agriculture (Section 5.4.5).

Barriers to technology transfer have been described and classified in many different ways. Reddy (1991) classifies barriers by actors, consumers, energy providers, etc.; and others (Hirst and Brown, 1990; Evans, 1991; Hirst, 1992) by the type of barrier, financing, pricing, etc. Technological and social changes offer new opportunities for the diffusion of GHG-mitigative technologies. Rapidly changing economies and institutional and social structures offer opportunities for locking into GHG-mitigative technologies that are likely to grow over the long term. Exploiting opportunities during a period of rapid change is typically easier than in a static environment. For example, the Internet revolution means that many aspects of society and the economy are being reshaped, offering opportunities to build environmental and sustainable development practices into the emerging paradigms. At the more micro-level, the beginning of an investment cycle for power supply systems and house purchase by individuals and families is a period when they are making major purchase decisions. Governments can influence these decisions through various regulations, financial incentives and information at such times to make the new investment less-GHG intensive. Synergies exist between GHG mitigation and other policy goals, e.g., reducing transport air pollution or conserving soils. Measures to address the latter offer opportunities for GHG mitigation also. While the chapter focuses broadly on both barriers and opportunities, Sections 5.3.1 and 5.3.8 specifically review the models of, and experience with, technological and social innovation and the opportunities offered for the diffusion of GHG-mitigative technologies. Synergies too are noted throughout, but particularly so in the sectoral sections 5.4.4 through 5.4.7.

The chapter focuses not only on the energy demand and supply sectors, which have a rich literature in this field, but also on the agriculture, forestry, and waste sectors. In the introductory sections below, a conceptual framework for understanding the role of opportunities and barriers, and a review of the two earlier Intergovernmental Panel on Climate Change (IPCC) reports that have dealt with this topic, namely the Second Assessment Report (SAR) and the Special Report on Technology Transfer (SRTT) are presented. Section 5.3 then discusses the generic opportunities and barriers that apply across all sectors, which is followed by a discussion of the prominent barriers and opportunities in appropriate sectors of the economy.

5.1.1 Summary of the Second Assessment Report – Barriers and Opportunities

The topic of barriers to the market penetration of environmentally sound technologies (ESTs) was treated in Section 1.5.3 ("Market failures and government responses") of the Working Group (WG) III SAR, and also in Chapter 8, Sections 8.2.3 ("Key factors affecting the magnitude of costs: Costs as a function of baselines and policy strategies"), and 8.4.3 ("The top-down vs. bottom-up modelling controversy: Some lessons from the energy field") respectively. The latter sections dealt with the discussion of the differences between top-down and bottom-up modelling when estimating the costs of strategies to reduce or control GHG emissions. The primary question raised in the discussion in the SAR may be summarized as: Given market prices, do firms fail to take advantage of all the energy efficiency opportunities available to them? Thus, a business investment decision, considering private costs, may not undertake all the available efficiency opportunities. Likewise, in the modelling sections, the discussion focused on the existence of the "no regrets" potential. Its existence implies that (1) market and/or institutional failures exist, and (2) cost-effective policies targeted to correct these can be identified and implemented. The SAR notes four categories of market imperfections that explain the above phenomena, and the policies that could be used to address them. A more detailed discussion of these topics is included in other sections of this chapter.

Information Dissemination
Acquiring information is costly, and markets, on their own, do not provide an efficient level of disclosure of information. Governments can amend this by providing information or instituting legislation and/or regulations that requires disclosure of information, e.g., requiring energy performance labels on household appliances (see Section 5.3.7 for further discussion on lack of information as a barrier).

Bureaucratic Structure and Limited Scope of Attention
Economic and organizational theory has emphasized that large organizations are not, in general, run by owners; that the managers, even with best-designed incentives, do not in general maximize the firm's market value; and that among the principal scarce factors within an organization are time and attention. Governments could provide information on energy efficiency that managers could access with ease, which may yield private returns higher than their marginal costs. (See Sections 5.3.5.2 and 5.4.3 for further discussion on barriers and opportunities in the industrial sector.)

Returns to Scale and Network Externalities
Technologies or projects may require large infrastructure or size in order to make them economic. The scale of such a project, e.g., a natural gas-based transportation system, may deter investment, although it may be cost-effective in comparison to a gasoline-based system at some higher future oil price (see Sections 5.3.5 for further discussion on network externalities).

Capital Market Imperfections
Studies of implicit discount rates have shown that households and firms behave as if they use rates substantially above the market rate for long-term government bonds. Firms use discount rates that reflect the riskiness of projects, and, as a result of imperfect information, households and firms often face rationing in capital markets for credit and equity. Economists emphasize that timing, risk, capital constraints, and information or lack thereof should be dealt with separately. A discount rate should reflect investment timing questions, risk should be treated by converting costs and benefits into certainty equivalents, and shadow pricing should address constraints on capital. Lack of information could be addressed through government intervention (see Sections 5.3.3 for further discussion on financing).

5.1.2 Special Report on Technology Transfer – Barriers and Opportunities

This IPCC Special Report was prepared in response to a request made by the UN Framework Convention on Climate Change through its Subsidiary Body for Scientific and Technological Advice (SBSTA) to provide input on the issue of "Development and assessment of methodological and technological aspects of transfer of technology". The focus of the report is on transfer of technology, and it describes actions that governments and other stakeholders can undertake to enhance technology transfer within and between countries. It emphasizes that governments have a key role to play in initiating and facilitating technology transfer, either directly or by creating an enabling environment for the private sector and community involvement.

While the technology transfer process can be complex and intertwined, certain stages can be identified. These include the identification of needs, choice of technology, assessment of conditions of transfer, agreement, and implementation. Evaluation and adjustment to local conditions, and replication are other important stages.

Barriers to the transfer of ESTs arise at each stage of the process. These vary according to the specific context from sector to sector and can manifest themselves differently in developed and developing countries, and in EITs. These barriers range from lack of information; insufficient human capabilities; political and economic barriers, such as lack of capital, high transaction costs, lack of full cost pricing, and trade and policy barriers; institutional and structural barriers; lack of understanding of local needs; business limitations, such as risk aversion in financial institutions; institutional limitations, such as insufficient legal protection; and inadequate environmental codes and standards.

The report further notes that there is no preset answer to enhancing technology transfer. The identification, analysis, and prioritization of barriers should be country based, and actions should be tailored to overcome specific barriers, interests, and influences of specific stakeholders in order to develop effective policy tools.

The thrust of the technology transfer report is on the identification of actions that governments may pursue to overcome barriers that slow or prevent the transfer of technology either within or across countries. This chapter of the TAR (Third Assessment Report) provides an in-depth discussion of the literature on barriers and opportunities, and provides a framework for differentiating between different types of potentials and barriers to technology penetration. The framework also helps in identifying the role of research, development and demonstration phases, and their linkage to the eventual market acceptance of technology. The chapter also discusses the opportunities for technology penetration, but it limits the discussion on policies and measures to sectoral interventions. A discussion of the broader policies and measures is found in Chapter 6.

5.2   Conceptual Framework for Understanding Barriers and Opportunities

The opportunity to mitigate GHG concentrations by removing or modifying barriers to the spread of technology may be viewed as an association between different types or categories of barriers and different concepts of the potential for GHG mitigation (*Figure 5.1*). Each concept of the potential represents a hypothetical projection that might be made today regarding the extent of GHG mitigation over time into the future. The bottom line, labelled "market potential" indicates the amount of GHG mitigation that might be expected to occur under forecast market conditions, with no changes in policy or implementation of measures whose primary purpose is the mitigation of GHGs.

At the other extreme, the "technical potential" describes the maximum amount of GHG mitigation achievable through technology diffusion. This is a hypothetical projection of the extent of GHG mitigation that could be achieved over time if all technically feasible technologies were used in all relevant applications, without regard to their cost or user acceptability.

By definition, it can be said that whatever physical, cultural, institutional, social, or human factors are preventing the progress from the market potential to the technical potential are "barriers" to the mitigation of GHG via technology diffusion. Since, however, the ultimate goal is to understand policy options for mitigation, it is useful to group these barriers in a way that facilitates understanding the kinds of policies that would be necessary to overcome them. As these different categories of barriers are created, there is a corresponding creation of intermediate conceptions of the potential for GHG mitigation. Starting at the bottom, it is possible to imagine addressing barriers (often referred to as "market failures") that relate to markets, public policies and other institutions that inhibit the diffusion of technologies that are (or are projected to be) cost-

effective for users without reference to any GHG benefits they may generate. Amelioration of this class of market and institutional imperfections would increase GHG mitigation towards the level that is labelled as the "economic potential". The economic potential represents the level of GHG mitigation that could be achieved if all technologies that are cost-effective from consumers' point of view were implemented. Because economic potential is evaluated from the consumer's point of view, cost-effectiveness would be evaluated using market prices and the private rate of time discounting, and also take into account consumers' preferences regarding the acceptability of the technologies' performance characteristics.[3]

Of course, elimination of all of these market and institutional barriers would not produce technology diffusion at the level of the technical potential. The remaining barriers, which define the gap between economic potential and technical potential, are usefully placed in two groups separated by a socioeconomic potential. The first group consists of barriers derived from people's preferences and other social and cultural barriers to the diffusion of new technology. That is, even if market and institutional barriers are removed, some GHG-mitigating technologies may not be widely used simply because people do not like them, are too poor to afford them, or because existing social and cultural forces operate against their acceptance. If, in addition to overcoming market and institutional barriers, this second group of barriers could be overcome, the "socioeconomic potential" would be achieved. Thus, the socioeconomic potential represents the level of GHG mitigation that would be achieved if all technologies that are cost effective (on the basis of a social rather than a private rate of discount) are implemented, without regard to existing concerns about their performance characteristics, and without regard to social and cultural obstacles to their use.

Finally, even if all market, institutional, social, and cultural barriers were removed, some technologies might not be widely used simply because they are too expensive. That is, the definition of socioeconomic potential includes the requirement that technologies be cost-effective. Elimination of this requirement would therefore allow a progression to the level of "technical potential", the maximum technologically feasible extent of GHG mitigation through technology diffusion.

An issue arises as to how to treat the relative environmental costs of different technologies within this framework. Because the purpose of the exercise is ultimately to identify opportunities for global climate change policies, the technology poten-

---

[3] The identification of "economic potential" with implementation of technologies that are cost-effective from the consumer's point of view adopts, in effect, the economist's view that economic potential corresponds to the elimination of market failures. Other analysts have used the phrase "economic potential" to incorporate a broader conception, similar to what is dubbed "socioeconomic potential" in this report (Jaffe and Stavins, 1994).

tials are defined without regard to GHG impacts. Costs and benefits associated with other environmental impacts would be part of the cost-effectiveness calculation underlying economic potential only insofar as existing environmental regulations or policies internalize these effects and thereby impose them on consumers. Broader impacts might be ignored by consumers, and hence not enter into the determination of economic potential, but they would be incorporated into a social cost-effectiveness calculation. Thus, to the extent that other environmental benefits make certain technologies socially cost-effective, even if they are not cost-effective from a consumer's point of view, the GHG benefits of diffusion of such technologies would be incorporated in the socioeconomic potential.

The technical potential can be illustrated with reference to the fuel cell as a power source for private vehicles. Current fuel cell technology, making use of hydrogen manufactured from natural gas, can offer GHG emission reductions of around 50%-60% relative to conventional vehicles. This gives some indication of the current technical potential for mitigation. It is imaginable that in the future, fuel cell vehicles using hydrogen or other fuels from non-fossil sources would have even lower GHG emissions, on a full fuel cycle basis (Michaelis, 1997c). Thus, the technical potential of fuel cells for GHG mitigation is significant, and is expected to improve over time, as shown in *Figure 5.1*, through scientific discovery and technological development. However, the Energy Technology Support Unit (ETSU, 1994) notes numerous challenges that would have to be overcome before such vehicles could enter widespread use and offer more substantial emission reductions. In other words, the current market potential is very small at best. The large gap between the market and technical potentials (at the present time) can be understood in terms of specific barriers. Some of these relate to technology performance and cost, while others have to do with fitting non-fossil fuels into the existing infrastructure. The need to improve the cost and performance of the technology would represent barriers separating the technical and socioeconomic potentials. To the extent that the diffusion of cost-effective fuel cells is or will be limited by rigidities in the existing infrastructure, these could be considered barriers separating the economic and socioeconomic potentials for this technology.

The economic potential can be similarly illustrated, for example, with reference to energy conservation opportunities in buildings. Engineering-based analysis in the United States and other countries indicates that measures such as replacing tungsten filament bulbs with compact fluorescent lamps (CFLs), insulating hot water tanks, and introducing more energy-efficient refrigerators, could reduce residential electricity by about 40% and deliver a net saving to consumers (IPCC, 1996). To the extent that achievement of these savings is limited by market and institutional imperfections (such as imperfect information or misplaced incentives), the savings they offer represent the economic potential of these technologies. But even if all of these imperfections were corrected, these technologies would not be used in all possible applications. Some people will not

use them because they find them inferior on aesthetic or performance grounds. Other potential users will judge that the high private discount rates they believe are appropriate to this kind of investment render the savings too small to justify the high up-front cost. If, in addition to overcoming market and institutional imperfections, these aspects of consumer preferences were ignored, the socioeconomic potential could then be identified. Finally, even this level of GHG mitigation is smaller than the technical potential, as illustrated in *Figure 5.1*, because many technologies that are available, such as rooftop solar photovoltaic electricity supplies, would not pay for themselves in energy savings even at the social discount rate.

Table 5.1 begins with the baseline level of GHG mitigation that could be achieved without policy intervention (market potential), and then examines in more detail the nature of the barriers and opportunities that are encountered as greater mitigation is pursued, i.e., move towards the technical potential in *Figure 5.1*. Identification of the nature of the barriers and opportunities that separate each of the levels is necessary in order to formulate policy responses to overcome the barriers. The barriers to the achievement of economic potential are market and other institutional failures in the markets for technology, and government policies that distort these markets. These include market failures related to information and capital markets, subsidies for energy use. and trade barriers that inhibit the import of energy-efficient technologies. In principle, policies can be designed to address each of these market or government failures.

Identification of the opportunities to achieve economic potential is important, because removal of these barriers in a cost-effective way would be desirable even if global climate change (GCC) were not a policy concern. That is, if policies can be devised to overcome market and institutional barriers to the use of cost-effective technologies with desirable performance characteristics, consumers would be better off even before any consideration of GCC benefits. The barriers to the achievement of socioeconomic potential include social and cultural constraints, as well as economic forces that cannot be characterized as imperfections of markets or of other institutions. Policies to mitigate the market and institutional imperfections separating market and economic potential constitute "no regrets" policies, i.e., policies that societies would not regret implementing no matter what is learned later about the severity of the GCC problem.

The barriers to the achievement of socioeconomic potential include social and cultural constraints, as well as economic forces that cannot be characterized as imperfections of markets or of other institutions. Other barriers to socioeconomic potential relate to consumer preferences, including attitudes towards uncertainty. Uncertainty about whether estimates of new technologies and cost savings will actually come to pass limits the adoption of new technologies; such hesitation in the face of uncertainty is completely rational given the irreversible nature of many energy-conservation investments (Hassett and

Table 5.1: *Taxonomy of barriers and opportunities*

Source of barrier and/or opportunity	Examples of market and/or institutional imperfections and opportunities[a]	Examples of social & cultural barriers and opportunities
Prices	Missing markets (market creation) Distorted prices (rationalization of prices)	
Financing	Financial market imperfections (sector reform or restructuring of economy) Constraints of official development assistance (ODA) (removing tied aid and/or better targeting of ODA)	Long time and high transaction costs for small projects (pooling of projects)
Trade and environment	Tariffs on imported equipment and restrictive regulations (rationalization of customs tariffs)	
Market structure and functioning		Circumstances requiring rapid payback (fuel subsides) Weaknesses of suppliers in market research (form associations to support market research)
Institutional frameworks	Transactions costs Inadequate property rights (improve land tenure) Misplaced incentives Distorted incentives	Institutional structure and design (restructuring of firms) National policy styles (shifting balance of authority) Lack of effective regulatory agencies (informal regulation)
Information provision	Public goods nature of information (increase public associations) Adoption externality (build demonstration projects)	
Social, cultural, and behavioural norms and aspirations		Inadequate consideration of human motivations and goals in climate mitigation (modify social behaviour) Individual habits (targeted advertising)

a: Remarks in parenthesis indicate opportunities, e.g., missing markets denote an opportunity for the creation of markets.

Metcalf, 1993, 1994). Even putting aside the effects of uncertainty, private decision makers may utilize discount rates to assess the value of future energy savings that are significantly higher than the discount rates applied in the engineering-economic calculations to indicate that particular technologies are cost-effective. Such higher discount rates make the energy savings less valuable and, hence, may lead to a conclusion that the technologies are not cost-effective for a particular user.

Socioeconomic potential also recognizes that the economic feasibility of particular technologies is constrained by social structures and cultural forces; it is possible to consider changing those structures because of GCC objectives. For example, if the land-use and transportation systems of the USA could be radically transformed, the potential for improvement of energy efficiency in the transportation sector would be much greater than anything that could be achieved taking those structures as given

(see Section 5.4 below). Hence, part of the gap between the economic and socioeconomic potential represents the savings that could result from changes in the structure of such systems.

The last set of barriers to achieving technical potential relate to the cost and performance of the technologies. These can be improved upon by solving scientific and technological problems, so policies to overcome this category of barriers could be aimed at fostering the research and development (R&D) process, either in the public or private sectors. In addition, because production costs typically fall as experience with a particular technology accumulates, policies that foster adoption of new technologies can, over time, produce cost reductions and performance improvements. The effect of such improvements would be to make the technologies more cost-effective and consumer-favoured, thus moving both the economic and socioeconomic potentials towards the level of the technical potential.

Figure 5.1 provides illustrative examples of the barriers that separate one potential from another. Actions to overcome these barriers need not necessarily take place in the order of the potentials. R&D could take place to approach the technical potential at the same time that institutional and subsidy reforms are being carried out to approach the socioeconomic and economic potentials respectively. While the figure denotes a hierarchy in terms of the potentials, there is no hierarchy in the interventions that might be pursued to overcome the barriers. Furthermore, an intervention may overcome more than one barrier that need not be in a hierarchical order either, e.g., the provision of information could address all categories of barriers.

Because some interventions may be more effective than others, the gaps between the various potentials are likely to be reduced to varying degrees as well. Thus, the gap between the socioeconomic and economic potential may completely disappear, and yet that between the economic and market potential may remain in place. This indicates that while the market potential has moved up, it still could be improved by removing what economists refer to as market failures.

5.3 Sources of Barriers and Opportunities

Barriers to climate change mitigation are inherent to the process of development. Sustainable development in a participatory framework can minimize these barriers, but the inequitable distribution of income and wealth forms a core feature of barriers to effective implementation of any type of intervention, and those related to climate change are no exception. The poor in any society bear a disproportionate burden of the impact of externalities. Climate change affects them more, because they often lack the infrastructure to withstand its impacts. The poor also pay more as a proportion of their income for energy services, and often tend to use traditional fuels secured outside the formal market system. They are not able to access subsidized fuels for instance, because they do not have the collateral to access these fuels and the equipment to use them. Appropriate ways of financing would be one way to overcome such barriers, provided they explicitly account for the non-existence of markets for some segments of society. The issue of segmentation is valid for firms as well. Small and medium-sized firms for instance face information and market-structure barriers that well-structured large firms can readily overcome with the resources at their disposal.

Lifestyles, behaviour, and consumption patterns all evolve as societies develop within their own socio-cultural contexts. With the advent of global communications these factors are being increasingly influenced by changes that are taking place in societies residing thousands of miles away. The communication channels may be viewed as an opportunity to influence the manner in which tomorrow's society might develop in countries where modern but resource-consumptive technologies and lifestyles have not taken root. Progress in achieving climate change mitigation will depend on how well the seeds of mitigative technological change can be planted and nurtured.

As a prelude to the more detailed sectoral discussion in Section 5.4, this section provides a general overview of the process of technological innovation, and the different sources of barriers to the diffusion of new technology and practices, as well as the policy opportunities that they represent. This section is organized by the following categories: prices, financing, trade and environment, market structure, institutional frameworks, information provision; and social, cultural and behavioural norms and aspirations. Within each of these areas, some of the barriers represent failures or imperfections in markets, policies, or other institutions that lie between the status-quo of the market potential and the possible achievement of the economic potential. Other barriers are aspects of institutions or social and cultural systems that economists may not characterize as market imperfections, but which nonetheless limit diffusion of GHG-efficient technology. These latter barriers separate the economic and socioeconomic potentials. Within each of the subsections below barriers and opportunities in both categories are discussed.

5.3.1 *Technological Innovation*

Many governments and firms have focused their strategies for GHG mitigation on encouraging technological innovation – various processes of research, experimentation, learning, and technology development. Innovation may lead to improvements in technology performance, reductions in GHG emissions per unit of service provided, or reductions in cost for low-GHG technology, all of which can contribute to GHG mitigation. Innovation can help to raise the technological, socio-political, economic, and market potentials for adoption of low-GHG technology, and for GHG mitigation. Identifying the barriers to, and opportunities for, technological innovation depends on understanding the innovation process. Since the IPCC SAR, there has been a rapid growth of interest in the theory of innovation, and in the development and application of models to evaluate climate mitigation policies that take account of endogenous technological change (Azar, 1996; Goulder and Mathai, 2000).

5.3.1.1 *The Innovation Process*

Until the 1980s, policy analysts generally viewed innovation as a linear process from R&D through to demonstration and deployment. Policies were focused on "science push" and "demand pull" for new technologies (OECD, 1992). Over the last twenty years there has been a growing recognition of the interconnectedness of the many processes involved in technological change, and the possibility of finding new insights or knowledge anywhere from the research lab to the customer service department.

Technological change can take many different forms including: (1) incremental improvements in existing technology; (2) radical innovation to introduce completely new technology; (3) changes in a system of linked technologies, and (4) changes in the "techno-economic paradigm" involving widespread reorganization of production and consumption patterns (Freeman and Perez, 1988). These four types of innovation have different dynamics. Thus, the first type is likely to occur continually through the accumulation of experience, selection of successful techniques and adaptation to a changing economic, legislative and socio-cultural context. The second and third types of technological change involve more positive creativity, being linked to new information in the form of a discovery, idea, or invention; or to a creative application of an existing invention. The fourth type, again, involves creativity but, because it involves a radical change in culture and markets, may also depend on these being "ripe" for change – on a general perception of a major challenge requiring a radical response.

Technology diffusion, the spread of existing technology through the population of potential users, can be distinguished from innovation – the first commercial application of a new technology. At a local level, however, there may be little difference between the two. Wallace (1995) notes the importance of an active and creative absorption process in the uptake of the new technology.

Technological change is a complex process. It occurs through a variety of interdependent mechanisms (Nelson *et al.*, 1967; Rosenberg, 1982; Dosi, 1988; OECD, 1992; Rosenberg, 1994; Lane and Maxfield, 1995), which can include:
- assessment of needs and potential markets;
- basic research: a search for new information;
- creative generation of new ideas;
- learning from experience;
- exchange of new information, ideas, and experience through the scientific and technical literature, patents, and a variety of other communication channels and networks including face-to-face contact and collaboration;
- experimentation to implement and test the new information and ideas;
- development of new technology;
- demonstration and market testing of new technology; and
- selection of successful technology, under the influence of the economic, social, legal, and physical context.

Because of the complexity of the technological innovation process, there are many different ways of looking at it. A variety of theories or models may be helpful, depending partly on specific circumstances.

From the perspective of neoclassical economics, innovation can be seen as the result of a process of investment in "knowledge capital", in the form of R&D to develop both formal and tacit knowledge (Griliches, 1979). The former includes the scientific literature and patents; the latter includes the skills and experience developed by those involved in developing new technology and can also be viewed as "human capital". Increasing capital, again, tends to feed into higher levels of economic output and improved efficiency. Sometimes this may contribute to GHG mitigation, but more often the improvement is in labour productivity, leading to increases in GHG emissions. In so-called "new growth" theory economic models (e.g., Grossman and Helpman, 1991, 1993), new knowledge may be assumed to result directly from R&D spending which, in turn, can be modelled as a result of the expected returns from the investment. In this framework, firms and research institutes are treated as rational investors in R&D. The size of their investment will depend on the opportunity cost of capital and the expected return from R&D. While new growth theory has generated useful insights into the sources of national differences in competitiveness at an aggregate or sectoral level, it is less useful for describing technology innovation for GHG mitigation.

In addition to R&D investment, knowledge capital can also be accumulated through the process of "learning by doing" (Arthur, 1994; Grubb, 2000). Empirical studies show that the cost of a generic technology such as solar photovoltaic cells tends to fall with the level of existing investment in that technology, including spending on R&D (Christiansson, 1995; Messner, 1996; Nakicenovic, 1996).

An alternative to the neoclassical investment approach to innovation is that pioneered by Nelson and Winter (1982), to view technological change from the perspective of the firm, as a stochastic process of search, imitation, experimentation, and learning (Winter *et al.*, 2000). Recent developments in agent-based modelling adopt this type of "evolutionary" framework, helping to bring out the role of information networks, the importance of existing experience, and also some of the spatial aspects of technology development and diffusion.

Finally, several analysts have adopted models of technology competition and diffusion analogous to those used to represent species competition and diffusion in ecosystems. Regularities have been found, for example, in the market succession of technology in energy supply, transport, and the iron and steel industry (Häfele *et al.*, 1982; Grübler and Nakicenovic, 1991; Nakicenovic, 1996). However, no approach can hope to foresee reliably the form of the next "wave" of technology in any of these sectors.

5.3.1.2 Barriers and Opportunities for GHG Mitigation through Technological Change

Barriers to GHG mitigation and opportunities for overcoming them arise throughout the innovation system. They relate both to the rate of technological change and its direction. The predominant concern of governments, firms, and researchers considering innovation policies has been to maximize the rate of technological change and its contribution to national competitiveness (e.g., Freeman, 1987; Dosi *et al.*, 1988; Grossman and

Helpman, 1991). Environmental concerns are usually recognized but are rarely a major priority for national systems for innovation. Indeed, there may even be a concern that paying more attention to innovation strategies about environmental objectives would be detrimental to competitiveness.

There may be many opportunities to find synergies between the goals of improving competitiveness and reducing GHG emissions. The most obvious of these opportunities are cases where GHG mitigation could reduce costs. A greater challenge for businesses and governments is to seize opportunities to create new markets for low-GHG-emitting technology. One case of a successful strategy is the Danish development of wind turbine technology (Kemp, 2000).

Communication – among firms, between firms and users, and between firms and universities or government labs – is an important contributor to technological change. Most innovations require some social or behavioural change on the part of technology users (Rosenberg, 1994). Product innovations, if they are noticeable by the user, demand a change in consumer behaviour and sometimes in consumer preferences (OECD, 1998a). Some product innovations – such as those that result in faster computers or more powerful cars – provide consumers with more of what they already want. Nevertheless, successful marketing may depend on consumer acceptance of the new technology. Other innovations, such as alternative fuel vehicles or compact fluorescent lights, depend on consumers accepting different performance characteristics or even redefining their preferences. While consumer preferences are often seen as barriers to technological change, some of the most successful firms are those that seize the opportunities they present, by working with their customers in the development of new technology and services (Lane and Maxfield, 1995).

One of the most obvious barriers to using innovation to address GHG emissions is the lack of incentives. Economic, regulatory, and social incentives for reducing GHG emissions will also act as incentives for innovation to find new means of mitigation. Another important type of barrier, which both slows technological change in general and tends to skew it in particular directions, is that posed by "lock-in" (see *Box 5.1*). The tendency for societies to lock in to particular clusters of technologies and patterns of development can prevent new, low-GHG emission technologies entering the market. Meanwhile, it is important to recognize when previously locked-in technology is beginning to change, so that the opportunity can be grasped to introduce low-emission technology.

Box 5.1. Lock-In

Schumpeter (1928) emphasized the effectiveness of the capitalist system in encouraging experiments and in selecting successes. This effectiveness can be ascribed partly to the capitalist's ability to invest in risky endeavours, trading off uncertainty against the size of the anticipated return. The competitive market system also introduces the element of "creative destruction" to the innovation process, analogous to natural selection, ensuring that an innovation that does not meet the needs of the market does not survive. Yet, despite their ability to select adequate technologies, markets sometimes "lock-in" to technologies and practices that are suboptimal because of increasing returns to scale, which block out any alternatives (Arthur, 1988, 1994). The QWERTY English keyboard layout is often mentioned as an example of an inefficient technology designed to solve a specific problem (to avoid keys sticking in mechanical typewriters) but which has become "locked in" (David, 1985). It has been claimed that alternative keyboard designs could double typing speeds, but these are not adopted because of the retraining costs that would be necessary for any change. Lock-in phenomena are familiar in the energy sector, with technologies and design standards in applications ranging from power stations to light bulbs and urban design to vehicles.

In many cases, a given technology helps to satisfy several different types of need. This is particularly evident in two of the most significant areas of energy use: cars and houses. Any individual may have a variety of potentially conflicting objectives when choosing a technology. This tendency of successful technologies to serve multiple needs contributes to lock-in by making it harder for competing innovations to replace them fully. Hence, many government attempts to introduce new, energy efficient or alternative fuel technology, especially in the case of the car, have failed because of a failure to meet all the needs satisfied by the incumbent technology. If alternative fuel vehicles have difficulty entering a market dominated by gasoline cars, alternatives to the car face even greater barriers. Owners have learned to associate their cars not only with personal mobility, but also with freedom, flexibility, fun, status, safety, a personal territory, and perhaps most powerful of all, a means of self-expression. Different owners may place emphasis on different needs. To succeed without some form of enforcement, any replacement must satisfy at least several of these needs better than the existing technology.

When a radical innovation does occur in a technology of fundamental importance, it may trigger an avalanche as a complex web of technologies and institutions require redevelopment (Schumpeter, 1935; Freeman and Perez, 1988). Such a shift may now be occurring with the spread of mobile information, communication, and networking technologies. Achieving substantial GHG mitigation may depend on recognizing when such transformations are occurring, and taking advantage of them.

5.3.1.3   The Context for Technological Change

The wider context plays an important role in shaping technological change and hence in determining the feasibility of GHG
mitigation. There are several important elements or dimensions
of the context for technological change:

- market conditions, including ease of entry for new
  firms and technologies; availability of capital; the
  degree of internalization of social and environmental
  concerns through taxes, subsidies, insurance, and other
  mechanisms; and the degree of competitiveness,
  including any oligopolistic practices or informal
  arrangements between government and the private sector;
- the legal system, including the system of intellectual
  property rights; the allocation (e.g., among firms or
  between the public and private sector) of liability for
  past and future environmental damage; freedom of
  speech and information; and ease of litigation;
- the physical infrastructure, including the design of
  cities and other settlements, transport systems, and utilities; and their flexibility in permitting the adoption of
  alternative technologies, lifestyles, and production systems;
- social and political structures, including the role of the
  public in decision-making; the location of power in
  institutional and social relationships; the presence of
  formal or informal alliances, for example involving
  government, industry, and the media; and the allocations of roles within households and communities;
- culture, including cultural diversity; the role of technology and material consumption in establishing individual identity, status, and social bonds; tendencies
  towards competition and co-operation, conformity, and
  distinction; and
- psychology, including awareness, understanding, and
  attitudes relating to climate change, its causes and
  potential impacts, and to changes in technology and
  lifestyles.

Of these dimensions, most attention has been paid in the literature, including the SAR, to the role of markets and legal systems. Existing market and legal incentives can pose barriers to
some kinds of technological change, as discussed in later sections of this chapter. Changes in the market and legislative context can also provide opportunities for innovation. For example, the need to address local pollution through government
regulations may stimulate innovation that can contribute to
GHG mitigation. Porter and Van der Linde (1995a) argued that
environmental regulation of industries could also promote their
competitiveness through accelerated innovation, although this
has been disputed by Palmer *et al.* (1995), who argues that
most evidence is that regulation, as historically practised, has
not fostered competitiveness, and has encouraged innovation
only narrowly aimed at regulatory compliance (Berman and
Bui, 1998; Xepapadeas and de Zeeuw, 1999).

The effects of physical infrastructure have been less studied,
being harder to measure than those of prices and regulations.
Infrastructure often acts as a constraint on changes in technology and behaviour: existing road systems and settlement patterns in many countries tend to encourage car dependency; the
existing supply networks for domestic and transport fuels make
it difficult for individual households or firms to adopt alternatives. In this chapter, the role of infrastructure is considered in
relation to buildings, transport, and energy supply (see
Sections 5.4.1 to 5.4.3).

The social capital passed on from generation to generation
offers an opportunity for diffusion of GHG mitigation technologies in traditional and modern societies alike. Societies in
which trust and civic co-operation are strong have significant
positive impact on productivity, especially human capital productivity, and provide stronger incentives to innovate and to
accumulate physical capital. More investment in consultation
and participation of the local population in decision making
about GHG mitigation technologies contribute both to information sharing, to building trust, and civic co-operation. The
former may contribute to changes in beliefs, norms, and values
if participants are convinced that they are better off after effect-
ing the change (Gibson *et al.*, 1998).

Reliance on market mechanisms alone, without an appropriate
institutional framework that performs a co-ordinating function
among sectors, is inadequate and may be destructive of social
capital. Policy attention to learning by doing, and network
externalities, together with policy stability and enforcement
favour the diffusion of GHG mitigation technologies.

Addressing the last three dimensions listed above thus involves
understanding human psychology, relationships, communities,
institutions, and the process through which social norms and
decisions are established. These aspects of climate mitigation
are addressed in Sections 5.3.6 and 5.3.8 of this Chapter.

5.3.2   Prices

Prices can have an important influence on the consumption of
resources and hence on GHG emissions. There is extensive literature on the use of prices to reflect environmental and other
social costs associated with resource use. If such costs were
fully reflected in prices, they would encourage producers and
consumers to adopt environmentally sustainable technologies
and practices. Where an adequate legal framework exists, it
should be possible in principle for those suffering the effects of
pollution or climate change to seek compensation from those
responsible. In practice, markets in environmental and social
damages function poorly, if at all, because transaction costs
(e.g., the costs for victims to identify polluters and seek compensation) are high compared with the environmental and
social costs suffered.

Where environmental and social costs are not reflected in markets (i.e., they are externalities), there are many ways in which governments can internalize them, notably through environmental regulations and taxes. However, governments have to balance a large number of objectives and the outcome may not be efficient in linking resource prices to GHG emissions. A variety of different types of government policy tend to reduce prices, in addition to the direct budgetary subsidies that are often introduced to support employment in particular sectors or to enable the poor to meet basic energy needs (OECD, 1997b). Examples include policies requiring electric utilities to provide universal, low-priced access to grid systems or even to maintain supplies when consumers fail to pay their bills (EBRD, 1999; World Bank, 1999). In India, electricity has historically been subsidized for residential consumers, serving as a disincentive for the adoption of efficient lighting and appliances (Alam *et al.*, 1998). When energy subsidies are reformed or removed, transitional or permanent supports are often required for some of the former recipients (OECD, 1997b). For example, in Russia, the introduction of long-run marginal cost electricity pricing has led to pensioners being unable to afford their electricity bills, requiring support that amounts to 20%-35% of local authority budgets (Gritsevich, 2000).

Government policies to address a wide range of environmental and social problems can encourage GHG mitigation by increasing the prices of carbon-intensive energy sources or decreasing the prices of non-carbon options. Such policies include pollution taxes and charges for the use of infrastructure and services, subsidies for renewable energy, and regulations requiring producers to sell electricity generated from low-carbon sources.

The developers of new technologies often seek to recover their investment in R&D through license fees for the use of their innovations. Such license fees may inhibit the adoption of the best available technology for GHG mitigation in developing countries.

Energy price expectations can have a strong influence on investments in low-GHG technology. Where energy prices fluctuate in unpredictable ways, investors may tend to delay investments in new technology, and be unwilling to adopt low-emission technology where this entails increased up-front costs. The next section discusses the effects of risk on investment.

A substantial literature has developed on the tendency of consumers and businesses to pay more attention to initial investments than operating costs, when considering technology choices (Hassett and Metcalf, 1995; Jaffe and Stavins, 1995). In the past, prices for some types of appliance, such as refrigerators, have tended to show little correlation with energy intensity within a given range of size and performance characteristics (Greening *et al.*, 1997). The prices of appliances and vehicles are influenced by many factors, not least their aesthetic features, and energy efficiency is usually a minor source

of variation. On the other hand, several governments have used taxation to introduce a price incentive for buying cars with smaller engines, lower fuel consumption, and to encourage the use of alternative fuel vehicles (IPCC, 1996; ECMT, 1997).

5.3.3   *Financing*

Many environmentally beneficial technologies require significant "up-front" investment. This investment will be typically offset, over time, by the environmental benefits, out-of-pocket cost savings, or financial revenues associated with the new technology. There are, however, many circumstances where users are unable to purchase equipment that is financially viable to them or beneficial to the society, simply because they do not have access to the private or government investment funds necessary to install the equipment. To the extent that private entities are not willing to provide funds to implement investments that are financially viable and in addition reduce GHG emissions, they constitute failures of capital and financial markets that must be overcome to reach the level of economic potential. In contrast to private financiers, who are primarily concerned about the risk-adjusted financial return, governments are expected to evaluate desirability of investments in a wider context of the well-being of the whole society, including harms and benefits that some entities impose on others. To the extent that governments are not willing to finance investments that are socially desirable thanks to climate and other environmental benefits, they constitute policy failures that prevents achievement of socioeconomic potential. All these market and policy failures are aggravated in developing countries and low income transition economies, where they interact with poverty and capital constraints.

Commercial Banks
Notwithstanding the significant potential as a supplier of investment capital for climate-friendly technology transfer, commercial banks thus far have not developed large portfolios of environmental loans (Delphi Int. Ltd. and Ecologic GMBH, 1997). Banks face high up-front cost of developing new, "green" financial products (e.g., energy-efficiency loans). To bear these costs is often perceived risky by the bankers, given uncertain and policy-dependent future market conditions. Relatively low capital requirements and the long-term cash-flow profile of many climate friendly investments, as well as high transaction costs of servicing large numbers of small and medium-sized projects, further reduce comparative attractiveness of this sector to the commercial banks (Berry, 1995). Technologies such as energy efficiency or public transport often have low collateral value compared to their traditional alternatives, making it difficult for the banks to use some financing instruments such as project finance.

Even if the size of the loan for manufacturing or distributing climate friendly technologies would justify the attention of bankers, the debt carrying capacity of such projects hinges upon the availability of financing for the end users, e.g., households to enable them to purchase those technologies. These down-stream projects most often require completely different financial products, which commercial banks are often not able to offer (e.g., micro-credits or grants to low income households with no assets).

Different energy producers and consumers have varying access to capital in financial markets, and at different rates of interest. In general, energy suppliers can obtain capital at lower interest rates than can energy consumers – thus, an "interest rate gap". Differences in these borrowing rates may reflect differences in the knowledge base of lenders about the likely performance of investments, as well as the financial risk of the potential borrower. At one extreme, electric and gas utilities are able to borrow money at low interest rates. At the other extreme, low-income households may have essentially no ability to borrow funds, resulting in an essentially infinite discount rate for valuing improvements in energy efficiency. The broader market for energy efficiency (including residential, commercial, and industrial consumers) faces interest rates available for efficiency purchases that are also much higher than the utility cost of capital (Hauseman, 1979; Ruderman *et al.*, 1987; Ross, 1990).

"Green" Financial Institutions
In response to the difficulties faced by the emerging environmental business sector in accessing traditional financing institutions, such as banks (Asad, 1997), a number of innovative approaches and specialized financial institutions have developed. These include environmental project finance (Stewart, 1993; Shaughnessy, 1995; Davis, 1996), green investment funds, leasing (Carter, 1996), environmental and ethical banks, environmental funds (OECD, 1999b), and energy service companies (ESCOs). Not clearly defined property rights to GHG emitting assets create obstacles to ESCOs and other similar institutions, that invest in the assets of third parties and rely on a contracts with owners to recuperate the return (WB and IFC, 1996). The growth of new "green" financial institutions hinges upon the long-term market growth prospects for the environmental business sector, which in turn depends fundamentally on the consistent and clear commitment by governments to climate policies (Delphi Int. Ltd. and Ecologic GMBH, 1997). Specific incentives, such as tax allowances, have been shown to stimulate the market penetration by green investment funds in some developed countries (e.g., The Netherlands).

In the last years of the decade sustainable forestry has started to attract private finance. Some new green financial institutions have worked towards capturing values of standing forests through innovative financial mechanisms. Sustainable forestry has provided attractive returns relative to stock markets. Forestry investment funds have typically achieved annualized returns in excess of 14% over the last decade. This was in excess of the returns on the S&P 500 index for the equivalent period (Ecosecurities, 1999). Forestry investments had lower volatility than stock markets, and could provide solid long-term returns. However, to the extent that these involve wood plantation where logging is an important part, the climate benefits are negligible. Managing forests and harvesting their products and services efficiently significantly improves financial return to the standing forests versus logging. The marketable goods and services of forests include pharmaceuticals (Simpson *et al.*, 1996), genetic resources (Rosenthal, 1997), and ecotourism (Panayotou, 1997). An important factor stimulating financial viability of sustainable forestry is the move of government, world business, and consumer demand towards confining wood procurement to environmentally sustainable sources.

Investors
Individual and institutional investors send important signals to companies in the pricing of new capital raised by the companies and in on-going valuation of quoted companies. They can also exert direct influence by using their rights as shareholders and owners. The key concern for investors is the relationship between environmental performance and investment performance. Many investors remain unconvinced that the present value of their portfolios may be affected by the future consequences of climate change. They also are not convinced that environmental performance contributes to good financial performance.

There is some empirical evidence, however, that investors do value environmental performance of firms. Dasgupta *et al.* (1998) showed that capital markets in Argentina, Chile, Mexico, and the Philippines reacted positively (increasing the firms' market value) to the announcement of rewards and explicit recognition of superior environmental performance. They found capital markets to react negatively (decreasing the firms' value) to citizens' complaints and to news of adverse environmental incidents (such as spills or violations of permits). Environmental regulators could harness market forces by introducing structured programmes to release firm-specific information about environmental performance, and empower communities and stakeholders through environmental education programmes. Lanoie *et al.* (1997) arrived at similar conclusions, drawing on evidence from American and Canadian studies.

Insurance Firms
The potential of the insurance sector lies in its ability to diversify its investment portfolio and to have its premium structure reflect environmental risks (Delphi Int. Ltd and Ecologic GMBH, 1997). The insurance industry may provide project finance and insurance for preventive infrastructure projects, thereby enhancing their access to finance. The insurance industry also provides strong financial incentives for loss prevention and mitigation to their clients and the public, e.g., by means of deductibles (UNEP, 1999). Some insurance companies have launched the "Insurance Industry Initiative for the Environment", in association with UNEP.

User Charges
Generation of revenues from the users of public infrastructure can be an important source of funds for financing GHG emissions reduction in the power and district heating sector and

other types of GHG emission-intensive infrastructure. Covering the costs of operation, maintenance, depreciation calculated according to the international accounting standards, and eventually debt service for investments is essential for the sustainability of infrastructure systems and important for attracting multilateral development banks (MDBs) and private finance (UNIDO, 1996; EBRD, 1999). In low-income countries this needs to take full account of affordability constraints. However, concern about the social impacts too often makes the governments reluctant to adopt higher tariff levels, even though evidence suggests consumers in many countries could afford and would be willing to pay more for improved service (Lovei, 1995; Gentry, 1997; AFDB, 1999).

Government-created Disincentives to Private Investment
Government policies may themselves be a source of risk to private investments, creating detrimental framework conditions for all, not only environmental, investments through unstable fiscal policy and a macroeconomic environment. This leads to high interest rates, elevated inflationary expectations, and fluctuating exchange rates. The traditional response to these problems through fiscal consolidation and tight monetary policies usually induces low liquidity in the enterprise and banking sector (EBRD, 1999). This liquidity constraint may be sharpened by obstacles to trade and bank credit, barriers to entry, especially for SMEs and foreign firms, barriers to foreign direct investments (FDIs) and to long-term foreign capital investments, all of which could otherwise relieve capital shortages (EBRD, 1997b; EBRD, 1998; EBRD, 1999). Weak governance, typically manifested by the lack of the rule of law, soft budget constraints, absence of competition in government procurement, and corruption, may foster a perverse microeconomic incentive structure that rewards private sector entities not for being competitive and efficient in using resources, but rather for "seeking rents" through friendly and not transparent relations with politicians (Gady and Ickes, 1998).

Governments sometimes introduce distortions directly to financial markets, constraining the private lending to investments. Imprudent government borrowing can raise interest rates and crowd out bank loans from the "real" sector of the economy (OECD, 1998b). Also, excessive subsidies to environmental investments may crowd out private sector financing (Peszko and Zylicz, 1998). The risk of lending for investments may additionally be increased by inadequate protection of creditors. This occurs when an underdeveloped legal and institutional system does not make it easy for creditors to seize collateral or initiate a turnover of management in the event of default.

Government-created Disincentives to Public Investments
Ill-designed taxation, as well as failures in budget planning and expenditure control may cause fiscal imbalances and high budget deficits, which contribute to high country sovereign risk, constrained access to foreign capital, and high cost of borrowing by the government. Increased nominal interest rates and related discount rates applied by the governments inhibit financing for most public environmental investments. Budget expenditure cuts usually involve ceilings for investment expenditures, while financing is made available for operation of existing technologies or infrastructure. This often leads to continuing operation of inefficient and polluting assets, even if their replacement through investment would bring a high rate of return.

A barrier to efficient use of government funds is poor management of public investment programmes and government budgets (OECD, 1998b). This is sometimes a result of an underdeveloped civil society, and absence of government accountability and transparency in budget preparation and implementation. Under these circumstances budgetary spending on environmental infrastructure and biodiversity tends to be neglected (OECD, 1999a; Partridge, 1996). An important opportunity to enhance government spending on climate friendly investments is through revising public sector expenditure choices (de Moor, 1997; Pieters, 1997). Many developing countries and the countries of the former Soviet Union could help both climate and economic development by phasing out ongoing subsidies to loss-making state owned, or even private enterprises.

Central and local governments have ample opportunities to create new mechanisms and new sources of finance for climate-related environmental investment (Tlaie and Biller, 1994; Pearce *et al.*, 1997). Budgetary resources can be used more cost-effectively (Lovei, 1995) and more creatively (Clements *et al.*, 1995) to leverage private capitalization of public environmental investments (World Bank, 1994; Partridge, 1996; UNIDO, 1996; Gentry, 1997; Peszko and Zylicz, 1998). Central governments can foster the use of economic instruments (tariffs, taxes, fees, etc.) to achieve environmental goals while generating budgetary revenues (Herber, 1997; Schlegelmilch, 1999). In the area of biodiversity pricing, instruments can result in a "double dividend". They can prevent the "tragedy of the commons" by limiting otherwise open access to vulnerable natural reserves. Prices also generate revenue to pay for the sustainable use of biodiversity resources and for afforestation. Successful examples of these government initiatives could be found in Latin America (Umana, 1996; Lopez, 1997), OECD countries (OECD, 1996) and Central and Eastern Europe .

Official Development Assistance
There is a mixed experience with donor aid programmes (Killick, 1997). Choice of beneficiary countries, sectors, and types of projects by the donor governments has often been driven by the geopolitical interests of donors rather than environmental or global priorities in the recipient countries. Bilateral aid is often a tool to support friendly regimes or strengthen the spheres of influence (Alesina and Dollar, 1998). Tied aid still dominates bilateral programmes, whereby the contracts are available only to firms from the donor country (Michaelowa, 1996).

Because of restrained competition tied aid may increase the costs of purchasing capital or providing services anywhere from 10% to 50%, and host governments are usually required to co-finance these projects. Some host governments have found themselves locked in the expensive, capital intensive, and inappropriate technologies that additionally created dependency for maintenance and spare parts. Tied aid may distort the efficiency of technology choice, and crowd out good technologies and viable business models (Graham and Hanlon, 1997). Tied aid has also had an impact on GHG emission reduction projects in the context of the Activities Implemented Jointly (AIJ) pilot phase (Michaelowa *et al.*, 1998).

Multilateral Development Banks
Sovereign guarantees required with most MDB lending involve host governments in making budgetary commitments that may be difficult to attain in many low income countries. Furthermore, strict adherence to sound banking principles (of not lower standards than in the highest-rated private banks) poses very high requirements for the internal financial viability of projects. It is not, clear, however that the MDBs can do otherwise. They can provide low cost lending only as a consequence of their high credit ratings. Maintenance of these high ratings requires very low exposure to default risk, which in turn depends on sovereign guarantees and sound financial parameters of a project.

Another problem with MDB loans is a longer time for and higher transaction costs of project preparation relative to the typical GHG emissions reduction project size. It usually takes 1.5-2 years and several hundred thousand US dollars to develop a project for financing. This can only be justified if the size of a project is minimum US$10-15 million. MDBs are trying to develop financial products that could reach small and medium-sized environmental projects (ADB, 1999). Trust funds and donor grants are used to lower project preparation costs. Smaller businesses are targeted trough intermediaries (local banks, leasing, ESCOs, or even NGOs) which "on-lend" MDB loans as a package of smaller financial products. Structural lending is used to finance multi-project programmes.

Most of the financing difficulties discussed above are most severe in developing countries, where they interact with poverty to severely constrain investment in GHG-efficient technology. Less developed capital and financial markets call for innovative financing to enable low-income households to afford GHG-mitigating technologies. This offers an important opportunity to integrate the broader objectives of development, equity, and sustainability (DES).

5.3.4   Trade and Environment

The barriers discussed in this section pertain to the whole economy of a country, and constitute a type of market failure. They inhibit the implementation of mitigation options indirectly by maintaining conditions in which investments in energy efficiency and fuel switching are ignored, undervalued, or considered too risky by economic actors.

High tariffs on imported goods or policies that constrain entry of imported products into the market can prevent new and GHG-efficient technology from entering the country. Since countries often rely on imports for high-efficiency equipment, duties can raise the price of imported equipment considerably. When both types of equipment are imported, the duty raises the price differential between the two.

An example of the limitations created by government regulation was a high import duty imposed on CFLs in Pakistan. When this duty was reduced from 125% to 25% in 1990, the price of CFLs dropped by almost half, and sales started to rise, leading to improved energy efficiency (US AID, 1996).

Government regulations that prohibit foreign firms from bidding on the construction of new industrial factories or power plants limit a country's access to new foreign technology. Conditions that constrain the entry of imported products, while beneficial in establishing a new industry or in achieving rapid expansion of an existing one, can also lead to the use of obsolete technology. The history of government intervention to address a severe paper shortage in India during the early 1970s illustrates this barrier. To address the shortage, the Indian government promoted the establishment of small paper mills that could be quickly set up (Datt and Sundharam, 1998). This led to the import of inexpensive energy-intensive and highly-polluting second-hand paper mills that were set up in many regions of the country. The inefficient mills grew to account for 50% of the country's paper production. Then, in 1988, the government removed the protection it had accorded the paper industry, which led to the shutdown of many of these small, inefficient plants. The elimination of government protection will in the long run increase GHG efficiency and economic productivity.

The transfer of modern technology takes place mainly through licensing of designs for local production, joint ventures, and export and/or import. Practices of transnational corporations, and policies of countries can inhibit these modes of technology transfer. Also, large fluctuations in exchange rates and inflation can inhibit capital flows. The fuel economy of motor vehicles across developing countries varies with the type of technology that is imported. Countries either import new (high fuel economy) or used (mostly lower fuel economy) motor vehicles, manufacture vehicles with outmoded low fuel-economy technology (Ambassadors in India or the VW Bug in Mexico and Brazil, the VW Jetta in China), and/or manufacture modern vehicles with some domestic components (Nissan in the Philippines, Maruti/Suzuki in India) (Sathaye and Walsh, 1992). Lack of suitable local firms to supply components and services, limited access to capital, and restrictions on repatriation of foreign exchange are some of the conditions that slow the introduction of modern efficient vehicles (Davidson, 2000, Section 8, Transportation).

There is not much empirical evidence for a relationship between trade and environmental regulation (Cropper and Oates, 1992; Rauscher, 1999) though there is a little more in the direction of the impact of trade on the environment (van Beers *et al.*, 1997). This lack of empirical relationship is caused by two reasons. First, it is most cost effective to use the same technology everywhere and, therefore, to operate everywhere according to the most stringent environmental regulations (Levinson, 1994). Second, the industry cost of environmental regulation is too small relative to other costs, such as labour, to weigh heavily in location decisions (Dean, 1992; Jaffe *et al*. 1995; Markusen, 1999; Steininger, 1999). In particular, there is little empirical evidence that developing countries tend to become pollution havens. This is because their production is primarily for the domestic market, their comparative advantage lies in less-polluting labor-intensive sectors, and weak environmental standards often go hand in hand with other factors that deter investment such as social capital weaknesses (Frederikson, 1999; Markusen, 1999). There is no empirical evidence of systematic FDI in polluting industries (Leonard, 1988). The environmental effects of trade liberalization seem to be highly country- and policy-specific (Frederikson, 1999).

There is also little evidence, both on theoretical and empirical grounds, of a "race to the bottom" when other countries use environmental standards to retaliate against trade measures. A globally optimal solution remains a combination of free-trade and co-operative environmental policies. This does not mean that, as environmental resources become scarcer, free trade may not generate negative environmental impacts under some circumstances as suggested by theoretical models (Copeland and Taylor, 1994; 1995). Little is known both theoretically and empirically about the links among trade, environment, and innovation (Carraro, 1994, Steininger, 1999). There is also little evidence, both theoretical and empirical, in favour of the Porter Hypothesis that stronger environmental regulation creates a long-term technological advantage (Jaffe, 1995; Ulph, 1997). Regulatory capture through which interest groups striving for protection against foreign competition lobby against environmental standards and for environmental tariffs is a possible barrier to diffusion of technology. Capture is less likely under a market-based instrument approach to environmental policy, which regulates polluting substances than under a command-and-control one, which regulates polluters. This is because the former raises the cost of lobbying and decreases the agency problem as the regulated group is larger and more heterogeneous under the former regime than under the latter and has no incentive to hide information from the regulator (Rauscher, 1999). Many international environmental agreements allow for trade sanctions. Though, in a less than efficient world, trade sanctions for environmental violations can be justified, the latter are discriminatory and may jeopardize diffusion of required technologies (Rauscher, 1999).

5.3.5 Market Structure and Functioning

Market failures related to pricing, information and institutional imperfections are discussed elsewhere in this section. In this subsection a variety of other barriers and opportunities related to the behaviour of market actors and the features of specific markets are considered. The majority of these opportunities and barriers affect the demand for higher energy efficiency, but in many developing and transitioning countries there are also problems on the supply side of markets.

In considering opportunities and barriers related to market behaviour and features, it is important to recognize that consumers (broadly defined to include households, firms, and other actors) and producers and/or providers in specific markets are in continual communication. In general, suppliers deliver what they think consumers want. But in markets characterized by a high degree of inertia or aversion to risk on the part of suppliers, there may be latent demand for higher levels of energy efficiency than are readily available in the market. Suppliers may not expend the effort to cultivate the demand for more efficient products or to develop marketing approaches to help overcome some of the barriers on the demand side (such as financing schemes).

The importance of particular barriers varies among specific markets. On the demand side, barriers tend to be greater with respect to households and small firms than with large companies, who are more able to evaluate investments. Similarly, in markets where the supply side is heavily comprised of small firms with low levels of technical, managerial, and marketing skills, the barriers tend to be higher.

Network Externalities

Some technologies operate in such a way that any given user's equipment interacts with the equipment of other users so as to create what economists call network externalities (David, 1985; Katz and Shapiro, 1986). For example, since vehicles must be refuelled, the attractiveness of vehicles using alternative fuels is very dependent on the availability of convenient sites for refuelling. Furthermore, the development of a rich infrastructure devoted to distributing any given fuel is, in turn, dependent on there being sufficient vehicles using that fuel to generate a large demand for that infrastructure. This need to create an interacting network of equipment and infrastructure can be a barrier to the diffusion of new technology, in that a potentially superior technology may have difficulty diffusing because of the lack of necessary infrastructure, while the diffusion of the infrastructure is impeded by the low diffusion of the new vehicles.

5.3.5.1 Demand Side of the Market

The diffusion of GHG-efficient technology may be limited by "irrational" or less-than-rational behaviour of households and firms. Such behaviour may be observed because of the way individuals process and act on whatever information they may have. The behaviour of an individual during the decision-making process may seem inconsistent with their goals. More or better information alone may be insufficient to change behaviour, which is strongly influenced by habit or custom (Brown and Macey, 1983).

Within organizations, various factors discourage or inhibit cost-effective decisions regarding new technology. For businesses, the priority of other investment opportunities (e.g., to maintain or expand market share and production capacity) may cause the firm to reject cost-effective GHG-efficiency investment opportunities. Where energy costs are a small component of total production costs, management may not provide sufficient support for energy efficiency investments. In addition, within a firm, no single party or department may have clear and explicit responsibility for managing energy costs.

Another facet of behaviour that is often cited as a barrier to energy efficiency investments is the demand for a rapid payback that may be either explicit or implicit in behaviour. To some degree, the so-called high discount rate applied by consumers could be seen as an aspect of "irrational" behaviour. However, the demand for a rapid payback is also related to particular features of energy-efficient products or services (such as uncertain performance), specific circumstances related to home and appliance ownership, the context in which these products are placed, or to macro-economic conditions, such as high inflation or uncertain future energy prices.

5.3.5.2   Supply Side of the Market

Limited Availability of Products or Services
This may result from decisions and practices of manufacturers and/or distributors. Firms that provide services related to energy efficiency may be few in number. Availability is typically lower (and prices are higher) in rural areas than in large cities. To some extent, limited availability of products and services is a "chicken and egg" problem, which tends to be most problematic in the early stages of market development for a more efficient product or service.

Weakness of Suppliers in Market Research
Firms may lack the resources or capability to do adequate market research, thereby inhibiting the development of new products or services for which there might be a demand.

Weakness of Suppliers in Product Development
Firms may be lacking in skills required for the development of new products, or in capital for investment in new production capacity. Gaining access to advanced designs and/or manufacturing techniques may also be a problem (related to international technology flows).

Weak Marketing Capabilities of Suppliers
Firms may lack the skills for adequate marketing of more efficient products or services.

5.3.6   Institutional Frameworks

Economic actors interact and organize themselves to generate growth and development through institutions (and policy making). While organizations are material entities possessing offices, personnel, equipment, budgets, and a legal character; institutions are systems of rules, decision-making procedures, and programmes that give rise to social practices, assign roles to participants in these practices, and guide their interactions. Organizations may administer institutions (Young, 1994). Institutions exhibit substantial continuity and offer narrow and infrequent windows of opportunity for reform (Aghion and Howitt, 1998; Rip *et al.*, 1998). Institutions operate in larger settings characterized by material conditions such as the nature of available technologies and the distribution of wealth, by cognitive conditions such as prevailing values, norms, and beliefs, and by transaction costs, costs of co-ordination, laws, etc. (Young, 1994; Coase, 1998). The market is a "set of institutions, expectations, and patterns of behaviour that enable voluntary exchanges" based on the willingness to pay of the parties to the exchange (Haddad, 2000). One major concern of the new institutional economics is the boundary between the market on which transactions are negotiated and organizations such as the firm (Simon, 1991).

On one level, all barriers can be considered institutional in origin, because markets, firms, governments, etc. are all institutions. In this section, however, the focus is on those barriers that derive from widespread or generic attributes of institutions. The distinctions are necessarily arbitrary, and some overlap between the discussion in this and other subsections is inevitable.

Institutions are a form of capital, social capital (Coleman, 1988). Social capital, like natural and human capital, is at the same time an input and an amenity. As an input, it enhances the benefits of investments in other factors and, thereby, shares the "shift" feature of technology (World Bank, 1997). Social capital is a public good and suffers, therefore, from underinvestment. Generally, weaknesses in social capital resulting from prevailing beliefs, norms, and values are an important generic barrier to the effectiveness of institutions. At the microeconomic level, social capital may be viewed as a social network, and as associated norms which may improve the functioning of markets and the productivity of the community for the benefit of the members of the association (Coleman, 1988; Putnam, 1993; World Bank, 1997; Young, 1999). At the macroeconomic level, social capital includes the political regime, the legal frameworks, and the government's role in the organization of production in order to improve macroeconomic performance as well as market efficiency (Olson, 1982; North, 1990). Institutions may remedy market failures due to asymmetric information through information sharing (Shah, 1991). Societies in which trust and civic co-operation are strong, a component of social capital, have significant positive impact on productivity and provide stronger incentives to innovate and to accumulate physical capital. Trust and civic co-operation tend to affect human capital productivity especially (Knack and Keefer, 1997).

5.3.6.1   Achievement of Economic Potential

High transaction costs and inadequate property rights
Substantial cost reductions may go unrealized when the trans-

action costs are high. Attempts to reduce transaction costs and to clarify property rights may yield substantial long-term gains. Uncertain property rights, especially as far as intellectual property rights are concerned, act to increase discount rates. Procurement routines which include energy consumption as a criterion, and accounting procedures which are adapted to the polluter-pays principle may need to be adopted to provide appropriate incentives for production units to reduce energy consumption. Intellectual property rights encourage foreign investment, but could also have a negative impact on the adaptation of existing technologies to local conditions (Blackman, 1999).

Demonstration projects, advertising campaigns, testing and certification of new technologies, subsidies to technological consulting services, and science parks are ways for governments to enhance the flow of information on new technologies. This information is bound to be imperfect, because firms have no incentive to supply information about new technologies to late adopters, and technology suppliers are more concerned about market share than about technology diffusion (Blackman, 1999).

Misplaced Incentives
In some situations, the incentives of the agent charged with purchasing a product or service are not aligned with those of the persons who would benefit from higher efficiency. An example is in rental housing where the tenant is responsible for the energy bill, so the landlord has little or no incentive to undertake energy efficiency improvements or acquire more efficient equipment. Other examples of misplaced incentives are present in contracts which pay fees to architects and technical advisors that are measured as a percentage of total project investment, and give rise to over-sizing and "gold-plating" without sufficient attention to the (energy) performance of the investments.

Inefficient Labour Markets
These may prevent the efficient movement of skilled workers among sectors. This may slow technology diffusion and therefore growth (Aghion and Howitt, 1998).

Co-ordination Problems between Technology-producing Sectors
Some technologies, dubbed "general purpose technologies" (or "GPTs", Bresnahan and Trajtenberg, 1995), are characterized by much initial scope for improvement, many varied uses, applicability across many diverse sectors, and strong complementarities among the uses in different sectors (Helpman, 1998). Development and diffusion of these technologies requires co-ordination between the sector or sectors producing the GPT and the application sectors, because rapid development of the GPT is dependent on improvements in the technologies in the application sectors, and vice versa.

Policy Uncertainty
Climate change uncertainties inhibit desirable investment in new technologies and long-term capital goods. The resulting uncertainty about energy prices, especially in the short-term, seems to be an important barrier. Therefore, policy uncertainty should not add to the incentive of holding off relatively irreversible investments. Policy stability is a virtue and institutions are patterns of routinized behaviour that stabilize perceptions, interpretations, and justifications (Giddens, 1984; O'Riordan *et al.*, 1998; Schmalensee, 1998). Lack of credibility of technology forcing policy is a form of policy uncertainty as is illustrated by the example of the 1970 US automobile emissions standards (*Box 5.2*) (Rip and Kemp, 1998). Another form of policy uncertainty results from a crisis by crisis government management style, or from the fact that issues are sometimes championed by individuals and die off when these individuals leave the political scene.

5.3.6.2 Achievement of Socioeconomic Potential

Vested Interests
Organizations provide not only public goods to their members

Box 5.2. United States Automobile Emission Standards

Federal controls on US automotive air pollution (carbon monoxide, oxides of hydrogen, and hydrocarbons), inspired by a technology forcing philosophy, were first applied to 1968 model cars in the US. In the following years, the standards were gradually made more stringent (White, 1982). The 1970 Clean Air Act imposed stringent nationally uniform emission limitations on new motor vehicles requiring 90% reductions over uncontrolled emissions by 1975-1976 with fines of up to US$10,000 per car and limited provision for deadline extensions. The ambitious standards established proved to be difficult to achieve. By 1976, it became clear that many air quality areas were not going to meet the deadline for implementing the ambient standards (Ashford *et al.*, 1985). The deadlines for achieving 90% reduction were repeatedly waived or statutorily postponed. The $10,000 fine was not credible (Stewart, 1981). The Clean Air Act was amended in 1977, and moved away from technology forcing by introducing market incentives such as innovation waivers (Ashford *et al.*, 1985). Empirical results show, however, with very little ambiguity, significantly lower emissions from vehicles for 1968 and after, especially for the years in which emissions were tightened (White, 1982). Therefore, compliance was achieved despite the fact that industry argued, or that compliance with the regulation was doubtful or thought to be impossible. Rapid diffusion of add-on catalysts and minor modifications to the standard combustion engine were achieved but basic changes in engine technologies did not materialize. Uncertainty about whether a deadline or a fine will be enforced gives the signal to industry that a technology developed may ultimately not be needed, and that adopting low-risk existing technologies *i.e.*, technology diffusion is the way to go (Stewart, 1981; Ashford *et al.*, 1985).

but also selective incentives, i.e., private goods. These selective incentives may be sufficient to maintain the organization even if the public good it once provided is no longer needed. Organizations that represent a narrow segment of society do not have an incentive to increase society's output, but rather to increase the share of output going to its members. These organizations are themselves barriers by being rent-seeking coalitions which reduce efficiency and output, and increase the political divisiveness of society. Rent-seeking coalitions interfere with an economy's capacity to change because of their slow decision-making processes, and because they increase the complexity of regulation and the role of government (Olson, 1982).

A major barrier to the diffusion of technical progress appears to lie in the existence of vested interests among economic agents specialized in the old technologies and who may, therefore, be tempted to collude and exert political pressure on governments to impose administrative procedures, taxes, trade barriers, and regulations in order to delay or even prevent the arrival of new innovations that might destroy their rents (Olson, 1982). The duration of the delay will depend in part on the design of political institutions and in part on technological characteristics (learning by doing and knowledge externalities), and on the balance of power between innovators and incumbents. The more learning by doing and the more positive knowledge externalities on the older technology, other things being equal, the lower the frequency of new innovations (Jovanovic and Nyarko, 1994; Krusell and Rios-Rull, 1996; Aghion and Howitt, 1998).

Firms' Institutional Structures
Firms' institutional structures shape their responses to technological opportunities and policies. Firms that tend to maximize stability and tend to rely on single–source internal analyses for information are the least likely to be first adopters of a new technology. On the other hand, firms that maximize profitability, and rely on multiple internal and external sources of information were most likely to experiment with a variety of technologies, but unlikely to commit themselves to a single fuel or process (Braid *et al.*, 1986; O'Riordan *et al.*, 1998). A vertically integrated firm may be slower to absorb information and respond to change than a firm where lateral transfers are possible ("smart workplaces"). An integrated firm also has less incentive to innovate than a decentralized one ("Arrow replacement effect"). On the other hand, as a lot of climate change innovation research is of an applied nature, research is more productive when it is carried out by the firm itself than when delegated to a research institution. Delegation of the research function to a specific entity within the firm increases the incentive to acquire information, but also increases the probability of getting a suboptimal innovation (Aghion and Howitt, 1998; DeCanio, 1998).

Inadequate Attention to Institutional Design
This lack of attention, for example, is especially connected to the institutional context ("national innovation system") for the

heuristic search which gives rise to a set of new findings, blueprints, artifacts ("selection environment"), and which may yield a protected space ("niche") in which a new product can survive more easily because of technology forcing. A national innovation system provides long-term goals, predictions of long–run outcomes, creation of an actors' network, adequate experimentation, and monitoring of outcomes, formulation of standards, tax and subsidies, etc. for alternative energy technologies (Freeman and Soete, 1997; Rip and Kemp, 1998).

National policy styles, as routinized institutional methods to deal with issues, in which the balance of authority is shifted from formal institutions toward informal networks and associations may help achieve economic potential. This shift favours the development of innovative policy formulation, and implementation (Wynne, 1993; O'Riordan *et al.*, 1998).

Lack of Effective Regulatory Agencies
Many developing countries have excellent constitutional and legal provisions for environmental protection but the latter are not enforced (O'Riordan *et al.*, 1998). However, "informal regulation" under community pressure from e.g., non-governmental organizations, trade unions, neighbourhood organizations, etc. may substitute for formal regulatory pressure (Pargal and Wheeler, 1996). Informal regulation is correlated with the adoption of clean technologies (Blackman and Bannister, 1998). Differences in regulatory costs between the old and the new technologies affect the rate of return on the new technology and the speed of diffusion of technologies (Millman and Prince, 1989; Ecchia and Mariotti, 1994).

Reliance on Market Mechanisms when Inappropriate
Organizations co-ordinate behaviour by promulgating standards and rules, and by offering certification that allows actors to formulate stable expectations about the environment and about the behaviour of other actors. Markets perform such functions incompletely or not at all. Thus, reliance on market mechanisms, to the exclusion of the development of organizations needed to perform standard-setting and other co-ordination functions limits the spread of new technology by increasing uncertainty and preventing the realization of co-ordination benefits.

5.3.7   Information Provision

Consumers of energy-using technologies cannot make good decisions regarding which technologies to employ unless they possess the appropriate information. The need for information creates three potential types of market failures with respect to energy-using technologies (Jaffe and Stavins, 1994).

Information as a Public Good
Generic information regarding the availability of different kinds of technologies and their performance characteristics may have the attributes of a "public good", and hence may be underprovided by the private market. This relates both to infor-

mation that consumers need to acquire about specific technologies, as well as to information that manufacturers need to acquire regarding the attributes and needs of consumers. This problem is exacerbated by the fact that even after a technology is in place and being used, it is often difficult to quantify the energy savings that resulted from its installation, since usage patterns and outside influences such as weather may have changed. Knowing that this uncertainty will prevail can itself inhibit technology diffusion, as internal or external advocates for a new technology may doubt that they will be able to justify investment decisions after the fact. Firms supplying products or services within a particular market learn from one another with respect to understanding the market and operating effectively in it. The processes and networks by which this learning takes place, such as professional associations, conferences, publications, and informal networks are often weak in developing countries and EITs.

Adoption Externalities

One of the mechanisms for the transmission of both generic information and application-specific information may be the process of technology adoption itself. That is, one way that a user learns about a new technology is by seeing it used or communicating with other agents that have used it. In this case, the adoption of the new technology by a given user creates a positive information externality, by lowering the cost for others to acquire useful information. This implies that the act of adoption has social benefits that exceed its private benefits, and hence will be inadequately undertaken by private agents.

Misplaced or "Split" Incentives

The third form of informational barrier arises in an institutional context in which investment decisions regarding energy technology must be made in an environment of "agency," that is, one economic agent must make an investment decision that affects the energy costs of some other agent. Examples include contractors who build for others, and tenant-landlord situations where investments are made by the landlord that reduce a tenants' energy costs (or vice versa). In these situations, the party making the investment can recover that investment from the party paying the energy costs only if the investor can credibly convince the consumer that the energy savings justify the investment. That may not happen, however, because information is costly to convey credibly.

The limitations that inadequate information places on decision-making depend on the context. Institutions play an important role both in transmitting information, and in determining the extent to which incentives exist to share and act on information. There is therefore a close relationship between informational and institutional barriers, as evidenced by the discussion of "misplaced incentives" in the previous paragraph and previous subsection. Finally, in many situations it is difficult to determine the extent to which apparently efficient decisions are limited by inadequate information, or whether instead the information is available but decision makers' bounded rationality limits their ability to utilize information effectively.

5.3.8 Social, Cultural, and Behavioural Norms and Aspirations

Perhaps the most significant barriers to GHG mitigation, and yet the greatest opportunities, are linked to social, cultural, and behavioural norms and aspirations. In particular, success in GHG mitigation may well depend on understanding the social, cultural, and psychological forces that shape consumption patterns.

5.3.8.1 Experience from Energy Efficiency Programmes

Conventional policy development is based on a model of human motivation that has been widely criticized (Stern, 1986; Jacobs, 1997; Jaeger *et al.*, 1998). People are assumed to be rational welfare-maximizers and to have fixed values, which, along with the information and means available to them, determine their behaviour. Practical analysis of energy efficiency and other GHG mitigation options often makes the narrower assumption that people are cost-minimizers (Komor and Wiggins, 1988). Such assumptions are undermined by experience with energy efficiency programmes. It has long been recognized that consumers do not necessarily act on their stated values (Maloney and Ward, 1973; Verhallen and van Raaij, 1981), and fail to take up measures that appear on paper to be economically worthwhile (Stern, 1986; Komor and Wiggins, 1988). Some of the reasons, such as energy price uncertainty and transaction costs, have been discussed elsewhere in this chapter and are consistent with the conventional view of consumers as "rational actors". Another important influence on behaviour is the source and quality of information on mitigation measures (the experiences of friends and family are trusted more than the advice of industry, retail sales staff or government) (Anderson and Claxton, 1982; Stern, 1986; Komor and Wiggins, 1988). It is much harder for the "rational actor paradigm" to accommodate features of human behaviour such as the gap between attitudes and action, the tendency to adopt behavioural routines rather than to optimize continually the limited number of variables that individuals typically take into account in their choices, and the tendency for people to rationalize their choices after the fact.

The gap between current practice and the economic potential has been characterized in this chapter as being caused by "barriers". However, Shove (1999) argues that the language of potentials, gaps, and barriers is itself an impediment to finding socially viable solutions for energy saving, and that new, more socially-sensitive approaches are needed to the analysis of measures, with researchers, industry actors, and policymakers working closely together. One of the greatest challenges for GHG mitigation strategies is that, for most people, neither energy saving nor GHG mitigation is a high priority (see for example, Gritsevich, 2000). Consumers' decisions about energy use are often motivated less by cost-minimization than by improving comfort and convenience (Wilhite *et al.*, 2000).

5.3.8.2 Drivers of Consumption

If energy use, GHG mitigation, and cost-minimization are peripheral interests in most people's everyday lives, it might be helpful to consider what does shape their consumption patterns. The influences on human behaviour are complex, and can be described and understood in many different ways. Insights can be found in several disciplines, including anthropology, biology, economics, mathematics, sociology, philosophy, and psychology. Michaelis (2000a) summarizes some of the different drivers of consumption patterns. They include

- demographic, economic, and technological change;
- resources, infrastructure, and time constraints;
- motivation, habit, need, and compulsion; and
- social structures, identities, discourse, and symbols

The first and second of these groups of influences are addressed elsewhere in the TAR and in this chapter, and will not be considered here. The current section focuses on the third and fourth groups. It draws partly on an IPCC expert meeting held in Karlsruhe in March 2000 (German Federal Ministry of Environment, 2000b). It also considers the insights to be gained from viewing behavioural change as an innovation process.

5.3.8.3 Human Need and Motivation

Human need is central to sustainable development as defined by the Brundtland Report: sustainable development is development that meets the needs of current generations without compromising the ability of future generations to meet their own needs (WCED, 1987). But the concept of human needs is controversial. The word "need" is used in many ways: as a strongly felt lack or want; as a positive motivation or desire; and as a necessary condition for something, such as survival, social acceptance, or health. The failure to distinguish these different meanings has confused efforts to agree on the morality of need-fulfilment (Michaelis, 2000b).

One major barrier to the success of many policies is the failure to take account of the full range of human motivations and goals. For example, an engineer may design an energy-efficient building that provides occupants with adequate shelter and warmth, but it may be hard to get people to live in it if it is in the wrong area, or lacks features normally associated with adequate social status. Similarly, public transport may provide fast, efficient mobility for certain trips, but young men may see car ownership as the only way to attract a girlfriend. Maslow (1954) explained motivation in terms of human needs, which he divided into categories: physiological needs, sense of belonging, esteem, and "self-actualization". He saw these categories as a hierarchy, arguing for example that we are only concerned about self-esteem when we have had enough to eat. While the idea of a hierarchy has been largely discredited (Douglas *et al.*, 1998), Maslow's categorization of needs continues to be widely used. Max-Neef (1991) proposed a more complex categorization of needs, divided into "having", "doing", "being", and "relating" needs, and emphasized the distinction between needs and "satisfiers".

While some consumption may respond to perceived needs, much is habitual. Habit formation is an important barrier to GHG mitigation as consumers may be unwilling or unable to change their behaviour or technology choices. The continuation of rising consumption levels has been widely observed and was noted by Jean-Jacques Rousseau in 1755 (Schor, 1998; Wilk, 1999). What was once luxury rapidly becomes habit, and then need. This is partly a social, as opposed to an individual psychological phenomenon, and will be discussed further in the next section. Often, we may try to use inappropriate satisfiers to meet particular needs (Max-Neef, 1991) – for example, eating in response to feelings of loneliness. Consumption of such ineffective satisfiers can become compulsive, especially when they give a short-term feeling of relief but fail to satisfy in the long term.

There may be opportunities for GHG mitigation in identifying where low-GHG-emitting behaviour can help to meet needs better than existing behaviour. Argyle (1987) finds from a review of several studies that human happiness is influenced mainly by health, the quality of family life, marriage, and friendships. Having meaningful work is also important. Absolute levels of material wealth are relatively unimportant: many studies have found that, once basic material and healthcare needs are met, happiness is largely independent of absolute income levels (Jackson and Marks, 1999; Inglehart, 2000), although relative income remains important as an indicator of social status. Efforts to promote low-GHG consumption patterns such as domestic energy conservation, cycling rather than relying on a car, living in higher density housing, or eating less meat might have the most success if they emphasize ancillary benefits in terms of improving health, family life, and community relationships rather than saving money.

Sen (1980, 1993) has developed a concept, related to human need, of the "capabilities" that individuals must have if they are to "flourish" or to live a good life. Individuals require different capabilities depending on their personal circumstances and the community they live in. While the good life is to some extent subjective, it is also socially defined. Some aspects of energy-using behaviour may be very hard to change because they play important roles in culture-specific ideals of the good life, varying from country to country. Wilhite *et al.*, (1996) describe the cultural significance of lighting and heating in Norway, and of bathing in Japan, suggesting that energy saving measures in these areas would need to be very sensitive to cultural requirements. They also observe that other aspects of household behaviour, such as washing clothes, are less culturally significant and may be easier to change. International differences in habitual behaviour in such areas might provide opportunities for encouraging change through information and education programmes emphasizing best practice.

Moisander (1998) describes how motivation is shaped by both broad values and attitudes, and by more specific priorities, and also how the ability to act depends on both personal capabilities or resources, and external factors or opportunities. Surveys of public attitudes in the United States find an increasing level of concern about climate change, and agreement that action is needed to save energy and protect the environment (Kempton *et al.*, 1992; Kempton, 1997). One of the challenges for individuals in acting on environmental values and attitudes is the need to reconcile divergent objectives. This is all the more difficult in the case of climate change, which is poorly understood by most people (Kempton, 1991, 1997; Lofstedt, 1992; Wilhite *et al.*, 1996). Moisander (1998) finds that being concerned about the environment provides some motivation for environmentally friendly behaviour. But identity (as a "green consumer") and internalized moral ideals or imperatives play a much stronger role. Identity and ethics, which play an important role in shaping consumption patterns, are largely social phenomena and will be discussed in more detail in the next two sections.

5.3.8.4   Social Structures and Identities

Most of the perspectives discussed in the last section treat the individual as a self-contained person with intrinsic motivations. While this is a dominant assumption in modern Western societies, in many cultures, individuals are understood primarily in relation to others, and behaviour is largely explained in terms of the social context (Hofstede, 1980; Cousins, 1989; Markus and Kitayama, 1991; Dittmar, 1992). In fact, the social and cultural context of the individual is important in all societies. It contributes to individuals' moral ideals and identity, to their areas of empowerment or constraint, and to the options they perceive to be open to them. Social and cultural influences are mediated through the use of discourse and symbolism and through the actions of others. Individuals often conform to the cultural norms of their community because of their needs for safety, sense of belonging, love, and esteem.

Social structures help to shape consumption, for example, through the association of objects and activities with status (Veblen, 1899; Hirsch, 1977) and class (Bourdieu, 1979). Social structures also allow some individuals to influence the consumption patterns of others. In many societies, women are mainly responsible for purchasing food and clothing for other household members, while men are more influential over large household expenditures (Grover *et al.*, 1999). Individuals within wider communities also influence each other's consumption patterns and habits in a wide variety of ways, depending on the social structure and their respective positions within it.

Much human behaviour can be understood as an expression of identity or self-definition (Meyer-Abich, 1997). In modern consumer societies, consumption patterns in particular are also used to establish and communicate identity. The combinations of goods people purchase help to confirm to themselves and express to others their personalities and values (Douglas and

Isherwood, 1979; Tomlinson, 1990), their membership of particular social groups or communities (Schor, 1998), and their relationship to their social and physical environment (Dittmar, 1992).

Some of the consumption choices that have the greatest effect on GHG emissions, such as car and house ownership and international travel, are also among the most significant means of establishing personal identity and group membership (Schor, 1998). Where such consumption patterns are closely connected to individual and collective identities, they may be particularly difficult to change, although the role of consumption in society is changing.

Some argue that, with urbanization, conspicuous consumption may have become more important as a form of status display – in small, close-knit communities it is unnecessary because everybody knows each other (Kempton and Layne, 1994). The status and group membership function of consumption has also been altered with the spread of television. Some viewers experience emotional attachments to TV characters as if they were real people; viewers also use the characters and situations they see as reference points for their own lives, helping to shape and reinforce their own values and identities (McQuail *et al.*, 1972). Those who watch a large amount of television increasingly compare themselves with the portrayed lifestyles of the super-rich, resulting in higher desired levels of consumption (Schor, 1998). While the media can pose a barrier to GHG mitigation by reinforcing current trends towards more GHG-intensive lifestyles, it may also offer opportunities. Raising awareness among media professionals of the need for GHG mitigation and the role of the media in shaping lifestyles and aspirations could be an effective way to encourage a wider cultural shift. The role of the media in GHG mitigation will be discussed further in the next section.

Ongoing developments in the media and communication technology could also generate barriers and opportunities for GHG mitigation. Many scenarios have been painted of the potential impacts of information and communication technology on society. The growth of Internet usage and other interactive communication forms are widely expected to stimulate economic development and technological innovation (Cairncross, 1997). However, they may also lead to increased social stratification, social exclusion, and a decline in trust and social solidarity (or social capital) (Castells, 1998). Such developments could have major implications for the feasibility of responding collectively to threats such as climate change. Fukuyama (1999) argues that, although social capital has declined in recent decades with the development of the information society, similar declines occurred during previous economic and technological upheavals and were followed by the creation of new institutions, leading to new heights of morality and social solidarity. Cairncross (1997) even suggests that free communication may lead to global peace. Slevin (2000) points to the development of personal web pages as a new, versatile, and sophisticated means of establishing personal identity. Inglehart

(1990) finds signs of the emergence of a new "postmaterial" culture that emphasizes networking and communication rather than possessions. However, Castells (1998) believes that more investment is needed in education and science if societies are to reap social benefits from new information and communication technologies and respond to environmental and other challenges.

5.3.8.5 *Discourse and Symbolism*

The spread of new communication technology may make it increasingly difficult for governments to exert a direct influence on social structure and culture. On the other hand, governments, along with the business community and NGOs, continue to have a substantial presence in the media and they all contribute to the shaping of the public discourse on climate change (see *Box 5.3*). Some of the essential features of that discourse are the differing views on the risks and uncertainties associated with GHG emissions; the costs and benefits of GHG mitigation; the allocation of blame for past and current emissions; and the rights of the victims of climate change to compensation. Disagreement on these various points poses an important barrier to GHG mitigation, especially where media presentation tends to emphasize controversy. There are many ways of helping to build of a common discourse, or narrative, about climate change, involving the various players taking all available opportunities to meet, discuss, and work together for common goals. An important example is the growing development of partnerships between transnational companies and environmental NGOs, for example, to develop accreditation schemes for green products or to design environmental strategies.

The linking of symbols to fundamental values may also be important in shaping behaviour. Ger *et al.* (1998) compare the symbolism of consumption patterns, based on interviews and observations in Denmark, Turkey, and Japan. They find that the symbolic attractions of resource-intensive consumption patterns are more powerful than those of more sustainable consumption patterns. The symbolic attachments are different depending on the country and the subculture within the country.

Narrative and symbols carried by the mass media form a large part of the means through which ideas, arguments, and values are transferred from the public to the private sphere, and ultimately may be integrated into individuals' consciousness and identity. Moisander (1998) has observed that consumption choices respond strongly to personal morals or ethics. It is in shaping ethics that the public narrative can play a particularly strong role.

5.3.8.6 *Ethics of GHG Mitigation: the Commons Dilemma*

There are several important ethical dilemmas both in the public discourse and in most people's minds regarding GHG mitigation. Essentially, they boil down to questions about human relationship with nature, about justice and equity between human beings, and about the nature of the "good life" (Michaelis, 2000b). In modern society, images of and narratives about the good life often emphasize individual independence and material well-being. These values may appear to conflict with messages about the interdependence of people around the world and the need to moderate the consumption of natural resources.

In addition to the perceived conflict with improving material wellbeing, ethical arguments for GHG mitigation face several barriers including the perceived weakness of the evidence that climate change is happening; the difficulty in understanding the risks associated with low-probability extreme weather events; the difficulty in tracing climate change impacts to particular emitters of GHGs; and the large physical and social distance between GHG emitters and victims of climate change (Pawlik, 1991). It seems that people are inclined to deny and remain passive about about those kinds of environmental nuisances and risks that they believe to be uncontrollable (Pawlik, 1990). From an institutional perspective, the "commons" dilemma charaterizes situations in which people are unable to co-operate to achieve collective benefits, because they are unable to change the rules affecting their perverse incentives; these incentives are themselves institution-dependent (Ostrom, 1990; Ostrom *et al.*, 1993). Current climate may be seen as an infrastructure which is used jointly by many people, which is subject to many decision makers, including some in the public sector, and whose benefits and costs are perceived differently by different people because these are borne by many people who do not take the protection decisions. Lack of clear limits on using up resources such as current climate generates costs (climate change) on all participants through unsustainable exploitation because GHG concentrations and, therefore, current climate are stocks like fish and timber. Complex institutional arrangements are required to overcome perverse incentives (Ostrom *et al.*, 1993). Commons dilemmas reflect persistent conflicts among (not between) many individuals (producers and consumers).

5.3.8.7 *The Need for Social Innovation*

Given the complexity of the social, cultural, and psychological drivers of human behaviour, there are no simple recipes for behaviour change. However, there are considerable opportunities to be grasped in taking advantage of the desire for change and the willingness to experiment and learn on the part of individuals, communities, and institutions.

There are many analogies between social and technological change: the two processes are closely linked, equally fundamental to the development of consumption patterns, and the processes behind the development and diffusion of behaviour patterns and cultures are similar to those of new technologies (Michaelis, 1997a, 1997b; Grübler, 1998). They include:

- Development and discovery of new narratives, ideas, symbols, concepts, behaviours, and lifestyles;

Table 5.2: *Strategies for risk management in social dilemmas and barriers to transformations of unsustainable behaviour (Vlek et al., 1999)*

Strategy	Method	Barrier
Provision of physical alternatives, (re)arrangements	Adjusting /depleting /changing behaviour options, enhancing efficacy	-Absence of physical or technical alternatives -Failure to identify, or disbelief in feasible alternatives -Unwillingness to make feasible alternatives available -Inability to utilize available alternatives
Regulation-and-enforcement	Enacting laws, rules; setting and/or enforcing standards, norms	-Absence of pertinent laws or regulations -Insufficient and/or ineffective law enforcement -Disbelief in effectiveness of law or regulation -Inability to abide by law or regulation
Financial-economic stimulation	Rewards and/or fines, taxes, subsidies, posting bonds	-Absence of financial incentives (rewards and punishments) -Inconsistency of financial incentive systems -Insufficient, ineffective financial incentives -Incentive systems justifying squandering ("I paid for it")
Provision of information, education, communication reduction strategies	About risk generation, types and levels of risk, others' perceptions and intentions, risk reduction strategies	-Lack of Knowledge (LoK) accumulating negative externalities -LoK about own causal role and possible contribution to solution -LoK about others' problem awareness and willingness to co-operate -Uninformed expectations about effects of proposed policies
Social modelling and support	Demonstrating co-operative behaviour, others' efficacy	-Absence of invisibility of model behaviour by opinion leaders -Fear of setting public examples and living by principles -Inability to understand and follow visible model behaviours -Failure of managers to provide needed social support
Organizational change	Resource privatization, sanctioning system, leadership institution, organization for self-regulation	-Too large organization, too much diffusion of responsibility -Organization form obscuring negative externalities -Inefficient organization requiring unnecessary energy, materials, and labour
Changing values and morality	Appeal to conscience, enhancing "altruism" towards others and future generations, reducing "here and now" selfishness	-Personal identity associated to material possessions and consumption -Importance of social superiority in spending capacity -View of "whole world as my playground" -Basic attitude biased against ("hostile") natural environment -Inability to feel responsibility for future generations

- Exchange of ideas, behaviours, etc., among firms, communities, government organizations, *etc.*;
- Experimentation with new ideas, behaviours, etc., possibly selecting those that could contribute to GHG mitigation and other policy objectives;
- Replication of successful ideas, behaviours, etc.; and
- Selection by the contextual framework of markets, laws, infrastructure, and culture.

Barriers and opportunities take various forms in association with each of the above processes. The willingness of some groups in society to take risks and to experiment provides an important opportunity for GHG mitigation. New values and behaviour patterns on the part of consumers (e.g., "ethical" or "green" consumption) can spread, encouraging producers to change production methods and management practice. The media plays an important role in the exchange of ideas and in shaping the way new ideas are viewed, whether as exciting new opportunities, as threats, or as eccentric oddities. Alliances among powerful groups can encourage or inhibit experimentation and the replication of successful ideas. And the government can play a key role in setting the contextual framework to encourage shifts in behaviour that would reduce GHG emissions, as well as in removing bureaucratic and regulatory barriers and providing support for local initiatives. Where the institutional structure and culture supports innovation, and where all contextual drivers point in the same direction,

changes in technology and behaviour can proceed very rapidly (Michaelis, 1998).

Box 5.3. Narratives about Climate Mitigation

Discourse or narrative – the written and spoken word – is one of the most important ways in which governments, businesses, NGOs, and the media influence each other and build agreement on policy directions. One of the most important barriers to GHG mitigation is the perception by some participants in national and international discourses that mitigation efforts might be costly, or might conflict with values such as individual freedom and equity. By analyzing these people's discourse, new opportunities may be identified for developing GHG mitigation measures that are consistent with their core values. It may also be possible to build new coalitions among institutions and actors, to seek mutually satisfactory GHG mitigation strategies.

Discourse and narrative can take many forms, including history, science, philosophy, folklore, and "common sense". Foucault (1961, 1975) has shown how narratives become an instrument for wielding power. MacIntyre (1985) offers a way of thinking about narrative as part of our cultural context or tradition, as something that we inhabit. Professional analysts, such as scientists and economists, are members of groups that define themselves by such traditions and have their own narratives about the world. Our narratives co-evolve with our notions of "the good", our understanding of our selves, our conception of society, our science (conception of nature), and our understanding of God or the spiritual dimension (Taylor, 1989; Latour, 1993). These understandings and conceptions are also central to our responses to climate change.

Analyzing discourses can provide essential insights into different people's assumptions and beliefs about the world. Thompson and Rayner (1998), Ney (2000) and Thompson (2000) have mapped out some of the essential features of the discourses that are used to describe and define positions on climate change. They focus in particular on two axes of the discourses: their view of nature and their conception of society. For example, some view the environment as robust, while others view it as fragile and vulnerable to human interference. Some believe that society works best through market-based institutions, while others believe that there should be more explicit emphasis on egalitarian, participatory approaches. Ney differentiates three main orientations: market-based, egalitarian, and contractarian or hierarchical. Some characteristics of these orientations are summarized in *Table 5.3*. Of these three, the market orientation clearly dominates international negotiations as well as the dialogue on climate change within many countries. It is also the source of the dominant discourse on climate mitigation policy within the IPCC.

Table 5.3: Discourses on climate change (adapted from Thompson and Rayner, 1998)

Discourse	Hierarchical	Market	Egalitarian
Myth of nature	Perverse, tolerant	Benign, robust	Ephemeral, fragile
Diagnosis of climate problem	Population	Pricing/market failure	Profligacy
Policy bias	Regulation	Libertarian	Egalitarian
Public consent to policy	Hypothetical	Revealed (voting)	Explicit (direct)
Intergenerational responsibility	Present>future	Present>future	Future>present

There are, in fact, many "axes" that can be used to map out discourses on climate change. Another important perspective is that of gender (Grover *et al.*, 1999; Hemmati, 2000). To some extent, the different axes can be correlated with those chosen by Ney, Rayner, and Thompson: feminist discourses have tended to align themselves with egalitarian discourses and in opposition to the hierarchical and market discourses as defined in *Table 5.3*.

While analyzing different positions can be a first step to resolving differences, something more is needed: we need to understand how the dialogues that underlie the climate debate have evolved over time, and might change in the future. In particular, we need to be more aware of the links between our scientific understanding of nature, our political and economic structures, and our ethics. Michaelis (2000) finds traces in the climate debate of a long-running process of development of alternative cultures or traditions in our society:

- The modern tradition, with roots in the 17th-18th century European Enlightenment, is built on a separation of humanity and nature, with its central aims of economic and technological progress and its commitment to finding "the good" in the everyday working life. This tradition is dominant in the words of government, business, and science. To a large extent, the different positions analyzed by Ney (2000), Thompson and Rayner (1998), and Thompson (2000) fall within the modern tradition. The climate debate within this tradition revolves around different ways of understanding nature and society.
- The romantic tradition, a reaction to the early Enlightenment in the late 18th and early 19th century, is committed to the emotional life of individuals, to romantic love and the family, and to an ideal harmony between humanity and nature. This tradition is dominant in the world

of entertainment, advertising, and individuals' private lives. It views climate change as a problem caused by the modern tradition, and tends to blame institutions such as businesses and governments which represent that tradition. However, narratives within the romantic tradition tend not to recognize the role of romanticism in shaping the consumption patterns for which industry produces.

- The humanist tradition, with much older roots going back to ancient Greece, is maintained by academic and intellectual circles in modern society, and is committed to the search for "the good life". Viewed from this tradition, the climate change problem appears to be caused by the failure of the modern and romantic traditions to understand human nature, and the nature of the good life. Less emphasis should be placed on material production and consumption, and more should be placed on developing family relationships, communities, civic involvement, and opportunities for learning and contemplation.

Writers such as MacIntyre (1985), Gare (1995), and Latour (1993) see little hope within the modern tradition for solving the problems of our time. MacIntyre advocates a revival of humanism. However, many social scientists have described the emergence of "postmodern" values, which recognize the multiplicity of valid traditions and narratives. This recognition sometimes leads to nihilism, but it could also be the basis for a renewed search for shared values and conceptions of the good life.

5.4 Sector- and Technology-specific Barriers and Opportunities

GHG emissions from some sectors are larger than those from others, and the importance of each GHG varies across sectors as well. Methane (CH_4) for instance is a much bigger contributor to emissions from agricultural activity than, for instance, from the industry sector. *Table 5.4* shows the carbon emissions from energy use in 1995. Emissions from electricity generation are allocated to the respective consuming sector. Carbon emissions from the industrial sector clearly constitute the largest share, while those from agricultural energy use form the smallest share. In terms of growth rates of carbon emissions, however, the fastest growing sectors are transport and buildings. With rapid urbanization promoting increased use of fossil fuels for habitation and mobility in many countries, the two sectors are likely to continue to grow faster than others will in the future.

Annual carbon emissions from land-use change were estimated in the IPCC Special Report on Land Use, Land-use Change and Forestry at 1.6 ±0.8GtC/yr for the period 1989 to 1998

(IPCC, 2000a). Tropical forests are estimated to be net emitters, but temperate and boreal forests are net sequesters of carbon. CH_4 emissions from livestock, rice paddies, biomass burning, and natural wetlands add up to $1.8GtC_{eq}$/yr with considerable uncertainty about these estimates. Below we describe the sector-specific barriers to and opportunities for reducing the sectoral GHG emissions.

5.4.1 Buildings

The buildings (residential and commercial) sector accounted for about a third of carbon emissions from fossil fuel combustion in 1995. Its share of the total emissions has increased faster than in other sectors (Price *et al.*, 1998). About half the emissions in this sector are from fuel use in the commercial sector, and the other half from the residential sector. Energy use in the sector is for cooking, space conditioning, water heating, and lighting and appliances. Aside from the use of modern energy, biomass use constitutes a significant portion of the energy supply, particularly in the developing world. The bulk of households in rural areas use biomass for cooking, and water

Table 5.4: Carbon emissions from fossil fuel combustion (MtC)

Sector	Carbon emissions and % share[1] 1995	Average annual growth rate (%)	
		1971 to 1990	1990 to 1995
Industry	2370 (43%)	1.7	0.4
Buildings			
-- Residential	1172 (21%)	1.8	1.0
-- Commercial	584 (10%)	2.2	1.0
Transport	1227 (22%)	2.6	2.4
Agriculture	223 (4%)	3.8	0.8
All sectors	*5577 (100%)*	*2.0*	*1.0*
-- Electricity generation[2]	1762 (32%)	2.3	1.7

[1] Emissions from energy use only; does not include feedstocks or carbon dioxide from calcination in cement production. Biomass = no emissions.

[2] Includes emissions only from fuels used for electricity generation. Other energy production and transformation activities are not included.

Source: Price *et al.*, 1999

and space heating. Much of the biomass (particularly for firewood, and charcoal combustion and charcoal production processes) in developing countries is used in an unsustainable fashion and results in additions to anthropogenic emissions (CEEEZ, 1998).

Barriers to the full realization of the opportunities for improving energy efficiency in this sector have been extensively studied. The key barriers are traditional customs, lack of skills, social barriers, misplaced incentives, lack of financing, market structure, administratively set prices, and imperfect information (Golove and Eto, 1996; Brown, 1997).

Traditional Customs
Lack of appreciation in the design and manufacture of energy-using devices can inhibit their penetration. In the case of improved biomass stoves it has been shown (ESD, 1995) that despite savings on household charcoal budgets, improved stove commercialization still remains a problem, because of inconsistent design and quality control in the manufacture of stoves. In some programmes (CEEEZ, 1998), field surveys showed that most users of improved cookstoves returned to traditional stoves, owing to a preference for speed in cooking with traditional stoves as compared to the former.

Lack of Skills
Insufficient skills in the manufacture of efficient appliances can slow or stop their diffusion. For example, dissemination of improved stoves could not be sustained (CEEEZ, 1998), because of various reasons, among them being increased production time arising from the complexity of the stove design. As a result, local producers switched to the production of familiar items, which were easy for them to manufacture.

Behaviour and Style
Despite the existence of demand-side management programmes, in most developed countries, and the availability of more technologically efficient household devices (such as air conditioners) in the market place, changes in behaviour and style (associated with a desire to increase dwelling size) tended to increase the demand for energy services (Wilhite *et al.*, 1996). Energy use for space heating increased in Norwegian homes from 1960 to 1990 thanks to a doubling of dwelling area per capita (Hille, 1997) in spite of more stringent building codes and the doubling of thermal efficiency.

Another example is space cooling in Japan, where air conditioners are technically very efficient, but space cooling demand is still increasing dramatically, because of changes in dwelling size, changing tastes, and modern building design which does not support natural cooling (Wilhite *et al.*, 1996). For most home owners, the lowest first cost is more important than a higher energy efficiency level when purchase decisions are made about an appliance or a home (Hassett and Metcalf, 1995).

Misplaced Incentives
These result between landlords and tenants with respect to acquisition of energy-efficient equipment for rental property. Where the tenant is responsible for the monthly cost of fuel and/or electricity, the landlord is prone to provide the least-first-cost equipment without regard to its monthly energy use. Fee structures for architects and designers are based on capital cost of the building. Designing an energy efficient heating, ventilation, and air-conditioning system costs more, and reduces the capital and operating costs of the building, both of which serve as a disincentive to architects for the design of energy-efficient structures (Lovins, 1992). Also, in the buildings sector compensation to architects and engineers based directly or indirectly on a percentage of the costs of the building provides perverse incentives.

Lack of Financing
This refers to the significant restrictions on capital availability for low-income households and small commercial businesses. Home mortgages for instance do not as a rule carry a lower interest rate for efficient homes, which have low annual energy costs. In case of switching to modern cooking stoves (electric, kerosene, or liquefied petroleum gas (LPG) for example) in rural areas of developing countries, the barriers result from household income, accessibility to modern fuels, the relative cost of traditional and modern fuels, and cooking habits (Soussan, 1987). For example, in view of both national and global benefits, use of low-cost electric stoves has been noted as a viable substitute for improved biomass cookstoves, as they can contribute effectively to preserve forests to enhance carbon sequestration (CEEEZ, 1998). Despite this realization, there has been a low level of switching from charcoal stoves to electric stoves. This is largely because of a lack of finance, resulting from low monthly income of which 35 % to 45 % is spent on fuel (CEEEZ, 1998).

Market Structure
This can imbue power to firms who may inhibit the introduction by competitors of energy-efficient equipment such as compact fluorescent lighting (Haddad, 1994). The design, construction and maintenance of buildings is largely fragmented. This is in part cause by the lack of integration and communication between sub-sectors, and in part a reflection of the diverse and large number of suppliers. This results in many instances of building design, insulation, and energy-using devices that do not exhibit high levels of energy efficiency (OTA, 1992). One response in Switzerland since 1978 has been to ensure that architects are fully integrated into the selection and construction of energy using devices in buildings (Jefferson, 2000).

Administratively Set Prices
These distort investment and the choice of energy forms and end-use equipment. Electricity has been historically subsidized to residential customers in India, and serves as a disincentive to faster penetration of efficient lighting and appliances (Alam *et al.*, 1998). In contrast to subsidies in the electricity industries of India, non-availability of subsidies in the commercial dissemination of improved cookstoves in Kenya has lead to dra-

Box 5.4. Commercial Dissemination of Improved Cookstoves in Kenya

One of the most successful improved cookstoves in Africa is the Kenya Ceramic Jiko (KCJ) (Karekezi, 1991). The KCJ was introduced in Kenya in 1982 and mainly targets urban populations who used charcoal.

The KCJ is produced and marketed through the informal sector. One of the key characteristics of this project was the ability to utilise existing production and distribution system for the traditional stove to produce and market the KCJ.

The most important factor to the successful commercialization of the KCJ is the conscious decision made by the project initiators not to provide subsidies. Although stove prices were initially high, the ensuing competition between producers reduced the price from as high as US$15.00 to a prices of US$2.50 in 1989 (Karekezi, 1991). Purchases made by high income groups in the earlier stove dissemination, however, effectively subsidized the stove development process thus making it available for lower income groups (Otiti, 1991).

matic improvements in the marketing and distribution of improved stoves as shown in *Box 5.4*.

Economic Pricing
Economic pricing in the electricity sector, particularly in developing countries and countries in transition, has been hampered by a lack of adherence to economic tariff setting (based on long-run marginal cost (LRMC)). Attempts to rigorously follow this concept, however, have resulted in social problems. For example, in Russia, a country in the process of transformation to a market economy, LRMC has led to pensioners not being able to afford their electricity bills, requiring subsidies amounting to 20%-35% of the budgets of local authorities (Gritsevich, 2000).

Imperfect Information
The lack of adequate and accurate information, and the limited ability of users to absorb it adds to the cost of its provision to consumers. Since energy costs are typically small on an individual basis, it is rational for consumers to ignore them in the face of information gathering and transaction costs. For instance, Sony was able to reduce the standby power loss in TVs from 7-8 watts to about 0.6 watt, a saving of US$5 per year per TV. One reason for consumers to not buy more efficient appliances, despite a label advertising this fact, is that consumers are wary or mistrustful because of past experience with advertised misinformation (Stern and Aronson, 1984). Kempton and Montgomery (1982) have shown that residential consumers systematically underestimate energy savings, because they lack the ability to use the information to calculate and compare savings with investment. Furthermore, Kempton and Layne (1994) liken today's energy bills to receiving a single monthly bill for all groceries purchased with no identification of the cost of individual items.

5.4.1.1 Opportunities, Programmes, and Policies to Remove Barriers

Technological and social changes bring about opportunities to improve the efficiency of buildings and appliances. A change in the production line for the manufacture of an appliance offers an opportunity for introducing new energy saving features in an appliance. Likewise, when buildings are sold, a city government may have the opportunity to intervene and have energy saving features installed prior to the registration of that sale. Targeting opportunities at a point where the stock is likely to turnover physically or contractually can reduce the perceived and actual cost to producers and consumers.

Governments have designed policies, programmes, and measures to tap these and other opportunities, and in the residential and commercial buildings sector they fall into nine general categories: voluntary programmes, building efficiency standards, equipment efficiency standards, state market transformation programmes, financing, government procurement, tax credits, accelerated R&D, and a carbon cap and trade system. The last three items are generic and are not dealt with in this section.

Voluntary programmes, such as Energy Star, which is operated by the United States Department of Energy (DOE) and Environmental Protection Agency (EPA), exist for both residential and commercial buildings, and appliances (Harris and Casey-McCabe, 1996). The Energy Star programme works with manufacturers to promote existing energy-efficient products, such as residential buildings, personal computers, TVs, etc., and to develop new ones. Manufacturers can affix an easily visible label to products that meet Energy Star minimum standards. These programmes also facilitate the exchange of information between end-users on their experience with energy-saving techniques.

Building efficiency standards focus primarily on the building shell and/or the HVAC (heating, ventilation, and air conditioning) system, and in commercial buildings also on lighting and water heating. Standards are being implemented in California and other states in the USA, and also in Singapore and Malaysia, and have been proposed or are on the books in Indonesia, the Philippines, and Mexico (Janda and Busch, 1994).

Equipment standards require that all new equipment meet minimum energy efficiency standards. Standards on household appliances and lighting have been in place in the US for over a decade and are expected to be tightened between 2000 and 2005 (McMahon and Turiel, 1997). About 30 developed and developing countries and EITs have voluntary or mandatory standards and labels in place on more than 40 household appliances (CLASP, 2000).

Demand-side management (DSM) programmes provide rebates, targeted delivery of efficient appliances and lighting to low-income households, information campaigns, and the like. These were pursued vigorously in some states in the USA. The deregulation of the US energy supply sector has reduced the emphasis on these programmes. Nevertheless, in several states that previously had these programmes, public benefit funds for energy efficiency have replaced the DSM programmes, and are typically charged to the electricity consumer on his electricity bill (Kushler and Witte, 2000).

Financing programmes spread the incremental investment costs over time and reduce the first cost impediment to adoption of energy-efficient technologies. For commercial buildings, ESCOs offer energy savings performance contracts that guarantee a fixed amount of savings and are paid through the cost savings.

Government procurement policies have accelerated the adoption of new technologies in the USA and Sweden. In the USA, federal regulations regarding procurement were amended in 1997 to limit purchases to equipment that falls in the top 25% of energy efficiency for similar products (McKane and Harris, 1996).

To effectively enhance dissemination of improved cookstoves, policies, and measures need to be put in place. The introduction of affordable credit financing is widely recognized in Africa as one of the effective measures, which will go a long way in removing the financing barrier. Assistance is still needed in some locations on the design, introduction of centralized small and medium-sized production centres, and marketing of energy efficient stoves, especially where biomass fuels are commercialized – typically as part of small enterprise development. Further research and development work is also essential to increase the efficiency of improved biomass stoves. For example, the British NGO, Energy for Sustainable Development (ESD) is financing and supporting a team of Ethiopian professionals working in household management and supply. It has achieved remarkable success in developing and commercializing two types of improved biomass cookstoves through an iterative approach of needs assessment, product design, redesign, and performance monitoring (Farinelli, 1999). The team consists of consumers, stove producers and stove installers, and pays attention to promotion, technical assistance, and quality production.

5.4.2   Transport

Carbon emissions from fossil fuel use in the transport sector are rising faster than those from any other sector (Price *et al.*, 1998). The transport modes responsible for most of the growth are car travel, road freight, and air transport.

Vehicular air pollution is a major environmental problem in many large urban centres in both developed and developing countries. Although urban air quality in developed countries has been controlled to some extent during, the past two decades, in many developing countries it is worsening and becoming a major threat to the health and welfare of people and the environment (UNEP, 1992).

Chapter 3 notes the existence of a range of technologies whose use in cars could substantially reduce emissions, including lightweight materials, gasoline direct injection engines, electric hybrid drive-trains, and fuel cell-electric drive-trains. Considerable and unexpected progress has been made in commercializing some of these technologies since the SAR. Chapter 3 also reviews studies that estimate the socioeconomic potential for energy efficiency improvements. The rapid emission growth from the sector, despite the considerable apparent mitigation potential, is mainly a result of a continuing increase in demand for mobility of people and goods. The energy intensity of personal travel is near-constant or increasing in many countries, with increasing use of sports utility vehicles and people carriers, and rising vehicle weight and power in most categories of vehicle (ECMT, 1997; Davis, 1999).

In addition to energy efficient technologies, IPCC (1996, Chapter 21) noted an extensive range of options for reducing GHG emissions, including the use of alternative fuels, public and non-motorized transport, and changes in transport and urban planning.

5.4.2.1   Barriers to Mitigation

IPCC (1996, Chapter 21) noted many reasons why GHG mitigation in the transport sector has proved difficult. Transport activity is closely interwoven with infrastructure, lifestyles, economic development, and patterns of industrial production. Partly because of these complex links, experts do not always agree on the best mitigation strategy. Climate change and energy saving is usually a minor factor in decisions and policy in the sector, and mitigation strategies may not be implemented if they seem to reduce the benefits provided by the transport system to individuals and firms. Appropriate mixes of policies need to be designed for local situations. And policies can be very slow to take effect because of the inertia of the infrastructure, technologies, and practices associated with the existing transport system.

Stated preference surveys in the United States have shown that consumers would prefer to purchase energy efficient cars, and would be prepared to pay US$400-600 for each litre/100km reduction in fuel consumption (Bunch *et al.*, 1993; US DOE, 1995). This is about the amount that would be expected from the fuel savings over the life of the car (Michaelis, 1996b). However, there is no evidence that this valuation of fuel economy is reflected in the car market. There may be several reasons. First, many vehicle purchasers have to work within budgets set by the size of loan they can obtain to buy a car, and such budgets are likely to be set independent of the amount they will have to spend on fuel. Where they have a number of high priorities in their vehicle choice such as comfort, size, safety, and performance, they will spend their budgets on those priorities rather than on energy efficient technologies that increase vehicle price. Second, vehicle manufacturers have no incentive to promote energy efficiency, and a strong interest in selling more sports utility vehicles and mini-vans where their profit margins are higher than for cars. The outcome can be viewed as a rational response to consumer preferences subject to a budget constraint, but it has been repeatedly noted in European government-industry discussions that marketing helps to shape those preferences (Dietz and Stern, 1993; Michaelis, 1996a).

Cars may also provide a good example of the principal-agent barrier. The first owner of a car may be more concerned with its status value and other aspects, and less concerned with cost minimization than subsequent owners. Secondhand owners' preferences for cost minimization do appear to be reflected in the secondhand car market, where more fuel efficient cars tend to be more expensive (Daly and Mayer, 1983; Kahn, 1986), reflecting perhaps half to three quarters of the value of fuel savings they will offer (Michaelis, 1996a). The lack of control of vehicle users over technology is exacerbated by the concentration of the global car industry in Annex I countries, and in a small number of transnational companies (IPCC, 2000b).

While information on the fuel efficiency of vehicles is widely available, it may not be easy to find or assimilate for the average purchaser. Labelling laws and information programmes have been introduced in many countries to overcome this information gap (ECMT, 1997). Nevertheless, the fuel economy information on labels is usually obtained in standard test cycles, the information from which may be inaccurate, underestimating consumption in real driving conditions by 10%-20% (IPCC, 1996).

Car technology is also a good example of "lock-in". A century of development has put the gasoline engine, and the infrastructure to maintain it and supply its fuel, in a virtually unassailable position. Technologies based on alternative fuels, batteries, or fuel cells will have to compete with gasoline engine performance and cost levels that continue to improve.

The phenomenon of lock-in can also be seen to apply to road transport more generally. Cars are preferred over other transport modes partly because of their intrinsic advantages in flexibility, convenience, comfort, and privacy. A car makes it possible to live in a suburban or rural area poorly served by public transport, taking advantage of low house prices and pleasant surroundings. However, there are also many sources of "positive returns to scale", strengthening the incentives for using cars as their prevalence grows.

As car fleets have grown, modern western societies, cultures, and economies are increasingly built around motorized road transport. Car-oriented culture has charged cars with significance as a means of freedom, mobility and safety, a symbol of personal status and identity, and as one of the most important products in the industrial economy. Car-oriented infrastructure and settlement planning makes it hard to use any alternative transport mode. Many attempts to encourage a shift in planning provision away from cars, toward public and non-motorized transport also fail because of the strength of links among trans-

port planners, construction firms and the financing institutions (e.g., Stenstadvold, 1995).

A second aspect of the lock-in to car transport is the result of economies of scale, and a century of R&D and learning from experience in car production. The real cost of owning and operating a car has declined over the last half century while public transport costs have risen. The declining number of people using buses, especially in rural areas, makes it uneconomic to operate services without subsidies. Falling bus and train occupancy levels also reduce their energy intensity advantage relative to cars, indeed, in some countries, trains consume more energy per passenger-km than cars (IPCC, 1996).

A third source of lock-in is linked to personal safety. With growing numbers of cars on the roads and declining numbers of pedestrians, the streets have become more dangerous. While travelling by car poses a higher risk of death or injury from accidents than travelling by bus or train, a car does offer protection from personal assault.

Because of the social and economic importance of transport, most governments provide budgetary subsides for construction and maintenance of transport infrastructure, and for transport services including many linked to car use (de Moor and Calamai, 1996; OECD, 1997b). Public finance for public and non-motorized transport has been generally less readily available than for road building since the 1950s. Other government instruments often support road transport, one example being planning laws that require off-street parking to be provided in new urban developments. It is the combination of policies and institutional relationships protecting road transport interests that poses the greatest barrier to change, rather than any single type of instrument (OECD, 1997b).

People have distorted perceptions of the relative convenience and cost of transport modes, usually justifying their habitual mode choices (Goodwin, 1985; OECD, 1997a). Bus users perceive trains as more expensive and less convenient than they really are, while train users have a similar misperception of buses. Car drivers believe that car use is cheaper and faster than it is.

GHG mitigation efforts in freight transport also face many barriers. The energy intensity of road freight can be reduced by improving fleet dispatching and routing, reducing the number of empty trips, and improving driving skills. While freight firms continue to make substantial efforts to minimize fuel use by trucks, speed, flexibility and responsiveness to customers is often a higher priority.

Moving freight by rail instead of by road can offer considerable energy savings in some countries (IPCC, 1996), mainly where long distances are involved and the freight can travel relatively slowly. However, nearly all freight movements must start and end by road, so that taking advantage of the low energy intensity of rail freight entails a loss of convenience as either containers must be loaded onto the train and unloaded for delivery, or trucks must be carried "piggy-back". Increasing rail freight depends on substantial investments in road-rail terminals. Meanwhile, it may be difficult for railways to operate efficiently with high levels of both passenger and freight traffic owing to the different operating patterns entailed.

5.4.2.2 *Opportunities for Mitigation*

Some of the most promising opportunities for GHG mitigation in the transport sector are linked to the growing need for action to address a wider range of concerns about the sector's social and environmental impacts. Several studies have evaluated environmental and social externalities associated with road transport (IPCC, 1996; ECMT, 1997; OECD, 1997a). Some have explored the effects of internalizing those costs through fuel taxes and other measures (EC, 1996; Michaelis, 1996b; ECMT, 1997). However, transport fuel taxes have proved very unpopular in some countries, especially where they are seen as revenue-raising measures (MVA, 1995), and may be an inefficient means of internalizing environmental costs other than those associated with carbon dioxide (CO_2) emissions. Charges on road users, including parking fees in many towns and tolls, especially on motorways, have been accepted where they are earmarked to cover the costs of road provision (Michaelis, 1997a). Several studies have explored the potential for adjusting the way existing road taxes, license fees, and insurance premiums are levied, and have found potential emissions reductions in the region of 10% in OECD countries (Wenzel, 1995; Michaelis, 1996b).

While it may be possible to adjust the price incentives in the transport sector, overcoming the many forms of inertia and lock-in is more difficult. Effective mitigation strategies would entail combinations of measures, just as the status quo is currently maintained by a combination of forces (IPCC, 1996). Often, the best opportunities for such concerted action arise at a local level, where the negative impacts of transport are most keenly felt (Michaelis, 1997a). There are several positive experiences of change, such as a Scottish example where a public consultation process led to a large shift in local government spending towards public transport (Macaulay *et al.*, 1993), initiatives to introduce toll rings around Norwegian cities, and the comprehensive transport strategies in Singapore (Ang, 1993), Curitiba (Rabinovitch, 1993), and other cities (IPCC, 1996).

Achieving the promise of new technology may depend on international co-operation to develop larger markets for low-GHG-emission vehicles through fiscal and regulatory measures and public purchasing. During high oil price periods, car importing countries have imposed restrictions and incentives on car importers to discourage the use of more energy-intensive cars. Agricultural surpluses and foreign exchange shortages have been important stimuli for technology development in the past, in particular in the case of the Brazilian ethanol programme.

While several studies have found that people living in denser and more compact cities rely less on cars (Armstrong, 1993), energy savings alone are unlikely to motivate the shift away from suburban sprawl to compact cities advocated by Newman and Kenworthy (1990). However, there is a growing concern to reverse the decline in the environment and in communities in city centres by moving away from zoning and car-based transport, and towards multi-function, high-density pedestrian zones. There is a considerable opportunity for GHG mitigation in linking to this concern. In particular, there is scope where infrastructure is developing rapidly to implement planning measures that encourage more sustainable transport patterns, avoiding the pollution, congestion, higher accident rates, and GHG emissions associated with cars.

5.4.3   *Industry*

Under perfect market conditions, all additional needs for energy services are provided by the lowest cost measures for increased energy supply or reduced energy demand. There is considerable evidence that energy efficiency investments that are lower in cost than the cost of marginal energy supply are not being made in real markets, suggesting that market barriers exist. A study of the industrial electric motor market in France has demonstrated the existence of barriers arising from decision-making practices, within an environment characterized by lack of information and split incentives (de Almeida, 1998). Barriers may exist at various points in the diffusion process of measures to reduce energy use and/or GHG emissions. The diffusion process depends on many factors such as capital cost, operating cost savings, information availability, network connections, imitation effects, and other factors (DeCanio and Laitner, 1997). All of these factors influence the probability of a firm adopting a given technology at a particular point in time. Barriers may take many forms in this process, and should be reviewed in the context of the industrial and business environment (e.g., multi-criteria optimization, firm size and structure, market structure, opportunity, and information routes). While barriers exist, it is important to note that ESTs and practices may also represent a strategic and competitive advantage through the development of new markets or new market opportunities, as shown by various authors (Porter and Van der Linde, 1995b; Reinhardt, 1999). This section focuses on barriers and opportunities in the industrial sector, and cites examples of successful approaches that have been used to remove barriers.

Decision-making Processes

In firms, decision-making processes are a function of its rules of procedure, business climate, corporate culture, managers' personalities, and perception of the firm's energy efficiency (DeCanio, 1993; OTA, 1993) and perceived risks of the investment, stressing the importance of firm structure, organization, and internal communication (Ramesohl, 1998). Energy awareness as a means to reduce production costs seems not to be a high priority in many firms, despite a number of excellent examples in industry worldwide. For example, Nelson (1994) reports on a (discontinued) successful programme at a major chemical company in the USA, which resulted in large energy savings with internal rates of return of over 100%. However, such programmes are only reported in a relatively small number of plants. A recent analysis of the Green Lights programme in the USA demonstrated the shortcomings in traditional decision-making processes, as investments in energy efficient lighting showed much higher paybacks than other investments. (DeCanio, 1998). These analyses demonstrate the need for a better understanding of the decision-making process, to be appropriately accounted in modelling and policy development.

Lack of Information
Cost-effective energy efficiency measures are often not undertaken as a result of lack of information on the part of the consumer, or a lack of confidence in the information, or high transaction costs for obtaining reliable information (Reddy, 1991; Sioshansi, 1991; OTA, 1993; Levine *et al.*, 1995). Information collection and processing consumes time and resources, which is especially difficult for small firms (Gruber and Brand, 1991; Velthuijsen, 1995). In many developing countries public capacity for information dissemination is especially lacking (TERI, 1997). The information gap concerns not only consumers of end-use equipment but all aspects of the market (Reddy, 1991). Many producers of end-use equipment have little knowledge of ways to make their products energy efficient, nor access to the technology for producing the improved products. Equipment suppliers may also lack the information, or ways to assess, evaluate, or disseminate the information. End-use providers are often unacquainted with efficient technology. In addition, there is a focus on market and production expansion, which may be more effective than efficiency improvements, to generate profit maximization. In the New Independent States (NIS) firms are more directed towards increasing competitiveness, although there are examples where firms have used energy efficiency as a means to reduce production costs (Gritsevich, 2000). Also, a lack of adequate management tools, techniques, and procedures to account for the economic benefits of efficiency improvements is an information barrier (see below). Finally, other policies and regulations may limit access to energy-efficient technologies. For example, import regulations for specific projects and industries in China (Fisher-Vanden, 1998) and India (Schumacher and Sathaye, 1999) limited or imposed high levies on the import of industrial technologies for some periods.

Limited Capital Availability
Energy efficiency investments are made to compete with other investment priorities, and many firms have high hurdle rates for energy efficiency investments because of limited capital availability. Capital rationing is often used within firms as an allocation means for investments, leading to even higher hurdle rates, especially for small projects with rates of return from 35% to 60%, much higher than the cost of capital (~15%) (Ross, 1986). In many developing countries cost of capital for domestic enterprises is generally in the range of up to 30%-

40%. Especially for SMEs capital availability may be a major hurdle in investing in energy efficiency improvement technologies because of limited access to banking and financing mechanisms. When energy prices do not reflect the real costs of energy (without subsidies or externalities) then consumers will necessarily underinvest in energy efficiency. Energy prices, and hence the profitability of an investment, are also subject to large fluctuations. The uncertainty about the energy price, especially in the short term, seems to be an important barrier (Velthuijsen, 1995). The uncertainties often lead to higher perceived risks, and therefore to more stringent investment criteria and a higher hurdle rate.

Lack of Skilled Personnel
A lack of skilled personnel, especially for SMEs, leads to difficulties installing new energy-efficient equipment compared to the simplicity of buying energy (Reddy, 1991; Velthuijsen, 1995). In many firms (especially with the current development toward "lean" firms) there is often a shortage of trained technical personnel, as most personnel are busy maintaining production (OTA, 1993). In most developing countries there is hardly any knowledge infrastructure available that is easily accessible for SMEs. Also, the position within the company hierarchy of energy or environmental managers may lead to less attention to energy efficiency, and reduced availability of human resources to evaluate and implement new measures.

In addition to the problems identified above, other important barriers include (1) the "invisibility" of energy efficiency measures and the difficulty of demonstrating and quantifying their impacts; (2) lack of inclusion of external costs of energy production and use in the price of energy, and (3) slow diffusion of innovative technology into markets (Fisher and Rothkopf, 1989; Levine *et al.*, 1994; Sanstad and Howarth, 1994). Regulation can contribute to more successful innovation (see above), but sometimes, indirectly, be a barrier to implementation of low GHG emitting practices. A specific example is industrial co-generation (CHP), which may be hindered by the lack of clear policies for buy-back of excess power, regulation for standby power, and wheeling of power to other users (*Box 5.5*). Co-generation in the Indian sugar industry was hindered by the lack of these regulations (WWF, 1996), while the existence of clear policies can be a driver for diffusion and expansion of industrial co-generation, as is evidenced by the development of industrial co-generation in the Netherlands (Blok, 1993). Finally, firms typically under-invest in R&D, despite the high paybacks (Nelson, 1982; Cohen and Noll, 1994), but recent analyses seem to suggest that public and private R&D funding for sustainable energy technologies is decreasing in developed countries (Kammen and Margolis, 1999).

Programmes and Policies for Technological Diffusion
A wide array of policies, to reduce the barriers or the perception of barriers has been used and tested in the industrial sector in developed countries (Worrell *et al.*, 1997), with varying success rates. With respect to technology diffusion policies there

is no single instrument to reduce barriers; instead, an integrated policy accounting for the characteristics of technologies, stakeholders, and countries addressed would be helpful.

Selection of technology is a crucial step in any technology transfer. Information programmes are designed to assist energy consumers in understanding and employing technologies and practices to use energy more efficiently. Information needs are strongly determined by the situation of the actor. Therefore, successful programmes should be tailored to meet these needs. Surveys in western Germany (Gruber and Brand, 1991) and the Netherlands (Velthuijsen, 1995) showed that trade literature, personal information from equipment manufacturers and exchange between colleagues are important information sources. In the United Kingdom, the "Best Practice" programme aims to improve information on energy efficient technologies, by demonstration projects and information dissemination. The programme objective is to stimulate energy savings worth US$5 for every US$1 invested (Collingwood and Goult, 1998). In developing countries technology information is more difficult to obtain. Energy audit programmes are a more targeted type of information transaction than simple advertising. Energy audit programmes exist in numerous developing countries, and limited information available from 11 different countries found that on average 56% of the recommended measures were implemented by audit recipients (Nadel *et al.*, 1991).

Environmental legislation can be a driving force in the adoption of new technologies, as evidenced by the case studies for India (TERI, 1997), and the process for uptake of environmental technologies in the USA (Clark, 1997). Market deregulation can lead to higher energy prices in developing countries (Worrell *et al.*, 1997), although efficiency gains may lead to lower prices for some consumers.

Direct subsidies and tax credits or other favourable tax treatments have been a traditional approach for promoting activities that are socially desirable. An example of a financial incentive programme that has had a large impact on energy efficiency is the energy conservation loan programme that China instituted in 1980. This loan programme is the largest energy efficiency investment programme ever undertaken by any developing country, and currently commits 7% to 8% of total energy investment to efficiency, primarily in heavy industry. The programme contributed to the remarkable decline in the energy intensity of China's economy. Since 1980 energy consumption has grown at an average rate of 4.8% per year (compared to 7.5% in the 1970s) while GDP has grown twice as fast (9.5% per year), mainly thanks to falling industrial sector energy intensity. Of the apparent intensity drop in industry in the 1980s, about 10% can be attributed directly to the efficiency investment programme (Sinton and Levine, 1994).

New approaches to industrial energy efficiency improvement in developed countries include voluntary agreements (VA). A VA generally is a contract between the government (or an other regulating agency) and a private company, association of companies or other institution. The content of the agreement may vary. The private partners may promise to attain certain energy efficiency improvement, emission reduction target, or at least try to do so. The government partner may promise to financially support this endeavour, or promise to refrain from other regulating activities. Many developed countries have adopted VAs directed at energy efficiency improvement or environmental pollution control (EEA, 1997; IEA, 1997; Börkey and Lévêque, 1998; OECD, 2000). There is a wide variety in VAs, ranging from public and consumer recognition for participation in a programme (e.g., Energy Star Program in the USA) to legally binding negotiated agreements (e.g., the Long-Term Agreements in the Netherlands). Voluntary agreements can have some apparent advantages above regulation, in that they may be easier and faster to implement, and may lead to more cost-effective solutions. Initial experiences with environmental VAs with respect to effectiveness and efficiency varied strongly, although only a few ex-post evaluations are available as most voluntary approaches are recent (EEA, 1997; Worrell *et al.*, 1997, Börkey and Lévêque, 1998). The Dutch long-term agreements on energy efficiency in industry have been evaluated favourably, and are expected to achieve the targets for most sectors (Universiteit Utrecht, 1997). The evaluation highlighted the need for more open and consistent mechanisms for reporting, target setting, and supportive policies. Preliminary evaluations show that VAs are most suitable for pro-active industries, a small number of participants, mature sectors with limited competition, and long-term targets (EEA, 1997). The evaluations also show that VAs are most effective if they include clear targets, a specified baseline, a clear monitoring and reporting mechanism, and if there are technical solutions available with relatively limited compliance costs (EEA, 1997). In some cases the result of a VA may come close to those of a regulation, *i.e.*, in the case of negotiated agreements as used in some European countries. Outside developed countries, also some NICs, e.g., Republic of Korea, consider the use of VAs (Kim, 1998), while the Global Semiconductor Partnership is an example of an international voluntary agreement to reduce PFC emissions.

5.4.4   Energy Supply

There are two primary types of options available for reducing emissions. One is to increase the efficiency of energy supply, and the second is to switch from carbon intensive fuels to low or no carbon content sources of energy. The two options face different categories of barriers and the most relevant are described in this section.

Energy Prices
Low prices are, in part, a consequence of direct and indirect subsidies to producers, and the non-inclusion of external costs in their production and use (Watson *et al.*, 1996; Harou *et al.*, 1998). It is common in the energy supply sector to find price policies (public or private) which do not reflect the "full costs". These full costs include environmental externalities, which, for

example, are not included in any coal transaction or gasoline prices in the United States. Producers and users of new energy technologies are not usually rewarded for the associated environmental benefits (World Bank, 1999).

Lack of Consistency in the Evaluation of Energy Costs

Closely related to the price barrier faced by clean fuels is the selective evaluation of energy costs from different energy sources. There is a need to make a comprehensive evaluation of all costs and benefits.

Lack of Adequate Financial Support

Multilateral development banks, public banks, and private banks generally do not offer soft credit, or programmes aimed specifically at energy technologies. This acts as a further barrier to capital-intensive energy projects. The absence, up until now, of specific programmes and an administrative process adapted to this type of project has resulted in high transaction costs and a lack of discussion of this key issue as a solution in the climate change problem. The role of a multilateral system could be especially important for the development of a hydropower programme, financing of regional interconnections, and developing small, sound environmental technologies for energy supply like mini hydro, solar, and wind.

Institutional Transformation and Reforms

Privately-owned generation, transmission, and distribution entities are playing increasingly large roles in electric utility systems worldwide. Many national power utility systems have been totally or partially privatized.

The liberalization of the power industry, which introduces competition within the generation segment, could have a significant impact on the viability of renewable sources. Some observers may argue that subsidies of any sort are antithetical to the concept of a deregulated market, and that the purpose of liberalization is precisely to eliminate such subsidies and market distortion. In competitive markets where the process is replaced by the market-driven decisions of generation companies subsidies to renewable sources may become less acceptable (Bouille, 1998).

Segmentation of the electricity chain may reduce the incentives for electricity companies, especially electricity distribution companies, to act on end-use efficiency (Poole *et al.*, 1995).

There are institutional and administrative difficulties associated with the development of technology transfer contracts. These are necessary to qualify regional construction companies as partners in any undertaking. There is a need for greater regional co-operation among developing countries in both research and development, and the development of an international commercial contracting network, to improve technology transfer.

Along with the institutional difficulties of technology transfer projects, high transaction and implementation costs act as bar-

riers as well. Often, cost estimations of new technologies do not include items related to transaction costs or items associated with technology penetration (policy implementation costs). Both transaction costs and policy implementation costs are additional expenses to technology transfer, limiting competitiveness and market potential.

Legal and Regulatory Framework

Many energy supply sources are subject to a lack of regulation other than for safety, inadequate tariffs for transport and distribution, and no incentives to increase efficiency. For example, there is often no penalty for natural gas flaring. This reduces the motivation for improving the efficiency of the supply chain of such sources.

If electric utility companies sell electricity within a regulatory system that allows them to recover all operating expenses, including taxes and a fair return for their investments, they will show no interest in increasing their efficiency. Within this system, utilities will be reimbursed the operational costs independent of the quality of the service offered (US DOE, 1996).

Distributed electricity generators often face a complex bureaucratic process for authorizing the construction and operation of co-generation facilities. Complicated terms of grid connection, as well as technical, economic, and institutional rules limit access to the grid for distributed generators (Verbruggen, 1990,1992, 1996).

Lack of Information

While lack of information on energy technology performance, technical, and economic characteristics is not a very significant barrier in the energy supply sector, this market failure is related to market transparency. The inability of the private market to provide generic information (no transparency), and the possibility that "in the field" operation of a technology may differ from controlled environment operation by a technology producer, both increase uncertainty and risk in an investment. These problems are extensions of the information barrier[4].

Developed countries generally have more capital and technological resources than do developing countries (World Bank, 1999). This can greatly affect the decision-making process in developing countries, as they may not have the newest knowledge to adequately assess new technology opportunities.

Decision-making Process and Behaviour

Many organizations are interested in using the most economically competitive technology, in terms of cost and availability of fuels, though not necessarily in terms of energy or carbon

---

[4] Any decision-making process is one where the decision maker "buys" information to reduce uncertainty and risk in order to make a "better" decision. Lack of information means, essentially, uncertainty. The lower the degree of information the higher the uncertainty and the barrier to adoption of a specific technology.

efficiency. The most competitive investments offer short pay-back periods, minimize overall investment, and receive an attractive rate of return. In such a framework, a relatively narrow range of technologies exists. Most of them are efficient in the economic sense but not necessarily in relation to GHG emissions reductions or avoidance. This represents a significant barrier to both developed and developing countries.

Co-generation as a distributed technology is an example of this type of barrier (*Box 5.5*). Another example of the "competitive" decision-making process as a barrier is typified in the case of Argentina, where systems with shorter payback periods (such as natural gas-fired systems) are favoured over others (*Box 5.6*). Changes similar to those described in *Box 5.6* are taking place in other developing and developed countries as well.

Box 5.5. Combined (Cooling) Heating and Power or Cogeneration

Co-generation is applied in utility district heating and in distributed on-site power units. Most barriers to on-site co-generation are the same barriers as the ones that impede the development of other types of distributed and/or independent power generation projects. The most important barriers are related to information, technology character, regulatory and energy policy.

Informational barriers
The significant technological advances of recent years (Major, 1995; Rohrer, 1996) are not spread widely enough. This barrier is the most stringent in developing countries and in small institutions and companies, especially when the latter have no technical background. When donors, international institutions, lending banks, etc. are not familiar with the co-generation technology, it will not be implemented by developing and transitional economies (Dadhich, 1996; Nielsen and Bernsen, 1996). Additionally, the economics of co-generation is relatively complex (Verbruggen *et al.*, 1992; Hoff *et al.*, 1996; Verbruggen, 1996). Optimization of co-generation projects requires extensive information about many determinants of profitability. This span of know-how makes its availability to small-scale independent projects exceptional. Finally, uncertainty about the main determinants like fuel prices, fuel availability, regulatory conditions, environmental legislation, contract terms with the power grid, etc. constitutes a significant barrier.

Decentralized character of the technology
Private investors impose high profitability standards on distributed generation projects. This payback gap is mainly due to a risk-averse attitude regarding non-core business activities. The distances to the energy grids (electricity, natural gas) limit the capacity or co-generation opportunities. Unequal treatment with respect to fuel supplies, authorization and licensing arrangements, and environmental and emissions regulation, constitutes an additional set of barriers that especially affect the small-scale distributed generation projects and add to the costs of the technology (COGEN Europe, 1997).

The terms of grid connection
In several countries, the position and attitude of the grid operator have been hostile towards distributed generation initiatives (Rüdig, 1986; Dufait, 1996). Incumbent power companies sometimes impose heavy regulations on producers or industries that file for a connection to the electricity grid, imposing technical prescriptions that cannot be set in standard packages. Tariff conditions are a particularly difficult issue, because the value of the kWh is dependent on time, place, quality, and reliability of supply, and differs for the three types of power flows that can be exchanged: surplus power that the co-generator delivers to the grid, shortage or make-up power bought by the co-generator at the grid, and back-up power (Verbruggen, 1990). Although there are widely accepted principles to fix the tariff for the different transactions, theoretical and practical difficulties in defining and measuring the costs constrain the development of contracts (Dismukes and Kleit, 1999). In many countries high tariffs on wheeling of electricity act as an additional barrier. In several countries the opportunities for small-scale distributed power generation are improving because grid connection is provided at neutral or even subsidized terms (the Netherlands and Japan; Blok and Farla, 1996).

Energy policy
Utility co-generation requires long-term planning from an integrated point of view (WEC, 1991). Very few nations own the intellectual and administrative capacity to realize an integrated energy policy plan that preserves the place for district heating and related co-generation. Some countries (e.g., Denmark) and international organizations have favoured the development of CHP (EC, 1997). Firm public policy and regulatory authority is necessary to install and safeguard harmonized conditions, transparancy and unbundling of the main power supply functions, and the position of independent players (Fox-Penner, 1990).

Box 5.6. Argentine Power Supply: Some Barriers Related to Institutional and Regulatory Topics

There is no doubt that the Argentine electric power system shows a trend towards the improvement of energy efficiency, both in final consumption as well in electricity-supply activities. Rising competition levels within the electricity industry are favouring efficiency in electricity generation. However, market trends show a rising dependency on natural gas to the detriment of the participation of non-GHG emitting technologies.

Several obstacles will have to be overcome to modify this trend. These are related to the following aspects.

Spot and contract prices. Within a context of falling prices at the spot market, distributors have been reluctant to long-term fixed price commitment. In fact, the indicator used to adjust the price at distribution level is the spot price. Should its supply be totally or partially contracted, the distributor cannot transfer to retail rates the costs of their contracts if they have, occasionally, a higher price than the spot. Long-term payback investment, with higher investment costs, major risks, and lower internal rate of return, are not favoured by a context based on spot prices.

Volatility of prices. A system with important hydro generation capacity shows variation depending on hydrological conditions. Dry and humid years represent important impacts on the income of hydroelectric generators and introduce an additional source of risk. This volatility could potentially increase if the interconnection with the Brazilian system becomes a reality in the short term. The Brazilian system is almost entirely supplied with hydroelectric generation, which has frequent surplus capacity. This surplus or non-firm energy, with zero value, could enter the Argentine wholesale market and introduce a fantastic volatility in the spot price market which would affect all generators.

Behaviour. Private investors are reluctant about options that imply higher risks, longer payback, lower internal rate of return, and high investment per unit of capacity. The decision-making process clearly shows this behaviour: all the new capacity installed after the privatization process is based on open and combined cycle thermal power plants using natural gas as fuel. In the past, Argentine public utilities, using lower discount rates, assuming higher risks, and making investments assisted by the multilateral financing system, developed an important hydropower system that represented near 50% of the supply. The new context offers lower opportunities for this "old" technology, and acts as a barrier to a more "costly" option from a private point of view.

Economics of the technologies. In the case of nuclear power plants, additional costs for waste treatment, plant decommissioning, and insurance reduce the competitiveness of this technology. In the case of hydroelectric stations, the payments of royalties, the need for insurance, and the transmission network expansion mechanism (payback in 15 years) increase the costs and decreases the possibilities of such technologies in the decision-making behaviour described above.

Uncertainty and risk aversion discourage long-term investments. Many forms of sustainable energy production require long-term investment. Most multilateral and international lending institutions are averse to technologically risky investments. As a result, both government and private entities may be reluctant to invest in high-tech projects that entail high capital costs (ECOSOC, 1994).

The lack of performance data for newer energy technologies often results in an unwillingness on the part of smaller firms to risk purchasing these more expensive technologies. While they may offer greater future savings than traditional technologies, the lack of test data prompts fears that reported energy savings may not materialize in practice.

The uncertainty inherent in new technologies leads investors to use high discount rates, which would make investments that are clearly cost-effective from a global perspective seem unattractive to private actors (Bouille, 1999). In the case of energy-efficiency investments, however, some may be for well-established technologies with low technological and economic risk.

Inclusion of renewable energy in a wholesale electricity market could affect price volatility for generators. Volatility is remarkably affected by hydroelectricity supply. Any mitigation action which increases the share of such a source in the electricity market will most likely contribute to further price volatility, increasing the level of risk for the actors.

Social and Cultural Constraints

The environmental impacts and risks of technologies, such as nuclear power and hydropower generation, may not be acceptable to many social groups. The real or perceived environmental risks of such technologies pose a significant barrier to their implementation (Bouille, 1998).

Cultural Aspects Related with Decentralized Systems in Rural Areas

There are cultural barriers that oppose the use of decentralized systems in rural areas. Renewable energy is often promoted in rural areas to reduce local environmental impacts, and accomplish social and welfare goals. While these technologies may be competitive, easy to operate, and adequate for the project needs, technology diffusion is often confronted with cultural barriers (Barnett, 1990).

In order to overcome cultural and social barriers, a project must take into account the needs of potential users of the project technology, and harmonize the diffusion strategy with local physical, human, and institutional resources. A project should also build local technical and institutional capabilities so the project may be fully realized (Barnett, 1990).

Capital Availability
There are substantial opportunities in developing countries for expansion of electricity supply. While the capacity being installed is improving in efficiency, this process is slowed by difficulties in accessing the necessary capital. Many ESTs require large up-front investments. In effect, the cost of pollution abatement is paid in advance. This is a serious obstacle for some technologies, particularly nuclear power generation and large hydropower schemes. These technologies also have other constraints, however. A reduction in nuclear unit size and/or improved safety and maintenance features could help to overcome this barrier.

Co-generation or combined production of power and heat is a much more efficient process than the production of each of these energy sources alone. Implementation of co-generation, however, faces barriers such as shortages of capital. There is also currently a lack of regulatory policies allowing commercialization of the excess electricity produced through access to existing grid systems (*Box 5.5*).

5.4.5 Agriculture

The Special Report on Land use, Land-use Change, and Forestry (IPCC, 2000a) estimated a significant potential for increasing carbon stocks in the agricultural sector. Improved management of cropland and grazing-lands, agroforestry, and rice paddies have the potential to sequester 398 MtC annually, and the conversion of cropland to agroforestry practices and grasslands can sequester an additional 428 MtC annually by 2010. These estimates are highly uncertain, however, and do not include the impact on the net emissions of methane (CH_4) or nitrous oxide (N_2O) from agricultural practices or wetlands and/or permafrost management.

CH_4 emissions from agriculture produce about eight per cent of the radiative forcing of all GHGs (Watson *et al.*, 1996). CH_4 from manure can be captured and used for fuel; emissions from ruminants can be reduced with better diets, feed additives, and breeding; and emissions from rice paddies can be mitigated by nutrient management, water management, altered tillage practices, cultivar selection, and other practices (Mosier *et al.*, 1998).

Many of the mitigation options to address these opportunities may provide multiple benefits to the farmer and society at large. Improving soil management for crop production, for instance, can also improve water relations, nutrient retention, and nutrient cycling capacity (Paustian *et al.*, 1998). Retiring surplus agricultural lands can result in improved water quality, reduced soil erosion, and increased wildlife habitat. As Izac (1997) points out, however, farmers, who will be the ultimate decision makers about which mitigation option to adopt, have shorter planning horizons than national or international beneficiaries, and many mitigation options ask them to bear costs up front while the benefits are longer term and to the society at large.

Furthermore, in order to realize these opportunities a very large proportion of farmers who pursue diverse agricultural practices will have to be convinced to adopt mitigation options. Economic, cultural, and institutional barriers exist which restrict the rate of adoption of such practices. Farmers who are accustomed to traditional practices may be reluctant to adopt new production systems. Crop price supports, scarcity of investment capital, and lack of economic incentives for addressing environmental externalities are some of the economic barriers. Limited applicability of mitigation options to different types of agriculture, negative effects on yield and soil fertility for rice production, and the increased skilled labour requirements are some of the other constraints. Among these barriers the especially critical ones are highlighted here.

Farm-level Adoption Constraints
Several generic constraints characterize the adoption of most new agricultural technology. These include small farm size, credit constraints, risk aversion, lack of access to information and human capital, inadequate rural infrastructure and tenurial arrangements, and unreliable supply of complementary inputs. Participatory arrangements that fully engage all the involved actors may help to overcome many of these barriers.

Government Subsidies
Subsidies for critical inputs to agriculture, such as fertilizers, water supply, and electricity and fuels, and to outputs in order to maintain stable agricultural systems and an equitable distribution of wealth can distort markets for these products. These types of subsidies prevail in both developed and developing countries. Low electricity prices in India, for example, provide a disincentive for the use of efficient pump sets, and encourage increased use of ground water, which depletes the water reservoirs. In the OECD, for example, high levels of farm subsidies have also contributed to the intensification of farm practices and often provide incentives to increase fertilizer use, livestock density, etc. (Storey, 1997).

Lack of National Human and Institutional Capacity and Information in the Developing Countries
Several of the Consultative Group on International Agricultural Research (CGIAR) systems are experiencing difficulty as their funding slows. The systems have not transferred capacity to national centres in the developing countries that they are expected to serve. The national centres also lack access to

information, and are not aware of technologies that suit their local conditions (IPCC, 2000b).

Intellectual Property Rights

To some extent the reduced public funding on new technologies has been replaced by the private sector's contribution. Private sector funding offers one approach to increasing investment for mitigation projects worldwide. Private plant breeding research has more than quadrupled in the USA in real terms between 1970 and 1990. Its international role is, however, controversial. Protection of intellectual property rights is weak, especially for commercially developed seed varieties (Deardorff, 1993; Frisvold and Condon, 1995, 1998; Knudson, 1998). On the other hand, hybridization will help to stimulate more investment from the private sector at the risk of increasing the farmers' dependency on the annual purchase of new seeds. There are also concerns that genetic resources that have not been considered as privately-owned intellectual property may get patented worldwide by private investors.

Several measures may be pursued to address the above barriers. These include

- The expansion of internationally supported credit and savings schemes, and price support, to assist rural people to manage the increased variability in their environment (Izac, 1997);
- Shifts in the allocation of international agricultural research for the semi-arid tropics towards water-use efficiency, irrigation design, irrigation management, and salinity, and the effect of increased CO_2 levels on tropical crops (Tiessen *et al.*, 1998);
- The improvement of food security and disaster early warning systems, through satellite imaging and analysis, national and regional buffer stocks, improved international responses to disasters, and linking disaster food-for-work schemes to adaptation projects (e.g., flood barricades);
- The development of institutional linkage between countries with high standards in certain technologies, for example flood control; and
- The rationalization of input and output prices of agricultural commodities taking DES issues into consideration which would lead to more efficient use of input resources.

5.4.6   Forestry

In addition to the several generic barriers that are discussed in Section 5.3, the forestry sector faces land use regulation and other macroeconomic policies that usually favour conversion to other land uses such as agriculture, cattle ranching, and urban industry. Insecure land tenure regimes, and tenure rights and subsidies favouring agriculture or livestock are among the most important barriers for ensuring sustainable management of forests as well as sustainability of carbon (C) abatement.

The Special Report on Land Use, Land-use Change and Forestry (IPCC, 2000a) notes significant opportunities for forestry and other land-use change activities to sequester carbon. Afforestation and reforestation activities could capture between 197 to 584MtC/yr in all countries under the IPCC "definitional" scenario between 2008 to 2012. The estimated deforestation, however, would negate this sequestration potential. Halting deforestation offers additional opportunity to reduce emissions. Forest management and agroforestry options offer a potential to capture another 700MtC/yr by 2010. Capturing these opportunities, however, entails significant hurdles of the types noted below.

Lack of Technical Capability

In many developing countries, the national and state forest departments play a predominant role in all aspects of forest protection, regeneration, and management. Currently lack of funding and technical capabilities in most tropical countries limit generation of information required for planning and implementation of forestry mitigation projects. Apart from a few exceptions, developing countries do not have adequate capacity to participate in international research projects and to adapt and transfer results of the research to the local level. Research on forests has not only suffered from a lack of resources; it has not been sufficiently interdisciplinary to provide an integrated view of forestry (FAO, 1997). However, the majority of the forestry research institutions do not function as R&D laboratories as they do in industry, and the main focus is on research and not technology development and dissemination. Unlike in the energy or transportation sectors, the technologies or even the management systems are going to be forest type or country specific.

Lack of Capacity for Monitoring Carbon Stocks

Forestry-sector GHG mitigation activities and joint implementation projects generally face a wide range of technical issues that challenge their credibility. The twin objectives of using forestry to mitigate climate change and managing forests sustainably do pose a challenge in monitoring and verifying benefits from carbon offset projects in the sector (Andrasko, 1997). While methods generally exist to monitor carbon stocks in vegetation, soils and products, operational systems that could be readily implemented for this purpose are lacking in all countries (IPCC, 2000a). Monitoring and verification are key elements in gaining the credibility needed to capture the potential benefits of forestry sector response options, particularly in reducing deforestation (Fearnside, 1997). While this is a generic barrier to deforestation reduction initiatives, it also represents an opportunity for transferring the technologies needed to monitor land-use change and carbon stocks and flows. Among the mitigation options, there is a higher degree of certainty on reforestation and/or afforestation, less on forest management, and even less on forest conservation.

Under the GEF-UNDP sponsored Asian Least-Cost Greenhouse Gas Abatement Strategy (ALGAS), the US Country Studies Program (Sathaye *et al.*, 1997a), and other forestry sector capacity building and analytical activities have identified mitigation options and technologies. Furthermore, the policies to promote technology transfer have been identified (e.g., regulations, financial incentives) and sometimes implemented (e.g., Mexico, Bolivia). Under the UNFCCC, each party is required to communicate a national inventory of GHG emissions by sources and sinks. A large portion of the parties has completed this task and is trying to understand forestry sector emissions and removals by sinks, which has improved dramatically. Many parties are taking steps to manage forest systems as C reservoirs (Kokorin, 1997; Sathaye *et al.*, 1997a).

As a result of the UNFCCC and Kyoto Protocol, many developing and transitional countries are developing National Climate Change Action Plans (NCCAPs) which incorporate forestry-sector mitigation and adaptation options (Benioff *et al.*, 1997). "No regrets" adaptation and mitigation options have been identified that are consistent with national sustainable development goals. Bulgaria, China, Hungary, Russia, Ukraine, Mexico, Nigeria, and Venezuela all have developed very specific forestry sector climate action plans.

The Russian Federation has a progressive forestry sector climate change action plan (Kokorin, 1997), although its implementation is uncertain under the current economic conditions. Based on current economic and climate change scenarios several mitigation and adaptation scenarios have emerged: (1) creating economic mechanisms to increase forestry sector effectiveness and efficiency in logged (removal) areas, (2) providing assistance for forestation in the Europe-Ural region, (3) promoting fire management and protection for central and northeastern Siberia, and (4) limiting clear-cut logging in southern Siberia. These steps are significant since Russia contains approximately 22% of the world's coniferous forests.

Forestry mitigation projects are likely to be largely funded by Annex I countries and implemented in non-Annex I countries and EITs. Technology, including management systems, is an integral part of all projects funded by bilateral or multilateral or commercial agencies. Thus, promotion of mitigation projects also automatically promotes the flow of technology from donor agencies or countries to host countries or agencies. In fact, technology transfer is already happening. Forestry sector options are of relatively low cost compared to those in the energy sector (Sathaye and Ravindranath, 1998). But there are some problems and uncertainties regarding the incremental C abated: its sustainability, measurement, verification, and certification. All forestry sector GHG mitigation projects must ensure that they meet accepted standards for sustainable forest management (Sathaye *et al.*, 1997b). Independent verification of C abatement would help to increase the credibility and funding of forestry-sector mitigation projects.

5.4.7   Waste Management

Waste management represents an important challenge for the reduction of GHG emissions. Waste is also a potential resource, much of which can be recycled and reused (CPCB, 1998). Residential and commercial waste may be differentiated from industrial waste, a component of the latter being toxic and requiring special treatment. In all cases, there are options for bulk reduction at source. Thus, waste management entails the three R's – Reduction, Recycling and Reclamation – for recovery of usable components either directly (example: chemical recovery in pulp and paper mills) or indirectly through processing of waste (example: CH_4 recovery from landfills and from distillery effluents).

Wastes of various kinds including energy, raw materials, effluents, emissions, and solid wastes are omnipresent in different walks of life (ESCAP 1992, Debruyn and Rensbergen, 1994; Doorn and Barlaz, 1995). Non-availability of appropriate technology is often perceived as a major impediment (Nyati, 1994; Narang *et al.*, 1998). However, there are cases to cite that even the proven technologies do not penetrate into society as rapidly as their potential would suggest (Reddy and Shrestha, 1998; Shrestha and Kamacharya, 1998).

5.4.7.1   Barriers to Mitigation

One of the major driving factors in waste management is the economic environment. Market forces favour waste utilization when there is a shortage of raw materials or their prices are high. Waste utilization is directly influenced by the economic incentive for recovery of usable materials (Vogel, 1998). Apart from market forces, the other barriers (Painuly and Reddy, 1996; Parikh *et al.*, 1996; Mohanty, 1997) in waste management relate to the following:

- Lack of enabling policy initiatives, an institutional mechanism, and information on opportunities for reduction, recycling, and reclamation of waste;
- Organizational problems in collection and transport of waste from dispersed sources for centralized processing and value addition; and,
- Lack of co-ordination among different interest groups, although there are several examples of successful initiatives taken through private sector and NGO efforts as well as business-to-business waste minimization and recycling programmes.

5.4.7.2   Programmes and Policies to Remove Barriers

To overcome the barriers and to exploit the opportunities in waste management, it is necessary to have a multi-pronged approach which includes the following components:

- Building up of database on availability of wastes, their characteristics, distribution, accessibility, current practices of utilization and/or disposal technologies and their economic viability;
- An institutional mechanism for technology transfer though a co-ordinated programme involving the R&D institutions, financing agencies, and industry (Schwarz, 1997); and

- Defining the role of stakeholders including local authorities, individual house holders, NGOs, industries, R&D institutions, and the government.

The efforts of local authorities in waste management could focus on: the separation and reclamation of wastes through seperate collection of reusable wastes for recovery; provision of reclamation centres where the public can deliver wastes; arrangements for separation and reclamation at disposal sites and transfer stations (de Uribarri, 1998); arrangements for waste disposal with by-product recovery; and landfilling of residuals. Local authorities may enlist the support of the public and individual householders as well as NGOs to store recoverable wastes separately or deliver these to the reclamation centres. Local authorities can also consult the industry on how wastes could be best ultilized to meet their raw material requirement. Industry can be encouraged to accept wastes as secondary raw materials (NWMC, 1990).

R&D institutions could play an important role in waste utilization by development and dissemination of viable technological alternatives including pilot scale demonstration, organizing technology transfer workshops, and dissemination of information to industries. Land use and industrial estate planners can internalize waste utilization and/or minimization concerns in the process of siting industrial plants (Datta, 1999). The possibilities of siting industrial activities in such a way that wastes from one unit could be used as raw material for another could be explored. The arrangement might reduce capital outlay and operating costs, and also facilitate transfer and processing of products and/or raw materials.

Governments may introduce fiscal and regulatory measures for reduction of wastes and promotion of waste utilization. These may include incentives to producers and users to accept reduced packaging, incentives to consumers to return reclaimable wastes, incentives to local authorities to support reclamation and/or waste utilization activities, incentives to industries using recovered materials, financial support to R&D activities, awards to individuals and/or organizations for waste utilization, and penalties for not adopting waste minimization and/or utilization practices.

Programmes for providing training and education on waste minimization and utilization with an interdisciplinary approach could be developed. Waste utilization as a profession has no fixed boundaries. Skills of psychology, economics, material sciences, process design, and ecology are but some of the many requirements for the trained professional.

Even the best planned, designed, and executed waste utilization programme would fail without the effective participation of the public. Education of the public on waste utilization issues, therefore, would play a vital role in ensuring the success of the programme. A public education programme would be aided by the identification of appropriate communication systems (AIT, 1997; ESCAP, 1997; Bhide, 1998).

5.5   Regional Aspects

There are many barriers and opportunities, from the ones described before, which have a particular relevance to developing countries and EITs. The issues of sustainable and equitable development resonate in these countries as they undergo a rapid transformation towards market-oriented systems that are immersed in a global economy. Institutionally, the transformation in developing countries is significant, but it is often confined to specific sectors, such as the deregulation of the energy sector. On the other hand, the socialist economies are undergoing a more radical shift of the whole economy. These global patterns of change provide an opportunity for introducing GHG mitigation technologies and practices that are consistent with DES goals. At the macro-level the change to a market economy and the liberalization and opening of markets to foreign investment provides an opportunity to make significant improvements in the GHG intensity of the economy. Similarly, the restructuring of the energy sectors also offers an opportunity to introduce demand management and low or no GHG-emitting energy sources. As the sections below note, however, a culture of energy subsidies, institutional intertia, fragmented capital markets, vested interests, etc. presents major barriers to the introduction of such technologies and practices. The developed countries face different types of barriers and opportunities that prevent or slow the penetration of GHG mitigation technologies. These barriers and opportunities are related to their more affluent lifestyles. The sections below emphasize situations in the three groups of countries that call for a more careful consideration of the barriers and opportunities they face.

5.5.1   *Developing Countries*

As a group, the developing countries are undergoing rapid urbanization, which leads to increased industrialization and motorization that has altered the manner in which people relate to their environment (Rabinovitch, 1992). Much of their technology stock is derived from developed countries, and increased globalization tends to expose even remote populations to socio-cultural patterns observed in the developed countries. Yet, the majority of the population in these countries lives in rural areas, and often in absolute poverty. These underlying attributes and phenomena create or emphasize barriers and opportunities that are particular to this group of countries.

Trade and Environment
A larger external debt and balance-of-payments (BoP) deficit is a reality in many developing countries. If a GHG mitigation technology has to be imported, it is likely to add to this debt and BoP deficit. Another barrier to the technology transfer process is the requirement in technology transfer contracts of "intellectual property rights" (IPR), which guarantee that private firms are compensated for sharing their technology. If IPR laws are not effectively enforced, there is little incentive for private firms to share their technology. However, patents and licensing fees can

be very expensive and in such situations, developing countries may prefer the lowest priced, albeit possibly less efficient technology alternatives (Srivastava and Dadhich, 1999).

Institutional Framework

Deregulation and privatization offer an opportunity for improving energy efficiency and reducing GHG emissions in the energy sector. Studies and scenario analyses show, however, a consequent increase in emissions resulting from low fuel prices, displacement of hydro and nuclear plants by cheaper fossil-fired capacity, and a change in attitudes and behaviour of the energy suppliers (Bouille, 1999).

Distorted Energy Prices

Energy price subsidies have been in place in many developing countries in the name of reducing the financial burden on the poor. This has spawned a culture of dependency on energy subsidies that is gradually diminishing (Jochem, 1999).

Finance

Lack of available capital and lack of finance at low interest rates is pervasive in developing countries. Together with the absence of standards or energy labeling schemes, these barriers support the proliferation of inefficient equipment and first-cost-minimization philosophy. Additionally, low incomes and poverty constrain access to adequate finance, and oblige the purchase of inexpensive and often GHG-intensive equipment (Bouille, 1999). Provision of special funds targeted to the poor and government financing of the first cost of equipment are ways to increase the provision of energy services.

Barriers

Information gap hindering proper technology selection, lack of adaptation and absorption capability, lack of access to state of the art technology, and the small scale of many projects (Jochem, 1999) are specific and important barriers in low income developing countries to effectively exploit the full potential benefit of technology transfer. Lack of information also slows the decision-making processes in developing countries.

5.5.2 Countries Undergoing Transition to a Market Economy in Central and Eastern Europe and the New Independent States

The collapse of communism in Central and Eastern Europe and the subsequent disintegration of the Soviet Union brought the region's serious environmental problems to the attention of the international community. Although the countries in this vast area of the world are remarkably diverse, central economic planning had created a common pattern of environmental problems which included wastefulness, pollution-intensive economic systems, ill-designed and resource heavy technologies, and perverse incentives encouraging increase of output rather than enhancing efficiency of resource use. A universal feature was also the world's highest energy and carbon intensity of economies.

A Soviet-type economy has left a legacy of acute health effects from local pollution. Having very scarce resources, the transition economies have so far focused mainly on mitigating local pollution rather than emissions of GHGs. However, wherever environmental policies were successful in the region, they have also brought important climate dividends. Some countries in the region (e.g., Poland) have introduced specific climate change mitigation policy instruments, such as charges on CO_2 and CH_4 emissions.

At the end of first decade of the transition to a market economy, contrasts between different countries in the region have outstripped bygone relative homogeneity. Central Europe and the Baltic countries have made a successful leap in economic reforms and restructuring, while countries of the former Soviet Union (so called New Independent States - NIS) continue to struggle with economic recession and political instability (EBRD, 1999). Recent empirical studies on the interrelationship between environmental improvement and economic development in transition economies undertaken by the World Bank, EBRD, and OECD have demonstrated that countries that were more successful in economic development and structural reforms have generally also been more successful in curbing emissions through targeted environmental policies. Aggregated GDP among advanced reforming countries has been gradually increasing, while emissions of main air pollutants have continued to decrease. Energy consumption has been stabilized and a switch away from coal has been recorded mainly in Poland and the Czech Republic causing GHG-intensity of GDP to decrease. In contrast, in the slower reforming countries in NIS, falling output, rather than economic restructuring or environmental protection efforts, appears to have been the main factor behind the decrease of energy use and emissions of pollutants, including GHGs (OECD, 1999a).

In the more advanced economies of the region, economic reforms have helped generate resources for investment in cleaner, more efficient technologies; reduced the share of energy- and GHG-intensive heavy industries in economic activity; and helped curb emissions as part of the shift towards more efficient production methods (OECD, 2000). In some sectors, however, the transition has brought greater climate pressures. For example, in those countries returning to economic growth, the use of motor vehicles for both passenger and freight transport has increased rapidly.

Energy Pricing and Subsidies

Virtually all countries in the region have embarked on the liberation of energy prices and elimination of energy subsidies. Significant successes in this field have been achieved in Central European and Baltic States. However, in NIS a sharp reduction of explicit subsidies has resulted in an almost immediate build up of hidden subsidies to energy producers and users, such as arrears and non-monetary forms of payments for energy (EBRD, 1999).

Finance and Income

Lack of adequate access to capital for GHG emission reduction technologies is perceived as a bottleneck in many countries in the region (World Bank, 1998). However, in CEE financial and capital markets are becoming mature enough to provide increasingly better access to credit for fuel switching or energy efficiency, especially given stable macroeconomic conditions and relatively high energy prices. In these countries the main bottleneck to environmental finance is not the lack of finance, but rather the lack of a "pull factor". Lack of implementation of the Polluter Pays Principle, and weak enforcement of the environmental and climate policy framework does not stimulate sufficient demand for investments that would bring mainly GHG reduction benefits, with little private financial return (OECD, 1999b). In NIS, however, the weak policy framework is aggravated by the overwhelming lack of liquidity both in the public and private sector. Limited financial resources, which are available to authorities have not always been used in a cost-effective way. Opportunities to leverage additional financing from public and private, domestic and foreign sources were also underutilized (OECD, 2000).

Institutional Aspects

The countries in the region have undergone a rapid deregulation and privatization on a short time scale that has no precedence in the history of the world. This process in the Baltic and Central European countries has generally led to increased resource efficiency and replacement of obsolete and GHG intensive technologies. However, in a number of countries of the former Soviet Union, particularly in Russia and Ukraine, the rapid pace of liberalization and privatization has not been matched by the development of institutions as well as a regulatory and incentive framework necessary to support a well-functioning market economy. Perverse incentives that had generated many of the environmental problems of centrally planned economies, such as rent seeking and lack of incentives for efficiency and restructuring, now undermine restructuring of already private enterprises (EBRD, 1999). But successful economic policies have not been a panacea for successful GHG-mitigation improvements. Targeted environmental policies and institutions in Central Europe were required to harness the positive forces of market reform, and ensure that enterprises and other economic actors improve their environmental performance which are still weak in the NIS (World Bank, 1998).

5.5.3 Developed Countries

Compared to the developing countries and those undergoing an economic transition, the GHG emissions in the developed countries originate increasingly from the energy used by households and other consumers for personal activities. Mitigation opportunities therefore lie increasingly in the area of personal transport, space conditioning, and other home use of energy, and in the energy used by the commercial sector, although opportunities exist in all sectors. Financial and income-related, social and behavioural, and institutional barriers thus become predominant in limiting the choice of mitigation technologies in these countries.

In the household sector, for instance, although a CFL offers a relatively short payback period, the large price differential between the CFL and an incandescent bulb poses a significant first-cost barrier to consumers. Most programmes to promote CFLs have focused on a subsidy to lower its first cost (Mills, 1993; Meyers, 1998). Raising the efficiency of other consumer appliances encounters barriers such as the relatively low energy cost, bundling of higher efficiency with other higher value attributes, and lack of information about energy consumption. Standards and labels are being implemented in several countries in order to overcome these barriers. While many communities and national governments have regulations for more efficient construction, rising affluence has increased the demand for homes with a larger floor area, which negates efficiency gains. Disincentives may also exist in the market structure, e.g., a building owner may not be interested in energy efficient designs if the user is responsible for paying for the energy used.

In the transport sector, manufacturers are producing cars that have more efficient engines and lower air resistance, but coupled with higher weight and more power (and other options), there has been little or no gain in vehicle fuel economy. Fuel economy is also not an important criteria in most purchasing decisions (see Section 5.4.3). The movement of households to suburban areas increases the distance traveled to work, and for leisure, and adds to a vehicle's fuel consumption. The lock-in of transport into motorized private transport is an important barrier to new efficient forms of mass transport, while the well-established gasoline-based infrastructure is a barrier to the introduction of new less GHG-intensive fuels and associated technologies.

Energy efficiency and GHG-intensity in industry still vary widely among and within developed countries, suggesting the existence of barriers. Decision makers do not have sufficient information to evaluate GHG mitigation opportunities. The relative high transaction costs reduce the changes of innovative technologies. Output growth is slow or stagnant in the large energy-intensive industries. The resulting slow stock turnover has slowed the penetration of new GHG mitigation technologies in these industries. As industries improve their labour productivity, concentration on a few core activities has led to a lack of skilled personnel to evaluate and implement new technology.

The energy supply sector is undergoing changes in the regulatory structure in almost all developed countries. These changes may not all be conducive to the goal of GHG mitigation. Increasing profitability through reduction of capital costs may lead to less efficient power generation options, and reduce the penetration rate of generally capital intensive renewable energy technologies. In general, grid operators (*i.e.*, utility companies) have put up high barriers against more efficient generation options like co-generation (CHP) through low buyback tariffs, high interconnection charges, or power quality demands (*Box 5.5*). Deregulation experiences have differed with respect to the treatment of co-generation and renewable energy.

5.6   Research Needs

The earlier chapters show a significant potential for GHG mitigation in energy and non-energy sectors. All types of barriers limit this potential. These barriers are specific to a technology, sector, and region, and they evolve over time. Research would be useful in several areas to collect data, establish databases, improve methods, and develop computerized models that would help decision makers to devise improved policies and measures to address these barriers:

- What is the quantitative global and regional market potential for different categories of mitigation technologies? Chapters 3 and 4 note the technical and socioeconomic potential but a parallel quantitative estimate for market potential is yet to be developed. Data and models that explicitly incorporate barriers to achieving the market potential would be helpful.

- What mix of barriers prevents the adoption of major mitigation technologies? Are social capital and related investment policies more or less important, and how might these vary across cultures and physical environments? What are the decision processes that foster technology transfer? Can technology transfer be managed

such as to support sustainable and equitable development? The IPCC-SRTT provides one framework for a technology transfer process. Models of processes that reflect "real world" decision-making are needed, however, in order to identify and elaborate on the barriers that prevent or slow the diffusion of mitigation technologies. The models would also need to take alternative development pathways into consideration. An improved understanding of technology transfer both within and across countries would be required since the actors and barriers tend to be very different.

- What is the appropriate role for stakeholders in the above decision-making processes? The roles of governments and other stakeholders change over time. This is particularly important in sectors where the social, cultural, institutional, and market context is changing rapidly. An identification of their emerging roles would help decision makers manage technology transfer better.

- Does market globalization favour or hamper the diffusion of mitigation technologies? Does environmental regulation confer to firms and nations a long-term technological advantage? Market globalization offers opportunity to plant seeds of mitigation technologies that are less GHG intensive, but it could also bring about proliferation of polluting technologies. It is important to understand the ongoing processes and to determine ways to assist the transfer of less GHG-intensive technologies.

References

ADB (Asian Development Bank), 1999: *Microfinance Development Strategy.* Asian Development Bank.

AFDB (African Development Bank), 1999: *African Development Report 1999.* African Development Bank.

Aghion, P. and P. Howitt, 1998: *Endogenous Growth Theory.* MIT Press, Cambridge, MA.

Alam, M., J. Sathaye, and D. Barnes, 1998: Urban Household Energy Use in India: Efficiency and Policy Implications. *Energy Policy,* **26**(11), 885-892.

Alesina, A., and A. D. Dollar, 1998: *Who gives foreign aid to whom and why?* National Bureau of Economic Research, Working Paper 6612.

Anderson, C. D., and J. D. Claxton, 1982: Barriers to Consumer Choice of Energy Efficient Product. *J. Consumer Research,* **9**, 163-168.

Andrasko, K., 1997: Forest management for Greenhouse gas benefits: Resolving monitoring issues across project and national boundaries. *Mitigation and Adaptation Strategies for Global Change,* **2**, 117-132.

Ang, B.W., 1993: An energy and environmentally sound urban transport system: the case of Singapore. *International Journal of Vehicle Design,* **14**(4).

Argyle, M., 1987: *The Psychology of Happiness,* Routledge, London and New York.

Armstrong, D.M., 1993: *Transport infrastructure, urban form and mode usage: an econometric analysis based on aggregate comparative data,* Regional Science Association Thirty-Third European Congress, Moscow, 24-27 August 1993, Northern Ireland Economic Research Centre, Belfast, N. Ireland.

Arthur, W.B., 1988: *Competing Technologies: An Overview.* In Dosi et al., 1988, 590-607.

Arthur, W.B., 1994: *Increasing Returns and Path Dependence in the Economy.* Michigan University Press, Ann Arbor.

Asad, M., 1997: Innovative global environmental management. *Environment Matters,* The World Bank, Winter/Spring issue, 12-13.

Ashford, N., C. Ayers, and R. Stone, 1985: Using Regulation to Change the Market for Innovation. *Harvard Environmental Law Review,* **9**, 419-466.

AIT (Asian Institute of Technology), 1997: *Energy, Efficient and Environmentally Sound Industrial Technologies.*

Azar, C., 1996: *Technological Change and the Long-Run Cost of Reducing CO_2 Emissions.* Working Paper 96/84/EPS, INSEAD Centre for the Management of Environmental Resources, Fontainebleau, France.

Barnett, A., 1990: The Diffusion of Energy Technology in the Rural Areas of Developing Countries: A Synthesis of Recent Experience. *World Development,* **18**, No. 4, pp 539-553.

Benioff, R., E. Ness, and J. Hirst, 1997: *National Climate Change Action Plans.* Interim Report for Developing and Transition Countries, U.S. Country Studies Program, Washington DC, 156 pp.

Berman, E., and L. Bui, 1998: *Environmental Regulation and Productivity: Evidence from Oil Refineries.* NBER WP No. 6776.

Berry, D., 1995: You've got to pay: Photovoltaics and transaction costs. *Electricity Journal* **8**(2), 42-49.

Bhide A. D., 1998: *Current Scenario on Management of urban solid waste.* Proceedings of the National workshop on Energy Recovery from Urban and Municipal Solid Waste, Ministry of Non-Conventional Energy Sources. Government of India.

Blackman, A., 1999: *The Economics of Technology Diffusion: Implications for Climate Policy in Developing Countries.* Resources for the Future, DP 99-42.

Blackman A., and G.J. Bannister, 1998: Community Pressure and Clean Technology in the Informal Sector: An Econometric Analysis of the Adoption of Propane by Traditional Mexican Brickmakers. *Journal of Environmental Economics and Management,* **35**(1), 1-21.

Blok, K., 1993: The Development of Industrial CHP in The Netherlands. *Energy Policy,* **2**(21), 158-175.

Blok, K., and J. Farla, 1996: The continuing story of CHP in the Netherlands. *International Journal of Global Energy Issues,* **8**(4), 349-361.

Bouille, D., 1999: Lineamientos para la regulación del uso eficiente de la Energia en Argentina. *Serie Medio Ambiente y Desarrollo,* 16, Naciones Unidas-CEPAL, Santiago.

Bouille, D., 1998: *Implications of Electric Power Sector Restructuring on Climate change Mitigation in Argentina.* IDEE/FB, Buenos Aires.

Bourdieu, P., 1979: *Distinction: A Social Critique of the Judgement of Taste.* Routledge, London. English translation: R. Nice, 1984.

Börkey, P., and F. Lévêque, 1998: *Voluntary Approaches for Environmental Protection in the European Union.* Working Document ENV/EPOC/GEEI(98)29/Final, OECD, Paris, France.

Braid, R. B., R. A. Cantor, and S. Rayner, 1986: Market Acceptance of New Nuclear Technologies. *Nuclear Power Options Viability Study* Vol. III, ORNL/TM–9780, /3, Oak Ridge National Laboratory, Oak Ridge, Tennessee.

Bresnahan, T., and M. Tratjenberg, 1995: General Purpose Technologies, Engines of Growth. *Journal of Econometrics,* **65**, 83–108.

Brown, M. A., and S. M. Macey, 1983: Residential energy conservation through repetitive household behaviors. *Environment and Behavior,* **15**, 123-141.

Brown, M. S., 1997: *Energy Efficient Buildings: Does the Market Place Work?* Proceedings of the 24th Annual Illinois Energy Conference, Chicago, Illinois, University of Illinois, 233-255.

Bunch, D. S., M. A. Bradley, T. F. Golob, R. Kitamura, and G. P. Occhiuzzo, 1993: Demand for clean-fuel vehicles in California: a discrete choice stated preference approach. *Transportation Research, A* **27A**(3), 237-253.

Cairncross, F., 1997: *The Death of Distance: How the Communications Revolution Will Change Our Lives.* Orion, London, 302 pp.

Carraro, C., 1994: *Trade, Innovation, Environment.* Kluwer, Dordrecht.

Carter, L. W., 1996: *Leasing in emerging markets: the IFCs experience in promoting leasing in emerging economies,* International Finance Corporation, Washington D.C.

Castells, M., 1998: *'Information Technology, Globalization and Social Development'.* UNRISD Conference on Information Technologies and Social Development, Palais des Nations, Geneva, 22-24 June 1998. http://www.unrisd.org/infotech/conferen/conf.htm.

CPCB (Central Pollution Control Board), 1998: Planning and design principles for land treatment of domestic and biodegradable industrial waste water. *Resource Recycling Series:* RERIS/3/1998/99, Government of India.

CEEEZ (Centre for Energy, Environment and Engineering), 1998: *Climate Change in Africa: Methodological Development, National Mitigation Analysis and Institution Capacity Building in Zambia.* ISBN 8-70550-2430.0.

Christiansson, L., 1995: *Diffusion and Learning Curves of Renewable Energy Technologies.* Working Paper WP-95-126, International Institute for Applied Systems Analysis, Laxenburg, Austria.

Clark, W.W., 1997: *The Role of Publicly-Funded Research and Publicly-Owned Technologies in the Transfer and Diffusion of Environmentally Sound Technologies, The Case Study of the United States of America.* Proceedings of the International Expert Meeting of CSD on the Role of Publicly-Funded Research and Publicly-Owned Technologies in the Transfer and Diffusion of Environmentally Sound Technologies. Ministry of Foreign Affairs, Republic of Korea, February 4-6, 1998.

CLASP, 2000: http://www.clasponline.org/standardlabel/programs/country.php3.

Clements, B., R. Hugouneng, and G. Schwartz, 1995: *Government Subsidies: Concepts, International Trends And Reform Options.* IMF working paper WP/95/91, International Monetary Fund, Washington D.C.

Coase, R., 1998: The New Institutional Economics, American Economic Association, *Papers and Proceedings,* **88(2)**, 72-74.

COGEN Europe, 1997: *The impact of liberalisation of the European Electricity Market on Cogeneration, Energy Efficiency, and the Environment.* COGEN Europe, Brussels, Belgium, 156 pp.

Cohen, L. R. and Noll, R. G., 1994: Privatizing Public Research. *Scientific American,* September, 72-77.

Coleman, J., 1988: Social Capital in the Creation of Human Capital. *American Journal of Sociology,* 94 (Supplement), S95-S120.

Collingwood, J., and D. Goult, 1998: *The UK Energy Efficiency Best Practice Programme: Evaluation Methods & Impact 1989 –1998.* Paper presented at Industrial Energy Efficiency Policies: Understanding Success and Failure, Utrecht, The Netherlands, 11-12 June 1998.

Copeland, B., and M. Taylor, 1994: North-South Trade and the Environment, *Quarterly Journal of Economics,* **109**, 755-87.

Copeland, B., and M, Taylor, 1995: Trade and Transboundary Pollution. *American Economic Review*, **85**, 716-36.

Cousins, S., 1989: Culture and selfhood in Japan and the US. *Journal of Personality and Social Psychology*, **56**, 124-131.

Cropper, M., and W. Oates, 1992: Environmental Economics: A Survey. *Journal of Economic Literature*, **30**, 675-740.

Dadhich, P.K., 1996: *Cogeneration. Policies, potential, and technologies.* Proceedings of Cogen India, **96**, 10-12 March 1996, Tata Energy Research Institute, 296 pp.

Daly, G.G., and T.H. Mayer, 1983: Reason and Rationality during Energy Crises. *Journal of Political Economy*, **91**(1), 168-181.

Dasgupta, S., B. Laplante, and N. Mamingi, 1998: *Capital Market Responses to Environmental Performance in Developing Countries*. World Bank Policy Research Working Paper nr. 1909.

Datt, R., and K.P.M. Sundharam, 1998: *Indian economy*. S. Chand and Company Ltd., New Delhi, India.

Datta, M., 1999: *Integrated Solid Waste Management: Industrial Solid Waste Management and Landfilling Practices*. Narosa Publishing House, New Delhi, India.

David, P., 1985: CLIO and the Economics of QWERTY, *American Economic Review Papers and Proceedings*, **75** (2): 332-337.

Davidson, O., 2000: Transportation. In *Methodological and Technological Issues in Technology Transfer. A Special Report of IPCC Working Group III*. B. Metz, O.R. Davidson, J.W. Martens, S.N.M. van Rooijen, and Laura Van Wie McGrory, (eds.), Cambridge University Press, 466 pp.

Davis, H.A., 1996: *Project Finance: Practical Case Studies*. Euromoney Books, London.

Davis, S.C., 1999: *Transportation Energy Data Book: Edition 19*. ORNL-6958, Oak Ridge National Laboratory, Oak Ridge, Tennessee.

De Almeida, E.L.F., 1998: Energy Efficiency and the Limits of Market Forces: The Example of the Electric Motor Market in France. *Energy Policy*, **26**, 643-653.

DeCanio, S., 1993: Barriers within firms to energy-efficient investments. *Energy Policy*, September 1993, 906-977.

DeCanio, S.J., 1998: The Efficiency Paradox: Bureaucratic and Organizational Barriers to Profitable Energy-Saving Investments. *Energy Policy*, **26**, 441-454.

DeCanio, S.J., and J.A. Laitner, 1997: Modeling Technological Change in Energy Demand Forecasting. *Technological Forecasting and Social Change*, **55**, 249-263.

De Moor, A., and P. Calamai, 1996: *Subsidising Unsustainable Development: Undermining the Earth with Public Funds*. Institute for Research on Public Expenditure, The Hague, The Netherlands, and Earth Council, San José, Costa Rica.

De Moor, A., 1997: *Key issues in subsidy policies and strategies for reform. Finance for Sustainable Development*. Proceedings of the Fourth Group Meeting on Financial Issues of Agenda 21, Santiago, Chile 1997, United Nations, New York, 285-313.

De Uribarri, C., 1998: From Transfer Stations to integrated waste Processing facilities. *International Solid Waste Association Year Book*, **97**(8), 61 pp.

Dean, J., P. Low, 1992: Trade and the Environment: *A Survey of the Literature. International Trade and the Environment*. World Bank Discussion Paper 159.

Deardorff, A., 1993: Should patent protection be extended to all developing countries? In *Multilateral Trading System: Analysis and Options for Change*. R.M. Stern (ed.)], Ann Arbor, University of Michigan Press, U.S.A., 435-448.

Debruyn. W., and J. Van Rensbergen, 1994: *Greenhouse Gas Emissions from Municipal and Industrial Wastes*, ENNE. RA9410, VITO, Energy Division, Belgium.

Delphi, 1997: *The role of financial institutions in achieving sustainable development*. Delphi Int. Ltd and Ecologic GMBH, Report to the European Commission.

Dietz, T., and P.C. Stern, 1993: *Individual Preferences, Contingent Valuation, and the Legitimation of Social Choice*. National Research Council, Washington, DC.

Dismukes, D.E., and A.N. Kleit, 1999: Cogeneration and Electric Power Industry Restructuring. *Resource and Energy Economics*, **21**(1999), 153-166.

Dittmar, H., 1992: *The Social Psychology of Material Possessions. To Have is To Be*. Harvester Wheatsheaf, Hemel Hempstead, England, ix, 250 pp.

Doorn, M., and M.A. Barlaz, 1995: *Estimate of Global Methane Emissions from landfills and Open Dumps*. Prepared for US EPA Office of Research and Development, February 1995, EPA-600/R-95-019.

Dosi, G., 1988: The Nature of the Innovation Process. In *Technical Change and Economic Theory*. G. Dosi, C. Freeman, R. Nelson, G. Silverberg, and L. Soete (eds.), 221-238.

Dosi, G., C. Freeman, R. Nelson, G. Silverberg and L. Soete (eds.), 1988: *Technical Change and Economic Theory*. Pinter, London and New York.

Douglas, M., and B. Isherwood, 1979: *The World of Goods*. Basic Books, New York, xi, 228 pp.

Douglas, M., D. Gasper, S. Ney and M. Thompson, 1998: Human Needs and Wants. In *Human Choice and Climate Change, Volume 1, The Societal Framework*. S. Rayner and E. L. Malone (eds.), Battelle Press, Columbus, Ohio, 195-263.

Dufait, N., 1996: Attitudes of the electric utilities towards CHP in the European Union. *International Journal of Global Energy Issues*, **8**(4), 329-337.

Ecchia, G., and M. Mariotti, 1994: *A Survey on Environmental Policy: Technological Innovation and Strategic Issues*. EEE Working Paper 44.94, Fondazione Eni Enrico Mattei, Milan, Italy.

ECMT (European Conference of Ministers of Transport), 1997: *CO_2 Emissions from Transport*. OECD, Paris.

EC (European Commission), 1997: *A Community strategy to promote combined head and power and to dismantle barriers to its development*. Communication from the Commission to the Council and the European Parliament, Brussels, Belgium, 18 pp.

Ecosecurities, 1999: *Forestry Investment*. http://www.ecosecurities.com.

ECOSOC, 1994: *Transfer of Environmentally Sound Technology, Cooperation and Capacity Building*. Report of the Secretary General of the Commission on Sustainable Development, intersessional ad hoc working group on technology transfer and cooperation, 23-25 February.

ESD (Energy for Sustainable Development), 1995: *A study of woody biomass derived energy supplies in Uganda*.

ESCAP, 1992: *Technology Planning for industrial Pollution Control*. United Nations Economic and Social Commission for Asia and the Pacific, Bangkok, Thailand.

ESCAP, 1997: Regional cooperation in climate change. Report of the Expert Group Meeting, May 1996. In *Waste Minimisation using Economic Instruments, the International Solid Waste Association Year Book*. G. Vogel, 44 pp.

ETSU (Energy Technology Support Unit), 1994: Appraisal of UK Energy Research, Development, Demonstration and Dissemination. *Transport*, Vol. 7, HMSO, London.

EBRD (European Bank for Reconstruction and Development), 1996: *Annual Report 1995 - Project Evaluation Lessons*. London.

EBRD (European Bank for Reconstruction and Development), 1997a: *Guide To Bankable Energy Efficiency Proposals*. P. Petit, London, January Draft Version.

EBRD (European Bank for Reconstruction and Development), 1997b: *Transition Report 1997*, London.

EBRD (European Bank for Reconstruction and Development), 1998: *Transition Report 1998*, London.

EBRD (European Bank for Reconstruction and Development), 1999: *Transition Report 1999*. London.

EC (European Commission), 1996: Towards fair and efficient pricing in transport. *Bulletin of the European Union*, Supplement 2/96, Office for Official Publications of the European Communities, Luxembourg.

ECMT (European Conference of Ministers of Transport), 1997: *CO_2 Emissions from Transport*. OECD, Paris.

EEA (European Environmental Agency), 1997: Environmental Agreements. *Environmental Issues Series No.3*, European Environmental Agency, Copenhagen, Denmark.

Evans, R., 1991: *Barriers to energy efficiency*. A report for the Energy Efficiency Office, London.

Farinelli, U., 1999: *Energy As a Tool for Sustainable Development for African Caribbean and Pacific Countries*. ISBN 92 1126 112 8.

Fearnside, P.M., 1997: Monitoring needs to transform Amazonian forest maintenance into a global warming mitigation option. *Mitigation and Adaptation Strategies for Global Change,* **2**(2-3), 285-302.

Fisher, A.C.R., and M. Rothkopf, 1989: Market failure and Energy Policy. *Energy Policy,* **17**, 397-406.

Fisher-Vanden, K., 1998: *Technological Diffusion in Chinas Iron and Steel Industry.* ENR-Discussion Paper E-98-26, John F. Kennedy School of Government, Harvard University, Cambridge, MA.

FAO (Food and Agriculture Organization), 1997: *Environmentally Sound Forest Harvesting: Testing the Applicability of the FAO Model Code in the Amazon in Brazil.* FAO Forest Harvesting Case Study 8, Food and Agriculture Organization of the United Nations, Rome, 78 pp.

Foucault, M., 1961: *Madness and Civilisation,* Routledge, London.

Foucault, M., 1975: *Discipline and Punish: the Birth of the Prison,* Penguin, London.

Fox-Penner, P.S., 1990: Cogeneration after PURPA: Energy Conservation and Industry Structure. *Journal of Law & Economics,* **33**, 517-552.

Frederikson, P., 1999: *Trade, Global Policy and the Environment.* Working Paper 402, World Bank.

Freeman, C., 1987: *Technology Policy and Economic Performance.* Pinter, London.

Freeman, C., and C. Perez, 1988: *Structural crises of adjustment: business cycles and investment behaviour.* In Dosi *et al.*, 38-66.

Freeman, C., and L. Soete, 1997: *The Economics of Industrial Innovation.* 3rd ed., MIT Press, Cambridge, MA.

Frisvold, G.B., and P.T. Condon, 1995: Technological Forecasting and Social Change. *An International Journal,* vol. 50, Number 1, Special Issue September 1995, pp 41-54.

Frisvold, G.B., and P.T. Condon, 1998: The Convention on Biological Diversity and Agriculture: Implications and Unresolved Debates. *World Development,* **26**(4), 551-570.

Fukuyama, F., 1999: *The Great Disruption.* Simon and Schuster, New York, 368 pp.

Gady C., and B.W. Ickes, 1998: *To Restructure or Not to Restructure: Informal Activities and Enterprise Behaviour in Transition.* The Davidson Institute Working Paper Series, No. 134, February.

Gare, A.E., 1995: *Postmodernism and the Environmental Crisis.* Routledge, London and New York.

Gentry, B., 1997: Making Private Investment Work for the Environment. In *Finance for Sustainable Development.* Proceedings of the Fourth Group Meeting on Financial Issues of Agenda 21, Santiago, Chile 1997, United Nations, New York, 341-402.

Ger, G., H. Wilhite, B. Halkier, J. Laessoe,, M. Godskesen, I. Røpke, 1998: *Symbolic meanings of high and low impact daily consumption practices in different cultures.* Working paper prepared for Second European Science Foundation, Workshop on Consumption, Everyday Life and Sustainability, Lancaster University, England. See http://www.lancs.ac.uk/users/scistud/esf/title.htm.

German Federal Ministry of Environment, 2000: Proceedings of the IPCC Expert Meeting on Conceptual Framework for Mitigation Assessment from the Perspective of Social Science, March 21-22, Karlsruhe, Germany, Kluwer (in press)

Gibson, C., E. Ostrom, and T. Ahn, 1998: *Scaling Issues in the Social Sciences.* Working Paper 1, International Human Dimensions Program.

Giddens, A., 1984: *The Constitution of Society.* Polity Press, Cambridge.

Goldemberg, J., 1996: *Energy, Environment and Development.* Earthscan, London

Golove, W., and J. Eto, 1996: *Market Barriers To Energy Efficiency: A Critical Reappraisal Of The Rationale For Public Policies To Promote Energy Efficiency.* LBL-38059.

Goodwin, P.B., 1985: Passenger Transport. In *Changes in Transport Users Motivations for Modal Choice.* European Conference of Ministers of Transport Round Table 68, OECD, Paris, 61-90.

Goulder, L.H., and K. Mathai, 2000: Optimal CO_2 Abatement in the Presence of Induced Technological Change. *Journal of Environmental Economics and Management,* **39**, 1-38.

Graham C., and M.O. Hanlon, 1997: Making foreign aid work. *Foreign Affairs,* July/August, 96-104.

Greening, L.A., A.H. Sanstad, and J.E. McMahon, 1997: Effects of Appliance Standards on Product Price and Attributes: An Hedonic Pricing Model. *The Journal of Regulatory Economics,* **11**(2), March, 181-194.

Griliches, Z., 1979: Issues in Assessing the Contribution of Research and Development to Productivity Growth. *Bell Journal of Economics,* **10**, 92-116.

Gritsevich, I., 2000: *Motivation and Decision Criteria in Private Households, Companies and Administration on Energy Efficiency in Russia.* Proceedings of the IPCC Expert Meeting on Conceptual Frameworks for Mitigation Assessment from the Perspective of Social Science, 21-22 March 2000, Karlsruhe, Germany.

Grossman, G. M., and E. Helpman, 1991: *Innovation and Growth in the Global Economy,* MIT Press, Cambridge, MA.

Grossman, G. M., and E. Helpman, 1993: *Endogenous Innovation in the Theory of Growth.* Working Paper no. 4527, National Bureau of Economic Research (NBER), Cambridge, MA.

Grover, S., C. Flenley, and M. Hemmati, 1999: *Gender and Sustainable Consumption,* Report for the United Nations Commission on Sustainable Development, UNED-UK, London.

Grübler, A., 1998: *Technology and Global Change,* Cambridge University Press, Cambridge, England.

Grübler, A., and N. Nakicenovic, 1991: *Evolution of Transport Systems: Past and Future.* RR-91-8, International Institute for Applied Systems Analysis, Laxenburg, Austria.

Grubb, M., 2000: Economic Dimensions of Technological and Global Responses to the Kyoto Protocol. *Journal of Economic Studies,* **27**(1/2), 111-125.

Gruber, E., and M. Brand, 1991: Promoting Energy Conservation in Small-And Medium-Sized Companies. *Energy Policy,* **19**, 279-287.

Haddad, B., 1994: Why Compact Fluorescent Lamps are not Ubiquitous: Industrial Organization, Incentives, and Social Convention. *ACEEE Summer Study on Energy Efficiency in Buildings,* American Council for an Energy Efficient Economy, Washington DC, **10**(10), 77-84.

Haddad, B., (2000): *Rivers of Gold: Designing Markets to Allocate Water in California.* Island Press, Washington DC.

Häfele, W., J. Anderer, A. McDonald, and N. Nakicenovic, 1982: *Energy in a Finite World: Paths to a Sustainable Future.* Ballinger, Cambridge, MA.

Harou, P.M., L.A. Bellu, and V. Cistulli, 1998: *Environmental Economics and Environmental Policy: A Workbook.* World Bank, Washington DC.

Harris, J., and N. Casey-McCabe, 1996: *Energy-Efficient Product Labeling: Market Impacts on Buyers and Sellers.* Proceedings of the 1996 ACEEE Summer Study on Energy Efficiency in Buildings, Washington, DC:

Hassett, K.A., and G.E. Metcalf, 1993: Energy Conservation Investment: Do Consumers Discount the Future Correctly? *Energy Policy,* **21**(6), 710-716.

Hassett, K.A., and G.E. Metcalf, 1995: Energy Tax Credits and Residential Conservation Investment: Evidence From Panel Data. *Journal of Public Economics,* **57**, 201-217.

Helpman, E., 1998: *General Purpose Technologies and Economic Growth.* MIT Press, Cambridge, MA.

Hemmati, M., 2000: *Women and Sustainable Development 2000 – 2002.* Prepared by the CSD NGO Women's Caucus for Discussion at the UN Commission on Sustainable Development Intersessional Working Group, 22 February - 3 March and Its 8th Session, 24 April - 5 May, http://www.earthsummit2002.org/wcaucus/csdngo.htm.

Herber, B., 1997: Innovative Financial Mechanisms for Sustainable Development: Overcoming the Political Obstacles to International Taxation. *Finance for Sustainable Development,* Proceedings of the Fourth Group Meeting on Financial Issues of Agenda 21, Santiago, Chile 1997, United Nations, New York, 455-485.

Hille, J., 1997: *Sustainable Norway, Local Agenda 21, Idehefte.* Idebanken, Oslo.

Hirsch, F., 1977: *Social Limits to Growth.* Routledge, London.

Hirst E., and M. Brown, 1990: Closing The Efficiency Gap: Barriers To Improving Energy Efficiency. *Resources, Conservation, and Recycling,* **3**, 267-281.

Hirst E., 1992: Making Energy Efficiency Happen. In *Technologies for A greenhouse Constrained Society.* M. Kuliasha, A. Zucker, and K. Ballew (eds.), Lewis Publishers MI.

Hoff, T., H.J. Wenger, and B.K. Farmer, 1996: Distributed Generation: an Alternative to Electric Utility Investments in System Capacity. *Energy policy*, **24**(2), 137-147.

Hofstede, G., 1980: *Culture's Consequences: International Differences in Work-related Values.* Sage, Beverly Hills.

Inglehart, R., 1990: *Culture Shift in Advanced Industrial Society.* Princeton University Press, Princeton, NJ.

Inglehart, R., 2000: Globalization and Postmodern Values. *The Washington Quarterly*, Winter 2000, 215-228.

IPIECA, 1999: *Technology Assessment in Climate Change Mitigation.* Report of the IPIECA Workshop, Paris, May.

IEA (International Energy Agency), 1997: *Voluntary Actions for Energy-Related CO_2 Abatement.* IEA/OECD, Paris, France.

IPCC (Intergovernmental Panel on Climate Change), 1996: *Climate Change 1995: Impacts, Adaptations and Mitigation of Climate Change: Scientific-Technical Analyses.* Contribution of Working Group II to the Second Assessment Report of the Intergovernmental Panel on Climate Change. R.T. Watson, M.C. Zinyowera, and R.H. Moss (eds.), Cambridge University Press, Cambridge, UK and New York, NY.

IPCC (Intergovernmental Panel on Climate Change), 2000a: *Land Use, Land-Use Change, and Forestry.* A Special Report of the IPCC. R.T. Watson, I.R. Noble, B. Bolin, R.H. Ravindranath, D.J. Verardo, and D.J. Dokken (eds.), Cambridge University Press, Cambridge, United Kingdom and New York, NY, USA.

IPCC (Intergovernmental Panel on Climate Change), 2000b: *Methodological and Technological Issues in Technology Transfer.* A special report of IPCC Working Group III. B. Metz, O.R. Davidson, J.W. Martens, S.N.M. van Rooijen, L. Van Wie McGrory, (eds.), Cambridge University Press, Cambridge, UK, and New York, NY.

Izac, B.A.M.N., 1997: Developing Policies for Soil Carbon Management in Tropical Regions. *Geoderma*, **79**, 261-276.

Jackson, T., and M. Marks, 1999: Consumption, Sustainable Welfare And Human Needs–with reference to UK expenditure patterns between 1954 and 1994. *Ecological Economics*, **28**, 421-441.

Jacobs, M., 1997: Sustainability And Markets: On The Neoclassical Model Of Environmental Economics. *New Political Economy*, **2**(3), 365-385.

Jaffe, A., S. Peterson and R. Stavins, 1995: Environmental Regulation and the Competitiveness of US Manufacturing. *Journal of Economic Literature*, **33**, 132-163.

Jaffe, A.B., and R.N. Stavins, 1994: The Energy-Efficiency Gap: What Does it Mean. *Energy Policy*, **22**(10), 804-810.

Jaffe, A.B., and R.N. Stavins, 1995: Dynamic incentives of environmental regulations: The effects of alternative policy instruments on technology diffusion. *Journal of Environmental Economics and Management*, **29**, 43-63.

Jaeger, C.C., O. Renn, E.A. Rosa, and T. Webler, 1998: Decision Analysis and Rational Action. In *Human Choice and Climate Change. Volume III. Tools for Policy Analysis.* S. Rayner, and E. L. Malone (eds.), Battelle Press, Columbus, Ohio, 141-215.

Janda, K.B., and J.F. Busch, 1994: Worldwide Status of Energy Standards for Buildings. *Energy*, **19**(1), 27-44.

Jefferson, M., 2000: Energy Policies for Sustainable Development. In *World Energy Assessment: Energy and the challenge of sustainability*, UNDP, New York, p. 425.

Jochem, E., 1999: *Energy End-Use Efficiency.* World Energy Assessment, UNDP.

Jovanovic, B., and Y. Nyarko, 1994: *The Bayesian Foundations of Learning by Doing.* NBER WP 4739.

Kahn, J.A., 1986: Gasoline prices and the used automobile market: a rational expectations asset price approach. *Quarterly Journal of Economics*, **101**(2), 323-339.

Kammen, D.M., and R. Margolis, 1999: Evidence of Under-Investment in Energy R&D in the United States and the Impact of Federal Policy. *Energy Policy*, **10**(27), 575-584.

Karekezi, S., 1991: The Development of Stoves and Their Effectiveness. *Renewable Energy Technology and Environment*, **1**, Pergamon Press 46, Oxford, United Kingdom.

Katz, M., and C. Shapiro, 1986: Technology Adoption in the Presence of Network Externalities. *Journal of Political Economy*, **94**, 822-841.

Kemp, R., 2000: Constructing Transition Paths through the Management of Niches. In *Path Creation and Dependence.* R. Garud and P. Karnoe (eds.), Lawrence Erlbaum Associates Publishers.

Kempton, W., 1991: Lay Perspectives On Global Climate Change. *Global Environmental Change: Human Policy Dimensions*, **1**, 183-208.

Kempton, W., 1997: How the Public Views Climate Change. *Environment*, **39**(9), 10-21.

Kempton, W., J.M. Darley, and P.C. Stern, 1992: Psychological Research for the New Energy Problems. *American Psychology*, **47**, 1213-1223.

Kempton W., and L. Montgomery, 1982: Folk Quantification of Energy. *Energy* **7**(10), 817-827.

Kempton W., and L. Layne, 1994: The Consumers Energy Analysis Environment. *Energy Policy*, **22**(10), 857-866.

Killick, T., 1997: What Future for Aid? In *Finance for Sustainable Development.* Proceedings of the Fourth Group Meeting on Financial Issues of Agenda 21, Santiago, Chile, United Nations, New York, 77-107.

Kim, J.I., 1998: Industry's Effort to Mitigate Global Warming: Options for Voluntary Agreements in Korea. In *Energy savings and CO_2 Mitigation Policy Analysis.* Proceedings of the Conference on Energy use in Manufacturing, Korea Energy Economics Institute, Korea Resource Economics Association, Seoul, Korea, 19-20 May 1998.

Knack, S., and P. Keefer, 1997: Does Social Capital have an Economic Payoff: A Cross Country Investigation. *Quarterly Journal of Economics*, **CXII**(4), 1251-1288.

Knudson, M., 1998: Agricultural Diversity: Do We Have The Resources To Meet Future Needs? In *Global Environmental Change: Assessing the Impacts.* G. Frisvold and B. Kuhn (eds.), Edward Elgar, Aldershot, UK.

Kokorin, A., 1997: Forest carbon sequestration scenarios and priorities for the russian federation action plan. *Energy Policy*, **56**, 407-421.

Komor, P.S., and L.L. Wiggins, 1988: Predicting conservation choice: beyond the cost-minimization assumption. *Energy*, **13**(8), 633-645.

Krusell, P., and J.V. Ríos–Rull, 1996: Vested Interests in a Positive Theory of Stagnation and Growth. *Review of Economic Studies*, **63**, 301–329.

Kushler M., and P. Witte, 2000: *A Review and Early Assessment of Public Benefit Policies Under Electric Restructuring - Volume 1: A State-By-State Catalog of Policies and Actions.* American Council for an Energy Efficient Economy, Washington DC, 87 pp.

Lane, D., and R. Maxfield, 1995: *Foresight, Complexity and Strategy.* Santa Fe Institute, New Mexico.

Lanoie, P., B. Laplante, and M. Roy, 1997: *Can Capital Markets Create Incentives for Pollution Control?* World Bank Policy Research Working Paper nr. 1753

Latour, B., 1993: *We Have Never Been Modern.* Prentice Hall, Harlow, England.

Leonard, J., 1988: *Pollution and the Struggle for World Product: Multinational Corporations, Environment and International Comparative Advantage.* Cambridge University Press.

Levine, M.D., E. Hirst, J.G. Koomey, J.E. McMahon, and A.H. Sanstad, 1994: *Energy Efficiency, Market Failures, and Government Policy.* Lawrence Berkeley Laboratory / Oak Ridge National Laboratory, Berkeley / Oak Ridge, USA.

Levine, M.D., J.G. Koomey, L.K. Price, H. Geller, and S. Nadel, 1995: Electricity and end-use efficiency: experience with technologies, markets, and policies throughout the world, *Energy*, **20**, 37-65.

Levinson, A., 1994: *Environmental Regulations and Industry Location: International and Domestic Evidence.* University of Wisconsin Working Paper.

Lofstedt, R., 1992: Lay perspectives concerning global climate change in Sweden. *Energy and Environment*, **18**(2), 161-175.

Lopez, R., 1997: Demand-based mechanisms to finance the green environment in Latin America. In *Finance for Sustainable Development.* Proceedings of the Fourth Group Meeting on Financial Issues of Agenda 21, Santiago, Chile 1997, United Nations, New York, 431-454.

Lovei, M., 1995: *Financing Pollution Abatement: Theory and Practice.* Environment Department Papers No. 028, The World Bank.

Lovins, A., 1992: *Energy-efficient buildings: Institutional barriers and opportunities.* E-source Rocky Mountain Institute.

Macaulay, A., D. Russell and D. MacKie, 1993: All Change! A Transport Strategy for Central Region: The Role of Workshops in Public Participation. In *Seminar F.* PTRC Transport, Highways and Planning Summer Annual Meeting, University of Manchester, England, 13-17 September, PTRC, London.

MacIntyre, A., 1985: *After Virtue.* Second Edition, Duckworth, London.

Major, G., 1995: *Learning from experiences with Small-scale Cogeneration.* Analyses Series No. 1, CADDET Centre for the Analysis and Dissemination of Demonstrated Energy Technologies, IEA International Energy Agency, OECD Organisation for Economic Co-operation and Development, Paris, 129 pp.

Maloney, M.P., and M.P. Ward, 1973: Ecology: lets hear it from the people: an objective scale for the measurement of ecological attitudes and knowledge. *The American Psychologist,* **28**, 583-586.

Markusen, J.R., 1999: Location Choice, Environmental Quality and Public Policy. In *Handbook of Environmental and Resource Economics.* C. van den Bergh, Edward Elgar, Aldershot, UK, 569-580.

Markus, H.R., and S. Kitayama, 1991: Culture and the self: implication for cognition, emotion and motivation. *Psychological Review,* **98**(2), 224-253.

Maslow, A., 1954: *Motivation and Personality.* Harper and Row, New York.

Max-Neef, M., 1991: *Human Scale Development - Conception, Application and Further Reflection.* Apex Press, London.

McKane, A.T., and J.P. Harris, 1996: *Changing Government Purchasing Practices: Promoting Energy Efficiency on a Budget.* Proceedings of the 1996 ACEEE Summer Study on Energy Efficiency in Buildings, August, 1996.

McMahon, J., and I. Turiel, 1997: Introduction to special issue devoted to appliance and lighting standards. *Energy and Buildings,* **26**(1), p. 1.

McQuail, D., J.G. Blumler and J. R. Brown, 1972: The television audience: a revised perspective. In Sociology of Mass Communications. D. McQuail (ed.), Penguin, Harmondsworth. Reproduced in part in *Media Studies: A Reader.* P. Marris and S. Thornham, (eds.), 1999, Edinburgh University Press, Edinburgh, 438-454.

Messner, S., 1996: Endogenized Technological learning in an Energy Systems Model. In *Journal of Evolutionary Economics.*

Metcalf, G., 1994: Economics and rational conservation policy. *Energy Policy,* **22**(10), 819-825.

Meyer-Abich, K.M., 1997: *Praktische Naturphilosophie – Erinnerung an einen vergessen Traum.* C.H. Beck, München, 520 pp.

Meyers, S, 1998: *Improving Energy Efficiency: Strategies for Supporting Sustained Market Evolution in Developing and Transitioning Countries.* Lawrence Berkeley National Laboratory, Berkeley, CA, LBNL-41460.

Michaelis, L., 1996a: *Sustainable Transport Policies: CO_2 Emissions from Road Vehicles, Policies and Measures for Common Action, Working Paper 1, Annex I.* Expert Group on the UNFCCC, OECD, Paris.

Michaelis, L., 1996b: Mitigation options in the transportation sector. In *Climate Change 1995: Impacts, Adaptations and Mitigation of Climate Change: Scientific-Technical Analyses.* R. Watson, M. Zinyowera and R. Moss (eds.), Cambridge University Press, Cambridge, 679-712.

Michaelis, L., 1997a: *Innovation in Transport Behaviour and Technology.* Working Paper 13, Annex I, Expert Group on the UNFCCC, OCDE/GD(**97**)79, OECD, Paris.

Michaelis, L., 1997b: Technical and Behavioural Change: Implications for Energy End-Use. In *Energy Modelling: Beyond Economics and Technology.* B. Giovannini and A. Baranzini (eds.), International Academy of the Environment, Geneva, and the Centre for Energy Studies of the University of Geneva.

Michaelis, L., 1997c: *Policies and Measures to Encourage Innovation in Transport Behaviour and Technology, Policies and Measures for Common Action.* Working Paper 13, Annex I, Expert Group on the UNFCCC, OECD, Paris.

Michaelis, L., 1998: Economic and technological development in climate scenarios. *Mitigation and Adaptation Strategies for Global Change,* **3**(2-4), 231-261.

Michaelis, L., 2000a: Drivers of consumption patterns. In *Towards Sustainable Consumption: A European Perspective.* B. Heap and J. Kent (eds.), The Royal Society, London, 75-84.

Michaelis, L., 2000b: *Ethics of Consumption.* Oxford Commission for Sustainable Consumption, document OCSC 2.1, Oxford Centre for the Environment, Ethics and Society, Oxford, 33 pp.

Michaelowa, A., K. Michaelowa, and S. Vaughan, 1998: Joint Implementation and trade policy. *Aussenwirtschaft,* **53**(4), 573-589.

Michaelowa, K., 1996: *Who Determines the Amount of Tied Aid: A Public-Choice Approach.* HWWA Discussion Paper No. 40, Hamburg Institute for Economic Research, Hamburg.

Millman, S., and R. Prince, 1989: Firm Incentives to Promote Technological Change in Pollution Control. *Journal of Environmental Economics and Management,* **17**, 247-265.

Mills E., 1993: Efficient lighting programs in Europe: Cost effectiveness, consumer response and market dynamics. *Energy,* **18**(2), 131-144.

Mohanty. B., 1997: *Technology, energy efficiency and environmental externalities in the pulp and paper industry.* Asian Institute of Technology, Bangkok, Thailand.

Moisander, J., 1998: *Motivation for Ecologically Oriented Consumer Behaviour.* Working paper prepared for Second European Science Foundation Workshop on Consumption, Everyday Life and Sustainability, Lancaster University, England. See http://www.lancs.ac.uk/users/scistud/ esf/title.htm.

Mosier, A.R., J.M. Duxbury, J.R. Freney, O. Heinemeyer, K. Minami, and D.E. Johnson, 1998: Mitigating agricultural emissions of methane. *Climatic Change,* **40**, 39-80.

Munasinghe, M., and R. Swart, 1999: An Overview of Climate Change and its Links with Development, Equity and Sustainability. In *Climate Change and its Linkages with Development, Equity and Sustainability.* IPCC Expert Meeting, Colombo, April 1999.

MVA, 1995: *Effects and Elasticities of Higher Fuel Prices.* Final Report to UK Department of Transport, The MVA Consultancy, Woking, Surrey, UK.

Nadel, S., V. Kothari, and S. Gopinath, 1991: *Opportunities for improving end-use electricity efficiency in India.* American Council for an Energy-Efficient Economy, Washington DC, USA.

Nakicenovic, N., 1996: Technological Change and Learning. In *Climate Change: Integrating Science, Economics, and Policy.* N. Nakicenovic, W.D. Nordhaus, R. Richels and F.L. Tol (eds.). International Institute for Applied Systems Analysis, Laxenburg, Austria, 271-294.

Narang, H.P., R.K. Gupta, R.C. Sharma, and S. Kaul, 1998: *Methane emission factors from biomass burning in traditional and improved cookstoves.* Proceedings of Convention and Symposium on Sustainable Biomass Production and Rural Development, Bio-Energy Society of India, March, 1998.

Nelson, R.R., and S.G. Winter, 1982: *An Evolutionary Theory of Economic Change.* Harvard University Press, Cambridge, MA.

Nelson, R.R. (ed.), 1982: *Government and Technical Progress.* Pergamon Press, New York, USA.

Nelson, R.R., M.J. Peck, and E.D. Kalachek, 1967: *Technology, Economic Growth and Public Policy.* The Brookings Institute, Washington DC.

Nelson, K., 1994: Finding and implementing projects that reduce waste. In *Industrial Ecology and Global Change.* R.H. Socolow, C. Andrews, F. Berkhout, and V. Thomas (eds.), Cambridge University Press, Cambridge, UK.

Newman, P., and M. Kenworthy, 1990: *Cities and Automobile Dependence.* Gower, London.

Nielsen, M., and E. Bernsen, 1996: Combined heat and power in Eastern Europe: potentials and barriers. *International Journal of Global Energy Issues,* **8**(4), 375-384.

North, D., 1990: *Institutions, Institutional Change, and Economic Performance.* Cambridge University Press.

NWMC (National Waste Management Council), 1990: *Circular No.17,* (1)/87-PL HSMD, Ministry of Environment and Forests, Government of India

Nyati, K.P., 1994: *Prospect, barriers and strategies- cleaner industrial production in developing countries.* Asia Pacific Tech Monitor, ESCAP APCTT, New Delhi, India, 10-14.

O'Riordan, T., C.L. Cooper, A. Jordan, S. Rayner, K.R. Richards, P. Runci, and S. Yoffe, 1998: Institutional Framework for Political Action. In *Human Choice and Climate Change.* S. Rayner and E.L. Malone, (eds.), Battelle Press.

OECD, 1992: *Technology and the Economy: the Key Relationships.* OECD, Paris.

OECD, 1996: *Saving Biological Diversity. Economic Incentives.* OECD, Paris.

OECD, 1997a: *Report of the OECD Policy Meeting on Sustainable Consumption and Individual Travel Behaviour.* Paris, 9-10 January 1997, OCDE/GD(**97**)144, OECD, Paris.

OECD, 1997b: *Reforming Energy and Transport Subsidies: Environmental and Economic Implications.* OECD, Paris.

OECD, 1998a: *Eco-Efficiency.* OECD, Paris.

OECD, 1998b: *Environmental Financing in CEE/NIS: Conclusions and Recommendations.* OECD, Paris.

OECD, 1999a: *Environment in the Transition to a Market Economy. Progress in Central and Eastern Europe and the New Independent States.* OECD, Paris.

OECD, 1999b: *Sourcebook on environmental funds in economies in transition.* OECD, Paris.

OECD, 2000: *Integrating Public Environmental Expenditure Management and Public Finance in Transition Economies.* OECD, Paris, CCNM/ENV/EAP(**2000**)90.

Olson, M., 1982: *The Rise and Decline of Nations: Economic Growth, Stagflation, and Social Rigidities.* Yale University Press.

Ostrom, E., 1990: *Governing the Commons.* Cambridge University Press.

Ostrom, E., L. Schroeder, and S. Wynne, 1993: *Institutional Incentives and Sustainable Development.* Westview Press, Boulder, Colorado, United States.

OTA (Office of Technology Assessment), 1992: *Energy Efficient Buildings: Does the Market Place Work?* Proceedings of the 24th Annual Illinois Energy Conference, University of Illinois, Chicago, Illinois, 76-78.

OTA (Office of Technology Assessment), 1993: *Industrial Energy Efficiency.* US Government Printing Office, Washington DC, USA.

Painuly, J.P., and B.S. Reddy, 1996: Electricity conservation programmes - Barriers in implementation. *Energy Sources*, **18**(3).

Palmer, K., W.E. Oats, and P.R. Portney, 1995: Tightening environmental standards: the benefit-cost or the no-cost paradigm? *Journal of Economic Perspectives*, **9**(4), 119-132.

Pargal, S., and D. Wheeler, 1996: Informal Regulation of Industrial Pollution in Developing Countries: Evidence from Indonesia. *Journal of Political Economy*, **104**(6), 1314-1324.

Parikh, J.P., B.S. Reddy, R. Banerjee and Koundinya, 1996: DSM Survey in India: Awareness, barriers and implementability. *Energy Sources*, **12**(10), 3 pp.

Partridge, W.L., 1996: Effective Financing of Environmentally Sustainable Development in Latin America and the Carrabien, In *Effective Financing of Environmentally Sustainable Development.* I. Serageldin, A. Sfeir-Younis (eds.), Environmentally Sustainable Development Proceedings Series No. 10, The World Bank, Washington DC.

Paustian, K., C.V. Cole, D. Sauerbeck, and N. Sampson, 1998: CO_2 mitigation by Agriculture: An Overview. *Climatic Change*, **40**, 135-162.

Pawlik, K. (ed.), 1996: Perception and Assessment of Global Environmental Conditions and Change (Page C). In *Human Dimensions of Global Environmental Change.* L. Reprot (ed.), Programme of the International Social Science Council, Barcelona, Spain.

Pawlik, K., 1991: The psychology of global environmental change: some basic data and an agenda for cooperative international research. *International Journal of Psychology*, **26**, 547-563.

Pearce, D., S. Ozdemiroglu, and Dobson, 1997: Replicating innovative financing mechanisms for sustainable development. In *Finance for Sustainable Development.* Proceedings of the Fourth Group Meeting on Financial Issues of Agenda 21, Santiago, Chile 1997, United Nations, New York, 405-430.

Peszko, G., and T. Zylicz, 1998: Environmental Financing in European Economies in Transition. *Environmental and Resource Economics*, **11**(3-4), 521-538.

Pieters, J., 1997: Subsidies and environment: on how subsidies and tax incentives may affect production decisions and the environment. In *Finance for Sustainable Development.* Proceedings of the Fourth Group Meeting on Financial Issues of Agenda 21, Santiago, Chile 1997, United Nations, New York, 315-339.

Porter, M.E., and C. van der Linde, 1995a: Green and Competitive: Ending the Stalemate. *Harvard Business Review* **5**(73), 120-123.

Porter, M.E., and C. van der Linde, 1995b: Toward a new conception of the environment-competitiveness relationship. *Journal of Economic Perspectives*, **9**(4), 97-118.

Price, L., L. Michaelis, E. Worrell, and M. Khrushch, 1998: Sectoral trends and driving forces of global energy use and greenhouse gas emissions. *Mitigation and Adaptation Strategies for Global Change*, **3**(2-4), 263-319.

Putnam, R., 1993: The Prosperous Community - Social Capital and Public Life. *American Prospect*, **13**, 35-42.

Rabinovitch, J., 1992: Curitiba: Toward Sustainable Urban Development. *Environment and Urbanization*, **4**(2), 62-73.

Rabinovitch, J., 1993: Urban public transport management in Curitiba, Brazil. *Industry and Environment*, Vol. 16, No. 1-2. United Nations Environment Programme, Paris.

Ramesohl, S., 1998: Successful Implementation of Energy Efficiency in Light Industry: A Socio-Economic Approach to Industrial Energy Policies. In *Industrial Energy Efficiency Policies: Understanding Success and Failure.* N. Martin, E. Worrell, A. Sandoval, J.W. Bode, and D. Phylipsen (eds.), Lawrence Berkeley National Laboratory, Berkeley, CA, USA. LBNL-42368.

Rauscher, M., 1999: Environmental Policies in Open Economies. *In Handbook of Environmental and Resource Economics.* C. van den Bergh, Edward Elgar, Aldershot, UK, 395-403.

Reddy, B.S., and R.M. Shrestha, 1998: Barriers to the adoption of efficient electricity technologies: a case study of India. *International Journal of Energy Research*, **22**, 257-270.

Reddy, A.K.N., 1991: Barriers to improvements in energy efficiency. *Energy Policy*. **19**, 953-961.

Reinhardt, F., 1999. Market Failure and the Environmental Policies of Firms Economic Rationales for Beyond Compliance Behavior. *Journal of Industrial Ecology* **1**(3), 9-21.

Rip, A., and R. Kemp, 1998: Human Choice and Climate Change. In *Technological Change.* S. Rayner and E.L. Malone (eds.), Battelle Press, 327-399.

Rohrer, A., 1996: Comparison of Combined Heat and Power Generation. *International Journal of Global Energy Issues*, **8**(4), 319-328.

Rosenberg, N., 1982: *Inside the Black Box: Technology and Economics.* Cambridge University Press, Cambridge, England.

Rosenberg, N., 1994: *Exploring the Black Box: Technology, Economics and History.* Cambridge University Press, Cambridge, England.

Rosenthal, J.P., 1997: Equitable Sharing of Biodiversity Benefits: Agreements on Genetic Resources. In: *Investing in Biological Diversity.* Proceedings of the Cairns Conference, OECD, Paris, 253-274.

Ross, M., 1990: Capital Budgeting Practices of Twelve Large Manufacturers. In *Advances in Business Financial Management*, Ph. Cooley (ed.), Dryden Press, Chicago, 157-170.

Ross, M.H., 1986: Capital budgeting practices of twelve large manufacturers. *Financial Management*, **Winter**, 15-22.

Ruderman, H., M.D. Levine, and J.E. McMahon, 1987: The Behavior of the Market for Energy Efficiency in Residential Appliances Including Heating and Cooling Equipment. *The Energy Journal*, **8**(1), 101-123.

Rüdig, W., 1986: Energy conservation and electricity utilities - A comparative analysis of organisational obstacles to CHP/DH. *Energy policy*, **4**, 104-116.

Sanstad, A.H., and R.B. Howarth, 1994: Normal markets, market imperfections and energy efficiency. *Energy Policy*, **22**, 811-818.

Sathaye, J., and M. Walsh, 1992: Transportation in Developing Nations: Managing the Institutional and Technological Transition to a Low-Emissions Future. In *Confronting Climate Change: Risks, Implications and Responses.* Cambridge University Press.

Sathaye, J., R.K. Dixon, and C. Rosenweig, 1997a: Climate Change Country Studies. *Applied Energy*, **56**(3/4), 225-235.

Sathaye, J., W. Makundi, B. Goldberg, M. Pinard, and C. Jepma, 1997b: International workshop on sustainable forestry management: monitoring and verification of greenhouse gases. In *Summary report, Mitigation and Adaptation Strategies for Global Change*, **2**, 91-99.

Sathaye, J., and N.H. Ravindranath, 1998: Climate change mitigation in the energy and forestry sectors of developing countries. *Annual Review Energy Environment*, **23**, 387-437.

Schlegelmilch, K. (ed.), 1999: *Green budget reform in Europe: countries at the forefront.* Springer-Verlag, Berlin Heidelberg.

Schmalensee, R., 1998: Greenhouse policy architectures and institutions. In *Economic and Policy Issues in Climate Change.* W.D. Nordhaus (ed.), Resources for the Future.

Schor, J., 1998: *The Overspent American.* Basic Books, New York.

Shah, R.K., 1991: Faillibility in Human Organizations and Political Systems. *Journal of Economic Perspectives,* **5**(2), 67-88.

Shove, E., 1999: Gaps, barriers and conceptual chasms: theories of technology transfer and energy in buildings. *Energy Policy,* **26**(15), 1105-1112.

Schumacher, K., and J. Sathaye, 1999: Energy and Productivity Growth in Indian Industries. In *Proceedings of the European Energy Conference on Technological Progress and Energy Challenges,* 30 September – 1 October 1999, Paris, France, 76-87.

Schumpeter, J., 1928: The Instability of Capitalism. *Economic Journal,* **XXXVIII**(151), 361-386.

Schumpeter, J., 1935: The analysis of economic change. *Review of Economic Statistics,* **May,** 2-10.

Schwarz, E.J., 1997: *Recycling networks - A building block towards a sustainable development.* International Solid Waste Association Year Book 1997/8, 183

Shaughnessy, H., 1995: *Project finance in Europe.* John Wiley & Sons, Chichester.

Slevin, J., 2000: *The Internet and Society.* Polity Press, Cambridge, England, 266 pp.

Shrestha, R.M., and B.K. Karmacharya, 1998: Testing of barriers to the adoption of energy efficient lamps in Nepal. *The Journal of Energy and Development,* **23**(1), International Research Center for Energy and Economic Development (ICEED).

Simon, H.A., 1991: Organizations and Markets. *Journal of Economic Perspectives,* **5**(2), 25-44.

Simpson, R.D., R.A. Sedjo, and J.W. Reid, 1996: Valuing Biodiversity for Use in Pharmaceutical Research. *Journal of Political Economy,* **104,** 1548-1570.

Sinton, J.E., and M.D. Levine, 1994: Changing Energy Intensity in Chinese Industry. *Energy Policy,* **22,** 239-255.

Sioshansi, F.P., 1991: The Myths and Facts of Energy Efficiency. *Energy Policy,* **19,** 231-243.

Soussan, J., 1987: Fuel Transitions within Households. Discussion paper No. 35. In *Discussion between traditional and commercial energy in the Third World.* W. Elkan, *et al.* (eds.), Survey Energy Economics Centre, University of Surrey, Guildford, Surrey, UK.

Srivastava L., and P. Dadhich, 1999: *Developing Economies, Capital Shortages and Transnational Corporation (TNCs) in Economic Impact of Mitigation Measures.* IPCC Expert Meeting, The Hague, May.

Stenstadvold, M., 1995: Institutional Constraints to Environmentally Sound Integrated Land-Use and Transport Policies: Experiences from the Norwegian Integrated Land-Use and Transport Planning Scheme. In *The 23rd European Transport Forum.* Coventry, England, 11-15 September, PTRC, London.

Stern, P.C., 1986: Blind Spots in Policy Analysis: What Economics Doesn't Say About Energy Use. *Journal of Policy Analysis and Management,* **5**(2), 200-227.

Stern P., and E. Aronson, 1984: *Energy use: The human dimension.* W.H. Freeman, New York.

Stewart, P., 1993: Project Finance. *International Financial Law Review* and *Euromoney Publications,* London, August.

Stewart, R., 1981: Regulation, Innovation and Administrative Law: A Conceptual Framework. *California Law Review,* **691,** 256-377.

Storey M., 1997: *The climate implications of agricultural policy reform. Policies and Measures for Common Action.* Working Paper 16, Annex I, Expert Group on the UNFCCC, OECD.

TERI (Tata Energy Research Institute), 1997: *Capacity building for technology transfer in the context of climate change.* TERI, New Delhi, India.

Taylor, C., 1989: *Sources of the Self: The Making of the Modern Identity.* Cambridge University Press.

Tomlinson, A., 1990: *Consumption, Identity and Style. Marketing, Meanings and the Packaging of Pleasure.* Routledge, London and New York.

Thompson, M., and S. Rayner, 1998: Cultural discourses. In *Human Choice and Climate Change. Volume 1. The Societal Framework.* S. Rayner and E. L. Malone (eds.), Battelle Press, Columbus, Ohio, 265-343.

Thompson, 2000: Consumption, Motivation and choice across scale: Consequences for selecting target groups. In **IPCC,** 2000b: *Society, Behavior, and Climate Change Mitigation.* E. Jochem, D. Bouille, and J. Sathaye (eds.), Proceedings of the IPCC Expert Meeting, Karlsruhe, Germany.

Tiessen H., C. Feller, E.V.S.B. Sampaio, and P. Garin, 1998: Carbon sequestration and turnover in semiarid savannas and dry forest. *Climatic Change,* **40,** 105-117.

Tlaie L., and D. Biller, 1994: *Successful environmental institutions: Lessons from Colombia and Curitiba Brazil.* Dissemination Note No.12, The World Bank, Washington DC.

Ulph, A., 1997: Environmental Policy and International Trade. In *New Directions in Economic Theory of the Environment.* C. Carraro and D. Siniscalco, Cambridge University Press, 147-192.

Umana, A., 1996: Financing biodiversity conservation programs. Effective financing of environmentally sustainable development. In *Environmentally Sustainable Development Proceedings.* I. Serageldin and A. Sfeir-Younis (eds.), Series No. 10, The World Bank, Washington D.C.

UNEP (United Nations Environment Programme), 1992: Health. In *The world environment 1972-1992: Two decades of challenge.* M.K. Tolba, O.A. El-Kholy, E. El-Hinnawi, M.W. Holdgate, D.F. McMichael, and R.E. Munn, Chapman & Hall, New York, 529-567.

UNEP (United Nations Environment Programme), 1999: *Insurance Industry Initiative.* UNEP Financial Services Initiatives, UNEP Economics and Trade Unit, Paris.

UNIDO (United Nations Industrial Development Organisation), 1996: *Guidelines for infrastructure development through build-operate-transfer (BOT) projects.* Vienna.

US AID, 1996: Strategies for financing energy efficiency. Washington D. C., US AID.

US DOE (Department of Energy), 1995: Effects of Feebates on Vehicle Fuel Economy, Carbon Dioxide Emissions, and Consumer Surplus. *Energy Efficiency in the U.S. Economy,* Technical Report Two DOE/PO-0031. United States Department of Energy, Washington, DC.

US DOE (Department of Energy), 1996: *Policies and measures for reducing energy related greenhouse gas emission. Lessons from Recent Literature.* Office of Policy and International Affairs, USDOE, Washington DC, July.

Universiteit Utrecht, 1997: *Afspraken Werken, Evaluatie Meerjarenafspraken over Energie-Efficiency.* Ministry of Economic Affairs, The Hague, The Netherlands.

van Beers, and J. Van den Bergh, 1997: An Empirical Multi-Country Analysis of the Impact of Environmental Regulations on foreign Trade Flows. *Kyklos,* **50,** 29-46.

Veblen, T., 1899: The Theory of the Leisure Class, 1993 edition, Dover. In *Transformation of Unsustainable Consumer Behaviors and Consumer Policies.* C. Vlek, L. Reisch, and G. Scherhorn, IHDP-IT Open Science Meeting, Discussion Paper 2, February 1999.

Velthuijsen, J.W., 1995: *Determinants of investment in energy conservation.* SEO, University of Amsterdam, The Netherlands.

Verbruggen, A., 1990: Pricing independent power production. *International Journal of Global Energy Issues,* **2**(1), 41-49.

Verbruggen, A., 1992: Combined Heat and Power. A real alternative when carefully implemented. *Energy Policy,* **20**(9), 884-892.

Verbruggen, A., 1996: An introduction to CHP issues. *International Journal of Global Energy Issues,* **8**(4), 301-318.

Verbruggen, A., M. Wiggin, N. Dufait, and A. Martens, 1992: The impact of CHP generation on CO_2 emissions. *Energy Policy,* **20**(12), 1207-1214.

Verhallen, T.M.M., and W.F. van Raaij, 1981: Household behaviour and the use of natural gas for home heating. *Journal of Consumer Research,* **8,** 253-257

Vlek, C., L. Reisch, and G. Scherhorn, 1999: *Transformation of unsustainable consumer behaviors and consumer policies.* IHDP-IT Open Science Meeting, Discussion Paper 2, February 1999.

Vogel, G., 1998: *Waste Minimisation using Economic Instruments.* The International Solid Waste Association Year Book, 44 pp.

Wallace, D., 1995: *Environmental Policy and Industrial Innovation: Strategies in Europe, the US and Japan*. Energy and Environment Programme, The Royal Institute of International Affairs, Earthscan, London.

Watson, R.T., M.C. Zinyowera, and R.H. Moss (eds.), 1996: *Climate Change 1995: Impacts, Adaptations and Mitigation of Climate Change: Scientific-Technical Analyses*. Cambridge University Press.

WCED (World Commission on Environment and Development), 1987: *Our Common Future*. Oxford University Press, Oxford and New York.

WEC (World Energy Council), 1991: *District Heating/Combined Heat and Power. Decisive factors for a successful use - as learnt from experiences*. WEC, London, 69 & 33 pp.

Wenzel, T., 1995: *Analysis of National Pay-As-You-Drive Insurance Systems and Other Variable Driving Charges*. Energy and Environment Division, Lawrence Berkeley Laboratory Paper, Berkeley, CA, LBL-37321.

White, L., 1982: US Automotive Emissions Controls: How Well Are They Working? *American Economic Review*, **75**(2), 332-335.

Wilhite, H., H. Nakagami, T. Masuda, Y. Yamaga, and H. Haneda, 1996: A cross-cultural analysis of household energy use behaviour in Japan and Norway. *Energy Policy*, **24**(9), 795-803.

Wilhite, H., E. Shove, L. Lutzenhiser, and W. Kempton, 2000: After twenty years of "demand side management" (DSM) we know a little about individual behaviour but next to nothing about energy demand. In *Society, Behavior, and Climate Change Mitigation*. E. Jochem, D. Bouille, and J. Sathaye (eds.), Proceedings of the IPCC Expert Meeting, Karlsruhe, Germany.

Wilk, R., 1999: *Towards a useful multigenic theory of consumption*. Paper presented at the Summer School of the European Council for an Energy Efficient Economy, Mandelieu, France.

Winter, S.G., Y.M. Kaniovski, and G. Dosi, 2000: Modeling industrial dynamics with innovative entrants. *Structural Change and Economic Dynamics*, **11**, 255-293.

WB and IFC (World Bank Group and International Finance Corporation), 1996: *Financing Private Infrastructure*. Washington DC.

World Bank, 1994: *Submission and Evaluation of Proposals for Private Power Generation Projects in Developing Countries*. Discussion Paper 250, P. Cordukes, Washington DC, September.

World Bank, 1997: *Expanding the Measure of Wealth*, World Bank.

World Bank, 1998: *Transition Toward a Healthier Environment. Environmental Issues and Challenges in the Newly Independent States*. Washington DC.

World Bank, 1999: *Fuel for Thought: Environmental Strategy for the Energy Sector*. Washington DC, July.

WWF (World Wildlife Fund), 1996: *Sustainable Energy Technology in the South*. A Report to WWF by Institute of Environmental Studies, Amsterdam, The Netherlands, and Tata Energy Research Institute, New Delhi, India.

Worrell, E., M. Levine, L. Price, N. Martin, R. van den Broek, and K. Blok, 1997: *Potentials and Policy Implications of Energy and Material Efficiency Improvement*. United Nations Division for Sustainable Development, New York.

Wynne, B., 1993: Implementation of Greenhouse Gas Reductions in the European Community: Institutional and Cultural Factors. *Global Environmental Change*, **3**(1).

Xepapadeas, A., and A. de Zeeuw, 1999: Environmental Policy and Competitiveness: the Porter Hypothesis and the Composition of Capital. *Journal of Environmental Economics and Management*, **37**(2), 101–128, 165-182.

Young, O.R. 1994: *International Governance: Protecting the Environment in a Stateless Society*. Cornell University Press.

Young, O.R. (ed.), 1999: *Science Plan for the Project on Institutional Dimensions of Global Environmental Change*. International Human Dimensions Programme of Global Environmental Change, http://www.uni-bonn.de/ihdp/wpo1.

6

Policies, Measures, and Instruments

Co-ordinating Lead Authors:
IGOR BASHMAKOV (RUSSIAN FEDERATION), CATRINUS JEPMA
(NETHERLANDS)

Lead Authors:
Peter Bohm (Sweden), Sujata Gupta (India), Erik Haites (Canada), Thomas Heller (USA), Juan-Pablo Montero (Chile), Alberto Pasco-Font (Peru), Robert Stavins (USA), John Turkson† (Ghana), Huaqing Xu (China), Mitsutsune Yamaguchi (Japan)

Contributing Authors:
Scott Barrett (UK), Andrew Dearing (UK), Bouwe Dijkstra (Netherlands), Ed Holt (USA), Nathaniel Keohane (USA), Shinya Murase (Japan), Toshio Nakada (Japan), William Pizer (USA), Farhana Yamin (Pakistan)

Review Editors:
Dilip Ahuja (India), Peter Wilcoxen (USA)

† John Turkson passed away in January 2000. He contributed actively to the drafting process.

CONTENTS

EXECUTIVE SUMMARY

The purpose of this chapter is to examine the major types of policies and measures that can be used to mitigate net concentrations of greenhouse gases (GHGs) in the atmosphere.[1] Alternative policy instruments are described and assessed in terms of specific criteria, on the basis of the most recent literature. Naturally, emphasis is on the instruments mentioned in the Kyoto Protocol (the Kyoto mechanisms), because they focus on achieving GHG emissions limits, and the extent of their envisaged international application is unprecedented. In addition to economic dimensions, political, economic, legal, and institutional elements are considered insofar as they are relevant to the discussion of policies and measures.

Any individual country can choose from a large set of possible policies, measures, and instruments to limit domestic GHG emissions. These can be categorized into market-based instruments (which include taxes on emissions, carbon, and/or energy, tradable permits, subsidies, and deposit–refund systems), regulatory instruments (which include non-tradable permits, technology and performance standards, product bans, and direct government spending, including research and development investment) and voluntary agreements (VAs) of which some fall in the category of market-based instruments. Likewise, a group of countries that wants to limit its collective GHG emissions could agree to implement one, or a mix, of instruments. These are (in arbitrary order) tradable quotas, Joint Implementation (JI), the Clean Development Mechanism (CDM), harmonized taxes on emissions, carbon, and/or energy, an international tax on emissions, carbon, and/or energy, non-tradable quotas, international technology and product standards, VAs, and direct international transfers of financial resources and technology.

Possible criteria for the assessment of policy instruments include environmental effectiveness, cost effectiveness, distribution considerations, administrative and political feasibility, government revenues, wider economic effects, wider environmental effects, and effects on changes in attitudes, awareness, learning, innovation, technical progress, and dissemination of technology. Each government may apply different weights to various criteria when evaluating policy options for GHG mitigation, depending on national and sector-level circumstances. Moreover, a government may apply different sets of weights to the criteria when evaluating national (domestic) versus international policy instruments.

---

[1] In keeping within the defined scope of Working Group III, policies and measures that can be used to reduce the costs of adaptation to climate change are not examined.

The economics literature on the choice of policies adopted emphasizes the importance of interest-group pressures, focusing on the demand for regulation. However, it has tended to neglect the "supply side" of the political equation, which is emphasized in the political science literature of the legislators and government and party officials who design and implement regulatory policy, and who ultimately decide which instruments or mix of instruments will be used. The point of compliance of alternative policy instruments, whether they are applied to fossil fuel users or manufacturers, for example, is likely to be politically crucial to the choice of policy instrument. And a key insight is that some forms of regulation actually benefit the regulated industry, for example, by limiting entry into the industry or by imposing higher costs on new entrants. A policy that imposes costs on industry as a whole might still be supported by firms who, as a consequence, would fare better than their competitors. Regulated firms, of course, are not the only group with a stake in regulation: opposing interest groups will fight for their own interests.

To develop reasonable assessments of the feasibility of implementing GHG mitigation policies in countries in the process of structural reform, it is important to understand this new policy context. Recent measures taken to liberalize energy markets were inspired mainly by desires to increase competition in energy and power markets, but they can have significant emissions implications also, through their impact on the production and technology pattern of energy and/or power supply. In the long run, the consumption pattern change might be more important than the sole implementation of climate change mitigation measures (e.g. see Chapter 2, the B1 scenario).

Market-based instruments–principally domestic taxes and domestic tradable permit systems–are attractive to governments in many cases because they are efficient; they are frequently introduced in concert with conventional regulatory measures. When implementing a domestic emissions tax, policymakers must consider the collection point, the tax base, the variation or uniformity among sectors, the association with trade, employment, revenue, and the exact form of the mechanism. Each of these can influence the appropriate design of a domestic emissions tax, and political or other concerns are likely to play a role also. For example, a tax levied on the energy content of fuels could be much more costly than a carbon tax for the equivalent emissions reduction, because an energy tax raises the price of all forms of energy, regardless of their contribution to carbon dioxide emissions. Yet, many nations may choose to use energy taxes for reasons other than cost-effectiveness, and much of the analysis in this chapter applies to energy taxes as well as to carbon taxes.

A country committed to a limit on its GHG emissions can also meet this limit by implementing a tradable permit system that directly or indirectly limits emissions of domestic sources. Like taxes, permit systems pose a number of design issues, including type of permit, sources included, point of compliance, and use of banking. To cover all sources with a single domestic permit regime is unlikely. The certainty provided by a tradable permit system that a given emission level for participating sources is achieved incurs the cost of uncertain permit prices (and hence compliance costs). To address this concern, a hybrid policy that caps compliance costs could be adopted, but the level of emissions would no longer be guaranteed.

For a variety of reasons, in most countries the management of GHG emissions will not be addressed with a single policy instrument, but with a portfolio of instruments. In addition to one or more market-based policies, a portfolio might include standards and other regulations, VAs, and information programmes:

- Energy-efficiency standards have reduced energy use in a growing number of countries. Standards may also help develop the administrative infrastructure needed to implement market-based policies. The main disadvantage of standards is that they can be inefficient, but efficiency can be improved if the standard focuses on the desired results and leaves as much flexibility as possible in the choice of how to achieve the results.

- VAs may take a variety of forms. Proponents of VAs point to low transaction costs and consensus elements, while sceptics emphasize the risk of free riding, and the risk that the private sector will not pursue real emissions reduction in the absence of monitoring and enforcement.

- Imperfect information is widely recognized as a key market failure that can have significant effects on improved energy efficiency, and hence emissions. Information instruments include environmental labelling, energy audits, and industrial reporting requirements, and information campaigns are marketing elements in many energy efficiency programmes.

A growing literature demonstrates theoretically, and with numerical simulation models, that the economics of addressing GHG reduction targets with domestic policy instruments depends strongly on the choice of those instruments. The interaction of abatement costs with the existing tax structure and, more generally, with existing factor prices is important. Policies that generate revenues can be coupled with policy measures that improve the efficiency of the tax structure.

Turning to international policies and measures, the Kyoto Protocol defines three international policy instruments, the so-called Kyoto mechanisms: international emissions trading (IET), JI, and CDM.[2] Each of these international policy instruments provides opportunities for Annex I Parties[3] to fulfil their

commitments cost-effectively. IET essentially allows Annex I Parties to exchange part of their assigned amounts (AAs). IET implies that countries with high marginal abatement costs (MACs) may acquire emissions reductions from countries with low MACs. Similarly, JI allows Annex I Parties to exchange emissions reduction units among themselves on a project-by-project basis. Under the CDM, Annex I Parties receive Certified Emissions Reduction (CERs)–on a project-by-project basis–for reductions accomplished in non-Annex I countries.

Economic analyses indicate that the Kyoto mechanisms could reduce significantly the overall cost of meeting the Kyoto emissions limitation commitments. However, to achieve the potential cost savings requires the adoption of domestic policies that allow the use the mechanisms to meet their national emissions limitation obligations. If domestic policies limit the use of the Kyoto mechanisms, or international rules that govern the mechanisms limit their use, the cost savings may be reduced.

In the case of JI, host governments have incentives to ensure that emission reduction units are issued only for real emission reductions, assuming that they face strong penalties for non-compliance with national emissions limitation commitments. In the case of CDM, a process for independent certification of emission reductions is crucial, because host governments do not have emissions limitation commitments and hence may have less incentive to ensure that certified emission reductions are issued for real emission reductions only. The main difficulty in implementing project-based mechanisms, both JI and CDM, is to determine the net additional emissions reductions (or sink enhancement) achieved. Various other aspects of these Kyoto mechanisms await further decision making, including monitoring and verification procedures, financial additionality (assurance that CDM projects do not displace traditional development-assistance flows) and other additionalities, and possi-

---

[2] The ability of two or more Annex I Parties to form a "bubble" under Article 4 of the Kyoto Protocol is sometimes classified as one of the flexibility mechanisms as well. This mechanism allows a one time redistribution of the emissions limitation commitments among the participants. Since such a redistribution is strictly a political decision this mechanism is not discussed here.

[3] Annex I Parties to the UNFCCC (as amended by decision 4/CP.3) include all 39 Parties (38 countries plus the European Economic Community) listed in Annex B of the Protocol that will have quantified emissions limitation or reduction commitments for the 2008 to 2012 commitment period, plus Turkey and Belarus, which are Parties to the Convention but not listed in Annex B of the Protocol. To be precise, one should refer to the commitments of Annex I Parties listed in Annex B of the Kyoto Protocol. To avoid confusion, the term Annex I countries is used throughout this chapter to refer to Annex I Parties listed in Annex B of the Protocol; Turkey and Belarus are understood to be included within this umbrella term, but not within the group of countries that will have limitation commitments.

ble means of standardizing methodologies for project baselines.

The extent to which developing country (non-Annex I) Parties effectively implement their commitments under the United Nations Framework Convention on Climate Change (UNFCCC; referred to as the Convention in this chapter) may depend on the effective implementation by developed country Parties of their commitments under the Convention related to the transfer of financial resources and technology. The transfer of environmentally sound technologies from developed to developing countries is now seen as a major element of global strategies to achieve sustainable development and climate stabilization.

Any international or domestic policy instrument can be effective only if accompanied by adequate systems of monitoring and enforcement. There is a linkage between compliance enforcement and the amount of international co-operation that will actually be sustained. Many multilateral environmental agreements address the need to co-ordinate restrictions on conduct taken in compliance with the obligations they impose and the expanding legal regime under the World Trade Organization (WTO) and General Agreement on Tariff and Trade (GATT) umbrella. Neither the UNFCCC nor the Kyoto Protocol provides for specific trade measures in response to non-compliance. But several domestic policies and measures that might be developed and implemented in conjunction with the Kyoto Protocol could conflict with WTO provisions. International differences in environmental regulation may have trade implications also.

One of the main concerns in environmental agreements (including the UNFCCC and the Kyoto Protocol) is with reaching wider participation. The literature on international environmental agreements predicts that participation will be incomplete, and so further incentives may be needed to increase participation.

6.1   Introduction

6.1.1   Introduction and Key Questions

The main purpose of this chapter is to discuss the various policies and measures in relation to the different criteria that can be used to assess them, on the basis of the most recent literature. There is obviously a relatively heavy focus on the Kyoto instruments, because they focus on climate policy, have been agreed since the IPCC Second Assessment Report (SAR; IPCC, 1996, Section 11.5), and the extent of their envisaged international application is unprecedented. Wherever feasible, political economic, legal, and institutional elements are discussed insofar as they are relevant to the implementation of policies and measures. To make both theoretical and practical points the chapter offers occasional examples of policy instrument application, but the effort in this regard is limited by the existing literature, which is weighted towards the experience of industrialized countries.[4]

The chapter does not systematically discuss policies and measures typically used to encourage sector-specific technologies; such policies and measures are described in Chapters 3, 4, and 5. The emphasis is on the general description and assessment of policies and measures.

6.1.2   Types of Policies, Measures, and Instruments

A country can choose from a large set of policies, measures, and instruments to limit domestic greenhouse gas (GHG) emissions or enhance sequestration by sinks. These include (in arbitrary order): (1) taxes on emissions, carbon, and/or energy, (2) tradable permits[5], (3) subsidies[6], (4) deposit–refund systems, (5) voluntary agreements (VAs), (6) non-tradable permits, (7) technology and performance standards, (8) product bans, and (9) direct government spending and investment. Definitions of these instruments are provided in *Box 6.1*. The first four are often called market-based instruments, although some VAs also fall into this category.

A group of countries that want to limit their collective GHG emissions could agree to implement one, or a mix, of instru-

---

[4] While an exhaustive review of in-country experiences with policy instruments is beyond the scope of this chapter, other recent works have focused on this issue (Panayotou, 1998; Huber *et al.*, 1999; Speck, 1999; Stavins, 2000).

[5] What makes a tradable permit a market-based instrument is the possibility of trading the permit, not the initial allocation of the permits (unless such allocation is through auction). The SAR adopted the convention of using "permits" for domestic trading systems and "quotas" for international trading systems. This convention is followed throughout the chapter.

[6] Sometimes taxes are combined with subsidies, known as "fee/rebate".

Box 6.1. Definitions of Selected National Greenhouse Gas Abatement Policy Instruments

- An emissions tax is a levy imposed by a government on each unit of emissions by a source subject to the tax. Since virtually all of the carbon in fossil fuels ultimately is emitted as CO_2, a levy on the carbon content of fossil fuels–a carbon tax–is equivalent to an emissions tax for emissions caused by fossil fuel combustion. An energy tax–a levy on the energy content of fuels–reduces the demand for energy and so reduces CO_2 emissions through fossil fuel use.

- A tradable permit (cap-and-trade) system establishes a limit on aggregate emissions by specified sources, requires each source to hold permits equal to its actual emissions, and allows permits to be traded among sources. This is different from a credit system, in which credits are created when a source reduces its emissions below a baseline equal to an estimate of what they would have been in the absence of the emissions reduction action. A source subject to an emissions-limitation commitment can use credits to meet its obligation.

- A subsidy is a direct payment from the government to an entity, or a tax reduction to that entity, for implementing a practice the government wishes to encourage. GHG emissions can be reduced by lowering existing subsidies that in effect raise emissions, such as subsidies to fossil fuel use, or by providing subsidies for practices that reduce emissions or enhance sinks (e.g., for insulation of buildings or planting trees).

- A deposit–refund system combines a deposit or fee (tax) on a commodity with a refund or rebate (subsidy) for implementation of a specified action.

- A VA is an agreement between a government authority and one or more private parties, as well as a unilateral commitment that is recognized by the public authority, to achieve environmental objectives or to improve environmental performance beyond compliance.

- A non-tradable permit system establishes a limit on the GHG emissions of each regulated source. Each source must keep its actual emissions below its own limit; trading among sources is not permitted.

- A technology or performance standard establishes minimum requirements for products or processes to reduce GHG emissions associated with the manufacture or use of the products or processes.

- A product ban prohibits the use of a specified product in a particular application, such as hydrofluorocarbons (HFCs) in refrigeration systems, that gives rise to GHG emissions.

- Direct government spending and investment involves government expenditures on research and development (R&D) measures to lower GHG emissions or enhance GHG sinks.

ments. These are (in arbitrary order):
- tradable quotas;
- Joint Implementation (JI);
- the Clean Development Mechanism (CDM);
- harmonized taxes on emissions, carbon, and/or energy;
- an international tax on emissions, carbon, and/or energy;

Box 6.2. Definitions of Selected International Greenhouse Gas Abatement Policy Instruments

- A tradable quota system establishes national emissions limits for each participating country and requires each country to hold quota equal to its actual emissions. Governments, and possibly legal entities, of participating countries are allowed to trade quotas. Emissions trading under Article 17 of the Kyoto Protocol is a tradable quota system based on the assigned amounts (AAs) calculated from the emissions reduction and limitation commitments listed in Annex B of the Protocol.
- JI allows the government of, or entities from, a country with a GHG emissions limit to contribute to the implementation of a project to reduce emissions, or enhance sinks, in another country with a national commitment and to receive emission reduction units (ERUs) equal to part, or all, of the emissions reduction achieved. The ERUs can be used by the investor country or another Annex I party to help meet its national emissions limitation commitment. Article 6 of the Kyoto Protocol establishes JI among Parties with emissions reduction and limitation commitments listed in Annex B of the Protocol.
- The CDM allows the government of, or entities from, a country with a GHG emissions limit to contribute to the implementation of a project to reduce emissions, or possibly enhance sinks, in a country with no national commitment and to receive CERs equal to part, or all, of the emissions reductions achieved. Article 12 of the Kyoto Protocol establishes the CDM to contribute to sustainable development of the host country and to help Annex I Parties meet their emissions reduction and limitation commitments.
- A harmonized tax on emissions, carbon, and/or energy commits participating countries to impose a tax at a common rate on the same sources.[7] Each country can retain the tax revenue it collects.
- An international tax on emissions, carbon, and/or energy is a tax imposed on specified sources in participating countries by an international agency. The revenue is distributed or used as specified by participant countries or the international agency.
- Non-tradable quotas impose a limit on the national GHG emissions of each participating country to be attained exclusively through domestic actions.
- International product and/or technology standards establish minimum requirements for the affected products and/or technologies in countries in which they are adopted. The standards reduce GHG emissions associated with the manufacture or use of the products and/or application of the technology.
- An international VA is an agreement between two or more governments and one or more entities to limit GHG emissions or to implement measures that will have this effect.
- Direct international transfers of financial resources and technology involve transfers of financial resources from a national government to the government or legal entity in another country, directly or via an international agency, with the objective of stimulating GHG emissions reduction or sink enhancement actions in the recipient country.

- non-tradable quotas;
- international technology and product standards;
- international VAs; and
- direct international transfers of financial resources and technology.

Box 6.2 defines some of the instruments most prominently discussed in the literature. The first five are often called market-based instruments, although VAs can fall into this category also.

6.1.3 Policy Developments since the Second Assessment Report

In December 1997, Parties to the United Nations Framework Convention on Climate Change[8] negotiated the Kyoto Protocol (UNFCCC, 1997). The Protocol established, for the first time, legally binding quantified emissions limitation and reduction commitments that cover the emissions of six GHGs from a wide range of sources for the period 2008 to 2012 for 38 countries and the European Economic Community (EEC; Annex I Parties). These commitments represent a 5.2% reduction from the 1990 emissions of the Annex I Parties, and a 10% to 20% reduction from their projected emissions during the 2008 to 2012 period.

Annex I Parties can meet their commitments through measures to reduce domestic emissions, specified actions to enhance domestic sinks, and co-operative action with other Parties under Articles 4, 6, 12, or 17. Article 4 allows a group of Annex I Parties to agree to reallocate their collective emissions reduction commitment and to fulfil this commitment jointly. Such an arrangement is commonly referred to as a "bubble". The members of the EEC are the only countries, to-date, to indicate that they are likely to establish one "bubble" to meet their commitments.

Article 6 defines JI for Annex I Parties, Article 12 establishes the CDM for projects in non-Annex I countries, and Article 17 allows emissions trading, a form of tradable quota, among Annex B Parties (see *Box 6.2*). The principles, modalities, rules, and guidelines for these three Kyoto Protocol mechanisms remain to be finalized. The Fourth Session of Conference of the Parties (CoP4) in Buenos Aires in November

---

[7] A harmonized tax does not necessarily require countries to impose a tax at the same rate, but to impose different rates across countries would not be cost-effective.

[8] That is, those countries that ratified the Convention, 186 countries as of September 2000.

1998 adopted a Plan of Action that includes development of these principles, modalities, rules, and guidelines for adoption at CoP6 at The Hague in November 2000.[9]

Annex I Parties have been implementing domestic policies to address their commitment under Article 4.2 of the Convention and evaluating possible policies to meet their more stringent commitments under the Protocol, taking into account the options afforded by the Kyoto mechanisms. Annex I Parties' national climate programmes are described in their National Communications, which are compiled by the UNFCCC Secretariat and subjected to external expert review under the Convention (UNFCCC, 1999, addenda 1-2).

Structural adjustment and energy sector reforms have been pursued in many countries. Although these are not GHG policies, they often have significant implications for GHG emissions, increasing or reducing emissions depending upon the circumstances (see Section 6.2).

6.1.4   *Criteria for Policy Choice*

Governments implement policies and measures to achieve particular objectives that they believe will not be achieved in the absence of government intervention, possibly because externalities or public goods are involved. Policies and measures can be generic, such as a general carbon tax or emissions trading, or sector-specific, such as a regulation applied to the construction sector, or a subsidy for green farming practices. The objective of this chapter is to assess different types of policies and measures, not to provide a complete list of these, so sector-specific policies and measures are discussed only in general terms.

Chapter 5 draws a distinction among five types of policy targets, each of which refers to a different interpretation (definition) of the concept of "barriers" to technological change: market potential, economic potential, socioeconomic potential, technological potential, and physical potential. Policies and measures can differ in the type of potential they aim to reach, but it is difficult to link specific policy instruments and specific potentials, because the potential achieved through virtually any policy instrument depends upon the "degree" to which that instrument is employed. For example, an emissions tax can be set at various levels; depending upon the level at which the emissions tax is set, it could have the effect (if perfectly implemented) of achieving any of the types of "potential" defined in Chapter 5.[10] For this, among other reasons, the prime focus in this section is on the possible criteria for policy instrument choice and evaluation.

Evaluation criteria are required both for the *ex-ante* choice of instruments and for the *ex-post* assessment of implementation and performance. Each government may apply different weights to the criteria when it evaluates GHG mitigation policy options.[11] Moreover, a government may apply different weights to the criteria when it evaluates national and international policy instruments, and the appropriateness of the criteria may vary depending on the degree of uncertainty about the pollution abatement cost and pollution damage functions. This general remark should be kept in mind when the various domestic and international policies, instruments, and measures discussed in this chapter are evaluated against the background of these criteria.

The criteria identified in SAR for the evaluation of policy options (Fischer *et al.*, 1998) are:

- Environmental effectiveness. How well does the policy achieve the environmental goal, such as a GHG emissions reduction target? How reliable is the instrument in achieving that target, does the instrument's effectiveness erode over time, and does the instrument create continual incentives to improve products or processes in ways that reduce emissions?
- Cost-effectiveness. Whether the policy achieves the environmental goal at the lowest cost, taking transaction, information, and enforcement costs into account.
- Distributional considerations. How the costs of achieving the environmental goal are distributed across groups within society, including future generations.
- Administrative and political feasibility. This includes considerations such as flexibility in the face of new

---

[9] The Plan of Action also includes work on the development and transfer of technologies, the financial mechanism, implementation of Articles 4.8 and 4.9 of UNFCCC, and preparations for the first session of the CoP serving as the meeting of the Parties to the Kyoto Protocol. This involves, *inter alia,* decisions on rules that govern sink enhancement activities under Articles 3.3 and 3.4 of the Kyoto Protocol.

[10] However, some concepts of "barriers" seem to imply combinations of instruments and levels or degrees of implementation. For example, if the problem is viewed as one of externalities, it is natural to use a tax on the relevant externality, with the tax set equal to the marginal social damages at the efficient level of control. On the other hand, although categories of "potential" refer to targets (ends), categories of policy instruments refer to the means of achieving those ends.

[11] The choice of weights is strongly influenced by many national and sector-level circumstances. These include government jurisdictional structure (e.g., sharing of government powers at various levels); geographical and climate profile (e.g., area size, regional weather patterns, heating degree days and temperature distribution, annual temperature variations, climate variability, latitude); economic setting (e.g., gross domestic product (GDP), GDP/capita, and GDP by sector); international trade patterns, such as percentage of energy-intensive exports; energy and natural resource base; demographics (e.g., population total and distribution, growth rate); land use and/or spatial patterns (e.g., distances driven/capita); industry and agriculture structure; building stock, and urban structure (e.g., home sizes); and environmental and/or health patterns (e.g., potential for highly variable climate change mitigation impacts across different national regions and urban areas).

knowledge, understandability to the general public, impacts on the competitiveness of different industries, and other government objectives (such as meeting fiscal targets and reducing emissions of pollutants).

The literature (e.g., OECD, 1997d) identifies some additional criteria, such as:

- Revenues raised in the case of market mechanisms, for instance, may constitute a second source of benefits from their use, over and above their direct environmental impact, depending on if and how the revenues are recycled.
- Wider economic effects include potential effects on variables such as inflation, competitiveness, employment, trade, and growth.
- Wider environmental effects, such as local air-quality improvement (usually referred to as the ancillary benefits).
- "Soft" effects, which relate to the impact of environmental policy instruments on changes in attitudes and awareness.
- Dynamic effects, which relate to the impact on learning, innovation, technical progress, and dissemination and transfer of technology.

The above lists of criteria guide the discussion of national and international policies and measures related to GHG abatement. However, the economics literature–particular theory development–focuses more on the cost-effectiveness criterion than on the other criteria mentioned, and there is a similar emphasis in this chapter, which is a review of the best available scientific literature. Wherever possible, literature on the potential equity impact of policies and measures is referred to. In addition, specific attention is paid to the political economy literature that describes policy choice (Section 6.1.5), the interactions of policy instruments with fiscal systems (Section 6.5.2), and the impacts on technological change (Section 6.5.3).

6.1.5 The Political Economy of National Instrument Choice

Some of the key lessons from the scholarly literature on political economy can be applied to instrument choice in climate policy at the national level. Since much of that scholarship focuses on policymaking in a limited set of developed nations, in particular in the USA, great care must be taken before applying any of these lessons to domestic politics generally.

6.1.5.1 Key Lessons from the Political Economy Literature

A useful starting-point is to view the policy process (at least in countries with strong legislatures) as analogous to a "political market" (Keohane et al., 1999). The demand side of such a "market" consists of the interest groups with a stake in the policy; in the environmental arena, such groups include regulated industries, producers of complementary products, environmen-

tal organizations, and (to a lesser extent) labour and consumer organizations. The supply side consists of the legislators and the administration involved in the design and implementation of the environmental policies and measures.

One key insight of this literature is that some forms of regulation can actually benefit the regulated industry, for example, by limiting entry into the industry or imposing higher costs on new entrants (Rasmusen and Zupan, 1991; Stigler, 1971). In the environmental arena, conventional regulation may provide firms with rents that result from reductions in output and raised prices as a consequence of regulation (Buchanan and Tullock, 1975; Maloney and McCormick, 1982). Stricter standards for new pollution sources benefit existing firms by raising barriers to entry (Nelson et al., 1993). Polluters' self-interest may also help explain the prevalence of tradable permits that have been allocated free ("grandfathered") when market-based instruments have been used. Permits allocated free to existing firms represent a transfer of rents from government to industry while auctioned permits and emissions taxes generally impose a heavier burden on polluters. Finally, VAs may be the preferred policy approach from industry's perspective, because these leave more of the initiative with the private sector (at least so it is perceived), which may enhance industry's chances of capturing rents.

Of course, it is important to recognize that industry may not act monolithically, since policies may have differential distributional impacts within a sector. A policy that imposes costs on industry as a whole might still be supported by firms that would fare better than their competitors. For example, firms that can achieve emissions reductions more cheaply may be more supportive of market-based schemes, such as tradable permits, than their higher-cost competitors (Kerr and Maré, 1997). In the realm of global environmental policy, the ban on ozone-depleting chlorofluorocarbons (CFCs) under the Montreal Protocol was, for instance, supported by those who expected to dominate the market for HFCs, then the leading substitute chemicals (Oye and Maxwell, 1995).

Regulated firms are not the only group with a stake in regulation; opposing interest groups will defend their own interests. Environmental groups, for example, tend to favour stringent targets, although many have opposed market-based instruments out of a philosophical concern that such policies give firms "licenses to pollute" or because of objections to attempts to quantify or monetize the environmental damages from pollution (Kelman, 1981; Hahn, 1989; Sandel, 1997). Some groups draw an ethical distinction between taxes and tradable permit systems, in which taxes are morally deficient because they put a price on emissions but set no upper limit on allowable pollution, while permits ensure a set level of emissions (Goodin, 1994). Other environmental groups support market-based policies in the hope that the resultant cost savings will make a higher level of environmental quality politically attainable, and possibly in part because of their own self-interest in distinguishing themselves from other environmental organizations (Svendsen,

1999). The US Clean Air Act defines permits as "limited autho-rizations to emit", to avoid limiting the ability to set lower emissions limits, which may also be a response to concerns of the environmental lobby that air should not become private property (Tietenberg, 1998). This indicates that the design of market-based instruments may be flexible enough to accommo-date ethical concerns without undermining effectiveness.

While the political economy literature emphasizes the impor-tance of preferences of interest groups, it has tended to neglect the "supply side" of the political equation: the legislators and government officials who ultimately design and implement regulatory policy. Government actors may have their own interests and preferences with respect to policy instruments:

- ideology or past experience may favour one instrument over another (Kneese and Schulze, 1975; Hahn and Stavins, 1991);
- legislators may prefer policies with (large but) hidden costs to those with (small but) visible ones (McCubbins and Sullivan, 1984; Hahn, 1987); and
- legislators responsible to local districts may emphasize distributional concerns over efficiency (Shepsle and Weingast, 1984).

Finally, the environmental administration may prefer direct regulation over market-based instruments, not only insofar as they are more familiar with it, but also because it gives them more control, and usually requires a relatively large adminis-trative capacity.

These political factors, however, vary widely among coun-tries. Whether or not a legislature exists, and if so whether in a parliamentary or presidential system, affects the support for particular policy instruments. Whether legislators are elected by district or by party list may affect the political support for different policy instruments as well. Factors such as the extent of interest-group organization and how groups interact with government are also critical–interest groups lobby legislators in some countries, sit on quasi-governmental decision-making bodies in others, are relegated to raising public awareness elsewhere, and in some countries are non-existent. Less tangi-ble cultural and historical factors can also be critical in influ-encing the choice of instrument. For example, a country's experience with free markets generally may influence whether or not it chooses to use market-based policy instruments for environmental protection (Keohane, 1998). Finally, there are clear political economy limitations of individually applied price, non-price, and regulatory policies that often lead to the linked or combined policy strategy that is observed in prac-tice.

6.1.5.2   Implications for Global Climate Change Policy

Since the political factors on the "supply side" are so hetero-geneous across nations, the focus here is on the demand for regulations, building on the literature reviewed above to draw conclusions about the likely preferences and positions of key interest groups involved in climate change policy. Five groups seem particularly important: environmental organizations (especially in the USA and Europe), producers of carbon-based fuels (e.g., coal and oil producers), large users of fuel (e.g., electric utilities), manufacturers of energy-using products (e.g., automobile manufacturers), and manufacturers of energy-effi-cient and GHG-abatement technologies (e.g., manufacturers of efficient lighting). Environmental organizations in the USA and Europe seem to be divided–some groups have embraced market-based policies such as emissions permits and carbon taxes, while others object to such policies being applied with-out restrictions. Some also object to the option of so-called exchanges of "hot air" (national quota surpluses not created by active policies).

The range of industry sectors with large stakes in global cli-mate policy suggests an important point: the various regulato-ry instruments that might be employed in climate change poli-cy would each act at different levels of regulation, creating dif-ferent points of compliance with very different implications for interest groups. Examples are:

- a system of tradable carbon permits (or a carbon tax, for that matter) imposed at the mine mouth, wellhead, or point-of-entry directly affects fuel producers (although the true economic incidence of the policy would be shared by downstream firms and consumers according to relative elasticities);
- a CO_2 tax, tradable emissions permit system, or emis-sions standard directly affects power plants; and
- energy-efficiency or fuel-efficiency standards directly affect manufacturers.

Industry groups–in particular, large producers and users of fuel–are also likely to focus their efforts on the allocation of carbon-reduction responsibilities, whatever the instrument. If a system of emissions standards is put into place, for example, existing firms will benefit if tighter standards are imposed on new sources, as has happened in a number of countries. Under an emissions tax, firms are likely to seek tax credits, differen-tial tax rates, or exemptions to relieve their tax burden. In a sys-tem of tradable permits, firms are likely to support the free allocation of permits to participants, rather than to sell them at auction or distribute them to the public (for subsequent sale to firms). For project-based mechanisms–CDM and JI–they would favour leaving much of the initiative with the private sector (Jepma and Van der Gaast, 1999). Industries that stand to profit from GHG abatement, including renewable energy sources, are likely supporters of climate policies (Michaelowa and Dutschke, 1999a, 1999b).

From a political standpoint, the success of such efforts at the distribution of the burdens (or rents) is likely to depend on the political saliency of climate change policy. Taxpayers and organized "public-interest groups" are likely to oppose alloca-tion schemes that benefit firms and/or benefit existing firms at the cost of the newcomers, thus reducing the scope for compe-tition. If such groups wield clout, and if public interest in climate policy is high, then mechanisms that appear to benefit polluters at the expense of the public are less likely to be implemented.

In contrast, some environmental organizations have not opposed the allocation of rents to industry, recognizing that free allocation of permits may be the most likely path to implementing emissions reduction in some countries. Such concessions on allocation of rents to the industry have allowed these groups to secure other goals in return, such as continuous emissions monitoring–the US Acid Rain Program is a good example (Kete, 1992; Svendsen, 1999). In summary, allocation schemes favourable to industry appear likely in practice, because the question of distribution is central to industry, including industries that will profit from climate policy, but it is only of secondary importance to environmental groups that do not support free allocation and to other groups that seek to reduce GHG emissions. In the US Acid Rain Program, for example, sulphur dioxide (SO_2) emissions allowances worth about US$5 billion per year were allocated free to electric utilities, in part because of interest group politics (Joskow and Schmalensee, 1988).

Although the "supply side" is heterogeneous across nations, it is likely that some governments will favour policies that raise revenue while others will be more concerned with the distribution of costs across sources, regardless of the revenue implications.

6.2 National Policies, Measures, and Instruments

Before policies and measures that aim to reduce, or remove barriers that hamper, GHG emissions or enhance sequestration by sinks are analyzed, it is necessary to understand the substantial impact that other policies (such as the structural reforms of trade liberalization and liberalization of energy markets) have had on GHG emissions in several developing countries, economies in transition (EITs), and some developed countries. These policies, sometimes coupled with macroeconomic, market-oriented reforms, set the framework in which more specific climate policies would be implemented. Therefore, to assess correctly the feasibility of any particular policy, it is important to understand this new policy context. The effect of these reforms on energy use and GHG emissions is not clear *a priori*. Impacts can differ widely among countries, depending on implementation strategies and the existence of other regulatory policies designed to prevent the undesired effects of free market operation in the presence of externalities, information, and co-ordination problems.

6.2.1 Non-Climate Policies with Impacts on Greenhouse Gas Emissions

6.2.1.1 Structural Reform Policies

During the 1990s, several countries, especially EITs and developing countries, implemented drastic market-oriented reforms that have had important effects on energy use and energy efficiency, and therefore on GHG emissions.[12] Most countries have undergone what has been called the first generation of structural reforms: trade liberalization, financial deregulation, tax reform, privatization of state-owned enterprises, and opening the capital account as part of a strategy to attract foreign investment. Some countries have also implemented macroeconomic stabilization packages that include fiscal discipline, independence of monetary policy from the public sector, and exchange rate unification.

The two largest countries in terms of population and coal reserves, China and India, have also started to reform their economic systems towards a more free-market orientation, although at a slower pace than many other countries. Since 1978, energy use in China has increased, on average, 4%/yr. However, the energy–output ratio in China fell 55% between 1978 and 1995.[13] Garbaccio *et al.* (1999), using input–output tables, found that most of this reduction arose from technical change, a result supported by other studies (Polenske and Lin, 1993; Sinton and Levine, 1994). An increase in energy-intensive imports has also led to decreased energy use per unit of GDP. Others have attributed the reduction to sectoral shifts in the composition of output (Smil, 1990; Kambara, 1992). As reform-induced changes aimed at increasing GDP may increase the use of energy, the net effect on GHG emissions of structural reform in China is an empirical problem that depends on the choice of development strategies, technologies, and complementary policies.

Future economic growth in all countries may be accompanied by increases in GHG emissions. Even if economic growth increases energy efficiency (both in terms of production and consumption), the scale effect may dominate and GHG emissions may rise, depending on the extent to which other policies and measures are implemented to curb emissions (Fisher-Vanden, 1999).

---

[12] For a description of the main reforms implemented in Latin America see Lora (1997) and for EITs see Chandler (2000).

[13] Energy-output ratios discussed here require a caveat about China's GDP statistics. China's energy consumption in 1997 and 1998 decreased 0.6% and 1.6%, respectively, and its energy intensity drastically declined 80% from 1985 to 1998. Many analysts consider statistics on Chinese GDP to be speculative (IEA, 1998b).

6.2.1.2   Price and Subsidy Policies

Price signals can only influence demand and supply if they actually reach economic agents and if those economic agents have the opportunity to respond to them. In Russia, energy intensity increased by 30% between 1990 and 1998, while energy prices also increased tremendously (IEA, 1997b, p. 50).[14] Experience shows that it takes time for economic agents to adjust their behaviour to new price signals, not only because of capital stock turnover, but also because consumers often do not have an accurate knowledge of their energy consumption, or the technical capacity to reduce it. Various types of energy market reforms and the pace of energy price reforms are designed to create and clear channels for market signals to work.

It is a difficult policy challenge, and therefore a time-consuming process, to bring prices into line with real costs. This is true both in developing countries, where the poor pay a high cost for low-quality energy services (or a low cost that is heavily subsidized) and in developed countries. Although data on energy subsidies are incomplete, partly because such support is difficult to identify and measure, some evidence indicates that subsidies on coal production, including transfers from both consumers and taxpayers, are declining in a number of OECD and developing countries. Recent data suggest that the total producer subsidy estimates for the coal production of Germany, UK, Spain, Belgium, and Japan, which amounted to over US$13 billion at the beginning of the 1990s, had declined to less than US$7 billion by 1996 (OECD, 1998a, 1998b). In addition, case studies in the energy supply sector identified the following areas for potential subsidy reforms: removal of coal-producer grants and price supports; reforming subsidies to investment in the energy supply industry; and regulatory reform to eliminate non-tariff barriers to the energy trade (OECD, 1997a, 1997b).

An IEA (1999b) analysis of fossil energy subsidies in China, Russia, India, Indonesia, Iran, South Africa, Venezuela, and Kazakhstan—which accounted for 27.5% of the world's total energy demand in 1997—claimed that removing such subsidies would lower CO_2 emissions by 16% in these countries, amounting to a 4.6% reduction in global emissions.[15]

The transport sector–to give an important example–is another sector that receives subsidies detrimental to the environment. Transport is indirectly subsidized through infrastructure financing and through tax benefits, which enhance the transport volume. According to Shelby *et al.* (1997), energy subsidies were higher than those to transportation for the OECD area. They also found for the USA that larger CO_2 savings could be achieved through reform of indirect rather than direct transport subsidies, such as free parking and supporting the highway infrastructure. Reform policies to internalize external the effects will, according to one study, probably lower sector-wide emissions by 10–15% (OECD, 1997c).[16] These findings are in line with the results from other work on internalizing the external cost of transportation (ECMT, 1998). The same studies also indicate that local communities can better carry out policy reform in the transport sector, because transport subsidies may originate at the local level and local communities are more likely to value other ancillary benefits through policy reform (OECD, 1997c; ECMT, 1998). The transport sector is only mentioned as an example, because it is responsible for a large share of the national emissions in many countries.[17]

6.2.1.3   Liberalization and Restructuring of Energy Markets

Liberalization of energy markets gives the suppliers greater freedom in the extraction, processing, generation, transportation, and distribution or supply of energy products and the consumers greater freedom to choose from different providers (WEC, 1998). In the electricity subsector, the separation of transmission from generation followed the realization that only transmission is a natural monopoly (Hunt and Shuttleworth, 1996). Recently, various measures have been taken to liberalize energy markets. The EU, for instance, adopted rules to liberalize its electricity market (IEA, 1997a), which became operational early in 1999 (although some EU countries, such as the UK, had started earlier). It is expected that this will be followed soon by rules regarding a liberalization of the natural

---

[14] The major reasons for such growth were: a shift from an energy-intensive industrial structure to an even more energy-intensive industrial structure through maintaining the competitive advantages of energy and raw materials production in parallel with a sharp reduction of production in less energy-intensive industries; reduced share of production-related energy consumption at the expense of heating, ventilation, and air conditioning (HVAC) related energy consumption; reduced GDP, industrial production, and industrial energy consumption with the background of a relatively stable energy consumption in the residential sector; lack of control and metering devices; non-payment problem, which appeared partly as a reaction to the sky-rocketing growth of energy prices; weak capital markets and high interest rates to attract capital for energy-efficiency improvement projects (see Bashmakov, 1998).

[15] The percentage reduction in energy consumption was calculated by adding the gross calorific value of the reductions of the different fuels under consideration and expressing the sum as a percentage of total primary energy supply (TPES). As the calculations in this study did not take into account the refinery sector (a 5% reduction in gasoline use can amount to a reduction in TPES of more than 5%), the number thus derived constitutes a lower bound to the true reductions in energy consumption. Some country experts strongly criticized the methodology and quantitative results of the study (Bashmakov, 2000).

[16] In this regard the time element could be crucial. In fact, during the time in which prices adjust, transport volumes may grow, but growth may be retarded. Additional research is needed to establish these findings.

[17] Other sectors, including electricity generation, mining, cement, agriculture, and forestry, can also involve significant GHG emissions but benefit from subsidies that increase emissions.

gas market.[18] In the USA, as a result of changes in policies at both federal and state levels, the generation and sales of electricity are being opened to competition. Liberalization of the energy markets in developing countries and EITs has, in many cases, been part of the macroeconomic restructuring in these countries. Both in Africa and Latin America, one of the main driving forces behind the reform of the power sector is to attract private capital to expand and improve the sector.

Although these policies are mainly inspired by the wish to increase competition in the energy and power markets, they can have, through their impact on the choice of production technology, significant emissions implications. Energy restructuring may include regulation of the transmission monopoly, environmental cost internalization, and system-benefit charges (SBCs; see *Boxes 6.6* and *6.7*). Several studies have examined the effects on GHG emissions of the restructuring of the electricity industry, but the issue is far from resolved. Indications are that the impacts can be either positive or negative (IEA, 1998b). The degrees of the environmental effects of liberalization of the electric utility industry are case specific and depend on pre-existing circumstances (e.g., fuel mix, vintage of plant, taxation schemes, and other factors). They also depend on such factors as national endowment of resources, the fuel mix, the vintage structure of generation capacity, scope for restructuring, and the size and speed of policy reform (OECD, 1999). In short, energy-sector structural reform cannot, in itself, guarantee a shift towards less carbon-intensive power generation.[19] On the whole, however, it may provide for a more economically driven behaviour that would be more responsive to price signals placed on GHG emissions.

Finally, the impacts of energy-sector structural reforms can be enhanced if appropriate additional policy measures are taken,[20] such as demand-side management (DSM). An example of the latter is the British Energy Savings Trust, which was set up 3 years after restructuring the UK energy markets, in 1992, to finance DSM programmes run by regional electric companies.

According to an IEA study (IEA, 1999a), in the UK energy sector the structural reforms in the electricity, coal, and gas supply sectors reduced the share of electricity generated from coal from 65% in 1990 to 35% in 1997. This resulted from closure of older coal-fired plants and the construction of combined cycle gas turbines. In countries where the electricity systems are largely based on non-fossil fuels, like Brazil, Norway, Sweden, and Switzerland, competition without environmental regulation may well lead to increased CO_2 emissions, as gas-fired power stations often will be the most economically attractive option for the development of new capacity.[21]

In Japan, after liberalization of the power-generation market several independent power producers entered it. However, around 85% of their fuels were coal and residual oil that, though inexpensive, emit more CO_2 per unit of power generated. With the liberalization of the retail market, adopted in 2000 for large power users, it is possible that the construction of an atomic power or liquefied natural gas (LNG) plant, both of which require a longer lead time and a huge investment, will become difficult. This may lead to adverse effects in terms of CO_2 emissions (Sagawa, 1998).

Several studies in the USA have tried to quantify the potential impacts of restructuring the electricity industry on GHG emissions (see Lee and Darani, 1995; Rosen *et al.*, 1995; US FERC, 1996; Palmer and Burtraw, 1997). The FERC study suggests that there would be no significant increase in total CO_2 and nitrogen oxides (NO_x) emissions. The other studies, however, suggest that the impact of a more open transmission grid on CO_2 and NO_x emissions could be substantial. A more recent study by the US Department of Energy's (DOE) Office of Policy found that the restructuring envisioned under the Comprehensive Electricity Competition Act (CECA) will lead to 145–220 megatonnes (Mt) less CO_2 emissions in 2010 than would have occurred in the absence of an explicit policy to reduce CO_2 emissions from the electricity sector (US DOE, 1999).[22]

There is a growth in literature that focuses on the impacts of liberalization and restructuring of energy markets on the key technologies of interest in the context of GHG reduction, such as energy efficiency, co-generation, and renewables.[23]

---

[18] An EU Directive on Natural Gas was adopted by the European Council of Ministers in May 1998 after publication in mid-1998; member states will have 2 years to implement the Directive.

[19] While it led to reduced emissions in some countries, such as the UK (Fowlie, 1999), it had the opposite effect in others, such as Australia.

[20] Another example is how liberalization of energy markets can reduce mitigation costs, especially when permit trading is not allowed, insofar as, for instance, electricity trade makes it easier to fulfil mitigation commitments (see also Hauch, 1999).

[21] In this regard, in Sweden the increase in carbon intensity was more the result of the political choice to phase out some nuclear power plants than of a link to the creation of an exchange. More generally, it may well be that the long-term impact of the international power exchange between Norway and Sweden will be that gas-fired power plants are added to the Nordic electricity system, causing coal-fired generation to decline. For some general information on the relationship between market deregulation and national mitigation commitments, see also Baron and Hou (1998).

[22] The DOE study incorporated policy proposals such as increasing the renewable-energy portfolio standards (RPSs) and removing barriers to the use of combined heat and power technologies where they are economical. In response to calls from environmentalists to reduce the potential impacts of restructuring the electricity industry, some countries initiated specific policies aimed at increasing the role of renewable energy in the electricity generation mix (Mitchell, 1995b, 1997; Wolsink, 1996; Wiser, 1997, 1999; Wiser and Pickle, 1997; Novem, 1998; Haddad and Jefferis, 1999; Wiser *et al.*, 1999).

6.2.2   Climate and Other Environmental Policies

Section 6.2.1 sets the general policy context in which any environmental policy will operate. This section focuses on specific policies to address climate change. The various policy instruments are assessed generically. In other words, there is not a sector-specific focus, because it is beyond the scope of this chapter. This may create some bias insofar as most sector-specific policies are technology oriented and of the command-and-control type.

6.2.2.1   Regulatory Standards

Regulatory environmental standards set either technology standards or performance standards, enforceable through fines and other penalties[24] (voluntary standards are discussed in Section 6.2.2.4). They may attach to a product, a line of products (e.g., US Corporate Average Fuel Economy (CAFE) standards), or the provision of a service (e.g., Japan requires that firms employ an energy manager).[25] In this chapter regulatory standards are distinguished from economic or market-based instruments (taxes and fees, permits, subsidies). Although all regulatory standards have consequences upon economic decision making, they differ from market-based instruments, which operate by directly changing relative prices rather than by specifying technology or performance outcomes.

Regulatory standards can be effective policies to address market failures and barriers associated with information, organization, and other transactions costs. They also are widely used to require actors to account for environmental externalities and, if continually modified to account for technical progress, they can provide dynamic innovation incentives (see Section 6.5.3). The principal sources of inefficiency associated with some regulatory standards derive from too narrow specifications of uniform behaviour in heterogeneous situations, weakness in controlling aggregate levels of pollution, and relatively more difficult application to products other than component or turnkey technologies. By requiring a certain level of performance without specifying how it should be achieved, performance standards generally reduce losses through inflexibility when compared to technology standards.

On the whole, energy efficiency standards have proved to be an effective energy conservation policy tool. Energy efficiency standards are widely used in over 50 nations and the number of standards is still growing.[26,27,28] For appliance standards enacted in the USA, cumulative energy savings in 1990 to 2010 are estimated at 24 etajoules (EJ), consumer life-cycle costs savings at US$46 billion, and emission reductions at about 400MtCO$_2$. For an early estimate, see McMahon (1992). The introduction of refrigerator and freezer standards in the EU is estimated to generate 300 TeraWatt hours (TWh) of cumulative electricity savings during 1995 to 2010 (Lebo and Szabo, 1996). Similar measures in Central and Eastern Europe are expected to save 60 TWh energy and to reduce emissions by 25 MtCO$_2$ (Bashmakov and Sorokina, 1996). In Japan, the law concerning the rational use of energy was strengthened on 1 April 1999 and is expected to reduce, in combination with the industries' voluntary actions plan, a maximum of 140 MtCO$_2$ in industry, transportation, and other sectors in total (Yamaguchi, 2000). Energy efficiency standards are especially effective in countries with high and growing appliance ownership and in countries in which consumers' energy awareness is low because of historically low energy prices.

The development of an effective regulatory standard requires national and, potentially, international, leadership to balance the interests of manufacturers, consumers, environmental non-government organizations (NGOs), and other interest groups, while creating sufficient societal support and incentives for successful implementation. While decisions to introduce regulatory standards are commonly made by legislatures, the development and implementation of standards over time is often left to a less transparent public administration. Although the enforcement and monitoring of all policy instruments is costly

---

[23] See Mitchell (1995b); Weinberg (1995); Boyle (1996); Lovins (1996); Nadel and Geller (1996); Owen (1996); Brown *et al.* (1998); Eyre (1998); Patterson (1999).

[24] There is no general agreement on terms by which regulatory standards are classified. In the USA, technology standards are often called command-and-control standards because they dictate particular technologies or best practices that limit the range of compliant behaviours. In other nations, command and control normally refers to all regulatory standards because they command behaviour and control compliance therewith.

[25] Mandatory standards are put in place by either specific legislation or government regulation. See, for instance, the Comprehensive National Energy Policy Act (USA, 1992), versus the Energy Conservation in Buildings Requirement for Thermal Performance and Heat–Water–Power Supply (Moscow City Government, 1999).

[26] In the USA in 1997 standards set for appliances are estimated to cover 75% and 84% of primary and delivered energy, respectively, in the residential sector. Similarly, it is estimated that standards covered 49% of both primary and delivered commercial energy use in 1997 (EIA, 1999).

[27] Technological progress provides a basis for regular updates of efficiency standards. In Russia, for instance, 1976 standards for refrigerators were improved by 50% in the 1980s and then in 1991 by an additional 50%. As a result, energy consumption of new units decreased by a factor of three (Bashmakov and Sorokina, 1996). The American Society of Heating, Refrigeration, and Air Conditioning Engineers updates its codes for residential and commercial buildings on average every 10 years.

[28] In France, successive building codes in the residential sector alone have generated 75% of the total energy savings over past 20 years. After building codes were set in 1974 they were made stricter in 1982, 1988, and 1998 (IEA, 1996, p.38).

and subject to failures, including discriminatory treatment and corruption, social science literature that examines the implementation of regulatory standards is more extensive.

Recent literature indicates that regulatory standards often precede market-based instruments and build institutional capacity in policy evaluation, monitoring, and enforcement (Legro *et al.*, 1999). This is especially true in developing countries that lack both trained personnel and the financial resources to implement market-based instruments.[29] Technology standards have provided the initial training ground for public officials unfamiliar with any approach to environmental regulation. Russell and Powell (1996) found that developing countries with a better institutional capacity developed through experience with regulatory standards generally are more successful in implementing market-based environmental policies than less well-equipped countries. Cole and Grossman (1998) suggest that when historical, technological, and institutional contexts are taken into account, technology standards are efficient in the initial stages of environmental policy development.

The use of regulatory standards to force the internalization of environmental costs has initial distributional consequences different from those of environmental taxes or subsidies.[30] Regulatory standards reduce economic benefits previously shared by consumers, capital, and labour only to the extent of compliance costs and/or output foregone. Unlike environmental taxes or auctioned permits, regulatory standards do not extract the value of environmental costs on inframarginal production that continues after the policy is mandated.

Regulatory standards may also be used to correct barriers that arise from information failures and can yield net benefits to society if the costs associated with the regulation are less than the losses due to informational barriers.

6.2.2.2 Emissions Taxes and Charges

An emission tax on GHG emissions requires domestic emitters to pay a fixed fee, or tax, for every tonne of CO_{2eq} of GHG released into the atmosphere. Such a fee would encourage reductions in GHG emissions in response to the increased price associated with those emissions. In particular, measures to reduce emissions that are less expensive than paying the tax would be undertaken.

Since every emitter faces a uniform tax on emissions per tonne of CO_{2eq} (if energy, equipment, and product markets are perfectly competitive) this would result in the least expensive reductions throughout the economy being undertaken first (IPCC, 1996, Section 11.5.1; Baumol and Oates, 1988). In the real world, markets, especially energy markets, deviate from this ideal, so an emissions tax may not maximize economic efficiency. Rather, the efficiency of an emissions tax should be compared with that of alternative policy measures. Criteria other than efficiency, such as distributional impacts, are likely to influence the design of the emissions tax where this is the chosen policy. Although equity considerations could be, in theory, better addressed through other redistribution mechanisms, in practice most energy and emissions taxes apply differential tax rates to different sources.

An emissions tax, unlike emissions trading, does not guarantee a particular level of emissions. Therefore, it may be necessary to adjust the tax level to meet an internationally agreed emissions commitment (depending on the structure of the international agreement; see Section 6.3). The main economic advantage of an emissions tax is that it limits the cost of the reduction programme by allowing emissions to rise if costs are unexpectedly high (IPCC, 1996, Section 11.2.3.1; see also Section 6.3.4.2).

An emissions tax needs to be adjusted for changes in external circumstances, like inflation, technological progress, and increases in emissions (Tietenberg, 2000). Inflation increases abatement costs, so to achieve a target emission reduction the tax rate needs to be adjusted for inflation. Fixed emissions charges in the transition economies of Eastern Europe, for example, have been significantly eroded by the high inflation (Bluffstone and Larson, 1997). Technological change generally has the opposite effect, reducing the cost of making emissions reductions. Thus, technological change generally increases the emissions reductions achieved by a fixed (real) tax rate. New sources increase emissions. If the tax is intended to achieve a given emissions limit, the tax rate will need to be increased to offset the impact of new sources (Tietenberg, 2000).

Implementation of a domestic emissions tax touches on many issues (Baron, 1996). Policymakers must consider the collection point, the tax base, the variation or uniformity among sectors, the association with trade, employment, revenue, or R&D policies, and the exact form of the mechanism (e.g., an emissions tax alone or in conjunction with other policy measures). Each of these can influence the appropriate design of a domestic emissions tax.

6.2.2.2.1 Collection Point and Tax Base

Since GHG emissions caused by the combustion of fossil fuels are closely related to the carbon content of the respective fuels, a tax on these emissions can be levied by taxing the carbon content of fossil fuels at any point in the product cycle of the

---

[29] This also applies to current climate policy. Under certain circumstances it is preferable to adopt a more intensive regulatory standards phase by financing capacity building and hands-on-experience in the flexible instruments for administrators in developing countries (Montero, 2000c). The Activities Implemented Jointly (AIJ) pilot phase might be considered a step in this direction.

[30] Regulatory standards reverse the distributional effects of efficient subsidies in that the incremental costs of regulation are borne not by producers, but by the subsidy-financing tax base or those consumers who must cross-subsidize the environmental goods (see Section 6.2.2.6).

fuel (EIA, 1998).[31, 32, 33] A producer–importer tax on the carbon content of fossil fuels, coupled with a crediting scheme for exports and non-combustion end-uses, closely replicates the effect of a direct emissions tax on end-users (CCAP, 1998). Further, by focusing on producers and importers rather than end-users, the number of regulated entities is dramatically reduced. Fewer regulated entities lead to substantially lower monitoring and enforcement costs. Modelling studies show that taxing fossil fuels on a basis other than carbon content–for example, energy content or value–also reduces CO_2 emissions, but usually at a higher cost for a given emissions reduction target (IPCC, 1996, Section 11.5.1).[34]

6.2.2.2.2 Association with Trade, Employment, Revenue, and Research and Development Policies

In an open economy, countries are often concerned about the impact of emissions taxes on tradable goods sectors (OECD, 1996a; IPCC, 1996, Section 11.6.4). In practice, therefore, current carbon taxes generally tend to have a lower rate on the tradable goods sectors, especially when they are energy intensive. When some trading partners do not undertake emissions reductions, for example, domestic emissions taxes on carbon-intensive tradable goods might simply shift production to countries without such taxes. One solution is corrective taxes on imports and exports (OECD, 1997d). If this option is not available (see Section 6.4.2), an emissions tax that is differentiated among various sectors in the economy may be preferred (Hoel, 1996). Another solution, which Böhringer and Rutherford (1997) find to be more efficient, is sector-specific wage subsidies to protect jobs in the carbon-intensive tradable goods sector.

Opposition to increased environmental regulation in general often centres on concerns that firms might relocate and/or people might lose their jobs (Rosewicz, 1990).[35] Emissions taxes are particularly vulnerable to this criticism since they require firms not only to pay abatement costs, but also taxes on their unabated emissions (Vollebergh *et al.*, 1997). Several recent papers, however, argue that emissions taxes are more cost-effective than direct regulation and may even lead to higher employment (Wellisch, 1995; Hoel, 1998). The intuition is that the right to emit pollution constitutes a rent. With mobile capital markets, part of that rent accrues (inefficiently) to owners of capital unless it is taxed (Schneider, 1998). By using the tax revenue to offset labour taxes, employment can be higher than in similarly designed policies using direct (technology) regulation (Hoel, 1998; see also Section 6.5.1). Chapters 8 and 9 refer to various sources corroborating the evidence that using emissions and/or energy taxes to reduce distortionary labour taxes tends to increase employment.

Even with an efficient outcome, the immediate profit losses to firms under an emissions tax might be considered "unfair" to firms in carbon-intensive industries. In that case, a portion of the tax revenue can be returned to firms (lump sum) to compensate them for lost profit without a loss of efficiency. Bovenberg and Goulder (1999) estimate that only 15% of the revenue from an emissions tax would need to be refunded to industry to maintain existing profit.

In addition to reducing emissions and raising revenue, a carbon tax also influences innovation. This occurs alongside any distinct R&D policies that are undertaken (see also Section 6.2.2.6). Early work in this area indicated that auctioned permits would provide the largest incentive to innovate, followed by emissions taxes and then permits allocated free (Milliman and Prince, 1989). More recent work demonstrates that with a large number of competitive firms and imperfect R&D markets, taxes may induce more innovation than auctioned permits, although the welfare effects remain ambiguous (Fischer *et al.*, 1998). The incentive for innovation is therefore a necessary design consideration (Grubb *et al.*, 1995; see Section 6.2.2.6). It has been suggested in this regard that the targetted recycling of emissions taxes that support renewable energy and energy efficiency activities may offer specific benefits (see Sections 6.2.2.6, 6.5.1, 6.5.2; IPCC, 1996).

In practice, both energy and carbon taxes have already been adopted as responses to commitments under the UNFCCC. The European Commission (EC), for instance, has issued several tax proposals designed to reduce emissions of CO_2 from fossil fuel use. For example, Finland, Netherlands, Denmark,

---

[31] Empirical work suggests that to focus on all six gases of the Kyoto Protocol (CO_2, CH_4, N_2O, SF_6, PFCs, HFCs) and not just the carbon content of fuels reduces compliance costs substantially (Reilly *et al.*, 1999).

[32] This assumes that "carbon removal and disposal" strategies (e.g., removing CO_2 from stack gases and sequestering them in geological formations or land use change involving afforestation and reforestation) receive payments equivalent to the tax rate per tonne CO_{2eq} sequestered. It also assumes that non-energy GHG emissions are also subject to the tax or to policies for which the marginal abatement cost is equal to the tax rate.

[33] One aspect that is also relevant is taxing net emissions versus gross emissions. Land use changes are included in the Kyoto Protocol. A national Computable General Equilibrium (CGE) model in which emissions from the use of timber and carbon accumulation in the forest are taken into account, thus calculating net emissions, is given by Pohjola (1999). If net emissions are taxed, Pohjola (1999) finds that the carbon tax needed to reduce net emissions by the same amount as emissions from fossil fuels is significantly lower.

[34] Energy taxes may be more efficient than taxes on carbon alone if there are negative externalities unrelated to CO_2 associated with the energy services delivered.

[35] A 1997 OECD study (OECD, 1997d) suggests that the evidence that more stringent environmental regulation is reflected in the pattern of international trade in goods produced by traditionally polluting activities is not yet clear. This conclusion could change, however, if energy taxation or any comparable measure is introduced at a large scale.

Sweden, and Norway all have energy taxes based in part on carbon content (Speck, 1999; see Section 6.1.3). Other countries that have recently introduced carbon or energy taxes to help achieve their climate change commitments include Slovenia, UK, Italy, Germany, and Switzerland. France is also considering increasing energy taxes on industry for the same purpose. None of these countries have been able to introduce a uniform carbon tax for all fuels in all sectors, because unilateral nature policies raise. In most cases for which an energy or carbon tax is implemented, the tax is implemented in combination with various forms of exemptions (e.g., rebates, VAs).

6.2.2.3 Tradable Permits

A country committed to a limit on its GHG emissions can meet this limit by implementing a tradable permit system that directly or indirectly limits emissions of the domestic sources covered by the commitment. The large number and diverse nature of the sources covered by national limits on GHG emissions raises issues of how to assign permit liability. If permit liability is imposed at the point of release to the atmosphere, a so-called "downstream" system, individual vehicle owners and households would have to participate.

Some emissions, such as HFCs, sulphur hexafluoride (SF_6), and energy-related CO_2, can be controlled indirectly, with a so-called "upstream" system, by limiting substances that ultimately result in GHG emissions (see, e.g., IPCC, 1996; Bohm, 1999).[36] Since energy-related CO_2 emissions are linked to the carbon content of fossil fuels, the system could be implemented by requiring fossil fuel producers and importers to hold permits equal to the carbon content of the fuels sold domestically.[37] Permit liability for energy-related CO_2 emissions could be imposed at any point in the fossil fuel distribution chain and at different points for different categories of sources, for example downstream for large industrial sources and on petroleum com-

panies for transportation fuels.[38] Industrial non-energy sources of GHG emissions also lend themselves, at least partially, to inclusion in a tradable permit system (Haites and Proestos, 2000).

Permits equal to the emissions limit are distributed (*gratis* or by auction, usually to permit-liable entities) and each permit-liable entity is required to hold permits equal to its actual GHG emissions or actual sales of regulated substances as appropriate. Permits may be traded, at least domestically and at least among permit-liable entities. Such a tradable permit system is well known from the literature to be cost-effective if transactions costs are not prohibitively high and if there are no significant imperfections in the permit market and other markets pertaining to the emitting activities (see IPCC, 1996, p. 417).[39]

Some sources of GHG emissions, such as methane emissions from livestock, as well as small sources, are very difficult to include in a tradable permit system because it is difficult to measure actual emissions (or an accurate proxy for actual emissions). In practice, then, the emissions cap for the tradable permit system is less than the national emissions limit and some sources need to be addressed by other policies.[40] For example, a government that takes part in an international agreement, such as the Kyoto Protocol, may establish an emissions cap for the tradable permit system on the basis of the initial national limit or the *ex post* limit, taking into account its net transfers under the Kyoto mechanisms.[41]

With a significant number of permit-liable entities it should be possible to establish market institutions that have low transac-

---

[36] HFCs and SF_6 are manufactured gases used in a variety of applications and ultimately escape to the atmosphere. Limiting sales of these gases in the country effectively limits the subsequent emissions. Cost-effectiveness then requires that the prices of the regulated substances rise to reflect the social marginal cost of abatement (see Section 6.4.1), so that the sources have the correct incentive to implement the appropriate abatement measures.

[37] Virtually all of the carbon content of fossil fuels is converted to CO_2 upon combustion. Thus, if there are no commercially viable CO_2 capture and sequestration technologies, the CO_2 emissions are closely related to the carbon content of the fuel. If CO_2 capture and sequestration is implemented, an upstream system based on the carbon content of fossil fuels could still be implemented, but it should be complemented by a system of credits for sequestered CO_2. Some fossil fuel is used as a feedstock for products that sequester the carbon for a relatively long time. An upstream system should include provisions to exempt the carbon sequestered in such products. A particular aspect that can be introduced is to specify a validity period of permits by establishing gradual devaluation and an expiration date. By introducing this dynamic incentives could be created.

[38] See NRTEE (1999) for a comprehensive overview of options for the design of a domestic GHG tradable permit system. Remember that the liable point may differ from the point of allocation. See Matsuo (1999) and Iwahashi (1998) on this.

[39] Tradable permits have been used to implement a cap on SO_2 emissions by electricity generators in the USA and on NO_x and SO_x emissions by large sources in the greater Los Angeles area (the Regional Clean Air Incentives Market (RECLAIM) Programme; e.g., Schmalensee *et al.* (1998); Stavins (1998a)).

[40] NRTEE (1999) suggests that coverage for different designs can range from 30% to over 90% of total emissions. Given a satisfactory solution to the monitoring problems, cost-effectiveness is improved by including as large a share of total emissions as possible in the tradable permit system.

[41] The national government could choose to be a net buyer or seller using the Kyoto mechanisms. To achieve compliance, the cap for the domestic trading system should reflect the national limit after adjustment for these international transfers. Whether the domestic cap is based on the initial commitment (no government transfers under the mechanisms) or the *ex post* limit, permit-liable entities could be allowed to acquire quotas under the Kyoto mechanisms, if allowed by the rules governing those mechanisms, for use towards compliance with their obligations under a domestic tradable permit system.

tion costs and that limit the scope for market power.[42] The only situation in which there might not be enough permit-liable entities is in a small country with an oligopolistic market for fossil fuels and an "upstream" trading system.[43] In particular, if an exchange institution is used, transaction costs are likely to be small and market power (the possibility of one or more market parties to manipulate market conditions in their favour, or to try to achieve such a result by taking speculative positions) is unlikely to have a noticeable influence on the transaction volume or final market prices (e.g., Smith and Williams, 1982; Carlén, 1999).[44] If the domestic tradable-permit system is integrated with an IET market (see Section 6.3.1)–which further increases cost-effectiveness–any remaining market power concerns are greatly diminished.

Some analysts argue that to allow entities, in addition to permit-liable participants, to participate in the market is desirable for several reasons. It allows the risks of changes in permit prices to be borne by the entities (e.g., private brokerage firms, traders, professional speculators, or arbitrators) best able to bear those risks. It may also improve intertemporal efficiency if other entities have relevant information not heeded by permit-liable participants. The behaviour of participants in the permit market might need to be supervised in the same manner as in other financial markets, regardless of whether they are permit-liable or not, to prevent abuses such as insider trading and efforts to manipulate the market.

Permit prices fluctuate, but this does not mean that prices of the products of permit-liable entities fluctuate to the same extent. Crude oil prices change daily, but the prices of various petroleum products, such as gasoline, are much more stable. Forward contracts and options are used to transfer the risks of price fluctuations to sources willing and able to bear those risks.[45] The same mechanisms are likely to be used by permit-liable entities to deal with the risks of fluctuations in permit prices.

The market value of the permits needed by a permit-liable entity is passed on to customers in the form of higher prices, to employees through lower wages, to shareholders through lower returns, and to suppliers through lower prices. To answer how the costs are shifted to these different groups requires a comprehensive model of the economy with accurate values for relevant price elasticities. Ultimately, the costs are borne by individuals, with the impact on a particular person reflecting his or her role as an employee, investor, and/or consumer of various products.[46]

Permits can be distributed to permit-liable entities (and/or others) *gratis* or by auction.[47] *Gratis* allocation requires a rule for distributing the permits among the recipients. Since the permits represent an asset transferred to the recipients it can be difficult to find a rule that is considered fair by all. An auction raises revenue. All of the revenue could be returned to permit-liable entities, but this needs to be done in a manner that leaves them with an economic incentive to reduce their emissions. The revenue could also be used for a variety of other purposes. Compensation could be provided to industries, whether or not they are permit-liable entities, or households that bear a disproportionate share of the impact. The revenue could also be used to reduce existing distortionary taxes and so reduce the net cost of the emission reduction policy (see Section 6.5.1). The introduction of an emissions trading programme, like the imposition of any new tax or regulation, imposes adjustment costs on the affected entities. This is true whether the permits are auctioned or distributed *gratis*. Moreover, some *gratis* allocation rules discriminate against new entrants (IPCC, 1996; Cramton and Kerr, 1998; Zhang, 2000).

---

[42] The number of participants in the trading programme could be small if a country chooses to make fossil-fuel producers and importers permit-liable and there are very few such firms. This implies that the domestic market for fossil fuels is not competitive. If the country created a competitive market for fossil fuels, the number of permit-liable entities would likely be sufficiently large to create a competitive market for permits as well. Sweden, which imports all its fossil fuel, has some 350 fossil fuel importers that are now liable to a carbon tax and that would be permit-liable should it choose to shift to a tradable permit system.

[43] Even under such circumstances a competitive permit market could be created by restructuring the fossil fuel market.

[44] Although the US SO_2 allowances are not traded on an exchange, over 9.5 million allowances were transferred between economically unrelated parties in 1998 and brokerage commissions for a simple transaction are approximately 1% of the sale price.

[45] For crude oil and natural gas, the options are exchange-traded contracts. Such transactions also occur in the SO_2 allowance market; they do not require exchange-listed contracts.

[46] This is true regardless of the domestic policy adopted to meet the GHG emissions limit. However, the total cost of meeting the limit, and the distribution of that cost, may differ with the policy adopted (see Section 6.5.1).

[47] Auctioned permits are equivalent to a tax, if adjusted with a similar frequency, and are designed to achieve an equal emissions reduction by the same sources. If, instead, tradable permits are allocated *gratis* to certain entities, the same distribution is obtained as in the tax case if the tax revenue is redistributed to these entities in the amount of the wealth of the permits otherwise allocated to them (IPCC, 1996, p. 410). To redistribute the tax revenue it is necessary to confirm the total amount of permits allocated. This means that the taxation system in combination with the revenue redistribution inevitably involves a key dimension of the permit trading system, so that the advantage of the taxation system in administrative costs diminishes significantly. If the scale of allocation for the permits in *gratis* is determined on the basis of historical factors, the allocation in *gratis* does not reduce efficiency in emissions reduction. Tax exemption and reduction, however, may reduce or even eliminate incentives for emissions reduction and depreciate the efficiency factor embraced in the taxation policies, because the scale of reduction or exemption is determined by the current emissions quantities.

Assuming compliance, permits are a more certain means than taxes of achieving quantified national emission limits. In addition, a tradable permit system with auctioned permits is more likely to provide the efficient price signal than a tax rate set by the government. However, the certainty of achieving the emissions levels provided by a tradable permit system incurs the cost of permit prices being uncertain. Some have argued in favour of introducing a trigger price into a permit trading system to meet this concern, namely the absence of an upper bound on the price and hence on compliance costs (See e.g., Kopp *et al.*, 1999a). When the permit price reaches the trigger, additional permits are sold by the government to prevent the price from rising further. Such a hybrid system fails to guarantee particular emissions levels, but does limit the economic cost of the programme for its users.[48]

6.2.2.4 Voluntary Agreements

No international definition of a VA is universally accepted (CEC, 1996; EEA, 1997; OECD, 1998a). VA is used here to mean an agreement between a government authority and one or more private parties, as well as a unilateral commitment that is recognized by the public authority, to achieve environmental objectives or to improve environmental performance beyond compliance.[49]

VAs may take a wide variety of different forms. The large-scale VAs in the field of GHG mitigation activities in Japan and the Netherlands are referred to in *Boxes 6.3 and 6.4*. For a description of the US "market transformation" type VA and the German VAs, see Mazurek (1998) and Storey *et al.* (1999), and Eichhammer and Jochem (1998), respectively. Sometimes these involve agreements between the government and a set of firms, but in other cases industry associations represent member firms. Sometimes the agreement only relates to general issues, such as R&D activities, reporting on emissions, or energy efficiency, but in other cases specific quantified targets, such as emissions targets, are agreed upon. A few VAs are legally binding once signed, but most are not.[50]

Although VAs are a relatively new environmental policy instrument, they are gaining popularity as a tool to cope with environmental issues. That in 1996 in the EU alone there existed more than 300 VAs at least suggests this type of policy measure is administratively and politically feasible, especially if it is used in a policy mix or in new policy areas (OECD, 1998a, p. 102). VAs are political feasible simply because most of the industries seem to prefer VAs over other tools (Dijkstra, 1998; Svendsen, 1999). VAs may precede more formal arrangements; the vast majority of GHG emissions reductions in the USA called for in the US Climate Change Action Plan come, for instance, from voluntary initiatives to increase energy efficiency. However, VAs may not be a satisfactory substitute for mandatory efficiency standards (Krause, 1996).

Sometimes the "voluntary" aspect of a VA is questioned, as the main motivation for industries to join the VA was to avoid the implementation of a carbon and/or energy tax and/or other mandatory policy (Torvanger and Skodvin. 1999, p. 28). Segerson and Miceli (1997) found that the level of abatement under a VA is closely related to the probability of regulatory action in the absence of an agreement.

Proponents of voluntary approaches point to the low transaction costs, the merits of the consensus elements in the approach, and the advantages of leaving the choice of abatement measures to the participants. Although free riding is a concern with VAs, the risk can be addressed through the proper design of the VA. Free riding can take place if firms that do not comply or participate benefit from the agreement while bearing no cost. Governments may encourage participation in VA programmes and discourage free riders by providing incentives such as permits to use labels and other marketing claims. As for possible abuse, some or all of the participants may use their initiating role in the process to create an agreement that benefits them, and hence obstruct real abatement progress. It could also involve introducing measures that benefit some firms, and reinforces their market dominance.

To assess the environmental effectiveness, the trade-off between how ambitious the objectives are and how well they are attained should be recognized. There is a suspicion that if the goals are too ambitious, they will not be attained. As most VAs are non-binding they may not attain ambitious goals (EEA, 1997; OECD, 1998a). VA objectives may be less stringent if environmental groups are left out off the negotiation process. Since VAs are a relatively new policy instrument to cope with environmental issues, it is too early to determine their effectiveness (OECD, 1998a, pp. 78–83).

From a methodological perspective, it is rather complex to assess the effectiveness of VAs because it is difficult to establish a counterfactual.[51]

Voluntary provisions also may accompany mandatory policies. The Substitution Provision of the US Acid Rain (SO_2 Emissions Trading) Program is the first example of a voluntary

---

[48] See Kopp *et al.* (1999b) for a discussion in the context of domestic US policy; Roberts and Spence (1976) provide a theoretical discussion.

[49] "For the purpose of this Communication Environmental Agreements ... can also take the form of unilateral commitments on the part of industry recognized by the public authorities" (CEC, 1996, p. 5).

[50] In some countries (e.g., Denmark) negotiated agreements are explicitly linked to favourable treatment under tax regimes.

[51] The issue of counterfactual baselines is revisited in Section 6.3.2.3 in the context of the Kyoto mechanisms.

Box 6.3. Keidanren Voluntary Action Plan on the Environment (See http://www.keidanren.or.jp/)

Keidanren (Japan Federation of Economic Organizations), the largest private and non-profit economic organization in Japan, announced the "Keidanren Appeal on the Environment" in 1996, in which concrete courses of action for measures to cope with global warming were specified. Following the Appeal, 37 trade associations set forth the "Keidanren Voluntary Action Plan on the Environment" in June 1997. Although the above action plan is a unilateral commitment on the part of the industries, it should be considered an environmental agreement.[52] In fact it constitutes a major component of the Japanese government's "Basic Principles for the Promotion of Measures Dealing with Global Warming"; a follow-up survey is to be conducted every year and reported to the government councils, including the Industrial Structure Council of the Ministry of International Trade and Industry, for third party review.

This action plan, which contributes to meeting the Japanese commitment under the Kyoto Protocol, has as its goal "to endeavour to reduce CO_2 emissions from the 28 industrial and energy-conversion sectors to below the levels of 1990 by 2010." Under a baseline (or business-as-usual) scenario these emissions are estimated to increase by 10%. The 28 sectors represent approximately 76% of CO_2 emissions generated by all industry and energy-conversion sectors in Japan, which in turn generated 42% of Japan's total CO_2 emissions in 1990.

Each participating business sector made a social commitment by setting a numerical target (in terms of: size of CO_2 or energy consumption; emissions or index of CO_2 emissions; or energy input per unit output), which was compiled and published by Keidanren. For example, the Japanese Iron and Steel Federation set a target of reducing energy consumption in 2010 by 10% from the 1990 level (57.22kt crude oil).

The second survey, presented just before CoP5, showed that CO_2 emissions in fiscal year 1998 were 126MtC, or 2.4% less than 1990 and 6% less than 1997 levels. Keidanren stressed that to meet the emissions goal it would:
- continue to make annual surveys of emissions by participating associations;
- intensify co-operation between the government and other sectors, such as transportation, households, *etc.*;
- promote the construction of new nuclear power plants; and
- explore positively the utilization of the Kyoto mechanisms.

compliance provision within an emissions trading regime.[53] Voluntary compliance was characterized by adverse selection; units that "opted in" to the programme tended either to have low emissions below their permitted allocations, or to have low costs of abatement (Montero, 1999). While the VA kept aggregate costs low, the adverse selection increased aggregate emissions (Montero, 1999). This inevitable trade-off between adverse selection and cost-savings means that the design of voluntary programmes will influence their net emissions impact (Montero, 2000a).

The OECD (1998a) noted that no empirical evidence is available on the cost-effectiveness of VAs. CEC (1996), however, argues that the flexibility of VAs provides room for industries to find the most efficient way to achieve the targets, which could be a major advantage. EEA (1997) recently concluded,

after analyzing six case studies of European VAs, that, while there was quantitative evidence for environmental improvement in most case studies, more sophisticated analysis would be necessary to distinguish between the effects of the VAs and those of other factors (EEA, 1997, pp. 84–85). In the same study it was recognized, however, that in five of the six cases the interviewed experts felt VAs incurred lower costs than alternative instruments.

OECD has indicated various conditions under which VAs can be implemented most effectively (EEA, 1997, p. 15; OECD, 1998a):
- clear targets are set prior to the agreement;
- the agreement specifies the baseline against which improvements will be measured;
- the agreement specifies reliable and clear monitoring and reporting mechanisms;
- technical solutions are available to reach the agreed target;
- costs of complying with the VA are limited and are relatively similar for all members of the target group; and
- third parties are involved in the design and application of VAs.

The EC, for instance, recommends prior consultation with interested parties, a binding form, quantified and staged objectives, the monitoring of results, and so on.

---

[52] This point of view is supported by the EC: "For the purpose of this Communication Environmental Agreements ... can also take the form of unilateral commitments on the part of industry recognized by the public authorities" (CEC, 1996, p. 5).

[53] The SO_2 emissions trading regime has been implemented in two phases. The first phase (beginning in 1995) imposed annual emissions caps (with trading) on the 263 dirtiest large electricity-generating units. The Substitution Provision allowed units regulated only by the second phase (beginning in 2000) to voluntarily "opt in" in the first phase. Owners of the first-phase plants could use these "substitution" units to lower the compliance costs.

Box 6.4. Voluntary Agreements in the Netherlands

In the early 1990s, the Dutch government entered into agreements with all energy-intensive industries to improve energy efficiency. The purpose was both to improve competitiveness by cutting energy costs and to reduce CO_2 emissions. This win–win situation is favoured by the Ministry of Economic Affairs, which was primarily responsible for the execution of the long-term agreement (LTA) policy. Efficiency is usually defined as the ratio of relevant physical output to primary energy consumed. The target for most sectors is to improve energy efficiency by 20% in 2000, compared to 1989. Most sectors were audited before entering into an agreement, to ensure that the efficiency improvement was feasible. The coverage of industrial energy consumption is high, almost 90% when non-energy consumption is excluded. There is a similar agreement with the horticultural greenhouse sector, which is the second largest energy-consuming sector after the chemical industry. An intermediate organization co-ordinates the annual monitoring and runs programmes for technological support and R&D. The government publishes results annually. It is expected that, on average, the 2000 efficiency target will be reached.[54] Based on interviews and analysis, 30%–50% of the efficiency improvement identified is implemented because of LTA and related supporting policies (Glasbergen *et al.*, 1997). The results for the LTA sectors in total manufacturing industry through 1996 are depicted in *Figure 6.1*, together with general statistics (Van Dril, 2000).

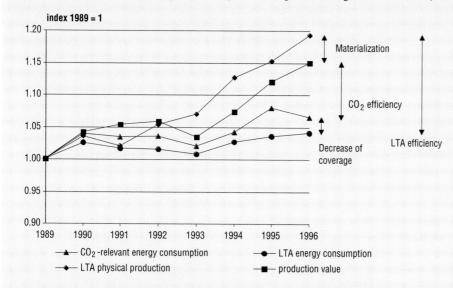

Figure 6.1: *Aggregated results of manufacturing industry LTAs and statistics.*

As a general observation, LTA results diverge from the actual average of the entire manufacturing sector. Both the energy and output indicators show significant deviations. The main explanations for the divergence are, first, that energy-intensive products such as primary materials have grown faster than average production value. In monitoring practice, there may be some bias towards adjusting for energy-intensive products, to avoid negative effects on efficiency results. A second explanation is that statistics on the chemical industries are unreliable and that no insight is provided by the entities responsible for monitoring. For example, no clear information is available on the share of non-energy consumption and its impact on CO_2 emissions.

6.2.2.5 Informational Instruments

As Chapter 5 shows, information drives decisions. Information gaps result in uncertainties, risks, and missed opportunities. Poor information is widely recognized as a barrier to improved energy efficiency or reduced emissions (Tietenberg and Wheeler, 2000). Markets are not always fully informed on the quality of information and application of decision-support technologies. In Russia, for instance, it is estimated that institutional barriers and information limitations result in only 2% of the market potential to improve energy efficiency actually being realized (Bashmakov, 1998).[55]

Reliable data are a prerequisite for decision-making. At the micro level, feasibility studies or business plans are used to explore opportunities to raise energy efficiency and energy productivity. They are based on metering and energy audits in specific situations. At the macro level, detailed statistical data on major aspects of energy consumption are the basis for development and evaluation of efficiency improvement policies, and

---

[54] Ministry of Economic Affairs, Netherlands (1999): Long-term agreements on energy efficiency, results (published annually).

[55] OECD/IEA (1997) includes 47 case studies of successful energy-efficiency improvement projects and policies.

their success or failure (Japan Energy Conservation Center, 1997). Comparisons between nations and companies and benchmarking on energy efficiency indicators also raise awareness and allow for better determination of efficiency potentials (see also OECD/IEA, 1997; Fenden, 1998, p. 203; Phylipsen *et al.*, 1998, p. 230; ADEME-European Commission, 1999). Also, improved accessibility to new technology information enhances technology transfer. Information-based policies can also be used to reveal low levels of performance.

Policy instruments to improve information are applied on three levels. First, they are used to raise awareness of climate issues. Governments communicate their targets and policy measures to the public. The information may influence preferences to contribute to GHG mitigation. Social marketing is becoming a crucial instrument in creating an appropriate social environment for GHG emissions reduction policies (Legro *et al.*, 1999). Second, governments stimulate research to analyze climate issues and create mitigation opportunities that can be widely applied. R&D generates new information on possibilities and determines the technical potential. Information on the economic situation (prices, taxes, interests rates, etc.) in turn constrains the technical potential to what is commercially feasible. Third, information instruments are used to help the implementation of measures. They can assist the public in making the right choices with respect to GHG mitigation.

There are several reasons for using instruments to further information on climate issues. First, climate change involves complex negative externalities, so the process of policymaking with regards to GHG reduction needs broad support and understanding. Second, information, once generated, can be widely used, which is regarded as a reason for collective funding of its collection, dissemination, and use. Many of the possible ways to reduce GHG emissions are similar all over the world. Markets for this information are not yet developed.

6.2.2.5.1 Education Programmes

Energy efficiency centres, government offices, utilities, equipment vendors, professional organizations and associations, educational channels, etc., deliver information on GHG reduction. Improved data and metering, energy audits and monitoring, workshops and exhibitions, campaigns in the mass media, education and training, efficiency and environmental labelling, publications and databases are all typical instruments used to enhance information dissemination.

Educational and training programmes may improve decision making and can have long-lasting effects. Consumer education is an important social marketing tool in implementing DSM programmes (see *Box 6.5*).

Information campaigns are used as marketing elements in most energy efficiency programmes. Typical examples of such campaigns are:
- publications and advertising;

Box 6.5. Public Education Component of Poland Efficient Lighting Project (OECD/IEA, 1997, p. 480)

The IFC/GEF Poland Efficient Lighting Project (PELP) was designed to reduce emissions of CO_2 and other GHGs emitted by Poland's electricity sector by stimulating the Polish consumer market for energy-efficient compact fluorescent lamps (CFLs). The public education component of PELP promoted the CFL subsidy programme to the public by providing consumer information on the benefits of energy-efficient lighting from a trustworthy, non-industry source. The generic PELP advertising bore the logos and endorsement of four respected Polish organizations. The PELP logo featured in advertisements and on participating products, was promoted as a symbol by which consumers could identify energy efficient, high-quality products. PELP organized high-level seminars for lighting professionals on technical and design aspects of energy-efficient lighting. Finally, to educate tomorrow's consumers on the benefits of CFLs, PELP created an energy-efficiency curriculum for schools, and sponsored an art and essay competition for schools on energy-efficient lighting.

- broadcasting of special programmes on television and radio;
- distribution of special brochures;
- creation of special easily accessible databases; and
- public awareness programmes, such as "Energy Conservation Day" and "Energy Conservation Month", which are implemented on a regular basis at the national level in Japan and South Korea.

Publication of books and periodicals on energy-efficient technologies and systems, and energy efficiency success stories, guidelines, and policies is another powerful information instrument.[56] Costs of information programmes vary according to their scale, coverage of specific groups of customers, and use of media.

6.2.2.5.2 Labelling

One instrument that is increasingly applied in the area of environmental policy is environmental and energy efficiency labelling. Labelling programmes can be mandatory or voluntary.[57] Mandatory energy efficiency labels have long been estab-

---

[56] The Russian Center for Energy Efficiency, for instance, since 1993 has published the "Energy Efficiency" quarterly bulletin. This stimulated regional energy-efficiency legislation, policies, and programmes in Russia. The bulletin is available on the Internet at http://www.glasnet.ru/~cenef.

[57] Mandatory labelling programmes are under implementation not only at a national level (e.g., *Energy Guide* in the USA), but also internationally, such as "SAVE" in the EU. Voluntary labelling programmes were also initially launched at the national level, such as "Blue Angel" in Germany and "Power Smart" and "EcoLogo" in Canada, but then some were internationalized (e.g. the originally US "Energy Star" programme for office equipment was introduced in Japan).

lished in the USA, Japan, and South Korea, and recently in the EU where they are part of appliance and automobile efficiency legislation. Labels and marketing may have a pervasive impact on consumers' behaviour and the introduction of clean technology. *Boxes 6.5* and *6.7* provide some examples of such developments to illustrate how these phenomena work in practice. The strengths of energy efficiency and environmental labelling are, first, that labels do not distort the market. Second, in many instances they are voluntary for both the producer and the consumer because the former is free to decide whether or not to join the system and the consumer is free to decide whether or not to buy the labelled product. Voluntary labels are a non-official instrument, and may be instituted without the usual delays associated with official policymaking. Third, labels are usually based on considerable information exchange among the various stakeholders, which may increase the overall acceptance of the instrument.

This is not to say, however, that energy efficiency and environmental labelling do not have weaknesses. If all products are labelled, the consumer must learn how to interpret the label (e.g., do higher numerical values indicate a better or worse product?). If products must meet a specified standard to qualify for a label, only part of the market will be covered by the labelled product. Competing labels for the same product or less reliable labels may easily undermine the trust of the consumers in the labelling instrument. This may turn out to be an inherent limitation.

In sum, environmental labels represent an important tool to create transparency in markets and thus give orientation to the consumer. The overall success of this instrument, however, will probably depend on the solution to the following dilemma: if applied too strictly, market coverage may be too low for the label to be effective; if applied too leniently, the environmental effectiveness may be limited.

6.2.2.6   *Subsidies and Other Incentives*

6.2.2.6.1 Environmental Subsidies

A subsidy for GHG emissions reduction pays entities a specific amount per tonne of CO_{2eq} for every tonne of GHG reduced or sequestered. Such a subsidy encourages implementation of measures to reduce emissions or enhance sequestration that are less costly than the subsidy.

Under certain circumstances, a uniform subsidy can lead to the same emissions reduction outcome as an equivalent uniform tax. In theory, in an industry with homogeneous firms, both taxes and subsidies (set at the same levels) yield exactly the same outcome in the short run. In general, a tax is more efficient than a subsidy because the subsidy can result in too many firms in the industry, and thus an inefficient amount of both pollution and goods associated with the pollution (Kolstad, 2000). This is always the case in the long run because a subsidy lowers the average cost of production, while the tax

Table 6.1: *Public expenditures as percentage of gross domestic expenditures on R&D (1985–1995) (OECD, 1998a)*

Country/ region	1985 Public % of total	1990 Public % of total	1995 Public % of total
Overall OECD	43.0	37.8	34.5
USA	50.3	43.8	36.1
Canada	48.9	44.3	37.7
EU	44.4	40.9	33.1
UK	42.2	35.5	33.3
France	52.9	48.3	–
Japan	21.0	–	22.4
Germany	37.6	33.9	37.1
South Korea	–	17.0	18.2
Czech Republic	–	30.6	34.9
India	88.5	87.3	84.6

increases the average cost of production. In the short run, it is also the case in an industry with heterogeneous firms. A subsidy may allow some firms to continue operating that would not continue in the case of a tax (those with average variable costs above prices). Besides, a subsidy requires that revenue be raised somewhere else in the economy, which can also produce dead-weight losses.

An emissions reduction subsidy, like an emissions tax, does not guarantee a particular level of emissions. Therefore, it may be necessary to adjust the subsidy level to meet an internationally agreed emissions commitment. In addition, criteria other than efficiency, such as distributional impacts, are likely to influence the design of the emissions subsidy (or the combination of subsidies and taxes in what is known as fee and/or rebate). The distributional and competitiveness impacts help explain why, in practice, some energy and emissions taxes are coupled with tax exemptions or subsidies. Also, the use of subsidies for environmental purposes may cause problems under WTO agreements on subsidies and countervailing measures.

6.2.2.6.2 Research and Development Policies

Technological progress is mainly achieved in the private sector, through learning by doing, incorporating new findings developed elsewhere into the production process, or through firms own R&D activities. A major, and generally increasing, part of funding of R&D expenditures is initiated by and in the private sector itself (*Table 6.1*). Government funding of R&D on energy has historically favoured nuclear and coal technologies (IEA, 1998a; OECD, 1998a). Research on renewable energy and energy-efficient technologies is gaining ground, but it is still a relatively small portion of R&D budgets in the OECD. This is important when assessing what governments can do to promote innovation. Perhaps governments can provide a reliable legal framework to protect research findings in the area of energy efficiency improvement from being copied elsewhere without compensation.[58]

6.2.2.6.3 Green Power

Green power policies establish mechanisms through which part of the electricity supply (whether in a regulated or competitive environment) must come from designated renewable energy sources. Regulatory policy mandates include set-asides for renewables, renewable portfolio standards (RPSs), and various kinds of subsidies created from SBCs or renewable energy funds. The cost of compliance for policy mandates is borne by all consumers. Despite this 100% participation, however, the policies may or may not be effective in stimulating renewable energy generation, depending on how aggressive they are and how they are implemented. Some examples are given in *Boxes 6.6* and *6.8*. To reduce the cost of compliance regulatory policy may be supplemented by tradable renewable energy certificates as described in *Box 6.7*.

Green power and green pricing programmes encourage consumers to voluntarily pay a higher price for electricity generated from "green" (environmentally friendly) energy sources. Green power products are offered by some suppliers where electricity markets have been liberalized, while green pricing is a green power option offered by the monopoly utility in jurisdictions where consumers are not yet permitted to choose their retail provider (Swezey and Bird, 2000). Green power marketing programmes are relatively new, dating from 1993, and are being implemented in Australia, Canada, Germany, Netherlands, Switzerland, UK, and USA (Markard, 1998; Crawford-Smith, 1999; Holt, 2000a, 2000b).

In the USA, about 30 green power products are being marketed by 15 retailers in competitive states and about 140 electric utilities offer a green pricing option that emphasizes wind or photovoltaics (Holt and Wiser, 1998; Holt, 2000b). Market penetration so far is low, a little over 1% on average, although it reaches as high as 4%–5% for a few utility programmes (Wortmann *et al.*, 1996; Holt, 2000a, 2000b). Of those who switch suppliers in competitive markets, some 20%–95% choose a green power product (the higher percentage results from significant renewable energy subsidies in California).

Wiser *et al.* (2000) assessed green power marketing programmes in the USA. They conclude that the collective impact of customer-driven demand for renewable energy has been modest to date, but that it is too early to draw definitive conclusions about the potential contribution of green power marketing in the long run.

In support of green power marketing and of policies that mandate renewable set-asides and RPSs, renewable energy certificates (also called credits, labels, or tags) may be traded sepa-

rately from green electricity. Whether renewable energy or other environmental attributes should remain with the purchaser of the underlying commodity, or be sold to different entities, is under debate. There are either plans for or limited experience with tradable certificates in Belgium, Denmark, Italy, Netherlands, UK, and USA, and it is likely to grow in importance (Benner, 2000; Rucker, 2000). Trading in renewable energy certificates promises greater liquidity and potentially lower costs to meet policy commitments and marketing claims. An example is given in *Box 6.7*.

6.2.2.6.4 Demand-side Management

Information programmes are often applied in combination with other initiatives (such as rebating in DSM programmes, energy audits, labelling, and regulation). In the US cumulative electric utility DSM spending to date is about US$15–20 billion. Close to 60% of utility customers are served by such programmes. Reductions in national electricity demand of 3%–4% percent were achieved with these programmes (Hadley and Hirst, 1995; Eto *et al.*, 1996). Studies on the efficiency of DSM programmes find that a large proportion of the reported conservation impacts are statistically observable after accounting for economic and weather effects (Parfomak and Lave, 1997).

With utility restructuring and the emergence of electricity generation competition, the rationale of utility resource acquisition has been greatly diminished. The new generation of programmes funded by SBCs emphasizes permanent market transformation effects aimed at technology manufacturers, including financial incentives paid directly to manufacturers, guaranteed minimum market sales for new energy efficient products, and competitive technology procurement programmes.

6.2.3 *Mixes of National Policy Instruments*

Section 6.2.2 discusses various policy instruments to manage GHG emissions in isolation. Various authors (e.g., Bernstein, 1993; Richards, 1998; Stavins, 1998b) argue that to select the best approach to attain the environmental goal, various cost and other aspects must be taken into account. These include production costs, cost differences across sources, transaction costs, monitoring and enforcement costs, implementation, administrative costs, and other socio-economic conditions idiosyncratic to each country. For these reasons, it can be anticipated that in most countries GHG emissions will be managed using a portfolio of policy instruments, rather than a single policy instrument. Furthermore, the portfolio of instruments is likely to differ from country to country. Using a portfolio of policy instruments enables a government to combine the strengths, while compensating for the weaknesses, of individual policy instruments, thus improving overall effectiveness and efficiency.

Under some conditions a combination of market-based and information policies and regulations can improve economic

---

[58] For some additional remarks see also Section 6.5.3.

Box 6.6. Examples of Policies to Promote Renewables in a Liberalized Power Market

Renewables Set-aside

The UK has been promoting wind and other renewable energy technologies through its Non Fossil Fuel Obligation (NFFO; see Mitchell, 1995a, 1997). The renewable NFFO sets aside a certain portion of the electricity market to be supplied by designated renewable energy technologies under a competitive bidding framework. Within each technology band (wind, biomass, landfills, solar, etc.) developers submit bids of proposed projects and the projects with the lowest cost/kWh price are awarded power purchase contracts. Regional electricity companies are mandated to purchase power from NFFO-awarded renewable electricity generators at a premium price. The companies are reimbursed for the difference between the NFFO premium price and the average monthly power pool purchasing price through the Fossil Fuel Levy (Mitchell, 1995a). The main weakness of the NFFO is that the implementation rate of approved projects is very low, because bids have such low cost/kWh that they do not allow the profitable operation of projects. Moreover, the intermittent character of NFFO rounds has precluded the development of a steady domestic market for renewable technologies (Michaelowa, 2000).

Renewable Portfolio Standard (RPS)

The RPS has received considerable attention in the USA. The RPS is similar to the NFFO concept in the UK, in that both are competitive least-cost mechanisms. Unlike NFFO there is no funding levy. Under RPS, all retail power suppliers are required to obtain a certain minimum percentage (e.g. 5%) of their electricity from specified renewable energy sources. Efficiency is obtained by allowing the market to determine the most cost-effective solution for each electricity retailer (Radar, 1996; Haddad and Jefferis, 1999). State legislatures and/or public utility commissions have approved various versions of RPS in several US states (Wiser, 1999b).

Production Subsidy and/or System-benefit Charges (SBCs)

Another support mechanism to promote renewables in a liberalized electricity market is a fee/kWh on all energy users to support renewable energy development. This charge is often referred to as SBCs (Haddad and Jefferis, 1999). In California, a total of US$540 million collected from 1998 up to 2002 from electricity customers is directed to support existing, new, and emerging renewable electricity generation technologies (California Assembly Bill, 1996, AB 1890, Ch. 854, Sec. 381). In addition, nine other states in the USA have established SBC policies under the restructuring of their electric utility industries to promote the use of renewables (Wiser, 1999b). Unlike in NFFO and RPS, there is no supplier obligation.

Box 6.7. Green Certificates for Wind Energy in the Netherlands

Support for wind energy in the Netherlands has included both R&D grants and a variety of market-stimulation mechanisms. These have included an integrated programme for wind energy, which provided subsidies of 35–40% of investment costs for newly built turbines. Electricity distribution companies raised environmental levies to purchase wind generated kiloWatt hours for high guaranteed prices (about US$0.07; Wolsink, 1996). More recently, however, significant changes have occurred in the Dutch renewable energy policy, reinforced by the liberalization of the electricity market. Most of the direct subsidies are now eliminated and other market-oriented and fiscal mechanisms introduced. One such mechanism is the green certificates market, which started in 1998. By law, local energy distribution companies must purchase renewable electricity from independent power generators. Distribution companies issue green certificates to the renewable generators equal to the amount of renewable kiloWatt hours sold to the grid. The renewable generator can then sell these green certificates on an open market to distribution companies that want to sell green-certified electricity (Schaeffer *et al.*, 1999). Green electricity is exempt from energy taxes, which will be raised to about US$0.06/kWh in 2001. The tax exemption makes wind energy competitive with electricity from conventional sources, and thus the subsidies are obsolete.

efficiency. Well-designed policies aimed at energy prices are economically most efficient when transaction costs are low and/or cannot be substantially reduced through market transformation policies. They also work best when the potential for technological learning by doing is small or known with reasonable certainty. Well-designed regulatory and incentives-based policies aimed at factors other than energy prices are economically most efficient when the transaction costs are large and can be substantially reduced at low administrative cost. They also work best when the potential for technological learning by doing is large. Virtually all end-use markets for energy efficiency suffer from high transaction costs and related market problems. Also, many energy efficiency and renewables technologies

exhibit large potentials for learning by doing. The most effective and economically efficient approach to achieve lower energy sector emissions is to apply market-based instruments, standards, and information policies in combination. Policies to administer energy price changes provide a uniform signal to all economic actors and overcome fragmentation. Standards and information policies can move the economy closer to the frontier of production possibilities, which raises total factor productivity.

Overriding non-economic reasons may also exist for combining different types of policy instruments to manage GHG mitigation. First, the number and diversity of sources is large and even

Box 6.8. Renewable Energy Policy in India

Renewable energy (RE) sources were first recognized and incorporated in official policy documents in the early 1970s. Several national-level programmes for RE technologies have been initiated, for example, the National Project for Biogas Development (NPBD) with a target of 1.5 million plants by 2001, a national programme for improved cook-stoves, a programme for mass demonstration of RE sources like wind, solar, biomass, *etc*. The Ministry of Non-conventional Energy Sources (MNES) co-ordinates and implements the RE policy at the national level with counter-part departments in the state governments. The Indian Renewable Energy Development Agency (IREDA) operates a revolving fund for development, promotion, and commercialization of RE through the provision of soft term financial assistance.

Under the New Strategy and Action Plan for RE, the following, two-pronged action plan was devised:
* High priority accorded to generation of grid-quality power from wind energy, small hydropower, bio-energy and solar energy.
* Rural energization programme is promoted through:
    - electrification of villages through photovoltaic and biomass gasifier power systems,
    - supply of solar lanterns to unelectrified households,
    - use of solar water heating systems,
    - rural energy programmes, for example, National Project on Biogas Development,
    - production of energy from agricultural waste, *etc*.

Currently, a three-fold strategy has been pursued by the government for promotion of RE sources through private sector involvement. These include:
* Providing budgetary resources by government for demonstration projects.
* Extending institutional finance from IREDA and other financial institutions for commercially viable projects, with private sector participation; and external assistance from international and bilateral agencies.
* Promoting private investment through fiscal incentives, tax holidays, depreciation allowance, facilities for wheeling and banking of power for the grid and remunerative returns for power provided to the grid. The emphasis has shifted from direct financial incentives (e.g., subsidies) to indirect fiscal incentives (e.g., low interest loans, financing packages for consumers, reduced tariff and taxes, viable power-purchase prices, *etc*.). Some fiscal incentives include: accelerated 100% depreciation on specified renewable energy based devices/projects, 100% tax deduction from profits and gains for first five years of operation, and 30% for the next five years for industrial undertakings set up for generation and/or distribution of power.

The new policy for RE tried to give a focus on commercialization and, market orientation and to encourage greater private sector involvement. Despite this there exists significant unexploited potential. The main barriers are: high initial and transaction costs, under-developed markets and market-support infrastructure for RE products, weak linkages between market development and R&D, product development not responsive to users' needs, and the pricing of conventional energy sources (TERI, 2000).

the most comprehensive instruments (an emissions tax or a tradable permit system) is not suitable for all of these sources. Second, the conditions needed to administer efficiently these comprehensive instruments (e.g., a manageable number of participants, but enough to create a competitive market for a tradable permit system) may reduce the scope of their application. Third, different policy instruments can be used to distribute the mitigation-cost burden across sources in ways that lessen opposition to the policy goal. Fourth, policy instruments have multiple impacts, so different instruments and sets of impacts are preferred for different sources. Finally, governments have frequently adopted a portfolio of policies, rather than a single policy instrument, to deal with complex environmental issues.

One important aspect in the policy analysis has been a shift of attention from the assessment of single policy instruments to questions of the optimal policy mix (OECD, 1996b). Assessing the performance of particular environmental policy instruments from historical evidence is difficult because these were often combined in policy packages, as was the case with the

phase-down of leaded petrol in a number of European countries. Econometric analysis has been employed to separate out the effects of individual policy instruments under such conditions, but this is not always possible (Katsoulacos and Xepapadeas, 1996; Boom, 1998).

6.3 International Policies, Measures, and Instruments

Although only Annex I Parties that have made commitments under the Kyoto Protocol's Annex B have quantified emissions limitations, all Parties have committed to take climate change considerations into account, to the extent feasible, in their relevant social, economic, and environmental policies and actions (UNFCCC, 1992, Article 4.1.f). It is recognized, however, that non-Annex I Parties' efforts to take actions that contribute to national development and GHG emissions reduction may be limited by capital constraints, lack of knowledge, or other factors. The UNFCCC and the Kyoto Protocol, therefore, include several provisions that can help overcome such barriers, such as the provision that:

- All Parties are "committed to promote and co-operate in the development, application and diffusion, including transfer, of technologies, practices and processes that control, reduce or prevent anthropogenic emissions of greenhouse gases not controlled by the Montreal Protocol in all relevant sectors, including the energy, transport, industry, agriculture, forestry and waste management sectors" (UNFCCC, 1992, Article 4.1.c).

- Parties agreed to establish a financial mechanism "for the provision of financial resources on a grant or concessional basis, including for the transfer of technology" (UNFCCC, 1992, Article 11.1).

Additionally,

- "The CDM created by the Kyoto Protocol creates an incentive for (entities in and governments of) Annex I Parties to assist the development and implementation of climate change mitigation projects that contribute to sustainable development in, and are approved by, a non-Annex I Party" (UNFCCC, 1997, Article 12).

This section discusses the three Kyoto mechanisms: international emissions trading (IET) (Article 17) in Section 6.3.1 (for some clarifying remarks, see also Section 6.1.3), and JI (Article 6) and the CDM (Article 12) in Section 6.3.2. Thereafter, the section deals with international transfers (Section 6.3.3) and with the various other international policies, measures, and instruments (Section 6.3.4).

6.3.1 International Emissions Trading

If the Kyoto Protocol comes into force Annex I Parties will have agreed to an allocation of AAs (here also called emission quotas) of GHG emissions for the first commitment period, 2008 to 2012. Article 17 of the Protocol allows them to trade part of these emission quotas among themselves in accordance with rules currently being negotiated.[59]

IET implies that countries with high marginal abatement costs (MACs) may acquire emission reductions from countries with low MACs. In principle, such trading tends to equalize MACs across these countries, so that an aggregate emissions reduction can be attained cost-effectively.[60] Parties have not yet decided whether IET based on Article 17 will be restricted to governments or whether legal entities also will be allowed to participate with the approval of their national governments. To

support compliance with their AAs after adjusting for trading, governments may use any of the domestic policy instruments discussed in Section 6.2 above.

Limiting all transactions to multilateral and potentially anonymous trade on an exchange would help IET move in the direction of becoming efficient and non-discriminatory. Bilateral trading cannot be relied upon to reveal to others the true full transaction prices (including undisclosed side-payments), which is required to give all participants equal access to gains from trade. Non-anonymous trading may eliminate transactions between Parties who are in conflict with each other, thus reducing market efficiency. Transparent, anonymous, and efficient trading would be possible on a continuous stock-exchange kind of market (Bohm, 1998). The scope for the exertion of market power is small on such markets, contributing to efficiency (Smith and Williams, 1982).

According to Article 17 in the Protocol, "any such trading shall be supplemental to domestic actions for the purpose of meeting quantified emission limitation and reduction commitments." How to implement this provision is still under debate.[61] A restriction on free IET as a result of binding supplementary requirements could prohibit equalization of the MACs across participating countries, and hence increase aggregate abatement costs.[62]

It has also been argued that constraints on the use of IET and the project-based Kyoto mechanisms (see also Section 6.3.2) might accelerate technological innovation in Annex I countries by increasing the relative price of alternative options for carbon mitigation. Limited analytical studies are inconclusive as to whether such constraints will induce significant innovation, but do suggest that they could reduce the flow of technology to other countries.

An initial quota allocation that turns out to exceed a baseline projection for a country's emissions–possibly relevant for some signatories of the Kyoto Protocol with substantial changes in political and economic systems since 1990–implies that sales of AA units (AAUs) will exceed emission reductions because of active climate mitigation policies, sometimes referred to as "hot air". Restricting trade of "hot air", as some Parties have proposed, would force larger reductions in emis-

---

[59] With sufficient incentives, some non-Annex I Parties may ask to join in IET, replacing their expected use of the CDM, which is discussed in Section 6.3.2.2 (Barros and Conte Grand, 1999; Bohm and Carlén, 2000; Montero *et al.*, 2000).

[60] For illustrations of the potential gains from IET, see Bohm (1997), Manne and Richels (1999), and Weyant and Hill (1999).

[61] Each of the mechanisms, not just IET, includes a so-called supplementarity provision, although the wording differs in the case of the CDM. Some Parties have proposed rules to address supplementarity that apply an overall limit on the use of the three mechanisms, rather than a separate limit for each mechanism.

[62] For a preliminary estimate of the cost implications of the EU proposal for the definition of supplementarity, see Baron *et al.* (1999), Bernstein *et al.* (1999), Criqui *et al.* (1999), and Ellerman and Wing (2000). See Woerdman (2000) and Michaelowa and Dutschke (1999a) on additional reasons for supplementarity.

sions by countries that would otherwise import emissions quotas during the first commitment period.[63] In addition, constraints on hot-air trading, other things being equal, would make the Protocol less beneficial for some countries with "hot air" allocations (Bohm, 1999).

Emissions trading creates a risk that sellers of AAUs might not undertake the emissions reductions that their sales require, in spite of the political costs of non-compliance and despite the sanctions to be instituted. Several options that provide Annex I Parties with an incentive to transfer only part of their AAUs that are surplus to their compliance needs are under consideration. Such options, called liability provisions, are discussed in Section 6.3.5.3. Liability provisions are intended to enhance environmental integrity and are also necessary for the functioning of the market.

6.3.2   Project-based Mechanisms (Joint Implementation and the Clean Development Mechanism)

Project-based mechanisms allow actions that reduce GHG emissions from, or enhance sinks beyond, what would otherwise occur to receive "credits" for the emissions mitigated; these credits can be used by Annex I Parties to help meet their emissions limitation commitments. These mechanisms include technology transfer and provide opportunities for mutual co-operation. JI involves emissions reduction or sink enhancement projects in Annex I countries. CDM involves emissions mitigation projects in non-Annex I countries.[64] Central to these mechanisms is the operational definition of what emissions would have been in the absence of the project; the baseline from which emission reductions (or sink enhancements) are measured. This section focuses on setting the baselines for crediting.

6.3.2.1 Joint Implementation (Article 6)

Article 6 of the Kyoto Protocol allows an Annex I country to contribute to the implementation of a project to reduce emissions (or enhance a sink) in another Annex I country and to receive emission reduction units (ERUs) equal to part or all of the emission reduction (sink enhancement) achieved. The ERUs received by the investor country can be used to help meet its national emissions limitation commitment.

In the case of JI, some analysts have suggested that an independent authority responsible for approving the project baseline is needed in addition to the Parties' approval of the project. Others argue that the host government has an incentive to ensure that ERUs are issued for real emission reductions only if the government is bound to strong and credible penalties for non-compliance (see also Section 6.3.5).

Numerous issues related to JI remain to be agreed, including:
- host and project eligibility;
- the possibility of awarding ERUs for emission reductions from JI projects prior to the start of the first commitment period (see Parkinson *et al.*, 1999);
- monitoring, verification, and reporting requirements;
- baseline updating frequency;
- ERU approval, registry, and trading conditions;
- supplementarity provisions; and
- incentives for compliance.

6.3.2.2   The Clean Development Mechanism (Article 12)

The purposes of the CDM are to assist non-Annex I Parties to achieve sustainable development and to contribute to the ultimate objective of the Convention while assisting compliance by Annex I Parties (UNFCCC, 1997, Article 12.2). The CDM allows a project to reduce emissions, or possibly to enhance sinks, in a country without a national commitment to generate certified emission reductions (CERs) equal to the reduction achieved.[65] Annex I Parties can use CERs to meet national emissions limitation commitments. In contrast to JI, for which there is little peer-reviewed literature, the literature is rapidly growing on the CDM (Goldemberg, 1998; Michaelowa and Dutschke, 1998; TERI, 1998; Hassing and Mendis, 1999; Jepma and van der Gaast, 1999; Haites and Yamin, 2000).

A process for independent review of the certification of the emission reductions achieved is necessary for the credibility of the CDM. Article 12.4 establishes an executive board for the CDM and Article 12.5 specifies that emission reductions must represent real, measurable, and long-term benefits related to the mitigation of climate change and be certified by designated operational entities. The certification process and the respective roles of the operational entities and the executive board remain to be defined, but they will be critical.

The host government must approve proposed CDM projects. As part of its approval process it will need to assess whether the proposed project contributes to sustainable development (Matsuo, 1998; Begg, *et al.*, 2000). Some Parties have proposed criteria or procedures that the host government be required to follow when determining whether a project contributes to sustainable development of the country (see also Thorne and La Rovere, 1999; Chadwick, *et al.*, 2000; Begg, *et al.*, 2000).

---

[63] The EU proposal to address supplementarity, for example, includes a provision that limits the transfers of quotas and thereby limits the trade of "hot air". Restricting trade of "hot air" allows these AAs to be banked for use or sale during future commitment periods, thus reducing the cost of compliance from what it otherwise would be during the future periods.

[64] Whether sink-enhancement projects are eligible under the CDM is still being negotiated.

---

[65] How CDM projects can be financed is still being negotiated. See Haites and Yamin (2000) for a summary of options.

Investments in CDM projects by Annex I governments could lead to a reduction in their official development assistance (ODA).[66] The effect of government investment in CDM projects on the level of ODA will be difficult to determine since the level of ODA in the absence of CDM projects is unobservable. However, historical figures compiled by the OECD Development Assistance Committee could be used to try to deal with this.

Article 12.8 specifies that a share of the proceeds from CDM projects will be used to cover administrative expenses and to assist developing country Parties that are particularly vulnerable to the adverse effects of climate change to meet the costs of adaptation. Articles 6 and 17 do not impose a comparable levy on JI projects or international transfers of AAUs, although a number of developing countries have proposed that the levy be applied to all three mechanisms.

CDM projects can begin to create CERs upon ratification of the Kyoto Protocol. The advantage is that it supports developing countries obtaining access to cleaner technologies earlier. It means that a supply of CERs should be available prior to the start of the 2008 to 2012 commitment period when they can be used by Annex I Parties.[67] Parkinson *et al.* (1999) argue that creation of CERs during 2000 to 2007, which are credited towards 2008 to 2012 compliance, increases the emissions trajectories of Annex I countries for 2000 to 2012. They estimate that increased Annex I emissions offset 30–60% of the CERs created during 2000 to 2012.

Some analysts argue that the CDM facilitates the transfer of CERs from low-cost emission reduction actions to Annex I investors when they might subsequently be needed by the host government to meet a future emissions limitation commitment. However, this assumes a fixed stock of emission reduction actions. In practice, the stock of possible emission reduction (or possibly sink enhancement) actions changes over time in response to turnover of the capital stock, technological change, and other developments. Rose *et al.* (1999) analyzes the optimal strategy for a host government given a dynamic stock of potential projects.[68]

Numerous issues related to implementation of the CDM remain to be negotiated, including:

- host and project eligibility;
- eligibility of sequestration actions;
- demonstrating contribution to sustainable development;
- project financing arrangements;
- monitoring, verification, and reporting requirements;
- baseline establishment;
- CER certification, registry, and trading conditions;
- the share of proceeds for administrative expenses and adaptation assistance;
- adaptation assistance fund administration;
- supplementarity provisions;
- executive board composition and responsibilities;
- process for designation of operational entities; and
- penalties for non-compliance.

6.3.2.3 Baselines

Credible project-based mechanisms under the Kyoto Protocol require the achieved net emission reduction (sink enhancement) to be determined.[69] The reduction is defined as the difference between what emissions (sequestration) would have been in the absence of the measure, the baseline, and actual emissions (sequestration). Thus, the baseline is an estimate of a situation that will never exist (Bohm, 1994; Jepma *et al.*, 1998; Kerr, 1998; Begg *et al.*, 1999).

Since the true baseline can never be observed, a baseline from which emission reductions are calculated may be estimated through reference to emissions from similar activities and technologies in the same country or other countries, or to actual emissions prior to project implementation.[70] Although this judgement is exercised through review by qualified, independent experts, possibly by stakeholders (such as environmental organizations), and by an entity with the final decision authority, the baseline will be an approximation of the counterfactual.[71] One way to reduce baseline uncertainty may be to limit the crediting period or to issue credits for only a fraction of the estimated emission reductions. However, this reduces the investors' interest in financing the projects.

Baseline determination requires a trade-off between the transaction costs of certification and the environmental costs of adverse selection, adjustments for increased emissions at other

---

[66] Therefore, some developing countries have proposed that a financial additionality requirement, which currently exists for the AIJ pilot phase and states that ODA funds should not be invested in such projects, should be extended to include the CDM. This view is not necessarily shared by all Parties.

[67] Estimates of the potential of the CDM are mounting (see, e.g., Austin *et al.*, 1998; Ellerman and Decaux, 1998; Zhang, 1999a), but the estimates are very sensitive to the rules applied for the CDM and the other Kyoto mechanisms.

[68] Contractual options to address this concern are available as well; the host government can insist on an option to acquire the right to future CERs from a project without cost at a specified future date, such as 2013.

---

[69] Note that the eligibility of sink enhancement projects under the CDM is still being negotiated.

[70] Harrison and Schatzki (2000) examine how baselines are established for several environmental and energy programmes in the USA.

[71] Parkinson *et al.* (2000) have estimated the range of uncertainty in estimates of emissions reduction because of the counterfactual nature of the baseline (based on a number of AIJ energy sector projects) to be between ±35% and ±60% depending on the project type.

Table 6.2: *Characteristics of activities implemented jointly projects*

Number of projects	94	Annex I countries: 68; non-Annex I countries: 26
Investors		Public sector: 61; private firms: 32
Project types	Renewable energy: 44%; energy efficiency: 38%; forestry or agriculture: 15%	
Project life (years)	16.5[a]	Range: 1 year to 60 years
Average emission reduction (tCO_{2eq})	1,658,320	Range: 13×10^6 to 57,467,271 (tCO_{2eq})
Average investment	US$6,298,065	Range: US$73,000 to US$130,000,000
Total investment	US$558,000,000[b]	
Average cost of emission reductions	Annex I: US$97/$tCO_{2eq}$; excluding "expensive" projects: US$26/$tCO_{2eq}$	
	Other: US$158/$tCO_{2eq}$; excluding "expensive" projects: US$9/$tCO_{2eq}$	

Source: Woerdman and van der Gaast, 1999.

[a] Average lifetime of projects considered.

[b] Total investment in all projects considered.

locations caused by the project (leakage), moral hazard, and changes over time in contextual economic, technological, and institutional conditions. Several options for baseline methodologies to try to deal with these trade-offs–including sectoral benchmarks, dynamic baselines, and selective eligibility of project types–are discussed in the literature (Chomitz, 1998; Hargrave *et al.*, 1998; Jepma, 1999; Michaelowa and Dutschke, 1999a; NEDO, 2000). In addition, numerous IEA/OECD and other studies have been published on standardization of baselines for specific sectors.[72]

Also several options for baseline determination have been proposed in the literature (Chomitz, 1998; Hargrave *et al.*, 1998; Jepma, 1999; Michaelowa and Dutschke, 1999a; NEDO, 2000). Several of these proposals try to deal with the issues of adjustment for increased emissions at other locations (leakage) and changes to the baseline over time.

Regardless of the method used to develop the project baseline, the partners involved in the project, excluding the JI host government, have an incentive to propose a baseline that yields as large a reduction as possible (Bohm, 1994; Wirl *et al.*, 1998).[73]

Baseline inflation would increase the number of credits created and raise the return to investors and/or the host firm or country. To minimize the risk of baseline inflation, an independent body with the authority to review certifications could be identified or created. In the case of the CDM the entity with the authority to make the final decision will be the Operating Entity, in accordance with the executive guidelines, or the Executive Board, or the CoP/MoP (Meeting of the Parties). In the case of JI the entity will be the host government.[74] The process adopted by the independent body would also determine the transaction costs involved in defining baselines.

6.3.2.4   Experience with Activities Implemented Jointly

Decision 1/CP5 of CoP1 in 1995 established a pilot phase for emissions reduction projects called Activities Implemented Jointly (AIJ). AIJ projects cannot create credits that can be used by Parties to meet commitments under the Convention or the Kyoto Protocol. This is a crucial difference between AIJ and JI or CDM projects. *Table 6.2* summarizes the characteristics of AIJ projects.

Dixon (1999) provides a comprehensive review of the experience with AIJ projects and the implications for JI and CDM projects, illustrating the valuable experiences gained in project baseline development and monitoring. However, several authors argue that AIJ projects may not be representative of

---

[72] For example, recent OECD/IEA baseline study references are: "Revised Framework for Baseline Guidelines", "Multi-Project Emission Baselines: Iron and Steel Case Study", "Multi-Project Emission Baselines: Final Cement Case Study", "Multi-Project Emission Baselines: Forestry Status Report", "Multi-Project Emission Baselines: Final Case Study on Energy Efficiency", Multi-Project Emission Baselines: Final Electricity Case Study", and "Case-Studies on Baselines for the Project-Based Mechanisms" (see http://www.oecd.org).

[73] Parties may not respond to these incentives, for instance, if such a response is incompatible with good business practices or would generate public criticism.

---

[74] The host government for JI projects has an incentive to minimize baseline inflation only if it faces effective penalties for non-compliance. Otherwise the benefits from the project could exceed the penalties because of non-compliance. If the penalties for non-compliance by Annex I Parties are weak or poorly enforced, JI projects could be subject to an international review process with authority to establish the quantity of ERUs issued and/or the ERUs could be incorporated into the liability provisions (see Section 6.3.5.3).

future JI and CDM projects (JIQ, 1998; Trexler, 1998; Woerdman and van der Gaast, 1999). Others suggest that AIJ projects provide limited guidance on how to establish baselines for emissions reduction or sequestration projects (Ellis, 1999; Lile *et al.*, 1999).

6.3.3 Direct International Transfers

The UNFCCC states that Annex II Parties (basically Annex I Parties except for the Parties in Central and Eastern Europe) shall provide new and additional financial resources, including the transfer of technology, needed by the developing country Parties to meet the agreed full incremental costs of implementing measures taken under the Convention and that are agreed between a developing country Party and the international entity or entities referred to in Article 11 of the Kyoto Protocol (UNFCCC, 1997, Article 11). So, the extent to which developing country Parties effectively implement their commitments under the Convention will depend on the effective implementation by developed country Parties of their commitments under the Convention related to financial resources and transfer of technology.

6.3.3.1 Financial Resources

Sustainable development requires increased investment, for which domestic and external financial resources are needed, particularly for developing countries (UN, 1992, Agenda 21, Chapter 34). In its Resolution 44/228 of 1989 giving a mandate to the convening of the UN Conference on Environment and Development (UNCED) in Rio de Janeiro, the UN General Assembly notes, *inter alia*: "that the largest part of current emission of pollutants into the environment originates in developed countries, and therefore recognizes that those countries have the main responsibility for combating such pollution", and that "new and additional financial resources will have to be channelled to developing countries in order to ensure their full participation in global efforts for environmental protection." Developed country Parties reaffirmed their commitments in the related provisions of the Kyoto Protocol. "The implementation of these existing commitments shall take into account the need for adequacy and predictability in the flow of funds and the importance of appropriate burden sharing among developed country Parties" (UNFCCC, 1997, Article 11).

Accordingly, Agenda 21 (UN, 1992, Chapter 33, especially its 15th Section) carries the consensus formulation that for developing countries: "ODA is a main source of external funding, and substantial new and additional funding for sustainable development and implementation of Agenda 21 will be required." In practice, however, there has been a clear trend of a continuing decline in ODA levels since UNCED. Total ODA dropped from 0.35% of total gross national product of the developed countries in 1991 to 0.29% in 1995, with further declines in 1996 and 1997 (OECD, 1998c). Some developed countries are contributing to solving the environmental problems that developing countries face with financial resources

other than ODA. For instance, the Japanese government is implementing the Green Aid Plan that aims to achieve both economic development and environmental protection in developing countries in Asia. Most developing countries maintain that a sufficient level of financial resources is key to effective implementation of Agenda 21 and is a priority issue to be resolved to enable the implementation of the global consensus reached at the UNCED.

6.3.3.2 Technology Transfer

The transfer of environmentally sound technologies from developed to developing countries has come to be seen as a major element of the global strategies to achieve sustainable development and climate change mitigation. Article 4.5 and other relevant provisions of the UNFCCC (UNFCCC, 1992) clearly define the nature and scope of the technology transfer, which includes environmentally sound and economically viable technologies and know-how conducive to mitigating and adapting to climate change. Technology transfer implemented through the financial mechanism of the UNFCCC is to be "on a grant or concessional basis", on non-commercial terms. The Parties included in Annex II "shall take all practicable steps to promote, facilitate and finance, as appropriate, the transfer of, or access to, environmentally sound technologies and know-how to other Parties, particularly developing country Parties, to enable them to implement the provisions of the Convention." Article 10, paragraph (c) of the Kyoto Protocol (UNFCCC, 1997) reiterated that all Parties shall: "co-operate in the promotion of effective modalities for the development, application and diffusion of, and take all practicable steps to promote, facilitate and finance, as appropriate, the transfer of, or access to, environmentally sound technologies, know-how, practices and processes pertinent to climate change, in particular to developing countries, including the formulation of polices and programmes for the effective transfer of environmentally sound technologies that are publicly owned or in the public domain and the creation of an enabling environment for the private sector, to promote and enhance the transfer of, and access to, environmentally sound technologies."

Three conditions have to be fulfilled for an effective transfer of technologies. First, the technology holder country must be willing to transfer the technology. Second, the technology must fit into the demand of the recipient country. Third, the transfer must be made at reasonable cost to the recipient. The IPCC Special Report on Technology Transfer (IPCC, 2000) identifies various important barriers that could impede environmental technology transfer, such as:

- lack of data, information, and knowledge, especially on "emerging" technologies;
- inadequate vision about and understanding of local needs and demands; and
- high transaction costs.

Some analysts argue with respect to the third item that the technology should be provided on favourable terms and therefore on non-commercial conditions, strictly separated from traditional technology transfers, and supported by government funding.

In fact, Agenda 21 (UN, 1992) states that "governments and international organizations should promote effective modalities for the access and transfer of environmentally sustainable technologies (ESTs) by means of activities, including the formulation of policies and programmes for the effective transfer of ESTs that are publicly owned or in the public domain." The major role of the government could be to supply EST research and develop funds to transfer publicly owned technology to developing countries. In this regard, the Commission on Sustainable Development, at its fifth session, concluded that: "a proportion of technology is held or owned by Governments and public institutions or results from publicly-funded research and development activities. The Government's control and influence over the technological knowledge produced in publicly-funded research and development institutions opens up a potential for the generation of publicly-owned technologies that could be made accessible to developing countries, and could be an important means for Governments to catalyze private sector technology transfer." In all countries the role of publicly funded R&D in the development of ESTs is significant. Through both policy and public funding, the public sector continues to be an important driver in the development of ESTs.

An additional role of the government is to make the legal provisions for the transfer of ESTs (including checking on abuse of restrictive business practices (Raekwon, 1997)). Good governance creates an enabling environment for private sector technology transfer within and across national boundaries. Although many ESTs are in common use and could be diffused through commercial channels, their spread is hampered by risks such as those arising from weak legal protection and inadequate regulation in developed and developing countries. However, many technologies that can mitigate emissions or contribute to adaptation to climate change have not yet been commercialized. Beyond an enabling environment, it will take extra efforts to enhance the transfer of those ESTs (IPCC, 2000). It should also be recognized that the effective transfer of ESTs requires substantial upgrading of the technological capacities in the developing countries (TERI, 1997) (see also Chapters 5 and 10).

6.3.4   *Other Policies and Instruments*

6.3.4.1   *Regulatory Instruments*

There are two ways to apply regulatory instruments internationally. One is to establish uniform standards for various products and processes for adoption by countries that participate in an international emission reduction agreement. There are several reasons why establishing uniform international standards for GHGs reduction is unlikely; for example, it is difficult to achieve agreement on the appropriate standards by affected interest groups in participating countries, and such an approach

would limit the domestic policy choices of individual countries. The second way is to adopt fixed national emission levels (non-tradable emission quotas) for participating countries. These national emission limits can be considered performance standards that each country must meet through domestic action. This leads to inefficiency because marginal emission abatement costs differ among countries (IPCC, 1996, p. 404).

6.3.4.2   *International and Harmonized (Domestic) Carbon Taxes*

An international carbon tax, payable to an international agency, or domestic carbon taxes harmonized across countries, offer potentially cost-effective means of obtaining CO_2 reductions (IPCC, 1996, 11.2.2.2).[75] By associating a uniform price with carbon emissions in every country, only reductions that cost less than the tax will be implemented, assuming that the tax is implemented perfectly. To provide a common price signal in all countries, the new carbon tax may need to be differentiated across countries to account for existing domestic fuel taxes and revenue constraints (Hoel, 1993). Providing a common price signal to all sources subject to the tax also requires that all countries refrain from policies that directly or indirectly offset the tax (such as subsidies or regulations).

The revenue raised by an international carbon tax must be redistributed or used in an agreed manner. It is likely to be difficult to obtain an agreement on the share of the revenue that each country should receive. Harmonized domestic taxes avoid this difficulty by letting each country keep the revenue it collects. In practice, it is difficult also to achieve agreement on minimum levels of harmonized carbon and/or energy taxes high enough to impact carbon emissions significantly. Political pressures to combine tax proposals with exemptions for specific sectors contribute to this difficulty and, if accepted, reduce the efficiency and effectiveness of the tax.

International or harmonized taxes provide greater certainty about the likely costs of an emissions reduction programme, compared with a similarly designed international emissions trading programme (Toman *et al.*, 1999). This advantage can also be obtained by a hybrid policy, consisting of domestic emissions trading programmes coupled with a harmonized "trigger price", at which countries would sell additional permits domestically (McKibbin and Wilcoxen, 2000). The hybrid policy sets an upper bound on the marginal cost of abatement (like a carbon tax), but otherwise operates like an emissions trading programme. For a discussion of the pros and cons of such a hybrid system, see Sections 6.2.2.2 and 6.2.2.3.

The two major concerns about international price-based policies are the emissions levels, and the feasibility of international agreement:

---

[75] To improve efficiency a tax should be applied to as many sources of GHG emissions as feasible.

- The first concern is that price-based policies (taxes or hybrid systems) fail to guarantee particular emissions levels if it is not possible to adjust the tax rate frequently to achieve emission reductions in accordance with the set targets. If one assumes, for instance, that taxes are the only instrument used to fulfil the Kyoto Protocol commitments, in practice they most likely cannot guarantee that emissions commitments will be fulfilled either in the aggregate and/or for individual countries.
- The second concern is that an international agreement involving international or domestically harmonized taxes may be more difficult to negotiate than one involving emissions quotas. Wiener (1998) argues that the voluntary assent nature of international agreements means that nations must be made better off to participate, unlike domestic policies for which individuals can be coerced. While in theory international or domestically harmonized taxes can be combined with side payments to compensate losers, in practice such side payments are difficult to negotiate and tend to introduce dynamic inefficiencies since individual firms (and countries) do not bear the full social cost of their activities (Mestelman, 1982; Baumol and Oates, 1988; Kohn, 1992).[76]

Cooper (1998) takes the opposite position, arguing that taxes are the more feasible international approach. He argues that because of their rising contribution to global emissions, the participation of developing countries is essential for the long-term success of a programme to stabilize GHG concentrations in the atmosphere. He argues that it may be impossible to forge an agreement between rich and poor countries on the allocation of future quotas. Instead, "mutually agreed-upon actions", such as nationally collected emission taxes, are the logical alternative.

6.3.4.3 Standardization of Measurement Procedures

Several efforts are underway to standardize measurement procedures. For example, in the automotive industry, manufacturers from Europe, Japan, and the USA, jointly with respective governments, are trying to harmonize exhaust emission measurement methods for heavy-duty diesel vehicles (such as actual running conditions, measurement equipment, and procedures) by 2006. If successful, the automobile manufacturers' association intends to ask their respective governments to mandate the outcome.

Other international standards are set by the Organization for International Standardization (ISO). The ISO has begun to establish international Environmental Management standards in its 14000 series. The first standard among them (ISO 14001, Environmental Management Standard or EMS) was published in 1996 (ISO, 1996).

ISO environmental standards are framework standards and do not set any performance standards. They are flexible to facilitate application by a wide variety of organizations throughout the world. An organization can select any environmental aspects (such as emissions to air and/or water, ozone depletion, climate change, etc.) it considers important for its activities. This means that the standards may be effective as tools to cope with global warming if they are utilized for that purpose. In December 1997, the Climate Technology Initiative (CTI) of the OECD and the ISO issued a Joint Statement concerning the potential contribution of international standards to climate change (ISO, 1998a). In 1998, ISO established a Climate Technology Task Force to review the application of the ISO 14000 series to climate change (ISO, 1998b).

In January 2000, ISO's Technical Management Board established an Ad Hoc Group on Climate Change (AHGCC) to develop a comprehensive ISO strategy for climate change. While ISO has not ratified a climate change strategy, the AHGCC has identified several areas in which the development and use of ISO standards may help facilitate implementation of the UNFCCC and its Kyoto Protocol, including (among others):

- codes of practice and guidelines for accreditation bodies and operational entities;
- CDM project validation, verification, and/or certification standards; and
- GHG measurement, monitoring, and reporting standards.

6.3.4.4 International Voluntary Agreements with Industry

Several voluntary initiatives that have an international impact have been identified. For instance, various multinational firms have undertaken voluntary actions to cope with climate change, including setting up emissions trading systems and engaging in trades.

A VA was concluded in July 1998 between the EC and the European Automobile Manufactures Association (ACEA). The EC subsequently negotiated a similar agreement with Japanese and Korean car manufacturers. The agreements are expected to reduce CO_2 emissions from new cars in 2008 by 25% below the 1995 level. Implementation is contingent on several preconditions, such as fuel quality improvement. The EC has been engaged in discussions with European industry associations regarding a possible VA on energy efficiency in televisions and videocassette recorders (EEA, 1997).

The United Nations Environment Program (UNEP) Statement by Financial Institutions on the Environment and Sustainable Development and the UNEP Insurance Initiative may be clas-

---

[76] Unlike side payments on a lump-sum basis, which remain efficient, side payments and/or subsidies determined by emission levels are not (because then the environmental impact of the original policy measure would be reduced). In the case of a tradable quota regime, the side payments take the form of more generous quota allocations, which are efficient, unless they are tied to emission levels.

sified as international VAs. Banks and insurance companies that sign these initiatives have to pay attention to environmental protection in their management and in their product selections and operations. These initiatives are not binding and no monitoring of conduct has been carried out. In addition, the territorial distribution of the signing banks and insurance companies is uneven; participation from developing countries and the USA is rare, and no Japanese banks signed the Financial Institutions Initiative.

Some domestic VAs may evolve as *de facto* international VAs. The Energy Star programme began in 1992 as a voluntary partnership between the US DOE, the US Environmental Protection Agency, product manufacturers, and others. Partners promote energy efficient products by labelling them with the Energy Star logo and educating consumers about the benefits of energy efficiency.[77] A similar programme has started in Japan, and several European governments and manufacturers are considering setting up similar programmes. No analyses of the costs and impacts of these programmes are available.

6.3.5   *International Climate Change Agreements: Participation, Compliance, and Liability*

6.3.5.1   *Participation*

One of the concerns in the economics literature on environmental agreements (including the UNFCCC and Kyoto Protocol) has been with increasing participation. The most obvious way in which international agreements seek to increase participation is by means of a minimum participation clause. This is an article that specifies the agreement will not be binding on any of its Parties until a large enough number of countries–and, sometimes, particular countries or types of countries–have ratified the agreement. The minimum participation clause effectively makes the obligations of each of its signatories a (non-linear) function of the total number of signatories.

The minimum participation clause can serve as a strategic device, but this need not always be the case. Suppose that the minimum participation level is given as k^+. Then, if the actual number of signatories is k, and $k < k^+ - 1$, accession by a non-signatory neither costs this country anything nor confers upon it any advantage. This is because the agreement would not yet be binding on this country. However, if $k = k^+ - 1$, then accession has a non-marginal effect on the environmental problem, for the accession will mean that *all* of the k^+ countries must undertake the measures prescribed by the treaty. One way to sustain full co-operation would be to set k^+ equal to the total number of countries affected by an environmental problem (i.e., all countries), while ensuring that every potentially participating country is better off with the agreement than without it. Obviously, the threat not to undertake any abatement for a

smaller value of k can be an important incentive for countries that consider joining the agreement to actually do so (because they believe that free-riding doesn't pay). It is therefore extremely important that this threat be credible. However, in the vast majority of cases it will not be (Hoel, 1993; Carraro and Siniscalco, 1993; Barrett, 1994).

More importantly, the actual number of Parties to an agreement usually exceeds the minimum participation level, which is another reason why the above threat mechanism cannot be used to deter free-riding. The minimum participation level clause may rather serve as a co-ordinating device than as an actual incentive to join the agreement.

The point, however, is that while agreements must offer some alternative means for deterring free-riding, often they do not. The literature on international environmental agreements therefore predicts that participation will be incomplete, and it often is. One of the few agreements that disproves this general rule is the 1987 Montreal Protocol on Substances that deplete the ozone layer (UNEP, 1987; revised and amended in 1990 and 1992). The revised Protocol contains provisions that control trade between Parties and non-Parties to the regime. Coupled with the financial resources available to developing countries Parties that are not available to non-Parties, the Party–non-Party trade provisions are widely cited as a major factor in explaining the near universal participation in the ozone regime (Rowlands, 1995). See also Chapter 10 for a further discussion on participation in international regimes.

6.3.5.2   *Compliance*

The bulk of environmental agreements cannot operate the financial "carrots" and/or trade restriction "sticks" illustrated by the ozone regime (Wiser, 1999a). The key question therefore becomes: how can compliance by all Parties be secured, given the consensual basis of international law and the reluctance of Parties to endow international bodies with legal authority to enforce the international commitments Parties have (freely) undertaken against them? The UNFCCC has near universal participation based on the traditional consensual approach buttressed by provisions that aim to facilitate developing country participation through the provision of financial and technological resources. The general nature of the commitments contained in the Convention would, in any case, prove difficult to enforce. These factors explain why Parties have not endowed the supreme body of the Convention, the CoP, with the authority to impose legally binding consequences on a Party in the event of non-compliance. Thus at present, no legal body exists to enforce compliance in the climate change context.

The quantified, legally binding commitments of the Kyoto Protocol pose a different challenge (Werksman, 1998). In the period after Kyoto, the majority of Parties signalled a clear desire to move towards a compliance system based on legally binding consequences, even though the compliance provisions of the Kyoto Protocol provide that legally binding conse-

---

[77] See http://www.epa.gov/appdstar/estar/

quences can only be adopted by means of a formal amendment to the Protocol. Be that as it may, UNFCCC negotiations on the institutions and procedures of a compliance system for the Protocol are well advanced.

Various suggestions have been put forward in the literature and by Parties for the kind of legally binding consequences deemed appropriate in the climate regime (Corfee Morlot, 1998; Wiser and Goldberg, 2000). These include the following (Grubb *et al.*, 1998; UNFCCC, 2000):

- allowing a "true-up" or grace period with opportunity to buy quotas;
- payment into a national or international compliance fund that would invest in quotas;
- issuing cautions and/or reports to motivate public pressure;
- suspending treaty privileges (such as voting or the right to nominate members for office);
- exclusion from access to the Kyoto mechanisms; and
- financial penalties and implementing trade sanctions.

As a result of the difficulties in agreeing any of these consequences, and their future enforceability, more attention has been paid to policy tools that prevent non-compliance. Again, suggestions in the literature and from the Parties focused on ensuring that emissions trading must be transparent at both the Party and entity level[78], and that emissions data, such as inventories, are publicly available. The idea being that Parties and/or firms may fear the reputation consequences of being identified as polluters. Furthermore, trading could be authorized only for eligible Parties or entities, namely those meeting some minimum standards on monitoring and reporting. Non-eligible Parties and/or entities could be suspended from the trading system.

Parties also could require that insurance be obtained for traded tonnes of emissions reductions. An extra quota reserve held for the premium payer could then be claimed if the traded tonnes fail to be verified as emission reductions. A similar proposal is to establish a "true-up" period or grace period (of some several months or years) after 2012; a party that is able to come into compliance at the end of this true-up period would be deemed to have complied with the agreement. Several other possibilities have been mentioned to enforce compliance with the Kyoto targets in a situation with IET.

6.3.5.3   Liability

Liability provisions prescribe how quotas transferred by a party that subsequently is not in compliance with its emissions limitation commitment are treated. Since the developing country hosts of CDM projects do not have emissions limitation commitments, this is not an issue for CERs once they have been certified and issued by the operational entity or the Executive Board. However, this does not deal with the question of what happens if the certification has not been undertaken to acceptable standards or if there are other significant irregularities in issuance procedures. Since both JI and IET involve only Parties with emissions limitation commitments, treatment of quotas traded using these mechanisms must be addressed if the issuer does not achieve compliance.

With regard to JI, Article 6.4 of the Kyoto Protocol specifies that if compliance by an Annex I country is questioned under Article 8, any ERUs acquired from that country cannot be used to meet the buyer's commitments under Article 3, until the question of non-compliance by the originating country is satisfactorily resolved (UNFCCC, 1997).

If the ERUs issued for JI projects are determined by an international review process, they reflect corresponding reductions of the host country's emissions and hence do not contribute to its non-compliance. However, if the decision on the quantity of ERUs issued is left to the host government and the penalties for non-compliance are weak or not effectively enforced, JI projects could contribute to non-compliance by the host country. Since any ERUs transferred must be deducted from the party's AA, they could be made subject to the liability provisions for IET.

Article 17 does not include any provisions to deal with quotas that have been transferred by a country that subsequently fails to meet its emissions limitation commitment. A number of options and variants have been proposed in the literature (Goldberg *et al.*, 1998; Grubb *et al.*, 1998; Haites, 1998; Baron, 1999; Zhang, 1999b). The proposals reflect various strategies, including seller and (its opposite) buyer liability, eligibility requirements for buyer and sellers, limits on the quantity of quota that can be sold, limiting sales to quantities surplus to estimated or actual compliance needs, or restoration of default. These approaches can be grouped into those that aim to prevent or limit the risk of non-compliance, and those designed to provide sufficient deterrence (either requiring the defaulting party to face the regimes' non-compliance system or else harnessing the market to discount quotas from those Parties considered to be most at risk).

These liability proposals differ in terms of their environmental effectiveness, impact on compliance costs of Annex I Parties, and market liquidity. The proposals can change the ratio of domestic reductions to purchased quotas used for compliance and the mix of quotas purchased. In this way they can change the distribution of costs across countries, including non-Annex I countries through the volume of CDM activity. In policy terms, it is likely that the most effective strategy would aim to combine one or more of them. Details of how this may be undertaken, as well as on how many of the proposals would be implemented in practice, are currently subject to international negotiations.[79]

---

[78] Given the wording of Article 17, the participation of legal entities in IET based on Article 17 would require an explicit decision by CoP/MoP (see also Section 6.3.1).

[79] For example, whether under a buyer liability system transferred quotas would be invalidated pro rata or in reverse chronological order.

6.4   Interrelations Between International and National Policies, Measures, and Instruments

6.4.1   *Relationship Between Domestic Policies and Kyoto Mechanisms*

It is important to consider ways in which international and national (domestic) policy instruments are likely to complement or conflict with one another in achieving GHG emissions reduction commitments at least cost. A substantial number of economic models suggest that use of the Kyoto mechanisms, established by Articles 6, 12, and 17 of the Kyoto Protocol (see Sections 6.3.1 and 6.3.2), combined with efficient domestic policies could significantly reduce the cost of meeting the emissions limitation commitments in the Protocol.[80] The results of these models rely on assumptions of perfect foresight or certainty over future levels of emissions and on fully efficient domestic mitigation policies in Annex I Parties. They also assume that developing countries will respond to the market signal given by the international market of CERs and generate CDM projects accordingly.[81] Moreover, these models implicitly assume that national economies are operating within an efficient market framework. However, when an inefficient market framework is assumed the conclusions may differ. This is an area in which further research is necessary.

Articles 6 and 12 of the Kyoto Protocol enable governments and entities of Annex I countries to support JI projects in other Annex I countries and CDM projects in non-Annex I countries, respectively, in return for emissions credits. Several countries have suggested the participation of legal entities in IET, although Article 17 (on IET) does not mention the participation of entities in IET other than Parties (see Australia *et al.* (1998) and United Kingdom of Great Britain and Northern Ireland (1998)).

The following discussion assumes that any supplementarity provisions are not binding.[82] In addition, MAC refers here to the marginal social abatement costs. Also, in this discussion it is assumed that the initial market is perfect and then how various factors influence this is assessed.

If IET under Article 17 is limited to Annex I governments, they would need to trade AAUs or introduce a domestic emissions trading scheme to equate their national MACs. Views differ as to whether national governments have the information to equate the national MACs. Experimental evidence indicates that governments have the necessary incentive when trading with other governments.[83] If both Annex I governments and legal entities are allowed to engage in IET under Article 17, this difference of views becomes academic as long as the domestic policies allow the legal entities to use the three Kyoto mechanisms as part of their compliance strategy. Government participation in the Kyoto mechanisms changes the AAs available for emissions by domestic sources.

For entities to equalize their MACs there must exist either a fully comprehensive domestic taxation system, which reflects the international price of AAUs, or open access to the international emissions market for sources of emissions and entities covered by domestic policies. In theory, several domestic policy regimes can be envisioned that would allow entities in Annex I countries to equalize their MAC so as to minimize the total cost of reduction. The implications for different types of domestic policy instruments are as follows (Dutschke and Michaelowa, 1998; Hahn and Stavins, 1999):[84]

- *Domestic tradable permits.* The domestic tradable permit programme must cover virtually all emissions sources, the cap must be set equal to the national AA after trading by the government, and the participants must be allowed to engage in international exchanges using the Kyoto mechanisms.[85] Participants would be allowed to use CERs and ERUs from CDM and JI projects towards compliance with their domestic obligations. The country could also host JI projects. Participants could also buy or sell AAUs under Article 17 if participation by legal entities was allowed.[86]
- *Domestic emissions and/or carbon tax.* The domestic emissions and/or carbon tax must cover virtually all emissions sources, and the tax must be set equal to or

---

[80] See Chapter 8 and Bernstein *et al.* (1999), Bollen *et al.* (1999), Cooper *et al.* (1999), Jacoby and Wing (1999), Kainuma *et al.* (1999), Kurosawa, *et al.* (1999), Manne and Richels (1999), McKibbin *et al.* (1999), Nordhaus and Boyer (1999), Tol (1999), Tulpulé *et al.* (1999), and Weyant and Hill (1999) for estimates of the cost savings resulting from various international quota trading arrangements. All the models assume efficient domestic policies, and international trading, within each region. The models typically have between four and 20 regions.

[81] See Baron and Lanza (2000) for a review of modelling results on the contribution of the Kyoto mechanisms.

[82] See Section 6.3.1 for a discussion of supplementarity.

[83] An experiment with emissions trading among government teams representing four Nordic countries revealed a trading efficiency of 97% (Bohm, 1997). Thus, their social MACs almost exactly equated at the national level.

[84] To explore the conditions under which different domestic policies can minimize costs with the aid of the Kyoto mechanisms, Hahn and Stavins (1999) examine pairs of countries with different combinations of domestic policies. The discussion here presumes that all Annex I Parties wish to implement domestic policies that enable all sources to equalize their MACs.

[85] A similar system could be based on voluntary agreements in which sources are allowed to trade emission reductions in the form of AAUs, ERUs, or CERs.

[86] The permits used in the domestic tradable permit system could be AAUs. Alternatively, the domestic permits could be freely exchangeable for AAUs.

less than the national marginal cost of abatement after trading by the government. Entities receive tax credits for CERs and ERUs, and for AAUs if participation by legal entities is allowed under Article 17. The country could also host JI projects.[87]

- *Non-tradable permits.* Virtually all sources are covered by non-tradable emissions limits, which allow the use of quotas to achieve compliance. The total emissions allowed under the permits must be equal to or less than national AAs after trading by the government. Entities could use purchased CERs and ERUs–and AAUs if participation by legal entities is allowed under Article 17–towards compliance with their emissions limits. The country could also host JI projects to reduce emissions below the emissions limits or to enhance sinks.

If sources are subject to regulations, design or performance standards, VAs, or taxes and at the same time there is no permit allocated to the source, and CERs, ERUs, or AAUs cannot be used for compliance, entities might still be allowed to trade them on the international market, provided that the volume sold does not exceed the volume of quotas acquired. Such domestic policies are unlikely to equate MACs across sources and so will not result in the lowest cost of compliance with the national emissions limitation commitment.

In practice, the combination of domestic policies and Kyoto mechanisms necessary to achieve cost-effectiveness may not be implemented for at least two reasons. First, use of the Kyoto mechanisms may be restricted in some countries, either because supplementarity restrictions are binding or because a national government that imposes an emissions tax may limit the use of the mechanisms towards compliance with tax liabilities to protect its revenue.

Second, it is difficult to cover all sources and relevant sinks with policies that provide an incentive to implement measures that equate MACs. Some sources are small and are excluded for administrative reasons. Other sources, such as methane emissions by livestock, are difficult to include in a trading or tax regime. Thus, the overall cost-effectiveness of the system will fall short of the theoretical ideal.

When part of the GHG emissions reduction needed to realize the Kyoto commitments offers net economic benefits to the national economy, the role of the Kyoto mechanisms changes significantly. Relative to a theoretical scheme of complete and perfect trading, a purely national mitigation strategy would still give rise to inefficiencies for individual countries or sources, as a result of differentials in MACs. However, the advantages that could be obtained from eliminating such inefficiencies through international mechanisms are more limited because of principal-agent problems.

Thus, if access to the international mechanisms is limited to governments, the Kyoto mechanisms are likely to be used only to reduce positive marginal domestic abatement costs. And, since measures with positive costs under a regime with restrictions make up only a fraction of total mitigation under an efficient domestic policy, the quantitative significance of the Kyoto mechanisms is greatly reduced. If access to the Kyoto mechanisms is given to individual sources, there arises the potential for a second principal-agent problem in that individual entities may mitigate in ways that minimize private costs but fail to minimize social costs in the national economy. In this case, both international efficiency and domestic efficiency are jeopardized.

6.4.2 Conflicts with International Environmental Regulation and Trade Law

Compatibility of environmental protection with free trade and/or investment has been important in both the environmental and trade fields. The Committee on Trade and Environment of the WTO has under discussion the relationships between the provisions of the multilateral trading system and trade measures for environmental purposes, including those pursuant to multilateral environmental agreements (MEAs). Also under discussion are the relationships between environmental policies and measures with significant trade effects and the provisions of the multilateral trading system. Some analysts suggest that the WTO is not an appropriate forum to resolve these questions and propose the establishment of a multilateral environmental organization for this purpose (Esty, 1994).

The UNFCCC is one of more than 200 multilateral and bilateral international environmental agreements (MEAs) whose compatibility with free trade and investment is debated (UNEP, 1983, 1991). More than 20 MEAs incorporate explicit trade measures.[88] Other MEAs address the need to co-ordinate restrictions on conduct taken in compliance with obligations they impose and the expanding regime of trade and investment law under the WTO/GATT umbrella.[89] UNFCCC Article 3.5 (UNFCCC, 1992), following GATT Article XX, stipulates that "Parties shall co-operate to promote a supportive and open international economic system that would lead to

---

[87] Only sources or sinks not subject to the tax are likely to be approved as JI projects, to reduce the risk of double counting.

[88] See Ward and Black (2000, p. 122). Some MEAs, like the 1973 Convention on International Trade in Endangered Species (CITES, 1973), the 1989 Basel Convention on the Control of Transboundary Movements of Hazardous Waste (UNEP, 1989), or the 1987 Montreal Protocol on Substances that Deplete the Ozone Layer Articles, 4.1, 4.2, and 4.3, restrict trade in polluting products or products that contain controlled substances (UNEP, 1987). Some, like Montreal Protocol Article 4.4, propose trade sanctions, but these are, however, not implemented, even on products manufactured with polluting substances.

[89] Agenda 21 (UN, 1992), Chapters 2.3, 2.11, 2.20 and 17; Principles 11 and 12 of the Rio Convention; Convention on Biodiversity, Article 22.1.

sustainable economic growth and development in all Parties ... Measures taken to combat climate change, including unilateral ones, should not constitute a means of arbitrary or unjustifiable discrimination or a disguised restriction on international trade."

There are no presently cited cases of trade claims against measures enacted in widely subscribed MEAs like the CITES or the Montreal Protocol. Neither the UNFCCC nor the Kyoto Protocol now provides for specific trade measures. The debate over conflicts between trade and MEAs stems from the prospect that trade-related measures might be enacted to limit trade in polluting products and in endangered species, or trade in goods created by means of polluting processes and production methods (PPMs). MEAs could also require general or specific trade measures to sanction non-Parties to the MEA or non-compliant MEA members.

IET under Article 17 of the Kyoto Protocol has raised questions of WTO compatibility. Early analysis concludes that the rules governing the transfer and mutual recognition of allowances are not covered by WTO because they are neither products nor services (Werksman, 1999).[90] However, several domestic policies and measures that may be taken in conjunction with the Kyoto Protocol might be considered to pose WTO problems, such as excessively restricting trade regulations, GATT-inconsistent border charges, or illegal subsidies.[91]

National programmes of permit distribution for emissions trading (see Section 6.2.2.3) or national environmental aid (subsidies) might benefit domestic firms or sectors over importers or foreign competitors (Black *et al.*, 2000).[92] In addition, a Party or group of Parties (as part of the national implementation pro-

grammes) might apply taxes or environmental policies and measures in a way that arguably discriminates against WTO trade partners. Environmental regulations, taxes, or voluntary measures could be challenged as indirect forms of protection that fall disproportionately on imported products. Recent cases suggest more cases could be argued under the agreement on Technical Barriers to Trade (TBT) rather than GATT.[93]

Trade-related environmental measures traditionally pose problems for the multilateral trading regime. Considerations of sovereignty favour the autonomy of WTO Parties to set health or environmental standards for all products, domestic or imported, consumed in their national territories. Each country has broad discretion to introduce its own policies and measures, including energy efficiency standards and import restrictions, to protect its environment and/or its people's health, subject to GATT Article 3 (national treatment). However, more debatable is whether GATT permits a government to place restrictions or bans on the import of goods or services, themselves not dangerous or polluting, that are produced outside its borders through PPMs that do not meet its national environmental regulations or standards. PPM issues may be characterized as "clean products produced through dirty processes". As MEAs increasingly utilize trade measures to prevent non-members from free-riding, the consistency of such trade measures with the relevant GATT articles (Article XX, in particular) has been questioned when they are based on the lack of corresponding PPM requirements in the exporting countries (Murase, 1995, 1996). At present, the relation between WTO-compatible environmental measures and MEAs remains unsettled. It is also unclear whether WTO law is neutral in its treatment of alternative trade-related measures (e.g., standards, taxes, and subsidies).

Prior to 1995, when GATT 1994 replaced GATT 1947 under the WTO agreement, six panel reports involved environmental issues related to trade measures under Article XX (Ahn, 1999). The Appellate Body under the revised WTO dispute settlement system has since decided two further cases.[94] While none of these disputes challenged the environmental objectives pur-

---

[90] Black *et al.* (2000) note that if emissions trading is treated as a financial service, there is no clear policy reason to exclude non-Parties to the Kyoto Protocol from trading in these markets.

[91] For example, a Party might impose equivalent product-specific energy-efficiency standards on domestic and imported refrigerators or automobiles. Or a Party might ban the domestic production and import of rice grown under methane intensive cultivation methods or of wood harvested under non-sustainable forestry practices. Alternatively, a party might impose a tax on the carbon content of domestic and imported fuels or the carbon consumed in the production of national and imported products. Finally, a Party might impose countervailing duties against imports from nations that do not force the internalization of GHG emissions costs on national producers.

[92] National energy policies have long been replete with distortionary subsidies (Black *et al.*, 2000, pp. 90–98). However, since even subsidies that encourage production below marginal factor costs have rarely been GATT challenged, it is unlikely that national policies that fail to internalize full environmental costs will be GATT illegal, unless they explicitly discriminate between national production and imports. Energy efficiency subsidies to internalize environmental benefits are, in principle, permissible under the GATT subsidies code.

[93] For example, when the EC concluded a voluntary agreement with ACEA to reduce CO$_2$ emissions from automobiles in February 1999, the Commission asked non-EU automobile manufactures to conclude the same kind of agreements, fearing that the European car manufacturers might lose international competitiveness. As a result, in October 1999, the Japanese Automobile Manufacturing Association's voluntary commitment to follow the same standards was approved by the EU. When the Japanese government enacted an amendment to strengthen fuel efficiencies of automobiles, both European and American governments expressed their concern, through formal TBT procedure, that it would become an invisible trade barrier for automobile export. Also, when the EC intended to propose a Directive on Waste Electrical and Electronic Equipment, both the US and Japanese governments expressed the same concern.

[94] Perkins (1999); see also United States -Standards for Reformulated and Conventional Gasoline (World Trade Organization, 1996).

sued by the governments concerned, all rulings found that the contested trade restrictions were in some respect discriminatory or unnecessarily trade restrictive. However, more recent rulings, including those of the Appellate Body, have narrowed or rejected earlier panel interpretations that had held PPMs either *per se* inconsistent with the intent of GATT or highly restricted by the terms of Article XX.

The GATT Panel rulings in the Tuna–Dolphin I dispute read Articles XXb and XXg so as to preclude provisional justification for extraterritorial PPMs as inherently arbitrary measures destructive to the system of international trade.[95] The panel in the Shrimp–Turtle dispute also explicitly held that the US shrimp embargo belonged to the class of measures (PPMs) that threatened the multilateral trading system and therefore violated the terms of the Chapeau of Article XX. However, the WTO Appellate Body overruled the Panel's view in the Shrimp–Turtle case. The WTO indicated implicitly that it does not categorically disallow the use of extrajurisdictional PPMs.[96] Although the import restrictions in question applied to shrimp harvesting practices and not to any characteristic of the shrimps themselves, the Appellate Body treated the measure as provisionally justifiable. It considered the legality of the specific restrictions, which were held to be invalid under the prohibition of the Chapeau of Article XX of discriminatory and arbitrary measures. The US embargo was ruled overly broad, its enforcement inflexible in considering the conservation effects of other nations' shrimping practices, disparate in its treatment of other nations, deficient in due process, and put into effect without sufficient good-faith efforts to secure wider multilateral acceptance of its exclusionary programme (Berger, 1999).

Although as yet there is no universally accepted interpretation of the Shrimp–Turtle Appellate Body decision, some analysts suggest the holding implies PPMs no longer violate WTO by their very nature (Ahn, 1999). Others argue such a conclusion is premature legally or has been insufficiently debated and tested in the scientific literature (Jackson, 2000). In either case, the ruling did not refer to important questions relevant to the interaction of WTO and the UNFCCC and/or Kyoto Protocol. It is unclear whether national PPMs need only be enacted by Parties to an MEA in their compliance programmes, or whether each particular PPM, its mode of application, and/or its sanction scheme are the subjects of multilateral accord. Nor is it certain how widely the multilateral agreement that supports the PPM must be subscribed to make it WTO compatible.[97]

Parties to MEAs might base national climate programmes on pollution taxes rather than product or PPM standards. WTO law does allow compensating charges or border adjustments to similar imported products to equalize the tax burden on domestic production. While direct taxes (wages, incomes) may not be compensated on imports or refunded on exports, certain indirect taxes, such as sales taxes or excises, may be adjusted at the border.[98] Indirect environmental taxes levied on a locally polluting product like imported fuel or gas guzzling automobiles, as long as not in excess of charges imposed on like domestic products, would be WTO consistent. Analogous indirect taxes, equal to domestic taxes, imposed on non-locally polluting imports produced through foreign process and production methods that were environmentally damaging have been approved in the GATT dispute settlement process.[99]

Nevertheless, some border charges on products manufactured through GHG-intensive PPMs might be WTO inconsistent. Although specific taxes on final products (e.g., fuels) and on "goods physically incorporated" into final products (e.g., a feedstock or catalyst) may be adjusted at the border, so-called hidden taxes on inputs, such as transport, machinery, advertising, or energy entirely consumed during production, have not been legally adjustable. Current practice is not fully symmetrical in its treatment of regulatory standards and taxes as environmental instruments.[100]

---

[95] United States–Restrictions on Imports of Tuna (World Trade Organization, 1994).

[96] See Shrimp–Turtle decision, paragraphs 121 and 187b (World Trade Organization, 1999); see also Perkins (1999, p. 119).

[97] Nor is there yet guidance whether trade restrictions could be enforced against a UNFCCC party with differentiated responsibilities under the MEA, even if WTO-legal restrictions against imports from non-Annex I Parties would confuse the meaning of "differentiated responsibilities".

[98] Economists have long noted the lack of precision of the categories direct and indirect taxes, as well as the dependence of the ability to pass on the incidence of taxes to consumers (indirect taxes) on market structure. However, the terms continue to be applied with reasonable ease in legal practice (Demaret and Stewardson, 1994, pp. 14–16).

[99] In the Superfund Tax case, US border charges on certain waste creating feedstock chemicals used as inputs in the processing of imported chemical derivative products were ruled to be legal. These border charges, equal to taxes imposed on similar US feedstock, were held valid even though there was no transboundary damage outside the nation of origin (World Trade Organization, 1988). A border tax on shrimp caught with turtle-unsafe methods and similar to a domestic tax on such products would seem to fall under this rule. Products that have been produced through differential production methods, like products that have different environmental qualities in themselves, have usually been considered to be not "like products" and therefore allowable objects of differential, non-discriminatory taxes (Demaret and Stewardson, 1994, pp. 34–41).

[100] A limited amendment to the treatment of energy taxes was made in the Uruguay 1994 Agreement on Subsidies and Countervailing Measures. Border adjustments were allowed for those nations that still imposed cumulative prior-stage indirect taxes on "energy, fuels, and oil used in the production process". This exception to the hidden tax rule was intended to cover only a limited set of nations (Demaret and Stewardson, 1994, pp. 29–30).

6.4.3   *International Co-ordination of Policy Packages*

When developing domestic policies to meet their emissions limitation commitments under the Kyoto Protocol, some Annex I Parties may wish, or be under pressure, to impose less stringent obligations on some industries to improve their competitiveness. The sensitivity of industry location to the stringency of environmental regulation is called "ecological dumping". International co-ordination of environmental policies may be needed to reach an economically efficient outcome in which it is impossible to make one country better off without making at least one other country worse off.

Under certain ideal conditions (e.g., perfect competition in all markets) there is theoretically no need for international policy co-ordination (Oates and Schwab, 1988). However, such conditions do not hold if there is imperfect competition in goods markets or unemployment (Rauscher, 1991, 1994; Barrett, 1994; Kennedy, 1994; A. Ulph, 1994; D. Ulph, 1994). If, and to what extent, international differences in environmental regulation have trade or even relocation implications obviously depends on a host of factors. These include country size, availability of alternatives, relative resource endowment, mobility of production factors, competition level, scope for innovation, possibility of border-tax adjustment, chances of retaliation, and redistribution of environmental tax revenues (OECD, 1996a).

Although it is clear that many factors affect the relationship between the stringency of pollution control policies (if implemented unilaterally) and net exports, some authors have carried out rather straightforward empirical tests on the relationship between the two variables. Han and Braden (1996) examined 19 US manufacturing industries between 1973 and 1990 with the help of regression analysis. They found the relationship between pollution abatement costs and net exports to be negative in most of the sample period, but diminishing over time (with elasticities close to zero in many industries). Van Beers and Van den Bergh (1997), using a gravity model of international trade and two measures of environmental stringency, did not find a significant relationship between environmental stringency and total exports for the "dirty" industries. However, when they focused on the non-resource based, and therefore more "mobile", industries only this relationship was significant.

Early empirical research on the impact of environmental policy on trade found little evidence of a measurable relationship, partly because of low environmental taxes and partly through data and statistical limitations. Therefore, many studies have concentrated on simulations of environmental tax regimes. From a survey of these studies, IPCC's SAR (IPCC, 1996) concluded that estimates of the effects of environmental policies (notably carbon taxes) on trade vary wildly, depending on model parameters (such as energy demand elasticities and assumptions regarding the substitutability of traded goods) and the policy scenario examined (extent of reduction in emissions and extent of international co-ordination).

Various partial equilibrium models have been designed to analyze ecological dumping, many using static or dynamic game theory. Early analyses used a Cournot setting, which models long-run competition among firms as a series of strategic capacity or output choices. The general conclusion from these early models is that the optimal tax (or any comparable domestic environmental policy instrument) would be set below marginal damage. As a consequence, environmental policies are designed to try to protect domestic industries. If producers collude, however, the incentive for governments to engage in ecological dumping is reduced (Ulph, 1993).

The ecological dumping conclusion could change completely if governments act strategically in setting taxes, and if there is Bertrand competition (firms compete by choosing the price to charge, rather than the quantity to produce) instead of Cournot competition (Eaton and Grossman, 1986; Barrett, 1994; Conrad, 1996; Ulph, 1996). If, however, producers act strategically or can collude, then the outcome in terms of ecological dumping is not straightforward. Quantity-based environmental regulation, if implemented unilaterally in a duopolistic case with a domestic and foreign supplier, might actually benefit domestic firms at the cost of domestic consumers (Kooiman, 1998). If both governments and producers act strategically, again, the incentive for governments to distort the environmental policy is less than when only governments acted strategically, so that the Bertrand outcome can be similar to the Cournot outcome (Ulph, 1996).

Ecological dumping also has been analyzed with the help of general equilibrium models of international trade involving externalities (Rauscher, 1994). It was shown that in a second-best world for several market structures–monopoly power of the exposed sector or oligopoly on an outside market (Elbers and Withagen, 1999)–ecological dumping might not (always) be beneficial from a welfare point of view. This is contrary to the conclusions of some of the earlier partial equilibrium models,

The most interesting case for analyzing policy co-ordination needs is that in which national commitments have been decided internationally, but individual Parties may, but need not, co-ordinate their national policies to fulfil their commitments. This would be the Kyoto Protocol case, after ratification. Hoel (1997) has addressed this case and argues that governments may tend to subsidize indirectly particular imperfectly competitive industries selling on the international market. To prevent this from happening, an argument can be made in favour of policy co-ordination, which is possible but not required in the Kyoto Protocol, except insofar as the Kyoto mechanisms are concerned.[101]

6.4.4   *Equity, Participation, and International Policy Instruments*

The participation of developing countries and EITs in the UNFCCC is important, since these countries are both large future emitters of carbon, and sources and potential sources of low-cost abatement investments (McKibbin and Wilcoxen, 2000). Since the participation of regions with low marginal abatement costs may be critical for aggregate cost and emissions reduction, encouraging their participation may require a serious consideration of the equity implication of that policy (Morrisette and Plantinga, 1991). Unlike efficiency, there is no universal consensus definition of equity by which policy instruments can be evaluated. Recent research on equity, however, analyzed the welfare impacts of climate policy alternatives to understand the participation incentives (for different countries and regions) of various policy instruments (Bohm and Larsen, 1994; Edmonds *et al.*, 1995; Rose *et al.*, 1998).

The types and structures of mechanisms adopted (such as uniform taxes, tradable quota, or individual non-tradable targets) affect the scope and timing of participation in some predictable ways (Edmonds *et al.*, 1995). For example, individual non-tradable targets based on the stabilization of national emissions would shift more than 80% of aggregate costs to non-OECD regions by 2020, making it unlikely that these regions would participate in such an agreement (Edmonds *et al.*, 1995).

Alternatively, with a common global carbon tax and full participation, the burden of abatement costs would be distributed unevenly across the world and would change with time. A large burden would fall on OECD and economies in transition in the early years, shifting to developing nations in later years (Edmonds *et al.*, 1995). Transition economies would thus be unlikely to participate in a common global carbon tax agreement. If such nations were to participate in the short run, growth and changing economic and political circumstances may increase the probability of their dropping out of a tax agreement when they face increasing net participation costs (Edmonds *et al.*, 1995).

The equity implications of a global tradable quota system depend on quota allocation. The portion of global abatement costs borne by a country or group of countries depends on its relative position in the quota market; net sellers of quota effectively receive income transfers from net buyers. *Table 6.3* describes the relative position of groups of countries in an international quota market, based on six possible initial allocations (Edmonds *et al.*, 1995).[102]

Of course, a country's participation in an allocation scheme depends on net costs (the sum of transfer payments associated with quota trade, plus direct mitigation costs), not just the direction of income transfers. However, that the direction of transfers may change over time, especially for China and the transition economies, complicates the incorporation of equity goals in quota system design (Edmonds *et al.*, 1995). Although quota allocation is referred to here, the analysis applies equivalently to redistributing international carbon tax revenues (Pezzey, 1992; Rose *et al.*, 1998).

Bohm and Larsen (1994) explore the participation implications of two of the more frequently discussed of the allocation schemes listed in *Table 6.3* (allocation by population and by GDP) for a quota regime covering Western Europe and Eastern Europe. Both of these allocations, and combinations thereof, lead to substantial losses by the Eastern European countries, making their participation unlikely (Bohm and Larsen, 1994). Given the aggregate cost-savings associated with their participation, an ideal allocation system would provide the minimum possible participation incentive to the Eastern European countries, while maximizing potential abatement cost savings to the western countries. The authors identify this lower bound in terms of eastern country quota-to-emissions ratios that would induce participation, ranging among countries from 0.85 to 0.91. This incentive scenario results in zero net gains (losses) to the eastern countries, and net costs to each western country of 0.09% of GDP. In the presence of wide disparity in current regional economic welfare, the perceived equity benefits of such a scenario may facilitate a more cost-effective agreement than any that might be achieved without Eastern European participation.[103]

If quota allocations are used to induce participation by transition economies and developing countries, the international wealth transfers that occur as a result may cause fluctuations in real exchange rates and international capital and trade flows (McKibbin and Wilcoxen, 1997a, 1997b). The magnitude of these fluctuations and the extent to which they could be problematic are uncertain. McKibbin and Wilcoxen (2000) suggest an alternative approach to the problem of equity versus participation incentive, which includes both short-run emissions quota and long-run emissions "endowments". In this approach, the price of emissions quota is set through international negotiation at regular intervals (they suggest every decade), and each country issues as many quotas as necessary to keep the price at the negotiated level. The price of emissions endowments, however, is flexible, and the quantity allowed per country is fixed. Each participating country's endowment prices reflect expected future prices of emissions quota.

---

[101] Hoel (1997) uses a simple model of a group of identical countries that interact through mobile real capital. Given the total stock of real capital for the group of countries as a whole, he demonstrates that if competition in the goods markets is imperfect or if unemployment exists, a lack of international policy co-ordination may lead to outcomes that are not Pareto optimal. However, he finds there is no need for policy co-ordination if after-tax wages are exogenous, but this seems to be a rather strong assumption.

[102] Rose *et al.* (1998) analyzed the welfare impacts of various tradable permit allocations and obtained results that are consistent with many of the results of Edmonds *et al.* (1995).

---

[103] GDP per capita in 1989 ranged from US$1,200–1,500 in Albania, Turkey, Tajikistan, and Uzbekistan to more than US$20,000 in Switzerland, Luxembourg, and the Scandinavian countries (Bohm and Larsen, 1994).

Table 6.3: Direction of income transfers in international emissions trading, six possible quota allocation schemes

| Tradable quota allocation | Anticipated position of participating countries, 2005-2095 | | | |
	OECD countries	EITs	China and other centrally-planned Asian countries	Rest of world
Grandfathering	Net sellers	Net sellers	Net buyers	Net buyers
Equal per-capita emissions	Net buyers	Net buyers	Net sellers early, Net buyers post-2035	
GDP-weighted emissions	Net sellers	Net effect small and ambiguous	Net buyers	Net effect small and ambiguous
GDP-adjusted grandfathering	Net buyers	Net sellers	Net effect small and ambiguous	Net effect small and ambiguous
No harm to developing nations	Net buyers	Net sellers early, net buyers post-2035	Net sellers	Net sellers
No harm to non-OECD nations	Net buyers	Net sellers early, net buyers post-2035	Net sellers	Net sellers

Source: Edmonds *et al.* (1995).

Notes: Under GDP-adjusted grandfathering, emissions rights have a baseline at current levels, adjusted for income growth. The "no harm" scenarios allocate sufficient quota to the relevant countries to cover their own emissions and to generate enough revenue to cover economic costs of protocol participation.

6.5   Key Considerations

This section deals with the most important aspects that could be considered in designing climate change policy.

6.5.1   *Price versus Quantity Instruments*

Optimal climate change policy–irrespective of whether it is national or international–under uncertainty and/or asymmetric information deviates from more typical analyses with best-guess parameter values and/or information symmetry, not only in terms of the stringency of policies, but also in terms of policy design (Weitzman, 1974). Depending on the degree of uncertainty and correlation between the marginal damage and MAC curves, taxes could be a better or inferior alternative to tradable permits (Watson and Ridker, 1984; Stavins, 1996).[104] Recent literature shows that taxes dominate quotas for the control of GHGs when the environmental damage function is rather flat (Hoel and Karp, 1998). Hoel (1998) and Pizer (1997b) point out that the lack of a clear, short-term threshold for severe climate damages favours the use of market-based

policies, like taxes, that limit cost uncertainty. In addition, there is mounting evidence that rigid emission limits are not appropriate in the short run under a weak emissions reduction regime (Newell and Pizer, 1998).

Recently, Pizer (1997a) argued that excluding uncertainty might lead to policy recommendations that are too lax. Ebert (1996) has argued that improving the information of the regulator is crucial, because decision makers always overestimate abatement costs if they neglect that firms possess an abatement option other than decreasing output–additional abatement technology.

To increase the effectiveness and efficiency of domestic GHG emissions reduction policies, it is argued that governments could adopt policies that take a comprehensive approach, stimulating the development of all kinds of new materials, materials substitution, product re-design, resource productivity, and waste management strategies that can reduce GHG emissions. Moreover, governments could set long-term GHG emissions reduction targets, since the optimal set of technical options at low GHG mitigation levels may not include options that are efficient at high GHG emissions reduction levels.

---

[104] Montero (2000b) finds that under incomplete enforcement tradable permits perform relatively better than taxes.

6.5.2 Interactions of Policy Instruments with Fiscal Systems

It is important to consider how the domestic policy instruments examined in this chapter may interact with existing fiscal systems, because such interactions can have significant effects on the overall costs of achieving specified GHG emissions reduction targets. A growing literature demonstrates theoretically, and with numerical simulation models, that the costs of addressing GHG targets with policy instruments of all kinds–command-and-control as well as market-based approaches–can be greater than anticipated because of the interaction of these policy instruments with existing domestic tax systems.[105] Domestic taxes on labour and investment income change the economic returns to labour and capital, and distort the efficient use of these resources.

The cost-increasing interaction reflects the impact that GHG policies can have on the functioning of labour and capital markets through their effects on real wages and the real return to capital (see, e.g., Parry *et al.*, 1999). By restricting the allowable GHG emissions, permits, regulations, or a carbon tax raise the costs of production and the price of output, thus reducing the real return to labour and capital, and exacerbating prior distortions in the labour and capital markets. Thus, to attain a given GHG emissions target, before or after use of IET and other Kyoto mechanisms, all the instruments have a cost-increasing "interaction effect".

For policies that raise revenue for the government, carbon taxes and auctioned permits, this is only part of the story, however. These revenues can be recycled to reduce existing distortionary taxes. Thus, to attain a given GHG emissions target, revenue-generating policy instruments have the advantage of a potential cost-reducing "revenue-recycling effect" as compared to the alternative, non-auctioned tradable permits or other non-revenue-generating instruments (Bohm, 1998). For a more complete theoretical discussion, see Chapter 7, and see Chapter 8 for the empirical results.

6.5.3 The Effects of Alternative Policy Instruments on Technological Change

In the long run, the development and widespread adoption of new technologies can greatly ameliorate what, in the short run, sometimes appear to be overwhelming conflicts between economic well being and environmental quality. Therefore, the effect of public policies on the development and spread of new technologies may be among the most important determinants of success or failure in environmental protection (Kneese and Schultze, 1975).

To achieve widespread benefits from a new technology, three steps are required (Schumpeter, 1942):
- invention, the development of a new technical idea;
- innovation, the incorporation of a new idea into a commercial product or process and the first marketplace implementation thereof; and
- diffusion, the typically gradual process by which improved products or processes become widely used.

Rates of invention, innovation, and technology diffusion are affected by opportunities that exist for firms and individuals to profit from investing in research, in commercial development, and in marketing and product development (Stoneman, 1983).

Governments often seek to influence each of these directly, by investment in public research, subsidies to research and technological development, dissemination of information, and other means (Mowery and Rosenberg, 1989). Policies with large economic impacts, such as those intended to address global climate change, can be designed to foster technological invention, innovation, and diffusion (Kemp and Soete, 1990). For the impact of R&D policies on technology development and transfer, see the IPCC Special Report on Technology Transfer (IPCC, 2000).

To examine the link between policy instruments and technological change, environmental policies can be characterized as market-based approaches, performance standards, technology standards, and voluntary agreements. All these forms of intervention have the potential to induce or force some amount of technological change, because by their very nature they induce or require firms to do things they would not otherwise do. Performance and technology standards can be explicitly designed to be "technology forcing", mandating performance levels that are not currently viewed as technologically feasible or mandating technologies that are not fully developed. The problem with this approach can be that while regulators typically assume that some amount of improvement over existing technology will always be feasible, it is impossible to know how much. Standards must either be made not very ambitious, or else run the risk of being ultimately unachievable, which leads to great political and economic disruption (Freeman and Haveman, 1972). However, in the case of obstructed technology, regulators know quite well the technology improvements that are feasible. Thus, although the problem of standards being either too low or too ambitious remains a possibility, it does not make standards inherently incapable of implementing some portion of the available technology base, and to do so cost-effectively on the basis of cost–benefit tests.[106]

---

[105] For the basic analysis and economic intuition of this literature, see Kolstad (2000, pp. 281–284).

[106] There is, however, an interesting example in which ambitious standards were finally achieved. New emission standards for passenger cars (the so-called "Muskie" standard), when first enacted in the USA in 1970, were thought to be too ambitious because no such technologies existed in the world. However, a technology breakthrough by two automobile manufacturers in Japan achieved the standard (Honma, 1978; OECD, 1978).

6.5.3.1   Theoretical Analyses

Most of the work in the environmental economics literature on the dynamic effects of policy instruments on technological change has been theoretical, rather than empirical, and the theoretical literature is considered first. The predominant theoretical framework involves what could be called the "discrete technology choice" model. In this, firms contemplate the use of a certain technology that reduces the marginal costs of pollution abatement and that has a known fixed cost (Downing and White, 1986; Jung *et al.*, 1996; Malueg, 1989; Milliman and Prince, 1989; Zerbe, 1970).

While some authors present this approach as a model of innovation, it is perhaps more useful as a model of adoption.[107] The adoption decision is one in which firms face a given technology with a known fixed cost and certain consequences, and must decide whether or not to use it; this corresponds precisely to the discrete technology choice model. Innovation, on the other hand, involves choices about research and development expenditures, with some uncertainty over the technology that will result and the costs of developing it. Models of innovation allow firms to choose their research and development expenditures, as in Magat (1978, 1979), or incorporate uncertainty over the outcome of research (Biglaiser and Horowitz, 1995; Biglaiser *et al.*, 1995).

Several researchers have found that the incentive to adopt new technologies is greater under market-based instruments than under direct regulation (Downing and White, 1986; Jung *et al.*, 1996; Milliman and Prince, 1989; Zerbe, 1970). This view is tempered by Malueg (1989), who points out that the adoption incentive under a freely allocated tradable permits system depends on whether a firm is a buyer or seller of permits. For permit buyers, the incentive is larger under a performance standard than under tradable permits.

Comparisons among market-based instruments are less consistent. Downing and White (1986), who consider the case of a single (sole) polluter, argue that taxes and tradable permit systems are essentially equivalent. On the other hand, Milliman and Prince (1989) find that auctioned permits provide the largest adoption incentive of any instrument, with emissions taxes and subsidies second, and freely allocated permits and direct controls last. Jung *et al.* (1996) consider heterogeneous firms, and model the "market-level incentive" created by various instruments. This measure is simply the aggregate cost savings to the industry as a whole from adopting the technology. Their rankings echo those of Milliman and Prince (1989).

On the basis of an analytical and numerical comparison of the welfare impacts of alternative policy instruments in the presence of endogenous technological change, Fischer *et al.* (1998) argue that the relative ranking of policy instruments depends critically on firms' ability to imitate innovations, innovation costs, environmental benefit functions, and the number of firms that produce emissions.[108] Finally, the study includes an explicit model of the final output market, and finds that it depends upon empirical values of the relevant parameters whether (auctioned) permits or taxes provide a stronger incentive to adopt an improved technology.

Finally, recent research investigates the combined effect of the pollution externality and the positive externality that results from learning-by-doing with mitigation technologies. Since the benefit from learning occurs after the learning has taken place, a dynamic analysis is needed. Some analyses shown that dynamic efficiency (discounted least cost, aggregated over time) requires that the incentive for emissions-mitigating innovations be set higher than the penalty on emissions, especially if account is taken of "leakage". This is in contrast with the conclusions of comparative static analysis upon which most environmental policy analysis is grounded (e.g., Baumol and Oates, 1988), under which the two incentives should be equal in all time periods (for a formal analysis, see Read (1999, 2000)).

6.5.3.2   Empirical Analyses

Empirical analyses[109] of the relative effects of alternative environmental policy instruments on the rate and direction of technological change are limited in number, but those available focus on technological change in energy efficiency, and thus are potentially of direct relevance to global climate policy. These studies can be considered within the three stages of technological change introduced above–invention, innovation, and diffusion. It is most illuminating, however, to consider the three stages in reverse order.

Beginning, then, with empirical analyses of the effects of environmental policy instruments on technology diffusion, Jaffe and Stavins (1995) conducted econometric analyses of the factors that affected the adoption of thermal insulation technologies in new residential construction in the USA from 1979 to 1988. They examined the dynamic effects of energy prices and technology adoption costs on average residential energy-efficient technologies in new home construction. The effects of

---

[107] Zerbe (1970) couches his research in terms of adoption. Downing and White (1986) frame their work in terms of innovation. Milliman and Prince (1989) use one model to discuss both diffusion and innovation, the latter being defined essentially as the initial use of the technology by an "innovating" firm. Malueg (1989) presents the same framework as a model of adoption. Jung *et al.* (1996) present their model as one of either adoption or innovation.

[108] Related to this is the finding of Parry (1998) that the welfare gain induced by an emissions tax is significantly greater than that induced by other policies only for very major innovations. Also related is Montero's (2000c) conclusion that the relative superiority of alternative policy instruments in terms of their effects on firm investments in R&D depends upon the nature of the underlying market structure. This is implied by Laffont and Tirole (1996).

[109] For further literature references, see Chapter 8.

energy prices can be interpreted as suggesting what the likely effects of taxes on energy use would be, and the effects of changes in adoption costs can be interpreted as indicating what the effects of technology-adoption subsidies would be. They found that the response of mean energy efficiency to energy price changes was positive and significant, both statistically and economically. Interestingly, they also found that equivalent percentage cost subsidies would have been about three times as effective as taxes in encouraging adoption, although standard financial analysis suggest they ought to be about equal in percentage terms. This finding does, however, offer confirmation for the conventional wisdom that technology adoption decisions are more sensitive to up-front cost considerations than to longer-term operating expenses.

In a study of residential conservation investment tax credits, Hassett and Metcalf (1995) also found that tax credit or deductions were many times more effective than "equivalent" changes in energy prices–about eight times as effective in their study. They speculate that one reason for this difference is that energy price movements may be perceived as temporary. The findings by Jaffe and Stavins (1995), and by Hasset and Metcalf (1995) are consistent with other analyses of the relative effectiveness of energy prices and technology market reforms in bringing about the adoption of lifecycle cost-saving technologies. Up-front subsidies can be more effective than energy price signals (see, e.g., Krause *et al.*, 1993; Howarth and Winslow, 1994; IPSEP, 1995; Eto *et al.*, 1996; Golove and Eto, 1995; IPCC, 1996, Executive Summary, p. 13). A disadvantage of such non-price policies relative to administered prices is that they have to be implemented on an "end-use by end-use" or "sector by sector" basis in a customized fashion. Also, an effective institutional and regulatory framework needs to be created and maintained to evaluate and ensure the continued cost-effectiveness of such policies.

This and other research on energy efficiency programmes also highlights a major difference in the way energy price signals and technology subsidies function. The technology adoption response to taxes may include a secondary increase in the demand for energy services. This secondary effect takes two forms: a direct effect that results from the increased utilization of energy-using equipment and capital stocks, and an indirect effect from increased disposable income. Studies of such demand effects suggest that the combined effects are generally not sufficient to offset more than a minor portion of emissions reductions.

In addition, technology subsidies and tax credits can require large public expenditures per unit of effect, since consumers who would have purchased the product even in the absence of the subsidy will still receive it.[110]

Some recent empirical studies suggest that the response of relevant technological change to energy price changes can be surprisingly swift. Typically, this is less than 5 years for much of the response in terms of patenting activity and the introduction of new model offerings (Jaffe and Stavins, 1995; Newell *et al.*, 1999; Poppe, 1999). Substantial diffusion can sometimes take longer, depending on the rate of retirement of previously installed equipment. The longevity of much energy-using equipment reinforces the importance of taking a longer-term view towards energy-efficiency improvements–on the order of decades.

An optimal set of policies would be designed in such a way as to achieve two outcomes simultaneously: release any obstructed emission and cost-reduction potentials from already available technologies through various market reforms that try to reduce market distortions (see IPCC, 2000), and induce the accelerated development of new technologies. This approach allows significant carbon abatement over the near-term by diffusing existing technologies, while at the same time preparing new technologies for the longer term.

6.6   Climate Policy Evaluation

Theoretically, it is unnecessary to monitor and evaluate national policies and programmes to see whether Annex I Parties fulfil their Kyoto commitments, provided national communications give a clear and reliable picture of the net impact of those actions on the net national GHG emissions and net uptake via sinks. Indeed, national inventories, usually updated on an annual basis, are the backbone of the monitoring system. Of course, governments might want to monitor the impact of their own policies for domestic assessment purposes. To meet the international commitments, such monitoring, however, would not be necessary if monitoring at the aggregate level were completely reliable. However, this may not be true. Evidence suggests that there can be a considerable margin of error in the national data provided to the UNFCCC Secretariat within the framework of the national communications.

Over the past 25 years an extensive literature, including programme evaluation, value-for-money audits, and comprehensive audits, has developed on the evaluation of government programmes. Much of this literature is specific to the type of programme–low-income housing, training, employment creation, policing, transit, energy efficiency, etc.–and has little relevance to the monitoring and evaluation of policies for climate change mitigation. However, the literature also includes numerous evaluations of energy-efficiency, DSM, emissions trading, environmental taxes, and other programmes that could provide useful insights into the design, monitoring, and evaluation of climate change policies.

---

[110] It may be possible to reduce the number of free-riders through subsidy programme design.

References

ADEME-European Commission, 1999: *Energy Efficiency Indicators: The European Experience*.ADEME, Paris.

Ahn, D., 1999: Environmental Disputes in the GATT/WTO: Before and After the US-Shrimp Case. *Michigan Journal of International Law, 20*, 819-870.

Austin, D., P. Faeth, R. Serôa da Motta, C.E.F. Young, C. Ferraz, Z. Ji, L. Junfeng, M. Pathak, and L. Srivastava, 1998: *Opportunities for financing sustainable development via the CDM: A Discussion Draft.* Tata Energy Research Institute, Instituto de Pesquisa Economica Aplicada, Institute of Environmental Economics at the Renmin University of China, and World Resources Institute, Washington, DC.

Australia, Canada, Iceland, Japan, New Zealand, Norway, Russian Federation, and the United States of America, June 1998: *Non-Paper on Principles, Modalities, Rules, and Guidelines for an International Emissions Trading Regime.* http://wwwmed.govt.nz/ers/environment/umbrellagroup/index.html

Barrett, S., 1994: Strategic environmental policy and international trade. *Journal of Public Economics, 54*, 325-338.

Baron, R. 1996: Economic/fiscal instruments: Taxation (i.e. carbon/energy), policies and measures for common action, Working Paper. Annex I Experts Group on the UNFCCC, OECD WPAPER No.4.

Baron, R., 1999: An assessment of liability rules for international GHG emissions trading, IEA Information Paper. International Energy Agency, Paris.

Baron, R., and J. Hou, 1998: Electricity trade, the Kyoto Protocol and emissions trading, IEA Information Paper for the Fourth Conference of the Parties to the UNFCCC. IEA, Paris.

Baron, R., and A. Lanza, 2000: Kyoto commitments: macro and micro insights on trading and the clean development mechanism. *Integrated Assessment, 1 (2),* 137-144.

Baron, R., M. Bosi, A. Lanza, and J. Pershing, 1999: *A Preliminary Analysis of the EU Proposal on the Kyoto Mechanisms.* International Energy Agency, Paris.

Barros, V., and M. Conte Grand, 1999: *El significado de una meta dinámica de reducción de emisiones de gases efecto invernadero: el caso Argentino (documento de trabajo).* Universidad del CEMA, Buenos Aires, Argentina.

Bashmakov, I., 1998: *Strengthening the Russian economy through climate change policies.* Proceedings, Dealing with Carbon Credits After Kyoto, May 28-29, Callantsoog, Netherlands.

Bashmakov, I., 2000: Energy Subsidies and "Right Prices". *Energy Efficiency, 27*, April-June 2000.

Bashmakov, I., and S. Sorokina, 1996: *Refrigerator energy efficiency standards in Eastern Europe.* Center for Energy Efficiency (CENEf), Moscow.

Baumol, W.J., and W.E. Oates, 1988: *The Theory of Environmental Policy, Second Edition.* Cambridge University Press, New York.

Begg, K., S.D. Parkinson, T. Jackson, P-E. Morthorst, and P. Bailey, 1999: *Overall Issues for Accounting for the Emissions Reductions of JI Projects.* Proceedings of Workshop on Baselines for the CDM, Tokyo, February 25-26, 1999, Global Industrial and Social Progress Research Institute (GIS-PRI).

Begg, K.G., S.D. Parkinson, Y. Mulugetta, and R. Wilkinson, 2000: An initial evaluation of CDM type projects in DC's. Final Report to DFID.

Benner, J., 2000: *Development of an International Green Certificate Trading System.* Proceedings of the Fifth National Green Power Marketing Conference, August 8, Denver, CO.

Berger, J.R., 1999: Unilateral Trade Measures to Conserve the World's Living Resources: An Environmental Breakthrough for the GATT in the WTO Sea Turtle Case. *Columbia Journal of Environmental Law, 24*, 355-411.

Bernstein, J., 1993: Alternative approaches to pollution control and waste management regulatory and economic instruments. Urban Management Program UNDP/UNCHS, World Bank, Washington DC.

Bernstein, P.M., W. Montgomery, T. Rutherford, and G. Yang, 1999: Effects of restrictions on international permit trading: The MS-MRT model. *The Energy Journal, Special Issue, The Costs of the Kyoto Protocol: A Multi-Model Evaluation,* **May**, 221-256.

Biglaiser, G., and J.K. Horowitz, 1995: Pollution regulation and incentives for pollution-control research. *Journal of Economic Management Strategy,* **3**, 663-684.

Biglaiser, G., J.K. Horowitz, and J.Quiggin, 1995: Dynamic pollution regulation. *Journal of Regulatory Economics,* **8**, 33-44.

Black, D., M. Grubb, and C Windham, 2000: *International Trade and Climate Change Policies.* The Royal Institute of International Affairs, London.

Bluffstone, R., and B.A. Larson, 1997: *Controlling Pollution in Transition Economies.* Edward Elgar, Cheltenham, UK.

Bohm, P., 1994: On the feasibility of Joint Implementation of carbon emissions reductions, *Climate Change: Policy Instruments and their Implications.* Proceedings of the Tsukuba Workshop of IPCC Working Group III, A. Amano (ed.), Center for Global Environmental Research, Environment Agency of Japan, Tsukuba, Japan.

Bohm, P., 1997: Joint Implementation as Emission Quota Trade: An Experiment Among Four Nordic Countries. *Nord,* **4**, Nordic Council of Ministers, Copenhagen. Reprinted (with the exception of Appendix 3) in *Pollution for Sale.* S. Sorrell, J. Skea (eds.), Edward Elgar, Cheltenham, UK, 1999.

Bohm, P., 1998: Public investment issues and efficient climate change policy, part 1. In *The Welfare State, Public Investment, and Growth.* H. Shibata and T. Ihori, (eds.), Springer Verlag, Tokyo.

Bohm, P., 1999: *International GHG emission trading with special reference to the Kyoto Protocol, TemaNord:506.* Nordic Council of Ministers, Copenhagen.

Bohm, P., and B. Larsen, 1994: Fairness in a tradable-permit treaty for carbon emissions reductions in Europe and the former Soviet Union. *Environmental and Resource Economics,* **4**, 219-239.

Bohm, P., and B. Carlén, 2000: Cost-effective approaches to attracting low-income countries to international emissions trading: Theory and experiment. In *Studies in Climate Change Policy: Theory and Experiments*, PhD thesis B. Carlén, Department of Economics, Stockholm University, Stockholm.

Böhringer, C., and T.F. Rutherford, 1997: Carbon taxes with exemptions in an open economy: A general equilibrium analysis of the German Tax Initiative. *Journal of Environmental Economics and Management, 32*(2), 189-203.

Bollen, J., A. Gielen, and H. Timmer, 1999: Clubs, ceilings and CDM: Macroeconomics of compliance with the Kyoto Protocol, *The Energy Journal, Special Issue, The Costs of the Kyoto Protocol: A Multi-Model Evaluation,* **May**, 177-206.

Boom, J.T., 1998: *Market performance and environmental policy: Four tradable permits scenarios, SOE.* Foundation for Economic Research of the University of Amsterdam.

Bovenberg, L., and L. Goulder, 1999: Neutralizing the adverse industry impacts of CO_2 abatement policies: What does it cost? Paper prepared for the FEEM-NBER Conference on Behavioural and Distributional Effects of Environmental Policy, June, Milan, Italy.

Boyle, S., 1996: DSM Progress and lessons in the Global Context. *Energy Policy,* 24(4), 345-360.

Brown, M.A., M.D. Levine, J.P. Romm, A.H. Rosenfeld, and J.G. Koomey, 1998: Engineering-Economic Studies of Energy Technologies to Reduce Greenhouse Gas Emissions: Opportunities and Challenges. *Annual Review Energy Environ*ment, **23**, 287-385.

Buchanan, J.M., and G. Tullock, 1975: Polluters, profits and political response: Direct controls vs. taxes. *American Economic Review,* 65(1), 139-147.

California Assembly Bill No. 1890, 1996: Chapter 854 signed September 23 1996.

Carlén, 1999: Large-country effects in international emissions trading: A laboratory test. Working Paper #15, Department of Economics, Stockholm University, Stockholm.

Carraro, C., and D. Siniscalco, (eds.), 1993: The European carbon tax: An economic assessment, Fondazione ENI, Enrico Mattei Series on Economics. *Energy and Environment.*

CCAP (Center for Clean Air Policy), 1998: US carbon emissions trading: Description of an upstream approach. [online: http://www.ccap.org].

CEC, 1996: Communication from the Commission to the Council and the European Parliament on Environmental Agreements, COM(96) 561 Final, 1996.

Chadwick, M., K. Begg, G. Haq, and T. Jackson, 2000: Environmental and Social aspects of JI -methodologies and case study results, Chapter 6. *Flexibility in Climate Policy: Making the Kyoto Mechanisms Work.* T. Jackson, K. Begg, S.D. Parkinson, (eds.), Earthscan, UK.

Chandler, W., 2000: *Energy and the environment in the transition economies: between cold war and global warming.* Westview Press, USA.

Chomitz, K., 1998: *Baselines for GHG reductions: Problems, precedents, solutions.* World Bank, Washington, DC.

Cole, D., and P.Z. Grossman, 1998: When is command-and-control efficient? Institutions, technology and the comparative efficiency of alternative regulatory regimes for environmental protection. Working Paper Series, Indiana University and Butler University, USA.

Conrad, K., 1996: Choosing emission taxes under international price competition. In *Environmental Policy and Market Structure.* C. Carraro, Y. Katsoulacos, and A. Xepapadeas, (eds.), 85-98.

Cooper, A., S. Livermore, V. Rossi, A. Wilson, and J. Walker, 1999: The economic implications of reducing carbon emissions: A cross-country quantitative investigation using the Oxford Global Macroeconomic and Energy Model. *The Energy Journal, Special Issue, The Costs of the Kyoto Protocol: A Multi-Model Evaluation,* **May**, 335-366.

Cooper, R., 1998: Toward a real global warming treaty. *Foreign Affairs,* **77**(2), 66-79.

Corfee Morlot, J., 1998: Ensuring Compliance with a Global Climate Change Agreement. OECD Information Paper, Paris.

Cramton, P., and S. Kerr, 1998: *Tradable carbon permit auctions – how and why to auction not grandfather.* University of Maryland/Resources for the Future, Washington, DC.

Crawford-Smith, C., 1999: *The Market for Green Power in Australia - Future Role and Implications.* World Renewable Energy Congress, Murdoch University, Perth, WA.

Criqui, P., S. Mima, and L. Viguier, 1999: Marginal Abatement Costs of CO_2 Emissions Reductions, Geographic Flexibility and Concrete Ceilings. An Assessment Using the POLES Model. *Energy Policy,* **27**, 585-601.

Demaret, P., and R. Stewardson, 1994: Border Tax Adjustments under the GATT and EC Law and General Implications for Environmental Taxes. *Journal of World Trade,* **28**(4), 5-65.

Dijkstra, B.R., 1998: *The Political Economy of Instrument Choice in Environmental Policy.* Groningen University, Groningen, The Netherlands.

Dixon, R.K., 1999: *The United Nations Framework Convention on Climate Change Activities Implemented Jointly Pilot: Experiences and Lessons Learned.* Kluwer, Dordrecht.

Downing, P.B., and L.J. White, 1986: Innovation in pollution control. *Journal of Environmental Economics and Management,* **13**, 18-29.

Dutschke, M., and A. Michaelowa, 1998: Issues and open questions of greenhouse gas emission trading under the Kyoto Protocol. Discussion paper 68, HWWA, Hamburg Institute for International Economics, Hamburg.

Eaton, J., and G.M. Grossman, 1986: Optimal trade and industrial policy under oligopoly. *Quarterly Journal of Economics,* **100**, 383-406.

Ebert, U., 1996: Naïve use of environmental instruments. In *Environmental Policy and Market Structure,* C. Carraro, Y. Katsoulacos, and A. Xepapadeas, (eds.), pp. 45-64.

ECMT, 1998: *Efficient Transport for Europe.* OECD Publications, Paris.

Edmonds, J., M. Wise, and D.W. Barns, 1995: Carbon coalitions: The cost and effectiveness of energy agreements to alter trajectories of atmospheric carbon dioxide emissions. *Energy Policy,* **23**(4/5), 309-335.

EEA, 1997: *Environmental Agreements - Environmental Effectiveness.* European Environment Agency, Copenhagen.

EIA, 1998: Emissions of GHGs in the United States 1997. Energy Information Administration, U.S. Department of Energy, DOE/EIA-0573(97), Washington, DC.

EIA, 1999: *Annual Energy Outlook 2000.* Energy Information Administration, U.S. Department of Energy, December 1999, Washington, DC.

Eichhammer, W., and Jochem, E., 1998: Voluntary Agreements for the reduction of CO_2 GHG emissions in Germany and their recent first evaluation. Paper prepared for CAVA Workshop, Gent, Belgium.

Elbers, C., and C. Withagen, 1999: General equilibrium models of environmental regulation and international trade. Center for Economic Research, paper no. 9924, Tilburg University, Tilburg, The Netherlands.

Ellerman, A.D.,and A. Decaux, 1998: *Analysis of post-Kyoto CO_2 emissions trading using marginal abatement curves.* Report 40, Joint Program on the Science and Policy of Global Change, Massachusetts Institute of Technology, Cambridge, MA.

Ellerman, A.D., and I.S. Wing, 2000: *Supplementarity: An Invitation to Monopsony?* MIT Joint Program on the Science and Policy of Climate Change. Report No. 59 (April, 2000).

Ellis, J., 1999: Experience with emission baselines under the AIJ pilot phase. OECD Information Paper, Paris.

Esty, D.C., 1994: *Greening the GATT: trade, environment and the future.* Institute for International Economics, Washington, DC.

Eto, J., R. Prahl, and J. Schlegel, 1996: *A Scoping Study on Energy Efficiency Market Transformation by California Utility DSM Programs,* LBNL-39058. Energy and Environment Division, Lawrence Berkeley Laboratory, Berkeley CA.

Eyre, N.J., 1998: A Golden Age or a False Dawn? Energy Efficiency in UK Competitive Energy Markets. *Energy Policy,* **26** (12), 963-972.

Fenden, P., 1998: Norwegian Industry's Network for Energy Conservation.Industrial Energy Efficiency Policies: Understanding Success and Failure. Workshop organized by International Network for Energy Demand Analysis in the Industrial Sector, June 1998, Utrecht, The Netherlands.

Fischer, C., I.W.H. Parry, and W. Pizer, 1998: Instrument choice for environmental protection when technological innovation is endogenous. Resources for the Future Discussion Paper 99-04 (revised), Washington, DC.

Fisher-Vanden, K., 1999: Structural change and technological diffusion in transition economies: Implications for energy use and carbon emissions in China. Unpublished PhD thesis, Harvard University, Cambridge, MA.

Fowlie, M., 1999: *The Environment and Competition in Electricity in the USA and the UK.* Oxford Institute for Energy Studies, Oxford, UK.

Freeman, A.M., III, and R.H. Haveman, 1972: Clean rhetoric and dirty water. *Public Interest,* **28**, 51-65.

Garbaccio, R., M. Ho, and D.W. Jorgenson, 1999: Why has the energy-output ratio fallen in China? *Energy Journal.*

Glasbergen, P., M.C. Das, P.P.J. Driessen, N. Habermehl, W.J.V. Vermeulen, K. Blok, J. Farla, and E. Korevaar., 1997: *Afspraken werken. Evaluatie Meerjarenafspraken over Energie-efficiency.* Ministry of Economic Affairs, Directorate-General for Energy, The Hague, 197 pp.

Goldberg, D., S. Porter, N. LaCasta, and E. Hillman, 1998: *Responsibility for non-compliance under the Kyoto Protocol's mechanisms for cooperative implementation.* Center for International Environmental Law and Euronatura, Washington, DC.

Goldemberg, J., 1998: *Issues & options: The Clean Development Mechanism.* United Nations Development Program, New York, NY.

Golove, W., and J. Eto 1995: *Market Barriers to Energy Efficiency: A Critical Reappraisal of Public Policies to Promote Energy Efficiency, LBNL-38059.* Energy and Environment Division, Lawrence Berkeley Laboratory, Berkeley, CA.

Goodin, R.E., 1994: Selling Environmental Indulgences. *Kyklos,* **47**(4), 573-596.

Grubb, M., T. Chapuis, and M.H. Duong, 1995: The economics of changing course: Implications of adaptability and inertia for optimal climate policy. *Energy Policy,* **23**(4/5), 417-432.

Grubb, M., T. Tietenberg, B. Swift, A. Michaelowa, Z. Zhang, and F.T. Joshua, 1998: *GHG Emissions Trading: Defining the Principles, Modalities, Rules and Guidelines for Verification, Reporting & Accountability.* United Nations Conference on Trade and Development, Geneva.

Haddad, B.M., and P. Jefferis, 1999: Forging consensus on National Renewables Policy: The Renewables Portfolio Standard and the National Public Benefit Trust Fund. *Electricity Journal,* **12**(2), 68-80.

Hadley, S., and E. Hirst, 1995: *Utility DSM Programs from 1989 through 1998: Continuation or Cross Roads? ORNL/CON-405.* Oak Ridge National Laboratory, Oak Ridge, TN.

Hahn, R.W., 1987: Jobs and environmental quality: Some implications for instrument choice. *Policy Sciences,* **20**, 289-306.

Hahn, R.W., 1989: Economic Prescriptions for Environmental Problems: How the Patient Followed the Doctor's Orders. *Journal of Economic Perspectives,* **3**(2), 95-114.

Hahn, R.W., and R.N. Stavins, 1991: Incentive-based environmental regulation: A new era from an old idea? *Ecology Law Quarterly,* **18**(1), 1-42.

Hahn, R.W., and R.N. Stavins, 1999: What has Kyoto wrought? The real architecture of international tradable permit markets, working paper. Cambridge, MA.

Haites, E., 1998: International emissions trading and compliance with GHG limitation commitments. Working Paper W70, International Academy of the Environment.

Haites, E., and A. Proestos, 2000: Suitability of non-energy GHGs for emissions trading. In *Non-CO₂ Greenhouse Gases: Scientific Understanding, Control and Implementation.* J. van Ham, A.P.M. Baede, L.A. Meyer, R. Ybema, (eds.), Kluwer Academic Publishers, Dordrecht, 417-424.

Haites, E., and F. Yamin, 2000: The Clean Development Mechanism: Proposals for its operation and governance. *Global Environmental Change,* **10**(1), 27-45.

Han, K., and J.B. Braden, 1996: Environment and trade: New evidence from US manufacturing. Unpublished research paper, Department of Economics, University of Illinois, and Department of Agriculture and Consumer Economic, University of Illinois at Urbana-Champaign, IL.

Hargrave, T., N. Helme, and I. Puhl, 1998: *Options for simplifying baseline setting for Joint Implementation and Clean Development Mechanism projects.* Center for Clean Air Policy (CCAP), Washington, DC.

Harrison, D., and T. Schatzki, 2000: *Certifying baselines for credit-based greenhouse gas trading programs: Lessons from experience with environmental and non-environmental programs.* EPRI, Palo Alto, CA.

Hassett, K.A., and G.E. Metcalf, 1995: Energy tax credits and residential conservation investment: Evidence from panel data. *Journal of Public Economics,* **57**, 201-217.

Hassing, P., and M. S. Mendis, 1999: An international market framework for CDM transactions. Paper presented at the UNFCCC technical workshop, Bonn, Germany.

Hauch, J., 1999: *Elephant – A Simulation Model for Environmental Regulation at Nordic Energy Markets.* Ph.D. thesis, Danish Economic Council, Copenhagen.

Hoel, M., 1993: Harmonization of carbon taxes in international climate agreements. *Environmental and Resource Economics,* **3**(3), 221-232.

Hoel, M., 1996: Should a carbon tax be differentiated across sectors? *Journal of Public Economics,* **59**(1), 17-32.

Hoel, M., 1997: Coordination of environmental policy for transboundary environmental problems. *Journal of Public Economics,* **66**, 199-224.

Hoel, M., 1998: Emission taxes versus other environmental policies. *Scandinavian Journal of Economics,* 100(1), 79-104.

Hoel, M., and L. Karp, 1998: Taxes versus quotas for a stock pollutant. Working Paper No. 885. Department of Agricultural and Resource Economics and Policy (CUDARE). University of Berkeley, California.

Holt, E., 2000a: *Green Power in Competitive Markets, 1999.* Electric Power Research Institute, TR-114210, Palo Alto, CA.

Holt, E., 2000b: *Green Pricing Update, 1999.* Electric Power Research Institute, TR-114211, Palo Alto, CA.

Holt, E., and R. Wiser, 1998: *Understanding Consumer Demand for Green Power.* National Wind Coordinating Committee.

Honma, S., 1978: Document 0.25: Was the Muskie standards successful? (in Japanese).

Howarth, R.B., and M. Winslow 1994: Energy Use and Climate Stabilization: Integrating Pricing and Regulatory Policies. *Energy: The International Journal.*

Huber, R.M., J. Ruitenbeek, and R. Serôa da Motta, 1999: Market-Based Instruments for Environmental Policymaking in Latin America and the Caribbean: Lessons from Eleven Countries. World Bank Discussion Paper No. 381, World Bank, Washington, DC.

Hunt, S., and G. Shuttleworth, 1996: *Competition and Choice in Electricity.* John Wiley & Sons, New York, NY.

IEA (International Energy Agency), 1997a: *The directive on the EU electricity market.* IEA, Paris.

IEA, 1997b: *Energy Efficiency Initiative,* Vol. 1. IEA, Paris.

IEA, 1998a: *Energy Policies in IEA Countries.* OECD Publications, Paris.

IEA, 1998b: *World Energy Outlook.* IEA, Paris.

IEA, 1999a: *Electricity Market Reform: An IEA Handbook.* IEA, Paris.

IEA, 1999b: *Looking at Energy Subsidies: Getting the Prices Right. World Energy Outlook, 1999 Insights,* IEA, Paris.

IPCC, 1996: *Climate Change 1995: Economic and Social Dimensions of Climate Change.* Cambridge University Press, Cambridge, UK

IPCC, 2000: *Special Report on Methodological and Technological Issues in Technology Transfer (SRTT).* Cambridge University Press, Cambridge, UK.

IPSEP, 1995: Cutting Carbon Emissions: Burden or Benefit? A Summary. In *Energy Policy in the Greenhouse, Volume II,* International Project for Sustainable Energy Paths, El Cerrito, CA.

Iwahashi, T., 1998: Basic Design of a Domestic System of Greenhouse Gas Emissions Trading. *Osaka Law Review,* **48**(3), 857-890.

Jackson, J.H., 2000: Comments on Shrimp/Turtle and the Product/Process Distinction, *European Journal International Law,* **11**(2).

Jacoby, H., and I.S. Wing, 1999: Adjustment time, capital malleability and policy cost. *The Energy Journal, Special Issue, The Costs of the Kyoto Protocol: A Multi-Model Evaluation,* **May**, 73-92.

Jaffe, A.B., and R.N. Stavins, 1995: Dynamic incentives of environmental regulations: The effects of alternative policy instruments on technological diffusion. *Journal of Environmental Economics and Management,* **29**, S43-S63.

Japan Energy Conservation Center, 1997: *Energy Conservation Handbook,* Japan.

Jepma, C.J., 1999: Determining a baseline for project co-operation under the Kyoto Protocol: a general overview. Paper presented at the GISPRI baseline workshop, 25-26 February, Tokyo, Japan.

Jepma, C.J., and W. van der Gaast, (eds.), 1999: *On the Compatibility of Flexibility Instrument.* Kluwer Academic Publishers, Dordrecht.

Jepma, C.J., W. van der Gaast, and E. Woerdman, 1998: *The compatibility of flexible instruments under the Kyoto Protocol.* Dutch National Research Program on Transboundary Global Air Pollution and Climate Change, Paterswolde, Netherlands.

JIQ, 1998: AIJ Pilot Phase: A Mile Wide but only an Inch Deep. *Joint Implementation Quarterly,* **4**(3).

Joskow, P.L., and R. Schmalensee, 1988: The political economy of market-based environmental policy: The U.S. acid rain program. *Journal of Law and Economics,* **41**, 37-83.

Jung, C., K. Krutilla, and R. Boyd, 1996: Incentives for advanced pollution abatement technology at the industry level: An evaluation of policy alternatives. *Journal of Environmental Economics and Management,* **30**, 95-111.

Kainuma, M., Y. Matsuoka, and T. Morita, 1999: Analysis of post-Kyoto scenarios: The Asian-Pacific integrated model. *The Energy Journal, Special Issue, The Costs of the Kyoto Protocol: A Multi-Model Evaluation,* **May**, 207-220.

Kambara, T., 1992: The energy situation in China. *China Quarterly,* **September**, 608-636.

Katsoulacos, Y., and A. Xepapadeas, 1996: Environmental innovation, spillovers and optimal policy rules. In *Environmental Policy and Market Structure,* C. Carraro, Y. Katsoulacos, and A. Xepapadeas, (eds.), pp. 143-150.

Kelman, S., 1981: Cost-Benefit Analysis: An Ethical Critique. *Regulation,* **5**(1).

Kemp, R., and L. Soete, 1990: Inside the "green box:" on the economics of technological change and the environment. *New Explorations in the Economics of Technological Change,* C. Freeman, and L. Soete, (eds.), Pinter, London.

Kennedy, P.W., 1994: Equilibrium pollution taxes in open economies with imperfect competition. *Journal of Environmental Economics and Management,* **27**, 49-63.

Keohane, N.O., 1998: The political economy of environmental policy in Venezuela. Report to the World Bank (Photocopy).

Keohane, N.O., R.L. Revesz, and R.N. Stavins, 1999: The positive political economy of instrument choice in environmental policy. In *Environmental and Public Economics: Essays in Honor of Wallace E. Oates,* A. Panagariya, P.R. Portney, and R.M. Schwab, (eds.), Cheltenham, UK, pp. 89-125.

Kerr, S., 1998: Enforcing Compliance: The Allocation of Liability in International GHG Emissions Trading and the Clean Development Mechanism. Resources for the Future Climate Issue Brief, No 15, October 1998, p4.

Kerr, S., and D. Maré, 1997: Efficient Regulation Through Tradable Permit Markets: The United States Lead Phasedown. Working Paper 96-06, Department of Agricultural and Resource Economics, University of Maryland, MD.

Kete, N., 1992: The U.S. acid rain control allowance trading system. In *Climate Change: Designing a Tradable Permit System*, T. Jones, J. Corfee-Morlot (eds.), Organization for Economic Cooperation and Development, Paris, pp. 69-93.

Kneese, A.V., and C.L. Schulze, 1975: *Pollution, Prices, and Public Policy*. Brookings Institution, Washington, DC.

Kohn, R.E., 1992: When subsidies for pollution abatement increase total emissions. *Southern Economic Journal, 59*, 77-87.

Kolstad, C.D., 2000: *Environmental Economics*. Oxford University Press, New York, USA and Oxford, UK.

Kooiman, J., 1998: *Topics in the economics of environmental regulation*. Thesis, Tinbergen Institute Research Series, no.182, 141-159, Amsterdam.

Kopp, R., R. Morgenstern, and W. Pizer, 1999a: *Something for everyone: A climate policy that both environmentalists and industry can live with*. Resources for the Future, Washington, DC.

Kopp, R., R. Morgenstern, W. Pizer, and M. Toman, 1999b: A proposal for credible early action in U.S. climate policy. *Weathervane*. Resources for the Future, Washington, DC. [online: http://www.weathervane.rff.org].

Krause, F., 1996: The policy effectiveness of mandatory energy efficiency standards and voluntary agreements: an illustrative case study. In Proceedings of the U.S. Climate Change Analysis Workshop, June 6-7, Springfield, VA., US Environmental Protection Agency, Washington, DC.

Krause, F., E. Haites, R. Howarth, and F. J. Koomey, 1993: Cutting Carbon Emissions: Burden or Benefit? In *Energy Policy in the Greenhouse, Vol. 2, Part 1*. International Project for Sustainable Energy Paths, El Cerrito, CA.

Krause, F., J. Koomey, and D. Olivier, 1998: Low Carbon Comfort: The Cost and Potential of Energy Efficiency in EU Buildings, *Energy Policy in the Greenhouse series, Vol. 2, Part 5*. International Project for Sustainable Energy Paths, El Cerrito, CA.

Kurosawa, A., H. Yagita, Z. Weisheng, K. Tokimatsu, and Y. Yanagisawa, 1999: Analysis of carbon emission stabilization targets and adaptation by integrated assessment model. *The Energy Journal, Special Issue, The Costs of the Kyoto Protocol: A Multi-Model Evaluation*, **May**, 157-176.

Laffont, J.J., and J. Tirole, 1996: Pollution Permits and Environmental Innovation? *Journal of Public Economics*, **62**, 127-140.

Lebo, B., and A. Szabo, 1996: *Preparing minimum energy efficiency standards for European appliances*. Proceedings of International Energy Conference on Use of Efficiency Standards in Energy Policy, Sophia-Antipolis, France, June, IEA/OECD.

Lee, H., and N. Darani, 1995: Electricity Restructuring and the Environment. CSIA Discussion Paper 95-13, Kennedy School of Government, Harvard University, Cambridge, MA.

Legro, S., I. Gritsevich, V. Berdin *et al.*, 1999: *Climate Change Policy and Programs in Russia: An Institutional Assessmen..*Advanced International Study Unit, PNNL, Washington, DC.

Lile, R., M. Powell, and M. Toman, 1999: Implementing the Clean Development Mechanism: Lessons from U.S. private-sector participation. In Activities Implemented Jointly, Discussion Paper 99-08, Resources for the Future, Washington, DC.

Lora, 1997: A Decade of Structural Reforms in Latin America: What Has Been Reformed and How to Measure It. Document presented at the Annual Meetings of the Inter-American Development Bank, Barcelona.

Lovins, A.B., 1996: Negawatts. Twelve transitions, eight improvements and one distraction. *Energy Policy, 24*(4), 331-344.

Magat, W.A., 1978: Pollution control and technological advance: A dynamic model of the firm. *Journal of Environmental Economics and Management, 5*, 1-25.

Magat, W.A., 1979: The effects of environmental regulation on innovation. *Law and Contemporary Problems*, **43**, 3-25.

Maloney, M.T., and R.E. McCormick, 1982: A positive theory of environmental quality regulation. *Journal of Law and Economics* **25**(4), 99-123.

Malueg, D.A., 1989: Emission credit trading and the incentive to adopt new pollution abatement technology, *Journal of Environmental Economics and Management*, **16**, 52-57.

Manne, A.S., and R. Richels, 1999: The Kyoto Protocol: A cost-effective strategy for meeting environmental objectives? *The Energy Journal, Special Issue, The Costs of the Kyoto Protocol: A Multi-Model Evaluation*, **May**, 1-24.

Matsuo, N., 1998: *How Is the CDM Compatible with Sustainable Development?* The Institute for Global Environmental Strategies (IGES), Japan.

Matsuo, N., 1999: Points and Proposals for the Emissions Trading Regime of Climate Change – For Designing Future System, September.

Markard, J., 1998: *Green Pricing – Welchen Beitrag können freiwillige Zahlungen von Stromkunden zur Förderung regenerativer Energien leisten?* Öko-Institute-Verlag, Freiburg.

McCubbins, M.D., and T. Sullivan, 1984: Constituency influences on legislative policy choice. *Quality and Quantity, 18*, 299-319.

McKibbin, W.J., and P.J. Wilcoxen, 1997a: Salvaging the Kyoto Climate Change Negotiations. Brookings Policy Brief no. 27, November 1997, The Brookings Institution, Washington, DC.

McKibbin, W.J., and P.J. Wilcoxen, 1997b: A Better Way to Slow Global Climate Change. Brookings Policy Brief no. 17, June 1997, The Brookings Institution, Washington, DC.

McKibbin, W.J., and P.J. Wilcoxen, 2000: The Next Step for Climate Change Policy. Brookings Background Paper #1 (January), Washington, DC.

McKibbin, W., M. Ross, R. Shackleton, and P. Wilcoxen, 1999: Emissions Trading, Capital Flows and the Kyoto Protocol. *The Energy Journal, Special Issue, The Costs of the Kyoto Protocol: A Multi-Model Evaluation*, **May**, 287-334.

McMahon, J., 1992: *Quantifying the benefits and costs of US Appliance energy performance standards*. Proceedings of International Energy Conference on Use of Efficiency Standards in Energy Policy, Sophia-Antipilis. France, 4-5 June 1992, IEA/OECD.

Mestelman, S., 1982: Production externalities and corrective subsidies: A general equilibrium analysis. *Journal of Environmental Economics and Management, 9*, 186-193.

Michaelowa, A., 2000: Klimapolitik in Großbritannien – Zufall oder gezieltes Handeln? *Zeitschrift für Umweltpolitik und Umweltrecht*.

Michaelowa, A., and M. Dutschke, 1998: Interest groups and efficient design of the Clean Development Mechanism under the Kyoto Protocol. *International Journal for Sustainable Development*, 1(1), 24-42.

Michaelowa, A., and M. Dutschke, 1999a: Economic and Political Aspects of Baselines in the CDM Context. In *Promoting development while limiting greenhouse gas emissions: trends and baselines*. J. Goldemberg, and W. Reid (eds.), New York, NY, pp. 115-134.

Michaelowa, A., and M. Dutschke, 1999b: Creation and sharing of credits through the Clean Development Mechanism under the Kyoto Protocol.. In *On the compatibility of flexible instruments*. C. Jepma, and W. van der Gaast (eds.), Kluwer, Dordrecht, pp. 47-64..

Milliman, S.R., and R. Prince, 1989: Firm incentives to promote technological change in pollution control. *Journal of Environmental Economics and Management, 17*, 247-265.

Ministry of Economic Affairs, 1997: Long term agreements on energy efficiency progress in 1995, Den Haag, Netherlands.

Mitchell, C., 1995a: Renewables NFFO. *Energy Policy, 23*(12).

Mitchell, C., 1995b: *Renewable Energy in the UK: Financing Options for the Future*. Council for the Protection of Rural England.

Mitchell, C., 1997: Renewable Non-Fossil Fuel Obligation - The Diffusion of Technology by Regulation. TIP Workshop on Regulation and Innovative Activities, Vienna, February 1997.

Montero, J.P., 1999: Voluntary Compliance with Market-Based Environmental Policy: Evidence from the U.S. Acid Rain Program. *Journal of Political Economy*, **107**(5), 998-1033.

Montero, J.P., 2000a: Optimal design of a phase-in emissions trading program. *Journal of Public Economics*, **75**, 273-291.

Montero, J.P. 2000b: Prices versus quantities with incomplete enforcement. *Journal of Public Economics,*

Montero, J.P., 2000c: Market structure and environmental innovation. Working Paper, Catholic University of Chile (January 21), place.

Montero, J.P., L. Cifuentes, and F. Soto, 2000: Participación voluntaria en políticas internacionales de cambio climático: Implicancias para Chile. *Estudios de Economía.*

Morrisette, P., and A. Plantinga, 1991: The Global Warming Issue: Viewpoints of Different Countries. *Resources*, **103**, 2-6.

Moscow City Government, 1999: *Energy Conservation in Buildings, Requirement for Thermal Performance and Heat-Water-Power Supply*, Moscow.

Mowery, D., and N. Rosenberg, 1989: *Technology and the Pursuit of Economic Growth.* Cambridge University Press, Cambridge, UK.

Murase, S., 1995: Perspectives from International Economic Law on Transnational Environmental Issues. *Recueil des Cours de L'Academie de Droit International de la Haye*, **253**, 283-431.

Murase, S., 1996: Unilateral Measures and the WTO Dispute Settlement. In *Asian Dragon and Green Trade: Environment, Economics and International Law.* S.S.C. Tay, and D.C. Esty (eds.), Times Academic Press, London, pp. 137-144.

Nadel, S., and Geller, H., 1996: Utility DSM. What have we learned? Where are we going? *Energy Policy*, **24**(4), 289-302.

NEDO, 2000: *A Report on Baseline Study Group.* Secretariat. New Energy and Industrial Technology Development Organization (NEDO) and Mitsubishi Research Institute, Tokyo.

Nelson, R.A., T. Tietenberg, and M.R. Donihue, 1993: Differential Environmental Regulation: Effects On Electric Utility Capital Turnover and Emissions. *Review of Economics and Statistics*, **75** (2), 368-373.

Newell, R., and W. Pizer, 1998: Regulating Stock Externalities under Uncertainty. Resources for the Future Discussion Paper 99-10.

Newell, R.G., A.B. Jaffe, and R.N. Stavins, 1999: The induced innovation hypothesis and energy-saving technological change. *Quarterly Journal of Economics* **114**(3), 941-975.

Nordhaus, W.D., and J.G. Boyer, 1999: Requiem for Kyoto: An Economic Analysis of the Kyoto Protocol. *The Energy Journal, Special Issue, The Costs of the Kyoto Protocol: A Multi-Model Evaluation*, **May**, 93-133.

Novem, 1998: Green Financing. The Netherlands Agency for Energy and Environment, January 28. [online: http://www.novem.org/netherl/green.html].

NRTEE, 1999: *Canada's Options for a Domestic GHG Emissions Trading Program.* National Round Table on the Environment and the Economy, Ottawa, 1999.

Oates, W., and R.M. Schawb, 1988: Economic competition among jurisdictions: Efficiency enhancing or distortion inducing? *Journal of Public Economics*, **35**, 333-354.

OECD, 1978: *Environmental Policies in Japan.* OECD, Paris.

OECD, 1996a: *Implementation Strategies for Environmental Taxes.* OECD, Paris.

OECD, 1996b: *Integrating Environment and Economy: progress in the 1990s.* OECD, Paris.

OECD, 1997a: *Environmental Taxes and Green Tax Reform.* OECD, Paris.

OECD, 1997b: Voluntary Agreements with Industry, Annex I Expert Group on the UNFCCC. Working Paper No.8, OCDE/GD997) 75, Paris [online: http://www.oecd.org/env/cc/freedocs.htm].

OECD, 1997c: *Reforming Energy and Transport Subsidies: Energy and Environmental Implications.* OECD, Paris.

OECD, 1997d: Economic/fiscal instruments: Competitiveness issues related to carbon/energy taxation, Annex I Expert Group on the UNFCCC. Working Paper No.14, OCDE/GD(97)190, OECD, Paris.

OECD, 1998a: *Improving the Environment through Reducing Subsidies, Part I: Summary and Policy Conclusions, and Part II: Analysis and Overview of Studies.* OECD, Paris.

OECD, 1998b: Voluntary Approaches for Environmental Policy in OECD Countries, 21 October, ENV/EPOC (98)6/REV1, OECD, Paris.

OECD, 1999: *Implementing Domestic Tradable Permits for Environmental Protection*, OECD, Paris.

OECD/IEA, 1997: *Energy Efficiency Initiative. Country Profiles & Case Studies with 47 case studies of successful energy efficiency improvement project and policies implementation to learn from 480.* OECD, Paris.

Owen, G., 1996: *A Market in Efficiency: Providing Energy Savings through Competition.* IPPR.

Oye, K.A., and J.H. Maxwell, 1995: Self-interest and environmental management. In *Local Commons and Global Interdependence: Heterogeneity and Cooperation in Two Domains.* R.O. Keohane, and E. Ostrom, (eds.), Sage Publications, London, pp. 191-221.

Palmer, K., and Burtraw, D., 1997: Electricity restructuring and regional air pollution. *Resource and Energy Economics*, **19**, 139-174.

Panayotou, T., 1998: *Instruments of Change: Motivating and Financing Sustainable Development.* Earthscan Publications and United Nations Environment Program, London.

Parfomak, P., and L. Lave, 1997: How Many Kilowatts are in a Negawatt?. Verifying Ex Post Estimates of Utility Conservation Impacts at the Regional Level. *The Energy Journal*, **17**(4).

Parkinson, S., K. Begg, P. Bailey, and T. Jackson. 1999: JI/CDM crediting under the Kyoto Protocol: does "interim period banking" help or hinder emissions reduction? *Energy Policy*, **27**, 129-136.

Parkinson, S.D., P. Bailey, K. Begg, and T. Jackson, 2000: Accounting for Emissions Reduction and Costs - methodology and case study results, Chapter 7. *Flexibility in Climate Policy: Making the Kyoto Mechanisms Work.* T. Jackson, K. Begg, and S.D. Parkinson, (eds.), Earthscan, UK.

Parry, I.W.H., 1998: Pollution regulation and the efficiency gains from technological innovation. *Journal of Regulatory Economics*, **14**, 229-254.

Parry, I.W.H., R. Williams, and L.H. Goulder, 1999: When Can Carbon Abatement Policies Increase Welfare? – The fundamental role of distorted factor markets. *Journal of Environmental Economics and Management*, **January**.

Patterson, W., 1999: *Transforming Electricity.* The Royal Institute of International Affairs/Earthscan, UK.

Perkins, N. 1999: Introductory Note, World Trade Organization: US-Import Prohibition of Certain Shrimp and Shrimp Products. *International Legal Materials*, **38**, 118-120.

Pezzey, J., 1992: The Symmetry Between Controlling Pollution by Price and by Quantity. *Canadian Journal of Economics*, **25**, 983-991.

Phylipsen, G., K. Blok, and E. Worrel, 1998: *Handbook on International Comparisons of Energy Efficiency in the Manufacturing Industry.* Utrecht, the Netherlands.

Pizer, W.A., 1997a: Optimal choice of policy instrument and stringency under uncertainty: the case of climate change. Resources for the future, report.

Pizer, W.A., 1997b: Prices vs quantities revisited: The case of climate change. Resources for the Future Discussion Paper 98-102.

Pohjola, J., 1999: Economy wide Effects of Reducing CO_2 Emissions: A Comparison between Net and Gross Emissions. *Journal of Forest Economics*, **5**(1), 139-168.

Polenske, K., and X. Lin, 1993: Conserving Energy to Reduce Carbon Dioxide Emissions in China. *Structural Change and Economic Dynamics*, **4**(2).

Poppe, D., 1999: Induced innovation and energy prices. Working Paper, University of Kansas, KS.

Radar, N., 1996: Renewables Portfolio Standard in California as Envisioned by the American Wind Energy Association, American Wind Energy Association (revised May 7, 1996). [online: http://www.igc.apc.org/awea/pol/rpsca.html].

Raekwon, Ch., 1997: Unexplored potential of the public-owned technology for promoting environmentally sound technology transfer. Korean Mission to the UN.

Rasmusen, E., and M. Zupan, 1991: Extending the economic theory of regulation to the form of policy. *Public Choice*, **72**(2-3), 275-96.

Rauscher, M., 1991: Foreign trade and the environment. In *Economics and the Environment: The International Dimension.* H. Siebert,(ed.), Mohr, Tübingen, Germany,

Rauscher, M., 1994: On Ecological Dumping. *Oxford Economic Papers*, **46**, 822-840.

Read, P., 1999: Comparative Static Analysis of Proportionate Abatement Obligations (PAO's) – A Market Based Instrument for Responding to Global Warming. *.NZ Econ Papers*, 33(1), 137-147.

Read, P., 2000: An Information Perspective on Dynamic Efficiency in Environmental Policy. *Information Economics and Policy*, 12(1), 47-68.

Reilly, J., R. Prinn, J. Harmisch, J. Fitzmaurice, H. Jacoby, D. Kicklighter, J. Melillo, P. Stone, A. Sokolov, and C. Wang, 1999: Multiple Gas Assessment of the Kyoto Protocol. *Nature*, **401**, 549-555.

Richards, K., 1998: Framing Environmental Policy Instrument Choice. Working Paper Series, Indiana University, USA.

Roberts, M. J., and M. Spence, 1976: Effluent charges and licenses under uncertainty. *Journal of Public Economics,* **5**, 193-208.

Rose, A., B. Stevens, J. Edmonds, and M. Wise, 1998: International Equity and Differentiation in Global Warming Policy. *Environmental and Resource Economics,* **12**, 25-51.

Rose, A., E. Bulte, and H. Folmer, 1999: Long-Run Implications for Developing Countries of Joint Implementation of GHG Mitigation. *Environmental and Resource Economics,* **14**, 19-31.

Rosen, R., T. Woolf, B. Dougherty, B. Biewald, and S. Bernow, 1995: Promoting Environmental Quality in a Restructured Electric Industry. Prepared for the National Association of Regulatory Utility Commissioners, Tellus Institute, Boston, MA.

Rosewicz, B., 1990: Americans Are Willing to Sacrifice to Reduce Pollution, They Say. *Wall Street Journal,* April 20, A1.

Rowlands, I., 1995: *The politics of global atmospheric change.* Manchester University Press, Manchester, UK.

Rucker, M., 2000: Linking International Green Power Markets: Lessons from the APX Green Tickets. Presented at Fifth National Green Power Marketing Conference, Denver, CO, August 8 2000.

Russell., C. and P. Powell, 1996: *Choosing Environmental Policy Tools Theoretical and Practical Considerations.* Inter American Development Bank, Washington DC.

Sagawa, N., 1998: *Energy in Japan.* The Institute of Energy Economics, March, Japan.

Sandel, M.J., 1997: It's Immoral to Buy the Right to Pollute [Editorial]. *New York Times,* December 15, A29.

Schaeffer, G., M. Boots, J. Martens, and M. Voogt, 1999: *Tradable Green Certificates, A New Market-Based Incentive Scheme for Renewable Energy: Introduction and Analysis.* Netherlands Energy Research Foundation ECN, Petten.

Schmalensee, R., P.L. Joskow, A.D. Ellerman, J.P. Montero, and E.M. Bailey, 1998: An Interim Evaluation of Sulfur Dioxide Emissions Trading. *Journal of Economic Perspectives,* **12**(3), 53-68.

Schneider, K., 1998: Comment on M. Hoel: Emission taxes versus other Environmental Policies. *Scandinavian Journal of Economics,* **100**(1), 105-108.

Schumpeter, J., 1942: *Capitalism, Socialism and Democracy.* Harper, New York, NY..

Segerson, K., and Miceli, T., 1997: Voluntary Approaches to Environmental Protection: The Role of Legislative Threats, Nota di lavoro 21.97, Fondazione Eni Enrico Mattei, Milan.

Shelby, M., R. Shackleton, M. Shealy, and A. Cristofaro, 1997: *The Climate Change Implications of Eliminating U.S. Energy (and Related) Subsidies.* US Environmental Protection Agency, Washington, DC.

Shepsle, K.A., and B.R. Weingast, 1984: Political solutions to market problems. *American Political Science Review,* **78**, 417-34.

Sinton, J.E., and M.D. Levine, 1994: Changing Energy Intensity in Chinese Industry. *Energy Policy,* **March**, 239-255.

Smil, V., 1990: *China's Energy.* Report prepared for the U.S. Congress. Office of Technology Assessment, Washington, DC.

Smith, V.L., and A.W. Williams, 1982: The Effects of Rent Asymmetries in Experimental Auction Markets. *Journal of Economic Behavior and Organization,* **3**(1), 99-116.

Speck, S., 1999: A database of environmental taxes and charges. European Union. (online: http://europa.eu.int/comm/dg11/enveco/).

Stavins, R.N., 1996: Correlated uncertainty and policy instrument choice. *Journal of Environmental Economics and Management,* **30**, 218-232.

Stavins, R.N., 1998a: What Can We Learn from the Grand Policy Experiment? Lessons from SO$_2$ Allowance Trading. *Journal of Economic Perspectives,* **12**(3), 69-88.

Stavins, R.N., 1998b: *Market Based Environmental Policies.* Kennedy School of Government, Harvard University, Harvard MA.

Stavins, R.N., 2000: Experience with Market-Based Environmental Policy Instruments. *The Handbook of Environmental Economics.* K. Göran Mäler, J. Vincent, (eds.), Elsevier, place.

Stigler, G.J., 1971: The theory of economic regulation. *Bell Journal of Economics and Management Science,* 2, 3-21.

Stoneman, P., 1983: *The Economic Analysis of Technological Change.* Oxford University Press, Oxford, UK.

Storey, M., G. Boyd, and J. Down, 1999: Voluntary Agreements with Industry. In *Voluntary approaches in Environmental Policy.* C. Corraro, and F. Leveque (eds.), Kluwer Academic Publishers, pp. 187-207.

Svendsen, G.T., 1999: U.S. interest groups prefer emission trading: a new perspective. *Public Choice,* **101**(1-2), 109-128.

Swezey, B., and L. Bird, 2000: *Green Power Marketing in the United States: A Status Report.* National Renewable Energy Laboratory, 5th editon. Golden, CO.

TERI, 1997: Capacity building for technology transfer in the context of climate change. TERI (Draft.), New Delhi, India.

TERI, 1998: *Clean Development Mechanism: Issues and Modalities.* TERI, New Delhi, India.

The People's Republic of China, 1997: National Report on Sustainable Development.

Thorne, S., and E.L. La Rovere, 1999: Criteria and indicators for the appraisal of Clean Development Mechanism Projects. Paper presented at COP5, Helio International.

Tietenberg, T., 1998: Ethical influences on the evolution of the US tradable permit approach to air pollution control. *Ecological Economics.* **24**, 241-257.

Tietenberg, T., 2000: *Environmental and Natural Resource Economics,* 5th ed. Addison-Wesley, Reading, MA.

Tietenberg, T., and D. Wheeler, 2000: Empowering the Community: Information Strategies for Pollution Control. In *Frontiers of Environmental Economics,* H. Folmer, (ed.).

Tol, R.S.J., 1999: Kyoto, Efficiency, and Cost-Effectiveness: Application of FUND, *The Energy Journal, Special Issue. The Costs of the Kyoto Protocol: A Multi-Model Evaluation,* **May**, 131-156.

Toman, M.A., R.D. Morgenstern, and J. Anderson, 1999: The economics of when flexibility in the design of GHG abatement policies. Resources for the Future Discussion Paper 99-38 (revised).

Torvanger, A., and Skodvin, T., 1999: *Implementing the Kyoto Protocol - The role of environmental agreements.* CICERO (Center for International Climate and Environmental Research).

Trexler, M., 1998: *The Role of Forestry as a Climate Change Mitigation Strategy.* U.S. Environmental Protection Agency, Washington DC.

Tulpulé, V., S. Brown, J. Lim, C. Polidano, H. Pant, and B.S. Fisher, 1999: The Kyoto Protocol: An Economic Analysis Using GTEM. *The Energy Journal, Special Issue, The Costs of the Kyoto Protocol. A Multi-Model Evaluation,* **May,** 257-286.

Ulph, A., 1993: Environmental Policy and International Trade When Governments and Producers Act Strategically. *Journal of Environmental Economics and Management.*

Ulph, A., 1994: Environmental policy, plant location and government protection. In *Trade, Innovation, Environment.* C. Carraro, C. (ed.), Kluwer, Dordrecht.

Ulph, A., 1996: Strategic environmental policy and international trade - the role of market conduct. In *Environmental Policy and Market Structure.* C. Carraro, Y. Katsoulacos, and A. Xepapadeas, (eds.), pp. 99-127.

Ulph, D., 1994: Strategic innovation and strategic environmental policy. In *Trade, Innovation, Environment.* C. Carraro (ed.), Kluwer, Dordrecht.

UNCTAD, 1998: *Greenhouse Gas Emissions Trading: Defining the Principles, Modalities, Rules and Guidelines for Verification, Reporting and Accountability.* United Nations, p.15., August.

UNEP, 1983; 1991: Multilateral Treaties in the Field of the Environment, Vol. 1 (1983), Vol. 2 (1991).

UNFCCC, 1992: *United Nations Framework Convention on Climate Change.* Climate Change Secretariat, Geneva, June.

UNFCCC, 1997: *The Kyoto Protocol to the Convention on Climate Change.* Climate Change Secretariat, Bonn, December.

UNFCCC, 1999: Procedures and mechanisms relating to compliance under the Kyoto Protocol. Submissions from Parties, Note by the secretariat, United Nations (FCCC/SB/1999/MISC.4 and addenda).

UNFCCC, 2000: Procedures and mechanisms relating to compliance under the Kyoto Protocol. Proposals by the Co-Chairman of the Joint Working Group on Compliance, 15 September.

United Kingdom of Great Britain and Northern Ireland, 1998: Non-Paper on Principles, Modalities, Rules and Guidelines for an International Emissions Trading Regime on behalf of the European Community and its Member States and Czech Republic, Slovakia, Croatia, Latvia, Switzerland, Slovenia, Poland and Bulgaria, June 8.

US Congressional Record, 1992: The Comprehensive National Energy Policy Act, Vol. 138, No. 142-Part V, Washington, DC.

US DOE, 1999: *Supporting Analysis of the Clinton Administration's Comprehensive Electricity Act (CECA)*. US DOE, May.

US FERC, 1996: Environmental Impact Statement, Promoting Wholesale Competition through Open Access Non-Discriminatory Transmission Services by Public Utilities and Recovery of Stranded Cost by Public Utilities and Transmitting Utilities, US FERC, April.

Van Beers, C. and Van den Bergh, 1997: An empirical multi-country analysis of the impact of environmental regulations on foreign trade flows. *Kyklos* **50**, 29-46.

Van Dril, A.W.N., 2000: *Testing Dutch LTA's: eco-efficiency can be improved.* ENER.

Vollebergh, H.R.J., J.L. de Vries and P.R. Koutstaal, 1997: Hybrid carbon incentive mechanisms and political acceptability. *Environmental and Resource Economics* **9**, 43-63.

Ward, H. and D. Black, 2000: Trade, Investment and the Environment, The Royal Institute of International Affairs, London.

Watson, W.D. and R.G. Ridker, 1984: Losses from effluent taxes and quotas under uncertainty. *Journal of Environmental Economics and Management* **11**, 310-316.

WEC, 1998: *Benefits and Deficiencies of Energy Sector Liberalization*. World Energy Council.

Weinberg, C.J., 1995: The Electricity Utility: Restructuring and Technology - A Path to Light or Dark. In *Profits in the Public Interest*. National Association of Regulatory Utility Commissioners. Washington DC.

Weitzman, M., 1974: Prices versus quantities. *Review of Economic Studies* **41**, 477-491.

Wellisch, D., 1995: Locational choices of firms and decentralized environmental policy with various instruments. *Journal of Urban Economics* **37**(3), 290-310.

Werksman, J., 1998: Compliance and the Kyoto Protocol: Building a Backbone into a "Flexible" Regime. *9 Yearbook International Environmental Law*, **48**.

Werksman, J., 1999: *GHG Emissions Trading and the WTO*. Foundation for International Law and Development, London.

Weyant, J.P., and J. Hill, 1999: Introduction and Overview. *The Energy Journal, Special Issue, The Costs of the Kyoto Protocol: A Multi-Model Evaluation*, **May**, vii-xiv.

Wiener, J., 1998: Global Environmental Regulation: Instrument Choice in Legal Context. *Yale Law Journal, 677*, 108.

Wirl, F., C. Huber, and I.O. Walker, 1998: Joint Implementation: Strategic Reactions and Possible Remedies. *Environmental and Resource Economics*, 12(2), 203-224.

Wiser, G., 1999: *Compliance Systems Under Multilateral Agreements*. Centre for International Environmental Law, Washington, DC., October.

Wiser, G., 2000: *Restoring the Balance, Using Remedial Measures to Avoid and Cure Non-Compliance under the Kyoto Protocol*, March.

Wiser, R.H., 1997: Renewable energy finance and project ownership: the impact of alternative development structures on the cost of wind power. *Energy Policy*, 25(1), 15-27.

Wiser, R. 1999. *Comparing State Portfolio Standards and System-Benefits Charges Under Restructuring*. LBNL, April.

Wiser, R., and S. Pickle, 1997: *Green Marketing, Renewables and Free Riders: Increasing Customer Demand for Public Good*. Lawrence Berkeley National Laboratory, Berkeley, CA, LBNL-40532, UC-1321, September.

Wiser, R., J. Fang, and A. Houston, 1999: *Green Power Marketing in Retail Competition: An Early Assessment*. The Topical Issues Brief Series sponsored by DOE's Office of Energy Efficiency and Renewable Energy, place.

Wiser, R., K. Porter, and M. Bolinger, 2000: *Comparing State Portfolio Standards and System-Benefits Charges Under Restructuring*. Lawrence Berkeley National Laboratory. Berkeley, CA.

Woerdman, E., 2000: *The EU Proposal on Supplementarity: a Questionnaire*. University of Groningen, Groningen, The Netherlands.

Woerdman, E., and W.P. van der Gaast, 1999: A Review of the Cost-Effectiveness of Activities Implemented Jointly: Projecting JI and CDM Credit Prices. ECOF Working Paper (First Draft), University of Groningen, Groningen, The Netherlands.

Wolsink, M., 1996: Dutch Wind Power Policy: Stagnating Implementation of Renewables. *Energy Policy*, 24(12).

World Trade Organization, 1999: *United States- Import Prohibition of Certain Shrimp and Shrimp Products (Shrimp-Turtle decision)*. Report of the Appellate Body (WT/DS58/AB/R), 12 October.

Wortmann, K., M. Klitzke, S. Lörx, R. Menges, 1996: Grüner Tarif. Klimaschutz durch freiwillige Kundenbeiträge zum Stromtarif. Akzeptanz, Umsetzung, Verwendung. Energiestiftung Schleswig-Holstein Studie 2, Kiel, Germany.

Yamaguchi, M. 2000. *Global Environmental Issues and Corporate Activity*. Iwanami Publishing Company. Tokyo (in Japanese).

Yamin, F., 1998: The Clean Development Mechanism and Adaptation, Working Paper, Foundation for International Environmental Law and Development, London, p. 38.

Yamin, F, J. Burniaux, and A. Nentjes, 2000: Kyoto Mechanisms: Key Issues for Policy-Makers for COP-6. *International Environmental Agreements: Politics, Law and Economics*, **1**(2).

Zerbe, R.O., 1970: Theoretical efficiency in pollution control. *Western Economic Journal*, **8**, 364-376.

Zhang, Z., 1999a: Estimating the Size of the Potential Market for All Three Flexibility Mechanisms under the Kyoto Protocol.

Zhang, Z., 1999b: Towards a Successful International GHG Emissions Trading System. *On the Compatibility of Flexible Instruments*. C.J. Jepma, W. van der Gaast (eds.), Kluwer Academic Publishers, Dordrecht, pp. 93-102.

Zhang, Z., 2000: Should the rules of allocating permits be harmonized? *Ecological Economics*.

7

Costing Methodologies

Co-ordinating Lead Authors:
ANIL MARKANDYA (UK), KIRSTEN HALSNAES (DENMARK)

Lead Authors:
Alessandro Lanza (Italy), Yuzuru Matsuoka (Japan), Shakespeare Maya (Zimbabwe), Jiahua Pan (China/Netherlands), Jason Shogren (USA), Ronaldo Seroa de Motta (Brazil), Tianzhu Zhang (China)

Contributing Author:
Tim Taylor (UK)

Review Editor:
Eberhard Jochem (Germany)

CONTENTS

EXECUTIVE SUMMARY

Using resources to mitigate greenhouse gases (GHGs) generates opportunity costs that should be considered to help guide reasonable policy decisions. Actions to abate GHG emissions or increase carbon sinks divert resources from other uses like health care and education. Assessing these costs should consider the total value that society attaches to the goods and services forgone because of the diversion of resources to climate protection. In some cases, the benefits of mitigation could exceed the costs, and thus society gains from mitigation.

This chapter addresses the methodological issues that arise in the estimation of the monetary costs of climate change. The focus is on the correct assessment of the costs of mitigation measures to reduce the emissions of GHGs. The assessment of costs and benefits should be based on a systematic analytical framework to ensure comparability of estimates and transparency of logic. One well-developed framework assesses costs as changes in social welfare based on individual values. These individual values are reflected by the willingness to pay (WTP) for environmental improvements or their willingness to accept (WTA) compensation. From these value measures can be derived measures such as the social surpluses gained or lost from a policy, the total resource costs, and opportunity costs.

While the underlying measures of welfare have limits and using monetary values remains controversial, the view is taken that the methods to "convert" non-market inputs into monetary terms provide useful information for policymakers. These methods should be pursued when and where appropriate. It is also considered useful to supplement this welfare-based cost methodology with a broader assessment that includes physical impacts when possible. In practice, the challenge is to develop a consistent and comprehensive definition of the key impacts to be measured. In this chapter the costing methodology is overviewed, and issues involved in using these methods addressed.

The costs of climate protection are affected by decisions on some key elements, the analytical structure, and the assumptions made. Among other key presumptions, these include the definition of the baseline, assumption about associated costs and benefits that arise in conjunction with GHG emission reduction policies, the flexibility available to find the carbon emissions of lowest cost, the possibility of no regret options, the discount rate, the assumption of the rate of autonomous technological change, and whether revenue is recycled.

First, defining the baseline is a key part of cost assessment. The baseline is the GHG emissions that would occur in the absence of climate change interventions. It helps determine how expensive GHG emissions reduction might be. The baseline rests on key assumptions about future economic policies at the macroeconomic and sectoral levels, including structure, resource intensity, relative prices, technology choice, and the rate of technology adoption. The baseline also depends on presumptions of future development patterns in the economy, like population growth, economic growth, and technological change.

Second, climate change policies may have a number of side-impacts on local and regional air pollution associated, and indirect effects on issues such as transportation, agriculture, land use practices, employment, and fuel security. These side-impacts can be negative as well as positive and the inclusion of the impacts then can tend to generate higher as well as lower climate change mitigation costs compared with studies that do not include such side-impacts.

Third, for a wide variety of options, the costs of mitigation depend on the regulatory framework adopted by national governments to reduce GHGs. The more flexibility allowed by the framework, the lower the costs of achieving a given reduction. More flexibility and more trading partners can reduce costs, as a firm can search out the lowest-cost alternative. The opposite is expected with inflexible rules and few trading partners.

Fourth, no regrets options are by definition actions to reduce GHG emissions that have negative net costs. Net costs are negative because these options generate direct or indirect benefits large enough to offset the costs to implement the options. The existence of no regrets potential implies that people choose not to exercise some carbon-reducing options because of relative prices and preferences, or that some markets and institutions do not behave perfectly. The presumption of effective policies that capture large no regrets options reduces costs.

Fifth, there are two approaches to discounting—an ethical or prescriptive approach based on what rates of discount should be applied, and a descriptive approach based on what rates of discount people (savers as well as investors) actually apply in their day-to-day decisions. For mitigation analysis, the country must base its decisions at least partly on discount rates that reflect the opportunity cost of capital. Rates that range from 4% to 6% would probably be justified in developed countries. The rate could be as high as 10%–12% in developing countries. It is more of a challenge to argue that climate change mitigation projects should face different rates, unless the mitigation project is of very long duration. Note that these rates do not reflect private rates of return, which typically must be greater to justify a project, at around 10%–25%.

Sixth, modellers account for the penetration of technological change over time through a technical coefficient called the "autonomous energy efficiency improvement" (AEEI). AEEI reflects the rate of change in the energy intensity (the ratio of energy to gross domestic product) holding energy prices constant. The presumed autonomous technological improvement in the energy intensity of an economy can lead to significant differences in the estimated costs of mitigation. As such, many observers view the choice of AEEI as crucial in setting the baseline in which to judge the costs of mitigation. The costs of mitigation are inversely related the AEEI–a greater AEEI the lower the costs to reach any given climate target. The costs decrease because people adopt low-carbon technology unrelated to changes in relative prices.

Other issues to be considered in the assessment of mitigation policies include the marginal cost of public funds, capital costs, and side effects. Policies such as carbon taxes or auctioned (tradable) carbon-emissions permits generate revenues that can be recycled to reduce other taxes that are likely to be distortionary. There has been considerable debate as to whether such revenue recycling might eliminate the economic costs of such mitigation policies. Theoretical studies indicate that this result can occur in economies with highly inefficient tax systems. Some empirical studies obtain the no-cost result, although many such studies do not. Tax recycling reflects several complicated assumptions in the baseline and policy case regarding the structure of the tax system and the overall policy framework, among others. Target setting and timing also affect cost estimates. Reduction targets defined as percentage reductions of future GHG emissions create significant uncertainty about GHG emission levels.

In addition, several issues on technology use in developing countries and economies in transition (EITs) warrant attention as critical determinants for climate change mitigation potential and related costs. These include current technological development levels, technology transfer issues, capacity for innovation and diffusion, barriers to efficient technology use, institutional structure, and human capacity aspects.

Equity is another issue in evaluating mitigation policies. The use of income weights is one approach to address equity. Under this system each dollar of costs imposed on a person with low income is given greater weight relative to the cost for a person with a high income. This method is, however, controversial and it is difficult to obtain agreement on the weights to be used. An alternative method is to report the distributional impacts separately. In this case it is important that all the key stakeholders are identified and the distributional effects on each reported. A third possibility is to use average damage estimates and apply these to all those impacted, irrespective of their actual WTP.

Given these presumptions on structure, the costs of climate protection can be modelled and assessed at three levels:

- Project level analysis estimates costs using "stand-alone" investments assumed to have minor secondary impacts on markets.
- Sector level analysis estimates costs using a "partial-equilibrium" model, in which other variables are presumed as given.
- Macroeconomic analysis estimates costs by considering how policies affect all sectors and markets, using various macroeconomic and general equilibrium models. The modeller confronts the trade-off between the level of detail in the cost assessment and complexity of the system. For example, a macroeconomic system tries to capture all direct and indirect impacts, with little detail on the impacts of specific smaller scale projects.

Modelling climate mitigation strategies can be done using several techniques, including input–output models, macroeconomic models, computable general equilibrium models, and models based on the energy sector. Hybrid models have also been developed to provide more detail on the structure of the economy and the energy sector. Two broad classes of integrated assessment models can be identified: policy optimization models and policy evaluation models. The appropriate use of these models depends on the subject of the evaluation and the availability of data.

Finally, the main categories of climate change mitigation policies include market-oriented, technology-oriented, voluntary, and research and development (R&D) policies. Climate change mitigation policies can include elements of two or more policy options. Economic models, for example, mainly assess market-oriented policies and in some cases technology policies, primarily those related to energy supply options. In contrast, engineering approaches mainly focus on supply and demand-side technology policies. Both approaches are relatively weak in the representation of R&D policies.

7.1 Introduction

7.1.1 Background and Structure of the Chapter

This chapter addresses the methodological issues that arise in the estimation of the monetary costs of climate change. The focus here is on the correct assessment of the costs of mitigation measures to reduce the emissions of greenhouse gases (GHGs). The other two areas in which cost issues arise are the estimation of the climate change impacts in monetary terms, and the assessment of measures to adapt to climate change. Working Group II (WGII) is charged with the responsibility to evaluate the impacts and adaptation measures. It is important, though, that much of the discussion in this chapter is relevant to these areas. The basic principles of cost estimation certainly apply in all three areas. Moreover, some of the key issues in cost estimation that arise in the assessment of impacts are also relevant to the estimation of the costs of mitigation. Hence, the relationship between the costs discussed by WGII and those discussed by WGIII is close.

The chapter begins by providing the background to this assessment report; by giving a summary of the Second Assessment Report (SAR) and of the developments in the literature since SAR (IPCC, 1996a, 1996b). Section 7.2 discusses the elements in any climate change cost estimation. It begins by setting out the decision-making framework for mitigation decisions. Unfortunately, this framework is complex, as it involves the application of different modelling techniques and assumptions. Important within the framework are issues of ancillary and co-benefits of climate change mitigation, evaluation techniques, the treatment of barrier removal and implementation costs, discounting, and the linkages between adaptation and mitigation. The conventional cost-effectiveness and the cost–benefit tools used for making decisions to reduce GHGs, or to select adaptation measures, provide only part of the information required by the decision maker. The extensions that are currently being discussed, and used in some cases, include the valuation of external effects, and considerations of equity and sustainability. The outline of the extended decision-making framework and its relationship to the cost methodology is discussed in Sections 7.2.1 to 7.2.5.

Section 7.3 discusses the critical assumptions made in the application of the methodology to climate change problems. The key issues are:
- different systems in which the cost analysis is carried out–project sector and macro level;
- determination of baselines;
- treatment of technological change;
- assessment of cost implications of including alternative GHG emission reduction options and carbon sinks; and
- treatment of uncertainty.

Section 7.4 covers the practical problems that arise in cost estimation, particularly relating to the linkages between the "micro" cost exercise and the broader "macro" picture. The

problems covered are:
- relationship to objectives of development, equity, and sustainability (DES);
- income and other macroeconomic effects of mitigation and adaptation policies;
- issues of spillovers;
- treatment of equity; and
- treatment of future costs and sustainability issues.

Section 7.5 considers the special issues that arise in the estimation of costs in developing countries and economies in transition (EITs).

Section 7.6 discusses the relationship between the cost assessment methodology and the models used to estimate mitigation costs. Issues discussed include classification of models (Section 7.6.2), top-down and bottom-up models (Section 7.6.3) integrated assessment models (IAMs; Section 7.6.4), categorization of climate change mitigation options (Section 7.6.5), and critical assumptions (Section 7.6.6).

The links between this chapter with others is as follows. Section 7.1 overlaps with Chapter 10, Section 7.2 with Chapter 6, and Section 7.6 with Chapters 8 and 9.

7.1.2 Summary of the Second Assessment Report on Cost Issues

IPCC's SAR published a separate volume on the economic and social dimensions of climate change (IPCC, 1996a). This report considered all aspects of climate change, including impacts, adaptation, and mitigation of climate change. The volume on economic and social dimensions was supplemented by a report from another working group of the IPCC that dealt with scientific and technical analyses of the impacts, adaptation, and mitigation of climate change (IPCC, 1996b). The Third Assessment Report (TAR) is structured in a different way. Impacts and adaptation are addressed together by one working group (WGII), and mitigation by another group (WGIII). All the technical areas, including scientific, engineering, economic, and social aspects of climate change impacts, adaptation, and mitigation, however, are integrated in the working groups.

The WGII SAR (IPCC, 1996b) reported a number of cost estimates for individual climate change mitigation technologies, but did not include specific subsections or extensive discussions on the cost assessment framework or methodological issues related to valuation issues. This section therefore only provides a short summary of the coverage of costing methodologies in the report of the social and economic dimensions by WGIII (IPCC, 1996a).

Costing methodologies were addressed as part of several chapters in the WGIII SAR (IPCC, 1996a). These included chapters on the decision-making framework, equity and social considerations, and intergenerational equity: discounting and economic efficiency. Furthermore, the report included two conceptual chapters on cost and methodologies, namely Chapter 5 (Applicability of Techniques of Cost–Benefit Analysis to Climate Change) and Chapter 8 (Estimating the Costs of Mitigating Greenhouse Gases). The first of these chapters included a general outline of analytical approaches applied to climate change cost assessment, with emphasis on cost–benefit analysis and further development of this framework to facilitate multi-attribute analysis. The analytical approaches presented were discussed in relation to different decision frameworks and valuation approaches.

Chapter 8 of the WGIII SAR (IPCC, 1996b) was a methodological introduction to a subsequent chapter on comparative assessments of the modelling results for mitigation costs. A taxonomy of the mitigation cost components applied in the models was presented, including the direct engineering and financial costs of specific technical measures, economic costs for a given sector, macroeconomic costs, and welfare costs. The importance of different assumptions, such as development patterns, technological change, and policy instruments, were then assessed in relation to cost concepts and modelling approaches. Some of the focal areas considered were "top-down" versus "bottom-up" models, double dividend issues and no regret options, long-term projections, and special issues related to mitigation-cost analysis for developing countries.

The WGIII SAR (IPCC, 1996a) also included an extensive review of the mitigation costs for different parts of the world based on top-down and bottom-up methodologies. The review, which was based on an assessment of several hundred studies, raised a number of important costing issues that are critical to the further development of cost concepts and models. These issues include, *inter alia,* model structure, assumptions on demographic and economic growth, availability and costs of technical options, timing of abatement policies, discount rate, and the effect of research and development (R&D).

7.1.3   *Progress since the Second Assessment Report*

A number of IPCC activities based on SAR have developed cost methodologies and applied them to the appraisal of specific policies. Some of the main activities are the IPCC Technical Paper on Technologies, policies, and measures for mitigating climate change (IPCC, 1996c) and the UNEP report on *Mitigation and Adaptation Cost Assessment Concepts, Methods and Appropriate Use*, which was developed on the basis of an IPCC workshop in June 1997 (Christensen *et al.*, 1998).

The IPCC Technical Paper (IPCC, 1996c) summarizes the information on mitigation technology costs provided by WGII SAR (IPCC, 1996b), and the chapter on policy instruments of WGIII SAR (IPCC, 1996b, Chapter 11). The aim of the Technical Paper was to provide a short overview of cost information to be used by climate change policymakers and by the Subsidiary Body for Scientific and Technical Advice (SBSTA) of the UN Convention on Climate Change.

The UNEP report (Christensen *et al.*, 1998) defines and clarifies mitigation and adaptation cost concepts to be used in the field of climate change based on WGIII SAR (IPCC, 1996a). The aim is to overcome some of the variations in the cost concepts that were presented in various chapters of SAR and to develop a generic overview of cost concepts that are easier to use for practitioners in the field. The report includes chapters on general mitigation and adaptation cost concepts, sectoral applications, macroeconomic analysis, and special issues in costing studies for developing countries, and concludes on the applicability of the various cost concepts in the formulation of national climate change policies and programmes.

During the TAR process, a crosscutting issues paper was prepared (Markandya and Halsnaes, 2000). The purpose was to provide a non-technical guide to the application of cost concepts in the analysis of climate change policies by any of the working groups involved in the TAR. Costs of mitigation, adaptation, or GHG emissions are likely to be estimated and their implications discussed in many parts of the TAR. It is essential, therefore, that a common understanding of the use of different cost concepts is employed. The crosscutting paper proposed a set of definitions for these concepts. The paper also identified categories of costs and their relevance in the climate change area. In this chapter the crosscutting issues paper is taken as the point of departure, the ideas are developed further and an elaboration of some of them provided. The purpose, however, is the same: to ensure a common understanding of commonly used cost terms, and the role of cost analysis within the broader decision-making framework for climate change policies.

After SAR, extensive debates arose regarding suitable costing methods to quantify the relative indirect economic impacts of various policies in distinct regions, with no consensus on the most suitable methods to be employed. However, a consensus is now beginning to emerge on how to quantify some ancillary benefits (OECD, 2000), and Chapters 8 and 9 herein**.** In preparation for TAR, Burtraw *et al.* (1999) provide a synthesis of methodological issues relevant to the assessment of ancillary costs and benefits of GHG mitigation policies. The magnitude of potential ancillary benefits depends upon the regulatory, demographic, technological, and environmental baselines. The magnitude and scope of potential benefits of GHG mitigation policies can be expected to be greater in cases in which higher emission baselines obtain and lower for cases in which regulatory and technological innovation have been more long standing (Morgenstern, 2000).

7.2   **Elements in Costing**

7.2.1   *Introduction*

This section addresses a number of key conceptual issues related to mitigation cost concepts, including definitions of private and social costs and methods to assess the side effects and equity aspects of mitigation policies. An overview is given of analytical approaches to assess mitigation costs, including a classi-fication and discussion of different modelling approaches and critical assumptions. The issue of ancillary and co-benefits of climate change mitigation is discussed. Valuation techniques are presented, as is the treatment of barrier removal and imple-mentation costs. A review of recent developments in the field of discounting is then presented and the section concludes with an investigation of the linkages between adaptation and mitigation.

7.2.2 Cost Estimation in the Context of the Decision-making Framework

Actions taken to abate GHG emissions or to increase the size of carbon sinks generally divert resources from other alterna-tive uses. The theoretically precise measure of the social costs of climate protection, therefore, is the total value that society places on the goods and services forgone as a result of the diversion of resources to climate protection. A social cost assessment should ideally consider all welfare changes that result from the changes in resources demanded and supplied by a given mitigation project or strategy in relation to a specific non-policy case (see Hazilla and Kopp, 1990). The assessment should include, as far as possible, all resource components and implementation costs. This means that both the benefits and the costs of a mitigation action should be included in the estima-tion. In some cases, the sum of all the benefits and costs asso-ciated with a mitigation action could be negative, meaning that society benefits from undertaking the mitigation action.

The conceptual foundation of all cost estimation is the value of the scarce resources to individuals. Thus, values are based on individual preferences, and the total value of any resource is the sum of the values of the different individuals involved in the use of the resource. This distinguishes this system of values from one based on "expert" preferences, or on the preferences of political leaders. It also distinguishes it from value systems based on ecological criteria, which give certain ecological goals a value in themselves, independent of what individuals might want, now or in the future.

The values, which are the foundation of the estimation of costs, are measured by the applied welfare economic concepts of the willingness to pay (WTP) of individuals to buy the resource, or by the individuals' willingness to accept (WTA) compensation to part with the resource. The WTP measure of value reflects the maximum people are willing to pay to live in a world with climate policy in force rather than not. WTA is the minimum compensation people would accept to live without this climate policy (e.g., Willig, 1976; Randall and Stoll, 1980; Hanemann, 1991; Shogren et al., 1994). The concepts of WTP and WTA therefore play a critical part in defining the social cost method.

WTP or WTA is most commonly approximated by the con-sumer and producer surplus as revealed in the demand and sup-ply schedules for the resources whose consumption and pro-duction is affected by the mitigation action. These measures

are standard economic tools of cost–benefit analysis (Hanley et al., 1997). In some cases, however, the resources that are affected do not have well-defined markets and hence lack iden-tifiable demand and supply schedules. Examples are changes in air quality, or changes in recreational use of forests. In such cases other methods of measuring WTP and WTA are required. These have been developed recently and can now provide cred-ible estimates for a range of non-marketed resources, though some debate remains over the application of such values to all policy-relevant impacts.

There is also a relationship between WTP and WTA and the conventional aggregate measures of economic activity such as gross domestic product (GDP). The classic paper on this is Weitzman (1976), which showed that GDP less depreciation of capital (or "net national product") is a measure of the net out-put that represents the income on the economy's capital stock when that economy is operating according to competitive mar-ket rules. However, a competitive economy is also one that maximizes the sum of consumer and producer surpluses. Hence GDP is closely linked to consumer and producer surplus maximization for commodities that operate through the market place. However, the relationship breaks down if competitive markets do not exist for all scarce resources. In this case, GDP changes do not fully reflect changes in social welfare.

A frequent criticism of this costing method is that it is inequitable, as it gives greater weight to the "well off". This is because, typically, a well-off person has a greater WTP or WTA than a less well-off person and hence the choices made reflect more the preferences of the better off. This criticism is valid, but there is no coherent and consistent method of valua-tion that can replace the existing one in its entirety. Concerns about, for example, equity can be addressed along with the basic cost estimation. The estimated costs are one piece of information in the decision-making process for climate change that can be supplemented with other information on other social objectives, for example impacts on key stakeholders and the meeting of poverty objectives.

7.2.2.1 Analytical Approaches

Cost assessment is an input into one or more of the rules for decision making, which are discussed in more detail in Chapter 10 of this report. Economic approaches to decision making include cost–benefit analysis, and cost-effectiveness analysis, and these approaches can be supplemented with multi-attribute analysis that facilitates an integrated assessment of economic impacts and other quantitative and non-quantitative informa-tion. These approaches are briefly described in *Box 7.1*.

It should be recognized that some types of impacts can be mea-sured in both monetary terms and physical terms. This applies, for example, to changes in air pollution as a result of the reduc-tions in GHGs.
There is a major difference between the economic approaches and multi-attribute analysis in how the various dimensions of

the assessment are summarized. The economic approaches seek to provide aggregates to single measures based on an economic welfare evaluation, while multi-attribute analysis does not provide an aggregation of the different dimensions of the analysis.

7.2.2.2   Cost Analysis and Development, Equity, and Sustainability Aspects

The underlying objective behind any cost assessment is to measure the change in human welfare generated as the result of a reallocation or change in use of resources. This implies the existence of a function in which welfare or "utility" depends on various factors such as the amounts of goods and services that the individual can access, different aspects of the individual's physical and spiritual environment, and his or her rights and liberties. Constructing a "utility function", representing social welfare, that is an aggregate measure of all such impacts for all individuals involves a number of complexities and controversial equity issues that have been intensively studied by economists (see, for example, Blackorby and Donaldson, 1988). However, the sum of the individual WTPs and WTAs can be taken as a measure of the social welfare, which finesses these difficulties to a considerable extent. There remain, however, issues that cannot be fully addressed in this WTP–WTA framework, most important of which are equity and sustainability.

Box 7.1. Decision-making Approaches

Cost–benefit analysis
This measures all negative and positive project impacts and resource uses in the form of monetary costs and benefits. Market prices are used as the basic valuation, as long as markets can be assumed to reflect "real" resource scarcities. In other cases the prices are adjusted to reflect the true resource costs of the action. Such adjusted prices are referred to as shadow prices (Squire and van der Tak, 1975; Ray, 1984).

Cost-effectiveness analysis
A special case of cost–benefit analysis in which all the costs of a portfolio of projects are assessed in relation to a policy goal. The policy goal in this case represents the benefits of the projects and all the other impacts are measured as positive or negative costs. The policy goal can, for example, be a specified goal of emissions reductions for GHGs. The result of the analysis can then be expressed as the costs (US$/t) of GHG emissions reductions (Sathaye *et al.*, 1993; Markandya *et al.*, 1998).

Multi-attribute analysis
The basic idea of multi-attribute analysis is to define a framework for integrating different decision parameters and values in a quantitative analysis without assigning monetary values to all parameters. Examples of parameters that can be controversial and very difficult to measure in monetary values are human health impacts, equity, and irreversible environmental damages (Keeney and Raiffa, 1993).

The above analysis of welfare focuses on the narrowly economic dimension. Even within this framework there are complexities that make a full assessment difficult. In addition, however, issues of DES need to be taken into account.[1]

A key question in broadening the analysis of costs to cover these dimensions is whether they can be measured in the same units as the costs (i.e., in money). The authors take the view that the methods to "convert" some of these other dimensions into monetary terms are useful and should be pursued. These are discussed further in Section 7.2.3. At the same time, there is some controversy about the measurement of equity, of environmental impacts and sustainability in monetary terms, as, for example, in the discussion on social cost-benefit analysis in Ray (1984).[2] This is because of disagreement about what values should be attached to physical and social changes that are of interest. Furthermore, it is generally accepted that not all these impacts can be put in monetary terms.[3] Hence it is important, indeed imperative, that the cost methodology be supplemented by a broader assessment of the impacts with physical values reported wherever possible. These questions are discussed further in Sections 7.3 and 7.4.

7.2.2.3   Ancillary Benefits and Costs and Co-benefits and Costs

The literature uses a number of terms to depict the associated benefits and costs that arise in conjunction with GHG mitigation policies. These include co-benefits, ancillary benefits, side benefits, secondary benefits, collateral benefits, and associated benefits. In the current discussion, the term "co-benefits" refers to the non-climate benefits of GHG mitigation policies that are explicitly incorporated into the initial creation of mitigation policies. Thus, the term co-benefits reflects that most policies designed to address GHG mitigation also have other, often at least equally important, rationales involved at the inception of these policies (e.g., related to objectives of development, sustainability, and equity). In contrast, the term ancillary benefits connotes those secondary or side effects of climate change mitigation policies on problems that arise subsequent to any proposed GHG mitigation policies. These include reductions in local and regional air pollution associated with the reduction of fossil fuels, and indirect effects on issues such as transportation, agriculture, land use practices, employment, and fuel security. Sometimes these benefits are referred to as "ancillary impacts", to reflect that in some cases the benefits may be negative. From the perspective of policies to abate local air pollution, GHG mitigation may be an ancillary benefit.

---

[1] Other issues that may need to be considered include incomplete information, perceptual biases, and learning.

[2] Indeed, many of the comments on earlier drafts of this chapter took different positions on this issue.

[3] For some impacts, such as those on "sustainability", the selection of physical indicators is also a matter of controversy.

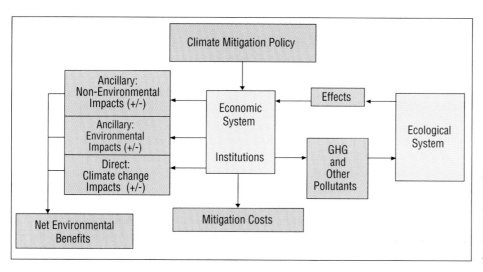

Figure 7.1: *Mechanism for the Generation of Ancillary Impacts. Please note that climate change impacts are not discussed in this report, but in the Third Assessment Report of Working Group II.*

Figure 7.1 illustrates the generation of ancillary benefits to GHG emission reduction policies.[4] These policies operate through the economic and institutional system within a country and lead to reductions in GHGs, changes in other pollutants, and mitigation costs. Changes in GHG emissions in turn lead to changes in air and water pollution, which ultimately extend throughout the environment and feed back into the economy. Then, depending on baseline conditions, technologies, and institutions, such as labour markets, tax systems, and existing environmental and other types of regulations (represented by "institutions" in the economic system box), these feedbacks may become:

- environmental impacts (such as the value of changes in conventional air or water pollution);
- non-environmental impacts (such as the value of employment effects); and,
- direct climate change impacts.

There appear to be three classes of literature regarding the costs and benefits of climate change mitigation:

(1) literature that primarily looks at climate change mitigation, but that recognizes there may be benefits in other areas;

(2) literature that primarily focuses on other areas, such as air pollution control, and recognizes there may be benefits in the area of climate mitigation; and

(3) literature that looks at the combination of policy objectives (climate change and other areas) and looks at the costs and benefits from an integrated perspective.

Each of these classes of literature may have their own preferred terms, and for class (3) it seems to be "co-benefits". TAR acknowledges the relevance of all three, yet specifically wants to make the case for an integrated approach, linking climate change mitigation to the achievement of sustainable development and other policy objectives. Therefore, in this report, the term "co-benefits" is used only when speaking generically about the issue because of the limited availability of literature. The term "ancillary benefits" is used when addressing class (1) and (2) literature. Class (1) literature appears to be the most extensive; it is this literature on the ancillary benefits of climate change mitigation that is primarily covered in this section.

The discussion of ancillary impacts and/or co-benefits and costs, and the estimation of these are closely related to the concept of external cost, which is discussed below.

7.2.2.4   Market Failures and External Cost

The term external cost or externality is used to define the costs that arise from any human activity when the agent responsible for the activity does not take full account of the impacts on others of his or her actions. Equally, when the impacts are positive and not accounted for in the actions of the agent responsible they are referred to as external benefits. Consider first the following example of external costs. Emissions of particulate pollution from a power station affect the health of people in the vicinity, but this is not often considered, or is given inadequate weight in private decision-making, as there is no market for such impacts. Such a phenomenon is referred to as an externality, and the costs it imposes are referred to as the external costs.

External costs are distinct from the costs that the emitters of the particulates take into account when determining their outputs, costs such as the prices of fuel, labour, transportation, and energy. Categories of costs that influence an individual's decision-making are referred to as private costs. The total cost to society is made up of both the external cost and the private cost, which together are defined as social cost:

Social Cost = External Cost + Private Cost

---

[4] Various additional interrelationships are omitted from this graphic. An example is that estimated health benefits might be lower if a GHG mitigation policy reduces temperature increases, thereby creating less ozone (O_3).

The private cost component is generally taken from the market prices of the inputs. Thus, if a project involves an investment of US$5 million, as estimated by the inputs of land, materials, labour and equipment, that figure is used as the private cost. That may not be the full cost, however, as far as the estimation of social cost is concerned. If, for example, the labour input is being paid more than its value in alternative employment, the private cost is higher than the social cost. Adjustments to private costs based on market prices to bring them into line with social costs are referred to as shadow pricing. A fuller discussion of shadow pricing is given in Ray (1984).

External costs typically arise when markets fail to provide a link between the person who creates the "externality" and the person who is affected by it, or more generally when property rights for the relevant resources are not well defined. If such rights were defined, market forces and/or bargaining arrangements would ensure that the benefits and costs of generating the external effect balanced properly. The failure to take into account external costs, however, may be a product not only of a lack of property rights, but also the result of a lack of full information and non-zero transaction costs.

7.2.2.5   Critical Assumptions in Studies of Ancillary Benefits and Co-benefits

Policies aimed at mitigating GHGs, as stated earlier, can yield other social benefits and costs (here called ancillary benefits and costs), and a number of empirical studies have made a preliminary attempt to assess these impacts. It is apparent that the actual magnitude of the ancillary benefits or co-benefits assessed critically depends on the scenario structure of the analysis, in particular on the assumptions about policy management in the baseline case (IPCC, 2000b; Krupnick *et al.,* 1996; Krupnick *et al.,* 2000).[5] This implies that whether a particular impact is included or not depends on the primary objective of the programme. Moreover, something that is seen as a GHG reduction programme from an international perspective may be seen, from a national perspective, as one in which local pollutants and GHGs are equally important.

A second point is that the economic accounting of ancillary benefits depends crucially on assumptions about the demographic characteristics, regulatory regime, and available technology and how these will evolve. For example, consider the case in which a government imposes a cap on emissions of sulphur. If a GHG mitigation programme is introduced it may reduce the associated amount of sulphur produced, but other activities may take up the slack and so result in no net change in emissions. Alternatively, consider the situation in which the government has a tax on emissions. If the tax is set equal to the marginal damage from sulphur, a small mitigation programme will not generate any direct benefits in terms of sulphur reduc-

tions (the value of the reductions is exactly matched by the loss of charge revenue). As a third example, consider the case in which the regulator has a plan to tighten the controls on local pollutants. Any GHG mitigation programme that reduces the levels of these emissions has then to be valued relative to the costs of achieving the dynamic baseline, and not in terms of the benefits of reduced emissions themselves. To sum up, the valuation of ancillary and/or co-benefits requires the policymaker to look not only at the external costs of the pollutants, but also at the net costs and benefits of measures being introduced to deal with them.

Externalities do not necessarily arise when there are effects on third parties. In some cases, these effects may already be recognized, or "internal", contained in the price of goods and services. Consider a stylized example, such as damages to vehicles in an automobile accident. If each driver is fully liable for damages to other vehicles and one can reliably assess fault and enforce liability, the damage in an accident would not be an externality because the party at fault would fully recognize the costs. Only if the drivers are not fully liable, or if fault cannot be established, or if liability is not enforceable is there a justification for treating the damage to vehicles in the example as an externality. The key idea is that such exceptions constitute a deviation from ideal institutions. In economic vocabulary, this is referred to as market failure. For damage to be considered an externality from the viewpoint of economic efficiency, some kind of failure in markets or other institutions that causes individuals to fail to take into account the social costs and benefits of their individual actions should be identifiable. From a practical perspective, it is also important that such failures result in an important misallocation of resources.

A full discussion of the empirical relevance of ancillary and/or co-benefits is provided in Chapters 8 and 9.

7.2.2.6   A Partial Taxonomy

A variety of effects may result from GHG policies that are secondary to the reduction in GHG emissions. Existing studies have identified mortality and morbidity benefits associated with collateral reductions in particulates, nitrogen oxides (NO_x), and sulphur dioxide (SO_2) from power plants and mobile sources as a major source of ancillary benefits. Reduced private vehicle use and substitution of mass transit will reduce air pollution and congestion and may also reduce transportation-related fatalities from accidents, although the size of this effect and the degree to which it counts as an ancillary benefit are unclear.[6] Substitution to mass transit may also involve additional costs, in terms of the opportunity cost of

---

[5] See Burtraw *et al.* (1999) and reviews by Burtraw and Toman (1997), Ekins (1996), and Pearce (2000).

[6] A major study in the early 1990s considered externalities throughout various fuel cycles for electricity generation in the USA. It concluded that of the highest-valued endpoints (among many specifically defined endpoints) were fatalities associated with the rail transport of coal and damage to roadway surfaces beyond those internalized in road fees (Lee *et al.*, 1995).

Table 7.1: *Ancillary Impacts*

Ancillary Impact	Expected sign
Reduction in particle pollution when fossil fuel use is reduced	(+)
Increases in urban air pollution when diesel vehicles are introduced to substitute gasoline	(−)
Increased availability of recreational sites when reforestation programmes are introduced	(+)
Increases in household air pollution relative to a baseline when electrification rates are reduced	(−)
Increases in technological efficiency when new technologies are adopted and unit costs fall	(+)
Increases in welfare with a shift to carbon taxation and a reduction in unemployment	(+)
Reductions in road-use related mortality when a shift from private to public transport takes place	(+)
Reductions in congestion with a shift from private to public transport	(+)
Decreases in employment when energy technologies that substitute the use of local fuels are introduced	(−)
Increases in employment that result from GHG projects in which there is an excess need for labour	(+)
Decline in employment because of decreased economic activity resulting from costs associated with GHG projects	(−)
Savings in household time in poor rural households when fuel wood use is replaced by biogas energy	(+)

time, and these ancillary impacts may also need to be considered. Additional areas that might be considered include improvements in ecosystem health (for instance, from reduction in nitrate deposition to estuaries), visibility improvements, reduced materials damages, and reduced crop damages.

At the same time, there may be ancillary costs of GHG mitigation, such as an increase in indoor air pollution associated with a switch from electricity to household energy sources (such as wood or lignite) or greater reliance on nuclear power with its attendant externalities. In developing countries pollution may rise if electrification slows as a result of policy-induced increases in electricity prices relative to other fuels (Markandya, 1994). A related cost stems from forgoing the benefits of electrification, which include increased productive efficiency and emergence of new technologies, to increases in literacy (Schurr, 1984). *Table 7.1* offers an illustrative set of examples of ancillary benefits (+) and costs (−). Under certain conditions, some of these observed impacts do not necessarily count as externalities from the standpoint of economic efficiency, depending on whether the market or institutions fail to account for these impacts in the incentives they provide for individual behaviour.

A taxonomy of the main externalities linked with the public health impacts of air pollution, which was developed in the social cost of electricity studies and is likely to be relevant to ancillary benefit estimation, is provided in *Table 7.2*.

7.2.3 Valuation Techniques for External Effects

The external effects described above cannot be valued directly from market data, because there are no "prices" for the resources associated with the external effects (such as clean air, or clean water). Hence indirect methods have to be adopted. Values have to be inferred from individuals' decisions in related markets, or from directly eliciting the WTP for the environmental good through questionnaires. Values of environmental

goods are broadly divided into use values and non-use values. The former comprises those values that result from some direct or indirect use to which the environment is put. Non-use values arise when individuals have a WTP for an environmental resource even when they make no use of it, or never will make any use of it, see Perman *et al.* (1999) for a discussion of this distinction.

The following methods have been developed and used in valuing environmental (and other) externalities. Further details can be found in several books (Hanley *et al.*, 1997; Bateman and Willis, 1999; Markandya *et al.*, 2000).

7.2.3.1 Impact Pathway Analysis

Impact pathway analysis measures the losses of goods and services affected by environmental impacts which are themselves (or their substitutes) priced in the market. To identify these losses, the effects of an action are traced from the release of pollutants and their dispersion in the ambient environment through to their impacts on natural resources and on humans. Based on the changes of market prices of these goods and services caused by the environmental impacts, demand schedules and the respective consumer surplus, measures can be estimated to reflect the welfare losses. This method has been used extensively to value the impacts of air pollution generated by electricity generation and transport (ExternE, 1995; 1997; 1999). Its main limitations are (a) the physical data on the linkages are not always quantified and those that are can be highly uncertain, (b) market prices are not available for all impacts, and (c) the more sophisticated analysis of price changes requires a level of modelling that is not always possible.

7.2.3.2 Property Prices or the Hedonic Method

Property prices vary according to the many attributes associated with them. House prices, for example, reflect size, commercial facilities, local infrastructure, and other attributes such as environmental quality of the house location. From statistical

Table 7.2: A Sample of externalities assessed in studies of electricity generation

	Health		Materials	Crops	Forests		Amenity[a]	Ecosystems
	Mortality	Morbidity			Timber	Other		
PM10	AM	AM	AM	NE	NE	NE	AM	NE
SO_2[b]	AM	AM	AM	AM	AM	AP	AM	AP
NO_x[b]	AM	AM	AM	AM	AM	NA	NE	AP
Ozone	AM	AM	AM	AM	NA	NA	NE	NE
Mercury and other heavy metals	NA	NA	NE	NE	NE	NE	NE	?
Routine operations[c]	AM	AM	NE	NE	NE	NE	NE	NE
Water pollutants[d]	NE	NE	NE	NE	NE	NE		AP
Noise	NE	NA	NE	NE	NE	NE	AM	NE

AM, assessed in monetary terms, at least in some studies. AP, assessed in physical terms and possibly partly in monetary terms. NA, not assessed, although they may be important. NE, no effect of significance is anticipated.

[a]   Effects of particulate matter less than 10 microns (PM_{10}), NO_x, and SO_2 on amenity arise with respect to visibility. In previous studies these have not been found to be significant in Europe, although they are important in the USA.

[b] SO_2 and NO_x include acid-deposition impacts.

[c]   Routine operations generate externalities through mining accidents, transport accidents, power-generation accidents, construction and dismantling accidents, and occupational health impacts. All these involve mortality and morbidity effects and are externalities to the extent that labour markets do not allow individuals to choose employment with different combinations of risk and reward.

[d]   Water pollution effects include impacts of mining (including solid wastes) on ground and surface water, power-plant emissions to water bodies, and acid deposition and its impacts on lakes and rivers (partly quantified).

Source: Developed from Markandya and Pavan (1999).

analyses of house prices, the contribution of environmental quality to house price variations can be assessed, which is an estimate of how much people are willing to pay for changes in environmental quality. This measure represents a use value for that environmental change from which a demand function can be estimated. The method has been used to value external effects such as noise, air quality, and visibility. The main limitation is that to work efficiently it requires the affected parties to be well informed about the impacts and markets, so that decisions about location can be made freely and easily. For examples of relevant studies see ExternE (1999), Palmquist (1991), and Zabel and Kiel (2000).

7.2.3.3   Contingent Valuation Method

By asking people directly how much they are willing to pay for a change in a provision of benefits from an environmental resource, a hypothetical market can be created in which a demand curve for ecological goods and services can be estimated. This method is the only one by which non-use values can be estimated, since hypothetical markets can be created for them. Since it is not based on revealed preferences, on which the other demand approaches are based, contingent valuation may incur in various biases, from strategic answers to lack of information. Such biases are currently well documented and techniques have been developed to reduce them. Contingent valuation methods have been used to value the use and non-use of sites of special significance, health effects (including changes in the risk of death), and damages to ecosystems (Bateman and Willis, 1999). Despite the considerable amount

of work on reducing the biases that arise because such data do not report actual transactions, this method arouses considerable scepticism among policymakers and its results are not always accepted.

Nevertheless, although such methods of valuation have problems, there is often no suitable alternative and they provide policymakers with important information for decision-making purposes. As suggested above, both physical impacts and values should be used in this process. In relation to climate change, the estimation of external effects arises primarily in the assessment of damages that result from such change, including those in agriculture, forests, energy use, recreation, and health. In relation to mitigation, the applications are primarily in valuing the impacts of O_3, NO_x, SO_x, particulate matter, and secondary particles. In adaptation, the valuation of external effects arises with respect to loss of land, changes to recreational facilities, and changes to agriculture.

7.2.3.4   Benefit Transfer

The valuation of improvements in environmental quality can be expensive. As research budgets are tight, economists explored the concept of "benefit transfer" as a cost-effective alternative to new non-market valuation studies (Desvousges *et al.*, 1992; McConnell, 1992). The term benefit transfer reflects its purpose: transfer the estimated economic value from one environmental good or site to another. Benefit transfer reduces the need to design and implement a new and potentially expensive valuation exercise for the second site. A general four stage process (Atkinson *et al.*, 1992):

- defines the purpose and desired precision of the benefit estimates;
- develops the transfer protocol for the question in hand;
- identifies existing studies that satisfy the protocol; and
- selects the appropriate statistical transfer method that allows for efficient extrapolation of economic data.

Consider the transfer of health risk estimates. For instance, an estimate of WTP for a given risk reduction from contaminated water in Wyoming could be transferred to a reduced risk of poor water quality in Mongolia, as long as the transfer protocol is satisfied. This protocol can be rather strict, however. For a health risk, the researcher must first specify the commodity. This includes defining the response (death or illness) and causal agent (e.g., chemical), as well as understanding the probability and severity of the risk and risk reduction methods, the temporal dimensions of the risk, whether the risk is voluntary or involuntary, and the exposure pathways and exposure levels. Once the risk is defined, the sample and site characteristics have to be classified, including socioeconomic and location particulars. Finally, the protocol has to address the market and exchange mechanisms that define the frame of how risk is reduced. Three elements are likely to matter–the set of risk reduction mechanisms (e.g., mitigation and adaptation options), the measure of value (e.g., WTP or WTA), and the exchange institution or "payment vehicle" (see Kask and Shogren, 1992).

7.2.4 *Implementation Costs and Barrier Removal*

All climate change policies necessitate some costs of implementation, that is costs of changes to existing rules and regulations, making sure that the necessary infrastructure is available, training and educating those who are to implement the policy as well those affected by the measures, etc. Unfortunately, such costs are not fully covered in conventional cost analyses. Implementation costs in this context are meant to reflect the more permanent institutional aspects of putting a programme into place and are different to those costs conventionally considered as transaction costs. The latter, by definition, are temporary transition costs. Considerable work needs to be done to quantify the institutional and other costs of programmes, so that the reported figures are a better representation of the true costs that will be incurred if the programmes considered in Chapter 6 are actually implemented. This section discusses the issues of implementation and the associated costs further.

Several economic and technical studies suggest that there is a large potential for climate change mitigation with no cost or very low cost (see the review on mitigation costing studies given in Chapters 8 and 9 of this report). Low mitigation costs, for example, may result from energy-efficiency improvements relating to end-use savings, as well as from the introduction of more efficient supply technologies. There is also potential for the introduction of renewable energy technologies with low

costs, such as wind turbines, biomass combustion, and solar water-heating systems. The implementation of such low-cost options in many cases implies that a number of current institutional failures and market barriers exist and that policies should be implemented to correct these.

Following this, mitigation cost assessment, in addition to the direct costs of the programmes, should consider implementations costs that arise in the following areas:

- financial market conditions;
- institutional and human capacities;
- information requirements;
- market size and opportunities for technology gain and learning; and
- economic incentives needed (grants, subsidies, and taxes).

Only some of these implementation conditions can be included in the formal cost assessment carried out for individual mitigation options. It is generally more complicated to design implementation programmes targeted to many individual actors (e.g., a demand-side management (DSM) scheme or a tradable carbon permits scheme) than those with centralized project planning (e.g., large-scale power sector changes). In this context it is important to distinguish between marginal and non-marginal projects, since the latter may well induce significant price effects.

Implementation policies can be separated into small "marginal" efforts (which create an incentive to change specific behaviour or introduce new technologies), and more general policy efforts, like economic instruments or general educational programmes (which work by changing the general market conditions and the capability of the actors).

Whether an implementation policy is "marginal" or "general" depends on the general market conditions, as well as on the whole design of policy instruments targeted towards climate change mitigation. Given a "general" environment in which energy and financial markets are efficient, competitive, and have little government intervention, and in which the institutional context is perceived as favourable for climate change mitigation programmes, the implementation policies need only take the form of information programmes, energy auditing, and other specific regulation efforts. However, if energy prices are heavily subsidized and financial markets are very limited, the implementation policy may require general price reforms, specific grants, and other institutional changes.

Implementation policies of the "marginal" sort can be integrated relatively easily into project or sector-level mitigation assessment. Implementation assessment includes the costs of different kinds of programmes for information, training, institution strengthening, and the introduction of technical standards. The most difficult part of such an assessment relates to the behaviour of the target groups. A detailed amount of information is needed on the behaviour of specific actors, including households and private companies, to design the most effective policy options.

It is difficult to integrate general implementation policies, like price changes, into specific project and sector assessments. For a DSM programme in the commercial lighting sector, implementation costs include information and training programmes, institutional capacity building, and sometimes also "costs" of changing the market conditions (prices and taxes). The costs of general changes in market prices and tax systems can only be assessed at the economy-wide level. The introduction of energy or carbon taxes or the removal of subsidies can cause significant structural effects that, again, change energy demand and technology choice. Thus, the proper full analysis of the implementation costs necessitates an economy-wide analysis that involves, for example, the use of computable general equilibrium (CGE) models and intersectoral macroeconomic models.

To a limited extent, such feedbacks can be integrated into a project- or sector-level mitigation-cost assessment by the use of shadow prices. These shadow prices reflect underlying social valuations of the use of different goods and services by different agents. By estimating them in a suitable manner some of the implementation costs, such as changes in government income or expenditure, or the higher value of foreign exchange, can be captured in the cost analysis. Importantly, however, implementation costs assessed using shadow prices do not pick up factors such as quantitative or physical constraints on the use and allocation of some resources, particularly financial ones.

A framework to assess implementation costs thus includes the costs of project or policy design, institutional and human capacity costs (management and training), information costs, and monitoring costs. The costs of resources involved should, in each case, be based on economic opportunity costs.

7.2.5 Discounting

The debate on discount rates is a long-standing one. As SAR notes (IPCC, 1996a, Chapter 4), there are two approaches to discounting; an ethical, or prescriptive, approach based on what rates of discount should be applied, and a descriptive approach based on what rates of discount people (savers as well as investors) actually apply in their day-to-day decisions. SAR notes that the former lead to relatively low rates of discount (around 2%–33% in real terms) and the latter to relatively higher rates (at least 6% and, in some cases, very much higher rates).

The ethical approach applies the so-called social rate of time discount, which is the sum of the rate of pure time-preference and the rate of increase of welfare derived from higher per capita incomes in the future. The descriptive approach takes into consideration the market rate of return to investments,

whereby conceptually funds can be invested in projects that earn such returns, with the proceeds being used to increase the consumption for future generations. Portney and Weyant (1999) provide a good overview of the literature on the issue of intergenerational equity and discounting.

For climate change the assessment of mitigation programmes and the analysis of impacts caused by climate change need to be distinguished. The choice of discount rates applied in cost assessment should depend on whether the perspective taken is the social or private case. The issues involved in the application of discount rates in this context are addressed below.

For mitigation effects, the country must base its decisions at least partly on discount rates that reflect the opportunity cost of capital. In developed countries rates around 4%–6% are probably justified. Rates of this level are in fact used for the appraisal of public sector projects in the European Union (EU) (Watts, 1999). In developing countries the rate could be as high as 10%–12%. The international banks use these rates, for example, in appraising investment projects in developing countries. It is more of a challenge, therefore, to argue that climate change mitigation projects should face different rates, unless the mitigation project is of very long duration. These rates do not reflect private rates of return, which typically need to be considerably higher to justify the project, potentially between 10% and 25%.

For climate change impacts, the long-term nature of the problem is the key issue. The benefits of reduced GHG emissions vary with the time of emissions reduction, with the atmospheric GHG concentration at the reduction time, and with the total GHG concentrations more than 100 years after the emissions reduction. These are very difficult to assess.

Any "realistic" discount rate used to discount the impacts of increased climate change impacts would render the damages, which occur over long periods of time, very small. With a horizon of around 200 years, a discount rate of 4% implies that damages of US\$1 at the end the period are valued at 0.04 cents today. At 8% the same damages are worth 0.00002 cents today. Hence, at discount rates in this range the damages associated with climate change become very small and even disappear (Cline, 1993).

A separate issue is that of the discount rate to be applied to carbon. In a mitigation cost study, should reductions of GHG in the future be valued less than reductions today? It could argued that this is the case, as the impacts of future reductions will be less. This is especially true of "sink" projects, some of which will yield carbon benefits well into the future. Most estimates of the cost of reductions in GHGs do not, however apply a discount rate to the carbon changes. Instead, they simply take the average amount of carbon stored or reduced over the project lifetime (referred to as flow summation) or take the amount of carbon stored or reduced per year (flow summation divided by the number of years). Both these methods are inferior to the application of a discount rate to allow for the greater benefit of present reductions over future reductions. The actual value, however remains a matter of disagreement, but the case for anything more than a very low rate is hard to make (Boscolo *et al.*, 1998).

More recent analysis on discounting now examines rates that vary with the time period considered. In surveys of individual trade-offs over time, Cropper *et al.* (1994) estimated a nominal rate of around 16.8%, based on a sophisticated questionnaire approach to valuing present versus future risks. Most importantly, however, these authors found evidence that respondents do not discount future lives saved at a constant exponential rate of discount. Rather, median rates seem to be decline over time (i.e., a rate is not constant over time but decreases as the time horizon lengthens). Using different econometric specifications that allow the discount rate to decline over time, Cropper *et al.* (1994) estimate that mean discount rates are greater for short time periods relative to long time horizons. For example, fitting their data to a hyperbolic function suggests that mean discount rate is 0.80 for 1 year and 0.08 for 100 years. While the pattern is consistent, the implied rates using linear discount rate functions are much larger: 34% for the initial period and about 12% for the last period.

Hyperbolic discounting implies that a person's relative evaluation of two payments depends on both the delay between the two payments and when this delay will occur–sooner or later. For instance, people often have an impulsive preference for immediate reward. Some people prefer to receive US$1000 today over US$1010 in a month's time, and yet they also prefer US$1010 in 21 months to US$1000 in 20 months, even though both choices involve a month's wait to obtain $10 more (see Lowenstein and Prelec, 1992). Theoretical support for hyperbolic discount rests on the idea that, while interest rates from financial instruments can be used to identify appropriate discount rates for time horizons of a few decades, they do not apply to future interest rates for far distant horizons. These will be determined by future opportunity sets created by many factors, such as economic growth. The fact that the scope of these future opportunity sets for the far distant future is not known adds another layer of uncertainty into climate policy, which tends to drive discount rates down.

Weitzman (1998) surveyed 1700 professional economists and found that (a) economists believe that lower rates should be applied to problems with long time horizons, such as that being discussed here, and (b) they distinguish between the immediate and, step by step, the far distant future. The discount rate implied by the analysis falls progressively, from 4% to 0%, as the perspective shifts from the immediate (up to 5 years hence) to the far distant future (beyond 300 years). Weitzman (1998) suggests the appropriate discount rate for long-lived projects is less than 2%. Finally, hyperbolic discounting has less support if it leads to time-inconsistent planning, as argued by Cropper and Laibson (1999). Time inconsistency arises when a policymaker has an incentive to deviate from a plan made with another person, say in the future, even when no new information has emerged. Policymakers of today try to commit future policymakers to a development path that is sustainable. But when the future actually arrives, these new policymakers deviate from the sustainable path and reallocate resources that are efficiently based on prevailing interest rates.

Finally the case is made for calculating all intertemporal effects with more than one rate. The arguments outlined above for different rates are unlikely to be resolved, given that they have been an issue since well before climate change. Hence it is good practice to calculate the costs for more than one rate to provide the policymaker with some guidance on how sensitive the results are to the choice of discount rate.[7] A lower rate based on the ethical considerations is, as noted above, around 3%.

7.2.6   *Adaptation and Mitigation Costs and the Linkages between Them*

Climate change puts society at risk. It is possible to prevent damages through mitigation and adaptation. Mitigation strategies against the risks of climate include curtailing GHG emissions to lower the likelihood that worse states of nature will occur. Adaptation strategies to climate risk include the changing of production and consumption decisions to reduce the severity of a worse state in the scenario if it does occur (Ehrlich and Becker, 1972; Crocker and Shogren, 1999). A portfolio of mitigation and adaptation actions jointly determines climate risks and the costs of reducing them. Since individuals in their private capacity have the liberty to undertake adaptation to climate change on their own accord, modellers and policymakers need to address these adaptive responses when choosing the optimal degree of public mitigation. If this is not the case, then policy actions are likely to be more expensive than they need be, with no additional reduction in climate risk (see, e.g., Schelling, 1992).

While most people appreciate that actions on adaptation affect the costs of mitigation, this obvious point is often not addressed in climate policymaking. Policy is fragmented–with mitigation being seen as addressing climate change and adaptation seen as a means of reacting to natural hazards. As a consequence, the estimated costs of each can be biased (see Kane and Shogren, 2000). Usually, mitigation and adaptation are modelled separately as a necessary simplification to gain traction on an immense and complex issue. One question that must be addressed is "How reasonable is this assumption?" Another is "What are the likely consequences of this assumption on the estimated costs of mitigation?"

---

[7] It is also useful to display graphically the time path of undiscounted costs, as discounting can obscure important information.

First, separability presupposes that the overall effectiveness and costs of mitigation do not depend on adaptation. However, for this assumption to hold, the implicit presumption is that climate risk is exogenous–a risk beyond people's private or collective ability to reduce. The necessary economic conditions for this to hold are rather restrictive. In particular, climate risk can be considered as exogenous only if markets are complete. A complete set of markets exists if people can contract to insure against all risks from each conceivable state of nature that might be realized (Marshall, 1976). Complete markets allow for perfect risk spreading and risk pooling such that the only remaining risk is outside the control of human actions (e.g., phases of the moon). However, markets for climate risk are notorious incomplete or non-existent because of the high cost of contracting (Chichilnisky and Heal, 1993). People make private and collective adaptation decisions through the markets that do exist and through collective policy actions. The economic circumstances that influence these choices matter to the level of risk, and addressing these conditions is essential for the successful estimation of costs. People choose to create and reduce risk. How people perceive risk, the relative costs and benefits of alternative risk reduction strategies, and relative wealth affect these choices.

Similar to income and substitution effects, adaptation can have two effects on the costs of mitigation. First, more adaptation can lower mitigation costs because policymakers choose to move to another point on the same mitigation cost curve - adaptation does not alter the marginal productivity of mitigation, it induces a shift along the cost curbe. Second, adaptation acting as a technical substitute or complement shifts the mitigation cost curve. For example, flood defences change land use and thereby change costs and prices in an area, which impacts on mitigation costs. Whether adaptation causes a shift along the mitigation cost curve or a shift of the entire curve itself, or both, then becomes a modelling question, and an empirical one to determine the magnitude of the shift along and to a new cost curve.

Second, sectoral work in agriculture, forestry, and coastal areas shows that cost estimates are sensitive to the inclusion of adaptation (see, e.g., Sohngen and Mendelsohn, 1997; Sohngen *et al.*, 1999). Greater climate variability, for instance, can influence how adaptation affects mitigation in agriculture. Increased levels of risk directly induce a nation to adapt more by switching its crop mix and crop varieties to those more tolerant of drier or wetter conditions, and by modifying its weed control strategies. The magnitude of this adaptation depends on how risk affects the perceived marginal productivity of mitigation (e.g., more or less effective soil sequestration per unit of area), and how mitigation and adaptation work with or against each other. Bouzaher *et al.* (1995), for example, estimate that winter cover crops can be used to increase soil organic carbon by expanding annual biomass production. They also show that conservation tillage, the Conservation Reserve Program, and the Wetlands Reserve Program can increase soil carbon by minimizing soil disturbance and targeting bottomland for hardwood trees. For non-climate risk, models that account for

mitigation and adaptation risk estimate that benefits are under-estimated by 50% when adaptation is ignored (e.g., Swallow, 1996).

Third, uncertainty in cost is affected by interaction of the technologies for risk reduction–mitigation and adaptation. By mitigation, humans reduce the odds that a deleterious event happens; by adaptation, they reduce the consequences when a damaging event actually does occur. For the most part, climate change literature contains models that deal with mitigation and adaptation separately. This is unfortunate, since significant interactions are likely to exist between how people choose to mitigate and adapt (Shogren and Crocker, 1999). These risk-reduction strategies probably complement or negate each other. Understanding the interaction between the two can help formulate better the analysis of mitigation costs. The benefits of mitigation will be lower if more people can adapt to the climate.

These results suggest that more it would be worthwhile to pay more attention to the interaction of mitigation and adaptation, and its empirical ramification. The challenge is to capture in a reasonable way the linkages between these sets of actions, and to establish how this interaction can impact the estimated costs of climate protection. Even if a complete empirical application of the portfolio of risk avoidance is currently unreachable, an understanding of which unmeasured links might be most valuable to decision makers in the future could indicate whether the costs of mitigation are being underestimated.

7.3   Analytical Structure and Critical Assumptions

7.3.1   *System Boundaries: Project, Sector, Macroeconomic*

Assessing climate change mitigation involves a comparison between a policy case and a non-policy case, otherwise referred to as a baseline case. The two should, as far as possible, be defined in a way that the assessment can include all major economic and social impacts of the policies, spillovers, and leakages, as well as GHG emission implications. In other words, the cases should be assessed in the context of a "system boundary" that include all major impacts. The system boundary can be a specific project, include one or more sectors, or the whole economy.

The project, sector, and macroeconomic levels can be defined as follows:
- *Project*. A project level analysis considers a "stand-alone" investment that is assumed not to have significant impacts on markets (both demand and supply) beyond the activity itself. The activity can be the implementation of specific technical facilities, infrastructure, demand-side regulations, information efforts, technical standards, etc. Methodological frameworks to assess the project level impacts include cost–benefit analysis, cost-effectiveness analysis, and lifecycle analysis.

- *Sector.* Sector level analysis considers sectoral policies in a "partial-equilibrium" context, for which other sectors and the macroeconomic variables are assumed to be as given. The policies can include economic instruments related to prices, trade, and financing, specific large-scale investment projects, and demand-side regulation efforts. Methodological frameworks for sectoral assessments include various partial equilibrium models and technical simulation models for the energy sector, agriculture, forestry, and the transportation sector.
- *Macroeconomic.* A macroeconomic analysis considers the impacts of policies across all sectors and markets. The policies include all sorts of economic policies, such as taxes, subsidies, monetary policies, specific investment projects, and technology and innovation policies. Methodological frameworks include various sorts of macroeconomic models such as general equilibrium models, Keynesian models, and Integrated Assessment Models (IAMs), among others.

A "trade-off" is expected between the details in the assessment and the complexity of the system considered. For example a project system boundary allows a rather detailed assessment of GHG emissions and economic and social impacts generated by a specific project or policy, but excludes sectoral and economy-wide impacts. Conversely, an economy-wide system boundary, in principle, allows all direct and indirect impacts to be included, but has little detail on the impacts of implementing specific projects.

The system boundaries may be selected on the basis of the specific scope of the study and the availability of analytical tools, such as models. Many studies have been organized, in practice, on the basis of the scope and structure of the modelling tools applied. For example, climate change mitigation studies for the energy sector were frequently structured according to traditional modelling approaches used in that sector, which are often rich in detail on technologies, but do not include market behaviour. In contrast, macroeconomic models are often rich in detail on market behaviour and price relationships, but do not explicitly include major GHG emitting sources and related technologies.

Project assessment methodologies are generally very rich in detail and include an assessment of various direct and indirect costs and benefits of the GHG reduction policy considered. The assessments are often conducted as very data-intensive exercises, in which various project assessment tools and expert judgements are combined. They require rather strong technical skills of the experts in the collection of data, to ensure consistency in the structure and results of the analysis.

A combination of different modelling approaches is required for an effective assessment of the options. For example, detailed project assessment has been combined with a more general analysis of sectoral impacts, and macroeconomic carbon tax studies have been combined with the sectoral modelling of larger technology investment programmes.

7.3.2   *Importance of Baselines*

7.3.2.1   *Development Patterns and Baseline Scenario Alternatives*

The baseline case, which by definition gives the emissions of GHGs in the absence of the climate change interventions being considered, is critical to the assessment of the costs of climate change mitigation. This is because the definition of the baseline scenario determines the potential for future GHG emissions reduction, as well as the costs of implementing these reduction policies. The baseline scenario also has a number of important implicit assumptions about future economic policies at the macroeconomic and sectoral levels, including sectoral structure, resource intensity, prices and thereby technology choice.

Macroeconomic issues that are particularly relevant to developing countries (such as instability of output, constrained capital, and foreign exchange) similarly have important implications on GHG emissions through impacts on energy sector investments and energy-intensive production sectors. These assumptions have important implications for the efficiency of policy instruments applied to climate change mitigation strategies and thereby for implementation costs, which are discussed in Section 7.2.3.

Economic policies have a number of direct and indirect impacts on GHG emitting sectors. It is generally expected that successful economic policies generate increased growth and the emissions intensity of the economy then depends on the mix of products produced as well as on the efficiency with which they are produced. Economic policies in some cases can imply a more efficient use of resources, which means that the GHG emission intensity per unit of economic output decreases. The tendency to increase GHG emissions alongside economic growth is expected to be particularly "strong" in countries that presently have low energy consumption. The challenge is to pursue a development pattern in which economic development is achieved alongside relatively low GHG emissions and other environmental impacts.

Many macroeconomic and sectoral policies have important consequences for future GHG emissions through the impacts on sectoral structure, resource intensity, prices, and thereby technology choice. Macroeconomic issues like constrained capital and foreign exchange can lead to low investments in the energy sector, to major energy-intensive production sectors, or to the high utilization of pollution-intensive domestic fuels. In the same way, uncertainty or macroeconomic instability has a tendency to slow down investments because of the risk perceptions of foreign and national investors, and because of high interest rates.

As noted, GHG emissions are interlinked with general economic development patterns and economic policies. These policies have an influence both on the baseline as well as on the effectiveness of the mitigation options, and thereby on GHG emission levels. It is useful to "decompose" the GHG emission/GDP intensity factor into subcomponents that explain the implicit resource components behind the GHG emissions. One way to achieve this for the energy sector is based on the so-called Kaya identity (Kaya, 1989):

$$\frac{GHG\ emissions}{GDP} = \frac{GHG\ emissions}{energy} \times \frac{energy}{GDP}$$

The first component of the identity, GHG emissions per energy unit, reflects the GHG emission intensity of energy consumption, which again reflects natural resource endowment and relative prices of the different energy sources. The second factor (energy consumption per GDP unit) reflects both the weight of energy-intensive processes in GDP and the efficiency of the resources used. The same approach can be used to assess GHG emission intensities of other sectors, such as agriculture, forestry, waste management, and industry.

Development may follow different paths in countries according to socioeconomic conditions, resources, national policies and priorities, and institutional issues. For instance, a rapidly growing economy develops a different composition of capital stock and energy use pattern compared with a slowly growing country. A nation following development policies that emphasize greater investments in infrastructure, such as efficient rail transport, renewable energy technologies, and energy efficiency improvements, exhibits a low GHG emission trajectory. However, a nation with substantial coal resources, scarce capital, and a low level of trade can be pushed towards a development path with high emissions.

7.3.2.2   Multiple Baseline Scenarios

The above discussion identifies a number of reasons why the establishment of a baseline case is very difficult and uncertain. There are some additional reasons why this is so. The difficulty in predicting the evolution of development patterns over the long term stems, in part, from a lack of knowledge about the dynamic linkages between technical choices and consumption patterns and, in turn, how these interact with economic signals and policies. Technology and consumption patterns are endogenous, their direction being determined at least partly by political decisions. There are also many general uncertainties that impact on the establishment of a baseline case, for example political and social changes.

The above considerations further emphasize the need for work on the basis of several alternative baseline scenarios characterized by different assumptions regarding development patterns and innovation. This allows the mitigation or adaptation assessments to create an estimate range for the costs associated with very different development paths. Indeed, the range of emission levels associated with alternative baseline scenarios could well be greater than the difference between a certain baseline and the corresponding active policy case.

In reality, this can only provide a partial insight into the costs of climate change. Despite the large disparities in cost estimates likely to arise through the use of multiple baselines, they do allow the future to be framed within a much wider analytical perspective. Using a number of different development patterns is of particular importance to developing countries. Since the major part of their infrastructure and energy systems is yet to be built, the spectrum for future development is wider than in industrialized countries. A baseline scenario approach that assumes current development trends to continue is therefore not very useful in these countries (IPCC, 1996a, Chapter 8).

The scenarios of the IPCC *Special Report on Emissions Scenarios* (IPCC, 2000a) show that alternative combinations of driving-force scenario variables can lead to similar levels and structure of energy use and land-use patterns. Hence for a given scenario outcome, for example in terms of GHG emissions, alternative pathways can lead to that outcome. The conclusion is therefore that one and only one development path does not exist and studies preferably should include multiple baseline scenarios that facilitate a sensitivity analysis of the key scenario variables and assess the consequence of different development patterns.

7.3.2.3   Baseline Scenario Concepts

The literature reports several different baseline scenario concepts, including (Sanstad and Howart, 1994; Halsnæs *et al.*, 1998; Sathaye and Ravindranath, 1998):

- efficient baseline case, which assumes that all resources are employed efficiently; and
- "business-as-usual" baseline case, which assumes that future development trends follow those of the past and no changes in policies will take place.

These different baseline scenario concepts represent different expectations about future GHG emission development trends, as well as different perspectives on the trade-offs between climate change mitigation policies and other policies. The costs of a given GHG emissions reduction policy depend in a very complicated way on numerous assumptions about future GHG emissions, the potential for emissions reductions, technological developments and penetration, resource costs, and markets.

The different GHG emission profiles of the alternative baseline-scenario approaches depend on a number of assumptions. These include economic growth, mix of products, GHG emissions, intensity of energy production and consumption, and other material use. A "business-as-usual" baseline case is often associated with high GHG emissions, particularly if current main GHG emission sources, such as the energy industry, run at low efficiency. Such a baseline case can reflect the continuation of current energy-subsidy policies (which implies relatively high energy consumption and thereby high GHG emissions) or various other market failures of particular importance for GHG emission intensive sectors, such as capital market constraints. An efficient baseline case that assumes properly

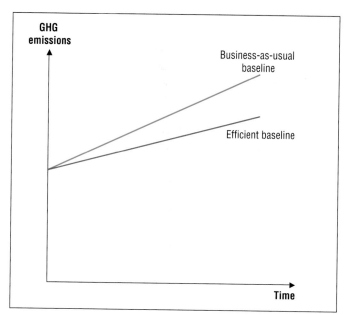

Figure 7.2: *Greenhouse gas emission profiles of different baseline case approaches.*

functioning markets, all other things being equal, can be expected to reflect relatively high energy efficiency and thereby lower GHG emissions than a business-as-usual baseline case. GHG emission profiles of the different baseline case approaches are illustrated in *Figure 7.2*.

The GHG emission reduction potential of a given policy is to be measured as the difference between the GHG emissions in the baseline case and the GHG emissions after the implementation of the policy. Clearly, this difference depends on both the baseline and the options chosen for the mitigation. High baseline-scenario GHG emissions based on a business-as-usual scenario approach in some cases can imply that the net mitigation costs measured per unit of GHG emission reduction are relatively low. Such a result, for example, can reflect that the mitigation scenario is assumed to imply a general efficiency improvement of the energy systems compared with the baseline wich both reduces GHG emission and generates fuel cost savings. The total costs of achieving a given GHG emission level (e.g., defined in relation to 1990 emissions), however, can be relatively high when the mitigation strategy is assessed in relation to a business-as-usual baseline scenario that has a large growth in GHG emissions. Conversely, GHG emission reduction costs per unit of emission can be relatively high in relation to an efficient baseline case, but total reduction costs of meeting a target can be low.

It is important to emphasize consistency and transparency in the definition of baselines, and in the reporting of any costs associated in moving from a given baseline case to a climate change policy case. Furthermore, when reporting the range of cost estimates for the different baselines, it is important also to provide information about the assumptions that underlie each baseline.

7.3.2.4 Specific Baseline Issues Related to International Co-operative Mechanisms for Greenhouse Gas Emission Reductions

The Kyoto Protocol of the United Nations Framework Convention on Climate Change (UNFCCC) includes a number of mechanisms for international co-operation about GHG emission reductions. The Protocol includes two project-based mechanisms, namely the clean development mechanism (CDM) and joint implementation (JI). The operational details of these two mechanisms are discussed in a number of studies which include a number of different arguments for baseline case approaches. A number of these arguments are subsequently referred and discussed.

A number of studies suggest the use of a so-called standard methodology for setting the baseline case for CDM and JI projects. Here, the baseline case serves as a metric for calculating GHG emission reductions that originate from the approved projects and the main issue is therefore to specify GHG emissions in the absence of the project. A number of specific complexities arise in relation to the definition of baseline cases for projects that do not include major new capital equipment, such as projects that include changes in operational practice, land use, land-use changes, and forestry projects.

Papers that evaluate alternative options for the baseline determination of CDM projects include Michaelowa and Dutschke (1998), Chomnitz (1999), Jepma (1999), Matsuo (1999), Parson and Fisher-Vanden (1999), and Harrison *et al.* (2000). These papers deal with various baseline issues including technology benchmarks, normative benchmarks that are politically chosen, and historical benchmarks based on GHG emission trends. Other important aspects considered include assumptions about baseline development over the timeframe of the CDM project.

He and Chen (1999) have suggested a set of criteria to establish baseline cases from a micro level perspective. In this approach, GHG emissions reduction projects are divided into three project categories:
- technology innovation, in which the GHG emission reduction project should be compared with existing technologies;
- new constructed plants, in which the GHG emission reduction project should be compared with alternative new advanced technologies; and
- technology substitution, in which the GHG emissions reduction project should be compared with a newly constructed existing plant.

A benchmark technology baseline to assess power-sector CDM projects could include assumptions about the efficiency and costs of power production technologies in a specific national or regional area, or could be based on international standards. The actual definition of baseline technologies will has major implications on the GHG emission reduction "performance" of the CDM project.

The choice of baseline case approach for CDM projects or JI projects might have major implications on the global cost effectiveness of climate change mitigation projects. A baseline scenario approach that uses internationally standardized technology data implies that the GHG emission reduction potential and related costs are estimated to be similar for projects implemented at quite different sites. Project host countries that have a relatively low GHG emission intensity from their power system compared with the international baseline standard have a relatively strong "market position" in this case, because the GHG emission reductions achieved with the particular CDM or JI project will be assessed to be relatively high. Project host countries with a relatively high GHG emission intensity compared with the international standard will tend to have a weaker market position than in the alternative approach, in which the baseline case reflects specific national GHG emissions. Baseline cases that underestimate the reductions from a particular project in this way result in fewer projects than is justified. This use of international benchmark technology standards can tend to imply a loss in the global cost-effectiveness of CDM or JI projects.

Another drawback to using a baseline case not related to the specific development context of the project host country is that it can be difficult to design the project such that it creates both global (GHG emission reduction) and local benefits (improvements in the local environment, employment, and income generation, and institutional strengthening). Such drawbacks, however, should be balanced against the expected decrease in transaction costs from using an international benchmark baseline case approach.

7.3.3   *Cost Implications of Different Scenario Approaches*

The costs of climate change mitigation policies are, by definition, a net incremental cost relative to a given scenario, which includes assumptions on both the baseline case and the policy case. The following section presents a taxonomy of baseline cases and policy scenario cases and discusses these in relation to cost assessments.

In Section 7.2 it is stated that cost assessments should include, in principle, all costs and benefits related to the policies as well as any ancillary benefits and costs. The actual determination of impacts related to the policies, however, is open to interpretation and discussion, and the actual selection of system boundaries for the cost assessment will reflect specific assumptions in the baseline as well as in the policy case scenario.

One way to evaluate the impact of different scenario structures on costs is to distinguish between the gross and the net costs of climate change mitigation policies. Gross costs are here defined to reflect all direct and indirect costs and benefits of the mitigation policy, when this policy is considered as the primary policy objective. Net costs are the gross costs corrected for side effects that result from potential synergies or trade-offs

between mitigation policies and general economic policies or non-GHG environmental policies. These side effects can be divided into three categories (IPCC, 1996a, Chapter 8):

- A double dividend related to recycling of the revenue of carbon taxes in such a way that it offsets distortionary taxes.
- Ancillary impacts, which can be synergies or trade-offs in cases in which the reduction of GHG emissions have joint impacts on other environmental policies (i.e., relating to local air pollution, urban congestion, or land and natural resource degradation). These are referred to as ancillary or co-benefits and are discussed in Section 7.2.2.
- Impacts on technological development and efficiency. These include specific incentives to develop and penetrate new technologies, technology learning, and reduction of current barriers to efficiency improvements in existing technical systems (part of these impacts are considered as part of the so called no regret potential, see Section 7.3.4.2 for a more detailed discussion).

7.3.3.1   *Double Dividend*

The potential for a double dividend arising from climate mitigation policies has been extensively studied during the 1990s. In addition to the primary aim of improving the environment (the first dividend), such policies, if conducted through revenue-raising instruments such as carbon taxes or auctioned emission permits, yield a second dividend, which can be set against the gross costs of these policies.

The literature demonstrates theoretically that the costs of addressing greenhouse targets with policy instruments of all kinds–command-and-control as well as market-based approaches–can be greater than otherwise anticipated, because of the interaction of these policy instruments with existing domestic tax systems.[8] Domestic taxes on labour and investment income change the economic returns to labour and capital and distort the efficient use of these resources.

The cost-increasing interaction reflects the impact that GHG policies can have on the functioning of labour and capital markets through their effects on real wages and the real return to capital.[9] By restricting the allowable GHG emissions, permits, regulations, or a carbon tax raise the costs of production and the prices of output, and thus reduce the real return to labour and capital. If government revenues are to remain unchanged, labour or capital tax rates have to be raised, exacerbating prior distortions in the labour and capital markets. Thus, to attain a given GHG emissions target, all instruments have a cost-increasing "interaction effect".

---

[8] For a very readable account of the basic analytics and economic intuition at the heart of this literature, see pages 281–284 in Kolstad (2000).

[9] See, for example, Parry *et al.* (1999).

For policies that raise revenue for the government (carbon taxes and auctioned permits), this is only part of the story, however. These revenues can be recycled to reduce existing distortionary taxes. Thus, to attain a given GHG emissions target, revenue-generating policy instruments have the advantage of a potential cost-reducing "revenue-recycling effect", as compared to the alternative, non-auctioned tradable permits or other instruments that do not generate revenue (Bohm, 1998). In a simple, stylized representation of the economy, Bovenberg *et al.* (1994) and Goulder (1995a, b) suggest that in only a few cases is the tax interaction effect fully offset by the revenue-recycling effect. In theoretical, numerical analyses, the "interaction effect" is found to be larger than the "revenue-recycling effect" (Parry *et al.*, 1999), which means that the introduction of an environmental policy, regardless of the policy instrument(s) used, has a net cost to the economy.[10] It is also true, however, that under some circumstances the (cost-reducing) "revenue-recycling effect" might exceed the (cost-increasing) "interaction effect". This could happen if, for example, the interaction effect was small, for example because of a sufficiently inelastic labour supply, or if some highly distortionary pre-existing taxes could be lowered.[11]

However, it is unclear whether the empirical findings of the interaction effect are due more to the assumptions invoked for tractable general equilibrium analysis than to real-world considerations (Kahn and Farmer, 1999).

In summary, all domestic GHG policies have an indirect economic cost from the interactions of the policy instruments with the fiscal system, but in the case of revenue-raising policies this cost is partly offset (or more than offset) if, for example, the revenue is used to reduce existing distortionary taxes. Whether these revenue-raising policies can reduce distortions in practice depends on whether revenues can be "recycled" to tax reduction. See Chapter 6 for the policy relevance of these estimated effects and Chapter 8 for model-based empirical studies.

7.3.3.2   Ancillary Impacts

The definition of ancillary impacts is given in Section 7.2.2.3. As noted there, these can be positive as well as negative. It is important to recognize that gross and net mitigation costs cannot be established as a simple summation of positive and negative impacts, because the latter are interlinked in a very complex way. Climate change mitigation costs (gross and well as net costs) are only valid in relation to a comprehensive specific scenario and policy assumption structure.

An example is transportation sector options that have an impact on both GHG emissions and urban air pollution control programmes. GHG emission control policies, like vehicle maintenance programmes, reduce both GHG emissions and other pollution, but another option, like the introduction of diesel trucks as a substitute for gasoline trucks, decreases GHG emissions but increases NO_x emissions and thereby local air pollution. The gross and net costs assessed for these programmes depend on specific baseline and policy case scenarios (specifically, the assumptions on urban air pollution control policies are critical).

It is important that assumptions about environmental control policies outside the specific area of GHG emissions reduction be carefully specified in relation to the baseline as well as to the policy case. If the baseline assumes that some environmental control policies are implemented in the time frame considered, the side effects of the GHG reduction policy in relation to these areas cover part of these environmental policy objectives. The mitigation costs then eventually offset part of the control cost in the baseline case. However, if the baseline case includes specific flue-gas cleaning systems on power plants to control SO_2 and NO_x emissions that are already installed, then investments in these plants are irreversible. In this case, the joint benefit of climate change mitigation programmes in the form of avoided control cost on the other emissions is low, while the public health ancillary benefits may be substantial (see also the discussion on ancillary and/or co-benefits in Section 7.2.2).

7.3.3.3   Technological Development and Efficiency Impacts

Assumptions about technological development and efficiency in the baseline and mitigation scenarios have a major impact on mitigation costs, in particular in bottom-up mitigation cost studies. Many of these studies structure the cost assessment around an estimation of the costs and other impacts of introducing technological options that imply lower GHG emissions. The existence and magnitude of a potential for technological efficiency improvements depends on expectations about technology innovation and penetration rates given consumer behaviour and relative prices. These assumptions are discussed in more detail in Section 7.3.4.

A number of cost studies assessed different parts of the three above-mentioned side effects. The double dividend is assessed predominantly in macroeconomic studies on the basis of fairly

---

[10] The environmental policy should yield environmental benefits, including some non-market benefits. A cost–benefit analysis of the proposed policy compares these benefits with the estimated cost of the policy.

[11] The term "strong double dividend" has been used in the literature for cases in which the revenue-recycling effect not only exceeds the interaction effect but also the direct (GDP) costs of reducing emissions, thus making revenue-generating environmental policy costless. A revenue-recycling effect this large presupposes that the original tax structure is seriously inefficient (e.g., that capital is highly overtaxed relative to labour). This in itself calls for a tax reform the benefits of which should not be ascribed to the introduction of a revenue-generating environmental policy, even if the two were made on one and the same occasion (see SAR, Chapter 11). In this perspective, the term "strong (or weak) double dividend" becomes redundant (see also Chapter 8).

detailed modelling representation of tax systems and specific labour market constraints that cover the short-to-medium term time horizon. Joint environmental impacts of climate change mitigation policies are examined in various studies, including macroeconomic studies, sectoral studies, and technology-specific engineering studies. Impacts of technological development and efficiency are basically addressed in all sorts of studies, sometimes explicitly but sometimes implicitly. The lack of an integrated treatment of all three issues is, *inter alia,* a consequence of the different approaches to the technology characterisation in top-down models (macroeconomic) and bottom-up models (technology- or policy-specific models), which are further explained and discussed in Section 7.6. A few studies exist, however, that attempt such an integration (see, e.g., Walz, 1999).

7.3.4   *Assumptions about Technology Options*

7.3.4.1   *Technological Uncertainty*

Costing climate change policy is an uncertain business. This uncertainty often manifests itself in the choice of technologies to mitigate and adapt to risks from climate change. Firms and nations can attempt to reduce risk by using more of the low-carbon technologies presently on the shelf or they can invent new ones. How quickly people will switch within the set of existing technologies with or without a change in relative energy prices is open to debate; how creative people are at inventing new technologies given relative prices is also a matter of discussion.

The key to addressing uncertainty is to capture a range of reasonable behaviours that underpins the choice to adopt existing or develop new low-carbon technology. Two key questions that should be addressed are:

- What explains the rate of adoption of existing low-carbon technologies given the relative price of energy?
- What explains the rate of invention of new low-carbon technologies given relative prices?

Which answers to these questions are accepted determines whether some weighted average of the estimates or a lower or upper estimate is used to guide policy.

For any given target and set of policy provisions, costs decline when consumers and firms have more plentiful low-cost substitutes for high-carbon technologies. Engineering studies suggest 20%-25% of existing carbon emissions could be eliminated (depending on how the electricity is generated) at low cost if people switched to new technologies, such as compact fluorescent light bulbs, improved thermal insulation, heating and cooling systems, and energy-efficient appliances. The critical issue is how this adoption of efficient technologies occurs in practice and which sort of regulation and economic instruments could eventually support this adoption. Chapter 5 of this report assesses the literature regarding technology adoption and regulation frameworks.

Many economists have emphasized that technological progress is driven by relative prices, and that people do not switch to new technologies unless prices induce them to switch. New efficient technologies, according to this argument, then are not taken up without a proper price signal. People are also perceived to behave as if their time horizons are short, perhaps reflecting their uncertainty about future energy prices and the reliability of the technology. Also, factors other than energy efficiency matter to consumers, such as a new technology's quality and features, and the time and effort required to learn about it and how it works. This issue has already been flagged in relation to technology adoption and implementation costs, but it also has an uncertainty element to it.

The different viewpoints on the origin of technological change appear in the assumed rate at which the energy-consuming capital can turnover without a change in relative energy prices. Modellers account for the penetration of technological change over time through a technical coefficient called the "autonomous energy efficiency improvement" (AEEI). The AEEI reflects the rate of change in energy intensity (the energy-to-GDP ratio) holding energy prices constant (see IPCC, 1996a, Chapter 8). The presumed autonomous technological improvement in the energy intensity of an economy can lead to significant differences in the estimated costs of mitigation. As such, many observers view the choice of AEEI as crucial in setting the baseline scenario against which to judge the costs of mitigation. The costs of mitigation are inversely related the AEEI– the greater the AEEI the lower the costs to reach any given climate target. The costs decrease because people adopt low-carbon technology of their own accord, with no change in relative prices.

Modellers have traditionally based the AEEI on historical rates of change, but now some are using higher values based on data from bottoms-up models and arguments about "announcement effects". For instance, some analysts have optimistically argued that the existence of the Kyoto Protocol will accelerate the implementation of energy efficient production methods to 2% per year or more. Policymakers and modellers continue to debate the validity of this assumption (see, e.g., Kram, 1998; Weyant, 1998). A range of AEEIs has been adopted in the modelling literature (see Chapter 8 for more details). The AEEI has ranged from 0.4% to 1.5% per year for all of the regions of the world, and has generated large differences in long-term project baselines (e.g., Manne and Richels, 1992). Edmonds and Barns' (1990) sensitivity study confirms the importance of the AEEI in affecting cost estimates. However, as noted by Dean and Hoeller (1992): "unfortunately there is relatively little backing in the economic literature for specific values of the AEEI ... the inability to tie it down to a much narrower range ... is a severe handicap, an uncertainty which needs to be recognized."

7.3.4.2   *No Regrets Options*

No regrets options are by definition GHG emissions reduction options that have negative net costs, because they generate direct or indirect benefits that are large enough to offset the costs of implementing the options. The costs and benefits included in the assessment, in principle, are all internal and external impacts of the options. External costs arise when markets fail to provide a link between those who create the "externality and those affected by it; more generally, when property rights for the relevant resources are not well defined. External costs can relate to environmental side-impacts, and distortions in markets for labour, land, energy resources, and various other areas. By convention, the benefits in an assessment of GHG emissions reduction costs do not include the impacts associated with avoided climate change damages. A broader definition could include the idea that a no regrets policy would, in hindsight, not preclude (e.g., by introducing lock-in effects or irreversibilities) even more beneficial outcomes, but this is not taken up in the mitigation literature. The no regret concept has, in practice, been used differently in costing studies, and has in most cases not included all the external costs and implementation costs associated with a given policy strategy.

The discussion of "no regrets" potential has triggered an extensive debate, which is particularly well covered in the SAR (IPCC 1996a, Chapters 8 and 9). The debate is summarized rather simply in graphical form in *Figure 7.3*.

Figure 7.3 illustrates the production frontier (F) of an economy that shows the trade-off between economic activity (Q) and emissions reduction (E). Each point on the curve shows the maximum level of emissions reduction for a given level of economic activity. The economy is producing composite goods, namely an aggregation of all goods and services Q and environmental quality E, which here represent GHG emissions. Given such an assumption it is possible to construct a curve F(Q,E) that represents the trade-off between Q and E. For a given economy at a given time, each point on F shows the maximum size of the economy for each level of GHG emissions, and therefore it shows the loss in economic output measured by Q associated with reductions in GHG emissions level E. If the economy is at a level below F then it is possible to increase the total production of Q and/or E. If O is taken as the starting point of the economy in *Figure 7.3* then all movements in the "triangle" OO'O" increase environmental quality E and/or economic output Q, but do not decrease either of these goods. Movements to positions outside this "triangle" imply a decrease in both economic activity Q and environmental quality E, or a trade-off in which one of these two goods decreases.

In estimating the costs, the crucial question is where the baseline scenario is located with respect to the efficient production frontier of the economy F. If the chosen baseline scenario assumes that the economy is located on the frontier, as in the efficient baseline case, there is a direct trade-off between economic activity and emissions reduction. Increased emissions reduction moves the economy along the frontier to the right. Economic activity is reduced and the costs of mitigation increase. If the economy is below the frontier, at a point such as O, there is a potential for combined GHG emissions reduction policies and improvements of the efficiency of resource use, implying a number of benefits associated with the policy.

Returning to the implications for the cost of climate change mitigation, it can be concluded that the no regrets issue reflects specific assumptions about the working and efficiency of the economy, especially the existence and stability of a social welfare function, based on a social cost concept. Importantly, the aggregate production frontier is uncertain, as it is dependent on the distribution of resources and is changed by technological development. Since it also involves the weighting of different goods and services by market valuations to form an aggregate, it is also affected by personal and social preferences that influence those valuations.

The critical question is how climate change mitigation policies can contribute to efficient and equitable development of the economy.

In this way it can be argued that the existence of a no regret potential implies:
- that market and institutions do not behave perfectly, because of market imperfections such as lack of information, distorted price signals, lack of competition, and/or institutional failures related to inadequate regulation, inadequate delineation of property rights, distortion-inducing fiscal systems, and limited financial markets;
- that it is possible to identify and implement policies that can correct these market and institutional failures without incurring costs larger than the benefits gained; and

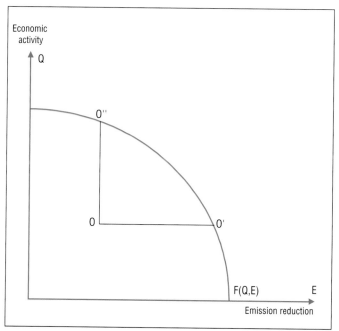

Figure 7.3: *Trade-off between emissions reduction and economic activity.*

- that a policy decision is made to eliminate selectively those failures that give rise to increased GHG emissions.

In other words, the existence of market and institutional failures that give rise to a no regrets potential is a necessary, but not a sufficient, condition for the potential implementation of these options. The actual implementation also requires the development of a policy strategy that is complex and comprehensive enough to address these market and institutional failures and barriers.

The costs that actually face private agents are different from the social costs, and therefore the market potential (as defined in Chapter 5) may be very different from the potential based on social costs. This implies that the actual implementation of no regrets options requires that it be possible to introduce policies that "narrow the gap" between the market potential and a potential estimated on the basis of social costs. Cameron *et al.* (1999) give a systematic overview of market failures and market barriers important to the implementation of no regrets options.

Returning to the implications for climate change mitigation cost, it can be concluded that the no regrets issues reflect specific assumptions about the location of the economy in relation to the efficient production frontier. Bottom-up studies have (in most cases on the basis of a specific assessment of production practices in main GHG emitting sectors, such as the energy sector) assumed that the economy in the baseline case operates below the optimal frontier and that mitigation policies imply an increased efficiency of technologies. The costs of implementing mitigation policies are then partly offset by direct and indirect benefits, which sometimes are large enough to generate a negative cost result. Top-down approaches, however, assume that the economy is efficient in the baseline case and mitigation policies therefore always imply a trade-off with other goods and thereby have a positive cost.

7.3.5 *Cost Implications of Alternative GHG Emission Reduction Options and Carbon Sinks*

For a wide variety of options, the costs of mitigation depend on what regulatory framework is adopted by national governments to reduce GHGs. In general, the more flexibility the framework allows, the lower the costs of achieving a given reduction. A stringent, inflexible carbon-mitigation policy induces greater economic burden than a loose, flexible policy. More flexibility and more trading partners can reduce costs. The opposite is expected with inflexible rules and few trading partners.

Flexibility can be measured as the ability to reduce carbon emissions at the lowest cost, either domestically or internationally, including "when and where" flexibility—which assumes a world emissions budget could be spent optimally over space and time to capture all potential intra- and intertemporal efficiencies. Providing a firm or nation with more flexibility to reach a given target and timetable also reduces costs.

The details as to how flexibility is achieved matter. Many advocates prefer emissions trading over carbon taxes because the quantity of carbon flowing into the atmosphere is fixed, thereby shifting risk from the environment to the economy in the form of price uncertainty. However, some suggestions on the design of emissions trading create relatively high transaction costs that would limit the cost savings of a trading system. Furthermore, the key issue of how the emissions rights should be allocated has yet to be resolved (IPCC, 1996a; Jepma and Munasinghe, 1998).

Another source of flexibility is to include carbon sinks in the policy framework. Recall that a carbon sink is a process that destroys or absorbs GHGs, such as the absorption of atmospheric carbon dioxide by terrestrial (e.g., trees) and oceanic biota. The main anthropogenic sink is tree planting and other forest management actions. Soils and other types of vegetation also provide a potential sink. It is estimated that forests around the world contain roughly about 1,146GtC in their vegetation and soil, with about twice as much in soil as in vegetation (See IPCC, 2000c). For the USA, forests are an important terrestrial sink, given that they cover about 750 million acres (about 300 million hectares). Land use changes in the USA have increased the uptake of carbon to an estimated 200MtC$_{eq}$.

A few studies found that carbon sequestration through sinks could cost as little as US\$25/tonne C in the USA for 150MtC$_{eq}$ (Stavins, 1999). But serious uncertainties remain about how to measure and account for estimates of net carbon. For example, how forest management activities affect soil carbon is unknown, and since forest soils contain over 50% of the total stored forest carbon in the USA, this difference can have a significant impact on estimates. And some researchers have shown that sinks are not as effective as predicted when the interaction of forest reserves and the timber market is accounted for. The more land that is set aside for carbon sinks, the quicker the cycle of harvesting on other forestland, and the less total net carbon sequestration. Some fear that these ambiguities about sinks could divert attention from first-order priorities to second-order technicalities (Jacoby *et al.*, 1998).

To sum up, flexibility in the regulatory framework can play a major role in reducing the costs of GHG emissions reduction. The extent to which particular instruments can be adopted, however, depends on resolving serious political differences as to how the burden of emissions reduction should be shared, between developed countries themselves, and between both developed and developing countries. It is important also not to underestimate the costs of implementing changes in regulatory policy (see Section 7.2.3), especially in developing countries. For some of the practical problems in using flexible instruments in such countries, see Seroa da Motta *et al.* (1999).

7.3.6 Uncertainty[12]

A thread that runs through much of the discussion of costs is that of uncertainty. The whole exercise of estimating mitigation costs is confounded by imprecise information about baselines, and the costs of mitigation and adaptation measures (especially future costs). It is critical that such uncertainties be recognized and conveyed to the policymakers in the most effective manner possible.

As discussed above, uncertainty about baselines is best dealt with by taking more than one baseline and reporting cost estimates for multiple baselines. Hence costs should not be given as single values, but as ranges based on the full set of plausible baselines.

Technological uncertainty is another key area. As noted in Section 7.3.4.1, the autonomous rate of improvement in the energy-to-GDP ratio that underlies almost all models of climate economics is a clear example of an exogenous parameter currently subject to uncertainty. This is not easy to overcome by endogenizing technical change, as practical models currently available have difficulties in dealing with endogenous technical change. Thus, the way firms develop new technologies is probably an issue surrounded by a greater uncertainty than uncertainty on the consumer side. There is a moderate degree of consensus in the literature on these issues. As with baselines, a scenario approach is essential and results have to be reported for both "optimistic" and "pessimistic" development paths.

Taking a different approach, the way consumers adopt existing lower carbon technologies and firms develop new ones can be viewed as key sources of uncertainty in costing methodologies. These assumptions are crucial, as different valuations are likely to affect the conclusions. However, the ways in which guidance and information about these two crucial issues are provided are radically different. Two different options are available from the consumer side. First, energy oriented macroeconometric models can provide a price elasticity to show how changes in the fuel mix are driven by relative prices. No specific direction of technological change can be derived from this class of model. However, differences in the results in terms of different energy structures (and different carbon impacts) could easily emerge. Second, engineering studies can provide some indications about available lower energy technologies to show the impact on energy demand and carbon emissions. Hence, from the point of view of uncertainty there is no *a priori* reason to choose between bottom-up and top-down models.

Finally there are uncertainties in the estimated costs as well as in the estimation of the ancillary benefits and/or co-benefits. As the literature on potential ancillary benefits is continues to develop, current estimates of the net social impacts of various mitigation policies are necessarily incomplete. Private cost figures are generally more certain than the external ones, but some imprecision remains. As with baselines, a scenario approach is recommended, with estimates prepared for a "low value", a "mid value", and a "high value". Uncertainty about the external costs is well recognized. As with the private costs, again a scenario approach that gives a range from low, through mid, to high values is recommended. In both cases the scenario approach provides a sensitivity analysis for the costing exercise.

In the crosscutting paper on uncertainty (Moss and Schneider, 2000), a number of scales are proposed to assess the level of imprecision in the reported impacts, costs, etc. One that has frequently been used for costing exercises is the three-point scale that seeks to evaluate the degree of confidence in a particular result using a scale of: low, medium, and high confidence levels. This has been expanded to a five-point scale, which asks the researcher to select one of the following:
- "very high confidence" (over 95% certain);
- "high confidence" (67%–95% certain);
- "medium confidence" (33%–67% certain);
- "low confidence" (5%–33% certain); and
- "very low confidence" (below 5% certain).

This has not been applied to cost estimates, but it would useful to establish whether it could be applied and, if so, whether it would provide policymakers with better guidance as to the reliability of the results.

7.4 Issues in Estimating Costs

7.4.1 Relationship between Mitigation Costs and Development, Equity, and Sustainability

A number of key concepts applied in cost assessment provide important insights about the DES aspects of mitigation policies without intending to be comprehensive in coverage. This section discusses a number of the important linkages between costing studies and DES approach.

Chapter 1 states that a system's capacity for mitigation depends on a number of characteristics that must be considered in the context of its unique position and aspirations including:
- a range of viable technical options;
- a range of viable policy instruments;
- resource availability and distribution; and
- human and social capital.

Each of these characteristics is interrelated with DES issues, but also has major impacts on mitigation costs. Thus, the interaction between DES aspects and mitigation costs is two-way.

---

[12] Uncertainty that is relevant to cost estimation arises from three sources. First, the intrinsic uncertainty of the climate system, second uncertainty about the impacts, and third uncertainty about the costs. This section only deals with the last of these. The broader issues are discussed in the crosscutting paper devoted exclusively to this topic (Moss and Schneider, 2000).

DES policies have, on the one hand, major implications for economic structure and viability of policy instruments, as well for man-made, natural, and social capital. Mitigation policies, on the other hand, have implications for the same DES issues. The focus of this section is on the second of these feedback mechanisms.

The DES implications of mitigation policies are different according to the geographical scale of the efforts. International as well as national large-scale mitigation efforts can potentially impose a large demand for exhaustible resources or can be thought to impose irreversible damages on environmental resources and these impacts should be reflected in mitigation studies. Mitigation policies also have long-term implications on future climate change and thereby on intergenerational equity. A number of issues related to how mitigation costing studies address intergenerational equity issues are discussed in Section 7.4.5.[13]

Climate change mitigation policies implemented at a national level will, in most cases, have implications for short-term economic and social development, local environmental quality, and intragenerational equity. Mitigation cost assessments that follow this line can address these impacts on the basis of a decision-making framework that includes a number of side-impacts to the GHG emissions reduction policy objective. The goal of such an assessment is to inform decision makers about how different policy objectives can be met efficiently, given priorities of equity and other policy constraints (natural resources, environmental objectives). A number of international studies have applied such a broad decision-making framework to the assessment of development implications of CDM projects (Austin *et al.*, 2000).

The following sections highlight a number of key linkages between mitigation costing issues and broader development impacts of the policies, including macroeconomic impacts, employment creation, inflation, marginal costs of public funds, capital availability, spillovers, and trade. This leads to discussion of a number of issues involved in an economic assessment of intergenerational equity aspects.

7.4.2   *Income and Other Macroeconomic Effects*

7.4.2.1   *Macroeconomic Indicators*

Major programmes of mitigation or adaptation, particularly those that involve the use of instruments such as energy and carbon taxes, cause changes in the values of key macroeconomic variables. These include growth in GDP, employment,

external account balance, and the rate of inflation. As part of the decision-making process, information on all these variables should be provided. Changes in GDP, however, have a special role in the analysis. As noted in Section 7.2.2, under certain circumstances GDP is a valid welfare measure of the value of the goods and services produced in an economy. In so far as this is the case, changes in GDP in real terms (i.e., adjusting for price changes) are also a valid measure of the costs of any mitigation policy. The major qualification is that prices should reflect social costs and that all activities that affect welfare should be included. To the extent that this is not the case a change in GDP is not an accurate measure of the costs of a programme. One common reason for divergence between GDP and welfare is the presence of external effects. Another is the failure to account for the economic value of leisure or household work. The macroeconomic models referred to in Section 7.6, and analyzed in detail in Chapter 8, do not report the costs of market-based programmes for GHG reduction at the microeconomic level, but do so in terms of conventional GDP.[14]

It must be recognized that the full set of adjustments to GDP measures needed to obtain a correct welfare measure of the costs is difficult to compute. If the policies have ancillary benefits and/or co-benefits, then the overall costs of the measures are less than any fall in GDP. This adjustment can be made (using the methods discussed in Section 7.2.3) to the GDP measure if the data on the ancillary benefits are collected. Other adjustments relate to changes in distributional effects and the shadow pricing of goods and services for which prices do not reflect social costs. Without a detailed microlevel analysis of which sectors are affected, however, these corrections are not possible. Hence it has to be recognized that GDP changes are less accurate as measures of the true costs of mitigation programmes, and that the use of multi-attribute and other similar analyses is even more important for the assessment of such programmes.

Several authors suggest the inclusion of more comprehensive welfare measures in macroeconomic studies to give a better reflection of social costs. The United Nations Commission for Sustainable Development (UNCSD) has developed a system for Green GDP accounting and a list of sustainable development indicators that can be used to include part of the social cost aspects in GDP measures (UNCSD, 1999). The indicators cover social, economic, environmental, and institutional DES aspects. A study by Håkonsen and Mathiesen (1997), based on a CGE model, assessed large differences in welfare implications of three mitigation policy cases, namely:

- case A, in which carbon tax revenue is recycled lump-sum to the household;
- case B, in which carbon tax revenue substitutes labour taxes; and

---

[13] To some extent DES impacts overlap with ancillary impacts. Examples are reductions in air pollution, changes in employment, etc. The concept of DES is, however, wider than that of ancillary benefits, covering issues of long-term equity, social and economic development, and sustainability.

---

[14] A study that summarizes the macroeconomic level costs of alternative climate change policies in Germany and the USA, including the employment impacts, is Jochem *et al.* (2000).

- case C, in which the model includes ancillary benefits related to local air pollution and the transport sector.

Sen (1999) presents a broader perspective on economic development and emphasizes that economic welfare is not the primary goal of development, but is rather an instrument to achieve the primary goal to enhance human freedom. Freedom, at the same time, is instrumental in achieving development. The studies should consider a broad range of development issues including impacts on economic opportunities, political freedoms, social facilities, transparency guarantees, and protective security.

7.4.2.2   The Marginal Costs of Public Funds

As noted in Section 7.2.4 shadow prices have to be applied to market prices when these prices do not reflect the true opportunity costs. Shadow prices have also been applied to the funds used to finance mitigation programmes. Public expenditures, regardless of the benefits they confer, impose a cost on society, which reflects the "marginal excess burden" of a tax policy. The marginal costs of public funds should include the impacts of eventually reduced distortions compared with existing tax systems, as well as administration costs, compliance costs, the excess burden of tax evasion, and avoidance costs incurred by the taxpayers. Slemrod and Yizhaki (1996) also suggest the distributional impacts of public funds collection be included.

The marginal costs of public funds are critically dependent on the dead-weight loss associated with distortionary taxation, which is dependent on the specific tax structure in place in the non-policy case. To evaluate the true social cost of the funds it is necessary to estimate or know the marginal cost of public funds, that is the cost per dollar of finance, which is greater by US$1 than the welfare cost of raising the tax revenue. In general there will not be one figure for this cost for the whole tax system. Each source of finance will have its own marginal cost. In general there will not be one figure for this cost for the whole tax system. Each source of finance will have its own marginal cost[15]. If such a correction is not made, mitigation policies underestimate the costs of reducing GHGs.

Håkonsen (1997) has surveyed the theoretical discussion of the marginal cost of public funds, and empirical estimates of the marginal costs have been made by the World Bank and others (Devarajan *et al.*, 1999, European Commission, 1998, Ruggeri, 1999).

Estimates tend to suggest that the marginal costs of public funds are larger in developing countries than in developed countries. Devarajan *et al.* (1999) estimates that these costs vary between US$0.48 and US$2.18 for developing countries and US$1.08 and US$1.56 for the USA. The European Commission uses a value of US$1.28 for the shadow price of public funds.

7.4.2.3   Employment

This section deals with the valuation of employment impacts on a project basis. If a project creates jobs, it benefits society to the extent that the person employed would otherwise not have been employed or would have been employed doing something of lower value. Conversely, if the project reduces employment there is a corresponding social cost. These benefits depend primarily on the period that a person is employed, what state support is offered during any period of unemployment, and what opportunities there are for informal activities that generate income in cash or kind. In addition, unemployment is known to create health problems, which have to be considered as part of the social cost.

A physical measure of the extent of the employment created is therefore an important task of any project assessment in an area where there is unemployment.[16] The data that have to be estimated are:
- number of persons to be employed in the projects;
- duration for which they are employed;
- present occupations of the individuals (including no formal occupation); and
- gender and age (if available).

This physical information can be used in the multi-attribute selection criteria discussed in Section 7.2.1 (*Box 7.1*). In addition, however, it is possible to place some money value on the employment, or to deduct from the payments made to the workers the value of the benefits of the reduced unemployment.

Before considering the framework for such an evaluation, it is important to set out the theoretical reasons for arguing that unemployment reduction has a social value. In neoclassic economic analysis, no social cost is normally associated with unemployment. The presumption is that the economy is effectively fully employed, and that any measured unemployment results from matching the changing demand for labour to a changing supply. In a well-functioning and stable market, individuals can anticipate periods when they will be out of work, as they leave one job and move to another. Consequently, the terms of labour employment contracts, as well as the terms of unemployment insurance, reflect the presence of such periods, and there is no cost to society from the existence of a pool of such unemployed workers. However, these conditions are far from the reality in most of the developing and some of the developed countries in which the GHG projects will be undertaken. Many of those presently unemployed have poor prospects of employment.

---

[15] It remains true, however, that if the system is optimally designed, the marginal costs of different fiscal instruments will be equalized.

[16] Account must also be taken of any divergence between the market price and the social value of the output derived from the labour, both in its pre-project stage and as a result of the project (for details, see Ray, 1984).

In these circumstances, therefore, it seems entirely appropriate to treat the welfare gain of those made employed as a social gain. For developed economies this welfare gain is calculated as follows (Kirkpatrick and MacArthur, 1990):

a. gain of net income as a result of a new job, after allowing for any unemployment benefit, informal employment, work-related expenses, etc.; minus

b. the value of the additional time that the person has at his or her disposal as a result of being unemployed and that is lost as a result of being employed; plus

c. the value of any health-related consequences of being unemployed that are no longer incurred.

To calculate the social benefits (the unemployment avoided as a result of the project), the welfare cost ((a) minus (b) plus (c)) has to be multiplied by the period of employment created by the project.[17] The above method can also be applied to obtain employment benefit estimates for projects in developing countries (see, e.g., Markandya, 1998).

7.4.2.4   Inflation

Price levels are always changing to reflect changes in the relative scarcity of inputs and other factors. However, when the overall cost of goods and services increases in a certain period, then the economy faces inflation. Two aspects of inflation need to be considered. First, for comparison at different points in time an adjustment should be made for any general increase in the price level, that is the comparisons should be made in real terms. The appropriate deflator is a matter of judgement, but it should be based on a basket of goods consumed by the relevant group of consumers in the country. Also, any such adjustments do not preclude the possibility of increases in "real prices". It is quite possible that the costs and benefits attached to some impacts increase slower (or faster) than the general price level.

The second issue relates to the welfare cost of any inflation generated by the mitigation or adaptation activities. One of the main causes of inflation is when a country incurs a fiscal deficit (i.e., public expenditures exceed tax revenues) that is financed by printing money. Such an increase in inflation is effectively a tax on money holdings, on assets denominated in nominal terms, and on those with fixed money incomes.

The distributional consequences of the inflation tax are germane to the decision-making process. There is no simple way, however, to estimate this welfare cost; doing so requires sophisticated measurements of losses in the consumption level that affect distinct income groups. Moreover, for most mitigation and adaptation measures, the increase in inflation is likely to be quite small. Hence, in the majority of cases it is sufficient to report any increase in inflation that results from the climate change policy and use that information as a direct element in the decision-making process.

7.4.2.5   Availability of Capital

The capital costs of mitigation and adaptation programmes may be underestimated if the true scarcity of capital is not reflected in the costs incurred by the parties that implement the programme. This can arise if capital is "rationed", that is the demand for investment projects exceeds the supply. In such a situation it is appropriate to apply a shadow price for capital, for the estimation of which the World Bank (1991) and others have made estimates. This adjustment is in addition to the adjustment for the marginal cost of public funds (Section 7.4.2.2). Moreover, when a shadow price for capital of greater than one is applied, it acts to ration capital when the discount rate applied is low.

The above discussion assumes that the capital allocated to the project is free to be used for any other project. What happens, however, if capital is not "fungible" in this sense, but is made available by a donor or third party for the specific purpose of implementing climate change programmes? In these circumstances the assessment of the programme from the national viewpoint differs from its assessment from the viewpoint of the third party. The national assessment could take the shadow price of capital as zero if it genuinely could not be used for any other purpose. If, however, there were a number of alternative projects to which the capital could be allocated, a comparison between them should be based on a shadow price of capital that reflects its scarcity relative to the investment opportunities available. The party providing the finance, on the other hand, will have its own set of alternative projects to which the capital could be allocated and it may apply its own shadow price. The important point is that the evaluation and ranking of projects from a domestic viewpoint may differ from their ranking from a donor perspective. When rankings differ, a compromise is usually reached, based on the relative bargaining strengths of the two parties.

7.4.3   Valuation of Spillover Costs and Benefits

In a world in which countries are linked by international trade, capital flows, and technology transfers GHG abatement by one country has welfare effects on others. In some cases these impacts, or spillovers, are positive and in others negative. Spillovers are a broad concept that has been used in relation to a number of different international inter-linkages between GHG emission reduction policies and impacts on industrial competitiveness, reallocation of industry, and a development and implementation of technologies. This section provides a short introduction to these main categories of spillovers as an introduction to Chapters 8 and 9 that include a review of economy-wide and sectoral studies on spillovers.

---

[17] Note that this method implies a social benefit for the employment that is likely to be much less than the product of the average earnings and hours worked.

7.4.3.1 *Industrial Competitiveness and Potential Reallocation of Industries*

GHG emission reduction policies potentially will have a major impact on industrial competitiveness because sub-sectors that have relatively high GHG emission intensity or have relatively high reduction costs potentially can lose in competitiveness.

The basic theoretical framework is that of a full employment, open economy, and no international capital mobility (Dixit and Norman, 1984). Within this model an emissions constraint shifts the production possibility frontier inwards, as long as the constraint requires some "no regret" measures to be undertaken. The spillover impact of this shift depends on whether the emissions reductions have a greater impact on the production of the export good, or on the import competing good. If it is the former, abatements turn the terms of trade in favour of the country that undertakes abatement and against the country that does not. In these circumstances the non-abating country suffers some welfare loss, while the abating country could be better or worse off, depending on the size of the shift in terms of trade relative to costs of abatement. Conversely, if emissions have a greater impact on the production of the import-competing good, the terms of trade move in favour of the non-abating country, which should have an increase in welfare. The analysis of industrial reallocation considered in the previous section becomes further complicated when international capital mobility is taken into account. Carbon constraints typically alter relative rates of return against abating and in favour of non-abating countries. A flow from the former to the latter is then likely, which shifts further inwards the production possibility frontier in the abating country. At the same time, it causes an outwards shift of the frontier in the non-abating country. Modelling capital flows is notoriously difficult, however, and no theoretical results can be obtained for the complex and empirically relevant cases. Hence the indisputable need to use simulation models and to undertake primary empirical research. The welfare impacts of changes in international capital flows are seldom reported. Progress depends on the further development of techniques such as decomposition analysis (Huff and Hertel, 1996)[18] and multiple simulations in which some variables are held constant to isolate their influence on the final outcome.

Seen from a more practical perspective the theoretical arguments about competitiveness and international capital flows have at least two versions of what happens without specific developing country targets: either domestic industry relocates abroad, or the demand for domestic energy-intensive goods declines and the trade balance deteriorates; or both occur.

Consider four factors that affect location or trade effects. First, do the non-tradable sectors account for a substantial share of carbon emissions? Second, are energy costs a small or large percentage of the total costs in key manufacturing sectors? Third, is the burden of meeting an emission reduction target partially borne by non-participating countries because of changes mediated through international trade? For example, developed nations could demand fewer exports from non-participating countries. This would shift the terms of trade against these countries, and they would bear some of the costs of reducing GHGs. Fourth, how do resources shift across sectors because of carbon policy? For instance, there could be a shift from the energy-intensive sector to the domestic goods sector that is non-energy intensive. The aggregate impact could be positive or negative depending on the potential returns from the non-energy intensive sector.

First, consider the "pollution havens" hypothesis, in which firms are tempted to relocate to or to build new plants in nations with lax environmental standards (see Dean, 1992; Summers, 1992; Esty, 1994; Jaffe *et al.*, 1994). Palmer *et al.* (1995) point out that the following must be considered:

- whether the cost of complying with environmental regulation is a small fraction of total cost;
- whether the differences between the developed nation's environmental regulations and those of most major trading partners are small or large; and
- whether the firms of the developed nation build state-of-the art facilities abroad regardless of the host nation's environmental regulations.

The evidence to date on pollution havens is not strong, although this may change in the future as international agreements on climate change come into force.

In the context of climate change, cost estimates must consider how carbon taxes affect trade flows in the short and long runs. The "leakage effect" reflects the extent to which cuts in domestic emissions are offset by shifts in production and therefore increases in emissions abroad. The empirical question is whether nations that are a net exporter in fossil fuel intensive products (e.g., steel) gain under Annex I-only carbon policies. Other developing nations might not gain because less capital will be available as the income in the developed nations drops, and it becomes more costly to import from developed nations the capital goods that promote growth (e.g., machinery and transportation equipment). See Chapters 8 and 9 for any empirical evidence on the magnitude of leakage.

7.4.3.2 *Technological Spillovers*

The theoretical discussion about spillovers emerging from impacts on industrial competitiveness and industrial reallocation is based on a comparative static framework. When extended to a dynamic context, the production possibility frontiers of industries are assumed to shift outwards in a way determined by technological change in different sectors as a reflection of

---

[18] Verikios and Hanlow (1999) illustrate in a comparative static framework how the welfare impacts of international capital mobility can be assessed using decomposition analysis.

an endogenous feedback from GHG emission reduction policies on technological change.

There are three routes by which technology policies in one country affect development in other countries or specific sectors. First, R&D may increase the knowledge base and this will be a general benefit for all the users of a technology. Second, increased market access for low-CO_2 technologies, through niche-markets or preferential buyback rates in one country may induce a generic improvement in technology in others. Third, domestic regulations on technology performance and standards, whether imposed or voluntary can create a strong signal for foreign industrial competitors. A paper by Goulder and Schneider (1999) similarly argues that climate change policies bias technical change towards emissions savings.

The possibility of a positive technological spillover from GHG emission reduction policies has not been taken into account in any of the global mitigation studies reviewed in Chapter 8. If this materializes, it could cause further complex shifts of the production possibility frontier, including an outwards shift in the production of the affected goods.

7.4.4   *Equity*

7.4.4.1   *Alternative Methods of Addressing Equity Concerns*

A key issue in evaluating climate change policies is their impact on intragenerational equity, in which one impact indicator is the income distributional consequences of the policies seen in a national context or across countries. Other related equity issues are the distributional impacts of avoided climate change damages that emerge as a result of mitigation policies, which is dealt with by the IPCC WGII TAR, and intergenerational equity, which is discussed in Section 7.2.4.

There are essentially two ways to deal with intragenerational equity. The first is not to deal with it at all in the benefit–cost analysis, but to report the distributional impacts separately. These can then be taken into account by policymakers as they see fit, or the information can be fed into a multi-criteria analysis that formalizes the ranking of projects with more than one indicator of their performance.

The second method of analysis is to use "income weights", so that impacts on individuals with low incomes are given greater weight than those on individuals with high incomes. Although a number of analysts do not support the use of such weights, some do and policymakers sometimes find an assessment that uses income weights useful. Hence they are included in this chapter.

The costs of different GHG programmes, as well as any related benefits, belong to individuals from different income classes. Economic cost–benefit analysis has developed a method of weighting the benefits and costs according to who is impacted. This is based on converting changes in income into changes in welfare, and assumes that an addition to the welfare of those on a lower income is worth more an addition of welfare to richer people. More specifically, a special form can be taken for the social welfare function, and a common one that has been adopted is that of Atkinson (1970). He assumes that social welfare is given by the function:

$$W = \sum_{i=1}^{N} \frac{A Y_i^{1-\varepsilon}}{1-\varepsilon}$$

where:
W is the social welfare function,
Y_i is the income of individual i,
ε is the elasticity of social marginal utility of income or inequality aversion parameter, and
A is a constant.

The social marginal utility of income is defined as:

$$\frac{\partial W}{\partial Y_i} = A Y_i^{-\varepsilon}$$

Taking per capita national income, $\bar{Y}$, as the numeraire, and giving it a value of one gives:

$$\frac{\partial W}{\partial Y_i} = A \bar{Y}^{-\varepsilon} = 1$$

and

$$\frac{\partial W / \partial Y_i}{\partial W / \partial \bar{Y}} = \left[\frac{\bar{Y}}{Y_i} \right]^{\varepsilon}$$

In this way the marginal social welfare impact of income changes by individuals is the elasticity of the ratio of the per capita income $\bar{Y}$ and the income of individual i, Y_i. The marginal social welfare impact of income changes by individual i also can be denoted as SMU_i, where SMU_i is the social marginal utility of a small amount of income going to individual i relative to income going to a person with the average *per capita income*. The values of SMU_i are, in fact, the weights to be attached to costs and benefits to groups relative to different cost and benefit components.

To apply the method, estimates of $\bar{Y}$ and ε are required. The literature contains estimates of the inequality aversion parameter (ε) in the range 1–2 (Murty *et al.* 1992; Stern, 1977). Some recent studies that estimate the value of ε for the Indian economy (Murty *et al.*, 1992) resulted in values in the range 1.75–2.0.[19]

7.4.4.2 The Use of Average Damages

A special case of the income distributional weights approach is to estimate the money value of impacts for different groups of individuals or countries and then apply the average damage to all individuals and countries. The best example of this is the value attached to changes in the risk of death. These risks are valued in terms of the statistical value of life, which caused much controversy in SAR (IPCC 1996a, Chapter 6). The "value of a statistical life" (VSL) converts individual WTP to reduce the risk of death into the value of a life saved, when it is not known which life that will be. For example, if each person in a community has a WTP of US$10 to reduce the risk of death by one in a hundred thousand, then the collective WTP of a group of 100,000 is US$1 million for a measure that would, on average, save one life. Hence, the figure of US$1 million is referred to as the VSL. This measure is one way of valuing changes in risks of mortality. Other ways include a "human capital" approach, which values the loss of income and multiplies it by the change in risk, or a "life years lost" approach, which takes the WTP for life years that could be lost as a result of changes in the survival probabilities an individual faces. Of these, the VSL has been used most commonly in recent years. The human capital approach is not well founded in terms of welfare and the life years lost approach is still being developed.

The VSL is generally lower in poor countries than in rich countries, but it is considered unacceptable by many analysts to impose different values for a policy that has to be international in scope and decided by the international community. In these circumstances, analysts use average VSL and apply it to all countries. Of course, such a value is not what individuals would pay for the reduction in risk, but it is an "equity adjusted" value, in which greater weight is given to the WTP of lower income groups. On the basis of EU and US VSLs and a weighting system that has some broad appeal in terms of government policies towards income distribution, Eyre *et al.* (1998) estimate the average world VSL at around 1 million Euros (approximately US$1 million at 1999 exchange rates).[20]

Formally, it can be shown that the use of average values for damages implies income weights based on an elasticity of one, which, as can be seen from above, is broadly consistent with government policies towards income redistribution (Fankhauser *et al.*, 1997; Eyre *et al.*, 1998). The advantage of this approach is that it addresses equity concerns while retaining a valuation of damages that is broadly consistent with the efficiency approach. Such an approach may be a way to reflect the equal value of lives as seen from a global policy perspective. National perspectives and opportunities should be addressed in another way.

7.4.5 Estimating Future Costs and Sustainability Implications

Mitigation policies that are large in scale can have significant long-term implications on future climate change and thereby have implications for intergenerational equity. The issue is to model future changes in ecological systems and economic welfare associated with different levels of climate change caused by specific mitigation efforts.

Climate change offers an imposing set of complications for the policymaker–global scope, wide regional variations, the potential for irreversible damages or costs, multiple GHGs, a very long planning horizon, and long time lags between emissions today and future impacts on ecosystem services. For the economist, to assess how these distant climate-induced changes in ecosystem services might affect the economic wellbeing of citizens in the far distant future is no less imposing.

The challenge rests in capturing accurately three general issues: (1) how climate change might affect ecological systems; (2) how these altered ecosystems might affect the demand for different market and non-market goods and services; and (3) how this demand change affects the welfare of our descendants. The first two issues can only be dealt with by broad scenario analyses that consider alternative development patterns for ecological systems and the interactions with man-made systems. The third issue can be addressed by applying assumptions about the preferences of future generations, which, for example, can be assumed to reflect the preferences of present generations.

Those who undertake studies of welfare losses brought about by climate change often focus on an assessment of the potential welfare losses suffered by future citizens through climate change. Typically, such an assessment is based on measuring the demand curve for people alive today under today's climate given the substitution possibilities implied by extant technologies and knowledge constraints that define today's opportunity set. Essentially, these analysts ask, "If the climate of the future enveloped us today, what would be our welfare loss?"

The question often not asked is this: "Does the opportunity set of today's citizens reflect, in any way, the opportunity faced by citizens in 2050 or 2100?" A welfare loss based on today's opportunity set may or may not be related to the potential climate-related loss in wellbeing to the citizens of the far distant future. Climate change triggers direct changes in the opportunity set and relative prices, and indirect changes in the adaptation of technology and supply. This is critical. More opportunities in the future will reduce the welfare loss; fewer opportunities could inflate the loss. The opportunities will depend on a complex mix of available substitutes, complementary recre-

---

[19] One reviewer has pointed out that different values of ε may be appropriate for costs and benefits.

[20] The parameter from which the weights are derived is called the elasticity of the marginal utility of income. The greater this elasticity, the greater the weight given to the WTP of poor households.

ational and non-recreational activities, relative prices, transaction costs, and preferences. These substitutes will be determined by the various different types of capital stock that contribute to human wellbeing, including man-made capital, human capital, natural capital, and social capital, as emphasized by the sustainability literature. For a more elaborate discussion on these issues see Chapter 1.

It is difficult to account for the opportunity sets of citizens in the far distant future and to predict the preferences of future generations, which adds a significant uncertainty to estimates of future damages from climate change. Climate change might affect household resources, human resource investment prices and levels, endowments, preferences, labour market opportunities, and natural environment, all of which influence our descendant's opportunity set–the basic materials needed for attainment in life. These risks indirectly modify our heirs' life chances by reducing and reallocating household resources or by constraining their choices or both. Our descendants may shift resources towards a sick child and away from recreation. Their children might have to forego the life experience of fishing the same river as their ancestors. Faced with these consequences, individuals today might be willing to pay to prevent risks that restrict our heir's opportunities. But this is a different question.

When considering future generations' opportunities the impacts of today's climate change investments on future generations' opportunities should also be considered. Investments might, for example, enhance the capacity of future generations to adapt to climate change, but at the same time they potentially displace other investments that could create other opportunities for future generations.

Two things are likely to be different in the future–the climate and our heirs' opportunities. Accounting for one change and not the other will not markedly advance our understanding of expected benefits. The question should be "How could these future effects be linked to existing models to value non-market effects?" For the most part, the valuation question is how to account for changes, both good and bad, of future opportunities. Accounting for these decisions probably requires a new model that focuses on the value of maintaining or enhancing the future's opportunities so as to maximize their life chances, whatever their preferences might be.

7.5   Specific Development Stages and Mitigation Costs (Including Economies in Transition)

Developing countries and EITs exhibit a number of special characteristics that should be reflected in mitigation cost studies. There is a need for further development of the methodologies and approaches that reflect these issues; this section introduces a number of distinct features for such economies and concludes with a number of suggestions for the expansion of studies and methodology development.

7.5.1   *Why Developing Countries Have Special Problems in Their Mitigation Strategies*

The term "developing countries" covers a wide variety of countries with distinct differences in their economic, political, social, and technological levels. The group of countries termed "least developing countries" have very little basic infrastructure, the "newly industrialized countries" have a structure closer to that of the developed countries, and others lie between these two extremes. Almost all developing countries have a relatively low level of GHG emissions per capita at present, but large countries like India, China, and Brazil will soon become very important in terms of their contribution to total global emissions. It is therefore important to understand how these countries might participate in globally cost-effective policies.

Mitigation costs in a country depend critically on the underlying technological and socioeconomic conditions. Studies that assess these costs make assumptions about current and future socioeconomic development patterns and the potential to implement climate change mitigation policies. Developing countries exhibit a number of specific complexities that are of major importance to costing studies. Data are limited, exchange processes are constrained, markets are incomplete, and a number of broader social development issues are potentially important for future GHG emissions, such as living conditions of the poor, gender issues, and institutional capacity needs. Some of these difficulties arise particularly in relation to land-use sectors, but can also be important in relation to the energy sector and transportation.

To sum up, a number of special issues related to technology use should be considered for developing countries as the critical determinants for their climate change mitigation potential and related costs. These include current technological development levels, technology transfer issues, capacity for innovation and diffusion, barriers to efficient technology use, institutional structure, human capacity aspects, and foreign exchange earnings.

The methodology of most current mitigation cost studies was developed on the basis of approaches originally designed for the market-based economies of developed countries. The application of these methodologies in a developing countries context typically poses special problems relating to data, sectoral coverage, activity projections, and assumptions about markets, behaviours, and policy instruments. A simplified application of these methodologies in developing countries can lead to a number of inaccuracies in mitigation studies:

- Major GHG emission sources and drivers for future emission can be overlooked. This is especially relevant for the land-use sectors.
- Mitigation studies may focus on specific technical options that are not consistent with national macroeconomic policy contexts and broader social and environmental policy priorities.
- The technical potential of specific options, for example electricity saving options, may be overestimated

because consumer behaviour and power market failures are not captured.

- The impacts of using different policy instruments cannot be assessed because the studies do not include any information on national institutional structure, taxes, and other regulation policies and various technology promotion programmes.

- Implementation issues, including institutional and human capacity aspects and local market development, are not represented.

7.5.2 Why Economies in Transition (EIT) Have Special Problems in Their Mitigation Strategies

Estimating the costs of mitigation for EITs presents its own challenges, which can be described as past, present, and future. In the recent past, prices were not the rationing mechanism of choice. The listed prices (where there were any) did not necessarily reflect the actual level of scarcity, since they were not set by supply and demand. As such, data based on listed prices from which to construct marginal abatement cost curves is sketchy at best, and completely missing at worst.

Today, problems still exist in the construction of such curves, in that each transition economy has its own unique mix of fee markets and state control. The newer sources of data reflect a mix of price and quantity rationing that needs to be better understood on a country-by-country basis.

Finally, using this data to estimate mitigation costs into the near or distant future depends on critical assumptions about how the political, legal, and economic institutions will evolve in these economies. Any estimates of mitigation costs into the twenty-first century made under the assumption that current institutions will be held constant are almost certainly not going to be correct. Hence, it is essential to devote a good deal of effort to develop scenarios of evolution for these institutions and their implications for economic development.

7.5.3 Development Projections

The establishment of long-term projections for GHG emissions is particularly complicated and uncertain for both developing countries and the EITs. These economies are often in a transition process in which important GHG emission sectors, such as the energy sector, industry, and transportation, are expected to play an increasing role. It is not possible, however, to project accurately the actual speed of this growth process and/or the GHG emission intensity of these future activities. Modelling tools and data are also very limited or even non-existent, and the only available information sources from which to generate GHG emission projections are often the official national development plans that cover a time horizon of 5–10 years only.

Changes in the structure of GDP have to be given careful consideration. One important aspect that could be integrated into the scenario development are the changes in economic structure and relative prices that emerge from structural adjustment programmes and other macroeconomic policies that many countries are currently undertaking. Another crucial issue, following that, will be the development of energy intensive and heavily polluting industrial activities, such as steel and aluminium production. As the recent shift of heavy industries from the developed towards the developing countries reaches its end, long-term economic output could come from services and other less energy-intensive activities. In EITs the issue is how fast and deep will the shift out of energy intensive industries be, and what will replace it.

The basic uncertainty of long-term GHG emission projections encourages analysts to use multiple baselines, each corresponding to a particular expectation of the future development pattern. Each development pattern may exhibit a unique emissions trajectory. A nation following development policies that emphasize greater investments in infrastructure, such as efficient rail transport, renewable energy technologies, and energy-efficiency improvements will exhibit a low emissions trajectory. However, a nation with substantial coal resources, scarce capital, and a low level of trade can be pushed towards a development path with high emissions.

The spatial distribution of the population and economic activities is still not settled in the developing countries. This raises the possibility of adopting urban and/or regional planning and industrial policies to strengthen small and medium cities and rural development, and thus reduce the extent of the rural exodus and the degree of demographic concentration in large cities. In the same way, technological choices can substantially decrease the energy demand and/or GDP elasticities. The preservation of a certain cultural diversity, as opposed to the trend towards a global uniformity of lifestyles, also favours less energy-intensive housing, transportation, leisure, and consumption patterns, at least in some cases. One example is related to development policies that avoid low urban population density coupled with long daily trips to work and large shopping centres by car.

It is a special challenge in costing studies to translate preferences for biological and cultural diversity into a useful value measure. The market does not price most of the services provided by biological or cultural diversity. Roughgarden (1995) argues that there is no need to quantify the benefits of these services, which are either so obvious or impossible to capture that measurement is unnecessary. Following this line of argument, "science" should dictate a target that could be used to establish a safe minimum standard–a level of preservation that guarantees survival of the species or culture in question (Ciriarcy-Wantrup, 1952). This minimum standard approach puts an infinite value on avoiding extinction. This view puts biological or cultural diversity beyond the reach of economic trade-offs, and the analyst attempts to find the least-cost solution to achieve some set standard.

However, Epstein (1995) argues that preservation without representation of benefits is unacceptable. It is suggested that hard evidence is needed to prove that the biological and cultural preservation benefits dominate those from development. It is then logical to compare the costs and benefits when resources are scarce, and an attempt should be made to balance the costs and benefits so that funds are allocated to their highest valued use.

Estimating the social value of biodiversity and culture is a major challenge. For biodiversity values there is no consensus as to the usefulness of the primary tool used to reveal the monetary value of these preferences–contingent valuation surveys. These public opinion surveys use a sequence of questions to put a monetary value on personal preferences. However, since people are responding to a survey rather than facing their own budget constraint and actually spending their own money, no market discipline exists to challenge their statements (Brown and Shogren, 1998).

The above possibilities of alternative development patterns highlight the technical feasibility of low carbon futures in the developing countries that are compatible with national objectives. However, the barriers to a more sustainable development in developing countries can hardly be underestimated, from financial constraints to cultural trends in both developed and developing countries, including the lack of appropriate institutional building. Any abatement-cost assessment relies on the implicit assumptions taken in the baseline or mitigation scenarios with regard to the probability of removing these barriers.

Since mitigation costs for different development patterns may vary substantially, one way to reflect this in mitigation cost analysis is to use a scenario-based range of mitigation costs rather than a single mitigation cost (see also Section 7.3.6).

7.5.4 Broadening the National Decision-making Framework

Although cost is a key component of the decision as to which policies to select, it is not the only consideration. Other factors enter the decision, such as the impacts of policies on different social groups in society, particularly the vulnerable groups, the benefits of GHG limitation in other spheres, such as reduced air pollution, and the impacts of the policies on broader concerns, such as sustainability. In developing countries these other factors are even more important than in developed countries. GHG limitation does not have as high a priority relative to other goals, such as poverty reduction, employment, etc., as it does in the wealthier countries. Indeed, it can be argued that the major focus of policy will be development, poverty alleviation, etc., and that GHG limitation will be an addendum to a programme designed to meet those needs. Accounting for the GHG component may change the detailed design of a policy or programme, rather than be the main issue that determines the policy.

Markandya (1998) developed a framework to expand the cost analysis with an assessment of the other impacts of climate change mitigation projects, such as employment, income distribution, environmental changes, and sustainability indicators. The suggestion is that monetary cost and benefit estimates be combined with physical indicators and qualitative information. These include the impacts of projects on vulnerable groups, on the environment more generally, and on sustainability in a broader sense.

Markandya and Boyd (1999) and Halsnæs and Markandya (1999) assessed the implications for cost-effectiveness of using an expanded cost-analysis framework compared with a focus on direct costs. They examined a number of case studies, including renewable energy options (biogas, solar water-heating systems, photovoltaic streetlights, and wind turbines), DSM programmes, and a number of transportation sector options. The expanded cost assessment includes a specific valuation for the welfare impacts of increased employment, local environmental improvements related to reduced non-GHG pollutants, and income distribution weights. The conclusion is that in a number of cases the application of an expanded cost-assessment framework has major implications for the cost-effectiveness ranking of mitigation projects compared with their ranking on direct costs alone. In particular, large differences in cost-effectiveness are seen for a biogas plant in Tanzania, for which combined social costs considered in the expanded framework go down to minus US$30/$tCO_2$ reduction compared with a purely financial cost of plus US$20/$tCO_2$. This cost difference reflects a positive welfare impact on presently unemployed low-income families and the time saved through reduced fuelwood collection. The case examples generally suggest that the combined social costs of mitigation policies in developing countries in particular will be lower than the purely financial costs, especially if the policies require presently unemployed labour and reduce the damages from local non-GHG pollutants. Similar studies for EITs reveal great large value of ancillary benefits in the form of reduced air pollution and increased employment, especially for carbon sink projects.

7.5.5 Addressing the Specific Characteristics of Markets and Other Exchange Processes in Developing Countries

Climate change studies focus on the cost assessment of activities through their presentation on the markets. The GHG emission sources considered, on this basis, are predominantly those represented in official economic and sectoral statistics, and the prices used to value the resources are derived on a market basis. Such information, however, is incomplete for developing countries for which markets are incomplete, property rights are not well established, and a significant part of the exchange process belongs to the informal economic sector. This section discusses the implications of these specific features for climate change studies.

GHG emissions in the energy and agriculture sectors are greatly influenced by present subsidies. Subsidy removal in the energy sector, if supported by improvements in managerial efficiency, could reduce CO_2 emissions and other pollutants by up to 40% in developing countries with very low or even negative costs (Anderson, 1994; Halsnæs, 1996). It should be recognized that general macroeconomic policies, such as structural adjustment programmes, already include a number of subsidy removal policies.

Most major markets in developing countries are characterized by supply constraints, but the labour market is an exception for unskilled labour is frequently in excess supply. Examples of such supply constraints are seen in the financial sector, power production, and infrastructure development. This results from high transaction costs that originate from weak market linkages, limited information, inadequate institutional set-ups, and policy distortions. Such market imperfections make it difficult to establish reliable parameters such as price elasticity of demand.

In many developing countries and EITs, commodity prices, including those of energy resources, are regulated and are not market determined. The consequent market distortions are often not adequately captured by models. There is therefore a need to apply some price-correcting rules to reflect social costs.

Traditional cost–benefit analysis suggests the use of shadow prices to correct for market distortions (see Section 7.2.3.1). Such a procedure is in line with the approach of CGE models. In both these approaches mitigation policies and related costs are assessed in relation to an "optimal resource allocation case", in which markets are in equilibrium and prices (and thereby cost) reflect resource scarcities. However, these conditions are far from those currently found in these countries, so studies should consider how a transformation to the optimal resource allocation case is likely to take place over a certain time frame. Developing countries are presently undergoing market-oriented economic reforms. However, the price distortions are only partially and gradually being remedied because of the high social costs associated with speedy reforms. The complexities in modelling this process cannot be underestimated, and it should therefore be recognized that only part of the transformation can be captured.

Integration of market transformation processes in cost studies should include an assessment of barrier removal policies. Such policies include efforts to strengthen the incentives for exchange (prices, capital markets, international capital, and donor assistance), to introduce new actors (institutional and human capacity efforts), and to reduce the risk of participation (legal framework, information, and general policy context of market regulation). Some of these policies can be reflected in cost studies, such as barrier removal policies that address market prices, capital markets, and technology transfers, while other areas like capacity building need to be addressed in a more qualitative way.

A number of important interrelationships and spillovers occur between the informal and formal sectors with regard to climate change mitigation policies. An example is the potential to introduce advanced production technologies in the energy and agriculture sectors that, on the one hand, use domestic resources (e.g., biomass) in a more sustainable way and, on the other, improve efficiency and create capacity in local companies and institutions. The impact of introducing policy instruments such as carbon taxes or energy subsidy removal also depends on potential substitutions to non-commercial wood fuels that might be unsustainable. Mitigation cost studies for developing countries should, as far as possible, include an assessment of energy consumption and biomass potential in the informal sector and apply assumptions about price relations and substitution elasticities between the formal and informal sectors. Similarly studies should consider the capacity of enterprises in both the formal and informal sectors to adapt and manage the advanced technologies that are suggested as cost-effective mitigation options in national programmes.

7.5.6 *Suggestions for Improvements in the Costing Study Approach Applied to Developing Countries and Economies in Transition*

Climate change studies in developing countries need to be strengthened in terms of methodology, data, and policy frameworks. Although a complete standardization of the methods is not possible, to achieve a meaningful comparison of results it is essential to use consistent methodologies, perspectives, and policy scenarios in different nations.

The following modifications to conventional approaches are suggested:

- Alternative development pathways should be analyzed with different patterns of investment in:
  - infrastructure (e.g., road versus rail and water);
  - irrigation (e.g., big dams versus small decentralized dams, surface irrigation versus ground water irrigation);
  - fuel mix (e.g., coal versus gas, unclean coal versus clean coal, renewable versus exhaustible energy sources);
  - employment; and
  - land-use policies (e.g., modern biomass production and afforestation).
- Macroeconomic studies should consider market transformation processes in the capital, labour, and power markets.
- In the less developed of the developing countries, informal and traditional sector transactions should be included in national macroeconomic statistics. The value of the unpaid work of household labour for non-commercial energy collection is quite significant and needs to be considered explicitly in economic analysis.
- Similarly, in such countries the traditional and informal sectors also account for an overwhelming proportion of

agriculture and land-use activities, employment, and household energy consumption; therefore, insofar as possible, these activities should be integrated into cost studies.

- Non-commercial energy sources, essentially traditional biomass, should be represented explicitly in the model as this has a crucial influence on both future energy flows and GHG emissions.
- The costs of removing market barriers should be considered explicitly.

In addition to paying attention to these factors, it is important to bear in mind that perhaps the most serious limitation of cost studies for developing countries is the paucity of data. Some mitigation studies have tried to circumvent data problems by making opaque assumptions or using estimates from data that relate to different circumstances. It is preferable to use simplified approaches that provide insights into basic development drivers, structures, and trade-offs than to use standardized international models in which the data and assumptions are duplicated from industrial countries.

7.6   Modelling and Cost Assessment

7.6.1   *Introduction*

The costs of climate policy are assessed by various analytical approaches, each with its own strengths and weaknesses. This section considers first the modelling options currently used to assess the costs of climate policy, and then the key assumptions that influence the range of cost estimates. The focus is on the general conceptual elements of cost assessment and on an evaluation of how model structures and input assumptions affect the range of cost estimates.

7.6.2   *Classification of Economic Models*

The models presented here are described and discussed in more detail in Chapter 8, in which a review of the main literature on these models is presented. However, it is useful to present an overview of the main modelling techniques applied in this kind of analysis here.

Input–Output Models
Input–output (IO) models describe the complex interrelationships among economic sectors using sets of simultaneous linear equations. The coefficients of equations are fixed, which means that factor substitution, technological change, and behavioural aspects related to climate change mitigation policies cannot be assessed. IO models take aggregate demand as given and provide considerable sectoral detail on how the demand is met. They are used when the sectoral consequences of mitigation or adaptation actions are of particular interest (Fankhauser and McCoy, 1995). The high level of sectoral disaggregation, however, requires strong restrictions that limit the validity of the model to short runs (5–15 years).

Macroeconomic (Keynesian or Effective Demand) Models
Macroeconomic models describe investments and consumption patterns in various sectors, and emphasize short-run dynamics associated with GHG emission reduction policies. Final demand remains the principal determinant of the size of the economy. The equilibrating mechanisms work through quantity adjustments, rather than price. Temporary disequilibria that result in underutilization of production capacity, unemployment, and current account imbalances are possible. Many macroeconomic models are available. They implicitly reflect past behaviour in that the driving equations are estimated using econometric techniques on time-series data. As a consequence, macroeconomic models are well suited to consider the economic effects of GHG emission reduction policies in the short- to medium-horizon.

Computable General Equilibrium Models
CGE models construct the behaviour of economic agents based on microeconomic principles. The models typically simulate markets for factors of production (e.g., labour, capital, energy), products, and foreign exchange, with equations that specify supply and demand behaviour. The models are solved for a set of wages, prices, and exchange rates to bring all of the markets into equilibrium. CGE models examine the economy in different states of equilibrium and so are not able to provide insight into the adjustment process. The parameters in CGE models are partly calibrated (i.e., they are selected to fit one year of data) and only partly statistically or econometrically determined (i.e., estimated from several years of data). Hence it is difficult to defend the validity of some of the parameter values.

Dynamic Energy Optimization Models
Dynamic energy optimization models, a class of energy sector models, can also be termed partial equilibrium models. These technology-oriented models minimize the total costs of the energy system, including all end-use sectors, over a 40–50 year horizon and thus compute a partial equilibrium for the energy markets. The costs include investment and operation costs of all sectors based on a detailed representation of factor costs and assumptions about GHG emission taxes. Early versions of these models assessed how energy demands can be met at least cost. Recent versions allow demand to respond to prices. Another development has established a link between aggregate macroeconomic demand and energy demand. Optimization models are useful to assess the dynamic aspects of GHG emissions reduction potential and costs. The rich technology information in the models is helpful to assess capital stock turnover and technology learning, which is endogenous in some models.

Integrated Energy-System Simulation Models
Integrated energy-system simulation models are bottom-up models that include a detailed representation of energy demand and supply technologies, which include end-use, conversion, and production technologies. Demand and technology development are driven by exogenous scenario assumptions often linked to technology vintage models and econometric forecasts. The demand sectors are generally disaggregated for

industrial subsectors and processes, residential and service categories, transport modes, etc. This allows development trends to be projected through technology development scenarios. The simulation models are best suited for short- to medium-term studies in which the detailed technology information helps explain a major part of energy needs.

Partial Forecasting Models
A wide variety of relatively simple techniques are used to forecast energy supply and demand, either for single time periods or with time development and varying degrees of dynamics and feedback. The main content is data on the technical characteristics of the energy system and related financial or direct cost.

Limits of Economic Models Taxonomy
The macroeconomic and CGE approaches can be further classified as "top-down" methodologies, while the technology-rich dynamic optimization/partial equilibrium, simulation, and partial forecasting approaches can be considered "bottom-up" approaches. It is also noted that the dynamic optimization/partial equilibrium, simulation, and partial forecasting approaches are sometimes collectively referred to as the family of engineering–economic models.

While useful, this taxonomy has its limits. First, differences in parameter values among the models within a given category may be more significant than the differences in model structure across categories. Second, many differences emerge between the theory underlying a particular model group and the actual models. Third, most models are hybrid constructions linked to provide greater detail on the structure of the economy and the energy sector (Hourcade *et al.*, 1998). A hybrid approach sheds light on both the economic and technological aspects of reducing energy-related CO_2 emissions, but it does have its drawbacks. Consistent results require that a hybrid approach remove all the inconsistencies across the linked models. This process is often cumbersome and time consuming.

7.6.3   Top-down and Bottom-up Models

Top-down and bottom-up models are the two basic approaches to examine the linkages between the economy and specific GHG emitting sectors such as the energy system. Top-down models evaluate the system from aggregate economic variables, whereas bottom-up models consider technological options or project-specific climate change mitigation policies. IPCC SAR on economic and social dimensions (IPCC, 1996a, Chapter 8) includes an extensive discussion on the differences between top-down and bottom-up models. It concluded that the differences between their results are rooted in a complex interplay among the differences in purpose, model structure, and input assumptions (IPCC, 1996a, Section 8.4.3).

In previous studies, bottom-up models tended to generate relatively low mitigation costs (negative in some cases), whereas top-down models suggested the opposite. Understanding why this range of costs arises requires exploration of the differences in the two modelling approaches.

The terms "top" and "bottom" are shorthand for aggregate and disaggregated models. The top-down label comes from the way modellers apply macroeconomic theory and econometric techniques to historical data on consumption, prices, incomes, and factor costs to model the final demand for goods and services, and the supply from main sectors (energy sector, transportation, agriculture, and industry). Some critics complain, however, that aggregate models applied to climate policy do not capture the needed sectoral details and complexity of demand and supply. They argue that energy sector models were used to explore the potential for a possible decoupling of economic growth and energy demand, which requires "bottom-up" or disaggregated analysis of energy technologies. Some of these energy sector technology data were, however, integrated in a number of top-down models, so the distinction is not that clearcut.

Macroeconomic models are often also detailed, but in a different way to bottom-up models. Top-down models account for various industrial sectors and household types, and many construct demand functions for household expenditures by summing "individual demand functions". Such functions can facilitate a reasonably detailed assessment of economic instruments and distributional impacts of climate change mitigation policies.

Another distinction between the top-down and bottom-up approaches is how behaviour is endogenized and extrapolated over the long run. Econometric relationships among aggregated variables are generally more reliable than those among disaggregated variables, and the behaviour of the models is more stable with such variables. It is therefore common to adopt high levels of aggregation for top-down models when they are applied to long time frames (e.g., beyond 10–15 years). The longer the period the greater the aggregation gap expected between top-down and bottom-up models.

Top-down models examine a broad equilibrium framework. This framework addresses the feedback between the energy sector and other economic sectors, and between the macroeconomic impacts of climate policies on the national and global scale. As such, early top-down models usually had minimal detail on the energy-consuming side of the economy. Specific technologies were not directly captured. In contrast, bottom-up models mimicked the specific technological options, especially for energy demand. Attention to the detailed workings of technologies required early modellers to pass over the feedbacks between the energy sector and the rest of the economy.

Top-down and bottom-up models also have different assumptions and expectations on the efficiency improvements from current and future technologies. Bottom-up models often focus on the engineering energy-gains evident at the microeconomic level and detailed analysis of the technical and economic dimensions of specific policy options. The sector-specific focus generates lower costs relative to the top-down model, which captures the costs caused by the greater production costs and lower investment in other sectors.

The basic difference is that each approach represents technology in a fundamentally different way. The bottom-up models capture technology in the engineering sense: a given technique related to energy consumption or supply, with a given technical performance and cost. In contrast, the technology term in top-down models, whatever the disaggregation, is represented by the shares of the purchase of a given input in intermediary consumption, in the production function, and in labour, capital, and other inputs. These shares constitute the basic ingredients of the economic description of a technology in which, depending on the choice of production function, the share elasticities represent the degree of substitutability among inputs.

7.6.4   Integrated Assessment Models

Researchers have also assessed the costs of climate protection by considering both the economic and biophysical systems, and the interactions between them. IAMs do this by combining key elements of biophysical and economic systems into one integrated system. They provide convenient frameworks to combine knowledge from a wide range of disciplines. These models strip down the laws of nature and human behaviour to their essentials to depict how increased GHGs in the atmosphere affect temperature, and how temperature change causes quantifiable economic losses. The models also contain enough detail about the drivers of energy use and energy–economy interactions to determine the economic costs of different constraints on CO_2 emissions (see, e.g., Shogren and Toman, 2000).

IAMs fall into two broad classes: policy optimization and policy evaluation models. Policy optimization models can be divided into three principal types:
- cost–benefit models, which try to balance the costs and benefits of climate policies;
- target-based models, which optimize responses, given targets for emission or climate change impacts; and
- uncertainty-based models, which deal with decision making under conditions of uncertainty.

Policy evaluation models include:
- deterministic projection models, in which each input and output takes on a single value; and
- stochastic projection models, in which at least some inputs and outputs take on a range of values.

Current integrated assessment research uses one or more of the following methods (Rotmans and Dowlatabadi, 1998):
- computer-aided IAMs to analyze the behaviour of complex systems;
- simulation gaming in which complex systems are represented by simpler ones with relevant behavioural similarity;
- scenarios as tools to explore a variety of possible images of the future; and
- qualitative integrated assessments based on a limited, heterogeneous data set, without using any models.

A review by Parson and Fisher-Vanden (1997) shows that IAMs have contributed to the establishment of important new insights to the policy debate, in particular regarding the evaluation of policies and responses, structuring knowledge, and prioritizing uncertainties. They have also contributed to the basic knowledge about the climate system as a whole. The review concludes that IAMs face two challenges, namely managing their relationship to research and disciplinary knowledge, and managing their relationship to other assessment processes and to policymaking.

7.6.5   Categorization of Climate Change Mitigation Options

An overview of how the different modelling approaches address the main categories of policies is given here in preparation for a discussion of the main assumptions behind study results. The main categories of climate change mitigation options include:

1. Market oriented policies:
   - taxes and subsidies;
   - emission charges;
   - tradable emission permits;
   - soft loans; and
   - market development and/or efforts to reduce transaction costs.

2. Technology oriented policies:
   - norms and standards;
   - effluent or user charges;
   - institutional capacity building; and
   - market development efforts (information, transaction cost coverage).

3. Voluntary policies:
   - ecolabelling; and
   - voluntary agreements.

4. R&D policies:
   - research programmes; and
   - innovation and demonstration.

5. Accompanying measures:
   - public awareness;
   - information distribution;
   - education;
   - transport; and
   - free consultancy services.

While climate policies can include elements of all four policies, most analytical approaches focus on a few of the options. Economic models, for instance, mainly assess market-oriented policies, and occasionally technology policies related to energy supply options. Engineering approaches primarily focus on supply- and demand-side technology policies. Both of these approaches have opportunities to expand their representation of R&D policies.

Table 7.3 shows the application of market-oriented, technology-oriented and voluntary climate policies in different analytical approaches. The schematic overview covers a large number of applications in global, regional, national, and local analyses. Chapters 8 and 9 of discuss the actual details and specific methods for different assessment levels. A few general conclusions on the representation of different climate policies in the analytical approaches are:

- Market-oriented policies can be examined by macroeconomic models, but only indirectly in technology-driven models through exogenous assumptions. Market descriptions, however, are often stylized representations in many macroeconomic models, which makes it difficult to address transaction costs.

- Technology-driven models can assess various technology-oriented policies. Exogenous assumptions on behaviour and preferences, however, need to be supplied to explain market development. This separation of technology data and market behaviour can make implementation cost-assessment difficult.

- It is a challenge to integrate market imperfections in CGE and partial equilibrium models, because these models tend to be structured around assumptions of efficient resource allocation. Recent work modelled labour market imperfections in such models (see, e.g., Welsch, 1996; Honkatukia, 1997; Cambridge Econometrics, 1998; European Commission, 1998).

- Key presumptions such as technological change, R&D policies and changes in consumer preferences are difficult to assess in both macroeconomic models and technology-driven models.

It is expected that the cost of climate change mitigation policies–all else being equal–decreases with the number of policy categories and options included in the analysis. This means that approaches that are either rich in detail (or facilitate great flexibility) in a number of policy areas can be expected to identify

Table 7.3: Application of climate change mitigation policies in different analytical approaches

	Market-oriented policies	Technology-oriented policies	Voluntary-oriented policies
Macroeconomic models			
IO models	All instruments difficulties with modelling of transaction costs	CGE: Exogenous assumptions; few examples with endogenous assumptions	Demand functions for ecological values
Keynesian			
CGE			
estimated			
calibrated			
Technology-driven simulation and/or scenario models	Exogenous	Exogenous, learning	Qualitative assumptions
Sectoral models			
Partial equilibrium	All instruments	Changes in capital stock	Exogenous demand function for ecological values
Technology-driven models optimization simulation	All instruments modelled through changes in capital stock	Exogenous assumptions on standards and R&D Leaning curves	Investments reflect future expectations on ecological values and policies
Project assessment approaches			
Cost–benefit analysis	All instruments	Exogenous technology data	Exogenous demand function for ecological values
Cost-effectiveness analyses	All instruments	Vintage models	
Technology assessment	No instruments		

relatively large mitigation potentials and relatively low costs compared with approaches that only address a few instruments or options.

A number of studies have assessed climate change mitigation costs given different regimes of global flexibility mechanism.[21] Climate change mitigation costs in these different policy regimes depend on the specific definition of the policy instrument, and on assumptions about market scale, competition, and restrictions. It is generally expected that climate change mitigation costs decrease with increasing supply of carbon-reduction projects.[22] Restrictions on this supply, or market imperfections in global markets for carbon-reduction projects, have a tendency to increase the "price" of the projects (Burniaux, 1998; Mensbrugge, 1998).

7.6.6   Key Assumptions of Importance to Costing Estimates

There are a number of sensitive issues in the debate about how to interpret cost estimates generated by different models, including assumptions about tax recycling, target setting, and international co-operative mechanisms.

7.6.6.1   Tax Recycling

Tax recycling issues revolve around two critical points concerning the interactions between existing tax systems and a tax system that integrates carbon taxes:

- Assumptions on the structure of the tax system in the baseline and mitigation cases, which include assumptions on tax substitution generated by the recycled revenue of carbon taxes. These baseline assumptions have to be projected into the future for a considerable period if the revenue recycling is to be calculated correctly.
- The total impact of the policy scenario that includes the recycling of carbon taxes, in terms of both distribution and compensation.

The net cost of climate policy depends on (1) the structure of the tax system prior to the introduction of the mitigation policy and (2) the nature of the mitigation policy (e.g., which sectors are covered, what tax instruments are employed, and the way that revenues are recycled). Estimates of the size of the effect are discussed in Chapter 8. This is closely related to the double-dividend literature, which is discussed in Section 7.3.3.1. As noted there, the welfare loss (or burden) of a given climate policy depends on the structure of existing taxes. The more dis-

torted the pre-existing tax the higher the welfare loss. This means that a carbon tax can result in either a totally increased burden (welfare loss of the whole tax system) or a double dividend (in which the total welfare loss of the tax system is lower because the carbon tax substitutes other "burdensome" taxes). In general, however, a larger benefit from a carbon tax is found in comparison with other instruments that meet the Kyoto Protocol targets (e.g., permits issued *gratis*) than is found in comparison between different methods of recycling.

7.6.6.2   Target Setting for Greenhouse Gas Emissions Reduction

The choice of targets and timing affects cost estimates. Emission reduction targets are related to baseline case assumptions, and can be defined in relation to a given base year, or in relation to expected future development trends. Targets defined relative to base-year levels are accurate in terms of the target for the future total GHG emissions, but the actual GHG emissions reduction effort that is required is uncertain because future emission levels are unknown. Reduction targets defined as percentage reductions of future GHG emissions create uncertainty as to the GHG emission levels.

Figure 7.4 illustrates the different target-setting principles. Target setting related to base-year emissions compares the GHG emissions level in the "dotted" line base-year emissions and GHG emissions reduction case 2. In contrast, target setting in relation to future GHG emissions compares the baseline case line and GHG emission reduction case 1.

Climate change damages are related to the accumulated stock of atmospheric GHG concentrations. As such, target setting for GHG reduction policies should reflect the long atmospheric lifetime of the gases. What matters is the accumulated GHG emissions over several decades and the "technically correct" GHG reduction targets imply that the targets were defined for a given time horizon.

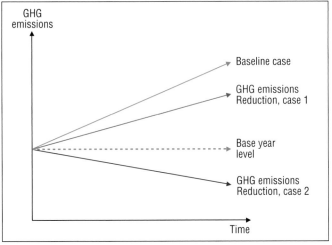

Figure 7.4: Baseline cases and target setting for GHG emissions reduction.

---

[21] As defined by the Articles 6, 12, and 17 of the Kyoto Protocol, these studies include JI between Annex I parties, CDM between Annex I and non-Annex I parties, and emissions trading between Annex I parties. See the discussions about these mechanisms in Chapter 6.

[22] Carbon-reduction projects are projects supplied by the potential host countries, where the policy is to be implemented.

Target setting over the relevant time horizon involves a number of technical challenges. The time dimension of the emissions reduction should reflect the dynamic aspects, such as time and path dependence of emissions and climate change damages. In addition, the costs of climate protection depend on the "when" and "where" flexibility of specific emission targets. Many of these time dimensions have been considered in IAMs.

The time flexibility involves several issues addressed in top-down and bottom-up models. Assumptions about technological change are, as discussed in Section 7.3.4, critical in the studies, and follow a simple "rule" that technological change over time expands the range of available GHG emissions reduction options. Technological change lowers the mitigation costs for a long-term target relative to a short-term target. It is emphasized that the mitigation costs and future technological change depend on GHG emissions reduction policies initiated and planned over the short- and long-term horizons. This reflects the point that technological change itself relates to R&D programmes and to current technology implementation.

The timing of mitigation policies also affects transition costs. From a short-term perspective, mitigation is constrained by the existing capital stock, infrastructure, and institutional structure related to technology. One key cost-determinant during the transition period is the turnover of capital stock. A time profile for mitigation that requires early retirement of capital stock increases the costs of achieving any target. This is predominantly an issue for developed countries, in which the capital stock and infrastructure are well developed.

7.6.6.3   *International Co-operative Mechanisms*

Mitigation costs vary across countries with different resource endowments, economic structure and development, institutional structure, and various other factors. These cost differences provide the opportunity to create and capture the gains from exchange that arise through international co-operative flexibility mechanisms. Mechanisms such as international carbon trading can facilitate collaborative emission reductions across countries and regions, and thereby minimize global control costs (see Chapter 6 for a detailed discussion on the issues involved in establishing such mechanisms).

The assumptions on international co-operative mechanisms include:
- Sectors and GHGs, which are included in the mechanisms.
- Specific constraints on countries and regions included in the trading regimes.
- Specific constraints on different co-operative mechanisms like those established by the Kyoto Protocol. The Protocol includes two project-based mechanisms: Article 6 on JI and Article 12 on CDM. Both JI and CDM aim to establish exchange institutions for projects to reduce GHG emissions. JI projects are between Annex I countries of the UNFCCC, and CDM projects are between countries with a reduction commitment specified in the Kyoto Protocol (termed Annex B countries) and countries without such a commitment. Another mechanism of the Kyoto Protocol is Article 17 that facilitates emissions trading among Annex B countries.
- Boundaries on GHG emissions trading markets, for example that set the minimum amount of domestic emission reductions for developed countries, specify a relationship between domestic GHG emissions reduction efforts and the GHG emissions reduction they can implement in collaboration with international partners.

Mitigation costs usually fall with greater flexibility for international emissions trading. This suggests that constraints on trading increase the costs of any emission target. Some critics point out that this argument does not address the potential positive impacts on technological development that can arise from implementing GHG emissions reduction policies domestically in developed countries, such as incentives for innovation and R&D.

7.6.6.4   *Critical Assumptions in the Energy Sector*

Table 7.4 provides an overview of the key assumptions behind mitigation cost studies for the energy sector. It is based on SAR (IPCC, 1996a, Chapter 8) and Halsnæs *et al.* (1998). Some of the new modelling areas that have important implications include assumptions on technology change, transaction costs and barrier removal policies, alternative demand projections (including lifestyle), and ancillary benefits. Similarly, assumptions related to climate change mitigation policies with major implications on costs include timing of the emissions reduction policies, and extent and function of global markets for emissions reduction projects.

The input assumptions are linked between the baseline case and the climate policy case in a complex way. There is the potential for many assumption combinations in baseline and mitigation scenarios, and the full set of assumptions in these two scenarios impacts the assessment of mitigation potential and related costs.

An OECD workshop in September 1998 (Mensbrugghe, 1998) concluded that the emissions reduction costs rely on baseline assumptions. Factors that lead to high cost estimates include high population and GDP growth rates, a relatively clean fuel mix, and relatively high energy costs. Among model parameters two areas were emphasized: the ability to substitute labour for energy, and the interfuel substitution elasticity. Low elasticities lead to high costs.

7.7   Conclusions on Further Needs for Research

It can be concluded generally that, since SAR (IPCC, 1996a, 1996b) was published, much progress has been achieved in the

Table 7.4: *Input assumptions used in energy sector mitigation studies*

Input assumptions	Meaning and relevance
Population	All else being equal, high growth increases GHG emissions.
Economic growth	Increased economic growth increases energy-using activities and also leads to increased investment, which speeds the turnover of energy-using equipment. Various assumptions on GHG emissions and resource intensities can be used for alternative scenarios.
Energy demand	
– structural change	Different sectors have different energy-intensities; structural change therefore has a major impact on overall energy use.
– technological change	This "energy-efficiency" variable influences the amount of primary energy needed to satisfy given energy services required by a given economic output.
– "lifestyle"	Explains structural changes in consumer behaviour.
Energy supply	
– technology availability and cost	Potential for fuel and technology substitution.
– backstop technology	The cost at which an infinite alternative supply of energy becomes available; this is the upper bound of cost estimates.
– learning	Technology costs related to time, market scale, and institutional capacity.
Price and income elasticities of energy demand	Relative changes in energy demand through changes in price or income, respectively; higher elasticities result in larger changes in energy use.
Transaction costs	Implementation, administration, scale of the activity.
Policy instruments and regulation	
– instruments	Economic versus regulatory measures.
– barriers	Implementation costs, including costs of overcoming barriers either in the form of institutional aspects or improvements in markets (including capacity building and institutional reforms); behavioural assumptions.
Existing tax systems and tax recycling	Recycling of carbon taxes; substitution of distortionary taxes decreases costs.
Ancillary benefits	Integration of local and regional environmental policies in most cases generates secondary benefits. Social policy goals, like income distribution and employment, can result in different policy rankings.

development of consistent and transparent approaches to assess climate change mitigation costs. This has facilitated understanding of the differences in mitigation cost results generated by different modelling approaches, based on different assumptions. A number of new research topics have been considered particularly important in the establishment of more information about globally efficient and fair climate change mitigation policies. These issues include a better understanding of the relationship between economic costs of climate change mitigation policies and the sustainable development implications in different parts of the world. Specifically, a number of key research issues for further work include:

- Development and application of methodological approaches for the integrated assessment of linkages between climate change mitigation costs and sustain-

able development, including development, environment, and social dimensions:
  - assessment of macroeconomic impacts using different welfare measures,
  - co-benefit studies, and
  - assessment of equity impacts (intragenerational equity impacts should be represented as detailed studies of distributional impacts, and can be integrated as formal decision criteria in policy assessments).
- Development of a framework for the assessment of intra- and intergenerational equity aspects of climate change mitigation studies.
- Integration of environmental impact assessments in climate change mitigation studies. This will require the

development of consistent methodological approaches and empirical studies.

- Establishment of approaches to conduct implementation cost analysis in both top-down and bottom-up models.
- Implementation cost studies that reflect financial market conditions, institutional and human capacities, information requirements, market size and opportunities for technology gain and learning, economic incentives, and policy instruments.
- Development of a systematic approach for reporting baseline assumptions and the costs of moving from one specific baseline case to a climate change mitigation policy.
- Further development of a consistent analytical structure and a format for reporting the main assumptions that underlie costing results, including:
  - main scenario drivers: economic growth, technological development, sectoral activity, and fuel prices;
  - behavioural assumptions;
  - flexibility of climate change mitigation policies, including timing of the reduction policies, GHG emissions included, and international co-operative mechanisms; and
  - assumptions about tax recycling options, side-impacts of climate change mitigation policies, and the potential implementation of no regrets options.

- Development of approaches to and conduct of studies for developing countries and EITs that better reflect the specific characteristics of these economies in implementing climate change mitigation policies. Some of the major research topics are:
  - assessment of alternative development patterns and their relationship to development, social, and environmental sustainability dimensions;
  - macroeconomic studies that consider structural adjustment policies and market transformation processes;
  - studies of the informal sector and implications for GHG emissions and reduction policies;
  - non-commercial energy use; and
  - specific implementation policy issues.
- Estimates of future costs and sustainability implications that both reflect how climate change might affect future ecosystems, how these altered ecosystems might affect the demand for different goods, and how this demand might affect the welfare of our descendants.

References

Anderson, D., 1994: Cost-effectiveness in Addressing the "CO_2 problem" with Special Reference to the Investments of the Global Environment Facility. *Annual Review of Energy and Environment*, **19**, 423-55.

Atkinson, A.B., 1970: On the Measurement of Inequality. *Journal of Economic Theory*, **2** (3), 244-63.

Atkinson, S., T. Crocker, and J. Shogren, 1992: Bayesian Exchangeability, Benefits Transfer, and Research Efficiency. *Water Resources Research*, **28**, 715-722.

Austin, D., P. Faeth, R. Seroa Da Motta, C. Ferraz, C.E.F.Young, Z. Ji, L. Junfeng, M. Pathak, L. Shrivastava, and S. Sharma, 2000: How Much Sustainable Development Can We Expect From The Clean Development Mechanism? World Resources Institute, Washington, DC.

Bateman, I., and K. Willis, (eds.), 1999: *Valuing Environmental Preferences: Theory and Practice of the Contingent Valuation Method in the US, EU and Developing Countries*. Oxford University Press, Oxford.

Blackorby, C., and D. Donaldson, 1988: Money Metric Utility: Harmless Normalization? *Journal of Economic Theory*, **46**, 120-129.

Bohm, P., 1998: Public Investment Issues and Efficient Climate Change Policy. In *The Welfare State, Public Investment, and Growth*, H. Shibata, T. Ihori, (eds.), Springer-Verlag, Tokyo.

Boscolo, M., J.R. Vincent, and T.Panayatou, 1998: Discounting Costs and Benefits in Carbon Sequestration Projects. Environment Discussion Papers - No. 41, Harvard Institute for International Development, Cambridge, MA.

Bouzaher, A., D. Holtkamp, R. Reese, and J. Shogren, 1995: Economic and Resource Impacts of Policies to Increase Organic Carbon in Agricultural Soils. In *Soil Management and Greenhouse Effects*. R. Lal, J. Kimble, E. Levine, B. Stewart, (eds.), CRC Lewis Publishers, Boca Raton, LA, 309-328.

Bovenberg, A., R. Lans, and A. de Mooij 1994: Environmental Levies and Distortionary Taxation. *American Economic Review*, **84**(4), 1085-1089.

Brown, G., and J. Shogren, 1998: Economics of the Endangered Species Act. *Journal of Economic Perspectives*, 3-20.

Burniaux, J., 1998: How Important is Market Power in Achieving Kyoto? An assessment based on the Green Model. In Economic Modelling of Climate Change, OECD Workshop Report, OECD, Paris.

Burtraw, D., and M.Toman, 1998: The Benefits of Reduced Air Pollutants in the U.S. from Greenhouse Gas Mitigation Policies. Discussion Paper 98-01-REV, Resources for the Future, Washington, DC.

Burtraw, D., A. Krupnick, K. Palmer, A. Paul, M. Toman, and C. Bloyd, 1999: Ancillary Benefits of Reduced Air Pollution in the US from Moderate Greenhouse Gas Mitigation Policies in the Electricity Sector. Discussion Paper 99-51, Resources for the Future, Washington, DC.

Cambridge Econometrics, 1998: Industrial Benefits from Environmental Tax Reform in the UK. Cambridge Econometrics Technical Report No. 1, 1998. Sustainable Economy Unit of Forum for the Future, United Kingdom.

Cameron, L.J., W.D. Montgomery, and H.L. Foster, 1999: The economics of strategies to reduce greenhouse gas emissions. *Energy Studies Review*, **9**(1) 63-73.

Chichilnisky, G., and G. Heal, 1993: Global Environmental Risk. *Journal of Economics Perspectives* **7**, 65-86.

Chomnitz, K., 1999: Baselines for greenhouse gas reductions: problems, precedents, solution. In Proceedings on CDM workshop on baseline for CDM, NEDO/GISPRI, (eds.), Tokyo, 23-87.

Christensen, J.M., K. Halsnæs, and J. Sathaye, (eds.), 1998: *Mitigation and Adaptation Cost Assessment: Concepts, Methods and Appropriate Use*. Risø National Laboratory, UNEP Collaborating Centre on Energy and Environment, Roskilde, Denmark.

Ciriacy-Wantrup, S., 1952: *Resource Conservation: Economics and Policies*. University of California, Berkeley, CA.

Cline, W.M. 1993: Give Greenhouse Abatement a Chance. *Finance and Development*, **March.**

Crocker, T., and J. Shogren, 1999: Endogenous Environmental Risk. In *The Handbook of Environmental and Natural Resource Economics*, J. van der Bergh, (ed.), Edward Elgar, Cheltenham, UK.

Cropper, M., and D. Laibson, 1999: The Implications of Hyperbolic Discounting for Project Evaluation. In P. Portney, J. Weyant, (eds.), *Discounting and Intergenerational Equity*, Resources for the Future, Washington, DC.

Cropper, M., S. Aydede, and P. Portney, 1994: Preferences for Life Saving Programs: How the Public Discounts Time and Age. *Journal of Risk and Uncertainty*, **8**, 243-265.

Dean, J. 1992: Trade and the Environment: A Survey of the Literature. P. Low, (ed.), International Trade and the Environment - World Bank Discussion Paper No. 159.

Dean, A., and P. Hoeller, 1992: Costs of reducing CO_2 emissions - Evidence from six global models. OECD Economic Studies **19**, (Winter).

Desvousges, W., M. Naughton, and R. Parsons, 1992: Benefits Transfer: Conceptual Problems in Estimating Water Quality Benefits. *Water Resources Research*, **28**, 675-683.

Devarajan, S., K.E. Thierfelder, and S. Shutiwart-Narueput, 1999: The Marginal Cost of Public Funds in Developing Countries. World Bank Working Paper.

Dixit, A., and V. Norman, 1984: Theory of International Trade: A Dual General Equilibrium Approach. Cambridge University Press, Cambridge, 339

Edmonds, J., and D.W. Barns, 1990: Estimating the Marginal Cost of Reducing Global Fossil Fuel CO_2 Emissions. PNL-SA- 18361, Pacific Northwest Laboratory, Washington, DC.

Ehrlich, I., and G.S. Becker, 1972: Market Insurance, Self-insurance and Self-protection. *Journal of Political Economy*, **80**, 623-648.

Ekins, P., 1996: How large a carbon tax is justified by the secondary benefits of CO_2 abatement? *Resource and Energy Economics*, **18**(2), 161-187.

Epstein, R., 1995: *Simple Rules for a Complex World*. Harvard University Press, Cambridge, MA.

Esty, D., 1994: Greening the GATT. Institute for International Economics, p. 159.

ExternE, 1995: *Externalities of Energy, Volume Three: Coal and Lignite*. Commission of the European Communities, DGXII, Luxembourg.

ExternE, 1997: Externalities of Fuel Cycles 'ExternE' Project: Results of National Implementation. Draft Final Report, Commission of the European Communities, DGXII, Brussels.

ExternE, 1999: *Externalities of Energy, Volume 7 : Methodology 1998 Update*. M. Holland, J. Berry, D. Forster, (eds.), Office for Official Publications of the European Communities, Luxembourg.

European Commission, 1998: Auto-Oil Cost Effectiveness Study. First Consolidated Report: Scope and Methodology, Commission of the European Communities, DGII, Brussels.

Eyre, N., T.E. Downing, R. Hoekstra, and R. Tol, 1998: Global Warming Damages. Report prepared under contract JOS3-CT95-0002, ExternE Programme of the European Commission, Brussels.

Fankhauser, S., and D. McCoy, 1995: Modelling the economic consequences of environmental policies. In *Principles of environmental and resource economics: A Guide to Decision Makers and Students*. H. Folmer, L. Gabel, J. Opschoor, (eds.), Edward Elgar, Aldershot, UK, 253-275.

Fankhauser, G.S., J. Smith, and R. Tol, 1997: The Costs of Adapting to Climate Change. Working Paper No. 13, Global Environmental Facility, Washington, DC.

Goulder, L.H., 1995a: Effects of carbon taxes in an economy with prior tax distortions: An intertemporal general equilibrium analysis. *Journal of Environmental Economics and Management*, **29**, 271-297.

Goulder, L.H., 1995b: Environmental taxation and the double dividend: A reader's guide. *International Tax and Public Finance*, **2**, 157-183.

Goulder, L.H., and S. Schneider, 1999: Induced technological change, crowding out, and the attractiveness of CO_2 emissions abatement. *Resource and Environmental Economics*, **21**(3-4), 211-253.

Håkonsen, L., 1997: An Investigation into Alternative Representations of the Average and Marginal Cost of Public Funds. Discussion Paper, Norwegian School of Economics and Business Administration.

Håkonsen, L., and L. Mathiesen, 1997: CO_2-Stabilization may be a "No-Regrets" Policy. *Environmental and Resource Economics*, **9**(2), 171-198.

Halsnæs, K., 1996: The economics of climate change mitigation in developing countries. *Energy Policy* (Special Issue), **24** (10/11).

Halsnæs, K., and A. Markandya, 1999: Comparative assessment of GHG limitation costs and ancillary benefits for developing countries. Paper presented to the IPCC Meeting on Costing Issues for Mitigation and Adaptation to Climate Change, GISPRI, Tokyo.

Halsnæs, K., D. Bouille, E. La Rovere, S. Karakezi, W. Makundi, V. Wanwarcharakul, and J.M. Callaway, 1998: Sectoral Assessment (Chapter 4). In *Mitigation and Adaptation Cost Assessment: Concepts, Methods and Appropriate Use.* J.M. Christensen, K. Halsnæs, J. Sathaye, (eds.), Risø National Laboratory, UNEP Collaborating Centre on Energy and Environment, Roskilde, Denmark.

Hanley, N., Shogren, J., and B. White, 1997: *Environmental Economics: In Theory and Practice.* Macmillan, Basingstoke, UK.

Hanemann, W. M., 1991: Willingness to Pay and Willingness to Accept: How Much Can They Differ? *American Economic Review,* **81,** 635-647.

Harrison, D.Jr., S.T. Schatzki, E. Haites, and T. Wilson, 2000: Critical Issues in International Greenhouse Gas Emissions Trading: Setting Baselines for Credit-Based Trading Programs – Lessons Learned from Relevant Experience. Electric Research Institute, Palo Alto, CA.

Hazilla, M., and R. Kopp, 1990: Social Costs of Environmental Quality Regulations. *Journal of Political Economy,* **98,** 853-873.He, J.K., and W.Y. Chen, 1999: Study on Assessment Method for GHG Mitigation Project. *Research of Environmental Sciences,* 12(2), 24-27.

Honkatukia, J., 1997: Are there Double Dividends for Finland? The Swedish Green Tax Commission Simulations for Finland. Helsinki School of Economics and Business Administration, Helsinki.

Hourcade, J., E. Haites, and T. Barker, 1998: Macroeconomic Cost Assessment (Chapter 6). In *Mitigation and Adaptation Cost Assessment: Concepts, Methods and Appropriate Use.* J.M. Christensen, K. Halsnæs, J. Sathaye, (eds.), Risø National Laboratory, UNEP Collaborating Centre on Energy and Environment, Roskilde, Denmark.

Huff, K., and T. Hertel, 1996: Decomposing Welfare Changes in the GTAP Model. GTAP Technical Paper No. 5, Purdue University, West Lafayette, Indiana, Unites States.

IPCC, 1996a: *Climate Change 1995: Economic and Social Dimensions of Climate Change.* J.P. Bruce, H. Lee, E. Haites, (eds.), Cambridge University Press, Cambridge.

IPCC, 1996b: *Climate Change 1995: Impacts, Adaptations and Mitigation of Climate Change: Scientific-Technical Analyses.* R.T. Watson, M.C. Zinyowera, R.H. Moss, (eds.), Cambridge University Press, Cambridge.

IPCC, 1996c: Technical Paper on Technologies, Policies and Measures for Mitigating Climate Change. IPCC, Geneva.

IPCC, 2000a: Special Report on Emission Scenarios. Main Text, IPCC Working Group III on Mitigation of Climate Change.

IPCC, 2000b: Expert Workshop on Assessing the Ancillary Benefits and Costs of Greenhouse Gas Mitigation Policies, March 27-29, Washington, DC.

IPCC, 2000c: Special Report on Land Use, Land Use Change, and Forestry. IPCC, Geneva.

Jacoby, H., R. Prinn, and R. Schmalensee, 1998: Kyoto's Unfinished Business. *Foreign Affairs,* July/August, 54-66.

Jaffe, A., S. Peterson, P. Portney, and R. Stavins, 1994: Environmental Regulation and International Competitiveness: What Does the Evidence Tell Us? Discussion Paper 94-08, Resources for the Future, Washington, DC.

Jepma, C., 1999: Determining a baseline for project co-operation under the Kyoto Protocol: a general overview. In Proceedings, CDM workshop - workshop on baseline for CDM, NEDO/GISPRI, (eds.), Tokyo.

Jepma, C., and M. Munasinghe, 1998: Climate Change Policy. Cambridge University Press. Cambridge.

Jochem, E., U. Kuntze, and M. Patel, 2000: Economic Effects of Climate Change Policy – Understanding and Emphasising the Costs and Benefits. Federal Ministry of Environment, Berlin.

Kahn, J.R. and A. Farmer, 1999: The double-dividend, second-best worlds and real-world environmental policy. *Ecological Economics,* 30 (3), 433-439.

Kask, S., and J.Shogren, 1994: Benefit Transfer Protocol for Long-term Health Risk Valuation: A Case of Surface Water Contamination. *Water Resources Research,* 30, 2813-2823.

Kane, S., and J. Shogren, 2000: Linking Adaptation and Mitigation in Climate Change Policy. *Climatic Change,* 45, No. 1, 75-101.

Kaya, Y., 1989: *Impact of carbon dioxide emissions on GDP growth: Interpretation of proposed scenarios.* Intergovernmental Panel on Climate Change /Response Strategies Working Group, IPCC, Geneva.

Keeney, R.L., and H. Raiffa, 1993: *Decisions with Multiple Objectives. Preferences and Value Tradeoffs.* Cambridge University Press, Cambridge.

Kirkpatrick, C.H., and J.D. MacArthur, 1990: Shadow Pricing Unemployed Labour in Developed Economies: An Approach at Estimation. *Project Appraisal,* **5** (2), 101-112.

Kolstad, C., 2000: *Environmental Economics.* Oxford University Press, New York, NY and Oxford.

Kram, J., 1998: The Costs of Greenhouse Gas Abatement. *Economics and Policy Issues in Climate Change.* W. Nordhaus, (ed.), Resources for the Future, Washington, DC, 167-189.

Krupnick, A.J., K. Harrison, E. Nickell, and M. Toman, 1996: The Value of Health Benefits from Ambient Air Quality Improvements in Central and Eastern Europe. *Environmental and Resource Economics,* 7 (4), 307-332.

Krupnick, A.J., D. Buttraw, and A. Markandya, 2000: The Ancillary Benefits and Costs of Climate Change Mitigation: A Conceptual Framework. Paper presented to the Expert Workshop on Assessing the Ancillary Benefits and Costs of Greenhouse Gas Mitigation Strategies, 27-29 March 2000, Washington, DC.

Lee, R., A.J. Krupnick, D. Burtraw, *et al.* 1995: *Estimating Externalities of Electric Fuel Cycles: Analytical Methods and Issues and Estimating Externalities of Coal Fuel Cycles.*McGraw-Hill/Utility Data Institute, Washington, DC.

Loewenstein, G., and D. Prelec, 1992: Anomalies In Intertemporal Choice - Evidence and an Interpretation. *Quarterly Journal of Economics,* **107,** 573-597.

Manne, A. S., and R.G. Richels, 1992: *Buying Greenhouse Insurance: The Economic Costs of CO_2 Emission Limits.* MIT Press, Cambridge, MA

Markandya, A. 1994: Measuring the External Costs of Fuel Cycles in Developing Countries. In *Social Costs of Energy,* O. Hohmeyer, R.L. Ottinger, (eds.), Springer-Verlag, Berlin.

Markandya, A. 1998: The Indirect Costs and Benefits of Greenhouse Gas Limitation. *Economics of GHG Limitations. Handbook Series.* UNEP Collaborating Centre on Energy and Environment, Risø National Laboratory, Denmark.

Markandya, A., and R. Boyd, 1999: The Indirect Costs and Benefits of Greenhouse Gas Limitation. Case Study for Mauritius. Economics of GHG Limitations. Country Study Series. UNEP Collaborating Centre on Energy and Environment, Risø National Laboratory, Denmark.

Markandya, A., and M. Pavan, 1999: *Green Accounting in Europe: Four Case Studies.* Kluwer Academic Publishers, Dordrecht.

Markandya, A., and K. Halsnæs, 2001: Costing Methodologies: A Guidance Note. In IPCC supporting material - Guidance Papers on the Cross Cutting Issues of the Third Assessment Report of the IPCC, IPCC, Geneva.

Markandya, A., K. Halsnæs, and I. Milborrow, 1998: Cost Analysis Principles. In *Mitigation and Adaptation Cost Assessment: Concepts, Methods and Appropriate Use.* K. Halsnæs, J. Sathaye, J. Christensen, (eds.), UNEP Collaborating Centre on Energy and Environment, Risø National Laboratory, Roskilde, Denmark.

Markandya, A., L. Bellù, V. Cistulli, and P. Harou, 2000: *Environmental Economics for Sustainable Growth: A Handbook for Policy Makers.* Edward Elgar, Cheltenham, UK (in press).

Marshall, J., 1976: Moral Hazard. *American Economic Review,* **66,** 680-690.

Matsuo, N., 1999: Baselines as the critical issue of CDM - possible pathways to standardisation In Proceedings CDM workshop - workshop on baseline for CDM, NEDO/GISPRI, (eds.), Tokyo, 9-22.

McConnell, K., 1992: Model Building and Judgment: Implications for Benefits Transfers with Travel Cost Models. *Water Resources Research,* **28,** 695-700.

Mensbrugghe, D. van der, 1998: Summary of OECD Workshop Report Economic Modelling of Climate Change. OECD, Paris.

Michaelowa, A., and M. Dutschke, 1998: Economic and Political Aspects of Baselines in the CDM Context. In *Promoting development while limiting greenhouse gas emissions: trends and baselines,* J. Goldemberg, W. Reid, (eds.), New York, NY, 115-134.

Moss, R.H., and S. Schneider, 2000: Uncertainties. In R. Pachauri, T. Taniguchi, and K. Tanaka (eds), Guidance Papers on the Cross Cutting Issues of the Third Assessment Report of the IPCC. IPCC Geneva.

Murty, M.N. *et al*, 1992: National Parameters for Investment Project Appraisal in India. Working paper no E/153/92, Institute of Economic Growth, University Enclave, Delhi.

OECD, 2000: *Ancillary benefits and costs of GHG mitigation*. Proceedings of an IPCC Co-sponsored Workshop, 27-29 March 2000, Washington, DC., OECD, Paris.

Palmer, K., W. Oates, and P. Portney, 1995: Tightening Environmental Standards: The Benefit-Cost of the No-Cost Paradigm? *Journal of Economic Perspectives,* **9**, 119-132.

Palmquist, R.B., 1991: Hedonic Methods (Chapter IV). In *Measuring the Demand for Environmental Quality*, J. Braden, C. Kolstad, (eds.), North-Holland Publ. Co. New York, NY.

Parry, I., W.H., Williams, C. Roberton III and L.H. Goulder, 1999: When Can Carbon Abatement Policies Increase Welfare? The Fundamental Role of Distorted Factor Markets. *Journal of Environmental Economics and Management,* **37** (1) , 52-84.

Parson, E.A., and K. Fisher-Vanden, 1997: Integrated Assessment Models of Global Climate Change. *Annual Review of Energy and the Environment,* **22**, 589-628.

Parson, E.A., and K. Fisher-Vanden, 1999: Joint Implementation of greenhouse gas abatement under the Kyoto Protocol's Clean Development Mechanism: Its scope and limits. *Policy Sciences,* **32**, 207-224.

Pearce, D., 2000: Policy Frameworks for the Ancillary Benefits of Climate Change Policies. Expert Workshop on Assessing the Ancillary Benefits and Costs of Greenhouse Gas Mitigation Policies, March 27-29, Washington, DC.

Perman, R., Y. Ma, J. McGilvary, and M. Common, 1999: *Natural Resource and Environmental Economics, 2nd edition*. Longman, Harlow.

Portney, P.R., and J. Weyant, (eds.), 1999: *Discounting and Intergenerational Equity*. Johns Hopkins University Press, Baltimore, MD.

Randall, A., and J. Stoll, 1980: Consumer's Surplus in Commodity Space. *American Economic Review,* **71**, 449-457.

Ray, A., 1984: *Cost Benefit Analysis: Issues and Methodologies*. Johns Hopkins University Press, Baltimore, MD.

Rotmans, J. and H. Dowlatabadi, 1998: Tools for Policy Analysis (Chapter 5). *Integrated Assessment Modelling in: Human Choice and Climate Change: an International Assessment, Vol. 3.*, S. Rayner, and E.L. Malone, (eds.), Batelle Press: Columbus, OH.

Roughgarden, J., 1995: Can Economics Protect Biodiversity? In *The Economics and Ecology of Biodiversity Decline*, T. Swanson, (ed.), Cambridge University Press, Cambridge, 149-154.

Ruggeri, G., 1999: The Marginal Cost of Public Funds in Closed and Open Economies. *Fiscal Studies,* **20**, 1-60.

Sanstad, A.H., and R. Howart, 1994: Normal Markets, Market Imperfections, and Energy Efficiency. *Energy Policy,* **22**(10).

Sathaye, J.A., and N.H. Ravindranath, 1998: Climate Change Mitigation in the Energy and Forestry Sectors of Developing Countries. *Annual Review of Energy and the Environment,* **23**, 387-437

Sathaye, J., R. Norgaard, and W. Makundi, 1993: A Conceptual Framework for the Evaluation of Cost-Effectiveness of Projects to Reduce GHG Emissions and Sequester Carbon. LBL-33859, Lawrence Berkely Laboratory, Berkeley, CA.

Schelling, T., 1992: Some Economics of Global Warming. *American Economic Review,* **82**, 1-14.

Schurr, S.H., 1984: Energy Use, Technological-Change and Productive Efficiency - An Economic-Historical Interpretation. *Annual Review of Energy,* **9**, 409-425.

Seroa da Motta, R., R. Huber, and J. Ruitenbeck, 1999: Market based instruments for environmental policy-making in Latin America and the Caribbean: lessons from eleven countries. *Environment and Development Economics,* **4** (2).

Sen, A., 1999: *Development As Freedom*. Alfred A. Knopp, New York, NY.

Shogren, J., and T. Crocker, 1999: Risk and its Consequences. *Journal of Environmental Economics and Management,* **37**, No. 1, 44-51

Shogren, J., and M. Toman, 2000: Climate Change. *Public Policies for Environmental Protection, 2nd edition*. P. Portney, R. Stavins, (eds.), Resources for the Future Press, Washington, DC.

Shogren, J., S. Shin, D. Hayes, and J. Kliebenstein, 1994: Resolving Differences in Willingness to Pay and Willingness to Accept. *American Economic Review,* **84**, 255-270.

Slemrod, J., and S. Yizhaki, 1996: The Costs of Taxation and the Marginal Efficiency Cost of Funds. *IMF Staff Papers,* **43**(1), 171-198.

Sohngen, B., and R. Mendelsohn 1997: Valuing the Impact of Large-Scale Ecological Change in a Market: The Effect of Climate Change on U.S. Timber. *American Economic Review,* **88,** 686-710.

Sohngen, B., R. Mendelsohn, and R. Sedjo 1999: Forest Management, Conservation, and Global Timber Markets. *American Journal of Agricultural Economics,* **81**, 1-13.

Squire, L., and H. van der Tak, 1975: Economic Analysis of Projects. Johns Hopkins University Press, Baltimore, MD.

Stavins, R., 1999: The Costs of Carbon Sequestration: A Revealed-Preference Approach. *American Economic Review,* **89** (4), 994-1009.

Stern, N., 1977: The Marginal Valuation of Income. In *Studies in Modern Economic Analysis*, M.J. Artis, A.R. Nobay, (eds.), Blackwell, Oxford.

Summers, L., 1992: Foreword. In International Trade and the Environment World Bank Discussion Paper No. 159, P. Low (ed.).

Swallow, S., 1996: Resource Capital Theory and Ecosystem Economics: Developing Nonrenewable Habitats with Heterogeneous Quality. *Southern Economic Journal,* **63**, 106-123.

UNCSD, 1999: United Nations Sustainable Development - Indicators of Sustainable Development (online: http://www.un.org/esa/sustdev/indisd).

Verikios, G., and K. Hanslow, 1999: Modelling the effects of implementing the Uruguay Round: A comparison using the GTAP model under alternative treatments of international capital mobility. Paper presented to the Second Annual Conference on Global Economic Analysis, Denmark, June 20-22.

Walz, R., 1999: Microeconomic Foundations of the Innovative Impact of a CO_2/Energy Surcharge. In *Schriftenreihe Innovative Wirkungen Umweltpolitischer Instrumente, Vol. 4* ('Regulation, Taxes and Innovations'), *Analytica Berlin*, R. Walz, U. Kuntze, (eds.) (in German)

Watts, W., 1999: Discounting and Sustainability. The European Commission, DGII, Brussels (photocopy).

Weitzman, M., 1976: On the Welfare Significance of National Product in a Dynamic Economy. *Quarterly Journal of Economics.* **9**, 156-162.

Weitzman, M., 1998: Gamma Discounting for Global Warming. Discussion Paper, Harvard University, Harvard, MA.

Welsch, H., 1996: Recycling of Carbon/Energy Taxes and the Labour Market. *Environmental and Resource Economics,* **8,** 141-155.

Weyant, J., 1998: The Costs of Carbon Emissions Reductions. *Economics and Policy Issues in Climate Change*, W. Nordhaus, (ed.), Resources for the Future, Washington, DC, 191-214.

Willig, R., 1976: Consumer's Surplus Without Apology. *American Economic Review,* **66**, 589-597

World Bank, 1991: *The Economic Analysis of Projects: A Practitioners Guide* (EDI Technical Materials), The World Bank, Washington, DC.

Zabel, J., and K. Kiel, 2000: Estimating the demand for air quality in four US cities. *Land Economics,* **76** (2), 174-194.

8

Global, Regional, and National Costs and Ancillary Benefits of Mitigation

---

Co-ordinating Lead Authors:
JEAN-CHARLES HOURCADE (FRANCE), PRIYADARSHI SHUKLA (INDIA)

Lead Authors:
Luis Cifuentes (Chile), Devra Davis (USA), Jae Edmonds (USA), Brian Fisher (Australia), Emeric Fortin (France), Alexander Golub (Russian Federation), Olav Hohmeyer (Germany), Alan Krupnick (USA), Snorre Kverndokk (Norway), Richard Loulou (Canada), Richard Richels (USA), Hector Segenovic (Argentina), Kenji Yamaji (Japan)

Contributing Authors:
Christoph Boehringer (Germany), Knut Einar Rosendahl (Norway), John Reilly (USA), Kirsten Halsnæs (Denmark), Ferenc Toth (Germany), ZhongXiang Zhang (Netherlands)

Review Editors:
Lorents Lorentsen (Norway), Oyvind Christopherson (Norway), Mordechai Shechter (Israel)

CONTENTS

EXECUTIVE SUMMARY

The United Nations Framework Convention on Climate Change (UNFCCC) has as its ultimate goal the "stabilization of greenhouse gas concentrations in the atmosphere at a level that will prevent dangerous anthropogenic interference with the climate system." Whereas mitigation costs play only a secondary role in establishing the target, they play a more important role in determining how the target is to be achieved. UNFCCC states that "policies and measures to deal with climate change should be cost-effective so as to ensure global benefits at the lowest possible costs." This chapter examines the literature on the costs of greenhouse gas mitigation policies at the national, regional, and global levels. The net welfare gains or losses are reported, including (when available) the ancillary benefits of mitigation policies. These studies employ the full range of analytical tools described in Chapter 7, from the technologically rich bottom-up models to more aggregate top-down models, which link the energy sector to the rest of the economy.

Models can also be distinguished through their level of geographical disaggregation. Global models, which divide the world into a limited number of regions, can provide important insights with regard to international emissions trade, capital flows, trade patterns, and the implications of alternative international regimes regarding contributions to mitigation by various regions of the globe. National models are more appropriate for examining the effectiveness of alternative fiscal policies in offsetting mitigation costs, the short-term effects of macro shocks on employment and inflation, and the implications of domestic burden-sharing rules for various sectors of the economy.

To cope with their wide range of diversity, the studies are grouped into three categories. The first two focus on the near-to-medium term. In one of these, the focus is exclusively on domestic policies. In the other, the domestic/international interface is explored. The third category focuses on the long-term goals of climate policy and explores cost-effective implementation strategies. That is, what is the least-cost emission reduction pathway for accomplishing a prescribed goal? The major conclusions are summarized below.

For any class of models, the emissions baseline is critically important in determining mitigation costs. It defines the size of the reduction required for meeting a particular target. The growth rate in carbon dioxide (CO_2) emissions is determined by:
- growth rate in gross domestic product (GDP);
- decline rate of energy use per unit of output, which depends on structural change in the economy and on technological development; and
- decline rate of CO_2 emissions per unit of energy use.

Much of the difference in cost projections can be explained by differences in these key variables.

Economic studies vary widely in their estimate of mitigation costs (both across and within countries). These differences can be traced to assumptions about economic growth, the cost and availability of existing and new technologies (both on the supply and demand side of the energy sector), resource endowments, the extent of "no regrets" options and the choice of policy instruments.

Virtually all analysts agree on the existence of "no regrets" options. Such options are typically assumed to be included in the reference (no policy) scenario by economic modellers. Even so, the overwhelming majority of emission baselines show that emissions continue to rise well into the future. This suggests that zero cost options are insufficient to reduce emissions in the absence of policy intervention.

Mitigation costs to meet a prescribed target will be lower if the tax revenues (or revenues from auctioned permits) are used to reduce existing distortionary taxes (the so-called "double dividend"). The preferred policy depends on the existing tax structure. Most European studies find that cutting payroll taxes is more efficient than other types of recycling. A significant number of these studies conclude that, within some range of abatement targets, the net costs of mitigation policies can be close to zero and even slightly negative. Conversely, in the USA, studies suggest that reducing taxes on capital is more efficient, but few models report negative costs.

Policies aimed at mitigating greenhouse gases can have positive and negative side effects (or ancillary benefits and costs, not taking into account benefits of avoided climate change) on society. Although this report overall emphasizes co-benefits of climate policies with other policies (to reflect the reality in many regions that measures are taken with multiple objectives rather than climate mitigation alone), the literature that focuses on climate mitigation uses the term "ancillary benefits" of specific climate mitigation measures. In spite of recent progress in methods development, it remains very challenging to develop quantitative estimates of the ancillary effects, benefits and costs of GHG mitigation policies. Despite these difficulties, in the short term, ancillary benefits of GHG policies under some circumstances can be a significant fraction of private (direct) mitigation costs. In some cases the magnitude of ancillary benefits of mitigation may be comparable to the costs of the mitigating measures, adding to the no regrets potential. The exact magnitude, scale and scope of these ancillary benefits and costs will vary with local geographical and baseline conditions. In some circumstances, where baseline conditions involve relatively low carbon emissions and population density, benefits may be low. For the studies reviewed here, the biggest share of the ancillary benefits is related to public health.

Mitigation costs are highly dependent on assumptions about trade in emission permits. Cost estimates are lowest when there would be full global trading. That is, when reductions are made where it is least expensive to do so regardless of their geographical location. Costs increase as the size of the emissions market contracts. In the case of Annex B trading only, the availability of excess assigned amount units in Russia and Ukraine can be critical in lowering the overall mitigation costs. Carbon trade provides some means for hedging against uncertainties regarding emissions' baselines and abatement costs. It also reduces the consequences of an inequitable allocation of assigned amounts.

It has long been recognized that international trade in emission quota can reduce mitigation costs. This will occur when countries with high domestic marginal abatement costs purchase emission quotas from countries with low marginal abatement costs. This is often referred to as "where flexibility". That is, allowing reductions to take place where it is cheapest to do so regardless of geographical location. It is important to note that where the reductions take place is independent upon who pays for the reductions. The chapter discusses the cost reductions from emission trading for Annex B and full global trading compared to a no-trading case. All of the models show significant gains as the size of the trading market is expanded. The difference among models is due in part to differences in their baseline, the cost and availability of low-cost substitutes on both the supply and demand sides of the energy sector, and the treatment of short-term macro shocks. In general, all calculated gross costs for the non-trading case are below 2% of GDP (which is assumed to have increased significantly in the period considered) and in most cases below 1%. Annex-B trading would generally decrease these costs to well below 1 % of GDP for OECD regions. The extent to which domestic policies relying on revenue recycling instruments can lower these figures is conditional upon the articulation of these policies and the design of trading systems.

Emissions constraints in Annex I countries are likely to have so-called "spillover" effects on non-Annex B countries. For example, Annex I emissions reductions result in lower oil demand, which in turn leads to a decline in the international price of oil. As a response, non-Annex I countries may increase their oil imports and emit more than they would otherwise. Oil-importing non-Annex I countries may benefit, whereas oil exporters may experience a decline in revenue.

A second example of spillover effects involves the location of carbon-intensive industries. A constraint on Annex I emissions reduces their competitiveness in the international marketplace. Recent studies suggest that there will be some industrial relocation abroad, with non-Annex I countries benefitting at the expense of Annex I countries. However, non-Annex I countries may be adversely affected by the decline in exports likely to accompany a decrease in economic activity in Annex I countries.

The cost estimates of stabilizing atmospheric CO_2 concentrations depend upon the concentration stabilization target, the emissions pathway to stabilization and the baseline scenario assumed. Unfortunately, the target is likely to remain the subject of intense scientific and political debate for some time. What is needed is a decision-making approach that explicitly incorporates this type of uncertainty and its sequential resolution over time. The desirable amount of hedging in the near term depends upon one's assessment of the stakes, the odds, and the costs of policy intervention. The risk premium–the amount that society is willing to pay to reduce risk–is ultimately a political decision that differs among countries.

The concentration of CO_2 in the atmosphere is determined more by cumulative rather than year-by-year emissions. A number of studies suggest that the choice of emissions pathway can be as important as the target itself in determining overall mitigation costs. A gradual near-term transition from the world's present energy system minimizes premature retirement of existing capital stock, provides time for technology development, and avoids premature lock-in to early versions of rapidly developing low-emission technology. On the other hand, more aggressive near-term action would decrease environmental risks associated with rapid climatic changes, stimulate more rapid deployment of existing low-emission technologies, provide strong near-term incentives to future technological changes that may help to avoid lock-in to carbon intensive technologies, and allow for later tightening of targets should that be deemed desirable in light of evolving scientific understanding.

8.1   Introduction

8.1.1   Summary of Mitigation Cost Analysis in the Second Assessment Report

Chapters 8 and 9 of Second Assessment Report (SAR) (IPCC, 1996) reviewed the literature on costs of greenhouse gas (GHG) mitigation prior to 1995. At that period, the debate was dominated by the differences in results from "bottom-up" (B-U) models and "top-down" (T-D) models. The former contain more details of technology and physical flows of energy, and the latter give more consideration to linkages between a given sector and a set of measures and macroeconomic parameters like gross domestic product (GDP) and final household consumption.

B-U models showed that energy efficiency gains of 10%–30% above baseline trends could be realized at negative to zero net costs over the next two or three decades. However, the costs of stabilizing emissions at 1990 levels reported by T-D analysis for Organization for Economic Co-operation and Development (OECD) countries were less optimistic, in the range –0.5% to 2% of GDP. SAR devoted much effort to explain the reasons for these differences and their meaning for policymakers. B-U models identify negative-cost mitigation potentials because of the difference between the best available techniques and those currently in use; the key question is then the extent to which market imperfections that inhibit access to these potentials can be removed cost-effectively by policy initiatives. T-D analyses focus on the overall macroeconomic effect of new incentive structures, such as carbon taxes or subsidies for energy efficiency; their results reflect a judgement on the capacity of non-price policies (market reforms, information, capacity building) to enhance the effectiveness of such signals to decarbonize the economy. The lesson is that, for a given abatement target, the content of the policy mix (carbon tax, carbon-energy tax, or auctioned emissions trading system) is as important as the assumptions regarding technology.

A second lesson of SAR is that, for both B-U and T-D models, the differences in cost assessment usually result from differences in the definition of baseline scenarios and in the time frame within which a given abatement target has to be met. Less often, they result from divergences in the costs of achieving this target from the same baseline scenario. This, in turn, relates to:
*   the structural features of the scenario (assumptions about population, the rate and structure of economic growth, consumption patterns, and technology development paths); and
*   its level of suboptimality (higher efficiency in the baseline scenario results in higher mitigation-costs estimates for a given target, while the existence of market failure, which enhances GHG emissions, or of fiscal distortions provides a possibility for economic and environmental double dividends).

The third lesson is that some of the determinants of costs are beyond the field of energy and environmental policy *stricto sensu*. This is why SAR emphasizes the importance of developing multiple baseline scenarios to support policymaking. This issue of the multiplicity of baseline scenarios is specifically important for the developing countries and countries with economies in transition. These regions were underinvestigated compared with the number of studies available for OECD countries.

8.1.2   Progress since the Second Assessment Report

Since SAR, the most important advance is the treatment of new topics related to linkages between national policies and the international framework of these policies in the context of the pre-Kyoto and post-Kyoto negotiation process. Of specific interest is the articulation between international emissions trading systems and domestic policies (taxes, domestic emissions trading, and standards). This link has been made in national models and global models that provide a description of relationships among various regions of the world. Some models represent solely GHG emissions trading, while others also incorporate energy flows, trade of other goods, and capital flows. In this context, while SAR discussed only the carbon leakage between abating and non-abating economies, an increasing number of studies have captured spillover effects (see *Box 8.1*) such as those triggered by trade effects and the modification of the capital flows.

A second evolution is the emergence of studies on the local and regional ancillary benefits of climate policies.

The third evolution is the development of studies on various abatement pathways towards given long-run concentration targets and on rules for emissions quota allocation among countries. These approaches, more dynamic in nature, capture the consequences of various abatement timetables on the behaviour of carbon prices and on the sharing of the overall burden among countries. They provide basic information about the equity of various designs of climate policies.

8.1.3   Coverage

This chapter covers studies on global assessments of the net cost of GHG mitigation policies irrespective of the avoided costs of climate change: total mitigation expenditure, and welfare gains or losses resulting from the economic feedbacks of mitigation policies and from their environmental co-benefits.

A specific effort has been devoted to ensure a balanced representation of global models and national models. Global models incorporate linkages between regions and countries; they cannot, however, represent very precisely the specific characteristics of each country, such as differences in national fiscal policies, in regional arrangements, and in socioeconomic constraints. Results from these models are widely diffused within the scientific community through publications in international journals, but are less utilized by national policymakers. The second type of study uses models the scope of which is limited to within the national frame. Results of such studies are more frequently reported in local languages and are more reflective of national debates and the specifics of the country in question. They incorporate linkages with the rest of the world economy, although in a more simplistic manner than the global models.

The results of these studies group into three large clusters. Section 8.2 reviews the studies that entail near-to-mid term impacts of domestic mitigation policies on factors such as GDP, welfare, income distribution, and social and environmental co-benefits at the local and national levels. Section 8.3 contains the results of mitigation studies that examine the interface between these domestic policies and the international context: international trade regimes, and spillover effects of the implementation of mitigation measures by a country or a block of nations on other countries. Section 8.4 reviews studies that focus on social, environmental, and economic impacts of alternate pathways to meet a range of concentration stabilization pathways beyond the Kyoto Protocol. They encompass a longer time horizon and do not incorporate details of macro economic policies, but highlight the question of technological change over the long run and the consequences of various sets of national targets.

8.2   Impacts of Domestic Policies

Evaluation of the economic impacts of domestic mitigation policies can no longer be made independently of the linkages between these policies and the international framework. However, it is important to disentangle the mechanisms that are themselves independent of the international regimes from those specifically driven by the interplay between these regimes and domestic policies. In addition, the existence of an international framework does not rule out the importance of domestic policies for addressing the specific problems of each country.

This section basically relies on national studies, including integrated economic regions such as the European Union (EU), but it also reports the results of multiregional studies for the concerned countries or region.

8.2.1   *Gross Aggregated Expenditures in Greenhouse Gas Abatements in Technology-rich Models*

In technology-rich B-U models and approaches, the cost of mitigation is constructed from the aggregation of technological and fuel costs. These include investments, operation and maintenance costs, and fuel procurement, but also included (and this is a recent trend) are revenues and costs from imports and

exports, and changes in consumer surplus that result from mitigation actions. In all the studies, it is customary to report the mitigation cost as the incremental cost of some policy scenario relative to that of a baseline scenario. The total cost of mitigation is usually presented as a total net present value (NPV) using a social discount rate selected exogenously (the NPV may be further transformed into an annualized equivalent). Many (but not all) report also the marginal cost of GHG abatement (in US$/tonne of CO_2-equivalent), which is the cost of the last tonne of GHG reduced. Chapter 7 discusses cost concepts and discount rates in more depth.

Current B-U analysis can be grouped in three categories:

- Engineering economics calculations performed technology-by-technology (Krause, 1995; LEAP (1995), Von Hippel and Granada (1993); UNEP, 1994a; Brown *et al.*, 1998; Conniffe *et al.*, 1997). The costs and reductions from the large number of actions are aggregated into whole-economy totals in these studies. Each technology (or other action on energy demand) is assessed independently via an accounting of its costs and savings (investment costs, operational and maintenance cost, fuel costs or savings, and emissions savings). Once these elements are estimated, a unit cost (per tonne of GHG reduction) is computed for each action. The unit costs are then sorted in ascending order, and thus the actions are ordered from least expensive to the most expensive, per tonne of abatement, to create a cost curve. This approach requires a very careful examination of the interactions between the various actions on the cost curve: it is often the case that the cost and GHG reduction attached to an action depends on those of other actions in the same economy. Although the simpler interactions are easily accounted for by careful analysis, there exist many other instances in which complex, multi-measure interactions are very difficult to evaluate without the help of a more complex model that captures the system's effects. As an example, consider simultaneously: (a) changing the mix of electricity generation, (b) increasing interprovincial trade of electricity, and (c) implementing actions to conserve electricity in several end-use sectors. As each of these three actions has an impact on the desirability and penetration of each other action, such a combination requires many iterations that assess the three types of action separately, before an accurate assessment of the full portfolio can be obtained.
- Integrated partial equilibrium models that facilitate the integration of multiple GHG reduction options and the aggregation of costs. To achieve this, the majority of B-U studies use the whole energy system (MARKAL, MARKAL-MACRO, MARKAL-MATTER, EFOM, MESSAGE, NEMS, PRIMES[1]). These models have the advantage of simultaneously computing the prices

---

[1] For references see *Table 8.1.*

Table 8.1: *List of the models referred to in this chapter*

Model	Region	Reference
ABARE-GTEM	USA/EU/Japan/CANZ	In Weyant, 1999
ADAM	Denmark	Andersen *et al.*, 1998
AIM	USA/EU/Japan/CANZ	In: Weyant, 1999
	Japan	Kainuma *et al.*, 1999; Kainuma *et al.*, 2000
	China	Jiang *et al.*, 1998
CETA	USA/EU/Japan/CANZ	In: Weyant, 1999
E3-ME	UK/EU/World	Barker 1997, 1998a, 1998b, 1998c, 1999
ELEPHANT	Denmark	Danish Economic Council, 1997; Hauch, 1999
ECOSMEC	Denmark	Gørtz *et al.*, 1999
ERIS		Kypreos et al, 2000
G-Cubed	USA/EU/Japan/CANZ	In: Weyant, 1999
GEM-E3	EU	Capros *et al.*, 1999c
GEM-E3	Sweden	Nilsson, 1999
GemWTrap	France/World	Bernard and Vielle, 1999a, 1999b, 1999c
GESMEC	Denmark	Frandsen *et al.*, 1995
GRAPE	USA/EU/Japan/CANZ	In: Weyant, 1999
IMACLIM	France	Hourcade *et al.*, 2000a
IPSEP	EU	Krause *et al.*, 1999
ISTUM	Canada	Jaccard *et al.*, 1996; Bailie *et al.*, 1998
MARKAL	World	Kypreos and Barreto, 1999
	Canada	Loulou and Kanudia, 1998, 1999a and 1999b; Loulou *et al.*, 2000
	Ontario (Canada)	Loulou and Lavigne, 1996
	Quebec, Ontario, Alberta	Kanudia and Loulou, 1998b; Kanudia and Loulou, 1998a; Loulou *et al.*, 1998
	Canada, USA, India	Kanudia and Loulou, 1998b
	EU	Gielen, 1999; Seebregts *et al.*, 1999a, 1999b; Ybema *et al.*, 1999
	Italy	Contaldi and Tosato, 1999
	Japan	Sato *et al.*, 1999
	India	Shukla, 1996
MARKAL-MACRO	World	Kypreos, 1998
	USA	Interagency Analytical Team, 1997
MARKAL-MATTER	EU	Gielen *et al.*, 1999b, 1999c
MARKAL and EFOM	EU	Gielen *et al.*, 1999a; Kram, 1999a. 1999b
	Belgium, Germany, Netherlands, Switzerland	Bahn *et al.*, 1998
	Switzerland, Colombia	Bahn *et al.*, 1999a
	Denmark, Norway, Sweden	Larsson *et al.*, 1998
	Denmark, Norway, Sweden, Finland	Unger and Alm, 1999
MARKAL Stochastic	Quebec	Kanudia and Loulou, 1998a
	Netherlands	Ybema *et al.*, 1998
	Switzerland	Bahn *et al.*, 1996
MEGERES	France	Beaumais and Schubert, 1994
MERGE3	USA/EU/Japan/CANZ	In: Weyant, 1999
MESSAGE	World	Messner, 1995
MISO and IKARUS	Germany	Jochem, 1998
MIT-EPPA	USA/EU/Japan/CANZ	In: Weyant, 1999
MobiDK	Denmark	Jensen, 1998
MS-MRT	USA/EU/Japan/CANZ	In: Weyant, 1999
MSG	Norway	Brendemoen and Vennemo, 1994
MSG-EE	Norway	Glomsrød *et al.*, 1992; Alfsen *et al.*, 1995; Aasness *et al.*, 1996; Johnsen *et al.*, 1996
MSG-6	Norway	Bye, 2000
MSG and MODAG	Norway	Aaserud, 1996
NEMS + E-E	USA	Brown *et al.*, 1998; Koomey *et al.*, 1998; Kydes, 1999
Oxford	USA/EU/Japan/CANZ	In: Weyant, 1999
POLES	USA, Canada, FSU, Japan, EU, Australia, New Zealand	Criqui and Kouvaritakis, 1997; Criqui *et al.*, 1999
PRIMES	Western Europe	Capros *et al.*, 1999a
RICE	USA/EU/Japan/CANZ	In: Weyant, 1999
SGM	USA/EU/Japan/CANZ	In: Weyant, 1999
SPIT	UK	Symons *et al.*, 1994
SPIT	Ireland	O'Donoghue, 1997
WorldScan	USA/EU/Japan/CANZ	In: Weyant, 1999

CANZ: Other OECD countries (Canada, Australia, and New Zealand); FSU: Former Soviet Union.

of energy and of end-use demand as an integral part of their routine. They are based on least-cost algorithms and/or equilibrium computation routines similar to those used in T-D approaches. They increasingly cover both the supply and demand sides, and include mechanisms to make economic demands responsive to the changing prices induced by carbon policies. Furthermore, many implementations of these models are multiregional, and represent explicitly the trading of energy forms and of some energy intensive materials.

- Simulations models (based on models such as ISTUM) that take into account the behaviour of economic agents when different from pure least cost. To accomplish this, economic agents (firms, consumers) are allowed to make investment decisions that are not guided solely by technical costs, but also by considerations of convenience, preference, and so on. Such models deviate from least-cost ones, and so they tend to produce larger abatement costs than least-cost models, all things being equal otherwise.

The boundaries between these three categories is somewhat blurred. For instance, NEMS and PRIMES do include behavioural treatment of some sectors, and MARKAL models use special penetration constraints to limit the penetration of new technologies in those sectors in which resistance to change has been empirically observed. Conversely, ISTUM has recently been enhanced to allow the iterative computation of a partial equilibrium (the new model is named CIMS).

Several studies go further: they are based on partial equilibrium models in which energy service demands are sensitive to prices. Therefore, even the quantities of energy services may increase or decrease in carbon scenarios, relative to the base case. For these models report not only the direct technical costs, but also the loss or gain in consumer surplus because of altered demands for energy services. The results of this new generation of partial equilibrium B-U models tend to be closer than those of other B-U models to the results of the general equilibrium T-D models, which are also discussed in this chapter. Loulou and Kanudia (1999) argue that, by making demands endogenous in B-U models, most of the side-effects of policy scenarios on the economy at large are captured. When a partial equilibrium model is used, the cost reported is the net loss of social surplus (NLSS), defined as the sum of losses of producers and consumers surpluses (see Chapter 7).

As is apparent from the results presented below, considerable variations exist in the reported costs of GHG abatement. Some of these differences result from the inclusion/exclusion of certain types of cost in the studies (*e.g.*, hidden costs and welfare losses), others from the methodologies used to aggregate the costs, others from the feedback between end-use demand and prices, and still others from genuine differences between the energy systems of the countries under study. However, the most significant cause of cost variations seems to lie not only

(see also Chapter 9) in methodological differences, but in the differences in assumptions. Finally, although most recent B-U results consider the abatement of a fairly complete basket of GHG emissions from all energy-related sources, a few essentially focus on CO_2 abatement only and/or on selected sectors, such as power generation. In this chapter, only results are reported that have sufficient scope to qualify as GHG abatement costs in most or all sectors of an economy.

To facilitate the exposition of the various results, the rest of this subsection is divided into four parts, as follows:
- studies that assume a large potential for efficiency gains, even in the absence of a carbon price;
- other B-U studies for Annex I countries or regions;
- Annex I studies that account for trade effects; and
- studies devoted to non-Annex I countries.

8.2.1.1 *National and Regional Cost Studies Assuming Large Potentials for Efficiency Gains (the Impact of No Regrets or Non-price Policies)*

An important part of climate policy debates is underpinned by a lasting controversy between believers and non-believers in the existence of a large untapped efficiency potential in the economy. If there, this potential could be realized at such a small societal cost that it would be more than compensated by cost savings that accrue from the efficiency improvements. Options that have a negative net social cost add up to an overall negative cost potential that may be quite large. *Figure 8.1* is a sketch of the successive marginal costs of abatement, as a function of GHG reduction relative to some baseline point A. The total cost is simply the area between the curve and the horizontal axis. From A to B, marginal abatement costs are negative, and from B onwards, they are positive. The debate revolves around the size of the total (negative) cost from A to B. The studies discussed in this subsection argue that the negative cost area is potentially quite significant, and compensates

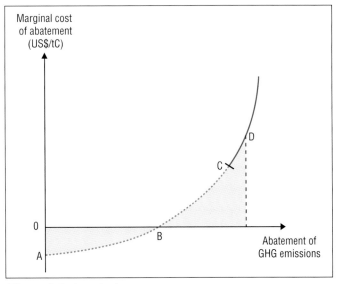

Figure 8.1: *A typical cost curve.*

to a large extent for the positive costs incurred after point B. Most other B-U studies analyzed in the next subsection do not even attempt to evaluate the relative positions of points A and B, since they optimize the system even in the absence of carbon constraint, and thus compute only the points beyond point B.

Krause (1995, 1996) identifies two main reasons why the negative cost area may be quite large: untapped potential for efficiency gains mainly in end-use technologies, both on the demand and on supply sides. Several major studies concretize this view in Europe as well as in North America (USA and Canada). For Europe, the monumental IPSEP reports (summarized in Krause *et al.* (1999)) conclude that emissions could be reduced by up to 50% below the 1990 level by 2030, at a negative overall cost. This involves the judicious implementation of technologies and practices in all sectors of the economy, and the application of a large number of government policies (incentives, efficiency standards, and educational). In the US, some of the 5-Lab studies (Brown *et al.*, 1997a, 2000, particularly the HE/LC scenario) indicate that the Kyoto reduction target could be reached at negative overall cost ranging from –US\$7 billion to –US\$34 billion. Another study based on the NEMS model (Koomey *et al.*, 1998) indicates that 60% of the Kyoto gap could be bridged with an overall increase in the US GDP. The latter study contrasts with another NEMS study (Energy Information Administration, 1998) that indicates GDP losses from 1.7% to 4.2% (depending on the extent of permit trading and sink options) for the USA to reach the Kyoto targets. Laitner (1997, 1999) further stresses the impact of efficient technologies on the aggregate cost of mitigation in the USA. In Canada, the MARKAL model was used with and without certain efficiency measures in various sectors (Loulou and Kanudia, 1998; Loulou *et al.*, 2000): the results show costs of Kyoto equal to US\$20 billion without the additional efficiency measures, versus –US\$26 billion when efficiency measures are included in the database. Again in Canada, the ISTUM model was used (Jaccard *et al.*, 1996, Bailie *et al.*, 1998) considering a set of pro-active options. For example, in the residential sector large emissions reductions of 17% to 25% relative to 1990 could be achieved as early as 2008 with many negative costs options, and beyond that level of reduction, the marginal costs is ranging from US\$25 to US\$89/tC.

As extensively discussed in SAR, many economists argue that the real magnitude of negative cost options is not so large if account is taken of:

- Transaction costs of removing market imperfections that inhibit the adoption of the best technologies and practices;
- Hidden costs, such as the risks of using a new technology (maintenance costs, quality of services);
- "Rebound effect" because, for example, an improvement in motor efficiency lowers the cost per kilometre driven and has the perverse effect of encouraging more trips; and
- Real preferences of consumers: options such as driving

habits and modal switches towards rail and mass transit are considered to entail negative costs. This does not consider enough the reality of consumers' behaviour preferences for flexibility and non-promiscuity in transportation modes, or even "symbolic" consumption (such as the preference for high-power cars even in countries with speed limits).

These arguments should not be used to refute the very existence of negative cost potentials. They indicate that the applicability of non-price policy measures apt to overcome barriers to the exploitation of these potentials must be given serious attention. Some empirical observations do confirm that active sectoral policies can result in significant efficiency gains, in demand-side management for electricity end-uses for example. However, the many sources of gaps between technical costs and economic costs cannot be ignored (see the taxonomy of Jaffe and Stavins, 1994). The few existing observations (Ostertag, 1999) suggest that the transaction costs may represent, in many cases, a large fraction of the costs of new technology, and there is always an uncertainty about the efficiency and the political acceptability of the policies suggested in the above studies. This issue is clearly exemplified by the set of studies carried out in the USA and collected under the name "5-LAB studies". In these, some scenarios produce positive incremental costs and others negative costs, depending on the aggressiveness with which efficiency measures are implemented (Interlaboratory Working Group, 1997; Brown *et al.*, 1998).

8.2.1.2 Bottom-up Costs Resulting from Carbon Pricing (Developed Countries)

Contrary to the studies discussed above, the partial equilibrium studies reviewed in this section do not report negative costs. This is because the least-cost algorithms employed, which are powerful to compute the incremental cost of the system with and without a carbon constraint (*i.e.*, point B in *Figure 8.1*), demand a set of somewhat arbitrary parameters to be calibrated in such a way that they calculate a suboptimal baseline; but such an operation demands resorting to a set of somewhat arbitary parameters and the results are less easy to interpret. This is why the B-U studies reported hereafter explore only the section of the cost curve with positive carbon prices (section CD in *Figure 8.1*).

It is very hard to encapsulate in a short presentation the many studies carried out with a B-U approach using a crosscutting, carbon-pricing instrument. *Figure 8.1* summarizes a number of these results, obtained with a variety of B-U models applied to a single Annex I country or region, ignoring the trade effects. Included are those studies that contain enough information to present the marginal abatement cost along with the level of GHG emission variation from 1990 (other studies that reported only the total abatement cost are discussed separately). In *Figure 8.2*, each point represents one particular reduction level (relative to 1990) and the corresponding marginal cost of reduction. Points that are linked together by a line correspond

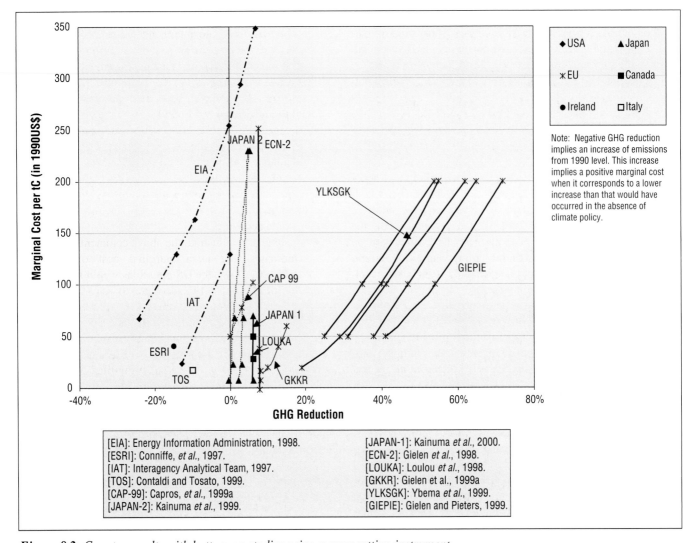

Figure 8.2: *Country results with bottom-up studies using a crosscutting instrument.*

to a multi-run study effected with the same model, but in which the amount of reduction was varied.

Evidently, *Figure 8.2* shows considerably discrepancies from study to study. These large variations are explained by a number of factors, some of which reflect the widely differing conditions that prevail in the countries studied, while others result from the modelling and scenario assumptions. These variations are discussed next, illustrated by examples from *Figure 8.2*.

8.2.1.2.1 Cost Discrepancies that Result from Specific Country Conditions

• *Energy endowment.* Countries that are richly endowed with fossil fuels find it generally less expensive to replace coal with gas, and thus have a greater potential than other countries to reduce emissions with readily available means. (This assumes that the change is not done very rapidly, so as to affect as little as possible the turnover rate of the existing investments.) For instance, this is the case for the USA (coal and oil products).[2] At

the other extreme of the spectrum some countries have fuelled their economy almost exclusively on hydropower, nuclear power, and some gas, and will thus find fewer opportunities to switch to less CO_2-intensive fuels. This occurs for Norway (hydro), France (nuclear and hydro), Japan (nuclear and some fossil fuels), Switzerland (hydro and nuclear), and to some extent Canada (hydro and nuclear).

• *Economic growth.* An economy with high growth rate faces the following dilemma. On the one hand, the growth allows for a rapid capital turnover, and thus many opportunities to install efficient or low-carbon technologies. On the other hand, the same economy requires more energy precisely because of its fast growth. The net result is that such countries have a tendency to decrease markedly their energy intensity

---

[2] However, although the direct cost of switching away from coal may be relatively low, the indirect costs (including the political cost) of disrupting the coal sector may be high.

(energy per GDP), but to increase significantly their total emissions. For such countries, a net reduction of GHG emissions below a base year's emissions is usually costly. Typically, many fast-developing countries in East and South East Asia are in this category. In the studies cited above, their emissions "reductions" are often computed relative to the baseline rather than to a fixed base year.

- *Energy intensity.* The degree of energy intensity of an economy acts in opposing directions when the economy wishes to achieve net emissions reductions: on the one hand, a high degree of energy intensity may occur because that the country has not yet implemented some efficiency measures implemented elsewhere. On the other hand, such an economy may have been built on energy, and may thus find it hard to veer to a different, less energy-intensive mode, in a short time. Its development path is somewhat frozen, at least in the short term. The higher the carbon intensity, the more important the time frame of abatement. Such a pattern is observable in North America; despite its rapid capital turnover in the industry, the large inertia in sectors such as transportation is a determining factor of high abatement cost when the required abatement implies short term actions on these sectors.

- *Other specific conditions.* For example, Germany faces a very special situation because of the reunification in 1990 to 1991. The East German part of the country emits much less now than in 1990, and the country as a whole is able to effect significant reductions at essentially zero or very low cost, up to a certain point, beyond which its marginal cost may well accelerate considerably.

8.2.1.2.2   Discrepancies in Results Due to Modelling and Scenario Assumptions

- *Policy assumptions.* The results summarized in *Figure 8.2* are based mostly on partial equilibrium models, which tend to approach general equilibrium computations, such as AIM, NEMS, MARKAL, MARKAL-MACRO, PRIMES, CIMS, etc. Some of these models allow evaluation of the impact on mitigation cost of the redistribution of the proceeds of a carbon tax (the results obtained with the AIM model for Japan (Kainuma *et al.*, 1999, 2000) show very clearly that suitable redistribution reduces the marginal cost of abatement).

- *Modelling differences.* Some models include partial economic feedbacks in the form of demand elasticities, as for example MARKAL (Loulou and Kanudia, 1999a), and for these models the abatement marginal costs are generally lower than when demands are fixed, because it becomes unnecessary to tap the most costly technological options. MARKAL-MACRO and NEMS include macroeconomic components in the computation of the equilibrium, and therefore qualify as general equilibrium models, albeit simplified ones. In addition, these two models include behavioural considerations in the calculation of the equilibrium, which tend to raise the cost of abatement, compared to least-cost models such as MARKAL.

- *Scenario variation.* The variety of scenarios used is quite large, as a result of varying some or all of the relevant elements. These include whether the technologies are allowed to penetrate freely or in a limited fashion (typically renewables, nuclear, and some new end-use technologies), the basket of GHG gases considered (CO_2 alone versus multigas studies), assumed economic growth, and sectoral scope (energy only *versus* whole economy).

- *Example.* To illustrate the above comments, *Figure 8.2* indicates that at a marginal cost of less than US$100/t$CO_{2eq}$, the US emissions would still be larger in 2010 than in 1990, according to the NEMS (Energy Information Administration, 1998) and MARKAL (Interagency Analytical Team, 1997) studies. Note that NEMS predicts higher marginal costs than MARKAL for the same emission level, as expected, since NEMS includes many behavioural considerations, whereas MARKAL is a least-cost model. Japan's emissions would be reduced by 1% to 8% (AIM studies (Kainuma *et al.*, 1999, 2000); Ireland's (Conniffe *et al.*, 1997) and Italy's (Contaldi and Tosato, 1999) emissions would also increase, whereas Canada's emissions would decrease by 6% (MARKAL study (Loulou *et al.*, 1998)). The several European studies show a wide range of reductions, from relatively small reductions (PRIMES study (Capros *et al.*, 1999a)) to medium or large reductions with the various MARKAL studies (Gielen and Pieters, 1999; Gielen *et al.*, 1999a; Ybema *et al.*, 1999). These large variations are mainly explained by the modelling and scenario assumptions: PRIMES marginal costs are expected to be larger than MARKAL's (just as NEMS costs were larger than MARKAL's in the US case). In addition, scenario assumptions vary across studies: the number of gases modelled, degree of efficiency of the instrument used across the EU countries, and availability of international permits trading.

Several studies are not represented in *Figure 8.2*, since only incremental or average costs were reported. For instance, a German study (Jochem, 1998) indicates reductions of 30% to 40% in 2010 at average costs ranging from US$12 to US$68/tCO_{2eq}. In Canada (Loulou and Lavigne, 1996), a measure of the impact of demand reduction is obtained by running MARKAL with and without elastic demands for energy services: the total cost is US$52 billion with fixed demands, and US$42 billion with elastic demands. Chung *et al.* (1997) arrive at much higher total costs for Canada, using a North American equilibrium level (the higher cost apparently results from fewer technological options than in MARKAL) A Swedish MARKAL study (Nystrom and Wene, 1999) find total cost of 210 billion Swedish krona for a stabilization scenario, against 640 billion Swedish krona for a 50% emissions reduction in 2010. This same study investigates the opportunity cost of a nuclear phase out, and evaluates a rebound effect on the demands of a 9% emissions reduction for Sweden.

8.2.1.3   Country Studies for Developing Countries

Several recent studies have been carried out as part of internationally co-ordinated country study programs conducted by the United Nation Environment Programme (UNEP) Collaborating Centre of Energy and Environment (UNEP, 1999a–1999g), and by the Asian Development Bank, United Nations Development Programme (UNDP), and the Global Environment Facility (ALGAS, 1999c–h). Summaries and analyses appear in Halsnaes and Markandia (1999). These recent studies supplement a number of earlier ALGAS studies of Egypt, Senegal, Thailand, Venezuela, Brazil, and Zimbabwe. The relevant results on aggregate cost are presented as individual country reports and summarized in ALGAS (1999) and in Sathaye *et al.* (1998). National study teams undertook the UNEP and ALGAS studies, using a variety of modelling approaches. The study results reported in *Table 8.2* are based primarily on energy sector options, which are supplemented with a number of options in the transportation sector, waste management, and from the land-use sectors. The GHG emissions reductions are defined as percentage reductions below baseline emissions in 2020 or 2030, or as accumulated GHG emission reductions over the timeframe of the analysis. These analyses are very useful to indicate the extent and cost of clean development mechanism (CDM) potentials in all countries studied.

The ALGAS cost curves show a total accumulated CO_2 emission reduction potential of between 10% and 25% of total emissions in the period 2000 to 2020. The marginal reduction cost is below US$25/t$CO_2$ (see *Table* 8.2) for a major part of this potential, and a large part of the potentials in many of the country studies are associated with very low costs which even in some cases are assessed to be negative. The magnitude of the potential for low cost options in the individual country cost curves depends on the number of options that have been included in the studies. Countries like Pakistan and Myanmar have included relatively many options and have also assessed a relatively large potential for low-cost emission reductions.

Most of the country studies have concluded that options like end-use energy efficiency improvements, electricity saving options in the residential and service sectors, and introduction of more efficient motors and boilers are among the most cost-effective GHG emission reduction options. The studies have included relatively few GHG emission reduction options related to conventional power supply.

The UNEP cost curves exhibit a number of interesting similarities across countries. All country cost curves have a large potential for low cost emission reductions in 2030, where 25% (and in some cases up to 30%) of the emission reduction can

be achieved at a cost below US$ 25/tCO_2 (See *Table 8.2*). The magnitude of this "low cost potential" is like in the ALGAS studies, influenced by the number of climate change mitigation options included in the study. Individual studies indicate that some of the countries like Ecuador and Botswana experience a very steep increase in GHG emission reduction costs when the reduction target approaches 25%. It must be noted that these country studies primarily have assessed end use energy efficiency options and a few renewable options and have not included major reduction options related to power supply which probably could have extended the low cost emission reduction area. The studies for Hungary and Vietnam estimate a relatively small emission reduction potential, which primarily can be explained by the specific focus in the studies on end use efficiency improvements and electricity savings that do not include all potential reduction areas in the countries.

The options in the low-cost part of the UNEP cost curves typically include energy efficiency improvements in household and industry, and a number of efficiency or fuel switching options for the transportation sector. The household options include electricity savings such as compact fluorescent light-

Table 8.2: Emission reduction potentials achievable at or less than US$25/t$CO_2$ for developing countries and two economies in transition

Annual reduction relative to reference case		
Country	**MtCO$_2$/yr**	**%**
Argentina (UNEP, 1999a)	–	11.5
Botswana (UNEP, 1999c)	2.87	15.4
China (ALGAS, 1999c)	606	12.7
Ecuador (UNEP, 1999b)	12.7	21.3
Estonia (UNEP, 1999g)	9.6	58.3
Hungary (UNEP, 1999f)	7.3	7.6
Philippines (ALGAS, 1999h)	15	6.2
South Korea(ALGAS, 1999d)	5.3	5.7
Zambia (UNEP, 1999d)	6.09	17.5
Brazil (UNEP, 1994)	–	29
Egypt (UNEP, 1994)	–	52
Senegal (UNEP, 1994)	–	50
Thailand (UNEP, 1994)	–	29
Venezuela (UNEP, 1994)	–	24
Zimbabwe (UNEP, 1994)	–	34
Cumulative reduction relative to reference case		
Country	**MtCO$_2$/yr**	**%**
Myanmar (ALGAS, 1999e)	44	23
Pakistan (ALGAS, 1999f)	1120	23.7
Thailand (ALGAS, 1999g)	431	4.2
Vietnam (UNEP, 1999e)	1016	13.4

bulbs (CFLs) and efficient electric appliances and, for Zambia, improved cooking stoves. A large number of end-use efficiency options have been assessed for electricity savings, transport efficiency improvements, and household cooking devices, but very few large scale power production facilities.

There are a number of similarities in the low cost GHG emission reduction options identified in the ALGAS and UNEP studies. Almost all studies have assessed efficient industrial boilers and motors to be attractive climate change mitigation options and this conclusion is in line with the conclusions of earlier UNEP studies (UNEP 1994b). A number of transportation options, in particular vehicle maintenance programmes and other efficiency improvement options, are also included in the low-cost options. Most of the studies have included a number of renewable energy technologies such as wind turbines, solar water thermal systems, photovoltaics, and bioelectricity. The more advanced of these technologies tend to have medium to high costs in relation to the above mentioned low-cost options. A detailed overview of the country study results is given in the individual country study reports (UNEP 1999a-g; ALGAS a-h, 1999).

Apart from the UNEP and ALGAS studies presented above, several additional independent studies were carried out for large countries with the help of equilibrium models. Examples are the ETO optimization model (for India, China, and Brazil), the MARKAL model for India, Nigeria, and Indonesia, and the AIM model for China. *Table 8.3* reports the marginal costs (or other cost in some cases) for the abatement levels considered in the studies (relative to baseline). Marginal costs vary from moderate to negative, depending on the country and model used, for emission reductions that are quite large in absolute terms compared to the baseline emissions.

These studies point out the interest of the same set of technologies for most of the countries, such as efficient lighting, efficient heating or air-conditioning (depending upon the region), transmission and distribution losses, and industrial boilers.

Importantly, it should be emphasized that in the way these studies are conducted, the potential for cheap abatement increases in proportion the baselines. In reality, this may not be the case because, in cases of rapid growth, an acceleration of the diffusion of efficient technologies is expected, which

Table 8.3: *Abatement costs for five large less-developed countries*

Country	China	India	Brazil	China	India	Nigeria	Indonesia
Reference	Wu *et al.* (1994)	Mongia *et al.* (1994)	La Rovere *et al.*, (1994)	Jiang *et al.* (1998)	Shukla (1996)	Adegbulugbe *et al.* (1997)	Adi *et al.* (1997)
Span of study	1990–2020	1990–2025	1990–2025	1990–2010	1990–2020	1990–2030	1990–2020
Emissions in 1990 ($MtCO_2$)	2411	422	264				
Emissions in final year, baseline ($MtCO_2$)	6133	3523	1446				
% change	154%	735%	447%	130%	650%		
Emissions in final year, mitigation ($MtCO_2$)	4632	2393	495				
% change	92%	467%	88%	53%	520%		
% change: mitigation versus baseline, final year	**–40%**	**–36%**	**–80%**	**–59%**	**–20%**	**–20%**	**–20%**
Marginal cost in final year (US\$/tCO_2)	**32**	**–16**	**–7**	**28**	**28**	**<30**	
Average cost in final year (US\$/tCO_2)						**<5**	
Annual cost in final year (billion US\$/yr)							**47**

would lower the magnitude of the negative cost potentials. A second caveat to be placed is that an increase of the GDP per capita is consistent with the increase of wages and purchasing power parities which would increase the cost of carbon imported from these countries through CDM projects.

8.2.1.4   Common Messages from Bottom-up Results

Clearly, the impact of policy scenarios has a large influence on abatement costs. Certain studies propose a series of public measures (regulatory and economic) that tap deep into the technical potential of low carbon and/or energy-efficient technologies. In many cases, such policies show low or negative costs. A comparison with least-cost approaches is difficult because these evaluate systematically both the baseline and the policy scenario as optimized systems and do not incorporate market or institutional imperfections in the current world. It would be of great interest to conduct a more systematic comparison of the results obtained via the various B-U approaches, so as to establish the true cause of the discrepancies in reported costs. A timid step in this direction is illustrated in Loulou and Kanudia (1999a).

This leads to a general discussion about the extent to which all these results suffer from a lack of representation of transaction costs, which are usually incurred in the process of switching technologies or fuels. This category of transaction cost encompasses many implementation difficulties that are very hard to capture numerically. The general conclusion from SAR (that costs computed using the B-U approach are usually on the low side compared to costs computed via econometric models, which assume a history-based behaviour of the economic agents) is no longer generally applicable, since some B-U models take a more behavioural approach. Models such as ISTUM, NEMS, PRIMES, or AIM implicitly acknowledge at least some transaction costs via various mechanisms, with the result that market share is not determined by visible (market-based) least-cost alone. Least-cost modellers (using MARKAL, EFOM, MESSAGE, ETO) also attempt to impose penetration bounds, or industry-specific discount rates, which approximately represent the unknown transaction costs and other manifestations of resistance to change exhibited by economic agents. In both cases these improvements result in partially eschewing the "sin" of optimism and blur the division between B-U and T-D models. While the former, indeed, tend to be less optimistic when they account for real behaviours, it is symmetrically arguable that the latter underestimate the possibility of altering these behaviours through judicious policies or better information. All this area still remains underworked.

A common message is the attention that must to be paid to the marginal cost curve. Despite the limitations and differences in results discussed above, B-U analyses convey important information that lies beyond the scope of T-D models, by computing both the total cost of policies and their marginal cost. Very often, indeed, the marginal abatement cost of a given target is high, although the average abatement cost is reasonably low, or even negative. This is because the initial reductions of GHG emissions may have a very low (or negative) cost, whereas additional reductions have, in general, a much higher marginal cost. This fact is captured in the curve representing marginal abatement cost versus reduction quantity, which starts with negative marginal costs, as illustrated in *Figure 8.1*. The initial portion of the curve (section A–B) exhibits negative cost options, which may add up to a significant portion of the reductions targeted by a given GHG scenario. As the reduction target increases (section B–C–D of the curve), the marginal cost becomes positive, and also eventually the total mitigation cost if the reduction target is large enough. But there is systematically a wedge between the marginal and total costs of abatement, and this wedge is all the more important as the macro-economic impacts of climate policies are driven in large part by the marginal costs (because the latter dictate the change in relative commodity prices). They are driven only modestly by the total amount of abatement expenditures.

A crucial, albeit indirect, message, is the importance of innovation: indeed, B-U models depend on a reasonable representation of emerging or future technologies. When this representation is deficient, the models present a pessimistic view of the costs of more drastic abatements in the long term. This issue is not one of the modelling paradigm, but rather of feeding the models with good estimates of technical progress. Some works are currently underway to make explicit the drivers of technical change, such as learning-by-doing (LBD) or uncertainty. These studies are discussed further in Section 8.4.

8.2.2   Domestic Policy Instruments and Net Mitigation Costs

Tapping the technical abatement potentials requires setting up new incentive structures (taxes, emissions trading, technical standards, voluntary agreements, subsidies) for production and consumption, i.e. climate policies. In the following, empirical models that measure net mitigation costs of climate policies are reviewed in order to disentangle the reasons why certain policy packages have similar or different outcomes in various countries. As a first step, the results are presented at an aggregated level; then the impact of measures meant to mitigate the sectoral and distributional consequences of climate policies is examined. Finally, in a third step, the ancillary benefits from the joint reduction of carbon emissions and other pollutants are considered to complete the picture.

8.2.2.1   Aggregate Assessment of Revenue-raising Instruments

Introducing a carbon tax (or auctioned tradable permits) provides an incentive to change the technology over the short and long term. Such policies generate tax revenues and the way these revenues are used has major impacts on the social costs of the climate policy. The reason is that these revenues are, in principle, available to offset some or all of the costs of the mitigation policy. When emission targets go beyond the negative cost potentials, there is a general agreement among economists (see Chapters 6 and 7) that if standards are used (or if emissions permits are allocated for free) the resultant social cost is higher than the total abatement expenditures. Producers pass part of the marginal abatement cost on to consumers through higher selling prices, which implies a loss of consumer surplus. If the elasticity of supply is quite high, this might lead to a net loss of producer surplus. However, if the elasticity of supply is fairly low, overall (or net) producer surplus can rise when policies cause a restriction in output, because the policy-generated rents per unit of production enjoyed by producers more than compensate for the net decrease in sales.

In the 1990s there was considerable interest in how revenue-neutral carbon taxes may mitigate this effect on the economy by enabling the government to cut the marginal rates of pre-existing taxes, such as income, payroll, and sales taxes. The possibility is a double dividend policy (Pearce, 1992), by both (1) improving the environment and (2) offsetting at least part of the welfare losses of climate policies by reducing the costs of the tax system (see the discussion in Chapter 7). The same mechanism occurs when nationally auctioned permits are used; for simplicity, the term carbon tax is used in the rest of this chapter, except when the distinction between these two instruments is necessary.

The starting point in a discussion of a double dividend is how expensive it is to raise government income, that is, how big is the marginal cost of funds (MCF). A high MCF gives more scope for a double dividend than a small MCF in the economy. This arises because the parameters that determine the magnitude of the double-dividend (see Chapter 7) are:

- direct cost to the regulated sector (sector's changes in production methods or installation of pollution-abatement equipment);
- tax-interaction effect (prices are increasing);
- revenue-recycling effect associated with using revenues to finance cuts in marginal tax rates.

When the revenues of carbon taxes are returned in a lump-sum fashion to households and firms, the tax-interaction effect is systematically higher than the revenue-recycling effect. Also the net cost of climate policy is higher than its gross cost (while lower than that with a no-tax policy, see A_1 and A_2 in *Figure 8.3*). However, it is possible to improve this result by targetting tax revenues to cuts in the most distortionary taxes; this can yield either a weak or a strong form of double dividend (Goulder, 1995a). The weak double dividend occurs as long as there is a revenue-recycling effect due to the swap between carbon taxes and the most distortionary taxes. Mitigation costs are systematically lower when revenues are recycled this way than when they are returned lump sum. The strong double dividend is more difficult to obtain. It requires that the (beneficial) revenue-recycling effect more than offset the combination of the primary cost and the tax-interaction effect. In this case, the net cost of abatement is negative (at least within some range). As discussed in Chapter 7, this is possible if, prior to the introduction of the mitigation policy, the tax system is already highly inefficient along non-environmental dimensions. In terms of *Figure 8.3*, the revenue-recycling effect is represented by the downward shift from curve A_1 to curve A_2 or A_3. If the shift is from A_1 to A_2, the weak double dividend occurs, but not the strong double dividend. If the shift is from A_1 to A_3, not only does the weak double dividend occur, but the strong double dividend is realized as well, since the net costs are negative within a range.

While the weak form of double dividend enjoys broad support from theoretical and numerical studies, the strong double dividend hypothesis is less broadly supported and more controversial. Indeed, reaching an economical dividend is impossible when the economy is at full employment and if all other taxation is optimal (abstracting for the environmental externality). Therefore, it may be argued that the double dividend accrues from the tax reform, independently of the climate policy. However, empirical models capture the fact that, in the real world, a carbon tax or auctioned emissions permits will not be implemented after the enforcement of an optimal fiscal reform. To the contrary, introducing a new tax may be a *sine qua non* condition to the fiscal reform. For a given carbon tax revenue, models help interpret the best way to recycle this revenue.

Specific features of the tax systems and markets of the production factors (labour, capital, and energy) ultimately determine the presence or absence of a strong double dividend. For example, a double dividend is likely if production factors are very distorted by prior taxation or specific market conditions, if there is a problem of trade-balance because of the import of fossil energy, or if consumer choice is highly distorted because of tax-deductible spending provisions (Parry and Bento, 2000).

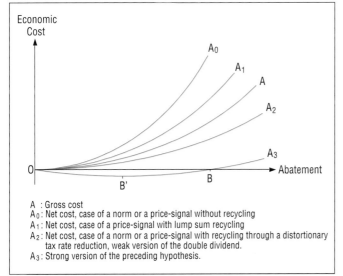

A : Gross cost
A_0: Net cost, case of a norm or a price-signal without recycling
A_1: Net cost, case of a price-signal with lump sum recycling
A_2: Net cost, case of a norm or a price-signal with recycling through a distortionary tax rate reduction, weak version of the double dividend.
A_3: Strong version of the preceding hypothesis.

Figure 8.3: *Carbon taxes and the costs of environmental policies.*

Table 8.4: *Energy Modelling Forum Results: carbon tax and GDP losses in 2010 with lump-sum recycling (in 1990 US$)*

Model	Carbon tax in 2010				GDP losses in 2010 (%)			
	USA	OECD-E	Japan	CANZ	USA	OECD-E	Japan	CANZ
ABARE-GTEM	322	665	645	425	1.96	0.94	0.72	1.96
AIM	153	198	234	147	0.45	0.31	0.25	0.59
CETA	168				1.93			
G-Cubed	76	227	97	157	0.42	1.50	0.57	1.83
GRAPE		204	304			0.81	0.19	
MERGE3	264	218	500	250	1.06	0.99	0.80	2.02
MIT-EPPA	193	276	501	247				
MS-MRT	236	179	402	213	1.88	0.63	1.20	1.83
Oxford	410	966	1074		1.78	2.08	1.88	
RICE	132	159	251	145	0.94	0.55	0.78	0.96
SGM	188	407	357	201				
WorldScan	85	20	122	46				

Source: Weyant (1999). The carbon tax required (either explicitly or implicitly) and the resultant GDP losses are calculated to comply with the prescribed limits under the Kyoto Protocol for four regions under a no trading case: the USA, OECD Europe (OECD-E), Japan, and Canada, Australia, and New Zealand (CANZ).

Empirical studies try to gauge the impact of these many determinants and to understand why the effects of a given recycling strategy (reducing payroll, personal income, corporate income, investment income, or expenditure taxes) differ from one country to another.

8.2.2.1.1 Net Economic Costs under Lump-sum Recycling

The simplest way to simulate the recycling of a carbon tax or of auctioned permits is through a lump-sum transfer. Such recycling does not correspond to any likely policy in the real world. However, these modelling experiments provide a useful benchmark to which other forms of recycling can be compared. In addition, they allow an easy intercountry comparison of the impacts of emissions constraint before the impacts of the many types of possible recycling policies are considered.

The comparative study carried out by the Energy Modeling Forum (EMF, Stanford University) is very useful in this respect: EMF-16 (1999) examined the costs of compliance with the Kyoto Protocol as calculated by more than a dozen modelling teams in the USA, Australia, Japan, and Europe (*Table 8.4*). Most of the models used are general equilibrium models. While not strictly comparable to the marginal technical abatement costs reported in Section 8.2.1, the magnitude of the carbon tax used in these models is determined again by the difference between the costs of marginal source of supply (including conservation) with and without the target. As in the B-U models, this parameter depends in turn on such factors as the size of the necessary emissions reductions, assumptions about the cost and availability of carbon-based and carbon-free technologies, the fossil fuel resource base, and short- and long-term price elasticity. Also important is the choice of base year: a model that provides 3 years to adapt to a constraint beginning

in 2010 shows higher marginal abatement costs than one that provides 8 years.

Figures 8.4-a to *8.4-d* show the incremental cost of reducing a ton of carbon for alternative levels of CO_2 reductions in the USA, OECD Europe, Japan, and Other OECD countries (CANZ) when all reductions are made domestically. Note there are two differences with the B-U studies:

- these numerical experiments do not consider negative cost abatement potentials and presume that if an action is economically justifiable in its own right, it will be undertaken independent of climate-related concerns; and

- because they incorporate demand elasticity and multiple macroeconomic feedback, these marginal cost curves do not behave as those found in the B-U studies.

A first conclusion that could be drawn from *Table 8.4* is that no strict correlations occur between the necessary carbon tax to reach a certain emission target and the GDP loss faced by a country. While the carbon tax in Japan is systematically higher than that for the USA, most studies conclude lower GDP losses in Japan than in the USA. In general, the carbon taxes are highest in OECD Europe and Japan, while the GDP losses are highest in the USA and Other OECD countries. This absence of strict correlation between marginal taxes and GDP losses is explained by the pre-existing energy supply, the structural economic features, and the pre-existing fiscal system. For instance, if a country relies more on renewable energy, and is specialized in low carbon-intensive industry, the impact of a given level of carbon tax will be lower. However, as the burden of emission reductions falls only on a few sectors, the carbon tax for a given target will be higher.

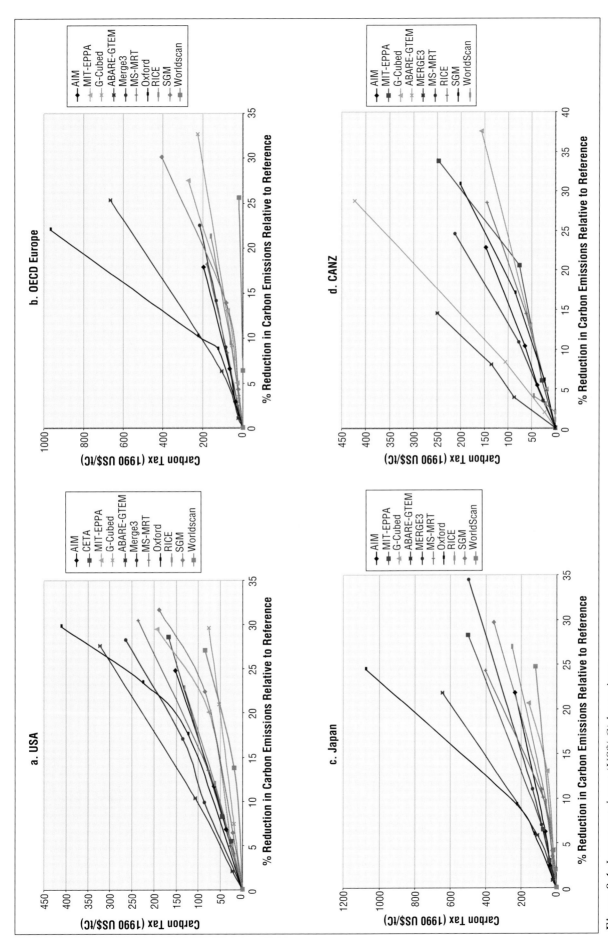

Figure 8.4: *Incremental cost (US$/tC) by regions.*

A second observation from this comparison is that the lump-sum recycling of tax revenue never gives a strong double dividend, which is in accordance with theory and is confirmed by country studies. This is the case in particular for Denmark (Frandsen *et al.*, 1995; Jensen, 1998; Gørtz *et al.*, 1999), France (Bernard and Vielle, 1999a), Finland (Jerkkola *et al.*, 1993; Nilsson, 1999), and Norway (Brendemoen and Vennemo, 1994; Johnsen *et al.*, 1996; Håkonsen and Mathiesen, 1997). These studies demonstrate welfare losses of the same order of magnitude as those of global models, ranging from 0.14% to 1.2% for various levels of emissions abatement ranging from 15% to 25% over a 10-year time period. Only a very few studies conclude to some strong form of double-dividend but do not explain the contradiction between this result and lessons from analytical works.

8.2.2.1.2 Carbon Taxes and Reducing Payroll Taxes

Figure 8.5 plots the range of the numerical findings for a wide set of countries. In comparison with the previous results, these findings are far more optimistic. This confirms the theoretical results that the gross costs of meeting given abatement targets can be significantly reduced by using the revenue of carbon taxes to finance cuts in the existing distortionary taxes, instead of returning the revenues to the economy in a lump-sum fashion. Only a few studies provide results that allow for a systematic assessment of the attractiveness of payroll recycling through comparing its welfare implication with that of lump-sum recycling. For Norway, Håkonsen and Mathiesen (1997) use a static computable general equilibrium (CGE) model to

compare lump-sum recycling to private households and a reduction in employers' social security contributions. Welfare is measured by a combined index of commodity and leisure consumption. When CO_2 emissions are reduced by 20% (*i.e.*, stabilizing emissions in 2000 at the 1990 level), welfare is reduced by 1% with lump-sum recycling, but only by 0.3% when tax income is used to reduce social security contributions. These authors also found ancillary benefits that decrease welfare losses even further (see Section 8.2.4 below).

In this report, it is impossible to identify all the sources of discrepancies in results across models. Only the differences between results concerning the USA and European economies are considered. These discrepancies arise because labour taxes represent one of the most important sources of distortion in European countries as a result of the pre-existing tax structure and of the type of labour-market regulation that prevails in these countries. Note that a systematic outcome of these studies is that an increase in employment is easier to obtain than an increase in total consumption or social welfare, which leads some authors to discuss the employment double dividend as distinct from the efficiency dividend.

While studies conclude that the swap between carbon and payroll taxes reduces the net burden of climate policies but does not avoid net welfare losses in the USA (Goulder 1995b; Jorgenson and Wilcoxen, 1995; Shackleton *et al.*, 1996), a strong double dividend often occurs in Europe. As suggested by theoretical analyses (Carraro and Soubeyran, 1996), these differences can be explained by the differences both in taxation

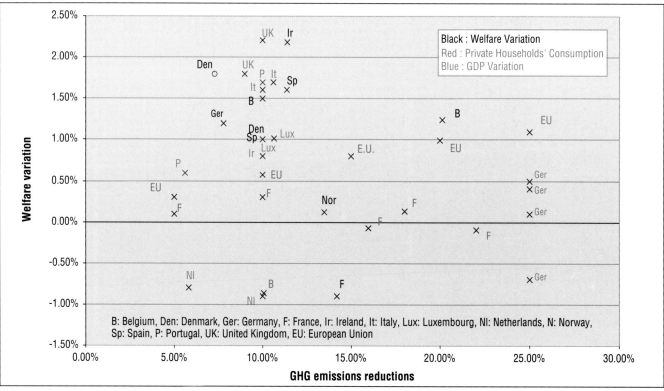

Figure 8.5: *Welfare variation with recycling through payroll taxes reduction.*

systems and in the rigidities of the labour markets. Capros *et al.* (1999b) demonstrate (*Figure 8.6*) that the increase of employment in the EU countries due to payroll tax reduction is far higher under the assumption of wage rigidities than under the assumption of a classic flexible labour market. In the same way, Bernard and Vielle (1999c) do not conclude to a strong double dividend in France, while Hourcade *et al.* (2000a) find a modest increase in total consumption of households (up to 0.2% for carbon taxes up to US$100/tC) because they incorporate structural unemployment. This is also why the E3ME model (Barker, 1999), econometrically driven and neo-Keynesian in nature, provides the most optimistic results; they indeed incorporate the rigidities of the real labour markets. It systematically finds a net increase in GDP in Europe (from 0.8% to 2.2%), except for the Netherlands, with a maximum in the UK. The DRI and LINK models, similar in nature to E3M3, do not find such a gain for the US economy, but a loss of 0.39%.

The magnitude of the double dividend for the European countries is lower in general equilibrium models than in Keynesian models: the welfare effects in different studies are between –1.35% and 0.57%. Even if these estimates cover different emission reduction levels for different time periods, they confirm the attractiveness of payroll recycling. In addition, it is remarkable that negative figures are found for small economies such as Belgium (Proost and van Regemorter, 1995) and Denmark (Andersen *et al.*, 1998) in the situation of a unilateral policy, which confirms the specific interest of these countries in international coordination.

The magnitude of the second dividend (the net economic benefit of tax recycling) is not independent of the abatement target. For a given fiscal system, it is determined by parameters for which sizes vary with the taxation levels (*e.g.*, the elasticity of decarbonization in the production sector and in household consumption, the crowding-out effect between carbon-saving technological change and non-biased technological change). Unfortunately, only a few studies report the range of taxes in which the double-dividend hypothesis holds. Hourcade *et al.* (2000a) found a curve similar to A_3 in *Figure 8.3*; after an optimum around US$100/tC, the double dividend tends to vanish in the same way. Håkonsen and Mathiesen (1997) found that tax recycling is actually welfare improving in the range of a 5% to 15% reduction in CO_2 emissions. Capros *et al.* (1999c) are more optimistic in this respect. They found that the final consumption of households in the EU is increases (about 1%) when the abatement target increases from 20% to 25%. The marginal increase is, however, lower than when the abatement target increases from 5% to 10%.

8.2.2.1.3 Other Forms of Taxes Reduction

Other forms of tax reductions, such as value-added tax (VAT), capital taxes, and other indirect taxes have also been studied in addition to recycling via the national debt and public deficit reductions.

Studies for the USA confirm that the nature of the existing fiscal system matters. While no study found a strong double div-

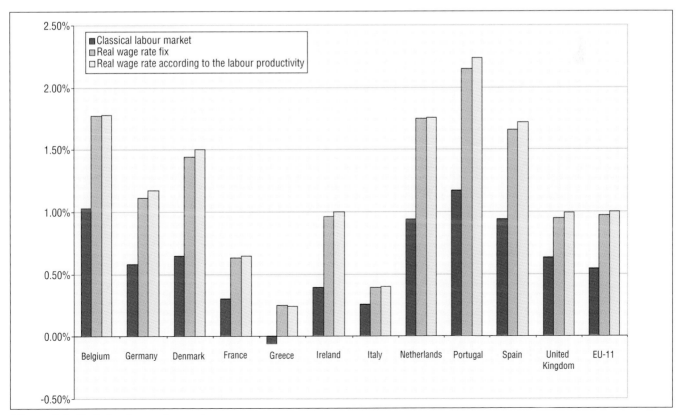

Figure 8.6: *Variation in employment.*

idend for the USA in the case of labour-tax recycling, the Jorgenson–Wilcoxen model supports this notion when recycling takes the form of a reduction in capital taxes (Shackleton, 1998). The pre-existing marginal distortions from taxes on capital are considerably larger than those from labour taxes. Consequently, according to Jorgenson (1997), if the revenues were rebated to consumers in the form of reduced taxes on wage and salary incomes, the cost would be reduced to 0.6%, or by a factor of three compared to the lump-sum recycling case. But if the taxes were rebated on capital income instead, the loss would turn into a gain (0.19%). This higher attractiveness of capital taxation recycling is not found in European countries, with the exception of the Newage model for Germany (Boehringer, 1997).

The other recycling modes have been scrutinized less systematically, but yield in general less favourable results than labour- and capital-taxation recycling. *Figure 8.7* synthetises these results. For Australia, McDougall and Dixon (1996) found that for all the scenarios in which energy taxes were used to offset reductions in payroll taxes, rises in GDP and employment were achieved. A decrease in GDP and employment resulted in the only scenario in which energy taxes were used to reduce the budget deficit. Fitz Gerald and McCoy (1992) found the same type of result for repayment of national debt in Ireland (1% GDP loss). These results are also confirmed in the German case, which is particularly interesting, because several models (Almon, 1991; Welsch and Hoster, 1995; Conrad and Schmidt, 1997; Boehringer *et al.*, 1997) simulate the same emission target (–25%) for the same year (2010) with different types of recycling. They generally conclude to a strong double dividend, and they find a significantly more pessimistic variation in welfare (–4.2% against –0.7% in Almon (1991), –0.03% against +0.1% in Conrad and Schmidt (1996)) when the revenues of the carbon tax are used to lower public deficit rather than reduce social contribution. The results are less clear concerning the relevance of recycling via a capital tax reduction in this country.

For France, Schubert and Beaumais (1998) found, for a carbon tax of US$140/tC, that these tax recycling schemes are less efficient in terms of welfare than recycling through payroll tax, because they trigger no mechanism that enhances employment and general activity. Bernard and Vielle (1999c) confirm this result for the same country. In a short-run analysis for Sweden, Brännlund and Gren (1999) found that private income remains almost unchanged if a reduction in VAT is implemented, because it compensates the regressive income effect of carbon taxation. Nevertheless, as the income increase in this study is relatively important compared to the changes in prices, taxes can be raised without altering consumer behaviour in any considerable way. But this balance may not be preserved in the case of higher carbon taxes.

There are few studies on mitigation costs and recycling for developing countries, but China is one exception. Zhang (1997, 1998) analyzed the implications of two scenarios under which China's CO_2 emissions in 2010 will be cut by 20% and 30% relative to the baseline. Gross National Product drops by 1.5% and

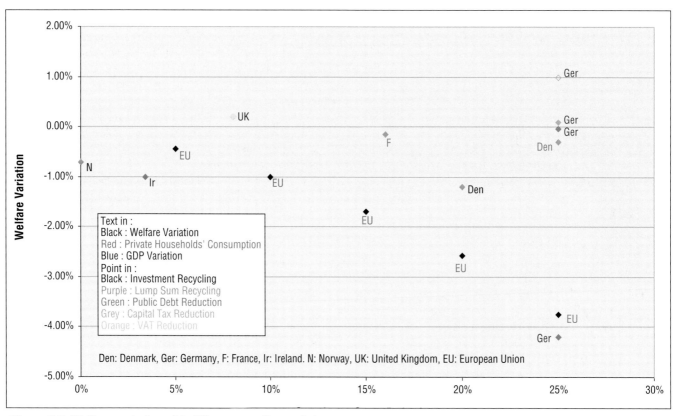

Figure 8.7: *Welfare variation with different recycling policies.*

2.8%, respectively, in 2010 relative to the baseline, and welfare, measured in Hicksian equivalent variation (defined in Chapter 7), drops by 1.1% and 1.8%. If part of the revenues raised by carbon taxes is recycled by equally reducing indirect taxes by 5% and 10%, respectively, for all sectors the welfare effect is markedly improved, and there may even be a gain. Garbaccio *et al.* (1998) report an even more optimistic view from their simulations on a dynamic CGE model for China. Uniform emissions reductions of 5%, 10%, and 15% from baseline were studied, and carbon tax revenues recycled by reducing all other taxes proportionally. In all of the alternative scenarios, a very small decline in GDP occurs in the first year of the simulation. However, in each case, GDP is increased in every year thereafter. The result arises through a shift from consumption to investment brought about indirectly through the imposition of the carbon tax. Thus, a double dividend may be achieved in China.

8.2.2.1.4 Conclusions: Interest and Limits of Aggregate Analysis

A lesson from this section is that, despite their great diversity, the findings of empirical models confirm the theoretical diagnosis. Revenue-raising instruments such as carbon taxes or auctioned emissions permits are, if properly utilized, the most efficient instrument for minimizing the aggregate welfare losses (or maximizing the welfare gains) of climate policies.

It should be noted however, that, even if the only one available study for China suggests that opportunities for revenue recycling exist in developing countries, no swapping generalization can be made at this stage. While theoretical modelling and empirical evidence suggest that such opportunities are available in many OECD countries, developing countries in many cases start from a different fiscal baseline (e.g., fewer entrenched distortionary payroll taxes). They also have other potentially underused tax bases that may become more developed as their economies grow at rates that typically exceed growth rates in OECD countries. In developing countries, direct welfare losses associated with a carbon tax may, therefore, reduce opportunities for mitigation within the fiscal reform policy envelope. At this stage, however, insufficient evidence exists either to confirm or to substantiate these hypotheses; studies to date have mainly concentrated on developed countries and their conclusions may not be directly transferable.

Beyond controversies about the capacity of government to warrant fiscal neutrality, that is the fact that the total fiscal burden remains unchanged, the adoption of carbon taxes or auctioned permits confronts the fact that their enforcement must be done in the heterogeneity of the real world, and can have very significant distributive implications:

- *Across economic sectors.* The carbon content of the steel, aluminium, cement, basic chemical, and transport industries are, indeed, four to five times higher per unit of value added than for the rest of industry. For unilateral initiatives, carbon taxes drastically impact the competitiveness of these sectors (with potential eco-

nomic shocks at the regional level); even with an internationally co-ordinated policy, their equity value will be lowered compared with the rest of industry.

- *Across households income groups.* Carbon taxation increases the relative prices of energy services such as heating, lighting, and transport. The resultant impact on welfare is then more negative for low income levels and people living in cold areas and in low density areas. It is also higher for high income groups and more beneficial for medium income groups in case of swap with other taxation.

Economic analysis can define the compensation necessary to offset these negative distributional effects but, in the real world, winners cannot (or are not willing to) compensate losers. This is especially relevant when the losers suffer heavy impacts and the winners enjoy only marginal gains, which leads to the so-called political mobilization bias (Olson, 1965; Keohane and Nye, 1998) when the losers are more ready to organize a lobbying and incur mobilization costs than the winners (Williamson, 1996). Under such circumstances, policies yielding the largest aggregate net benefits may prove very difficult to enforce. Economic models provide no answer to this issue, but can try to frame the debate by providing the stakeholders with appropriate information. This is the objective of Sections 8.2.2.2 and 8.2.2.3.

8.2.2.2 Mitigating Sectoral Implications: Tax Exemptions, Grandfathered Emission Permits, and Voluntary Agreements

In all countries in which CO_2 taxes have been introduced, some sectors are exempt, or the tax is differentiated across sectors (see, e.g., ECON, 1997). Typically, households pay the full tax rate, whereas export-oriented industries pay either nothing or a symbolic rate.[3] Very few countries have actually implemented a CO_2 tax, and (unsurprisingly) tax exemptions are more systematically analyzed in these countries, such as the Scandinavian countries. Concerns about the sectoral implications of revenue-raising policies have led to four types of responses being studied:

- exemption of the most carbon-intensive activities;
- differentiating the carbon tax across sectors;
- compensation subsidies; and
- government's free provision of emissions permits to firms on a grandfathering basis or on the basis of voluntary agreements on sectoral objectives.

8.2.2.2.1 Tax Exemption

Lessons from the few modelling exercises suggest that the efficiency cost for the whole economy of offsetting the sectoral

---

[3] Some exceptions occur. For instance, in Norway emissions of CO_2 from oil and gas production have traditionally been charged the maximum rate.

impacts of carbon taxes through tax exemptions are very high. Böhringer and Rutherford (1997) show for Germany that exemptions to energy- and export-intensive industries increase the costs of meeting a 30% CO_2 reduction target by more than 20%. Jensen (1998) has similar findings for Denmark with respect to a unilateral reduction of CO_2 emissions by 20% (Jensen, 1998). To exempt six production sectors that emit 15% of Denmark's total emissions implies significantly greater welfare costs (equivalent variation) than full taxation to meet the same abatement target. Namely, welfare loss of 1.9% and a carbon tax on the non-exempted sectors of US$70/t$CO_2$, against a welfare loss of 1.2% and a carbon price of US$40/t$CO_2$ in the no-exemption case (uniform taxes). A similar result is found in Hill (1999) for Sweden: the welfare costs of using exemptions are more than 2.5 times higher than in the uniform carbon tax case for a 10% emission decrease. The high costs of tax exemption are also confirmed by a US study (Babiker *et al.*, 2000).

8.2.2.2.2   Tax Differentiation

Tax differentiation is studied in a CGE model for Sweden in Bergman (1995), who compares its effect with a uniform tax for given emission targets. The tax rate applicable to the industrial firms is set to one-quarter of the tax rate for non-industrial firms and households. The GDP loss increases slightly compared to the uniform tax, but it is still quite small. However, the purchasing power of the aggregated incomes of labour and capital is significantly reduced. Consequently, tax differentiation does not seem to have as much of an adverse effect as full tax exemption. The reason is that all sectors pay a carbon tax when taxes are differentiated, while this is not the case for tax exemptions. Thus, the burden on sectors that pay the highest carbon tax is not that large, and hence results in lower welfare losses.

8.2.2.2.3   Compensating or Subsidizing Mitigation Measures

Böhringer and Rutherford (1997) as well as Hill (1999) envisage labour subsidies used to keep a given employment target. They conclude that – compared to tax exemptions for energy- and export-intensive industries – a uniform carbon tax cum wage subsidy achieves an identical level of national emission reduction and employment at a fraction of the costs.

A second option is a special case of voluntary agreements. In most of the literature, voluntary agreements result from negotiations on emission levels between public authorities and firms adversely impacted by environment policies. Carraro and Galeotti (1995) examined another form of voluntary agreement for European countries: firms receive financial benefits if they have engaged in environmental research and development (R&D) spending. This option is justified because economic tools may be inefficient in reaching the optimal R&D level, even in a pure and perfect market competition (Laffont and Tirole, 1993). According to this study, a strong double dividend could occur in all European countries except Belgium and the

UK, even if the impact on employment is weak. One of the reasons for this double dividend is the technical progress induced by this policy.

8.2.2.2.4   Free Allocation of Emissions Permits

Annex B countries are currently considering the creation of a market for GHG emissions on the basis of grandfathered quotas or of quotas delivered in function of voluntary agreements of sectors to given emissions targets (see Chapter 6). This option does not generate revenue, but (contrary to tax exemptions) implies participation of the carbon-intensive industry to climate policy and does not transfer the full burden to households and the rest of industry. However, the welfare impacts are systematically found to be less favourable than under a full revenue-neutral taxation. Jensen (1998) found a welfare loss of 1.4% in Denmark and a permit price of US$110/t$CO_2$, while a uniform tax to meet the same –20% target is only US$40 and the resultant welfare loss is 1.2%. Bye and Nyborg (1999) investigated the effects on welfare (total discounted utility) of both uniform taxes and tradable permits issued freely compared to the current system of tax exemptions. To keep total tax revenues unchanged for the government, the payroll tax is adjusted accordingly. They found that a permit system gives a welfare loss of 0.03% compared to the current system, while with uniform taxes there is a gain of similar size. The main reason is that payroll taxes must increase to maintain the budget balance when carbon taxes are not used. There are similar findings for the USA. Parry *et al.* (1999) show that the net economic impact (after accounting for environmental benefits but not without climate benefits) of carbon abatement is positive when permits are auctioned, but switches to negative when permits are grandfathered.

Other allocation rules have been tested, but do not improve the result compared with grandfathered permits. For Denmark, Jensen and Rasmussen (1998) examined the aggregate welfare loss (equivalent variation) of an emission target of 80% of 1990 levels from 1999 to 2040; they found 0.1% with a permit auction, 2.0% for grandfathered permits, and 2.1% when the permits are given to firms in the proportion of market shares and sectoral emissions, similar to an output subsidy.

Such results are obtained because, in the case of free delivered permits, the interactions with the tax system occur without the compensating effect of tax-revenue recycled, as in the cases of environmental taxes and auctioned permits. Studies by Parry (1997), Goulder *et al.* (1997), Parry *et al.* (1999), and Goulder *et al.* (1999) show that the costs of quotas or marketable permits are higher if there are prior taxes on the production factors concerned than in if there are no such taxes. Quotas or permits tend indeed to raise the costs of production and the prices of output. This reduces the real return to labour and capital, and thereby exacerbates prior distortions in relevant markets and decrease the overall efficiency of the economy.

Bovenberg and Goulder (2001) found that avoiding adverse impacts on the profits and equity values in fossil fuel industries involves a relatively small efficiency cost for the economy. This arises because CO_2 abatement policies have the potential to generate revenues that are very large relative to the potential loss of profit for these industries. By enabling firms to retain only a very small fraction of these potential revenues, the government can protect the firms' profit and equity values. Thus, the government needs to grandfather only a small percentage of CO_2 emissions permits or, similarly, must exempt only a small fraction of emissions from the base of a carbon tax. This policy involves a small sacrifice of potential government revenue. Such revenue has an efficiency value because it can finance cuts in pre-existing distortionary taxes. These authors also found a very large difference between preserving firms' profits and preserving their tax payments. Offsetting producers' carbon tax payments on a dollar-for-dollar basis (through cuts in corporate tax rates, for example) substantially overcompensates firms, raising their profit and equity values significantly relative to the situation prior to the environmental regulation. This reflects that producers shift onto consumers most of the burden from a carbon tax. The efficiency costs of such policies are far greater than the costs of policies that do not overcompensate firms.

8.2.2.2.5 Conclusions

The costs of meeting the Kyoto targets are very sensitive to the type of recycling used for the revenue of carbon taxes or auctioned permits. In general, however, modelling results show that the sum of the positive revenue-recycling effect and the negative tax-interaction effect of a carbon tax or auctioned emission permits is roughly zero. Thus, in some analyses the sum is positive, while in others it is negative. In economies with an especially distortive tax system (as in several European analyses), the sum may be positive and hence confirm the strong double-dividend hypothesis. In economies with fewer distortions, such as in various models of the US economy, the sum is negative. Another conclusion is that even with no strong double-dividend effect, a country fares considerably better with a revenue-recycling policy than with one that is not revenue recycling, like grandfathered quotas. Analyses of the US economy found that revenue recycling reduces the cost of regulation by about 30%–50% for a certain range of targets, while European analyses report cost savings that are even higher than 100%.

However, at this stage insufficient evidence exists either to confirm or to substantiate these results in the context of developing countries. Studies to date have concentrated on developed countries and, while these studies are comprehensive and rigorous, their conclusions may not be directly transferable. It can be argued that, in developing countries, direct welfare losses typically associated with specific factor taxes (such as a carbon tax) may have fewer opportunities for mitigation within the fiscal-reform policy envelope. Nevertheless, the complex linkages between formal and informal sectors of the economy may show this intuition to be incorrect; the only existing study for China reviewed here suggests that this may be the case but further research is needed to confirm this more generally.

8.2.2.3 The Distributional Effects of Mitigation

A policy that leads to an efficiency gain may not improve overall welfare if some people are in a worse position than before, and vice versa. Notably, if there is a wish to reduce the income differences in a society, the effect on the income distribution should be taken into account in the assessment.

An evaluation of the distributional incidence of higher energy prices is significantly conditional upon the indicator used. Distributional impacts appear to be higher when additional costs are measured in terms of percentage of total household expenditures rather than income, and higher if current income is considered instead of lifetime income. Lifetime income is relevant in the sense that households can borrow or save, and also move between different income classes. According to Poterba (1991), a person had only a 41% chance of being in the same quintile of income distribution in 1971 and in 1978. This percentage rises to 54% if the person initially belongs to the poorest quintile. However, current income is relevant to studies on the short-term intergeneration impact of a new tax. For example, an elderly person is more adversely affected by new taxes on expenditures than are those on an income, even if subsequent generations pay the same lifetime tax bill under each factor influencing the macroeconomy.

International competition limits the ability of firms to pass the tax onto prices, thus reducing the size of the indirect distributional effect. In the same way, the degree of production factor substitution determines the extent to which the tax changes prices. Moreover, as the substitution is generally supposed to be limited in the short term, but increasing as existing plants are replaced, the distributional effect of an environmental tax changes over time. Last, but not least, the distributional effect depends basically on the utilization of the tax revenue.

Two British studies looked at distributional effects of climate policies. Barker and Johnstone (1993) investigated the distributional effects of a carbon–energy tax. Revenues are recycled through an energy efficiency programme and compared to lump-sum transfers. The results show that the burden of a carbon–energy tax falls most heavily on low-income groups. At the same time, for these low-income groups the potential gains to be realized by increasing energy efficiency are higher to offset this regressive outcome. Symons *et al.* (1994) investigated other various assumptions of revenue recycling for the UK, and found that to introduce a carbon tax without recycling or with recycling through VAT or petrol excise-duty reductions is significantly regressive. Conversely, recycling the carbon tax by a combination of VAT rate reductions and benefits reforms directed towards poorer households results in favourable distributive effects.

The conclusion is similar for other countries. For Ireland, O'Donoghue (1997) found that carbon taxation is generally regressive, but that recycling the carbon tax through a fixed basic income for all individuals allows the distributional effects to become almost neutral. For Norway, Brendemoen and Vennemo (1994) concluded that a global carbon tax of US$325/tC in 2000 and US$700/tC in 2025 (1987 prices) has no significant impact on the regional distribution of welfare. For Australia, Cornwell and Creedy (1996) found that a carbon tax only affecting households (the input–output matrix is constant in their model) is clearly regressive, but can become neutral if adequate recycling is implemented. In addition, the distributional differences across income are not affected much. On the other hand, Aasness *et al.* (1996) conclude for the same country that poor households are less favourably affected than rich households, because of smaller budget shares on consumer goods (which imply relatively more CO_2 emissions) in the rich households. Harrison and Kriström (1999) studied the general equilibrium effects of a scenario in Sweden in which the existing carbon taxes increase by 100% and labour taxes are reduced to maintain constant governmental revenue, but without removing the existing exemptions from carbon taxes. All households lose from this carbon tax (with tax exemptions) increase. They point out that the distributional effects are very dependent on the size of the household (the more affected being those with children). In a study for 11 EU member states, Barker and Kohler (1998) examined emission reductions of 10% below baseline by 2010. They concluded that the changes would be weakly regressive for nearly all the member states if revenues are used to reduce employers' taxes, and strongly progressive if they are returned lump-sum to households.

In summary, most studies show that the distributional effects of a carbon tax are regressive unless the tax revenues are used either directly or indirectly in favour of the low-income groups (see also Poterba, 1991; Barker, 1993; Hamilton and Cameron, 1994; OECD, 1995; Cornwell and Creedy, 1996; Oliveira Martins and Sturn, 1998; Smith, 1998; Fortin, 1999). This undesirable effect can be totally or partially compensated by a revenue-recycling policy if the climate policy is implemented through carbon taxes or auctioned permits.

Three other issues of distributional effects, not dealt with here, are industry sector impacts, regional effects, and how people are affected by environmental damage. For instance, a tax on CO_2 emissions obviously leads to very different effects in energy-intensive industries than in sectors producing labour-intensive services (see Chapter 9). In addition, the poor household generally lives in the most polluted area and then benefits first from the amelioration of air quality induced by GHG reduction policy (see Section 8.2.4).

8.2.3   The Impact of Considering Multiple Gases and Carbon Sinks

The overwhelming majority of T-D mitigation studies concentrate upon CO_2 abatement from fossil fuel consumption, while an increasing number of B-U studies tend to incorporate all the GHG emissions from the energy sector, but still not include emissions from the agricultural sector and sequestration. However, the Kyoto Protocol also includes methane (CH_4),

nitrous oxide (N_2O), perfluorocarbons, hydrofluorocarbons, and sulphur hexafluoride (SF_6) as gases subject to control. The Protocol also allows credit for carbon sinks that result from direct, human-induced afforestation and reforestation measures taken after 1990. This may have significant impacts on abatement costs.

A recent study (Reilly *et al.*, 1999) estimated the mitigation costs for the USA and included consideration of all of these gases and forest sinks. The study assumes that the Kyoto Protocol is ratified in the USA and implemented with a cap and trade policy. The analysis considers the effects of including the other gases in the Kyoto Protocol in terms of the effect on allowable emissions, reference emissions, the required reduction, and the cost of control.

For the USA, the authors estimate that base year (1990) emissions were 1,654MtC_{eq}, converting non-CO_2 gases to carbon equivalent units using 100-year global warming potential indices (GWPs) as prescribed in the Kyoto Protocol. This compares with 1,362MtC_{eq} for carbon emissions alone. The result is that allowable emissions are 1,539MtC_{eq} in the multigas case compared with 1,267MtC_{eq} if other gases had not been included in the agreement.

The authors also projected emissions of other gases to grow substantially through 2010 in the absence of GHG control policies, so that total emissions in the reference case reach 2,188MtC_{eq} compared with 1,838MtC_{eq} of carbon only. The combination of these factors means that the required reduction is 650MtC_{eq} in the multigas case compared with 571MtC_{eq} if only carbon is subject to control. To analyze the impact of including the other gases in the Kyoto Protocol the authors consider three policy cases:

*   *Case 1, fossil CO_2 target and control.* This case includes only CO_2 in determining the allowable emissions under the Kyoto Protocol and includes only emissions reductions of CO_2, unlike the requirements in the Kyoto Protocol that require consideration of multiple gases.
*   *Case 2, multigas target with control on CO_2 emissions only.* This case is constructed with the multigas target (expressed as carbon equivalents using GWPs) as described in the Kyoto Protocol, but only carbon emissions from fossil fuels are controlled.
*   *Case 3, multigas target and controls.* The multigas Kyoto target applies and the Parties seek the least-cost control across all gases and carbon sinks.

Case 1 is thus comparable to many other studies that only consider CO_2 and provides an approximate ability to normalize results with other studies. For Case 1 the resultant carbon price is US$187 in 1985 price (US$269 in 1997 price). Case 2 illustrates that, if the USA does not adopt measures that take advantage of abatement options in other gases and sinks, the cost could be significantly higher (US$229 in 1985 price or US$330 in 1997 price). In 1997 US$, the total cost in terms of reduced output is estimated to be US$54 billion for Case 1, US$66 billion in Case 2, and US$40 billion in Case 3.

By comparison with Case 1, the introduction of all gases and the forest sink results in a 20% decline in the carbon price to US$150 (1985 price, US$216 in 1997 price).

Cases 2 and 3 are comparable in the sense that they nominally achieve the same reduction in GHGs (when weighted using 100-year GWPs). Thus, for a comparable control level, the multigas control strategy is estimated to reduce US total costs by nearly 40%.

The Reilly *et al.* (1999) study did not conduct sensitivity analyses of the control costs, but noted the wide range of uncertainties in any costs estimates. Both base year inventories and future emissions of other GHGs are uncertain, more so than for CO_2 emissions from fossil fuels. Moreover, some thought will be required to include other GHGs and sinks within a flexible market mechanism such as a cap and trade system. Measuring and monitoring emissions of other GHGs and sinks could add to the cost of controlling them and so reduce the abatement potential.

8.2.4   *Ancillary Benefits*

"Co-benefits"[4] are the benefits from policy options implemented for various reasons at the same time, acknowledging that most policies resulting in GHG mitigation also have other, often at least equally important, rationales. "Ancillary benefits" are the monetized secondary, or side benefits of mitigation policies on problems such as reductions in local air pollution associated with the reduction of fossil fuels, and possibly indirect effects on congestion, land quality, employment, and fuel security. These are sometimes referred to as "ancillary impacts" to reflect that these impacts may be either positive or negative. *Figure 8.8* shows the conceptual framework for analyzing ancillary and co-benefits and costs. The figure shows that climate and social/environmental benefits can be direct benefits, ancillary benefits, or co-benefits, depending on the objectives of the policies.

The term co-benefits is used in this report despite its limited literature because it shows the case for an integrated approach, linking climate change mitigation to the achievement of sustainable development. However, there appear to be three classes of literature regarding the impacts of climate change mitigation: (1) literature that primarily looks at climate change mitigation, but that recognizes there may be benefits in other areas (illustrated in the top panel of *Figure 8.8*); (2) literature that primarily focuses on other areas, such as air pollution mitigation, and recognizes there may be "ancillary benefits" in the

area of climate mitigation (illustrated in the centre panel of *Figure 8.8*), (3) literature that looks at the combination of policy objectives and examines the costs and benefits from an integrated perspective (illustrated in the bottom panel of *Figure 8.8*). In this report, the term "co-benefits" is used when speaking generically about this latter perspective and when reviewing class (3) literature. The term "ancillary benefits" is used when addressing class (1) and (2) literature. This section covers primarily class (1) literature, which is the most extensive.

Very few economic modelling studies that examine the impacts on economic welfare of various GHG abatement policies explicitly consider their ancillary consequences, *i.e.* effects which would not have occurred in the absence of specific GHG policies. These range from public health benefits through reduced air pollution to reduced CH_4 from animal farms, and through impacts on biodiversity, materials, or land use (see Rothman, 2000).

Existing studies provide evaluations of net ancillary benefits ranging from a small fraction of GHG mitigation costs to more than offsetting them (see Burtraw *et al.*, 1999, and reviews by Pearce, 2000; Burtraw and Toman, 1997; and Ekins, 1996). Such variation in estimates is not surprising because the underlying features differ by sectors considered and the geographic area being studied; but this variation also reflects the lack of agreement on the definition, reach, and size of these impacts and on the methodologies to estimate them. This literature is growing, particularly with respect to the impacts on public health[5,6], so a critical review of it is given in this section. Most of the studies reviewed focus on public health, which is the largest quantifiable impact; therefore this assessment also focuses on it. Ancillary impacts to specific economic sectors are reviewed in Chapter 9.

Most of the key ancillary benefits quantified to date are relatively short term and 'local', that is affecting the communities relatively close to the sources of the emissions changes. In both these respects ancillary benefits can be thought of as offsetting all or part of the welfare losses associated with the costs of reducing GHGs. In this regard the best measure of ancillary

---

[4] See Chapter 7 for a more formal definition of ancillary and co-benefits and costs of GHG mitigation.

[5] A number of possibly important side benefits are not amenable as yet to either quantitative or economic analyses (e.g., ecosystem damages, biodiversity loss).

[6] In SAR, IPCC estimated that, for European countries and the USA, benefits such as reduced air pollution could offset between 30% and 100% of the abatement costs (IPCC, 1996, p. 218). These estimates were controversial and not supported by a standardized methodology. After SAR, extensive debates arose regarding suitable costing methods to quantify the relative economic impacts of various policies in distinct regions, with as yet no consensus on the most suitable methods to be employed (Grubb et al., 1999). However, a consensus is now beginning to emerge on how to quantify some ancillary benefits. See OECD, Proceedings from Workshop on the Ancillary Benefits and Costs of Climate Change Mitigation (OECD, 2000).

impacts may be the percentage (or absolute) variation in welfare loss from considering a carbon tax (or other instrument) that does not include direct climate-mitigation benefits. Few studies provide such estimates (Dessus and O'Connor (1999), is an exception).

Other metrics in the literature help to shed light on the size and uncertainty associated with ancillary impacts estimates. The first normalizes ancillary benefits with carbon reductions, that is, ancillary benefits per tonne of carbon reduced (e.g., Burtraw *et al.*, 1999). The second is the average ancillary benefits per tonne as a fraction of the carbon tax. This latter measure is useful because it has some linkage to the net benefits question. Private marginal carbon mitigation costs are equalized to the tax in the models in the literature. Given that average mitigation costs are less than (or equal to) marginal costs, if the met-

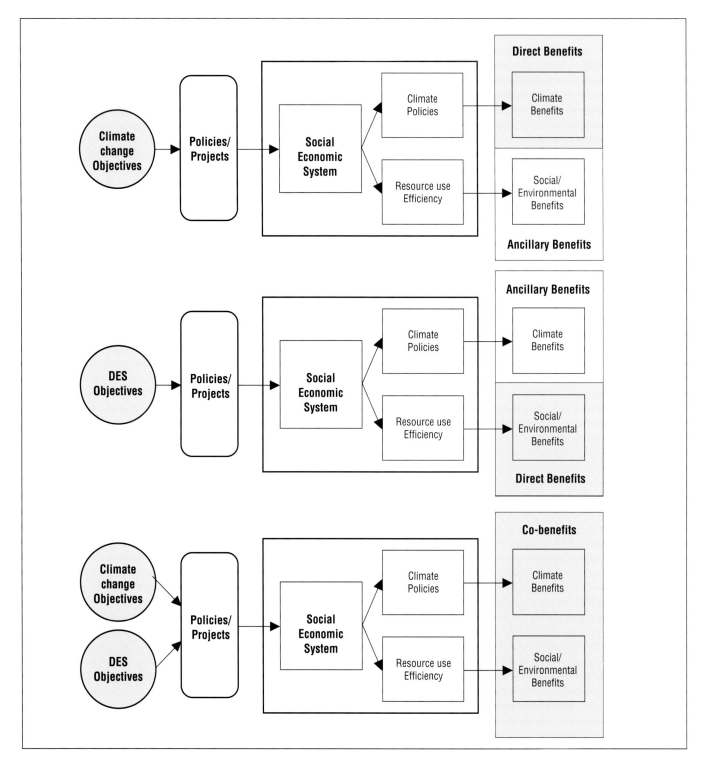

Figure 8.8: *Conceptual framework for analyzing ancillary and co-benefits and costs.*

ric equals more than one then the carbon policy modelled has net private benefits even without counting the direct climate benefits. If (average) ancillary benefits are lower than private marginal costs no claim for net benefits can be made. However, the lower the fraction, the less likely the policy will have net benefits.

A few important caveats are, however, in order. The most important is that the relevant cost measure in the above fraction is *social* not *private* cost. In this case, marginal *social* costs are likely to exceed marginal *private* costs because of tax interaction effects. Thus, ancillary benefits may not exceed social marginal cost even if the former exceeds private marginal cost (equal to the carbon tax). Second, ancillary benefit (cost) measures need to measure social welfare gains (or losses) if they are to be comparable to losses on the mitigation side. But, many measures of ancillary benefits understate social welfare gains and other benefits remain unmonetized or even unquantified, while in other cases, the ancillary benefits overstate welfare gains (say by counting all traffic fatality reductions as external benefits). Thus, *reported* ancillary benefits can under- or overstate actual ancillary benefits. If this indicator is greater than one, then the carbon policy has net private benefits even without counting the direct climate benefits.

The section reviews some of the recent studies estimating ancillary benefits of GHG mitigation policies. The studies are briefly described and examined for the credibility of their methods and estimates.

8.2.4.1 The Evaluation of the Ancillary Public Health Impacts

Studies estimating ancillary public health impacts from climate policies were examined, relying on three surveys of this literature (Ekins, 1996; Burtraw *et al.*, 1999; Kverndokk and Rosendahl, 2000) and on summaries of the older literature, supplemented by some of the newer studies. *Table 8.5* provides a description of each study, as well as the estimates of ancillary benefits per tonne of carbon. *Table 8.6* summarizes the modelling choices of the studies reviewed.

The Burtraw *et al.* (1999) review of US ancillary benefit studies of public health impacts linked to mitigation policies applied to the electricity sector came to several important conclusions:

- Estimates from early studies of ancillary benefits tended to exceed later ones because of the former's use of more crude and less disaggregate modelling.
- Studies that did not factor into the baseline the reductions in conventional pollutants required under the 1990 Clean Air Act estimated benefits an order of magnitude larger than the studies that did include the 1990 Clean Air Act in the baseline. Analyzing Ekins (1996), Burtraw *et al.* (1999) found that whether the Second Sulphur Protocol is added to the baseline or not can alter the estimate of ancillary benefit by over 30%.

- Some studies did not consider the "bounceback" effect (*i.e.*, the offsetting increase in conventional pollutants) when a less carbon-intensive technology is substituted for a more intensive one in response to a carbon mitigation policy.
- Ancillary benefit estimates are very sensitive to assumptions about the mortality risk coefficient and the value of statistical life (VSL). Routine values used in the literature can lead to a difference of 300% in ancillary benefit estimates.
- Burtraw *et al.* (1999) and earlier studies to reconcile US and European estimates for the social costs of fuel cycles found that population density differences between Europe and the USA account for 2 to 3 times larger benefit estimates in Europe. Also, the fact that much East Coast US pollution is blown out to sea while European pollution is blown inland can account for large ancillary benefit differences.
- With a cap on SO_2 emissions, abatement cost savings are considered ancillary benefits of a carbon policy unless the reductions are so large that the cap becomes non-binding. When this happens, with SO_2 effects on mortality being as large as they appear to be, ancillary benefits increase in a discontinuous and rapid fashion, as the health benefits begin to be counted.

Kverndokk and Rosendahl (2000) review much of the recent ancillary benefit literature in the Nordic countries, UK, and Ireland, concluding that benefits are of the same order of magnitude as gross (*i.e.*, private) mitigation costs. They also conclude that the benefits should be viewed as highly uncertain, because of the use of simplistic tools and transfers of dose–response and valuation functions from studies done in other countries. They point out that most of the Norwegian studies use expert judgement instead of established dose–response functions and estimates of national damages per tonne rather than distinguishing where emissions changes occur and exposures are reduced. Also, they point out that large differences in ancillary benefits per tonne across several Norwegian studies can be attributed to differences in energy demand and energy substitution elasticities. If energy production is reduced rather than switched to less carbon-intensive fuels, ancillary benefits will be far larger. Kverndokk and Rosendahl (2000) point out also that studies that feed environmental benefits back into the economic model add significantly to ancillary benefits.

8.2.4.2 Summarizing the Ancillary Benefit Estimates

8.2.4.2.1 Presentation of the Studies

Figure 8.9 summarizes the ancillary benefits per tonne of carbon from 15 studies, along with available confidence intervals around the mid estimate. Multiple entries for a study on the Figure result from modelling of multiple policy scenarios. Most of the studies focus solely on public health impacts.

Table 8.5: Scenarios and results of studies on ancillary benefit reviewed

Study	Area and sectors	Scenarios (1996 US$)	Average ancillary benefits (US$/tC; 1996 US$)	Key pollutants	Major endpoints
Dessus and O'Connor, 1999	Chile (benefits in Santiago only)	Tax of US$67 (10% carbon reduction) Tax of US$157 (20% carbon reduction) Tax of US$284 (30% carbon reduction)	251 254 267	Seven air pollutants	Health—morbidity and mortality, IQ (from lead reduction)
Cifuentes *et al.*, 2000	Santiago, Chile	Energy efficiency	62	SO_2, NO_x, CO, NMHC Indirect estimations for PM_{10} and resuspended dust	Health
Garbaccio *et al.*, 2000	China – 29 sectors (4 energy)	Tax of US$1/tC Tax of US$2/tC	52 52	PM_{10}, SO_2	Health
Wang and Smith, 1999	China – power and household sectors	Supply-side energy efficiency improvement Least-cost per unit global-warming-reduction fuel substitution Least-cost per unit human-air-pollution-exposure-reduction fuel substitution		PM, SO_2	Health
Aunan *et al.* 2000, Kanudia and Loulou , 1998a	Hungary	Energy Conservation Programme	508	TSP, SO_2, NO_x, CO, VOC, CO_2, CH_4, N_2O	Health effects; materials damage; vegetation damage
Brendemoen and Vennemo, 1994	Norway	Tax US$840/tC	246	SO_2, NO_x, CO, VOC, CO_2, CH_4, N_2O, Particulates	Indirect: health costs; lost recreational value from lakes and forests; ; corrosion Direct: traffic noise, road maintenance, congestion, accidents
Barker and Rosendahl, 2000	Western Europe (19 regions)	Tax US$161/tC	153	SO_2, NO_x, PM_{10}	Human and animal health and welfare, materials, buildings and other physical capital, vegetation

(continued)

Table 8.5: continued

Study	Area and sectors	Scenarios (1996 US$)	Average ancillary benefits (US$/tC; 1996 US$)	Key pollutants	Major endpoints
Scheraga and Leary, 1993	USA	US$144/tC	41	TSP, PM_{10}, SO_x, NO_x, CO, VOC, CO_2, Pb	Health – morbidity and mortality
Boyd *et al.*, 1995	USA	US$9/tC	40	Pb, PM, SO_x, SO_4, O_3	Health, visibility
Abt Associates and Pechan-Avanti Group, 1999	USA	Tax US$30/tC Tax US$67/tC	8 68	Criteria pollutants	Health – mortality and illness; Visibility and household soiling (materials damage)
Burtraw *et al.*, 1999	USA	Tax US$10/tC Tax US$25/tC Tax US$50/tC	3 2 2	SO_2, NO_x	Health

NMHC, non-methane hydrocarbons; PM, particulate matter; PM_{10}, particulate matter less than 10 microns; TSP, total suspended particulate; VOC, volatile organic compounds; IQ, intelligence quotient

Table 8.6: Modelling choices of studies on valuation of ancillary benefits reviewed[21]

Study	Baseline (as of 2010)	Economic modelling	Air pollution modelling	Valuation	Uncertainty treatment
Dessus and O'Connor, 1999	4.5%/yr economic growth; AEEI: 1% Energy consumption: 3.6% PM: 1% Pb: 4.1% CO: 4.8%	Dynamic CGE	Assumed proportionality between emissions and ambient concentrations	Benefits transfer used: PPP of 80% US VSL: $2.1 mill. VCB: $0.2 mill. IQ loss: $2500/point	Sensitivity tests on WTP and energy substitution elasticities
Cifuentes *et al.*, 2000	For AP control, considers implementation of Santiago Decontamination Plan (1998 to 2011)	No economic modelling Only measures with private, non-positive costs considered	Two models for changes in $PM_{2.5}$ concentrations: (1) Box model, which relates SO_2 and CO_2 to $PM_{2.5}$ (2) Simple model assumes proportionality between $PM_{2.5}$ concentrations apportioned to dust, SO_2, NO_x, and primary PM emissions. Models derived with Santiago-specific data and applied to nation	Benefits transfer from US values, using ratio of income/capita Uses original value for mortality decreased by 1 standard deviation VSL = US$407,000 in 2000	Parameter uncertainty through Monte Carlo simulation. Reports centre value and 95% CI
Garbaccio *et al.*, 2000	1995 to 2040 5.9% annual GDP growth rate; carbon doubles in 15 years; PM grows at a bit more than 1%/yr	Dynamic CGE model; 29 sectors; Trend to US energy/ consumption patterns; Labour perfectly mobile; Reduce other taxes; Two-tier economy explicit.	Emissions/concentration coefficients from Lvovsky and Hughs (1998); three stack heights	Valuation coefficients from Lvovsky and Hughs (1998); VSL: US$3.6 million (1995) to RMB 82,700 Yuan (RMB 8.3 yuan = $1) in 2010 (income elasticity = 1). 5%/yr increase in VCB to US$72,000	Sensitivity analysis
Wang and Smith, 1999		No economic modelling	Gaussian plume	Benefit transfer using PPP. VSL = US$123,700, 1/24 of US value	

[21] AEEI, Autonomous Energy Efficiency Improvement; $PM_{10, 2.5}$, particulate matter less than 10 or 2.5 microns, respectively; CGE, Computable General Equilibrium Model; PPP, Purchasing Power Parity; W TP, Willingness To Pay; AP: air pollution; CAA: Clean Air Act; NAAQS: National Ambient Air Quality Standards; SIP: State Implementation Plan; CRF: conenrtration-response function; CL: confidence level; VSLY and VSL: Value of Statistical Life Year, Value of Statistical Life; RIA: Regulatory Impact Analysis; VCB, value of a case of bronchitis.

(continued)

Table 8.6: continued

Study	Baseline (as of 2010)	Economic modelling	Air pollution modelling	Valuation	Uncertainty treatment
Aunan *et al.*, 2000	Assumes status quo emissions scenario	Two analyses: bottom-up approach and macroeconomic modelling	Assumes proportionality between emissions and concentrations	Benefit transfer of US and European values using "relative income" = wage ratios of 0.16	Explicit consideration through Monte Carlo simulation Reports centre value and low, high
Brendemoen and Vennemo, 1994	2025 rather than 2010 2%/yr economic growth 1% increase in energy prices 1%–1.5% increase in electricity and fuel demand CO_2 grows 1.2% until 2000, and 2% thereafter.	Dynamic CGE		Health costs of studies reviewed based on expert panel recommendations Contingent valuation used for recreational values	Assume independent and uniform distributions
Barker and Rosendahl, 2000	SO_2, NO_x, PM expected to fall by about 71%, 46%, 11% from 1994 to 2010	E3ME Econometric Model for Europe		US\$/emissions coefficients by country from EXTERNE: €1,500/t NO_x for ozone (€1= \$1); NO_x and SO_2 coefficients are about equivalent, ranging from about €2,000/t to €16,000/t; PM effects are larger (2,000–25,000) Uses VSLY rather than VSL: €100,000 (1990)	
Scheraga and Leary, 1993	1990 to 2010 7% growth rate carbon emissions Range for criteria Pollutants 1%–7%/yr	Dynamic CGE		US\$/emissions coefficients	
Boyd *et al.*, 1995	Static CGE				

(continued)

Table 8.6: continued

Study	Baseline (as of 2010)	Economic modelling	Air pollution modelling	Valuation	Uncertainty treatment
Abt Associates and Pechan-Avanti Group, 1999	2010 baseline scenarios – 2010 CAA baseline emission database for all sectors, plus at least partial attainment of the new NAAQS assumed. Benefits include coming closer to attainment of these standards for areas that would not reach them otherwise. Includes NO_x SIP call	Static CGE		From Criteria Air Pollutant Modelling System (used in USEPA Regulatory Impact Analysis and elsewhere)	SO_2 sensitivity – SO_2 emissions may not go beyond Title IV requirements – NO_x sensitivity – NO_x SIP call reductions not included in final SIP call rule
Burtraw *et al.*, 1999	Incorporates SO_2 trading and NO_x SIP call in baseline	Dynamic regionally specific electricity sector simulation model with transmission constraints. The model calculates market equilibrium by season and time of day for three customer classes at the regional level, with power trading between regions.	NO_x and SO_2. Account for conversion of NO_x to nitrate particulates	Tracking and Analysis Framework: the numbers used to value these effects are similar to those used in recent Regulatory Impact Analysis by the USEPA.	Monte Carlo simulation for CRF and valuation stages.

Box. 8.1. Global Public Health Effects of Greenhouse Gas Mitigation Policies

It is useful to estimate ancillary benefits through quantitative indicators, even if they are not monetized (Pearce, 2000). One such global scale effort was produced by the WHO/WRI/EPA Working Group on Public Health and Fossil Fuel Combustion on the range of avoidable deaths that could arise between 2000 and 2020 under current policies, and under the scenario proposed by the EU in 1995. This EU Scenario assumed that by 2010 GHG emissions would be 15% below 1990 levels for Annex I countries, and 10% below projected emissions for 2010 for non-Annex I countries (Davis, 1997; Working Group on Public Health and Fossil Fuel Combustion, 1997). The total change in carbon emissions was estimated globally, based on a source–receptor matrix for four specific sectors (industry, transport, household, and energy) that was adjusted for local temperature and humidity. Applied to nine regions and adjusted for temperature and humidity, this matrix yielded changes in projected fuel types and formed the basis for calculating total emissions of particulates. Mortality tied with particulates was calculated based on best estimates (Borja-Aburto *et al.*, 1997, 1998; Pereira *et al.*, 1998; Gold *et al.*, 1999; Braga *et al.*, 1999; Linn *et al.*, 2000).

The report included a sensitivity analysis of the range of deaths, predicting that by 2020, 700,000 avoidable deaths (90% Confidence Interval, 385,000–1,034,000) will occur annually as a result of additional particulate matter (PM) exposure under the baseline forecasts when compared with the climate policy scenario. From 2000 to 2020, the cumulative impact on public health related to the difference in PM exposure could reach 8 million deaths globally (90% CI, 4.4–11.9 million). In the USA alone, the number of annual deaths from PM exposure in 2020 (without control policy) would equal in magnitude deaths associated with human immunodeficiency diseases or all liver diseases in 1995. "The mortality estimates are indicative of the magnitude of the potential health benefits of the climate-policy scenario examined and are not precise predictions of avoidable deaths. While characterized by considerable uncertainty, the short-term public-health impacts of reduced PM exposure associated with greenhouse-gas reductions are likely to be substantial even under the most conservative set of assumptions."

The framework for this assessment is described in more detail in Abt Associates (1997); Pechan and Associates (1997).

From *Figure 8.9*, it can be observed that:
- midpoint estimates are mostly less than US$100/tC, but range from less than US$2 up to almost US$500/tC;
- US estimates are the lowest while estimates from one study for Chile and several for Norway are the highest (the latter includes a broader range of benefits);
- significant divergence in estimates occurs across studies for the same country; and
- uncertainty bounds are quite large for most of the studies that report them.

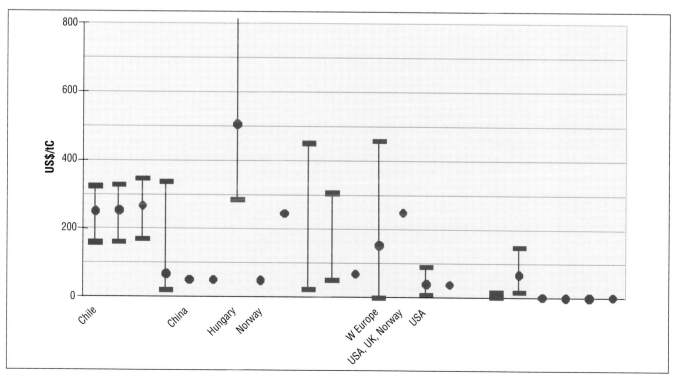

Figure 8.9: *Summary of ancillary benefits estimates in 1996 US$/tC.*

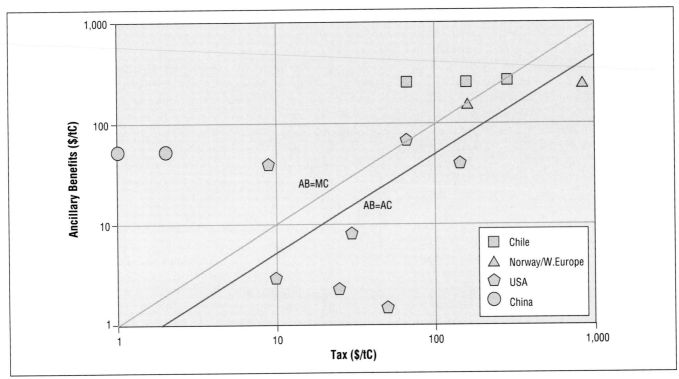

Figure 8.10: *Ancillary benefits in 1996 US$/tC versus levels of carbon tax.*

Figure 8.10 provides ancillary benefits per tonne estimates related to the size of the carbon tax (in 1996 US$/tC). Points on the diagonal line AB = MC indicate that marginal private mitigation costs (MC) equate to the tax. Some points fall on this line; more appear above it than below, with the Norwegian/Western Europe and the US studies split. If the damage (benefit) function is linear, then average benefits equate marginal benefits. Thus, points on the diagonal imply that the carbon tax is "quasi-optimal" (Dessus and O'Connor, 1999), in that the tax is optimal without considering either the direct climate mitigation benefits or any social costs over private costs (such as deadweight losses from the tax interaction effect). Alternatively, it can be assumed that the private mitigation cost function is quadratic (Total Cost=b(X^2)), where X is carbon reduction. In this case, the tax rate equals marginal private mitigation cost and average private mitigation cost is half marginal private mitigation cost. The heavy diagonal line equates ancillary benefits to average private mitigation cost. Points above this line imply there are net benefits to carbon policy, with the same important caveats as above. More points appear above the corresponding line (AB=AC) on the graph than above the AB=MC line.

In the general case, a larger carbon tax should lead to progressively smaller carbon reductions (if the marginal abatement cost curve is upward sloping). For all but one study (Abt Associates and Pechan-Avanti Group, 1999), the ratio of ancillary benefits to the tax rate does fall. As for the change in ancillary benefits per tonne of carbon, Burtraw *et al.* (1999) show this ratio falling dramatically in percentage terms with higher carbon taxes. In contrast, Dessus and O'Connor (1999) show it

rising slightly, and in the Abt study the ratio of benefits to the tax rate rises dramatically (Abt Associates and Pechan-Avanti Group, 1999). This last result reflects that this analysis treated the SO_2 cap as non-binding considerably below the higher tax-rate modelled. In addition, this study treated the National Ambient Air Quality Standards as a cap, with abatement below these "caps" treated as benefits, but reductions above these caps treated as saving abatement costs.

8.2.4.2.2   Evaluation of the Studies

Almost all the studies analyze the effects of a GHG reduction policy through a tax on carbon. The ranges of the tax are from modest levels (RMB Y9/tC[7]) in 2010 for Garbaccio *et al.* (2000), US$10/tC for Burtraw *et al.* (1999); up to high levels (US$254/tC for Dessus and O'Connor (1999), and US$840/tC for Brendemoen and Vennemo (1994). The US studies employ relatively modest taxes, between US$10/tC and US$67/tC. Only two studies consider alternative programmes: Aunan *et al.* (2000) considers a National Efficiency Programme, and Cifuentes *et al.* (2000) considers energy efficiency improvements. The level of abatement considered by these two studies is relatively modest, however.

Baseline
An analysis of ancillary benefits requires a time line and a clear definition of the key constituents of the baseline against which the prospective scenario can be measured, including the eco-

---

[7] The exchange rate is US$1 vs Renminbi yuan (RMB Y) 8.3.

nomic, demographic, regulatory[8], environmental[9], and technological conditions, and their implications for emissions or other inputs to an ancillary benefit calculation.

The importance of the baseline is evident in a review of previous studies for the USA in Burtraw *et al.* (1999). Assessments varied in their estimates of ancillary benefits, chiefly because they employed different assumptions regarding the regulatory baselines, that is the 1990 US Clean Air Act Amendments and, especially, the tradable permit programme for SO_2. Among these baseline parameters, the most critical are the spatial location of emissions relevant to potentially exposed populations, regulatory conditions, and available technologies (Morgenstern, 2000). The importance of the location of emission reductions and exposed populations means that highly disaggregated models are the preferred tools of analysis. This may conflict with other goals for the analysis of GHG mitigation strategies. For example, large CGE models, which are used for cost estimation, operate at a different scale than the more localized models relevant to estimating ancillary benefits.

Economic Modelling

Most of the studies in *Table 8.5* use static or dynamic CGE models (one uses an econometric model) that provide T-D and sectorally aggregate estimates of ancillary benefits and/or costs. The Burtraw *et al.* (1999) model stands out for the location specificity of its economic model (although only for the electricity sector), which permits more credible modelling of population exposure reductions than that from spatially aggregate models. Another specific feature is its detailed representation of investment choices and their dependence on other factors covered in the model. Finally, several studies do not use an economic model. Instead, they follow a B-U approach, positing some increase in energy efficiency or reduction in carbon and estimating the ancillary benefits that would result, at a reasonably detailed spatial level. Such studies suffer from not accounting for behavioural adjustments, such as energy substitutions, which could alter their estimates of ancillary benefits considerably. The high ratio of ancillary benefits to the carbon tax for Garbaccio *et al.* (1999) appears to arise from very optimistic assumptions about energy substitution elasticities.

Emissions and Environmental Media Modelling

All the studies in *Table 8.5* account for the most important pollutant affecting public health – particulates. Most, however, do not consider secondary particulate formation from SO_2 and NO_x, or do so in a very simplistic manner. In a developing country, direct particulate emissions are likely to be a large fraction of particulate mass, making the lack of attention to secondary products less important. In developed countries, however, secondary products are likely to be far more important than primary particulates. Omitting these products could bias ancillary benefit estimates downwards; using proportionality assumptions or other simple approaches raises uncertainties and may carry biases. Only one study considered lead emissions (Dessus and O'Connoer, 1999); few address ozone.

The Abt study (Abt Associates and Pechan-Avanti Group, 1999) is the most comprehensive in its modelling of secondary particulate formation and dispersion. It found that 12 urban areas in the USA would come into compliance with the recently promulgated standard for particulate matter less than 2.5 microns ($PM_{2.5}$)[10] for a carbon tax of US\$67 (US\$1996). Without this tax, these areas would not be able to meet the new standard. With there being at best sparse information on the actual $PM_{2.5}$ concentrations in US urban areas, these estimates should be viewed as highly speculative.

Health Effects Modelling

Three recent studies (Hagler-Bailly 1995; Lee *et al.*, 1995; European Commission, 1999) developed methods that set the stage for much of the recent estimates of ancillary benefits. However, studies that draw on this literature, but reduce its information to coefficients that link emissions directly to health effects (or values) ignore spatial and demographic heterogeneity. This is particularly so when such coefficients are generated for one country or region and then directly applied to another, without taking into account local conditions. In the absence of country-specific information, transfer of risk information may be made between countries, with appropriate caveats to take into account underlying differences in health status, access to care, and other important factors (see *Box 8.2*).

Most of the studies rely on concentration–response functions from the literature on health, and apply them using a standard methodology (Ostro, 1996; EPA, 1999). The most important health effects are premature mortality and chronic respiratory effects.

Aside from differences in the base rates of the effects[11], due to local characteristics such as the age distribution of the population and health care services, other factors help explain the different outcomes of the studies. First, some use PM_{10}, while

---

[8] For example, if they are implemented, the recent proposed tightening of the US standards for ozone and particulates and associated improvements over time imply that benefits from reductions in the criteria air pollutants that result from climate policies will be smaller in the future than if carried out now.

[9] Some environmental effects exhibit thresholds or non-linearities that imply benefits do not move directly with reductions in local and regional pollutants. Acidification is an interesting example because damage may result only after critical load thresholds are violated. On the other hand, recovery may not occur with a reduction in conventional pollutants until some new threshold is achieved or after a significant time lag.

[10] The new US $PM_{2.5}$ standard and the tighter ozone standard have been remanded to the EPA by the D.C. Court of Appeals and aspects of the case are currently being heard by the US Supreme Court. Thus, these standards are not yet in effect (November 2000).

Box 8.2. The Impact of Air Pollution on Health Differs by Country

For any society, deaths at earlier ages result in more productive years of life lost than for those that occur at later ages. One study in Delhi, India, found that children under 5 and adults over 65 years of age are not at risk from air pollution, because other causes of death (notably infectious diseases) predominated in those who survive to reach these age groups (Cropper *et al.*, 1997). However, people between 15 and 45 years of age are at increased risk of death from air pollution relative to those in developed countries. Since the population distribution in India includes many more people in these middle age groups, the net impact on the country from air pollution measured in terms of years of life lost is similar to that of a developed country.

others use fine particles ($PM_{2.5}$), or serveral components of them (sulphates and nitrates). When the individual components of $PM_{2.5}$ are used, the implicit assumption is that their risk is similar to that of $PM_{2.5}$. To date, this has not been verified (especially for nitrates, the secondary particulate product from NO_x emissions). Second, studies that look at age groups separately generally report higher impacts (Aunan *et al.* (2000), for example, used a steeper dose–response coefficient for people older than 65 years of age than that used by other studies). Very few consider the chronic effects on mortality, derived from cohort studies (*e.g.*, Pope *et al.*, 1995) (Abt Associates and Pechan-Avanti Group, 1999 is one, while others consider it for their "high estimate" only). Use of the latter results in estimates of death three times larger than use of the time series studies. Also, few studies consider effects on child mortality. Finally, different studies consider different health endpoints, which is important for reconciling morbidity estimates.

Valuation of Effects

The most important monetary benefit is related to mortality risk reductions, which can be expressed in terms of the VSL (see Chapter 7). The VSL should ideally be indigenously estimated (Krupnick *et al.*, 2000)[12] but almost of the studies build on a consensus on the appropriate values to use (Davis *et al.*, 2000), given the state of research on valuation (albeit concentrated in the UK and USA).

A major difference in the treatment of values across the studies is whether these values are adjusted for different income levels and increased for future income growth. Adjustments that assume an income elasticity of willingness to pay (WTP) of 1.0 are inconsistent with the admittedly thin literature. A number of studies found elasticities in the 0.2-0.6 range based on income differentials *within* a country. Such elasticities, when applied to transfers *among* countries, yield quite high values. Most of the developing country ancillary benefit studies reported in *Table 8.6* use an income elasticity of 1.0. The US Science Advisory Board has endorsed the idea of making adjustments for future income growth within a country.

The state of the art of the valuation of air pollution-related mortality effects is currently in ferment, with serious questions being raised about the inappropriateness of basing such valuation on labour market studies. Ad hoc adjustments for the shorter life span of those thought to be most affected by air pollution (the elderly and ill) have been made but more credible estimates of willingness to pay await new research. Such efforts are more likely to lower such estimates relative to current estimates than raise them (see Davis *et al.*, 2000 and Krupnick *et al.*, 2000).

Environmental Externalities

All the studies, except those in the USA, assume that improvements in public health count as externalities and, hence, as ancillary benefits. As noted in Krupnick *et al.* (2000), this assumption may not always hold. Burtraw *et al.* (1999) and Abt Associates and Pechan-Avanti Group (1999) count the abatement cost savings from reducing SO_2 emissions in response to a carbon tax because SO_2 emissions are capped in the USA. Similar adjustments are not made for SO_2 and other pollutant taxation in Europe. Moreover, not all ancillary benefits are necessarily externalities. In some cases, these effects may be already "internalized" in the price of goods and services: for example, where accident insurance against road fatalities exists, much of this effect is already accounted for through purchasing insurance and the penalties for failure to obtain it.

Treatment of Uncertainty

The uncertainty that surrounds the estimates of benefits is no less than that associated with mitigation costs, extending from physical modelling, through valuation, to modelling choices. Several of the studies use Monte Carlo simulation, but others use less sophisticated sensitivity analyses to characterize uncertainties.

Allowance for Ancillary Costs

None of the studies reviewed in this assessment reported estimates of ancillary costs. Some studies, such as Burtraw *et al.*

---

[11] Most of the concentration-response functions for health effects of air pollution are based on relative risks models, which give the percentage increase in the number of health effects due to a change in air pollution concentration. This percentage change needs to be applied to the base rate of the effects (i.e. the number of effects observed without change in air pollution). For example, for the non-accidental mortality in the USA, this base rate is about 800/100,000.

[12] Where there is a lack of local information on willingness to pay, one option is to use studies from developed countries and "adjust" the estimates for local conditions. This procedure is called benefit transfer: "an application of monetary values from a particular valuation study to an alternative or secondary policy-decision setting, often in another geographic area than the one where the original study was performed" (Navrud, 1994). The problems of such transfers are discussed in greater detail in Davis *et al.* (2000).

(1999), discuss the bounce-back effect associated with energy substitution to natural gas and other less carbon-intensive fuels. However, even these studies, not surprisingly, estimate positive net ancillary benefits from GHG mitigation policies. The issue is whether the models were designed to capture ancillary costs. In general, our conclusion is no, except for fossil fuel substitution in the power and transport sectors. From an energy substitution perspective, substitution to nuclear power or hydropower does not generate reported ancillary costs because these ancillary effects are not present in the studies. Other sources of ancillary costs were also left out of the modelling exercise, either because of model boundaries or through making some standard modelling choices. All the studies examined effects on one country or region, and therefore do not consider the leakage effect. None of the studies considered health linkages that might result from slower income and employment growth following the implementation of a GHG mitigation policy.

8.2.4.3   Why Do Studies for the Same Country Differ?

It is enlightening to consider why estimates of ancillary benefits (or costs) for two different studies of the same country differ.

In the case of Chile, Dessus and O'Connor (1999) estimate benefits of about US$250/tC, as compared to US$62/tC in Cifuentes *et al.* (2000). Half of the Dessus and O'Connor (1999) benefits are attributable to effects on intelligence quotient (IQ) associated with reduced lead exposure, an endpoint not considered by Cifuentes *et al.* (2000) and by most studies. The large lead–IQ effect seems to be at variance with US and European studies that consider this and more conventional endpoints. Also, the VSL used by Dessus and O'Connor (1999) is more than twice as large as that used by Cifuentes *et al.* (2000; US$2.1 million versus US$0.78 million by the year 2020). These choices were driven by alternative benefit transfer approaches: Dessus and O'Connor (1999) used 1992 purchasing power parity to transfer a mid estimate of US VSL, while Cifuentes *et al.* (2000) used 1995 per capita income differences and the exchange rate to transfer a lower bound US VSL. This comparison illustrates that the choice of benefit transfer approach in estimating ancillary benefits dominates by far the modelling choices (Dessus and O'Connor (1999) used a T-D model while Cifuentes *et al.* (2000) used a B-U approach).

For the USA, Burtraw *et al.* (1999) found that for a US$25 carbon tax, the ancillary benefits per tonne are US$2.30, while Abt Associates and Pechan-Avanti Group (1999) found that for a slightly larger tax (US$30), the ancillary benefits per tonne are US$8. For a US$50/tC tax, Burtraw *et al.* (1999) found ancillary benefits of only US$1.50/tC, while for an even larger tax (US$67), Abt Associates and Pechan-Avanti Group (1999) estimated the ancillary benefits to be US$68/tC. These differences are explained by:

- The effect of a unit change in particulate nitrates (derived from NO_x emissions) on the mortality rate

which in Burtraw *et al.* (1999) are about one-third of those used by Abt Associates and Pechan-Avanti Group (1999).

- The value of statistical life used to value mortality risk reductions (about 35% lower in Burtraw *et al.* (1999) who adjust the VSL for the effects of pollution on older people rather than on those of average age).
- Sectors included (Burtraw *et al.*, 1999) are restricted to the electricity sector by 2010, and NO_x emissions per unit carbon are projected to be lower for this sector than in the general US economy.
- Effect of carbon tax on SO_2 emissions (Abt Associates and Pechan-Avanti Group, 1999) finds that the US$67 carbon tax is large enough to bring SO_2 emissions significantly under an SO_2 cap 60% lower than the current cap. It also cuts NO_x emissions enough to bring significant numbers of non-attainment areas under the national ambient standards. Burtraw *et al.* (1999) does not find such a large effect.
- Baseline emissions (Burtraw *et al.*, 1999) do not account for new, tighter ozone and PM standards, but Abt Associates and Pechan-Avanti Group (1999) do (while assuming only partial attainment of the standards). This baseline assumption leaves lower emissions of conventional pollutants to be controlled in the Abt Associates and Pechan-Avanti Group (1999) study than in the Burtraw *et al.* (1999) study.

8.2.4.4   Conclusions

The diffusion of methods and key studies to estimate health effects and their monetization has contributed to a reasonable degree of standardization in the literature. However, some of the differences in estimates result from different assumptions and/or methodologies used to estimate them:

- Selection of concentration–response functions, such as use of time series rather than the cohort mortality studies.
- Consideration of more and/or different endpoints, such as considering the lead effects on IQ.
- Use of different assumptions to perform benefit transfers across countries and across time. For example, considering per capita income as opposed to purchasing power parity to perform the unit value transfer; choice of the income elasticity value.
- Defining the baseline differently: most of the literature on ancillary benefits systematically treats only government regulations with respect to environmental policies. In contrast, other regulatory policy baseline issues, such as those relating to energy, transportation, and health, are generally ignored, as have baseline issues that are associated to technology, demography, and the natural resource base.

Therefore, although the standard methodology is generally accepted and applied, a number of assumptions or judgements can lead to estimates of ancillary benefits in terms of US$/tC for a given country that differ by more than an order of magnitude. The least standardized, least transparent and most uncertain component for modelling ancillary benefits is the link from emissions to atmospheric concentrations, particularly in light of the importance of secondary particulates to public health.

Also, the above review reveals implicitly the lack of studies estimating non-health effects from GHG mitigation policies (damages from traffic crashes, the effects of air pollution on materials, and air pollution effects on crops losses, which have been shown to be quite high in some regions). Depending upon the GHG mitigation policies selected, some of this damage could well be reduced, but the nature of this relationship remains a speculative matter. More information can be found in sectoral studies reviewed in Chapter 9, but no comprehensive evaluation can be derived from them.

For all these reasons, it remains very challenging to arrive at quantitative estimates of the ancillary benefits of GHG mitigation policies. Despite the difficulties, it can be said that the ancillary benefits related to public health accrue over the short term, and under some circumstances can be a significant fraction of private (direct) mitigation costs. With respect to this category of impacts alone mortality tends to dominate. The exact magnitude, scale, and scope of these ancillary benefits varies with local geographical and baseline conditions; if the baseline scenario assumes a rapid decrease in non-GHG pollutant emissions, benefits may be low, especially in low density areas. Net ancillary costs (i.e., where the ancillary benefits are less than ancillary costs) may occur under certain conditions, but the models reviewed here are generally not designed to capture these effects. While most of the studies assessed above address ancillary benefits of explicit climate mitigation measures, it should be noted that in many cases, these ancillary benefits can be expected to be as least as important as climate mitigation for decision making. Hence, the terms co-benefits is also used in this report. Therefore, there is a strong need for more research in the area of integrated policies addressing climate mitigation alongside other environmental, social or economic objectives.

8.3   Interface between Domestic Policies and International Regimes

For every country, the costs of achieving a given level of abatement will be dramatically affected by the interface between its domestic policy and international regimes. Since a co-ordination on the basis of simple reporting mechanisms has not be adopted from the outset because it would not have been stringent enough for UNFCCC objectives, some studies were devoted to clarifying the differences between the two main tools for co-ordinating climate policies: country emissions quotas or agreed carbon taxes.

Theoretically, both solutions are equivalent in a world with complete information (the optimal quota leads to the same marginal abatement cost as the optimal level). However, Pizer (1997), building on a seminal work by Weitzman (1974), demonstrated that this is not the case if uncertainties about climate damages and GHG abatement costs are considered. Indeed, welfare losses due to an error of anticipation are not the same in these two approaches, depending upon whether the steepness of marginal abatement cost curve is higher or lower than the steepness of the damage curve. If the marginal abatement cost curve is steeper, then it is preferable to agree on a pre-determined level of taxation because if this level is either too low or too high, the resulting welfare losses trough climate impacts will not be dramatic. This is the case in most modelling efforts as long as there is no large probability of dramatic non-linearity in climate systems over the middle term. This policy conclusion can be reverted if one considers a high level of risk-aversion to catastrophic events (which makes the damage curve steeper), or a large proportion of "no regret" policies (which make the mitigation cost curve flatter). The main message, however, is that in a tax harmonization approach, the costs of complying with commitments on climate policies are known in advance (but the outcome is not predictable), while in a quota approach the outcome is observable but there is an uncertainty about the resultant costs. In this respect, emissions trading is logically a companion tool to a system of emissions quotas, to hedge against the distributional implications of surprises regarding abatement costs and emissions baselines.

After the Berlin Mandate (1995), a quota co-ordination approach was implicitly adopted and the focus of analysis was placed on linkages between emissions trading regimes and national policies. Contrary to the preceding period, very few works were devoted to the case of co-ordinated carbon taxes. Hourcade *et al.*, (2000a) confirmed that, because of the existing uneven distribution of income, discrepancies in pre-existing taxation levels, and differences in national energy and carbon intensities, a uniform carbon tax would result in very differentiated losses in welfare across countries, unless appropriate compensation transfers operated. However, a differentiated taxation does not minimize total abatement expenditures (rich countries would have to tap more expensive abatement potentials) and creates distortions in international competition. The suggested solution, a uniform tax for carbon-intensive industry exposed to international competition and a differentiated taxation for households, has to be at least adapted to the Kyoto framework which does not preclude the use of carbon taxes but changes the condition of their applicability. However, the underlying issue of how to minimize abatement expenditures while guaranteeing a fair distribution of welfare costs still remains.

Under the Kyoto framework, the interface between domestic policies and the international regime passes through three main channels: the impact of international emissions permit trading (under Article 17), international trading in project-related credits (under Articles 6 and 12 (Read, 1999)) on abatement costs, and spillover effects across economies through commercial and capital flows.

Table 8.7: *Energy Modelling Forum main results; marginal abatement costs (in 1990 US$/tC; 2010 Kyoto target)*

Model	No trading				Annex I trading	Global trading
	USA	OECD-E	Japan	CANZ		
ABARE-GTEM	322	665	645	425	106	23
AIM	153	198	234	147	65	38
CETA	168				46	26
Fund					14	10
G-Cubed	76	227	97	157	53	20
GRAPE		204	304		70	44
MERGE3	264	218	500	250	135	86
MIT-EPPA	193	276	501	247	76	
MS-MRT	236	179	402	213	77	27
Oxford	410	966	1074		224	123
RICE	132	159	251	145	62	18
SGM	188	407	357	201	84	22
WorldScan	85	20	122	46	20	5
Administration	154				43	18
EIA	251				110	57
POLES	135.8	135.3	194.6	131.4	52.9	18.4

Source: cited in Weyant, 1999; Council of Economic Advisors, 1998; EIA (Energy Information Administration), 1998; Criqui *et al.*, 1999.

8.3.1   *International Emissions Quota Trading Regimes*

8.3.1.1   *"Where Flexibility"*

Table 8.7 synthesizes marginal abatement costs for the USA, Japan, OECD-Europe, and the rest of the OECD (CANZ) calculated by 13 world T-D models co-ordinated by the Energy Modeling Forum. It also includes the results obtained with the POLES model, which provides a multiregional partial equilibrium analysis of the energy sector, and two other studies of the economic impacts of Kyoto conducted by the US Government, the Administration's Economic Analysis (Council of Economic Advisors, 1998), and a study by the Energy Information Administration (1998). These results cannot be directly compared with those of the B-U analysis reported in Section 8.2.1.1, because

they incorporate feedback on energy demand, oil prices, and macroeconomic equilibrium. They give, however, an idea of the assumptions on technical abatement potentials retained for each region in these exercises, the main difference with B-U analysis being that these exercises do not explicitly consider negative cost potentials (they are implicit in most optimistic baselines).

Despite the wide discrepancies in results across models, the robust information is that, in most models, marginal abatement costs appear to be higher in Japan than in the OECD-Europe. CANZ and the USA have comparable results, approximately two-thirds the European one, and much lower than in Japan.

This means that Kyoto targets are likely to be unequitable. This risk is confirmed by uncertainty analyses based on existing

Table 8.8: *Energy Modelling Forum main results; GDP loss in 2010 (in % of GDP; 2010 Kyoto target)*

Model	No trading				Annex I trading				Global trading			
	USA	OECD-E	Japan	CANZ	USA	OECD-E	Japan	CANZ	USA	OECD-E	Japan	CANZ
ABARE-GTEM	1.96	0.94	0.72	1.96	0.47	0.13	0.05	0.23	0.09	0.03	0.01	0.04
AIM	0.45	0.31	0.25	0.59	0.31	0.17	0.13	0.36	0.20	0.08	0.01	0.35
CETA	1.93				0.67				0.43			
G-CUBED	0.42	1.50	0.57	1.83	0.24	0.61	0.45	0.72	0.06	0.26	0.14	0.32
GRAPE		0.81	0.19			0.81	0.10			0.54	0.05	
MERGE3	1.06	0.99	0.80	2.02	0.51	0.47	0.19	1.14	0.20	0.20	0.01	0.67
MS-MRT	1.88	0.63	1.20	1.83	0.91	0.13	0.22	0.88	0.29	0.03	0.02	0.32
Oxford	1.78	2.08	1.88		1.03	0.73	0.52		0.66	0.47	0.33	
RICE	0.94	0.55	0.78	0.96	0.56	0.28	0.30	0.54	0.19	0.09	0.09	0.19

models which provide a pretty wide range of outcomes that can be interpreted as covering the uncertainties prevailing in the real world. This can be shown in the results of domestic cost of carbon: from US$85 to US$410 in the USA, US$20 to US$966 for the OECD-Europe, US$122 to US$1074 for Japan, US$46 to US$423 for CANZ. The variance remains significant if the extreme values:

- from US$76 to 236/tC for the USA if one excludes GTEM, Merge 3, and Oxford;
- from US$159 to US$276/tC for the OECD-Europe and from US$145 to US$250 for CANZ if one excludes Worldscan, GTEM, and Oxford; and
- a continuum from US$122 to US$645/tC for Japan if Oxford is excluded.

In terms of GDP losses, the ranking of impacts differs because of the various pre-existing structures of the economy and of the energy supply and demand in various countries and because these studies do not consider the domestic policies targeted to tackle these pre-existing conditions; the GDP losses are from 0.45% to 1.96% for the USA, from 0.31 to 2.08 for the EU, from 0.25 to 1.88 for Japan. This variation is reduced under emissions trading; 0.31 to 1.03 for the USA, 0.13 to 0.73 for the OECD-Europe, from 0.05 to 0. 52 for Japan.

This discrepancy in results reflects differences in judgements about parameters such as technical potentials, emissions baselines, how the revenues of permits are recycled, and how near-term shocks are represented. Another important source of uncertainty is the feedback of the carbon constraint on the demand for oil; a drop in oil prices requires indeed higher prices of carbon to meet a given target since the signals not conveyed by oil prices as to be passed through price of carbon which leads to a totally different incremental cost of the carbon constraint.

These uncertainties about mitigation costs are reflected in the net welfare losses. The preceding discussion in Section 8.2 demonstrated the many sources of a wedge between total abatement costs and welfare losses, including the double dividend from fiscal reforms and the very structures of the economy (share of carbon intensive activities) and of the energy system.

The wide range of cost assessments, far from resulting from purely modelling artefacts, help to capture the range of possible responses of real economies to emissions constraints and to appreciate the magnitude of uncertainties that governments have to face.[13] They demonstrate that without emissions trad-ing, the Kyoto targets lead to a misallocation of resources, a non-equitable burden-sharing (notwithstanding its mitigation through double-dividend domestic policies analyzed in Section 8.2.2.1) and distortions in international competition. Even in the most optimistic models regarding abatement costs such as Worldscan, trading offers the potential for countries with high domestic marginal abatement costs to purchase emissions permits in countries with low marginal abatement costs and hence a way of minimizing total abatement costs and of hedging against risks of a too high and unequitable burden.

The full global trading scenarios presented in *Tables 8.7* and *8.8* assume non-restricted trade within Annex I and ideal CDM implementation that can exploit all cost effective options in developing countries with unlimited trading. Beyond the fact that the price of carbon is drastically reduced, it is remarkable that the variance of results is far lower than in the no-trade scenarios (between US$15/tC and US$86/tC). Uncertainty about costs persists, but this lesser variance arises because uncertainty is higher on each regional cost curve than on the aggregation of the same regional cost curves, which is exploited in the case of full trading.

In the case of Annex I trading (without considering the CDM) the price of permits ranges from US$20 to US$224/tC instead of US$15 to US$86/tC in the full trade case, which represents a far greater variance. This is mainly from the amount of so called "hot air"[14] retained in simulations. Some countries in Eastern Europe and the former Soviet Union have had a decline in emissions in the 1990s, resulting from the economic dislocations associated with restructuring. As a result, their emissions during the first commitment period are projected to be lower than their negotiated target. If trading is allowed within Annex I, these excess emissions quota may be sold to countries in need of such credits. Hence, the assumption regarding the availability of "hot air" is important. This, of course, will be governed in part by the rate of economic recovery, but also by the role of energy efficiency improvements and fuel switching during the restructuring process.

The main lessons from the above studies using T-D approaches (namely that trade has a marked, beneficial effect on costs of meeting mitigation targets), are confirmed by a series of recent studies using B-U approaches. These provide a more detailed information on the potentials for CDM projects. The MARKAL, MARKAL-MACRO, and MESSAGE models have been adapted and expanded to facilitate such multicountry studies. In North America, Kanudia and Loulou (1998) report MARKAL results for a three-country Kyoto study (Canada, USA, India). The total cost of Kyoto for Canada and the USA amounts to some US$720 billion with no trade, versus US$670

---

[13] This is exemplified by two others studies of the economic impacts of Kyoto conducted by the US Government. It is remarkable that GDP losses span from virtually zero to 3.5% and are correlated with the level of marginal abatement cost. The EIA assessment is the highest because it accounts for near-term shocks, such as inflationary impacts of higher energy prices (requiring higher interest rates, which dampen the investment), and for a 5-year delay in the response of agents). The EIA estimates rise to 4.2% when the non-CO_2 gases and carbon sinks are excluded from the analysis.

[14] Hot air: a few countries, notably those with economies in transition, have assigned amount units that appear to be well in excess of their anticipated emissions (as a result of economic downturn). This excess is referred to as "hot air".

billion when North American emissions and electricity trading is unimpeded, and only US$340 billion when India is added to the permit trading. MARKAL studies in the Nordic states (see Larsson *et al.* (1998) for Denmark, Sweden, and Norway, and Unger and Alm (1999) for the same plus Finland) show the considerable value of trading electricity and GHG permits within the region when severe GHG reductions are sought. Another MARKAL study computes the net savings of trading GHG permits between Belgium, Switzerland, Germany, and the Netherlands (Bahn *et al.*, 1998) at about 15% of the total Kyoto cost without trading. Another study (Bahn *et al.*, 1999a) shows that Switzerland's Kyoto cost may be reduced drastically if it engages in CDM projects with Columbia, in which case the marginal cost of CO_2 drops to US$12/tC. This type of B-U analysis has also been extended to the computation of a global equilibrium between Switzerland, Sweden, and the Netherlands, using MARKAL-MACRO (Bahn *et al.*, 1999b), with the conclusion that GDP losses resulting from a Kyoto target are 0.2% to 0.3% smaller with trade than without. More ambitious current research aims at building worldwide B-U models based on MARKAL (Loulou and Kanudia, 1999a) or on MARKAL-MACRO (Kypreos, 1998).

8.3.1.2 Impacts of Caps on the Use of Trading

From the above results, it is seen that all OECD countries have an interest in making the market as large as possible. Some Parties to the UNFCCC, however, have suggested that the supplementarity conditions of Articles 6.1.d, 12 and 17 of the Kyoto Protocol be translated into quantitative limits placed on the extent that Annex I countries can satisfy their obligations through the purchase of emission quotas. The rationale for the supplementarity condition is that, if the price of permits were too low, this would discourage domestic action on structural variables (infrastructure, transportation) or on innovation apt to modify the emissions trends over the long run. These measures are very often liable to high transaction costs and governments may prefer to import additional emissions permits instead of adopting such measures. In other words, minimization of the costs of achieving Kyoto targets may not guarantee minimization of the costs of climate policies over the long run; this is the case when the inertia of technical systems is considered (Ha-Duong *et al.*, 1999) and when one accounts for the long term benefits of inducing technical change through abatements in the first period (Glueck and Schleicher, 1995).

Some works have studied the consequences of enforcing the supplementarity condition through quantitative limits: one of the EMF scenarios imposed a constraint on the extent to which a region could satisfy its obligations through the purchase of emission quota (the limit was one-third).

However, the models cannot deliver any response without an assumption about *ex ante* limits on carbon trading, resulting into a stable duopoly between Russia and Ukraine or into a monopsony (Ellerman and Sue Wing, 2000). In the first case, the price of carbon will be higher than in a non-restricted mar-

ket, and most of the additional burden will fall on countries in which the marginal cost curve is high because they have a lesser potential for cheap abatement. This is typically the case for Japan and most of the European countries (Hourcade *et al.*, 2000b). The other possibility is for the market power to be controlled by the carbon-importing countries; in this case, the risk is that all or most of the trading will be of "hot air" at a very low price. Which of these alternatives will be realized cannot be predicted but, in both cases, quantitative limits to trade lead to outcomes that contradict the very objective of the supplementarity condition. Criqui *et al.* (1999) assessed the order of magnitude at stake with the POLES model, and examined a scenario in which the carbon tax is US$60/tC with unrestricted trade. They found that the carbon prices under the concrete ceiling conditions proposed by the EU fall to zero (with no market left for the developing countries) if the market power is held by the buyers. Alternatively, the carbon prices increase up to US$150/tC if the market power is held by the sellers, this risk being increased in the case of caps on hot air trading which increases the monopolistic power of Russia and Ukraine. Böhringer (2000) assesses the economic implications of the EU cap proposal within competitive permit markets. He concludes that part of the efficiency gains from unrestricted permit trade could be used to pay for higher abatement targets of Annex-B countries which assure the same environmental effectiveness as compared to restricted permit trade but still leaves countries better off in welfare terms.

8.3.1.3 The Double Bubble

Here the case of the "double bubble" is examined, in which countries belonging to the EU have a collective target, making use of the flexibilty to shift emission quota within the group and the remaining Annex I countries trade among themselves to reach their individual targets.

Figure 8.11 shows the incremental value of carbon emission for the two groups and compares them with that of full Annex I trading. Notice that for the USA, the tax is lower in the case of the "double bubble" than with Annex I trading. The reason is that without the EU bidding for the Russian "hot air", the demand for emission quotas falls as does its price. The EU on the other hand is disadvantaged under such a scenario. With their access to low cost emission quotas limited, the incremental value rises.

8.3.2 Spillover Effects: Economic Effects of Measures in Countries on Other Countries

In a world in which economies are linked by international trade and capital flows, abatement by one economy induces spillover effects and has welfare impacts on other economies. It matters to understand the conditions under which both abating and non-abating economies will experience positive or negative impacts from the policy adopted in other groups of countries; it matters also to understand the results of these spillover

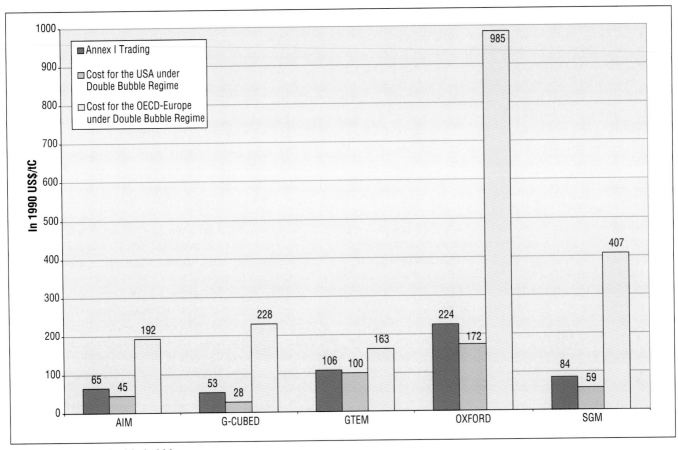

Figure 8.11: *The double bubble.*

effects in terms of carbon leakage. Chapter 7 provides the basic concepts of such an analysis and here some brief comments are added to explain the strengths and weaknesses on the results found by modelling exercises.

In static terms, without international capital mobility, the welfare costs of abatement for an open economy can be decomposed into two components (Dixit and Norman, 1984):

- costs that would be incurred if the economy were closed; and
- changes in the terms of trade, which are the first transmission mechanism for spillover effects (see Chapter 7).

If they require to go beyond "no regrets" potentials, binding emissions constraint comes to increasing the cost of carbon-intensive products and, if emissions arise from the production of its export goods, the abating economy benefits from better terms of trade. If, indeed, the importing economy cannot produce a perfect substitute easily, it will sell the same product at a higher price and increase its purchasing power of imported goods. The non-abating economy will symmetrically suffer a welfare loss because of more expensive imports, while the net result for the abating economy depends on the size of improvement in the terms of trade relative to the production costs of abatement. The welfare impacts are more important in the economies that are very dependent on foreign trade.

In the real world, an emission constraint simultaneously affects both export and import goods, but this does change the nature of the mechanism. Increased production of emission-intensive goods in non-Annex I regions is stimulated by both increased non-Annex I consumption and increased exports to Annex I regions. The net relative balance between these parameters is influenced by the extent to which Annex I emission constraints fall on export competing industries (when the country is specialized in such industries) as opposed to import-competing industries (when it imports carbon intensive goods). If a constraint predominantly affects export industries, it encourages increased non-Annex I production for internal consumption. If the constraint predominantly affects import-competing industries, increased non-Annex I production is mainly exported to Annex I regions. Emissions leakage is beneficial to non-Annex I economies only in the second case, since it is associated with an improvement in their terms of trade, whereas their terms of trade deteriorate in the first case.

Another factor that affects the increase of emission-intensive goods in non-Annex I regions is the effect of Annex I abatement on the intermediate demand for fossil fuels. As discussed above, Annex I abatement will reduce fossil fuel prices. Lower prices for fossil fuels will encourage the production of more emission-intensive goods and the use of more emission-intensive production techniques in non-Annex I regions.[15]

So far, it was assumed that changes in the production structures in both Annex I and Non-Annex I countries result only in changes in final demand and in price structures. The introduction of international capital mobility complicates the analysis since, in addition to production costs and changes in the terms of trade, carbon constraints alter the relative rates of return in the abating and non-abating countries. If capital flows from the first country to the second in response to these changes, there will be a further restriction of the production frontier (the set of possible productive combinations) for the abating economy and an outward shift for the other economy. Factor rewards in both countries are also affected. Part of the income from foreign investment accrues to the home economy and subtracts from income in the foreign economy; abating economies are affected by changes in income and factor prices that result from changes in international capital flows, with symmetric gains for non-abating economies.

No theoretical results for complex and empirically relevant cases can be obtained as to the extent that international capital mobility modifies the conclusions of the static analysis of the role of the trade effects. However, modelling results are seldom reported on the welfare impact of changes in international capital flows, although McKibbin *et al.* (1999) emphasize the macroeconomic repercussions. It is still, indeed, impossible to derive clear conclusions about the role of these changes, because of the methodological difficulties in interpreting the results from complex CGE models. It is usually conceded that modelling international capital flows is one of the more contentious issues; technically indeed, such a modelling relies on equalizing rates of return on capital across countries, but, because this makes capital flows too reactive, various "ad hoc" devices are used to obtain less irrealistic outcomes. Differences in the riskiness of rates of returns are clearly relevant to explain most of the real behaviours, but how this can be "best" dealt with in a deterministic model is an open question. Progress depends on the further development of techniques.[16] It depends also on progress in theoretical and empirical analyses to capture more effectively how the exchange rate of currencies reacts to external payment deficits. This depends on the level of confidence on the future economic expansion of each country and how monetary policies (including the determination of

the public discount rate) employed to mitigate adverse impacts can change the return to capital in a country relative to other countries.

Models reviewed in this section have in common features that must be clearly borne in mind when interpreting the results:

- They assume perfect competition in all industries.
- Most of them use the so-called Armington specification that identical goods produced in different countries are imperfect substitutes: it is known that the results may then be sensitive to the particular commodity and chosen regional aggregation models (Lloyd, 1994).
- All of the models, apart from the G-cubed model of McKibbin and Wilcoxen (1995), are long-term growth models with international trade, without explicitly modelled financial markets that affect the macroeconomic adjustment.[17]
- Emission reductions involve only carbon dioxide.[18]
- The bias in technological change is unaffected by the emissions constraints and the production possibilities frontier always lies below the unconstrained frontier. Under such a hypothesis, the aggregate impact is unlikely to be positive, but some economies may benefit from favourable changes in their terms of trade and from changes in international capital flows.

Simulation studies covered in this report were conducted prior to and after the negotiation of the Kyoto Protocol. Pre-Kyoto studies consider more stringent emissions reduction targets for Annex I regions than the average 5.2% actually adopted under the Protocol. The major findings are that Annex B abatement would result in welfare losses for most non-Annex I regions under the more stringent targets. The magnitude of these losses is reduced under the less stringent Kyoto targets. Some non-Annex I regions that would experience a welfare loss under the more stringent targets experience a mild welfare gain under the less stringent targets.

Studies using a variety of more stringent pre-Kyoto targets include Coppel and Lee (1995; the GREEN model), Jacoby *et al.* (1997; the EPPA model), Brown *et al.*1997b) and Donovan *et al* (1997; the GTEM model), and Harrison and Rutherford (1999; the IIAM model). The last two models are based on the Global Trade Analysis Project (GTAP) database (Hertel, 1997).

In these studies, most non-Annex I countries suffer deterioration in their terms of trade and also welfare losses. Since the analysis at the region or country level depends on the type of

---

[15] To the extent that increased non-Annex I emissions result from more emission-intensive production techniques and increased production of emission-intensive goods for internal consumption, policies to control emission leakage by curbing the imports of emission-intensive goods into Annex I regions are likely to be counterproductive. Curbing imports may restrict substitution options in Annex I economies, requiring further cuts in output and exports that would stimulate greater non-Annex I emission-intensive production.

[16] These include techniques such as decomposition analysis (Huff and Hertel, 1996) and multiple simulations, in which some variables are held constant to isolate their influence on the final results. Verikios and Hanslow (1999) employed such a framework to asses the welfare impacts of international capital mobility.

---

[17] In the G-cubed model such a mechanism is superimposed on the structure of a long-term growth model.

[18] It is evident from simulations with the GTEM model (Brown *et al.*, 1999) that somewhat different results may be obtained if emission reductions involve a least-cost mix of the different GHGs identified under the Kyoto Protocol.

aggregation, it is difficult to give a comprehensive list of exceptions. The reasons for these exceptions are, however, easy to explain. Brazil and South Korea are, in many models, found to enjoy welfare gains from Annex I abatement policies because, unlike other non-Annex I regions, they are net importers of fossil fuels and have a high relative dependence on exports of iron and steel and non-ferrous metal products. In addition, in Brazil these products are far less intensive in fossil energy than in many other economies.[19] Brazil gains from lower prices for fossil fuel imports and higher prices for exports of iron and steel and non-ferrous metal products. Conversely, non-Annex I regions with the greatest dependence on fossil fuel exports, such as the Middle East and Indonesia, suffer the greatest deterioration. Non-Annex I regions that are net importers of manufacture goods that are fossil-fuel intensive also suffer a deterioration even if they benefit from lower oil prices.

One of the most important conclusions is that a number of those among non-Annex I regions that experienced a welfare loss under the pre-Kyoto targets experience a welfare gain under the Kyoto targets. For example, in the GREEN model, India and the Dynamic Asian Economies experienced a loss in real income in the pre-Kyoto simulation (Coppel and Lee, 1995). They experience a mild gain in real income under simulations of the Kyoto Protocol that involve varying degrees of policy co-ordination among the non-Annex I regions (van der Mensbrugghe, 1998). In pre-Kyoto simulations of the GTEM model (Brown *et al.*, 1997b; Donovan *et al.*, 1997), Chinese Taipei, India, Brazil, and the Rest of America were all found to experience welfare losses; with Kyoto targets (Tulpulé *et al.*, 1999) these regions experience mild welfare gains.

There is one key reason why some regions that experienced welfare losses under the more stringent targets experience mild gains in welfare under the Kyoto targets: the changing balance between substitution and output reduction with the level of abatement. GDP losses or the required level of a carbon tax for Annex I regions are, indeed, an increasing function of the level of abatement and the milder Kyoto targets are expected to be achieved with a greater reliance on substitution relative to output reduction than the more stringent targets.

A fairly similar regional pattern of non-Annex I welfare changes is found in simulations of Kyoto targets in a number of studies in which comparable pre-Kyoto target simulations are not available. These studies include Kainuma *et al.* (1999; the AIM model drawing on the GTAP database), McKibbin *et al.* (1999; the G-Cubed model), Bernstein *et al.* 1999; MS-MRT, drawing on the GTAP database), and Brown *et al.*, (1999; the multigas (CO_2, CH_4, and NO_x) version of GTEM) and Böhringer and Rutherford (2001).

---

[19] These calculations are based on version 3 of the GTAP database after reconciliation with energy data, mainly from the International Energy Agency.

8.3.2.1   *Impact of Emissions Trading*

All of the above studies considered various forms of emissions trading for Annex I economies. It was universally found that most non-Annex I economies that suffered welfare losses under uniform independent abatement also suffer smaller welfare losses under emissions trading. This is also the case in all of the studies for which results on movements in the terms of trade are published (Coppel and Lee 1995; Harrison and Rutherford, 1999).

Why are overall welfare losses to non-Annex I regions reduced by emissions trading? A key point is that because the marginal and average cost of abatement for the aggregate Annex I is lower under emissions trading than under uniform abatement, a higher GDP is achieved for a given reduction in emissions. This means that the reduction in emissions is achieved through a heavier reliance on substitution relative to output reduction (substitution involves the substitution of less emission-intensive for more emission-intensive Annex I produced inputs). The heavier reliance on substitution means that there is a less severe decline in fossil fuel prices and a lower increase in the price of manufactured goods that are fossil-fuel intensive. There is also less increase in non-Annex I exports of fossil-fuel intensive manufactured goods to Annex I regions under emissions trading than independent abatement. However, these increased exports divert resources from activities in which the original non-Annex I comparative advantage was higher and the overall result is less beneficial to most non-Annex I economies.

Some non-Annex I economies that experience welfare gains under independent abatement also experience smaller gains under emissions trading; however, the aggregate effect of emissions trading is found to be positive for non-Annex I economies: those that suffer welfare losses under independent abatement suffer smaller losses under emissions trading.

To summarize, despite a number of identifiable numerical discrepancies, there is agreement that the mixed pattern of gains and losses under the Kyoto targets results in a more positive aggregate outcome than under the assumed and more stringent pre-Kyoto targets. Similarities in the regions that are identified as gainers and losers are also quite marked. Oil-importing economies that rely on energy-intensive exports are gainers (and more so if the exports' carbon intensity is low), economies that rely on oil exports experience losses, and the results are more unstable for economies between these two extremes.

8.3.2.2   *Effects of Emission Leakage on Global Emissions Pathways*

As discussed above, a reduction in Annex I emissions tends to increase non-Annex I emissions, reducing the environmental effectiveness of Annex I abatement. Emissions leakage is measured as the increase in non-Annex I emissions divided by the reduction in Annex I emissions.

A number of multiregional models have been used to estimate carbon leakage rates (Martin *et al.* 1992; Pezzey 1992; Oliveira-Martins *et al.* 1992; Manne and Oliveira-Martins, 1994; Edmonds *et al.*, 1995; Golombek *et al.*, 1995; Jacoby *et al.* 1997; Brown *et al.* 1999). In SAR (IPCC, 1996, p. 425) a high variance in estimates of emission leakage rates was noted; they ranged from close to zero (Martin *et al.* (1992) using the GREEN model) to 70% (Pezzey (1992) using the Whalley–Wigle model). In subsequent years, some reduction in this variance has occurred, in the range 5%–20%. This may in part arise from the development of a number of new models based on reasonably similar assumptions and data sources, and does not necessarily reflect more widespread agreement about appropriate behavioural assumptions. However, because emission leakage is an increasing function of the stringency of the abatement strategy, this may also be because carbon leakage is a less serious problem under the Kyoto targets than under the targets considered previously.

Technically, there is a clear correlation between the sign and magnitude of spillover effects analyzed above and the magnitude of carbon leakage. It is important, however, to recognize those parameters that have a critical influence on results:

- The assumed degree of substitutability between imports and domestic production. This is why models based on the Armington assumption that imports and domestic production are imperfect substitutes produces lower estimates of emission leakage than models based on the assumption of perfect substitutability.
- The ease of substitution among technologies with different emissions intensities in the electricity and the iron and steel industries in Annex I regions.
- The assumed degree of competitiveness in the world oil market; this issue is considered in Section 8.3.2.3.
- The existence of an international carbon-trading system: for a given abatement strategy, emission leakage is lower under emissions trading than under independent abatement. This conclusion flows logically from the discussion above on movements in terms of trade. Greater Annex I output reduction under independent abatement stimulates greater emission-intensive production in non-Annex I regions, through both higher prices for emission intensive products and lower prices for fossil fuels. Support for the above conclusions on the impact of emissions trading is found in ABARE-DFAT (1995), Brown *et al.* (1997b), Hinchy *et al.* (1998), Brown *et al.* (1999), McKibbin *et al.* (1999), Kainuma *et al.* (1999), and Bernstein *et al.* (1999).

8.3.2.3 Effects of Possible Organization of Petroleum Exporting Countries (OPEC) Response

In the preceding discussion, a competitive equilibrium in the world economy was assumed. However, OPEC may be able to exercise a degree of monopoly power over the supply of oil. The issue has been raised in the literature as to the possible nature of an OPEC response to reduced demand for oil as a result of Annex I abatement. If in the short term OPEC were to reduce production to maintain prices in the face of lower demand, the time path for Annex I carbon taxes may need to be modified. See also Chapter 9.

A number of theoretical papers examined how a carbon tax might alter the optimal timing of extraction of given reserves of oil and, symmetrically, how significantly the potential supply response could alter the optimal time path of the price of carbon tax (Sinclair, 1992; Ulph and Ulph, 1994; Farzin and Tahovonen, 1996; Hoel and Kverndokk, 1996; Tahvonen, 1997). However, the severity of the potential problem depends on a number of key parameter values and implementation issues. Although it has been assumed that OPEC can "Granger cause" the world price of oil (Güllen, 1996), there is some question about the degree of cartel discipline that could be maintained in the face of falling demand (Berg *et al.*, 1997a). Any breakdown in the cartel would tend to increase the supply of oil on the market, which in the short term may require a higher carbon tax to meet a given abatement target. On the other hand, Bråten and Golombek (1998) suggest that implementing an Annex I climate change agreement might be seen by OPEC members as a hostile act and could strengthen the resolve to maintain cartel discipline. The OPEC response is likely to be related to the size of its potential loss in revenue to OPEC and these potential losses would be smaller under Annex I emissions trading than under independent abatement.

A number of empirical studies have tried to assess the significance of the potential OPEC response within a game theoretic framework. To do so, Berg *et al.* (1997b) resorted to a Cournot–Nash dynamic game in which parameter values are based on empirical estimates. They also identify (non-OPEC) "fringe" oil producers and other fossil fuel sources. A scenario is examined in which a carbon tax is maintained at a level of US$10 per barrel of oil. Initially, OPEC cuts back on production to try to maintain price, but this is partly offset by increased production by the fringe. Bråten and Golombek (1998) derive a similar pattern of OPEC response in a static model. Berg *et al.* (1997b) found that the optimal OPEC policy is not heavily influenced by intertemporal optimization in shifting supplies from one time period to another to maximize discounted net revenue.

If OPEC acts as a cartel, the extent of emissions leakage in response to Annex I abatement may be reduced (Berg *et al.*, 1997b), because the resultant higher price for oil reduces the incentives for increased emission-intensive activity in non-Annex I regions. Lindholt (1999) examined the Kyoto Protocol in an enhanced version of the same model and assumed that an efficient tradable permit scheme is established between Annex B countries. Whether or not OPEC acts as a cartel does not affect the shape of the time path of permit prices, only their level according to Lindholt (1999). A permit price of US$14/tCO$_2$ would be required in 2010 if OPEC acts as a cartel, whereas it would be US$24/tCO$_2$ in a competitive oil market. The lower permit price when OPEC acts as a cartel stems from OPEC cutting back production to maintain a higher oil price, which slows the growth in emissions in Annex B countries.

These studies mentioned demonstrate that whether or not OPEC acts a cartel will have a modest effect on the loss of wealth to OPEC and other oil producers and the level of permit prices in Annex B regions. A natural extension of this research would be to trace through all the ramifications of cartel behaviour by OPEC in the more complex CGE models discussed in this section.

8.3.2.4   *Technological Transfers and Positive Spillovers*

In a dynamic context, a progressive outward shift in the production possibilities frontier occurs over time as a result of technical change. A strand of literature (Goulder and Schneider, 1999) argues that climate policies will bias technical change towards emissions savings. In that case, there will be an outwards shift in the production possibilities frontier at some points, and an inwards shift at other points relative to the baseline.

One potentially important related issue not captured in the above models is that cleaner technologies, developed in response to abatement measures in industrialized countries, tend to diffuse internationally. The question is to what extent this will offset the negative aspects of leakage noted above and to amplify positive spillover. Grubb (2000) presents a simplified model, which represents this spillover effect in terms of its impact on emissions per unit GDP (intensities). The results suggest that, because the impact of cleaner technologies is cumulative and global, this effect tends to dominate over time, provided the connection between industrialized and developing country emission intensities is significant (higher than 0.1 on a scale where 0 represents an absence of connection and 1 a complete convergence of intensities by 2100). At this stage, empirical analysis is still lacking to derive a robust conclusion from this result. A recent work by Mielnik and Goldemberg (2000) suggests that the potential for technological leap-frogging in developing countries is important, but to what extent climate mitigation in Annex B accelerates this leap-frogging is still unclear. However, this demonstrates that the trickling down of technical change across countries deserves more attention in modelling exercises, all the more so since theoretically it (see Chapter 10 of this report) demonstrates that technological spillovers may be a major stabilizing force of any climate coalition.

8.4   Social, Environmental, and Economic Impacts of Alternative Pathways for Meeting a Range of Concentration Stabilization Pathways

The appropriate timing of mitigation pathways depends upon many factors including the economic characteristics of different pathways, the uncertainties about the ultimate objective, and the risks and damages implied by different rates and levels

of atmospheric change. This section focuses upon the mitigation costs of different pathways towards a predetermined concentration ceiling. No policy conclusion should be derived from it before reading Chapter 10, which discusses mitigation timing in the wider context of uncertainties, risks and impacts.

8.4.1   *Alternative Pathways for Stabilization Concentrations*

A given concentration ceiling can be achieved through a variety of emission pathways. This is illustrated in *Figure 8.12*. The top panel shows alternative concentration profiles for stabilization at 350-750ppmv. The bottom panel shows the corresponding emission trajectories. In each case, two different routes to stabilization are shown: the IPCC Working Group I profiles (from IPCC, 1995) and Wigley, Richels and Edmonds (WRE) profiles (from Wigley *et al.*, 1996).

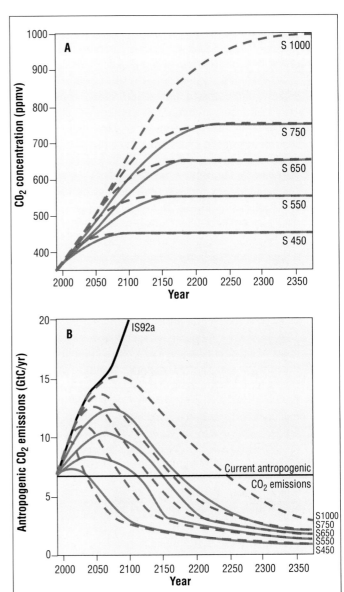

Figure 8.12: *Alternative pathways to stabilization.*

The choice of emission pathways can be thought of as a carbon budget allocation problem. To a first approximation, a concentration target defines an allowable amount of carbon to be emitted into the atmosphere between now and some date in the future. The issue is how best to allocate this budget over time. A number of modellers have attempted to address this issue. Unfortunately, to model stabilization costs is a daunting task. It is difficult enough to forecast the evolution of the energy and economic system to 2010. Projections over a century or more are necessary, but must be treated with considerable caution. They provide useful information, but their value lies not in the specific numbers but in the insights.

This section examines how mitigation costs might vary both with the stabilization level and with the pathway to stabilization. Also discussed are key assumptions that influence mitigation cost projections. Important, this discussion begins with the assumption that the stabilization ceiling is known with certainty and neglects the costs of different damages associated with different pathways (discussed in Chapter 10). Here, the challenge is to identify the least-cost mitigation pathway to stay within the prescribed ceiling. In Chapter 10, the issue of decision-making under uncertainty is discussed regarding the ultimate target and impacts of different pathways. Decision making under uncertainty requires indeed examining symmetrically the costs of accelerating the abatement in case of negative surprises about damages of climate change and adopting a prudent near-term hedging strategy. That is, one that balances the

risks of acting too slowly to reduce emissions with the risks of acting too aggressively.

8.4.2 Studies of the Costs of Alternative Pathways for Stabilizing Concentrations at a Given Level

Some insight into the characteristics of the least-cost mitigation pathway can be obtained from two EMF studies (EMF-14, 1997; EMF-16, 1999) and from Chapter 2 in the SRES mitigation scenarios (IPCC, 2000). In the first EMF study, modellers compared mitigation costs associated with stabilizing concentrations at 550ppmv using the WGI and WRE profiles (see *Figure 8.12*), Note that the WGI pathway entails lower emissions in the early years, with less rapid reductions later on. The WRE pathway allows for a more gradual near-term transition away from carbon-venting fuels. *Figure 8.13* shows that in these models the more gradual near-term transition of the two examined results in lower mitigation costs.

The above experiment compares mitigation costs for two emission pathways for stabilizing concentrations at 550ppmv. It does not identify the least-cost mitigation pathway, however. This was done in the subsequent EMF (1997) study. The results are presented in *Figure 8.14*. In these studies the least-cost mitigation pathway tends to follow the models reference case in the early years with sharper reductions later on.

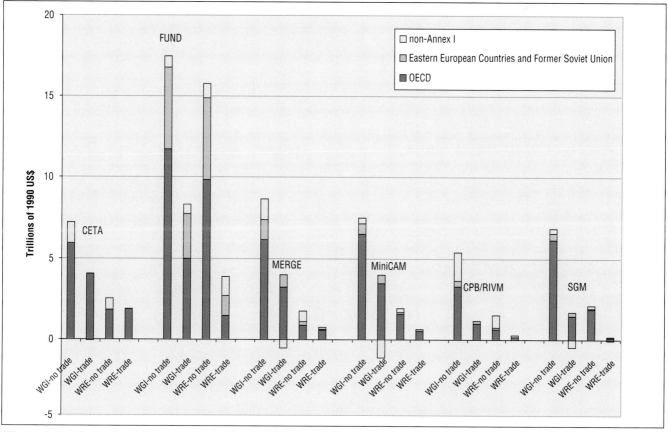

Figure 8.13: *Costs of stabilizing concentrations at 550ppmv; discounted to 1990 at 5%.*

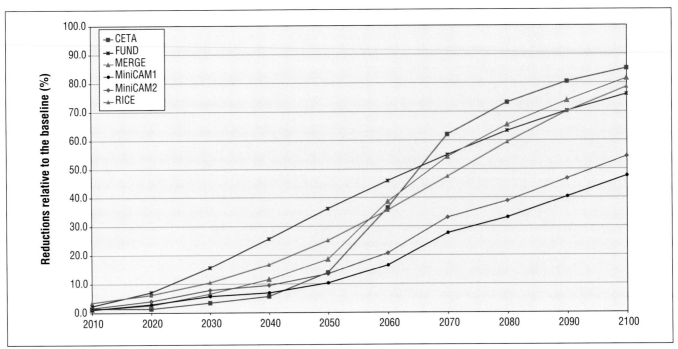

Figure 8.14: *Rate of departure from the baseline corresponding to least-cost mitigation pathway for a 550ppmv stabilization target.*

The selection of a 550ppmv target was purely arbitrary and not meant to imply an optimal concentrations target. Given the present lack of consensus on what constitutes "dangerous" interference with the climate system, three models in the EMF-16 study examined how mitigation costs are projected to vary under alternative targets. The results are summarized in *Figure*

8.15. As would be expected, mitigation costs increase with more stringent stabilization targets.

In Chapter 2, nine modelling groups reported scenario scenario results using different baseline scenarios. An analysis focused on the results of stabilizing the SRES A1B scenario at 550 and

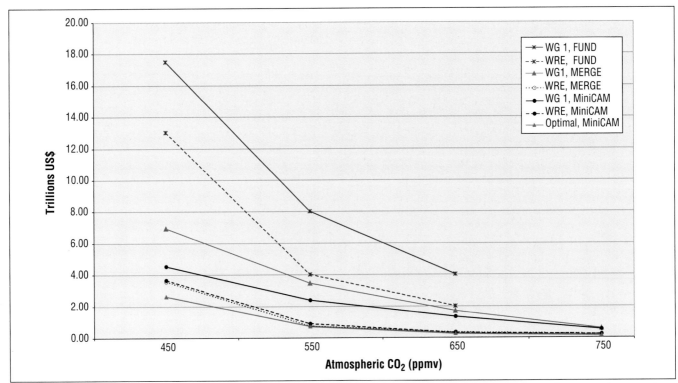

Figure 8.15: *Relationship between present discounted costs for stabilizing the concentrations of CO_2 in the atmosphere at alternative levels.*

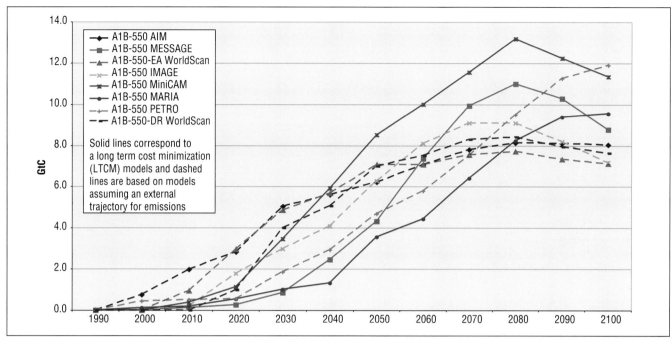

Figure 8.16: *Reductions in carbon emissions for the SRES A1B 550 case.*

450ppmv provides additional insight into the relationship between mitigation and baseline emissions. For the 550ppmv case, there are eight relevant trajectories (see *Figure 8.16*) giving the carbon reductions necessary to achieve a stabilization level of 550ppmv, where the models which impose a long-term cost minimization (LTCM) are represented as solid lines, and the models which use an external trajectory as the basis for their mitigation strategy are presented as dashed lines. The first impression of *Figure 8.16* is that even given common assumptions about GDP, population, and final energy use, and a com-

mon stabilization goal, there is still a lot of difference in the model results. A preliminary examination suggests that, in contrast to the non-optimization model results, a common characteristic among the LTCM models is that the near-term emissions pathways departs only gradually from the baseline.

Figure 8.17 clarifies the results by converting the absolute reduction to a percent reduction basis and averages them for the two classes of models. LTCM models show clearly a more gradual departure from the emissions baseline. *Figure 8.17*

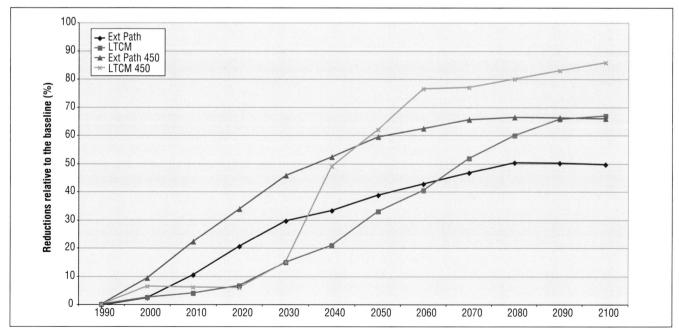

Figure 8.17: *Time path of emissions reductions.*

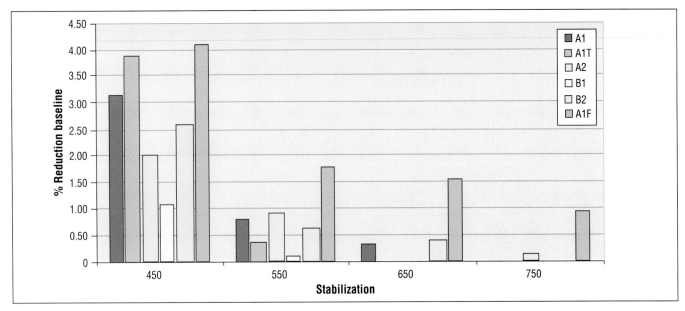

Figure 8.18: *Global average GDP reduction in 2050 for alternative stabilization targets and six SRES reference scenarios.*

also gives comparable results for the four cases with a 450ppmv target. The LTCM show a very similar decoupling until 2030, when this decoupling increases rapidly, and exceeds the other models by 2050, earlier in the 450ppmv case than in the 550ppmv case.

8.4.3 Economywide Impact of CO_2 Stabilization in the Post-SRES Scenarios

The economy-wide impact of stabilizing atmospheric CO_2 concentrations was assessed based on 42 post-SRES stabilization scenarios developed using the AIM, ASF, MARIA, MiniCAM, MESSAGE, and World SCAN models. These scenarios were developed by applying various mitigation policies and measures to the six illustrative scenarios (baselines) presented in the Special Report on Emission Scenarios (SRES).

The economy-wide impact of CO_2 stabilization was assessed on the basis of the difference in GDP in baseline scenarios and corresponding stabilization cases in a given year. This difference is expressed in percent (reflecting a relative GDP loss) and is positive when GDP in a baseline scenario is larger than in a stabilization case and is negative when GDP in a stabilization scenario is larger. Such an approach to measuring effects of stabilization was selected since it better reflects the societal burden of emission stabilization than absolute changes in GDP. For example, a 1% reduction in the 2100 GDP of the SRES A1 world is equal to about US$5.5 trillion and is larger in absolute terms than a 2% or US$5.0 trillion reduction in the 2100 GDP of the poor A2 world. Nonetheless, the relative level of effort in the latter case would be much greater. It should be also emphasized here that the GDP reduction itself represents a very crude indicator of economic consequences of the CO_2 stabilization. For example, most of the stabilization scenarios

reviewed here have not rigorously accounted for the economic effects of introducing new low-emission technologies, new revenue rising instruments or adequate inter-regional financial and technology transfers, all elements which contribute to lower the costs as explained in the rest of the chapter.

The average GDP reduction in most of the stabilization scenarios reviewed here is under 3% of the baseline value (the maximum reduction across all the stabilization scenarios reached 6.1% in a given year). At the same time, some scenarios (especially in the A1T group) showed an increase in GDP compared to the baseline due to apparent positive economic feedbacks of technology development and transfer. The GDP reduction (averaged across storylines and stabilization levels) is lowest in 2020 (0.99%), reaches a maximum in 2050 (1.45%) and declines by 2100 (1.30%). However, in the scenario groups with the highest baseline emissions (A2 and A1FI), the size of the GDP reduction increases throughout the modelling period.

Due to their relatively small scale when compared to absolute GDP levels, GDP reductions in the post-SRES stabilization scenarios do not lead to significant declines in GDP growth rates over this century. For example, the annual 1990-2100 GDP growth rate across all the stabilization scenarios was reduced on average by only 0.003% per year, with a maximum reduction reaching 0.06% per year.

Figure 8.18 shows the relationship between the relative GDP reduction, the scenario group, and the stabilization level in 2050. The reduction in GDP tends to increase with the stringency of the stabilization target. But the costs are very sensitive to the choice of baseline scenario. The maximum relative reduction occurs in the A1FI scenario group, followed by the other A1 scenario groups and the A2 group, while the mini-

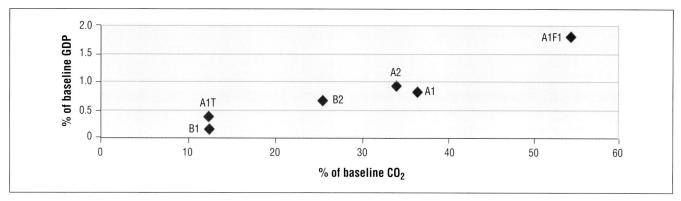

Figure 8.19: *Average GDP and CO$_2$ reductions in 550ppmv stabilization scenarios: year 2050 (labels identify different scenario groups).*

mum reduction occurs in the B1 group[20]. By 2100, the situation slightly changes with GDP reductions in the A2 scenario group becoming relatively more pronounced.

Differences in relative GDP reductions in different scenario groups are explained by the magnitude of corresponding CO$_2$ emission reductions needed to achieve a particular stabilization level. The emission reduction is apparently the largest in the A1FI scenario group, which is also associated with the largest GDP loss (*Figure 8.19*). Meanwhile, the smallest relative GDP loss occurs in the A1T and B1 groups, which have low baseline emissions and accordingly require the smallest reductions to reach the CO$_2$ stabilization.

Regional GDP reduction patterns in the post-SRES stabilization scenarios are also generally explained by corresponding required reductions in CO$_2$ emissions, which are determined by baseline emissions, the stabilization levels, assumptions about emissions trading mechanisms and about the relative contribution of regions to global CO$_2$ emissions; reduction and associated financial and technology transfer. In most of the baseline (SRES) scenarios starting from 2020, absolute CO$_2$ emissions in developing (non-Annex I) regions remain larger for the rest of the 21st century than in the industrialized (Annex I) regions.

8.4.4 Reasons why Energy-economy Models Tend to Favour Gradual Departure from Baseline in the Near-term

There are several reasons why the models tend to favour a more gradual departure from their reference path if used to study a pre-determined concentration level. First, energy using and energy producing capital stock (e.g., power plants, buildings, and transport) are typically long lived. The current system was put into place on the basis of a particular set of expectations about the future. Large emission reductions in the near term require accelerated replacement, which is apt to be costly.

There is more opportunity for reducing emissions cheaply at the point of capital stock turnover.

Second, the models suggest that currently there are insufficient low-cost substitutes, on both the supply and demand sides of the energy sector, for deep near-term cuts in carbon emissions. With the anticipated improvements in the efficiency of energy supply, transformation, and end-use technologies, such reductions should be less expensive in the future.

Third, because of positive returns on capital, future reductions can be made with a smaller commitment of today's resources. For example, assume a net real rate of return on capital of 5% per year. Further, suppose that it costs US$50 to remove a tonne of carbon, regardless of the year in which the reduction occurs. To remove the tonne today would cost US$50. Alternatively, it only needs US$19 to be invested today to provide the resources to remove a tonne of carbon in 2020.

Finally, for higher near-term emissions, the size of the carbon budget (to meet a prescribed emission target) is higher, reflecting that the products of early emissions have a longer time to be removed from the atmosphere, and because the higher concentrations give higher oceanic and terrestrial sinks.

The fact that the least-cost mitigation pathway tends to follow the baseline in the early years has been misconstrued by some analysts as an argument for inaction. Wigley *et al.* (1996) note that this is far from the case: First, all stabilization targets still require future capital stock to be less carbon-intensive than under a business as usual (BAU) scenario. As most energy production and use technologies are long-lived, this has implications for current investment decisions. Second, new supply options typically take many years to enter the marketplace. To ensure sufficient quantities of low-cost, low-carbon substitutes in the future requires a sustained commitment to research, development and demonstration today. Third, any no regrets measures for reducing emissions should be adopted immediately. Lastly, it is clear that one cannot go on deferring emission reductions indefinitely, and that the need for substantial reductions in emissions is sooner the lower the concentration target.

---

[20] Please note that only one scenario was available for the GDP loss data for A1T-450, A1T-550, A1FI-450, A1FI-650, A2-450, and B2-650.

8.4.5 Critical Factors Affecting the Timing of Emissions Reductions: The Role of Technological Change

As pointed out by Grubb (1997), there are several key assumptions imbedded in the energy-economy models that influence the shape of the least-cost mitigation pathway. For a pre-determined target, these relate to the determinants of technical change; capital stock turnover and the inertia in the energy system; discounting; and, the carbon cycle. When the target is uncertain, they include in addition the probability attached to each target and risk aversion (see Chapter 10) which tend to favour a more aggressive departure from current trends.

The discount rate will not be discussed because it is less important in cost-efficiency frameworks (when the target is predetermined) than in a cost-benefit one (when the discount rate reduces the weight of future environmental impacts, see Chapter 10). Neither are the very few studies discussed which try to assess different benefits in terms of environmental co-benefits of reducing GHG emissions presented. Wigley *et al.* (1996), for example, show pathway-related differentials up to 0.2°C in global mean temperature and 4cm in global mean sea-level (by 2100) for the WGI and WRE 550 stabilization pathways. See Chapter 10 for an elaboration of these timing issues. This part will rather insist on the key features of technical change that are numerically of utmost importance.

To the extent that the cost of reducing emissions is lower in the future than at present, the overall cost of stabilizing the CO_2 concentration is less if emissions mitigation is shifted towards the future. This shift occurs in all models. The extent of the shift that minimizes the cost of limiting the concentration of atmospheric CO_2 depends, at least in part, on the treatment of technological change. Without technological change, the problem is simple and the results of Hotelling (1939) apply. With endogenous technological change, the problem becomes more complex.

This discussion of the determinants of technological change must begin with the acknowledgement that no adequate theory of endogenous technological change exists at present. Many researchers have contributed to the field, but the present state of understanding is such that present knowledge is partial and not necessarily fully consistent. Although no complete theory of technological change exists, two elements have been identified and explored in the literature: induced technological change (ITC) and learning-by-doing (LBD). Work by Ha-Duong *et al.* (1997), Grubb *et al.* (1995), Grubb (1997), and Kypreos and Barreto (1999) examined the implication of ITC, LBD, and inertia within the context of uncertainty and an imperative to preserve the option of concentration ceilings such as 450ppmv. They conclude that emissions mitigation can be shifted from the future towards the present under appropriate circumstances.

Goulder and Mathai (1998) also explore how the effect on timing depends on the source of technological change. When the channel for technological change is R&D, ITC makes it preferable to concentrate more abatement efforts in the future. The reason is that technological change lowers the costs of future abatement relative to current abatement, making it more cost-effective to place more emphasis on future abatement. However, when the channel for technological change is LBD, the presence of ITC acts in two opposite directions. On the one hand, ITC makes future abatement less costly but, on the other hand, there is an added value to current abatement because such abatement contributes to experience or learning and helps reduce the costs of future abatement. Which of these two effects dominates depends on the particular nature of assumptions and firms. In recent years, there has been a good deal of discussion about the potential for ITC (e.g., Anderson *et al.*, 1999). Proponents argue that such changes might substantially lower, and perhaps even eliminate, the costs of CO_2 abatement policies. These discussions have exposed very divergent views as to whether technological change can be induced at no cost, or whether a resource cost is involved. For example, in a 1995 article, Porter and van der Linde (1995) contend that properly designed regulation can trigger innovation that may partially or more than fully offset the costs of compliance. Indeed, they argue that firms can actually benefit from more stringent regulation than that faced by their competitors in other countries. However, in an accompanying article, a strongly contrary view is put forward by Palmer *et al.* (1995). Examining available data, they found that such offsets pale in comparison to expenditures for pollution abatement and control.

8.4.5.1 ITC through Dedicated R&D

Including R&D driven ITC in climate mitigation models leads to ambiguous results in terms of time profile and tax level in a cost-benefit framework In a cost-effectiveness framework, the optimal tax is lower in the case with R&D driven ITC and has to be set up early even if the effective resulting abatement shifts from the near-term to the more distant future. If there are market failures in the R&D market (e.g., knowledge spillover), then subsidies for R&D are justified as it enhances social welfare and raises the abatement level (Goulder and Schneider, 1999; Weyant and Olavson, 1999; Goulder and Mathai, 2000).

However, R&D driven-ITC can reduce the gross costs of a carbon tax under special circumstances. Specifically, if R&D has been substantially over-allocated towards the fossil fuel industries prior to the imposition of a carbon tax, the carbon tax can reduce this allocative inefficiency and, as a result, its costs can be quite low or even negative. A substantial prior misallocation towards carbon-intensive industries could occur if there were prior subsidies towards R&D in the fossil fuel industries (with no comparable subsidies in other industries), or if there were substantial positive spillovers from R&D in non-carbon industries (with no comparable spillovers in the fossil fuel industries). Under other plausible initial conditions, however, R&D driven-ITC raises, rather than lowers, the net social costs of a given carbon tax because of the crowding out of R&D from other sectors; to put it clearly the tax level for a given abatement is lower than under the hypothesis of exogenous technical change and part of this decrease is offset when all the general equilibrium effects are accounted.

The same model has been employed to compare the costs of achieving a given abatement target through carbon taxes and R&D subsidies (Schneider and Goulder, 1997; Goulder and Schneider, 1999). If there are no spillovers to R&D, the least-cost way to reach a given abatement target is through a carbon tax alone. The carbon tax best targets the externality from the combustion of fossil fuels related to climate change, and thus is the most cost-effective. However, if there are spillovers to R&D, the least-cost way to achieve a given abatement target is through the combination of a carbon tax and R&D subsidy. If spillovers are present, there is a market failure in the R&D market as well as a (climate change related) market failure associated with the use of carbon. Two instruments (the R&D subsidy and the carbon tax) are needed to address the two distinct market failures most efficiently. In general, a R&D subsidy by itself does not offer the least-cost approach to reducing carbon emissions. Results from this model are highly sensitive to assumptions about the nature and extent of knowledge spillovers. Further empirical work that sheds light on these spillovers would have considerable value.

8.4.5.2 *Learning by Doing (LBD)*

LBD as a source of technical change was first emphasized by Arrow (1962). Nakicenovic (1996) discussed the importance of LBD in energy technology, and Messner (1995) endogenizes the learning process in energy models. LBD is a happy consequence of those investments in which learning is a result of cumulative experience with new technologies. LBD typically refers to reductions in production cost, in which learning takes place on the shop floor through day-to-day operations, not in the R&D laboratory. The LBD component of change is significant too. Kline and Rosenberg (1986) discuss industry studies that indicate that LBD-type improvements to processes in some cases contribute more to technological progress than the initial process development itself.

LBD models use the installed capacity or cumulative use as an indicator of accumulating knowledge in each sector. The abatement costs are represented by the specific investment costs in US\$/kWh. The models are global and therefore the diffusion process is not represented. The optimization problems are non-convex, which raises a difficult computational problem to find an optimum. However, pioneering work at the International Institute for Applied Systems Analysis (IIASA) on the MESSAGE model and additional developments based on models like MARKAL and ERIS; (MATSSON), Kypreos and Barreto (1999), Seebregts *et al.* (1999a), (SKFB), Tseng *et al.* (1999), and Kypreos *et al.* (2000) demonstrate progress in this direction. They show that several technologies are likely to play a prominent role in reducing the cost of abatement, if ITC is indeed taken into account when computing the equilibrium. A problem with modelling endogenous technological change is

that the traditional baseline scenario versus optimal policy run argumentation is not feasible. This follows directly from the path dependence. The most important results are: greater consistency of model results with the observed developments of technological change;

- new technologies first appear in niche markets with rising market shares;
- the time of breakthrough of new technologies can be influenced by policy measures (taxes and R&D) if they are strong enough;
- identification of key technologies, like photovoltaic modules or fuel cells, for public R&D investments is difficult; and
- technological lock-in effects depend on costs.

The most important conclusion for the timing of a mitigation policy is that early emissions-reduction measures are preferable when LBD is considered. This is confirmed unambiguously by a macroeconomic modelling study (van der Zwaan *et al.*, 1999/2000) which finds also lower levels of carbon taxes than those usually advocated.

These findings must be tempered by the fact that the models are not only highly non-linear systems, and therefore potentially sensitive to input assumptions, but also the quantitative values employed by modellers are typically drawn from successful historical examples. Furthermore, the empirical foundations of LBD are drawn from observations of the relationship between cumulative deployment and/or investment in new technology and cost. This relationship is equally consistent with the hypothesis that a third factor reduced costs, in turn leading to increases in demand. The authors restrict their findings to more qualitative assertions, because of the limitations of current models (Messner, 1997; Grübler and Messner, 1998; Barreto and Kypreos, 1999; Seebregts *et al.*, 1999a, 1999b). The research so far has been limited to energy system models and ignored other forms of endogenous, complex changes that are important for emissions, like changes in lifestyles and social institutions.

8.4.5.3 *The Distinction Between Action and Abatement*

The key message from this discussion about technical change is that a clear distinction has to be made between the timing of action and the timing of abatement. As a result of inertia in technological innovation, short-term action is required to abate more in the future, but a given amount of abatement at a given point in time is not a good measure of the effort. The necessity of this distinction is reinforced by the consideration of inertia in capital stocks. Mitigation costs are influenced by assumptions about the lifespan of existing plants and equipment (e.g., power plants, housing, and transport). Energy-related capital stock is typically long lived and premature retirement is apt to be costly. For example, an effort to change the transportation infrastructure will not reduce carbon emissions significantly for two decades or more. Hence, a drastic departure from the current trend is impossible without high social costs and a delay of action in this sector will require higher abatement costs in the more flexible sectors to meet a given target. Lecocq *et al.* (1999) found that these costs would be increased by 18% in 2020 for a 550ppmv target and by 150% for a 450ppmv target.

This irreversibility built into technological change is far more critical when the uncertainty about the ultimate target is considered. In this case indeed, many of the parameters that legitimize the postponing of abatement play in the opposite direction. If indeed the concentration constraints turn out to be lower than anticipated, there may be a need for abrupt reduction in emissions and premature retirement of equipment. In other words, even if the permanent costs of an option (in case of perfect expectation) are lower than those of an alternative option, it may be the case that its transition costs are higher because of inertia. For example, two ideal transportation systems can be envisaged, one relying on gasoline, the other on electric cars and railways, both with comparable costs in a stabilized situation; however, a brutal transition from the first system to the second may be economically disruptive and politically unsustainable. These issues are examined in more depth in Chapter 10 because the selection of the ultimate target depends upon the decision-making framework and upon the nature of the damage functions. But, it matters here to insist on the fact that the more inertia is built into the technical system, and the less processes of learning by doing and induced technical change have operated, the more costly corrections of trajectories in hedging strategies will be, for example, moving from a 550ppmv concentration goal to 450ppmv (Ha-Duong *et al.*, 1997; see also Grubb *et al.*, 1995; Grubb, 1997). This possibility of switching from one objective to another is supported by current material regarding climate damages, in particular (Tol, 1996) if the rate of change is considered in the analysis and the delay between symptoms and the response by society (see Chapter 10).

References

Aaserud, M., 1996: Costs and Benefits of Climate Policies: An Integrated Economy-Energy-Environment Model Approach for Norway. In *Modelling the Economy and the Environment.* B. Madsen, C. Jensen-Butler, J.B. Mortensen, A.M. Bruun Christensen,(eds.), Springer, Berlin.

Aasness, J., T. Bye, and H. T. Mysen, 1996: Welfare effects of emission taxes in Norway. *Energy Economics,* **18**(4), 335-46.

ABARE-DFAT, 1995: *Global Climate Change. Economic Dimensions of a Global Policy Response Beyond 2000.* ABARE, Canberra, Australia.

Abt Associates, 1997: *International PM Mortality Impacts in 2020 from Greenhouse Gas Policies.* Memorandum, Environmental Research Area, USA.

Abt Associates, and Pechan-Avanti Group, 1999: *Co-Control Benefits of Greenhouse Gas Control Policies.* Prepared for the Office of Policy, U.S. Environmental Protection Agency, Contract No. 68-W4-0029.

Adegbulugbe, A.O., F.I. Ibitoye, W.O. Siyanbola, G.A. Oladosu, J.F.K. Akiubonu, *et al.,* 1997: GHG Emission Mitigation in Nigeria. In *Global Climate Change Mitigation Assessment: Results from 14 transitioning and developing countries,* US Country Studies Program, Washington, DC.

Adi, A.C., C.L. Malik, and A. Nurrohim, 1997: Mitigation of CO_2 from Indonesia's Energy System. *Applied Energy,* **56**(3-4), 253-263.

Alfsen, K., H. Birkelund and M. Aaserud, 1995: Impacts of an EC Carbon/Energy Tax and Deregulating Thermal Power Supply on CO_2, SO_2 and NO_x Emissions. *Environmental and Resource Economics* **5**, 165-189

ALGAS (Asian Least-Cost Greenhouse Gas Abatement Strategy), 1999a: *Summary Report.* Asian Development Bank, Manila.

ALGAS, 1999b: *Profiles of Investment and Technical Assistance.* Greenhouse Gas Abatement Projects. Asian Development Bank, Manila.

ALGAS, 1999c: *People's Republic of China.*

1999d. *Republic of Korea.*

1999e. *Myanmar.*

1999f. *Pakistan.*

1999g. *Philippines.*

1999h. *Thailand.* Asian Development Bank, Manila.

Almon, C., 1991: The INFORUM Approach to Interindustry Modelling. *Economic Systems Research,* **3**, 1-7.

Andersen, F.M., H.K. Jacobsen, P.E Morthorst, A. Olsen, M. Rasmussen, T. Thomsen, and P. Trier, 1998: EMMA: En energi- og miljørelateret satellit-model til ADAM (EMMA an energy and environmental satellite model related to ADAM). *Nationaløkonomisk Tidsskrift,* **136**(3), 333-349.

Anderson, J., R. Morgenstern, and M. Toman, 1999: *The Rationale for Flexibility in the Design of Greenhouse Gas Abatement Policies: A Review of Economic Issues.* Resources for the Future, Washington, DC.

Arrow, K., 1962: The Economic Implication of Learning by Doing. *Review of Economic Studies,* **29**, 155-173.

Aunan, K., H.A. Aaheim, and H.M. Seip, 2000: Reduced Damage to Health and Environment from Energy Saving in Hungary. Expert Workshop on Assessing the Ancillary Benefits and Costs of Greenhouse Gas Mitigation Policies, March 27-29, Washington, DC.

Babiker, M., M. E. Bautista, H. D. Jacoby, and J. M. Reilly, 2000: Effects of Differentiating Climate Policy by Sector: A U.S. Example. In *Sectoral Economic Costs and Benefits of GHG Mitigation,* Proceedings of the IPCC Expert Meeting, February 14-15, 2000, L. Bernstein, J. Pan, (eds.), Eisensach, Germany.

Bahn, O., E. Fragnière, and S. Kypreos, 1996: Swiss Energy Taxation Options to Curb CO_2 Emissions. *Conference Proceedings,* The 1996 European Environment Conference, West Yorkshire, UK, pp.1-6.

Bahn, O., A. Haurie, S. Kypreos, and J.-P. Vial, 1998: Advanced Mathematical Programming Modeling to Assess the Benefits from International CO_2 Abatement Cooperation. *Environmental Modeling and Assessment,* **3**, 107-115.

Bahn, O., A. Cadena, and S. Kypreos, 1999a: Joint Implementation of CO_2 Emission Reduction Measures between Switzerland and Colombia. *International Journal of Environment and Pollution,* **11**, 1-12.

Bahn, O., S. Kypreos, B. Bueler, and H. J. Luthi, 1999b: Modelling an International Market of CO_2 Emission Permits. *International Journal of Global Energy Issues,* **12**(1-6), 283-291.

Bailie, A., B. Sadownik, A. Taylor, M. Nanduri, R. Murphy, J. Nyboer, M. Jaccard, and A. Pape, 1998: *Cost Curve Estimation for Reducing CO_2 Emissions in Canada: An Analysis by Province and Sector.* Office of Energy Efficiency, Natural Resources, Canada.

Barker, T., 1993: Secondary Benefits of Greenhouse Gas Abatement: The Effects of a UK Carbon/Energy Tax on Air Pollution. Energy-Environment-Economy Modelling. Discussion Paper N°4, ESRC Research Project on Policy Options for Sustainable Energy Use in a General Model of the Economy.

Barker, T., 1997. Taxing pollution instead of jobs: Towards more employment without more inflation through fiscal reform in the UK. *Ecotaxation,* T. O'Riordan, (ed.), Earthscan Publ., London.

Barker, T., 1998a: Policy Evaluation in a Multi-sectoral Framework. *European-US Conference on Post-Kyoto Strategies,* Stanford University, Stanford, CA.

Barker, T, 1998b: The Effects on Competitiveness of Coordinated *versus* Unilateral Fiscal Policies Reducing GHG Emissions in the EU: An Assessment of a 10% Reduction by 2010 Using the E3ME Model. *Energy Policy,* **26**(14), 1083-1098.

Barker, T. 1998c: The Use of Energy-Environment-Economy Models to Inform Greenhouse Gas Mitigation Policy, Special Section.In *Impact Assessment and Project Appraisal: Reflections on the Use of Models in Policy Making,* S. Shackley, (ed.), **16**(2), June 1998, pp. 133-131.

Barker, T., 1999: Achieving a 10% cut in Europe's CO_2 Emissions using Additional Excise Duties: Coordinated, Uncoordinated and Unilateral Action Using the Econometric Model E3ME. *Economic Systems Research,* **11** (4), 401-421.

Barker, T., and N. Johnstone, 1993: Equity and Efficiency in Policies to Reduce Carbon Emissions in the Domestic Sector . *Energy & Environment,* **4** (4), 335-361.

Barker, T., and J. Kohler, 1998: Equity and Ecotax Reform in the EU: Achieving a 10% Reduction in CO_2 Emissions Using Excise Duties. *Fiscal Studies,* **19** (4), 375-402.

Barker, T., and K.E. Rosendahl, 2000: Ancillary Benefits of GHG Mitigation in Europe: SO_2, NO_x and PM_{10} reductions from policies to meet Kyoto targets using the E3ME model and Externe valuations. *Expert Workshop on Assessing the Ancillary Benefits and Costs of Greenhouse Gas Mitigation Policies,* March 27-29, Washington, DC.

Berg, E., S. Kverndokk, and K. Rosendahl, 1997a: Gains from Cartelisation in the Oil Market. *Energy Policy,* **25**(13), 1075-1091.

Berg, E., S. Kverndokk,, and K. E. Rosendahl, 1997b: Market Power, International CO_2 Taxation and Petroleum Wealth. *The Energy Journal,* **18**(4), 33-71.

Bergman, L.,1995: Sectoral Differentiation as a Substitute for International Coordination of Carbon Taxes: A Case Study of Sweden. In *Environmental Policy with Political and Economic Integration.* J. Braden , T. Ulen, (eds.), Edward Elgar, Cheltenham, UK.

Bernard, A.L., and M. Vielle. 1999a: Efficient Allocation of a Global Environment Cost between Countries : Tradable Permits VERSUS Taxes or Tradable Permits AND Taxes ? An appraisal with a World General Equilibrium Model. Working Paper, Ministère de l'équipement des transports et du logement, France.

Bernard, A.L., and M. Vielle. 1999b: The Pure Economics of Tradable Pollution Permits. Working paper, Ministère de l'équipement des transports et du logement, France.

Bernard, A.L., and M. Vielle. 1999c. Rapport à la Mission Interministérielle sur l'Effet de Serre relatif aux évaluation du Protocole de Kyoto effectuées avec le modèle GEMINI-E3 (Report to the MIES on the greenhouse effect: An evaluation of the Kyoto Protocol with the GEMINI-E3 Model, Ministère de l'équipement des transports et du logement, Commissariat à l'énergie atomique, France.

Bernstein, P., D. Montgomery, and T. Rutherford, 1999: Global Impacts of the Kyoto Agreement: Results from the MS-MRT model. *Resource and Energy Economics,* **21**(3-4), 375-413.

Böhringer, C. , 1997: NEWAGE - Modellinstrumentarium zur gesamtwirtschaftlichen Analyse von Energie- und Umweltpolitiken, In *Energiemodelle in der Bundesrepublik Deutschland - Stand der Entwicklung*, S. Molt, U. Fahl, (ed.), Jülich, Germany, pp. 99-121.

Böhringer, C. , 2000: Cooling Down Hot Air - A Global CGE Analysis of Post-Kyoto Carbon Abatement Strategies, *Energy Polic,* **28,** 779-789.

Böhringer, C., and T.F. Rutherford, 1997: Carbon Taxes with Exemptions in an Open Economy - A General Equilibrium Analysis of the German Tax Initiative. *Journal of Environmental Economics and Management,* **32,** 189-203.

Böhringer, C., T.F. Rutherford, 1999: Decomposing General Equilibrium Effects of Policy Intervention in Multi-Regional Trade Models - Methods and Sample Application. ZEW Discussion Paper 99-36, Mannheim, Germany.

Böhringer, C., T.F. Rutherford, 2001: World Economic Impacts of the Kyoto Protocol. In *Internalization of the Economy, Environmental Problems and New Policy Options*, P.J.J. Welfens, R. Hillebrand, A. Ulph, (eds.), Springer, Heidelberg/New York.

Böhringer, C., *et al.*, 1997: Volkswirtschaftliche Effekte einer Umstrukturierung des deutschen Steuersystems unter besonderer Berücksichtigung von Umweltsteuern. Forschungsbericht des Instituts für Energiewirtschaft und Rationale Energieanwendung, Universität Stuttgart, Band 37, Stuttgart.

Borja-Aburto, V.H., D.P. Loomis, S.I. Bangdiwala, C.M. Shy, and R.A. Rascon-Pacheco, 1997: Ozone, Suspended Particulates and Daily Mortality in Mexico City. *American Journal of Epidemiology* **145**(3), 258-68.

Borja-Aburto, V.H., M. Castillejos, D. R. Gold, S. Bierzwinski, and D. Loomis, 1998: Mortality and Ambient fine Particles in Southwest Mexico City, 1993-1995. *Environmental Health Perspectives,* **106**, 849-855.

Bovenberg, A. L., and L.H. Goulder, 2001: Neutralizing the Adverse Industry Impacts of CO2 Abatement Policies: What Does It Cost? In *Behavioral and Distributional Effects of Environmental Policies*, C. Carraro , G. Metcalf, (eds.), University of Chicago Press, Chicago/London.

Boyd, R., K. Krutilla and W.K. Viscusi, 1995: Energy Taxation as a Policy Instrument to Reduce CO_2 Emissions: A Net Benefit Analysis. *Journal of Environmental Economics and Management,* **29**(1), 1-25.

Braga, A., G. Conceicao, L. Pereira, H. Kishi, J. Pereira, and M. Andrade, 1999: Air Pollution and Pediatric Hospital Admissions in Sao Paulo. *Journal of Environmental Medicine* **1**, 95-102.

Brännlund, R., and I.-M. Gren, 1999: Green Taxes in Sweden: A Partial Equilibrium Analysis of the Carbon Tax and the Tax on Nitrogen in Fertilizers. In *Green Taxes – Economic Theory and Empirical Evidence from Scandinavia*. R. Brännlund, I-M. Gren, (eds.), Edward Elgar, Cheltenham, UK.

Bråten, J., and R. Golombek, 1998: OPEC's Response to International Climate Agreements. *Environmental and Resource Economics,* **12**(4), 425-42.

Brendemoen, A., and H. Vennemo, 1994: A Climate Treaty and the Norwegian Economy: A CGE Assessment. *The Energy Journal,* **15**(1), 77-93.

Brown *et al., 1997a: Scenarios of U.S. Carbon Reductions: Potential Impacts of Energy Technologies by 2010 and Beyond*. Prepared by ORNL, LBNL, PNNL, NREL, ANL for the U.S. Department of Energy, Washington, DC.

Brown, S., D. Donovan, B. Fisher, K. Hanslow, M. Hinchy, M. Matthewson, C. Polidano, V. Tulpulé, and S. Wear, 1997b: The Economic Impact of International Climate Change Policy. ABARE Research Report 97.4, Canberra, Australia.

Brown, M., M. Levine, J. Romm, A. Rosenfeld, and J. Koomey, 1998: Engineering-Economics Studies of Energy Technologies to reduce GHG emissions: Opportunities and Challenges. *Annual Review of Energy and the Environment,* **23**, 287-385.

Brown, S., D. Kennedy, C. Polidano, K. Woffenden, G. Jakeman, B. Graham, F. Jotzo, and B. Fisher, 1999: Economic Impacts of the Kyoto Protocol: Accounting for the Three Major Greenhouse Gases. ABARE Research Report 99.6, Canberra, Australia.

Brown *et al.* 2000: *A Clean Energy Future for the US*, prepared by ORNL, LBNL, PNNL, NREL, ANL for the U.S. Department of Energy, Washington, DC.

Burtraw, D., and M. Toman, 1997: The Benefits of Reduced Air Pollutants in the U.S. from Greenhouse Gas Mitigation Policies. Discussion Paper 98-01-REV, Resources for the Future, USA.

Burtraw, D., A. Krupnick, K. Palmer, A. Paul, M. Toman, and C. Bloyd, 1999: *Ancillary Benefits of Reduced Air Pollution in the US from Moderate Greenhouse Gas Mitigation Policies in the Electricity Sector*. Discussion Paper 99-51, Resources for the Future, USA.

Bye, B., 2000: Environmental Tax Reform and Producer Foresight: An Intertemporal Computable General Equilibrium Analysis. *Journal of Policy Modeling,* **22**,(6), 719-752.

Bye, B., and K. Nyborg, 1999: The Welfare Effects of Carbon Policies: Grandfathered Quotas versus Differentiated Taxes. Discussion Papers 261, Statistics Norway, Norway.

Capros, P., L. Mantzos, L. Vouyoukas, and D. Petrellis, 1999a: European Energy and CO_2 Emissions Trends to 2020. Paper presented at IEA/EMF/IIASA Energy Modelling Meeting, Paris, June 1999, 1-18.

Capros, P., L. Mantzos, P. Criqui, Kouvaritakis, Sorai, Scrattenkolzer, and L. Vouyoukas, 1999b: *Climate Technology Strategy 1: Controling Greenhouse Gases. Policy and Technology Options*. Springer Verlag eds., Berlin.

Capros, P., T. Georgakopoulos, D. Van Regemorter, S. Proost, T.F.N Schmidt, K. Conrad, and L. Vouyoukas, 1999c: *Climate Technology Strategy: Controlling Greenhouse Gases 2: The Macroeconomic Costs and Benefits of Reducing Greenhouse Gases Emissions in the European Union*. Springer Verlag eds., Berlin.

Carraro, C., and M. Galeotti, 1995: Voluntary agreements: A macro-economic assessment. Dipartimento di Scienze Economiche, Università di Bergamo, Italy, June.

Carraro, C., and A. Soubeyran, 1996: Environmental Taxation and Employment in a Multi-sector General Equilibrium Model. In *Environmental Fiscal Reform and Unemployment*. Kluwer Academmic Publishers, London.

Chung, W., Y. June Wu, and J. D. Fuller, 1997: Dynamic Energy Environment Equilibrium Model for the Assessment of CO_2 Emission in Canada and the USA. *Energy Economics*, **19**, 103-124.

Cifuentes, L., J.J. Prieto, *et al.,* 1999: Valuation of mortalitiy risk reductions at present and at an advanced age: Preliminary results from a contingent valuation study. Annual Meeting of the Latin American and Caribean Economic Association, Santiago, Chile.

Cifuentes, L., E. Sauma, H. Jorquera, and F. Soto, 2000: Co-controls Benefits Analysis for Chile: Preliminary estimation of the potential co-control benefits for Chile. COP-5: Progress Report (Revised Version), L.A. Cifuentes, E. Sauma, H.Jorquera, F. Soto, (eds.), 12 Nov 1999.

Conniffe, D., J. Fitz Gerald, S. Scott, and F. Shortall, 1997: *The Costs to Ireland of GHG Abatement*. ESRI (Economic and Social Research Institute), Dublin. See also: B.O.Cleirgh, 1998: *Environmental Resources Management*, Economic Research Institute; *Limitation and Reduction of CO_2 and other Greenhouse Gas Emissions in Ireland*. Government of Ireland, Dublin.

Conrad, K., and T. F. N. Schmidt, 1997: Double Dividend of Climate Protection and the Role of International Policy Coordination in the E.U: An Applied General Equilibrium Model Analysis with the GEM-E3 Model. ZEW Discussion Paper n° 97-26.

Contaldi, M., and G.C. Tosato, 1999: *Carbon Tax and its Mitigation Effect through Technology Substitution: An Evaluation for Italy*. Proceedings of 22nd IAEE Annual International Conference, Rome, June 9-12, 1999.

Coppel, J., and H. Lee, 1995: Model Simulations: Assumptions and Results. In *Global Warming: Economic Dimensions and Policy Responses*, OECD, pp. 74-84.

Cornwell, A., and J. Creedy 1996: Carbon Taxation, Prices and Inequality in Australia, *Fiscal Studies*, **17**(3), 21-38.

Council of Economic Advisors, 1998: The Kyoto Protocol and the President's Policies to Address Climate Change: Administration Economic Analysis. Washington, DC.

Criqui, P. and N. Kouvaritakis, 1997: Les coûts pour le secteur énergétique de la réduction des émissions de CO_2: une évaluation avec le modèle POLES. To be published in les actes du Colloque de Fontevraud: *Quel environnement au XXIème siècle? Environnement, maîtrise du long-terme et démocratie*, 20 pp.

Criqui, P., S. Mima, and L. Viguer, 1999: Marginal Abatement Costs of CO_2 emission Reductions, Geographical Flexibility and Concrete Ceilings: An Assessment Using the POLES Model. *Energy Policy*, **27**(10), 585-602.

Cropper, M., N. Simon, A. Alberini, S. Arora, and P.K. Sharma, 1997: The Health Benefits of Air Pollution Control in Delhi. *American Journal of Agricultural Economics*, **79**(5), 1625-1629.

Danish Economic Council, 1997: *Dansk økonomi, efterår 1997* (Danish Economy, Autumn 1997). Danish Economic Council, Copenhagen.

Davis, D., 1997: The Hidden Benefits of Climate Policy: Reducing Fossil Fuel use Saves Lives Now. *Environmental Health*, Notes 1-6.

Davis, D.L., A. Krupnick, and G. Thurston, 2000: The Ancillary Health Benefits and Costs of GHG Mitigation: Scope, Scale, and Credibility. Expert Workshop on Assessing the Ancillary Benefits and Costs of Greenhouse Gas Mitigation Policies, March 27-29, Washington, DC.

Dessus, S., and D. O'Connor, 1999: Climate Policy Without Tears: CGE-Based Ancillary Benefits Estimates for Chile. OECD Development Centre.

Dixit, A., and V. Norman, 1984: *Theory of International Trade: A Dual General Equilibrium Approach*. Cambridge University Press, Cambridge.

Donovan, D., K. Schneider, G. Tessema, and B. Fisher, 1997: International Climate Change Policy: Impacts on Developing Countries. ABARE Research Report 97.8, Canberra, Australia.

ECON, 1997: Avgiftenes rolle i klimapolitikken (The role of taxes in the climate policy). ECON report n°60/97, ECON Centre for Economic Analysis, Oslo.

Edmonds, J., M. Wise, and D. Barns, 1995: Carbon Coalitions: The Cost and Effectiveness of Energy Agreements to Alter Trajectories of Atmospheric Carbon Dioxide Emissions. *Energy Policy*, **23**(4/5), 309-336.

Ekins, P., 1996: How Large a Carbon Tax is justified by the Secondary Benefits of CO_2 Abatement. *Resource and Energy Economics*, **18**(2), 161-187.

Ellerman, A.D., and I. Sue Wing, 2000: *Supplementarity: A Invitation for Monopsony*, Report n°59, Joint Program on the Science and Policy of Global Change, MIT, Cambridge, MA.

EMF-14 (Energy Modelling Forum) Working Group, 1997: *Integrated Assessment of Climate Change*. The EMF 14 Working Group, Stanford Energy Modeling Forum, Stanford University, Stanford, CA.

EMF-16 (Energy Modelling Forum), Working Group 1999: *Economic and Energy System Impacts of the Kyoto Protocol: Results from the Energy Modeling Forum Study*. The EMF 16 Working Group, Stanford Energy Modeling Forum, Stanford University, Stanford, CA, July 1999.

EIA (Energy Information Administration), 1998: *Impacts of the Kyoto Protocol on U.S. Energy Markets and Economic Activity*. Washington, US Department of Energy.

EPA (US Environmental Protection Agency), 1999: *The Benefits and Costs of the Clean Air Act, 1990 to 2010*. EPA-410-R-99-001

European Commission, 1999: ExternE: Externalities of Energy. European Commission, DGXII, Science, Research and Development, JOULE.

Farzin, Y., and O. Tahvonen, 1996: Global Carbon Cycle and the Optimal Time Path of a Carbon Tax. *Oxford Economic Papers*, **48**(4), 515-36.

Fitz Gerald, J., and D. McCoy, 1992: The Macroeconomic Implications for Ireland (Chapter 5). In The Economic Effects of Carbon Taxes. Policy Research Series, Paper No. 14, J. Fitz Gerald, D. McCoy, (eds.), Economic and Social Research Institute, Dublin.

Frandsen, S.E., J. V. Hansen, and P. Trier, 1995: *GESMEC - En generel ligevægtsmodel for Danmark. Dokumentation og anvendelser* (GESMEC – An applied general equilibrium model for Denmark – Documentation and applications). The Secretariat of the Economic Council, Copenhagen.

Fortin, E., 1999. Etude des effets redistributifs d'une taxe sur le carbone et l'énergie à l'aide du modèle IMACLIM (Distributive impacts of a Carbon Tax: A Study with the IMACLIM Model). Internal Document, CIRED, France.

Garbaccio, R.F., M.S. Ho and D.W. Jorgenson, 1999: Controlling Carbon Emissions in China. *Environment and Development Economics*, **4**(4 Oct), 493-518

Garbaccio, R.F., M.S. Ho and D.W. Jorgenson, 2000: The Health Benefits of Controlling Carbon Emissions in China. Expert Workshop on Assessing the Ancillary Benefits and Costs of Greenhouse Gas Mitigation Policies, March 27-29, Washington, DC.

Gielen, D.J., 1999: The Impact of GHG Emission Reduction on the Western European Materials System. In *Material Technologies for Greenhouse Gas Emission Reduction*, D.J.Gielen, (ed), Workshop Proceedings, April 1998, Utrecht. Netherlands Energy Research Foundation ECN, Petten /Dutch National Research Programme on Global Air Pollution and Climate Change, The Netherlands, 3-16.

Gielen, D.J. and J.H.M. Pieters, 1999: *Instruments and Technologies for Climate Change Policy: An Integrated Energy and Materials Systems Modeling Approach*. Paris, OECD, Environment Directorate, Environment Policy Committee, ENV/EPOC/GEEI(99)15.

Gielen, D.J., P.R. Koutstaal, T. Kram, and S.N.M. van Rooijen, 1999a: Post-Kyoto: Effects on the Climate Policy of the European Union. In *Dealing with Uncertainty Together*. Summary of Annex VI (1996-1998). T. Kram (Project Head), International Energy Agency, Energy Technology Systems Analysis Programme (ETSAP).

Gielen, D., S. Franssen, A. Seebregts, and T. Kram, 1999b: Markal for Policy Instrument Assessment - The OECD TOG Project. Paper prepared for the IEA/ETSAP Annex VII Workshop, Washington, DC, 1-12.

Gielen, D., T. Kram, and H. Brezet, 1999c: Integrated Energy and Materials Scenarios for Greenhouse Gas Emission Mitigation. Paper prepared for the IEA/DOE/EPA Workshop Technologies to Reduce GHG Emissions: Engineering-Economic Analysis of Conserved Energy and Carbon, Washington, DC, 1-15.

Glomsrød, S., H. Vennemo, and T. Johnsen, 1992: Stabilisation of Emissions of CO_2: A Computable General Equilibrium Assessment. *The Scandinavian Journal of Economics*, **94**(1), 53-69.

Glueck, H., and S. Schleicher, 1995: Endogenous Technological Progress Induced by CO_2 Reductions Policies. *Environmental and Ressources Economics*, **5**, 151-153.

Gold, D., A. Damakosh, A. Pope, D. Dockery, W. McDonnell, P Serrano, S. Retama, and M. Castillejos, 1999: Particulate and Ozone Pollutant Effect on the Respiratory Function of Children in Southwest Mexico City. *Epidemiology*, **10**(1), 8-17.

Golombek, R., C. Hagem, and M. Hoel, 1995: Efficient Incomplete International Climate Agreements. *Resource-and-Energy-Economics*, **17**(1), 25-46.

Gørtz, M., J.V., Hansen, and M. Larsen, 1999: CO_2-skatter, dobbelt-dividende og konkurrence i energisektoren. Anvendelser af den danske AGL-model ECOSMEC (CO_2 taxes, double dividend and competition in the energy sector: Applications of the Danish CGE model ECOSMEC). *Arbejdpapir*, **1**.

Goulder, L.H., 1995a: Effects of Carbon Taxes in an Economy with Prior Tax Distortions: An Intertemporal General Equilibrium Analysis. *Journal of Environmental Economics and Management*, **29**, 271-297.

Goulder, L.H., 1995b: Environmental Taxation and The Double Dividend: A Reader's Guide.*International Tax and Public Finance*, **2**, 157-183.

Goulder, L. H., and S. H. Schneider, 1999: Induced Technological Change and the Attractiveness of CO_2 Abatement Policies. *Resource and Energy Economics*, **21**, 211-253.

Goulder, L. H. and K. Mathai, 2000: Optimal CO_2 Abatement in the Presence of Induced Technological Change. *Journal of Environmental Economics and Management*, **39**, 1-38.

Goulder, L.H., I.W.H. Parry, and D. Burtraw, 1997: Revenue-Raising *versus* Other Approaches to Environmental Protection: The Critical Significance of Pre-Existing Tax Distortions. *RAND Journal of Economics*, **Winter**.

Goulder, L.H., I.W.H. Parry, C, Roberton, C. Williams III, and D. Burtraw, 1999: The Cost-Effectiveness of Alternative Instruments for Environmental Protection in a Second-best Setting. *Journal of Public Economics*, **72** (3), 329-360.

Grubb, M., 1997. Technologies, Energy Systems and the Timing of CO_2 Abatement: An Overview of Economic Issues. *Energy Policy*, **25**, 159-172.

Grubb, M., 2000: Economic Dimensions of Technological and Global Responses to the Kyoto Protocol. *Journal of Economic Studies*, **27**(1/2), 111-125

Grubb, M., T. Chapuis, and M. Ha Duong, 1995: The Economics of Changing Course: Implications of Adaptability and Inertia for Optimal Climate Policy. *Energy Policy*, **23**(4/5).

Grubb, M., C. Vrolijk, and D. Brack, 1999: *The Kyoto Protocol, a guide and assessment*. The Royal Institute of International Affairs and Earthscan Publ., London.

Grübler, A., and S. Messner, 1998: Technological Change and the timing of Mitigation Measures. *Energy Economics*, **20**, 495-512.

Güllen, G., 1996: Is OPEC a Cartel? Evidence from Co-integration and Causality Tests. *Energy Journal*, **17**(2), 43-57.

Ha-Duong, M., M. Grubb, and J-C. Hourcade, 1997: Influence of Socio-economic Inertia and Uncertainty on Optimal CO_2 Emission Abatement. *Nature*, **390**, 270-273.

Ha-Duong, M., J-C. Hourcade and F. Lecocq, 1999: Dynamic Consistency Problems behind the Kyoto Protocol. *Int.J. Environment and Pollution*, **11**(4), 426-446.

Hagler Bailly, 1995: Final Report prepared by Hagler Bailly Consulting, Inc., Boulder, Colorado, under subcontract to ICF Incorporated, Fairfax, VI for U.S. Environmental Protection Agency, Acid Rain Division.

Håkonsen, L., and L. Mathiesen, 1997: CO_2-Stabilization May Be a 'No-Regrets' Policy. *Environmental and Resource Economics*, **9**(2), 171-98.

Halsnaes, K., and A. Markandya, 1999: Comparative Assessment of GHG Limitation Costs and Ancillary Benefits in Country Studies for DCs and EITs. Working Paper, UNEP Collaborating Centre on Energy and Environment, Denmark, June, 1-20.

Hamilton, K., and C. Cameron, 1994: Simulating the Distributional Effects of a Carbon Tax. *Canadian Public Policy*, **20**, 385-399.

Harrison, G. W., and B. Kriström, 1999: General Equilibrium Effects of Increasing Carbon Taxes in Sweden. In Green Taxes - Economic Theory and Empirical Evidence from Scandinavia. R. Brännlund, I-M. Gren, (eds.), Edward Elgar, Cheltenham, UK.

Harrison, G.W., and T. Rutherford, 1999: Burden sharing, joint implementation and carbon coalitions. In *International Agreements on Climate Change*, C. Carraro, (ed.), Kluwer Academic Publishers, Dordrecht.

Hauch, J., 1999: *Elephant – A Simulation Model for Environmental Regulation at Nordic Energy Markets*. PhD thesis, Danish Economic Council, Copenhagen.

Hertel, T., (ed.), 1997: *Global Trade Analysis: Modeling and Applications*. Cambridge University Press, Cambridge, 403 pp.

Hill, M. 1999: Green Tax Reforms in Sweden: The Second Dividend and the Cost of Tax Exemptions. Beijer Discussion Paper Series No. 119, Beijer Institute of Ecological Economics, The Royal Swedish Academy of Sciences, Stockholm.

Hinchy, M., K. Hanslow, and B. Fisher, 1998: Gains from International Emissions Trading in a General Equilibrium Framework. ABARE Working Paper.

Hoel, M., and S. Kverndokk, 1996: Depletion of fossil fuels and the impacts of global warming. *Resource and Energy Economics*, **18**(2), 115-136.

Hotelling, H., 1931: The Economics of Exhaustible Resources. *Journal of Political Economy*, **39**, 137-175.

Hourcade, J-C., K. Heloui, and F. Ghersi. 2000a: Les déterminants du double dividende d'écotaxes : Rôle du changement technique et du risque d'embauche (The Determinants of Double Dividend: The Importance of Technical Change and Employment Risk). *Economie et Prévisions*, **143-144**, 47-68.

Hourcade, J-C., P. Courtois, and T. Lepesant, 2000b: Socio-economic of Policy Formation and Choices. In *Climate Change and European Leadership*. J. Gupta, M. Grubb. (eds), Kluwer Academic Publishers, Dordrecht.

Huff, K., and T. Hertel, 1996: Decomposing Welfare Changes in the GTAP Model. GTAP Technical Paper No. 5, Purdue University, USA, 80 pp.

IAT (Interagency Analytical Team), 1997: *Climate Change Economic Analysis*. Technical Annex. Markal-Macro Modeling Analysis Supporting the Interagency Analytical Team: Methodology, Technology Assumptions and Results.

Interlaboratory Working Group, 1997: *Scenarios of US Carbon Reductions: Potential Impacts of Energy efficient and Low Carbon Technologies by 2010 and beyond*. Oakridge and Lawrence Berkeley National Laboratories, CA, ORNL-444 and LBNL-40533.

IPCC (Intergovernmental Panel on Climate Change), 1995: *Climate Change 1994: Radiative Forcing of Climate Change and An Evaluation of the IPCC IS92 Emission Scenarios*. J.T. Houghton *et al.*, (eds.), Cambridge University Press. Cambridge.

IPCC (Intergovernmental Panel on Climate Change), 1996: Climate Change 1995. Economic and Social Dimensions of Climate Change. Contribution of Working Group III to the Second Assessment Report of the Intergovernmental Panel on Climate Change. Cambridge University Press, Cambridge.

IPCC, 2000: *Emissions scenarios*. A special report of Working Group III of the Intergovernmental Panel on Climate Change, Cambridge University Press, Cambridge.

Jaccard, M., A. Bailie, and J. Nyboer, 1996: CO_2 Emission Reduction Costs in the Residential Sector: Behavioral Parameters in a Bottom-up Simulation Model. *The Energy Journal*, **17**(4), 107-134.

Jacoby, H., R. Eckaus, D. Ellerman, R. Prinn, D. Reiner, and Z. Yang, 1997: CO_2 Emission Limits: Economic Adjustments and the Distribution of Burdens. *The Energy Journal*, **18**(3), 31-58.

Jaffe, B., and R. N. Stavins, 1994: The Energy-Efficiency gap. What does it mean?. *Energy Policy*, **22**(10) 804-810.

Jensen, J., 1998: Carbon Abatement Policies with Assistance to Energy Intensive Industry. Working paper No 2/98, The MobiDK Project, The Ministry of Business and Industry, Copenhagen.

Jensen, J., and T. N. Rasmussen, 1998: *Allocation of CO_2 Emission Permits: A General Equilibrium Analysis of Policy Instruments*. The MobiDK Project, Ministry of Business and Industry, Copenhagen.

Jerkkola, J., J. , and J. Pohjola, 1993: A CGE Model for Finnish Environmental and Energy Policy Analysis: Effects of Stabilizing CO_2 Emissions. Discussion papers No. 5, Helsinki School of Economics, Helsinki.

Jiang, K., X. Hu, Y. Matsuoka, and T. Morita, 1998: Energy Technology Changes and CO_2 Emission Scenarios in China. *Environmental Economics and Policy Studies*, **1**(2), 141-160.

Jochem, E., 1998: Policy Scenarios 2005 and 2020 – Using Results of Bottom-Up Models for Policy Design in Germany. Working Paper, Fraunhofer Institute for Systems and Innovation Research, Karlsruhe, Germany, pp. 1-4.

Johnsen, T.A., B. M. Larsen, and H. T. Mysen, 1996: Economic impacts of a CO_2 tax. In *MSG-EE: An Applied General Equilibrium Model for Energy and Environmental Analyses*. K. H. Alfsen, T. Bye, E. Holmøy, (eds.), Social and Economic Studies 96, Statistics Norway.

Jorgenson, D.W., 1997: How Economics Can Inform the Climate Change Debate. AEI Conference Summaries, May 1997, American Enterprise Institute for Public Policy Research. (http://www.aei.org/cs/cs7755.htm.).

Jorgenson, D.W., and P.J. Wilcoxen, 1995: Reducing U.S. Carbon Emissions: An Econometric General Equilibrium Assessment. In *Reducing Global Carbon Dioxide Emissions: Costs and Policy Options*. D. Gaskins, J. Weyant, (eds.), Stanford University Press, Stanford, CA (forthcoming).

Kainuma, M., Y. Matsuoka, T. Morita, and G. Hibino, 1999: Development of an End-Use Model for Analyzing Policy Options to Reduce Greenhouse Gas Emissions. *IEEE Transactions on Systems, Man, and Cybernetics – Part C: Applications and Reviews*, **29**(3 –August), 317-323.

Kainuma, M., Y. Matsuoka, and T. Morita, 2000: The AIM/End-Use Model and its Application to Forecast Japanese Carbon Dioxide Emissions. *European Journal of Operational Research*.

Kanudia, A., and R. Loulou 1998a: Robust Responses to Climate Change via Stochastic MARKAL: The Case of Québec. *European Journal of Operations Research*, **106**, 15-30.

Keohane, R. and J.S. Nye, 1998: Power and Interdependence in the Information Age. *Foreign-Affairs*; **77**(5), pp. 81-94.

Kline, S., and N. Rosenberg, 1986: An Overview of Innovation. In *The Positive Sum Strategy: Harnessing Technology For Economic Growth*. R. Landau, N. Rosenberg (eds.), National Academy Press, Washington, DC, pp. 275-305.

Koomey, J., C. Richey, S. Laitner, R. Markel, and C. Marnay, 1998: *Technology and GHG Amissions: An Integrated Analysis using the LBNL-NEMS Model*. Technical Report LBNL-42054, Lawrence Berkeley National Laboratory, Berkeley, CA.

Kram, T., 1999a: Energy Technology Systems Analysis Programme-ETSAP: History, the ETSAP Kyoto Statement and Post-Kyoto Analysis. In *Energy Models for Decision Support: New Challenges and Possible Solution,* Proceedings of conference in Berlin, May 4-5, 1998, Laege and Schaumann, (eds.), OECD, ETSAP, pp. 23-42.

Kram, T. (Project Head), 1999b: Dealing with Uncertainty Together. Summary of Annex VI (1996-1998). International Energy Agency, Energy Technology Systems Analysis Programme (ETSAP), 83 pp.

Krause, F., 1995: Cutting Carbon Emissions: Burden or Benefit? Executive Summary. In *Energy Policy in the Greenhouse*, IPSEP.

Krause, F., 1996: The Costs of Mitigating Carbon Emissions: A Review of Methods and Findings from European Studies. Energy Policy, Special Issue on the Second UN IPCC Assessment Report, Working Group III, 24 (10/11), 899-915.

Krause, F., J. Koomey, and D. Olivier, 1999: Cutting Carbon Emissions while Saving Money: Low Risk Strategies for the European Union. Executive Summary.In Energy Policy in the Greenhouse, Vol. II, Part 2, IPSEP, International Project for Sustainable Energy Paths, El Cerrito, CA, November, 33 pp.

Krupnick, A., D. Burtraw, and A. Markandya, 2000: The Ancillary Benefits and Costs of Climate Change Mitigation: A conceptual Framework. Expert Workshop on Assessing the Ancillary Benefits and Costs of Greenhouse Gas Mitigation Policies, March 27-29, Washington, DC.

Kverndokk, S., and K. E. Rosendahl, 2000: CO_2 Mitigation Costs And Ancillary Benefits in the Nordic Countries, the UK and Ireland: A survey. Memorandum 34/2000, Department of Economics, University of Oslo, Oslo.

Kydes, A.S., 1999: Energy Intensity and Carbon Emission Responses to Technological Change: The U.S. Outlook. *The Energy Journal*, **20**(3), 93-121.

Kypreos, S., 1998: The Global MARKAL-MACRO Trade Model. In *Energy Models for Decision Support: New Challenges and Possible Solutions*, Conference Proceedings, Berlin, May 4-5, 1998, Laege and Schaumann, (eds),. Paris, OECD, ETSAP, pp. 99-112.

Kypreos, S., and L. Barreto, 1999: *A Simple Global Electricity MARKAL Model with Endogenous Learning*. Villigen, Paul Scherrer Institute, General Energy Research Department-ENE, Energy Modeling Group, 24 pp.

Kypreos, S., Barreto, L., Capros, P. and Messner, S., 2000: ERIS: A model prototype with endogenous technological change, *International Journal of Global Energy Issues*, **14**(1,2,3,4):374-397.

Laffont, J.J., and J. Tirole, 1993: *A Theory of Incentives in Procurement and Regulation*. MIT Press, Cambridge, MA.

Laitner, J.A., 1997: *The Benefits of Technology Investment as a Climate Change Policy*. Proceedings of the 18th North American Conference of the USAEE/IAEE, September 7-10, San Francisco, CA.

Laitner, J.A., 1999: *Modeling the Influences of Technology Policy, Organizational Structure, and Market Characteristics on the Economic Evaluation of Climate Change Policies*. Proceedings of ACEEE Industrial Summer Study, Saratoga Springs, NY. June.

La Rovere, E.L., L.F.L. Legey, and J.D.G. Miguez, 1994: Alternative Energy Strategies for Abatement of Carbon Emissions in Brazil. *Energy Policy*, **22**(11), pp. 914-924.

Larsson, T., P.E. Grohnheit and, F. Unander, 1998: Common Action and Electricity Trade in Northern Europe. *International Transactions in Operational Research*, **5**(1), 3-11.

LEAP, 1995: *Use Guide for Version 95 of the LEAP Model*. Tellus Institute, Boston Centre, Boston, MA.

Lecocq, F., J-C. Hourcade, and T. Lepesant, 1999: Equity, Uncertainty and the Robustness of Entitlement Rules. IIASA, EMF, IEA Conference on Energy Modelling, Paris, June 16-18.

Lee, R., A.J. Krupnick, D. Burtraw, *et al.*, 1995: *Estimating Externalities of Electric Fuel Cycles: Analytical Methods and Issues and Estimating Externalities of Coal Fuel Cycles*. McGraw-Hill/Utility Data Institute, Washington, DC.

Lindholt, L., 1999: Beyond Kyoto: CO_2 Permit Prices and the Markets for Fossil Fuels. Discussion Paper 258, Statistics Norway, Oslo, 45pp.

Linn, W. S., Y. Szlachcic, H. Gong, Jr., P. L. Kinney and K. T. Berhane, 2000. Air pollution and Daily Hospital Admissions in Metropolitan Los Angeles. *Occup Environ Med* **57**(7), pp 477-483.

Lloyd, P. 1994: Aggregation by industry in high-dimensional models. *Review of International Economics*, **2**(2), 97-111.

Loulou, R., and D. Lavigne, 1996: MARKAL Model with Elastic Demands: Application to GHG Emission Control. In *Operations Research and Environmental Engineering*. C. Carraro, A. Haurie (eds.), Kluwer Academic Publishers, Dordrecht, pp. 201-220.

Loulou. R., and A. Kanudia, 1998: The Kyoto Protocol, Inter-Provincial Cooperation and Energy Trading: A Systems Analysis with Integrated MARKAL Models. *Energy Studies Review*.

Loulou, R., and A. Kanudia, 1999a. *The Regional and Global Economic Analysis of GHG Mitigation Issue via Technology-Rich Models: A North American Example*. International Modelling Conference, Paris, 15-17 June 1999, IEA/EMF/IIASA, 27 pp.

Loulou, R., and A. Kanudia, 1999b: Minimax Regret Strategies for Greenhouse Gas Abatement: Methodology and Application. *Operations Research Letters*.

Loulou, R., A. Kanudia, and D. Lavigne, 1998: GHG Abatement in Central Canada with Inter-provincial Cooperation. *Energy Studies Review*, **8**(2), 120-129 (This volume is dated 1996, but appeared in January 1998).

Loulou, R., A. Kanudia, M. Labriet, K. Vaillancourt, and M. Margolick, 2000: *Integration of GHG Abatement Options for Canada with the MARKAL Model*. Report prepared for the National Process on Climate Change and the Analysis and Modelling Group, Canadian Government, 232 pp.

Lvovsky, K. and G. Hughes, 1997: An Approach to Projecting Ambient Concentrations of SO_2 and PM-10. Unpublished Annex 3.2 to World Bank (1997).

Manne, A.S., and J. Oliveira-Martins, 1994: Comparisons of model structure and policy scenarios: GREEN and 12RT. Draft, Annex to the WP1 Paper on Policy Response to the Threat of Global Warming, OECD Model Comparison Project (II), OECD, Paris.

Martin, J.B., J-M. Burniaux, and O. Martins, 1992: The Costs of International Agreements to Reduce CO_2 Emissions: Evidence from Green. *OECD Journal Studies*, **19** (Winter).

McDougall, R., and P. Dixon, 1996: Analysing the Economy-wide Effects of an Energy Tax: Results for Australia from the ORANI-E Model. In *Greenhouse: Coping with Climate Change*. W. Bouma, G. Pearman, M. Manning, (eds.), CSIRO Publishing, Melbourne, Australia.

McKibbin, W., and P. Wilcoxen, 1995: The Theoretical and Empirical Structure of the G-Cubed Model. Brookings: Discussion Papers. *International Economics*, **118.**

McKibbin, W., M. Ross, R. Shackleton, and P. Wilcoxen, 1999: Emissions Trading, Capital Flows and the Kyoto Protocol. *The Energy Journal* (Special Issue), 287-333.

Messner, S., 1995. *Endogenized Technological Learning in an Energy Systems Model*. I.I.A.S.A. Working Paper WP-95-114, Laxenburg, Austria.

Messner, S., 1997: Endogenized Technological Learning in an Energy Systems Model. *Journal of Evolutionary-Economics*; **7**(3), pp. 291-313.

Mielnik, O., and J. Goldemberg, 2000: Converging to a Common Pattern of Energy Use in Developing and Industrialised Countries. *Energy Policy,* **23**(8), 503-508.

Mongia, N., Sathaye, J., Mongia, P., 1994: Energy Use and Carbon Implications in India. *Energy Policy*, **22**(11), pp. 894-906.

Morgenstern, R.D., 2000: Baseline Issues in the Estimation of the Ancillary Benefits of Greenhouse Gas Mitigation Policies. Expert Workshop on Assessing the Ancillary Benefits and Costs of Greenhouse Gas Mitigation Policies, March 27-29, Washington, DC.

Nakicenovic, N., 1997: Technological Change and Learning. *Perspectives in Energy*, Quarterly of Moscow International Energy Club and the International Academy of Energy, **4**(2), 173-189.

Navrud, S. 1994: Economic valuation of the external costs of fuel cycles. Testing the benefit transfer approach.. In *Models for Integrated Electricity Resource Planning*, A.T. Almeida, (ed.), Kluwer Academic Publishers, Dordrecht, pp. 49-66.

Nilsson, C., 1999: *Unilateral versus Multilateral Carbon Dioxide Tax Implementations - A Numerical Analysis with the European Model - GEM-E3*. Preliminary Version. National Institute of Economic Research, Stockholm.

Nystrom, I., and C. Wene, 1999: Energy-Economy Linking in MARKAL-MACRO: Interplay of Nuclear, Conservation and CO_2 Policies in Sweden. Working Paper. Department of Energy Conversion, Chalmers University of Technology, Goteborg, Sweden.

O'Donoghue, C., 1997: Carbon Dioxide, Energy Taxes and Household Income. Working paper 90, The Economic and Social Research Institute, Dublin.

OECD (Organisation for Economic Co-operation and Development), 1995: *Climate Change, Economic Instruments and Income Distribution*, OECD, Paris.

OECD, 2000: *Ancillary benefits and Costs of Greenhouse Gas Mitigation*. Proceedings of an IPCC Co-sponsored Workshop, 27-29 March 2000, Washington, DC.

Office of the President of the United States, 1998: *The Kyoto Protocol and the President's Policies to Address Climate Change: Administration Economic Analysis*, July 1998.

Oliveira-Martins, J. and P. Sturn, 1998: *Efficiency and Distribution in Computable Models of Carbon Emission Abatement*. Economics Department Working Papers (OECD).

Oliveira-Martins, J., J-M. Burniaux, and J.P. Martin, 1992: Trade and The Effectiveness of Unilateral CO_2 -Abatement Policies: Evidence from GREEN. *OECD Economic Studies*(19), 123-140.

Olson, M., 1965: *Logic of collective action*. Harvard University Press, Cambridge, MA.

Ostertag, K., 1999: Transaction Costs of Raising Energy Efficiency. Working Paper, Fraunhofer Institute for Systems and Innovation Research (ISI)/Centre International de Recherche sur l'Environnement et le Development (CIRED), Karlsruhe, Germany/Nogent sur Marne, France, February 1999, pp. 1-15.

Ostro, B.D. 1996: *A Methodology for Estimating Air Pollution Health Effects*. Office of Global and Integrated Environmental Health, World Health Organization, Geneva.

Palmer, K., W. Oats, and P. Portney, 1995: Tightening Environmental Standards: the Benefit-Cost or the No Cost Paradigm. *Journal of Economic Perspectives*, **9**(4), 119-132.

Parry, I.W.H., 1997: Environmental Taxes and Quotas in the Presence of Distorting Taxes in Factor Markets. *Resource and Energy Economics* **19**, 203-220.

Parry, I.W.H., and A.M. Bento, 2000: Tax-Deductions, Environmental Policy and the 'Double Dividend' Hypothesis. *Journal of Environmental Economics and Management*, **39** (1), 67-96.

Parry, I.W.H., R. Williams, and L.H. Goulder, 1999: When Can Carbon Abatement Policies Increase Welfare? The Fundamental Role of Distorted Factor Markets. *Journal of Environmental Economics and Management*, **January**.

Pearce, D.,1992: The Secondary Benefits Of Greenhouse Gas Control. Working Paper GEC 92-12, CSERGE, University College London and University of East Anglia, UK.

Pearce, D., 2000: Policy Frameworks for the Ancillary Benefits of Climate Change Policies. Expert Workshop on Assessing the Ancillary Benefits and Costs of Greenhouse Gas Mitigation Policies, March 27-29, Washington, DC.

Pechan, E.H. & Associates, 1997. Estimating Global Energy Policy Effects on Ambient Particulate Matter Concentrations. Pechan Report No. 10.001/2116.

Pereira, L., D. Loomis, G. Conceicao, A. Braga, R. Arcas, H.kishi, J. Songer, G. Bohm, and P. Saldiva, 1998: Association between Air Pollution and Intrauterine Mortality in Sao Paulo Brazil. *Environmental Health Perspectives*, **106**, 325-329.

Pezzey, J., 1992: Analysis of Unilateral CO_2 Control in the European Community and OECD. *The Energy Journal*, **13**(3), 159-71.

Pizer, W., 1997: Prices *versus* Quantities revisited: the Case of Climate Change. *RFF Discussion Paper 98-02*. Resources for the Future, Washington, DC.

Pope III, C. A., M. J. Thun, *et al.*, 1995: Particulate Air Pollution as a Predictor of Mortality in a Prospective Study of U.S. adults. *American Journal of Respiratory and Critical Care Medicine*, **151**, 669-674.

Porter, M., and C. van der Linde, 1995: Towards a New Conception of the Environment-Competitiveness Relationship. *Journal of Economic Perspectives*, **9**(4), 97-118.

Poterba, J., 1991: Tax Policy to Combat Global Warming : On Designing a Carbon Tax. In *Global Warming : Economic Policy Responses*. R. Dornbush, J.M. Poterba, (eds.), MIT Press, Cambridge, MA.

Proost, S., and D van Regemorter, 1995: The Double Dividend and the role of Inequality Aversion and Macroeconomic Regimes. *International Taxes and Public Finances*, **2**, 217-219.

Read, P., 1999: Cooperative Implementation After Kyoto: Joint Implementation and the Need for Commercialized Offsets Trading (Chapter 12). In *On the Compatibility of Flexible Instruments*. C. Jepma, W. van der Gaast, (eds.), Kluwer Academic Publishers, Dordrecht.

Reilly, J., R. Prinn, J. Harnisch, J. Fitzmaurice, H. Jacoby, D. Kicklighter, J. Melillo, P. Stone, A. Sokolov, and C. Wang. 1999: Multi-Gas Assessment of the Kyoto Protocol. *Nature*, **401**, 549-555.

Rothman, D.S., 2000: Estimating Ancillary Impacts, Benefits, and Costs on Ecosystems from Proposed GHG Mitigation Policies. Expert Workshop on Assessing the Ancillary Benefits and Costs of Greenhouse Gas Mitigation Policies, March 27-29, Washington, DC.

Sathaye, J., and N.H. Ravindranath, 1998: Climate Change Mitigation in the Energy and Forestry sectors of developing countries. *Annual Review of Energy and Environment*, **23**, 387-437.

Sato, O., M. Shimoda, K. Tatemasu, and Y. Tadokoro, 1999: Roles of Nuclear Energy in Japan's Future Energy System. English summary of the JAERI technical paper 99-015, March.

Scheraga, J.D. and N.A. Leary, 1993: *Costs and Side Benefits of using Energy Taxes to Mitigate Global Climate Change*. Proceedings 1993 National Tax Journal, pp. 133-138.

Schubert, K. , and O. Beaumais, 1998: Fiscalité, marché de droits à polluer et équilibre général: application à la France. In *L'environnement , une nouvelle dimension de l'analyse économique*. K. Schubert, P. Zagamé, (eds), Vuibert, France

Schneider, S.H., and L.H. Goulder, 1997: Achieving Carbon Dioxide Emissions Reductions at Low Cost. *Nature*, **389**, 4 September.

Seebregts, A.J., T. Kram, G.J. Schaeffer, and A.J.M. Bos, 1999a: Modelling Technological Progress in a MARKAL Model for Western Europe including Clusters of Technologies. European IAEE/AEE Conference: Technological Progress and the Energy Challenge, Paris, 30 Sept -1 Oct 1999, 16 pp.

Seebregts, A., T. Kram, G. Schaeffer, A. Stoffer, S. Kypreos, L. Barreto, S. Messner and L. Schrattenholzer, 1999b: Endogenous Technological Change in Energy System Models - A Synthesis of Experience with ERIS, MARKAL, and MESSAGE. Working Paper, ECN-C-99-025, Netherlands Energy Research Foundation, Petten, The Netherlands, April, pp. 1-29.

Shackleton, R., M. Shelby, A. Cristofaro, R. Brinner, J. Yanchar, L. Goulder, D. Jorgenson, P. Wilcoxen, and P. Pauly, 1996: The Efficiency Value of Carbon Tax Revenues. In *Reducing Global Carbon Dioxide Emissions: Costs and Policy Options*. D. Gaskin, J. Weyant, (eds.), Energy Modelling Forum, Stanford, CA.

Shackleton, R., 1998: The Potential Effects of International Carbon Emission Permit Trading Under the Kyoto Protocol. OECD Experts Workshop on Climate Change and Economic Modelling : Background Analysis for the Kyoto Protocol, Room Document n°8.

Shukla, P.R., 1996: The Modelling of Policy Options for Greenhouse Gas Mitigation in India. *Ambio*, **24**(4), 240-248.

Sinclair, P., 1992: High does nothing and rising is worse: Carbon Taxes should keep Declining to Cut Harmful Emissions. *Manchester School of Economic and Social Studies*, **60**(1), 41-52.

Smith, S., 1998: Distibutional Incidence of Environmental Taxes on Energy and Carbon: A Review of Policy Issues. Paper presented at Colloque du Ministere de l'Environnnement et de l'Amenagement du Territoire, Toulouse, 13 May 1998.

Symons, E., J. Proops, and P. Gay, 1994: Carbon Taxes, Consumer Demand and Carbon Dioxide Emissions: A Simulation Analysis for the UK. *Fiscal Studies*, **15**(2), 9-43.

Tahvonen, O., 1997: Fossil fuels, stock externalities and backstop technology. *Canadian Journal of Economics*, **30**(4), 855-874.

Tseng, P., J. Lee, S. Kypreos, and L. Barreto, 1999: Technology Learning and the Role of Renewable Energy in Reducing Carbon Emissions. International Workshop on Technologies to Reduce Greenhouse Gas Emissions: Engineering-Economic Analysis of Conserved Energy and Carbon, Washington, May, 1999.

Tulpulé, V., S. Brown, J. Lim, C. Polidano, H. Pant, and B. Fisher, 1999: An Economic Assessment of the Kyoto Protocol Using the Global Trade and Environment Model. *The Energy Journal* (Special Issue) 257-285.

Ulph, A., and D. Ulph, 1994. The Optimal Time Path of a Carbon Tax. *Oxford Economic Papers*, **46**(5), 857-868.

UNEP, 1994a: *UNEP GHG Abatement Costing Studies*. Collaborating Centre on Energy and Environment, Risø National Laboratory, Denmark.

UNEP, 1994b: *UNEP Greenhouse Gas Abatement Costing Studies. Part One: Main Report*. UNEP Collaborating Centre on Energy and Environment, Risø National Laboratory, Denmark.

UNEP, 1999a: *Economics of Greenhouse Gas Limitations. Country Study Series Argentina.*
1999b. *Economics of Greenhouse Gas Limitations. Country Study Series Ecuador*
1999c. *Economics of Greenhouse Gas Limitations. Country Study Series Botswana*
1999d. *Economics of Greenhouse Gas Limitations. Country Study Series Zambia*
1999e. *Economics of Greenhouse Gas Limitations. Country Study Series Vietnam.*
1999f. *Economics of Greenhouse Gas Limitations. Country Study Series Hung*ary.
1999g. *Economics of Greenhouse Gas Limitations. Country Study Series Estonia.*
UNEP Collaborating Centre on Energy and Environment, Risø National Laboratory, Denmark.

Unger, T., and L. Alm, 1999: Electricity and Emission – Permits trade as a Means of Curbing CO_2 Emissions in the Nordic Countries. Working Paper, Department of Energy Conversion, Chalmers University of Technology, Goteborg, Sweden.

Van der Mensbrugghe, D., 1998: A (Preliminary) Analysis of the Kyoto Protocol: Using the OECD GREEN Model. In *Economic Modelling of Climate Change*, OECD, Paris, pp. 173-204.

Verikios, G. and K. Hanslow, 1999*: Modelling the Effects of Implementing the Uruguay Round: A Comparison Using the GTAP Model under Alternative Treatments of International Capital Mobility.* Paper presented to the Second Annual Conference on Global Economic Analysis, Denmark, June 20-22, 1999.

Von Hippel, D., and B. Granada, 1993: *Application of the LEAP/EDB Energy/Environment Planning System*. SEI, Boston, MA..

Wang, X., and K. Smith, 1999: *Near-term Health Benefits of Greenhouse Gas Reductions: A Proposed Assessment Method and Application in Two Energy Sectors of China*. World Health Organization, Geneva.

Weitzman, M.L., 1974: Prices *versus* Quantities. *Review of Economic Studies*, **41**(4), 477-491.

Welsch, H. and F. Hoster, 1995: A General Equilibrium Analysis of European Carbon/Energy Taxation – Model Structure and Macroeconomic Results. *Zeitschrift fuer Wirtschafts- und Sozialwissenschaften* (ZWS), **115**, 275-303.

Weyant, J., 1999: The Costs of the Kyoto Protocol: A Multi-Model Evaluation. Special Issue of the *Energy Journal*. John Weyant (ed.).

Weyant, J., and T. Olavson, 1999: Issues in Modeling Induced Technological Change in Energy, Environment, and Climate Policy. *Journal of Environmental Management and Assessment*, **1**, 67-85.

Wigley, T., R. Richels, and J. Edmonds, 1996: Economic and Environmental Choices in the Stabilization of Atmospheric CO_2 Concentrations. *Nature*, **379**, 240-243.

Williamson, O.,1996: *The Mechanisms of Governance*. Oxford University Press, New York, USA and Oxford, UK.

Working Group on Public Health and Fossil Fuel Combustion, 1997: Short-term Improvements in Public Health from Global Climate Policies on Fossil Fuel Combustion: An Interim Report. *Lancet*, 1341-1348.

Wu, Z., J. He, A., Zhang, Q. Xu, S. Zhang, and J. Sathaye, 1994: A Macro Assessment of Technology Options for CO_2 Mitigation in China's Energy System. *Energy Policy*, **22**(11), pp. 907-913.

Ybema, J.R., M. Menkveld and T. Kram, 1998: Securing Flexibility in the Energy System to meet Future CO_2 Reduction: An Inventory of hedging options.Third Modelling Seminar on Uncertainty and Policy Choices to Meet UNFCCC Objectives, Paris, October 16-17, 1997, Netherlands Energy Research Foundation ECN, Petten.

Ybema, J.R., P. Lako, I. Kok, E. Schol, D.J. Gielen, and T. Kram, 1999: *Scenarios for Western Europe on Long Term Abatement of CO_2 Emissions*. Netherlands Energy Research Foundation ECN, Petten/Dutch National Research Programme on Global Air Pollution and Climate Change, 109 pp.

Zhang, Z.X., 1997: *The Economics of Energy Policy in China: Implications for Global Climate Change, New Horizons in Environmental Economics Series*. Edward Elgar, Cheltenham, UK.

Zhang, Z.X., 1998: Macroeconomic Effects of CO_2 Emission Limits: A Computable General Equilibrium Analysis for China. *Journal of Policy Modeling*, **20**(2), 213-250.

9

Sector Costs and Ancillary Benefits of Mitigation

Co-ordinating Lead Authors:
TERRY BARKER (UK), LEENA SRIVASTAVA (INDIA)

Lead Authors:
Majid Al-Moneef (Saudi-Arabia), Lenny Bernstein (USA), Patrick Criqui (France), Devra Davis (USA), Stephen Lennon (South Africa), Junfeng Li (China), Julio Torres Martinez (Cuba), Shunsuke Mori (Japan)

Contributing Authors:
Lee Ann Kozak (USA), Denise Mauzerall (USA), Gina Roos (South Africa), William Rhodes (USA), Sudhir Sharma (India), Sharmila B. Srikanth (India), Clive Turner (South Africa), Marcelo Villena (Chile)

Review Editor:
Tomihiro Taniguchi (Japan)

CONTENTS

EXECUTIVE SUMMARY

Policies adopted to mitigate global warming will have implications for specific sectors, such as the coal industry, the oil and gas industry, electricity, manufacturing, transportation and households. A sectoral assessment helps to put the costs in perspective, to identify the potential losers, and the extent and location of the losses, as well as to identify the sectors that may benefit. However, it is worth noting that the available literature to make this assessment is limited: there are few comprehensive studies of the sectoral effects of mitigation, compared with those on the macro gross domestic product (GDP) effects, and they tend to be for Annex B countries and regions.

There is a fundamental problem for mitigation policies. It is well established that, compared to the situation for potential gainers, the potential sectoral losers are easier to identify, and their losses are likely to be more immediate, more concentrated, and more certain. The potential sectoral gainers (apart from the renewables sector and perhaps the natural gas sector) can only expect a small, diffused, and rather uncertain gain, spread over a long period. Indeed many of those who may gain do not exist, being future generations and industries yet to develop.

It is also well established that the overall effects on GDP of mitigation policies and measures, whether positive or negative, conceal large differences between sectors. In general, the energy intensity and the carbon intensity of the economies will decline. The coal and perhaps the oil industries are expected to lose substantial proportions of output relative to those in the reference scenarios, but other sectors may increase their outputs yet by much smaller proportions. Energy-intensive sectors, such as heavy chemicals, iron and steel, and mineral products, will face higher costs, accelerated technical or organizational change, or loss of output (again relative to the reference scenario) depending on their energy use and the policies adopted for mitigation. Other industries, including renewables and services, can be expected to benefit in the long term from the availability of financial and other resources that would otherwise have been taken up in fossil fuel production. They may also benefit from reductions in tax burdens, if taxes are used for mitigation, and the revenues recycled as reductions in employer or corporate or other taxes.

Within this broad picture, certain sectors will be substantially affected by mitigation. The coal industry, producing the most carbon-intensive of products, faces almost inevitable decline in the long term relative to the baseline projection. However, technologies still under development, such as carbon dioxide (CO_2) sequestration from coal-burning plants and in-situ gasification, could play a future role in maintaining the output of coal whilst reducing CO_2 and other emissions. The oil industry also faces a potential relative decline, although this may be moderated by (1) lack of substitutes for oil in transportation and (2) substitution away from solid fuels towards liquid fuels in electricity generation. Modelling studies suggest that mitigation policies may have the least impact on oil, the most impact on coal, with the impact on gas somewhere between; these findings are established but incomplete. The high variation across studies for the effects of mitigation on gas demand is associated with the importance of its availability in different locations, its specific demand patterns, and the potential for gas to replace coal in power generation.

Particularly large effects on the coal sector are expected from policies such as the removal of fossil fuel subsidies or the restructuring of energy taxes so as to tax the carbon content rather than the energy content of fuels. It is a well-established finding that removal of the subsidies would result in substantial reductions in greenhouse gas (GHG) emissions, as well as stimulating economic growth. However, the effects in specific countries depend heavily on the type of subsidy removed and the commercial viability of alternative energy sources, including imported coal; and there may be adverse distributional effects.

There is a wide range of estimates for the impact of implementation of the Kyoto Protocol on the oil market using global models and stylized policies. All studies show net growth in both oil production and revenue to at least 2020 with or without mitigation. They show that implementation leads to a fall in oil-exporting countries' revenues, GDP, income or welfare, but significantly less impact on the real price of oil than has resulted from market fluctuations over the past 30 years. Of the studies surveyed, the largest fall in the Organization of Petroleum Exporting Countries (OPEC) revenues is a 25% reduction in 2010 below the baseline projection, assuming no permit trading and implying a 17% fall in oil prices; the reduction in OPEC revenues becomes just over 7% with Annex B trading.

However, the studies typically do not consider some or all of the following factors that could lessen the impact on oil production and trade. They usually do not include policies and measures for non-CO_2 GHGs or non-energy sources of GHGs, offsets from sinks, and actions under the Kyoto Protocol related to funding, insurance, and the transfer of technology. In addition, the studies typically do not include other policies and effects that can reduce the total cost of mitigation, such as the use of tax revenues to reduce tax burdens, ancillary environmental benefits of reductions in fossil fuel use, and induced technical change from mitigation policies. As a result, the studies may tend to overstate the overall costs of achieving Kyoto targets.

The very likely direct costs for fossil fuel consumption are accompanied by very likely environmental and public health benefits associated with a reduction in the extraction and burning of the fuels. These benefits come from a reduction in the damages caused by these activities, especially the reduction in the emissions of pollutants that are associated with combustion, such as sulphur dioxide (SO_2), nitrogen oxides (NO_x), carbon monoxide (CO) and other chemicals, and particulate matter. This will improve local and regional air and water quality, and thereby lessen damage to human, animal and plant health and the ecosystem. If all the pollutants associated with GHG emissions are removed by new technologies or end-of-pipe abatement (for example, flue gas desulphurization on a power station combined with removal of all other non-GHG pollutants), then this ancillary benefit will no longer exist. But removal of all pollutants is limited at present and it is expensive, especially for small-scale emissions from dwellings and cars.

Industries concerned directly with mitigation are likely to benefit from action. These include renewable electricity, producers of mitigation equipment (incorporating energy- and carbon-saving technologies), agriculture and forestry producing energy crops, research services producing energy and carbon-saving research and development (R&D). The extent and nature of the benefits will vary with the policies followed. Some mitigation policies can lead to overall economic benefits, implying that the gains from many sectors will outweigh the losses for coal and other fossil fuels, and energy-intensive industries. In contrast, other less well-designed policies can lead to overall losses.

These results come from different approaches and models. A proper interpretation of the results requires an understanding of the methods adopted and the underlying assumptions of the models and studies. Large differences in results can arise from the use of different reference scenarios or baselines. The characteristics of the baseline can also markedly affect the quantitative results of modelling mitigation policy. For example, if air quality is assumed to be satisfactory in the baseline, then the potential for air-quality ancillary benefits in any GHG mitigation scenario is ruled out by assumption. Even with similar or the same baseline assumptions, the studies yield different results. As regards the costs of mitigation, these differences appear to be largely a result of different approaches and assumptions, with the most important being the type of model adopted. Bottom-up engineering models assuming new technological opportunities tend to show benefits from mitigation. Top-down, general equilibrium models appear to show lower costs than top-down, time-series econometric models. The main assumptions leading to lower costs in the models are that:

- new flexible instruments, such as emission trading and joint implementation, are adopted;
- revenues from taxes or permit sales are returned to the economy by reducing burdensome taxes; and
- anacillary benefits, especially from reduced air pollution, are included in the results.

Finally, long-term technological progress and diffusion are largely given in the top-down models; different assumptions or a more integrated, dynamic treatment could have major effects on the results.

It is worth placing the task faced by mitigation policy in an historical perspective. CO_2 emissions have tended to grow more slowly than GDP in a number of countries over the last 40 years. The reasons for such trends vary but include:

- a shift away from coal and oil and towards nuclear and gas as the source of energy;
- improvements in energy efficiency by industry and households; and
- a shift from heavy manufacturing towards more service and information-based economic activity.

These trends will be encouraged and strengthened by mitigation policies.

9.1 Introduction and Progress since the Second Assessment Report

In the Second Assessment Report (SAR) and in the literature, the benefits and costs of mitigation have largely been measured in terms of macro concepts such as gross domestic product (GDP) or total welfare; sectoral effects have not been considered as a central issue. This chapter considers these sectoral implications. For a definition of co-benefits and ancillary benefits and costs, see Chapter 7; for the macroeconomic effects of mitigation policies, see Chapter 8.

The definitions of sectors adopted in this chapter is that of the UN System of National Accounts (1993 ISIC). This is an internationally agreed set of definitions, conventions, and accounts which includes the division of the macro economy into industrial sectors, such as manufacturing. The data for sectoral economic models are usually arranged according to these accounts, and the results of the models reported below (in as much as they provide a comprehensive sectoral disaggregation of the macroeconomic effects) will follow these definitions. However, the energy sector is further subdivided in this chapter, since the mitigation effects are so important and distinct for the component industries, namely coal, oil and gas, and electricity.

When assessing the sectoral responses to mitigation policies and measures, a distinction can be made between commercial firms (partnerships or corporations) and persons (such as car drivers and home-owners) as decision makers. Firms are generally expected to be more price-responsive in their fuel use, because of better access to capital, information, and technologies, while persons generally value lifestyles more highly in their fuel use decisions. Although "sectors" are largely taken to be industrial contributors to GDP, households and private motorists are also responsible for large amounts of greenhouse gas (GHG) emissions and are also covered in this chapter.

The effects of mitigation can be divided into the effects in the sector or region that undertake the mitigation policies and measures and the further, consequential effects, or spillovers, on other sectors or regions. More investment in energy-efficient equipment or in technology to develop a renewable source of energy may lead to technological spillovers on other sectors. Such spillovers are considered below.

This chapter continues with reviews of results from multisectoral studies (9.2.1), followed by those on each major sector in turn (coal, petroleum and gas, non-fossil-based energy, agriculture and forestry, manufacturing, construction, transport, service industries, and households in Sections 9.2.2 to 9.2.11). Section 9.3 reviews the literature on sectoral spillover effects of mitigation in one country or region on the rest of the world. Ancillary benefits associated with particular sectors or with sectoral mitigation policies are covered in Sections 9.2.2 to 9.2.11. Section 9.4 considers why the macro and sectoral studies come to different conclusions. Section 9.5 suggests areas for further research.

9.2 Economic, Social, and Environmental Impacts of Policies and Measures on Prices, Economic Output, Employment, Competitiveness, and Trade Relations at the Sector and Sub-sector Levels

Studies of the impact of mitigation policies on sectors can be divided into those which adopt a general approach and cover all the sectors of the economy in question, and those which concentrate on one sector or group of sectors, leaving aside indirect effects on the rest of the economy. The general studies are discussed in 9.2.1, and the sector studies are considered in the sections that follow.

The studies can also be arranged according to the methodology of the analysis:
(1) top-down studies, that capture general effects on the economy and tend to consider price-driven policies such as carbon taxes rather than technology policies;
(2) bottom-up studies that do not consider general effects but examine technology-driven options[1]; and
(3) financial cost-benefit analyses of individual mitigation measures, which do not include impacts on social factors, but sometimes do include the ancillary benefits (e.g., ADB-GEF-UNDP, 1998a).

The general studies tend to be top-down, although there have been major comprehensive bottom-up studies (e.g., Krause *et al.*, 1992). Many of the individual sector studies are bottom-up or cost-benefit. The top-down and bottom-up methodologies are compared in section 9.4.1.1.

9.2.1 *Impacts from Multisectoral Studies*

These studies tend to use large-scale models as a framework for the analysis. Important differences between the studies arise from the type of model being used (computable general equilibrium (CGE) or econometric), the method chosen for the recycling of any tax revenues, and the treatment of the world oil market. Two topics, the effects of carbon taxes (and more recently traded emission permits) and the removal of energy subsidies, have been assessed in some detail.

9.2.1.1 *Effects of Carbon Taxes and Auctioned Emission Permits*

Table 9.1 gives some details of studies of mitigation policies for which sectoral effects are available. These are all at a country or world-region level (e.g., the European Union). The table also shows the outcomes of different policies on carbon dioxide (CO_2) emissions, GDP and sectoral outputs. For some studies a range of outcomes is shown, corresponding to the range published for GDP depending on some critical assumption,

---

[1] US National Academy of Sciences (1992) reviews a number of studies on this debate.

Table 9.1: Some multisectoral studies of carbon dioxide mitigation

Region or reference country	China Garbaccio *et al.* (1999)	EU-6 DRI (1994)	EU-11 Barker (1999)	New Zealand Bertram *et al.* (1993)	UK Cambridge Econometrics (1998)	USA CRA and DRI (1994)	USA Jorgenson *et al.* (1999)	USA McKibbin *et al.* (1999)
Funding body	US Dept of Energy	EC	EC	NZ Min of Environment	FFF-FOE	Electric Power Research Institute		US EPA
Model		DRI-models	E3ME	ESSAM	MDM-E3	DRI	JWS	G-cubed
Model type	Static CGE	Macro	Macro	CGE	Macro	Macro	Dynamic CGE	Dynamic CGE
Policy	Carbon tax	Carbon tax	Carbon tax	Carbon & energy taxes	Carbon tax	Carbon Tax	Emission permits	Emission permits
Recycling mode	All other taxes	Employer taxes	Employer taxes	Corporate tax	Employer taxes	Lump-sum	Personal income	Lump-sum
Industries	29	20-30	30	28	49	About 100	35	12
Fuel types	4	17	11	4	10	4	4	5
Period	1992 to 2032	1992 to 2010	1970 to 2010	1987 to 1997	1960 to 2010	1990 to 2010	1996 to 2020	1996 to 2020
Effect year	2032	2010	2010	1996/97	2010	2010	2020	2010
Model run	15%	INT	Mult-coord.	324	C72F11	$100/tC	Personal	Unilateral US
CO$_2$	-15%	-15%	-10%	-46%	-4.4%	-15.3%	-31%	-29.6%
GDP	+1%	+0.9%	+1.4%	+4.6%	+0.1%	-2.3%	+0.6%	-0.7%
Output: coal	-19%	Energy -7%	-8%	-24%	0%	-25%	-52%	-40%
: refined oil	-2		-17	-22	-0	-6	-4	-16
: gas			-4	-41	-4	-18	-25	-14
: electricity	+3 (year 1)		-3	-17	-1	-17	-12	-6
: agriculture	+0 (year 1)	-7	+3	+4	+0		+4	-1
: forestry	..	..	..	+5	..		..	-1
: food, etc.	+0 (year 1)	Manufac-turing +1	+2	+3	+0		+5	Nondur-ables -1
: chemicals	+1 (year 1)		+2	+6	-0		-0	..
: steel	+1 (year 1)		+1	-26	-1	-5	-3	Durables -1
: construction	+1 (year 1)	..	+1	+0	+0		+1	..
: transport	+1 (year 1)	-2	+0	+5	+0	-4	+1	-2
: services	+0 (year 1)	+1	+1	+6	+0	-2	+3	-0
: consumer's expenditure	+0.8%			+6.7%	+0.1%	-1.9%	+0.7%	-0.4%

Notes: (1) "Multisectoral models" are defined as those in which GDP is divided into production sectors. Definitions of sectors differ between studies.

(2) .. denotes not available or not reported.

such as the method chosen to recycle government revenues. The effects are shown as differences from the reference scenario or the base in the final year of the projection. Note that the macroeconomic results of these studies are covered in Chapter 8.

Several conclusions are well established in this literature.

(1) The nature of the recycling of revenues from new taxes or permit schemes is critical to the sectoral effects (and the overall GDP effects - see Chapters 7 and 8 for a detailed discussion of the recycling literature). In some of the studies (e.g. Garbaccio *et al.*, 1999, 2000), GDP is increased above the reference scenario when rates for some burdensome tax are reduced. Those studies that report reductions in GDP do not always provide a range of recycling options, suggesting that policy packages that increase GDP have not been explored.

(2) Reductions in fossil fuel output below the reference case will not impact all fossil fuels equally. Fuels have different costs and price sensitivities, they respond differently to mitigation policies, energy-efficiency technology is fuel and combustion device specific, and reductions in demand can affect imports differently from output. Large effects on gas output are discussed below in section 9.2.3.2.

(3) In most instances the relative decline in output does not imply an absolute decline of the sector; rather it implies a decline in its rate of growth. This is particularly true for the oil sector, where under present technology there is a captive market in the use of oil for personal transportation, which is expected to increase substantially over the foreseeable future (this is not shown in *Table 9.1*, but reflected in the literature).

(4) The sectoral results suggest that agriculture usually benefits[2]. The effects on manufacturing are mixed and the reasons for these results are explored below. Finally, the service sectors generally increase their output as a result of the policy shifts; since services are such a large proportion of GDP, if the overall economy has higher output this usually implies that services have higher output.

It is worth placing these results and the tasks faced by mitigation policy in an historical perspective. CO_2 emissions have tended to grow more slowly than GDP in a number of countries over the last 40 years (Proops *et al.*, 1993; Price *et al.*, 1998; Baumert *et al.*, 1999). The reasons for such trends vary but include:

- a shift away from coal and oil, and towards nuclear and gas as the sources of energy;
- improvements in energy efficiency by industry and households; and

- a shift from heavy manufacturing towards more service and information-based economic activity.

These trends will be encouraged and strengthened by mitigation policies.

9.2.1.2 Reducing Subsidies in the Energy Sector

Empirical and theoretical studies indicate that no regrets policies can result from the removal of subsidies from fossil fuels or from electricity that relies on fossil fuels. The UN Framework Convention on Climate Change (UNFCCC article 4.2e (ii)) calls for Annex I Parties "to identify and periodically review its own policies and practices which encourage ... [greater emissions] than would otherwise occur". The Kyoto Protocol calls for such Parties to "implement ... measures ... such as ... progressive reduction or phasing out of market imperfections, fiscal incentives, tax and duty exemptions and subsidies in all greenhouse gas emitting sectors that run counter to the objective of the Convention ...".

The extent of the impact of reducing subsidies will depend on the specific characteristic of each country, the type of subsidy involved, and the international co-ordination to implement similar measures. Most countries introduce subsidies in order to accomplish several policy objectives. In the case of energy, these are usually in order to:

- secure domestic energy supplies;
- ensure that power supply is sufficient to meet demand;
- provide access to energy for low-income households;
- maintain or slow the loss of employment in mining communities; and
- retain the international competitiveness of domestic industry.

Coal subsidies have encouraged high production of coal in a number of industrial countries and high coal consumption in numerous developing and transition economies (OECD, 1997c). For example, a complete measure of the total support to producers can be estimated in the form of the producer subsidy equivalent (PSE), which has been calculated annually by the International Energy Agency (IEA) for several countries since 1988 (IEA, 1998b). DRI (1994) used revised versions of the IEA's coal PSE estimates (shown in *Table 9.2*) to model the effects of removing subsidies. These subsidies tend to increase GHG emissions and more general pollution.

In recent years many countries have changed their energy policy, from a focus on energy self-sufficiency, to broader policy objectives, oriented towards encouraging economic efficiency and taking into account environmental problems. Subsidies are currently under review by many countries, and in some cases reforms have already taken place. Nevertheless, large subsidies remain in both Annex I and non-Annex I countries.

In theoretical terms, polluting activities, such as coal mining and coal burning, could be taxed in order to achieve economic efficiency. Economic theory indicates that the optimal policy

---

[2] The reason for the major reduction of –7% in the DRI (1994) results for EU–6 agricultural net output is that the scenario contains a wide range of environmental policies in addition to climate change policies, and many of these impinge heavily on agriculture.

Table 9.2: *Producer subsidy equivalents for coal production in OECD countries in 1993*

	PSE per tonne US$/tce	Total PSE MUS$	Budgetary support	Price support Mtce	Subsidized production
France	43	428	100%	0%	10.0
Germany	109	6688	40%	60%	61.5
Japan	161	1034	12%	88%	6.4
Spain	84	856	37%	63%	10.2
Turkey	143	416	100%	0%	2.9
UK	15	873	2%	98%	57.4
US	0	0	0	0	0

Note: PSE is Producer Subsidy Equivalent; tce = tonne of coal-equivalent; Mtce = million tce. 1 tce = 29.308 GJ
Source: OECD (1997c).

would be to replace those production and consumption subsidies with optimal taxes. According to global studies, even without adding new taxes, removing the subsidies and trade barriers at a sectoral level would create a win-win situation, improving efficiency and reducing the environmental damage (Burniaux *et al.*, 1992; Hoeller and Coppel, 1992; Larson and Shah, 1992, 1995; Anderson and McKibbin, 1997). It is a well-established finding that removal of these subsidies would result in substantial reductions in GHG emissions, as well as stimulating economic growth. Local studies also indicate that removing support to the production and use of coal and other fossil fuels can result in substantial reductions in CO_2 emissions in the main coal-using countries, at the same time as reducing the cost of electricity production (DRI, 1994; Shelby *et al.*, 1994; Golub and Gurvich, 1996; Michaelis, 1996; OECD, 1997c, Appendix A). *Table 9.3* is a review of the quantitative results of these case studies, along with the global studies. Note, however, that these analyses adopt different methodologies, so that the figures are not directly comparable.

In spite of these results, it is not wise to generalize about the environmental and economic effects of removing subsidies in the energy industry (OECD, 1997c). For example, the effect of removing subsidies to coal producers depends heavily on the type of subsidy removed and the availability and economics of alternative energy sources, including imported coal. Removing some electricity sector subsidies may have very little effect on GHG emissions or may even increase emissions, for example, when subsidies to electricity supply industry investment are supporting the use of less polluting energy sources. Finally, there may be cases where removing a subsidy to an energy-intensive industry in one country would lead to a shift in production to other countries with lower costs or environmental standards, resulting in a net increase in global GHG emissions (OECD, 1997c). The issue of carbon leakage is addressed in greater detail in Chapter 8.

9.2.1.3   *Sectoral Impacts of the Kyoto Mechanisms*

The effects of the Kyoto Mechanisms at the sectoral level are complex. The available studies have looked at the effect of international emissions trading, but there have been no comprehensive studies on the sectoral effects of the Clean Development Mechanism (CDM) or joint implementation (JI). Countries buying assigned amount units (AAUs), or funding CDM and JI projects, may have less need to reduce fossil fuel consumption. Therefore, the sectors in these countries that depend on fossil fuel production or use may experience smaller economic impacts (Brown *et al.*, 1999). This would also reduce the impact on fossil fuel producers, both at the domestic and international level. However, countries selling credits, or hosting JI and CDM projects, will have to generate these AAUs or Certified Emission Reductions (CERs) through either reduction of GHG emissions or enhancement of sinks. The economic impact on sectors within those countries will vary depending on the source of the credits. Some sectors will benefit, while others may see reduced rates of growth. Until the rules for implementation of the Kyoto mechanisms have been decided, sectoral impacts of their use will remain speculative.

9.2.2   *Coal*

Coal remains one of the major global and long-term energy resources and is likely to continue being so as long as economically exploitable reserves are widely available. Though its relative importance has declined in industrialized countries during the last century, mainly as a result of the advent of oil and gas, 36% of world electricity is generated from coal and 70% of world steel is produced using coal and coke. Global hard coal production in 1998 was about 3,750Mt, mostly used to generate electricity, with reserves estimated at in excess of 1000 billion tonnes (WCI, 1999; IEA, 1998b, 1999). The dependence on coal use in electricity generation in developing countries is expected to continue. Depending on the efficiency of this power generation and the degree of substitution for

Table 9.3: *Summary results from case studies on energy subsidy removal (note that subsidies are defined in various ways and are not comparable)*

Study	Subsidy or group of subsidies removed	Monetary equivalent of distortion (US$ million, various years, 1988 –to 1995)	Decrease in annual CO_2 emissions relative to reference scenarios resulting from reforms by 2010 million tonnes	Other economic effects of removing subsidies
Larsen and Shah (1995)	Global price subsidies to consumers of fossil fuels (difference between domestic and world prices)[b]	215,000	1366[a]	Enhanced economic growth.
GREEN	Global price subsidies to consumers of fossil fuels (difference between domestic and world prices)[b]	235,000	1,800 in 2000 1,5000 in 2050	Enhanced economic growth in most regions, largest in CIS. Improved terms-of-trade for non-OECD countries.
DRI (1994)	Coal PSEs in Europe and Japan	5,800	10 (DRI estimate) >50 (OECD estimate)	Job loss in coal industry, increased coal trade.
Böhringer	Coal in Germany	6,700	NQ	Nearly 1% GDP increase. Job loss in coal industry, increased coal trade. Cost of using subsidies to maintain jobs is 94–145,000 DM per job/year. Reduces cost of meeting CO_2 target.
Australia	State procurement/planning	133	0.3	Reduces cost of meeting CO_2 target.
	Barriers to gas and electricity trade	1,400	0.8	Reduces cost of meeting CO_2 target.
	Below-market cost financing	NQ	NQ	
Italy	Net budgetary subsidies to the electricity supply industry (ESI)	4,000	12.5	Reduces cost of meeting CO_2 target/makes CO_2 tax more effective.
	VAT below market rate	300	0.6	
	Subsidies to capital	1,500	3.3	
	Excise tax exemption for fossil fuels use by ESI	700	5.9	
	Total net and cross-subsidies	10,000	19.2	
Norway	Barriers to trade	NQ	8 for Nordic region	
Russia	Direct subsidies and price control for coal	3,600	120	1% drop in employment
	Price control/debt forgiveness for electricity consumers	6,000	(about half caused by shift from coal to other fuels, half to reduced final energy demand)	(but note that model included no subsidy recycling mechanism).
UK	Grants and price supports for coal and nuclear producers	2,500	0 to 40	
	VAT on electricity below general rate	1,200	0.2	
USA	DFI (1993) analysis of federal subsidies	8,500[c]	10	
	DJA (1994) analysis of federal subsidies	15,400[c]	64	GNP increased 0.2% if revenue used to reduce capital taxes.

Source: OECD (1997c)

a   The model used is comparative static: emission reduction is calculated using mostly 1991 market data.

b   This measure of "subsidies" is a crude one, and does not necessarily indicate the existence of any particular government policy.

c   The two studies analyze different sets of energy supports and use slightly different estimates for some of them: these figures are not a reliable indication of total US federal energy subsidies. See Appendix A, Table 14, OECD (1997c) for details. Results are sensitive to assumptions regarding the future structure of the US electricity supply industry.

NQ = not quantified

direct coal combustion, fuel substitution can assist in reducing GHG emissions, for example when electrification reduces coal use by households (see Held *et al.*, 1996; Shackleton *et al.*, 1996; and Lennon *et al.*, 1994 for a discussion of the South African electrification programme).

The Special Report on Emissions Scenarios (Nakicenovic *et al.*, 2000) suggests that there is a very large range in the global primary energy demand expected to come from coal even in the absence of additional climate change policy initiatives. For example, in 2100, scenario A2 has a coal demand of some 900EJ, but scenario B1 has only 44EJ (the 1990 level is estimated to be 85-100EJ).

GHG mitigation is expected to lead to a decline in coal output relative to a reference case, especially in Annex B countries. Indeed the process may have already started; recent trends in coal consumption indicate a 4% reduction in OECD countries and a 12.5% increase in the rest of the world in 1997 versus 1987 (WCI, 1999). The process may lead to higher costs, especially if the change is rapid, but there are also substantial ancillary benefits. Chapter 3 discusses the wide variety of mitigation options that exist for the production and use of coal. These involve reducing emissions directly from the coal mining process, replacing coal with other energy sources or reducing coal utilization (directly through efficiency of coal combustion or indirectly via the more efficient use of secondary energy supplies).

Some of the options detailed in Chapter 3 could represent a "win-win" situation for GHG mitigation and the coal sector. For example, GHG mitigation can be achieved by reducing the coal sector's own energy consumption, beneficiation and coalbed CH_4 recovery, whilst maintaining coal production. Other options have clear, but often non-quantifiable, costs and/or ancillary benefits attached to them. The study Asia least-cost GHG abatement strategy (ALGAS)-India (ADB-GEF-UNDP, 1998a) reports that Indian CO_2 abatement would be primarily achieved by fuel switching and, to some extent, by a shift to more expensive but more efficient technologies. The most affected sector is coal as its consumption is modelled to decrease in power generation, followed by the industrial and residential sectors. The study concludes that this could lead to a significant reduction in labour employment in the coal sector. For China, using a dynamic linear programming model, Rose *et al.* (1996) find that CO_2 emissions may be reduced substantially by conserving energy and switching away from coal, without hindering future economic development.

9.2.2.1   *Costs for the Coal Sector of Mitigation Options*

Apart from the direct loss of output there are numerous other costs for the coal sector associated with mitigation. These costs relate mainly to the impact of the long-term reduction in coal consumption and hence coal production. In the short to medium term, these impacts will be moderate as global coal consumption is anticipated to continue to increase, albeit at a lower

rate. Whilst limited work has been undertaken in this area, macro impacts identified by the IEA (1997a and 1999) and the WCI (1999) include:

- reduced economic activity in coal-producing countries owing to reduced coal sales;
- job losses in the coal mining, coal transport, and coal processing sectors – especially in developing countries with high employment per unit of output;
- potential for the "stranding" of coal mining assets as well as coal processing assets;
- closure of coal mines, which are very expensive to re-open;
- higher trade deficits caused by reductions in coal exports from developing countries;
- reduction in national energy security resulting from an increased reliance on imported energy sources where local energy options are primarily coal based;
- negative impacts of mine closure on communities where the mine is the major employer; and
- possible slowdown of economic growth during the transition from coal to other energy sources in countries with a heavy reliance on coal.

Kamat *et al.* (1999) modelled the impact of a carbon tax on the economy of a geographically defined coal-based region, namely the Susquehanna River Basin in the USA. Their results indicated that maintaining 1990 emissions with a carbon tax of about US$17 per tonne of carbon could have a minor impact on the economy as a whole, however, the negative impacts on the energy sector could be considerable. In this regard the model indicates a decrease in total output of the coal sector of approximately 58%. Exports are also severely affected with resultant production cutbacks and job losses.

At the global level, Bartsch and Müller (2000) report results that suggest a significant reduction in the OECD's demand for coal under a Kyoto-style scenario against a baseline scenario. Coal demand is modelled to fall by 4.4mtoe[3] per day from this baseline in 2010 and 2020. Knapp (2000) indicates a substantial potential for relocation of the steel industry from Annex B countries to the rest of the world as coal becomes more expensive. Whilst compromising overall emission reduction objectives, this could be viewed as a positive equity contribution with economic benefits for non-Annex B countries. Knapp also indicates that the reduction in coal exports to Annex B countries for thermal power generation will severely impact some coal-exporting countries. In particular Colombia, Indonesia, and South Africa will incur substantial losses in export income with attendant job and revenue losses. These costs might, to an extent, be reduced through the use of the Kyoto CDM and technological innovation. The CDM might, for example, be used to transfer highly efficient clean coal technology to non-Annex B countries, as well as promote economic diversification to less

---

[3] mtoe means million tonnes oil equivalent; 1 tonne oil equivalent (toe) equals 45.37 GJ.

energy-intensive economic activity and the relocation of energy-intensive industries. To achieve full benefits the latter would have to be accompanied by efficiency improvements through the application of state of the art technology.

Pershing (2000) notes that internal economic growth could offset the negative export impacts within 5 years for Colombia and Indonesia, but not for South Africa. In this regard he reports that South Africa could feel the greatest impacts of the major non-Annex B coal-exporting countries. In particular, he forecasts revenue losses for Indonesia and South Africa as being as high as 1% and 4% of gross national product (GNP) respectively. Dunn (2000) reports that the coal industry has been shedding jobs for several years now and this trend is likely to continue in the coal industry as GHG mitigation actions take effect. Pershing (2000), however, suggests that such impacts may not materialize as a result of the implementation of the Climate Convention or Kyoto Protocol commitments. For example, most projections are based on the use of macroeconomic models - most of which do not take into account fossil fuel distribution effects at the national level, or the use of CO_2 sinks or non-CO_2 GHG mitigation options. Pershing also suggests that some of these impacts may be offset by other aspects of future energy and development paths. For example, in a world in which climate change mitigation policies have been taken, investment in non-conventional oil supply might be deferred - lowering the impacts on conventional fuel exporters.

9.2.2.2 Ancillary benefits for Coal Production and Use of Mitigation Options

The main ancillary benefits associated with reduction in coal burning, namely public health impacts, are considered in Chapter 8. However, there are also some ancillary benefits of mitigation directly affecting the coal industry. Mitigation could increase energy efficiency in coal utilization (Tunnah *et al.*, 1994; Li *et al.*, 1995). The uptake of new, high efficiency, clean coal technologies (IEA, 1998b) could lead to enhanced skills levels and technological capacity in developing nations. Further benefits include increased productivity as a consequence of increased market pressures, as well as the extension of the life of coal reserves. The costs of adjustment will be much lower if policies for new coal production also encourage clean-coal technology. Mitigation also may favour coal production in non-Annex B countries as a result of the migration of energy-intensive industries to developing countries (carbon leakage), although estimates of the scale of such leakage are highly dependent on the assumptions made in the models (Bernstein and Pan, 2000). There are also potential benefits in enhancing research and development (R&D) in the coal industry, especially in finding alternative and non-emitting applications for coal (IEA, 1999).

9.2.3 Petroleum and Natural Gas

Petroleum and natural gas are discussed in a single section, because they are often produced in the same countries and marketed by the same companies. In terms of value, petroleum is the largest single commodity traded on the world market. Coal, by comparison, is typically used in the country in which it is produced. Approximately 55% of the oil produced worldwide is exported, compared with 20% for gas and 12% for hard coal. The three fuels have quite different patterns of demand and different carbon contents per unit of useful energy.

9.2.3.1 Petroleum

Global production of crude oil in 1998 totalled 3,516Mt (approx. 147EJ). In 1997, 56% of oil was consumed in the transport sector, up from 42% in 1973 (IEA, 1997b). The emission scenarios in the Intergovernmental Panel on Climate Change (IPCC) Special Report (Nakicenovic *et al.*, 2000) show a wide range in demand for oil in 2100, from 0.5EJ in the A2 marker scenario to 248EJ in the illustrative scenario A1FI. Cumulative oil use between 1990 and 2100 in scenario A1FI is 29.6ZJ, about 200 times 1998 production, which is close to the combined conventional and unconventional resource base known today (see Chapter 3).

Oil is exported by more than 40 countries worldwide with 11 of which are members of the Organization of Petroleum Exporting Countries (OPEC). OPEC accounts for 76% of world crude oil reserves, 41% of world production and 55% of world exports (BP Amoco, 1999). On the other hand, around 54% of the world's downstream refining capacities are in the OECD, which controls 30% of the world's crude production. The petroleum industry is divided into two sectors, the "upstream" which involves finding and producing crude oil, and the "downstream" which involves refining crude oil into petroleum products and marketing those products to end-users. The distinction between OPEC and/or non-OPEC and upstream and/or downstream aspects of the market and industry is useful in assessing the impact of mitigation on prices, output and wealth.

9.2.3.1.1 The Global Oil Market

The market for crude oil is global, and a reduction in demand will affect all exporters via the price mechanism. However, the national economic impact of reduced demand varies greatly depending on the actual cost of production of crude oil and the degree to which the economies of individual producer countries are dependent on oil exports. It should be noted that the cost of production for crude oil can be very different from the market price, which includes royalties paid to government, transportation costs, and profit. Low-cost producers will be able to tolerate declines in the price of crude oil better than high-cost producers will. The more dependent a country is on oil and gas exports, the more its economy will be impacted if the value of these exports decreases.

Table 9.4: Costs of Kyoto Protocol implementation for oil exporting region/countries[a]

Model[b]	Without trading[c]	With Annex-I trading	With "global trading"
G-Cubed	-25% oil revenue	-13% oil revenue	-7% oil revenue
GREEN	-3% real income	"substantially reduced loss"	N/a[d]
GTEM	0.2% GDP loss	<0.05% GDP loss	N/a
MS-MRT	1.39% welfare loss	1.15% welfare loss	0.36% welfare loss
OPEC Model	-17% OPEC Revenue	-10% OPEC revenue	-8% OPEC revenue
CLIMOX	N/a	-10% some oil exporters' revenues	N/a

Source: Pershing (2000)

a   The definition of oil exporting country varies: for G-Cubed and the OPEC model it is the OPEC countries, for GREEN it is a group of oil exporting countries, for GTEM it is Mexico and Indonesia, for MS-MRT it is OPEC + Mexico, and for CLIMOX it is West Asian and North African oil exporters.

b   The models are all considering the global economy to 2010 with mitigation according to the Kyoto Protocol targets (usually in the models, applied to CO_2 mitigation by 2010 rather than GHG emissions for 2008 to 2012) achieved by imposing a carbon tax or auctioned emission permits with revenues recycled through lump-sum payments to consumers; no ancillary benefits, such as reductions in local air pollution damages, are taken into account in the results. See Weyant (1999).

c   "Trading" denotes trading in emission permits between countries.

d   N/a denotes "not available".

Different top-down models have been used to study the effects of CO_2 abatement on the oil market.[4] Few macroeconomic models have explicitly examined the economic impact of CO_2 abatement on energy-exporting countries. Most of the models (OECD's computable general equilibrium model (GREEN), OPEC's world energy model (OWEM), the IEA model, the international integrated assessment model (IIAM), and Whalley and Wigle's model (WW)) cover different world geographic regions or country groupings.

Wit (1995) surveys such models and concludes that they should be treated with caution, as hardly any of the global models have been constructed primarily to examine the economic impact of CO_2-abatement policies on energy exporters. The sensitivity of the parameters used in the surveyed models is high, which underlines the uncertainties with regard to the results. In three of the models (OWEM, GREEN, and WW) the CO_2-abatement policies would result in the energy exporters suffering the greatest welfare losses. (See Chapter 7 for a discussion on welfare losses.) The cumulative losses of a 1990 CO_2 emissions stabilization target range between 3% to 12% of GDP for energy exporting countries by 2010.

Pershing (2000) also surveys a number of model results for impacts of implementation of the Kyoto Protocol on oil export-ing countries (*Table 9.4*). Direct comparison of the model results is difficult, because each model uses a different measure of impact, and many use different groups of countries in their definition of oil exporters. However, the studies all show that use of the flexibility mechanisms will reduce the economic cost to oil producers.

These and other studies show a wide range of estimates for the impact of GHG mitigation policies on oil production and revenue. Much of these differences are attributable to the assumptions made about: the availability of conventional oil reserves, the degree of mitigation required, the use of emission trading, control of GHGs other than CO_2, and the use of carbon sinks. However, all studies show net growth in both oil production and revenue to at least 2020. As Pershing (2000) points out, these studies show significantly less impact on the real price of oil than has resulted from market fluctuations over the past 30 years. This feature (well-established) is illustrated in *Figure 9.1*. This figure shows the projection of real oil prices to 2010 from the IEA's 1998 World Energy Outlook (IEA, 1998b) and the effect of implementing the Kyoto Protocol from the G-cubed study (McKibbin *et al.*, 1999, p. 326), the study which shows the largest fall in OPEC revenues in *Table 9.4*. The 25% loss in OPEC revenues in the non-trading scenario implies a 17% fall in oil prices shown for 2010 in the figure; this is reduced to a fall of just over 7% with Annex B trading.

Many of the studies addressing the impact of CO_2 mitigation on oil producers are worth describing in more detail. Rosendahl (1996) uses a competitive dynamic model of the oil market with oil as an exhaustible resource. A constant unit cost of extraction and fixed amount of the oil resource are assumed in analyzing the impact of constant unit carbon tax. The model finds that a US$12/barrel carbon tax would reduce global oil wealth (defined as the net real value of accumulated oil pro-

---

[4] With the exception of Bartsch and Mueller (2000), all of the economic studies discussed in this section assume adequate supplies of conventional crude oil to 2020 and beyond, the generally accepted position on the availability of this resource. However, there are analysts who predict oil supply shortages before that date (Campbell, 1997). If such shortages were to develop, oil use, and therefore CO_2 emissions, would decline without the imposition of GHG mitigation policies. See Chapter 3 for an assessment of the literature on fossil fuel reserves and resources.

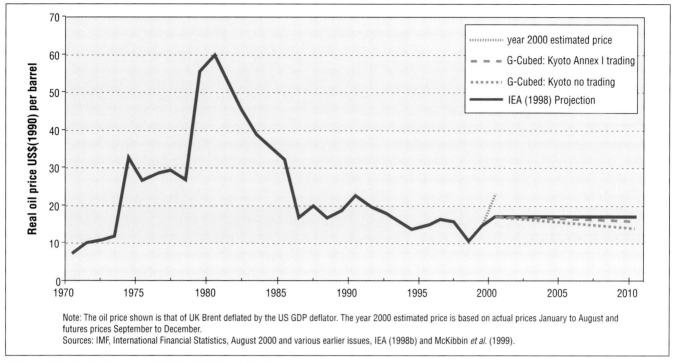

Note: The oil price shown is that of UK Brent deflated by the US GDP deflator. The year 2000 estimated price is based on actual prices January to August and futures prices September to December.
Sources: IMF, International Financial Statistics, August 2000 and various earlier issues, IEA (1998b) and McKibbin *et al.* (1999).

Figure 9.1: *Real oil prices 1970 to 2010 and the Kyoto target.*

duction) by 33%–42% and non-OPEC oil wealth 40%–54%, with the lower figure reflecting an assumed low price elasticity of -0.5[5]. Doubling the carbon tax would reduce global oil wealth by 58%–74% and non-OPEC wealth by 70%–96%. The average producer could lose about 10% of their wealth at the low tax assumption (US$3/barrel) and around two-thirds of their wealth at the high tax assumption (US$24/barrel). The marginal carbon tax increase would reduce the producers' price, or the resource rent, by 33%–50%, and increase the consumer price by 50%–67% of the tax increase.

Berg *et al.* (1997) examine the effect of a global CO_2 tax on the global oil, gas and coal markets using Statistics Norway's PETRO model. They use an optimizing, intertemporal equilibrium model with three demand regions (OECD-Europe, rest-OECD and non-OECD) and two supply regions (OPEC and the competitive fringe). They find that in the first 40 years starting in 1995, given a US$10/barrel oil-equivalent (boe[6]) carbon tax and assuming OPEC exercised market power, OPEC's oil wealth would be reduced by 20% and non-OPEC's oil wealth

by 8%, and the tax revenue would be collected by consuming countries. The tax reduces CO_2 emissions by 20% below the baseline levels over the first 50 years, and then eliminates fossil fuel combustion altogether by 2110; this comes about through an assumption of a falling real price for backstop (carbon-free) technology. Lindholt (1999) follows up this study, looking at the implications of a CO_2 tradable auctioned permit scheme to meet (a) Kyoto-style targets interpreted as CO_2 (not GHG) targets and (b) a global reduction below 1990 levels of 5.2% with emissions held constant 2010 to 2100. OPEC's production in (a) is reduced by 10% relative to the baseline in 2010 (the permit price is US$6.2/boe) rising sharply after 2040 before falling to zero in 2070, but oil prices are maintained to 2010; in (b) the reduction is 22% by 2010. Again for 2010 most of the permit revenues go to Annex B countries.

Donovan *et al.* (1997) model the economic impact of two CO_2 emission reduction scenarios compared to a reference case with no limits on emissions. In their less stringent scenario, Annex B countries stabilize their CO_2 emissions from fossil fuels at 1990 levels by 2010; in their more stringent scenario, they reduce their emissions to 15% below 1990 levels by 2010. Their model projections show oil use in 2010 reduced by 3.7% and 5.9% respectively in the less severe and more severe scenarios. These relatively small reductions are attributed to the fact that most oil is used in the transport sector where there is relatively little opportunity for substitution. In 2010, in the less stringent case, the value of oil exports from non-Annex B to Annex B countries declines by about 8%.

Jacoby *et al.* (1997) use an emissions predictions and policy analysis (EPPA) model, a CGE model derived from the OECD

[5] Estimates of the price elasticity of crude oil demand for the long and short terms differ across regions and sectors. A survey (Huntington, 1991) of inferred price elasticities from 11 world models found the OECD short-term elasticity to be –0.06 to –0.20 (average –0.12) and the long-term elasticity to be –0.35 to –0.80 (average –0.47). The corresponding estimates for the non-OECD short-term elasticity are –0.04 to –0.14 (average –0.11) and for the long term –0.17 to –0.54 (average –0.30). The world average elasticity in the short term is –0.10 and in the long term –0.38.

[6] 1 barrel oil equivalent (boe) equals 6.12GJ.

GREEN model. The world is divided into 12 trading regions, 8 production sectors, including 5 energy sectors, and 4 consumption sectors, as well as government and investment sectors. Model results show that when a quantitative emissions reduction is applied to the OECD region, all other regions suffer welfare loss from the reduction in economic activity and energy use in OECD, as well as the associated adjustment in prices of energy and the consequences on international trade. The welfare losses in energy exporting countries are no greater than that in other regions owing to the influence of backstop technologies on crude oil price and the net oil exports of the energy exporters. The OECD's lower production of the backstop (carbon-intensive heavy oil) leads to increased demand for crude oil; in addition the oil price is higher. Without the backstop technology constraint, the energy exporters suffer the largest welfare loss of all the regions.

Ghanem *et al.* (1998) use OWEM, an econometric model, to analyse the potential impacts of the Kyoto Protocol on OPEC members to 2020. The reference case for this study assumes a real oil price of US$17.4/barrel (1997$) in 2000, growing at 1.5%/year in real terms after that, and an average autonomous energy efficiency improvement of 1%/year, with higher rates in China and the economies in transition (EIT). The world economy is assumed to grow at 3.3%/year from 2000 to 2020. OPEC's production in 2020 is projected at 51.6M barrels/day (crude + natural gas liquids), and its share of world oil production is projected at 51.2%. Two scenarios are examined: firm oil prices, i.e., remaining at the reference level, and soft oil prices, US$16.9/barrel from 2000 to 2020. World oil demand in 2020 is projected at 100.7M barrels/day in the reference case, dropping to 81.1M barrels/day (OPEC at 33M barrels/day) in the firm oil price case and 83.6M barrels/day (OPEC at 37.8) in the soft price case. In the firm oil price case, OPEC has a cumulative loss in revenue compared to the base case of 20.6% (US$659bn), declining to 17.9% with trading. The loss rises to 27.2% (US$870bn) when oil prices are lower.

Brown *et al.* (1999) use the global trade and environment model (GTEM), a general equilibrium model of the world's economy, to evaluate the impact of the Kyoto Protocol's commitments, with and without unrestricted international emissions trading. The study does not consider the enhancement of sinks or the use of the CDM as mitigation policies. GTEM seeks the minimum cost for mitigation of 3 GHGs (CO_2, methane (CH_4) and nitrous oxide (N_2O)) and up to 54 economic sectors, covering 45 countries. GTEM results show that trading significantly reduces the losses in oil production in 2010 for all countries or regions reported. Because of the many assumptions that have to be made and the sector-specific impacts of emissions trading, only low confidence can be assigned to specific numerical results, but the benefits of unrestricted international emissions trading for oil producing countries has been confirmed in many studies.

Bartsch and Müller (2000) use CLIMOX, a global economic-environmental simulation model based on GREEN and GTAP,

to simulate the effects of two Kyoto scenarios on the global oil market. The "most likely" scenario assumes the implementation of the Protocol and its extension to the year 2020, with the policy instruments relying heavily on CO_2 trading permits among Annex B countries. Oil production declines by 3% in 2010 and 5% in 2020, and global oil revenues fall by an average of 12%. The model assumes supplies of conventional oil peaking in 2015 and a CH_4 leakage tax raising the price of natural gas.[7] The "global compromise" scenario assumes a global agreement to be achieved after 2012 incorporating all countries with world oil demand falling by 8% in 2020. Oil revenues fall by 19% in 2020 and by 32% in the absence of international emissions trading.

A number of studies (Kassler and Paterson, 1997; Ghasemzadeh, 2000; Pershing, 2000) have considered how impacts on oil producing countries might be alleviated. Options include: use of emissions trading and the CDM; removal of subsidies for fossil fuels that distort market behaviour; energy tax restructuring according to carbon content; increased use of natural gas, since many oil exporters are also major gas exporters; and efforts to diversify the economies of oil exporting countries.

Finally, Pershing (2000) points out that studies of the impact of GHG mitigation policies on the oil industry typically do not consider some or all of the following policies and measures that could lessen the impact on oil exporters:

- policies and measures for non-CO_2 GHGs or non-energy sources of all GHGs;
- offsets from sinks;
- industry restructuring (e.g., from energy producer to supplier of energy services);
- the use of OPEC's market power; and
- actions (e.g., of Annex B parties) related to funding, insurance, and the transfer of technology.

In addition, the studies typically do not include the following policies and effects that can reduce the total cost of mitigation:

- the use of tax revenues to reduce tax burdens or finance other mitigation measures;
- ancillary environmental benefits of reductions in fossil fuel use; and
- induced technical change from mitigation policies.

As a result the studies may tend to overstate both the costs to oil exporting countries and overall costs.

9.2.3.1.2 The US Oil Market

The US Energy Information Agency (EIA, 1998), using NEMS, an energy-economy model of the US, projects that implementation of the Kyoto Protocol would lower US petroleum consumption by 13% in 2010, and lower world oil price by 16% relative to a reference case price of US$20.77/ barrel.

---

[7] Methane has a Greenhouse Warming Potential (GWP) of 21 for a 100 year time horizon, making even small leaks significant contributors to potential impacts on climate.

Laitner *et al.* (1998) argue that an innovation-led climate strategy would be beneficial to the US economy and manufacturing. However, they project a loss of 36,000 jobs in the US oil and gas extracting industry (11% of 1996 employment) and of US$8.7bn (1993$) in contribution to GDP (about 18% of the 1996 level) (US Department of Commerce, 2000; US Bureau of Labor Statistics, 2000). Losses in the petroleum refining industry are smaller, namely 1000 jobs (1% of 1996 employment) and US$0.5bn in contribution to GDP (about 2% of the 1996 level).

Sutherland (1998) reports on a study of the impact of high energy prices on six energy-intensive industries, including petroleum refining. Prices of refined petroleum products are increased in two steps: US$75/tC in 2005 and US$150/tC in 2010. The mechanism of the price increase is not described; thus there is no discussion of who receives the revenues or how they are handled. The study finds that these price increases reduce the US demand for refined products by about 20%. The cost of other energy sources is also increased, which along with decreased demand, raises the cost of refining in OECD countries and intensifies the on-going shift of refining capacity from OECD to non-OECD countries. Shifting refining capacity to non-OECD countries reduces employment in, and increases imports by, OECD countries. Reductions in fuel use results in reductions in the emissions of local air pollutants.

9.2.3.2 Natural Gas

Global production of natural gas in 1998 totalled 2379bn cubic meters (approx. 93 EJ). In 1997, 45% of natural gas was consumed by industry, including for electric power generation, while 51% was consumed in other sectors, which include residential, commercial, agriculture, public service, and unspecified uses (IEA, 1998b). The emission scenarios in the IPCC Special Report on Emission Scenarios all show increased demand for natural gas in 2100, ranging from 127EJ in the B1 marker scenario to 578EJ in the A1FI illustrative scenario (Nakicenovic *et al.*, 2000). These scenarios are baseline scenarios, which do not include policies to limit GHG emissions.

World gas demand has grown by 3.2%/year over the past 25 years, compared to 1.6%/year for oil and 0.6%/year for coal. Most of the growth has been in power generation where it grew by 5.2%/year. This growth has increased in recent years in response to a variety of technological advantages and policy actions to reduce local air pollutants, particularly sulphur oxides (SO_x), a trend that is expected to continue through 2010, independent of policies to reduce GHG emissions (IEA, 1998b). IEA projects that demand for natural gas will grow at 2.6%/year from 1995 to 2020, 1.7%/year in OECD countries and 3.5%/year in non-OECD countries.

The IEA's projections to 2020 show that, while there is considerable further scope for switching from coal or oil to natural gas in OECD countries, the contribution of fuel switching to the further growth of gas demand in these countries is likely to

be more modest than in the past (IEA, 1998b). It is unlikely that there will be any significant switch from oil to natural gas in the transport sector during this period. Residential use of natural gas for space and water heating is reaching saturation. But it is uncertain whether natural gas demand for electricity generation will increase or decrease.

Natural gas has the lowest carbon content of the fossil fuels, and it is generally assumed that its use will increase as the result of efforts to reduce CO_2 emissions. Because of this and the possibilities for substitution in the power generation sector away from coal, Ferriter (1997) shows an increasing demand for natural gas in the two carbon tax scenarios and the efficiency-driven scenario compared to the reference case. Switching towards natural gas - especially high efficiency combined cycle and co-generation - is likely to be a very important part of reaching Kyoto targets in some countries. However, other studies (IEA, 1998b; IWG, 1997; EIA, 1998) conclude that the emissions limits set by the Kyoto Protocol will require reductions in total use of electricity and replacement of older generating capacity with non-fossil fuel units, either renewables or nuclear, decreasing the demand for natural gas.

Another uncertainty is the growth in demand from gas in non-Annex B countries. The IEA projects rapid growth in the use of natural gas in many of the non-Annex B countries e.g., 6.5%/year in China, 5.8%/year in South and East Asia, and 4.9%/year in Latin America. Bartsch and Müller (2000) also see a significant growth in gas demand in China and India to 2020, but Stern (2000) questions whether the investments in the necessary infrastructure can be made. The Kyoto Protocol's provisions on JI and the CDM could lead to further growth of natural gas use in EIT and developing nations. However, until the details of these mechanisms are agreed, it will be difficult to estimate their impact on natural gas demand.

Recent general modelling studies by Donovan *et al.* (1997) and Bernstein *et al.* (1999) suggest that, in Annex B countries, policies to reduce GHGs may have the least impact on the demand for oil, the most impact on the demand for coal, with the impact on the demand for natural gas falling in the mid-range. These results are different from recent trends, which show natural gas usage growing faster than use of either coal or oil, and can be explained as follows.

- Current technology and infrastructure will not allow much switching from oil to non-fossil fuel alternatives in the transport sector, the largest user of oil, before about 2020.
- The electric utility sector, the largest user of coal, can switch to natural gas, but the rate of switching will be limited by regional natural gas availability.
- Given the above considerations, modelling studies suggest that Annex B countries are likely to meet their Kyoto Protocol commitments by reducing overall energy use, which is likely to result in a reduction in natural gas demand.

Table 9.5: Changes in carbon dioxide emissions and gas demand from the reference case in alternative emissions abatement studies

	Change in CO_2 emissions (%)	Change in natural gas demand (%)	Ratio of changes in gas demand to changes in CO_2 emissions[d]	Year	Region
DRI (1992)	-11.7	-7.2	0.62	2005	EC
Hoeller *et al.* (1991)	-49.2	-27.4	0.56	2000	World
Bossier and De Rous (1992)	-8.2	3.0	-0.37	1999	Belgium
Proost and Van Regemorter (1992)	-28.8	15.3	-0.55	2005	Belgium
Burniaux *et al.*(1991)	-53.6	0.0	0.0	2020	World
Barker (1995)	-12.8	-6.2	0.48	2005	UK
Ghanem *et al.*(1998)	-30.7	-20.1	0.65	2010	World
Baron (1996)[a]	-8.5[b]	-4.0	0.47	2000	USA
Birkelund *et al.* (1994)	-10.7	-8.0	0.75	2010	EU
Bernow *et al.* (1997)	-17.8	-5.4	0.30	2015	Minnesota
Gregory *et al.* (1992)	-8.4	-5.2	0.62	2005	UK
WEC (1993) Scenario B	-11.1	0.0	0.0	2020	World
Kratena and Schleicher (1998)	-29.0	-36.4	1.26	2005	Austria
Mitsubishi Research Institute (1998)	-11.3[c]	9.2	-0.81	2010	OECD

a   Citing a study by US Congressional Budget Office (CBO)

b   Estimated.

c   Change in fossil fuel demand.

d   Median ratio (Column 3):   0.47

    Mean ratio: 0.26

    Std.dev.of ratio: 0.64

Given the agreement in the modelling studies and the logic that can be used to support the conclusions, this finding is established, but incomplete.

The GHG mitigation benefits of using natural gas depend on minimizing losses in its use. CH_4, the chief constituent of natural gas, is a GHG, and will be emitted to the atmosphere in natural gas leaks, most of which occur in older, low pressure distribution systems. CH_4 losses also are often a by-product of coal production. A full comparison of the benefits of switching from coal to natural gas, a step often included in mitigation strategies, requires a lifecycle analysis of CO_2 and CH_4 emissions for both fuels.

Brown *et al.* (1999) used GTEM, a general equilibrium model described above, to evaluate the impact of the Kyoto Protocol's commitments, with and without unrestricted international emissions trading, on the production of natural gas. They found the effect of emissions trading on projected natural gas production is mixed, with some countries seeing higher production rates and others, lower production rates. Because of the many assumptions that have to be made and the sector-specific impacts of emissions trading, only low confidence can be assigned to specific numerical results.

Table 9.5 summarizes a number of global economic modelling studies which project the impact of measures to mitigate CO_2

emissions on the demand for natural gas, expressed as the ratio in change in gas demand to the change in CO_2 emissions. The results are highly variable; the mean ratio is 0.14 with a standard deviation of 0.88. *Table 9.5* shows that some studies have pointed towards stronger gas demand of CO_2-abatement measures compared to the reference cases.

Longer term, natural gas would be the easiest of the fossil fuels to convert to hydrogen. This would significantly increase demand for natural gas. For technical details see Chapter 3.

9.2.3.3   Ancillary Benefits of GHG Mitigation in the Oil and Gas Industry

If, as projected, GHG mitigation policies reduce the growth in demand for crude oil they will result in several ancillary benefits: the rate of depletion of oil reserves will be slowed; and air and water pollution impacts associated with oil production, refining and consumption will be reduced, as will oil spills. Reduced growth in demand for natural gas will have similar benefits: slower rate of depletion of this natural resource, less air and water pollution associated with this industry, and less potential for natural gas explosions.

Table 9.6: *Projected nuclear energy capacity (MW)*

Country	1997	2007	2010
Japan	45248	49572	54672
South Korea	10316	19716	22716
China	2100	9670	11670
Taiwan, China	5148	7848	7848
India	1845	3990	4320
Pakistan	139	600	600
North Korea	0	2000	2000
Total	64796	93396	103826

Source: Hagan (1998)

9.2.4 Non-fossil Energy

This section covers the effects of mitigation on non-fossil-fuel-based energy production and use (electricity and biomass), and the ancillary benefits and costs associated with mitigation using non-fossil energy.

9.2.4.1 Electricity Use and Production Fuel Mix

World electricity demand in 1998 was 12.6bn MWh, about 60% of which (7.5bn MWh) was consumed in the industrialized countries (EIA, 2000a). Fossil fuels used for electricity generation account for about one third of the CO_2 emissions from the energy sector worldwide (EIA, 2000b). Globally, about 60% of all electricity is produced with fossil fuels. However, the fraction of electricity generated from fossil fuels varies across countries, from as little as 1% in Norway to 95% in the Middle East, and 97% in Poland (EIA, 2000a). Nuclear reactors are producing electricity with a global capacity of around 351GWe (IAEA, 1997), with each having an average of nearly 800MWe of installed capacity. Half of this total is concentrated in three countries: the USA with 25%, and France and Japan with 12.5% each (IAEA, 1997, pp. 10-11).

Recent projections show that electricity use will grow 37% to 16.8bn MWh by 2010, and 76% to 21.6bn MWh by 2020. About two thirds of this growth will occur outside the developed countries (EIA, 2000b). The IPCC Special Report on Emissions Scenarios (SRES) projections (Nakicenovic *et al.*, 2000) are similar, with worldwide electricity demand projected to more than double between 1990 and 2020 in scenarios A1B, A1F1 and B1, and to double between 1990 and 2020 in scenarios A2 and B2. Beyond 2020, the growth in electricity demand projected in the scenarios diverges. A1B shows the highest growth, more than 20 times between 1990 and 2100, while B1 shows the lowest growth, slightly less than 6 times between 1990 and 2100.

Much of this new power will be generated with fossil fuels. Globally, use of gas for electricity generation is projected to more than double by 2020. Global use of coal for generation is projected to grow by more than 50%, with about 90% of the projected increase occurring in the developing countries. In Asia, nuclear power is still expected to increase to meet the increasing electric power demand mainly because of resource constraint issues (Aoyama, 1997; Matsuo, 1997). *Table 9.6* shows estimates of nuclear electrical generating capacity by region to 2010.

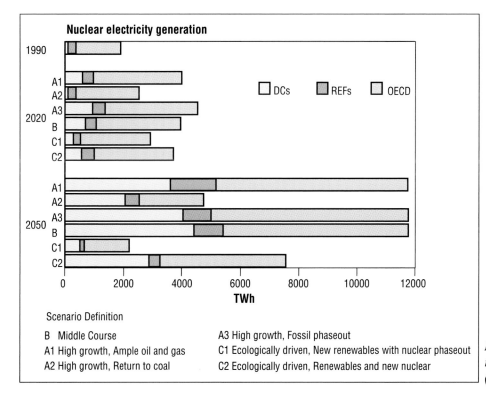

Scenario Definition

B   Middle Course
A1 High growth, Ample oil and gas
A2 High growth, Return to coal

A3 High growth, Fossil phaseout
C1 Ecologically driven, New renewables with nuclear phaseout
C2 Ecologically driven, Renewables and new nuclear

Figure 9.2: *Projection of world nuclear capacity to 2050 in TWh (Nakicenovic et al., 1998).*

Table 9.7: *Change in shares (percentage points) of alternative energy sources in electricity generation under stabilization relative to the baseline in 2010*

	Coal	Oil	Gas	Nuclear[a]	Renewables
United States	-18.1	-0.6	- 1.6	+14.1	+ 6.3
European Union	-21.2	-1.0	+ 1.7	+16.3	+ 4.2
Japan	-10.8	-8.0	- 8.2	+18.3	+ 8.6
Canada	-12.4	-1.0	- 0.3	+ 2.9	+10.8
Australia	-50.5	+2.2	+ 3.0	0.0	+45.4
New Zealand	- 2.4	-0.1	-14.0	0.0	+16.5

Source: ABARE, 1995

a These results do not take into account any barriers to the expansion of nuclear power in the USA, Canada, the EU, and Japan.

Uncertainty is reflected in the wide range in the long-term projections for nuclear energy capacity. The World Energy Council (Nakicenovic *et al.*, 1998; http://www.iiasa.ac.at/cgi-bin/ecs/bookdyn/bookcnt.py) projects a range of 2,227 to 11,840 TWh in 2050 under six possible future energy scenarios as shown in *Figure 9.2*.

9.2.4.2   Impacts of Mitigation on the Electricity Sector

Given the extensive use of fossil fuel in the production of electricity, it is not surprising that a variety of proposals have been put forth to mitigate GHG emissions in this sector. Many countries have proposed renewable technologies as one solution for GHG mitigation (Comisión Nacional de Energía, 1993; SDPC *et al.*, 1996; Piscitello and Bogach, 1997; European Commission, 1997). In some European countries such as Sweden and Austria, carbon taxes have been introduced. In Japan, nuclear power is planned to supply 480TWh in 2010, or 17.4% of total primary energy supply, to help meet the Kyoto target (Fujime, 1998). In contrast, in Sweden, a policy under debate to phase out nuclear power and restrict CO_2 emissions to 1990 levels by other means would result in significantly higher electricity prices (Anderson and Haden, 1997)

In general, mitigation policies work through two routes. First, they either mandate or directly provide incentives for increased use of zero-emitting technologies (such as nuclear, hydro, and other renewables) and lower-GHG-emitting generation technologies (such as combined cycle natural gas). Or, second, indirectly they drive their increased use by more flexible approaches that place a tax on or require a permit for emission of GHGs. Either way, the result will be a shift in the mix of fuels used to generate electricity towards increased use of the zero- and lower-emitting generation technologies, and away from the higher-emitting fossil fuels (Criqui *et al.*, 2000).

Quantitative analyses of these impacts are somewhat limited. *Table 9.1* presents published results from multisectoral models. Other multi-regional models used to assess the impacts of GHG reduction policies appear to have the capability to quantify these impacts on the electricity sector (Bernstein *et al.*, 1999; Cooper

et al., 1999; Kainuma *et al.*, 1999a, b and c; Kurosawa *et al.*, 1999; MacCracken *et al.*, 1999; McKibbin *et al.*, 1999; Tulpule *et al.*, 1999). However, the focus of the studies conducted with these models has generally been on broader economy-wide impacts, and many do not report results for the electricity sector. McKibbin *et al.* (1999) reported the price and quantity impacts on electric utilities if the USA unilaterally implements its Kyoto commitments. Under this scenario, electricity prices in the USA increase 7.2% in 2010 and 12.6% in 2020, while demand drops 6.2% and 9.5% in those years, respectively. The Australian Bureau of Agricultural and Resource Economics (ABARE, 1995) reported shifts in fuel share for Annex B under a policy where this group of countries stabilizes emissions at 1990 levels by 2000. They show that the share of coal in the generation of electricity for most Annex B countries would drop by 10% to 50%, with the combined shares for nuclear and renewables increasing 14 to 46%.[8] (See *Table 9.7* for detailed results.) They note that such a policy may require substantial structural changes in the industry and are likely to involve significant costs, but do not elaborate or quantify.

There are a number of analyses for the USA only that report detailed impacts on the electricity sector. Charles River Associates (CRA) and Data Resources International (DRI) (1994) assessed the potential impact of carbon taxes of US$50, $100, and $200 per tonne carbon, phased in to these levels over 1995 to 2000. By 2010, imposition of such taxes has increased prices of electricity by 13%, 27%, and 55% for the US$50, $100, and $200 tax, while sales dropped 8%, 14%, and 74%, respectively.

More recently, a group of studies assessing the impacts of the Kyoto Protocol on the US have reported electricity sector impacts (EIA, 1998; WEFA, 1998; DRI, 1998). These studies all use a flexible mechanism, such as tradable emissions per-

---

[8] They note that their results should only be viewed as indicative of the broad direction of the magnitude of impacts, and that they do not account for any barriers to the expansion of nuclear in the USA, Canada, the EU, and Japan.

mits, as the implementation policy. Taken together, the studies reflect a range of assumptions about the level of emissions reductions that would need to come from the domestic energy sector. The range of results for the EIA study for 2010 is summarized here, however, the results from all three studies are generally consistent. Key impacts in 2010, all of which increase as emissions reduction requirements increase, include the following.

- Electricity prices were projected to increase 20% to 86% above baseline levels.
- Electricity demand was projected to decrease 4% to 17% below baseline levels.
- Prices of natural gas were projected to increase by 35% to 206% over the baseline levels. Prices of coal for electricity production were projected to increase to about 2.5 to 9 times the baseline levels. And, despite a 7% to 40% decrease in fossil generation, fossil fuel expenditures increase 81% to 238% over baseline levels.
- About 9% to 43% of total generation will shift away from coal relative to the baseline. The large shift over this limited time period would reflect significant structural changes and potentially large stranded costs. Roughly half of this is replaced by natural gas generation, while most of the remainder is not replaced as a result of reduced demand. Renewable generation beyond baseline levels generally does not enter the mix until at least 2020.

None of the studies quantify the potential stranded costs associated with the premature retirement of existing generation.

9.2.4.3 Ancillary Benefits Associated with Mitigation in the Electricity Industry

The ancillary benefits expected from the increased use of new generating technologies adopted to achieve GHG mitigation would be sales and employment growth for those who manufacture and construct the new generation facilities. There could also be income and employment growth in the production of fuels for this new generation. The ancillary benefits associated with use of non-fossil energy for thermal applications would be similar.

Ancillary benefits of increased use of renewable sources have been described by several experts (Brower, 1992; Johansson *et al.*, 1993; Pimental *et al.*, 1994; Miyamoto, 1997). These include:

- further social and economic development, such as enhanced employment opportunities in rural areas, which can help reduce rural poverty and decrease the pressures to migrate to urban areas;
- land restoration activities such as improvement of degraded lands and associated positive impacts on farm economics, new rural development opportunities, prevention of erosion, habitats for wildlife;
- reduced emissions, in certain instances, of local pollutants;
- potential for fuel diversity; and
- elimination of the need for costly disposal of waste materials, such as crop residues and household refuse.

9.2.4.4 Ancillary Costs Associated with Mitigation in the Electricity Industry

There are also ancillary costs associated with actions to mitigate GHGs in the electricity sector. The growth experienced by those who benefit from mitigation would be offset by a decline in sales and employment for those who would have produced and constructed the facilities that would have been built without the mitigation activity. Similarly, there will be a loss of income and jobs for those that would have provided the fuel for those facilities no longer being built (i.e., the coal industry). The specifics of the mitigation policy and action will effect whether the net effect of this shifting of economic activity will be positive or negative.

There are also environmental issues associated with some of the renewable technologies. For example, concern has been raised about the ecological impacts of intensive cultivation of biomass for energy, the loss of land and other negative impacts of hydro electricity development, and the noise, visual interference, and potential for killing birds associated with wind generation (Brower, 1992; Pimental *et al.*, 1994; IEA, 1997a; Miyamoto, 1997; IEA, 1998a).

Nuclear power might be expected to increase substantially as a result of GHG mitigation policies, because power from nuclear fuel produces negligible GHGs. The construction of nuclear power stations, however, does lead to GHG emissions, but over the lifecycle of the plant these are much lower than those from comparable fossil fuel stations.

In spite of the advantages, nuclear power is not seen as the solution to the global warming problem in many countries. The main issues are (1) the high costs compared to alternative combined cycle gas turbines, (2) public concerns about operating safety and waste disposal, (3) safety of radioactive waste management and recycling of nuclear fuel, (4) the risks of nuclear fuel storage and transportation, and (5) nuclear weapon proliferation (Hagan, 1998). Whether the full potential for nuclear power development to reduce GHGs can be realized will be determined by political and public responses and safety management.

9.2.5 Agriculture and Forestry

The sectoral effects of mitigation on agriculture and forestry are described in detail in Chapter 4. This section covers ancillary benefits for agriculture.

9.2.5.1 Ancillary Benefits for Agriculture from Reduced Air Pollution

GHG mitigation strategies that also reduce emissions of ozone precursors, i.e., volatile organic compounds (VOCs) and nitrogen oxides (NO_x), may have ancillary benefits for agriculture. Elevated concentrations of tropospheric ozone (O_3) are damaging to vegetation and to human health (EPA, 1997). GHG mitigation strategies which increase efficiency in energy use or increase the penetration of non-fossil-fuel energy are likely to reduce NO_x emissions (the limiting precursor for O_3 formation in non-urban areas) and hence O_3 concentrations in agricultural regions.

Studies of the adverse impacts of O_3 on agriculture were first conducted in the United States in the 1960s, with major studies in the 1980s (EPA, 1997; Preston *et al.*, 1988) and later in Europe (U.K. DoE, 1997) and Japan (Kobayashi, 1997). These studies indicate that, for many crop species, it is well established that elevated O_3 concentrations result in a substantial reduction in yield. The US Environmental Protection Agency (EPA) funded the National Crop Loss Assessment Network (NCLAN) from 1980 to 1986, which developed O_3 dose-plant response relationships for economically important crop species (Heck *et al.*, 1984a and b). Results of this study are shown in *Figure 9.3*. The basic NCLAN methodology was used in 9 countries in Europe between 1987 to 1991 on a variety of crops including wheat, barley, beans, and pasture for the European Crop Loss Assessment Network (EUROCLAN) programme. EUROCLAN found yield reductions to be highly correlated with cumulative exposure to O_3 above a threshold of 30-40 parts per billion (ppb) (Fuhrer, 1995).

The World Health Organization (WHO) uses the AOT 40 standard to describe an acceptable O_3 exposure for crops. AOT 40 is defined as the accumulated hourly O_3 concentrations above 40 ppb (80 mg/m^3) during daylight hours between May and July. Acumulative exposure less than 6000 mg/m^3.h is necessary to prevent an excess of 5% crop yield loss (European Environment Agency, 1999). Observations indicate that this limit is exceeded in most of Europe with the exception of the northern parts of Scandinavia and the UK (European Environment Agency, 1999). Median summer afternoon O_3 concentrations in the majority of the eastern and southwestern United States presently exceed 50 ppb (Fiore *et al.*, 1998). As shown in *Figure 9.3* these concentrations will result in yield reductions in excess of 10% for several crops. IPCC Working Group (WG)I (Chapter 4) predicts that, if emissions follow their SRES A2 scenario, by 2100 background O_3 levels near the surface at northern mid-latitudes will rise to nearly 80ppb.

(However, scenario B1 has only small increases in O_3 emissions.) At the higher O_3 concentrations the yield of soybeans may decrease by 40%, and the yield of corn and wheat may decrease by 25% relative to crop yields at pre-industrial O_3 levels. Within a crop species, the sensitivity of individual cultivars to O_3 can vary (EPA, 1997), and it is possible that more resistant strains could be utilized. However, this would impose an additional constraint on agriculture.

An economic assessment of the impact of O_3 on US agriculture, based on data from the NCLAN study, found that when O_3 is reduced by 25% in all regions, the economic benefits are approximately US$1.9billion (bn) (1982 dollars) (Adams *et al.*, 1989). Conversely, a 25% increase in O_3 pollution resulted in costs of US$2.1bn (Adams *et al.*, 1985). Two recent studies found that crop production may be substantially reduced in the future in China owing to elevated O_3 concentrations (Chameides *et al.*, 1999; Aunan *et al.*, 2000, forthcoming). China's concerns about food security may make GHG mitigation strategies that reduce surface O_3 concentrations more attractive than those that do not.

9.2.5.2   Ancillary Benefits from Carbon Sequestration

Chapter 3 considers new technologies for using biomass, such as sugar cane, to replace fossil fuels. Such mitigation may have considerable associated benefits, particularly for sustainable development in creating new employment (see 9.2.10.4 below).

Alig *et al.* (1997) through modelling alternative carbon flux scenarios using the forestry and agricultural sector optimization model (FASOM) estimated the welfare effects of carbon sequestration for the US. They estimate total social welfare costs to range from US$20.7bn to $50.8bn. In the case of the agricultural sector, the consumer's surplus decreases in all scenarios.

9.2.6   Manufacturing

The effects of GHG mitigation on manufacturing sectors are likely to be very mixed, depending on the use of carbon-based fuels as inputs, and the ability of the producer to adapt produc-

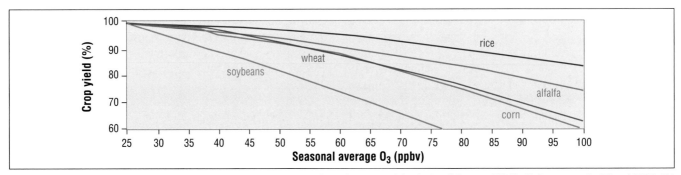

Figure 9.3: *Effect of seasonal O_3 concentrations on agricultural crop yields derived from a Weibull Parametric Fit of NCLAN O_3 exposure and yield data (Adams et al., 1989, adapted by Chameides et al., 1999)*

tion techniques and to pass on increases in costs to customers. Different manufacturing processes and technologies use carbon-based inputs in very different amounts in relation to output. High carbon-intensive sectors (using the UN System of National Accounts 1993 ISIC, p. 594-5) include basic metals (aluminium, steel), other non-metallic mineral products (cement, bricks, glass) and some chemicals (bulk chemicals). Low carbon-intensive sectors include office machinery (electronics) and other chemicals (pharmaceuticals). Several large sectors (food, textiles, machinery, and vehicles) are somewhere between these extremes.

If the Kyoto Protocol is ratified, manufacturing sectors in Annex I countries can expect to face mitigation policies to meet national targets. The possible options for the firms are basically: (1) energy conservation (adoption of more efficient technologies), (2) shift to products with lower carbon intensities, (3) accept extra taxation or emission permits and the possible effect on profits and/or product sales and (4) shift production abroad as foreign direct investment or joint ventures.

Generally, adopting these options will create ancillary benefits and costs. Speculative ancillary benefits include:
- the adoption of energy conservation technologies that mitigates local air pollution similar to the case of transportation sector;
- the accumulation of scientific and technological knowledge that contributes to the development of new products and processes; and
- the internationalization of manufacturing that stimulates technology transfer to developing regions and greater equity in wealth distribution.

If production is transferred to non-Annex B countries, ancillary costs include:
- losses in Annex B manufacturing employment; and
- increases in non-Annex B emissions.

Thus, the assessment of the effects of climate policy on manufacturing could take into account the interactions between sectors and economies. Multisectoral and multi-regional models have been used to evaluate them (see 9.2.6.1 below).

9.2.6.1 Effects on Manufacturing from Multisectoral Top-down Studies

Manufacturing sectors show mixed results in the multisectoral studies (see *Table 9.1* above). Reflecting industrial and financial globalization, recent studies tend to involve international trade on both goods and capital. In the main example, McKibbin *et al.* (1999) evaluate the potential sectoral impacts of the Kyoto Protocol using the G-Cubed model, mainly focusing on the real and the financial trading structure. In case of unilateral action by the USA, the effects on manufacturing industries show at most 1.4% and 1.2% decrease in quantity and price, respectively, although the effects on the energy industry sectors are large, e.g., 56% down in the coal mining industry with a 375.6% price increase in 2020.

9.2.6.2 Mitigation and Manufacturing Employment

Some mitigation policies would increase output and employment in the energy equipment industries. In 2010 under an innovation scenario for GHG mitigation of 10% relative to 1990, GDP for the US is projected to increase by 0.02% (Laitner *et al.* 1998). Wage and salary earnings are shown to rise in 2010 by 0.3% and employment (jobs) by 0.4%. From another perspective, these net job gains might be all provided by new small manufacturing plants in the USA; in that case, the redirected investments in energy-efficient and low-carbon technologies would produce additional employment equivalent to the jobs supported by about 6200 small manufacturing plants that open in the year 2010. While these impacts are small in relation to the larger economy, it is because the scale of investment is also relatively small. The anticipated extra energy-efficiency and renewable-energy investments in the year 2010 is less than 3% of US total investment in that year.

9.2.6.3 Mitigation Measures and Technology Strategy

Some bottom-up studies assess the relationships between climate policy and technological strategy. For instance, the implementation of a carbon tax or subsidies will strongly affect the investment decisions in manufacturing sectors. Kainuma *et al.* (1999a) assess how a subsidy affects the adoption of energy conservation technologies to meet Kyoto targets using the AIM/ENDUSE model of Japan. *Figure 9.4* shows the contribution of technologies undertaken by firms to reduce carbon emissions in 2010. Subsidies of US$30/tC compare with a carbon tax of US$300/tC to meet Kyoto targets without subsidies (using a rate of 100yen/$) (Kainuma *et al.*, 1997, 1999a).

For developing countries, environmental policy is often linked to technological improvement. Jiang *et al.* (1998) assess the potential for CO_2 emission reductions in China based on advanced energy-saving technology options under various tax and subsidy measures. For example, they consider the adoption of advanced coking-oven systems by the iron and steel industry in China. Without changes in policy only 15.9% of existing ovens will be replaced by advanced ones by 2010. With a carbon tax, but without a subsidy, the replacement share rises to 62%. With a tax and subsidies for energy-saving technologies, the share rises to 100%, i.e. the advanced ovens will be fully adopted by the industry. They also mention that taxes with subsidies do not give the best solutions for other sectors. They conclude that a carbon tax with subsidy could have reduced CO_2 emissions by 110Mt of carbon-equivalent in 2000 and 360Mt in 2010, from the baseline case of 980Mt in 2000 and 1380Mt in 2010.

Energy-saving technologies across the sectors, such as material and thermal recycling have a large potential to reduce carbon emission. Yoshioka *et al.* (1993) employed input-output analysis to evaluate the potential contribution of blast-furnace cement to reduce CO_2 emissions: improved technology for the utilization of 1 tonne of blast furnace slag to produce cement

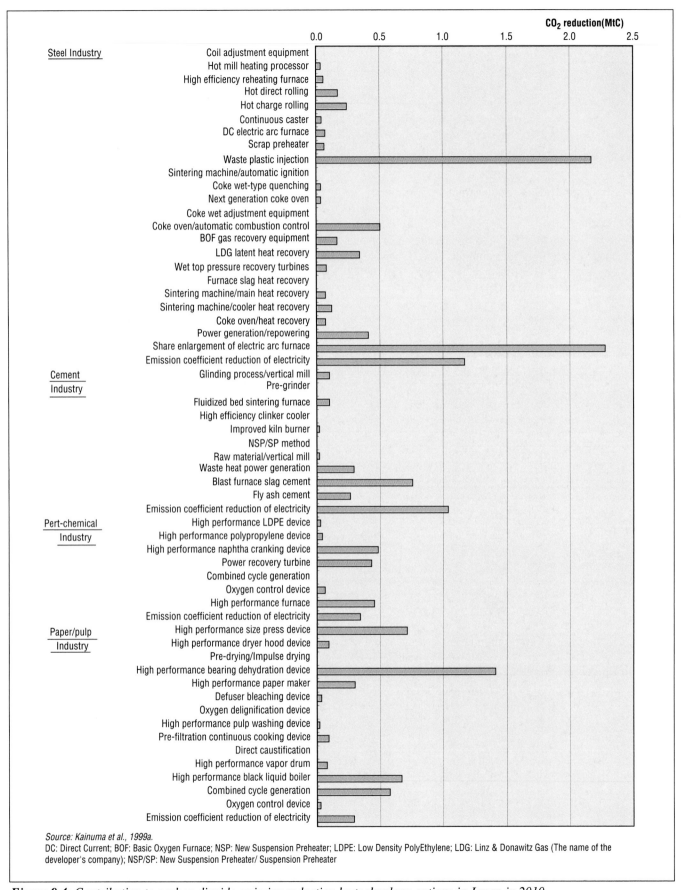

Figure 9.4: *Contribution to carbon dioxide emission reduction by technology options in Japan in 2010.*

could reduce CO_2 by 0.85 tonnes. In the same manner, Ikeda *et al.* (1995) estimated that the utilization of by-products in the steel and iron industry, and of steel scrap could reduce CO_2 emissions by 2.4% in Japan in 1990.

9.2.7 Construction

This section is concerned with the impact of mitigation on the construction industry, rather than with the options for mitigating energy use in buildings, which are considered in Chapter 3. One of the main products of the construction sector are buildings which require energy for a number of services such as lighting, space heating and cooling, and electricity for equipment. Energy consumption in buildings reaches nearly one-third of total primary energy consumption in the US, and hence their importance for GHG emission reductions. Mitigation will lead to changes in the materials used, and in design and heat control, all tending to increase the quantity (output) and improve the quality of buildings. Most renewable energy investments, such as hydropower and electricity from biomass, also require inputs from the construction sector.

Multisectoral modelling suggests that carbon tax and permit policies will have little impact on construction output and employment; this finding is established in the literature, but incomplete. *Table 9.1* shows that according to three different macroeconomic models (Garbaccio *et al.* 1999; Jorgenson *et al.*, 1999; and Barker, 1999) construction will increase its output by about +1%. Two other models in the same table (Bertram *et al.*, 1993; Cambridge Econometrics, 1998) find 0% variation in the construction output.

9.2.8 Transport

Transport energy use has been growing steadily worldwide, with the largest increases occurring in Asia, the Middle East and North Africa, and it is projected to grow more rapidly than energy use in other sectors through at least until 2020 (Michaelis and Davidson, 1996; IEA, 1997b; Schafer, 1998; Nakicenovic *et al.*, 2000). There are few options available to reduce transport energy use which do not involve significant economic, social or political costs. Governments presently find it difficult to implement measures to reduce overall demand for mobility (IEA, 1997b). Singapore is an exception to this general rule as a result of a comprehensive set of policies dating from 1975 to limit traffic (Michaelowa, 1996).

Almost all transport energy is supplied from oil, and the growing demand for transport seems inconsistent with macroeconomic studies that project decreased demand for oil as the result of GHG mitigation policies. Further research is needed to resolve this apparent inconsistency (Bernstein and Pan, 2000).

Local concerns, traffic congestion and air pollution, are currently the key drivers for transport policy (Bernstein and Pan,

2000). Measures to reduce traffic congestion also reduce CO_2 emissions, since they involve either reducing the number of vehicles on the road or increasing the average speed and fuel efficiency at which vehicles travel through urban areas. Policies to reduce traffic congestion include: improvements in mass transit, incentives for car pooling, and fees for entering city centres (Bose, 2000), as well as employer-based transport management, parking management, park-and-ride programmes, and road use pricing. One approach has been to assess the external (social) costs of transport, including contribution to global warming, as a guide to the level of taxes or user charges by transport modes that would internalize these costs, and hence improve the efficiency of the system (ECMT, 1998).

An "information society" based on a digital information network is sometimes projected to replace a substantial proportion of physical travel. However, historical data show that the telegraph and telephone did not affect the steady growth of transportation in France (see *Figure 9.5*). Mokhtarian *et al.* (1995) conclude that telecommuting, one aspect of the information society, does reduce transportation energy use. However, the reductions are smaller than often assumed, because they are partially offset by increased household energy use, and because some telecommuters do so only for part of their working day. Care must be taken in extrapolating future reductions from the limited case studies currently available; the behaviour of early-adapters may be different from that of later telecommuters. In the medium term, macro view, information technologies appear to be complementary to transportation (Gruebler, 1998); but in the longer term an "information society" could significantly replace travel and associated impacts, although this remains speculative.

9.2.8.1 Aviation

In 1999, in response to a request from the International Civil Aviation Organization (ICAO), the IPCC prepared a Special Report, *Aviation and the Global Atmosphere*, which included a comprehensive review of the potential impacts of aviation on the climate system (Penner *et al.*, 1999). The demand for air travel, as measured in revenue passenger-kilometres, is projected to grow by 5%/year for the next 15 years, but improvements in efficiency and operations are projected to hold the growth in CO_2 emissions to 3%/year. Aircraft also emit water vapour, NO_x, SO_x and soot, and trigger the formation of condensation trails (contrails) and may increase cirrus cloudiness – all of which contribute to climate change (Penner *et al.*, 1999).

Penner *et al.* present several growth scenarios for aviation that provide a basis for sensitivity analysis for climate modelling. These scenarios, which assume the scope for switching from air travel to other modes of travel is limited, show radiative forcing resulting from subsonic aircraft emissions growing from the 1992 level of $0.05 Wm^{-2}$ to between 0.13 and $0.56 Wm^{-2}$ by 2050. The scenario with economic growth equal to the IS92a

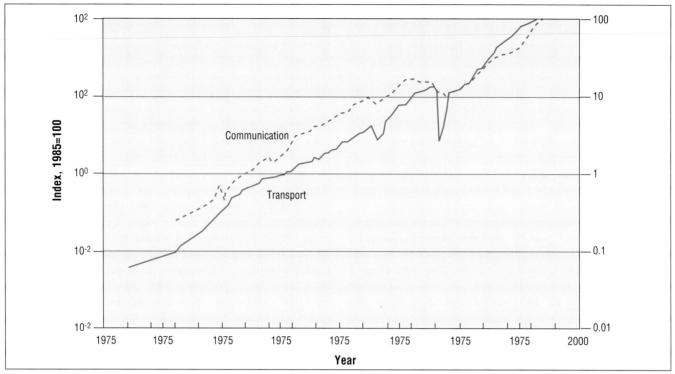

Figure 9.5: *Historical trends in transport and communication volume indices for France (Source: Gruebler, 1998: Chapter 7).*

reference scenario indicates that aviation may contribute 0.19 Wm^{-2}, or about 5% of anthropogenic radiative forcing, by 2050. More supersonic aircraft would substantially increase this contribution, although there is considerable uncertainty whether any such fleet will be developed. The growth scenarios do not consider air space and infrastructure limitations; however, recent experience in both Europe and North America indicates that the air traffic system is reaching saturation. Penner *et al.* assume that by 2050 all currently identified improvements in aircraft efficiency and operations will be implemented. However, turnover time in the aviation industry is long. Individual aircraft will be operated by commercial airlines for 25 years or more, and a successful product, including its derivatives, will be produced for possibly 25 years or longer. Thus, the overall life of an aircraft type can exceed 50 years.

Penner *et al.* (1999) conclude:
"Although improvements in aircraft and engine technology and in the efficiency of the air traffic control system will bring environmental benefits, these will not fully offset the effects of increased emissions from the projected growth in aviation. Policy options to reduce emissions further include more stringent aircraft emissions regulations, removal of subsidies and incentives that have negative environmental consequences, market-based options such as environmental levies (charges and taxes) and emissions trading, voluntary agreements, research programmes, and substitution of aviation by rail and coach. Most of these options would lead to increased airline costs and fares. Some of these approaches have not been fully investigated or tested in aviation and their outcomes are uncertain (Penner *et al.*, 1999, p. 11)."

The need for further research in this area is explored at the end of the chapter.

9.2.8.2   *Passenger Cars*

Chapter 3, section 3.4 discusses the status of low-GHG-emission technology for passenger cars. This section will discuss the effects of mitigation policies on the use of this technology and more generally on the use of passenger cars.

Government policies aimed at reducing passenger car fuel use, such as the US corporate average fuel economy (CAFE) standards, and the high tax placed on gasoline in many countries, have been in place for many years. These policies have been driven by two considerations: the cost of importing crude oil, and/or the desire to improve local environmental quality. The auto industry has responded to these policies with the introduction of successive generations of technology to improve passenger car efficiency. However, total passenger car fuel use has increased steadily as improvements in vehicle efficiency have been overwhelmed by increases in car sizes and car traffic. The number of passenger cars in use worldwide has risen from 193 to 477 million between 1970 and 1995, and total kilometres travelled have risen from 2.6 to 7.0 trillion vehicle-kilometres between 1970 and 1995 (OECD, 1997b). While growth in passenger car numbers has slowed in OECD countries, it is expected to continue to rise at a rapid rate in the rest of the world. Passenger car numbers in China are expected to increase 20-fold from 1995 to 2015 (Dargay and Gately, 1997).

Because gasoline is already taxed at a very high level in many countries, and the cost of fuel is a small portion of the total cost of driving, even fairly substantial increases in the cost of the fuel (as a GHG mitigation policy) may have little impact on vehicle use. The net present cost to the consumer of a tax equivalent to US$300/tC is approximately 5% of the capital cost of a typical new vehicle, assuming an initial cost of US$20,000, 12,000 km/year, and a 10-year life (IEA, 1997b). Furthermore, the users of company and/or government-provided cars may not be responsive to the increase in fuel cost at all, a typical case of principal-agent problem (see Chapter 3 and Chapter 6).

Initiatives to improve fuel economy continue, often with the express intention of reducing GHG emissions. European car manufacturers have voluntarily agreed to reduce the fuel consumption of new cars by 20% by 2010. In 1993, US car manufacturers entered into a partnership for a new generation vehicle (PNGV) with the US government aimed at developing a passenger car with triple the current fuel economy (to about 80 miles per gallon), by 2004, with no increase in cost or loss of performance compared with current vehicles. The incremental costs of these vehicles have been estimated to be as low as $2,500/car (DeCicco and Mark, 1997) to as high as more than $6,000/car (Duleep, 1997; OTA, 1995). Since these vehicles will be designed to meet the emissions standards anticipated to be in effect when they are produced, no ancillary local air pollution benefit is expected.

However, much of the increase in fuel efficiency may be taken up in increased demand for fuel if the lower operating costs are translated into increased ownership and use of vehicles. In addition, Dowlatabadi *et al.* (1996) find that increasing fuel economy to 60 miles per gallon had little beneficial effect on urban ozone concentration, and could decrease the safety of passenger cars unless offsetting steps were taken. Wang *et al.* (1998) estimate the capital investment required in the USA through to 2030 for fuel production and distribution to be (1) US$100bn (1995$) or less if the fuel for PNGV cars is reformulated gasoline or diesel, ethanol, methanol, liquefied petroleum gas (LPG), or liquefied natural gas (LNG); (2) approximately US$150bn for di-methyl ether; and (3) in the order of US$500bn for hydrogen. No estimate was made of the cost of applying this technology outside the USA.

The Australian Bureau of Transport and Communications Economics (BTCE, 1996) examine the social costs of 16 measures to reduce GHG emissions from the transport sector. In the longer term, five of these measures: (1) metropolitan road user charges, (2) reduced urban public transport fares, (3) city-wide parking charges, (4) labelling of new cars to inform buyers of their fuel efficiency, and (5) shifting inter-capital freight from road to rail were found to be "no regrets" options, i.e., they had zero or negative costs to society as a whole. Together these measures could reduce emissions from the Australian transport sector by about 5% to 10% of total projected emissions. A carbon tax on motor fuels and accelerated introduction of fuel-

saving technology for commercial vehicles are no regrets measures if applied at a low level, but incurred positive social costs if applied more broadly. Planting trees to offset transport emissions, scrapping older cars, and accelerating the introduction of energy efficiency technology for passenger cars and aircraft are found to be low-to-medium cost measures. Scrapping older commercial vehicles, compulsory tuning of passenger car engines twice a year, resurfacing highways, and increasing the use of ethanol as a motor fuel are found to be high cost measures.

Many parts of the developing world are faced with severe environmental problems caused in part by a rapid growth in the use of personal vehicles (scooters, motor cycles, mopeds, and cars). Many of these vehicles are old and poorly maintained, use two-stroke engines, and operate on inadequate road systems. The result is traffic congestion, greater fuel consumption, and noise and air pollution that degrade the urban environment. Bose (1998) finds that improving public transportation to meet as much as 80% of travel demand, and promoting cleaner fuels and improved engine technologies (i.e., phasing out two-stroke engines, increasing the share of cars equipped with three-way catalytic converters, using unleaded gasoline, electric vehicles, and vehicles fuelled with compressed natural gas) in six Indian cities can significantly reduce both emissions and fuel consumption. Total fuel savings for the six cities is 0.83mtoe (see footnote 3) in 2010 to 2011, and automotive emissions are reduced 30%–80% compared with a baseline case.

9.2.8.3 Freight Trucks, Rail, and Shipping

Freight transportation has been growing rapidly as a result of the growth of international merchandise trade, which has surpassed the growth in the world economy over the last two decades (IEA, 1997b). EIA (1998) consider the impacts of carbon fees to reduce US carbon emissions to 3% below 1990 levels, the amount estimated by the US administration as necessary to meet its Kyoto Protocol commitments when reductions in the emissions of other gases were taken into account. These fees raise the cost of diesel fuel by US$0.68/gallon, but result in only a 4.9% reduction in US freight truck travel, most of which is a result of lower economic activity. US rail transport is projected to decline by 32%, largely as the result of a 71% reduction in the demand for coal. The cost of marine fuel is projected to rise by US$0.84/gallon, nearly twice the reference price, but domestic shipping is projected to decline by only 10% (EIA, 1998).

9.2.8.4 Co-benefits from Reduced Road Traffic

Nations may choose to include GHG mitigation along with improvements in urban air quality and other traffic-related damages as objectives for policies designed specifically to reduce road traffic. The policies have co-benefits in terms of:
- reduced air emissions associated with less fuel use (e.g., Ross, 1999), and therefore consequent reductions in the damages caused by these emissions;

- reduced congestion;
- fewer traffic crashes;
- less noise; and
- less road damage.

The co-benefits from less noise and road damage are only likely to be large for substantial levels of mitigation (see ECMT, 1998 for valuations of these benefits for some European countries).

9.2.8.4.1   Air Pollution Associated with Road Traffic

There are likely to be substantial GHG co-benefits from some policies mainly aimed at reducing air pollution; these are mostly considered in Chapter 8.

Today 3 out of 4 of the world's highly dense megacities are in the rapidly developing countries, where traffic congestion is often high, involving highly polluting and inefficient vehicle fleets (WRR, 1998). Because of this, reducing traffic and congestion will also lower potential exposures to known hazards from the burning of road fuels, especially to those living near to congested roadways. Children are at high risk from the damaging neurological effects of pollution. A recent report from the WHO and the European Environment Agency estimates that 21,000 deaths annually are tied with air pollution from traffic in Central Europe (WHO, 1999).

The total of health damage costs from road traffic is significant. A recent study jointly produced by agencies of the Swiss, French and Austrian ministries of health, environment, and economy estimates that the annual number of deaths linked to traffic based pollution in these countries exceeds those that occur because of traffic crashes alone. This study uses a willingness-to-pay approach to economically evaluate traffic-related air pollution health effects. In all three countries, the total air pollution related health costs are US$49.7bn[9], with $26.7bn coming from road traffic-related pollution. As a percentage of GDP, such costs in these countries range from 1.1%-5.8% (Sommer *et al.*, 1999). A recent study from Sao Paulo (Miraglia *et al.*, forthcoming), estimated that by 2020, 35300 avoidable deaths from air pollution will occur if current trends in transportation continue and about 150,000 children will be admitted to the hospital or visit the emergency room.

9.2.8.4.2   Road Congestion

The research done on the ancillary benefits of GHG mitigation policies on road transport suggests that the value of the consequent reduction in congestion may be one of the most significant of such benefits (Barker *et al.*, 1993). Traffic congestion also contributes to increased exposure to pollutants by passengers during periods of congestion, with levels inside private vehicles found to be 2 to 8 times those in the surrounding air (Fernandez-Bremauntz and Ashmore, 1995). Action to reduce

this congestion can be expected to lower risks associated with such exposures, as well as lessen public health impacts of associated pollutants more generally.

9.2.8.4.3   Road Traffic Crashes

Section 9.2.8.2 lists several options for transport policies to mitigate GHGs. Some of these options, such as expanded reliance on mass transit and shifts away from individual passenger vehicles, can be expected to decrease the number of traffic crashes. The total number of damages resulting from crashes is substantial. With respect to traffic deaths and disabilities, the World Bank reports that traffic crashes are already the leading cause of death for young males and the 5th leading cause of death for young females worldwide. About 75% of all deaths occur in developing countries, although they have less than 1/4 of all vehicles. If present trends continue to 2020, one fourth of all health costs in developing countries may be spent on treating road injuries alone (Ross, 1999). However, policies that encourage the use of smaller vehicles could have increased death and injuries caused by traffic crashes (Dowlatabadi *et al.*, 1996).

The extent to which these total damages may be affected by various climate policies remains unknown, but is likely to be nontrivial and to vary in developed and developing countries. For instance, shifting travel from personal vehicles to mass transportation for large populations in the megacities of Sao Paulo and Shanghai (MacKenzie, 1997) has been projected to yield two sets of ancillary benefits:

(1) less net GHG emissions from transport, and
(2) lower incidence of traffic-accident-related morbidity and mortality.

9.2.9   Services

Since services employ more people and since they are much more employment-intensive than energy and manufacturing, employment usually increases as a result of GHG mitigation. However, the effects are small and diffused, and there is hardly any literature on specific sectoral effects for the service industries apart from the multisectoral studies reviewed in section 9.2.1 above.

9.2.10   Households

The impact of mitigation on households comes directly through changes in the technology and price of household use of energy for heat, light, and power, and indirectly through macroeconomic effects, particularly on the income of households and the employment of their members. An important ancillary benefit for households is the potential improvement in quality of indoor, local, and regional air.

Most studies analyze the effect of mitigation strategies on GDP, which is often taken as an indicator of welfare. However,

---

[9] The original calculations are in euro and a rate of US$1=€1 has been used for conversion.

this measure does not capture the effects on the distribution of income between households. There are some studies that look at private consumption and other constructed indices of welfare, but these are few in number. This literature concentrates more on the developed economies, as these are the countries that would be taking actions first to reduce the emissions. The effect on developing economies is indirect through the trade effects and energy price effect.

9.2.10.1 Distributional Effects of Mitigation

These are mainly discussed in Chapter 8 (section 8.2.2.3). There are a number of studies on the domestic income distributional effects of carbon taxation, mostly for developed countries (Johnson *et al.,* 1990; Chandler and Nicholls, 1990; Poterba, 1991; Bertram *et al.,* 1993; Hamilton and Cameron, 1994; Symons *et al.,* 1994; Cornwall and Creedy, 1996). These studies show a regressive effect of carbon taxes, but a progressive effect if revenues are returned to disadvantaged groups. As the share of household expenditure on energy and the dependence on high-carbon fuels of the lower income groups is high, the impact of a carbon tax would be disproportionately higher on these lower income groups (Goldemberg and Johansson, 1995; Yamasaki and Tominaga, 1997). Barker and Kòhler (1998) review a number of studies on impact of carbon taxation on households. Their analysis of an EU carbon tax indicates that taxation on domestic energy is regressive and taxation on road fuels is weakly progressive. They also show that revenues recycled through employer taxes could increase disposable income for all income groups in the study.

9.2.10.2 Electricity and Demand-side Management

A number of studies point out that power sector deregulation and competition will improve the efficiency of operations as well as management, which will result in a reduction in electricity rate charged to the end users (Hsu and Tchen, 1997). Demand-side management (DSM) instituted by electric utilities would increase electricity prices, but could lead to a reduction in total bills to participating customers (Hirst and Hadley, 1995), although the increased electricity prices could deter companies from using these measures in a competitive market. Parker (1995) indicated that DSM measures could lead to job creation from production and installation of equipment.

9.2.10.3 Effects of Improvement in Energy Efficiency

Improvements in the efficiency of energy production may have substantial impacts on households. Bashmakov (1993) reports a reduction in energy bills for end users and a substantial reduction in environmental costs for Russia. The study also reports that every rouble invested in energy efficiency generates 5 times more jobs than investments in energy production. On the other hand, Gaj *et al.* (1997) report a high social cost of economic transition caused by macroeconomic reforms, which indirectly reduce GHG emissions, because employment in non-competitive sectors is high in Poland.

9.2.10.4 Ancillary Benefits for Households

A major ancillary benefit of GHG mitigation is reductions in the emissions of local air pollutants. Glomsro *et al.* (1990) have indicated that improved health conditions as a consequence of improved air quality, etc., could offset roughly two-thirds of the calculated GDP loss arising out of policies to reduce emissions. Alfsen *et al.* (1995) indicate a 6 to 10% reduction in SO_2 and NO_x emissions by the year 2000 as a result of an energy tax of US$3/barrel in 1993 and increasing by US$1 in each subsequent year to 2000.

Transport sector mitigation could imply substantial price increases with associated negative political, economic, and social implications, such as hardship for low-income rural motorists without access to public transport (Koopman and Denis, 1995; Dargay and Gately, 1997). But the option of using public transport could benefit the lower income sections of society, especially in developing countries, along with associated reduction in emission of CO, NO_x and SO_2 (Bose and Srinivasachary, 1997). Lower fuel use by road transportation could have substantial health benefits in urban areas (Pearce, 1996; Zaim, 1997).

Some of the indirect benefits of GHG mitigation of fuel switching and efficient devices in the household sectors, typically in developing countries, include:

- improved indoor air quality;
- higher quality of life (simplifying household chores, better hygiene, and easier cleaning);
- reduced fuel demand with economic and time-saving benefits to the household (one study in Tanzania reported that women using wood as fuel spend 12 hours a week to collect it (Gopalan and Saksena, 1999));
- increased sustainability of local natural resources; and
- reductions in the adverse effects of biomass use on human health (WHO, 1992).

These points are particularly relevant in the case of biomass-burning stoves (Sathaye and Tyler, 1991; Smith 1996). Gopalan and Saksena (1999) report that the level of exposure to key pollutants in rural households can be 10 to 100 times higher than the health-related guidelines of the WHO.

The results of a study on potential fuelwood use in 2020 for Austria, Finland, France, Portugal, and Sweden reveal that upstream emissions from fuel extraction are generally higher for fossil fuels than biofuels (Schwaiger and Schlamadinger, 1998). However, some research indicates that local negative environmental implications may be greater for use of wood than fossil fuel (Radetzki, 1997). An associated impact of increased diversion of land for growing wood would be on agriculture production and hence the commodity prices (Alig *et al.*, 1997). The economic benefits of afforestation also include benefits from increase in supply of non-timber forest products (Mors, 1991; ADB-GEF-UNDP, 1998a). These options in developing countries would greatly increase the wood supply and address the forest degradation issue but via-

Table 9.8: *Typology of potential international spillovers from mitigation strategies*

Spillovers / Policies and Measures	Benefits from technology improvement	Impacts on energy industries activity and prices	Impacts on energy intensive industries	Resource transfers to sectors
Public R&D policies	Increase in the scientific knowledge base			
"Market access" policies for new technologies	Increase in know-how through experience, learning by doing			
Standards, subsidies voluntary agreements	New cleaner industry/ product performance standards			
Carbon taxes	Price-induced technical change and technology diffusion	Reduction of activity in fossil fuel industries. Lower international prices, negative impacts for exporters, positive for importers, possibility of a "rebound effect"	Carbon leakages, positive impacts for activity, negative for envir. in receiving country	
Energy subsidy removal			Reduced distorsions in industrial competition	
Harmonized carbon taxes				
Domestic emission trading			Distorsion in competition if differentiated schemes (grandfathered vs. auctioned)	
Joint Implementation, Clean Development Mechanism				Technology transfer
International emission trading				Net gain when permit price is superior (not equal) to average reduction costs

bility is an important issue as incomes are too low in rural areas for sizeable numbers of the population to buy wood.

Mitigation strategies in rural domestic energy use range from use of more efficient appliances, installation of PV solar, fuel-switching and use of bio-gas (ADB-GEF-UNDP, 1998a). Such strategies for developing countries are constrained by high capital costs (Biswas and Lucas, 1997). The ancillary benefits of lower use of traditional biomass are decreased deforestation, and lower loss of crop-nutrient from the system through use of agricultural residue as fuel (Bala, 1997). The ancillary environmental benefits that are associated with such strategies do not form a major factor in energy decisions of the household (Aacher and Kammen, 1996); it is the cost that is the important factor. And some mitigation measures at home, such as reduction of air leaks, tend to worsen indoor air quality (Turiel, 1985).

9.2.10.5 The Asia Least-cost Greenhouse Gas Abatement Strategy Studies

ALGAS was a regional technical assistance project of the Asian Development Bank (ADB) which enabled 12 Asian countries:

*   to prepare an inventory of anthropogenic emissions and sinks of GHGs;

*   to evaluate the costs and effectiveness of measures available to reduce GHG emissions or enhance sinks; and

*   to develop national action plan policy responses that will be required to implement the measures that are identified.

The ALGAS country reports highlight the forestry sector options: forest protection and reforestation will have both socio-economic benefits and environmental benefits. These forestry options will increase rural incomes, increase equity of income, and increase the availability of biomass (ADB-GEF-UNDP, 1998b, c, d and f). These studies also emphasize that the forestry options would reduce the pressure on forested land and have indirect benefits of reducing soil erosion in hilly terrain. However, some of the studies (ADB-GEF-UNDP 1998c and e) indicate that these changes are short term and do not have a significant effect.

The ALGAS-Bangladesh (ADB-GEF-UNDP, 1998d) study also reports that the options in the agricultural sector of reducing CH_4 emission from paddy fields and enteric fermentation in animals have direct benefits in terms of increased incomes, and also improve foodgrain production and availability.

9.3 International Spillovers from Mitigation Strategies

International spillovers[10] arise when mitigation in one country has an impact on sectors in other countries. The main factors are:

(1) improvement in the performance or reduction in the cost of low-carbon technologies;
(2) changes in the international prices, exports and outputs of fossil fuels, especially oil; and
(3) relocation of energy-intensive industries.

Table 9.8 shows how different policies and measures may give rise to such spillovers. These effects may be included in the design and assessment of policies, particularly in the search for internationally equitable strategies. Chapter 8 considers the macro aspects of spillovers; this section considers the sectoral aspects.

9.3.1 *Technology Policies*

In the sectoral perspective of this chapter, it appears that there are three routes by which technology policies in one country affect sectoral development in others (see Section 8.3.2.5 for a global perspective). First, R&D may increase the knowledge base and this will benefit every country. Second, increased "market access" for low-CO_2 technologies, through niche-markets or preferential buyback rates in one country may induce a generic improvement in technology in others. *Box 9.3.1* explains how this process can be modelled. Third, domestic regulations on performance and standards, whether imposed or voluntary can create a strong signal for foreign industrial competitors (Gruber *et al.*, 1997). For example, the ratio of emission standards for carbon monoxide, hydrocarbons, and NO_x for automobiles in the EU relative to those in the US has been reduced from a factor of more than 3 in the seventies to a factor 1.5 to 2 in the nineties (Anderson, 1990; IFP, 1998).

9.3.2 *Tax and Subsidy Policies*

Spillover effects from tax and subsidy policies for mitigation are less direct. The global economic impacts of the policies are examined, both in a theoretical and in a modelling perspective, in Chapter 8 (8.3.2.1 to 8.3.2.4). Their impacts on sectors are also analyzed in section 9.2 above. The sectoral effects of these policies can be summarized as follows.

---

[10] Spillover effects can be defined as interdependencies between countries, sectors or firms that take the form of technological synergies and flows of stimuli and constraints that do not entirely correspond to commodity flows (Dosi, 1988). The concept originates in the literature on technical change, in order to account for the non-appropriability of scientific and technological knowledge, which reduces the incentive to private R&D and thus motivates public investment in R&D (Arrow, 1962).

Box 9.1. International Technological Spillovers in the National Energy Modelling System Model of the US Energy Sector

The rate of international spillovers largely depends on the nature of the technology, the degree of internalization of the market, and the competitive structure of the industry. The NEMS model of the US energy sector is one of the rare models explicitly incorporating spillover effects. It is assumed, based on historical experience, that power plant development outside the US will also help to decrease costs in the USA. Thus, one unit installed abroad is incorporated in the experience curve, but only up to a fraction of the same unit in the USA. The corresponding factor (from 0 to 1) depends on the proximity of the country and firm developing this power plant. It gives the measure of the expected international spillover rate (NEMS model documentation, DOE-EIA; see Kydes, 1999).

(1) They will reduce the demand for carbon-based fuels, and thus introduce a downward pressure on their prices e.g., in the world price of crude oil;
(2) They may reduce the industrial competitiveness of sectors with higher costs in the mitigating country, raising competitiveness and hence market shares for sectors in other countries;
(3) They may create an incentive to industrial relocation and thus give rise to "carbon leakages";
(4) However, they may also stimulate the development of alternative technological solutions.

The effects of carbon taxation on international competitiveness are reviewed by Ekins and Speck (1998) and Barker and Johnstone (1998). Clearly, a carbon tax will raise the cost of production of some sectors of the economy, causing some consumers to switch from their products to the products of the sectors in other countries, changing international trade. National losses (and/or gains) for price competitiveness will be the net sum of the sectors' losses (and/or gains) for price competitiveness. The outcome for a particular sector will depend on the policy instruments used, how any tax revenue has been recycled, and whether the exchange rate has adjusted to compensate at the national level. The conclusions from these surveys are that the reported effects on international competitiveness are very small, and that at the firm and sector level, given well-designed policies, there will not be significant loss of competitiveness from tax-based policies to achieve targets similar to those of the Kyoto Protocol.

These conclusions are confirmed by later studies, although in general the effects of environmental taxation in one country on sectors in other countries are not well covered by the literature. Using an econometric model (E3ME), Barker (1998a) assesses policies reducing CO_2 emissions in 11 EU member states at the level of 30 industries and 17 fuel users, comparing unilateral with co-ordinated policies. The carbon tax reduces imports of oil and increases imports of carbon-intensive products. However, the results for trade are negligible.

Table 9.9: *Effects on sectoral output of Japan (in per cent) of an ad-valorem fuel tax*

	Change of output (%)		
Sector	**Japan only**	**OECD**	**World**
Agriculture	0.0998	0.0646	-0.0295
Forestry	0.1744	0.2044	0.0687
Mining	0.0488	0.1311	0.1415
Oil and coal	-0.3983	-0.1212	0.6689
Chemistry	-0.5143	-0.3929	-0.3884
Metal	-0.1619	-0.1032	0.0126
Other manufacture	-0.0604	-0.0065	-0.0500
Elec. water, gas	-0.3081	-0.3145	-0.3080
Transportation	0.0548	0.0480	0.0364
other services	0.0349	0.0376	0.0364
Capital goods	0.0007	0.0797	0.1078

Source: Ban (1998).

Ban (1998) assesses the effects of an *ad-valorem* tax on coal (20%), oil (10%), and gas (10%) using an applied general equilibrium model (GTAP) with 12 world regions and 14 industry sectors. He has three taxation cases, (a) Japan only, (b) OECD only, and (c) the world, with revenues used to increase government expenditure. The results are all shown against a reference case for 1992. *Table 9.9* shows the effects on the industrial output in Japan: the effects are very small when the tax is for Japan only, but they are even smaller when the taxation is at the OECD or world level, illustrating the size of the competitiveness effects. These results depend critically on the assumptions adopted as Ban points out.

There are other aspects to spillovers not well captured in existing models. As energy efficiency is generally higher in Annex B countries than in the rest of the world, some studies suggest that relocation of industry to developing regions would increase global CO_2 emissions (e.g., Shinozaki *et al.*, 1998). However, this conclusion would be altered if the relocated industry used up-to-date technologies rather than the average technology in developing countries. The international diffusion of improved technologies in response to CO_2 constraints is not captured in existing models and would tend to counteract the negative environmental aspects of spillovers.

9.4   Why Studies Differ

This section consolidates the explanations for the different findings in both the macro studies reviewed in Chapter 8 and the sectoral studies in this chapter. It extends and complements the methodological discussion in the SAR (Hourcade *et al.*, 1996, pp. 282-92), particularly in the role of assumptions leading to differing results.

In assessing the economy-wide effects of mitigation, considerable use has been made of top-down models (macroeconomic,

general equilibrium, and energy-engineering), while specific sectoral studies use both top-down and engineering-economic bottom-up models. Critical differences in the results come from the type of model used, and its basic assumptions. Repetto and Austin (1997), in a meta-analysis of model results on the costs of mitigation for the USA, show that 80% of predicted impacts come from choice of assumptions. They find that four assumptions are critical in leading to lower costs of mitigation. These are that:

- the economy responds efficiently to policy changes at least in the long run;
- international joint implementation is achieved;
- revenues from taxes or permit sales are returned to the economy through reducing burdensome taxes; and
- any co-benefits from reduced air pollution are fully included.

They conclude that under reasonable assumptions, the predicted economic impacts from the models for the USA in stabilizing CO_2 emissions at 1990 levels through to 2020 would be neutral or even favourable.

Most early studies are focused on the costs, rather than on the benefits of mitigation[11]. More recently, top-down modellers have studied the impact of using the revenues collected from carbon taxes (or from auctions of carbon permits) to correct economic distortions in some sectors of the economy (typically to reduce taxes on labour, taxes on incomes and profits, or taxes on investment).

---

[11] More formally, the studies impose taxes on the carbon content of energy as a factor of production (with labour and capital as other factors) in a production function; depending on the precise assumptions chosen this has the inevitable implication that output and GDP will fall. See Boero *et al.* (1991), Hoeller *et al.* (1991), Cline (1992), Ekins (1995), and Mabey *et al.* (1997) for surveys of the assumptions and results of the modelling in this area.

Table 9.10: *A comparison of top-down and bottom-up modelling methodologies*

Treatment	Top-down	Bottom-up
Concepts and terms	Economics-based	Engineering-based
Treatment of capital	Homogeneous and abstract concept	Precise description of capital equipment
Treatment of technical change	Trends rates (usually exogenous)	Menu of technical options
Motive force in the models	Responses of economic groups via income and price elasticities	Responses of agents via discount rates
Perception of the market in the model	Perfect markets are usually assumed	Market imperfections and barriers
Potential efficiency improvements	Usually low with assumption of all negative cost opportunities utilized	Opportunities for no regrets actions identified

Source: Bryden *et al.* (1995)

9.4.1   The Influence of Methods

9.4.1.1   Top-down and Bottom-up Modelling

The adoption of top-down or bottom-up methods makes a significant difference to the results of mitigation studies (see 8.2.1 and 8.2.2 for discussion and results). In top-down studies the behaviours of the economy, the energy system, and their constituent sectors are analyzed using aggregate data. In bottom-up studies, specific actions and technologies are modelled at the level of the energy-using, GHG-emitting equipment, such as power-generating stations or vehicle engines, and policy outcomes are added up to find overall results. The top-down approach leads easily to a consideration of the effects of mitigation on different broad sectors of the economy (not just the energy and capital goods sectors), so that the literature on these effects tends to be dominated by this approach.

Table 9.10 compares the methodologies. They have a fundamentally different treatment of capital equipment and markets. Top-down studies have tended to suggest that mitigation policies have economic costs because markets are assumed to operate efficiently and any policy that impairs this efficiency will be costly. Bottom-up studies tend to suggest that mitigation can yield financial and economic benefits, depending on the adoption of best-available technologies and the development of new technologies. Some hybrid models include both approaches (see Laroui and van Leeuwen, 1995, for an example).

9.4.1.2   General Equilibrium and Time-series Econometric Modelling

There are two main types of macroeconomic models used for medium- and long-term economic projections[12]: resource allocation models (i.e. CGE) and time-series econometric models.

Their main differences being the assumptions made about the real measured economy, aggregation, dynamics, equilibrium, empirical basis, and time horizons, among others.

The main characteristic of CGE models is that they have an explicit specification of the behaviour of all relevant economic agents in the economy. In the mitigation applications they have usually adopted assumptions of optimizing rationality, free market pricing, constant returns to scale, many firms and suppliers of factors, and perfect competition in order to provide a market-clearing equilibrium in all markets. Econometric models have relied more on time-series data methods to estimate their parameters rather than consensus estimates drawn from the literature. Results from these models are explained not only by their assumptions but also by the quality and coverage of their data. It is usually argued that CGE models are more suitable for describing long-run steady-state behaviour, while econometric models are more suitable for forecasting the short-run. However, the models have increasingly incorporated long-run theory and formal econometric methods, and several now include a mix of characteristics, from both resource allocation and econometric models; see Jorgenson and Wilcoxen (1993), McKibbin and Wilcoxen (1993, 1995), Barker and Gardiner (1996), Barker (1998b) and McKibbin *et al.* (1999).

9.4.2   The Role of Assumptions

9.4.2.1   Baseline

A critical point for the results of any modelling is the definition of the baseline (or reference or business-as-usual) scenario. The SRES (Nakicenovic *et al.*, 2000) explores multiple scenarios using six models and identifies 40 scenarios divided into 6 scenario groups. As OECD (1998) points out, among the key factors and assumptions underlying reference scenarios are:

- population and productivity growth rates;
- (autonomous) improvements in energy efficiency;
- adoption of regulations e.g., those requiring improvements in air quality; if air quality is assumed to be satisfactory in the baseline, then the potential for air qual-

---

[12] See Shoven and Whalley (1992), Dervis *et al.* (1982), Jorgenson (1995a, 1995b), Holden *et al.* (1994), Barro and Sala-i-Martin (1995) for different methods of long-term modelling.

ity ancillary benefits in any GHG mitigation scenario is ruled out by assumption;

- developments in the relative price of fossil fuels; some of the underlying factors are supply-side issues, for example oil and gas reserves, development of gas distribution networks, the relative abundance of coal; energy policies also play a role, particularly tax and subsidy policies;
- technological change, such as the spread of combined cycle gas turbines;
- supply of non-fossil fuel based electricity generation (nuclear and hydro); and
- the availability of competitively priced new sources of energy, so-called backstop fuels, for example solar, wind, biomass, tar sands.

Differences in the reference scenarios lead to differences in the effects of mitigation policies. Most notably, a reference scenario with a high growth in GHG emissions implies that all the mitigation scenarios associated with that reference case may require much stronger policies to achieve stabilization.

Nevertheless, even if reference scenarios were exactly the same, there are other reasons for changes in model results. Model specification and, more importantly, differences in model parameters also play a significant role in determining the results.

9.4.2.2   Costs and Availability of Technology

If any fuel becomes perfectly elastic in supply at a given price (i.e., the backstop technology), the overall price of energy will be determined independently of the level of demand, which will then become the critical determinant of mitigation costs. Hence, the assumption of a backstop technology strongly determines mitigation costs. Models without a backstop technology will tend to estimate higher economic impacts of a carbon tax, because they rely completely on conventional fuels, so that the tax rate has to rise indefinitely to keep carbon concentrations constant, to offset the effects of economic growth.

9.4.2.3   Endogenous Technological Change

The treatment of technology change is crucial in the macroeconomic modelling of mitigation. The usual means of incorporating technical progress in CGE models is through the use of time trends, as exogenous variables constant across sectors and over time. These trends give the date of the solution. Technical progress usually enters the models via two parameters: (i) autonomous energy efficiency (AEEI) (if technical progress produces savings of energy, then the value share of energy of total costs will be reduced); and (ii) as changes in total factor productivity. The implication of this treatment is that technological progress in the models is assumed to be invariant to the mitigation policies being considered. If in fact the policies lead to improvements in technology, then the costs may be lower then the models suggest.

9.4.2.4   Price Elasticity

In assessing the effects of mitigation, estimations of price-induced substitution possibilities between fuels and between aggregate energy and other inputs can be crucial for model outcomes. All such substitutions become greater as the time for adjustment increases. The problem is that estimates of substitution elasticities are usually highly sensitive to model specification and choice of sample period. There is little agreement on the order of magnitude of some of the substitution elasticities, or even whether they should be positive or negative, e.g., there is debate whether capital and energy are complements or substitutes. If energy and capital are complements, then increasing the price of energy will reduce the demand in production for both energy and capital, reducing both investment and growth. Most CGE models consider very different possibilities of substitution, for example WW, Global 2100, and Nordhaus's DICE/RICE models assume capital and labour as substitutes, while GREEN assumes capital and energy as direct substitutes.

9.4.2.5   Degree of Aggregation

There are many different products, skills, equipment, and production processes; many important features are missed when they are necessarily lumped into composite variables and functions. A basic difference among models and their results is the level of aggregation. Indeed, in practice, different goods have different energy requirements in production, and therefore any changes in consumption and production patterns will affect them differently. Hence, a highly aggregated model will miss some potentially major interactions between output and energy use, which is precisely the purpose of the analysis. For example, sectoral disaggregation allows the modelling of a shift towards less energy-intensive sectors, which might reduce the share of energy in total inputs. In the same way, when a carbon tax is introduced, it could reduce the estimated costs of abatement by allowing substitution effects of energy-intensive goods by less energy-intensive goods.

9.4.2.6   Treatment of Returns to Scale

Constant returns to scale represent a common assumption on the economic modelling of climate change. However, in practice, economies of scale seem to be the rule rather than the exception. Indeed, there are several reasons for economies of scale, see Pratten (1988), and Buchanan and Yoon (1994). For example, many electricity-generating stations benefit from economies of scale, utilizing a common pool of resources including fuel supply, equipment maintenance, voltage transformers, and connection to the grid. In spite of this fact, the impact of the effects of increasing returns and imperfect competition (IC) in the modelling of climate-change strategies has consistently been neglected in the literature. Most of the global models, if not all, assume explicitly perfect competition, for example, see DICE/RICE, G-Cubed, Global 2100, GREEN, GTEM, WorldScan, and WW.

9.4.2.7 Treatment of Environmental Damages

Most models are not able to incorporate the benefits of preventing climate change (or of the costs of doing nothing). Instead, modellers have only considered the economic impact of meeting some emission standard, which implicitly assumes (in the base situation) that climate change would have no economic impacts. Nevertheless, the potential costs caused by climate change are likely to be huge (even though some favourable effects are also expected) regarding: loss of human well-being, damage to property including agriculture and forestry, ecosystem loss, and risk of disaster, see Nordhaus (1991), Cline (1992, Chapter 3), Fankhauser (1995), Fankhauser and Tol (1995). This situation has been caused to some extent by two factors, the difficulties of economists in valuing environmental impacts, and the scientific uncertainty of predicting the physical effects of climate change[13].

9.4.2.8 Recycling of Tax Revenues

Carbon taxes will generate significant tax revenues. The effects of these revenues in the economy will depend on how this money is recycled into the economy (in practical terms, some mechanism for recycling is always needed in order to avoid a general deflationary impact). If it is assumed that revenues will not be fully recycled, the models tend to find that any carbon tax will reduce GDP. Usually, modellers have tried to separate the economic impacts arising from this environmental policy from those arising from a tax cut, assuming that revenues will be returned in the form of lump-sum rebates (an unrealistic assumption). The alternative is to assume that revenues collected from the carbon tax are used in correcting economic distortions in the economy, e.g., taxation of employment, which would benefit society not only by correcting the externality but also by reducing the costs of the distorting taxes (the so-called "double dividend"). Obviously, if the benefits from reducing existing taxes on labour are incorporated into the modelling, the projected economic impacts can be substantially more optimistic than if a lump-sum revenue recycling is assumed, although the size of the effect depends on model specification[14].

9.4.2.9 International Environmental Policy

Environmental policy to reduce climate change will be economically efficient when the incremental cost of emission reductions is equal in all complying countries. A way of achieving cost savings in the abatement policy is to allow emission sources to contract with each other to meet required emission reductions. In this sense, flexible instruments such as international emissions trading and JI are more efficient than a situation in which each country has to achieve its own emission reduction[15]. Usually, international emissions trading is modelled as if all countries set the same carbon tax rate, so that cost-effective emission reductions are advantageous to undertake in whatever country they arise. Hence, if models consider economic instruments for environmental regulation, the overall cost of controlling emissions should be lower as a consequence of cost savings in the control produced by these instruments[16].

It is important to point out that this kind of modelling implicitly assumes an ideal scenario. However, in practice some problems arise with the basic theory, involving the operation and design of the market. Some important considerations here are:

- the degree of competition in the market (i.e., that neither buyers nor sellers have sufficient weight to influence the price of the permit);
- high transaction costs derived from inadequate information;
- fairness in allocating the emission permits (auctioning versus "grandfathering"); and
- the institutional and administrative costs of implementing the system (are the costs negligible?)[17].

9.5 Areas for Further Research

The literature on sectoral economic costs and benefits is limited and additional research would be beneficial in all areas.

---

[13] In the same way, models should incorporate other benefits of limiting GHG emissions, but again the complexity of modelling and valuing of these benefits are substantial. The ancillary benefits associated with the abatement policies usually include reductions in damages from other pollutants jointly produced with GHGs (see Chapter 8 and Barker, 1993; Barker *et al.* 1993 and OECD 2000) but also include the conservation of biological diversity.

[14] Nevertheless, in general the research on double dividend of environmental taxes has resulted in conflicting and confusing conclusions. See Bohm (1997) for a clear statement of the issues and O'Riordan (1997) for several reviews of theoretical and practical evidence of the dividends from environmental taxation.

[15] In general terms, from the economists' point of view, environmental regulation should rely on economic instruments instead of command–and–control policies, considering the cost-effective allocation of the control responsibility of the former ones, which have proven to be efficient in simple settings, see Bohm and Russell (1985), Baumol and Oates (1988), Montgomery (1972).

[16] See Tietenberg (1990), Barrett *et al.* (1992), Barrett (1991), Rose and Tietenberg (1993). See also the studies reported to the OECD expert workshop on Climate Change and Economic Modelling, September 1998.

[17] In terms of the Kyoto Protocol, for example, a specific problem of modelling IET is the possibility that target emissions will be below the base-year emissions. In the same way, variations from the full unrestricted trading systems may change cost estimations. Two clear variations are: the definition of trading entities (i.e. bubbles), and the limits of the amount of trading.

Specific issues identified in this chapter (not in order of priority) include:

- Additional research on the impacts of climate change policies on the fossil fuel industries is needed. Questions include:
  - the apparent anomaly between studies indicating significant decreases in the demand for oil in Annex B countries, and studies indicating significant increases in the demand for transportation fuels, the major user of oil;
  - whether in the medium term (10 to 30 years) reserves of conventional oil are limited, which would soften the impact of climate change policies, or whether they are plentiful; and
  - whether the demand for natural gas will decrease as a result of a general decrease in the demand for fossil fuels, or increase, as a result of fuel switching from higher carbon content fuels and growth in demand in non-Annex B countries.

- The impacts of climate change policies on the financial industries have not been analyzed. IPCC (2001) details the potential impacts, positive and negative, of climate change on the financial industries, but there appears to be no literature evaluating the degree to which mitigation policies would affect these impacts.

- The applicability of existing climate policies, and their impacts on the aviation industry and the shipping industry have not been adequately studied. Further analysis is needed to determine the efficiency, effectiveness, and equity of various policy options, particularly involving taxation, on limiting GHG emissions from aviation and shipping. This would include the difficulties involved in changing the current treaty structure to allow for the taxation of aviation fuel. The International Maritime Organization is studying GHG emissions from shipping. The International Civil Aviation Organization is currently analyzing policy options for aviation and is expected to complete its evaluation by September, 2001.

- Further study would be helpful to determine the degree to which employment growth in the industries that would benefit from climate change policies (e.g., renewable energy) would offset the decrease in employment in industries that would suffer as the result of climate change policies (e.g., fossil fuels). These studies could also consider frictional unemployment.

- More generally, an assessment is needed of how sectoral costs of mitigation can be minimized and distributed more equitably, both at the national and the global levels. Babiker *et al.* (2000) found that macroeconomic costs for the US increased when climate change policies excluded one or more economic sectors. However, this study did not indicate the benefits, if any, to the protected sector.

- More research is needed on the ancillary and co-benefits of GHG mitigation and other objectives of transport policies (reductions in air pollution, lower levels of traffic congestion, fewer road crashes).

References

Aacher, R.N., and D.M. Kammen, 1996: The quiet (energy) revolution: Analysing the dissemination of PV Power System in Kenya. *Energy Policy*, **24**(1), 85-111.

ABARE, 1995: *Global Climate Change: Economic Dimensions of a Cooperative International Policy Response Beyond 2000*. Canberra, Australia.

Adams, R.M., J.D. Glyer, S.L. Johnson, and B.A. McCarl, 1989: A reassessment of the economic effects of ozone on United States agriculture. *Journal of Air Pollution Control Association*, **39**(7), 960-968.

Adams, R.M., S.A. Hamilton, and B.A. McCarl, 1985: An assessment of the economic effects of ozone on U.S. agriculture. *Journal of Air Pollution Control Association*, **35**, 938-943.

ADB-GEF-UNDP, 1998a: *Asia Least Cost Greenhouse Gas Abatement Strategy: India country report*. Asian Development Bank, Manila, The Philippines.

ADB-GEF-UNDP, 1998b: *Asia Least Cost Greenhouse Gas Abatement Strategy: Country report Thailand*. Asian Development Bank, Manila, The Philippines.

ADB-GEF-UNDP, 1998c: *Asia Least Cost Greenhouse Gas Abatement Strategy: Country report Republic of Korea*. Asian Development Bank, Manila, The Philippines.

ADB-GEF-UNDP, 1998d: *Asia Least Cost Greenhouse Gas Abatement Strategy: Country report of Bangladesh*. Asian Development Bank, Manila, The Philippines.

ADB-GEF-UNDP, 1998e: *Asia Least Cost Greenhouse Gas Abatement Strategy: Pakistan country report*. Asian Development Bank, Manila, The Philippines.

ADB-GEF-UNDP, 1998f: *Asia Least Cost Greenhouse Gas Abatement Strategy: The Philippines country report*. Asian Development Bank, Manila, The Philippines.

Alfsen, K.H., H. Birkelund, and M Aaserud, 1995: Impacts of EC carbon/energy tax and deregulatory thermal power supply on CO_2, SO_2 and NO_x emissions. *Environmental and Resource Economics*, **5**(2), 165-189.

Alig, R., D. Adams, B.A. McCarl, J.M. Callaway, and S. Winnett, 1997: Assessing effects of mitigation strategies for global climate change with an inter-temporal model of the US forest and agriculture sector. *Environmental and Resource Economics*, **9**(3), 259-274.

Anderson, R., 1990: *Reducing emissions from older vehicles*. American Petroleum Institute, Research Study 053, Washington DC, August 1990.

Anderson, B., and E. Haden, 1997: Power production and the price of electricity: an analysis of a phase out of Swedish nuclear power. *Energy Policy*, **25**(13), 1051-1064.

Anderson, K., and W.J. McKibbin, 1997: Reducing Coal Subsidies and Trade Barriers: Their Contribution to Greenhouse Gas Abatement. *Brookings Discussion Papers in International Economics*, No. 135.

Aoyama, T., 1997: Energy Perspectives in Asia – The Future of Nuclear Power Generation. *Energy in Japan*, No.145, 42-50.

Arrow, K.J., 1962: The Economic Implications of Learning by Doing. *Review of Economic Studies*, **29**, 155-173.

Aunan, K., T. Berntsen, and H. Seip, 2000: Surface ozone in China and its possible impact on agricultural crop yields. *Ambio*, **26**(6), September.

Babiker M., M. Bautista, H. Jacoby and J. Reilly, 2000: Effects of differentiating climate policy by sector: a U.S. example. In *Sectoral Economic Costs and Benefits of GHG Mitigation*. L. Bernstein and J. Pan (eds.), Proceedings of an IPCC Expert Meeting, 14 - 15 February 2000, Technical Support Unit, IPCC Working Group III.

Bala, B.K., 1997: Computer modelling of the rural energy system and of CO_2 emissions for Bangladesh. *Energy* **22**(10), 999-1004.

Ban, K., 1998: Applied General Equilibrium Analysis of Current Global Issues - APEC, FDI, New Regionalism and Environment. *The Economic Analysis*, No.156, Economic Research Institute, Economic Planning Agency (in Japanese, internally reviewed).

Barker, T., 1993: Secondary benefits of greenhouse gas abatement: The effects of a UK carbon/energy tax on air pollution. *Energy-Environment-Economy Modelling Discussion Paper No. 4*, Department of Applied Economics, University of Cambridge.

Barker, T., N. Johnstone, and T. O'Shea, 1993: The CEC Carbon/Energy Tax and Secondary Transport-Related Benefits. *Energy-Environment-Economy Modelling Discussion Paper No. 5*, Department of Applied Economics, University of Cambridge.

Barker, T., 1999: Achieving a 10% cut in Europe's CO_2 emissions using additional excise duties: coordinated, uncoordinated and unilateral action using the econometric model E3ME. *Economic Systems Research* **11** (4), 401-421.

Barker, T., and B. Gardiner, 1996: Employment, wage formation and pricing in the European Union: empirical modelling of environmental tax reform. In *Environmental Fiscal Reform and Unemployment*. C. Carraro and D. Siniscalculo (eds), Kluwer, Dordrecht, 229-272.

Barker, T., and J. K(hler, 1998: Equity and ecotax reform in the EU: achieving a 10% reduction in CO_2 emissions using excise duties. *Fiscal Studies*, **14** (4), 375-402.

Barker, T., and N. Johnstone, 1998: International competitiveness and carbon taxation. In *International Competitiveness and Environmental Policies*. T. Barker and J. K(hler (eds.), Edward Elgar, Cheltenham, UK, 71-127.

Barker, T., 1995: Taxing pollution instead of employment: greenhouse gas abatement through fiscal policy in the UK. *Energy and Environment*, **6** (1), 1-28.

Barker, T., 1998a: The effects on competitiveness of coordinated versus unilateral fiscal policies reducing GHG emissions in the EU: an assessment of a 10% reduction by 2010 using the E3ME model. *Energy Policy* **26** (14), 1083-1098.

Barker, T., 1998b: Large-scale energy-environment-economy modelling of the European Union. In *Applied Economics and Public Policy*. I. Begg and B. Henry (eds.), Cambridge University Press.

Barrett, S., 1991: Economic Instruments for Climate Change Policy. In *Responding to Climate Change: Selected Economic Issues*. OECD, Paris.

Barrett, S., M. Grubb, K. Roland, A. Rose, R. Sandor, and T. Tietenberg, 1992: *Tradeable Carbon Emission Entitlements: A Study of Tradeable Carbon Emission Entitlement Systems*. UNCTAD, Geneva.

Barro, R. J., and X. Sala-i-Martin, 1995: *Economic Growth*. McGraw-Hill, Inc.

Bartsch U., and B. Müller, 2000: Impacts of the Kyoto protocol on fossil fuels. In *Sectoral Economic Costs and Benefits of GHG Mitigation*. L. Bernstein and J. Pan, Proceedings of an IPCC Expert Meeting, 14 - 15 February 2000, Technical Support Unit, IPCC Working Group III.

Bashmakov, I., 1993: *Costs and Benefits of CO_2 reduction in Russia*. In *Costs, impacts and benefits of CO_2 mitigation*. Y. Kaya, N. Nakicenovic, W.D. Nordhaus, and F.L. Toth (eds.), Proceedings of the workshop held on 28-30 September 1992 at International Institute for Applied Systems Analysis, Laxenburg, Austria, 453-474.

Baumert, K., R. Bandhari, and N. Kete, 1999: *What Would a Developing Country Commitment to Climate Policy Look Like?* World Resources Institute, Climate Note, Washington, DC.

Baumol, W., and W. Oates, 1988: *The Theory of Environmental Policy*. Cambridge University Press, Cambridge.

Berg, E., S. Kverndokk, and K. E. Rosendahl, 1997: Market power, international CO_2 taxation and oil wealth. *Energy Journal*, **18**, 33-71.

Bernstein, L., and J. Pan (eds.), 2000: *Sectoral Economic Costs and Benefits of GHG Mitigation*. Proceedings of an IPCC Expert Meeting, 14 - 15 February 2000, Technical Support Unit, IPCC Working Group III.

Bernstein, P., W.D.Montgomery, T. Rutherford, and G. Yang, 1999: Effects of restrictions on international permit trading: the MS-MRT model. *Energy Journal*, Special Issue, 221-256.

Bertram, G., A. Stroombergen, and S. Terry, 1993: *Energy and Carbon Taxes Reform Options and Impacts*. Prepared for the Ministry of the Environment, New Zealand by Simon Terry and Associates and BERL, Wellington, New Zealand.

Biswas, W.K., and N.J.D. Lucas, 1997: Economic viability of biogas technology in a Bangladesh village, *Energy*, **22**(8), 763-770.

Boero, G., R. Clarke, and L.A. Winters, 1991: The Macroeconomic Consequences of Controlling Greenhouse Gases: A Survey. *Environmental Economics Research Series*. Department of the Environment, London.

Bohm, P., 1997: Environmental taxation and the double dividend: fact or fallacy. In *Ecotaxation*. T. O'Riordan (ed.), Earthscan, London.

Bohm, P., and C.S. Russell, 1985: Comparative Analysis of Alternative Policy Instruments. In *Handbook of Natural Resource and Energy Economics, vol. 1*. A.V. Kneese and J.L. Sweeney (eds.).

Bose, R., 2000: Mitigating GHG Emissions from the Transport Sector in Developing Nations: Synergies explored in local and global environmental agenda. In *Sectoral Economic Costs and Benefits of GHG Mitigation*. L. Bernstein and J. Pan (eds.), Proceedings of an IPCC Expert Meeting, 14 - 15 February 2000, Technical Support Unit, IPCC Working Group III.

Bose, R.K, 1998: Automotive energy use and emission control: a simulation model to analyse transport strategies for Indian metropolises. *Energy Policy*, **26**(13), 1001-1016.

Bose, R.K., and V Srinivasachary, 1997: Policies to reduce energy and environmental emissions in the transport sector. A case study of Delhi City. *Energy Policy*, **25**(14-15), 1137-1150.

BP, 2000: *Statistical Review of World Energy*. BP-Amoco, London, http://www.bp.com/worldenergy/

Brower, M.L., 1992: *Cool Energy: Renewable Solutions to Environmental Problems*. MIT press.

Brown, S., D. Kennedy, C. Polidano, K. Woffenden, G. Jakeman, B. Graham, F. Jotzo, and B.S. Fisher, 1999: *Economic impacts of the Kyoto Protocol: Accounting for the three major greenhouse gases*. ABARE Research Report 99.6, Canberra, Australia.

Bryden, C., N. Johnstone, and C. Hargreaves 1995: The integration of a bottom-up model of the electricity supply industry with a top-down model of the economy in the UK. In *Top-down or Bottom-up Modelling? An application to CO_2 abatement*. F. Laroui and M. J. van Leeuwen (eds.), Foundation for Economic Research of the University of Amsterdam (SEO), 75-92

BTCE, 1996: *Transport and Greenhouse: Costs and options for reducing emissions*. Bureau of Transport and Communications Economics Report 94, Australian Government Publishing Service, Canberra, Australia.

Buchanan. J.M., and Y.J. Yoon, 1994: *The Return to Increasing Returns*. The University of Michigan Press, Michigan.

Burniaux, J-M., J.P. Martin, G. Nicoletti, and J.O. Martins, 1991*: The Costs of Policies to Reduce Global Emissions of CO_2: Initial Simulation Results with GREEN*. OECD Department of Economics and Statistics, Working Paper No. 103, OECD/GD(91)115, Resource Allocation Division, Paris.

Burniaux, J-M., J. Martin, and J. Oliveira-Martins, 1992: The effect of existing distortions in energy markets on the costs of policies to reduce CO_2 emissions. *OECD Economic Studies*, **19**, 141-165.

Cambridge Econometrics, 1998: *Industrial Benefits from Environmental Tax Reform*. A report to Forum for the Future and Friends of the Earth, Technical Report No. 1, Forum for the Future, London.

Campbell, C.J., 1997: *The Coming Oil Crisis*. Multi-Science Publishing and Petroconsultants, Brentwood, Essex, UK.

Chameides, W. L., X. Li, X. Tang, X. Zhou, C. Luo, C. Kiang, J. St. John, R.D. Saylor, S.C. Liu, K.S. Lam, T. Wang, and F. Giorgi, 1999: Is ozone pollution affecting crop yields in China? *Geophysical Research Letters*, **26**(7), 867-870.

Chandler, W., and A.K. Nichols, 1990: Assessing carbon emission control strategies: A carbon tax or a gasoline tax*? Policy Paper No. 3*, American Council for an Energy-Efficient Economy (ACEEE), Washington DC.

CRA (Charles River Associates) and DRI/McGraw-Hill, 1994: *Economic Impacts of Carbon Taxes*. Report prepared for the Electric Power Research Institute, Palo Alto California, USA.

Cline, W.R., 1992: *The Economics of Global Warming*. Institute for International Economics, Washington, DC.

Comisión Nacional de Energía, 1993: *Programa de Desarrollo de las Fuentes Nacionales de Energía*. Cuba.

Cooper, A., S. Livermore, V. Rossi, A. Wilson, and J. Walker, 1999: The economic implications of reducing carbon emissions: a cross-country quantitative investigation using the Oxford Global Macroeconomic and Energy Model. *Energy Journal*, Special Issue, 335-366.

Cornwall, A., and J. Creedy, 1996: Carbon taxation, prices and inequality in Australia. *Fiscal Studies*, **17**(3), 21-38.

Criqui, P., N. Kouvaritakis, and L. Schrattenholzer, 2000: The impacts of carbon constraints on power generation and renewable energy technologies. In *Sectoral Economic Costs and Benefits of GHG Mitigation*. L. Bernstein and J. Pan, Proceedings of an IPCC Expert Meeting, 14 - 15 February 2000, Technical Support Unit, IPCC Working Group III.

Dargay, J., and D. Gately. 1997: Vehicle ownership to 2015: Implication for energy use and emissions. *Energy Policy,* **25** (14-15), 1121-27.

DeCicco, J., and J. Mark, 1997: Meeting the energy and climate challenge for transportation. *Energy Policy*, **26**, 395-412.

Dervis, K., J. de Melo, and S. Robinson, 1982: *General Equilibrium Models for Development Policy*. Cambridge University Press.

Donovan, D., K. Schneider, G.A. Tessema, and B.S. Fisher, 1997: *International Climate Policy: Impacts on Developing Countries*. Research Report 97-8, Australian Bureau of Agricultural & Resource Economics, Australia.

Dosi, G., 1988: Sources, procedures and microeconomic effects of innovation. *Journal of Economic Literature*, **36**, 1126-71.

Dowlatabadi, H., Lave and Russell, 1996: A free lunch at higher CAFÉ: A review of economic, environmental and social benefits. *Energy Policy*, **24**, 253-264.

DRI (Data Resources Incorporated, McGrawHill), 1992: *Impact of a Package of EC measures to Control CO_2 Emissions on European Industry*. Final Report prepared for the European Commission.

DRI (Data Resources Incorporated, McGrawHill), 1994: *The Energy, Environment and Economic Effects of Phasing Out Coal Subsidies in OECD Countries*. Report under contract to OECD, Paris.

DRI (Standard & Poor), 1998: *The Impact of Meeting the Kyoto Protocol on Energy Markets and the Economy*.

Duleep, K.G., 1997: Evolutionary and revolutionary technologies for improving fuel economy. In *Transport, Energy and Environment: How Far Can Technology Take Us?* J. DeCicco, and M. Delucchi (eds.), Presented at the Conference on Sustainable Transportation Energy Strategies, Asilomar, CA, USA, August 1995, American Council for an Energy-Efficient Economy, Washington DC, USA.

Dunn, S., 2000: Climate policy and job impacts: recent assessments and the case of coal. In *Sectoral Economic Costs and Benefits of GHG Mitigation*. L. Bernstein and J. Pan (eds.), Proceedings of an IPCC Expert Meeting, 14 - 15 February 2000, Technical Support Unit, IPCC Working Group III.

ECMT (European Conference of Ministers of Transport), 1998: *Efficient Transport for Europe: Policies for Internalisation of External Costs*. OECD Publications Service, Paris.

EIA, 1998: *Impacts of the Kyoto Protocol on U.S. Energy Markets and Economic Activity*. Report SR/OIAF/98-03, U.S. Energy Information Agency, Washington DC, USA.

EIA, 2000a: *International Energy Annual, 1998*. Washington DC.

EIA, 2000b: *International Energy Outlook 2000*. Washington DC.

Ekins, P., 1995: Rethinking the Costs Related to Global Warming: A Survey of the Issues. *Environmental and Resource Economics*, **6**, 231-277.

Ekins, P., and S. Speck, 1998: The impacts of environmental policy on competitiveness: theory and evidence. In *International Competitiveness and Environmental Policies*. T. Barker and J. Köhler (eds.), Edward Elgar, Cheltenham, UK, 33-70.

EPA, 1997: *The Benefits and Costs of the Clean Air Act, 1970 to 1990*. Report prepared for U.S. Congress, U.S. Environmental Protection Agency (EPA), Washington DC.

European Commission, 1997: *White Paper of EU Commission, Strategy and Action Plan.*.

European Environment Agency, 1999: *Environment in the European Union at the turn of the century*. Copenhagen, Denmark.

Fankhauser, S., 1995: *Valuing Climate Change: The Economics of the Greenhouse*. Earthscan, London.

Fankhauser, S., and R. Tol, 1995: *Recent Advancements in the Economic Assessment of Climate Change Costs*. CSERGE Working Paper GEC 95-31, London.

Fernandez-Bremauntz, A.A., and M.R. Ashmore, 1995: Exposure of commuters to carbon monoxide in Mexico City – II, Comparison of in-vehicle and fixed-site concentrations. *Journal of Exposure Analysis and Environmental Epidemiology*. **5**(4), 497-510.

Ferriter, J., 1997: The effects of CO_2 reduction policies on energy markets. In *Environment, Energy and Economy*. Y. Kaya and K. Yokobori (eds.), United Nations University Press, New York, NY, USA.

Fiore, A.M., D.J. Jacob, J. Logan, and J. Yin, 1998: Long-term trends in ground level ozone over the contiguous United States, 1980-1995. *Journal of Geophysical Research*, **103**, 1471-1480.

Fuhrer, J., 1995: The critical level for ozone to protect agricultural crops - an assessment of data from European open-top chamber experiments. In *Critical levels for ozone: a UNECE workshop report.* FAC Report no. 16, Swiss Federal Research Station for Agricultural Chemistry and Environmental Hygiene, Liebefeld-Bern, 42-57.

Fujime, K, 1998: Japan's Latest Long-term Energy Supply and Demand Outlook – From Kyoto Protocol's Perspectives. *Energy in Japan*, No. 153, 1-9

Gaj, H., M. Sadowski, and S. Legro, 1997: *Climate Change Mitigation: Case Studies from Poland.* Pacific North West Laboratory, Washington DC.

Garbaccio, R.F., M.S. Ho, and D.W. Jorgenson, 1999: Controlling carbon emissions in China. *Environment and Development Economics*, 4(4), 493-518

Ghanem, S., R. Lounnas, D. Ghasemzadeh, and G. Brennand, 1998: Oil and energy outlook to 2020: implications of the Kyoto Protocol. *OPEC Review*, **22**, 31-58.

Ghasemzadeh, D., 2000: Comments on "Impacts of the Kyoto Protocol on Fossil Fuels". In *Sectoral Economic Costs and Benefits of GHG Mitigation.* L. Bernstein and J. Pan (eds.), Proceedings of an IPCC Expert Meeting, 14 - 15 February 2000, Technical Support Unit, IPCC Working Group III.

Glomsro, D. S., H. Vennemo, and T. Johnsen, 1990: *Stabilization of emissions of CO_2: A computable general equilibrium assessment.* Central Bureau of Statistics, Discussion paper No. 48, Oslo, Norway.

Goldemberg, J., and T.B. Johansson, 1995: *Energy as an Instrument for Socio-Economic Development.* UNDP, New York.

Golub, A., and E. Gurvich, 1996: *The Implications of Subsidies Elimination in the Power Sector of Russia.* Report under contract to OECD, Paris.

Gopalan, H.N.B., and S. Saksena (eds.), 1999: *Domestic Environment and Health of Women and Children.* United Nations Environmental Programme and Tata Energy Research Institute, New Delhi, India.

Gruber E., K. Ostertag, and E. Bush, 1997: Consumption target values for electrical appliances in Switzerland. *Energy and Environment*, **8** (2), 105-113.

Gruebler, A., 1998: *Technology and Global Change*, Cambridge University Press.

Hagan, R., 1998: The future of nuclear power in Asia. *Pacific and Asian Journal of Energy*, 8(1), 9-22.

Hamilton, K., and C. Cameron, 1994: Simulating the distributional effects of a carbon tax. *Canadian Public Policy, 20*, 385-399.

Heck, W.W., W.W. Cure, J.O. Rawlings, L.J. Zaragoza, A.S. Heagle, H.E. Heggestad, R.J. Kohut, L.W. Kress, and P.J. Temple, 1984a: Assessing Impacts of Ozone on Agricultural Crops: I. Overview. *Journal of Air Pollution Control Association.* **34**, 729-735.

Heck, W.W., W.W. Cure, J.O. Rawlings, L.J. Zaragoza, A.S. Heagle, H.E. Heggestad, R.J. Kohut, L.W. Kress, and P.J. Temple, 1984b: Assessing Impacts of Ozone on Agricultural Crops: II. Crop Yield Functions and Alternative Exposure Statistics. *Journal of Air Pollution Control Association*, **34**, 810-817.

Held G., A.D. Surridge, G.R. Tosen, C.R. Turner, and R.D. Walmsley (eds.), 1996: *Air pollution and its impacts on the South African Highveld.* Environmental Scientific Association.

Hirst, E., and S. Hadley, 1995: Effect of electricity utility DSM Programmes on electricity prices. *Energy Studies Review*, 7(1), 1-10.

Hoeller, P., and J. Coppel, 1992: *Energy taxation and price distortions in fossil fuel markets: some implications for climate change policy.* OECD Economic Department Working Paper 110, Paris.

Hoeller, P., A. Dean, and J., Nicolaisen, 1991: Macroeconomic Implications of Reducing Greenhouse Gas Emissions: A Survey of empirical Studies. *OECD Economics Studies*, **16**, 46-78.

Hourcade, J. C., R. Richels, and J. Robinson 1996: Estimating the costs of mitigating greenhouse gases.In *Climate Change 1995: Economic and Social Dimensions of Climate Change.* Contribution of Working Group III to the Second Assessment Report of the Intergovernmental Panel on Climate Change, Cambridge University Press, Cambridge, UK.

Holden, K., D.A. Peel, and J.L. Thompson, 1994: *Economic forecasting: an introduction.* Cambridge University Press.

Hsu, G. J. Y., and T. Chem, 1997: The return of the electric Power industry in Taiwan. *Energy Policy*, 25 (11), 951-957.

Huntington, H.G., 1991: Inferred demand and supply elasticities from a comparison of world oil models. In *International Energy Economics*. T.Sterner (ed.), Lexington, Detroit, 239-261

IAEA, 1997: *Nuclear Power Reactors in the World.* April 1997 Edition, Reference Data Series No. 2, IAEA, Vienna.

IEA, 1997a: *Renewable energy policy in IEA countries.* International Energy Agency, Paris, France.

IEA, 1997b: *Transport, Energy and Climate Change.* International Energy Agency, Paris, France.

IEA, 1998a: *Biomass: Data, Analysis and Trends.* International Energy Agency, Paris, France.

IEA, 1998b: *World Energy Outlook.* 1998 Edition, International Energy Agency, Paris, France.

IEA, 1999: *Coal Information 1998.* 1999 edition, International Energy Agency, Paris, France.

Ikeda, A., M. Ishikawa, M. Suga, Y. Hujii, and K. Yoshioka, 1995: Application of Input-Output Table for Environmental Analysis (7) – Simulations on steel-scrap, blast furnace slag and fly-ash utilization. *Innovation and IO Technique – Business Journal of PAPAIOS*, 6(2), 39-61 (in Japanese)

IPCC (Intergovernmental Panel on Climate Change), 2001: Climate Change: Impacts and Adaptation. Cambridge Unviersity Press, Cambridge, United Kingdom.

IWG (Interlaboratory Working Group), 1997: *Scenarios of U.S. Carbon Reductions: Potential impacts of energy technologies by 2010 and beyond.* Office of Energy Efficiency and Renewable Technologies, U.S. Department of Energy, Washington DC, USA.

Jacoby, H.D, R.S. Eckaus, A.D. Ellerman, R.G. Prinn, D.M. Reiner, and Z. Yang, 1997: CO_2 Emissions Limits: Economic Adjustment and the Distribution of Burdens. *Energy Journal*, 18(3), 31-58.

Jiang,K., X. Hu, Y. Matsuoka, and T. Morita, 1998: Energy technology changes and CO_2 emission scenarios in China. *Environmental Economics and Policy Studies*, 141-160.

Johansson, T., H. Kelly, A.K.N. Reddy, and R.H. Williams (eds.), 1993: *Renewable Energy, Sources for Fuels and Electricity.* Island Press. ISBN 1-55963-138-4.

Johnson, P., S. McKay, and S. Smith, 1990: *The Distributional Consequences of Environmental Taxes.* Institute for Fiscal Studies, Commentary No. 23, London.

Jorgenson, D.W., 1995a: *Productivity, International Comparisons of Economic Growth.* MIT Press, Cambridge, MA.

Jorgenson, D.W., 1995b: *Productivity, Postwar US Economic Growth.* MIT Press, Cambridge, MA.

Jorgenson, D.W., and P.J. Wilcoxen, 1993: Reducing US Carbon Emissions: An Econometric General Equilibrium Assessment. *Resource and Energy Economics, 15*, 7-25.

Jorgenson, D.W., R.J. Goettle, D.T. Slesnick, and P.J. Wilcoxen, 1999: Carbon mitigation, permit trading and revenue recycling. *Harvard-Japan Project on Energy and the Environment 1998-99*, March 1999.

Kainuma, M., Y. Matsuoka, T. Morita, and G. Hibino, 1999a: *Development of an End-Use Model for Analyzing Policy Options to Reduce Greenhouse Gas Emissions.* IEEE Transactions on Systems, Man and Cybernetics, Part C, *29*(3), 317-324.

Kainuma, M., Y. Matsuoka, and T. Morita, 1999b: Analysis of Post-Kyoto Scenarios: The Asian-Pacific Integrated Model. *Energy Journal*, Special Issue, 207-220.

Kainuma, M, Y. Matsuoka, and T. Morita, 1999c: *The AIM/ENDUSE Model and Case Studies in Japan.* A Joint Meeting of the International Energy Agency, The Energy Modeling Forum and the International Energy Workshop, 16-18 June 1999, IEA Paris, France.

Kainuma, M., Y. Matsuoka, and T. Morita, 1997: The AIM model and simulations. In *Key Technology Policies to Reduce CO_2 Emissions in Japan, Tokyo.* H. Tsuchiya, Y. Matsuoka, A. van Wijk, and G.J.M. Phylipsen (eds.), WWF Japan. 39-57.

Kamat, R., A. Rose, and D. Abler, 1999: The impact of a carbon tax on the Susquehanna River Basin economy. *Energy Economics*, *21*(3), 363 – 384.

Kassler, P., and M. Paterson, 1997: *Energy Exporters and Climate Change.* The Royal Institute of International Affairs, London.

Knapp, R. 2000: Discussion on coal. In *Sectoral Economic Costs and Benefits of GHG Mitigation.* L. Bernstein and J. Pan (eds), Proceedings of an IPCC Expert Meeting, 14 - 15 February 2000, Technical Support Unit, IPCC Working Group III.

Kobayashi, K., 1997: Variation in the Relationship Between Ozone Exposure and Crop Yield as Derived from Simple Models of Crop Growth and Ozone Impact. *Atmospheric Engineering,* **31**(3), 703-714.

Koopman, G.J., and C. Denis, 1995: *EUCARS: A Partial Equilibrium European Car Emissions Simulations Model, Version 2.* European Commission, Technical Report, DGII.

Krause, F., W. Back, and J. Koomey, 1992: *Energy Policy in the Greenhouse.* John Wiley.

Kurosawa, A., H. Yagita, W. Zhou, K. Tokimatsu, Y. Yanagisawa, 1999: Analysis of Carbon Emission Stabilization Targets and Adaptation by Integrated Assessment Model. *Energy Journal,* Special Issue, 157-176.

Kydes, A.S., 1999: *Modeling Technology Learning in the National Energy Modeling System.* Report: EIA/DOE-0607 (99), on: http://www.eia.doe.gov/oiaf/issues/technology.html.

Laitner, S., S. Bernow, and J. DeCicco, 1998: Employment and other macroeconomic benefits of an innovation-led climate strategy for the United States. *Energy Policy,* **26** (5), 425-432.

Laroui, F., and M.J. van Leeuwen (eds.), 1995: *Top-down or Bottom-up Modelling? An application to CO₂ abatement.* Foundation for Economic Research of the University of Amsterdam (SEO), ISBN 90-6733-101-5.

Larsen, B.. and A. Shah, 1995: World Fossil Fuel Subsidies and Global Carbon Emissions. In *Public Economics and the Environment in an Imperfect World.* L. Bovenberg and S. Cnossen (eds.), Kluwer, Dordrecht.

Lennon, S.J., G.R. Tosen, M.J. Morris, C.R. Turner, and P. Terblanche, 1994: *Synthesis report on the environmental impacts of electrification.* National Electrification Forum, Working Group for the End-Use of Energy and the Environment, Synthesis Report WGEU05, Pretoria.

Li, J., T. Johnson, Zhou Changyi, R.P. Taylor, et al., 1995: Energy Demand in China: Overview Report. *Issues and Options in Greenhouse Gas Emissions Control.* Subreport No. 2, The World Bank, Washington DC, p. 17

Lindholt, L., 1999: *Beyond Kyoto: CO₂ permit prices and the markets for fossil fuels.* Discussion Papers No. 258, Statistics Norway, Oslo, August.

Mabey, N., S. Hall, C. Smith, and S. Gupta, 1997: *Argument in the Greenhouse. The International Economics of Controlling Global Warming.* Routledge, London.

MacCracken, C., J. Edmonds, S.H. Kim, and R. Sands, 1999: The Economics of the Kyoto Protocol. *Energy Journal,* Special Issue, 25-72.

MacKenzie, J.J., 1997: Climate Protection and the National Interest: The Links among Climate Change, Air Pollution, and Energy Security. *Climate Protection and the National Interest,* World Resources Institute.

Matsuo, N., 1997: Japan's Optimal Power Source Mix in the Long Run and Nuclear Power Generation – A Study on Perspectives of Nuclear Power Generation. *Energy in Japan,* **145,** 33-41.

McKibbin, W., M. Ross, R. Shackleton, and P. Wilcoxen, 1999: Emissions trading, capital flows and the Kyoto Protocol. *Energy Journal,* Special Issue, 287-334.

McKibbin, W.J., and P.J. Wilcoxen, 1993: The global consequences of regional environment policies: an integrated macroeconomic, multi-sectoral approach. In *Costs, Impacts and Benefits of CO₂ Mitigation,* Y. Kaya, N. Nakicenovic, W.D. Nordhaus, and F.L. Toth (eds.), Proceedings of the workshop held on 28-30 September 1992 at IIASA Laxenburg, Austria, 247 – 272.

McKibbin, W.J., and P.J. Wilcoxen, 1995: The theoretical and empirical structure of the G-Cubed model. *Brookings Discussion Papers in International Economics,* No. 118, Brookings Institution, Washington DC.

Michaelis, L., and O. Davidson, 1996: GHG mitigation in the transport sector. *Energy Policy,* **24**(10-11), 969-984.

Michaelis, L., 1996: OECD *Project on Environmental Implications of Energy and Transport Subsidies: Case Study on Electricity in the U.K..* OECD, Paris.

Michaelowa, A., 1996: Economic growth in South East Asia and its consequences for the environment. *Intereconomics,* **31**(6): 307-312.

Miraglia, S., L. Pereira, A. Braga, P. Saldiva, and G. Bohm (in press): *Air Pollution Health Costs in Sao Paolo City, Brazil.*

Miyamoto, K. (ed.), 1997: Renewable biological systems for alternative sustainable energy production. *FAO Agricultural Series Bulletin 128.*

Mokhtarian, P.L., S.L. Handy, and I. Salomon, 1995: Methodological issues in the estimation of the travel, energy, and air quality impacts of telecommuting. *Transportation Research A,* **29**(4), 283-302.

Montgomery, W.D., 1972: Markets in Licenses and Efficient Pollution Control Programs. *Journal of Economic Theory* **5,** 395-418.

Mors, M., 1991: The Economics of Policies to Stabilise or Reduce Greenhouse Gas Emissions: The Case of CO₂. *Economic Papers No 87,* Commission of European Communities, Brussels, Belgium.

Nakicenovic, N., A. Grubler, and A. McDonald (eds.), 1998: *Global Energy Perspectives.* Cambridge University Press, ISBN 0-521-64569-7.

Nakicenovic, N., J. Alcamo, G. Davis, B. de Vries, J. Fenhann, S. Gaffin, K. Gregory, A. Grübler, T-Y. Jung, T. Kram, E. Lebre La Rovere, L. Michaelis, S. Mori, T. Morita, W. Pepper, H. Pitcher, L. Price, K. Riahi, A. Rehrl, H-H. Rogner, A. Sankovski, M. Schlesinger, P. Shukla, S. Smith, R. Swart, S. van Rooijen, N. Victor, D. Zhou, 2000: *Special Report on Emissions Scenarios.* A Special Report of Working Group III of the Intergovernmental Panel on Climate Change, Cambridge University Press, Cambridge, UK.

Nordhaus, W.D., 1991: To slow or not to slow: the economics of the greenhouse effect. *Economic Journal,* **101,** 920-937.

OECD (Organization for Economic Cooperation and Development), 1997a: *Enhancing the market development of energy technology, a survey of eight technologies.* Paris, France.

OECD, 1997b: *OECD Environmental Data 1997.* Paris, France.

OECD, 1997c: *Reforming Coal and Electricity Subsidies.* Annex I Expert Group on the United Nations Framework Convention on Climate Change, Working Paper No. 2, OCDE/GD (97)70, Paris, France.

OECD, 1998: *Economic Modelling of Climate Change.* OECD Workshop Report, Paris, France.

OECD, 2000: *Proceedings of an IPCC/OECD Workshop on Assessing the Ancillary Benefits and Costs of Greenhouse Gas Mitigation Strategies.* Held in Washington DC, March 27-29, 2000.

O'Riordan, T., 1997: *Ecotaxation.* Earthscan, London.

OTA (Office of Technology Assessment), 1995: *Advanced Automotive Technology: Visions of a Super-Efficient Family Car.* Report OTA-ETI-638. U.S. Congress, Office of Technology Assessment, Washington DC, USA.

Parker, J.C., 1995: Including Economic externalities in DSM planning. *Energy Studies Review,* **7**(1), 65-76.

Pearce, D., 1996: Economic valuation of health damage from air pollution in the developing world. *Energy Policy,* **24**(7), 627-630.

Penner, J.E., D.H. Lister, D.J. Griggs, D.J. Dokken, and M. McFarland (eds.), 1999: *Aviation and the Global Atmosphere.* Intergovernmental Panel on Climate Change, Geneva, Switzerland.

Pershing J., 2000: Fossil fuel implications of climate change mitigation responses. In *Sectoral Economic Costs and Benefits of GHG Mitigation.* L. Bernstein and J. Pan (eds), Proceedings of an IPCC Expert Meeting, 14 - 15 February 2000, Technical Support Unit, IPCC Working Group III.

Pimental, D., G. Rodrigues, T. Wane, R. Abrams, K. Goldberg, H. Staecker, E. Ma, L. Brueckner, L. Trovato, C. Chow, U. Govindarajulu, and S. Boerke, 1994: Renewable energy: economic and environmental issues. *Bioscience,* **44**(8), September.

Piscitello, E.S., and V.S. Bogach, 1997: *Financial Incentives for Renewable Energy.* Proceedings of an International Workshop, 17-21 February 1997, Amsterdam, The Netherlands, World Bank Discussion Paper No. 391,

Poterba, J., 1991: Tax policy to combat global warming: on designing a carbon tax. In *Global Warming: Economic Policy Responses.* R. Dornbusch and J. M. Poterba (eds.), MIT, Cambridge, Massachusetts.

Pratten, C.F., 1988: A survey of the economies of scale. In *Research on the Cost of Non-Europe, Basic Findings.* Vol. 2, Studies on the Economics of Integration, Commission of the European Communities, Brussels, 2.1-2.153.

Preston, E., and D. Tingey, 1988: The NCLAN Program For Crop Loss Assessment. In *Assessment of Crop Loss From Air Pollutants.* W.W. Heck, O.C. Taylor, and D.T. Tingey (eds.), Elsevier Scinece Publishers Ltd, London, 45-60.

Price, L.K., L. Michaelis, E. Worrel, and M. Krushch, 1998: Sectoral trends and driving forces of global energy use and greenhouse gas emissions. *Mitigation and Adaptation Strategies for Global Change*, **3**, 261-319.

Proops, J.L.R., M. Faber, and G. Wagenhals, 1993: *Reducing CO$_2$ Emissions - A Comparative Input-Output Study for Germany and the UK*, Springer-Verlag, Berlin.

Radetzki, M., 1997: The economics of biomass in industrialized countries – an overview. *Energy Policy*, **25**(67), 545-554.

Repetto, R., and D. Austin, 1997: *The Costs of Climate Protection: A Guide for the Perplexed*. World Resources Institute.

Rose, A., and T.H. Tietenberg, 1993: An International System of Tradeable CO$_2$ Entitlements: Implications for Economic Development. *Journal of Environment & Development*, **2**, 348-383.

Rose, A., J. Benavides, D. Lim, and O. Frias, 1996: Global warming policy, energy, and the Chinese economy. *Resource and Energy Economics*, **18**(1), 31-63.

Rosendahl, K. E., 1996: Carbon taxes and the impacts on oil wealth. *Journal of Energy Finance and Development*, **1**(2), 223-234.

Ross, A., 1999: *Road Accidents: A Global Problem Requiring Urgent Action*. Roads & Highways Topic Note, RH-2, World Bank.

Sathaye, J., and Tyler, S. 1991: Transitions in household energy use in urban China, India, The Philippines, Thailand, and Hong Kong. *Annual Review of Energy and the Environment*, **16**, 295-335.

Schafer, A., 1998: The global demand for motorized mobility. *Transportation Research A*, **32**(6), 455-477.

Schwaiger, H., and B. Schlamadinger. 1998: The potential of fuelwood to reduce greenhouse gas emissions in Europe. In *Biomass and Bioenergy*, **15** (4/5), 369-379, Bohlin, Wisniewski and Wisniewski (eds.).

SDPC (State Development Planning Commission), 1996: *National strategy for new and renewable energy development plan for 1996 to 2010 in China*. Beijing, China.

Shackleton, L.Y., S.J. Lennon, and G.R. Tosen (eds.), 1996: *Global climate change and South Africa*. Environmental Scientific Association, Cleveland, South Africa.

Shelby, M., A. Cristofaro, B. Shackleton, and B. Schillo, 1994: *The Climate Change Implications of Eliminating U.S. Energy (and Related) Subsidies*. US EPA, Washington DC. Report for OECD, Paris.

Shinozaki, M., Y. Wake, and K. Yoshioka, 1998: Input-output analysis on Japan-China environmental issues (4). *Innovation and IO Techniques – Business Journal of PAPAIOS*, **8** (3), 40-49.

Shoven, J.B., and J. Whalley, 1992: *Applying General Equilibrium*, Cambridge University Press.

Smith, K.R., 1996: Indoor air pollution in developing countries: growing evidence of its role in the global disease burden. In *Proceedings of Indoor Air' 96: The 7th International Conference on Indoor Air Quality and Climate*. Institute of Public Health, Tokyo.

Sommer, H., R. Seethaler, O. Chanel, M. Herry, S. Masson, and J.C. Vergnaud, 1999: *Health Costs due to Road Traffic-Related Air Pollution: An impact assessment project of Austria, France, and Switzerland*. Economic Evaluation, Technical Report on Economy, World Health Organization.

Stern, J., 2000: Comments on "Impacts of the Kyoto Protocol on Fossil Fuels". In *Sectoral Economic Costs and Benefits of GHG Mitigation*. L. Bernstein and J. Pan (eds.), Proceedings of an IPCC Expert Meeting, 14-15 February 2000, Technical Support Unit, IPCC Working Group III.

Sutherland, R.J., 1998: The impact of potential climate change commitments on six industries in the United States. *Energy Policy*, **26**, 765-776.

Symons, E., J. Proops, and P. Gay, 1994: Carbon taxes, consumer demand and carbon dioxide emissions: a simulation analysis for the UK. *Fiscal Studies*, **15**(2), 19-43.

Tietenberg, T.H., 1990: Economic Instruments for Environmental Regulation. *Oxford Review of Economic Policy*, **6**, 17-33.

Tulpule, V., S. Brown, J. Lim, C. Polidano, H. Pant, and B. Fisher, 1999: The Kyoto Protocol: an economic analysis using GTEM. *Energy Journal*, Special Issue, 257-286.

Tunnah, B., S. Wang, and F. Liu, 1994: *Energy Efficiency in China: Technical and Sectoral Analysis*. World Bank Publications.

Turiel, I., 1985: *Indoor Air Quality and Human Health*. Stanford University Press, Stanford.

UK Department of Environment, 1997: Effects of ozone on vegetation, materials and human health. *Ozone in the United Kingdom*. Fourth Report of the Photochemical Oxidants Review Group, United Kingdom.

US Bureau of Labor Statistics, 2000: *Nonfarm Payroll Statistics*. From the National Current Employment Statistics Home Page http://stats.bls.gov/ceshome.htm.

US Department of Commerce, 2000: *Survey of Current Business*, Table 2.

US National Academy of Sciences, 1992: *Policy Implications of Greenhouse Warming: Mitigation, Adaptation, and the Science Base*, Washington DC.

Wang, J., M. Mintz, M. Singh, K. Storck, A. Vyas, and L. Johnson, 1998: *Assessment of PNGV Fuels Infrastructure - Phase 2 Report: Additional Capital Needs and Fuel-Cycle Energy and Emission Impacts*. Argonne National Laboratory Report ANL/ESD/37, Chicago, IL, USA.

WCI (World Coal Institute), 1999: *Coal – power for progress*. WCI, January 1999.

WEC (World Energy Council), 1993: *Energy for Tomorrow's World*. Kogan Page, St. Martin's Press, New York.

WEFA, Inc., 1998. *Global Warming: The High Cost of the Kyoto Protocol*. Burlington, Massachusetts.

WHO (World Health Organization), 1992: *Climate Change and Human Health*. A.J. McMichael, A. Haines, R. Slooff, and S. Kovats (eds.), Horley Studios, Redhill, United Kingdom.

WHO, 1999: Transport, Environment, and Health. *Third Ministerial Conference on Environment and Health*, London, 16-19 June 1999.

Wit, G. de, 1995: *Employment effects of ecological tax reform*. Center for Energy Conservation and Environmental Technology, The Hague.

Yamasaki, E., and N. Tominaga, 1997: Evolution of an ageing society and effect on residential energy demand. *Energy Policy*, **25** (11), 903-912.

Yoshioka, K., H. Hayami, A. Ikeda, K. Hujiwara, and M. Suga, 1993: Application of Input-Output Table for Environmental Analysis (4) – Promotion of Blast Furnace Cement. *Innovation and IO Technique – Business Journal of PAPAIOS*, **4**(3-4), 21-28 (in Japanese).

Zaim, K.K., 1997: Estimation of health and economic benefits of air pollution abatement for Turkey 1990 and 1993. *Energy Policy*, **25**(13), 1093-1097.

10

Decision-making Frameworks

Co-ordinating Lead Authors:
FERENC L. TOTH (GERMANY), MARK MWANDOSYA (TANZANIA)

Lead Authors:
Carlo Carraro (Italy), John Christensen (Denmark), Jae Edmonds (USA), Brian Flannery (USA), Carlos Gay-Garcia (Mexico), Hoesung Lee (South Korea), Klaus Michael Meyer-Abich (Germany), Elena Nikitina (Russian Federation), Atiq Rahman (Bangladesh), Richard Richels (USA), Ye Ruqiu (China), Arturo Villavicencio (Ecuador/Denmark), Yoko Wake (Japan), John Weyant (USA)

Contributing Authors:
John Byrne (USA), Robert Lempert (USA), Ina Meyer (Germany), Arild Underdal (Norway)

Review Editors:
Jonathan Pershing (USA), Mordechai Shechter (Israel)

CONTENTS

EXECUTIVE SUMMARY

Scope for and New Developments in Analyses for Climate Change Decisions

Climate change is profoundly different from most other environmental problems with which humanity has grappled. A combination of several features gives the climate problem its unique feature, which include:

- public good issues that arise from the concentration of greenhouse gases (GHGs) in the atmosphere (and require collective global action);
- the multiplicity of decision makers (ranging from global decision-making frameworks (DMFs) down to the micro-level of firms and individuals);
- the heterogeneity of emissions; and
- the consequences of emissions around the world.

Moreover, the long-term nature of climate change originates because it is the concentration of GHGs that is important, rather than annual emissions; this feature raises the thorny issues of intergenerational transfers of wealth and environmental good and bad outcomes. Next, human activities associated with climate change are so widespread that narrowly defined technological solutions are impossible and the interactions of climate policy with other broad socioeconomic policies are strong. Finally, large uncertainties or in some areas even ignorance characterize many aspects of the problem and require a risk management approach to be adopted in all DMFs that deal with climate change.

Experiments with cost–benefit models framed as a Bayesian decision-analysis problem show that optimal near-term (next two decades) emission paths diverge only modestly with perfect foresight and even with hedging for low-probability, high-consequence scenarios. Cost-effectiveness analyses seek the lowest cost that will achieve an environmental target by equalizing the marginal costs of mitigation across space and time. Long-term cost-effectiveness studies estimate the costs to stabilize atmospheric carbon dioxide (CO_2) concentrations at different levels. While there is a moderate increase in the costs when passing from a 750 ppmv to a 550 ppmv concentration stabilization level, there is a larger increase in costs passing from 550 ppmv to 450 ppmv unless the emissions in the baseline scenario are very low. The total costs of stabilizing atmospheric carbon concentrations are very dependent on the baseline scenario: for example, for scenarios focusing on the local and regional aspects of sustainable development costs are lower than for other scenarios. Rather than seeking a single optimal path, the tolerable windows or safe landing approaches seek to delineate the complete array of possible emission paths that satisfy externally defined climate impact and emission cost constraints. Results indicate that a delay in near-term effective emission reductions can drastically reduce the future range of options for relatively tight climate change targets. Less tight targets offer more near-term flexibility.

International Regimes and Policy Options

Different mitigation policy options include the timing of responses to climate change, the choice between mitigation and adaptation responses, the role of technological innovation and diffusion, the choice between domestic action and the adoption of international mechanisms, the combination of climate change mitigation with actions towards other environmental or socio-economic objectives, and others. The costs and benefits of these crucially depend on the characteristics of the international agreement on climate change that is adopted. In particular, they depend upon two main features of the international regime: the number of signatories, and the size of their quantitative commitment to control GHG emissions. The number of signatories depends on how equitably the commitments of the participants are shared. Cost-effectiveness (minimizing costs by maximizing participation) and equity (the allocation of emissions limitation commitments) are therefore strongly linked.

There is therefore a three-way relationship between the design of the international regime, the cost-efficiency of climate policies, and the equity of the consequent economic outcomes. Thus, it is crucial to design the international regime in a way that increases both its efficiency and its equity. The literature presents different strategies to optimize an international regime. For example, countries can be encouraged to participate in an international group committed to specific emissions limits and targets if the equity (and therefore efficiency) is increased by a larger agreement. This may include measures like an appropriate distribution of targets over time, the linkage of the climate debate with other issues ("issue linkage"), the use of financial transfers to affected countries ("side payments"), and technology transfer agreements.

Linkages to National and Local Sustainable Development Choices

Government structures involved in the decision-making process vary considerably among countries. Institutional articulation remains one of the critical factors affecting the consolidation of an effective decision-making process related to sustainable development. Even if rules and regulations exist to assign competence, tasks, and responsibilities among the institutions involved, a considerable gap exists between what might be desirable and what, for the most part, is practised. In this context, policies related to sustainable development are no longer seen as a hierarchical, government-controlled chain of commands, but rather as an open process in which the principles of "good governance"–transparency, participation, pluralism, and accountability–become the key elements of the decision-making process.

A critical requirement of sustainable development is the capacity to design policy measures that exploit potential synergies between national economic growth objectives and environmentally focused policies without hindering development and in accordance with national strategies. As also discussed in Chapters 1 and 2, climate change mitigation strategies offer a clear example of co-ordinated and harmonized policies that take advantage of the synergies between the implementation of mitigation options and broader objectives. The potential linkages between climate change mitigation issues and economic and social aspects have also brought an important shift in the focus of mitigation analysis literature. The three perspectives introduced in Chapter 1 (cost-effectiveness, equity and sustainability) illustrate this shift and broaden the array of options, for example by including options for institutional and behavioural changes. From being confined to project-by-project or sector-based approaches, analyses and studies are increasingly concerned with the use of broader policy issues as mechanisms to reduce GHG emissions. Thus the alternative energy paths of low carbon futures in developing countries can be compatible with national objectives. Although environmental concerns, and climate change issues in particular, were not explicitly addressed by macroeconomic and sectoral policies, analyzed country cases show clear synergies between reform policies and environmental improvements. It is also important to underline that for the elements that make up policies at different levels to operate in a mutually reinforcing manner, the creation of appropriate communication and information channels should be given special attention.

The private sector has played an important role in the development and transfer of energy efficiency technologies, which reduce the emission of GHGs, and it is becoming increasingly active in developing and transferring renewable energy technologies. Large enterprises can establish research and development (R&D) institutions on their own or jointly with other research centres to provide support to technological innovation and the integration of production and research. On the supply side, the government can play an important role in R&D and creating an enabling environment for technology transfer. While introducing technologies to mitigate and adapt to climate change, the developing countries should consider that the introduction of such technology could generate economic benefits and promote sustainable development. In many cases of technology transfer, much attention was paid to the introduction of technologies and a high cost was paid to procure expen-

sive technological facilities, but less effort was applied to the digestion, absorption, and innovation of the introduced technologies. Information can play a guiding role in technology transfer. To enable sound decision-making, the up-to-date information on the current status of technology research and development, the technical and economic evaluation of technologies, and the sources of technologies should be available.

Key Policy-relevant Scientific Questions

Answers to policy-relevant scientific questions need to draw on the vast material presented in this volume. These questions are concerned with the best possible current action with a view to a huge array of possible futures. Decisions need to be made regarding the short-term balance of various types of actions (mitigation, adaptation, information acquisition), their timing (in absolute terms and relative to each other), their location (of mainly mitigation activities), the character and content of international agreements, and the mode and broader policy context of implementation.

Striking the appropriate balance between mitigation and adaptation will be a tedious process. The need for, extent, and costs of adaptation measures in any region will be determined by the magnitude and nature of the regional climate change driven by shifts in global climate. How global climate change unfolds will be determined by the total amount of GHG emissions that, in turn, reflects nations' willingness to undertake mitigation measures. Balancing mitigation and adaptation efforts largely depends on how mitigation costs are related to net damages (primary or gross damage minus damage averted through adaptation plus costs of adaptation). Both mitigation costs and net damages, in turn, depend on some crucial baseline assumptions: economic development and baseline emissions largely determine emission reduction costs, while development and institutions influence vulnerability and adaptive capacity.

Options to mitigate climate change include actual emission reductions and CO_2 sequestration, investments in developing technologies that will make future reductions cheap relative to their current costs, institutional and regulatory changes to modify current decisions that distort in favour of GHG-emitting action, and others. Their relative weight in an optimal near-term portfolio of mitigation actions crucially depends on the assumptions behind the various mitigation-cost estimates and about the preconditions and future availability of inexpensive technologies. Estimates of costs of drastic near-term reductions tend to be high, but the proper way to encourage technological development remains heavily debated.

In principle, costs of near-term emission reductions could be reduced by using international flexibility mechanisms to realize reductions where they are least expensive. While there is a broad agreement on the cost-reducing effect of international flexibility mechanisms, there are concerns about their implications for incentives to technological development as well as about the political (domestic and international) and practical pitfalls of their implementation. In addition to costs, climate change impacts and mitigation efforts raise a whole array of equity issues.

Much of the debate about climate change mitigation revolves around the broader issue of development and the unequal distribution of wealth among countries of the world. Views diverge widely. Is climate change an opportunity to solve the problems of sustainable development and global distribution of wealth? Or would broadening the scope of the already complex and controversial issue of climate change run the risk of neither solving the climate problem nor improving prospects for sustainable development? This reports takes the view that by taking into account the broader perspective of sustainable development the portfolio of mitigation policies is enhanced. A central issue in linking development and climate concerns is technological transfer that could help less-developed countries speed-up their development and control GHG emissions at relatively low costs. Opportunities are ample, but barriers are significant also.

10.1   Introduction

10.1.1   Chapter Overview

The preceding chapters in this volume assess the scientific literature on specific aspects of climate change economics and policy. This chapter is intended to synthesize the most important policy-relevant scientific results by taking several cuts across the material. This chapter begins with a presentation of the special features of climate change in the context of how they affect decision-making in different frameworks. This is followed by a list of analytical frameworks adopted by scientists to provide advice to decision makers and by an overview of the most important new developments since the Second Assessment Report (SAR). This section closes with notes on decision-making processes and implications of uncertainty for the robustness of choices.

Section 10.2 presents an assessment of key insights from the economics and political science literature into international regimes and policy options. The chief issue addressed in the section is how international institutions for addressing climate change, such as the United Nations Framework Convention on Climate Change (UNFCCC), are simultaneously shaped by and influence national policy choice.

Section 10.3 considers the problem of local and national climate policy formulation in the broader context of sustainable development objectives. The interactions of development and environmental policy objectives, particularly as they affect non-Annex I nations, are discussed.

Section 10.4 looks at a series of policy-relevant scientific questions related to global and international climate policy in more detail. It focuses on what has been learned from work that examined decision making at the global scale. While much of this literature is also cognizant of the regional decisions that accumulate to determine global aggregates, it is united by a global focus, common to all of the work discussed in the section. It explores what is known about costs and benefits of actions, the timing and composition of policy responses, and the influence of equity and fairness considerations on policy. Finally, some concluding remarks and an outline of future tasks are presented in the closing section.

The long tradition of using the terms decision analysis (and frameworks) and decision making (and frameworks) largely interchangeably, and both meaning scientific inquiries to serve decision makers, has resulted in some confusion in the case of climate change. With a view to the political sensitivity of the issue, it is important to clarify the terminology here at the beginning of this chapter. Toth (2000) proposes a simple scheme to make a clear distinction to recognize the fine borderline between a policy-relevant scientific assessment and policy making proper. Climate change decision-making and decision analysis intended to support it can be structured in three major domains: decision making *per se* (the act of for-

mulating decisions), decision analysis (aimed at providing information for decision makers), and process analysis (investigating procedures of decision making). The last two are sometimes difficult to separate and they overlap in certain areas, but the distinction is still useful.

DMFs relevant to the climate problem have several levels. They stretch from global and supranational fora through national and regional institutions down to the micro-level of companies, families, and individuals. At each level, it is useful to distinguish two parts of these DMFs: institutions that provide the boundary conditions (jurisdictions, procedural rules, the body of earlier agreements, etc.) and processes that fall within these frameworks (negotiations, lobbying, persuasion). At the global level, for example, UNFCCC provides the institutional part and negotiations represent the process part of the DMF.

To keep the term comprehensive and flexible, decision-analysis frameworks (DAFs) are defined as analytical techniques aimed at synthesizing available information from many (broader or narrower) segments of the climate problem to help policymakers assess the consequences of various decision options within their own jurisdictions. DAFs organize climate-relevant information in a suitable framework, apply a decision criterion (based on some paradigms or theories), and identify options that are better than others under the assumptions that characterize the analytical framework and the application at hand. A broad range of DAFs has been used to provide substantial information for the various DMFs involved in climate decisions at various levels. The most important ones are depicted later in this section.

The third domain is process-analysis frameworks (PAFs), which involve assessments of the decision-making process and provide guidance for decision making in two main areas. The first is concerned with institutional framework design, that is how to build policy regimes that address the problem effectively (Victor *et al.*, 1998; Young, 1999). The second looks at procedures of decision making at various levels. The bulk of the literature on climate change addresses global regime-building in framework analysis and international negotiations in procedure analysis (Kremenyuk, 1991). Pertinent lessons from this literature are assessed in Section 10.2.

The objective in this chapter is to provide a critical appraisal of policy-oriented analyses and to summarize the emerging insights in a form that allows policymakers to make informed judgements within the various DMFs. It is clearly not intended to inflict any particular position upon the policymakers.

10.1.2   Scope of the Problem

Climate change is a problem that is inherently different from other environmental problems with which humanity has grappled, because the assumption that prior experience with other air-pollution problems is a good model upon which to base climate policy responses fails at many levels. At least six unique features characterize the issue.

10.1.2.1 The Problem Is Global

Public goods issues

Traditional environmental air-pollution problems have been amenable to local solutions. The dirty air in a North American city is of no direct consequence to a city in New Zealand. With climate change it is the emissions of all sources in all nations that determine the concentration of greenhouse gases (GHGs) in the atmosphere. As a consequence, the climate change problem is inherently a public goods problem. That is, the climate that everyone enjoys is the product of everyone's behaviour. No single individual or nation can determine the composition of the world's atmosphere. Any individuals' or nations' actions to address the climate change issue, even the largest emitting nation acting alone, can have only a small effect. As a consequence, individuals and nations acting independently will provide, together, fewer resources than all individuals and nations would if they acted in concert. This characteristic provides an important motivation for collective, global action.

Multiplicity of decision makers

Multiplicity of decision makers also implies that there are limits to collective actions. Decisions by actors at a wide range of levels–global governmental organizations, nation states, regional governments, private individuals, multinational firms, local enterprises–all matter. The global nature of the problem also implies that the full breadth of human social structures is encompassed. This in turn implies that a diversity of policy responses is needed. Policy responses that are effective and appropriate in one social context may be completely inappropriate in another.

Heterogeneity

Emissions and consequences are also heterogeneous around the world. This exacerbates the basic public goods nature of the problem. Countries are distributed across a spectrum of high emitters to low emitters and high impacts to low impacts. Nations with high emissions and low expected impacts have a high potential to control concentrations, but little incentive. On the other hand, nations with low emissions and high impacts have great incentive to control emissions, but little capability. While side payments could, in principle, resolve this dilemma, transaction costs may be significant and the present income distributions may lead to unacceptable outcomes. Furthermore, most of the people who will be directly affected by the problem have not been born yet, which limits their ability to negotiate. Both emissions and the capability to mitigate carbon emissions to the atmosphere are unevenly distributed around the world. A dozen countries control 95% of conventional carbon-based energy resources–conventional oil, conventional gas, and coal. Unconventional resources–deep gas, methane hydrates, and shales–while presently expensive relative to conventional fuels, have an unknown distribution in potentially

vast quantities. Fifteen nations emit more than 75% of the world's annual carbon emissions.

10.1.2.2 The Problem Is Long Term

It is concentrations not emissions that matter

Climate change is related to the concentration of GHGs and not to any individual year's emissions. Carbon dioxide (CO_2) concentrations are closely related to the net accumulation of emissions over long periods of time. That is, it is the sum of emissions over time that determines the atmospheric concentration. Any individual year's emissions are only marginally important[1]. Average residence times for GHGs can range up to thousands of years for some of the anthropogenic species. Strategies to control net emissions must account for long periods of time in a meaningful way. The ultimate objective of the UNFCCC is the "stabilization of GHG concentrations in the atmosphere at a level that would prevent dangerous anthropogenic interference with the climate system" (UNFCCC 1993, Article 2).

Intergenerational transfers are inevitable

The consequences of climate change will be visited primarily on those who are alive in the future. The present generation has inherited its atmosphere and associated climate from its ancestors. While individuals and governments make many decisions that affect future generations, most of these decisions are undertaken inadvertently. It is impossible to avoid the intergenerational wealth-transfer issue when addressing the climate problem. That most of the affected parties are not present to participate in the decision-making process raises complicated ethical questions. The implications of their absence are not immediately obviously. Future generations have a stake both in the environmental resources, such as climate, that they inherit, and in other wealth that is passed down to them. Sacrifices that are made by the present generation for the good of its descendants will alter the composition of wealth (e.g., environmental versus material) that is transferred from the present to the future, as well as the magnitude of the transfer. As climate change is anticipated to be greater in the future than it is at present, those who live in the future will reap most of the benefits that accrue to near-term actions to limit emissions. Intergenerational asymmetry can lead to a form of public goods problem in which the willingness to undertake emissions mitigation in the near-term may be less than would have been the case if the decision makers lived infinitely. Also implied is a greater sensitivity to emission-limitation costs than would be the case if the present generation lived to benefit from its emissions-mitigation actions.

To limit the concentration of atmospheric carbon dioxide, global carbon emissions must eventually peak and then decline
This result follows from the nature of the carbon cycle, as it is

---

[1] Clearly, the time path of emissions over a longer time period does affect the rate of change of concentrations and associated climate changes.

presently understood. While non-CO_2 GHGs with relatively short life times, such as methane (CH_4), have an atmospheric concentration that is stable with a stable rate of annual emission, CO_2 does not. The cumulative net introduction of carbon emissions from terrestrial reservoirs, such as fossil fuels or biological carbon, through (for example) energy production and use or land-use change, determines the long-term, steady state, atmospheric CO_2 concentration. Carbon cycle models require net emissions to asymptotically approach zero, though the process can take centuries. Most, but not all, emissions scenarios anticipate that, in the absence of a concern for climate change, future GHG emissions will continue to rise rather than fall (IPCC, 2000b). Where reference emissions scenarios exhibit increasing emissions over time, most of the emissions mitigation required to stabilize the concentration of carbon must occur in the future, with the deviation from the profile required for stabilization growing with time.

While emissions limitation is a policy response, it is not the only policy response available to decision makers
In addition to emissions limitations, policymakers have a wide array of other tools at their disposal including knowledge gathering, research and development of technologies to reduce emissions and enhancing the resilience of societies experiencing climate change. The optimal and actual mix of policy responses will vary over time.

10.1.2.3   Associated Human Activities Are Pervasive

Control of greenhouse gas concentrations implies eventual limitations on energy-related emissions
Energy is the single largest source of GHG emissions. It is responsible for approximately 80% of net carbon emissions to the atmosphere. While net emissions of carbon are associated with fossil fuel combustion, the carbon-to-energy ratio varies between high-carbon fuels, such as coal, and low-carbon fuels, such as CH_4, approximately by a factor of two. Technologies such as hydroelectric power, nuclear fission, wind power, and solar power are generally treated as if they have little or no direct carbon emissions, though this may not be the case. For example, CH_4 may be released in the process of creating a hydroelectric facility and carbon may be released in the manufacture of cement used in nuclear power reactors. Technologies do exist that can biologically sequester or physically remove and store carbon. Thus, in principle, controlling energy-related carbon emissions is possible for several sources of carbon emissions without foregoing fossil fuel use. These technologies are discussed in Chapter 3.

Narrowly defined technological solutions are unavailable, but a broad development and deployment of technology is key to controlling the cost of emissions limitation
Emissions of GHGs are associated with an extraordinary array of human activities. CO_2 emissions are associated with the combustion of fossil fuels and changes in land-use. They are thus affected by activities that range from, for example, household heating and cooling to commercial lighting and appli-

ances, to the transportation of goods and provision of services, to the manufacture of materials, to the growth and harvest of crops, and to the generation of electric power. As a consequence, GHG emissions are greatly affected by other exogenous and non-climate-policy factors. Narrowly defined technological solutions, such as were available to address the problem of stratospheric ozone depletion, are impossible for the climate issue. While no single technology provides a complete solution to the problem of controlling emissions of GHGs, a significant set of existing, emerging, and potential technologies is available to mitigation climate change, as discussed in Chapters 2, 3 and 4[2].

Policy interactions will be significant
Future emissions depend to a large degree on the rate and direction of technological developments in a broad array of human endeavours. For example, China's policies to stabilize its population size, taken for reasons unrelated to climate change, will have a profound effect on Chinese emissions of GHGs to the atmosphere. Policies to control non-GHG air pollutants can greatly affect GHG emissions. For example, measures to substitute natural gas and non-carbon-emitting energy forms, such as solar and nuclear power, for coal in electricity generation to control local and regional air pollution can affect GHG emissions as well. On the other hand, some policies that reduce local air pollution, such as scrubbing power plants for sulphur, can reduce power-plant efficiency and increase GHG emissions.

10.1.2.4   Uncertainty Is Pervasive

There are many uncertainties regarding the magnitude of future climate change, its consequences and the costs, benefits and implementation barriers of possible solutions. Future emissions to the atmosphere are inherently uncertain and can only be explored on the basis of scenarios. The change in concentration of GHGs that would result from a given emission rate is much less uncertain. But the timing, extent, and distribution of climate change and sea level rise for a given concentration of GHGs is not well known due to limitations in modelling climate change at the regional level. The impacts of climate change on ecosystems and humanity is known with limited certainty. The potential for an unspecified, low-probability, but catastrophic turn of events haunts the problem.

While uncertainties are great, they are not distributed evenly throughout the problem. The cost implications of emissions mitigation are better known than the more distant (in time) potential benefits from mitigation. In part this is because of temporal proximity, but it is also because most of the costs

---

[2] See also, for example, Energy Innovations, 1997; Interlaboratory Working Group on Energy Efficient and Low-Carbon Technologies, 1997, 2000; Koomey *et al.*, 1998; Bernow *et al.*, 1999; Edmonds *et al.*, 1999; Geller *et al.*, 1999; Laitner, 1999; Laitner *et al.*, 1999; PCAST, 1999; Hanson and Laitner, 2000; Kim *et al.*, 2000.

associated with emissions mitigation pass through markets, whereas many of the benefits do not. Some uncertainties will remain unresolved regardless of the decisions made. This follows directly from the fact that there is only one observed history. All the other potential histories are counterfactual, and therefore constructs from analytical tools that are limited in their veracity. In decision making terms the problem of climate change mitigation requires decision making under uncertainty. Given the long lead times of mitigation action, fully resolving uncertainties would make an adequate response infeasible.

10.1.2.5 The Consequences Are Potentially Irreversible

Many global biogeochemical processes have long time scales. Sea level changes as a consequence of changes in mean global temperature can take more than 1000 years to play out. Similarly, changes in the concentration of GHGs can rise rapidly, but decline slowly. And, even if concentrations can be reduced, the nature of the climate system is such that it might not return to the same climatic state associated with an earlier concentration.

10.1.2.6 The Global Institutions Needed to Address the Issue Are only Partially Formed

The UNFCCC has been ratified by more than 170 parties and entered into force in 1994. It provides the institutional foundations upon which international climate change negotiations occur. It sets as its ultimate objective the stabilization of the concentration of GHGs in the atmosphere at levels that prevent dangerous anthropogenic interferences with the climate. However, the UNFCCC establishes a process and does not create the institutions for implementing the objective. The objective has not yet been quantified. The term "dangerous" is left open to interpretation by the parties.

The Kyoto Protocol of December 1997, described in Chapter 1, represents a further important step in the international regime formation under the UNFCCC. The Kyoto Protocol has brought a number of new elements and broadened the context of the decision-making process regarding implementation of climate change policy. Ultimately, further institutional development is needed for the UNFCCC to meet its final objective.

10.1.3 Tools of Analysis and their Summary in the Second Assessment Report

10.1.3.1 Tools of Analysis

A wide variety of tools have been applied to the climate problem. These are enumerated and briefly described in *Table 10.1*. In general, these tools help decision makers in several ways–choose a policy strategy, understand the implications of alternative policy strategies, understand the joint interactions of multiple, individual policy strategies. The tool can be employed by either a single decision maker or by stakeholder

groups. Their quantitative nature and their ability to incorporate the global, long-term diversity of relevant human activities, the uncertainty, and the irreversibility characteristics of the problem mean that decision frameworks have been broadly applied to the climate problem. This approach has several special cases, which have themselves received broad attention, including cost–benefit analysis and cost-effectiveness analysis. We review progress in these areas later in this section, after the SAR and the "tolerable window and/or safe landing" (TWSL) work.

Other tools have also been employed or have the potential to be employed to help illuminate decision making. These include game theory, portfolio theory, public finance, culture theory, and simulation exercises, and are discussed in the body of the chapter.

10.1.3.2 Summary of the Second Assessment Report

SAR divided its discussion of DMFs into four sections–an introduction, a discussion of the context of decision making, a discussion of the tools for decision analysis, and concluded by considering the implications for national decision-making in the context of the UNFCCC. The chapter began by discussing the features of climate change that distinguish it from other environmental problems. It then described decision analysis and the present state-of-the-art.

Decision analysis uses quantitative techniques to identify the "best" choice from among a range of alternatives. Model-based decision analysis tools are often used as part of interactive techniques in which stakeholders structure problems and encode judgements explicitly in subjective-preference scales. It makes the major trade-offs explicit. Although decision analysis can generate an explicit value as a basis for choice, it is based on a range of relevant monetary and non-monetary criteria. It is used to explore the decision and to generate improved options that are well balanced in the major objectives and that are robust with respect to different futures. A review of the real world limitations of quantitative decision models and the consistency of their theoretical assumptions with climate change decision-making highlighted the following points:

- There is no single decision maker in climate change. As a result of differences in values and objectives, parties that participate in a collective decision-making process do not apply the same criteria to the choice of alternatives. Consequently, decision analysis cannot yield a universally preferred solution.
- Decision analysis requires a consistent utility valuation of decision outcomes. In climate change, many decision outcomes are difficult to value.
- Decision analysis may help keep the information content of the climate change problem within the cognitive limits of decision makers. Without the structure of decision analysis, climate change information becomes cognitively unmanageable, which limits the ability of decision makers to analyze the outcomes of alternative

Table 10.1: Decision-making frameworks: compatibility with decision-making principles, and applicability at geopolitical levels and in climate policy domains

Decision analysis frameworks	Description of the tool	Applicability to problem characteristics	Comment
Decision analysis	Decision analysis is a formal quantitative technique for identifying "best" choices from a range of alternatives. Decision analysis requires the development of explicit influence structures that specify a complete set of decision choices, possible outcomes, and outcome values. Uncertainty is incorporated directly in the analysis by assigning probabilities to individual outcomes.	G, L, H, U, IR	Virtues include quantification of results, reproducibility of analysis, ability to incorporate the full dimensionality of the climate problem explicitly. Limitations include the assumptions of: 1. A single decision-maker, with well-ordered preferences, who is expected to be present throughout the period of analysis. 2. The number of alternatives is finite and therefore limited in practice. 3. Outcomes must be comparable—implying the need for aggregation to a single set of common units, e.g. US\$, lives, utility. 4. Rationality. 5. Uncertainties are quantifiable.
Cost–benefit analysis	Estimates of the costs and benefits for selected decision variables are derived. The "best" outcome is the one with the highest net benefits.	G, L, H, U, IR	1. This is a special case of general decision analysis. 2. Requires an explicit mechanism for valuing costs and benefits across time.
Cost-effectiveness analysis	Accepts specific performance goals as given exogenously, then minimizes the cost to achieve the desired performance.	G, L, H, U, IR	1. This is a special case of general decision analysis. 2. Requires an explicit mechanism for valuing costs and benefits across time. 3. Provides no information about the selection process. For example, analysis might accept a fixed CO_2 concentration ceiling as specified exogenously, but cannot comment on the desirability of that choice.
Tolerable windows and/or Safe landing approach	Accepts specific performance goals as inequalities given exogenously, then enumerates paths that are consistent with the goals.	G, L, H, U, IR	1. Provides no information about the selection process. For example, analysis might accept a fixed CO_2 concentration ceiling as specified exogenously, but cannot comment on the desirability of that choice. 2. The analysis does not provide a "best" path.
Game theory	Provides information about the implication of multiple decision-makers' choices, taking into account expectations that each has of their own actions on others, and others' actions on them.	G, IN	Technique is descriptive rather than prescriptive. It offers information about potential outcomes within a specific context.

(continued)

Table 10.1: continued

Decision analysis frameworks	Description of the tool	Applicability to problem characteristics	Comment
Portfolio theory	Concerned with creating under a budget constraint an optimal composition of assets characterized by different returns and different levels of risks. Decision options (portfolio elements) are represented by a probability distribution of expected returns while risks are estimated on the basis of the variability of expected returns, and only these two factors determine the decision makers' utility function. The decision rule is to choose the efficient portfolio compared to which no other portfolio offers higher expected return at the same or lower level of risk or lower risk with the same (or higher) expected return.	G, L, H, IN	Application to climate change problem has been limited.
Public finance theory	Encompasses a variety of research techniques including the theory of the second best.	IN	Examines trade-offs between efficiency and other criteria.
Ethical and cultural prescriptive rules	Concerned primarily with the implications of alternative social organizations. Has had limited application to the climate problem.	IN	1. Used to consider explicitly the interactions between policy instrument choice and social structure. 2. It is non-quantitative.
Policy exercises, focus groups, and simulation gaming	Includes a suite of research activities that have been used to assist in the decision-making process. In general, groups examine potential outcomes by playing a role in a simulated decision-making environment.	G, IN	1. Results are generally not reproducible. 2. Computer models may be used to assist in the exercise. 3. Much of the value is pedagogical.

Notes: G = global; LT = long-term; H = pervasive human activities; U = uncertainty; IR = irreversible; IN = relevant to institutional framing.

actions rationally. Quantitative comparisons among decision options (and their attributes) are implied by choices between options (the concept of "revealed preference" in economics). Better decisions are made when these quantitative comparisons are explicit rather than implicit.

- The treatment of uncertainty in decision analysis is quite powerful, but the probabilities of uncertain decision outcomes must be quantifiable. In climate change, objective probabilities have not been established for many of the outcomes. In real-world applications subjective probabilities are used.

- The large uncertainties and differences between parties may mean there can be no "globally" optimal climate-change strategy; nevertheless, the factors that affect the optimal strategies for single decision makers still have relevance to individual parties.

The lack of an individual decision maker, utility problems, and incomplete information suggest that decision analysis cannot replace the political process for international climate-change decision-making. Although elements of the technique have considerable value in framing the decision problem and identifying its critical features, decision analysis cannot identify globally optimal choices for climate change abatement. Decision analysis suffers fewer problems when used by individual countries to identify optimal national policies.

The UNFCCC establishes a collective decision-making process within which the parties negotiate future actions. Although some features of the decision-making process are set out in the Convention, many are still undecided. It becomes important, then, to examine negotiation and compromise as the primary basis for climate change decisions under the Convention. Important factors that affect negotiated decisions include the following:

- Excessive knowledge requirements in negotiated environmental decisions may impede a collective rational choice. This difficulty could be reduced by making the negotiation process itself more manageable through the use of tools like stakeholder analysis or by splitting accords into more easily managed clusters of agreements.

- In the face of long-term uncertainties, sequential decision-making allows actions to be better matched to outcomes by incorporating additional information over time. Sequential decision-making also minimizes harmful strategic behaviour among multiple decision makers.

- Improved information about uncertain outcomes may have very high economic value, especially if that information can create future decision options.

- There are currently no effective mechanisms for sharing the risks related to climate change and their associated economic burdens. International risk sharing could yield substantial benefits for global economic and social welfare.

The Convention is, first and foremost, a framework for collective decision making by sovereign states. Given this collective decision mechanism and the uncertainties inherent in the climate problem, several recommendations emerge:

- decisions for actions under the UNFCCC are rather being taken sequentially to benefit from the gradual reduction in uncertainties;

- countries may implement a portfolio of mitigation, adaptation, and research measures;

- they may adjust this portfolio continuously in response to new knowledge (the value of better information is potentially very large); and

- efficient distribution of the risks of losses related to climate change may warrant new insurance mechanisms.

10.1.4 Progress since the Second Assessment Report on Decision Analytical Frameworks

Much work has been conducted since SAR. Work has focused on a wide array of issues ranging from that which explores the tools of analysis to that which employs those tools to shed light on the problem of climate change. Researchers such as De Canio (1997), De Canio and Laitner (1997), De Canio *et al.* (2000a, 2000b, 2000c), Laitner and Hogan (2000), Laitner *et al.* (2000), Peters and Brassel (2000), and Sanstad *et al.* (2000a, 2000b) have focused on integrated assessment, endogenous technological change, and behavioural, social, and organizational phenomena (discussed in Chapter 8). Work has also continued to examine the problems of cost–benefit, cost-effectiveness, and the interaction of uncertainty with decision making. New approaches have also been developed, including, for example, tolerable windows and safe landing.

10.1.4.1 Decision-making under Uncertainty

Work has continued in the development of tools to understand the influence of uncertainty on decision making. The initial work examined in SAR explores the problem of emissions-mitigation objectives under a cost-effectiveness framework, but the interaction between concentration limits and the date at which uncertainty is resolved influences the results. This interaction occurs because in decision analysis no option can ever be foreclosed before the date at which uncertainty is hypothesized to be resolved. Any concentration ceiling implies a cumulative emissions limit. Thus preserving the option to stay below any arbitrary limit means adopting a hedging strategy. Grubb (1997) characterizes the problem thus: "If we delay action in the belief that we are aiming at a 500ppmv target, for example, then after a couple of decades it may be simply too late to be able to stabilize at 400ppmv, however urgent the problem then turns out to be; and even stabilization at 450ppmv might by then involve radical changes of direction that could prove economically very disruptive."

The core of the issue is the interplay between inertia and uncertainty; without inertia any trajectory could indeed be corrected at no cost, but as inertia is important, changing course may be very costly. Fortunately, the Convention embodies the dynamic nature of the decision problem in drafting climate as an ongoing process, not a "once and for all" event. The UNFCCC (1993) requires periodic reviews "in light of the best scientific information on climate change and its impacts, as well as relevant technical, social and economic information."

Such a sequential decision-making process aims to identify short-term strategies in the face of long-term uncertainties. The next several decades will offer many opportunities for learning and mid-course corrections. The relevant question is not "what is the best course of action for the next 100 years", but rather "what is the best course for the near-term given the long-term objective?"

There have been several attempts to frame the issue. *Figure 10.1* reports the results of an analysis by Ha-Duong *et al.* (1997). The authors use their model of the Dynamics of Inertia and Adaptability for integrated assessment of climate-change Mitigation (DIAM) to determine the least-cost emission pathway given an uncertain concentration target. A defining feature of their model is an inertia parameter that accounts for the time scale of change in the global energy system. In their analysis they assign equal probability to a target of 450, 550, and 650ppmv. The solid 550ppmv line corresponds to the optimal pathway when the target is known to be 550ppmv from the outset. The analysis shows the optimal hedging strategy when uncertainty is not resolved until 2020. The authors note that "our results show that abatement over the next few years is economically valuable if there is a significant probability of having to stay below ceilings that would be otherwise reached within the characteristic time scales of the systems producing greenhouse gases."

The degree of near-term hedging in the above analysis is sensitive to the date of resolution of uncertainty, the inertia in the energy system, and assumes that the ultimate concentration target (once it has been agreed) must be met at all costs. The last stems directly from the formulation of the problem as one of

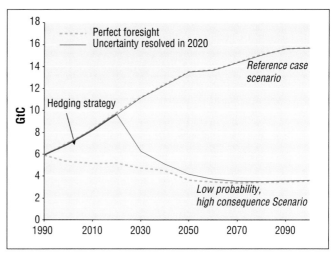

Figure 10.2: *Optimal hedging strategy for low probability, high consequence scenario using a cost-benefits optimization approach.*

finding the least-cost mitigation pathway in the face of uncertainty. Since a future political decision on a 450ppmv target cannot be excluded, decisions prior to 2020 must be such that they do not preclude the achievement of such a target.

One way to avoid the bias inherent in the framing of the emissions control problem under uncertainty is to reframe the problem as a decision tree structure within the context of cost–benefit analysis rather than cost-effectiveness analysis. This was the approach taken by the seven models used in an Energy Modeling Forum (EMF; Manne, 1995) exercise on climate change decision making under uncertainty (Weyant, 1997). The study focused on hedging strategies for low probability, high consequence scenarios in which uncertainty was not resolved until 2020. Two parameters were varied: the mean temperature sensitivity factor and the cost of damages associated with global warming. The unfavourable cases were defined as the top 5% of each of these two distributions. Two surveys of expert opinion were used to choose the distribution of these variables. For the opinion survey on climate sensitivity, see Morgan and Keith (1995), and for warming damages, see Nordhaus (1994b). *Figure 10.2* (Manne and Richels, 1995; Manne 1995) shows what happens when the unfavourable case has a probability of 0.5 and the expected case a probability of 0.95 (the two parameter values assumed for the unfavourable case are shown in the surveys cited above as being in the upper 5% of each of the distributions of the two key parameters, i.e., climate sensitivity and climate damages). The dashed lines show what happens if perfect foresight is available and can make today's decisions in the full knowledge of which of these outcomes will occur. The solid lines indicate the average results from an economically efficient hedging strategy. The analysis takes into account both the costs and benefits of emissions abatement. With a cost-benefit analysis, costs and benefits are balanced at the margin. Seven EMF modelling teams have confirmed these results (Weyant, 1997). The reason for so little hedging is the low probability of the extreme outcome,

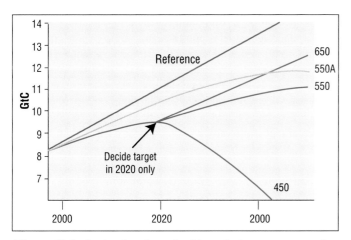

Figure 10.1: *Optimal carbon dioxide emissions strategy, using a cost-effectiveness approach.*

that is 0.25%. If one were to increases this probability, the desired degree of hedging would increase accordingly.

Another parameter for stochastic cost–benefit analysis is the importance of non-linearity in the impacts and the date at which some threshold is likely to occur. Peck and Tiesberg (1993) observed that optimal policies were more sensitive to uncertainty in the damage-function power parameter than to uncertainty in the scale parameter. Ha-Duong *et al.* (1999) confirm this view and demonstrate that introducing thresholds in the damage function leads to more significant decoupling from current emissions trends for a given probability distribution.

Ultimately, as recognized in the IPCC (1996c) one should try and assess the option value of the information incorporated in alternative emissions pathways, that is the capacity of society to adapt to any new information. As pointed out by Ulph and Ulph (1997), the environmental irreversibility has to be balanced against the technological irreversibility, including the crowding out between forms of technical progress. Ha-Duong (1998) finds, comparing Working Group I (WGI) and Wigley, Richels, Edmonds (WRE) strategies, that the magnitude of the value of information is significant compared with the opportunity costs of abatement. On the basis of nine scenarios he found that the information value of acting soon is, for most of them, higher than that of acting later, if low and high damages are assumed equally probable.

Whatever the approach, the basic message is quite similar. First the costs and benefits of quick action have to be balanced against those of delayed action; second, to assume that the concentration target is known with certainty is an over-simplification of the decision problem. What is needed is an approach that explicitly incorporates uncertainty and its sequential resolution over time. The desirable amount of hedging should depend upon assessment of the stakes, the odds, and the costs of policy measures. The risk premium – the amount that society is willing to pay to reduce risk – ultimately is a political decision that differs among countries.

Uncertainty also affects the choice of policy instrument. In principle many mechanisms can be employed to limit emissions, including, voluntary agreements among domestic and international parties, regulation, taxes, subsidies, and quotas or tradable permits (see Chapter 6). Economists have focused on the potential role of taxes and quotas because these tools hold potential for cost minimization. Although both instruments are equivalent in a world with complete information (the optimal quota leads to the same marginal abatement cost as the optimal tax level), Pizer (1999), building upon a seminal work by Weizman (1974), demonstrated that this is not the case if uncertainties about climate damage and GHG abatement costs are considered.

Indeed, welfare losses that result from imperfect foresight depend on whether the steepness of the marginal abatement cost curve is higher or lower than that of the damage curve.

Hence the finding that a co-ordination through price is preferable as long as the probability of dramatic non-linearity in climate systems is not large over the middle term. This policy conclusion can be reverted if the transaction costs of adopting co-ordinated taxation, high level of risk-aversion to catastrophic events, or a large amount of "no regrets" policies are considered. The main message, however, is that in a tax co-ordination approach costs are observable (while the outcome is not predictable), but in a quota approach the outcome is observable although there is an uncertainty about the resultant costs. In this respect, emissions trading is logically a companion tool for a system of emissions quotas, to hedge against the distributional implications of surprises regarding abatement costs and emissions baselines.

10.1.4.2   *Cost-effectiveness Analysis*

There is an increased interest in cost-minimizing paths that lead to alternative, stable steady-state concentrations of GHGs in the atmosphere. This interest stems from the objective of the UNFCCC–to stabilize the concentration of GHGs. Work has focused primarily on the problem of stabilizing the concentration of CO_2. The focus on CO_2 reflects the importance placed on this gas by the Intergovernmental Panel on Climate Change Working Group I (IPCC WGI) and the distinctive characteristic of CO_2. As CO_2 does not have an atmospheric sink, the net emissions to the atmosphere must eventually decline indefinitely to maintain any steady-state concentration (IPCC, 1996a). In contrast, GHGs such as CH_4 and N_2O, with atmospheric sinks, have steady-state concentrations associated with steady-state emissions. Cost-effective paths depend on many factors including reference emissions, technical options for emissions limitation, the timing and rate of change of the availability of options, the discount rate, and assumed control mechanisms and their efficiency. The analysis conducted to date generally does not take into account that long-term emissions mitigation must take place against a background of climate change that affects both the nature and composition of economic activity and the carbon cycle.

Both Manne and Richels (1997) and Edmonds *et al.* (1997) examined the relationship between steady-state concentrations of CO_2 and associated minimum costs. Both papers computed the minimum cost of honouring a concentration ceiling. All cost calculations assumed that all activities throughout the world pursued emissions mitigation based on a common marginal cost of carbon emissions mitigation. While real-world implementation strategies are likely to be less efficient, the choice of a cost-effective assumption for each period provides a unique benchmark for comparison purposes. Several assumptions regarding cost-effectiveness over time were examined. The two studies examined three cases:

- global emissions limited to a trajectory prescribed by IPCC (1995), labelled WGI;
- global emissions limited to a trajectory prescribed by Wigley *et al.* (1996), labelled WRE; and
- a model-determined minimum-cost emissions path.

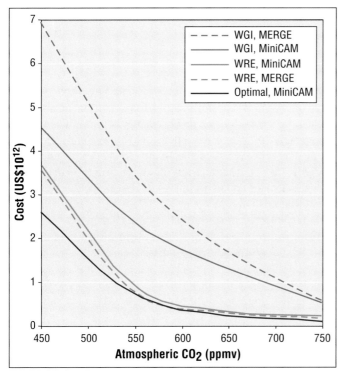

Figure 10.3: Relationship between present discounted costs for stabilizing the concentration of carbon dioxide in the atmosphere at alternative levels from two studies. Costs are discounted at 5%/yr over the time period 1990 to 2100. Sources: Manne and Richels (1997) and Edmonds et al. (1997).

Costs were discounted over time at 5%/yr over the period 1990 to 2100. The results are displayed in *Figure 10.3*.

Costs are roughly an order of magnitude greater for concentration ceilings of 450ppmv than for the 750ppmv ceiling between WGI, WRE, and optimal global emissions constraints. Furthermore, costs decline sharply as the constraint is relaxed from 450ppmv to 550ppmv. Relaxation of the constraint from 650ppmv to 750ppmv reduces costs, but at a more modest rate. As discussed in Chapters 2, 7 and 8, it should be noted that the total costs of stabilizing atmospheric carbon concentrations are very dependent on the baseline scenario: for example, for scenarios focusing on the local and regional aspects of sustainable development costs are lower than for other scenarios.

Progress has also been made in examining the time path of the value of a tonne of carbon when the cost of stabilizing the concentration of CO_2 is minimized. Peck and Wan (1996) demonstrated that the results of Hotelling (1931) could be applied to the problem of minimizing the cost to stabilize the concentration of CO_2 and generalized. They show that to minimize present discounted cost, the value of a tonne of carbon should rise at the rate of interest (discount rate). This theorem ensures that the marginal cost of emissions mitigation across both space and time is equal after taking into account that carbon is naturally removed from the system. Thus, the initial marginal costs

should be relatively modest, but should rise steadily (at the rate of interest plus the rate of carbon removal, approximately 1%/yr). The rise in marginal cost continues until it reaches the marginal cost of a "backstop" technology, one capable of providing effectively unlimited emissions mitigation at a constant marginal cost.

All cost-effective policies minimize the cost of stabilization by equalizing the marginal cost of mitigation across time and space, that is, in all regions, across all human activities, and across all generations, except to the extent that non-linearities, non-convexities, and corner solutions exist. The implementation of real-world regimes to control net emissions to the atmosphere is likely to be inefficient to some degree for a number of reasons, including, for example, the problems of "free riding"; cheating; in some cases considerations of fairness and equity; and monitoring, compliance, and transactions costs.

Some work has been undertaken to compare potential policy regimes with respect to cost-effectiveness. For example, Chapter 8 shows the difference in emissions mitigation requirements between various potential implementations of the Kyoto Protocol and more cost-effective paths. Edmonds and Wise (1998) examined the cost effectiveness of a strategy that sought to minimize the costs of monitoring and verification, and premature retirement of capital stocks, while simultaneously addressing concerns about fairness and equity. They considered a hypothetical protocol that focused on new investments in energy technology. They assumed that Annex I nations required new emissions sources to be carbon-neutral after a prescribed date. Existing sources were treated as new after a fixed period following their initial deployment. Non-Annex I nations remained unencumbered until their incomes reached levels comparable to those in Annex I nations. The authors concluded that the regulatory regime could stabilize the concentration of GHGs, and that the level at which concentrations stabilized is determined by the initial date of obligations. The hypothetical protocol is economically inefficient, however. That is, it does not minimize the cost of achieving a concentration limit. The authors compare the hypothetical protocol, which uses a technology regulation to limit emissions, with an alternative cap-and-trade regime that achieved the same emissions path. Costs in the hypothetical protocol were approximately 30% greater than in those in the alternative cap-and-trade regime.

Jacoby *et al.* (1998) also considered the problem of accession to the Kyoto Protocol. They reject the idea that there is such a thing as inter-temporal cost-effectiveness in the context of a century-scale problem. Rather, they begin with the proposition that a continuous process of negotiation and re-negotiation is required. They analyze a system of obligations based on per-capita income that can lead to the stabilization of concentrations of GHGs.

A substantial body of work has considered the implication of technology development and deployment on the cost of meeting alternative emissions-mitigation obligations. This line of investigation has a long tradition extending back to, for example, Cheng *et al.* (1985). These are discussed in Chapter 8[3]. Recent studies, for example by Dooley *et al.* (1999), Edmonds and Wise (1999), Grübler *et al.* (1999), PCAST (1999), Schock *et al.* (1999), and Weyant and Olavson (1999), have explored the potential role of a variety of technologies in both the near term and the longer term. The principal conclusion of this body of investigation is that the cost of emissions mitigation depends crucially on the ability to develop and deploy new technology. The value of successful technology deployment appears to be large with the value depending on the magnitude and timing of emissions mitigation and on anticipated reference scenario progress.

10.1.4.3   *Tolerable Windows and Safe Landing Approaches*

Considerable work since the SAR has explored the implications for global emissions of GHGs of a set of constraints on a variety of associated phenomena. This vein of research is referred to as the tolerable windows and/or safe landing (TWSL) approach. See, for example, Alcamo and Kreileman (1996a, 1996b) and Swart *et al.* (1998) for early work on the safe landing approach and Toth *et al.* (1997) for early work on the tolerable windows approach. The approach seeks to limit the emissions time-paths with implications for the near term and long term. While the tolerable windows and safe landing analyses differ somewhat in the detail of their implementation, they are similar in approach. We consider the safe landing approach first. In a multimodel exercise four constraints on emissions trajectories are considered: temperature change since 1990, maximum decadal rate of temperature change, sea level rise between 1990 and 2100, and maximum rate of sea level change. In addition, a limit on the rate of reduction of emissions is set.

Criteria	Low	Medium	High
Change in temperature from 1990	1.0°C	1.5°C	2.0°C
Decadal change in temperature	0.10°C	0.15°C	0.20°C
Change in sea level	20cm	30cm	40cm
Decadal change in sea level	2cm	3cm	4cm
Maximum reduction in emissions	2%	3%	4%

The safe landing interval is the range of emissions, given in CO_2 equivalent emissions (C_{eq}), in 2010. This range is 7.6–11.9GtC_{eq}; 1990 emissions were 7.10GtC, and approximately 9.8GtC_{eq}, equivalent, defined in terms of CO_2, CH_4, and N_2O only (Pitcher, 1999). Emissions for Annex I nations can be derived by subtracting the anticipated non-Annex I emissions from the global total.

Results from the analysis depend strongly on the constraints and model sensitivities. The tolerable windows approach (Toth *et al.*, 1997, 1999; Bruckner *et al.*, 1999; Petschel-Held *et al.*, 1999) is formulated as a type of extended and generalized cost–benefit analysis for which two kinds of normative inputs are required. First, with the help of climate-impact response functions that depict reactions of climate-sensitive socioeconomic and natural systems to climate change forcing, social actors can specify their willingness to accept a certain amount of climate change in their own jurisdiction. Second, the same social actors reveal their willingness to pay for climate change mitigation in terms of acceptable burden-sharing principles and implementation schemes internationally, as well as in terms of tolerable utility, consumption, or Gross Domestic Product (GDP) loss in their own jurisdiction. An integrated climate-economy model (e.g., Integrated Assessment of Climate Protection Strategies - ICLIPS) can then determine whether there exists a corridor of emission paths over time that keeps the climate system within the permitted domain.

If the corridor does not exist, a willingness to accept more climate change can be specified (e.g., as a result of resource transfers to increase the adaptive capacity in the most constraining region or sector on the impact side). Alternatively, willingness to pay for emission reductions can be increased or more cost-reducing flexibility instruments can be allowed on the mitigation side. If the corridor does exist, it can be perceived as the room to manoeuvre for global climate policy over the long term. The tolerable windows approach leaves the specification of climate-change mitigation regimes up to decision makers involved in climate-change policy making at the global and national levels. The primary goal of the ICLIPS integrated assessment model (IAM) is to determine the implications of different equity principles in burden sharing and of various implementation mechanisms on the existence and shape of the emission corridor. Nevertheless, the model can also produce cost-effective emission paths.

The German Advisory Council on Global Change (WBGU) proposed two climate change constraints based on geohistorical arguments: the tolerable magnitude of climate change is set to 2°C compared to the pre-industrial era[4] and the rate of temperature increase should not exceed 0.2 °C per decade. On the cost side, it is assumed that to reduce GHG emissions at a rate faster than 4%/yr would be economically too painful to implement. These constraints are used to illustrate the application of the tolerable windows approach. The results presented here are based on an extended atmospheric chemistry–climate model. In addition to CO_2, the model also includes CH_4, N_2O, chloro-

---

[3] See also Edmonds *et al.* (1994, 1996, 1997, 1999), Grübler and Nakicenovic (1994), Christiansson (1995), Shukla (1995), Goulder (1996), Energy Innovations (1997), Interlaboratory Working Group on Energy-Efficient and Low-Carbon Technologies (1997, 2000), Mattsson (1997), Grübler and Messner (1998), Koomey *et al.* (1998), Yamaji (1998), Bernow *et al.* (1999), Geller *et al.* (1999), Laitner (1999), Laitner *et al.* (1999), Lako *et al.* (1999), Hanson and Laitner (2000), and Kim *et al.* (2000).

[4] This 2 degree centigrade limit has also been adopted by the European Union as its provisional target for stabilizing greenhouse gas concentrations in the atmosphere under UNFCCC Article 2.

fluorocarbons (CFCs), and aerosols. One simplifying assumption is that all GHG emissions are reduced at the same rate, except for CFCs, which follow the IPCC IS92a scenario paths. For simplicity, energy-related global CO_2 emissions are presented in *Figure 10.4*.

Figure 10.4(a) presents the basic emission corridor for the WBGU window. It follows from the mathematical formulation of the model that at least one permitted emission path passes through any arbitrary point in the corridor. However, not every arbitrary path within the corridor is necessarily a permitted path. If emissions follow the upper boundary of the corridor in the first few decades after 1995, for example, this would entail a sharp turnaround and persistent emission reductions at the maximum annual rate (4%/yr) for many decades to come.

How do near-term emissions affect the available flexibility over the long-term? The scenario presented in *Figure 10.4(b)* shows this. Here it is assumed simply that CO_2 emissions follow the baseline path according to the IPCC IS92a scenario until 2010. The result is a much narrower corridor: it implies that the likelihood of a fast turnaround of emissions and persistent reductions at relatively higher rates (3%–4%/yr) is significantly higher.

The next analysis illustrates the implications of a fairness principle for the Annex I emission corridor. The assumption is that GHG emissions by non-Annex I countries follow the baseline path and these countries start emission reductions only when their per capita emissions reach those of Annex I levels on the basis of their 1992 populations. The resultant Annex I corridor is presented in *Figure 10.4(c)*. Obviously, the result is a relatively narrow corridor.

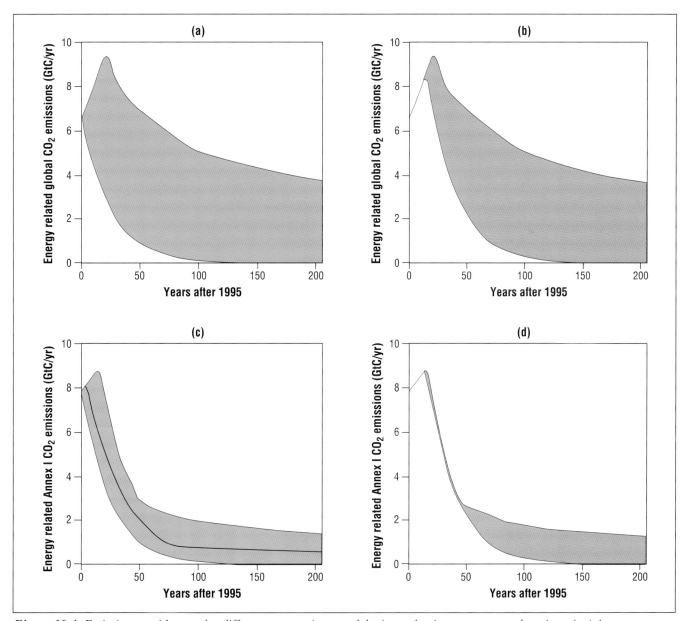

Figure 10.4: *Emission corridors under different assumptions on delaying reduction measures and equity principles.*

Figure 10.4(d) shows the resultant emission corridor if the above two assumptions about future emissions are combined. This implies that the world community follows the baseline emission path until 2010 and reduction obligations will be distributed between the Annex I and non-Annex I countries according to the case in *Figure 10.4 (c)*. The result for Annex I countries emissions through the first half of the 21st century looks like a straightjacket rather than an emission corridor with ample choice.

Importantly, Annex I corridors in *Figures 10.4(c)* and *10.4(d)* reflect the rigid implementation of emission quotas that result from the specified equity principle. No cost divergence is considered between Annex I and non-Annex I. The difference between *Figures 10.4(a)* and *10.4(c)* corridors indicates the potential to reduce abatement costs if Annex I countries are allowed to "buy" part of the non-Annex I corridor. The economic value of this transaction is the subject of many detailed energy-economy models (see Section 10.4).

It is clear that all these emission corridors are associated with the global climate window as specified by the Council. It is beyond the scope of this analysis to discuss arguments for and against whether the 2°C increase in global mean temperature above the pre-industrial level and the rate of temperature increase at no more than 0.2°C per decade are preferred or realistic propositions. The objective for the tolerable windows approach is to provide an assessment framework that can help test any climate protection proposal formulated through selected climate attributes. The computed emission corridors, nevertheless, can assist in deciding the magnitude and urgency of the policy measures associated with them, and/or trigger rethinking the originally proposed climate change targets. The presented example also shows how equity concerns can be analyzed in the tolerable windows approach, albeit in a terse form.

10.1.4.4   Computational, Multiscenario Simulation Approaches

Computational, multiscenario simulation is a new analytic approach to the assessment of climate change policy. Bankes (1993), Lempert *et al.* (1996), and Laitner and Hogan (2000) have employed this approach, as have Morgan and Dowlatabadi (1996), van Asselt and Rotmans (1997), and, to some extent, Yohe (1996). Also, the IPCC Special Report on Emissions Scenarios (IPCC, 2000b) presented a large set of very different baseline scenarios. The basic idea is to use computer simulation models to construct a range of a large number of fundamentally different scenarios of the future and, instead of aggregating the results using a probabilistic weighting, make policy arguments from comparisons of fundamentally different, alternative cases. These methods are most useful under conditions of deep uncertainty. For example, when we do not have reliable information or widespread agreement among the stakeholders about the system model, the prior probability distributions on the parameters of the system

model, and/or the loss function to use in evaluating alternative outcomes (Lempert and Schlesinger, 2000).

These multiscenario simulation approaches offer the promise of a powerful synthesis between the narrative, process-oriented methods of scenario-based planning (Schwartz, 1996; van der Heijden, 1996) and quantitative tools such as decision analysis, game theory, and portfolio analysis. From the quantitative methods, multiscenario simulation draws systematic methods of handling large quantities of data and normative descriptions of good decisions. From scenario-planning, multiscenario simulation draws the insight that multiple views of the future are crucial to allow groups to transmit and receive information about highly uncertain futures. Also scenario planning shows that groups can often agree on actions to take in the face of deep uncertainty without agreeing on the reasons for these actions (Lempert and Schlesinger, 2000). For instance, multiscenario simulation can adopt a meaningful cost–benefit framework for climate change, but at the same time acknowledge the deep uncertainty and differing values among stakeholders. These make it impossible to fully quantify the costs and benefits or to assign widely accepted probabilities to many of the key outcomes of interest. Such computational, multiscenario simulations are enabled by new computer technology–primarily large quantities of inexpensive memory; fast, networked processors; and powerful visualization tools–and are only just becoming available.

10.1.5   Robust Decision-making

Uncertainty is a feature that pervades discussions on climate change issues. IPCC SAR covered main areas of uncertainties, especially those related to:

- atmospheric concentrations of GHGs and their impact on meteorological phenomena (IPCC, 1996a);
- the potential of technological options and the relationships between climate change and the dynamics of natural systems (IPCC, 1996b); and
- socio-economic dimensions of climate change (IPCC, 1996c).

Several sections in this report (1.5; 2.2; 7.2; 10.1) review new and complementary perspectives that facilitate a better understanding of the tensions between the limited capacity to predict and the urgent need to act in a situation faced with high stakes of risk.

The implications of uncertainty are global in scale and long-term in their impact; quantitative data for baselines and the consequences of climate change are inadequate for decision making. In recent years, researchers and policymakers have become increasingly concerned about the high levels of inherent uncertainty, and the potentially severe consequences of decisions that have to be made.

Conventional frameworks for decision making on climate change policies presume that relevant aspects of the contextual environment are to some extent predictable; therefore uncertainty can be reduced to provide decision makers with appropriate information within appropriate time frames.

This *anticipatory management approach* is based on the premise that it is possible to predict and anticipate the consequences of decisions and hence to make a proper decision once all the necessary information is gathered to make a scientific forecast. The prevailing image is that "given enough information and powerful enough computers it is possible to predict with certainty, in a quantitative form, which in turn makes it possible to control natural systems" (Tognetti, 1999).

Anticipatory approaches have successfully managed a wide range of decision problems in which the relative uncertainties are reducible, and the stakes or outcomes associated with the decisions to be made are modest (Kay *et al.*, 1999). A number of uncertainty analysis techniques, such as Monte Carlo sampling, Bayesian methods, and fuzzy set theory, have been designed to perform sensitivity and uncertainty analysis related to the quality and appropriateness of the data used as inputs to models. However, these techniques, suitable for addressing technical uncertainties, ignore those uncertainties that arise from an incomplete analysis of the climate change phenomena, or from numerical approximations used in their mathematical representations (modelling uncertainties), as well as uncertainties that arise from omissions through lack of knowledge (epistemological uncertainties). Current methods thus give decision makers limited information regarding the magnitude and sources of the underlying uncertainties and fail to provide them with straightforward information as input to the decision-making process (Rotmans and de Vries, 1997).

The management of uncertainties is not just an academic issue but an urgent task for climate change policy formulation and action. Various vested interests may inhibit, delay, or distort public debate with the result that "procrastination is as real a policy option as any other, and indeed one that is traditionally favoured in bureaucracies; and inadequate information is the best excuse for delay" (Funtowicz and Ravetz, 1990).

Funtowicz and Ravetz have proposed a highly articulated and operational scheme for dealing with the problems of uncertainty and quality of scientific information in the policy context. By displaying qualifying categories of the information–numeral, unit, spread, assessment, and pedigree (NUSAP)–the NUSAP scheme provides a framework for the inquiry and elicitation required to evaluate information quality. By such means it is possible to convey alternative interpretations of the meaning and quality of crucial quantitative information with greater quality and coherence, and thus reduce distortion of its meaning.

In recent years a good deal of analytical work has addressed problem-solving strategies for different circumstances characterized by the inherent uncertainties in the situation and the severity of consequences that arise from the decision to be made. Adaptive management approaches to decision making start by accepting uncertainty as an inherent property of complex systems. The issue here is not the problem of a "deterministic version of scientific uncertainty"—a temporary matter of imprecision which will be eradicated when enough research has been devoted to the questions (Wynne, 1994). The starting point is the acknowledgement that uncertainty emerges not only from the long time-scales involved and/or the ability of models to predict long-term events, but mainly from the endemic uncertainty, indeterminacy, and ignorance related to the co-evolution of natural and social systems. Furthermore, these methods stress the relevance of values, ethical and social, and thus introduce the need for public discourse and debate (Westra, 1997).

A central concern in adaptive approaches is with the plurality of value systems and how multiple perspectives can inform the decision process. Various attempts have been made to incorporate a variety of perspectives in relation to uncertainty and to make uncertainty more explicit by expressing it in terms of risk.

Parallel modelling (Visser *et al.*, 2000) and computational, multiscenario simulation (Lempert *et al.*, 1996; Morgan and Dowlatabadi, 1996; van Asselt and Rotmans, 1997) are emerging approaches based on the idea that multiple views of the future are crucial to allow groups to transmit and receive information about highly uncertain futures.

Rather than aggregate different scenario or model results using probabilistic weights or using computer resources to increase the resolution of a single best-estimate model, analysts use simulation models to construct different scenarios to compare different, alternative policy options based on their robustness across the scenarios. Valuation is thus reframed as a process in which uncertainty is not banished, but is managed, and values are not presupposed but are made explicit.

The analysis of multiple and diverse perspectives as a source of uncertainty has been addressed by van Asselt and Rotmans (1997) within the framework of the Tool to Assess Regional and Global Environmental and health Targets for Sustainability (TARGETS) IAM. The authors introduce the idea of model routes–a chain of biased interpretations of the crucial uncertainties in the model–to analyze differences in future projections as the outcome of divergent views and valuations, instead of merely of low, high, and medium values. The approach distinguishes two dimensions of perspectives: (1) a world view, which entails a coherent view of how the world functions, and (2) a management style, that is policy preferences and strategies. By combining stereotypical views of nature and humanity as well as ethical attitudes with different management styles, the approach enables the analysis of "utopias" that result when views match the strategies and "dystopias" that result when they do not.

Dystopias are useful with respect to communicating the role of uncertainty and its consequences for decision making. They indicate the risk of decision making in uncertain conditions by showing to what kind of future the chosen strategies might lead, in the event that the adopted worldview fails to describe the reality adequately.

Another promising avenue for managing uncertainties is the exploratory modelling methods (Lempert *et al.*, 1996; Lempert *et al.*, 2000; Lempert and Schlesinger, 2000; and Robalino and Lempert, 2000) is discussed in Section 10.1.4.1.

Robustness is not a new concept, but it is just recently, under the pressure of global environmental problems and the acceleration of change, that such approaches have grown in formalization and sophistication (Rosenhead, 1990). Rooted in Savage's maximizing the minimum regret (1954), Simon's ideas of satisfying strategies (1959 a and b), and Lindbolm's incremental policies (1959), the search for robust strategies as a formal decision-making criterion has grown during the 1990s. However, it has always been more difficult to implement robustness, as opposed to optimization, within an analytical method, except for in very special cases. A new development over the past few years is that it is now becoming possible to implement robustness as an analytical criterion using simulation models of the type relevant to climate change policy. In conclusion, multiscenario simulation approaches, like multiple-model routes, exploratory models, or parallel modelling, show that uncertainty is no longer a theoretical scientific concept, but a notion that might be usefully deployed by decision makers in arriving at their decisions (van Asselt and Rotmans, 1997).

This ability to analytically address robustness is closely tied to the idea of adaptive decision strategies, that is, strategies that can change over time in response to observations of the climate and economic systems. (Adaptive decision strategies differ from sequential strategies in that in the former information is endogenous, that is, the type, rate, and quality of information gained depends on both the unfolding scenario and policy choices whereas information is exogenous in the sequential strategies.) Adaptive decision strategies are closely tied to the concept of robustness, because such strategies are most useful in situations of deep uncertainty—where robustness, as opposed to optimization, appears to be the best decision-making criterion.

10.2   International Regimes and Policy Options

10.2.1   *Introduction*

Previous chapters provide some answers to the most relevant policy questions related to the climate change problem. Issues such as the timing of optimal responses to climate change, the role of technological innovation and diffusion, the choice between domestic action and the adoption of "Kyoto mechanisms", the importance of co- and ancillary benefits, etc., have been analyzed from different perspectives. However, it is important to notice that the costs and benefits of all the above options crucially depend on the characteristics of the international agreement on climate change that is adopted. In particular, they depend upon two main features of the international regime: the number of signatories, and the size of their quantitative commitment to control GHG emissions.

It is therefore impossible to assess the costs and benefits of the Kyoto Protocol or of other potential agreements on climate change independently of the number of signatories of the agreement and of their abatement targets and/or policy commitment. However, the number of signatories is endogenous and depends on the abatement targets and mitigation polices adopted in various countries. Hence the weakness of most of the available literature on costs and benefits of climate change policies, which widely neglects the full interdependency between policies, costs–benefits, and signatories (more generally, the structure of the international agreements). For example, studies analyze the costs of implementing the Kyoto Protocol either through a set of domestic policies and measures or through a system of international tradable permits, with a fixed number of signatories. But the adoption of either policy crucially affects the number of signatories, which can be larger or lower under policies and measures than under tradable permits. And the number (and identity) of signatories crucially affects the costs and benefits of different agreements.

Therefore, this section aims to provide an analysis of the effectiveness of climate policies by focusing on the link between policy options on the one hand and the structure of the agreements and international regimes on the other. Some of the most important theoretical results are reviewed first, and then the existing literature is revisited to see which information it provides on the interdependencies described above. In particular, can such an analysis show whether there exist the conditions for an agreement on climate change to be signed by all or almost all world countries (see Carraro, 1998; Carraro and Siniscalco, 1998; and Barrett, 1999 for a theoretical analysis of these conditions)? Also, would it show which countries can play a leadership role with respect to achieving the largest possible coalition by proposing strategies, measures, and institutions that help expand the number of countries that commit to control their emissions (see Grubb and Gupta, 2000)? Notice that in this way we also analyze which strategies can be proposed to reduce the costs of mitigation policies. But this is a quite different approach to those analyzed in the previous sections of this chapter and in Chapters 8 and 9. The reason is that here a country's goal is not to identify a new climate friendly technology or an adequate redistribution of costs across sectors. Now the goal is to affect other countries' behaviour to increase the number of those that share the burden, and to share the burden more equitably.

The equity issue is also very important to understand which countries are going to reduce and/or control[5] their emissions.

---

[5] When using the word "emission reduction", reduction with respect to the baseline scenario is meant here. As a consequence, emissions in some countries can increase with respect to their 1990 level or other baselines.

As a consequence, given what is said above, equity is crucial to assess adequately the costs of emission reductions at the global and country level. It has been argued that some countries are allowed to reduce emissions less than other countries, both within (Kram, 1998) and outside the European Union (EU) bubble (Bosello and Roson, 1999; Metz, 1999; Rose and Stevens, 1999). Even when applying the Kyoto mechanisms, some countries will benefit from the agreements more than other ones (Nordhaus and Boyer, 1999). It has also been argued that some countries can exploit their monopolistic power in a future trading system (Burniaux, 1998). All these remarks address the problem of optimal burden-sharing (the distribution of costs) of climate change control. This problem is strictly related to the features of an international agreement on climate for two main reasons. First, increasing the number of participating countries reduces the direct costs for each signatory; second, an agreement in which the burden is equitably shared is more likely to be signed by a large number of countries (Convery, 1999). Therefore, equity and the structure of the international agreement (number and identity of signatories) are strictly linked. However, the number of signatories affects and is affected by costs. Hence, equity and efficiency cannot be separated.

These remarks reinforce the previous basic statement. An analysis of the costs and benefits of different policy options, and of the distribution of these costs and benefits across countries, cannot be done independently of an analysis of the likely features of the prevailing international regime (i.e., of the incentives that lead countries to sign an international agreement to control GHG emissions and to set quantitative emission targets).

Notice that an analysis of the features of climate international agreements and of their repercussions on the choice of different policy options (and vice versa) must take into account:

- basic features of the climate problem recalled in Section 10.1, and particularly the public-good nature of GHG abatement in the absence of a supranational authority;
- scenarios that describe the future evolution of economic and environmental climate-related variables;
- economic incentives for countries to sign an international agreement on climate change control, that is under what conditions, in terms of the number of countries, damaging effects of free-riding (leakage), structure of costs and benefits, can a coalition (*i.e.*, a group of signatories of the international agreement) emerge?[6]
- the political and institutional dimension of an international climate agreement, its history, the possibility of monitoring and sanctioning deviations, the links with other agreements.

This section is devoted to the analysis of the above issues and also aims to provide a framework to understand how future negotiations on climate change can evolve, and how costs and benefits of climate policies are modified by these possible evolutions.

10.2.2 Coalition Formation

If the goal is to understand which international regime is likely to emerge to control GHG emissions, game theory is certainly the best tool. Indeed, game theory has been used extensively to analyze the possibility of coalition formation in the presence of free riding (i.e., when parties have to agree on the provision of a public good). Early contributions (see Hardin and Baden, 1977) characterized the environmental game among countries as a prisoners' dilemma, inevitably leading to the so-called "tragedy" of the common property goods. However, in the real world, at the same time, many international environmental agreements on the commons were signed, often involving subgroups of negotiating countries and sometimes involving economic and technological transfers and other links to other policies (trade, technological co-operation, etc.). It was therefore necessary to develop new models to help understand the logic of coalition formation in the presence of spillovers, and the possibility to increase welfare by means of appropriate mechanisms and strategies. These new models were developed in the 1990s within a non-co-operative game-theory framework, and provide interesting indications on the likely outcomes of climate negotiations.

Consider first the case in which countries negotiate on a single worldwide agreement. Most papers in the game-theory literature on coalition formation applied to environmental agreements (Hoel, 1991, 1992; Carraro and Siniscalco, 1992, 1993; Barrett, 1994, 1997b; Heal, 1994; Parson and Zeckhauser, 1995) propose the following conclusions:

- the presence of asymmetries[7] across countries and the incentive to free-ride makes the existence of global self-enforcing agreements, that is agreements which are profitable to all signatories and stable, quite unlikely (Carraro and Siniscalco, 1993);
- when self-enforcing international environmental agreements exist, they are signed by a limited number of countries (Hoel, 1991, 1994; Carraro and Siniscalco, 1992; Barrett, 1994); and
- when the number of signatories is large, the difference between the co-operative behaviour adopted by the coalition and the non-co-operative one is very small (Barrett, 1997b; Hammitt and Adams, 1996).

---

[6] In the case of climate negotiations, possible coalitions are Annex B Parties of Kyoto Protocol, the Umbrella Group, UNFCCC Parties, *etc.*

[7] Herein countries are symmetric when they share the same production technologies, consumers' preferences, institutions, *etc.*, namely when their payoff (welfare) functions are identical.

The results are robust with respect to different specifications of countries' welfare function, and with respect to the burden-sharing rule[8] used in the asymmetric case (Barrett, 1997a; Botteon and Carraro, 1997a). They suggest that the attempt to negotiate effective emission reductions is unlikely to lead to a coalition formed by all or by almost all countries, unless more complex policy strategies, in which environmental policy interacts with other policy measures, are adopted.[9] This is why in the game-theoretic environmental economics literature two main sets of instruments are proposed to expand environmental coalitions, that is to increase the number of signatories of an environmental agreement. These instruments are "economic and technological transfers" and "issue linkage". The potential of these instruments is analyzed in Section 10.2.5, which deals with partial agreements and ways to broaden them.

Consider the case in which countries are free to sign the agreement proposed by a group of countries or to propose themselves a different agreement to the same or to other countries (Carraro, 1998). This may lead to the formation of multiple climate agreements, as happens with trade blocs (Bloch, 1997; Yi, 1997; Carraro and Moriconi, 1998). The multiplicity of coalitions may allow region-specific agreements in which the characteristics of countries in the region are better reflected by the contents of the agreement. Even in this case, game theory provides a clear analysis of the outcome of climate negotiations. Despite the large number of equilibrium concepts,[10] some conclusions seem to be quite robust:

- the equilibrium coalition structure is not formed by a single coalition, but usually by many coalitions;
- the grand coalition, in which all countries sign the same environmental agreement, is unlikely to be in equilibrium; and
- coalitions of different sizes may emerge at the equilibrium (even when countries are symmetric).

---

[8] In the asymmetric case, the rule chosen to divide the gains from co-operation among the countries in the coalition (usually called burden-sharing rule) plays a crucial role because it affects the likelihood that each country decides to sign the agreement. The burden-sharing rule is usually taken from co-operative game theory and Nash's and Shapley's is the most used. In contrast, in the symmetric case different rules lead to the same outcome (equal shares).

[9] Surveys of the above literature are proposed in Barrett (1997b), Tulkens (1998), and Carraro (1999a).

[10] Unfortunately, game theory is far from achieving a well-defined non-co-operative theory of coalition formation under the above general assumptions and definitions. Several stability concepts can be used, but these unfortunately provide different equilibrium coalition structures. Among these are the concepts of equilibrium binding agreements (Ray and Vohra, 1997), α-stability and β-stability (Hart and Kurz, 1983), sequential stability (Bloch, 1997), open-membership stability (Yi, 1997), and far-sighted stability (Chew, 1994; Mariotti, 1997).

The specific results on the size of the coalitions depend on the model structure and, in particular, on the slope of countries' reaction functions (i.e., on the presence of carbon leakage). If there is no or little leakage and countries are symmetric, then the Nash equilibrium of the multicoalition game is characterized by many small coalitions, each one satisfying the properties of internal and external stability (this result is shown in Carraro and Moriconi, 1998).

The remaining question is therefore a policy one. Is a country's welfare larger when one or when several coalitions form? And what happens with environmental effectiveness? The answer is still uncertain, both because theory provides examples in which a single agreement is preferred, at least from an environmental viewpoint, to many small regional agreements (and vice versa), and because empirical studies have not yet convincingly addressed this issue. Moreover, the conclusion crucially depends on the choice of the equilibrium concept and on the size of leakage.

The consequence of the results discussed above is that the structure of the international environmental agreements is a crucial dimension of the negotiating process. If all countries negotiate on a single agreement, the incentives to sign are lower than those that characterize a multiple-agreement negotiating process. But at the equilibrium, the environmental benefit (quality) may be higher.

Can more precise conclusions be made on the likely coalition(s) that can emerge at the equilibrium? Can existing studies be used, albeit not in their design, to address the above issues, and to increase our understanding of the implications of different policy strategies? In the next section, the aim is to provide, at least partially, a synthesis, by exploring the outcomes of the combinations of different coalition structures (international regimes) and of different policy options (with focus on different degrees of adoption of emissions trading and other Kyoto mechanisms). *Table 10.2* summarizes the main combinations for which impact is explored. The papers indicated in each cell are examples and do not cover the literature in total.

10.2.3   *No Participation*

No participation constitutes the benchmark for evaluating the costs and benefits of policies designed to control GHG emissions under alternative coalition structures. It is usually named the baseline (or business as usual) scenario, because it identifies the values of the main environmental and economic variables when no coalition forms and no action, unilateral or co-operative, is adopted (IPCC SAR (IPCC, 1995) is a good example of this approach). The construction of the baseline scenario is very important to assess both the profitability and the stability (i.e., whether it is self-enforcing) of a coalition. A coalition is profitable when welfare after the coalition is formed is larger than in the no participation case. A coalition is self-enforcing if there are no incentives to leave or enter the

Table 10.2: Coalition structures and policy options

Coalition structure \ Policy options	Domestic measures only	Co-ordinated carbon tax	Flexibility mechanisms with ceilings	Free flexibility mechanisms	Flexibility mechanisms with banking	Flexibility mechanisms with R&D	Flexibility mechanisms with monopoly power
No participation	IPCC (1995)						
Unilateral participation	Jorgensen et al. (1993) Barrett (1992)						
EU only	Carraro and Siniscalco (1992)	Bosello and Carraro (1999) Barker (1999)					
OECD only	Burniaux et al. (1992)	Capros (1998)		Harrison and Rutherford (1999) Holtsmark (1998) Capros (1998)			
Annex-1 countries	McKibbin et al. (1998)	Mensbrugghe (1998) Buonnano	Ellerman et al. (1998) Holtsmark (1998) et al. (1999) Manne and Richels (1998)	Ellerman et al.(1998) Grubb and Vrolijk (1998) Westkog (1999) McKibbin et al. (1998) Manne and Richels (1999a, 1999b) Mensbrugghe (1998) Nordhaus and Boyer (1999) Shackleton (1998)	Bosello and Roson (1999) (1999)	Nordhaus (1997) Buonnano et al.	Burniaux (1998) Ellerman et al. (1998)
Double umbrella				McKibbin et al. (1998) Shackleton (1998)			Ellerman et al. (1998)
All countries	Nordhaus and Yang (1996)		Ellerman et al. (1998) Buonnano et al. (1999)	Bohm (1999) Ellerman et al. (1998) Manne and Richels (1998; 1999a, 1999b) Nordhaus and Boyer (1999) Shackleton (1998)	Bosello and Roson (1998) Westkog (1999)	Nordhaus (1997) Buonnano et al. (1999)	

coalition. The baseline scenario crucially affects these incentives also. If the no participation case is such that emissions decline and the target can be achieved easily through small emission reductions, then the incentives to join the coalition (sign the agreement) are much higher, so a coalition with many countries is more likely to form (Barrett, 1997b). Symmetrically, if large emission reductions are necessary, abatement becomes more costly, and incentives to free-ride increase, which further increases the costs for co-operating countries (particularly if leakage is high).

A careful definition of the no participation case is therefore very relevant to assess the likelihood of large coalitions and thus the efficiency of a climate agreement. But it is also very relevant in terms of equity. When the burden of emissions abatement has to be shared equitably, it is important to distribute emissions targets with reference to the baseline scenario. Each country therefore has an incentive to pretend that its own baseline scenario implies larger emissions than is actually true (Grubb, 1998; Bohm, 1999). In this way, the actual cost for the country would be lower. An optimistic scenario in which predicted emissions are lower than "true" emissions (as measured ex-post) leads countries to agree on low emission-reduction targets, but forces countries to more reductions later and to pay abatement costs larger than expected. A pessimistic scenario makes the agreement more difficult because larger emission reductions have to be agreed, but countries find themselves in a better situation and pay lower costs ex-post. Hence, if a country succeeds in convincing the others that its own baseline emissions are larger than the "true" ones, then this country achieves relative benefits in terms of less-stringent emission targets and lower abatement costs.

The definition of a baseline scenario has therefore a strategic dimension and can hardly be defined as an "objective" evaluation of future economic and environmental cycles and trends. It is therefore important to collect, as in Chapter 2, the largest amount of information from different sources and to identify the scenario more as an average of much scattered information, rather than as a subjective analysis of likely future events. This may reduce the likelihood of strategic definitions of the baseline scenario and may partly prevent the consequent impacts on the equilibrium coalition and on the assessment of costs and benefits of climate policies.

10.2.4   Unilateral Participation

An extensive literature analyses the costs and sometimes the benefits of introducing policies to control GHG emissions in a single country (Hoel, 1991; Bucholz and Konrad, 1994; Porter and Van Linde, 1995; Hoel and Schneider, 1997; Endres and Finus, 1998). Given the arguments proposed in the Introduction and the results summarized in Section 10.3.1, this type of exercise may seem unreasonable. There are, however, two main justifications for undertaking it. The first is that domestic abatement costs (related to domestic policies and

measures) hardly depend on the coalition structure. Indeed, only if leakage is large, and if climate policies have a large impact on trade and financial flows, are the costs of domestic abatement policies significantly affected by the size of the coalitions and by the agreed emission targets. Hence, it may be useful to compute the costs of unilateral participation as a benchmark case, which identifies costs that can be reduced only when coalition forms and the Kyoto mechanisms are implemented among signatory countries. Notice the importance of a careful assessment of leakage and of trade and financial repercussions of climate policies (McKibbin *et al.*, 1998). Notice also that the above arguments concern the costs but not the benefits of climate policies. Indeed, the climate benefits of unilateral participation are likely to be zero or almost zero for all or almost all countries (a possible exception is the USA), given the global nature of the climate problem (Hoel, 1991; Bucholz and Konrad, 1994; Endres and Finus, 1998).

A second reason to assess the cost of a unilateral participation is that it could identify a series of low cost (or no cost) options (so called low hanging fruits or no regrets actions) that could be implemented independently of the formation of a climate coalition. It could also help identify policy mixes that help restructure the fiscal system and public regulatory and incentive schemes in such a way that emission abatement costs are more than compensated by other economic (non-environmental) benefits (the so-called double dividends).[11]

There are also cases in which unilateral actions have been analyzed from a very specific viewpoint. Examples are:
*   Bucholz and Konrad (1994) analyze the detrimental effect of pre-negotiation actions (more bargaining power can be achieved by unilaterally increasing emissions before negotiating);
*   Endres and Finus (1998) examine the negative effects on negotiations of a higher environmental consciousness in one country;
*   Hoel (1991) analyzes the costs of unilateral actions;
*   Hoel and Schneider (1997) analyze the role of social norms; and
*   Porter and Van Linde (1995) focus on the advantage of being a leader by adopting emission reductions before the other countries.

10.2.5   Partial Agreements

The case of partial agreements is most often analyzed in recent empirical literature, for two reasons. First, as shown in Section 10.2.1, theory suggests that a partial coalition forms at the equilibrium. Hence, the climate problem is neither a "tragedy of commons" nor a situation in which there are clear incentives to co-operative emission control. Second, the history of international environmental negotiations is a history of partial

---

[11] See Chapters 7 and 8, and Goulder (1995), Bovenberg (1997), for surveys of this literature.

agreements that are slowly broadened as more and more countries decide to join the group of signatories. In the case of climate, in particular, the Kyoto agreement can be seen as a first partial climate agreement. Therefore, many papers have dealt with the costs and benefits of the Kyoto agreement and with the possible strategies to increase the number of countries that commit themselves to emission control targets (see the papers gathered in OECD (1998), and in Carraro (1999b, 2000); see also Burniaux (1998), Capros (1998), Ellerman *et al.* (1998), Grubb and Vrolijk (1998), Holtsmark (1998), Manne and Richels (1998), Mensbrugghe (1998), Carraro, 1999c), Nordhaus and Boyer (1999), and the surveys by Metz (1999) and Convery (1999)).

Two remarks are important. First, even if most recent analyses deal with the Kyoto agreement, there are studies that try to compute the optimal coalition structures, in terms of both participation and targets, independently of the decisions taken in Kyoto (a recent attempt is in Nordhaus and Boyer (1999)). Usually the conclusion derived from these papers is that Kyoto is neither economically nor environmentally optimal. However, the notion of optimality is not very useful when analyzing coalition formation. Indeed, what matters is the notion of the stability of a coalition. This identifies which countries have an incentive to join the coalition (sign the agreement) for different membership and institutional rules, baseline scenarios, abatement costs (and therefore climate policies, including the degree of adoption of Kyoto mechanisms), and environmental benefits (and therefore impacts, adaptation costs, etc.).

Second, the Kyoto agreement can theoretically be interpreted as a partial (Carraro, 1998) or as a global agreement (Chander *et al.*, 1999). It is interpreted as a global agreement when all countries are seen as committed to emission targets. Those in Annex B are committed to emission targets with respect to 1990, the other ones are "committed" to emissions levels that evolve as in the baseline scenario. This second interpretation is nothing more than a "technical" interpretation, which is useful to show that:

- optimal emissions targets are not necessary because the same optimal outcome can be achieved through an international, unrestricted emissions-trading scheme among all countries (Chander *et al.*, 1999); and
- the resultant outcome can be profitable to all countries if an appropriate economic and technological transfer scheme is adopted (Markusen 1975; Chander and Tulkens, 1995, 1997; Germain *et al.*, 1997).

As a consequence, even a "partial", suboptimal agreement like Kyoto can be transformed into a "global" optimal agreement (see Section 10.3.5).

Away from this ideal world of perfectly competitive and international market mechanisms, are the analyses of coalitions that, like the coalition formed by Annex I countries of the UNFCCC or Annex B countries of the Kyoto Protocol, are par-

tial (formed by a subgroup of the negotiating countries). In this context, four questions need to be answered:

(a) Are these partial coalitions effective?
(b) Are they too costly for the signatory countries?
(c) Can partial coalitions be enlarged by providing incentives for other countries to join? and
(d) Is there a distribution of emission targets and/or of abatement costs such as to increase the size of partial coalitions and hence the effectiveness of a climate agreement?

(a) The answer to the first question depends on two main factors: the baseline scenario and the degree of leakage. If the baseline scenario is very ambitious and leakage is high, then countries find it difficult to undertake large emissions reductions (decreasing returns of scale in emission abatement are usually assumed), and also their effort is offset by the leakage effect (the increased emissions by free-riding countries). Hence, a partial coalition is effective whenever there is no or little leakage, high pollution levels characterize the baseline scenario, and signatory countries contribute a large share of the total emissions.

(b) For the second question, many studies try to assess the cost for Annex I countries of achieving given emissions targets under alternative policy options. These policy options include:

- the timing of the mitigation responses (see the special issue of *Energy Economics* edited by Carraro and Hourcade (1999));
- the degree of adoption of the Kyoto mechanisms and their features, such as banking (see the papers in OECD (1998) and Carraro (1999d));
- the role of complementary industrial policies, mainly designed to foster innovation (see Nordhaus, 1997; Schneider and Goulder, 1997; Kopp *et al.*, 1998; Buonnano *et al.*, 1999); and
- the effects of uncertainty about climate impacts or abatement costs (Carraro and Hourcade, 1999).

The main result can be summarized as follows. Despite their high variability, all the studies show that the Kyoto mechanisms sensibly reduce the costs of compliance, whatever the coalition structure. Hence, emissions trading, and more generally the application of the Kyoto mechanisms, can reduce overall mitigation costs without reducing the effectiveness of the climate policy. Chapters 6 and 8 give an extensive overview of relevant studies.[12]

---

[12] For example, Shogren (1999) notes that "it is estimated that any agreement without the cost flexibility provided by trading will at least double the USA costs, … the key is to distribute emissions internationally so as to minimise the costs of climate policy". Manne and Richels (1999b) state that losses in 2010 are two and one-half times higher with the constraint on the purchase of carbon emission rights; international co-operation through trade is essential if we are to reduce mitigation costs (see also Glomstrod *et al.*, 1992; Burniaux, 1998; Capros, 1998; Ellerman *et al.*, 1998; Mensbrugghe, 1998; Hourcade *et al.*, 1999; Nordhaus, 1999; Rose and Stevens, 1999; Tol, 1999a, 1999b).

If assuming an even broader type of flexibility than incorporated in the Kyoto mechanisms (banking and borrowing and international emissions trading (IET) among all countries) then compliance costs are further lowered. This result is shown in Bosello and Roson (1999) for banking, Westskog (1999) for banking and borrowing, Manne and Richels (1999a, 1999b), McKibbin *et al.* (1998), and many others for IET among all countries. If in addition the incentives to innovation provided by the Kyoto mechanisms are taken into account, then compliance costs are even lower (Buonnano *et al.* 1999).

However, all the above papers also show that the size of the coalition crucially affects the size of the benefits that derive from the adoption of the Kyoto mechanisms. The larger the number of participating countries, and the higher the variability of marginal abatement costs across them, the larger the benefits from emissions trading and the clean development mechanism (CDM). Hence, to reduce abatement costs and increase environmental benefits, policies, rules, and institutions should be designed to achieve the largest possible coalition.

(c) The third question, how to broaden a climate coalition, is often related to the issue of links between a climate agreement and other international agreements. Indeed, two types of policy options, based respectively on economic and technological transfers and on issue linkage, are often proposed as the way to achieve larger climate coalitions. These policies imply that links must be established between different multilateral agreements (e.g., agreements on both climate and free trade or technological co-operation).

First, consider economic and technological transfers. It is quite natural to propose these transfers to compensate those countries that may lose by signing the environmental agreement. In other words, a redistribution mechanism among signatories, from gainers to losers, may provide the basic requirement for a self-enforcing agreement to exist, that is the profitability of the agreement for all signatories. Therefore, if well designed, economic and technological transfers can guarantee that no country refuses to sign the agreement because it is not profitable. Moreover, Chander and Tulkens (1995, 1997) and Chander *et al.* (1999) show economic and technological transfers exist such that not only is each country better off within a global coalition than it is with no coalition at all (the no participation case), but also it is better off within a global coalition than it is in any subcoalition, provided the remaining countries behave non-co-operatively (see also Markusen 1975; Germain *et al.*, 1997). This result is important because it implies that no country or group of countries has an incentive to exclude other countries from the environmental coalition, that is the grand coalition is optimal (but it may not be stable).

Economic and technological transfers play a major role also with respect to the stability issue (Carraro and Siniscalco, 1993; Petrakis and Xepapadeas, 1996; Schmidt, 1997). Indeed, it is not sufficient to guarantee the profitability of the environmental agreement. Incentives to free-ride also need to be off-

set. The possibility of using self-financed economic and technological transfers to stabilize environmental agreements is analysed in Carraro and Siniscalco (1993) and Hoel (1994), which show that these transfers may be successful only if associated with a certain degree of commitment. For example, when countries are fairly symmetric, only if a group of countries is committed to co-operation can another group of uncommitted countries be induced to sign the agreement by a system of economic and technological transfers (Carraro and Siniscalco, 1993).[13] This gives developed countries the responsibility to lead the expansion of the coalition. However, the amount of resources that would be necessary to induce large developing countries to join the agreement may be such that some developed countries perceive the economic costs of a climate agreement to be larger than its environmental benefit. In this case, the transfer mechanism would undermine the existence of the leader coalition and would therefore be ineffective. This is why countries in the leader coalition must be strongly committed to co-operation on emission control.

Another general conclusion emerges from the analysis carried out in Carraro and Siniscalco (1993): both the existence of stable coalitions and the possibilities of expanding them depend on the pattern of interdependence among countries. If there is leakage (i.e., a non-co-operating country expands its emissions when the coalition restricts them, thus offsetting the effort of the co-operating countries), then environmental benefits from co-operation are low, the incentive to free-ride is high, and conditions for economic and technological transfers to be effective are unlikely to be met. If, on the contrary, there is no leakage (i.e., the free-riders simply enjoy the cleaner environment without paying for it, but do not offset the emission reduction by the co-operating countries), then environmental benefits are larger, free-riding is less profitable, and transfers may achieve their goal to expand the coalition.

A second policy strategy aimed at expanding the number of signatories to a climate agreement is based on the idea of designing a negotiation mechanism in which countries do not bargain only on GHG reductions, but also on another interrelated (economic) issue. For example, Barrett (1995, 1997c) and Kirchgässner and Mohr (1996) propose to link climate negotiations to negotiations on trade liberalization, Carraro and Siniscalco (1995, 1997) and Katsoulacos (1997) propose to link them to negotiations on R&D co-operation, and Mohr (1995) proposes a link to international debt.

Again we must distinguish the profitability from the stability problem. The idea of "issue linkage" was originally proposed by Folmer *et al.* (1993) and Cesar and De Zeeuw (1996) to solve the problem of asymmetries among countries. The intuition is that some countries gain on a given issue, whereas other countries gain on a second one. By "linking" the two issues it

---

[13] This condition is less stringent when countries are asymmetric (see Botteon and Carraro, 1997a).

may be possible that the agreement in which the countries decide to co-operate on both issues is profitable to all of them. The idea of "issue linkage" can also be used to achieve the stability goal. If countries that do not sign a climate agreement do not enjoy the benefits that arise from signing simultaneously other multilateral agreements (e.g., those on technological co-operation), then incentive for all countries to sign the linked agreement is strong.

This approach is likely to function when the negotiation on an issue with excludable benefits (a "club good" in economic words) is linked to the climate negotiation (which, if successful, typically provides a public good, that is a non-excludable benefit). An example could be the linkage of environmental negotiations with negotiations on technological co-operation whose benefits are largely shared among the signatories whenever innovation spillovers to non-signatories are low (see Carraro and Siniscalco, 1997).[14]

Therefore, issue linkage may be a powerful tool to address the enlargement issue. If the developed countries (USA, EU, and Japan above all) increase their financial and technological support to developing countries, and also make this support conditional on the achievement of given environmental targets, then other countries are likely to be induced to join the environmental coalition (i.e., to sign a treaty in which they commit themselves to adequate emission reductions).[15]

(d) The final question concerns the link between equity[16] and the size of a climate agreement and, as a consequence, between equity and the agreement's environmental effectiveness. It has been shown that the use of different criteria to share the cost of a given emission target crucially affects the size of the equilibrium coalitions, that is the number and identity of signatory countries (Barrett, 1997a; Botteon and Carraro, 1997a, 1997b; Eyckmans, 2000). For the Kyoto Protocol, Convery (1999) argues that without assigning generous emission targets to Russia and Ukraine, these countries would not have signed the agreement. Eyckmans (2000) proves the same conclusion by simulating different equilibrium climate coalitions with the Regional Integrated Model of Climate and the Economy (RICE). Indeed, without implementing the Kyoto mechanisms, Russia and Ukraine have an incentive not to ratify the protocol, whereas with joint implementation (JI) and trading, and with the possibility of exchanging excess GHG emissions, all countries find it profitable to ratify the protocol. Bosello *et al.*

(2000), using again the RICE model, confirm the same results and analyze different distributional rules in terms of their impacts on the equilibrium climate coalitions. They show that the Kyoto Protocol could be sustained and possibly expanded by adopting a more equitable sharing of the emission reduction commitments.

10.2.6 Global Agreements

The difficulty of achieving a global agreement on climate change underlined in the previous sections depends on four main factors:

- *The heterogeneity of countries with respect to the causes of climate change, the impacts, and the mitigation and adaptation costs.* This factor mainly influences the profitability of the decision to sign a climate agreement. Some countries may lose when signing the agreement, even when environmental benefits are fully accounted for. As shown by Chander and Tulkens (1995, 1997), there always exists a system of economic and technological transfers that may make all countries gain. But this again raises the equity problem and the related burden-sharing issue. Equity may have a large impact on the existence and size of a climate coalition. As previously argued, and as argued by many policymakers and scientists, the way in which the burden of controlling emissions is shared across countries crucially affects a country's decision to join a coalition. On the one hand, if the burden is not equitably shared, some countries may not find it profitable to sign the agreement. Profitability depends on two main factors: (1) the distribution of costs within the coalition and (2) the size of the coalition. It is possible that there exists a minimum size of the coalition above which it becomes profitable. And these two factors are strictly interdependent. On the other hand, equity also affects free-riding incentives. As in Section 10.2.5, in some cases it may be reasonable for some countries to transfer resources to other countries to induce them to join the coalition on which they would otherwise free-ride. In this case, the final outcome is not equitable–free-riders would gain more than countries in the starting coalition–but it may be environmentally and economically efficient.

- *The strong incentives to free-ride on the global agreement and the lack of related sanctions.* When all countries agree to control emissions, a defecting country achieves the whole benefit, because its incidence on global emission is marginal (with a few exceptions) and pays no cost. Hence, a defection with respect to a large coalition is the optimal strategy if there are no sanctions. However, credible sanctions are difficult to design (Barrett, 1994). Emissions themselves are hardly a credible sanction, because countries are unlikely to sustain self-damaging policies. Moreover, in this case,

---

[14] An extension to the case of structurally asymmetric countries is provided in Botteon and Carraro (1997b), whereas information asymmetries are accounted for in Katsoulacos (1997).

[15] It is, however, important to keep in mind the negative impact that such linkages may have on the (perceived) fairness of the envisaged enlarged regime: there are possible linkages that could easily be perceived as "blackmail" on part of the Parties with strategic advantages.

[16] For equity principles see Chapter 1, Section 1.3.

asymmetries play a double role: some countries may not gain from signing the environmental agreement, whereas some countries, even when gaining from environmental co-operation, may lose from carrying out the economic sanctions (Barrett, 1997c; Schmidt, 1997).

- *The absence of environmental leadership.* The process of achieving a global agreement can be a sequential one (Carraro and Siniscalco, 1993), in which case a group of countries take the leadership, start to reduce and/or control emissions and implement strategies such as to induce other countries to follow.[17] The presence of low-cost climate policies and equitable burden-sharing (Schmidt, 1997) are again important elements for the formation of an initial profitable coalition. As said, our definition of profitability accounts for the environmental benefits of emission control. Hence, benefits should be increased by increasing the number of countries that control emissions, but abatement costs should be minimized by exploiting all possible opportunities (including emissions trading). This is a prerequisite to achieving a strong leader coalition that can exert its leadership through the design of better negotiation rules, the implementation of transfer mechanisms, and the credibility of international-issue linkages. A preliminary model of the effects of leadership is given in Jacoby *et al.* (1998), who show how and when developing countries may join a leader coalition formed by Annex I countries.

- *The focus on a single international climate agreement.* As explained in Section 10.3.1, if countries may join different coalitions, which means that several agreements can be signed by groups of countries in the same way as countries form trade blocs, then the likelihood that all or almost all countries set emission reduction targets increases (Yi and Shin, 1994; Bloch, 1997; Carraro, 1997, 1998). The outcome of negotiations in which more agreements can be signed is usually a situation with several small environmental blocs (Carraro and Moriconi, 1998), but this can be considered a step in the right direction. If all or almost all countries set emission reduction targets within their own bloc (e.g., regional environmental agreements are signed), then, in a subsequent phase, negotiations among blocs may lead to more ambitious emission reductions.

Despite the warning that global agreements may be difficult to reach, many articles analyze the costs of agreements in which all countries participate, in one form or another (see, e.g., Capros, 1998, Ellerman *et al.*, 1998; Manne and Richels, 1998; Shackleton, 1998; Bosello and Roson, 1999; Nordhaus and Boyer, 1999). The weakest form, discussed in Section 10.2.4, is that in which a few countries commit to emission reductions, but all accept trade emissions in a single international market.

The strongest form is that in which a central planner is assumed to set optimal emissions levels for all world countries. This optimal solution is often proposed as a benchmark for actual negotiations and was often analyzed before Kyoto (see the collection of papers in Carraro (1999d)).

More interesting is the attempt made by Peck and Teisberg (1999) to model the negotiations between developed and developing countries to achieve a global agreement. This paper shows the potential for the achievement of co-operation to be achieved–the Pareto frontier is small, but not empty–but does not analyse the incentives to actually sign the agreement. However, the paper suggests a research direction that at least helps to identify the optimal emission reductions that are profitable for all negotiating countries.

The conclusions that can be derived from this type of empirical analyses are similar to those already mentioned for partial agreements. In the scenario in which baseline emissions are lower, it is easier to achieve a global agreement because lower emissions reductions are necessary (Barrett, 1997b) and consequently abatement costs are lower. Optimal emissions targets are such that they equalize marginal abatement costs. This optimal, cost-minimizing solution can also be achieved through an unconstrained emissions-trading system (Chander *et al.*, 1999). Hence, either emissions targets are optimally set, or countries are allowed to trade emissions for any given set of targets through which a global consensus can be achieved. Of course, these two options have different impacts on equity. As shown by Bosello and Roson (1999), starting from the Kyoto targets, international unconstrained emissions trading among all countries achieves optimality, but reduces equity.

10.2.7 Political Science Perspectives

Game theory and other rational-choice approaches are used frequently in political science. However, political science research considers political processes in more detail and their findings complement the results presented above, at least on three major issues. Although these extensions have important implications for the conclusions here, the basic insights remain the same.

While game theory analysis usually models states as *unitary* actors, much political science research conceives of states as complex political systems. The behaviour of a complex actor can be seen as a function of three main determinants: the internal configuration of preferences, the internal distribution of influence and power, and the nature of political institutions (which specify the decision rules). Domestic decision-making processes often produce outcomes that differ significantly from those that maximize the net national welfare. Particularly relevant in this context are three findings that illustrate systematic biases.

First, in "baseline" circumstances, the measures that are most easily adopted and implemented are those that offer tangible

---

[17] See Carraro (1999b) and Grubb (1999) for a more detailed analysis.

benefits to a specific sector of the economy or organized segments of society, while costs are widely dispersed throughout society (Underdal, 1998). For most conventional environmental-protection measures, costs are concentrated while benefits are indeterminate or widely dispersed, which indicates that—unless the issue really mobilizes the general public—the odds favour opponents to the measures, particularly in the implementation phase.

Second, (environmental) damage that hits the "social centre" of society tends to generate more political energy than damage that affects the social periphery only. This bias is stronger the more skewed the distribution of economic and political resources. This suggests, for example, that damage suffered primarily by poor farming communities in developing countries generates a less vigorous political response than damage that hits the infrastructure of the "modern" sectors of the economy (e.g., as a consequence of extreme weather events).

Third, domestic political processes often generate political "friction" that limits the scope for international package deals and compensatory arrangements. Only compensation that benefits the domestic actor(s) who are blocking a particular solution–or more powerful actors–will be fully effective. Only a subset of the compensatory arrangements that make sense in terms of economic criteria will pass the test of political feasibility. These issues of national DMFs are explored in Section 10.1.

Most of the research reviewed above examines climate change policy in isolation, on its own merits only. In the real world, new issues enter a policy space that is already crowded by other problems competing for attention. In such an environment, the priority given to a particular issue and the chances that a particular option will be adopted depend on how well it combines with other salient concerns. As we have seen in, for example, the acid rain case, policy confluence and synergy can make a significant difference for some of the parties. However, although the causal mechanism itself is well understood, it is triggered by circumstances that occur more or less at random. Thus, the aggregate net impact in terms of the climate change regime cannot be predicted (even if issue linkage, as seen above, may be a powerful strategy).

The conventional assumption in game theory analysis is that each party aims to maximize its own welfare, defined—when dealing with environmental problems—in terms of damage and abatement costs. Political science research modifies this assumption in three different directions.

First, it introduces a distinction between the "basic game" itself (i.e., the system of activities to be regulated) and the "policy game" through which decisions about regulations are made. The policy game generates its own stakes; certain kinds of behaviour—notably behaviour that meets the expectations of domestic "clients" and important others—are rewarded, while moves that violate these expectations are punished. Governments also consider such political stakes. Where such

stakes exist, a political scientist expects government behaviour to deviate to some degree from what national economic interests indicate. Such deviance may go both ways; the wish to placate politically important domestic "clients" most often leads to a more restrictive policy, while the momentum generated towards the end of a successful international conference can lead a lone "laggard" to go the extra mile to accommodate the majority.

Second, political science emphasizes (more and more) the relevance of "social norms", "social learning", and the operation of "social roles" in regime processes (Young, 1999). These approaches recognize that all international environmental regimes are "social institutions" that develop particular (social) dynamics and induce behavioural consequences: the matter of *social norms* refers to behaviour that roots in considerations of legitimacy or authoritativeness. Actors, who regard the rules of regimes as legitimate, often comply without engaging in detailed calculations of the costs and benefits (of their doing so). One important effect of international regimes is that they initiate *social learning processes*. Already, the start of global negotiating generally has resulted in the generation of new facts, ideas, and perspectives that reduce uncertainty and lead to changes in the prevailing discourses, values, and actual behaviour of actors. The operation of *social roles* refers to the observation that actors regularly take on new roles under the terms of institutional arrangements that shape identities and interests.

Third, norms of fairness are assumed to serve as (1) frameworks of soft constraints upon the pursuit of self-interest, and (2) as decision premises in situations in which interests provide no clear guidance. Studying international negotiations we can observe some rather general norms that are frequently invoked and very rarely disputed—at least on principled grounds. These norms seem to constitute a soft core of widely, though probably not universally, accepted ideas about distributive fairness. This core is described in summary fashion below.

The default option in international co-operation is the norm that all parties shall have *equal* obligations, usually defined in relative terms. The principle of equal obligations has a firm normative basis if all parties involved are roughly equal in all relevant respects. This condition is never met in global negotiations, although it usually applies to subgroups. When the range of variance exceeds a certain threshold, attention shifts to some notion of *equity*. The common denominator for equity norms is that costs and/or benefits be distributed in (rough) proportion to actor scores on the dimension(s) that led the parties to think about differentiation in the first place. Several such dimensions can be identified, but in international co-operation attention focuses primarily on two. One is the role that each party played in causing the problem or providing the good in question, the other refers to the consequences that a particular obligation or project would have for the various parties involved. This gives a matrix with four key principles (see *Table 10.3*). In a global setting, however, the range of variance in terms of criteria such as "guilt" or "capacity" is most often so large that even the

Table 10.3: *Key principles of equity in the political science context*

Focus on ↓	Object to be distributed	
	Costs (obligations)	**Benefits**
Cause of current state of affairs	"Guilt" or responsibility (for causing the problem)	Contribution (to solving the problem or providing the good)
Consequences for actors	Capacity (ability to pay)	Need

notion of soft proportionality leads to "unfair" burdens upon the poorest countries. When the latter threshold is reached, attention tends to shift to the simple principle of *exemption*; more precisely, exemption from any substantive obligation for which a party is not (fully) compensated.

This leaves a somewhat complex and elastic framework, but the bottom line is clear enough. A global agreement has to be at least roughly consistent with (1) the general pattern of differentiation outlined in the preceding paragraph, and (2) the combined implications of the equity principles of "guilt", "capacity", and "need" (i.e., implications that can be derived from all three principles).

These points are important to consider in the design of international environmental regimes. Political scientists focus on

sociopolitical dimensions and processes that current game-theory models neglect or are unable to capture adequately. Nevertheless, the policy-relevant conclusions from game theory remain valid and useful for the policy process.

10.2.8 Implementation and Compliance

Since SAR, political science analysis in the field of effectiveness and implementation of international environmental agreements has focused on the process of implementation. That is, how intent is translated into action to solve international environmental problems and what are the real effects of these efforts (Sand, 1992; Haas *et al.*, 1993; Young, 1994, 1999; Brown Weiss and Jacobson, 1998; Victor *et al.*, 1998). Analysts distinguish "implementation" and "compliance"

Box 10.1. Definitions of Political Science Terms

Implementation
Implementation refers to the actions (legislation or regulations, judicial decrees, or other actions) that governments take to translate international accords into domestic law and policy (Jacobson and Brown Weiss, 1995; Underdal, 1998; Brown Weiss, 1999). It includes those events and activities that occur after authoritative public policy directives have been issued, such as the effort to administer the substantive impacts on people and events (Mazmanian and Sabatier, 1983). It is important to distinguish between the legal implementation of international commitments (in national law) and the effective implementation (measures that induce changes in the behaviour of target groups; see Zürn, 1996).

Compliance
Compliance is a matter of whether and to what extent countries do adhere to the provisions of the accord (Jacobson and Brown Weiss, 1995; Underdal, 1998). The concept of compliance includes implementation, but it is generally broader. Compliance focuses not only on whether implementing measures are in effect, but also on whether there is compliance with the implementing actions. Compliance measures the degree to which the actors whose behaviour is targetted by the agreement (whether they be local government units, corporations, organizations, or individuals) conform to the implementing measures and obligations (Brown Weiss, 1999).

Effectiveness
Effectiveness measures the degree to which international environmental accords lead to changes of behaviour that help to solve environmental problems, that is the extent to which the commitment has actually influenced behaviour in a way that advances the goals that inspired the commitment (Victor *et al.*, 1998).

Enforcement
Enforcement refers to the actions taken once violations occur. It is customarily associated with the availability of formal dispute settlement procedures and with penalties, sanctions, or other coercive measures to induce compliance with obligations. Enforcement is part of the compliance process (Brown Weiss, 1999).

(Chayes and Chayes, 1993, 1995; Mitchel, 1994; Jacobson and Brown Weiss, 1995; Cameron *et al.*, 1996; Underdal, 1998; Victor *et al.*, 1998; Brown Weiss, 1999). See *Box 10.1.* for the definition of political science terms.

Although compliance is an important matter for the outcome of an agreement, it has to be distinguished from the effectiveness of the accord (Underdal, 1998; Victor *et al.*, 1998; Brown Weiss, 1999; Young, 1999). This refers to the extent to which the commitment actually influences behaviour in a way that advances the goals that inspired the commitment.

Discussions are underway on how to enforce international commitments, that is to make parties to the international treaties conform with their international obligations through application of various tools (penalties, sanctions, *etc.*). Some researchers argue that enforcement might be especially difficult in international systems and, thus, is often unlikely unless a party persistently fails to comply.[18] Besides, non-compliance is frequently the product of incomplete planning and miscalculation rather than a wilful act (Victor *et al.*, 1998). Thus, enforcement is often contrasted to the management of non-compliance and implementation failures (non-compliance is a problem to be solved, not an action to be punished), which includes greater transparency, non-adversarial forms of dispute resolution, technical and economic assistance, persuasion, and negotiation (Haas *et al.*, 1993; Chayes and Chayes, 1995; Sand, 1995; Downs *et al.*, 1996; Zürn, 1996; Peterson, 1997; Victor *et al.*, 1998; Vogel and Kessler, 1998). However, there are also good reasons to consider coercive "enforcement" techniques–in cases of severe violations they may be more effective. In this debate, standard solutions do not exist and a mixed approach seems to be reasonable.

The challenge today is how decisions regarding compliance and implementation of the UNFCCC and its Kyoto Protocol should be undertaken to make these international mechanisms more effective in solving the problems of both combatting global climate change and changes in the behaviour of the targets (Victor and Salt, 1995; O'Riordan and Jäger, 1996; O'Riordan, 1997; Soroos, 1997; Grubb *et al.*, 1999). Two crucial aspects of decision-making regarding implementation of the international climate change regime are:
- how national governments have translated international commitments into national rules and policies, and promote changes in behaviour of stakeholders; and
- how international institutions have aided monitoring of implementation and compliance, adherence to commitments, and adjustment of international rules by the parties.

Inadequate attention to implementation at both national and international levels is largely the reason why many interna-

tional agreements have fallen short of their promise (Victor *et al.*, 1998). Moreover, as the policy agenda has grown more demanding, international agreements play an ever greater role in affecting and co-ordinating the behaviour of national governments that have undertaken the international obligations and became responsible for meeting them. These agreements also influence the activities and responses of non-state actors (such as firms, individuals, scientists, interest groups, consumer and environmental groups), whose activities are affected by the international treaties after national governments adopt rules and policies for domestic implementation of the international regime. The importance of implementation has increased, and climate change is a good example of this. The stakeholders come to play an increasing role in design and implementation of the treaties (Michaelowa, 1998b; Victor *et al.*, 1998; De La Vega Navarro, 2000), and to involve them more broadly makes this process more effective.

Success or failure in the implementation of international environmental agreements depends to a large extent on how they are implemented in countries, once the parties to the agreements have returned back home. The process of domestic implementation of international environmental arrangements is very important to the overall effectiveness of the treaty. Results of attempts to develop co-operative solutions to international environmental problems are found in the domestic setting of the decision-making (Hanf and Underdahl, 1995; Hanf and Underdal, 2000). Indeed, to understand what is likely to happen at the international level, it is necessary to examine the underlying factors and processes, structures, and values at the national level (Kawashima, 1997; Kotov *et al.*, 1997; Kawashima, 2000). These determine the manner in which national positions on negotiating international agreements are arrived at and the ultimate agreements are then carried out. In turn, the expectation is that what has happened or is happening in various international arenas influences these domestic processes and decisions within individual countries; thus, national–international linkages within the decision-making process are very strong.

The decision-making and policy-making processes pertaining to international co-operation in the environmental field may be represented as a sequence of three interrelated phases (Hanf and Underdal, 1995):
- formation of national preferences and policy positions for international negotiation;
- translation of national preferences into international collective action; and
- implementation of international agreements at the national level.

The first two phases are analyzed from both economic and political science perspectives. As for the third phase, studies demonstrate that there are no standard decisions or standard implementation processes for the international environmental regimes. Even countries with similar political, economic, and social systems adopt different approaches, and within countries

---

[18] This is why in Sections 10.2.2–10.2.6 the focus is on self-enforcing agreements.

the implementation process varies markedly among different sectors. It is expected that implementation of the climate change international regime will illustrate this conclusion, and the canvas for the decision-making process will be extremely intricate and complex.

The literature on compliance and implementation indicates that a variance in the extent to which parties to international agreements fulfil their obligations, and that the extent of national compliance also varies across international regimes (Jacobson and Brown Weiss, 1995; Downs *et al.*, 1996; Brown Weiss, 1999; Young, 1999).

Signing (and ratification) of an international agreement constitutes no guarantee that it will be implemented effectively and complied with. Nor does the refusal to sign an agreement necessarily mean that an actor will act contrary to its terms. Moreover, an actor may comply with some provisions (e.g., procedural obligations), but not with others (e.g., substantive rules that require major behavioural change), and meet some obligations partially (for example, by reducing emissions, but less and/or later than required by the agreement). It is necessary to note that rule-consistent behaviour may not always be induced by the treaty, or necessarily result from the existence of a particular agreement. For one thing, some international agreements do not require that all actors change their behaviour—some actors may already behave as prescribed by regime rules (Brown Weiss and Jacobson, 1998). Moreover, in some cases for which behavioural change is prescribed, the required change may come about without any deliberate effort to meet the obligation, and compliance without implementation occurs. For example, the recent sharp economic recession in Russia (more than a 50% decrease in Gross National Product (GNP) during the 1990s) resulted in sufficient pollutant-emission reduction from industrial and other activities to meet (and even to "over-comply with") the domestic targets set by a number of international agreements. Little effort was required on the part of the government or non-government actors to honour these commitments (Kotov and Nikitina, 1996, 1998). A number of recent research efforts conclude that most domestic behavioural change can be attributed to many exogenous factors, and is not induced directly by the international regime, but this change contributes to compliance with the regime goals and targets (Levy *et al.*, 1995). These specifics affect seriously decision-making patterns regarding the implementation of the climate change regime.

Recent research brings together several important paths that can influence the variance in the extent to which parties fulfil their obligations (Haas *et al.*, 1993; O'Riordan and Jäger, 1996; Zürn, 1996; Peterson, 1997; Victor *et al.*, 1998). Enhancement of a contractual environment refers to the high relevance of an institution's transparency and credibility. An effective design introduces a shared set of norms and rules, provides information about membership and compliance, and helps to reduce transaction costs. Concern building describes the potential influence of institutions on actors' beliefs and ideas. These actors create, collect, and disseminate scientific knowledge and serve as centres for social learning processes. Capacity building refers to asymmetries across countries and their restrictive effects on an international commitment and to the possibilities of overcoming them. International institutions can manage transfers of cognitive, administrative, and material capacities to enable states to agree and comply with obligations. A broadly accepted management of non-compliance in many cases can lead to more effective solutions to defection. This approach of flexible responses covers various instruments, such as dispute resolution, interacting measures of assistance and persuasion, incentives, and greater transparency. Participation by "target groups" (e.g., regulated industries) and other non-governmental organizations (NGOs) reduces uncertainty, leads to more realistic agreements, and helps to ensure that countries put them into practice.

10.2.9 *Monitoring, Reporting, and Verification*

Studies confirm that in the past compliance of nearly all governments with their binding international environmental obligations has been quite high. However, this often reflects that the commitments were fairly trivial, in many cases simply codifying rather than changing behaviour (Brown Weiss, 1999; Victor and Skolnikoff, 1999). But the effectiveness of these commitments in reducing environmental problems was also low. Incentives to cheat were few and the need for strict monitoring and enforcement was low. As efforts to tackle environmental problems intensified, as in the case of climate change, countries' commitments became more demanding and stringent, the costs and complexity of implementation increased, and thus the incentives to cheat have grown. For this reason, stricter monitoring and enforcement are increasingly essential to ensure that these commitments are implemented fully (Sand, 1996; Victor and Skolnikoff, 1999). The historical record of high compliance without much monitoring and enforcement is a poor indicator of what will be needed for more effective international environmental protection in the future.

Although systematic reviews of implementation are commonplace in many national regulatory programmes (Lykke, 1993), the systematic monitoring, assessment, and handling of implementation failures by international institutions is relatively rare (General Accounting Office, 1992). Nonetheless, efforts to provide such review are growing, and today formal mechanisms for implementation review exist in nearly every recent international environmental agreement. Such mechanisms are incorporated into the UNFCCC structure as well (Victor and Salt, 1995). In addition, many informal mechanisms to review implementation and handle cases of non-compliance often operate in tandem with the formal mechanisms. Together, these formal and informal mechanisms are termed by some researchers as "systems for implementation review".

An implementation review process is especially vital when decisions are undertaken regarding complex and uncertain problems on the international environmental agenda. Such problems as global warming are still poorly understood, and involve a large number of stakeholders. Since regulation of the many diffused actors is often complex, governments cannot be sure in advance whether their efforts to put international commitments into practice will be successful. Moreover, some governments may intentionally violate their international obligations. Thus, there is a need to review implementation and handle problems that arise. Implementation review can also make it easier to identify problems with existing agreements, which can aid the process of renegotiation and adjustment. However, until recently implementation review has neither been the topic of much research nor high on the policy agenda.

International agreements that include procedures for gathering and reviewing information on implementation and handling implementation problems, as for the UNFCCC, are more likely to be effective than those in which little effort is given to developing the functions of implementation review (Zürn, 1996; Victor *et al.*, 1998). Agreements contain prescriptions for the governments to report regularly the data on their emissions and implementation measures. This has made parties more accountable for the implementation of their commitments, helped to direct assistance that facilitates compliance, and provided information and assessments that make it easier to adjust agreements over time.

Within the decision-making process regarding UNFCCC implementation, today more attention is given to assessing national emissions, policy, and measures. The process of compiling GHG emission inventories is well underway. Parties to the UNFCCC are obliged to compile and submit national communications on how they are implementing the convention (Green, 1995). These reports include inventories of GHG emissions, reports on policies and measures that the parties have adopted to try to stabilize or reduce emissions, and (eventually) an account of the extent to which emission abatement has been successful. Since 1991, IPCC and the Organization for Economic Co-operation and Development (OECD) have built effective guidelines for inventory reporting. All parties to the convention must use this system of reporting to the UNFCCC regarding emissions by sources and removals by sinks. Within this framework the governments are actively contributing to the international reporting process in submitting their national reports. Experts regard data reported by them as the backbone of the IPCC international system, while the EU is also engaging its own system – Coordination-Information-Air (CORINAIR). Without good data, systems of implementation review work poorly or not at all (Lanchbery, 1998). This system was intended to be applicable to all countries and for the main emissions sector (that for energy-related CO_2 emissions), and it makes use of energy flow statistics of the type that most developed countries collect routinely. Special methodologies and guidelines have been elaborated to convert the national inventory systems reasonably well, certainly for energy-related emissions, into the IPCC format. In this and in other respects it is what is known largely as a "top down" system.

Nevertheless, much work needs to be done and decisions undertaken in this area soon. Assessments indicate the poor quality of data. National data of the member states for major GHGs are not comparable, accurate, or reliable outside the energy sector (Lanchbery, 1997): emission figures given in national inventories are often of poor quality (in many subsectors, no estimates are made at all by some countries). This is not surprising given the rapid development of the climate issue and the requirement for reliable inventories. However, it impedes significantly the simplest reviews of implementation.

At the moment the reporting process may not be transparent enough. Further decisions could be undertaken, both at national and international levels, to improve its effectiveness. That is, to improve and develop further the compilation methodologies, increase the transparency of the compilation process and its reliability, and more work is needed that is specifically directed towards obtaining information for inventories, rather than purely for scientific purposes. It is crucial that inventories of GHGs are accurate, reliable, and comprehensive. Otherwise, it is not possible to determine the state of the emissions, where they originate, and how they are changing.

As the climate change regime develops after Kyoto, the issues of emissions measurement and verification, including the release and absorption of carbon from changes in land use, rice cultivation, and forest management, will become even more important. And it represents one of the toughest challenges for the scientific community. By adding three additional gases and sinks, the Kyoto Protocol fulfils its ambition to be more complete, but at the same time it makes compliance more difficult, and it complicates monitoring and verification (Corfee-Morlot and Schwengels, 1994). In particular, it raises the need for further modelling, a comprehensive new analysis, and better inventories.

Targets agreed by Kyoto are challenging. However, to implement the commitments and to meet the targets it is necessary to reach a common understanding of what they mean. Forecasts from different sources are often not comparable. For example, data from the International Energy Agency (IEA) are different from national figures. Even different ministries in the same country, let alone in other countries, use different assumptions, which significantly hampers the comparability of data. Assumptions on burden sharing and cost-effectiveness analysis become more difficult or even arbitrary. Thus, one of the first decisions regarding the steps of Kyoto Protocol implementation should be to make the data and assumptions to be used more consistent.

Verification and monitoring mechanisms are of particular importance to implement flexible mechanisms. Without a clear definition, measurement, and inventory of emissions and emissions reduction, binding targets cannot be achieved and flexible mechanisms cannot be realized, as is stressed in various parts of the Kyoto Protocol. Baseline calculations, monitoring, and verification play a crucial role in measuring emissions reductions that result from JI and CDM projects, and thereby ensure that these projects are based on real environment improvements (Jepma and Munasinghe, 1998). Decisions and agreement among parties is urgently needed on firm rules based on accepted methodologies (e.g., benchmarking). The same applies to emissions trading: rules that govern emissions-trading markets must be simple and transparent. Particularly important are rules on the total number of permits available in the market, the permit tenure (their duration), eligibility criteria, the method of initial permit allocation, and the monitoring mechanism.

How to make the verification and enforcement system more effective? Several suggestions in this respect refer to environmental agreements in general (Green, 1993). Different coercive measures, such as trade sanctions and other penalties, may be needed in cases of severe violations. To date, practice shows that sanctions have been used rarely, but when applied they have often been effective (Victor, 1995). A looming challenge is to determine when and how sanctions can be made compatible with international trade rules. Potential conflicts between the sanctions that have sometimes been vital to international environmental co-operation and the free-trade rules that discourage sanctions have not been tested or settled. There are also suggestions to use, for reluctant countries, various compensations for the costs of implementing the treaties (compensation for national reporting testifies to this approach). Other suggestions include bilateral funding programmes. Several funding programmes have been undertaken to support the compilation and reporting of national inventories by the developing countries and countries with economies in transition.

Regimes that elicit the most co-operation have at their disposal more powerful carrots and sticks with which to enforce international obligations. Such tools are increasingly being used, and they work—especially when the sanction is to withdraw financial assistance. The threat of cutting off finance has brought swift compliance. The combination of soft management backed by strict enforcement when necessary has been effective. The most flagrant violations have been deterred and reversed only when strong incentives, including threats of trade sanctions, have been applied (Chayes and Chayes, 1995).

Such market-based mechanisms as GHG emissions trading also may be regarded as a tool to make the UNFCCC implementation easier and less costly for many developed countries. The Kyoto Protocol envisions creating a system of internationally tradable emissions rights that can be used to lower the cost of cutting emissions of GHGs. The international use of market-based incentives is virgin territory. There are no direct historical precedents, and there is much to be learned about the institutions that will be needed to enable the successful international use of market-based systems.

10.3 Local and/or National Sustainable Development Choices and Addressing Climate Change

10.3.1 Introduction

Chapter 1 presented three perspectives on climate change mitigation: cost-effectiveness, equity, and sustainability. The first perspective dominates much of the assessments reviewed in the previous chapters and sections. It is also dominant in the scientific literature on climate change mitigation. As discussed in Sections 1.3 and 1.4, other key perspectives are relevant for mitigation assessment as well: equity and sustainability. This is especially relevant for the assessment of mitigative capacity at local and national levels, and certainly for incorporating climate change mitigation policies into national development agendas.

Decision making related to climate change is a crucial aspect of making decisions about sustainable development, simply because climate change is one of the most important symptoms of "unsustainability". Climate change could undermine economic activities, social welfare, and equity in an unprecedented manner, in particular both intra- and intergenerational equity is likely to be worsened. Now it is widely recognized that global environmental problems and the ability to meet human needs are linked through a set of physical, chemical, and biological processes that have an impact on global hydrological cycles, affect the boundaries and functioning of ecological systems, and accelerate land degradation and desertification.

Despite the close links, climate change and sustainable development have been pursued as largely separate discourses. The sustainable development research community has not generally considered how the impacts of changing climate may affect efforts to develop more sustainable societies. Conversely, methodological and substantive arguments associated with sustainable development are still absent in climate change discourse. It is difficult to generalize about sustainable development policies and choices. Sustainable development implies and requires diversity, flexibility, and innovation. Policy choices are meant to introduce changes in technological patterns of natural resource use, production and consumption, structural changes in the production systems, spatial distribution of population and economic activities, and behavioural patterns. Moreover, the process of integrating and internalising climate change and sustainable development policies into national development agendas requires new problem-solving strategies and decision-making approaches in which uncertainties need to be managed to produce robust choices.

In this section the dual structure of linkages between sustainable development and climate change is discussed. The existence of positive synergistic effects is reviewed, as is how specific strategies, especially those related to lifestyle options and technology-transfer policies, could reinforce potential synergies. Finally, the emergence of new and innovative decision frameworks, in which extended peer community participation is essential to incorporate into the decision process both the plurality of different legitimate perspectives and the management of irreducible uncertainties in knowledge and ethics, is examined.

10.3.2 *Development Choices and the Potential for Synergy*

Chapter 1 provides a concise overview of sustainable development as a context for climate change mitigation policy. As argued there, the concept of sustainable development defies objective interpretation or operational implementation. However, it is precisely the diversity of interpretations that "makes up the biggest advantage of the concept: it is sufficiently rich and flexible to refract the full diversity of human interests, values and aspirations" (Raskin *et al.*, 1998). So nearly everyone can agree that sustainable development is a good thing, and consensus has become possible over broad policy areas in which previously people could not agree. Or, in the words of O'Riordan (1993), "sustainable development may be a chimera. It may mark all kinds of contradictions. It may be ambiguously interpreted by all manners of people for all manners of reasons. But as an ideal it is nowadays as persistent a political concept as are democracy, justice and liberty."

Now, sustainability is perceived as an irreducible, holistic concept in which economic, social, and environmental issues are interdependent dimensions that must be approached in a unified framework. However, the interpretation and valuation of these dimensions give rise to a diversity of approaches. Different disciplines have their own conceptual framework, which translates into different variables, different pathways, and different normative judgements. Economists stress the goal to maximize the net welfare of economic activities, while maintaining or increasing the stock of economic, ecological, and sociocultural assets over time. The social approach tends to highlight questions of inequality and poverty reduction, and environmentalists the questions of natural resource management and ecosystems' resilience (Rotmans, 1997). Apart from the weight placed on each of the critical dimensions, the important conclusion from this ongoing debate is that achieving sustainable economic development, conserving environmental resources, and alleviating poverty and economic injustice are compatible and mutually reinforcing goals in many circumstances.

While the overall literature on sustainable development is very large, the literature that focuses on concrete policies to make operational the concept of sustainable development is, however, much smaller. This asymmetric coverage of the guidance and the operational principles for managing a sustainable development path constitutes a non-negligible barrier to an effective decision-making process, since policymakers lack concise and relevant information that would allow them to assess alternative development choices.

10.3.2.1 *Decision-making Process Related to Sustainable Development*

Actions that steer the course of society and its economic and governmental organizations are largely tasks of making decisions and solving problems. This requires choosing issues that require attention, setting goals, finding or designing suitable courses of action, and evaluating and choosing among alternative actions. The first three of these activities—fixing agendas, setting goals, and designing actions—are usually called problem solving; the last, evaluating and choosing, is usually called decision making (Simon *et al.*, 1986). Except for trivial cases, decision making generally involves complicated processes of setting actions and dynamic factors that begins with the identification of a stimulus for action and ends with a specific commitment to action (Mintzberg, 1994). The complexity of the decision-making process related to sustainable development becomes even more problematic simply because the difference between the present state and a desired state is not clearly perceived, so "we have a better understanding on what is unsustainable rather than what is sustainable" (Fricker, 1998).

Much of the ambiguity arises from the lack of measurements that could provide policymakers with essential information on the alternative choices at stake, on how these choices affect clear and recognizable social, economic, and environmental critical issues. Such measurements could also provide a basis for evaluating policymakers' performance in achieving goals and targets. Management requires measurement and now, as never before, government institutions and the international community are concerned with establishing the means to assess and report on progress towards sustainable development. "If we genuinely embrace sustainable development, we must have some idea if the path we are going on is heading towards it or away from it. There is no way we can know that unless we know what it is we are trying to achieve—i.e. what sustainable development means—and unless we have indicators that tell us whether we are on or off a sustainable development path" (Pearce, 1998). Therefore, indicators are indispensable to make the concept of sustainable development operational. They are particularly useful for decision making because they help (Hardi and Barg, 1997):

- understand what sustainable development means in operational terms (in this sense, measurement and indicators are explanatory tools, which translate the concepts of sustainable development into practical terms);
- make policy choices to move towards sustainable development (measurement indicators create linkages between everyday activities, and sustainability indicators provide a sense of direction for decision makers when they choose among policy alternatives, that is they are planning tools); and
- decide how successful efforts to meet sustainable development goals and objectives have been (in this sense measurement and indicators are performance assessment tools).

The past few years have witnessed a rapidly increasing interest in the construction of sustainable development indicators to assess the significance of sustainability concerns in economic analysis and policy. Different analytical frameworks have been suggested to identify, develop, and communicate indicators of sustainability. Hardi and Barg (1997), in an extensive survey of ongoing work on measuring sustainability, discuss the advantages and limitations of different approaches from the viewpoint of their practical applicability. The main differences among frameworks are (1) the ways and means by which they identify measurable dimensions, and select and group the issues to be measured; and (2) the concepts by which they justify the identification and selection procedure. Some of the major frameworks are briefly summarized below.

One of the most prominent is the Human Development Index developed by the United Nations Development Programme (UNDP) to ranks a country's performance on the criteria of human development, instead of solely the economic performance. Though the index was not developed as a sustainable development index, recent efforts have been made to supplement it with an environmental dimension to encompass explicitly the multiple dimensions of sustainability. Integrated environmental–economic accounting is a framework that is rapidly gaining prominence. The basic idea of this approach is to establish links between the conventional circular production–consumption economic accounting to the natural support system through the extraction of resources in one direction and the discharge of residuals in the other (Tietenberg, 1996). Another framework that is attracting a high level of interest is the multiple capital approach. This approach recognizes that a country's wealth is the combination of economic, environmental, and social capital and these dimensions of capitals should be preserved, enriched, or substituted if consumed. The World Bank's *Measure of the Wealth of Nations* (World Bank, 1997) is the most notable application of this framework. The concept of genuine savings is introduced in the World Bank approach to measure the true rate of saving of a nation after accounting for the depreciation of produced assets, the depletion of natural resources, investments in human capital, and the value of global damages from carbon emissions. Lastly, the Pressure–State–Response framework (OECD,1993; UNCSD, 1996) focuses on the causal relationships between stress-generating human activities, changes in the state of the natural and social environment, and society responses to these changes through environmental, general economic, and sectoral policies.

Different sets of thematic indicators are devised for use at different scales. The broadest scale is the international or global level. In this context, global conventions and protocols, such as the climate, biodiversity, desertification, and ozone agreements, are extremely important. It is becoming increasingly clear that unless specifically tailored indicators are developed and monitored, the implementation of these conventions is not possible. Both the secretariats of the conventions and international agencies are working intensely not only to identify and develop appropriate indicators, but also, most importantly, to give them acceptability in the eyes of the international community (Gallopin, 1997).

At the national level, several important steps to make operational the concept of sustainable development have being undertaken. Different sets of thematic indicators are being used for each of the major issues in national environmental policy, reflecting differences in national endowment, level of development, and cultural traditions, as well as the heterogeneity within countries. The indicators generally cover every aspect of pollution control, nature conservation, resource depletion, social welfare, health, education, employment, waste management, etc.—in short, a compendium of all the components of traditional development goals and conventional policy debate. Hence, factors that distinguish sustainable development from traditional development tend to be submerged under a sea of age-old problems that are made no more readily soluble by bearing the name sustainable development (George, 1999). The point is that current definitions of indicators and the use of terminology are particularly confusing and some clarity and consensus is required about the definition of what an indicator is, as well as in the definition of related concepts such as threshold, index, target, and standard. This consensus cannot be based solely on political agreement; logical and epistemological soundness is also necessary (Gallopin, 1997).

It is recognized (Hardi and Barg, 1997) that much work remains to be done. Some approaches lack causal linkages or they tend to over simplify interlinkages and relations among issues; others focus on the measurement of those segments of sustainable development that can be expressed in monetary terms; in some cases detailed calculations of indicators are highly technical and difficult to handle. Fresh initiatives oriented to capture complex interlinkages in the interactions between human activity and the environment, especially those related to pressure–state–response causalities, have been undertaken in recent years (Meadows, 1998; Bossel, 1999). Undoubtedly, all these efforts are needed to provide decision makers with information and operational criteria to assess current situations and evaluate strategic decisions. Furthermore, these efforts hold the additional promise of treating environmental problems within a framework that the key institutions and agencies in any government will understand.

10.3.2.2   *Technological and Policy Options and Choices*

It is clear from the preceding discussion that governments' commitments to sustainable development require indicators by which decision makers can evaluate their performance in achieving specific goals and targets. Furthermore, such indicators are essential, first to capture the complex interlinkages between the basic building blocks of sustainable development (environment, economic activity, and the social fabric), and second to balance the unavoidable trade-offs between the main policy issues related to each of these blocks (development, equity, and sustainability).

It is difficult to generalize about sustainable development policies and choices. Sustainability implies and requires diversity, flexibility, and innovation. Thus, there cannot be one "rightful" path of sustainable development that leads finally to a blissful state of sustainability (Bossel, 1998). Depending upon differences among individual countries (size, level of industrialization, cultural values, etc.) as well as on the heterogeneity within countries, policy choices are meant to introduce changes in:

- *Technological patterns of natural resource use, production of goods and services, and final consumption.* These encompass individual technological options and choices as well as overall technological systems. Sustainable development on a global scale requires radical technological changes focused on the efficient use of materials and energy for the sufficient coverage of needs, and with minimum impact on the environment, society, and future. This is of particular importance in developing countries, in which a major part of the infrastructure needed can avoid past practices and move more rapidly towards technologies that use resources in a more sustainable way, recycle more wastes and products, and handle residual wastes in a more acceptable manner. As discussed in Chapter 3, the range of opportunities is extensive enough to cope with different development styles and national circumstances, but what is even more important, economic potential increases as result of the continuous process of technological change and innovation. A number of technologies that less than 10 years ago were at the laboratory-prototype stage are now available in the markets. Issues on barriers and opportunities for technology development, transfer, and diffusion at the national level are discussed in Chapter 5 and Section 10.3.3 below.

- *Structural changes in the production system.* Economic growth continues to be a widely pursued objective of most governments and, therefore, policy decisions on development patterns may have direct impacts on both raw material and the energy content of production. Structural changes towards services or a low energy-intensity industrial base may or may not affect the overall level of economic activity, but could have significant impacts on the energy content of goods and services.

- *Spatial distribution patterns of population and economic activities.* Country-wide policies on the geographical distribution of human settlements and productive activities impact on sustainable development at three levels: on the evolution of land uses, on mobility needs and transport requirements, and on the energy requirements. These factors are of utmost relevance for most developing countries, in which spatial distributions of the population and of economic activities are not yet settled. Therefore, these countries are in a position to adopt urban and/or regional planning and industrial policies directed towards a more balanced use of their geographical space.

- *Behavioural patterns that determine the evolution of lifestyles.* Consumption behaviours, and individual choices in general, have a critical influence on sustainable development. After all, sustainability is a global project that requires big and small daily contributions from almost everybody (Bossel, 1998). Personal opportunities and freedom of choices are embedded in cultures and habits, but these are also shaped and supported by the products and services provided by the economic system, as well as by the organization and administration at all levels. Within the boundaries of individual freedom, government policies can discourage unsound consumption styles and encourage more sustainable social behaviour through the adoption of financial incentives (subsidies), disincentives (taxes), legal constraints, and the provision of wider choices of infrastructure and services. This point is elaborated further in Section 10.3.2.3.

The set of specific policies, measures, and instruments to mitigate climate change and consequently promote sustainable development is quite large. These include generic policies oriented to induce changes in the behaviour of economic agents, or control and regulatory measures to achieve specific targets at the sectoral level. A comprehensive discussion of various aspects of different types of policies and measures is presented in Chapter 6. Here it is important to note, first, that sustainability issues cannot be addressed by single isolated measures, but they require a whole set of integrated and mutually reinforced policies. Second, weights assigned to different policies depend on individual countries according to their national circumstances and specific priorities. Third, the cause–effect reaction in the process of policy implementation is not linear. Except in trivial cases, policies tend to disrupt existing patterns, social systems create and respond to changes within themselves through feedback loops, and new patterns emerge as social, economic, and environmental aspects interact in the process of convergence towards the desired goals.

10.3.2.3 Choices and Decisions Related to Lifestyles

There are two reasons why lifestyles are an issue of climate policy. First, consumption patterns are an important factor in climate change since they have become an essential element of lifestyles in developed countries. If, for instance, people changed their preferences from cars to bicycles, this would alleviate climate change and decrease mitigation costs considerably. Second, many promising domains for substantial environmental improvements through technological change also require changes in lifestyle. With respect to traffic, for instance, to reach sustainability beyond that of increases in efficiency requires changes in the modal split and ultimately in urban planning (Deutscher Bundestag, 1994). Yet lifestyles have been subjected to far less systematic investigation than technology (Duchin, 1998, p. 51). In SAR they were not discussed at all.

The concept of lifestyle (*Lebensführung*) was introduced by Weber (1922). Lifestyle denotes a set of basic attitudes, values, and patterns of behaviour that are common to a social group, including patterns of consumption or anticonsumption. It seemed for a while that a change from environmentally less benign to more benign consumption patterns had emerged by itself (Inglehart, 1971, 1977) in the 1970s. What really happened, however, was not a switch from one coherent and dominant set of values to another, but an end of coherence through a pluralization of values (Mitchell, 1983; Reusswig, 1994; Douglas *et al.*, 1998). Current lifestyles reflect this patchwork of values. Some of these, however, are environmentally more benign than others. The idea of promoting transfers from the latter to the former must take into account that lifestyles are not just a matter of behaving this or that way, but are basically an expression of people's self-esteem (see below). Lifestyles, therefore, are based on ideas with respect to the individual's identity. To this extent the issue is not only that individuals need to change their behaviour, but that they need to change themselves. This tends to be underestimated in policy considerations, but must be accounted for when such changes become relevant with respect to climate change. Otherwise discrepancies between people's environmental consciousness and behaviour are deplored but not understood.

10.3.2.3.1 Lifestyles as an Expression of Identity

As far as an individual's behaviour can be explained in terms of economic rationality, changes in lifestyles would seem to be a matter of changing relative prices of commodities by economic policy. In general, however, the rationality of human behaviour is beyond economic rationality. Examples from India as well as from the USA are referred to by Douglas *et al.* (1998), who note that the majority of lifestyles are "not economically rational, but they are still culturally rational". Therefore "the social and cultural dimensions of human needs and wants must be included in the theoretical approaches."

In cultural anthropology human behaviour is interpreted in terms of finding one's place within the social universe by relating oneself to others (not only to the proverbial Joneses next door), that is by setting up distinctions in the community. In doing so commodities are a means of discrimination. They "constitute the visible part of culture as the tip of the iceberg which is the whole of the social processes" (Røpke, 1999). Of course, many goods satisfy needs as well, but they do even this because of their social capacity to make sense in the individual's social context. This explanation of human consumption behaviour—as advanced by Douglas and Isherwood (1979)—seems to be considerably more comprehensive than a purely economic one. However, even if in their account "human beings are conceived of as social, ... they are just as unpleasant pursuers of their own interests as they are in economics" (Røpke, 1999).

Goods, however, make sense not only with respect to others. It has been observed in marketing research that people since the 1960s have gradually passed from buying goods like food, clothing, or housing to basically buying personality, the hard-

ware commodities being part of that (Tomlinson, 1990). In doing so an individual relates to him- or herself rather than to others. Consumers by now are "engaged in an ongoing enterprise of self-creation, ... a 'cultural project' ... the purpose of which is to complete the self" by consumption (McCracken, 1988, p. 88). As far as consumption is responsible for climate change, this means that people in developed countries (and their fellow consumers in less-developed countries) aim for self-realization at the expense of others.

The general rule is that human behaviour expresses one's implicit or explicit self-definition (Meyer-Abich, 1997). Moisander (in press) points out that this project of identity is not limited to the paradigm of the rational, autonomous, and self-certain individual. In the consumer society "The ways in which people relate to their possessions can be seen as reflections of how they view themselves and relate to their social and physical environment" (Dittmar, 1992, p. 125). They express who we are, even if they do so not necessarily in a consistent way. The "social life of things" (Appadurai, 1986) animates all kinds of commodities.

All this seems to imply that any attempt to change lifestyles intentionally is bound to fail. Intercultural experience, however, shows that "the Western conception of the person as a bounded, unique, more or less integrated motivational and cognitive universe ... is ... a rather peculiar idea within the context of the world's cultures" (Geertz, 1979, p. 229). Although in the Western world even the modern state is supposed to have been established by an agreement of independent, or decontextualized individuals, the question "Who am I?" in other cultures is generally answered by reference to the contexts in which one belongs. That is, to dependencies, and not by independence claims with respect to oneself. Western people tend to believe that they are what they are just for themselves, as if everybody had only his or her first name, but even in Europe the idea of individual salvation after death, for instance, did not develop before the 12th century (Ariès, 1977). In contrast, intercultural studies have shown that traditional Asian, African, Latin American, and even Southern European concepts of self indicate an interdependent identity (Cousins 1989; Markus and Kitayama, 1991). This means "that behaviour is seen as context-bound and aimed towards a harmonious fit with the expectations and evaluations of others, who are continuously involved in one's definition of self" (Dittmar, 1992, p. 190). The barrier of consumption-based identity at the expense of others might, therefore, be overcome by contextualizing the Western self in intercultural communication. Section 10.3.2.3.2. gives some indications of how this could be fostered politically.

10.3.2.3.2 Policies and Options for Change

Environmental education
Although political attention is always in a process of change, public awareness of environmental disruption in general and climate change in particular is at a fairly high level in many

developed countries, and has been rising over a long period. This consciousness is generally ahead of the corresponding behaviour, yet (apart from governmental action):

- in many countries citizens initiatives offer bottom-up solutions for alternative consumption patterns (Georg, 1999); and
- environmental behaviour tends to cover consciousness in low-cost situations, that is if the costs and inconvenience are not much higher for the environmentally more benign solution (see Diekmann and Preisendörfer (1992) for Germany and Switzerland).

Interestingly enough, these low-cost limits appear mainly to be a matter of equity—not to pay too much more than one's fellow citizens—because environmental legislation (for everybody) is accepted beyond those limits. Politically, environmental consciousness can be promoted by environmental education. This includes primary schools as well as high schools, adult education, and particularly lecturer's education at the college and university level. Environmental education could be more effective than it has proved so far if it recognized that human behaviour hinges on lifestyle or self-awareness.

Decreasing marginal satisfaction with rising private material consumption

John Stuart Mill's idea that in affluent societies people might prefer other forms of satisfaction to ever-increasing consumption of purchased commodities is not prominent in contemporary economics, but has not completely disappeared from economic thought (see Harrod, 1958; Hirsch, 1977; Xenos, 1989). Now there are indications that the marginal utility of those commodities is steadily decreasing with rising consumption (see Scherhorn, 1994; Inglehart, 1996). For instance, although consumption in the USA has doubled since 1957, it is reported that the average US citizen considers his or her happiness to have decreased since then (UNDP, 1998). Sanne (1998) reports from Sweden that 87% of the people have a car, but 14% of these do not "need" it while only 47% consider it to be "necessary". Similarly, 52% have a dishwasher, but 30% of these do not "need" it and only 12% deem it "necessary". The mismatch between economic consumption and the satisfaction of human needs is shown by the Index of Sustainable Economic Welfare. This ran parallel to GNP up to the 1970s, but rapidly departed after that (Nordhaus and Tobin, 1972; Daly and Cobb, 1989; Jackson and Marks, 1999). If these discrepancies became a political issue the personal relevance of consumption patterns could decrease. Politically, the introduction of a suitable index of welfare and stimulation of public dialogue on the goals of economic action could foster this.

New emphasis on immaterial and common goods

Goods may be either common or private and either material or immaterial, so four combinations arise:

- private material goods (e.g., house or car);
- private immaterial goods (e.g., well-being or creativity);

- common material goods (e.g., shared cars or household appliances); and
- common immaterial goods (e.g., environmental quality or collective actions, such as the liberation of a creek, common attendance of public facilities, *etc.*).

Among the four combinations, so far economic analysis of consumption has been based mainly on only one, private material goods. Since marginal satisfaction with these is decreasing, the neglected combinations have been reconsidered recently (Scherhorn, 1997). This is particularly relevant with respect to climate change, because immaterial and common goods (irrespective of their material basis) are or stimulate social activities that promote the integrity of society. They either foster the natural environment or endanger it much less than private and material goods generally do in terms of production, consumption, and waste management. Material goods are not an end in themselves, so that their real utility is different from their material reality. Politically, public education in consumer's behaviour can promote the awareness that identities can be expressed by a broader plurality of goods than just material and private goods.

New deals in collective action

Climate is a common public good and the debate as to what extent "commons" can be appreciated in market-based economies is ongoing. Much of the discussion originally derives from Olson's (1965) argument that if people were rational egoists in the sense of liberal economics, individual rationality must lead to collective irrationality in large informal groups, because free riders could not be excluded. As non-irrational collective actions exist, human behaviour cannot only be motivated by rational egoism, the problem posed by Olson's analysis was to identify these other motivations. Udéhn (1993) summarized the subsequent discussion comprehensively. The main outcome confirms that human co-operation generally cannot be explained economically, but only by taking into account social or personal commitments. Sen (1976/77, 1985) noted that such commitments can replace economic "rationality", or utility maximization. He also argued, that commitments are related to a person's "identity" (discussed above as the key to lifestyles). These identities, however, are not fixed once and for all but develop through social intercourse. Correspondingly, one of the most consistent results of the debate on collective action is that "co-operation increases dramatically if people are allowed to communicate before being subjected to a social dilemma" (Udéhn, 1993; see Dawes, 1980; Orbell *et al.*, 1984; van de Kragt *et al.*, 1986). This may be expected in market behaviour as well, so that environmental commitments can overcome price incentives. By co-operation in the common interest there is also "reason to believe that appeals to the full set of motivations and behaviours—accompanied by an analysis of bold options—can encourage lifestyle decisions that reduce pressures on the environment" (Duchin, in press).

Environmental legislation

Democratic governments cannot go far beyond public consciousness in environmental legislation. To the extent that people deliberately pay higher prices for environmentally more benign goods, as has been discussed above, governments can increase this threshold step by step.

Creative democracy

Better understanding (as the Olson debate has shown) can lead to the perception of common interests, but such understanding does not necessarily come about by itself. Its promotion is rather an objective of "creative democracy" (Burns and Ueberhorst, 1988). Generally, this is again a matter of education, particularly of political education. "Education and productive employment ... would be worth while policy goals in relation to global climate change" (Douglas *et al.*, 1998). Education implies formal and informal processes of creativity as well as receptivity, so that not only schools and universities are to be addressed here. For instance, in Germany most cities have their special Agenda 21 program, supported by citizens' movements and by an Agenda 21 office in local government. This stimulates learning by doing. The administration is in charge here not simply to implement sustainability locally, but to promote understanding as well as co-operation in the global interest of sustainability. A broad public dialogue can also encourage public confidence in making changes for the advantage of nature, the developing countries, and future generations. For behavioural change, "arguably the most important obstacle is the difficulty of imagining new scripts and removing the obstacles to actually living them" (Duchin, in press). If that public dialogue, therefore, is mainly concerned with such "new scripts", even the discrepancy between environmental consciousness and behaviour might finally disappear, so that people as consumers would no longer lag behind themselves as citizens.

Research needs

Lifestyle research is neglected compared to technology research, even where technological and lifestyle changes are linked. Particularly, nature-saving lifestyles and the general process of self- and world-constitution through goods are enormously understudied (see McCracken, 1988).

10.3.2.4   *Interaction of Climate Policy with other Objectives*

The linkages between the social, economic, environmental, and political dimensions of sustainable development call for policies that can serve multiple objectives, and requires that a balance be struck when objectives conflict. These linkages are often mutually reinforcing in the long run, but may sometimes be contradictory in the short term (OECD, 1999c). In this regard, a critical requirement of sustainable development is a capacity to design policy measures that, without hindering development and remaining consistent with national strategies, could exploit potential synergies between national economic growth objectives and environmentally focused policies. Climate change mitigation strategies offer a clear example of how co-ordinated and harmonized policies can take advantage of the synergies between the implementation of mitigation options and broader objectives.

Over the past years, of the policy options to mitigate climate change, technological options to limit or reduce GHG emissions have received by far the most attention. Chapters 3 and 4 provide a comprehensive review of technologies and practices to mitigate climate change. Energy efficiency improvements (including energy conservation), switches to low carbon-content fuels, use of renewable energy sources, and the introduction of more advanced non-conventional energy technologies are expected to have significant impacts on curbing actual GHG emission tendencies. Similarly, the adoption of new technologies and practices in agriculture and forestry activities, as well as the adoption of clean production processes, could make substantial contributions to the GHG mitigation effort. Depending on the specific context in which they are applied, these options may entail ancillary benefits, and in some cases are worth undertaking whether or not there are climate-related reasons for doing so.

The potential linkages between climate change mitigation issues and economic and social aspects have also brought an important shift in the focus of mitigation analysis literature. From being confined to project-by-project or sector-based approaches, analyses and studies are increasingly concerned with broader policy issues as mechanisms to reduce the increase of GHG emissions. Fresh methodological developments (UNEP, 1998) broaden climate change mitigation policies by incorporating distributional impacts, negative side effects, and the appropriate choice of instruments and institutional constraints, among others. This provides a somewhat different slant on the focus of climate change mitigation policies. More emphasis is now given to exploit mutually reinforcing links among individual actions, to take advantage of the potential interactions of mitigation options with other objectives, and to supplement individual mechanisms with economic instruments of wider scope.

10.3.2.5   *Synergies, Trade-offs, and No Regrets*

The existence of ancillary benefits and synergies in implementing mitigation options has been addressed in a preliminary way in IPCC (1996c). These issues are discussed in detail in Chapters 7, 8 and 9. Some relevant findings are highlighted here. The adoption of more sustainable agricultural practices in Africa (Sokona *et al.*, 1999) illustrates clearly the mutually reinforcing effects of climate change mitigation, environmental protection, and economic benefits. In fact, the introduction or expansion of agroforestry and organic agriculture (i.e., methods that intensify agricultural production while using less input), can improve food security and at the same time reduce GHG emissions. In agroforestry systems, trees are planted to delineate plots of land, and further to fix nitrogen, causing the nutrients lower in the soil to rise up. The trees also prevent soil erosion, supply firewood and animal fodder, and constitute a source of income. Organic farming improves the fertility of the soil through the addition of organic matter. The damage and diseases caused by insects are virtually eliminated through the technique of "growing in corridors" and other holistic methods. Costly inputs are not used at all or are kept to a minimum, and the system is flexible. In addition, these methods restore and maintain carbon levels in the soil. Hence, if practised on a large scale, they could transform soils from carbon sources into carbon sinks.

Energy efficiency improvements and energy conservation are other issues of economic and strategic concern. In developing countries, energy demand (for electricity in particular) continues to grow at a rate that is often hard to keep up with. The adoption of environmentally sound technologies (ESTs) for both energy production and energy consumption would enable these countries to lower the pressure on energy investments, reduce public investments (in some cases by up to one-third (World Bank, 1994)), improve export competitiveness, enlarge energy reserves, and also avoid a large increase in GHG emissions. Thus the alternative energy paths of low-carbon futures in developing countries can be compatible with national objectives. Such paths could prevent energy and/or GDP intensities from following the growth path of the developed world, in which energy demand and GDP elasticity first increased with successive stages of industrialization, but since have sharply decreased.

A large number of similar synergy effects can be found in industry, transportation, and human settlement patterns. For example, more decentralized development patterns based on a stronger role for small- and medium-sized cities can decrease the rural exodus, reduce needs for transportation, and allow the use of modern technologies (biotechnology, solar energy, wind, and small-scale hydropower) to tap the large reserves of natural resources. Building upon the lending experiences of World Bank operations and sector programmes in a number of countries, Warford *et al.* (1996) provide evidence for the positive linkages between economic policies and the environment. Although environmental concerns, and climate change issues in particular, were not explicitly addressed by macroeconomic and sectoral policies, the country cases analyzed show clear synergies between reform policies and environmental improvements. In some cases when adverse side effects do occur, the remedy is not to reverse the reform policies, but rather to introduce specific complementary measures that address the negative effects.

Finally, it is important to underline that for the elements that constitute policies at different levels to operate in a mutually reinforcing manner, the creation of appropriate communication and information channels should be given special attention. The topic of establishing effective and stable flows of communication among different stakeholders is seldom addressed in connection with climate change mitigation. This is mainly because policies related to climate change tend to treat mitigation options as isolated projects, each falling into a narrow area in which potential synergies may be ignored or misunderstood. As result, environmental policies risk resulting poorly structured interventions, with a limited scope of influence, and an overestimated cost-effectiveness (Eskeland and Xie, 1998).

Greater synergies could be achieved if agencies with global and local agendas did business together, through effective linkage mechanisms that allow co-ordination and support in implementing tasks or functions that belong to different subsystems and involve different actors.

10.3.2.6 Links to other Conventions

Awareness of the complex system of interrelated cause-and-effect chains among climate, biodiversity, desertification, water, and forestry has been growing in recent years. Now it is widely recognized that global environmental problems and the ability to meet human needs are linked through a set of physical, chemical, and biological processes. Climate change, for example, alters the global hydrological cycle, affects the boundaries and functioning of ecological systems, and accelerates land degradation and desertification (*Figure 10.5*). These negative impacts in turn reinforce each other through feedback loops, which results in a serious threat to land productivity, food supply, fresh-water availability, and biological diversity, particularly in vulnerable regions (Watson *et al.*, 1998).

Global environmental problems are addressed by a range of individual instruments and conventions—UN Convention on Climate Change, Convention on Biological Diversity, Convention to Combat Desertification, and Forestry Principles. Each of the instruments focuses on a specific issue and has its own defined objectives and commitments, with the exception of Forestry Principles, which has no binding legal agreement. A great deal of interaction exists among the environmental issues that these instruments address, and there is also a significant overlap in the implementation of the instruments. They contain similar requirements concerning (UNDP, 1997):

- common, shared, or co-ordinated governmental and civil institutions to enact the general objectives;
- formulation of strategies and action plans as a framework for country-level implementation;
- collection of data and processing information;
- new and strengthened capacities for both human resources and institutional structures; and
- reporting obligations.

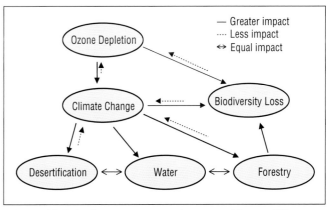

Figure 10.5: *Linkages among environmental issues (Watson et al., 1998).*

Table 10.4: Overlapping requirements of the Parties to the Rio agreements (UNDP, 1997)

	Climate change	Biological diversity	Desertification	Forestry principles
National inventories	Article 4(b)			Principle 12(a)
National and regional action plans	Article 4(b)	"Strategies" Articles 6(a), 6(b)	Articles 9, 10	Principles 3(a), 5(a), 6(b), 8(d), 8(h), 9(c) Article 4(b) and Intergovernmental Panel on Forests (IPF) Proposals for Action
Legislation	Preamble	Article 8(k)	Article 5(e)	Principles 8(f), 13(d)
Research	Article 5	Article 12(b)	Articles 17, 19(b)	Principle 12(a)
Public education	Article 6	Article 13	Articles 5(d), 19	Principle 12(d)
Environmental impact Assessment	Article 4(i), 4(d)	Article 14		Principle 8(h)
Public participation	Article 6(i)(a)(iii)	Article 9	Article 19(4)	Principle 2(d)
Exchange information	Article 7	Article 17	Article 16	Principles 2(c), 11, 12(c)
Training	Article 6	Article 12(a)	Article 19	Principles 3(a), 11, 12(b)
Reports	Article 12	Article 26		
Examine obligations; assess implementation	Article 7(e)	Article 23		Principle 12(a)
Report steps to Conference of the Parties (CoP)	Article 12	Article 26	Article 26	

Table 10.4 summarizes the actions and commitments of the parties under the different instruments. The requirements represent a significant burden, especially on developing countries, in terms of human and financial resources. *Table 10.4* illustrates the wide scope for overlaps between the instruments and the risks that their implementation will lead to a duplicative effort.

It is recognized (UNDP, 1997) that global conventions and instruments "can be more efficiently implemented through a greater understanding of the commonalities and overlaps between them and a co-ordinated and harmonized approach to their implementation at the local, national, and international levels. In other words, creating synergy among the instruments and their requirements". Indeed, linkages between instruments provide opportunities to implement them in a mutual-reinforcing manner, avoiding duplication. At least three clusters of activities are likely to gain advantage from potential synergies in implementing the conventions: the development and strengthening of organizational structures, capacity-building interventions, and data collection and information processing.

Implementing the conventions involves the participation of institutional structures with different responsibilities and concerns, their policy agendas are generally limited in scope, and frequently their immediate objectives diverge. Further, environmental issues are in general broadly diffused through different government agencies, endowed with uneven resources in terms of both authority and material resources. This institutional fragmentation, especially in developing countries, results in a lack of co-ordination and duplication of activities in areas where common organizational procedures, flows of information, and a set of coherent individual institutional actions are required for effective policy actions. Reporting to the respective conferences of the parties, setting up appropriate legislation, and formulation and periodical updating of national action plans are stipulated in the conventions. These need to move towards convergence on overlapping issues, seeking consensus, and agree on policy frameworks within which the ultimate goals can be achieved with greater effectiveness. Therefore, the opportunities for synergies can be exploited by enhancing and strengthening linkage mechanisms between institutions, either at the implementation of specific tasks or functions or through the creation of more formal and permanent links between different actors. The types of linkage mechanisms that might be most successful will depend on institutional, political, and economic factors in each specific case.

Concerning capacity building, the conventions and instruments emphasize the dimension of human resource development as a basic condition for addressing the crucial questions related to the evaluation and implementation of policy options. Here, the potential for synergies is considerable since different instruments focus on enhancing the cross-transfer of professional skills to bridge the gap between academic specialization and the job functions of professionals involved in multidisciplinary issues. A variety of complementary and overlapping areas exists in seminars, courses, and workshops on planning tools and methods, policy analysis, and shared fields, reflecting the training needs under each convention.

Data collection and management, analysis and processing of the information, and dissemination are the core of the conventions and instruments. This requires information systems to be set up so that information can be transposed into proper database structures to enable its archiving, retrieval, expansion, and application. Even though each convention addresses a specific set of problems, considerable overlap exists in the data requirements. Information on land uses, forestry, agriculture, infrastructure, and population, among other areas, is common data needed across the instruments. Taking advantage of synergy in information systems avoids redundancy and dispersion in data collection and management, especially in developing countries where the technical competence and expenditure required are beyond the capacity of local agencies.

At the international level the institutions responsible for the various instruments can also support synergy at the national level by co-ordinating among themselves and helping to ensure that participant countries are not burdened by conflicting directives or timing in reporting requirements (UNDP, 1997). Moreover, the scope for linkages among international bodies of scientific expertise, established under different conventions, is evident.

10.3.3 Technology Transfer

Technology transfer has broadly been discussed in the IPCC *Special Report on Technology Transfer* (IPCC, 2000a). The report provides a framework for analysis of the complex and multifaceted nature of the technology transfer process, emphasizing the sustainable development perspective. It examines broad trends of technology transfer in recent years, explores the international political context, discusses policy tools for overcoming key barriers and creating enabling environments, and provides an overview of financing and partnerships. The report also includes sectoral perspectives on the transfer of adaptation and mitigation technologies. These perspectives are illustrated by a wide variety of case studies. This section highlights the main findings of the IPCC report, especially those

issues related to the role that the main stakeholders must play in the formulation and implementation of policies that facilitate technology transfer.

10.3.3.1 The View of Technology Transfer

The effectiveness of measures to mitigate or adapt to climate change depends to a great extent on technological innovation and the diffusion of technologies. The transfer and/or diffusion of ESTs across and within countries is now considered a major element of global strategies to achieve climate stabilization and support sustainable development. At the same time, it is recognized that transferred technologies must meet the needs and priorities of specific local circumstances.

The term technology transfer is interpreted by some as a one-time transaction that maintains the dependency of the recipient. Some analysts therefore prefer the notion of technology co-operation or technology diffusion, which is seen by them as reflecting a process of technical change brought about by dispersed and uncoordinated decisions over time. Others still may see technology transfer as a two-way learning process that might more appropriately be called technology communication. According to the definition used by IPCC (2000a), "technology transfer encompasses the broad set of processes that cover the flows of knowledge, experience, and equipment for mitigating and adapting to climate change among different stakeholders. These include governments, international organizations, private sector entities, financial institutions, NGOs and research and/or education institutions. It comprises the process of learning to understand, utilize, and replicate the technology, including the capacity to choose it, adapt it to local conditions, and integrate it with indigenous technologies." Technology transfer will therefore be used as a broad term including all aspects mentioned above.

While technology transfer is now a common feature of all sectors of human activity, some features are unique to the area of climate change, including:

- scale, both in terms of geography, which may involve all countries of the world, and the number of technologies, which could easily run into the thousands;
- number of persons that might benefit from the success of these efforts, since the whole world is expected to be the beneficiary; and
- payback periods for the R&D expenditures, which may be too long to be of interest to the private sector.

These features determine technology transfer activities that could be evaluated at several levels—international, macro- or national, sector-specific, and project-specific levels—and that could follow different pathways according the interactions among the stakeholders involved in the transfer process. Each pathway represents different types of flows of knowledge, moneys, goods, and services among different sets of stakeholders. Each one has very different implications for the learning that occurs and, ultimately, the degree of technology-as-knowledge transfer that takes places beyond the simple hardware transfers.

10.3.3.2   *Technology Transfer: International Aspects*

10.3.3.2.1 International Technology Transfer Policy

The legal, economic, and political issues that surround technology transfer have invariably found their place in every international agreement that has anything to do with social, economic, and environmental topics. The Montreal Protocol on Substances that Deplete the Ozone Layer includes several provisions for technology transfer. The Multilateral Fund under the Protocol is a key factor that has facilitated technology transfer to developing countries to comply with the Protocol commitments. Several of the Rio Declaration principles address requirements for states to exchange scientific and technological knowledge and to promote a supportive and open international economic system for the development, adaptation diffusion, and transfer of ESTs. Chapter 34 of Agenda 21, devoted to technology transfer, supports these principles with more detailed proposals for action. The extent to which these proposals have been implemented varies, and debate continues within the UN Commission on Sustainable Development. The Convention on Biological Diversity specifically addresses access to and transfer of technology relevant to the conservation and the sustainable use of biological diversity, including biotechnology.

The UNFCCC requires the parties to the Convention "to promote and cooperate in the development, application, diffusion, including transfer, of technologies, practices and processes that control, reduce or prevent anthropogenic emissions of greenhouse gases" (Article 4.1.c). The Convention calls developed country parties to take all practical steps to promote, facilitate, and finance, as appropriate, the transfer of, as well as access to, ESTs to other parties, particularly to developing country parties. The importance of technology transfer is also recognized in the Kyoto Protocol of the UNFCC. As further discussed in Section 10.3.3.2.2, flexible mechanisms under the Kyoto Protocol provide strong incentives for technology transfer.

Foreign direct investment (FDI) has proved an effective channel of international technology transfer. Levels of FDI, commercial lending and equity investment all increased dramatically during the 1990s, to the point where official development assistance (ODA) became less than one quarter of the total foreign finance available to developing countries by mid-decade (IPCC, 2000a). The growing role of FDI in technology transfer is supported by various domestic and international developments, including the liberalization of markets, development of stronger domestic legal and financial systems, and tariff reductions under the Uruguay Round of the General Agreement on Tariffs and Trade (GATT). In this context, issues related to intellectual property rights, in particular the Agreement on Trade Related Aspects of Intellectual Property, will play a prominent role in shaping both the flows and intensity of technology transfer in the future. To function effectively, trade and investment require proper enabling frameworks. These include a stable economic system, transparent and equitable legal and/or financial structures, sound environmental laws, uniform non-discriminatory enforcement procedures, respect for local culture, safe and secure environment for workers and/or contractors, and removal of unnecessary barriers to the movement of personnel and materials.

Beyond the issues concerning property rights and the process of opening national economies, changes in the features of new technologies (systematic character, the important role of users, increasing knowledge intensity) have significant implications for technology transfer policy.[19] In particular, many ESTs are still in the early stages of their development and have a comparatively short track record, so private actors may be unwilling to accept the extra risks or costs involved in utilizing new technologies. In general, the spread of proven ESTs that should diffuse through commercial transactions may be limited because of existing barriers. Barriers to the transfer of ESTs arise at each stage of the process, as discussed in detail in Chapter 5. These vary according to the specific context, for example from sector to sector, and can manifest themselves differently in developed countries, developing countries, and countries with economies in transition. Some of the key barriers are summarized in *Box 10.2*. For the success of technology transfer, the parties concerned need to make common efforts to overcome the barriers and create opportunities for the transfer and/or diffusion of technology (Verhoosel, 1997). At present there is no easy answer for overcoming barriers. Measures to be taken depend on the specific barriers and the interests of different stakeholders and are discussed in Chapters 5 and 6.

To improve the enabling environment for technology transfer and diffusion, governments could consider a number of actions such as:

- Enact measures, including regulations, taxes, codes, standards, and removal of subsidies, to internalize the full environmental and social costs and reduce unfair commercial risks.
- Reform legal systems. Uncertain, slow, and expensive enforcement of contracts by national courts or international arbitration, and insecure property rights can discourage investment.
- Reform administrative law to reduce regulatory risk and ensure that public regulation is acceptable to stakeholders and subject to independent review.
- Protect intellectual property rights and licenses, and ensure the active use of patents.
- Encourage financial reforms, competitive national capital markets, and international capita flows that support FDI. Governments can expand financial lending for ESTs through regulation that allows the design of spe-

---

[19] An overview of international technology transfer mechanisms is given in Radosevic (1999).

Box 10.2. Main Barriers to Transfer of Environmentally Sound Technologies (IPCC, 2000a)

- Lack of full-cost pricing, which internalizes environmental and social cost.
- Poor macroeconomic conditions, which include underdeveloped financial sector, trade barriers (high tariffs and/or quantity controls), high or uncertain inflation or interest rates, uncertain stability of tax and tariff policies, investment risk.
- Risk of change from existing technology to application of new technology, especially risk aversion and business practices in financial institutions.
- Lack of data, information, knowledge, and awareness regarding the availability, characteristics, costs, and benefits of ESTs, especially in the case of "emerging" technologies.
- Lack of markets for ESTs because of lack of confidence in economic or technical viability.
- High transaction costs of obtaining information, negotiating, contracting, and enforcing contracts.
- Lack of vision about and understanding of local needs and demands.
- Low private sector involvement because of lack of access to capital, in particular inadequate financial strength of smaller firms to manufacture, purchase, and install new ESTs.
- Insufficient human and institutional capabilities.
- Lack of supporting legal institutions and frameworks, including codes and standards for the evaluation and implementation of ESTs.
- Low, often subsidized, conventional energy prices that result in disincentives to adopt energy-saving measures and renewable energy technology.

cialized credit instruments, capital pools, and energy service companies.
- Simplify and make transparent program and project approval procedures and public procurement requirements.
- Promote competitive markets, liberalize trade policies, and make investment policies transparent.
- Encourage national markets for ESTs to facilitate economies of scale and other cost-reducing practices.

10.3.3.2.2 Mechanisms for Technology Transfer

Technology intermediaries are needed to reduce barriers to technology transfer associated with information, management, technology, and financing. These operate between users and suppliers of technology and help to create links within networks and systems (through bridging between institutions), and encourage interaction between the system. They also assist with undertaking research, evaluation, and dissemination tasks. ODA programs mechanisms for technology transfer under the UNFCCC, and multilateral development banks (MDBs) can play a significant role in strengthening national institutional and organizational structures for technological development and innovation.

The 1990s have seen broad changes in the types and magnitude of international financial flows that drive technology transfer (IPCC, 2000a; French, 1998). ODA decreased and fell below the committed levels (OECD, 1999a). However, it plays an important role in technology transfer, especially for the sectors and areas that are commercially less attractive to FDI, such as forestry, public health, agriculture, and coastal zone management (OECD, 1997). Moreover, ODA is still critical for the poorest countries, particularly when it is aimed at development capacities to acquire, adapt, and use foreign technologies.

MDBs have become aware of the role they can play in helping to mobilize private capital to meet the needs of sustainable development and the environment, and of the potential to use financial innovation to encourage environmental projects and initiatives. The World Bank has developed a number of initiatives with potential support for environmental technology transfer. An important new initiative is the Carbon Investment Fund, which will provide additional finance for CO_2 mitigating projects in return for carbon offsets. Other MDBs, such as the regional development banks in Africa, Asia, Latin American, and the Caribbean region, can also play an important role in developing systematic approaches to create enabling environments for technology transfer, including South–South technology transfer.

The Global Environment Facility (GEF), the financial mechanisms of the UNFCCC, is a key multilateral institution for the transfer of ESTs (Anderson, 1997). Although this is of a modest scale in terms of total investment and mainstream investment flows, GEF-supported projects are especially significant for renewable energy technologies, such as wind, solar thermal, solar photovoltaic home systems, and geothermal[20].

The Kyoto Protocol mechanisms, in particular the project based Joint Implementation (JI) and Clean Development Mechanism (CDM), can increase technology transfer. CDM and JI can provide financial incentives for ESTs and influence technology choice. As voluntary mechanisms, they require cooperation among developed and between developed and developing country parties, as well as between governments, private

---

[20] From 1991 to 1998, GEF approved grants of in total US$610 million for 61 energy efficiency and renewable energy projects in 38 countries. An additional US$180 million in grants has been approved for climate change projects (IPCC, 2000a).

sector entities, and community organizations. Project-based crediting can lead to tangible investments and to the development of local capacity to maintain the performance of these investments. These investments could incrementally assist developing countries to achieve multiple sustainable development objectives, such as economic development, improvement of local environmental quality, minimization of risk to human health by local pollutants, and reduction of GHGs. Much about the design and governance of the CDM, however, remains to be resolved. There is a need to design simple, unambiguous rules that ensure environmental performance in the context of sustainable development, while also favouring investment. The multilateral oversight and governance provisions of the CDM, and the project-basis transactions, will raise the transaction cost of investment in CDM projects as compared to the cost of mitigation through other means. Chapter 6 discusses these aspects in more detail.

10.3.3.3   *Technology Transfer: National Aspects*

10.3.3.3.1 Research and Development: Supply Side

Research and development (R&D) is a process of forming new ideas and transforming them into products and services. Technology capacity at both the assessment and replication stages of the technology transfer process have to be underpinned by R&D. Central to this process are national systems of innovation and international co-operation between public research institutions and private-sector entities in R&D. Governments have been investing for three decades in R&D for ESTs in the energy sector. There may be a case for seeing whether results from this process have been used and disseminated sufficiently. Developing countries' R&D efforts are often adaptive, following externally developed technology, which suggests the need for additional resources to develop indigenous innovative capacity. The activities at all stages of technological development and implementation are necessary to attain short-term and long-term technical results (Elliot and Pye, 1998). In the field of climate change, R&D of mitigation and adaptation technologies can reduce the costs of implementation of mitigation and adaptation measures, and provides decision makers with viable alternatives in the formulation of response strategies to climate change.

The process of technological innovation includes not only research and development, but also innovation in the design of products, technological processes, and manufacturing, and innovation in management and market exploration. The private sector has played an important role in the development of energy-efficiency technologies, and is becoming increasingly active in developing renewable-energy technologies (Forsyth, 1999). The bulk of R&D and technology transfer in the energy sector is mainly driven by oil, natural gas, and power supply companies. Other energy supply technologies, such as coal, nuclear, and renewable sources, are often dependent on governments to preserve or increase their presence in the market. Governments can play an important role in R&D as follows:

- establish a National System of Innovation—institutional and organizational structures to support technological development and innovation[21];
- build and strengthen scientific and technical educational institutions and modify the form or operation of technology networks;
- guide the advancement in science and technology and the direction of investment through industrial and technological policies, and provide suggestions and consultation to enterprises;
- encourage enterprises to increase investment in R&D of ESTs through effective policies and create a favourable environment for the innovative activities of enterprises;
- make efforts to increase R&D investments through the governmental budget to accelerate the formation of diversified investment and financing systems, including different kinds of loans;
- give policy support to R&D to encourage the development of innovative technologies and products in the field of climate change, including preferential tax policies, import and export tax policies, and government procurement policies;
- develop modalities for the transfer of public owned or supported ESTs;
- provide funds for licensing of patented ESTs entities to encourage the private-sector to transfer ESTs they own to developing countries.

10.3.3.3.2 Technology Transfer: Demand Side

With the tendency towards globalization and closer integration of most countries in the world economy, countries generally have two sources of technologies: they can either develop their own technologies or procure technologies from other countries, adopting and developing them to fit the specific circumstances (Ding, 1998). When technology transfer is carried out between developed and developing countries, it is important to build up a mutual understanding. Developing countries not only need technologies relevant to climate change mitigation, but also those that are able to generate economic benefits and promote social and economic development, recognizing differences in social aspects (such as tradition and customs).

Besides the common problems of climate change, many developing countries are facing the challenges of poverty eradication and economic development. Technology development and technology transfer are effective mechanisms in alleviating these problems. The introduction of technologies will help to reduce the cost and shorten the time of technological development. For developing countries, the transfer and diffusion of appropriate technologies plays a key role in taking measures to mitigate and adapt to climate change, while pursuing the goal of sustainable development (Xu, 1998).

---

[21] National Systems of Innovation also have a broader role in creating an enabling environment for the transfer of ESTs (see IPCC, 2000a).

The scope of technology transfer should not be limited to the technology itself. The enabling environment of the technology should also be included. If technology transfer is to bring about economic and social benefits, local capacities to handle, operate, replicate, and improve the technology on a continuous basis must be taken into account, as well as the institutional and organizational circumstances. There is little developmental benefit in a technological initiative that remains confined within a very narrow sphere of influence with scarce possibilities of replication on a significant scale and without decreasing reliance on assistance from abroad. Technology transfer needs to build up strong links between:

- its operational context (tools, machines, equipment, processes);
- the organizational environment (management organization, product operation, and technology infrastructure); and
- knowledge (experience, skills, vocational training, advanced training).

In many cases of technology transfer, much attention was paid to the accelerated introduction of technologies and a high cost was paid to procure expensive technological facilities. However, less attention was paid to the digestion, absorption, and innovation of the introduced technologies, in other words, to the supportive base for technological development. In some cases, technology transfer to increase energy efficiencies achieved a one-off step of efficiency improvement, but disregarded that reversion to previous patterns of efficiency must be prevented and failed to ensure the basis for a continuing and self-sustained path of improving efficiency in the future. In conclusion, the process of development, application, and dissemination of technologies, and their accelerated commercialization, is not simply a technical programme. It concerns a wide range of issues, including policy formulation, personnel training, fund raising, and standardization; in general, an array of interlinked factors are related to the sustainability and replication of technological innovation. (Yang and Xu, 1998; Zhang, 1998; Xu, 1999).

10.3.3.3.3 Capacity Building

Human and institutional capacity building is required at all stages in the process of technology transfer. Much of the focus on capacity building has been on enhancing scientific and technical skills, capabilities, and institutions in developing countries, as a pre-condition for assessing, adapting, managing, and developing technologies (UNCTAD, 1995,1996).

Successful technology transfer depends to a great extent on the quality of human resources of the recipients. In general, developing countries lack qualified technical personnel and institutions. Therefore, it is important to build up local competence and an infrastructure that can adapt and "internalize" technology into the local specific conditions and traditions. The potential users of new technologies should learn to use the technologies. The process of learning includes demonstration, training,

and technical assistance. The local research community could be strengthened so that it can absorb the new science and technology into the local cultural and social fabric. Together with technology transfer, assistance could be provided to train technical personnel.

Information can play a guiding role in technology transfer. Decision making on technology transfer requires information on the current status of technology, research, and development, technical and economic evaluations of the technologies, and knowledge of the commercial sources of technologies. The establishment of information systems is an important component of institutional capacity building. These systems must also include information on technology assessment services, consultants, financial institutions, lawyers and accountants, and technical experts. Local government, industrial associations, NGOs, and communities can work together in the development of these kinds of systems.

In general, small and medium-size enterprises (SMEs) lack the capability and resources to access all the information necessary to make appropriate decisions. Technical support centres could be set up to provide technical assistance to the SMEs. Several developed countries and international organizations have already developed schemes of this type, with significant success. Electronic information networks can accelerate the exchange of information and therefore should be used more extensively. For example, the Greenhouse Gas Technology Information Exchange (GREENTIE), an initiative of the IEA and the government of Japan, aims to combine voluntary action by governments with incentives for private dissemination of technological information (Forsyth, 1998).

10.3.4 Decision-making Frameworks for Sustainable Development and Climate Change

Decision making related to climate change is a crucial part of making decisions about sustainable development simply because climate change is one the most important symptoms of "unsustainability". Indeed, global warming poses a significant potential threat to future development activities and the economic well being of a large number of human beings. Climate change could also undermine social welfare and equity in an unprecedented manner. In particular, both intra- and intergenerational equities are likely to be worsened. Lastly, increasing anthropogenic emissions and accumulations of GHGs might significantly perturb a critical global subsystem—the atmosphere. Policymakers routinely make macro-level decisions that influence both climate change mitigation and adaptation, but are of a broader scope than strategies specifically related to climate change. These decisions relate to economic development, environmental sustainability, and social equity issues—which invariably have a much higher priority in national agendas than does climate change (Munasinghe, 2000). In this context, economic–environmental–social interactions could be identified and analyzed and effective sustainable development policies formulated by linking and articulating them explicitly with climate change policies.

10.3.4.1 Forms of Decision-making

Despite the close links, climate change and sustainable development have been pursued as largely separate discourses. "The sustainable development research community has not generally considered how the impacts of a changing climate may affect efforts to develop more sustainable societies. Global warming is acknowledged as a problem, but is typically leaped over in an effort to push governments towards specific policy responses. Conversely, the concept of sustainable development and the methodological and substantive arguments associated with it are notably absent in the climate change literature" (Cohen *et al.*, 1998). Despite the strong synergies between policies oriented to climate change and national development objectives, different ways of thinking in approaching the two problems lead to different social practices and decision-making procedures, which makes it difficult to establish strong working linkages between them.

The main point here is that climate change and sustainable development are rooted in very different disciplines, which results in distinct conceptual frameworks and policy assessments. The dominant natural science approach to climate change has constructed it as an environmental problem, which can be identified and managed objectively by means of scientific rationality. This formulation has resulted in a number of "value neutral" decision-making approaches and methods that represent only the technical dimension of a much more complex set of decision-making problems (Jaeger *et al.*, 1998). These are not especially helpful in deciding how to respond politically, because they ignore the human dimensions of the problem and the difficult and locally differentiated politics of responding to it. In contrast, the human-centred sustainable development approach to environmental problems is more politically and geographically sensitive, but it is analytically vague. This makes it difficult to define or implement in practice (Cohen *et al.*, 1998).

This distinction does not simply apply to the formalities, but has rather practical consequences on the systems of rules, decision-making procedures, social practices, and role of stakeholders—the institutional arrangements that determine the processes of problem solving and decision making. Different disciplinary perspectives of climate change and sustainable development can be associated with two major streams of institutional arrangements models, characterized as collective-action and social-practice models (Clark, 1998). A collective-action model, which reflects the mainstream thinking of climate policy literature, embodies the rational actor paradigm. Social actors are coherent identities that possess well-defined preference structures and seek to maximize payoffs through a process of weighting the benefits and costs associated with alternative choices in situations that involve strategic interaction. According to this view, "climate change can be decom-

posed into a conceptually simple (if still practically challenging) problem, for which a rational solution can be constructed and implemented within the existing framework of political power and technical expertise" (Jaeger *et al.*, 1998). The role of government institutions, as the relevant actors in the decision-making process, is to co-ordinate regulation through policy instruments to prevent individualistic behaviour from producing outcomes that are worse for all participants than the feasible alternatives under optimal, rational choices (Clark, 1998; Young, 1998).

By contrast, sustainable development is closer to the idea of institutions as arrangements that engender patterned practices, which play a role in shaping the identity of participants and feature the articulation of normative discourses, the emergence of informal communities, and the encouragement of social learning. This category of social-practice institutional arrangements (Young, 1998) directs attention to processes through which actors become enmeshed in complex social practices. These subsequently influence their behaviour through the de facto engagement in belief systems and normative preferences, rather than through conscious decisions about compliance with regulatory rules. From this point of view, control, legitimacy, credibility, and appropriate decision-making processes become crucial issues in the construction of sustainable development practices.

With such dissimilar discourses it is not surprising that climate change and sustainable development have been pursued as two separate agendas for the purposes of policy formulation and action. Moreover, while these issues have achieved a high level of public interest and visibility, climate change is the issue that so far has formally been accepted for serious consideration in government agendas. Sustainable development has not yet been able to translate its ideals into concrete objectives for problem solving and decision making. In this context, scientists are confronted with the urgent task of "reforming the relationship between science research and policymaking" (Rayner and Malone, 1998b). This task implies a twofold effort. First, the sustainable development discourse needs greater analytical and intellectual rigor (methods, indicators, etc.) to make the concept advance from theory to practice. Second, the climate change discourse needs to be aware of both the restrictive set of assumptions that underlie the tools and methods applied in the analysis, and the social and political implications of the scientific constructions of climate change (Cohen *et al.*, 1998).

Over recent years a good deal of analytical work has addressed the problem in both directions. Various approaches have been explored to transcend the limits of the standard views of rationality in dealing with issues of uncertainty, complexity, and the contextual influences of human valuation and decision making. Jaeger *et al.* (1998) provide a useful discussion on the various attempts to create new interfaces between scientific rationality and a pluralistic society. As these authors remark, "a common theme emerges: the emphasis on cultural and social realities, which cannot be reduced to individual choices." Now, it is recognized that sustainable development and climate change deal not only with complex and poorly defined goals, but also that the values at stake are plural and conflicting, and even the very nature of each problem is successively transformed in the course of exploration. Problems are no longer seen as external constraints to the social progress, but as issues inherent to the structure itself of societies, so even the idea itself of finding a solution no longer applies, since problems are not solved but managed (Tognetti, 1999). Therefore, the process of integrating and internalizing climate change and sustainable development into national agendas requires new problem-solving strategies in which decision making takes on a new complexion. It becomes a task of finding a partially undiscovered landscape rather than charting a scientific course to an end point. Decisions must be made about which of the systemic possibilities to promote and which to discourage, how to deal with uncertainties, and what risks to take considering irreversible changes and potential bifurcation points. These decisions must be informed by science, but in the end they are an expression of human ethics and preferences, and of the sociopolitical context in which they are made (Kay *et al.*, 1999).

In a seminal work, Funtowicz and Ravetz (1993, 1994) provide a fruitful approach to problem-solving and decision-making strategies in terms of uncertainties in knowledge and complexities in ethics. Contrary to the traditional view that science is value-free, these authors claim that in any real problem of environmental management, scientific facts and sociopolitical values are inseparable dimensions. According to the degree of both uncertainty and decision stakes, Funtowicz and Ravetz distinguish three sorts of problem-solving strategies: applied science, professional consultancy, and post-normal science.

When system uncertainties and decision stakes are both low, applied science is able to manage problems by means of standard routines and procedures. Here, problems are regarded as objective states that exist independently of values and perceptions. The existence of one best solution is assumed, and the task of analysts and decision makers is to work out the optimal strategies by searching for the maximum utility among a number of options (Mintzberg, 1994). Since consensus on the problem definition and values at stake are assumed, the proposed solutions speak for themselves, and the implementation just requires their translation from the technical language of scientists to the pragmatic language of policymakers.

Professional consultancy deals with problems for which uncertainties cannot be managed at a technical level, because of the more complex aspects of the problem and because the decision stakes are also more complex, involving both stakeholders and natural systems. In response to the public demand for more inclusive processes, problems are treated as risks, and if techniques and procedures from the applied science are required, judgement becomes a key element in the decision process.

A third sort of problem-solving strategy emerges when uncertainties are of either an epistemological or ethical nature, or when the decision stakes reflect conflicting purposes among stakeholders. In this case, the "puzzle-solving exercises of normal science" are no longer applicable to the resolution of policy issues of risks and the environment. What is required here is an approach that allows:

- management of irreducible uncertainties in knowledge and ethics;
- plurality of different legitimate perspectives; and
- extension of the peer community to all those with a stake in the dialogue of the issue.

These are the elements of an emerging type of problem-solving and decision making known as post-normal science (Funtowicz and Ravetz, 1993).

The main contribution of post-normal science to policy analysis is to assert that when science is applied to policy issues, it cannot provide certainty for policy recommendations; and the conflicting values in any decision process cannot be ignored, even in the problem-solving work itself (Jaeger *et al.*, 1998; Rayner and Malone, 1998b). The epistemological analysis of the approach shows that the insertion of technocratic discourse into a broader social discourse and participation is not only possible, but also necessary to improve the substantive quality of decisions. At the practical level, the post-normal concept lays out a DMF for articulating new institutional arrangements in which power sharing between conventional decision-making agents and extended peer communities is a key element (De Marchi and Ravetz, 1999; Healy, 1999). This is not merely motivated by a "democratic sentiment", but by the conviction that the resultant decisions, although not necessarily economically the most efficient, will turn out to be better decisions, judged by a broad range of competing social criteria (Rayner and Malone, 2000).

10.3.4.2 Public and Private Decision Making

Decision analysis largely addresses both sustainable development and climate change at their most aggregated level as government policy. The implicit assumption of the government as a single decision maker has resulted in scant attention (even neglect) being paid to how government policies and decisions are connected to lower hierarchical levels at which policies must be implemented. This issue raises two interconnected questions: the first concerns the view of the government as a homogeneous and unitary decision-making actor, and the second relates to the links of government policies to everyday decisions by concerned stakeholders.

Regarding the first question, government structures involved in the decision-making process vary considerably among countries. Some governments have established interministerial committees to co-ordinate sustainable development policies, including climate change strategies, while others have assigned responsibilities to more formal permanent commissions or even to a ministry created specifically to handle sustainable development policies. With many different institutions involved in sustainable development issues, considerable confusion often exists regarding who has the responsibility for policy formulation, where the authority for making day-to-day decisions resides within the government, and how channels of communication and decision making should be achieved between the different actors involved. Institutional articulation remains one of the critical factors affecting the consolidation of an effective decision-making process related to sustainable development. Even if there exist rules and regulations that assign competence, tasks, and responsibilities among the institutions involved, a considerable gap exists between what might be desirable and what, for the most part, is practised.

Concerning the interface between macro-policies and the real decision-making levels, the situation is no more encouraging. It is true that sustainable development and climate change are primarily the responsibility of the government system simply because national economy-wide policies have widespread effects on the regulation of societal processes. As discussed above (Section 10.3.2.2), government policies shape structural changes in the production systems, affect the spatial distribution patterns of population and economic activities, influence behavioural patterns of the population, and regulate interaction with the environment and resource-base system. However, as recognized (Jaeger *et al.*, 1998; Rayner and Malone, 2000) all too often, especially in developing countries, the levers of state power have a small impact on or even no connection with the local level, at which policies must be implemented by ordinary people living in face-to-face communities.

Recent tendencies at different levels are emerging as appropriate responses to increase the legitimacy and competence of local communities, associations, movements, and NGOs in the public decision-making process. Increasing concern of local populations directly affected by environmental problems, together with current tendencies towards decentralization and weakening of authoritarian practices, especially in many developing countries, have opened a new political scenario for a more active participation of civil society in the public policy formulation and decision process. Present trends towards reassigning the setting of rules from government to the markets, together with the process of transferring the provision of services from the public sector to private ownership, have redefined the roles of social stakeholders. Within this context, sustainable development policies are no longer seen as a hierarchically, government-controlled chain of commands, but as an open process in which the principles of "good governance"—transparency, participation, pluralism, and accountability—are becoming the key elements of the decision-making process.

Public involvement in decision making is not a completely new phenomenon. For instance, traditional participatory mechanisms, such as public hearings, notice and comment procedures, and advisory committees, have been practised extensively by US government agencies (Beierle, 1998). However, it is only lately that participatory forms of decision making have acquired legitimacy and prominence in environmental issues, mainly because of their complexity, uncertainty, large temporal and spatial scales, and irreversibility (van den Hove, 2000). As discussed in Section 10.3.4.3, innovative mechanisms such as regulatory negotiations, mediations, stakeholder consultation, collaborative decision-making techniques, community-based methods, and others, are currently being applied by governments, institutions, and local administrations, as well as by intergovernmental organizations. Rayner and Malone (2000) conclude that, whether policy innovation and behavioural change are led locally or nationally, "they will be marked by a process of institutional learning that either moves presently peripheral concerns about climate change to the core of people's daily concerns or, at least, palpably and convincingly links climate policies to these everyday concerns."

10.3.4.3 Participatory Forms of Decision-making

A substantial body of work on participatory approaches to the decision making process has emerged in the 1990s. Theoretical roots of this resurgence originate in the Frankfurt School of Critical Theory and, more concretely, in Habermas' ideas of discursive ethics (Habermas, 1979; O'Hara, 1996). Discursive ethics views rationality as a social construction, inseparably linked to and informed by the human experience of a social, cultural, and ecological life world, which constitutes the context of human experience. It presupposes no norms other than the acceptance of a reasoned, reflective, and practical potential for discourse: that is, the mutual recognition and acceptance of others as "response-able" subjects (O'Hara, 1996). The main contribution of discursive ethics is to offer a conceptual framework for making visible the hidden normative assumptions, behaviours, and motivations that influence de facto decision-making and valuation processes.

Despite the resurgence of interest in public participation, no widely accepted consistent method has emerged to evaluate the success of individual processes or the desirability of many participatory methods. Diverse perspectives together with country-specific conditions favour different forms of participation. In most developed societies, participatory discourse has been motivated by public concerns on the rigid and constraining forms of technocratic decision-making practices, and their institutionalized forms of bureaucracy and social control. Following Beierle (1998), divergent models of the role of civil society in decision making arise from differences of view on the nature of democracy. A managerial perspective acknowledges public preferences as vital to the managerial role of democratic institutions in identifying and pursuing the common good, but public participation in decision making conveys the threat of self-interested strategic behaviour. Under a pluralistic perspective there is no objective "common good", but a relative common good that arises out of the free deliberation and negotiation among organized interest groups. The role of the government is arbitration among these groups. Lastly, a popular perspective calls for the direct participation of citizens as a mechanism to instil democratic values in citizens and strengthen the body politic. Each view provides different forms of participation: the managerial perspective may favour information flow mechanisms, such as surveys or the provision of right-to-know information; the pluralist perspective prefers stakeholder mediation; and the popular perspective favours citizen advisory groups (Beierle, 1998).

Participatory forms in decision making carry a distinct connotation in developing countries. They are rooted in the idea of grassroots participation, promoted by international development aid agencies since 1990 (UNDP, 1992). The concept is far from new, but in recent years it has received a different connotation. Before, participation was considered as an extension of partnership between governmental institutions and development institutions at the operational level. The scheme was oriented mainly to relieve the state of some of its executorial responsibilities without any effective form of decisional decentralization (Lazarev, 1994). Participatory development as it is envisaged today aims to renew these ideas of partnership, but to give due recognition to the role of local populations by letting them generate, share, and analyze information, establish priorities, specify objectives, and develop tactics (World Bank, 1996). It is viewed as a social learning process within which stakeholders, by generating and internalizing their own aspirations, themselves enable a social change process.

Impoverished and marginalized areas in developing countries have been the main targets for promoting participatory forms of decision making. The rationale is straightforward: these segments of population are generally the less educated and less organized, they are more difficult to reach, and the institutions that serve them are often weak. A range of participatory methods better adapted to work at the field level have been designed to engage and enable the poor to become active actors in development programmes. These include workshop-based and community-based methods for collaborative decision making, methods for stakeholder consultation, and methods for incorporating participation and social analysis into project design. Based on a World Bank (1996) survey of participatory methods, *Table 10.5* summarizes some of relevant participatory tools.

Involving citizens in the decision-making process is not an easy task. It requires careful planning, thoughtful preparation, and flexibility to change procedures on the demand of affected stakeholders. The selection of a supportive and conducive structure for public discourse is essential, not only to gain public acceptance, but also to take advantage of the full potential to articulate well-balanced decisions (Renn *et al.*, 1993). Setting aside technical aspects and contextual differences, participatory forms of decision making are viewed as proper mechanisms to achieve broader social goals (Beierle, 1998). These are to inform and educate the public, incorporate public values, assumptions, and preferences into decision making, increase the substantive quality of decisions, foster trust in institutions, and reduce conflict among stakeholders.

10.4 Policy-relevant Scientific Questions in Climate Change Response

10.4.1 Introduction

In this section a selected set of key policy-relevant scientific questions is examined in some detail. It surveys new developments and new results to foster our ability to make critical choices in climate policy, such as striking the right balance between mitigation and adaptation, the timing and location of actions, the costs of actions, and options to reduce and share them. After a brief discussion of the broad climate policy portfolio, the focus is on mitigation questions. The issues involved in these policy responses are structured as follows.

What should the response be? What are the most important factors to consider in crafting a short- to medium-term portfolio of mitigation and adaptation actions, and in acquiring information to resolve the large uncertainties? Drawing largely on IAMs, Section 10.4.2 takes a closer look at the first two components.

When should the response be made? The relationship between the timing of various types of mitigation responses, their costs, and their social, economic, technological, and environmental implications, raises a broad array of policy issues. The most important insights are summarized in Section 10.4.3.

Where should the response take place? Closely related to the timing issue, the location of mitigation responses is a multifacetted concern also. While the environmental value of a given amount of unreleased GHG is equal wherever its abatement takes place, there are far-reaching implications of whether and to what extent nations are allowed to use international flexibility instruments. The questions range from cost and efficiency concerns, to incentives for technological development, to implementation and verification problems. Section 10.4.4 summarizes some of the aspects.

Who should pay for the response? The location of the mitigation action can largely be separated from the question of who carries the costs. Numerous guidelines have been proposed for burden sharing. They range from historical responsibility, to various equity principles, to efficiency and international competitiveness concerns. Some fundamental points are reviewed in Section 10.4.5.

Towards what objective should the response be targeted? Current analyses of climate change impacts, adaptations, and mitigation normally cover the range between 450 and 850ppmv CO_2-equivalent concentration or an increase of between 1°C and 6°C in the global mean temperature. Completing the circle that started with the discussion of how the costs and benefits of balancing mitigation and adaptation activities influence the choice of the climate and/or GHG stabilization target, the issue of high versus low levels of stabilization is raised again in Section 10.4.6.

Table 10.5: *Participatory methods and tools (World Bank, 1996)*

Method	Tools
Collaborative decision making: Workshop-based methods	*Appreciation–influence–control (AIC)* AIC encourages stakeholders to consider the social, political, and cultural factors, along with the technical and economic aspects, that influence a given project or policy. Activities focus on building appreciation through listening, influence through dialogue, and control through action. *Objectives-oriented project planning (ZOPP)* The purpose of ZOPP is to undertake participatory, objectives-oriented planning that spans the life of the project or policy work, while building stakeholder commitment and capacity with a series of workshops. *TeamUP* TeamUp builds on ZOPP, but emphasizes team building. It enables teams to undertake participatory, objectives-oriented planning and action, while fostering a "learning-by-doing" atmosphere.
Collaborative decision making: Community-based methods	*Participatory rural appraisal (PRA)* PRA is a label given to a growing family of participatory approaches and methods that emphasize local knowledge and enable people to undertake their own appraisal, analysis, and planning. It enables development practitioners, government officials, and local people to work together in context-appropriate programmes. *SARAR* The purpose of this participatory method is to (a) provide a multisectoral, multilevel approach to team building through training, (b) encourage participants to learn from local experiences rather than from external experts, and (c) empower people at the community and agency levels to initiate action.
Methods for stakeholder consultation	*Beneficiary assessment (BA)* BA's general purposes are to (a) undertake systematic listening to "give voice" to poor and other hard-to-reach beneficiaries, thereby highlighting constraints to beneficiary participation, and (b) obtain feedback on development interventions. *Systematic client consultation (SCC)* SCC refers to a group of methods used by the World Bank to improve communication among Bank staff, direct and indirect beneficiaries and stakeholders of bank-financed projects, government agencies, and service providers, so that projects and policies are more demand-driven.
Methods for social analysis	*Social assessment (SA)* Objectives of SA are to (a) identify key stakeholders and establish the appropriate framework for their participation, (b) ensure that project objectives and incentives for change are appropriate and acceptable to beneficiaries, (c) assess social impacts and risks, and (d) minimize or mitigate adverse impacts. *Gender analysis (GA)* GA focuses on understanding and documenting the differences in gender roles, activities, needs, and opportunities in a given context. It highlights the different roles and learned behaviour of men and women based on gender attributes, which vary across culture, class, ethnicity, income, education, and time.

10.4.2 What Should the Response Be? The Relationship between Adaptation and Mitigation

The principal objective of mitigation activities is to reduce the amount of anthropogenic CO_2 and other GHG emissions in order to slow down and thus delay climate change. Ultimately, this is to achieve "stabilization of GHG concentrations in the atmosphere at a level that would prevent dangerous anthropogenic interference with the climate system" (UNFCCC, 1993, Article 2). In contrast, climate change adaptation aims to reduce adverse consequences of climate change and to enhance positive impacts, through private action and/or public measures (*Box 10.3*). Adaptation activities include behavioural, institutional, and technological adjustments. They capture a

Box 10.3. Mitigation and Adaptation

Mitigation consists of activities that aim to reduce GHG emissions directly or indirectly (e.g., by changing behavioural patterns, or by developing and diffusing relevant technologies), by capturing GHGs before they are emitted to the atmosphere or sequestering GHGs already in the atmosphere by enhancing their sinks.

Adaptation is defined as adjustments in human and natural systems, in response to actual or expected climate stimuli or their effects, that moderate harm or exploit beneficial opportunities (see IPCC, 2001b)

wide array of potential strategies, such as coastal protection, establishing corridors for migrating species, searching for drought-resistant crops, altering planting patterns, forest management, as well as personal savings or insurance that may cover the damage expected by individuals (Toman and Bierbaum, 1996). Adaptation is a central theme of WGII (IPCC, 2001b).

Whereas mitigation deals with the causes of climate change, adaptation tackles the consequences. As a result, the distribution of benefits from mitigation and adaptation policies is fundamentally different in terms of damage avoided. Mitigation will have only a long-term global impact on climate change damage, while adaptation options usually generate a positive effect in a shorter term. Adaptation activities mainly benefit those who implement them, while gains from mitigation activities accrue also to those who have not invested into the abatement policies. Mitigation is plagued by the free-rider problem and might create severe problems for decision making as opposed to adaptation, in which free-riding is much more limited. Hence, the output of mitigation activities can be viewed as a global public good, while the output of adaptation measures is either a private good in the case of autonomous adaptation or a regional or national public good in the case of public strategies (Callaway *et al.*, 1998; Leary, 1999). Mitigation policies at the global scale are efficient only if all major emitters implement their accepted reduction commitments. In contrast, most adaptation policies are carried out by those for whom averted damage exceeds the respective costs (Jepma and Munasinghe, 1998).

What adaptation and mitigation actions have in common is that they both avoid climate change damages. So far the debate about climate change policy has been dominated by emission reduction activities. The strong bias towards mitigation schemes has resulted in a relatively poor incorporation of adaptive response strategies into climate change analysis, although methods on how to evaluate and assess adaptive response strategies have already been elaborated (Feenstra *et al.*, 1998; Parry and Carter, 1998). The reasons for this are diverse. Adaptation has been associated with an attitude of fatalism and acceptance. Putting too much emphasis on adaptation strategies might raise the notions that mitigation efforts have little effect, that climate change is inevitable, and/or that mitigation measures are unnecessary. Approaching the climate issue from the adaptive side might inhibit concerted rational action by governments, as adaptation measures are conducted and

rewarded locally. Consequently, there is no incentive to participate in international negotiations if a country considers itself to be able to adapt fully to climate change (Pielke, 1998).

Emission reduction is recognized as attacking the immediate cause. However, the political and scientific discussion would certainly gain by broadening it beyond the issue of mitigation, if only because past emissions of GHGs together with their long atmospheric lifetime leave the earth with unavoidable adverse climate change impacts, irrespective of current mitigation actions (see Smith, 1997; Jepma and Munasinghe, 1998; Rayner and Malone, 1998b).

Even if mitigation efforts do succeed, adaptation strategies are considered to be reasonable because ancillary benefits independent of climate change might result (Pielke, 1998). Exploring adaptation strategies and the way in which people in "homes, factories, and fields" can be empowered institutionally and technologically to change their practice of living may, next to the generation of short-term benefits, contribute to substantial emission reductions, as well as to the development of strategies to cope with general aspects of global change and thus improve their ability to flexible (re)action. Hence, bottom-up analysis can be viewed as a necessary complementary tool in examining climate impacts with respect to top-down schemes employed in the derivation of national GHG emission reduction targets conducted by expert groups (Rayner and Malone, 1998a).

It is recognized increasingly that the impacts of global climate change are not determined solely by the physical characteristics of events. They also depend on the society's ability to adapt to changing patterns of the geophysical environment, as indicated by the analyses of documented impacts of historical climate hazards (Meyer *et al.*, 1998). Larger damages and higher losses of life are caused by extreme weather events in poor regions compared to similar events in affluent regions. Thus damages are not only a function of climatic change patterns. They are strongly influenced by economic, institutional, and technical development, which determine the capacity to adapt to changing patterns, as well as by exogenous shifts in socioeconomic conditions, such as population growth (Tol and Fankhauser, 1998).

The challenge is to find the right balance of adaptation and mitigation measures that represents an effective and complementary response strategy to climate change. For this purpose it is important to recognize the potential economic trade-off between mitigation and adaptation strategies. This trade-off entails the use of scarce resources in mitigation activities, like restructuring a nation's energy system, versus adaptation strategies, like protection against changing flood and/or drought patterns or sea-level rise. More generally, the trade-off implies greater or lesser stresses from climate change to be adapted to, depending on the level of mitigation effort. The question then is to what extent governments should focus on mitigation and adaptation strategies, recognizing that adaptation and mitigation decisions would generally not be made by the same entities. This implies that the search for the best possible combination of adaptation and mitigation strategies is a complex process.

Several approaches from different angles are possible to answer this question. From an economic point of view, the task is to compare the marginal costs and benefits of both strategies, and—in an optimization framework—minimize the overall welfare loss or macroeconomic costs. Following the heuristic principle of precaution would imply precautionary investments in both mitigation and adaptation to hedge against the uncertainties involved in climate change. However, there is little guidance in the discussion of the precautionary approach regarding how to operationalize critical levels of GHG emissions. Furthermore, the success of climate change policy depends on institutional structures and constraints that need to be analyzed with respect to the feasibility of mitigation and/or adaptation strategies.

10.4.2.1 *Economic Considerations*

From a global optimization perspective, the aim of coping with climate change is to determine the optimal scope and amount of adaptation and mitigation measures and thus to minimize the resultant global welfare loss. In this context, the quantity of adaptation depends on the level of mitigation, but the perceived level and costs of adaptation influence the level of mitigation. The task is then to set the share of mitigation and adaptation costs within the overall costs, which include the residual damage costs (Fankhauser, 1996; Jepma and Munasinghe, 1998). In the IAMs, which use a cost–benefit framework, the optimal mitigation and adaptation levels are theoretically resolved by comparing the marginal costs of further action with the marginal benefits of avoided damage. Many uncertainties characterize this framework, such as sector- and country-specific damage functions, and adaptation options and their costs are largely unknown, especially in developing countries. Assumptions and data behind the mitigation cost functions differ widely as well, as explained in previous chapters.

Integrated studies do not yet explicitly report adaptation costs and possible secondary benefits of adaptation strategies. In fact, they take into account individual market adjustments driven by changes in relative prices and changing consumption, investment, and production decisions to balance the private marginal benefits and costs (private adaptation; Callaway *et*

al., 1998). However, most IAMs do not balance the marginal costs of controlling GHG emissions against those of adapting explicitly to any level of climate change. Tol and Fankhauser (1998) give an overview of IAMs and their treatment of adaptation strategies (*Table 10.6*). Tol *et al.* (1998) approximate that about 7%–25% of the estimated global damage costs may be attributed to adaptation activities.

Another observation is that adaptation options are typically analyzed for a given amount of climate change independent of mitigation considerations (Fankhauser, 1996). Here the aim is to find the amount of adaptation necessary to minimize the net damage that results from a given level of climate change. Analysts often include predetermined adaptation options in an ad-hoc manner, and so there is a tendency to underestimate adaptive capacity. These analyses have been widely carried out and are reasonably well understood in the field of agricultural and coastal impacts, at least in some developed countries (Fankhauser, 1995a; Yohe *et al.* 1996; Mendelsohn and Schlesinger, 1997; Mendelsohn and Neumann, 1999).

In general, the integrated analysis of adaptation options is a rather complex process because all socioeconomic trends affect the vulnerability to climate change and vulnerability determines the optimal level of adaptation. Even without adaptation, impact assessments vary depending on the socioeconomic development projected for the future. Studies that examine the avoided damage under different emissions reduction targets (i.e., different costs of mitigation) and compare them with the costs of adaptation options are yet to be developed.

Giving policy advice on the basis of the efficiency concept within the IAM framework is often difficult, partly because IAMs capture only some elements of the potential coping strategies and are, thus, biased towards mitigation activities, and partly because damage estimates still have a rather low confidence (Tol, 1999a, 1999b). Nevertheless, IAMs are a useful tool in exploring the implications of new types of policies. They help to manage scientific knowledge and give insights into the major driving forces for present and future development with respect to social, economic, and ecological structures (Rotmans and Dowlatabadi, 1998).

The critical aspect of the efficiency approach is that it is only partially optimal because the level of climate change, which depends on the level of mitigation, is exogenous (Callaway *et al.*, 1998). Thus, this approach does not take into account that the emissions causing climate change are the result of externalities and therefore are not optimal. From this perspective, to correct the emissions' externality through mitigation is the first answer. However, the need for adaptation measures remains valid because of the adverse climate change effects that are already unavoidable. Strategies that incorporate both mitigation and adaptation are likely to be more efficient for limiting the damages of climate change than strategies that pursue only one or the other form of action.

Table 10.6: *Adaptation in integrated assessment models*

Model	Adaptation	Source
DICE	Not explicitly considered	Nordhaus (1994b)
RICE	Not explicitly considered	Nordhaus and Yang (1996)
		Nordhaus and Boyer (1999)
CONNECTICUT	Not explicitly considered	Yohe *et al.* (1996)
SLICE	Not explicitly considered	Kolstad (1994)
AIM	Not explicitly considered	Morita *et al.* (1997)
MERGE 2, 3	Not explicitly considered	Manne (1995)
CETA	Not explicitly considered	Peck and Teisberg (1992)
CETA revised		
IMAGE 2.1	Land allocation: expansion or contraction and intensification or extensification	Alcamo (1994)
CSERGE(M)	Not explicitly considered	Maddison (1995)
CSERGE(F)	Not explicitly considered	Fankhauser (1995a, b)
FUND 1.5	Induced adaptation	Tol (1996)
PAGE 95	Adaptation as policy variable	Plambeck and Hope (1996)
MARIA	Not explicitly considered	Mori and Takahaashi (1997)
ICAM 2.0, 2.5	Induced adaptation	Dowlatabadi and Morgan (1995)
MiniCAM 2.0	Induced adaptation	Edmonds *et al.* (1994)
PGCAM	Induced adaptation	Edmonds *et al.* (1994)
DIAM	Not explicitly considered	Grubb *et al.* (1995)
FARM	Production practices in agriculture and forestry, land, water, labour and capital allocation	Darwin *et al.* (1996) Darwin (1999)

AIM: Asian-Pacific Integrated Model
CETA: Carbon Emission Trajectory Assessment
CONNECTICUT: Connecticut
CSERGE: Centre for Social and Economic Research on the Global Environmnet
DIAM: Dynamics of Inertia and Adaptability for integrated assessment of climate-change Mitigation
DICE: Dynamic Integrated Model of Climate and the Economy
FARM: Future Agriculture Resource Model
FUND: Framework for Uncertainty, Negotiation and Distribution
ICAM: Integrated Climate Assessment Model
IMAGE: Integrated Model to Assess the Greenhouse Effect
MARIA: Multiregional Approach for Resource and Industry Allocation
MERGE: Model for Evaluating the Regional and Global Effects of greenhouse gas reduction policies.
MiniCAM: Mini Climate Assessment Model
PAGE: Policy Analysis for the Greenhouse Effect
PGCAM: Process Global Climate Assessment Model
RICE: Regional Integrated Model of Climate and the Economy
SLICE: Stochastic Learning Integrated Model of Climate and the Economy

Also, the efficiency criterion is often criticized because economic efficiency is not necessarily the only aim decision makers, economic agents, and governments wants to pursue, and it does not account for ecological systems and subsistence agriculture entirely outside the market sector. Distributional aspects of burden-sharing schemes and culturally determined risk preferences also play an important role in resource-allocation decisions.

10.4.2.2 *Precautionary Considerations*

In decision making, the precautionary principle is considered when possibly dangerous, irreversible, or catastrophic effects are identified, but scientific evaluation of the potential damage is not sufficiently certain, and actions to prevent these potential adverse effects need to be justified (Jonas, 1985; O'Riordan and Cameron, 1994; CEC, 2000). The precautionary principle implies an emphasis on the need to prevent such adverse effects. It thus acknowledges societal risk preferences, which are, plausibly, that humankind would rather be risk averse than risk neutral or risk seeking if one considers, for instance, future climate-induced loss of GNP (Pearce, 1994; Jaeger *et al.*, 1998). Hence, attitudes towards risk play a key role in decision making under uncertainty. However, one might also favour prevention to cure even where one is certain about the damage.

With the precautionary principle, uncertainty about the damage to be incurred does not serve as an argument to delay action. In the face of great uncertainty, a precautionary approach might even result in a more stringent emission-reductions target and/or adaptational response (Cantor and Yohe, 1998).

The evaluation of uncertainty and the necessary precaution is plagued with complex pitfalls. These include the global scale, long time lags between forcing and response, the impossibility to test experimentally before the facts arise, and the low frequency variability with the periods involved being longer than the length of most records (Moss and Schneider, 2000). Some of these uncertainty aspects may be irreducible in principle, and hence decision makers will have to continue to take action under significant uncertainty, so the problem of climate change evolves as a subject of risk management in which strategies are formulated as new knowledge arises (Jaeger *et al.*, 1998).

Aspects of uncertainty are associated with each link of the causal chain of climate change, beginning with GHG emissions, covering damage caused by climate change, followed by a set of mitigation and adaptation measures (Jepma and Munasinghe, 1998). In particular, damage-function estimates are prone to low confidence as they involve uncertainty in both natural and socioeconomic systems. To quantify the impact of climate change on flora and fauna needs consideration of many effects because of the complexity of the biological and ecological systems. Similarly, the manner in which humans adapt to climate change is not well known, socioeconomic modules are still at a stage of low disaggregation, and damage as a function of vulnerability, adaptation and time-dependency is poorly understood (Tol *et al.*, 1998; Tol, 1999a, 1999b).

However, following the precautionary principle, uncertainty is not an argument for delaying action, as the UNFCCC acknowledges in Article 3.3: parties should "take precautionary measures to anticipate, prevent or minimize the causes of climate change and mitigate its adverse effects. Where there are threats of serious or irreversible damage, lack of full scientific certainty should not be used as a reason for postponing such measures..." (UNFCCC, 1993). Pursuing this principle, mitigation and adaptation measures are to be implemented before full information is available and uncertainties regarding the scope and timing of climate change are resolved. Yet, the question of timing and extent of mitigation and/or adaptation policies remains unquantified by the precautionary principle (Portney, 1998).

10.4.2.3 Institutional Considerations

In contrast to the single-actor paradigm, which assumes that society can be identified with a unique optimizing decision maker, GHG emissions are, in fact, controlled by a multitude of individual agents and multiple decision makers that influence the transformation of individual to collective actions. Thus far, decision analysis has strongly emphasized the most aggregated level of government policy and neglected the mul-

tidimensionality of decision-making institutions (Jaeger *et al.*, 1998).

Although there are many country-specific differences in the relationships between national, regional, and local governments, most analysts consider local authorities to be the salient political actors. In addition to acting on their own, local governments serve as an interface between citizens and the nation state, and they are in regular contact with members of the community. O'Riordan *et al.* (1998) suggest that, as the need for more effective climate policy emerges, it might be useful to broaden the national response strategy to incorporate the local levels and so stimulate the very effective informal institutional dynamics of individuals and households. The rise in the number of informal networks of co-operation dispersed via schools, universities, religious communities and other social groups is regarded as an important step towards including climate change awareness into people's everyday concerns. This is of great importance, as the individual costs of contributing to climate change are less than the consequent social costs, and thus individual agents generate a changing climate that is socially suboptimal. Becoming aware of the gap between individual and social rationality is assumed to stimulate effective mitigation and adapation measures.

Striking the appropriate balance between mitigation and adaptation will be a tedious process. The need for, and extent and costs of adaptation measures in any region will be determined by the magnitude and nature of regional climate change as a local manifestation of global climate change. How global climate change unfolds will be determined by the total amount of GHG emissions that, in turn, reflects nations' willingness to undertake mitigation measures. Toth (unpublished) points out that balancing mitigation and adaptation efforts largely depends on how mitigation costs are related to net damages (primary or gross damage minus damage averted through adaptation plus costs of adaptation). Both mitigation costs and net damages, in turn, depend on some crucial baseline assumptions: economic development and baseline emissions largely determine emission reduction costs, while development and institutions influence adaptive capacity.

Different levels of globally agreed limits for climate change (or for atmospheric GHG concentrations, as frequently discussed), entail different balances of mitigation costs and net damages for individual nations. Considering the uncertainties involved and future learning, climate stabilization will inevitably be an iterative process. Nation states will determine their own national targets based on their own exposure and their sensitivity to other countries' exposure to climate change. The global target emerges from consolidating national targets, possibly involving side payments, in global negotiations. Simultaneously, agreement on burden sharing and the agreed global target determines national costs. Compared to the expected net damages associated with the global target, nation states might reconsider their own national targets, especially as new information becomes available on global and regional patterns and impacts of climate change. This becomes the starting point for the next round of negotiations. It follows from the above that establishing the "magic number" (i.e., the upper limit for global climate change or GHG concentration in the atmosphere) will be a long policy process, hopefully helped by improving science.

Mitigation and adaptation decisions related to anthropogenically induced climate change differ. Mitigation decisions involve many countries, disperse benefits globally over decades to centuries (with some near-term ancillary benefits), are driven by public policy action, based on information available today, and the relevant regulation will require rigorous enforcement. In contrast, adaptation decisions involve a shorter time span between outlays and returns, related costs and benefits accrue locally and their implementation involves local public policies and private adaptation of the affected social agents, both based on improving information. Local mitigation and adaptive capacities vary significantly across regions and over time. A portfolio of mitigation and adaptation policies will depend on local or national priorities and preferred approaches in combination with international responsibilities.

10.4.3 *When Should the Response Be Made? Factors Influencing the Relationships between the Near-term and Long-term Mitigation Portfolio*

A broad range of mitigation responses can be conceived. However, the bulk of attention, in both the analytical and policy arenas, has been devoted to reducing the emission of GHGs from anthropogenic sources and to removing the CO_2 (the most important GHG) already in the atmosphere by enhancing the biophysical processes that capture it. The timing of these efforts depends partly on the climatic constraints to be observed and on the costs of these actions, which are subject to change over time. Even with an exact knowledge of the timing and consequences of the future impacts of climate change, policymakers will still be faced with difficult choices regarding the implementation of response options. This is because the costs, availability, and associated impacts of future mitigation options are uncertain, and the choices involve trade-offs with important competing environmental and other social objectives. Chapter 8 (Section 8.1.4) discusses the costs of different pathways towards a fixed stabilization objective, and notes factors which would favour a larger proportion of preparatory activities relative to mitigation per se as well as factors that favour early mitigation. This section considers the wider context relating to climate change risks and damages.

Inertia and Uncertainty
Various attempts have been made over the past few years to explore these questions. Arguments that favour a larger fraction of preparatory activities (developing technologies, building institutions, and the like), rather than emission reductions in the near-term mitigation portfolio, include losses from the early retirement of installed capital stock, technological devel-

opment, the optimal allocation of resources over time (discounting effect), and the carbon cycle premium (Wigley *et al.*, 1996). See Chapter 8 for a detailed discussion. *Table 10.7* summarizes the most important arguments brought forward in favour of modest and stringent emissions reduction in the near term.

In addition to those emphasized by Wigley *et al.* (1996; see above), other arguments are proposed that support less stringent near-term emission reductions as well. Most refer to the significant inertia in economic systems. The first argument below is related to the economic lifetime of already installed capital stock. The second points to the possibility of low-cost mitigation technologies becoming available in the future.

Wigley *et al.* (1996) refer to the inertia of the capital stock. Researchers also identified other fields of inertia such as technological developments and lifestyles. The essential point of inertia in economic structures and processes is that it incurs costs to deviate from it and these costs rise with the speed of deviation. Such changes are often irreversible. The costs stem from premature retirement of the capital stock, sectoral unemployment, switching cost of existing capital, and rising prices of scarce investment goods. Emissions reduction in the present influences the marginal abatement cost in the future. The inertia of technological development arises from the path dependence. The capital stock can be divided into three parts. First, end-use equipment with a relatively short lifetime can be replaced within a few years. Second, infrastructure, buildings, and production processes can be replaced in up to 50 years. Structures of urban form and urban land-use can only be changed over 100 years. The demand and supply of goods and services in these three domains are interrelated in a complex way (Grubb *et al.*, 1995; Grubb, 1997; Jaccard *et al.*, 1997).

Technological Change
In the debate on weaker versus stronger early mitigation, the modelling of technological change and the resultant costs of the available mitigation technologies at any given time has far more influence when there is explicit consideration of damages from climate change. Many models assume an exogenous aggregated trend parameter, the rate of autonomous energy efficiency improvement. Other authors indicate phenomena such as inertia, lock-in, and the diversity of factors that affect the rate of technological development and diffusion. Energy technologies are changing and improved versions of existing technologies are becoming available, even without policy intervention. Modest early deployment of rapidly improving technologies allows learning-curve cost reductions, without premature lock-in to existing, low-productivity technology. Both the development of radically advanced technologies require investment in basic research and incremental improvements in existing technologies (e.g., learning by doing) is needed. Not only will new energy-system technologies be required to stabilize concentrations of CO_2, but also a host of peripheral technologies to distribute, maintain, transport, and store new fuels. On the other hand, endogenous (market-induced) change

Table 10.7: *Balancing the near-term mitigation portfolio*

Issue	Favouring modest early abatement	Favouring stringent early abatement
Technology development	• Energy technologies are changing and improved versions of existing technologies are becoming available, even without policy intervention. • Modest early deployment of rapidly improving technologies allows learning-curve cost reductions, without premature lock-in to existing, low-productivity technology. • The development of radically advanced technologies will require investment in basic research.	• Availability of low-cost measures may have substantial impact on emissions rajectories. • Endogenous (market-induced) change could accelerate development of low-cost solutions (learning-by-doing). • Clustering effects highlight the importance of moving to lower emission trajectories. • Induces early switch of corporate energy R&D from fossil frontier developments to low carbon technologies.
Capital stock and inertia	• Beginning with initially modest emissions limits avoids premature retirement of existing capital stocks and takes advantage of the natural rate of capital stock turnover. • It also reduces the switching cost of existing capital and prevents rising prices of investments caused by crowding out effects.	• Exploit more fully natural stock turnover by influencing new investments from the present onwards. • By limiting emissions to levels consistent with low CO_2 concentrations, preserves an option to limit CO_2 concentrations to low levels using current technology. • Reduces the risks from uncertainties in stabilization constraints and hence the risk of being forced into very rapid reductions that would require premature capital retirement later.
Social effects and inertia	• Gradual emission reduction reduces the extent of induced sectoral unemployment by giving more time to retrain the workforce and for structural shifts in the labour market and education. • Reduces welfare losses associated with the need for fast changes in people's lifestyles and living arrangements.	• Especially if lower stabilization targets would be ultimately required, stronger early action reduces the maximum rate of emissions abatement required subsequently and reduces associated transitional problems, disruption and the welfare losses associated with the need for faster later changes in people's lifestyles and living arrangements.
Discounting and intergenerational equity	• Reduces the present value of future abatement costs (*ceteris paribus*), but possibly reduces future relative costs by furnishing cheap technologies and increasing future income levels.	• Reduces impacts and (*ceteris paribus*) reduces their present value.
Carbon cycle and radiative change	• Small increase in near-term, transient CO_2 concentration. • More early emissions absorbed, thus enabling higher total carbon emissions this century under a given stabilization constraint (to be compensated by lower emissions thereafter).	• Small decrease in near-term, transient CO_2 concentration. • Reduces peak rates in temperature change.
Climate change impacts	• Little evidence about damages from multi-decade episodes of relatively rapid change in the past.	• Avoids possibly higher damages caused by faster rates of climate change.

could accelerate development of low-cost solutions and induces an early switch of corporate energy R&D from fossil frontier developments to low carbon technologies. Chapter 8 presents a discussion on induced technological change.

Intergenerational Equity
Assuming that current GHG emissions are too high from a sustainability point of view, it might be unfair of the current generation to decide to take the benefits related to emissions for themselves and that future generations should carry the burden of reductions. This argument on intergenerational equity is often emphasized to support early emission reduction.

Representation of Damages

An important implication of the debate on spiky versus smooth stabilization paths[22] is that relatively high emissions in the near term, especially for higher stabilization targets, may produce faster rates of climate change in the early 22nd century. There is little reliable information on what kind and how much risk this would pose to some ecosystems and socioeconomic sectors. Nevertheless, it should be kept in mind that paths towards the same ultimate environmental objectives might involve different environmental impacts for several decades.

This line of research investigates whether the choice of emission paths that lead to the same concentration target makes a difference in damages. In nearly all IAMs, climate damages depend on the magnitude of temperature change, but not on its rate. Experts point out that, because of the difficulties or higher costs of adaptation in some impact sectors, net damages could be higher for a relatively faster climate change. Moreover, some large-scale geophysical systems, like ocean circulation, may also be sensitive to the rate of temperature change (see IPCC, 2001a). To explore the former issue, Tol (1996; 1998a) used the Climate Framework for Uncertainty, Negotiations and Distribution (FUND) model for different damage functions and conducted an extensive sensitivity analysis with respect to the discounting rate, the power of the damage function, the optimal temperature for the level variant, and memory of damages in the rate variant. The results are ambiguous, but the flat path (early mitigation) is preferable in a significantly larger number of cases. This entails, for example, early reductions for discount rates of 0% and 3%, but not for 10%. If the optimal temperature increase exceeds pre-industrial levels by 3°C, early reduction is not required. Tol finds that impact costs for the spiky path are typically less than 20% higher than those of the smooth path. However, the differences are larger when impact costs also depend on the rate of change, when the exponent of the damage function is higher, and when the rate variant includes memory of damages.

Uncertainty with Respect to the Stabilization Target

Looking beyond the question of optimizing the emission path towards a specific concentration level, the main problem is that we do not know today what will be the desirable stabilization target. This kind of uncertainty, the expectation that it will be resolved over time, and the sequential nature of making mitigation decisions supports the arguments for timing mitigation actions in a context of uncertainty, raising various issues including that of inertia. From this perspective, it may be wise to prepare the ground now for possibly deep and fast emission reductions if resolution of the uncertainties would indicate that drastic climate protection measures are necessary, rather than

rush towards an uncertain target now by taking an expensive path.

Some models focus on the problem of near-term mitigation measures under long-term uncertain concentration targets, when the capital stock is inert. In these models, the equation for mitigation costs incorporates—beside the common permanent costs—an additional term to represent the transition costs, which are indicators for the inertia of the system. The costs in this field are typically calculated by comparing paths of immediate and of delayed reduction (usually 20 years). The latter is an approximation to the spiky path. In models that incorporate only one production sector, costs depend on the inertia of the system, the delay of reduction measures, and the concentration targets. For a concentration target of 450ppmv, mitigation costs may rise by 70% if the inertia is high (50 years characteristic time), compared to the lower (20 years) inertia of an increase by 25%–32%. The transition costs are more important than the permanent costs until 2050, with a maximum of 1.4% of Gross World Product (GWP) in 2040 and decline to zero until 2070. With respect to uncertain concentration targets, the results are most sensitive to inertia. Emissions reduction are double those of corresponding cases with a certain concentration target; for example, 9%–14% compared to 3%–7% in 2020 under a 550ppmv concentration goal (Ha-Duong *et al.*, 1997; see also Grubb *et al.*, 1995; Grubb, 1997). In a sectorally disaggregated model with two sectors of different inertias, the abatement levels are roughly the same, but the cost burden lies primarily in the more flexible sector. The costs are higher and the differences are more distinctive in the delayed cases compared to the immediate control cases. The sensitivity decreases with the concentration target. Analysts, however, warn that such models and results are still preliminary.

Possibilities to Reduce Near-term Costs

Chapters 8 discusses various possibilities to capture low-cost options, such as revenue recycling, integration of climate with non-climate policies to achieve ancillary benefits, and the availability of no regrets options. Such possibilities would be in favour of near term actions. For example, revenue recycling has been proposed as one instrument to reduce the costs of, and thus in support of, near-term emission reductions. In addition to the environmental considerations, this argument relates to the numerous distortions from taxes and subsidies in virtually all countries. Economy-wide effects of carbon taxes, and especially the double-dividend issue, are highly debated. Much enthusiasm has been given to "green taxes", such as the carbon tax, which might reduce the inefficiency of the current tax systems and lead to environmental improvements. Recent analyses show that the ultimate fiscal effect of substituting carbon taxes for other distortionary taxes is roughly neutral (Nordhaus, 1998), but the actual sign depends on the original size of the distortions in the economy. It may be positive in economies with highly distorted tax systems and hence confirm the double dividend hypothesis (see Chapter 8). It is likely to be negative in economies with less pre-existing distortions. In either case, revenue-recycling policies dominate other

---

[22] In a "spiky" emissions trajectory the emissions follow the baseline before declining sharply in order to meet a concentration stabilization target, while in a "smooth" trajectory the emissions diverge from the baseline early, allowing a more gradual decline of emissions eventually.

measures considerably. Goulder (1995) and others report about 30%–50% reductions in the cost of regulation, and many European studies find cost reductions over 100%.

10.4.4. Where Should the Response Take Place? The Relationship between Domestic Mitigation and the Use of International Mechanisms

Inquiries into the options and costs to reduce GHGs, especially CO_2, emissions indicate that the costs of reductions vary substantially across sectors in any given national economy and, perhaps even more significantly, across countries. This implies that for uniformly mixing pollutants like GHGs the costs of achieving any given level of environmental protection could be reduced if emission reductions were undertaken at locations where the associated costs are lowest. The concept has become known as "where-flexibility" in the climate policy literature. An institutional setting is required to exploit the opportunities of where flexibility, which involves a great variety of private and public decision-makers who originate from different cultures, represent different constituencies (if any), and live in systems with different social norms. Where-flexibility entails linkages both to other international agreements (GATT, Second European Sulphur Protocol, etc.) and to the legal systems of individual nations. As a result of its effects on relative prices, the choice between the international or domestic strategy also affects technological change.

In principle, two different mechanisms achieve where-flexibility: allowances and credit baseline. In the case of allowances, each participant starts with an initial endowment of pollution rights distributed by the government or through an auction. Emission rights must cover each unit of emission. This system has the character of emissions trading. Under a credit-baseline system, each participant has a baseline (i.e., a counterfactual, hypothetical emission trend) at the country, sector, or project level. Some measures are undertaken to reduce emissions. The difference between the baseline and the factual emissions is credited by an institutional body and can be traded. This system has the character of emissions reduction production.

The Kyoto Protocol contains three instruments to make use of where-flexibility: IET embodies the allowance system, while CDM and JI reflect the credit-baseline system. The Kyoto Protocol on IET allows Annex I parties with commitments listed in Annex B to trade emission allowances during the commitment period. As for JI, Article 6 declares that Annex I parties with commitments listed in Annex B are allowed to transfer or acquire emission reduction units that result from projects during the commitment period (the reduction units are specific to countries; these parties have national baselines). Finally, CDM as defined in Article 12 implies that, starting in 2000, Annex I parties listed in Annex B are allowed to acquire certified emission reductions from projects within the jurisdictions of non-Annex I parties.

Three general principles operate behind these arrangements:

- first, voluntarism indicates the freedom of contracting, i.e., the quantity-price combinations of exchange;
- second, supplementarity signifies the responsibility of Annex I parties to fulfil part of their commitments within their own jurisdictions; and
- third, additionality means that projects in CDM and JI have to be additional relative to the course of events in their absence (i.e., it must be decided what would happen anyway and what constitutes an additional project).

Ample attention has been paid to formulate principles for the design of where-flexibility instruments at the national and international level. The principles in the literature (Michaelowa, 1995; Watt *et al.*, 1995; Carter *et al.*, 1996; Matsuo, 1998, Matsuo *et al.*, 1998; OECD, 1998; Ott, 1998; EC, 1999) include:

- Environmental effectiveness. All units traded should be backed by sound data and verifiable emissions reductions; the use of the mechanisms is a means to achieve emission commitments agreed under the Protocol and the mechanisms should be designed to improve environmental performance and compliance with these commitments.
- Economic efficiency. This includes the cost-effectiveness of the emission reductions required by the Protocol, and over the longer term helping the community of nations to address climate change in a least-cost manner. It also requires the mechanisms to be administratively feasible, such that they do not impose excessive transaction costs on market actors. Economic efficiency will also improve if the market for trading and crediting is accessible to a wide range of potential players.
- Equity. While the main issue of equity under the Kyoto Protocol is the determination of assigned amounts or emission targets, the design of the implementation mechanism must also be perceived as equitable. Implementation of the mechanisms should not give an unfair advantage to any party or group of parties to the disadvantage of others (procedural equity). It should also allow new entrants over time.
- Credibility. Only a credible market mechanism should be used by the parties and will be accepted by the public. A mechanism of low credibility might be a source of various coalition formations at the negotiations and might undermine the will to comply with the commitments.

In creating a regime for flexible instruments, perhaps the most important lesson about multilateral agreements of the past two decades is that large and apparently "perfect" constructions have rarely been implemented quickly. Quite the contrary, the most successful examples of international regime building are based on a "piecemeal" approach, that is the stepwise evolution of political and legal mechanisms (Ott, 1998). For current DMFs, this might lead to a strategy with several phases that bring together the national and international levels at the speed of progress in international regime building (Holtsmark and Hagem, 1998). This would involve a two-stage game for IET (Ott, 1998), with a "twin cycle system" for JI (Heller, 1995) that focuses on the learning process in creating an international regime for JI.

There are some new and important factors to consider in the design of the instruments (see Ott, 1998). The economic and ecological dimensions of climate change and its mitigation affect different constituencies, sectors, and cultural values of the parties. Stakeholders range from states and international organizations to private companies and NGOs. Incentives motivate both private and governmental participants to report the highest possible baselines of GHG emissions to secure the largest amount of certified reduction credits. However, other processes create the opposite incentives.

The implementation of these instruments can be seen as a further step towards a more flexible and market-oriented policy in international environmental policy, and as an extension of national instruments to the international level. At the national level, some experience has already been accumulated with emissions trading, such as the Emission Trading Program under the US Acid Rain Program, the Los Angeles Regional Clean Air Incentives Markets, and the Norwegian Sulphur Trading programme introduced in 1999. Actual experience is much thinner internationally. Examples of the possibility for emission trading include the Montreal Protocol intended to curb CFC emissions that deplete the ozone layer and the United Nations Economic Commission for Europe (UNECE) Sulphur Protocol.

Many plausible arguments support the use of international mechanisms, as outlined below.

Static Eficiency
This argument is related to the positive allocative effects caused by trade. The argument is fundamentally dependent on the assumption of differences in the marginal reduction costs between countries in a well-defined market. This might lead to gains arising from trade for both sides. Trade reduces the overall costs of compliance with any specified set of internationally accepted reduction targets. Lile *et al.* (1999) and Edmonds *et al.* (1996) find the rationale for Annex I countries is that reduction costs in developing countries are much lower than their own. Ellerman *et al.* (1998) and Holtsmark and Hagem (1998) arrive at similar conclusions. However, some bottom-up, project-based country studies that quantify national mitigation-cost curves and the consequences (e.g., Jackson (1995); EC 1999)) identify lower mitigation costs in developed countries and thus smaller cost differences internationally. *Table 10.8* gives studies on the costs of Kyoto targets under different flexibility arrangements.

Willingness to Accept Deeper Reduction Goals
This argument is related to the reduction of the overall cost of compliance. If the reduction costs are lowered by the use of international mechanisms, then the nations might be willing to accept deeper GHG-reduction commitments. This argument does not hold in countries with potential hot air when a country's baseline (or projected future) emission is expected to be lower than its entitlement, so that a marketable good (emission permit) is created without the need for any effective reduction effort whatsoever. Nevertheless, members of the so-called "umbrella group", including the USA, Japan, Australia, Russia, and others, have made it clear that the level of commitment they accepted in Kyoto was contingent on the unfettered use of flexible mechanisms. In this sense they have already incorporated the willingness to accept deeper emission reductions in their existing commitments for the first budget period.

Complementarity to Other Goals
By using CDM and/or JI, climate protection can serve other goals such as accelerating socioeconomic and technological development, reducing regional and local pollution, and fostering integration and international understanding (Sun Rich, 1996). (See Chapter 1 for an extensive discussion of climate change in the context of sustainable development and Section 10.3 above on linkages to other issues and international agreements.)

Motivation for Private Institutions
Under JI and CDM, private institutions, such as enterprises and NGOs, are likely to be engaged, with the bulk of reduction measures probably taking place in the private sector. This might lead to a further reduction of mitigation costs, because private institutions tend to operate at higher efficiency than state bureaucracies do.

Technology Transfer
JI or CDM is often only possible if technology is transferred from rich and energy-efficient regions to poor and energy-inefficient countries. This might have the favourable effect that developing countries "leap-frog" over the inefficient development stages previously passed through by the developed countries.

Domestic versus International Strategies
Arguments to support a domestic strategy are often formulated as a critique to an international strategy. Some general criticisms focus on the question whether or not an international strategy is an adequate instrument to achieve the ultimate goal of the UNFCCC, that is stabilizing the GHG concentration. Bush and Harvey (1997) emphasize two key requirements: to sharply constrain GHG emissions in developing countries and to achieve significant GHG reduction in developed countries. The most frequent arguments in support of the domestic strategy entail the following.

Dynamic Efficiency
Technological and social innovation are dynamic processes that can be accelerated through pressure from commitments in the Kyoto Protocol. The international strategy allows devel-

Table 10.8: *Studies on costs of Kyoto targets under different grades of where-flexibility*

In this table the essential results (welfare implications) are presented with respect to achieving the Kyoto targets under different institutional settings for the use of flexibility instruments.

The following features are summarized in the table:

1. Reference and year of publication; footnote says where the reference is available
2. Welfare measure and scenario; the scenarios are not all with respect to the Kyoto-targets and the welfare measure is not the same in all studies
3. Different grades of "where-flexibility"
4. Comments on the study summarizes the most important features of the models used for the study.

Study	Welfare measure and scenario	No trade	Different grades							Comments
			Double-bubble[23]	Less than A1 trade[24]	x%-Cap[25]	Annex 1 trade	More than A1 trade[26]	A1 trade + CDM	Full/global trade	
McKibbin, Shackelton and Wilcoxen (1999)	%-change of GNP per region in 2010; not Kyoto: Stab. at 1990 level	USA: -.5 Jap: -.1 Aus: -2.2 OtherOECD (OOECD): -1.2		*OECD* USA: -.5 Jap: -.3 Aus: -1.5 OOECD: -.9 gain $90 bil[27] US-unilateral reduction: -.6						G-Cubed(Global General Equilibrium Growth Model), fossil fuels only; with monetary effects, no 'no-regret', 8 regions, 12 sectors, no terms of trade, capital flows, 1995$
McKibbin, Shackelton and Wilcoxen (1998)	%-change of GNP per region in 2010; not Kyoto: Stab. at 1990 level	USA:-.3 Jap:-.8 Aus:-2.5 OOECD:-1.4 Chi:0 LDC:+2.7	USA:+.1 Jap:-.1 Aus:-1.3 OOECD:-1.2 Chi:0 LDC:+1.8			USA:-.2 Jap:-.2 Aus:-2 OOECD:-.5 Chi:0 LDC:+1.4			USA:0 Jap:0 Aus:-7 OOECD:-.2 Chi:-.5 LDC:+.2[28]	
McKibbin, Ross, Shackelton and Wilcoxen (1999)	%-change of GNP per Region in 2010; not Kyoto: Stab. at 1990 level	USA:-.6 Jap:-.5 Aus:-1.6 OOECD:-1.3 Chi:-.1 LDC:+.7				USA:-.5 Jap:-.4 Aus:-.8 OOECD:-.6 Chi:-.1 LDC:+.7			USA:-.2 Jap:-.1 Aus:-.4 OOECD:-.2 Chi:-.4 LDC:0	
Bernstein, Montgomery, Rutherford and Yang (1999)	Hicks Equivalent Variation (HEV), %-change from baseline per region, Protocol costs	USA:-.5 Jap: -.6 EU15: -.4 OOE: -1. SEA: -.2 OAS: .1 Chi: +.3 FSU: -.4 MPC: -1.4 ROW: -.1				US: -.25 Jap: -.25 EU15: -.25 OOE: -.75 SEA: 0 OAS: 0 Chi: +.3 FSU: +4.4 MPC: -1.1 ROW: -.1			US: -.2 Jap: 0 EU15: -.1 OOE: -.4 SEA: +.25 OAS: +.2 Chi: +.25 FSU: -.4 MPC: -.3 ROW: 0	MS-MRT (Multi-sectoral, multi-regional General Equilibrium Model) (GEM), terms-of-trade, capital flows, leakage, 10 regions, 5 energy, 4 non-energy sectors, 1995$

[23] Tulpulé (1998) defines the double-bubble as EU + EE; McKibbin, Ross, Shackelton and Wilcox (1999) defines it as European-OECD.

[24] If less participants than the Annex 1 countries are assumed (e.g. without USA).

[25] Special Case EU-Cap: purchases or sales of emissions by annex B countries may not exceed 5% of the weighted average of base year emissions and the assigned Kyoto emission budget.

[26] If more participants than only the Annex 1 countries are assumed (e.g. India, China).

[27] 2010 – 20 discounted at 5%.

[28] non-Annex 1 countries accept commitments consistent with the baseline emissions.

(continued)

Table 10.8: continued

Study	Welfare measure and scenario	No trade	Double-bubble[23]	Less than A1 trade[24]	Different grades x%-Cap[25]	Annex 1 trade	More than A1 trade[26]	A1 trade + CDM	Full/global trade	Comments
Bernstein, Montgomery and Rutherford (1999)	HEV	US:-.56 Jap:-.64 EU:-.45 OOE:-.92 SEA:-.18 OAS:-.1 Chi:+.34 FSU:-.42 MPC:-1.39 ROW:-.1			US:-.43 Jap:-.52 EU:-.33 OOE:-.78 SEA:-.13 OAS:-.08 Chi:-.31 FSU:+.05 MPC:-1.26 ROW:-.08[29] US:-.34 Jap:-.31 EU:-.25 OOE:-.71 SEA:-.07 OAS:-.05 Chi:+.25 FSU:+2.18 MPC:-1.17 ROW:-.08 US:-.35, Jap:-.24 EU:-.24 OOE:-.75 SEA:-.05 OAS:-.02 Chi:+.22 FSU:+4.18 MPC:-1.15 ROW:-.08 US:-.43 Jap:-.30 EU:-.30 OOE:-.80 SEA:-.08 OAS:-.05 Chi:+.23 FSU:+4.57 MPC:-1.15 ROW:-.08 US:-.39 Jap:-.24 EU:-.25 OOE:-.82 SEA:-.02 OAS:.03 Chi:+.26 FSU:+4.27 MPC:-1.23 ROW:-.03	US:-.36 Jap:-.23 EU:-.25 OOE:-.76 SEA:-.04 OAS:-.01 Chi:+.22 FSU:+4.44 MPC:-1.15 ROW:-.08		US:-.32 Jap:-.18 EU:-.20 OOE:-.67 SEA:+.06 OAS:+.09 Chi:+.55 FSU:+3.47 MPC:-.92 ROW:+.01[30]	US:-.14 Jap:-.03 EU:-.05 OOE:-.3 SEA:+.25 OAS:+.19 Chi:+.34 FSU:+.48 MPC:-.36 ROW:+.03	MS-MRT GEM. GTAP4 Database, 10 regions, 3 fuels and electrical sectors, 2 goods Annex-I – B30: 30% ceiling for buyers; this restricts imports of TP Annex-I – B50: 50% ceiling for buyers; this restricts imports of TP Annex-I – S50: 50% ceiling for sellers; this restricts exports of TP No Hot Air: The QUELRCS[31] for FSU and EE are set to their baseline values

[29] Annex I - B10: 10% ceiling for buyers; this restricts imports of TP's.

[30] limit of permit sales by each A1-region to 15% of its total sales under unrestricted global trading.

[31] Quantified Emission Limitation or Reduction Commitments.

(continued)

Table 10.8: continued

Study	Welfare measure and scenario	No trade	Double-bubble[23]	Less than A1 trade[24]	Different grades x%-Cap[25]	Annex 1 trade	More than A1 trade[26]	A1 trade + CDM	Full/global trade	Comments
Cooper et al. (1999)	Change in potential output (GDP) in 2010 in %	US:-2.5 (-1.8), Ca:-3.9, Jp:-1.8 (-1.9), Ge:-2.2(-2.4), Fr:-2.2(-2.2), It:-2.3, UK:-1.9, EU:-2.2, Chi:+1.6, Rus:+0.9	US:-1.(-.5), Ca:-1., Jap:-3(-.3), Ger:-2.2(-4.9), Fr:-2.2(-2.2), It:-2.3, UK:-1.8, EU:-2.2, Chi:+.7, Rus:-1.(+2.5 income)			US:-1.4(-.8), Ca:-1.2, Jap:-.5(-.5), Ger:-.8(-.9), Fr:-.6(-.5), It:-.7, UK:-1., EU:-.7, Chi:+.6, Rus:-1.4 (+4.income)				Oxford Model, Energybased macroeconomic GEM, 6 fuels, 4 sectors, 22 regions
Tulpulé (1998, 1999)	%-loss of GNP	Annex 1: -1.2 Non-Annex I: 0	Annex I: -.3 Non-Annex I: 0			Annex I: -.3 Non-Annex I: 0				GTEM, GTAP3-database, 19 regions, 16 tradeables, S(y)=j(i), transport costs for trade
Brown et al. (1999)	% loss of GNP, global	-.8				-.2				GTEM, GTAP 4 –database, 18 regions, 23 tradeables, 3 GHG GWP:1,21,310; technological change rising with GHG-penalty, ref-scen = no policy, 1992 $
Böhringer (1999) (CO_2-emissions in bill. t CO_2)	HEV, % of BAU income	-.2 (28.51)			-.15 (28.74)	-.04 (29.03)				global GEM, GTAP4 and OECD/IEA-data, 7 sectors, 11 Regions
Manne and Richels (1997)	consumption loss 1990-2100 discounted with 5% to 1990 in $ WG1 10% cut in 2010 WRE	8.7 tril 4.4 tril 1.8 tril				5.9 tril 2.7 tril 0.9 tril			3.2 tril 1.9 tril 0.8 tril	
Manne and Richels (1998)	annual US-Costs, 1990 bill. $ in 2010 (2020)	85 (102)						51 (77), 15 % of potentials available due to complexities of flexibility instruments	23 (45)	MERGE 3.0 (model for evaluating the regional and global effects of GHG reduction policies), 9 regions, leakage, Global Trade (GT) non-AI restricted by baseline bounds,
Manne and Richels (1999b)	Kyoto Forever => K-constraints through 21st c. for AI and baseline for NonAI (no leakage); US GDP loss in bil $	87			(a) 55 (b) 61 1/3 limitation of satisfying Kyoto obligations by AI-countries			49 15 % of potential available	23	(a) buyers market - sellers are price takers (b) sellers market – buyers are price takers
Capros (1998)	costs only for regions and not Kyoto targets (15%), EU			* with 15%						POLES for world level, extensive assessment for EU

(continued)

Table 10.8: continued

Study	Welfare measure and scenario	Different grades								Comments
		No trade	Double-bubble[23]	Less than A1 trade[24]	x%-Cap[25]	Annex 1 trade	More than A1 trade[26]	A1 trade + CDM	Full/global trade	
Holtsmark (1998)	% loss of 1990 GDP per country	USA:+.49 Ca:+.51 EU[32]:-.05 DK:+.59 Fin:0 Swe:0 Nor:+1.36 Rus:+.17 OthEIT:-.07 AuNZ:+.24 Jap:+.09 N-AI:+.91				USA:+.48 Ca:+.5 EU:-.07 DK:+.37 Fin:0 Swe:0 Nor:+1.22 Rus:+.16 OthEIT:-.07 AuNZ:+.17 Jap:+.08 N-AI:+.46			USA:+.4 Ca:+.42 EU:-.04 DK:+.25 Fin:0 Swe:0 Nor:+1.26 Rus:-.4 OthEIT:-.27 AuNZ:+.17 Jap:+.07 N-AI:+.42	partial & static ACT, strategic OPEC, 3 regional gas markets, multi GHG, national tax with marginal excess burden to account for double dividend hypothesis, no "no regrets", terms of trade
Holtsmark and Hagem (1998)	Costs in % of 1990 GDP per country, EU targets are differentiated as agreed in June 1998	USA:+.29 Ca:+.46 EU:-.12 DK:+.53 Fin:-.12 Swe:+.08 Nor:+1.23 Rus:-.18 OthEIT:-.15 AuNZ:+.06 Jap:-.08: N-AI:+.13				USA:+.27 Ca:+.42 EU:+.15 DK:+.23 Fin:-.11 Swe:+.02 Nor:+1.18 Rus:-.17 OthEIT:-.14 AuNZ:-.36 Jap:-.07 N-AI:+.11			USA:+.18 Ca:+.3 EU:-.11 DK:+.1 Fin:-.1 Swe:-.1 Nor:+1.23 Rus:-.33 OthEIT:+.33 AuNZ:-.12 Jap:-.07 N-AI:+.08	partial & static ACT, strategic OPEC, 3 regional gas markets, multi GHG, national tax to account for double dividend hypothesis, no "no regrets", terms of trade
Sands et al. (1998)	US Costs; non-Kyoto-Targets, 4 policy scenarios: e.g.: 1990 + 10% (M90+1)	M90:+.2 M90-1:+.07 M90-1:+.43 M95:+.07				M90:-.18			M90:-.09	SGM, IEA and Government-data, 7 regions, 9 prod sectors, input sectors, determines global C-tax, hot air
Kainuma et al. (1998), (1999)	% GDP loss per region; Asia	USA:+.4 Jap:+.25 EU:+.3 EEFSU:+.2 Chi:+.2 Ind:-.25 Kor:-.55 MEA:+.15	USA:+.3 Jap:+.05 EU:+.4 EEFSU:-2.2 Chi:-.15 Ind:-.15 Kor:-.35 MEA:+.8			USA:+.3 Jap:+.15 EU:+.3 EEFSU:-3.6 Chi:+.2 Ind:-.15 Kor:-.25 MEA:+.65	USA:+.3 Jap:+.1 EU:+.15 EEFSU:-2.7 Chi:+.4 Ind:-.15 Kor:-.35 MEA:+.6[33]		USA:+.2 Jap:0 EU:+.05 EEFSU:-1.7 Chi:-.37 Ind:-.2 Kor:-.3 MEA:+.9	AIM, IEA, GTAP and IMF-data, 21 regions, 7 energy-, 4 non-energy goods,
Mensbrugghe (1998)	% GDP loss in bill. 1985 $	+.7				+.2			+.1	GREEN- A multi-sector, multi-region dynamic general equilibrium model for quantifying the costs of curbing CO_2 emissions: 12 regions, 4 activities, autonomous energy efficiency improvements (AEEI), backstops, terms of trades

(continued)

[32] EU targets are differentiated as agreed in June 1998.
[33] Annex 1 + China and India

Table 10.8: continued

Study	Welfare measure and scenario	No trade	Double-bubble[23]	Less than A1 trade[24]	Different grades x%-Cap[25]	Annex 1 trade	More than A1 trade[26]	A1 trade + CDM	Full/global trade	Comments
Richels *et al.* (1996)	20% below 1990 emissions until 2010 and the stab. emissions at his level; Annex 1 countries; trill. 1990 US$, OECD, GDP loss CETA EPPA MERGE MiniCAM	OECD:+2 NOECD:+.3 OECD: +5.2 NOECD:+1.1 OECD:+2 NOECD:+.3 OECD:+1.5 NOECD:-.1							OECD:+1.3 NOECD:-.7 OECD: +2 NOECD:-.5 OECD:+1.1 NOECD:0 OECD:+.5 NOECD:-.2	
Nordhaus and Boyer (1999)	RICE-98; GDP loss in 1990$	US: 84 bil. Jap: 38 bil. EU: 65 bil. CANZ: 19 bil.				US: 60 bil. Jp: 19 bil. EU: 36 bil. CANZ: 11 bil.			US: 21 bil. Jap: 6 bil. EU:15 bil. CANZ: 4 bil.	RICE-98; 13 regions, no trade in goods, IAM;
Tol (1999a)	consumption loss 1990-2100, discounted with 5% to 1990, in trill. $; target:Kyoto forever	No-AI: -15 OECD: 2.75 EEfSU: .75	No-AI: -.2 OECD: 2.25 EEfSU: .6		lim. purch.[34] No-AI:-.15 OECD:2.45 EEfSU: .5 lim.sale:[36] No-AI: .8 OECD: 1.9 EEfSU: -.15 lim. both: No-AI: .65 OECD: 2.15 EEfSU: -.15	No-AI: -0.15 OECD: 2.1 EEfSU: .6	No-AI: .5 OECD: 1.25 EEfSU: .45[35]		No-AI: .45 OECD: .9 EEfSU: .35	Framework for Uncertainty, Negotiation and Distribution (FUND)-IAM; 9 regions, dynamic damage
Kurosawa *et al.* (1999)	GDP loss in bill. 1990 $ in 2010	US: Jap: EU: 90 CANZ:				US: Jap:10 EU: 90 CANZ:			US: Jap: EU: 63 CANZ:	GRAPE-IAM, 10 regions
Criqui *et al.* (1999)	Total Cost (GDP)in 1990 $ in 2010	56419 Mio. (0.11)			only for US and Asian countries (0.03)	16 583 Mio.			5808 Mio. (0.01)	POLES
Ellerman *et al.* (1998)	Total Cost in bill. 1985 US$	+120				+88			+11	Emissions Prediction and Policy Assessment (EPPA)-Model
Zhang (1998)	Reductions in total abatement costs compared to the no-trade			[37]USA:81 Jap:91 EU:2.3 OOECD: 33.5 OECD:82.4	[38]USA:63.7 Jap:71.9 EU:39.2 OOECD:70.8 OECD:66 [39]USA:81.1 Jap:77.4 EU:19.1 OOECD:68.8 OECD:79.6				USA:85.2 Jap:93.1 EU:0.2 OOECD:45.3 OECD:86.5	12 regions, 6 GHG

34 Maximum Emissions are 110% of allotment (= commitment); *i.e.* restricts the quantity bought from abrought   35 Annex 1 plus Asia   36 90% of allotment must be emitted => this restricts the quantity sold to others
37 No Hot Air: Trading in Hot-Air is not allowed, indicating that any effectuated trading in GHG emissions must represent real emission reductions   38 EU-Caps
39 50% reduction from BAU scenario: The Maximum allowed acquisition from all three flexibility mechanisms are limited to 50% of the difference between the projected baseline emissions and the Kyoto targets in 2010

oped countries to lower this pressure and, as a result, less innovation would occur. Instead of innovation, inefficient technology would be exported to developing countries. This argument is strongly related to the endogenous growth argument. The problem from the scientific point of view is that so far no model with Learning by Doing (LBD) includes regional disaggregation and trade.

Missing No Regrets Options

This argument is related to the political, social, and economic barriers to making use of no regrets potentials in developed countries. The possibility to fulfil their commitments by using CDM and/or JI might be more favourable for developed countries than to explore and utilize no regrets opportunities. The potential of no-regrets in developing countries affects the principle of additionality and is therefore a problem in accepting CDM and JI projects (Rentz, 1998).

Implementation

This argument embraces two problem areas: implementation of an institutional framework for instruments at the national and international levels, and compliance with the Kyoto Protocol targets. On the institutional side, there are several impediments to building a strong regime for functional international instruments. These are related to both institutional problems and market imperfections.

The production character of CDM and JI requires a fixed baseline to be defined. The baseline provides an incentive to cheat by setting it unrealistically high so that the efficacy of the instrument decreases because no real reduction takes place. The possibility to cheat stems from the intricacies of fixing the baseline, which usually arise through vague guidelines (political issue) and the general problems of forecasting (technical issue; see Michaelowa, (1995; 1998b). Begg *et al.* (1999) and Parkinson *et al.* (2001) examine the uncertainty associated with baseline construction and propose it be managed through conservative esimates, use of monitored data and verification, and either baseline revision or limited crediting lifetimes. Parson and Fisher-Vanden (1999) argue that the opportunity for self-serving manipulation of project baselines will vary markedly among project types, and suggest a likely bias in any project-based JI system (such as the CDM) towards project types most resistant to baseline manipulation—retrofits and technological carbon management. They also propose a hybrid domestic–international system for project certification to limit the scope for cheating.

In this context, an optimal baseline strategy is required that (first of all) takes into account the high volume of projects that will be needed for the flexible mechanisms to achieve their main objective of an (overall) environmental effectiveness. The balance is likely to be achieved by optimizing baseline stringency and minimizing project complexity (as long as the ability to determine "what would have happened otherwise" is not compromised). The reasoning is that a higher number of effective projects will be more beneficial for the environment

(in terms of GHG reductions) than a lower number of individually very effective projects (OECD, 1999a). A related delicate balance should be reached between the requirement for rigorous monitoring and reporting efforts (to ensure environmental effectiveness) and the need to obtain cost-effective and predictable emission benefits via simple procedures that encourage such projects (OECD, 1999b).

The next problem is that each implemented project—especially a large one—affects the baseline in other parts of the economy. Jackson (1995) warns that this might implicate a multilevel system of baselines in the overall economy, sectors, and projects. Clearly, projects have to be monitored. Thus the political problem of creating guidelines for monitoring and the technical issue of registration arise. Furthermore, credible enforcement or penalization against cheating or non-fulfilling parties needs to be established, probably in the form of a special body for ascertainment (Janssen, 1999). A reliable basis of international law for contracting is essential, especially among private actors. Even if all these conditions are fulfilled, the problem of corruption might make the instrument inefficient or flawed outright. With respect to such problems Barrett (1998) raises the question whether the Kyoto Protocol will be implemented at all by any parties if every party believes that other parties do not obey the rules and follow their own commitments.

Turning to the second problem area, the efficient allocation of markets can be distorted by transaction costs associated with searching for partners, and the costs of contracting and negotiations (see Stavins, 1995). Price distortions result when large nations and corporations exercise market power (Hahn, 1984; Hagem and Westskog, 1998), or asymmetric information distribution between partners in JI or CDM projects is exploited by one of them (Hagem, 1996). It is well known that each of these deviations from the ideal world of competitive markets might lead to inefficient allocations. Other factors at work include the initial distribution of property rights, which might also reflect equity considerations.

Corruption and Other Host Internal Problems

Corruption is an important problem in several countries on both sides of the JI and CDM relationships. The negative consequence of corruption is that institutional settings are undermined, especially when hard currency is involved. In many developing countries with weak democratic institutions, politicians have strong incentives to maximize the financial flows and to ignore potential negative consequences. Heller (1995) points out that higher financial inflows from donor countries might result in shrinking domestic environmental budgets, so that no real emissions reduction occurs.

Balance Between Domestic and International Strategies

It is apparent from this section that the relationship between domestic mitigation and the use of international instruments remains an intricate one. Work by Hahn and Stavins (1995) indicates that the link between domestic implementation and international mechanisms may seriously limit the ultimate economic potential of emissions trading. Most economic studies of trading assume that trade occurs whenever there is the potential to lower compliance costs. However, to the extent that some countries implement their domestic strategies through regulatory and tax measures, emissions permits obtained through international transactions may have limited or no value in these countries. Moreover, Hahn and Stavins (1995) point out that domestic legislators may be concerned about the significant financial transfers implied by emissions trading and act to keep funds within their own borders. In another relationship, Montgomery (1997) raises the possibility that domestic legislators may try to impose trade barriers in an effort to limit the competitiveness consequence implied by the loss of capital and jobs that may accompany efforts to limit emissions.

In summary, the literature on where-flexibility reveals abundant opportunities to reduce the costs of emission reductions, but also raises concerns about the implementation. However, concerns as to whether the flexibility mechanisms of the Kyoto Protocol can be implemented because of the possibility that some parties may be corrupt or may cheat are universal concerns. They apply not only to the countries involved in flexibility mechanisms, but also to countries that take on any emissions reduction commitments (although flexibility instruments are particularly sensitive to cheating). Given the various ways proposed to reduce the risks of their misuse, the Kyoto mechanisms offer the double advantage of reducing the costs of climate change mitigation and fostering non-climate objectives as well.

10.4.5   Who Should Pay for the Response? Mitigation by Countries and Sectors: Equity and Cost-effectiveness Considerations

Equity and efficiency considerations in the context of decision making that addresses global climate change are important for various reasons, including ethical concerns, effectiveness, sustainable development, and implementation of UNFCCC itself (Munasinghe, 1998; see also Chapter 1). Principles of justice and fairness[40] are important in themselves, in all types of human interactions, and play a major role in practically all modern international agreements, including the UN Charter; they emphasize the basic equality of all humans (Jepma and Munasinghe, 1998).

---

[40] The terms "justice" and "fairness" are often used as synonyms, however, there are debates on the different notions of the terms. Following Albin (1995), justice means distributive justice, in the sense of a general standard for allocating collective benefits and burdens among the members of a community at local, national, or global levels. Principles of justice exist prior to and independently of any phenomenon to be judged. Fairness consists of individual perceptions of what is reasonable under the circumstances, often in reference to how a principle of justice regarded as pertinent should be applied.

Some authors argue that equitable decisions generally carry greater legitimacy and encourage parties with differing interests to co-operate better in carrying out mutually agreed actions. One of the major obstacles to reaching a comprehensive agreement on global warming—setting GHG emission limitation targets for individual countries—involves parties that act as a "veto" because they regard particular arrangements as unfair or unjust. Decisions that are widely accepted as equitable are likely to be implemented with greater willingness than those enforced under conditions of mistrust (Rowlands, 1997). Others find little evidence that fairness matters much. Victor (1999) examines the relationships between fairness and the compliance with international environmental agreements through the lessons learned about implementation and effectiveness of numerous earlier treaties. His conclusion is that equity concerns matter little in the success of negotiating and implementing such agreements. Even for cases in which fairness seems to play some role, willingness to pay had a stronger role. Victor argues that if parties to an agreement take the trouble to deviate from the simplest across-the-board commitments, then many criteria need to be considered in negotiating commitments. Fairness might be one criterion, but is probably not the most important.

In a broader context, equity and fairness are important elements of the social dimension, while efficiency is a crucial factor in the economic dimension of sustainable development. The impetus of sustainable development provides a crucial reason for finding efficient and equitable solutions to the problem of global warming, especially with regard to future generations. The UNFCCC recognizes these two principles in Article 3.1.

Equity principles apply to both procedural and consequential issues (Banuri *et al.*, 1996). Procedural issues concern the process of how decisions are made. The two aspects of procedural equity involve the effective participation in decision-making processes and the process itself, which should be the principle of equal treatment before the law. In this context, reference is made to Coase's model of social cost (1960) in that he assumes a situation of equal bargaining power among participants and equal distribution of the costs of making the bargain with respect to the internalization of externalities. The philosophical notion of procedural equity is the "ideal speech situation" (Habermas, 1981), a situation in which dialogue and decision making are free from inappropriate constraints such as barriers to the acquisition of knowledge or financial resources. Transfer of these concepts to climate change negotiations requires consideration of the influence of scientific information, human resources, institutional capacities, and financial assets on the bargaining, and a redistribution of these among participants to create procedural equity.

Consequential equity deals with the outcome of decision making, and with the distribution of costs and benefits of preventing global warming (including future emission rights) and of coping with the climate change impacts and adaptation. The consequential decisions have implications for burden sharing

Table 10.9: *Equity principles and burden-sharing rules*

Equity principle	Interpretation	General operational rule
Egalitarian	Every individual has an equal right to pollute or to be protected from pollution	Allow or reduce emissions in proportion to population
Sovereignty	All nations have an equal right to pollute or to be protected from pollution; current state of emissions constitutes a status quo ("grand-fathering")	Proportional reduction of emissions to given or existing emission levels' or equal percentage of emission reductions
Polluter pays	Welfare losses corresponding to gains by emissions (eventually including historical emissions)	Share abatement costs across countries in proportion to emission levels
Ability to pay	Mitigation costs vary directly with national economic well-being	Equalize abatement costs across nations (costs as proportion of GDP equal for each nation)
Horizontal	All countries with similar features have similar emissions rights and burden-sharing responsibilities	Equalize net welfare change across nations–net cost of abatement as a proportion of GDP
Vertical	Welfare losses vary positively with national economic well-being, welfare gains vary inversely with GDP	Progressively share net welfare change across nations, net gains inversely and net losses positively correlated with per capita GDP
Utilitarian	Achieving the greatest good (happiness) for the greatest number	Maximize net present value of the sum of individuals utility (maximize social welfare).
Compensation	No nation should be made worse off	Compensate net losing nations
Rawls' maximin	The welfare of the worst-off nations should be maximized	Maximize the net benefit to the poorest nations
Market justice	The market is just	Allocate emissions permits to the highest bidder
Consensus equity	The negotiation process is fair	Seek a political solution to emissions reduction
Convergence	Equalize per capita emissions	Converge to an upper boundary of emissions
Environmental	The environment receives preferential treatment	Maximize environmental values and cut back emissions accordingly

among and within countries (intragenerational and spatial distribution) and between present and future generations (intergenerational and temporal distribution). While actions to mitigate climate change have to be paid for by the present generation, benefits in the sense of avoided losses will affect generations to come. This involves discounting future benefits to a net present value (Portney and Weyant, 1999). However, most of the potentially affected parties are not present to participate in the decision making, so that the current generation has to discuss equity issues within climate change.

In total, four kinds of questions frame the issue of justice in climate change (Shue, 1993), of which the third (procedural equity) provides the basis for a just process in determining the other three kinds of allocations.

1. What is a fair allocation of the costs of preventing the global warming that is still avoidable?
2. What is a fair allocation of the costs of coping with the social consequences of the global warming that will not be avoided?
3. What background allocation of wealth would allow international bargaining about issues like (1) and (2) to be a fair process?
4. What is a fair allocation of emissions of GHGs over the long term and during the transition to the long-term allocation?

To answer these questions scientists have developed typologies for the various distributional equity principles; these are understood to be general concepts of distributive justice and fairness, which often overlap. Associated burden-sharing rules, on the other hand, represent an operational function to generate a specific scheme to reduce GHG emissions or to bear the costs of climate change impacts. *Table 10.9* gives an overview of general equity principles and accompanying operational rules (Thompson and Rayner, 1998).

Major devices to determine the order of equity principles are the following: Rose *et al.* (1998) distinguish between "allocation-based", "outcome-based", and "process-based" principles. The first group focuses on the initial allocation of property rights of GHG emissions, such as the egalitarian, sovereignty, polluter pays, and ability-to-pay principles. The second group of principles examines the outcome in terms of welfare changes[41] caused by emissions reduction efforts, such as the horizontal, vertical, compensation, and utilitarian principles. The third category recognizes the libertarian, political consensus, and Rawls' maximin as guiding principles to the process of emission allocation. Shue (1993) divides principles of justice into "fault-based" and "no-fault" principles. The ability-to-pay, for example, is no-fault in the sense that guilt is irrelevant to the assignment of responsibility to pay. The richest should pay the highest rates no matter how they acquired what they own. In contrast, the polluter-pays principle, an economic principle that polluters should bear the cost of abatement without subsidy (Rayner *et al.*, 1999), is based upon fault or, alternatively, upon an amoral rationale of causal responsibility, or simply that the assignment of burden creates an incentive to not pollute. Thus, fault need not be a moral issue. Rowlands (1997) differentiates, among other things, according to aspects of historical difference (if any). The classification is based on whether past usage has established present and future rights, be it the same (grandfathering) or be it a correction for injustices from the past (natural debt). Agarwal and Narain (2000) outline the concept of contraction and convergence. This is the entitlement of GHG emissions budgets in terms of future emissions rights. Such a global future emissions budget is based on a global upper limit of atmospheric concentration of CO_2, for instance 450ppmv (contraction). This budget is then distributed as entitlements to emit CO_2 in the future, and all countries will agree to converge on a per-capita emission entitlement (convergence). Level of contraction and timing of convergence are subject to negotiations with respect to the precautionary principle.

The Kyoto Protocol endorses the principle of differentiation among countries (between Annex B and non-Annex B) and within Annex B countries for emissions reduction targets. However, details of the form of JI and the endowment of GHG emissions rights remain to be established. Also, future negotiations to determine national targets after 2012, as well as the question of commitments for developing countries, need to be discussed. Accordingly, several proposals for the differentiation of national GHG reduction targets, as well as multiform modelling exercises to explore the consequences of the different proposals, have been published recently. An overview is given in *Table 10.10*.

The variety of equity principles reflects the diverse expectations of fairness that people use to judge policy processes and the corresponding outcomes. The demand for fairness arises from the existence of communities (social solidarity) and from publicly shared expectations of the conduct of community relations. As communities pursue manifold ways of organizing institutional structures and social relations, there are different perceptions of what is equitable and fair (Rayner *et al.*, 1999; Rayner and Malone, 2000). Distinct moral principles generate conflicting debates on how to share the burdens, even though there might be equally legitimate and justified claims. Therefore, it is very difficult to achieve a worldwide consensus on just one justice principle. One way of reaching an accord might be to set up a combination of the diverse equity-based distribution proposals (Müller, 1998). Even if agreement on a particular first principle is reached, the question of how reductions for each country should be generated would remain unresolved because of the different reference bases against which equity citeria could be applied, such as population, land area, GDP per capita, or emissions per capita. With respect to the spatial distribution of GHG emissions limitation burdens, should the burdens be laid more on the production or on the consumption side of CO_2 emissions and what are the accompanying effects in terms of intragenerational equity (Rose and Stevens, 1998)? In summary, manifold equity principles and different accompanying operational rules exist; these might best be applied as a combination to respect more than just one equity position and thus enhance political feasibility.

However, there is a strong bias towards the principle of efficiency and its underlying utilitarian maxim. Also, it is important to recognize that self-interest plays a crucial role in voting for a specific operation rule, and that self-interests or, alternatively, particular preferences are at the core of economic considerations. Closely related to the concept of preferences is that of willingness-to-pay. Developed countries usually have a much higher willingness-to-pay in terms of solving environmental problems. This is partly because willingness-to-pay depends on the ability to pay. Consequently, it seems reasonable that developed countries bear the primary burden involved in mitigating climate change (Victor, 1999), as endorsed in the Kyoto Protocol. Hence, economics in terms of efficiency is a major aspect when negotiating emissions-limitation commitments.

The problem of distributing emissions-limitation quotas is not solved by economic principles either, because emissions trading yields Pareto efficiency irrespective of the initial distribution of emission permits. Where-flexibility in emissions reduction follows Coase (1960), who addresses the assignment of property rights as an efficient solution to market failure. Under the assumptions of perfect competition, a marketable emissions permit scheme with full trading will be cost-effective no matter how the permits are distributed. It will lead to an equalizing of the marginal costs of emissions reduction across all sources (Nordhaus, 1994a) and generate the same costs no matter which burden-sharing rule is applied. Hence, there is no efficiency–equity trade-off and no obstacle to considering equity issues within climate change while emphasizing cost-

---

[41] The comparison and aggregation of welfare in terms of monetary units such as GDP across different countries is a controversial issue. Attempts have been made to incorporate equity considerations through weighting the welfare changes, giving attention to the unequal distribution of wealth among developed and developing countries (Tol *et al.*, 1999).

Table 10.10: *Selected studies of applied equity principles and burden-sharing rules*

Reference	Subject of investigation	Geographical mapping	Results			
			Numerical results*			
Torvanger and Godal (1999)	Emission limitations that could occur if burdens were to follow the	Countries in Baltic Sea Region	Sov.	Egal.	Abil.	
			all			
	• Sovereignty pinciple	Denmark	−6	18	−14	
	• Egalitarian principle (to fulfil the	Estonia	−6	−37	−4	
	Kyoto Protocol)	Finland	−6	27	−15	
	• Ability-to-pay principle (assuming	Germany	−6	8	−12	
	no increase in emissions)	Iceland	−6	45	−13	
		Latvia	−6	23	−4	
		Lithuania	−6	19	−3	
		Norway	−6	29	−13	
		Poland	−6	15	−1	
		Russia	−6	112	−14	
		Sweden	−6	−20	−4	
			* changes compared to 1990 levels, in per cent			
Rose *et al.* (1998)		Global, 9 Regions	Sov.	Egal.	Hor.	Vert.
	• Sovereignty	USA	8.2	67.7	9.5	17.3
	• Egalitarian	Can, W. Europe	5.6	29.8	7.0	3.3
	• Horizontal	Other OECD	1.5	12.5	3.8	8.2
	• Vertical	EEFSU	6.2	55.9	4.1	1.1
		China	3.9	−25.4	1.2	0.0
		Middle East	1.0	0.3	1.3	0.6
		Africa	0.8	−36.3	0.8	0.0
		Latin America	1.3	−10.6	1.3	0.1
		Southeast Asia	2.1	−63.3	1.6	0.0
	EEFSU: Eastern Europe and Former Soviet Union		* net cost impacts in the year 2005, in billions of 1990 US$			
OECD/IEA (1994)	Emission limitations following 10% reduction in world emissions according to	Global, 10 Regions		Egal.	Hor.	Vert.
		North America		11	2.5	12
		West/North Europe		7	2	12
	• Egalitarian	Pacific OECD		21	3	52
	• Horizontal	Central/E. Europe		25	39	6
	• Vertical	Former SU		11	8	4
		East Asia		8	14	6
		China		3	23	2
		Middle East		23	24	13
		Latin America		7	12	5
		Africa		5	24	3
			* in per cent			

(continued)

Table 10.10: *continued*

Reference	Subject of investigation	Geographical mapping	Main features
Elzen *et al.* (1999, 2000) FAIR model (Framework to Assess International Regimes for burden sharing)	• the Brazilian proposal (revised and original approach), as application of polluter-pays principle • Brazilian methodology for estimating historical emissions • Triptych approach 　Phylipsen *et al.* (1998) 　Blok *et al.* (1997) 　Sector oriented	Analysis extended to global scale	• only allocation-based criteria • accounting for historical emissions and/or a per-capita approach favour developing countries • inclusion of all GHG and land use emissions favours developed countries • energy-related CO_2 emissions may still increase because of high growth in non-Annex I emissions, especially in industrial sector • energy efficiency plays a major role in emissions reduction if combined with global diffusion of technology
Byrne *et al.* (1998)	Proposal for egalitarian principle on the basis of 1989 population	140 countries Four income groups	• achieving economic parity in 2050 • increase in CO_2 emissions for low-income countries • reduction in CO_2 emission for upper-income countries
Ringuis *et al.* (1998)	Horizontal: equal weight, CO_2/capita, CO_2/unit GDP, GDP/capita, GDP, CO_2	OECD	• none of the rules in which it is possible to allocate costs among countries and into economic and social drivers equalizes costs across the OECD
Rowlands (1997)	Historical (reactive and proactive) Equality Efficiency	OECD	• twin-track strategy: short term flat-rate approach, long-term differentiated approach

Note: EEFSU= Eastern Europe and Former Soviet Union

effectiveness. Equity rules play an important role when determining the initial distribution of emissions allowances, or the compensation schemes, as cost-effectiveness might result in a disproportionately high level of burden to certain groups of countries. Attempts can be made to provide resource transfers to compensate for the disadvantaged (Biermann, 1997). Usually, it is assumed that mitigation costs are relatively higher in developed countries. Thus, trading reduces the costs to developed countries and provide a transfer to developing countries. Yet, the magnitude of side payments needs to be considered when evaluating alternative burden-sharing rules, because they often generate rather high transaction and/or administrative costs (Burniaux, 1999). If, however, use of the flexibility mechanisms is restricted and equalization of marginal abatement costs throughout the countries cannot be fulfilled, the choice of burden-sharing rule matters with respect to the aggregate abatement costs. Furthermore, emissions trading is usually perceived to take place in a perfect market with parties having equal opportunities of involvement. Agarwal and Narain

(1991) see an advantage for developed nations who have stronger market capacities.

Montgomery (1997) points out that it is not only international negotiators who must consider equity, but also domestic legislators. In an attempt to limit the competitiveness consequences implied by the loss of capital and jobs that may accompany efforts to limit domestic emissions, legislators may act to impose trade barriers. This is another aspect of the need to link international equity and negotiations to the fairness concerns in domestic implementations.

This section shows that equity, opportunities for cost-effectiveness, and flexibility are among the main criteria that a burden sharing rule should satisfy. While it is clear that Pareto optimality is a broadly accepted efficiency principle, there is no agreement on a best equity principle. Therefore theories of justice do not generate one best solution for the international allocation of emissions permits. It appears more important to emphasize negotiating principles that are widely accepted, regarded as equitable, and politically feasible. Beckerman and Pasek (1995), for instance, propose to minimize the proportionate loss of welfare in any voluntary agreement for public goods and lay a smaller burden on the poorest participants.

Much of the debate about equity in climate change mitigation deals with social, economic, and political issues, including international economic development and the unequal distribution of wealth within and among countries. Views diverge widely. Is climate change an opportunity to solve the large problems of sustainable development and global distribution of wealth? Or would broadening the scope for the anyway complex and controversial issue of climate change run the risk of neither solving the climate problem nor improving prospects for sustainable development. Helm (1999) presents an analysis of fair sharing of GHG limitation burdens by separating the climate issue from the dispute about the global welfare distribution. In contrast, Rayner and Malone (2000) pursue a holistic approach to equity and address climate change as an arena in which to debate a wide variety of economic and political issues. In this context, equity is perceived as a basis for generating social capital, which is necessary, together with economic, natural, and intellectual capital, for sustainability.

10.4.6 Towards what Objective Should the Response Be Targetted? High versus Low Stabilization Levels–Insights on Mitigation

In a rational world, the ultimate level of climate and thus GHG concentration stabilization would emerge from a political process in which the global community would weigh mitigation costs and the averted damages associated with different levels of stabilization. Also weighed would be the risks of triggering systemic changes in large geophysical systems, like ocean circulation, or other irreversible impacts. In reality, the political process will inevitably be influenced by the distribution of positive and negative effects of climate change, as well as by the costs of mitigation among countries, largely determined by how risks, costs, environmental values, and development aspirations are weighed in different regions and cultures. This process will be strongly influenced by new scientific and technical knowledge and by experience gained in making and implementing policy. The climate change literature contains a diversity of arguments as to why either a low level or a relatively high level of stabilization is desirable (IPCC, 2001b).

Given the large uncertainties that characterize each component of the climate change problem, it is impossible to establish a globally acceptable level of stabilized GHG concentrations today. Studies discussed in this section and summarized in *Table 10.11* support the obvious expectations that lower stabilization targets involve exponentially higher mitigation costs and relatively more ambitious near-term emissions reductions, but, as reported by WGII (IPCC, 2001b), lower targets induce significantly smaller biological and geophysical impacts and thus induce smaller damages and adaptation costs.

10.4.7 Emerging Conclusions with Respect to Policy-relevant Scientific Questions

Looking at the dilemmas covered in previous sections, the following conclusions emerge:

- a carefully crafted portfolio of mitigation, adaptation, and learning activities appears to be appropriate over the next few decades to hedge against the risk of intolerable magnitudes and/or rates of climate change (impact side) and against the need to undertake painfully drastic emission reductions if the resolution of uncertainties reveals that climate change and its impacts might imply high risks;
- the nature of the climate change problem requires that mitigation action at any level needs to start in the near term, as well as the development of appropriate adaptation strategies;
- emission reduction is an important form of mitigation, but the mitigation portfolio includes a broad range of other activities, including investments to develop low-cost non-carbon energy, and to improve energy efficiency and carbon management technologies to make future CO_2 mitigation inexpensive;
- timing and composition of mitigation measures (investment in technological development or immediate emission reductions) is highly controversial because of the technological features of energy systems, and the range of uncertainties involved with, for example, their impacts of climate change;
- international flexibility instruments help reduce the costs of emission reductions, but they raise a series of implementation and verification issues that need to be balanced against the cost savings;
- while there is a broad consensus on Pareto optimality as an efficiency principle, there is no agreement on the best equity principle for burden sharing. Efficiency and equity are important concerns in negotiating emissions limitation schemes, and they are not mutually exclusive. Therefore, equity will play an important role in determining the distribution of emissions allowances and/or within compensation schemes that follow emissions trading resulting in a disproportionately high level of burden to certain countries. Finally, it is more important to rely on politically feasible burden-sharing rules than to select one specific equity principle.

Finally, a series of potential large-scale geophysical transformations that might exert a major influence on the desired level of stabilization have been identified and examined more closely in recent years. These imply thresholds that humanity might decide not to cross because the potential impacts or even the associated risks are considered to be unacceptably high. Little is know about these thresholds today. Most recent results and

Table 10.11: Selected studies on global mitigation costs for different stabilization targets

Study	Scenarios & dimensions	450ppmv	550ppmv	650ppmv	750ppmv	850ppmv	Notes
Nordhaus and Boyer (1999) **RICE-98**	billion 1990 US$ net impacts on global welfare difference from base (=0) mitigation reduction in cl.damage		335.00 459.00 794.00				discounted (d) back to 1950 IAM
Valverde and Webster (1999) **MIT-EPPA 1.6**	in billion 1985 US$ 500a global emissions path all nations equal % abatement OECD: no trade OECD: trade Non-OECD: no trade Non-OECD: trade Global: no trade Global trade		(500ppmv) 272.50 272.40 216.30 216.20 488.90 488.60				1985-2100, 5 year time steps d= 5%
Manne and Richels (1999b) **MERGE 3.0**	billion of 1990 US$ Kyoto followed by arbitrary reduc. Kyoto followed by least-cost least cost		2400.00 900.00 650.00				consumption loss through 2100 d=5% to 1990
Tol (1999c), FUND	percentage of world income, median		2.1%				average annual income loss 9 regions, 5 sectors in 2100
Ha-Duong et al. (1997) **DIAM**	percentage of 1990 GWP inertia of 50 years	1.1%					d=3% average annual costs period 2000–21000
Lecocq et al. (1998) **STARTS**	percentage abatement costs as consumption loss compared to BAU in A: flexible sector B: rigid sector		 1,5% 0.4%				average annual costs period 2000-2100 differential inertia in sectors

(continued)

Table 10.11: continued

Study	Scenarios & dimensions	450ppmv	550ppmv	650ppmv	750ppmv	850ppmv	Notes
Yohe and Wallace (1996) Connecticut	percent of GWP as costs in 1990, no benefit side one percentage point is ~210 billion US$		16.40 to 16.69	5.94 to 7.09	2.84 to 4.24	1.59 to 3.05	d, expected present value 7 scenarios
Dowlatabadi (1998) ICAM-3	percentage of GDP as costs mitigation costs within 9 scenarios with different technical change in energy sector		0.05 to 0.48				period: 2000–20025 sequential learning framework mitigation costs for the USA and Canada only
Richels a. Edmonds (1995) Global 2100 & ERB	percentage of GWP as costs Sc:500a: follow BAU through 2010 Sc: 500b: between 500a & stab. Sc: Emission stabilization at 1990 level	(400) GI: 0.9%; ERB:1.1%	(500) GI 2100: 0.6 %; ERB: 0.7% GI 2100: 0.9 %; ERB: 0.95% GI 2100: 1.15 %; ERB: 1.1% GI:0.6%; ERB: 0.7%				d=5% stabilization in 2100 Global 2100 Manne/Richels ERB: Edmonds-Reilly-Barns (1992)
Plambeck, Hope (1996) PAGE95	in trillion US$ BAU + 100 GtC BAU	2.50 2.20					d= 5 %, 1990-2200 further scenarios not listed here, e.g., non-linear etc.
Yohe and Jacobsen (1999) Connecticut	trillion 1990 US$ annual control costs of 7 sc Minimum cost: Sc3 to Sc7 Cost with Kyoto: Sc3 to Sc7 Minimum cost minus 10 % emis. Cost with Kyoto minus 10 %	A: 10.13 to 44.40 A: 10.47 to 47.04 B: 10.13 to 44.40 B: 10.40 to 46.77	A: 2.11 to 16.12 A: 2.12 to 16.19 B: 2.11 to 16.12 B: 2.13 to 16.16	A: 0.36 to 7.24 A: 0.40 to 7.26 B: 0.36 to 7.24 B: 0.42 to 7.28			d=3 % through 2100 (Ramsey) cost study in terms of deadweight loss, no opt A: alt. sink specifications B: alt. emissions targets for 2010
Manne (1995) MERGE	trillion 1990 US$ global damage benefits of stab.as reduced dam. costs of stab.	(415 ppmv) 1.90 2.50 18.50					d= 5 % to 1990

(continued)

Table 10.11: continued

Study	Scenarios & dimensions	450ppmv	550ppmv	650ppmv	750ppmv	850ppmv	Notes
Manne and Richels (1997) MERGE 3.0	trillion 1990 US$						
	WGI: w/o where flex.	14.20	9.00	5.00	3.00		d=5% to 1990
	WGI: with where flex	7.00	4.00	2.00	1.2		1990-2100
	WRE: w/o where flex	5.50	2.00	1.00	1.00		non-market and
	WRE: with where flex	3.50	1.00	0.6	0.50		market damages
	least cost: with where flex		0.60				
	WGI: Annex-1-trade		5.90				
	10% cut in 2010: A-1-trade		2.30				
	WRE: A-1-trade		0.90				
Tol (1999d), FUND 1.6	trillion net present costs in US$						
	WGI: no trade	32.50	17.5	10.50			5% through 2050
	WGI: trade	17.50	8.0	4.00			damage per year
	WRE: no trade	34.00	16.0	10.00			in billion US$: 216
	WRE: trade	13.00	4.0	2.00			
Tol (1999a) FUND 1.6	in trillion US$		below 550				
	Minimum Cost		2.4				d= 5% per year to 1990
	Min. Cost meeting Kyoto, trade		3.1				consumption losses p.a.
	Min. Cost meeting Kyoto		3.7				period 1990-2200
	2 % reduction, interm. trade		4.0				
	meeting Kyoto, trade		4.4				
	meeting Kyoto, no trade		14.6				

the implications of the possibility of such thresholds are summarized in Chapter 19 of WGII (IPCC, 2001b). Nevertheless, currently estimated "danger zones" are in the domain of high stabilization levels for most threshold events.

Considering the special combination of features of the climate problem listed at the beginning of this chapter, it is obvious that no "once forever" solution exists. Making long-term commitments in any area where retraction is possible is problematic. Making decisions that entail long-term and possibly irreversible consequences due to long delays, inertia and similar system properties is even more difficult, especially under severe uncertainties. Therefore, as emphasized in this chapter, the most promising approach to climate policy is sequential decision-making. This process involves a regular reassessment of the long-term climate risks (net damages from a given magnitude of climate change) and their management objectives (climate or GHG concentration stabilization) in the light of newly available information. Short-term strategies are then crafted so that both GHG emissions and the underlying socioeconomic processes (resource use, technologies) evolve in a direction which makes future course corrections in any direction the least expensive. The current structure of the international climate regime is formulated in this vein: the UNFCCC provides some, albeit vague, guidelines for long term stabilization objectives while short-term goals are settled in and implemented under protocols for each budget period.

The analytical tools to support the above decision-making processes need to handle this double feature. They should provide policymakers with guidance to set long-term targets and to formulate short-term policies and measures. Some models take a long-term view to explore deep future impacts of climate change, but this must not be interpreted as suggesting optimal strategies for the next 50-100-200 years. Other models explore what are the most promising near-term policies and how to implement them. Similarly, many studies and models reviewed in this chapter consider the world as a whole or broken down into a few regions, at best. Others take a more detailed look at subnational and regional aspects. They shed light on the smaller scale implications of climate change and its management strategies, often in the context of other social concerns characterizing the country or region. Our assessment has found a healthy diversity of DAFs along both the long-term-short term and the global-local axes. Nevertheless, the analytical capacity and thus quotable results are still badly missing in most developing countries. This is probably the most severe problem to be solved by the time the world community will prepare its next climate change assessment report.

References

Agarwal, A., and S. Narain, 1991: *Global warming in an unequal world: a case of environmental colonialism.* Centre for Science and Environment, New Delhi.

Agarwal, A., and S. Narain, 2000: How Poor Nations Can Help To Save The World. In *Climate Change and Its Linkages with Development, Equity, and Sustainability,* Proceedings of the IPCC Expert Meeting held in Colombo, Sri Lanka, 27-29 April 1999, M. Munasinghe, R. Swart, (eds.), LIFE, RIVM, World Bank, pp. 191-215.

Albin, C., 1995: Rethinking Justice and Fairness: the Case of Acid Rain Emission Reduction. *International Affairs,* **21**(2), 119-143.

Alcamo, J., 1994: *IMAGE 2.0 - Integrated Modeling of Global Climate Change.* Kluwer, Dordrecht.

Alcamo, J., and E. Kreileman, 1996a: *The Global Climate System: Near Term Action for Long Term Protection.* Background Report prepared for the Workshop on Quantified Emissions Limitation Reduction Objectives (QUELROS) at the Third Meeting of the Ad Hoc Group on the Berlin Mandate of the Framework Convention on Climate Change, Report 481508001, National Institute of Public Health and the Environment (RIVM), Bilthoven, The Netherlands.

Alcamo, J., and E. Kreileman, 1996b: Emissions Scenarios and Global Climate Protection. *Global Environmental Change,* **6**(4), 305-334.

Anderson, D., 1997: Renewable Energy Technology and Policy for Development. *Annual Review of Energy and Environment,* **22**, 187-215.

Appadurai, A., 1986: *The Social Life of Things. Commodities in Cultural Perspective.* Cambridge University Press, Cambridge/London/ New York.

Ariès, P., 1977: *L'homme devant la mort.* Ed. du Seuil, Pari.

Bankes, S. C., 1993: Exploratory Modeling for Policy Analysis. *Operations Research,* **41**(3), 435-449.

Banuri, T., K. Göran-Mäler, M. Grubb, H. K. Jacobson, and F. Yamin, 1996: Equity and Social Considerations. In IPCC, Climate Change 1995: Economic and Social Dimensions of Climate Change, Contribution of Working Group III to the Second Assessment Report of the IPCC, Cambridge University Press, Cambridge, 79-124.

Barker, T., 1999: Achieving a 10% Cut in Europe's Carbon Dioxide Emissions Using Additional Excise Duties: Coordinated, Uncoordinated and Unilateral Action Using the Econometric Model E3ME. *Economic Systems Research,* **11**(4), 401-442.

Barrett, S., 1992: Reaching a CO_2-emission limitation agreement for the Community: implications for equity and cost-effectiveness. *European Economy,* Special Edition, **1**(1), 3-23.

Barrett, S., 1994: Self-Enforcing International Environmental Agreements. *Oxford Economic Papers,* **46** (October), 878-894.

Barrett, S., 1995: *Trade Restrictions in International Environmental Agreements.* London Business School, London.

Barrett, S., 1997a: Heterogeneous International Environmental Agreements. In *International Environmental Agreements: Strategic Policy Issues.* C. Carraro, (ed.), E. Elgar, Cheltenham, UK.

Barrett, S., 1997b: Towards a Theory of International Co-operation. In *New Directions in the Economic Theory of the Environment.* C. Carraro, D. Siniscalco, (eds.), Cambridge University Press, Cambridge.

Barrett, S., 1997c: The Strategy of Trade Sanctions in International Environmental Agreements. *Resources and Energy Economics,* **19**(4), 345-361.

Barrett, S., 1998: Political Economy of the Kyoto Protocol. *Oxford Review of Economic Policy,* **14**(4), 20-39.

Barrett, S., 1999: Montreal vs. Kyoto International Cooperation and the Global Environment. In *Global Public Goods: International Cooperation in the 21st Century.* United Nations Development Program.

Beckerman, W., and J. Pasek, 1995: The equitable international allocation of tradable carbon emission permits. *Global Environmental Change,* **5**(5), 405-413.

Begg, K., S. D. Parkinson, T. Jackson, P-E. Morthorst, and P. Bailey, 1999: *Overall Issues for Accounting for the Emissions Reductions of JI Projects.* Proceedings of Workshop on Baselines for the CDM, February 25-26. Global Industrial and Social Progress Research Institute (GISPRI), Tokyo.

Beierle, T., 1998: Public Participation in Environmental Decisions: An Evaluation Framework Using Social Goals. *Discussion Paper 99-06,* Resources for the Future, Washington DC.

Bernow, S., K. Cory, W. Dougherty, M. Duckworth, S. Kartha, and M. Ruth, 1999: *America's Global Warming Solutions.* Worldwildlife Fund, Washington, DC.

Bernstein, P. M., W. D. Montgomery, and T. F. Rutherford, 1999: Global Impacts of the Kyoto Agreement: Results from the MS-MRT Model. *Resource and Energy Economics,* **21**(3/4), 375-413.

Bernstein, P. M., W. D. Montgomery, T. F. Rutherford, and G.-F. Yang, 1999: Effects of Restrictions on International Permit Trading: The MS-MRT Model. *Energy Journal,* Kyoto Special Issue, 221-256.

Biermann, F., 1997: Financing environmental policies in the south: Experience from the multilateral ozone fund. *International Environmental Affairs,* **9**(3), 179-218.

Bloch, F., 1997: Non-co-operative Models of Coalition Formation in Games with Spillovers. In *New Directions in the Economic Theory of the Environment.* C. Carraro, D. Siniscalco, (eds.), Cambridge University Press, Cambridge, 311-352.

Bohm, P., 1999: *International Greenhouse Gas Emission Trading.* Nordic Council of Ministers, Tema Nord 1999, Stockolm.

Böhringer, C., 1999: Cooling down Hot Air. *Discussion Paper No. 99-43,* Center for European Economic Research, Mannheim, Germany.

Bosello, F., and C. Carraro, 1999: Recycling Energy Taxes. Impacts on Disaggregated Labour Market. *FEEM Nota di Lavoro No. 79.99,* Milan, 79–99.

Bosello, F., and R. Roson, 1999: Carbon Emissions Trading and Equity in International Agreements. Paper presented at the 3rd FEEM-IDEI-INRA Conference on Environmental and Resource Economics, June 14-16, Toulouse.

Bosello, F., C. Carraro, L. Kernevez, D. Raggi, and R. Roson, 2000: *Equity, Stability and Efficiency of Climate Agreements.* Fondazione ENI E. Mattei, Milan.

Bossel, H., 1998: *Earth at Crossroads: Paths to a Sustainable Future.* Cambridge University Press, UK.

Bossel, H., 1999: *Indicators for Sustainable Development: Theroy, Method, Applications.* International Institute for Sustainable Development, Winnipeg, Canada.

Botteon, M., and C. Carraro, 1997a: Burden-Sharing and Coalition Stability in Environmental Negotiations with Asymmetric Countries. In *International Environmental Agreements: Strategic Policy Issues.* C. Carraro, (ed.), E. Elgar, Cheltenham, UK.

Botteon, M., and C. Carraro, 1997b: Strategies for Environmental Negotiations: Issue Linkage with Heterogeneous Countries. In *Game Theory and the Global Environment.* H.Folmer, N.Hanley, (eds.), E. Elgar, Cheltenham, UK.

Bovenberg, L. A., 1997: Environmental Policy, Distortionary Labour Taxation and Employment: Pollution Taxes and the Double Dividend. In *New Directions in the Economic Theory of the Environment.* C. Carraro, D. Siniscalco, (eds.), Cambridge University Press, Cambridge, 69-104.

Brown, S., D. Kennedy, C. Polidano, K. Woffenden, G. Jakeman, B. Graham, F. Jotzo, B., and S. Fisher, 1999: Modeling the Economic Impacts of the Kyoto Protocol. Accounting for the three major greenhouse gases. ABARE Research Report 99.6, Canberra., Australia.

Brown Weiss, E., 1999: Understanding Compliance with international environmental agreements: The Bakers`s Dozen Myths. *University of Richmond Law Review,* **32**, 1555-1585.

Brown Weiss, E., and H. K. Jacobson, (eds.), 1998: *Engaging Countries:Strengthening Compliance with International Environmental Accords.* MIT Press, Cambridge, MA/London, UK.

Bruckner, T., G. Petschel-Held, F. L. Toth, H.-M. Füssel, M. Leimbach, and H.-J. Schellnhuber, 1999: Climate Change Decision-Support and the Tolerable Windows Approach. *Environmental Modeling and Assessment,* **4**, 217-234.

Buchholz, W., and K. A. Konrad, 1994: Global Environmental Programs and the Strategy Choices of Technology. *Journal of Economics,* **60**(3), 379-387.

Buonnano, P., E. Castelnuovo, C. Carraro, and M. Galeotti, 1999: Efficiency and Equity of Emission Trading with Endogenous Environmental Technical Change. In *Efficiency and Equity of Climate Change Policy.* C. Carraro, (ed.), Kluwer Academic Publishers, Dordrecht.

Burniaux, J. M., 1998: *How Important is Market Power in Achieving Kyoto? An Assessment Based on the GREEN Model.* Proceedings of the Workshop on Economic Modelling and Climate Change, 17-18 September 1998, OECD, Paris.

Burniaux, J-M, 1999: Burden Sharing Rules in Post-Kyoto Strategies: A General Equilibrium Evaluation based on The Green Model. Working Paper, presented at the EMF/IEA meeting, Paris, June 1999, OECD, Paris.

Burniaux, J-M., J-P. Martin, G. Nicolletti, and J. O. Martins, 1992: The Cost of International Agreements to Reduce CO2 Emissions. *European Economy*, Special Edition, **1**, 271-298.

Burns, T.R., and R. Ueberhorst, 1988: *Creative Democracy: Systematic Conflict Resolution and Policymaking in a World of High Science and Technology.* Praeger, New York, NY.

Bush, E. J., and L. D. D. Harvey, 1997: Joint Implementation and the Ultimate Objective of the United Nations Framework on Climate Change. *Global Environmental Change*, **7**(3), 265-285.

Byrne, J., Y.-D. Wang, H. Lee, and J.-D. Kim, 1998: An equity- and sustainability based policy response to global climate change. *Energy Policy*, **26**(4), 335–343.

Callaway, J.M., L.O. Naess, and L. Ringuis, 1998: Adaptation Costs: A Framework and Methods. In *Mitigation and Adaptation Cost Asessment, Concepts, Methods and Appropriate Use.* UNEP Collaborating Centre on Energy and Environment, Riso National Laboratory, Denmark, 97-119.

Cameron, J., J.Wersman, and P. Roderick (eds.), 1996: *Improving Compliance with International Environmental Law.* Earthscan, London.

Cantor, R, and G. Yohe, 1998: Economic activity and analysis. *Human Choice & Climate Change Tools for Policy Analysis*, **3**, 1-103.

Capros, P., 1998: *Economic and Energy System Implications of European CO2 Mitigation Strategy: Synthesis of Results from Model Based Analysis.* Proceedings of the Workshop on Economic Modelling and Climate Change, 17-18 September 1998, OECD, Paris, 27-48.

Carraro, C., (ed.), 1997. *International Environmental Agreements: Strategic Policy Issues.* E. Elgar, Cheltenham, UK.

Carraro, C., 1998: *Beyond Kyoto: A Game-Theoretic Perspective.* Proceedings of the Workshop on Economic Modelling and Climate Change, 17-18 September 1998, OECD, Paris.

Carraro, C., 1999a: Environmental Conflict, Bargaining and Cooperation. In *Handbook of Natural Resources and the Environment.* J. van den Bergh, (ed.), E. Elgar, Cheltenham, UK.

Carraro, C., 1999b: The Economics of International Coalition Formation and EU Leadership. In *European Leadership of Climate Change Regimes.* M. Grubb, J. Gupta, (eds.), Kluwer Academic Publ., Dordrecht.

Carraro, C., 1999c: The Structure of International Environmental Agreements. In *International Environmental Agreements on Climate Change.* C.Carraro, (ed.), Kluwer Academic Publ., Dordrecht, 9-25.

Carraro, C., (ed.), 1999d: *International Environmental Agreements on Climate Change.* Kluwer Academic Publ., Dordrecht.

Carraro, C., (ed.), 2000: *Efficiency and Equity of Climate Change Policy.* Kluwer Academic Publ., Dordrecht.

Carraro, C., and D. Siniscalco, 1992: The International Protection of the Environment: Voluntary Agreements among Sovereign Countries. In *The Economics of Transnational Commons.* P. Dasgupta, K.G. Maler, (eds.), Clarendon Press, Oxford.

Carraro, C., and D. Siniscalco, 1993: Strategies for the International Protection of the Environment. *Journal of Public Economics*, **52**, 309-328.

Carraro, C. and D. Siniscalco, 1995: Policy Coordination for Sustainability: Commitments, Transfers, and Linked Negotiations. In *The Economics of Sustainable Development.* I. Goldin and A. Winters, (eds.), Cambridge University Press, Cambridge.

Carraro, C., and D. Siniscalco, 1997: R&D Cooperation and the Stability of International Environmental Agreements. In *International Environmental Agreements: Strategic Policy Issues.* C. Carraro, (ed.), E. Elgar, Cheltenham, UK.

Carraro, C. and F. Moriconi, 1998: Endogenous Formation of Environmental Coalitions. Paper presented at the 3rd Coalition Theory Network Workshop on Coalition Formation: Applications to Economic Issues, 8-10 January 1998, Venice, Italy.

Carraro, C. and D. Siniscalco, 1998: International Environmental Agreements: Incentives and Political Economy. *European Economic Review*, **42**(3-5), 561-572.

Carraro, C., and J. C. Hourcade (eds.), 1999: Optimal Timing of Climate Change Policies. *Energy Economics*, Special Issue, Elsevier, Amsterdam.

Carter, L., K. Andrasko, and W. v. d. Gaast, 1996: *Technical Issues in JI/AIJ Projects: A Survey and Potential Responses.* Proceedings of the Workshop on New Partnerships to Reduce the Buildup of Greenhouse Gases, 9-31 October 1996, San Jose, Costa Rica, UNEP.

CEC, 2000: Commission of the European Communities: Communication from the Commission. on the Recautionary Principle, COM(2000) 1, Brussels.

Cesar, H., and A. De Zeeuw, 1996: Issue Linkage in Global Environmental Problems. In *Economic Policy for the Environment and Natural Resources.* A. Xepapadeas, (ed.), E. Elgar, Cheltenham, UK.

Chander, P., and H. Tulkens, 1995: A Core-Theoretical Solution for the Design of Cooperative Agreements on Trans-Frontier Pollution. *International Tax and Public Finance*, **2**(2), 279-294.

Chander, P., and H. Tulkens, 1997: The Core of an Economy with Multilateral Environmental Externalities. *International Journal of Game Theory*, **26**, 379-401.

Chander, P., H. Tulkens, J. P. Van Ypersele, and S. Willems, 1999: The Kyoto Protocol: an Economic and Game Theoretic Interpretation. CLIMNEG working paper No. 12, CORE, UCL, Louvain, Belgium.

Chayes, A., and A. H. Chayes, 1993: On Compliance. *International Organization*, **47**(Spring), 175 – 207.

Chayes, A., and A. H. Chayes, 1995: *The New Sovereignty: Compliance with Treaties of International Regulatory Regimes.* Harvard University Press, Cambridge, MA.

Cheng, H. C., M. Steinberg, and M. Beller, 1985: *Effects of Energy Technology on Global CO2 Emissions.* TR030, DOE/NBB-0076, National Technical Information Service, U.S. Department of Commerce, Springfield, VA.

Chew, M. S., 1994: Farsighted Coalitional Stability. *Journal of Economic Theory*, **54**, 299-325.

Christiansson, L., 1995: Diffusion and Learning Curves of Renewable Energy Technologies. Working Paper WP-95-126, International Institute for Applied Systems Analysis (IIASA), Laxenburg, Austria.

Clark, W. R., 1998: Agents and Structures: Two views of preferences, two views of institutions. *International Studies Quarterly*, **42**(2), 245-269.

Coase, R., 1960: The problem of social cost. *Journal of Law and Economics*, **1**(October), 1-44.

Cohen, S., D. Demeritt, J. Robinson, and D. Rothman, 1998: Climate change and sustainable development: towards dialogue. *Global Environmental Change*, **8**(4), 341-371.

Convery, F. J., 1999: Blueprints for a Climate Policy: Based on Selected Scientific Contributions. In *Integrating Climate Policies in a European Environment.* C. Carraro, (ed.), Special Issue of Integrated Assessment, Baltzer Publ.

Cooper, A., S. Livermore, V. Rossi, A. Wilson, and J. Walker, 1999: The Economic Implications of Reducing Carbon Emissions: A Cross-Country Quantitative Investigation using the Oxford Global Macroeconomic and Energy Model. *Energy Journal*, Kyoto Special Issue, 335 - 365.

Corfee-Morlot, J. and P. Schwengels, 1994: *Greenhouse Gas Verification - Why, How and How Much?: proceedings of a workshp Bonn, April 28-29.* W. Katscher, G. Stein, J. Lanchbery, J. Salt, (eds.), KFA Forschungszentrum, Jülich, Germany.

Cousins, S., 1989: Culture and Selfhood in Japan and the US. *Journal of Personality and Social Psychology*, **56**(1), 124-131.

Criqui, P., S. Mima, and L. Viguier, 1999: Marginal Abatement Costs of CO2 Emission Reductions, Geographical Flexibility and Concrete Ceilings: an Assessment using the POLES Model. *Energy Policy*, **27**(10), 585-601.

Daly, H., and J. Cobb, 1989: *For the Common Good — Redirecting the Economy Towards Community, the Environment and Sustainable Development.* Beacon Press, Boston, MA, 482 pp.

Darwin, R., 1999: A FARMER's View of the Ricardian Approach to Measuring Agricultural Effects of Climatic Change. *Climatic Change*, **41**(3/4), 371-411.

Darwin, R., M. Tsingas, J. Lewandrowski, and A. Raneses, 1996: Land use and cover in ecological economics. *Ecological Economics*, **17**(3), 157-181.

Dawes, R. M., 1980: Social Dilemmas. *Annual Review of Psychology*, **31**, 169-193.

De Canio, S. J. and J. A. Laitner, 1997: Modeling Technological Change in Energy Demand Forecasting: A Generalized Approach. *Technological Forecasting and Social Change*, **55**(3), 249-263.

De Canio, S. J., C. Dibble, and K. Amir-Atefi, 2000a: The Importance of Organizational Structure for the Adoption of Innovations. *Management Science*, **46** (10), 1285-1299.

De Canio, S. J.., C. Dibble, and K. Amir-Atefi, 2000b: Organizational Structure and the Behavior of Firms: Implications for Integrated Assessment. *Climatic Change*, **48**, (2/3), 487-514.

De Canio, S. J., W. E. Watkins, G. Mitchell, K. Amir-Atefi, and C. Dibble, 2000c: Complexity in Organizations: Consequences for Climate Policy Analysis. In *Advances in the Economics of Environmental Resources*. R. B. Howarth , D. C. Hall, (eds.), JAI Press Inc., Greenwich, CT.

De La Vega Navarro, A., 2000: *Energy markets, oil companies and climate change issues.* Proceedings of the IPCC Havana Meeting on Development, Equity and Sustainability.

De Marchi, B., and J. Ravetz, 1999: Risk management and governance: a post-normal science approach. *Futures*, **31**(7), 743-757.

Den Elzen, M. G. J., M. Berk, M. Shaeffer, J. Olivier, C. Hendriks, and B. Metz, 1999: The Brazilian Proposal and other Options for International Burden Sharing: an evaluation of methodological and policy aspects using the FAIR model. Report 728001011, RIVM, Bilthoven.

Den Elzen, M. G. J., M. Berk, A. Farber, and R. Oostenrijk, 2000: FAIR 1.0 – An interactive model to explore options for differentiation of future commitments under the climate change convention, Report 728001012, RIVM, Bilthoven, The Netherlands.

Deutscher Bundestag, 1994:*Mobility and Climate. Developing Environmentally Sound Transport Policy Concepts.* Economica Verlag, Bonn.

Diekmann, A., and P. Preisendörfer, 1992: Persönliches Umweltverhalten – Diskrepanzen zwischen Anspruch und Wirklichkeit. *Kölner Zeitschrift für Soziologie und Sozialpsychologie*, **44**(1), 226-251.

Ding, X., 1998: On the relationship of scientific progress and economic development in the country. *The World Science and Technology Research and Development*, **19**(5), 78-81 (in Chinese).

Dittmar, H., 1992: *The Social Psychology of Material Possessions. To Have Is To Be.* Harvester Wheatsheaf, Hemel Hempstead, UK.

Dooley, J. J., J. A. Edmonds, and M. A. Wise, 1999 : The Role of Carbon Capture & Sequestration in a Long-Term Technology Strategy of Atmospheric Stabilization. In *Greenhouse Gas Control Technologies.* Eiasson, B., P. Riemer, A. Wokaun, (eds), Pergamon Press, 857-861.

Douglas, M., and B. Isherwood, 1979: *The World of Goods.* Basic Books, New York, NY.

Douglas, M., D. Gasper, S. Ney, and M. Thompson, 1998: Human Needs and Wants. In *Human Choice and Climate Change. Vol. I: The societal framework.* S. Rayner, E. L. Malone, (eds.), Battelle Press, Columbus, OH, 195-263.

Dowlatabadi, H., 1998: Sensitivity of Climate change mitigation estimates to assumptions about technical change. *Energy Economics*, **20**(5/6), 473-493.

Dowlatabadi, H., and M. G. Morgan, 1995: *Integrated assessment climate assessment model 2.0, technical Documentation.* Department of Engineering and Public Policy, Carnegie Mellon University.

Downs, G. W., D. M. Rocke, and P. Barsoom, 1996: Is the Good News about Compliance Good News about Cooperation? *International Organization*, **50**(3), 379-406.

Duchin, F., 1998: *Structural Economics. Measuring Change in Technology, Lifestyles, and the Environment.* Island Press, Washington, DC.

Duchin, F.: Environmentally Significant Consumption. In *Environmental/Ecoogical Sciences*. B.L. Turner II, (ed.); *International Encyclopedia of the Social and Behavioral Sciences.* N. J. Smelser, P. B. Baltes, (eds.), Elsevier Science, New York, NY (in press).

EC, 1999: *Accounting and Accreditation of Activities Implemented Jointly.* Final Report prepared for DG XII of the European Commision under the Environment and Climate Programme 1994-1998, Contract No. ENV4-CT96-0210 (DGXII-ESCY).

Edmonds, J., and M. Wise, 1998: The Economics of Climate Change: Building Backstop Technologies And Policies To Implement The Framework Convention On Climate Change. *Energy & Environment*, **9**(4), 383-397.

Edmonds, J., and M. Wise, 1999: Exploring A Technology Strategy for Stabilizing Atmospheric CO_2. In *International Environmental Agreements on Climate Change.* C. Carraro, (ed.), Kluwer Academic Publ., Dordrecht, 131-154.

Edmonds, J., M. Wise, and C. MacCracken, 1994: Advanced energy technologies and climate change: An analysis using the global change assessment model (GCAM). PNL-Report, PNL-9798, UC-402, Pacific Northwest Laboratory, Richland/Washington, DC.

Edmonds, J., M. Wise, H. Pitcher, R. Richels, T. M. L. Wigley, and C. MacCracken, 1996: An Integrated Assessment of Climate Change and the Accelerated Introduction of Advanced Energy Technologies: An Application of MiniCAM 1.0. *Mitigation and Adaptation Strategies for Global Change*, **1**(4), 311-339.

Edmonds, J., M. Wise, and J. Dooley, 1997: Atmospheric Stabilization and the Role of Energy Technology. In *Climate Change Policy, Risk Prioritization and U.S. Economic Growth.* C. E. Walker, M. A. Bloomfield, M. Thorning, (eds.), American Council for Capital Formation, Washington DC, 71-94

Edmonds, J., J.Dooley, and S.H. Kim, 1999: Long-Term Energy Technology: Needs and Opportunities for Stabilizing Atmospheric CO_2 Concentrations. In *Climate Change Policy: Practical Strategies to Promote Economic Growth and Environmental Quality, Monograph series on tax, trade and environmental policies and U.S. economic growth.* American Council for Capital Formation, Center for Policy Research, Washington, DC, 81-97.

Ellerman, A. D., H. D. Jacboy, and A. Decaux, 1998: The Effects on Developing Countries of the Kyoto Protocol and CO_2 Emissions Trading. Report No. 41, MIT Joint Project on the Science and Policy of Global Change, Cambridge, MA.

Ellerman, A. D., H. D. Jacoby, and A. Decaux, 1999: The Effects on Developing Countries of the Kyoto Protocol, and CO_2 Emissions Trading. Paper released under the "Joint Program on the Science and Policy of Global Change", MIT.

Elliott, R. N., and M. Pye, 1998: Investing in Industrial Innovation: A Response to Climate Change. *Energy Policy*, **26**(5), 413-423.

Endres, E. A., and M. Finus, 1998: Playing your Better Global Emissions Game: Does it Help to be Green? *Swiss Journal of Economics and Statistics*, **134**(1), 21-40.

Energy Innovations, 1997: *Energy Innovations: A Prosperous Path to a Clean Environment.* Alliance to Save Energy, American Council for an Energy-Efficient Economy, Natural Resources Defense Council, Tellus Institute and Union of Concerned Scientists, Washington, DC.

Eskeland, G. S., and J. Xie, 1998: Acting Globally while Thinking Locally: Is the Global Environment Protected by Transport Emission Control Programs? Policy Research Working Paper No. 1975, The World Bank.

Eyckmans, J., 2000: On the Farsighted Stability of the Kyoto Protocol: Some Simulation Results. Paper presented at the 5th Coalition Theory Network Workshop, Barcelona, 20-21 Jan. 2000.

Fankhauser, S., 1995a: *Valuing Climate Change – The Economics of the Greenhouse.* Earthscan, London.

Fankhauser, S., 1995b: Protections vs. retreat: the economic costs of sea level rise. *Environment and Planning A*, **27**, 299-317.

Fankhauser, S., 1996: The Potential Costs of Climate Adaptation. In *Adapting to Climate Change: Assessments and Issues.* J. B. Smith, N. Bhatti, G. Menzhulin, R. Benioff, M. I. Budyko, M. Campos, B. Jallow, F. Rijsberman, (eds.), Springer, Berlin, 80 - 96.

Feenstra, J. F., I. Burton, J. B. Smith, and R. S. J. Tol, (eds.), 1998: *Handbook on Methods for Climate Change Impact Assessment and Adaptation Strategies, Version 2.0.* UNEP and Free University (Vrije Universiteit), October 1998.

Folmer, H., P. Van Mouche, and S. E. Ragland, 1993: Interconnected Games and International Environmental Problems. *Environment and Resources Economics*, **3**(4), 313-335.

Forsyth, T., 1998: Technology Transfer and Climate Change Debate. *Environment,* **40**(9).

Forsyth, T., 1999: *International Investment and Climate Change: Energy Technology for Developing Countries.* Royal Institute of International Affairs/Earthscan, London.

French, H., 1998: Making Private Capital Flows to Developing Countries Environmental Sustainable; Policy Challenge. *Natural Resource Forum,* **22**(2).

Fricker, A., 1998: Measuring up to Sustainability. *Futures,* **30**(4), 367-375.

Funtowicz, S.O., and J.Ravetz, 1990: *Uncertainty and Quality in Science for Policy.* Kluwer Academic Press, Dordrech.

Funtowicz, S. O., and J. Ravetz, 1993: Science for the Post-Normal Age. *Futures,* **September,** 739-755.

Funtowicz, S.O., and J. Ravetz, 1994: The Worth of a Songbird: Ecological Economics as a Post-normal Science. *Ecological Economics,* **10**, 197-207.

Gallopin, C.G., 1997: Indicators and Their Use: Information for Decision Making. In *Sustainability Indicators.* B. Moldan, S. Billharz, R. Matravers, (eds.), John Wiley & Sons, 13-17.

Geertz, C., 1979: From the Native's Point of View. On the Nature of Anthropological Understanding. In *Interpretive Social Science: A reader.* P. Rabinow, W.M. Sullivan, (eds.), University of California Press, Berkeley, CA, 225-241.

Geller, H., S. Bernow, and W. Dougherty, 1999: *Meeting America's Kyoto Protocol Target: Policies and Impacts.* American Council for an Energy-Efficient Economy, Washington DC.

General Accounting Office, 1992: International Environment: International Agreements Are Not Well Monitored. GAO/RCED-92-43, Washington DC.

George, C., 1999: Testing for Sustainable Development through Environmental Assessment. *Environmental Impact Assessment Review,* **19**, 175-200.

Georg, S., 1999: The Social Shaping of Household Consumption. *Ecological Economics,* **28**(3), 455-466.

Germain, M., P.Toint, and H. Tulkens, 1997: Financial transfers to Ensure Co-operation International Optimality in Stock-Pollutant Abatement.CORE Discussion Paper No. 9701, Center for Operations Research and Econometrics, Université de Louvain, Belgium.

Glomsrød, S., H. Vennemo, and L. Johnsen, 1992: Stabilisation of Emissions of CO2: A Computable General Equilibrium Assessment. *The Scandinavian Journal of Economics,* **94**(1), 53–69.

Goulder, L.H., 1995: Environmental Taxation and the Double Dividend: A Reader's Guide. *International Tax and Public Finance,* **2**, 157-183.

Goulder, L. H., 1996: Notes on the Implications of Induced Technological Change for the Attractiveness and Timing of Carbon Abatement. Paper prepared for the Workshop on Timing the Abatement of Greenhouse Gas Emissions, June 17-18, 1996, Paris, Department of Economics, Stanford University, CA.

Green, O., 1993: International Environmental Regimes: Verification and Implementation Review. *Environmental Politics,* **2**(4).

Green, O., 1995: Environmental Regimes: Effectiveness and Implementation Review. In *The Environment and International Relations.* J. Vogler, M. Imber, (eds.), Routledge, London.

Grubb, M., 1997: Technologies, Energy Systems and the Timing of CO$_2$ Emissions Abatement An Overview of Economic Issues. *Energy Policy,* **25**(2), 159-172.

Grubb, M., 1998: International Emissions Trading Under the Kyoto Protocol: Core Issues in Implementation. Photocopy, RIIA, London.

Grubb, M., 1999: Corrupting the Climate? Economic Theory and the Politics of Kyoto. Text of Valedictory Lecture, The Royal Institute on International Affairs, London.

Grubb, M., and C. Vrolijk, 1998: The Kyoto Protocol: Specific Commitments and Flexibility Mechanisms. RIIA, *EEP Climate Change Briefing, No. 11.*

Grubb, M. and J. Gupta (eds.), 2000: *European Leadership of Climate Change Regimes,* Kluwer Academic Publ., Dordrecht.

Grubb, M. J., M. Ha Duong, and T. Chapuis, 1995: The economics of changing course. *Energy Policy,* **23**(4/5), 417-432.

Grubb, M., C. Vrolijk, and D. Brack, 1999*: The Kyoto Protocol: A Guide and Assessment.* RIIA/Earthscan, London.

Grübler, A., and N. Nakicenovic, 1994: *International Burden Sharing in Greenhouse Gas Reduction,* RR-94-9. International Institute for Applied Systems Analysis, Laxenburg, Austria.

Grübler, A., and S. Messner, 1998: Technological Change and the Timing of Mitigation Measures. *Energy Economics,* **20**(5/6), 495-512.

Grübler, A., N. Nakicenovic, and D. Victor, 1999: Dynamics of Energy Technologies and Global Change. *Energy Policy,* **27**(5), 247-280.

Haas, P. M., R. O. Keohane, and M. A. Levy (eds.), 1993: *Institutions for the Earth: Sources of Effective International Environmental Protection.* MIT Press, Cambridge, MA.

Habermas, J., 1979: *Communication and the Evolution of Society.* Beacon Press, Boston, MA.

Habermas, J., 1981: *Theory of communicative action.* Beacon Press, Boston, MA.

Ha-Duong, M., 1998: Quasi-Option Value and Climate Policy Choice. *Energy Economics,* **20**(5/6), 599-620.

Ha-Duong, M., M. J. Grubb, and J.-C. Hourcade, 1997: Influence of socioe-conomic interia and uncertainty on optimal CO$_2$-emission abatement. *Nature,* **390**, 270-273.

Ha-Duong, M., J-C. Hourcade and F. Lecocq, 1999: Dynamic Consistency Problems behind the Kyoto Protocol. *Int.J. Environment and Pollution,* **11**(4), 426-446.

Hagem, C. 1996: Joint Implementation under Asymmetric Information and Strategic Behavior. *Environmental and Resource Economics,* **8**(4), 431 - 447.

Hagem, C., and Westskog, 1998: The Design of a Tradeable Quota System under Market Imperfections. *Journal of Environmental Economics and Management,* **36**(1), 89-107

Hahn, R.W., 1984: Market Power and Transferable Property Rights. *Quarterly Journal of Economics,* **99**, 753-765.

Hahn, R., and R. Stavins, 1995: Trading in greenhouse permits: a critical examination of design and implementation issues. In: *Shaping National Responses to Climate Change: A Post-Rio Policy Guide.* H. Lee ,(ed.), Island Press, Cambridge, 177-217.

Hammitt and Adams, 1996: The Value of International Cooperation for Abating Global Climate. *Resource and Energy Economics,* **18**(3), 219-241.

Hanf, K., and A. Underdal, 1995: Domesticating International Commitments: Linking National and International Decision-Making In *The International Politics of Environmental Management.* A. Underdal, (ed.), Kluwer, Dordrecht, 101- 127.

Hanf, K., and A. Underdal (eds.), 2000*: The Domestic Basis of International Environmental Agreements.* Ashgate, Aldershot, UK.

Hanson, D., and S. Laitner, 2000: *An Economic Growth Model of Investment, Energy Savings, and CO$_2$ Reductions: An Integrated Analysis of Policies that Increase Investments in Advanced Efficient/Low-Carbon Technologies.* Proceedings of the 93rd Annual Air & Waste Management Conference, Salt Lake City, UT.

Hardi, P., and S. Barg, 1997: *Measuring Sustainable Development: Review of Current Practice.* Occasional Paper No. 17, Internatiuonal Institute for Sustainable Development, Winnipeg, Canada.

Hardin, G., and J. Baden, 1977: *Managing the Commons.* Freeman & Co, New York.

Harrison, G. W., and T. F. Rutherford, 1999: Burden Sharing, Joint Implementation, and Carbon Coalition. In *International Environmental Agreements on Climate Change.* C. Carraro, (ed.), Kluwer Academic Publ., Dordrecht, pp. 77-108.

Harrod, R. F., 1958: The Possibility of Economic Satiety — Use of Economic Growth for Improving the Quality of Education and Leisure. In *Problems of United States Economic Development, Vol. 1.* Committee for Economic Development, New York, NY, pp. 207-213.

Hart, S., and M. Kurz, 1983: Endogenous Formation of Coalitions. *Econometrica,* **51**(4), 1047-1064.

Heal, G., 1994: The Formation of Environmental Coalitions. In *Trade, Innovation, Environment.* C. Carraro, (ed.), Kluwer Academic Publ., Dordrecht, pp. 301-332.

Healy, S., 1999: Extended peer communities and the ascendance of post-nor-mal politics. *Futures,* **31**(7), 655-669.

Heller, T. C., 1995: Joint Implementation and the Path to a Climate Change Rate. Jean Monnet Papers No. 23, The Robert Schuman Center at the European University Institue, San Domenico, Italy.

Helm, C., 1999: Applying Fairness Criteria to the Allocation in Climate Protection Burdens: An Economic Perspective. In *Fair weather? Equity Concerns in Climate Change.* F. L. Tóth, (ed.), Earthscan Publications, London, 80-94.

Hirsch, F., 1977: *Social Limits to Growth..* Routledge, London.

Hoel, M., 1991: Global Environmental Problems: The Effects of Unilateral Actions Taken by One Country. *Journal of Environmental Economics and Management,* **20**(1), 55-70.

Hoel, M., 1992: International Environmental Conventions: the Case of Uniform Reductions of Emissions. *Environmental and Resource Economics,* **2**, 141-159.

Hoel, M., 1994: Efficient Climate Policy in the Presence of Free-Riders. *Journal of Environmental Economics and Management,* **27**(3), 259-274.

Hoel, M., and K. Schneider, 1997: Incentives to Participate in an International Environmental Agreement. *Environment and Resources Economics,* **9**, 154-170.

Holtsmark, B. J., 1998: *From the Kyoto Protocol to the Fossil Fuel Markets: An Analysis of the Costs of Implementation and Gains from Emission Trading taking benefits from Revenue Recycling into Account.* Proceedings of the Workshop on Economic Modelling and Climate Change, 17-18 September 1998, OECD, Paris, 123-138.

Holtsmark, B. J., and C. Hagem, 1998: Emission Trading under the Kyoto Protocol. Report No. 1998:1, CICERO, Oslo.

Hotelling, H., 1931: The Economics of Exhaustible Resources. *Journal of Political Economy,* **39**(2), 137-175.

Hourcade, J. C., M. Ha-Duong, and F. Lecoq, 1999: Dynamic Consistency Problems behind the Kyoto Protocol. *International Journal of Environment and Pollution,* **11** (Special Issue on Methodologies and Issues for Integrated Environmental Assessment).

Inglehart, R., 1971: The Silent Revolution in Europe: Intergenerational Change in Post-Industrial Societies. *American Political Science Review,* **65**(4), 991-1017.

Inglehart, R., 1977: *The Silent Revolution. Changing Values and Political Stiles Among Western Publics.* Princeton University Press, Princeton, NJ.

Inglehart, R., 1996: The Diminishing Utility of Economic Growth: From Maximizing Security toward Maximizing Subjective Well-Being. *Critical Review,* **10**(4), 509-533.

Interlaboratory Working Group on Energy-Efficient and Low-Carbon Technologies, 1997: Scenarios of U. S. Clean Carbon Reductions: Potential Impacts of Energy Technologies by 2010 and Beyond. U.S. Department of Energy, Office of Energy Efficiency and Renewable Energy, Washington DC.

Interlaboratory Working Group on Energy-Efficient and Low-Carbon Technologies, 2000: *Scenarios for a Clean Energy Future.* U.S. Department of Energy, Office of Energy Efficiency and Renewable Energy, Washington DC.

IPCC (Intergovernmental Panelon Climate Change), 1995. *Climate Change 1994: Radiative Forcing of Climate Change and An Evaluation of the IPCC IS92 Emissions Scenarios.* J. T. Houghton, L. G. M. Filho, J. Bruce, H. Lee, B. A. Callander, E. Haites, N. Harris, K. Maskell, (eds.), Cambridge University Press, Cambridge,

IPCC (Intergovernmental Panel on Climate Change), 1996a: *Climate Change 1995.* The Science of Climate Change. The Contribution of Working Group I to the Second Assessment Report of the Intergovernmental Panel on Climate Change. J. P. Houghton, L. G. Meira Filho, B. A. Callendar, A. Kattenberg, and K. Maskell, (eds.), Cambridge University Press, Cambridge.

IPCC (Intergovernmental Panel on Climate Change), 1996b: Climate Change 1995. Impacts, Adaptation, and Mitigation of Climate Change: Scientific-Technical Analysis. The Contribution of Working Group II to the Second Assessment Report of the Intergovernmental Panel on Climate Change. R. T. Watson, M. C. Zinyowera, R. H. Moss, (eds.), Cambridge University Press, Cambridge.

IPCC (Intergovernmental Panel on Climate Change), 1996c: Climate Change 1995. Economic and Social Dimensions of Climate Change. The Contribution of Working Group III to the Second Assessment Report of the Intergovernmental Panel on Climate Change. J. P. Bruce, H. Lee, E. F. Haites, (eds.), Cambridge University Press, Cambridge.

IPCC (Intergovernmental Panel on Climate Change), 2000a: *Special Report on Technology Transfer.* Cambridge University Press, Cambridge, UK.

IPCC (Intergovernmental Panel on Climate Change), 2000b: *Special Report on Emission Scenarios.* Cambridge University Press, Cambridge, UK.

IPCC, 2001a: Climate Change: The Scientific Basis. Contribution of Working Group I to the IPCC Third Assessment Report (TAR), Cambridge University Press. Cambridge, UK (in press).

IPCC, 2001b: Climate Change 2001, Impacts, Adaptation, and Vulnerability. Contribution of Working Group II to the IPCC Third Assessment Report (TAR) Cambridge University Press. Cambridge, UK (in press).

Jaccard, M., L. Failing, and T. Berry, 1997: From Equipment to Infrastructure: Community Energy Management and Greenhouse Gas Emission Reduction. *Energy Policy,* **25**(13), 1065-1074.

Jackson, T., 1995: Joint Implementation and Cost-Effectiveness under the Framework Convention on Climate Change. *Energy Policy,* **23**(2), 117-138.

Jackson, T., and N. Marks, 1999: Consumption, Sustainable Welfare and Human Needs — with Reference to UK Expenditure Patterns Between 1954 and 1994. *Ecological Economics,* **28**(3), 421-441.

Jacobson, H. K., and E. Brown Weiss, 1995: Strengthening Compliance with International Environmental Accords. *Global Governance,* **1**(2), 119-148.

Jacoby, H. D., R. Schmalensee, and I. S. Wing, 1998: *Toward a Useful Architecture for Climate Change Negotiations.* Joint Program on the Science and Policy of Global Change, Massachusetts Institute of Technology, Cambridge, MA.

Jaeger, C. C., O. Renn, E. A. Rosa, and T. Webler, 1998: Decision analysis and rational action. In *Human Choice & Climate Change, Tools for Policy Analysis, Vol. 3.* S. Rayner, E. Malone, (eds.), Battelle Press, Columbus, OH, 141-216.

Janssen, J., 1999: (Self-)Enforcement of Joint Implementation and Clean Development Mechanism Contracts. Paper presented at the First World Conference of Environmental and Resource Economists, June 1998, Venice.

Jepma, C. J., and M. Munasinghe, 1998: *Climate Change Policy: Facts, Issues, and Analysis.* Cambridge University Press, Cambridge.

Jonas, H., 1985: *The Imperative of Responsibility.* University of Chicago Press, Chicago, IL.

Jorgensen, D., W. Slesnick, and P. J. Wilcoxen, 1993: Carbon Taxes and Economic Welfare. Brooking Papers of Economic Activity, 393-431.

Kainuma, M., Y. Matsuoka, and T. Morita, 1998: *Analysis of Post-Kyoto Scenarios: The Asian Pacific Integrated Model.* Proceedings of the Workshop on Economic Modelling and Climate Change, 17-18 September 1998, OECD, Paris, 161-172.

Kainuma, M., Y. Matsuoka, and T. Morita, 1999: Analysis of Post-Kyoto Scenarios: The Asian Pacific Integrated Model. *Energy Journal,* Kyoto Special Issue, 207-220.

Katsoulacos, Y., 1997: R&D Spillovers, R&D Co-operation, Innovation and International Environmental Agreements. In *International Environmental Agreements: Strategic Policy Issues.* C. Carraro, (ed.), E. Elgar, Cheltenham, UK.

Kawashima, Y., 1997: Japan's Decision-Making About Climate Change Problems: Comparative Study of Decisions in 1990 and 1997. *Environmental Economics and Policy Studies,* **3**(1).

Kawashima, Y., 2000: Comparative Analysis of Decision-Making Process of the Developed Countries Towards CO_2 Emission Reduction Target. *International Environmental Affairs,* **9**.

Kay, J., H. Regier, M. Boyle, and F. George, 1999: An Ecosystem Approach for Sustainability: Addressing the Challenge of Complexity. *Futures,* **31**(7), 721-742.

Kim, S. H., C. MacCracken, and J. Edmomds, 2000: Solar Energy Technologies and Stabilizing Atmospheric CO_2 Concentrations. *Progress in Photovoltaics Research and Application,.* **8**(1), 3-15.

Kirchgässner, G., and E. Mohr, 1996: Trade Restrictions as Viable Means of Enforcing Compliance with International Environmental Law: An Economic Assessment. In *Enforcing International Environmental standards: Economics Mechanisms as Viable Means?* W. Rüdiger, (ed.), Springer Verlag, Berlin.

Kolstad, C. D., 1994: George Bush versus Al Gore - Irreversibilities in greenhouse gas accumulation and emission control investment. *Energy Policy*, **22**(9), 772-778.

Koomey, J. G., R. C. Cooper, J. A. Laitner, R. J. Markel, and C. Marnay, 1998: *Technology and Greenhouse Gas Emissions: An Integrated Scenario Analysis Using the LBNL–NEMS Model.* LBNL–42054 and EPA 430–R–98–021, Lawrence Berkeley National Laboratory and US Environmental Protection Agency.

Kopp, R. J., W. Harrington, R. D. Morgenstern, W. A. Pizer, and J. S. Shih, 1998: *Diffusion of New Technologies: A Microeconomic Analysis of Firm Decision Making at the Plant Level.* Resources for the Future, Washington, DC.

Kotov, V., and E. Nikitina, 1996: To Reduce or to Produce? Problems of Implementation of the Climate Change Convention in Russia. In *Verification 1996: Arms Control, Peacekeeping and Environment.* J. B. Poole, R.Guthrie, (eds.). Westview Press, Boulder, CO.

Kotov, V., and E. Nikitina, 1998: Implementation and Effectiveness of the Acid Rain Regime in Russia. In *The Implementation and Effectiveness of International Environmental Commitments.* D. G. Victor, K. Raustiala, E. B. Skolnikoff, (eds.), IIASA, Laxenburg, Austria and MIT Press, Cambridge, MA.

Kotov V., E. Nikitina, A. Roginko, O. S. Stokke, D. G. Victor, and R. Hjorth, 1997: Implementation of International Environmental Commitments in Countries in Transition. *MOCT-MOST Economic Policy in transition Economies,* **7**(2), 103 – 127.

Kram, T., 1998: *The Energy Technology Sytems Analysis Program: History, the ETSAP Kyoto Statement and Post-Kyoto Analysis.* Proceedings of the Workshop on Economic Modelling and Climate Change, 17-18 September 1998, OECD, Paris.

Kurosawa, A., H. Yagita, Z. Weisheng, K. Tokimatsu, and Y. Yanagisawa, 1999: Analysis of Carbon Emission Stabilization Targets and Adaptation by Integrated Assessment Models. *Energy Journal,* Kyoto Special Issue, 157-175.

Laitner, J. A. ("Skip"), 1999: The Economic Effects of Climate Policies to Increase Investments in Cost-Effective, Energy-Efficient Technologies. Presented at the 74th Annual WEA International Conference, San Diego, CA.

Laitner, J. A., and K. Hogan, 2000: *Solving for Multiple Objectives: The Use of the Goal Programming Model to Evaluate Energy Policy Options.* Proceedings of the ACEEE Energy Efficiency in Buildings Summer Study, August, American Council for an Energy Efficient Economy, Washington DC.

Laitner, J.A., K. Hogan, and D. Hanson, 1999: *Technology and Greenhouse Gas Emissions: an Integrated Analysis of Policies that Increase Investments in Cost Effective Energy–Efficient Technologies.* Proceedings of the Electric Utilities Environmental Conference, January, Tucson, AZ.

Laitner, J. A., S. De Canio, and I. Peters, 2000: Incorporating Behavioral, Social, and Organizational Phenomena in the Assessment of Climate Change Mitigation Options. An invited paper for the IPCC Expert Meeting on Conceptual Frameworks for Mitigation Assessment from the Perspective of Social Science, March 21-22, Karlsruhe, Germany, (revised June 2000).

Lako, P. (ed.), J. R. Ybema, and A. J. Seebregts, 1999: *Long-term Scenarios and the Role of Fusion Power—Synopsis of SEO Studies, Conclusion, and Recommendations.* ECN-C-98-095, Netherlands Energy Research Foundation (ECN), Petten, The Netherlands.

Lanchbery, J., 1997: What to Expect from Kyoto. *Environment,* **39**(9).

Lanchbery, J.,1998: Greenhouse gas inventories: national reporting process andimplementation review mechanisms in the EU. *Research on the Socio-Economic Aspects of Environmental Change (1992-1996).* European Communities, Laxenburg, Austria,

Lazarev, G., 1994: *People, Power and Ecology.* The Macmillan Press Ltd., London.

Leary, N., 1999: A framework for benefit-cost analysis of adaptation to climate change and climate variability. *Mitigation and Adaptation Strategies for Global Change,* **4**(3/4), 307-318.

Lecocq, F., J.-C. Hourcade, and M. Ha-Duong, 1998: Decision Making under Uncertainty and Inertia Constraints: Sectoral Implications of the When Flexibility. *Energy Economics,* **20**(5/6), 539-555.

Lempert, R. J., and M. E. Schlesinger, 2000: Robust Strategies for Abating Climate Change. *Climatic Change,* **45**(3/4), 387-401.

Lempert, R. J., M. E. Schlesinger, and S. C. Bankes, 1996: When We Don't Know the Costs or the Benefits: Adaptive Strategies for Abating Climate Change. *Climate Change,* **33**(2), 235-274.

Lempert, R. J., M. E. Schlesinger, S. C. Bankes, and N. G. Andronova, 2000: The Impact of Variability on Near-Term Climate-Change Policy Choices. *Climate Change,* 45 (3/4), 387-401.

Levy, M. A, O. R. Young, and M. Zürn, 1995: The Study of International Regimes, *European Journal of International Relations,* **1**(3), 267- 331.

Lile, R., M. Powell, and M. Toman, 1999: Implementing the Clean Development Mechanism : Lessons from U.S. Private-Sector Participation in Activities Implemented Jointly. *Discussion Paper 99-08,* Resources For The Future, November 1998.

Lindbolm, C. E., 1959: The Science of Muddling Through. *Public Administration Review,* **19**(1), 79-88.

Lykke, E. (ed.), 1993: *Achieving Environmental Goals: The Concept and Practice of Environmental Performance Review.* Belhaven, London.

Maddison, D., 1995: A cost-benefit analysis of slowing climate change. *Energy Policy,* **23**(4/5), 337-346.

Manne, A. S., 1995: Hedging Strategies for Global Carbon Dioxide Abatement: A Summary of Poll Results, EMF-14 Subgroup – Analysis for Decisions under Uncertainty. Stanford University, Stanford, CA.

Manne, A. S., and R. Richels, 1995: The Greenhouse Debate: Economic Efficiency, Burden Sharing, and Hedging Strategies. *Energy Journal* **16**(4), 1-37.

Manne, A. S., and R. G. Richels, 1997: On Stabilizing CO_2 Concentrations - Cost-Effective Emission Reduction Strategies. *Environmental Modeling & Assessment,* **2**(4), 251-266.

Manne, A. S., and R. G. Richels, 1998: *The Kyoto Protocol: A Cost Effective Strategy for Meeting Environmental Objectives.* Proceedings of the Workshop on Economic Modelling and Climate Change, 17-18 September 1998, OECD, Paris.

Manne, A. S., and R. G. Richels, 1999a: On Stabilizing CO2 Concentrations – Cost-Effective Emission Reduction Strategies. In *International Environmental Agreements on Climate Change.* C. Carraro, (ed.), Kluwer Academic Publ., Dordrecht.

Manne, A. S., and R. G. Richels, 1999b: The Kyoto Protocol: A Cost-Effective Strategy for Meeting Environmental Objectives? *Energy Journal,* Kyoto Special Issue, 1-25.

Manne, A. S., R. Mendelsohn, and R. G. Richels, 1995: MERGE - A model for evaluating regional and global effects of GHG reduction policies. *Energy Policy,* **23**(1), 17-34.

Mariotti, M., 1997: A Model of Agreements in Strategic Form Games. *Journal of Economic Theory,* **74**(1), 196-217.

Markus, H. R., and S. Kitayama, 1991: Culture and the self: implications for cognition, emotion, and motivation. *Psychological Review,* **98**(2), 224-253.

Markusen, G. R., 1975: Co-operation Control of International Pollution and Common Property Resources. *Quarterly Journal of Economics,* **89**, 618-632.

Matsuo, N., 1998: Points and proposals for the emission trading regime of climate change – For designing future systems (Version 2). IGES, Shonan Village, Japan.

Matsuo, N., A. Maruyama, M. Hamamoto, M. Nakada, and K. Enoki, K., 1998: *Issues and Options in the Design of the Clean Development Mechanism, Version 1.0.* IGES, Shonan Village, Japan.

Mattsson, N., 1997: Internalizing Technological Development in Energy Sysstems Models. ISRN CTH-EST-R-97/3-SE, Report 1997:3, Thesis for the Degree of Licentiate of Engineering, Chalmers University of Technology, Goteborg, Sweden.

McCracken, G., 1988: *Culture and Consumption. New Approaches to the Symbolic Character of Consumer Goods and Activities.* Indiana University Press, Bloomington and Indianapolis, IN.

McKibbin, W.J., L. Ross, R. Shackelton, and P.G.J. Wilcoxen, 1998: *The Potential Effects of International Carbon Emissions Permit Trading under the Kyoto Protocol*. Proceedings of the Workshop on Economic Modelling and Climate Change, 17-18 September 1998, OECD, Paris, 49-79.

McKibbin, W. J., R. Shackelton, and P. G. J. Wilcoxen, 1999: What to expect from an International System of Tradeable Permits for Carbon Emissions. *Resource and Energy Economics*, **21**(3/4), 319-346.

Meadows, D., 1998: *Indicators on Information Systems for Sustainable Development*. A Report to the Balaton Group, September 1998.

Mendelsohn, R., and M. E. Schlesinger, 1997: *Climate Response Functions*. Yale University, New Haven CT/University of Urbana, Champain, IL

Mendelsohn, R., and J. Neumann (eds.), 1999: *The Impacts of Climate Change on the U.S. Economy*. Cambridge University Press, Cambridge.

Mensbrugghe, D., 1998: *A (Preliminary) Analysis of the Kyoto Protocol: Using the OECD GREEN Model*. Proceedings of the Workshop on Economic Modelling and Climate Change, 17-18 September 1998, OECD, Paris, 173-204.

Metz, B., 1999: International Equity in Climate Change Policy. In *Integrating Climate Policies in a European Environment, special issue of Integrated Assessment*. C. Carraro, (ed.), Baltzer Publ., 111 – 126.

Meyer, W. B., K. W. Butzer, T. E. Downing, B. L. Turner II, G. W. Wenzel, and J. L. Wescoata, 1998: Reasoning by analogy - Human Choice & Climate Change. In *Tools for Policy Analysis*. S. Rayner, E. L. Malone, (eds.), Battelle Press, Columbus, OH, 217-289.

Meyer-Abich, K. M., 1997: *Praktische Naturphilosophie — Erinnerung an einen vergessenen Traum*. C. H. Beck, Munich.

Michaelowa, A., 1995: Joint Implementation of Greenhouse Gas Reductions under Consideration of Fiscal and Regulatory Incentives. In *BMWi Studienreihe, No.88*, Bonn.

Michaelowa, A., 1998a: Climate policy and Interest Groups - A Public Choice Analysis. *Intereconomics*, **33**(6), 251 – 259.

Michaelowa, A., 1998b: Joint Implementation - the Baseline Issue. Economic and Political Aspects. *Global Environmental Change*, **8**(1), 81 - 92.

Mintzberg, H., 1994: *The Raise and Fall of Strategic Planning*. Prentice Hall, Hemel Hempstead, UK.

Mitchel, R. B., 1994: *International Oil Pollution at Sea: Environmental Policy and Treaty Compliance*. MIT Press, Cambridge, MA.

Mitchell, A., 1983: *The Nine American Lifestyles. How we are and where we're going*. MacMillan, New York, NY.

Mohr, E., 1995: International Environmental Permits Trade and Debt: The Consequences of Country Sovereignty and Cross Default Policies. *Review of International Economics*, **3**(1), 1-19.

Moisander, J., 2001: Group Identity, Personal Ethic, and Sustainable Development Suggesting New Directions for Social Marketing Research. In *Society, Behavior, and Climate Change Mitigation*. Proceedings of the IPCC Expert Meeting. Conceptual Frameworks for Mitigation Assessment from the Perspective of Social Science. E. Jochem, D. Bouille, J. Sathaye (eds.), Kluwer, Dordrecht (in press).

Montgomery, W.D. 1997: Economic perspectives on agreements to limit carbon emissions. In *Setting the stage: Canada's negotiating position on climate change*, Climate Change Conference, 4-5 September 1997, Calgary, Canada, Canadian Energy Research Institut (CERI), Report Number Conf-9709126.

Morgan, M. G., and D. W. Keith, 1995: Subjective Judgement by Climate Experts. *Environmental Science and Technology*, **29**, 468-476.

Morgan, M. G., and H. Dowlatabadi, 1996: Learning from Integrated Assessments of Climate Change. *Climatic Change*, **34**(3), 337-68.

Mori, S., and M. Takahaashi, 1997: An integrated assessment model for the evaluation of new energy technologies and food production - An extension of multiregional approach for resource and industry allocation model. *International Journal of Global Energy Issues*, **11**(1-4), 1-17.

Morita, T., M. Kainuma, K. Masuda, H. Harasawa, K. Takahashi, Y. Matsuoka, J. Sun, Z. Li, F. Zhou, X. Hu, K. Jiang, P. R. Shukla, V. K. Sharma, T. Y. Jung, D. K. Lee, D. Hilman, M. F. Helmy, G. Yoshida, G. Hibino, and H. Ishii, 1997: *Asian-Pacific Integrated Model AIM*. National Institute for Environmental Studies, Tsukuba, Japan.

Moss, R. H., and S. H. Schneider, 2000: Uncertainties in the IPCC TAR: Recommendations to Lead Authors to More Consistent Assessment and Reporting. IPCC Supporting Material, Guidance Papers On The Cross Cutting Issues Of The Third Assessment Report of the IPCC, R.Pachauri, T. Taniguchi, K. Tanaka, (eds.), 33–51.

Müller, B., 1998: *Justice in Global Warming Negotiations: how to achieve a procedurally fair compromise*. Oxford Institute of Energy Studies, Oxford.

Munasinghe, M., 1998: Climate change decision-making: science, policy and economics. *International Journal of Environment and Pollution*, **10**(2), 188-239.

Munasinghe, M., 2000: Development, Equity and Sustainability in the Context of Climate Change. In *Climate Change and its linkages with Development, Equity and Sustainability*. M. Munasinghe, R. Swart, (eds.), IPCC.

Nordhaus, W. D., 1994a: Locational competition and the environment. Should Countries Harmonize Their Environmental Policies? Cowles Foundation Discussion Paper 1079, Cowles Foundation, Yale University, New Haven, CT.

Nordhaus, W. D., 1994b: *Managing the Global Commons: Economics of the greenhouse effect*. Cambridge, MIT Press, Cambridge, MA.

Nordhaus, W. D., 1997: Modelling Induced Innovation in Climate-Change Policy: Theory and Estimates in the R&DICE Model. Paper prepared for the Workshop on Induced Technological Change, June 1997, IIASA., Laxenburg, Austria.

Nordhaus, W. D., 1998: Assessing the Economics of Climate Change, An Introduction: InEconomics And Policy Issues In *Climate Change*, W.D. Nordhaus, (ed.), Resources for the Future, Washington, DC, 1-21.

Nordhaus, W. D., 1999: Requiem for Kyoto: An Economic Analysis of the Kyoto Protocol. *Energy Journal*, Kyoto Special Issue, 93-130.

Nordhaus, W. D., and J. Tobin, 1972: Is Growth Obsolete? In *Economic Growth. Fiftieth Anniversary Colloquium V, National Bureau of Economic Research*. Columbia University Press, New York, NY, 1-80.

Nordhaus, W. D., and Z. Yang, 1996: RICE: A Regional Dynamic General Equilibrium model of Alternative Climate Change Strategies. *The American Economic Review*, **86**(4), 726-741.

Nordhaus, W. D., and J. Boyer, 1999: Roll the DICE Again: The Economics of Global Warming. Draft Version, 28.1.1999, Yale University, New Haven, CT.

O'Hara, S. U., 1996: Discursive ethics in ecosystems valuation and environmental policy. *Ecological Economics*, **16**(2), 95-107.

O'Riordan, T., 1993: The Politics of Sustainability. In *Sustainable Environmental Economics and Management: Principles and Practices*, R. K. Turner (ed.), Belhamen Press, London.

O'Riordan, T., 1997: Climate Change Economic and Social Dimension. *Environment*, **39**(9).

O'Riordan, T., and J. Cameron, 1994: *Interpreting the Precautionary Principle*. Earthscan Publications, London.

O'Riordan, T. and J. Jaeger (eds.), 1996: *Politics of Climate Change: A European Perspective*. Routledge, London, UK/New York, USA.

O'Riordan, T., C. L. Cooper, S. R. Jordan, K. R. Richards, P. Runci, and S. Yoffe, 1998: Institutional frameworks for political action. In *Human Choice & Climate Change, The Societal Framework*. S. Rayner, E. Malone, (eds.), Battelle Press, Columbus, OH, 345-440.

OECD, 1993: Environmental Indicators: Basic Concepts and Terminology. Background Paper No. 1: Group on the State of the Environment, Environment Directorate, OECD, Paris.

OECD, 1997: *Technology Co-operation in Support of Cleaner Technology in Developing Countries*. Synthesis Report, Paris.

OECD, 1998: Proceedings of the OECD Workshop on 'Climate Change and Economic Modelling. Background Analysis for the Kyoto Protocol', Paris, 17-18 Sept, 1998,.

OECD, 1999a: Financial Flows to Developing Countries in 1998. Rise in Aid; Sharp Fall in Private Flows. News Release (10 June), OECD, Paris.

OECD, 1999b: *Technology and Environment; Towards Policy Integration*. DSTI/STP(99)19/Final, OECD, Paris.

OECD, 1999c: *The Three-Year Project on Sustainable Development*. A Progress Report, Paris.

OECD/IEA, 1994: *World Economic Outlook, 1994 edition*.

Olson, M., 1965: *The Logic of Collective Action. Public Goods and the Theory of Groups*, 2nd ed. Harvard University Press, Cambridge, MA.

Orbell, J. M., P. Schwartz-Shea, and R. T. Simmons, 1984: Do Cooperators Exit More Readily than Defectors? *The American Political Science Review*, **78**(1), 147-162.

Ott, H. E., 1998: Operationalizing 'Joint Implemenation'. Organizational and Institutional Aspects of a New Instrument in International Climate Policy. *Global Environmental Change*, **8**, 11-47.

Parkinson, S. D., P. Bailey, K. Begg, and T. Jackson, 2001: Uncertainty and Sensitivity Analysis: methodology and case study results. In *Flexibility in Climate Policy: making the Kyoto mechanisms work*, T. Jackson, K. Begg, S. D. Parkinson, (eds.), Earthscan, London, 137-161.

Parry, M., and J. Carter, 1998: *Climate Impact and Adaptation Assessment*. Earthscan, London.

Parson, E. A., and R. Zeckhauser, 1995: Cooperation in the Unbalanced Commons. In *Barriers to Conflict Resolution*, K. Arrow, R. Mnookin, L. Ross, A. Tversky, R. Wilson, (eds.), Norton, New York, NY, 212-234.

Parson, E. A., and K. Fisher-Vanden, 1999: Joint implementation of greenhouse gas abatement under the Kyoto Protocol's "clean development mechanism": its scope and limits. *Policy Sciences*, **32**(3), 207-224.

PCAST (President's Committee of Advisors on Science and Technology), 1999: *Powerful Partnerships: The Federal Role in International Cooperation on Energy Innovation*. The White House, Washington, DC.

Pearce, D., 1994: The Precautionary Principle and Economic Analysis. In *Interpreting the Precautionary Principle*. T. O'Riordan, J. Cameron, (eds.), Earthscan, London, 132-151.

Pearce, D., 1998: Measuring Sustainable Development. In *Sustainable Development Indicators – Proceedings of an OECD Workshop, Oct. 1998*, Paris.

Peck, S. C., and T. J. Teisberg, 1992: CETA: A model for carbon emissions trajectory assessment. *Energy Journal*, **13**(1), 55-177.

Peck, S. C., and T. J. Teisberg, 1993: CO_2 Emissions Control: Comparing Policy Instruments. *Energy Policy*, **21**(3), 222-230.

Peck, S. C., and T. J. Teisberg, 1999: CO_2 Concentrations Limits, the Costs and Benefits of Control and the Potential for International Agreements. In *International Environmental Agreements on Climate Change*. C. Carraro, (ed.), Kluwer Academic Publ., Dordrecht.

Peck, S. C., and Y. H. Wan, 1996: Analytic Solutions of Simple Greenhouse Gas Emission Models. In *Economics of Atmospheric Pollution* (Chapter 6), E.C. Van Ierland, K. Gorka (eds.), Springer Verlag, Berlin.

Peters, I., and K-H. Brassel, 2000: Integrating Computable General Equilibrium Models and Multi-Agent Systems – Why and How. In *2000 AI, Simulation and Planning in High Autonomy Systems*, H. S. Sarjoughian, F. E. Cellier, M. M. Marefat, J. W. Rozenblit, (eds.), Simulation Councils, Inc.

Peterson, M. J., 1997: International Organizations and the implementation of Environmental Regimes. In *Global Governance. Drawing Insights from the Environmental Experience*. O. Young, R. Oran, (eds.), MIT Press, Cambridge, MA, 115 – 151.

Petrakis, E. A., and A. Xepapadeas, 1996: Environmental Consciousness and Moral Hazard in International Agreements to Protect the Environment. *Journal of Public Economics*, **60**(1), 95-110.

Petschel-Held, G., H.-J. Schellnhuber, T. Bruckner, F. L. Toth, and K. Hasselmann, 1999: The Tolerable Windows Approach: Theoretical and Methodological Foundations, *Climatic Change*, **41**(3/4), 303-331.

Pielke, R. A. J., 1998: Rethinking the role of adaptation in climate policy. *Global Environmental Change*, **8**(2), 159 - 170.

Pitcher, H., 1999: *The Only Safe Landing is in the Hedge*. Pacific Northwest National Laboratory, Washington, DC.

Pizer, W, 1999: Choosing Price or Quantity Controls for Greenhouse Gases. *Climate Issues Brief No. 1 7*, Resources for the Future. Washington, DC.

Plambeck, E. L., and C. Hope, 1996: PAGE95: An updated valuation of the impacts of global warming. *Energy Policy*, **24**(9), 783-793.

Porter, M. E., and C. Van Linde, 1995: Toward a New Conception of Environment-Competitiveness Relationship. *Journal of Economic Perspectives*, **9**(4), 97-118.

Portney, P. R., 1998: Applicability of Cost-Benefit Analysis to Climate Change. In *Economics and Policy Issues in Climate Change*. W. D. Nordhaus, (ed.), Resource for the Future, Washington, DC, 111-127.

Portney, P. R., and Weyant, J. P., (ed.), 1999: *Discounting and Intergenerational Equity*, Resources for the Future, Washington.

Radosevic, S., 1999: *International Technology Transfer and Catch-up in Economic Development*. E. Elgar, Cheltenham, UK.

Raskin, P., G. Gallopin, P. Gutman, A. Hammond, and R. Swart, 1998: *Bending the Curve: Toward Global Sustainability*. SEI/UNEP, Pole Series Report No. 8, UNEP/DEIA/TR.98-4.

Ray, D., and R. Vohra, 1997: Equilibrium Binding Agreements. *Journal of Economic Theory*, **73**(1), 30-78.

Rayner, S., and E. L. Malone, 1998a: The challenge of climate change to the social sciences. In *Human Choice & Climate Change: What Have We Learned, Vol. 4*. Battelle Press, Columbus, OH.

Rayner, S., and E. L. Malone, 1998b: *Human Choice & Climate Change. Ten Suggestions For Policymakers-Guidelines from an International Social Science Assessment*. Battelle Press, Columbus, OH.

Rayner, S., and E. Malone, 2000: Climate Change, Poverty and Intragenerational Equity: the national Level. In *Climate Change and Its Linkages with Development, Equity, and Sustainability*. M. Munasinghe, R. Swart, (ed.), LIFE/RIVM/ World Bank, 215-243.

Rayner, S., E. L. Malone, and M. Thompson, 1999: Equity Issues and Integrated assessment. In *Fair weather? Equity concerns in climate change*, F. L. Tóth, (ed.), 11-44.

Renn, O., T. Webler, H. Rakel, P. Dienel, and B. Johnson, 1993: Public participation in decision making: a three-step procedure. *Policy Sciences*, **26**, 189-214.

Rentz, H., 1998: Joint Implementation and the Question of 'Additionality' - a Proposal for a Pragmatic Approach to Identify Possible Joint Implementation Projects. *Energy Policy*, **26**(4), 275-279.

Reusswig, F., 1994: *Lebensstile und Ökologie. Gesellschaftliche Pluralisierung und alltagsökologische Entwicklung unter besonderer Berücksichtigung des Energiebereichs*. Institut für sozial-ökologische Forschung, Frankfurt am Main.

Richels, R., and J. A. Edmonds, 1995: The economics of stabilizing atmospheric CO_2 concentrations. *Energy Policy*, **23** (4/5), 373-378.

Richels, R., J. A Edmonds, H. Gruenspecht, and T. Wigley, 1996: The Berlin Mandate: The Design of Cost-Effective Mitigation Strategies. In *Climate Change: Integrating Science, Economics, and Policies*, N. Nakicenovic, W. D. Nordhaus, R. Richels, F. L. Toth, (eds.), IIASA, Laxenburg, Austria, 229-248.

Ringius, L., A. Torvanger, and B. Holtsmark, 1998: Can multi-criteria rules fairly distribute climate burdens? OECD results from three burden sharing rules. *Energy Policy*, **26**(10), 777-793.

Robalino, D.A., and R.J. Lempert, 2000: Carrots and Sticks for New Technology: Abating Greenhouse Gas Emissions in a Heterogeneous and Uncertain World. *Integrated Assessment*, **1** (1), 1-19.

Røpke, I., 1999: The Dynamics of Willingness to Consume. *Ecological Economics*, **28**(3), 399-420.

Rose, A., and B. Stevens, 1998: Will a global warming agreement be fair to developing countries?. *International Journal of Environment and Pollution*, **9**(2/3), 157-178.

Rose, A. and B. Stevens, 1999: A Dynamic Analysis of Fairness in Global Warming Policy: Kyoto, Buenos Aires and Beyond. In *Efficiency and Equity of Climate Change Policy*. C. Carraro, (ed.), Kluwer Academic Publ., Dordrecht, 329 – 362.

Rose, A, B. Stevens, J. A. Edmonds, and M. Wise 1998: International Equity and Differentiation in Global Warming Policy. *Environmental and Resource Economics*, **12**, 25–51.

Rosenhead, J., 1990: Robustness Analysis: Keeping your Options Open. In *Rational Analysis for a Problematic World*. J. Rosenhead, (ed.), John Wiley and Sons, 193-218.

Rotmans, J., 1997: Indicators for Sustainable Development. In *Perspectives on Global Change*, J. Rotmans, B. de Vries, (eds.), Cambridge University Press, Cambridge.

Rotmans, J., and H. Dowlatabadi, 1998: Integrated assessment modeling. In *Human Choice & Climate Change Tools for Policy Analysis*. S. Rayner, E. Malone, (eds.), Battelle Press, Columbus, OH, 291-377.

Rowlands, I. H., 1997: International fairness and justice in addressing global climate change. *Environmental Politics*, **6**(3), 1-30.

Sand, P. H. (ed.), 1992: *The effectiveness of International Environmental Agreements.*Grotius Publishers, Cambridge.

Sand, P. H., 1995: *Principles of International Environmental Law, Vol. 1: Frameworks, Standards and Implementation.* Manchester University Press, Manchester, UK.

Sand, P. H., 1996: Institution-Building to Assist Compliance with International Environmental Law: Perspectives. *Zeitschrift für auslandisches öffentliches Recht und Völkerrecht*, **56**(3), 754 – 795.

Sands, R. D., J. A. Edmonds, S. H. Kim, C. N MacCracken, and M. A. Wise, 1998: *The Cost of Mitigating United States Carbon Emissions in the Post-2000 Period.* Proceedings of the Workshop on Economic Modelling and Climate Change, 17-18 September 1998, OECD, Paris, 139-160.

Sanne, Chr., 1998: The (Im)possibility of Sustainable Lifestyles - Can We Trust the Public Opinion and Plan for Reduced Consumption? Urban Research Program Working Paper No. 63. Research School of Social Sciences, Australian National University, Canberra.

Sanstad, A., S. De Canio, and G. Boyd, 2000a: *Estimating the Macroeconomic Effects of The Clean Energy Future Policy Scenarios.* Proceedings of the 93rd Annual Air & Waste Management Conference, June, Salt Lake City, UT.

Sanstad, A. H., S. J. De Canio, R. Howarth, S. Schneider, and S. Thompson, 2000b: *New Directions in the Economics and Integrated Assessment of Global Climate Change.* Pew Center on Global Climate Change, Washington DC.

Savage, L. J., 1954: *The Foundations of Statistics.* Wiley, New York.

Scherhorn, G., 1994: Die Wachstumsillusion im Konsumverhalten. In *Geld & Wachstum. Zur Philosophie und Praxis des Geldes.* H.C. Binswanger, P. von Flotow, (eds.), Weitbrecht, Stuttgart and Wien, 213-230.

Scherhorn, G., 1997: Das Ganze der Güter. In *Vom Baum der Erkenntnis zum Baum des Lebens – Ganzheitliches Denken der Natur in Wissenschaft und Wirtschaft.* K. M. Meyer-Abich, (ed.), C. H. Beck, Munich, 162-251.

Schmidt, C., 1997: *Enforcement and Cost-effectiveness of International Agreements: The Role of Side Payments.* Diskussionbeträge des Sonderforschungbereich 178, Serie II, NR 350, Universität Konstanz, Germany.

Schneider, S.H., and L.H. Goulder, 1997: Achieving Low-cost Emissions Targets: *Nature*, **389**, 13-14.

Schock, R. N., W. Fulkerson, M. L. Brown, R. L. San Martin, D. L. Greene, and J. Edmonds, 1999: *How Much is Energy R&D Worth As Insurance?* UCRL-JC-131205. Lawrence Livermore National Laboratory, **March**.

Schwartz, P., 1996: *The Art of the Long View.* Currency-Doubleday, New York, NY.

Sen, A., 1976/77: Rational Fools. A Critique of the Behavioural Foundations of Economic Theory. *Philosophy and Public Affairs*, **6**, 317-344.

Sen, A., 1985: Goals, Commitment and Identity. *Journal of Law, Economics, and Organization*, **1**(2), 341-355.

Shackleton, T., 1998: HANDOUT: The Potential Effects of International Carbon Emissions Mitigation Under the Kyoto Protocol: What we have learned from the G-Cubed Model.. In Proceedings of the OECD Workshop on Climate Change and Economic Modelling. Background Analysis for the Kyoto Protocol, 17-18 Sept, 1998, OECD, Paris.

Shogren, J., 1999: Benefits and Costs of Kyoto. In *Efficiency and Equity of Climate Change Policy.* C. Carraro, (ed.), Kluwer Academic Publ., Dordrecht, 279 – 289.

Shue, H., 1993: Subsistence Emissions and Luxury Emissions. *Law & Policy*, **15**(1), 39-59.

Shukla, P. R., 1995: Greenhouse Gas Models and Abatement Costs for Developing Nations. *Energy Policy*, **23**(8), 677-687.

Simon, H. A., 1959a: *Administrative Behaviour.* MacMillan, New York, NY.

Simon, H. A., 1959b: Theories of Decision-Making in Economics and Behavioral Science. *American Economic Review*, **49**, 254-283.

Simon, H., G. Dantzig, R. Hogarth, C. Piott, H. Raiffa, T. Schelling, K. A. Shepsle, R. Thaier, A. Tversky, and S. Winter, 1986: Decision Making and Problem Solving. In *Report of the Research Briefing Panel on Decision Making and Problem Solving.* National Academy Press, Washington DC.

Smith, J. B., 1997: Setting priorities for adapting to climate change. *Global Environmental Change*, **7**(3), 251-164.

Sokona, Y., S. Humphreys, and J. P. Thomas, 1999: *Sustainable Development: a Centrepiece of the Kyoto Protocol – An African Perspective.* ENDA Tiers Monde, Dakar, Senegal.

Soroos, M. S., 1997: *The Endangered Atmosphere: Preserving a Global Commons.*University of South Carolina Press, Columbia, SC.

Stavins, R. N., 1995: Transaction Costs and Tradeable Permits. *Journal of Environmental Economics and Management*, **29**(2), 133-148.

Sun Rich, C., 1996: Northeast Asia Regional Cooperation: Proposal for China-Japan Joint Implementation of Carbon Emissions Reduction. IGCC Policy Paper No. 29, University of California Institute on Global Conflict and Cooperation, La Jolla, CA.

Swart, R., M. Berk, M. Janssen, E. Kreileman, and R. Leemans, 1998: The safe landing approach: riks and trade-offs in climate change. In *Global change scenarios of the 21st century - Results from the IMAGE 2.1. Model.* J. Alcamo, R. Leemans, E. Kreileman,(eds.), Pergamon/Elsevier Science, Oxford, 193-218.

Thompson, M., and S. Rayner, 1998: Cultural discourses. In *Human Choice & Climate Change, The Societal Framework, Vol. 1.* S. Rayner, E. Malone, (ed.), Battelle Press, Columbus, OH, 195-264.

Tietenberg, T., 1996: *Environmental and Natural Resource Economic.* Harper Collins, New York, NY, 416 pp.

Tognetti, S., 1999: Science in a Double-Bind: Gregory Bateson and the Origins of Post-Normal Science. *Futures*, **31**(7), 689-703.

Tol, R. S. J., 1996: *The Climate Framework for Uncertainty, Negotiation and Distribution (FUND), Version 1.5.* In An Institute on the Economics of the Climate Resource, K.A. Miller and R.K. Parkin (eds.), University Corporation for Atmospheric Research, Boulder, CO, pp. 471-496.

Tol, R. S. J., 1998a: On the Difference in Impact of two almost identical Climate Change Scenarios. *Energy Policy*, **26**(1), 13-20.

Tol, R. S. J., 1999a: Kyoto, Efficiency, and Cost-Effectiveness: Applications of FUND. *Energy Journal*, **Kyoto Special Issue**, 131-156.

Tol, R. S. J., 1999b: *New estimates of the damage costs of climate change - Part I: Benchmark estimates.* Free University, Amsterdam, 20 pp.

Tol, R. S. J., 1999c: Safe policies in an uncertain climate: an application of FUND. *Global Environmental Change*, **9**(3), 221-232.

Tol, R. S. J., 1999d: Spatial and Temporal Efficiency in Climate Policy: Application of FUND. *Environmental and Resource Economics*, **14**(1), 33-49.

Tol, R. S. J., and S. Fankhauser, 1998: On the representation of impact in integrated assessment models of climate change. *Environmental Modeling and Assessment*, **3**, 63-74.

Tol, R. S. J., S. Fankhauser, and J. B. Smith, 1998: The scope for adaptation to climate change: what can we learn from the impact literature? *Global Environmental Change*, **8**(2), 109-123.

Tol, R. S. J., S. Fankhauser, and D. W. Pearce, 1999: Empirical and Ethical Arguments in Climate Change Impact Valuation and Aggregation. In *Fair weather? Equity concerns in climate change.* F. L. Tóth, (ed.), Earthscan, London, pp. 65-79.

Toman, M., and R. Bierbaum, 1996: An Overview of Adaptation to Climate Change. In *Adapting to Climate Change: An International Perspective.* J. B. Smith, N. Bhatti, G. V. Menzhulin, R. Benioff, M. Campos, B. Jallow, F. Rijsberman, M. I. Budyko, R. K. Dixon, (eds.), Springer, New York, NY, pp. 5-26.

Tomlinson, A., 1990: *Consumption, Identity, and Style. Marketing, Meanings, and the Packaging of Pleasure.* Routledge, London, UK/New York, USA, 244 pp.

Torvanger, A., and O. Godal, 1999: A survey of differentiation methods for national greenhouse gas reduction targets. CICERO Report 1999:5, Oslo, 52 pp.

Toth, F. L., 1999: Development, Equity and Sustainability Concerns in Climate Change Decisions. In *Climate Change and its Linkages with Development, Equity and Sustainability.* M. Munasinghe, R. Swart, (eds.), IPCC, Geneva, pp. 263-288.

Toth, F. L., 2000: Decision Analysis Frameworks. In *Guidace Papers on the Cross Cutting Issues of the Third Assessment Report of the IPCC.* R. Pachauri, T. Taniguchi, K. Tanaka, (eds.), IPCC, Geneva, pp. 53-68.

Toth, F. L., T. Bruckner, H.-M. Füssel, M. Leimbach, G. Petschel-Held, and H.-J. Schellnhuber, 1997: The tolerable windows approach to integrated assessments. In *Climate Change and Integrated Assessment Models (IAMs)*

- *Bridging the Gaps, Center for Global Environmental Research.* O. K. Cameron, K. Fukuwatari, T. Morita, (eds.), National Institute for Environmental Studies, Tsukuba, Japan, pp. 401-430.

Toth, F. L., G. Petschel-Held, and T. Bruckner, 1999: Climate change and integrated assessments: The tolerable windows approach. In *Goals and Economic Instruments for the Achievement of Global Warming Mitigation in Europe.* J. Hacker , A. Pelchen, (eds.), Kluwer, Dordrecht, pp. 55-78.

Toth, F. L.: *The relevance and use of cost assessment in climate change decisions.* Paper presented at IPCC Expert Meeting on Costing Issues, 29 June-1 July 1999, Tokyo, Japan (unpublished).

Tulkens, H., 1998: Cooperation versus Free-riding in International Environmental Affairs: Two Approaches. In *Game Theory and the Global Environment.* H. Folmer, N. Hanley, (eds.), E. Elgar, Cheltenham, UK, pp. 30 – 44.

Tulpulé, V., S. Brown, J. Lim, C. Polidano, H. Pant, and B. S. Fisher, 1998: *An Economic Assessment of the Kyoto Protocol using the Global Trade and Environment Model.* Proceedings of the Workshop on Economic Modelling and Climate Change, 17-18 September 1998, OECD, Paris, pp. 99-121.

Tulpulé, V., S. Brown, J. Lim, C. Polidano, H. Pant, and B. S. Fisher, 1999: The Kyoto Protocol: An Economic Analysis Using GTEM. *Energy Journal*, **Kyoto Special Issue**, 257-285.

Udéhn, L., 1993: Twenty-five Years with "The Logic of Collective Action". *Acta Sociologica*, **36**, 239-261.

UNCSD (United Nations Commission on Sustainable Development), 1996: *Indicators for Sustainable Development: Framework and Methodologies.* United Nations, New York, NY.

UNCTAD (United Nations Conference on Trade and Development), 1995. *Technological Capacity-Building and Technology Partnership: Field Findings, Country Experiences and Programs.* UN, New York /Geneva

UNCTAD,1996: *Fostering Technological Dynamics: Evolution of Thought on Technology Development Process and Competitiveness: A Review of Literature*, UN, Geneva.

Underdal, A., 1998: Explaining Compliance and Defection: Three Models. *European Journal of International Relations*, **4**(1).

UNDP (United Nations Development Program), 1992: *Grassroot Participation, Defining New Realities and Operationalizing New Strategies.* UNDP Discussion Paper Series, New York.

UNDP (United Nations Development Program), 1997: *Synergies in national Implementation.* Sustainable Energy and Environmental Division, 62 pp.

UNDP (United Nations Development Program), 1998: *Human Development Report.* Oxford University Press, New York, USA and Oxford, UK, 240 pp.

UNEP, 1998: *Mitigation and Adaptation Cost Assessment: Concepts, Methods and Appropriate Use.* UNEP Collaborating Centre on Energy and Environment, Risø National Laboratory, Denmark, 169 pp.

UNFCCC, 1993: *Framework Convention on Climate Change: Articles.* United Nations, New York, NY.

Valverde, A., Jr. L. J., and M. D. Webster, 1999: Stabilizing atmospheric CO_2 concentrations: technical, political, and economic dimensions. *Energy Policy*, **27**(10), 613-622.

Van Asselt, M. and J. Rotmans, 1997: Uncertainties in perspective. In *Perspectives on Global Change.* J. Rotmans, B. de Vries, (eds.), Cambridge University Press, UK..

Van de Kragt, A.J.C., R.M. Dawes, and J.M. Orbell with S.R. Braver and L.A. Wilson II, 1986: Doing Well and Doing Good as Ways of Resolving Social Dilemmas. In *Experimental Social Dilemmas.* H.A.M. Wilke, D.M. Messick, C.G. Rutte (eds.). Verlag Peter Lang, Frankfurt am Main, pp. 177-203.

Van den Hove, S., 2000: Participatory approaches to environmental policy-making: the European Commission Climate Policy Process as a case study. *Ecological Economics*, **33**, 457-472.

Van der Heijden, K., 1996: *Scenarios: The Art of Strategic Conversation.* John Wiley & Sons, West Sussex, UK, 305 pp.

Verhoosel, G., 1997: International Transfer of Environmental Sound Technology; The New Dimension of an Old Stumbling Block. *Environment Policy Law*, **27**(6).

Victor, D. G., 1995: *The Montreal Protocol's Noncompliance Procedure: Lessons for Making Other International Environmental Regimes More Effective.* IIASA, Laxenburg, Austria, 53 pp.

Victor, D. G., 1999: The Regulation of Greenhouse Gases: Does Fairness Matter? In Fair weather? Equity concerns in climate change. F. L. Tóth, (ed.), Earthscan Publications, pp. 193-207.

Victor, D. G., and J. E. Salt, 1995: Keeping the Climate Treaty Relevant. *Nature*, **373**, 280 – 282.

Victor, D.G., and E. B. Skolnikoff, 1999: Translating Intent into Action: Implementing Environmental Commitments, *Environment*, **41**(2), 16-20 & 39-44.

Victor, D. G., K. Raustiala, and E. B. Skolnikoff (eds.), 1998: *The Implementation and Effectiveness of International Environmental Commitments: Theory and Practice.* IIASA, Laxenburg, and MIT Press, Cambridge, MA/London, UK, 460 pp.

Visser, H., R. J. M. Folkert, J. Hoekstra, and J. J. de Wolff, 2000: Identifying Key Sources of uncertainty in Climate Change Projections. *Climatic Change*, **45**(3/4), 421-457.

Vogel, D. and T. Kessler, 1998: How Compliance Happens and Doesn't Happen Domestically. *Engaging Countries: Strengthening Compliance with International Environmental Accords.* E. Brown Weiss and H. K. Jacobson, (eds.), MIT Press, Cambridge, MA/ London, UK

Warford, J. J., M. Munasinghe, and W. Cruz, 1996: *The Greening of Economic Policy Reform.* Economic Development Institute, The World Bank, 187 pp.

Watson, R., J. Dixon, S. Hamburg, A. Janetos, and R. Moss, 1998: *Protecting our Planet, Securing our Future.* UNEP, U.S. NASA, The World Bank, 95 pp.

Watt, E., J. Sathaye, O. d. Buen, O. Masera, I. A. Gelil, N. H. Ravindranath, D. Zhou, J. Li, and D. Intarapravich, 1995: *The Institutional Needs of Joint Implementation Projects.* Lawrence Berkeley Laboratory, 81 pp.

Weber, M., 1922: *Wirtschaft und Gesellschaft. Grundriss der verstehenden Soziologie. Vol. 2. 4th ed..* J. Winckelmann, (ed.), J. C. B. Mohr (Paul Siebeck), Tübingen, Germany, pp. 387-1033.

Weizman, M, 1974: Prices versus Quantities, *Review of Economic Studies*, **41**, 477-491.

Westskog, H., 1999: Paper presented at the EEA Annual Conference, Santiago de Compostela, 26-28 September 1999.

Westra, L., 1997: Post-Normal Science, the Precautionary Principle and the Ethics of Integrity. *Foundations of Science*, **2**, 237-262.

Weyant, J., 1997: Incorporating Uncertainty in Integrated Assessment of Climate Change, Elements of Climate Change 1996, Aspen Global Change Institute, Aspen, CO, pp. 268-263.

Weyant, J. P., and T. Olavson, 1999: Issues in Modeling Induced Technological Change in Energy, Environment, and Climate Policy. *Environmental Modeling and Assessment*, **4**(2/3), 67-86.

Wigley, T. M. L., R. Richels, and J. A. Edmonds, 1996: Economic and Environmental Choices in the Stabilization of Atmospheric CO_2 Concentrations. *Nature*, **379**, 240-243.

World Bank, 1994: *World Development Report, 1994.* Oxford University Press, New York.

World Bank, 1996: *The World Bank Participation Sourcebook. Environment Department Papers No.1.,* Washington, DC, 204 pp.

World Bank, 1997: *Expanding de Measurement of Wealth, Indicators of Environmentally Sustainable Development.* Washington, DC., 110 pp.

Wynne, B., 1994: Scientific knowledge and the global environment. In *Social theory and the global environment*, M. Redclift, T. Benton, (eds.), Routledge, London.

Xenos, N., 1989: *Scarcity and Modernity.* Routledge, London, 122 pp.

Xu, G., 1999: Some issues concerning the implementation of the strategy of developing the country relying on science and education. *China Soft Science*, **1**, 8-17 (in Chinese)

Xu, Z., 1998: Appropriate Technology and Sustainable Development. *China Soft Science*, **8**, 79-82 (in Chinese)

Yamaji, K., 1998: A Study of the Role of End-of-pipe Technologies in Reducing CO_2 Emissions. *Waste Management*, **17**(5/6), 295-302.

Yang, F., and Q. Xu, 1998: A study on the Green Technology Innovation of Enterprises. *China Soft Science*, **3**, 47-51. (in Chinese)

Yi, S., 1997: Stable Coalition Structures with Externalities. *Games and Economic Behaviour*, **20**, 201-223.

Yi, S., and H. Shin, 1994: *Endogenous Formation of Coalition in Oligopoly: I. Theory.* Photocopy, Dartmouth College, 24 pp.

Yohe, G. W., 1996: Climate change Policies, the Distribution of Income, and U.S. Living Standards, In *Climate Change Policy, Risk Prioritization and U.S. Economic Growth*, C.E. Walker, M.A.Bloomfield, M. Thorning, (eds.), American Council for Capital Formation, Washington DC. 11-54.

Yohe, G., and R. Wallace, 1996: Near Term Mitigation Policy for Global Change Under Uncertainty: Minimizing the Expected Cost of Meeting Unknown Concentration Thresholds. *Environmental Modeling and Assessment*, **1**(1/2), 47-57.

Yohe, G., and M. Jacobsen, 1999: Meeting Concentration Targets in the Post-Kyoto World: Does Kyoto Further A Least Cost Strategy? *Mitigation and Adaptation Strategies for Global Change*, **4**(1), 1-23.

Yohe, G. W., J. Neumann, P. Marshall, and H. Ameden, 1996: The economic cost of sea level rise on US coastal properties. *Climatic Change*, **32**(4), 387-410.

Young, O., 1994: *International Governance*. Cornell University Press, Ithaca, NY, 221 pp.

Young, O., 1998: *Institutional Dimensions of Global Environmental Change*. International Human Dimensions Programme, Report No. 9, IHDP.

Young, O. (ed.), 1999: *The Effectiveness of International Environmental Regimes: Causal Connections and Behavioral Mechanisms*. MIT Press, Cambridge, MA, 364 pp.

Zhang, H., 1998: Social Environment and Conditions for the Advancement of Science and Technology. *China Soft Science*, **2,** 85-89 (in Chinese).

Zürn, M., 1996: Die Implementation internationaler Umweltregime und positive Integration. *MPIFG Discussion paper 3*, 29 pp.

CLIMATE CHANGE 2001: MITIGATION

APPENDICES

I

List of Authors and Reviewers

A. List of Co-ordinating Lead Authors, Lead Authors, Contributing Authors, and Review Editors[1]

F. Ackerman	Tufts University, United States of America
A.O. Adegbulugbe	Obafemi Awolowo University, Nigeria
R.S. Agarwal	Indian Institute of Technology Delhi, India
D. Ahuja	Nat. Institute Advance Studies, India
G. Akumu	Climate Network Africa (CAN), Kenya
J. Alcamo	University of Kassel, Germany
M.A. Al-Moneef	King Saud University, Saudi Arabia
E. Alsema	University of Utrecht, Netherlands
M. J. Apps	Natural Resources Canada, Canada
H. Audus	International Energy Agency (IEA, Greenhouse Gas R&D Programme), United Kingdom
T. Banuri	Stockholm Environment Institute Boston, USA (Pakistan)
T. Barker	University of Cambridge, United Kingdom
R. Baron	International Energy Agency, France
S Barrett	London Business School, United Kingdom
I.A. Bashmakov	Tsentr po Effektivnomu Ispolzovanlu Energii, Russian Federation
J. de Beer	Ecofys, Netherlands
S. Bernow	Tellus Institute, United States of America
L.S. Bernstein	International Petroleum Industry Environmental Conservation Association (IPIECA), United States of America
D.K. Biswas	Ministry of Environment & Forests, India
K. Blok	Ecofys, Netherlands
C. Bohringer	Institute of Energy, Economics and the Rational Use of Energy, Germany
P. Bohm	Stockholm University, Sweden
R.K. Bose	Tata Energy Research Institute (TERI), India
D.H. Bouille	Instituto de Economía Energética asociado a Fundación Bariloche (IDEE), Argentina
J. Byrne	University of Delaware, United States of America
E. Calvo	Comision Nac.de Cambio Clim., Consejo Nac. del Amb., Peru
P. Capros	University of Athens, Greece
C. Carraro	University of Venice, Italy
C. Cerri	Universidade de Sao Paolo, Brazil
M.J. Chadwick	United Kingdom
R. Christ	IPCC Secretariat, Switzerland (Austria)
J.M. Christensen	Risoe National Laboratory, Denmark
O. Christophersen	Ministry of Environment, Norway
L. A. Cifuentes	Catholic University of Chile, Chile
R. Costanza	University of Maryland, United States of America
P.J. Crabbe	University of Ottowa, Canada
P. Criqui	Institut d'Economie et de Politique de l'Energie; Centre National de la Recherche Scientifique (IEPE-CNRS), France
O.R. Davidson	University of Cape Town, South Africa (Sierra Leone)
G.R. Davis	Shell International, United Kingdom
D.L. Davis	Carnegie Mellon University, United States of America
J. Davison	IEA Greenhouse Gas R&D Programma, United Kingdom
A. Dearing	World Business Council for Sustainable Development (WBCSD), Switzerland (UK)
E. Demiraj Bruci	Hydrometeorological Institute, Albania
B. Dijkstra	University of Heidelberg, Germany (Netherlands)
T. Downing	University of Oxford, United Kingdom (USA)
J. Edmonds	Pacific Northwest National Laboratory, United States of America
B.S. Fisher	Abare, Australia
B.P. Flannery	Exxon Research and Engineering Co., United States of America
E. Fortin	Congrès International des Réseaux Electriques de Distribution (CIRED), France
P. Freund	International Energy Agency (IEA, Greenhouse Gas R&D Programme), United Kingdom

---

[1] Country in parenthesis: citizenship of contributor.

T. Fujimori	Japan Forest Technical Ass., Japan
C. Gay-Garcia	Centro de Ciencias de la Atmosfere, Universidad Nacional Autonoma de Mexico (UNAM), Mexico
L.J. Geng	United Nations Development Programme (UNDP), Peru
A. Golub	Center for Environmental Economics, Russian Federation
L..H. Goulder	Stanford University, United States of America
D.L. Greene	Oak Ridge National Laboratory, United States of America
K. Gregory	Centre for Business and the Environment, United Kingdom
M. Grubb	Imperial College of Science, Technology and Medicine, United Kingdom
S. Gupta	Tata Energy Research Institute (TERI), India
E.F. Haites	Margaree Consultants Inc., Canada
D.O. Hall	King's College London/University of London, United Kingdom
K. Halsnaes	Riso National Laboratory, Denmark
J. Harnisch	Ecofys, Germany
T.C. Heller	Stanford University, United States of America
D. Herbert	University of British Colombia, Canada
O. Hohmeyer	University of Flensburg, Germany
E. Holt	Climate Technology Initiative, United States of America
J.C. Hourcade	Congrès International des Réseaux Électriques de Distribution (CIRED); Centre National de la Recherche Scientifique (CNRS); École des Hautes Études en Sciences Sociales (EHESS), France
H. Imura	Kyushu University, Japan
H. Ishitani	Graduate School of Engineering, University of Tokyo, Japan
A. B. Jaffe	Brandeis University, United States of America
G. Jannuzzi	Universidade Estadual de Campinas (UNICAMP), Brazil
H.H. Janzen	Agriculture and Agri-Food Canada, Canada
T. Jaszay	Technical University of Budapest, Hungary
C.J. Jepma	University of Groningen, Netherlands
M. Jefferson	Global Energy & Environmental Consultants, United Kingdom
K. Jiang	Energy Research Institute, China
E. Jochem	Frauhofer Institute for Systems and Innovation Research, Germany
R.O. Jones	American Petroleum Institute, United States of America
S. Kartha	Tellus Institute, United States of America
T. Kashiwagi	Tokyo University of Agriculture & Technology, Japan
P.E. Kauppi	University of Helsinki, Finland
D. Keith	Carnegie Mellon University, United States of America
N. Keohane	Harvard University, United States of America
A.M. Khan	International Atomic Energy Agency, Austria
H.S. Kheshgi	Exxon Research & Engineering Company, United States of America
A, Khosla	Development Alternatives, India
A. Kollmus	Tufts University Boston, USA (Switzerland)
L. A. Kozak	Southern Company Services, United States of America
O.N. Krankina	Oregon State University, United States of America (Russian Federation)
A.J. Krupnick	Resources for the Future, United States of America
L. Kuijpers	Technical University Eindhoven, Netherlands
S. Kverndokk	Frischsenteret / Frisch Centre, Norway
A. Lanza	International Energy Agency, France (Italy)
E. Lebre la Rovere	Federal University of Rio de Janeiro, Brazil
H. Lee	Korea Energy Economics Institute, South Korea
R. Lempert	Organization for Research and Development (RAND), United States of America
S.J. Lennon	ESKOM Engineering, South Africa
M. Levine	Lawrence Berkeley National laboratory, United States of America
C. Li	Complex Systems Research Center, United States of America
J. Li	Energy Research Institute, China
J. Liski	European Forest Institute, Finland
L. Lorentsen	Ministry of Finance, Norway
R. Loulou	McGill University, Canada

W.R. Makundi	Lawrence Berkeley Laboratory/University of California, United States of America (Tanzania)
A.J.G. Manders	Centraal Planbureau (CPB), Netherlands
A. Markandya	University of Bath, United Kingdom
G. Marland	Oak Ridge Nat. Laboratory, United States of America
O. Masera Ceruti	Instituto de Ecologia, Universidad Nacional Autonoma de Mexico (UNAM), Mexico
Y. Matsuoka	Kyoto University, Japan
D. L. Mauzerall	Princeton University, United States of America
R. S. Maya	Southern Center for Energy and Environment, Zimbabwe
M. McFarland	DuPont Fluoroproducts, United States of America
B. Metz	National Institute for Public Health and the Environment (RIVM), Netherlands
I. Meyer	Potsdam Institute For Climate Impact Research, Germany
K.M. Meyer-Abich	Universitaet Essen, Germany
L. Michaelis	Oxford Centre for the Environment Ethics & Society (OCEES), United Kingdom
E. Mills	Lawrence Berkeley National Laboratory, United States of America
K. Minami	National Institute of Agro-Environmental Sciences, Japan
K. Minato	Japan Automobile Research Institute (JARI), Japan
J.F.B. Mitchell	Hadley Centre, United Kingdom
J. P. Montero	Catholic University of Chile, Chile
W.R. Moomaw	The Fletcher School of Law and Diplomacy, United States of America
D. Moorcroft	World Business Council for Sustainable Development, Switzerland
J. R. Moreira	Biomass Users Network (BUN), Brazil
S. Mori	Science University of Tokyo, Japan
T. Morita	National Institute for Environmental Studies, Japan
M. Munasinghe	University of Colombo, Sri Lanka
S. Murase	Sophia University, Japan
M.J. Mwandosya	Center for Energy, Environment, Science & Technology, Tanzania
G.-J. Nabuurs	ALTERRA Research Institute of the Green Environment, Netherlands
A. Najam	Boston University, United States of America (Pakistan)
T. Nakata	Mitsui & Co. Ltd., Japan
N. Nakicenovic	International Institute for Applied Systems Analysis (IIASA), Austria
E. Nikitina	Russian Academy of Science, Russian Federation
J.B. Opschoor	Institute for Social Studies, Netherlands
R.K. Pachauri	Tata Energy Research Institute (TERI), India
J. Pan	Chinese Academy of Social Sciences/TSU Working Group III IPCC, Netherlands (China)
J.K. Parikh	Indira Gandhi Institute of Development Research (IGIDR), India
K. Parikh	Indira Gandhi Institute of Development Research (IGIDR), India
A. Pasco-Font	Grade, Peru
J. Pershing	International Energy Agency, France (USA)
G Peszko	Organization for Economic Cooperation and Development (OECD), France (Poland)
R. Pichs-Madruga	Centro de Investigaciones de Economía Mundial (CIEM), Cuba
L. Pinguelli Rosa	Federal University Rio de Janeiro, Brazil
H.M. Pitcher	Pacific Northwest National Laboratories, United States of America
W. Pizer	Resources for the Future, United States of America
S. E. Plotkin	Argonne National Laboratory, United States of America
N. S. Prasad	Centre For Wind Energy Technology (C-WET), India
L. K. Price	Lawrence Berkeley National Laboratory, United States of America
A. Rahman	Bangladesh Centre for Advanced Studies (BCAS), Bangladesh
A Rana	National Institute for Environmental Studies, Japan (India)
S. Rand	United States Environmental Protection Agency (US EPA), United States of America
P. D. Raskin	Stockholm Environment Institute Boston, United States of America
N. H. Ravindranath	Indian Institute of Science, India
S. Rayner	Colombia University, United States of America
W. Razali Wan Mohd	Forest Research Institute Malaysia (FRIM), Malaysia
J. Reilly	Department of Agriculture, United States of America
W.J. Rhodes	United States Environmental Protection Agency (US EPA), United States of America
K. Riahi	International Institute for Applied Systems Analysis (IIASA), Austria (Iran)

R.G. Richels	Electric Power Research Institute (EPRI), United States of America
J.B. Robinson	University of British Columbia, Canada
R.A. Roehrl	International Institute for Applied Systems Analysis (IIASA), Austria
H.H. Rogner	International Atomic Energy Agency, Austria
G. Roos	ESKOM Engineering, South Africa
K.E. Rosendahl	Statistics Norway, Norway
W. Sachs	Wuppertal Institute, Germany
A. Sagar	Harvard University, United States of America (India)
P.H.N. Saldiva	Harvard School of Public Health, United States of America
D. A. Sankovski	ICF Consulting, United States of America
J. A. Sathaye	Lawrence Berkeley Nat. Laboratory, United States of America
A. Schafer	Massachusetts Institute of Technology, United States of America
R.A. Sedjo	Resources for the Future, United States of America
H. Sejenovich	Fundacion Barilloche, Argentina
R. Seroa da Motta	Instituto de Pesquisa Econômica Aplicada (IPEA), Brazil
R. Sharma	United Nations Environment Programme (UNEP), Kenya (India)
S. Sharma	Tata Energy Research Institute (TERI), India
M. Shechter	University of Haifa, Israel
J. Shogren	University of Wyoming, United States of America
S. B. Shrikanth	Tata Energy Research Institute (TERI), India
P. R. Shukla	Indian Inst. of Management, India
R.E.H. Sims	Massey University, New Zealand
V. I. Sokolov	Russian Academy of Sciences, Russian Federation
B. Solberg	Norwegian Forest Research Institute, Norway
L. Srivastava	Tata Energy Research Institute (TERI), India
R.N. Stavins	Harvard University, United States of America
R.T.M. Sutamihardja	Ministry of Environment, Indonesia
R Swart	National Institute for Public Health and the Environment (RIVM), Netherlands
K. Tanaka	Global Industrial and Social Progress Research Institute (GISPRI), Japan
T. Taniguchi	Global Industrial and Social Progress Research Institute (GISPRI), Japan
M.T. Tolmasquim	Cidade Universitaria (Federal University of Rio de Janeiro), Brazil
J. Torres Martinez	Ministerio de Ciencia, Cuba
T. Taylor	University of Bath, United Kingdom
F. L. Toth	Potsdam Institute for Climate Impact Research, Germany
J. Turkson	RISOE National Laboratory, Denmark (Ghana)
C.R. Turner	ESKOM Engineering, South Africa
A. Underdal	Centre for Advanced Study, Norway
A. Verbruggen	University of Antwerp, UFSIA, Belgium
D.G. Victor	Council on Foreign Relations, United States of America
A. Villavicencio	RISOE National Laboratory, Denmark (Ecuador)
M.J. Villena	Cambridge University, United Kingdom (Chile)
H.J.M. de Vries	National Institute for Public Health and the Environment (RIVM), Netherlands
Y. Wake	Keio University, Japan
A. C. Walter	State University of Campinas, Brazil
J.P. Weyant	Stanford University, United States of America
P. Wilcoxen	University of Texas, United States of America
J.J. Wise	Mobil Oil Corporation, United States of America
E. Worrell	Lawrence Berkeley National Laboratory, United States of America (Netherlands)
H. Xu	Energy Research Institute, China
M. Yamaguchi	Keio University, Japan
K. Yamaji	University of Tokyo, Japan
F. D. Yamba	University of Zambia (Lusaka), Zambia
F. Yamin	University of London, United Kingdom (Pakistan)
J. R. Ybema	ECN Policy Studies, Netherlands
R. Ye	State Environmental Protection Administration, China
G. Yohe	Wesleyan University, United States of America
T. Zhang	Tsinghua University, China

Z. Zhang	University of Groningen, Netherlands
P. Zhou	Energy, Environment, Computer and Geophysical Applications Group (EECG Consultants), Botswana
D. Zhou	Energy Research Institute, China
F. Zhou	Energy Research Institute, China

B. List of Reviewers

Armenia

| K. Ter-Ghazaryan | Ministry of Nature Protection |

Australia

| I. Jones | University of Sydney |
| A. Fuller | Australian Government |

Austria

B. Amon	University of Agricultural Sciences
G. Klaassen	International Institute for Applied Systems Analysis
K. Radunsky	Federal Environment Agency
L. Schrattenholzer	International Institute for Applied Systems Analysis
D.M. Tripold	Federal Ministry for Agriculture, Forestry, Environment and Water Management

Belgium

P. Boeckx	Gent University
J. Franklin	Solvay Research and Technology
D. Seed	BNFL

Benin

| E. Ahlonsou | Service Meteorologique National |

Brazil

J. Goldemberg	University of Sao Paolo
R. Schaeffer	Federal University of Rio de Janeiro
M. Silvia Muylaert	Federal University of Rio de Janeiro
M. Tolmasquim	Federal University of Rio de Janeiro
S. Trindade	SE_T International Ltd.

Canada

R. Desjardins	Agriculture and Agri-Food Canada
P. Edwards	Meteorological Service of Canada
M. Everell	Government of Canada
L. Gagnon	Hydro-Quebec

R. Grafton	University of Ottawa
P. Hall	Canadian Forest Service
B. Jacques	Environment Canada
N. Macaluso	Environment Canada
D. MacDonald	Alberta Department of Environment
J. Masterton	Government of Canada
J. Robinson	University of British Columbia
W. Smith	Environment Canada
J. Stone	Environment Canada

China

E. Lin	Chinese Academy of Agricultural Sciences
H. Xu	Energy Research Institute of SDPC
T. Teng	Chinese Academy of Social Sciences
Y. Xu	Tsinghua University
T. Zhang	Tsinghua University

Cuba

M. Castellanos	Ministry of Science, Technology and Environment
R. Pichs-Madruga	Centre for World Economy Studies/ Vice-Chair IPCC WG III
J. Torres	Ministry of Science, Technology and Environment
G. Trueba	Ministry of Science, Technology and Environment

Finland

P. Hakkila	VTT Energy
R. Korhonen	VTT Energy
P. Kortelainen	Finnish Environment Institute
R. Pipatti	VTT Energy
I. Savolainen	VTT Energy

France

N. Campbell	Elf-Atochem SA
J-Y. Caneill	Electricité de France
H. Connor-Lajambe	Helio International
C. Cros	Agency for the Environment and Energy Resources
M. Darras	Gaz de France
B. Lesaffre	Ministère de l'Aménagement du Territoire et de l'Environnement
P. Meunier	Interministerial Task Force on Climate Change
A. Michaelowa	Hamburg Institute for International Economics
M. Petit	Ecole Polytechnique

Germany

H-J. Ahlgrimm	Federal Agricultural Research Center
P. Burschel	Technical University Munich
A. Freibauer	University of Stuttgart
E. Jochem	Fraunhofer Institut für Systemtechnik und Innovationsforschung (FhG – ISI)
H. Kohl	Federal Ministry for the Environment, Nature Conservation and Nuclear Safety

K. Kristof	Wuppertal Institute for Climate Environment Energy
M. Lindner	Potsdam Institute for Climate Impact Research
K. Lochte	Baltic Sea Research Institute Warnemünde
P. Radgen	Fraunhofer Institute Systems and Innovations Research
S. Ramesohl	Wuppertal Institute for Climate Environment Energy
B. Schärer	Umweltbundesamt
S. Sicars	Sitec
F. Toth	Potsdam Institute for Climate Impact Research
M. Treber	Germanwatch e.V.
M. Weber	Technical University Munich

India

K. Chatterjee	Climate Change Centre
M. Murty	Institute of Economic Growth

Italy

M. Contaldi	National Environmental Protection Agency
A. Raudner	National Environmental Protection Agency

Japan

A. Amano	Kwansei Gakuin University
K. Asakura	Central Research Institute of Electric Power Industry
S. Baba	Ministry of Foreign Affairs
Y. Fujii	University of Tokyo
N. Goto	University of Tokyo
T. Hakamata	National Institute of Agro-Environmental Sciences
H. Hasuike	The Institute of Applied Energy
Y. Hosoya	The Tokyo Electric Power Company
T. Imai	The Kansai Electric Power Company
A. Inaba	National Institute for Resources and Environment
M. Inoue	Global Environmental Affairs Office, Ministry of Economy, Trade and Industry
Y. Ishida	Global Industrial and Social Progress Research Institute
T. Jung	Institute for Global Environmental Strategies
H. Kazuno	Climate Change Division, Ministry of Foreign Affairs
M. Kokitsu	Global Industrial and Social Progress Research Institute
H. Kuraya	Office of Research and Information, Ministry of the Environment
N. Matsuo	Global Industrial and Social Progress Research Institute / Institute for Global Environmental Strategies
K. Nakane	Hiroshima University
I. Nouchi	National Institute of Agro-Environmental Sciences
Y. Ogawa	Institute of Energy Economics
A. Rana	National Institute for Environmental Studies
I. Sadamori	Global Industrial and Social Progress Research Institute
T. Saito	Senshu University
S. Sato	Chubu Electric Power Company
R. Shimizu	Global Environment Division, Ministry of Foreign Affairs
N. Soto	University of Tokyo
K. Tanaka	Global Industrial and Social Progress Research Institute
Y. Tanaka	Ministry of Agriculture, Forestry and Fishery
Y. Uchiyama	University of Tsukuba
H. Watanabe	Global Industrial and Social Progress Research Institute

K. Yamaji	University of Tokyo
K. Yamazaki	Chubu Electric Power Company
S. Yokoyama	National Institute for Resources and Environment

Kenya

G. Akumu	Climate Network Africa
E. Kimuri	Ministry of Tourism, Trade and Industry
P. Mbuthi	Ministry of Energy
J. Ng'ang'a	University of Nairobi
J. Njihia	Consultant
W. Nyakwada	Kenya Meteorological Department
P. Oballa	KEFRI
K. Senelwa	Moi University

Korea

| J-S. Lim | Kwangwoon University |

Mexico

N. Montes	National Autonomous University of Mexico
C. Sheinbaum Pardo	National Autonomous University of Mexico
A. de la Vega Navarro	National Autonomous University of Mexico

Netherlands

G. Addink	Utrecht University
A. Baede	Royal Netherlands Meteorological Institute
M. Beeldman	Netherlands Energy Research Foundation
M. Berk	National Institute for Public Health and the Environment
H. Bersee	Ministry of Housing, Spatial Planning and the Environment
M. Davidson	Centre for Energy Conservation and Environmental Technology Delft
D. Gielen	Netherlands Energy Research Foundation
B. Hare	Greenpeace International
J. van der Jagt	Utrecht University
K. Jardine	Greenpeace International
T. Kram	Netherlands Energy Research Foundation
K. Mallon	Greenpeace International
B. Metz	IPCC WGIII Co-chair
L. Meyer	Ministry of Housing, Spatial Planning and the Environment
J. Pan	IPCC WGIII TSU
S. van Rooijen	Netherlands Energy Research Foundation
J. van Soest	Centre for Energy Conservation and Environmental Technology Delft
R. Swart	IPCC WGIII TSU
R. Tol	Vrije Universiteit Amsterdam
A. Webb	Netherlands Agency for Energy and the Environment
E. Woerdman	University of Groningen
R. Ybema	Netherlands Energy Research Foundation

New Zealand

J. Barnett	University of Canterbury
S. Kerr	Motu Economic Research
P. MacLaren	New Zealand Forest Research Institute Ltd.
D. Payton	Ministry of Foreign Affairs and Trade
H. Plume	Ministry for the Environment
P. Read	Massey University
A. Reisinger	Ministry for the Environment
R. Sims	Massey University
M. Storey	Agriculture New Zealand
A. Stroombergen	Infometrics Consulting

Nigeria

G. Ayoola	University of Agriculture

Norway

K. Alfsen	Center for International Climate and Environmental Research Oslo (CICERO)
T. Asphjell	The Norwegian Pollution Control Authority (SFT)
K. Brekke	Statistics Norway
O. Christophersen	Ministry of Environment
T. Gulowsen	Greenpeace Nordic
P. Haugan	University of Bergen
H. Kolshus	Centre for International Climate and Environmental Research Oslo (CICERO)
H. Leffertstra	The Norwegian Pollution Control Authority (SFT)
L. Mathiesen	Norwegian School of Economics and Business Administration
S. Mylona	The Norwegian Pollution Control Authority (SFT)
P. Neksa	Foundation for Scientific and Industrial Research
M. Pettersen	The Norwegian Pollution Control Authority (SFT)
J. Petterson	Foundation for Scientific and Industrial Research
A. Thorvik	Statoil
A. Torvanger	Centre for International Climate and Environmental Research Oslo (CICERO)

Poland

H. Gay	EnergSys Ltd.

Portugal

S. Dessai	Euronatura
P. Martins Barata	Euronatura
E. de Oliveira Fernandes	University of Porto

Saudi Arabia

M. Al-Sabban	Government of Saudi Arabia

Singapore

J. Ruitenbeek	Economy and Environment Program for Southeast Asia

Slovenia

A. Kranjc	Ministry of Environment, Hydrometeorological Institute
Z. Stojic	EkoNova

South Africa

G. Downes	ESKOM
W. Poulton	ESKOM
C. Turner	ESKOM
M. Veeran-Rambharos	Generation Environmental Management
B. Vrede	Eskom
M. de Wit	CSIR Environmentek

Spain

F. Hernández	Consejo Superior de Investigaciones Científicas, Institute of Economics and Geography

Sweden

U. Dethlefsen	Vattenfall Utveckling AB
L. Eidenstein	Vattenfall Utveckling AB
C. Ekström	Vattenfall Utveckling AB
S-O. Ericson	Vattenfall Utveckling AB
H. Fernqvist	Volvo Car Corporation
K. Maunsbach	Vattenfall Utveckling AB
B. Svensson	Vattenfall Utveckling AB

Switzerland

C. Albrecht	Swiss Secretariat for Economic Affairs
O. Bahn	Paul Scherrer Institute
J. Romero	Office fédéral de l'environnement, des forêts et du paysage

Ukraine

V. Demkin	Kyiv Mohyla Academy

United Kingdom

H. Audus	IEA Greenhouse Gas R&D Programme
P. Ashford	Caleb Management Services
K. Begg	University of Surrey
W. Bjerke	International Primary Aluminium Institute
S. Boehmer-Christiansen	University of Hull

D. Colbourne	Calor Gas Ltd
E. Cornish	The Uranium Institute
P. Costa	EcoSecurities Ltd.
J. Davidson	IEA Greenhouse Gas R&D Programme
N. Eyre	Energy Saving Trust
P. Freund	IEA Greenhouse Gas R&D Programme
J. Gale	IEA Greenhouse Gas R&D Programme
J. Grant	International Petroleum Industry Environmental Conservation Association
M. Grubb	Imperial College
M. Harley	English Nature / Countryside Council for Wales / Scottish Natural Heritage
S. Holloway	British Geological Survey
R. Knapp	World Coal Institute
G. Leach	Stockholm Environment Institute
N. Mabey	World Wide Fund for Nature
F. MacGuire	Friends of the Earth England, Wales and Northern Ireland
A. McCulloch	University of Bristol
M. Mann	Department of Environment, Transport and the Regions
A. Meyer	Global Commons Institute
B. Müller	Oxford Institute for Energy Studies
A. Murphy	Shell International
S. Parkinson	University of Surrey
J. Skea	Policy Studies Institute
M. Tight	University of Leeds
D. Warrilow	Department of the Environment, Transport and the Regions

USA

F. Ackerman	Tufts University
R. Alig	Forestry Sciences Laboratory
S. Archer	National Science and Technology Center
K. Arrow	Stanford University
C. Artusio	Office for Science and Technology Policy
P. Backlund	Office for Science and Technology Policy
S. Baldwin	Office for Science and Technology Policy
W. Barbour	Environmental Protection Agency
D. Bassett	US Department of Energy
D. Bauer	National Research Council
S. Bernow	Tellus Institute
G. Blomquist	University of Kentucky
E. Boedecker	US Department of Energy
E. Boes	National Renewable Energy Laboratory
G. Boyd	Argonne National Laboratory
M. Brown	Oak Ridge National Laboratory
S. Brown	Winrock International
C. Campbell	Petroplan Inc.
R. Caton	Alchemy Consulting Inc.
F. de la Chesnaye	US Environmental Protection Agency
D. Clark	University of California
B. DeAngelo	Environmental Protection Agency
S. DeCanio	University of California
M. Delmas	University of California
J. Dowd	US Department of Energy
R. Downs	Office of Science and Technology Policy
L. Drake	Massachusetts Institute of Technology
R. Eckaus	Massachusetts Institute of Technology
A. Farrell	Carnegie Mellon University

K. Fisher-Vanden	Dartmouth College
B. Flannery	Exxon Research and Engineering Co.
L. Flejzor	US Department of State
R. Fleagle	University of Washington
J. Frankel	Harvard University
K. Friedman	US Department of Energy
R. Friedman	Heinz Center
G. Frisvold	University of Arizona
W. Fulkerson	University of Tennessee
W. Gore	Boehringer Ingelheim Pharmaceuticals Inc.
J. Gowdy	Rensselaer Polytechnic Institute
K. Green	US Department of Transportation
L. Greening	Consultant
B. Haddad	University of California
S. Hadley	Oak Ridge National Laboratory
J. Hammitt	Harvard School of Public Health
D. Hanson	Argonne National Laboratory
M. Hanson	Energy Center of Wisconsin
D. Harrison	National Economic Research Associates
H. Herzog	Massachusetts Institute of Technology
E. Holt	US Department of Energy
J. Hrubovcak	US Department of Agriculture
H. Jacoby	Massachusetts Institute of Technology
J. Johnston	ExxonMobil Research and Engineering Company
P. Karpoff	US Department of Energy
J. Kerstetter	Washington State University
G. Kelly	Global Climate Coalition
H. Khesghi	ExxonMobil Research and Engineering Company
R. Kopp	Resources for the Future
L. Kozak	Southern Company Services
F. Krause	International Project for Sustainable Energy Paths
S. Laitner	Environmental Protection Agency
H. Lee	Harvard University
P. Lydon	Berkeley University
T. Lyon	Indiana University
T. Marx	General Motors Corporation
B. McCarl	Texas A&M University
S. McDonald	Pacific Northwest National Laboratory
B. McNutt	US Department of Energy
R. Mendelsohn	Yale School of Forestry and Environmental Studies
H. Miller	Air Transportation Associaton of America, Inc.
J. Miotke	Director Office of Global Change, US Department of State
J. Moore	TA Engineering Inc.
R. Morgenstern	Resources for the Future
A. Mosier	US Department of Agriculture
T. Muir	Office of Science and Technology Policy
B. Murry	Center for Economics Research
R. Newell	Resources for the Future
A. Nicholls	Pacific Northwest National Laboratory
M. Offutt	White House Office of Science and Technology Policy
W. Orr	Prescott College, Nasa Program
P. O'Rourke	Sparber & Associates
W. Pizer	Resources for the Future
S. Plotkin	Argonne National Laboratory
R. Prince	US Department of Energy
P. Quinlan	Office of Science and Technology Policy
R. Randall	The Rainforest Regeneration Institute

J. Reilly	Massachusetts Institute of Technology
A. Rose	Penn State University
M. Rose	University of Michigan
N. Rosenberg	Pacific Northwest National Laboratory
M. Ross	University of Michigan
D. Rothman	Columbia University
M. Ruth	Boston University
A. Sanstad	Lawrence Berkeley National Laboratory
S. Schneider	Stanford University
K. Segerson	University of Connecticut
A. Serchuk	Center for Renewable Energy and Sustainable Technology
W. Shadis	TA Engineering Inc.
M. Sheehan	Osterberg & Sheehan
J. Sheffield	Oak Ridge National Laboratory / University of Tennessee
J. Shiller	Ford Motor Company
W. Short	National Renewable Energy Laboratory
J. Shortle	Penn State University
T. Siddiqi	Global Environment and Energy in the 21st century
L. Silverman	US Department of Energy
K. Skog	USDA Forest Products Laboratory
K. Smith	North Carolina State University
W. Smith	US Environmental Protection Agency
A. Solomon	Executive Office of the President
J. Solomon	Praxair Inc.
T. Terry	US Department of Energy
J. Tester	Massachusetts Institute of Technology
T. Tietenberg	Colby College
M. Toman	Resources for the Future
R. Tuccillo	Executive Office of the President
Th. Vanderspurt	ExxonMobil Research and Engineering Company
C. Walker	US Agency for International Development
M. Walsh	Oak Ridge National Laboratory
E. Watts	US Department of Energy
L. Weber	Office of Science and Technology Policy
M. Weiss	Massachusetts Institute of Technology
H. Wesoky	Federal Aviation Administration
N. Young	Air Transportation Association of America, Inc.

Venezuela

L. Pérez	Ministerio del Ambiente y de los Recursos Naturales / Ministerio de Energia y Minas

IGO/NGO

A. Alexiou	United Nations Educational, Scientific and Cultural Organisation
R. Baron	International Energy Agency
L. Bernstein	International Petroleum Industry Environmental Conservation Association
G. Brennand	Organisation of the Petroleum Exporting Countries
J. Corfee Morlot	Organisation for Economic Cooperation and Development
J. Crayston	International Civil Aviation Organisation
D. Ghasemzadeh	Organisation of the Petroleum Exporting Countries
J. Grant	Int'l Petroleum Industry Environmental Conservation Association
B. Hare	Greenpeace International
V. Kagramanian	International Atomic Energy Agency
A. Khan	International Atomic Energy Agency

K. Mallon	Greenpeace International
D. Mansell-Moullin	International Petroleum Industry Environmental Conservation Association
D. O'Connor	Organisation for Economic Co-operation and Development
L. Schipper	International Energy Agency
W.-J. Schmidt-Küster	FORATOM – European Atomic Forum
J. Wise	International Petroleum Industry Environmental Conservation Association
F. Unander	International Energy Agency
D. Wallace	International Energy Agency

II

Glossary

Glossary[1]

AAs
See *assigned amounts*.

AAU
See *assigned amount unit*.

Activities Implemented Jointly (AIJ)
The pilot phase for *joint implementation*, as defined in Article 4.2(a) of the *United Nations Framework Convention on Climate Change*, that allows for project activity among developed countries (and their companies) and between developed and developing countries (and their companies). AIJ is intended to allow Parties to the *United Nations Framework Convention on Climate Change* to gain experience in jointly implemented project activities. There is no crediting for AIJ activity during the pilot phase. A decision remains to be taken on the future of AIJ projects and how they may relate to the Kyoto Mechanisms. As a simple form of tradable permits, AIJ and other market-based schemes represent important potential mechanisms for stimulating additional resource flows for the global environmental good. See also *Clean Development Mechanism*, and *emissions trading*.

Adaptation
Adjustment in natural or human systems to a new or changing environment. Adaptation to *climate change* refers to adjustment in natural or human systems in response to actual or expected climatic stimuli or their effects, which moderates harm or exploits beneficial opportunities. Various types of adaptation can be distinguished, including anticipatory and reactive adaptation, private and public adaptation, and autonomous and planned adaptation.

Additionality
Reduction in *emissions* by *sources* or enhancement of removals by *sinks* that is additional to any that would occur in the absence of a *Joint Implementation* or a *Clean Development Mechanism* project activity as defined in the *Kyoto Protocol* Articles on *Joint Implementation* and the *Clean Development Mechanism*. This definition may be further broadened to include financial, investment, and *technology* additionality. Under *financial additionality*, the project activity funding shall be additional to existing Global Environmental Facility, other financial commitments of Parties included in Annex I, Official Development Assistance, and other systems of co-operation. Under *investment additionality*, the *value* of the *Emissions Reduction Unit /Certified Emission Reduction Unit* shall significantly improve the financial and/or commercial viability of the project activity. Under *technology additionality*, the technology used for the project activity shall be the best available for the circumstances of the host Party.

Administrative costs
The costs of activities of the project or sectoral activity directly related and limited to its short-term implementation. They include the costs of planning, training, administration, monitoring, etc.

Afforestation
Planting of new forests on lands that historically have not contained forests[2]. See also *Deforestation* and *Deforestation.*

AIJ
See *Activities Implemented Jointly.*

Alliance of Small Island States (AOSIS)
The group was formed during the Second World Climate Conference in 1990 and comprises small island and low-lying coastal developing countries that are particularly vulnerable to the adverse consequences of *climate change*, such as sea level rise, coral bleaching, and the increased frequency and intensity of tropical storms. With more than 35 states from the Atlantic, Caribbean, Indian Ocean, Mediterranean, and Pacific, AOSIS share common objectives on environmental and sustainable development matters in the *UNFCCC* (*United Nations Framework Convention on Climate Change*) process.

Alternative development paths
Refer to a variety of possible scenarios for societal values and consumption and production patterns in all countries, including but not limited to a continuation of today's trends. In this Report, these paths do not include additional climate initiatives which means that no scenarios are included that explicitly assume implementation of the **UNFCCC** or the emission targets of the **Kyoto Protocol,** but do include assumptions about other policies that influence greenhouse gas emissions indirectly.

Alternative energy
Energy derived from non-fossil fuel sources.

Ancillary benefits
The ancillary, or side effects, of policies aimed exclusively at *climate change mitigation*. Such policies have an impact not only on *greenhouse gas emissions*, but also on resource use efficiency, like reduction in emissions of local and regional air pollutants associated with fossil fuel use, and on issues such as transportation, agriculture, *land-use* practices, employment, and fuel security. Sometimes these benefits are referred to as "ancillary impacts" to reflect that in some cases the benefits may be negative. From the perspective of policies directed at

---

[1] The terms that are independent entries in this glossary are highlighted in **bold and italics** in text as cross-references.

[2] For a discussion of the term *forest* and related terms such as *afforestation*, *reforestation*, and *deforestation (ARD)*: see the IPCC Special Report on Land Use, Land-Use Change and Forestry, Cambridge University Press, 2000.

abating local air pollution, ***greenhouse gas mitigation*** may also be considered an ancillary benefit, but these relationships are not considered in this assessment. See also ***co-benefits***.

Anthropogenic emissions
Emissions of ***greenhouse gases***, *greenhouse gas* precursors, and aerosols associated with human activities. These include burning of ***fossil fuels*** for energy, ***deforestation*** and ***land-use*** changes that result in net increase in emissions.

Annex I countries/Parties
Group of countries included in Annex I (as amended in 1998) to the ***United Nations Framework Convention on Climate Change***, including all the developed countries in the Organisation of Economic Co-operation and Development, and ***Economies in transition***. By default, the other countries are referred to as ***Non-Annex I countries***. Under Articles 4.2 (a) and 4.2 (b) of the Convention, Annex I countries commit themselves specifically to the aim of returning individually or jointly to their 1990 levels of ***greenhouse gas emissions*** by the year 2000. See also ***Annex II***, ***Annex B***, and ***Non-Annex B countries***.

Annex II countries
Group of countries included in Annex II to the ***United Nations Framework Convention on Climate Change***, including all developed countries in the Organisation of Economic Co-operation and Development. Under Article 4.2 (g) of the Convention, these countries are expected to provide financial resources to assist developing countries to comply with their obligations, such as preparing national reports. Annex II countries are also expected to promote the transfer of environmentally sound technologies to developing countries. See also ***Annex I***, ***Annex B***, ***Non-Annex I***, and ***Non-Annex B countries/Parties***.

Annex B countries/Parties
Group of countries included in Annex B in the ***Kyoto Protocol*** that have agreed to a target for their ***greenhouse gas emissions***, including all the ***Annex I countries*** (as amended in 1998) but Turkey and Belarus. See also ***Annex II***, ***Non-Annex I***, and ***Non-Annex B countries/Parties***.

AOSIS
See ***Alliance of Small Island States***.

Assigned amounts (AAs)
Under the ***Kyoto Protocol***, the total amount of ***greenhouse gas emissions*** that each ***Annex B country*** has agreed that its emissions will not exceed in the first commitment period (2008 to 2012) is the assigned amount. This is calculated by multiplying the country's total *greenhouse gas* emissions in 1990 by five (for the 5-year commitment period) and then by the percentage it agreed to as listed in Annex B of the Kyoto Protocol (e.g., 92% for the European Union; 93% for the USA).

Assigned amount unit (AAU)
Equal to 1 tonne (metric ton) of ***CO_2-equivalent emissions*** calculated using the ***Global Warming Potential***.

Average cost
Total cost divided by the number of units of the item for which the cost is being assessed. With ***greenhouse gases***, for example, it would be the total cost of a programme divided by the physical quantity of ***emissions*** avoided.

Banking
According to the ***Kyoto Protocol*** [Article 3 (13)], Parties included in Annex I to the ***United Nations Framework Convention on Climate Change*** may save excess ***emissions*** allowances or credits from the first commitment period for use in subsequent commitment periods (post-2012).

Barrier
A barrier is any obstacle to reaching a potential that can be overcome by a policy, programme, or measure.

Barrier removal costs
The costs of activities aimed at correcting market failures directly or at reducing the transactions costs in the public and/or private sector. Examples include costs of improving institutional capacity, reducing risk and ***uncertainty***, facilitating market transactions, and enforcing regulatory policies.

Baseline
A non-intervention ***scenario*** used as a base in the analysis of intervention scenarios.

Benefit transfer
An application of monetary values from a particular valuation study to an alternative or secondary policy-decision setting, often in a geographic area other than the one in which the original study was performed.

Biofuel
A fuel produced from dry organic matter or combustible oils produced by plants. Examples of biofuel include alcohol (from fermented sugar), black liquor from the paper manufacturing process, wood, and soybean oil.

Biological options
Biological options for mitigation of climate change involves one or more of the three strategies: *conservation* - conserving an existing carbon ***pool***, and thereby preventing ***emissions*** to the atmosphere; *sequestration* - increasing the size of existing carbon pools, and thereby extracting carbon dioxide from the atmosphere; and *substitution* - substituting biological products for ***fossil fuels*** or energy-intensive products, thereby reducing carbon dioxide emissions.

Biomass

The total mass of living organisms in a given area or volume; recently dead plant material is often included as dead biomass. Biomass can be used for fuel directly by burning it (e.g., wood), or indirectly by fermentation to alcohol (e.g., sugar) or extraction of combustible oils (e.g., soybeans).

Bottom-up models

A modelling approach that includes technological and engineering details in the analysis. See also *top-down models*.

Bubble

Article 4 of the *Kyoto Protocol* allows a group of countries to meet their target listed in *Annex B* jointly by aggregating their total *emissions* under one "bubble" and sharing the burden. The European Union nations intend to aggregate and share their emissions commitments under one bubble.

Cap

See *emissions cap*.

Capital costs

Costs associated with capital or investment expenditure on land, plant, equipment, and inventories. Unlike labour and operating costs, capital costs are independent of the level of output for a given capacity of production.

Capacity building

In the context of *climate change*, capacity building is a process of developing the technical skills and institutional capability in developing countries and *Economies in transition* to enable them to participate in all aspects of *adaptation* to, *mitigation* of, and research on climate change, and the implementation of the *Kyoto Mechanisms*, etc.

Carbon cycle

The term used to describe the flow of carbon in various forms (e.g., as *carbon dioxide*) through the atmosphere, ocean, terrestrial biosphere, and lithosphere.

Carbon dioxide (CO$_2$)

A naturally occurring gas, and also a by-product of burning *fossil fuels* and *biomass*, as well as *land-use* changes and other industrial processes. It is the principal anthropogenic *greenhouse gas* that affects the earth's radiative balance. It is the reference gas against which other *greenhouse gases* are measured and therefore has a *Global Warming Potential* of 1.

Carbon dioxide fertilization

The enhancement of the growth of plants as a result of increased atmospheric carbon dioxide concentration. Depending on their mechanism of photosynthesis, certain types of plants are more sensitive to changes in atmospheric carbon dioxide concentration. In particular, plants that produce a three-carbon compound (C$_3$) during photosynthesis; including most trees and agricultural crops such as rice, wheat, soybeans, potatoes and vegetables, generally show a larger response than plants that produce a four-carbon compound (C$_4$) during photosynthesis; mainly of tropical origin, including grasses and the agriculturally important crops maize, sugar cane, millet and sorghum.

Carbon leakage

See *leakage*.

Carbon tax

See *emissions tax*.

CDM

See *Clean Development Mechanism*.

CER

See *certified emission reduction*.

Certified emission reduction (CER)

Equal to 1 tonne (metric ton) of *CO$_2$-equivalent emissions* reduced or sequestered through a *Clean Development Mechanism* project, calculated using *Global Warming Potentials*. See also *emissions reduction units*.

CFCs

See *chlorofluorocarbons*.

CH$_4$

See *methane*.

Chlorofluorocarbons (CFCs)

Greenhouse gases covered under the 1987 Montreal Protocol and used for refrigeration, air conditioning, packaging, insulation, solvents, or aerosol propellants. Since they are not destroyed in the lower atmosphere, CFCs drift into the upper atmosphere where, given suitable conditions, they break down *ozone*. These gases are being replaced by other compounds, including hydrochlorofluorocarbons and *hydrofluorocarbons*, which are *greenhouse gases* covered under the *Kyoto Protocol*.

Clean Development Mechanism (CDM)

Defined in Article 12 of the *Kyoto Protocol*, the Clean Development Mechanism is intended to meet two objectives: (1) to assist Parties not included in Annex I in achieving sustainable development and in contributing to the ultimate objective of the convention; and (2) to assist Parties included in Annex I in achieving compliance with their quantified emission limitation and reduction commitments. *Certified emission reductions* from Clean Development Mechanism projects undertaken in *non-Annex I countries* that limit or reduce *greenhouse gas emissions*, when certified by operational entities designated by *Conference of the Parties/Meeting of the Parties*, can be accrued to the investor (government or industry) from Parties in *Annex B*. A share of the proceeds from the certified project activities is used to cover administrative expenses as well as to assist developing country Parties that are particularly vulnerable to the adverse effects of *climate change* to meet the costs of *adaptation*.

Climate change

Climate change refers to a statistically significant variation in either the mean state of the climate or in its variability, persisting for an extended period (typically decades or longer). Climate change may result from natural internal processes or external forcings, or to persistent anthropogenic changes in the composition of the atmosphere or in *land use*. Note that *United Nations Framework Convention on Climate Change*, in its Article 1, defines "climate change" as "a change of climate which is attributed directly or indirectly to human activity that alters the composition of the global atmosphere and which is in addition to natural climate variability observed over comparable time periods". *United Nations Framework Convention on Climate Change* thus makes a distinction between "climate change" attributable to human activities altering the atmospheric composition, and "climate variability" attributable to natural causes.

Climate Convention

See *United Nations Framework Convention on Climate Change*.

CO_2

See *carbon dioxide*.

CO_2-equivalent

The concentration of *carbon dioxide* that would cause the same amount of *radiative forcing* as the given mixture of carbon dioxide and other *greenhouse gases*.

Co-benefits

The benefits of policies that are implemented for various reasons at the same time – including *climate change mitigation* – acknowledging that most policies designed to address *greenhouse gas mitigation* also have other, often at least equally important, rationales (e.g., related to objectives of development, sustainability, and equity). The term co-impact is also used in a more generic sense to cover both the positive and negative side of the benefits. See also *ancillary benefits*.

Co-generation

The use of waste heat from electric generation, such as exhaust from gas turbines, for either industrial purposes or district heating.

Commercialization

Sequence of actions necessary to achieve market entry and general market competitiveness of new technologies, processes, and products.

Compliance

See *implementation*.

Conference of the Parties (CoP)

The supreme body of the *United Nations Framework Convention on Climate Change*, comprising countries that have ratified or acceded to the Framework Convention on Climate Change. The first session of the *Conference of the Parties* (CoP-1) was held in Berlin in 1995, followed by CoP-2 in Geneva 1996, CoP-3 in Kyoto 1997, CoP-4 in Buenos Aires, CoP-5 in Bonn, and CoP-6 in The Hague. See also *CoP/MoP* and *Meeting of the Parties*.

Consumer surplus

A measure of the *value* of consumption beyond the price paid for a good or service.

CoP

See *Conference of the Parties*.

CoP/MoP

The *Conference of the Parties* of the *United Nations Framework Convention on Climate Change* will serve as the *Meeting of the Parties (MoP)* the supreme body of the *Kyoto Protocol*, but only Parties to the Kyoto Protocol may participate in deliberations and make decisions. Until the Protocol enters into force, *MoP* cannot meet.

Cost-effective

A criterion that specifies that a *technology* or measure delivers a good or service at equal or lower cost than current practice, or the least-cost alternative for the achievement of a given target.

Deforestation

Conversion of forest to non-forest[3].

Demand-side management

Policies and programmes designed for a specific purpose to influence consumer demand for goods and/or services. In the energy sector, for instance, it refers to policies and programmes designed to reduce consumer demand for electricity and other energy sources. It helps to reduce *greenhouse gas emissions*.

Dematerialization

The process by which economic activity is decoupled from matter–energy throughput, through processes such as eco-efficient production or *industrial ecology*, allowing environmental impact to fall per unit of economic activity.

Deposit–refund system

Combines a deposit or fee (tax) on a commodity with a refund or rebate (*subsidy*) for implementation of a specified action. See also *emissions tax*.

Desertification

Land degradation in arid, semi-arid, and dry sub-humid areas resulting from various factors, including climatic variations and human activities. Further, the United Nations Convention to Combat Desertification (UNCCD) defines land degradation as a reduction or loss, in arid, semi-arid, and dry sub-humid

---

[3] See footnote 2.

areas, of the biological or economic productivity and complexity of rain-fed cropland, irrigated cropland, or range, pasture, forest, and woodlands resulting from land uses or from a process or combination of processes, including processes arising from human activities and habitation patterns, such as: (i) soil erosion caused by wind and/or water; (ii) deterioration of the physical, chemical and biological or economic properties of soil; and (iii) long-term loss of natural vegetation.

Double dividend

The effect that revenue-generating instruments, such as a *carbon tax* or auctioned (tradable) carbon emission permits, can (1) limit or reduce *greenhouse gas emissions* and (2) offset at least part of the potential welfare losses of climate policies through recycling the revenue in the economy to reduce other taxes likely to be distortionary. In a world with involuntary unemployment, the **climate change** policy adopted may have an effect (a positive or negative "third dividend") on employment. Weak double dividend occurs as long as there is a revenue-recycling effect; that is, as long as revenues are recycled through reductions in the marginal rates of distortionary taxes. Strong double dividend requires that the (beneficial) revenue recycling effect more than offset the combination of the primary cost and in this case, the net cost of abatement is negative. See also **interaction effects.**

Economic potential

Economic potential is the portion of **technological potential** for *greenhouse gas emissions* reductions or **energy efficiency** improvements that could be achieved cost-effectively through the creation of markets, reduction of market failures, increased financial and technological transfers. The achievement of economic potential requires additional **policies and measures** to break down **market barriers**. See also **market potential, socio-economic potential**, and **technological potential.**

Economies in transition (EITs)

Countries with national economies in the process of changing from a planned economic system to a market economy.

Ecosystem

A system of interacting living organisms and their physical environment. The boundaries of what can be called an ecosystem are somewhat arbitrary, depending on the focus of interest or study. Thus, the extent of an ecosystem may range from very small spatial scales to, ultimately, the entire earth.

Ecotax

See *emissions tax*

EITs

See **economies in transition.**

Emissions

In the **climate change** context, emissions refer to the release of **greenhouse gases** and/or their precursors and aerosols into the atmosphere over a specified area and period of time.

Emissions cap

A mandated restraint, in a scheduled timeframe, that puts a "ceiling" on the total amount of anthropogenic **greenhouse gas emissions** that can be released into the atmosphere. The **Kyoto Protocol** mandates caps on the *greenhouse gas* emissions released by **Annex B countries/Parties**.

Emissions factor

An emissions factor is the coefficient that relates actual **emissions** to activity data as a standard rate of emission per unit of activity.

Emissions permit

An emissions permit is the non-transferable or tradable allocation of entitlements by a government to an individual firm to emit a specified amount of a substance.

Emissions quota

The portion or share of total allowable *emissions* assigned to a country or group of countries within a framework of maximum total emissions and mandatory allocations of resources.

Emissions reduction unit (ERU)

Equal to 1 tonne (metric ton) of **carbon dioxide emissions** reduced or sequestered arising from a **Joint Implementation** (defined in Article 6 of the **Kyoto Protocol**) project, calculated using **Global Warming Potential**. See also **certified emission reduction** and **emissions trading**.

Emission standard

A level of emission that under law or voluntary agreement may not be exceeded.

Emissions tax

Levy imposed by a government on each unit of CO_2**-equivalent emissions** by a **source** subject to the tax. Since virtually all of the carbon in **fossil fuels** is ultimately emitted as carbon dioxide, a levy on the carbon content of fossil fuels – a *carbon tax* – is equivalent to an emissions tax for emissions caused by to fossil fuel combustion. An *energy tax* – a levy on the energy content of fuels – reduces demand for energy and so reduces carbon dioxide emissions from fossil fuel use. An *ecotax* is designated for the purpose of influencing human behaviour (specifically economic behaviour) to follow an ecologically benign path. International emissions/carbon/energy tax is a tax imposed on specified sources in participating countries by an international agency. The revenue is distributed or used as specified by participating countries or the international agency.

Emissions trading

A market-based approach to achieving environmental objectives that allows those reducing **greenhouse gas emissions** below what is required to use or trade the excess reductions to offset emissions at another source inside or outside the country. In general, trading can occur at the intracompany, domestic, and international levels. The Second Assessment Report by the Intergovernmental Panel on Climate Change adopted the convention of using "permits" for domestic trading systems and "quotas" for international trading systems. Emissions trading under Article 17 of the **Kyoto Protocol** is a *tradable quota system* based on the **assigned amounts** calculated from the emission reduction and limitation commitments listed in Annex B of the Protocol. See also **certified emission reduction** and **Clean Development Mechanism**.

Energy conversion
See **energy transformation**.

Energy efficiency
Ratio of energy output of a conversion process or of a system to its energy input.

Energy intensity
Energy intensity is the ratio of energy consumption to economic or physical output. At the national level, energy intensity is the ratio of total domestic **primary energy** consumption or **final energy** consumption to **Gross Domestic Product** or physical output.

Energy service
The application of useful energy to tasks desired by the consumer such as transportation, a warm room, or light.

Energy Tax
See **emissions tax.**

Energy transformation
The change from one form of energy, such as the energy embodied in **fossil fuels**, to another, such as electricity.

Equivalent CO$_2$
See **CO$_2$-equivalent.**

ERU
See **emissions reduction unit.**

Externality
See **external cost.**

External cost
Used to define the costs arising from any human activity, when the agent responsible for the activity does not take full account of the impacts on others of his or her actions. Equally, when the impacts are positive and not accounted for in the actions of the agent responsible they are referred to as *external benefits*. **Emissions** of particulate pollution from a power station affect the health of people in the vicinity, but this is not often considered, or is given inadequate weight, in private decision making and there is no market for such impacts. Such a phenomenon is referred to as an *externality*, and the costs it imposes are referred to as the external costs.

FCCC
See **United Nations Framework Convention on Climate Change.**

Final energy
Energy supplied that is available to the consumer to be converted into usable energy (e.g., electricity at the wall outlet).

Flexibility mechanisms
See **Kyoto Mechanisms**.

Forest
A vegetation type dominated by trees. Many definitions of the term *forest* are in use throughout the world, reflecting wide differences in bio-geophysical conditions, social structure, and economics[4]. See also **afforestation**, **deforestation** and **reforestation**.

Fossil fuels
Carbon-based fuels from fossil carbon deposits, including coal, oil, and natural gas.

Fuel switching
Policy designed to reduce **carbon dioxide emissions** by switching to lower carbon-content fuels, such as from coal to natural gas.

Full-cost pricing
The pricing of commercial goods – such as electric power – that includes in the final prices faced by the end user not only the private costs of inputs, but also the costs of **externalities** created by their production and use.

G77/China
See **Group of 77 and China**.

GDP
See **Gross Domestic Product**.

General equilibrium analysis
General equilibrium analysis is an approach that considers simultaneously all the markets and feedback effects among these markets in an economy leading to market clearance. See also **market equilibrium**.

Geo-engineering
Efforts to stabilise the climate system by directly managing the energy balance of the earth, thereby overcoming the enhanced **greenhouse effect**.

GHG
See **greenhouse gas**.

---

[4] See footnote 2.

Global warming
Global warming is an observed or projected increase in global average temperature.

Global Warming Potential (GWP)
An index, describing the radiative characteristics of well-mixed *greenhouse gases*, that represents the combined effect of the differing times these gases remain in the atmosphere and their relative effectiveness in absorbing outgoing infrared radiation. This index approximates the time-integrated warming effect of a unit mass of a given *greenhouse gas* in today's atmosphere, relative to that of *carbon dioxide*. Note that *GWP* also stands for *Gross World Product*.

GNP
See **Gross National Product**.

GPP
See **Gross Primary Production.**

Greenhouse effect
Greenhouse gases effectively absorb infrared radiation emitted by the earth's surface, by the atmosphere itself from these same gases, and by clouds. Atmospheric radiation is emitted to all sides, including downwards to the earth's surface. Thus, *greenhouse gases* trap heat within the surface–troposphere system. This is called the natural greenhouse effect. Atmospheric radiation is strongly coupled to the temperature of the level at which it is emitted. In the troposphere the temperature generally decreases with height. Effectively, infrared radiation emitted to space originates from an altitude with a temperature of, on average, $-19°C$, in balance with the net incoming solar radiation. However, the earth's surface is kept at a much higher temperature of on average $+14°C$. An increase in the concentration of *greenhouse gases* leads to an increased infrared opacity of the atmosphere, and therefore to an effective radiation into space from a higher altitude at a lower temperature. This causes a *radiative forcing*, an imbalance that can only be compensated for by an increase in the temperature of the surface–troposphere system. This is the enhanced greenhouse effect.

Greenhouse gas (GHG)
Greenhouse gases are those gaseous constituents of the atmosphere, both natural and anthropogenic, that absorb and emit radiation at specific wavelengths within the spectrum of infrared radiation emitted by the earth's surface, the atmosphere, and clouds. This property causes the *greenhouse effect*. Water vapour (H_2O), *carbon dioxide*, *nitrous oxide*, *methane* and *ozone* (O_3) are the primary *greenhouse gas*es in the earth's atmosphere. Moreover, there are a number of entirely human-made *greenhouse gas*es in the atmosphere, such as the halocarbons and other chlorine- and bromine-containing substances, dealt with under the *Montreal protocol*. Beside carbon dioxide, *nitrous oxide* and *methane*, the *Kyoto Protocol* deals with the *greenhouse gas*es **sulphur hexafluoride**, **hydrofluorocarbons**, and **perfluorocarbons**.

Gross World Product (GWP)
An aggregation of the *Gross Domestic Products* of the world. Note that *GWP* also stands for *Global Warming Potential*.

Gross Domestic Product (GDP)
The sum of gross *value added*, at purchasers' prices, by all resident and non-resident producers in the economy, plus any taxes and minus any subsidies not included in the value of the products in a country or a geographic region for a given period of time, normally 1 year. It is calculated without deducting for depreciation of fabricated assets or depletion and degradation of natural resources

Gross National Product (GNP)
GNP is a measure of national income. It measures *value added* from domestic and foreign sources claimed by residents. GNP comprises *Gross Domestic Product* plus net receipts of primary income from non-resident income.

Gross Primary Production (GPP)
The amount of carbon fixed from the atmosphere through photosynthesis.

Group of 77 and China (G77/China)
Originally 77, now more than 130 developing countries that act as a major negotiating bloc in the *UNFCCC (United Nations Framework Convention on Climate Change)* process. G77/China is also referred to as *non-Annex I countries* in the context of the *United Nations Framework Convention on Climate Change*.

GWP
See *Global Warming Potential, Gross World Product*.

Harmonized emissions/carbon/energy tax
Commits participating countries to impose a tax at a common rate on the same *sources*. Each country can retain the tax revenue it collects. A harmonized tax would not necessarily require countries to impose a tax at the same rate, but imposing different rates across countries would not be *cost-effective*. See also *emissions tax.*

HFCs
See *hydrofluorocarbons*.

Hydrofluorocarbons (HFCs)
Among the six *greenhouse gases* to be curbed under the *Kyoto Protocol*. They are produced commercially as a substitute for *chlorofluorocarbons*. HFCs largely are used in refrigeration and semiconductor manufacturing. Their *Global Warming Potentials* range from 1300 to 11,700.

IEA
See *International Energy Agency*.

IGO
See *Intergovernmental Organization*.

Implementation

Implementation refers to the actions (legislation or regulations, judicial decrees, or other actions) that governments take to translate international accords into domestic law and policy. It includes those events and activities that occur after the issuing of authoritative public policy directives, which include the effort to administer and the substantive impacts on people and events. It is important to distinguish between the legal implementation of international commitments (in national law) and the effective implementation (measures that induce changes in the behaviour of target groups). *Compliance* is a matter of whether and to what extent countries do adhere to the provisions of the accord. Compliance focuses not only on whether implementing measures are in effect, but also on whether there is compliance with the implementing actions. Compliance measures the degree to which the actors whose behaviour is targeted by the agreement, whether they be local government units, corporations, organizations, or individuals, conform to the implementing measures and obligations.

Implementation costs

Costs involved in the implementation of *mitigation* options. These costs are associated with the necessary institutional changes, information requirements, market size, *opportunities* for *technology* gain and learning, and economic incentives needed (grants, subsidies, and taxes).

Income elasticity

The percentage change in the quantity of demand for a good or service, given a 1% change in income.

Industrial ecology

The set of relationships of a particular industry with its environment; often refers to the conscious planning of industrial processes so as to minimize their negative interference with the surrounding environment (e.g., by heat and materials cascading).

Industrialization

The conversion of a society from one based on manual labour to one based on the application of mechanical devices.

Inertia

Property by which matter continues in its existing state of rest or uniform motion in a straight line, unless that state is changed by external force. In the context of *climate change mitigation*, it is associated with different forms of capital (e.g., physical man-made capital, natural capital, and social non-physical capital, including institutions, regulations, and norms).

Infrastructure

The basic installations and facilities upon which the operation and growth of a community depend, such as roads, schools, electric, gas and water utilities, transportation, and communications systems.

Integrated assessment

A method of analysis that combines results and models from the physical, biological, economic, and social sciences, and the interactions between these components, in a consistent framework to evaluate the status and the consequences of environmental change and the policy responses to it.

Interaction effect

The result or consequence of the interaction of *climate change* policy instruments with existing domestic tax systems, including both cost-increasing tax interaction and cost-reducing revenue-recycling effect. The former reflects the impact that *greenhouse gas* policies can have on the functioning of labour and capital markets through their effects on real wages and the real return to capital. By restricting the allowable *greenhouse gas emissions*, permits, regulations, or a *carbon tax* raise the costs of production and the prices of output, thus reducing the real return to labour and capital. For policies that raise revenue for the government, carbon taxes and auctioned permits, the revenues can be recycled to reduce existing distortionary taxes. See also *double dividend*.

Intergovernmental Organization (IGO)

Organizations constituted of governments. Examples include the World Bank, the Organization of Economic Co-operation and Development (OECD), the International Civil Aviation Organization (ICAO), the Intergovernmental Panel on Climate Change (IPCC), and other UN and regional organizations. The *Climate Convention* allows accreditation of these IGOs to attend the negotiating sessions.

International emissions/carbon/energy tax

See *emissions tax*.

International Energy Agency (IEA)

Paris-based energy forum established in 1974. It is linked with the Organization for Economic Co-operation and Development (OECD) to enable member countries to take joint measures to meet oil supply emergencies, to share energy information, to co-ordinate their energy policies, and to co-operate in the development of rational energy programmes.

International product and/or technology standards

See *Standards*.

JI

See *Joint Implementation*.

Joint Implementation (JI)

A market-based implementation mechanism defined in Article 6 of the *Kyoto Protocol*, allowing *Annex I countries* or companies from these countries to implement projects jointly that limit or reduce *emissions*, or enhance *sinks*, and to share the *Emissions Reduction Units*. JI activity is also permitted in Article 4.2(a) of the *United Nations Framework Convention on Climate Change*. See also *Activities Implemented Jointly* and *Kyoto Mechanisms*.

Known technological options
Refer to technologies that exist in operation or pilot plant stage today. It does not include any new technologies that will require drastic technological breakthroughs.

Kyoto Mechanisms
Economic mechanisms based on market principles that Parties to the *Kyoto Protocol* can use in an attempt to lessen the potential economic impacts of *greenhouse gas* emission-reduction requirements. They include *Joint Implementation* (Article 6), the *Clean Development Mechanism* (Article 12), and *Emissions Trading* (Article 17).

Kyoto Protocol
The Kyoto Protocol to the *United Nations Framework Convention on Climate Change* was adopted at the Third Session of the *Conference of the Parties* (COP) to the United Nations Framework Convention on Climate Change in 1997 in Kyoto, Japan. It contains legally binding commitments, in addition to those included in the UNFCCC. Countries included in Annex B of the Protocol (most OECD countries and countries with *Economies in transition*) agreed to reduce their anthropogenic *greenhouse gas emissions* (*carbon dioxide, methane, nitrous oxide, hydrofluorocarbons, perfluorocarbons*, and *sulphur hexafluoride*) by at least 5% below 1990 levels in the commitment period 2008 to 2012. The Kyoto Protocol has not yet entered into force (November 2000).

Land use
The total of arrangements, activities, and inputs undertaken in a certain land-cover type (a set of human actions). The social and economic purposes for which land is managed (e.g., grazing, timber extraction, and conservation).

Leakage
The part of *emissions* reductions in *Annex B countries* that may be offset by an increase of the emission in the non-constrained countries above their *baseline* levels. This can occur through (1) relocation of energy-intensive production in non-constrained regions; (2) increased consumption of *fossil fuels* in these regions through decline in the international price of oil and gas triggered by lower demand for these energies; and (3) changes in incomes (and thus in energy demand) because of better terms of trade. Leakage also refers to the situation in which a carbon *sequestration* activity (e.g., tree planting) on one piece of land inadvertently, directly or indirectly, triggers an activity, which in whole or part, counteracts the carbon effects of the initial activity.

Macroeconomic costs
Usually measured as changes in *Gross Domestic Product* or growth in *Gross Domestic Product*, or as loss of "welfare" or loss of consumption.

Marginal cost pricing
The pricing of commercial goods and services such that the price equals the additional cost that arises from the expansion of production by one additional unit.

Market barriers
In the context of *mitigation* of *climate change*, conditions that prevent or impede the diffusion of *cost-effective* technologies or practices that would mitigate *greenhouse gas emissions*.

Market-based incentives
Measures intended to use price mechanisms (e.g., taxes and tradable permits) to reduce *greenhouse gas emissions*.

Market equilibrium
The point at which demand for goods and services equals the supply; often described in terms of the level of prices, determined in a competitive market, that "clears" the market.

Market penetration
Market penetration is the share of a given market that is provided by a particular good or service at a given time.

Market potential
The portion of the economic potential for *greenhouse gas emissions* reductions or *energy efficiency* improvements that could be achieved under forecast market conditions, assuming no new *policies and measures*. See also *economic potential*, *socio-economic potential*, and *technological potential*.

Methane (CH_4)
Methane is one of the six *greenhouse gases* to be mitigated under the *Kyoto Protocol*.

Methane recovery
Method by which *methane emissions*, for example from coal mines or waste sites, are captured and then reused either as a fuel, or for some other economic purpose (e.g., reinjection in oil or gas reserves).

Meeting of the Parties (to the Kyoto Protocol) (MoP)
Conference of the Parties to the *United Nations Framework Convention on Climate Change* serving as the meeting of the Parties to the *Kyoto Protocol*. It is the supreme body of the Kyoto Protocol.

Mitigation
An anthropogenic intervention to reduce the *sources* or enhance the *sinks* of *greenhouse gases*. See also *biological options, geo-engineering*.

Mitigative capacity
The social, political, and economic structures and conditions that are required for effective *mitigation*.

Montreal Protocol
The Montreal Protocol on Substances that Deplete the *Ozone Layer* was adopted in Montreal in 1987, and subsequently adjusted and amended in London (1990), Copenhagen (1992), Vienna (1995), Montreal (1997) and Beijing (1999). It controls the consumption and production of chlorine- and bromine-containing chemicals that destroy stratospheric ozone, such as

chlorofluorocarbons, methyl chloroform, carbon tetrachloride, and many others.

MoP
See *Meeting of the Parties* (to the Kyoto Protocol).

N_2O
See **nitrous oxide**.

National Action Plans
Plans submitted to the ***Conference of the Parties*** by Parties outlining the steps that they have adopted to limit their anthropogenic ***greenhouse gas emissions***. Countries must submit these plans as a condition of participating in the ***United Nations Framework Convention on Climate Change*** and, subsequently, must communicate their progress to the *Conference of the Parties* regularly. The National Action Plans form part of the National Communications, which include the national inventory of *greenhouse gas* ***sources*** and ***sinks***.

Nitrous oxide (N_2O)
One of the six ***greenhouse gases*** to be curbed under the ***Kyoto Protocol***.

Non-Annex I Parties/Countries
The countries that have ratified or acceded to the ***United Nations Framework Convention on Climate Change*** that are not included in Annex I of the ***Climate Convention***.

Non-Annex B countries/Parties
The countries that are not included in Annex B in the ***Kyoto Protocol***.

No regrets options
See **no regrets policy**.

No regrets policy
One that would generate net social benefits whether or not there is climate change. *No regrets opportunities* for ***greenhouse gas emissions*** reduction are defined as those options whose benefits such as reduced energy costs and reduced emissions of local/regional pollutants equal or exceed their costs to society, excluding the benefits of avoided climate change. *No regrets potential* is defined as the gap between the ***market potential*** and the ***socio-economic potential***.

No regrets potential
See **no regrets policy**.

Optimal policy
A policy is assumed to be "optimal" if marginal abatement costs are equalized across countries, thereby minimizing ***total costs***.

Opportunity
An opportunity is a situation or circumstance to decrease the gap between the ***market potential*** of any ***technology*** or practice and the ***economic potential, socio-economic potential,*** or ***technological potential***.

Opportunity cost
Opportunity cost is the cost of an economic activity forgone by the choice of another activity.

Ozone
Ozone, the triatomic form of oxygen (O_3), is a gaseous atmospheric constituent. In the troposphere it is created both naturally and by photochemical reactions involving gases resulting from human activities ("smog"). Tropospheric ozone acts as a ***greenhouse gas***. In the stratosphere it is created by the interaction between solar ultraviolet radiation and molecular oxygen (O_2). Stratospheric ozone plays a decisive role in the stratospheric radiative balance. Its concentration is highest in the ozone layer.

PAMs
See ***Policies and Measures***.

Pareto criterion / Pareto optimum
A requirement or status that an individual's welfare could not be further improved without making others in the society worse off.

Pareto improvement
The opportunity that one individual's welfare can be improved without making the welfare of the rest of society worse off.

Performance criteria
See ***standards***.

Perfluorocarbons (PFCs)
Among the six ***greenhouse gases*** to be abated under the ***Kyoto Protocol***. These are by-products of aluminium smelting and uranium enrichment. They also replace ***chlorofluorocarbons*** in manufacturing semiconductors. The ***Global Warming Potential*** of PFCs is 6500–9200 times that of ***carbon dioxide***.

PFCs
See ***perfluorocarbons***.

Policies and Measures (PAMs)
In ***United Nations Framework Convention on Climate Change*** parlance, **policies** are actions that can be taken and/or mandated by a government–often in conjunction with business and industry within its own country, as well as with other countries–to accelerate the application and use of measures to curb ***greenhouse gas emissions***. **Measures** are technologies, processes, and practices used to implement policies, which, if employed, would reduce *greenhouse gas* emissions below anticipated future levels. Examples might include carbon or other energy taxes, standardized fuel efficiency ***standards*** for automobiles, etc. "Common and co-ordinated" or "harmonized" policies refer to those adopted jointly by Parties.

Pool
See *reservoir*.

PPP
See *Purchasing Power Parity*. It also stands for polluter-pays-principle.

Precautionary Principle
A provision under Article 3 of the *United Nations Framework Convention on Climate Change*, stipulating that the Parties should take precautionary measures to anticipate, prevent or minimize the causes of *climate change* and mitigate its adverse effects. Where there are threats of serious or irreversible damage, lack of full scientific certainty should not be used as a reason for postponing such measures, taking into account that *policies and measures* to deal with climate change should be *cost-effective* so as to ensure global benefits at the lowest possible cost.

Present value cost
The sum of all costs over all time periods, with future costs discounted.

Price elasticity
The responsiveness of demand to the cost for a good or service; specifically, the percentage change in the quantity consumed of a good or service for a 1% change in the price for that good or service.

Primary energy
Energy embodied in natural resources (e.g., coal, crude oil, sunlight, uranium) that has not undergone any anthropogenic conversion or transformation.

"Primary market" and "secondary market" trading
In commodities and financial exchanges, buyers and sellers who trade directly with each other constitute the "primary market", while buying and selling through the exchange facilities represent the "secondary market".

Private costs
Categories of costs influencing an individual's decision-making are referred to as private costs. See also *social cost, external cost,* and *total cost.*

Producer surplus
Returns beyond the cost of production that provide compensation for owners of skills or assets that are scarce (e.g., agriculturally productive land). See also *consumer surplus.*

Project costs
Project costs are all the financial costs of a project such as capital, labour, and operating costs.

Purchasing Power Parity (PPP)
Estimates of *Gross Domestic Product* based on the purchasing power of currencies rather than on current exchange rates. Such estimates are a blend of extrapolated and regression-based numbers, using the results of the International Comparison Program. PPP estimates tend to lower per capita *Gross Domestic Product*s in industrialized countries and raise per capita *Gross Domestic Product*s in developing countries. *PPP* is also an acronym for polluter-pays-principle.

QELRCs
See *quantified emission limitation or reduction commitments*.

Quantified emission limitation or reduction commitments (QELRCs)
The *greenhouse gas emissions* reduction commitments, in percentage terms relevant to base year or period, made by developed countries listed in Annex B of the *Kyoto Protocol*. See also *targets and timetables*.

Radiative forcing
Radiative forcing is the change in the net vertical irradiance (expressed in Watts per square meter: Wm^{-2}) at the tropopause due to an internal change or a change in the external forcing of the climate system, such as, for example, a change in the concentration of *carbon dioxide* or the output of the Sun. Usually radiative forcing is computed after allowing for stratospheric temperatures to readjust to radiative equilibrium, but with all tropospheric properties held fixed at their unperturbed values. Radiative forcing is called *instantaneous* if no change in stratospheric temperature is accounted for.

Rebound effect
Occurs because, for example, an improvement in motor efficiency lowers the cost per kilometre driven; it has the perverse effect of encouraging more trips.

Reforestation
Planting of forests on lands that have previously contained forests but that have been converted to some other use[5]. See also *afforestation* and *deforestation.*

Regulatory measures
Rules or codes enacted by governments that mandate product specifications or process performance characteristics. See also *standards.*

Renewables
Energy sources that are, within a short timeframe relative to the earth's natural cycles, sustainable, and include non-carbon technologies such as solar energy, hydropower, and wind, as well as carbon neutral technologies such as *biomass.*

---

[5] See also footnote 2.

Research, development, and demonstration

Scientific and/or technical research and development of new production processes or products, coupled with analysis and measures that provide information to potential users regarding the application of the new product or process; demonstration tests, and feasibility of applying these products processes via pilot plants and other pre-commercial applications.

Reserves

Refer to those occurrences that are identified and measured as economically and technically recoverable with current technologies and prices. See also *resources*.

Reservoir

A component of the climate system, other than the atmosphere, which has the capacity to store, accumulate or release a substance of concern, e.g. carbon, a *greenhouse gas* or a precursor. Oceans, soils, and forests are examples of reservoirs of carbon. *Pool* is an equivalent term (note that the definition of pool often includes the atmosphere). The absolute quantity of substance of concern, held within a reservoir at a specified time, is called the *stock*.

Resources

Resources are those occurrences with less certain geological and/or economic characteristics, but which are considered potentially recoverable with foreseeable technological and economic developments.

Resource base

Resource base includes both *reserves* and *resources*.

Revenue recycling

See *interaction effect*.

Safe landing approach

See *tolerable windows approach*.

Scenario

A plausible and often simplified description of how the future may develop, based on a coherent and internally consistent set of assumptions about key driving forces (e.g., rate of *technology* change, prices) and relationships. Note that scenarios are neither predictions nor forecasts.

Sequestration

The process of increasing the carbon content of a carbon *reservoir* other than the atmosphere. Biological approaches to sequestration include direct removal of *carbon dioxide* from the atmosphere through *land-use* change, *afforestation, reforestation*, and practices that enhance soil carbon in agriculture. Physical approaches include separation and disposal of *carbon dioxide* from flue gases or from processing *fossil fuels* to produce hydrogen- (H_2) and carbon dioxide-rich fractions and long-term storage underground in depleted oil and gas reservoirs, coal seams, and saline aquifers.

SF$_6$

See *sulphur hexafluoride*.

Sinks

Any process or activity or mechanism that removes a *greenhouse gas*, an aerosol, or a precursor of a *greenhouse gas* or aerosol from the atmosphere.

Social costs

The social cost of an activity includes the *value* of all the resources used in its provision. Some of these are priced and others are not. Non-priced resources are referred to as *externalities*. It is the sum of the costs of these externalities and the priced resources that makes up the social cost. See also *private cost, external cost,* and *total cost*.

Socio-economic potential

The socio-economic potential represents the level of GHG mitigation that would be approached by overcoming social and cultural obstacles to the use of technologies that are cost-effective. See also *economic potential*, *market potential*, and *technology potential*.

Source

A source is any process, activity or mechanism that releases a *greenhouse gas*, an aerosol, or a precursor of a *greenhouse gas* or aerosol into the atmosphere.

Spillover effect

The economic effects of domestic or sectoral *mitigation* measures on other countries or sectors. In this report, no assessment is made on environmental spillover effects. Spillover effects can be positive or negative and include effects on trade, carbon *leakage*, transfer, and diffusion of environmentally sound *technology* and other issues.

Stabilization

The achievement of stabilization of atmospheric concentrations of one or more *greenhouse gases* (e.g., *carbon dioxide* or a *CO_2-equivalent* basket of *greenhouse gases*).

Stabilization analysis

In this report this refers to analyses or *scenarios* that address the *stabilization* of the concentration of *greenhouse gases*.

Stabilization scenarios

See *stabilization analysis*.

Stakeholders

Person or entity holding grants, concessions, or any other type of *value* or interest that would be affected by a particular action or policy.

Standards

Set of rules or codes mandating or defining product performance (e.g., grades, dimensions, characteristics, test methods, and rules for use). *International product and/or technology or performance standards* establish minimum requirements for affected products and/or technologies in countries where they are adopted. The standards reduce *greenhouse gas emissions* associated with the manufacture or use of the products and/or application of the technology. See also *emissions standards*, *regulatory measures*.

Stock
See *reservoir*.

Storyline
A narrative description of a *scenario* (or a family of scenarios) that highlights the main scenario characteristics, relationships between key driving forces, and the dynamics of the scenarios.

Structural change
Changes, for example, in the relative share of *Gross Domestic Product* produced by the industrial, agricultural, or services sectors of an economy; or more generally, systems transformations whereby some components are either replaced or potentially substituted by other ones.

Subsidy
Direct payment from the government to an entity, or a tax reduction to that entity, for implementing a practice the government wishes to encourage. *Greenhouse gas emissions* can be reduced by lowering existing subsidies that have the effect of raising emissions, such as subsidies to *fossil fuel* use, or by providing subsidies for practices that reduce emissions or enhance *sinks* (e.g., for insulation of buildings or planting trees).

Sulphur hexafluoride (SF$_6$)
One of the six *greenhouse gases* to be curbed under the *Kyoto Protocol*. It is largely used in heavy industry to insulate high-voltage equipment and to assist in the manufacturing of cable-cooling systems. Its *Global Warming Potential* is 23,900.

Supplementarity
The *Kyoto Protocol* states that *emissions trading* and *Joint Implementation* activities are to be supplemental to domestic actions (e.g., energy taxes, fuel efficiency *standards*, etc.) taken by developed countries to reduce their *greenhouse gas emissions*. Under some proposed definitions of supplementarity (e.g., a concrete ceiling on level of use), developed countries could be restricted in their use of the *Kyoto mechanisms* to achieve their reduction targets. This is a subject for further negotiation and clarification by the parties.

Targets and timetables
A target is the reduction of a specific percentage of *greenhouse gas emissions* from a *baseline* date (e.g., "below 1990 levels") to be achieved by a set date, or timetable (e.g., 2008 to 2012). For example, under the *Kyoto Protocol's* formula, the European Union has agreed to reduce its *greenhouse gas* emissions by 8% below 1990 levels by the 2008 to 2012 commitment period. These targets and timetables are, in effect, an *emissions cap* on the total amount of *greenhouse gas* emissions that can be emit-ted by a country or region in a given time period. See also *quantified emission limitation or reduction commitments*.

Tax-interaction effect
See *interaction effect*.

Technological potential
The amount by which it is possible to reduce *greenhouse gas emissions* or improve *energy efficiency* by implementing a *technology* or practice that has already been demonstrated. See also *economic potential*, *market potential*, and *socio-economic potential*.

Technology
A piece of equipment or a technique for performing a particular activity.

Technology or performance standard
See *standard*.

Technology transfer
The broad set of processes that cover the exchange of knowledge, money, and goods among different *stakeholders* that lead to the spreading of *technology* for adapting to or mitigating *climate change*. As a generic concept, the term is used to encompass both diffusion of technologies and technological co-operation across and within countries.

Tolerable windows approach
These approaches analyse *greenhouse gas emissions* as they would be constrained by adopting a long-term climate - rather than *greenhouse gas* concentration *stabilization* - target (e.g., expressed in terms of temperature or sea level changes or the rate of such changes). The main objective of these approaches is to evaluate the implications of such long-term targets for short- or medium-term "tolerable" ranges of global *greenhouse gas* emissions. Also referred to as safe landing approaches.

Top-down models
The terms "top-down" and "bottom-up" are shorthand for aggregate and disaggregated models. The top-down label derives from how modellers applied macroeconomic theory and econometric techniques to historical data on consumption, prices, incomes, and factor costs to model final demand for goods and services, and supply from main sectors, like the energy sector, transportation, agriculture, and industry. Therefore, top-down models evaluate the system from aggregate economic variables, as compared to *bottom-up models* that consider technological options or project specific *climate change mitigation* policies. Some technology data were, however, integrated into top-down analysis and so the distinction is not that clear-cut.

Total cost
All items of cost added together. The total cost to society is made up of both the *external cost* and the *private cost*, which together are defined as *social cost*.

Trace gas

A minor constituent of the atmosphere. The most important trace gases that contribute to the **greenhouse effect** are, *inter alia*, **carbon dioxide**, **ozone**, **methane**, **nitrous oxide**, **perfluorocarbons**, **chlorofluorocarbons**, **hydrofluorocarbons**, **sulphur hexafluoride**, methyl chloride, and water vapour.

Tradable quota system

*See **emissions trading**.*

Trade effects

Economic impacts of changes in the purchasing power of a bundle of exported goods of a country for bundles of goods imported from its trade partners. Climate policies change the relative production costs and may change terms of trade substantially enough to change the ultimate economic balance.

Umbrella Group

A set of largely non-European developed countries who occasionally act as a negotiating bloc on specific issues.

United Nations Framework Convention on Climate Change (UNFCCC)

The Convention was adopted on 9 May 1992 in New York and signed at the 1992 Earth Summit in Rio de Janeiro by more than 150 countries and the European Economic Community. Its ultimate objective is the "stabilization of greenhouse gas concentrations in the atmosphere at a level that would prevent dangerous anthropogenic interference with the climate system". It contains commitments for all Parties. Under the Convention Parties included in **Annex I** aim to return greenhouse gas emission not controlled by the **Montreal Protocol** to 1990 levels by the year 2000. The convention entered in force in March 1994. See also **Conference of the Parties** and **Kyoto Protocol**.

Uncertainty

An expression of the degree to which a value (e.g., the future state of the climate system) is unknown. Uncertainty can result from lack of information or from disagreement about what is known or even knowable. It may have many types of sources, from quantifiable errors in the data to ambiguously defined concepts or terminology, or uncertain projections of human behaviour. Uncertainty can therefore be represented by quantitative measures (e.g., a range of values calculated by various models) or by qualitative statements (e.g., reflecting the judgement of a team of experts).

UNFCCC

See **United Nations Framework Convention on Climate Change**.

Value added

The net output of a sector after adding up all outputs and subtracting intermediate inputs.

Value

Worth, desirability, or utility based on individual preferences. The total value of any resource is the sum of the values of the different individuals involved in the use of the resource. The values, which are the foundation of the estimation of costs, are measured in terms of the willingness to pay (WTP) by individuals to receive the resource or by the willingness of individuals to accept payment (WTA) to part with the resource.

Vision

Picture of a future world, usually a desired future world.

Voluntary agreement

An agreement between a government authority and one or more private parties, as well as a unilateral commitment that is recognized by the public authority, to achieve environmental objectives or to improve environmental performance beyond compliance.

Voluntary measures

Measures to reduce **greenhouse gas emissions** that are adopted by firms or other actors in the absence of government mandates. Voluntary measures help make climate-friendly products or processes more readily available or encourage consumers to incorporate environmental **values** in their market choices.

III

Acronyms, Abbreviations, and
Chemical Compounds

Acronyms, Abbreviations, and Chemical Compounds

AAUs	Assigned Amount Units	FGD	Flue Gas Desulphurization
ABWR	Advanced Boiling Water Reactor	GATT	General Agreement on Trade and Tariff
ACEA	European Automobile Manufacturer's Association	GDP	Gross Domestic Product
		GEF	Global Environment Facility
ADB	Asian Development Bank	GHGs	Greenhouse Gases
AEEI	Autonomous Energy Efficiency Improvement	GNP	Gross National Product
		GWP	Global Warming Potential / Gross World Product
AIJ	Activity Implemented Jointly		
ALGAS	Asia-Least-Cost Greenhouse Gas Abatement Strategy	H_2O	Water vapour
		HC	Hydrocarbons
ARD	Afforestation, Reforestation and Deforestation	HCFC	Hydrochlorofluorocarbon
		HDI	Human Development Index
ASF	Atmospheric Stabilization Framework	HFCs	Hydrofluorocarbons (hydrogenated Fluorocarbons)
BAU	Business-As-Usual		
BIGCC	Biomass Integrated Gasification Combined Cycle	HFE	Hydrofluoroethers
		HVAC	Heating, Ventilation and Air Conditioning
BOP	Balance-Of-Payments	IA	Integrated Assessment
BWR	Boiling Water Reactor	IAEA	International Atomic Energy Agency
C	Carbon	IAMs	Integrated Assessment Models
C_2F_6	Perfluoroethane / Hexafluoroethane	ICAO	International Civil Aviation Organization
CAC	Command and control	ICE	Internal Combustion Engine
CAFE	Corporate Average Fuel Economy	IEA	International Energy Agency
CANZ	Canada, Australia and New Zealand	IET	International Emissions Trading
CBA	Cost Benefit Analysis	IGCC	Integrated Gasification Combined Cycle
CCGT	Combined Cycle Gas Turbine	IGCCS	Integrated Gasification Combined Cycle or Supercritical
CDM	Clean Development Mechanism		
CEA	Cost-Effectiveness Analysis	IMO	International Maritime Organization
CERs	Certified Emission Reduction	IPCC	Intergovernmental Panel on Climate Change
CF_4	Perfluoromethane / Tetrafluoromethane	IPR	Intellectual Property Rights
CFCs	Chlorofluorocarbons	IS92	IPCC 1992 Scenario
CFL	Compact Fluorescent Lamps	ISIC	International Standard Industrial Classification
CGE	Computable General Equilibrium		
CH_4	Methane	ISO	International Standardization Organization
CHP	Combined Heat and Power	IUCN	International Union for the Conservation of Nature and Natural Resources
CO	Carbon-monoxide		
CO_2	Carbon-dioxide	JI	Joint Implementation
COP	Conference of Parties	LESS	Low CO_2 – emitting Energy Supply System
CSD	Commission for Sustainable Development	LNG	Liquid Natural Gas
DCs	Developing Countries	LPG	Liquefied Petroleum Gas
DES	Development, Equity, and Sustainability	LWR	Light Water Reactor
DMF	Decision Making Framework	MAC	Marginal Abatement Cost
DSM	Demand Side Management	MDB	Multilateral Development Banks
EBRD	European Bank for Reconstruction and Development	MEA	Multilateral Environmental Agreements
		MNCs	Multinational Corporation
EEA	European Environmental Agency	N	Nitrogen (element)
EITs	Economies In Transition	N_2	Nitrogen (gas)
EMS	Environmental Management Standard	N_2O	Nitrous oxide
ERUs	Emission Reduction Units	Na_3AlF_6	Cryolite
ESCOs	Energy Service Companies	NACE	Nomenclature des Activites dans la Communaute Europienne (Index of Business Activities in the European Union)
ESTs	Environmentally Sound Technologies		
EU	European Union		
FAO	United Nations Food and Agricultural Organization	NGOs	Non-Governmental Organizations
		NH_3	Ammonia
FBC	Fluid Bed Combustion	NH_4^+	Ammonium ion
FDI	Foreign Direct Investments	NICs	Newly Industrialized Countries

NMHC	Non-Methane Hydrocarbon
NMVOCs	Non-Methane Volatile Organic Compounds
NO	Nitric oxide
NO_2	Nitrogen dioxide
NO_x	The sum of NO and NO_2
O_2	Oxygen
O_3	Ozone
ODA	Official Development Assistance
ODS	Ozone Depleting Substances
OECD	Organization for Economic Co-operation and Development
OPEC	Organization of Petroleum Exporting Countries
PEM	Proton exchange membrane
PFC	Perfluorocarbon
PPM	Processes and Production Method or Parts Per Million
PPP	Purchasing Power Parity or Polluter Pays Principle
PV	Photo Voltaic
PWR	Pressurized Water Reactor
QELRCs	Quantified Emission Limitation or Reduction Commitments
R&D	Research and Development
SAR	Second Assessment Report of the IPCC
SBSTA	Subsidiary Body for Scientific and Technological Advice
SF_6	Sulfur hexafluoride
SMEs	Small and Medium Sized Enterprises
SO_2	Sulphur dioxide
SO_x	Sulphur oxides
SPM	Summary for Policymakers
SRES	Special Report on Emissions Scenarios
SRLULUCF	Special Report on Land-Use, Land-Use Change and Forestry
SRTT	Special Report on Methodological and Technological Issues in Technology Transfer
TAR	Third Assessment Report
TPES	Total Primary Energy Supply
UNCED	United Nations Conference on Environment and Development
UNDP	United Nations Development Programme
UNEP	United Nations Environment Programme
UNFCCC	United Nations Framework Convention on Climate Change
VA	Voluntary Agreements or Value - Added
VAT	Value Added Tax
VOC	Volatile organic compound
WCED	World Commission on Environment and Development
WEC	World Energy Council
WG I	Working Group One of the IPCC
WG II	Working Group Two of the IPCC
WG III	Working Group Three of the IPCC
WHO	World Health Organization
WTA	Willingness to Accept compensation
WTO	World Trade Organization
WTP	Willingness to Pay
WWF	World Wide Fund for Nature

IV

Units, Conversion Factors, and
GDP Deflators

Units

SI (Systeme Internationale) Units

Physical Quanitty	Name of Unit	Symbol
length	metre	m
mass	kilogram	kg
time	second	s
thermodynamic temperature	kelvin	K
amount of substance	mole	mol

Fraction	Prefix	Symbol	Multiple	Prefix	Symbol
10^{-1}	deci	d	10	deca	da
10^{-2}	cent	c	10^2	hecto	h
10^{-3}	milli	m	10^3	kilo	k
10^{-6}	micro	μ	10^6	mega	M
10^{-9}	nano	n	10^9	giga	G
10^{-12}	pico	p	10^{12}	tera	T
10^{-15}	femto	f	10^{15}	peta	P
			10^{18}	exa	E
			10^{21}	zeta	Z

Special Names and Symbols for Certain SI-Derived Units

Physical Quantity	Name of SI Unit	Symbol for SI Unit	Definition of Unit
force	newton	N	$kg\ m\ s^{-2}$
pressure	pascal	Pa	$kg\ m^{-1}\ s^{-2}\ (=N\ m^{-2})$
energy	joule	J	$kg\ m^2\ s^{-2}$
power	watt	W	$kg\ m^2\ s^{-3}\ (=J\ s^{-1})$
frequency	hertz	Hz	s^{-1} (cycles per second)

Decimal Fractions and Multiples of SI Units Having Special Names

Physical Quantity	Name of Unit	Symbol for Unit	Definition of Unit
length	ångstrom	Å	$10^{-10}\ m = 10^{-8}\ cm$
length	micron	μm	$10^{-6}\ m$
area	hectare	ha	$10^4\ m^2$
force	dyne	dyn	$10^{-5}\ N$
pressure	bar	bar	$10^5\ N\ m^{-2} = 10^5\ Pa$
pressure	millibar	mb	$10^2\ N\ m^{-2} = 1\ hPa$
mass	tonne	t	$10^3\ kg$
mass	gram	g	$10^{-3}\ kg$
column density	Dobson units	DU	2.687×10^{16} molecules cm^{-2}
Stream function	Sverdrup	Sv	$10^6\ m^3\ s^{-1}$

Non-SI Units

°C	degree Celsius (0 °C = 273 K approximately)
	Temperature differences are also given in °C (=K) rather than the more correct form of "Celsius degrees".
ppmv	parts per million (10^6) by volume
ppbv	parts per billion (10^9) by volume
pptv	parts per trillion (10^{12}) by volume
yr	year
Btu	British Themal Unit
MWe	megawatts of electricity
tce	tonnes of coal equivalent
toe	tonnes of oil equivalent
boe	barrels of oil equivalent

The units of mass adopted in this report are generally those which have come into common usage and have deliberately not been harmonized, e.g.,

kt	kilotonnes (10^3 tonnes)
GtC	gigatonnes of carbon (1 GtC = (10^9 tonnes C = 3.67 Gt carbon dioxide)
PgC	petagrams of carbon (1 PgC = 1 GtC)
MtN	megatonnes (10^6 tonnes) of nitrogen
TgC	teragrams of carbon (1 TgC = 1 MtC)
$TgCH_4$	teragrams of methane
TgN	teragrams of nitrogen
TgS	teragrams of sulphur

Conversion Factors[1]

C - CO_2 Conversion Factor
$C/CO_2 = 1/3.67$

General Conversion Factors for Energy

To:	TJ	Gcal	Mtoe	MBtu	GWh
From:	*multiply by:*				
TJ	1	238.8	2.388×10^{-5}	947.8	0.2778
Gcal	4.1868×10^{-3}	1	10^{-7}	3.968	1.163×10^{-3}
Mtoe	4.1868×10^4	10^7	1	3.968×10^7	11630
Mbtu	1.0551×10^{-3}	0.252	2.52×10^{-8}	1	2.391×10^{-4}
GWh	3.6	860	8.6×10^{-5}	3412	1

---

[1] Energy related conversion factors are taken from *World Energy Outlook 2000*, International Energy Agency, Paris.

Conversion Factors for Mass

To:	kg	t	lt	st	lb
From:	*multiply by:*				
kilogram (kg)	1	0.001	9.84×10^{-4}	1.102×10^{-3}	2.2046
tonne (t)	1000	1	0.984	1.1023	2204.6
long ton (lt)	1016	1.016	1	1.120	2240.0
short ton (st)	907.2	0.9072	0.893	1	2000.0
Pound (lb)	0.454	4.54×10^{-4}	4.46×10^{-4}	5.0×10^{-4}	1

Conversion Factors for Volume

To:	gal US	gal UK	bbl	ft³	l	m³
From:	*multiply by:*					
US Gallon (gal)	1	0.8327	0.02381	0.1337	3.785	0.0038
UK Gallon (gal)	1.201	1	0.02859	0.1605	4.546	0.0045
Barrel (bbl)	42.0	34.97	1	5.615	159.0	0.159
Cubic foot (ft³)	7.48	6.229	0.1781	1	28.3	0.0283
Litre (l)	0.2642	0.220	0.0063	0.0353	1	0.001
Cubic metre (m³)	264.2	220.0	6.289	35.3147	1000.0	1

Specific Net Calorific Values

Crude Oil*

	toe/tonne
Saudi Arabia	1.0160
United States	1.0286
Former USSR	1.0050
Iran	1.0190
Venezuela	1.0685
Mexico	1.0115
Norway	1.0260
People's Rep. of China	1.0000
United Kingdom	1.0415
UAE	1.0180

* for selected countries

Petroleum Products*

	toe/tonne
Refinery gas	1.150
LPG	1.130
Ethane	1.130
Motor Gasoline	1.070
Jet Fuel	1.065
Kerosene	1.045
Naphtha	1.075
Gas/Diesel Oil	1.035
Fuel Oil	0.960
Other Products	0.960

* selected products – average
  values

Coal*

	toe/tonne
Peoples's Rep. of China	0.500
United States	0.646
India	0.477
South Africa	0.564
Australia	0.597
Russia	0.444
Poland	0.543
Kazakhstan	0.444
Ukraine	0.516
Germany	0.604

* steam coal production for selected
  countries

Gross Caloric Values

Natural Gas*

	kJ/m³
Russia	37579
United States	38416
Canada	38130
Netherlands	38220
United Kingdom	39518
Indonesia	40600
Algeria	42000
Uzbekistan	37889
Saudi Arabia	38000
Norway	40460

* for selected countries (production).
Note: to calculate the net heat content, the gross heat content is multiplied by 0.9.

Conventions for Electricity

Figures for electricity production, trade and final consumption are calculated using the energy content of the electricity (i.e. at a rate of 1TWh = 0.086Mtoe). Hydro-electricity production (excluding pumped storage) and electricity produced by other non-thermal means (wind, tide, photovoltaic, *etc.*) are accounted for similarly using 1TWh = 0.086 Mtoe. However, the primary energy equivalent of nuclear electricity is calculated from the gross generation by assuming a 33% conversion efficiency, i.e. 1TWh = (0.086 / 0.33) Mtoe. In the case of electricity produced from geothermal heat, if the actual geothermal efficiency is not known, then the primary equivalent is calculated assuming an efficiency of 10%, so 1TWh = (0.086 / 0.1) Mtoe.

GDP Deflators and Changes in Consumer Prices

(Per cent)

	1982-1991	1992-2001	1992	1993	1994	1995	1996	1997	1998	1999	2000	2001
GDP deflators												
Advanced economies	**4.8**	**2.0**	**3.2**	**2.7**	**2.2**	**2.2**	**1.8**	**1.7**	**1.4**	**1.0**	**1.5**	**1.9**
United States	3.7	2.0	2.4	2.4	2.1	2.2	1.9	1.9	1.2	1.5	2.0	2.3
Japan	5.8	2.5	4.3	3.5	2.7	3.0	2.5	1.9	2.0	1.6	1.7	1.7
European Union	1.8	-	1.7	0.6	0.2	-0.6	-1.4	0.3	0.3	-0.9	-0.8	0.9
Other advanced economies	8.7	2.4	3.8	3.8	3.3	3.4	2.9	2.1	1.5	0.3	1.3	2.2
Consumer prices												
Advanced economies	**4.9**	**2.3**	**3.5**	**3.1**	**2.6**	**2.6**	**2.4**	**2.1**	**1.5**	**1.4**	**1.9**	**2.0**
United States	4.1	2.5	3.0	3.0	2.6	2.8	2.9	2.3	1.6	2.2	2.5	2.5
European Union	5.7	2.5	4.6	3.8	3.0	2.9	2.5	1.8	1.4	1.4	1.8	1.8
Japan	1.9	0.7	1.7	1.2	0.7	-0.1	0.1	1.7	0.6	-0.3	0.1	0.9
Other advanced economies	8.8	2.8	3.8	3.4	3.3	3.8	3.2	2.4	2.6	1.0	2.5	2.4
Developing countries	**45.7**	**20.3**	**36.1**	**49.8**	**55.1**	**22.9**	**15.1**	**9.5**	**10.1**	**6.5**	**5.7**	**4.7**
Regional groups												
Africa	19.6	24.4	47.1	38.7	54.8	35.5	30.0	13.6	9.2	11.0	9.6	6.1
Asia	9.7	7.6	8.6	10.8	16.0	13.2	8.2	4.7	7.6	2.5	2.6	3.0
Middle East and Europe	21.2	24.7	26.5	26.6	33.3	38.9	26.6	25.3	26.0	20.3	16.2	9.4
Western Hemisphere	166.9	47.4	109.1	202.6	202.5	34.4	21.4	13.0	9.8	8.8	7.7	6.4
Analytical groups												
By source of export earnings												
Fuel	13.7	21.4	22.1	26.2	31.8	43.2	31.9	16.1	15.6	12.0	10.5	8.8
Nonfuel	51.2	20.3	38.0	53.0	58.0	20.8	13.5	8.9	9.6	6.0	5.2	4.3
By external financing source												
Net creditor countries	2.8	3.6	4.3	5.5	4.0	5.8	3.9	1.9	1.8	1.4	3.3	4.1
Net debtor countries	47.7	20.9	37.4	51.6	57.2	23.5	15.5	9.8	10.4	6.7	5.8	4.7
Official financing	34.3	24.0	59.3	37.4	64.8	30.9	22.4	11.2	8.2	10.4	7.6	4.4
Private financing	54.6	21.0	38.0	57.1	61.4	21.4	13.9	9.2	10.0	5.7	5.1	4.3
Diversified financing	22.5	19.2	24.6	28.5	26.2	33.0	26.1	13.3	12.5	11.5	10.7	8.6
Net debtor countries by debt-servicing experience												
Countries with arrears and/or rescheduling during 1994-1998	100.1	49.8	113.6	204.3	219.9	38.7	19.8	10.4	16.6	11.6	8.1	6.0
Other net debtor countries	27.5	11.0	14.0	14.1	18.6	18.0	13.9	9.6	8.3	5.0	5.0	4.3
Countries in transition	**15.5**	**118.4**	**788.9**	**634.3**	**273.3**	**133.5**	**42.4**	**27.3**	**21.8**	**43.7**	**19.5**	**14.2**
Central and eastern Europe	...	74.8	278.3	366.8	150.4	72.2	32.1	38.4	18.7	20.5	19.4	12.3
Excluding Belarus and Ukraine	...	34.0	104.8	85.1	47.5	24.8	23.3	41.4	17.0	10.9	10.7	7.1
Russia	...	156.1	1,734.7	874.7	307.4	197.4	47.6	14.7	27.7	85.9	20.5	15.9
Transcaucasus and Central Asia	...	193.8	949.2	1,428.7	1,800.7	265.4	80.8	33.0	13.1	15.5	16.3	17.9
Memorandum												
Median inflation rate												
Advanced economies	5.4	2.2	3.2	3.0	2.4	2.4	2.1	1.7	1.6	1.4	2.1	2.0
Developing countries	9.5	7.0	9.9	9.3	10.6	10.1	7.1	6.3	5.7	4.0	4.0	3.6
Countries in transition	11.9	155.2	839.1	472.3	131.6	39.2	24.1	14.8	10.0	8.1	7.9	5.2

Source: IMF (2000) *World Economic Outlook*, International Monetary Fund, Washington DC.

V

List of Annex I, Annex II, and Annex B Countries

List of Annex I Countries, UNFCCC

Australia
Austria
Belarus a/
Belgium
Bulgaria a/
Canada
Croatia*
Czech Republic a/ *
Denmark
European Union
Estonia a/
Finland
France
Germany
Greece
Hungary a/
Iceland
Ireland
Italy
Japan
Latvia a/
Liechtenstein*
Lithuania a/
Luxembourg
Monaco*
Netherlands
New Zealand
Norway
Poland a/
Portugal
Romania a/
Russian Federation a/
Slovakia a/*
Slovenia a/*
Spain
Sweden
Switzerland
Turkey
Ukraine a/
United Kingdom of Great Britain and Northern Ireland
United States of America

List of Annex II Countries, UNFCCC

Australia
Austria
Belgium
Canada
Denmark
European Union
Finland
France
Germany
Greece
Iceland
Ireland
Italy
Japan
Luxembourg
Netherlands
New Zealand
Norway
Portugal
Spain
Sweden
Switzerland
Turkey
United Kingdom of Great Britain and Northern Ireland
United States of America

Note:	Party included in Annex I means a Party included in Annex I to the Convention, as may be amended, or a Party which has made a notification under Article 4, paragraph 2(g), of the Convention.
a/	Countries that are undergoing the process of transition to a market economy.
*	Countries added to Annex I by an amendment that entered into force on 13 August 1998, pursuant to Decision 4/CP.3 adopted at CoP 3.
Source:	Annex I to the United Nations Framework Convention on Climate Change, p. 29.
	Annex II to the United Nations Framework Convention on Climate Change, p. 30.

List of Annex B Countries, Kyoto Protocol

Party	Quantified emission limitation or reduction commitment (percentage of base year or period)
Australia	108
Austria	92
Belgium	92
Bulgaria*	92
Canada	94
Croatia*	95
Czech Republic*	92
Denmark	92
Estonia*	92
European Community	92
Finland	92
France	92
Germany	92
Greece	92
Hungary*	94
Iceland	110
Ireland	92
Italy	92
Japan	94
Latvia*	92
Liechtenstein	92
Lithuania*	92
Luxembourg	92
Monaco	92
Netherlands	92
New Zealand	100
Norway	101
Poland*	94
Portugal	92
Romania*	92
Russian Federation*	100
Slovakia*	92
Slovenia*	92
Spain	92
Sweden	92
Switzerland	92
Ukraine*	100
United Kingdom of Great Britain and Northern Ireland	92
United States of America	93

* Countries that are undergoing the process of transition to a market economy.
Source: Annex B to the Kyoto Protocol to the Convention on Climate Change, p.28.

VI

List of Major IPCC Reports

Climate Change—The IPCC Scientific Assessment
The 1990 Report of the IPCC Scientific Assessment Working Group (also in Chinese, French, Russian, and Spanish)

Climate Change—The IPCC Impacts Assessment
The 1990 Report of the IPCC Impacts Assessment Working Group (also in Chinese, French, Russian, and Spanish)

Climate Change—The IPCC Response Strategies
The 1990 Report of the IPCC Response Strategies Working Group (also in Chinese, French, Russian, and Spanish)

Emissions Scenarios
Prepared for the IPCC Response Strategies Working Group, 1990

Assessment of the Vulnerability of Coastal Areas to Sea Level Rise–A Common Methodology
1991 (also in Arabic and French)

Climate Change 1992—The Supplementary Report to the IPCC Scientific Assessment
The 1992 Report of the IPCC Scientific Assessment Working Group

Climate Change 1992—The Supplementary Report to the IPCC Impacts Assessment
The 1992 Report of the IPCC Impacts Assessment Working Group

Climate Change: The IPCC 1990 and 1992 Assessments
IPCC First Assessment Report Overview and Policymaker Summaries, and 1992 IPCC Supplement

Global Climate Change and the Rising Challenge of the Sea
Coastal Zone Management Subgroup of the IPCC Response Strategies Working Group, 1992

Report of the IPCC Country Studies Workshop, 1992

Preliminary Guidelines for Assessing Impacts of Climate Change, 1992

IPCC Guidelines for National Greenhouse Gas Inventories
Three volumes, 1994 (also in French, Russian, and Spanish)

IPCC Technical Guidelines for Assessing Climate Change Impacts and Adaptations
1995 (also in Arabic, Chinese, French, Russian, and Spanish)

Climate Change 1994—Radiative Forcing of Climate Change and an Evaluation of the IPCC IS92 Emission Scenarios, 1995

Climate Change 1995—The Science of Climate Change – Contribution of Working Group I to the Second Assessment Report, 1996

Climate Change 1995—Impacts, Adaptations, and Mitigation of Climate Change: Scientific-Technical Analyses – Contribution of Working Group II to the Second Assessment Report, 1996

Climate Change 1995—Economic and Social Dimensions of Climate Change – Contribution of Working Group III to the Second Assessment Report, 1996

Climate Change 1995—IPCC Second Assessment Synthesis of Scientific-Technical Information Relevant to Interpreting Article 2 of the UN Framework Convention on Climate Change
1996 (also in Arabic, Chinese, French, Russian, and Spanish)

Technologies, Policies, and Measures for Mitigating Climate Change – IPCC Technical Paper I
1996 (also in French and Spanish)

An Introduction to Simple Climate Models used in the IPCC Second Assessment Report – IPCC Technical Paper II
1997 (also in French and Spanish)

Stabilization of Atmospheric Greenhouse Gases: Physical, Biological and Socio-economic Implications – IPCC Technical Paper III
1997 (also in French and Spanish)

Implications of Proposed CO_2 Emissions Limitations – IPCC Technical Paper IV
1997 (also in French and Spanish)

The Regional Impacts of Climate Change: An Assessment of Vulnerability – IPCC Special Report, 1998

Aviation and the Global Atmosphere - IPCC Special Report, 1999

Land Use, Land Use Changes and Forestry - IPCC Special Report, 2000

Methodological and Technological Issues in Technology Transfer - IPCC Special Report, 2000

Emissions Scenarios - IPCC Special Report, 2000

Climate Change 2001: The Scientific Basis, 2001

Climate Change 2001: Impacts, Adaptation, and Vulnerability, 2001

Climate Change 2001: Mitigation, 2001.

ENQUIRIES: IPCC Secretariat, c/o World Meteorological Organization, 7 bis, Avenue de la Paix, Case Postale 2300, 1211 Geneva 2, Switzerland

VII

Index

---

[1] This index refers to the main occurrences of key terms in the chapters of this Report and the Technical Summary, excluding the Summary for Policymakers, Executive Summaries, reference lists and title pages; numbers in bold refer to the main discussion on the term in the report.